Understanding G protein-coupled receptors and their role in the CNS

(Molecular and Cellular Neurobiology)

Understanding G protein-coupled receptors and their role in the CNS

(Molecular and Cellular Neurobiology)

Edited by

Menelas N. Pangalos
Neurology CEDD, GlaxoSmithKline,
New Frontiers Science Park North,
Harlow, Essex, UK

and

Ceri H. Davies
Neurology CEDD, GlaxoSmithKline,
New Frontiers Science Park North,
Harlow, Essex, UK

OXFORD
UNIVERSITY PRESS

Great Clarendon Street, Oxford OX2 6DP

Oxford University Press is a department of the University of Oxford.
It furthers the University's objective of excellence in research, scholarship,
and education by publishing worldwide in

Oxford New York
Auckland Bangkok Buenos Aires Cape Town Chennai
Dar es Salaam Delhi Hong Kong Istanbul Karachi Kolkata
Kuala Lumpur Madrid Melbourne Mexico City Mumbai Nairobi
São Paulo Shanghai Taipei Tokyo Toronto

Oxford is a registered trade mark of Oxford University Press
in the UK and in certain other countries

Published in the United States
by Oxford University Press Inc., New York

A catalogue record for this title is available from the British Library

Library of Congress Cataloging in Publication Data
(Data available)

ISBN 0 19 850916 2 (Hbk)

10 9 8 7 6 5 4 3 2 1

Typeset by Newgen Imaging Systems (P) Ltd., Chennai, India
Printed in Great Britain
on acid-free paper by T. J. International Ltd, Padstow

Dedication

This book is dedicated to the memory of our friend and colleague Dr Gary Price. As a colleague we valued your knowledge, wisdom, and enthusiasm. As our dear friend we loved you for your humour, honesty, and lust for life. Gary we will miss you.

Preface

The development of transmembrane signalling mechanisms was an important evolutionary step that enabled the cells of multicellular organisms to communicate with each other, as well as their environment. Within the different classes of transmembrane receptors, the G protein-coupled receptors (GPCRs) are amongst the oldest signal transduction proteins, being present in yeast, plants, protozoa, and metazoa. Gene prediction studies based on the recently sequenced human genome estimate that more than 2% of the 30,000 human genes code for GPCRs <http://www.gpcr.org>. Given their large number and sequence diversity, it is not surprising that these proteins transduce a highly diverse set of signals including lights, odorants, cations, amines, amino acids, peptides, lipids, sugars, and large proteins. Most commonly they are activated by the binding of endogenously secreted ligands. For example, the vast majority of endogenous hormones and neuroactive peptides utilize GPCRs to alter the physiology of their target cells as do many small-molecule neurotransmitters such as acetylcholine, GABA, and glutamate. Irrespective of the nature of the activating ligand, an established dogma for GPCR-medicated signal transduction is that the ligand-receptor interaction drives a conformational change of the GPCR, sensed on the inside of the cell by a guanine-nucleotide binding protein (G protein).

Mechanistically, activated GPCRs initiate the disassembly of these G proteins from a tightly associated, inactive $G\alpha\beta\gamma$ subunit heterotrimer to free $G\alpha$ and $G\beta\gamma$ subunits. This occurs through the acceleration of the rate-limiting dissociation of GDP from inactive $G\alpha$ subunits followed by replacement with GTP. Once dissociated both the $G\alpha$-GTP subunit and the $G\beta\gamma$ dimeric subunit are capable of initiating a cascade of intercellular signal transduction events. These can include the activation or inhibition of numerous biochemical cascades, as well as the opening or closing of different ion channels. Such changes in cellular physiology ultimately lead to alterations in gene transcription and fluctuations of transmembrane voltage.

One of the major obstacles in identifying the key steps and components involved in the GPCR mediated regulation of cell physiology has been the elaborate nature of many GPCR signalling cascades. Further complexity has recently been added by the demonstration that GPCRs can function not only as individual proteins, but as multimetric complexes. In addition, different G protein conformations have been identified that allow GPCRs to promiscuously couple to more than one G protein subtype. As such, the integration of numerous cellular properties strongly influence the specificity, diversity, amplification, and time course of GPCR activation in a cell. In this respect, it is now also evident that GPCRs interact not only with G proteins, but with novel receptor-associated proteins that are able to mediate and modify their functions. It is this high level of complexity that confers upon GPCRs their role in so many pathophysiological processes. As a consequence, this superfamily represents one of the most important drug target families for the pharmaceutical industry, with 33% of the best selling prescription drugs in the year 2000 interacting with GPCRs (Wood Mackenzie's, 2000).

In compiling this book our intention was three fold. First, we wanted to dissect the generic features of GPCR biology ranging from receptor classification, structure, regulation and signal transduction all the way to the use of bioinformatics in GPCR research. Secondly, we wanted to provide a thorough grounding in the biology and pharmacology of the principle members of the three GPCR subfamilies that make up the GPCR superfamily. Finally in doing this, our third intention was to bring together into one readable and accessible volume the current views and thoughts of some of the world's leading GPCR scientists. We hope that this book will act as a useful starting place for scientists at all stages in their careers, interested in or embarking upon GPCR research in the central nervous system.

MENELAS PANGALOS AND CERI DAVIES

Contents

List of Contributors

Alexander, S. P. H.
University of Nottingham Medical School
Physiology and Pharmacology Department
Queen's Medical Centre
Nottingham, NG7 2UH

Bartfai, T.
Department of Neuropharmacology
Harold L. Dorris Neurological Research Center
The Scripps Research Institute
10550 N. Torrey Pines Rd
La Jolla, CA, USA

Bockaert, J.
CNRS UPR-9023
Laboratory 'Mécanismes Moléculaires des
Communications Cellulaires'
141 rue de la Cardonille
34094 Montpellier Cedex5, France

Bowery, N. G.
Department of Pharmacology
The Medical School
University of Birmingham
Edgbaston, Birmingham, B15 2TT

Buckley, N. J.
Schools of Biochemistry & Molecular Biology and
Biomedical Sciences
University of Leeds
Leeds LS2 9JT

Burnstock, G.
Autonomic Neuroscience Institute
Royal Free and University College Medical School
Royal Free Campus
Rowland Hill Street
Hampstead, London NW3 2PF

Cesselin, F.
INSERM U 288 Neuropsychopharmacology
Faculté de Médecine Pitié-Salpêtrière
91, Boulevard de l'Hôpital
75634 PARIS Cedex 13
France

Challiss, R. A. J.
Department of Cell Physiology & Pharmacology
University of Leicester
Maurice Shock Medical Sciences Building

University Road,
Leicester LE1 9HN

Clapham, D. E.
Department of Cardiology
Howard Hughes Medical Institute
Children's Hospital Boston
320 Longwood Ave.
Boston, MA 02115, USA

Davies, C. H.
Neurology CEDD
GlaxoSmithKline
New Frontiers Science Park North
Harlow, Essex UK CM19 5AW

Duckworth, D. M.
GlaxoSmithKline
New Frontiers Science Park
Third Avenue
Harlow, Essex CM19 5AW

Dumont, Y.
Douglas Hospital Research Centre
Dept. Psychiatry
McGill University
6875 LaSalle Blvd.
Verdun, QC, Canada, H4H 1R3

Ellis, J.
Departments of Psychiatry and Pharmacology
Penn State University College of Medicine
Hershey, Pennsylvania USA

Feniuk, W.
Glaxo Institute of Applied Pharmacology
University of Cambridge
Department of Pharmacology
Tennis Court Road
Cambridge CB2 1QJ

Fidock, M.
GPCR Group
Discovery Biology
Pfizer Global Research and Development
Sandwich, Kent CT13 9NJ

Fredholm, B. B.
Department of Physiology and Pharmacology
Karolinska Institutet S-171 77 Stockholm
Sweden

Garriga-Canut, M.
Harvard University
Children's Hospital

Gazi, L.
School of Animal and Microbial Sciences
University of Reading, Reading
RG6 6AJ

Gether, U.
Molecular Neuropharmacology Group
Department of Pharmacology
The Panum Institute
University of Copenhagen
DK-2200 Copenhagen, Denmark

Hamon, M.
INSERM U 288
Neuropsychopharmacology
Faculté de Médecine Pitié-Salpêtrière
91, Boulevard de l'Hôpital
75634 PARIS Cedex 13, France

Hannon, J. P.
Nervous System Research
WSJ-386/745, Novartis Pharma AG
CH-4002 Basel, Switzerland

Hay, D. W. P.
Respiratory, Inflammation & Respiratory Pathogens
Centre of Excellence for Drug Discovery
GlaxoSmithKline, King of Prussia
PA 19406, USA

Hepler, J. R.
Department of Pharmacology
Emory University School of Medicine
5009 Rollins Research Center
1510 Clifton Road
Atlanta GA 30322-3090, USA

Hieble, J. P.
Cardiovascular & Urology Centre of Excellence for
Drug Discovery
GlaxoSmithKline, King of Prussia
PA 19406, USA

Ho, M. K. C.
Department of Biochemistry
The Molecular Neuroscience Center and
The Biotechnology Research Institute
Hong Kong University of Science and Technology
Clearwater Bay
Kowloon, Hong Kong, China

Hough, L. B.
Center for Neuropharmacology and Neuroscience
Albany Medical College
Albany
NY 12308, USA

Hoyer, D.
Nervous System Research
WSJ-386/745, Novartis Pharma AG
CH-4002 Basel, Switzerland

Humphrey, P. P. A.
Glaxo Institute of Applied Pharmacology
University of Cambridge
Department of Pharmacology
Tennis Court Road
Cambridge CB2 1QJ

Kahl, U.
Department of Neuropharmacology
Harold L. Dorris Neurological Research Center
The Scripps Research Institute
10550 N. Torrey Pines Rd
La Jolla, CA, USA

Kenakin, T.
GlaxoSmithKline Research and Development
5 Moore Drive
Research Triangle Park
NC 27709, USA

Kendall, D. A.
University of Nottingham Medical School
Physiology and Pharmacology Department
Queen's Medical Centre, Nottingham
NG7 2UH

King, B. F.
Department of Physiology
Royal Free and University College Medical School
Royal Free Campus
Rowland Hill Street, Hampstead
London NW3 2PF

Langel, Ü.
Department of Neurochemistry and
Neurotoxicology
Stockholm University
SE-106 91 Stockholm
Sweden

Leurs, R.
Leiden/Amsterdam Center for Drug Research
Department of Pharmacochemistry
Vrije University Amsterdam
Amsterdam 1081 HV
The Netherlands

Medhurst, A. D.
Neurology Centre of Excellence for Drug Discovery
GlaxoSmithKline
Third Avenue, Harlow Essex
CM19 5AW

Michalovich, D.
Inpharmatica
60 Charlotte St
London W1T 2NU

Middlemiss, D. N.
Psychiatry Centre of Excellence for Drug Discovery
GlaxoSmithKline
New Frontiers Science Park
Third Avenue, Harlow
Essex, CM19 5AW

Nahorski, S. R.
Department of Cell Physiology & Pharmacology
University of Leicester, Maurice Shock Medical
Sciences Building
University Road
Leicester, LE1 9HN

Noble, F.
Département de Pharmacochimie Moléculaire et
Structurale
INSERM U266, CNRS UMR 8600 UFR des
Sciences Pharmaceutiques et biologiques
4 Avenue de l'Observatoire,
75270 Paris Cedex 06, France

Nunn, C.
Nervous System Research
WSJ-386/745, Novartis Pharma AG
CH-4002 Basel, Switzerland

Pangalos, M. N.
Neurology CEDD, GlaxoSmithKline
New Frontiers Science Park North
Harlow, Essex CM19 5AW

Perrin, M. H.
The Clayton Foundation Laboratories for
Peptide Biology
The Salk Institute for Biological Studies
10010 N.Torrey Pines Road, La Jolla

California 92037
USA

Pin, J-P.
CNRS UPR-9023
Laboratory "Mécanismes Moléculaires des
Communications Cellulaires"
141 rue de la Cardonille
34094 Montpellier Cedex5,
France

Price, G. W.
Psychiatry Centre of Excellence for Drug Discovery
GlaxoSmithKline, New Frontiers Science Park
Third Avenue, Harlow
Essex, CM19 5AW

Quirion, R.
Douglas Hospital Research Centre
Dept. Psychiatry, McGill University
6875 LaSalle Blvd., Verdun
QC, Canada, H4H 1R3

Rasmussen, S. G. F.
Molecular Neuropharmacology Group
Department of Pharmacology
The Panum Institute
University of Copenhagen
DK-2200 Copenhagen
Denmark

Redrobe, J. P.
Douglas Hospital Research Centre
Dept. Psychiatry, McGill University
6875 LaSalle Blvd.
Verdun, QC
Canada, H4H 1R3

Riccardi, D.
School of Biological Sciences
University of Manchester
Manchester

Roberson, C.
Department of Cardiology
Howard Hughes Medical Institute
Children's Hospital Boston
320 Longwood Ave.
Boston, MA 02115, USA

Roberts, C.
Psychiatry Centre of Excellence for Drug Discovery
GlaxoSmithKline
New Frontiers Science Park
Third Avenue, Harlow
Essex, CM19 5AW

Roques, B. P.
Département de Pharmacochimie
Moléculaire et Structurale INSERM U266
CNRS UMR 8600
UFR des Sciences Pharmaceutiques
et biologiques 4 Avenue de l'Observatoire
75270 Paris Cedex 06, France

Ruffolo, R. R. Jr.
Wyeth-Ayerst Research
P.O. Box 42528,
Philadelphia, PA 19101
USA

Saugstad, J. A.
Robert S. Dow Neurobiology Laboratories
Legacy Research
1225 N.E. 2nd Avenue
Portland, OR 97232, USA

Schöneberg, T.
Institut für Pharmakologie
Universitätsklinikum Benjamin Franklin
Freie Universität Berlin
Thielallee 69-73
14195 Berlin
Germany

Selmer, I-S.
Glaxo Institute of Applied Pharmacology
University of Cambridge
Department of Pharmacology
Tennis Court Road
Cambridge CB2 1QJ

Shioda, S.
Department of Anatomy
Showa University School of Medicine and
CREST of JST

1-5-8 Hatanodai Shinagawa-ku
Tokyo 142-8555, Japan
Department of Medicine
Tulane University Health Science Center
New Orleans
LA 70112, USA

Strange, P. G.
School of Animal and Microbial Sciences
University of Reading
Reading, RG6 6AJ

Vale, W. W.
The Clayton Foundation laboratories for
Peptide Biology
The Salk Institute for Biological Studies
10010 N.Torrey Pines Road
La Jolla, California 92037 USA

Ward, D. T.
School of Biological Sciences
University of Manchester
Manchester

Waschek, J. A.
Department of Psychiatry and Mental Retardation
Research Center
University of California at Los Angeles
68-225 Neuropsychiatric Institute
760 Westwood Plaza
Los Angeles, CA 90024, USA

Wong, Y. H.
Department of Biochemistry
The Molecular Neuroscience Center and
The Biotechnology Research Institute
Hong Kong University of Science and Technology
Clearwater Bay, Kowloon
Hong Kong, China

List of Abbreviations

5-HT	5-hydroxytryptamine
Aβ	β-amyloid
AC	adenylyl cyclase
ACh	acetylcholine
ACTH	adrenocorticotropic hormone
ADH	autosomal dominant hyperparathyroidism
ADHD	attention deficit hyperactivity disorder
AP-2	adaptor protein-2
APC	adenomatous polyposis coli
ApoE	apolipoprotein E
APP	amyloid precursor protein
AS	alternative splice variant
AreN	arcuate nucleus
AT1aR	angiotensin type 1a receptors
AVP	arginine-vasopressin
β2-ARs	β2-adrenergic receptors
BAC	bacterial artificial chromosome
BDNF	brain-derived neurotrophic factor
BLAST	basic local alignment search tool
BNTX	7-benylidenenaltrexone
BRET	bioluminescence resonance energy transfer
Btk	Burton's tyrosine kinase
CT	calcitonin
CaM	calmodulin
cAMP	cyclic AMP
CAM	constitutively active mutant
CBP	CREB-binding protein
CC	coiled coil
C-CAM	Clocinnamox
CCK	cholecystokinin
cDNA	complementary DNA
CHO	chinese hamster ovary
CK1α	casein kinase 1α
CMA	correlated mutation analysis
CNS	central nervous system
Cort BP1	cortactin-binding protein 1

CRE	cAMP response element
CREB	cAMP responsive element binding protein
CRF	corticotropin releasing factor
CRLR	calcitonin receptor-like-receptor
CSF	cerebrospinal fluid
CST	cortistatins
cTAL	cortical thick ascending limb
CTC	cubic ternary complex
CTX	cholera toxin
DAG	1,2-diacylglycerol
DALCE	Tyr-D-Ala^2-Gly-Phe-Leu-Cys
DALDA	[D-Arg^2, Lys^4]dermorphin-(1-4)-amide
DAS	dense alignmemt surface
DCT	distal convoluted tubule
DDBJ	DNA data bank of Japan
DEP	dishevelled, eg1-10, pleckstrin
DOB	(-)2,5-dimethoxy-4-iodoamphetamine
DOI	2,5-dimethoxy-4-iodophenylisopropylene
DPDPE	Tyr-D-Pen-Gly-Phe-D-Pen
DR	dose-ratio
DRG	dorsal root ganglion
EBI	European bioinformatics institute
ECD	extracellular domains
EEG	electroencephalograph
EGFR	epidermal growth factor receptor
EMBL	European molecular biology laboratory
EMSA	electromobility shift assays
EPR	electron paramagnetic resonance spectroscopy
ER	endoplasmic reticulum
ERK	extracellular signal-regulated protein kinase
EST	expressed sequence tag
ETC	extended ternary complex
FACs	fluorescent activated cell sorting
FAK	focal adhesion kinase
FHH	familial hypocalciuric hypercalcaemia
FRET	fluorescence resonance energy transfer
FSH	follicle-stimulating hormone
FTIR	fourier transform infrared resonance spectroscopy
GAERS	genetic absence epilepsy rats
GALP	galanin-like peptide
GAP	GTPase-activating protein
GEF	guanine nucleotide exchange factor
GFP	green fluorescent protein
GGL	G protein gamma-like
GH	growth hormone
GHB	γ-hydroxybutyrate

GHRH	growth hormone releasing hormone
GIRK	G protein-gated inwardly rectifying potassium channel
GMAP	galanin message associated peptide
GnRH	gonadotropin-releasing hormone
GNTI	5'-guanidinonaltrindole
GPCR	G protein coupled receptor
GRK	G protein-coupled receptor kinase
GSS	genomic survey sequence
H1a	homer 1a
H1b	homer 1b
HA tag	hemagglutinin tag
HA	histamine
HEK293	human embryonic kidney 293
HPLC	high performance liquid chromatography
HTGS	high throughput genomic sequence
iCV	intracerebroventricular
IDdb	Investigational drugs database
iGluR	ionotropic glutamate receptor
IMCD	inner medullary collecting duct
IP_3	inositol 1,4,5-trisphosphate
ISO	isoproterenol
IUPHAR	International union of pharmacology
JNK	c-Jun N-terminal Kinase
KA	kainic acid
kb	kilobases
KSHV	Kaposi's sarcoma-associated herpes virus
LC	locus coereleus
LCR	Locus control region
LGN	lateral geniculate nucleus
LIVBP	leucine/isoleucine/valine-binding protein
LRR	leucine-rich repeat
LTB_4	leucotriene
LTD	long-term depression
LTP	long-term potentiation
MAPK	mitogen-activated protein kinase
MGI	mouse genome informatics
mGluR	metabotropic glutamate receptor
MS	multiple sclerosis
mSoS	mammalian 'Son-of-Sevenless'
mth	methuselah
MTSEA	methanethiosulfonate ethylammonium
N/OFQ	nociceptin/orphanin FQ
NA	noradrenaline
NAPE	N-acylphosphatidylethanolamine
NCBI	National Center for Biotechnology Information
NCC	nonselective cation channels

NF-AT	nuclear factor of activated T cells
NF-κB	nuclear factor kappa B
NGF	nerve growth factor
NHE3	Na^+/H^+ exchanger type 3
NHERF	Na^+/H^+ exchanger regulatory factor
NIH	National Institutes of Health
NKA	neurokinin A
NKB	neurokinin B
NMR	nuclear magnetic resonance
NPK	neuropeptide K
NPY	neuropeptide Y
NRSE	neuron restrictive silence element
NSF	N-ethyl maleimide fusion protein
NSHPT	neonatal severe hyperparathyroidism
NTB	naltriben
OGFr	opioid growth factor receptor
OP	opioid peptides
ORL-1	opioid Receptor-Like -1
PACAP	pituitary adenylate cyclase-activating polypeptide
PAF	platelet activating factor
PAR-2	Protease-activated receptor type 2
PBP	periplasmic amino acid binding protein
PCR	polymerase chain reaction
PeVN	periventricular nucleus
PH	pleckstrin homology
PIP_2	phosphatidylinositol 4,5-bisphosphate
PIR-PSD	protein information resource-protein sequence database
PK	protein kinase
PKA	cyclic AMP-dependent protein kinase
PKC	protein kinase C
PL	phospholipase
PLC-β	phospholipase C-β
PMSF	phenylmethylsulphonylfluoride
PP	pancreatic polypeptde
PPT-A	preprotachyinin A
PPT-B	preprotachyinin B
PSD	post-synaptic density
PSI-BLAST	position specific iterated BLAST
PT	proximal tubule
PTH	parathyroid hormone
PTX	pertussis toxin
PVA	paraventricular nucleus
PVN	paraventricular nucleus
PYY	peptide YY
RAMP	receptor activity modifying protein
REM	rapid eye movement

rER	rough endoplasmic reticulum
R-SAT	receptor selection and amplification technology
SAM	sterile alpha motif
SAR	structure-activity relationship
SCAM	substituted cysteine assessibility method
SCG	superior cervical ganglion
SFO	sub-fornical organ
SH2	src-homology domain 2
SH3	src-homology domain 3
SNPs	single nucleotide polymorphisms
SNV	stearyl-Nle17-VIP
SP	substance P
SPR	surface plasmon resonance
SRIF	somatostatin receptor inducing factor
sst2A	somatostatin 2A
SSTRIP	somatostatin receptor interacting protein
STN	subthalamic nucleus
TAPS	dermorphin tetrapeptide analogues
THC	TIGR human consensus
TIS	transcription initiation site
TM	transmembrane
TMD	transmembrane domain
TrEMBL	translated EMBL
TRP	transient receptor potential
VDCC	voltage-gated Ca^{2+} channels
VIP	vasoactive intestinal peptide
VTA	ventral tegmental area

Part 1

The GPCR superfamily

Chapter 1

GPCR superfamily and its structural characterization

Torsten Schöneberg

1.1 Classification of GPCRs

The diversity of receptor groups within the GPCR superfamily is the result of a long evolutionary process. It has been suggested that serotonin (5-HT) receptors have existed for more than 750 million years (Peroutka and Howell 1994). The tendency towards protein diversification depends upon gene duplications and the continuous accumulation of mutations. The maintenance of vital functions in organisms, however, strictly requires enough structural conservation to ensure the functionality of the corresponding proteins. Despite the remarkable structural variety of natural GPCR agonists, hydropathy analysis, biochemical, and immunological data suggest that all GPCRs share a common molecular architecture consisting of seven transmembrane domains (TMDs) connected by three extra- and three intracellular loops. Interestingly, this global architecture is maintained in all GPCRs discovered so far, despite a rather low amino acid sequence homology.

Molecular cloning studies and genome data analysis have revealed at least 1200 members of the GPCR superfamily in the human genome. About 40–60% out of all human GPCRs have orthologs in other species including more distantly related organisms such as *Caenorhabditis elegans* and *Drosophila melanogaster*. The GPCR superfamily comprises at least three families which share little sequence homology among each other (Fig. 1.1). To date, about 190 GPCRs have been assigned to an agonist or potential ligands. More than 900 are olfactory GPCRs. The remaining GPCRs are so called 'orphan' receptors or pseudogenes in man.

1.1.1 Family 1 GPCRs

About 90% out of all GPCRs are grouped into the rhodopsin-like family (Family 1, also known as Type 1 or Family A, see Part 2). Classifying these proteins is a difficult problem because the mean pairwise amino acid identity is 17% for Family 1 receptors. Based on structural similarities Family 1 GPCR can be subdivided into at least four subfamilies.

Subfamily I contains a subset of olfactory, adenosine, melanocortin peptides, and lysosphingolipid receptors. This subgroup also includes cannabinoid receptors, which bind the endogenous cannabinoid anandamide.

Subfamily II recruits receptors for acetylcholine, catecholamine, and indoleamine ligands.

Family 1

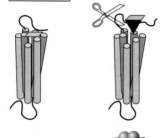

Amine
acetylcholine, catecholamine, histamine, serotonin

Peptide
angiotensin, bombesin, bradykinin, C5a anaphylatoxin, APJ-like, Fmet-leu-phe, interleukin-8, chemokine, CCK, endothelin, melanocortin, neuropeptide Y, neurotensin, opioid, somatostatin, tachykinin, thrombin, melatonin vasopressin, galanin, orexin, urotensin II, GnRH, TRH, GRH

Hormone protein
folicle stimulating hormone, lutropin-choriogonadotropic hormone, thyrotropin

(Rhod)opsin

Lipids and Phospholipids
prostaglandin, prostacyclin, thromboxane, platelet-activating factor, Lysophingolipid, LPA, cannabinoids

Olfactory

Viral receptors (probably peptide receptors)

Family 2

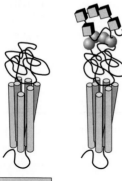

Peptide
calcitonin, corticotropin releasing factor, gastric inhibitory peptide, glucagon, growth hormone-releasing hormone, parathyroid hormone, PACAP, secretin, vasoactive intestinal polypeptide

Protein(?)
Latrotoxin
CD55

Family 3

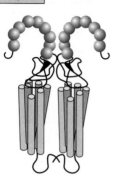

Amino acid
glutamate
GABA

Calcium
extracellular calcium-sensing receptor

Putative pheromone receptors

Subfamily III can be further divided into two groups, the vertebrate opsins and the neuro-peptide receptors such as receptors for tachykinin, cholecystokinin, neuropeptide Y, and bombesin.

Subfamily IV is a collection of receptors for autocrine, paracrine, and endocrine factors which include peptides (vasopressin, angiotensine, opioids, somatostatin, chemokine), glyco-protein hormones (LH/CG, TSH, FSH), eicosanoids, proteases, nucleotides, and platelet activating factor (PAF).

Members of these GPCR subfamily are also involved in immune chemotaxis by interacting with formylated peptides and C5a. Large extracellular domains (ectodomain) are rare in GPCRs of Family 1. One example comes from the glycoprotein hormone receptors in which the ectodomain is composed of leucine-rich repeats. Recent studies indicated the evolution of an expanding group of homologous leucine-rich repeat-containing GPCRs also in Family 1 (Kudo *et al.* 2000).

1.1.2 Family 2 GPCRs

The secretin receptor family or the glucagon/VIP/calcitonin receptor family (Family 2) com-prises receptors for secretin, glucagon, VIP, calcitonin, PTH, PTHrP, glucagon-like peptide 1, gastric inhibitory polypeptide, growth hormone-releasing hormone, corticotropin-releasing factor, and pituitary adenylate cyclase-activating peptide (see Part 3). The GPCR Family 2, the second largest family, recruits about 60 members and is characterized not only by the lack of the structural signature sequences present in the Family 1, but also by the presence of a large N-terminal ectodomain. The long and complex disulfide-bonded amino-terminal ecto-domain of these receptors plays an important role in agonist binding. Six cysteine residues within the N terminus, two cysteine residues connecting the e1 and e2 loops and about a dozen residues within the TMD core are well conserved among members of this family (Ulrich *et al.* 1998). Typically these receptors couple to more than one G protein, with the G_s/adenylyl cyclase pathway predominant and calcium signalling is also observed.

Recently, cloned GPCRs such as the α-latrotoxin receptor and Ig-Hepta reveal sequences with similarity to Family 2 GPCRs within the TMD region, but are unusual in so far as they contain large and complex extracellular domains those forming a subfamily within Family 2. The first reports of sequences related to Family 2 GPCRs followed the isolation of cDNA clones encoding EMR1 (EGF-module-containing mucin-like hormone receptor 1), F4/80, and CD97. The ectodomains are composed of various protein domains such as EGF domains, cadherin repeats, and thrombospondin type-1 repeats (reviewed in Stacey *et al.* 2000). Although the physiological functions of this subfamily are largely unknown, the acquisition of such extracellular domains leads to the possibility that members of this subfamily possess cell migration and adhesion properties. The *Drosophila* mutant *methuselah*

Fig. 1.1 Classification of GPCR based on the chemical nature of their ligands. Three main families can be distinguished when comparing their amino acid sequences. Except for the global structure of seven TMDs GPCRs from different families share no sequence similarity. The categorization shown here is in respect to their ligand preference and does not reflect the phylogenetic ancestry. Family 1 recruits receptors activated by ligands diverse as biogenic amines, proteases, light, and glycoprotein hormones. Family 2 comprises receptors for peptides and proteins. Dimerization appears to be essential for function of Family 3 receptors.

(*mth*) was identified from a screen for single gene mutations that extended average lifespan. The protein affected by this mutation is closely related to GPCRs of the secretin receptor family. The recently resolved 2.3 Å-resolution crystal structure of the *mth* ectodomain shows a three-domain architecture (West *et al.* 2001).

1.1.3 Family 3 GPCRs

Family 3 recruits about two dozens GPCRs such as metabotropic glutamate receptors (mGluR), the calcium-sensing receptor, and GABA$_B$ receptors, but also potential taste, pheromone, and olfactory receptors. Like in Family 2, these receptors possess large ectodomains responsible for ligand binding. The homology of the ectodomain of mGluR to the leucine/isoleucine/valine-binding protein (LIVBP) and other bacterial periplasmic binding proteins that mediate the transport of amino acids in prokaryotes is obvious. Pioneering studies with GABA$_B$ receptors implicate that GPCR may exist as dimers and that dimerization may potentially play important roles in the function of these receptors (see Chapter 4.3). Recently, crystal structures of the extracellular ligand-binding region of mGluR1—in a complex with glutamate and in two non-liganded forms—have been resolved showing disulphide-linked homodimers (Kunishima *et al.* 2000). There was also first evidence that G proteins might not always be necessary for recruitment of signalling molecules to the receptor. Several metabotropic receptors, including members of the mGluR subfamily have proline-rich intracellular domains, suitable for direct interaction with SH2- and SH3-containing proteins. Such direct protein–protein contacts form the structural basis of potential G protein-independent signalling cascades.

Family 3 GPCRs are defined as a group of receptors comprising at least three different subfamilies that share $\geq$20% amino acid identity over their seven membrane-spanning regions. Subfamily I includes the metabotropic glutamate receptors, mGluRs 1-8, which are receptors for the excitatory neurotransmitter glutamate and are widely expressed in the central nervous system (see Chapters 4.2 and 4.3). Subfamily II contains at least two types of receptors: the calcium-sensing receptor and a recently discovered, multigene subfamily of putative pheromone receptors. Subfamily III includes a subfamily of receptors, the GABA$_B$ receptors, that bind and are activated by the inhibitory neurotransmitter GABA.

1.1.4 Miscellaneous GPCRs

Three additional families encompass pheromone receptors, *Dictyostelium* cAMP receptors and proteins of the frizzled/smoothened receptor group. These receptors display a heptahelical transmembrane organisation, but for most of these receptors the terminology 'G protein-coupled receptor' is controversially discussed because the relevance of G protein coupling for receptor's signal transduction was not or not convincingly demonstrated.

The 'orphan' GPCRs are cloned GPCRs that bind unknown ligands. More than 200 'orphan' GPCRs, not including the olfactory GPCRs, have been discovered so far. In most cases, the extent of sequence homology is insufficient to assign these 'orphan' receptors to a particular receptor subfamily. Once the sequence of a GPCR is known, understanding the function of the encoded protein becomes a task of paramount importance. Consequently, reverse molecular pharmacological and functional genomic strategies are being employed to identify the activating ligands of the cloned receptors. The reverse molecular pharmacological methodology includes expression of orphan GPCRs in mammalian cells and screening these cells for a functional response to cognate or surrogate agonists present in biological extract preparations or peptide and compound libraries (Debouck and Metcalf 2000). Many new transmitter/receptor systems have been discovered recently and their physiological functions and potential relevance in human disease are currently being analysed (see Chapters 9 and 10).

GPCRs are not only encoded by eukaryotic genes but also by viral genes. To date, 18 putative GPCRs have been identified within herpes and pox viruses. Involvement of GPCRs in the pathophysiologic role of viruses has been impressively demonstrated for the Kaposi's sarcoma-associated herpes virus (KSHV) receptor and so-called UL78 gene family found in the cytomegalovirus (Oliveira and Shenk 2001). The KSHV-GPCR, closely related to chemokine and interleukin receptors, was found to agonist-independently activate the $G_{q/11}$/PLC signal transduction pathway (Arvanitakis *et al.* 1997).

1.2 Pharmacological diversity of GPCRs

In the post-genome (proteome) era long established views of receptor pharmacology are changing since modifications of GPCR structure and function can contribute to the pharmacological diversity found for products of a single GPCR gene. The changes in GPCR structure and function can occur at different levels. For example, gene duplication events that have led to multiple receptor subtypes are the cause of a considerable functional diversity found in one transmitter system. Moreover, tissue-specific splicing, RNA editing, and variations in post-translational modifications can multiply the products derived from a single GPCR gene (Fig. 1.2 and also see Chapter 2). In addition, tissue- or cell-specific expression of effectors can modify the ligand preference and signal transduction capabilities of GPCRs, so

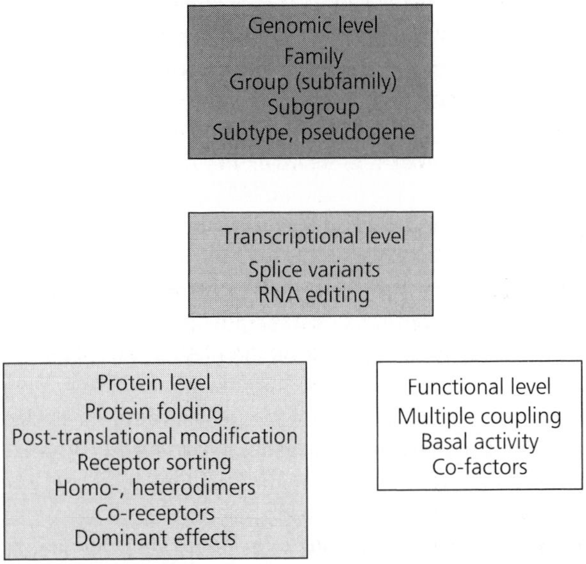

Genomic level
Family
Group (subfamily)
Subgroup
Subtype, pseudogene

Transcriptional level
Splice variants
RNA editing

Protein level
Protein folding
Post-translational modification
Receptor sorting
Homo-, heterodimers
Co-receptors
Dominant effects

Functional level
Multiple coupling
Basal activity
Co-factors

Fig. 1.2 Molecular basis of the functional diversity of GPCRs. Functional diversity of GPCRs is determined at different levels. The molecular mechanisms underlying this phenomenon include gene duplication, RNA and post-translational modifications as well as interaction with various co-factors and other GPCR molecules.

that specificity and function of a given GPCR can vary when expressed in a different cellular environment (Chapter 7).

1.2.1 Splice variants

Recent molecular characterization of cloned protein genes draws attention to alternative splicing as a source of structural and functional diversity. The amino acid sequences of most GPCRs are encoded by intronless single-copy genes (regarding the coding region). However, a number of GPCR genes show an exon/intron assembly of their coding regions, as described for rhodopsin, some amine and peptide receptors, and GPCRs with a large extra-cellular domain (glycoprotein hormone receptors, many Family 2 GPCRs). The existence of introns in GPCR genes provides the potential for additional diversity by virtue of alternative splicing events which may generate distinct receptor isoforms. For example, pharmacolo-gical and molecular biological studies have resulted in the cloning of cDNAs encoding four EP prostanoid receptors. The cloning of these receptors has revealed further heterogeneity due to alternative mRNA splicing. Specifically, eight human EP_3 receptor isoforms have been identified which differ only in their C termini (Pierce and Regan 1998). It should be noted that a tissue-specific occurrence of distinct splice variants has been described, for example, for the PACAP receptor (Chatterjee et al. 1996) and the corticotropin releasing factor (CRF) receptor (Ardati et al. 1999).

The genomic intron/exon structure and even the number of subtypes of a given GPCR are not necessarily conserved among species. Extensive studies on opsin genes have shown that introns in the coding region can appear and disappear during evolution. For example, human rhodopsin is encoded by four exons, but in some fish species the coding region for rhodopsin is intronless (Venkatesh et al. 1999). At least in principle, this implicates species-specific differences in the isoform pattern generated by alternative splicing events.

Two functional types of GPCR splice variants can be distinguished. First, usage of an alternative splice site can generate a functional receptor as demonstrated for a large num-ber of GPCRs such as the $mGluR_1$ (Prezeau et al. 1996) and the D_2 dopamine receptor (Seeman et al. 2000). In most cases, the divergence between receptor isoforms is limited to the C-terminal tail, a region involved in internalization, down-regulation and interactions with other proteins. As shown for the prostanoid EP_3 and $5-HT_4$ receptor isoforms, altern-ative splice products can vary in their basal activity when expressed in vitro. The extent of constitutive activity was found to be reversally correlated with the length of the C-terminal portion of the splice variants (Jin et al. 1997; Claeysen et al. 1999). Second, an improper splicing event can produce a non-functional receptor protein that may display dominant negative effects on the wild-type receptor. It has been demonstrated that the expression of a truncated isoform of the gonadotropin-releasing hormone (GnRH) receptor can decrease the signalling efficacy of the full-length receptor by reducing its cell surface expression levels (Grosse et al. 1997). This dominant negative effect was highly specific for the GnRH receptor and was probably due to heterocomplex formation between the two proteins. One may speculate that co-expression of truncated receptor isoforms may modulate the gon-adotropes' responsiveness to GnRH and thus contribute to the fine tuning of gonadotropin release in vivo. Similarly, the ability of an EP_1 receptor isoform to inhibit signalling by EP_1 as well as EP_4 receptors can be explained by complex formation between these different receptors (Okuda-Ashitaka et al. 1996). Impaired insertion of the wild-type receptor into the plasma membrane is a common mechanism underlying the dominant negative effects of co-expressed mutant or truncated receptors. It was demonstrated that a naturally occurring allele coding for a truncated CCR5 chemokine receptor which functions as a co-receptor for infection by primary M-tropic HIV-1 strains exerts a dominant negative effect on the viral env protein-mediated cell fusion (Samson *et al.* 1996). It was later shown that the truncated receptor complexes with the wild-type CCR5 and that this interaction retains CCR5 in the endoplasmic recticulum (ER) resulting in reduced cell surface expression (Benkirane *et al.* 1997). In addition, defective intracellular transport due to the formation of misfolded complexes between wild-type and mutant rhodopsin in the ER is held responsible for the dominant effect of one mutant allele in case of retinal degeneration in *Drosophila* (Colley *et al.* 1995). It should be noted that dominant negative effects of mutant GPCRs on wild-type receptor function can also be caused by other mechanisms such as titrating G proteins away from the wild-type receptor (Leavitt *et al.* 1999). Thus, expression of truncated or modified receptor proteins may highlight a novel principle of specific modulation of GPCR function.

1.2.2 RNA editing

RNA editing is a co- or post-transcriptional process in which select nucleotide sequences in RNA are altered from that originally encoded in the genome. Double-stranded RNA-specific adenosine deaminases convert adenosine residues to inosine in messenger RNA precursors (pre-mRNA). Their main physiological substrates are pre-mRNAs. Extensive analysis of cDNAs from 5-HT$_{2C}$ receptor reveals post-transcriptional modifications indicative of adenosine-to-inosine RNA editing (Burns *et al.* 1997). RNA transcripts encoding the 5-HT$_{2C}$ receptor undergo adenosine-to-inosine RNA editing events at up to five specific sites. Interestingly, reduced G protein-coupling efficiency for the edited isoforms is primarily due to silencing of the constitutive activity of the non-edited 5-HT$_{2C}$ receptor (Niswender *et al.* 1999). No further example of modified GPCR functions by mRNA editing has been reported yet.

1.2.3 Pseudogenes

A pseudogene is a sequence which is present in the genome of a given population and is typically characterized by close similarities to one or more paralogous genes, yet is non-functional. This lack of function is a result of either failure of transcription or translation, or production of a protein that does not have the same functional repertoire as the protein encoded by the normal paralog gene. Initial interpretation of the sequence data from human genome indicated that 10–20% of the coding sequences were pseudogenes (Dunham *et al.* 1999; Venter *et al.* 2001). There is no exact information about the number of GPCR pseudogenes but careful estimation suggests that approximately 17% of all GPCR coding sequences in the human genome are functionally inactive. The majority of vertebrate pseudogenes are a result of retrotransposition of transcripts derived from genes that encode functional proteins. By contrast, human olfactory receptor pseudogenes have been distributed to most of the human chromosomes by duplication of genomic DNA. For more than 900 identified olfactory receptors the sequence of at least 63% is disrupted by what appears to be a random process of pseudogene formation (Glusman *et al.* 2001). Consideration of how pseudogenes are formed suggests that most are unlikely to be transcribed, but pseudogene transcripts can nevertheless be identified. For example, the human 5-HT$_7$ receptor pseudogene transcripts can be identified in tissues such as kidney and liver, whereas the functional 5-HT$_7$ receptor transcripts cannot be detected (Olsen and Schechter 1999). The functional relevance of GPCR pseudogene transcripts remains unclear.

1.2.4 Single nucleotide polymorphisms

A polymorphic locus is one whose alleles or variants are such that the most common variant among them occurs with less than 99% frequency in the population at large, for example, if the locus is biallelic, the rarer allele must occur with a frequency greater than 1% in the population. The availability of a reference sequence of the human genome provides the basis for studying the nature of sequence variation, particularly single nucleotide polymorphisms (SNPs), in human populations. SNPs occur at a frequency of approximately 0.5–1 SNP/kb throughout the genome when the sequence of individuals is compared (Mullikin *et al.* 2000; The genome international sequencing consortium 2001; Venter *et al.* 2001). SNP typing is a powerful tool for genetic analysis because sequence variants are responsible for the genetic component of individuality, disease susceptibility, and drug response. The latter point will have an important impact on drug design, therapeutic regimes and side effects. This field just starts to develop in GPCR research. To date, a large number of variants in GPCR genes has been identified (for review see Rana *et al.* 2001). For example, the most prevalent SNP in the μ opioid receptor gene is a nucleotide substitution at position 118, displaying an allelic frequency of approximately 10%. The resulting amino acid change at a putative N-glycosylation site (N40D) of the μ opioid receptor increases the binding affinity and the potency of β-endorphin and may have implications for normal physiology, therapeutics, and vulnerability to develop addictive diseases (Bond *et al.* 1998). With growing efforts in identifying SNPs, it is likely that mutation-induced alterations in receptor function will become a major focus of future drug-development efforts (Kopin *et al.* 2000; Brodde *et al.* 2001).

1.3 GPCR structure and architecture

Currently, a high-resolution structure of an entire GPCR is available only for bovine rhodopsin (Palczewski *et al.* 2000), because of the difficulties inherent in producing, purifying, and crystallizing other GPCRs. As shown in Fig. 1.3, the mostly α-helical TMDs are arranged in a closely packed bundle forming the transmembrane receptor core. The N terminus of the polypeptide is located in the extracellular space, whereas the C terminus shows an intracellular localization. The seven transmembrane helices are connected by six alternating intracellular (i1-i3) and extracellular (e1-e3) loops. As predicted by nuclear magnetic resonance (NMR), in site-directed spin-labelling and mutagenesis studies most TMDs show a cytosolic α-helical extension (Altenbach *et al.* 1996; Farahbakhsh *et al.* 1995; Schulz *et al.* 2000*b*; Yeagle *et al.* 1997). The current rhodopsin model still lacks detailed structural information of some parts of the connecting loops and the C terminus.

1.3.1 Helical arrangement and helical structure

Like other polytopic membrane proteins, GPCRs are partially buried in the non-polar environment of the lipid bilayer by forming a compact bundle of transmembrane helices. The correct orientation and integration of the polypeptide chain is guided by a complex translocation apparatus residing in the ER. Two different folding stages can be distinguished following an initial translocation of the receptor N terminus into the ER lumen. In stage I,

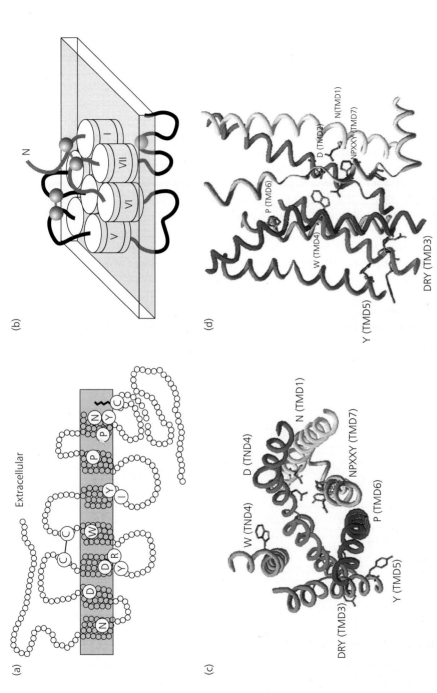

Fig. 1.3 Structure of Family 1 GPCRs. A two-dimensional model of Family 1 GPCRs (a) and its organization in the plasma membrane (b) are shown. The ring-like arranged seven TMDs assemble in a counterclockwise fashion as viewed from the extracellular surface. Some of the highly conserved residues are shown in enlarged circles (a). The e1 and the e2 loops, in some cases also the N terminus and the e3 loop, are linked by disulfide bonds (b). The recent crystal structure of rhodopsin (2.8 Å resolution) shows the orientation of the TMDs relative to each other (Palczewski *et al.* 2000). Positions of key residues are indicated in the three-dimensional rhodopsin structure viewed from extracellular (c) or laterally (d). (See Plate 1.)

hydrophobic α-helices are established across the lipid bilayer, and protein folding is predominantly driven by the hydrophobic effect. The TMDs adopt a secondary structure in order to minimize the polar surface area exposed to the lipid environment with the result that hydrophobic amino acids face the lipid bilayer and that more hydrophilic amino acid residues are orientated towards the core crevice of the TMD bundle. In stage II, a functional tertiary structure is formed by establishing specific helix–helix interactions, leading to the tightly packed, ring-like structure of the TMD bundle.

In the early stage of GPCR structure/function analysis, investigators used the structure of bacteriorhodopsin, a prokaryotic ion pump with structural similarities to the GPCR superfamily, as a scaffold for topographical models of the transmembrane core of GPCRs (Baldwin 1994; see Chapter 3 for more detail). The identification of specific interhelical contact sites was required to provide information about the relative orientation of the different helices towards each other. To determine the structural determinants which actually contribute to specific helix–helix interactions, chimeric receptors were generated. Studies with chimeric muscarinic receptors provided the first experimental evidence as to how TMD1 and TMD7 are oriented relative to each other and also strongly suggested that the TMD helices in muscarinic receptors are arranged in a counterclockwise fashion as viewed from the extracellular membrane surface (Liu *et al.* 1995). Functional analysis of artificial metal ion-binding sites (Elling *et al.* 1995) and disulfide bonds (Farrens *et al.* 1996) as well as spectroscopic approaches (Beck *et al.* 1998) allowed the identification of distinct amino acid residues that are involved in helix–helix contacts and the relative orientation of single TMDs to each other. Finally, the proposed helical arrangement was confirmed by structural data from rhodopsin crystallography highlighting the feasibility of mutagenesis approaches. Interestingly, the overall α-helical character of TMDs is often disrupted by non-α-helical components, such as intrahelical kinks (often due to residues other than proline), 3_{10}-helices and π-helices (Riek *et al.* 2001).

1.3.2 Conserved structural features

Most of the current knowledge about structure/function relationships of GPCRs is based on studies with rhodopsin and other members of the Family 1 of GPCRs and the mechanics of receptor activation are covered in more detail in Chapter 3. Only a few critical amino acids have been preserved during evolution of the rhodopsin-like GPCR family (Fig. 1.3). Despite an evolutionary conservation mutational alteration of some conserved amino acid residues does not always have the same functional consequence.

The majority of Family 1 and also Family 2 GPCRs contains a conserved pair of extracellular cysteine residues linking the first and second extracellular loops via a disulfide bond. Numerous functional analyses of mutant GPCRs in which the cysteine residues were replaced by other amino acids have shown that this disulfide bond may be critical for receptor signalling (Wess 1997). But in some receptors this disulfide bond is required to maintain more distinct functions. Systematic mutagenesis studies of the conserved cysteine residues in several GPCRs showed that disruption of the disulfide bond does not influence the receptor's ability to activate G protein, but interferes with high affinity ligand binding and receptor trafficking (Le Gouill *et al.* 1997; Perlmann *et al.* 1995; Schulz *et al.* 2000a; Zeng *et al.* 1999). Although the disruption of the disulfide bond, the receptor core structure appears to remain intact, allowing receptor function. Consistent with this notion, some GPCRs, for example, receptors for sphingosine 1-phosphate and lysophosphatidic acid, lack the conserved extracellular

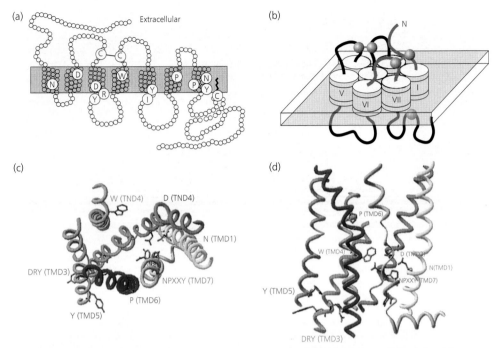

Plate 1 Structure of Family 1 GPCRs. A two-dimensional model of Family 1 GPCRs (a) and its organization in the plasma membrane (b) are shown. The ring-like arranged seven TMDs assemble in a counterclockwise fashion as viewed from the extracellular surface. Some of the highly conserved residues are shown in enlarged circles (a). The e1 and the e2 loops, in some cases also the N terminus and the e3 loop, are linked by disulfide bonds (b). The recent crystal structure of rhodopsin (2.8 Å resolution) shows the orientation of the TMDs relative to each other (Palczewski et al. 2000). Positions of key residues are indicated in the three-dimensional rhodopsin structure viewed from extracellular (c) or laterally (d).

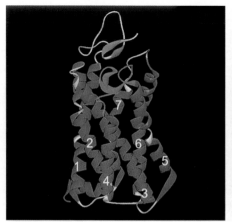
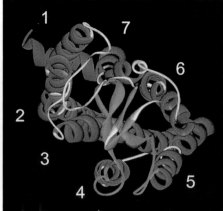

Plate 2 The high-resolution structure of rhodopsin seen from the side (*left panel*) and from the top (*right panel*) (Palczewski *et al.* 2000). The seven transmembrane helices (indicated by numbers) are organized in a counterclockwise fashion (as seen from the extracellular side in the right panel). A buried ligand binding crevice containing the covalently attached chromophore, retinal, is formed between the seven helices. The second extracellular loop 2 (ECL2, show in green) dives into the transmembrane domain and forms a plug in the binding crevice.

Plate 3 Characterization of agonist-induced conformational changes at the cytoplasmic side of TM6 in the β_2 adrenergic receptor. His269$^{6.31}$, Lys270$^{6.32}$, Ala271$^{6.33}$, and Leu272$^{6.34}$ at the bottom of TM6 (Fig. 3.2) were one by one mutated to cysteines in a background mutant with a reduced number of reactive cysteines (2AR-Cys-min) (Jensen *et al.* 2001). Panels a, b, c, e, and f show time course experiments where fluorescence intensity is measured over time in response to 3×10^{-4} M (-)isoproterenol (ISO) followed by 10^{-4} M(-)alprenolol (ALP). Panel d shows an extracellular view of a receptor model with an illustrative set of the preferred conformations of the IANBD side chain covalently attached to the four substituted cysteines (color-coded). Note that IANBD attached to Cys271$^{6.33}$ (yellow) and Cys272$^{6.34}$ (purple) are facing the interior of the TM helix bundle, while IANBD attached to Cys269$^{6.31}$ (blue) and Cys270$^{6.32}$ (red) are oriented towards the lipid membrane. Panel G shows the proposed conformations of the inactive and active states of the β2AR. The inactive conformation of the receptor (left panel) is characterized by a highly kinked TM6 helix (blue) with the cytoplasmic end in close proximity to TM3 and the helix bundle. An illustrative set of the preferred conformations of the IANBD moiety covalently attached to the four substituted cysteines at the cytoplasmic side of TM6 is shown. The hypothetical active conformation (right panel) of the receptor in which the cytoplasmic side of TM6 is moved away arbitrarily from the helix bundle and upwards towards the hydrophobic region, marked by straight lines. This putative rearrangement of TM6 moves all four IANBD labelled residues upwards and outwards allowing them to penetrate further into the more hydrophobic region of the membrane/detergent micelles and away from the more hydrophilic polar headgroups as well as from the predicted more hydrophilic interior of the receptor protein. The movement can explain the observed shift for all four IANBD labelled cysteines towards a less polar environment upon receptor activation. Note that the movement of the cytoplasmic part of TM6 is shown to occur around the conserved proline kink but could as well involve a rigid body movement of the entire helix. However, our previous simulation of the TM6 helix indicated the possibility that the kink in the TM6 helix induced by Pro287$^{6.50}$could behave as a flexible hinge, which can modulate the movement of the cytoplasmic side of TM6 helix relative to the extracellular region.

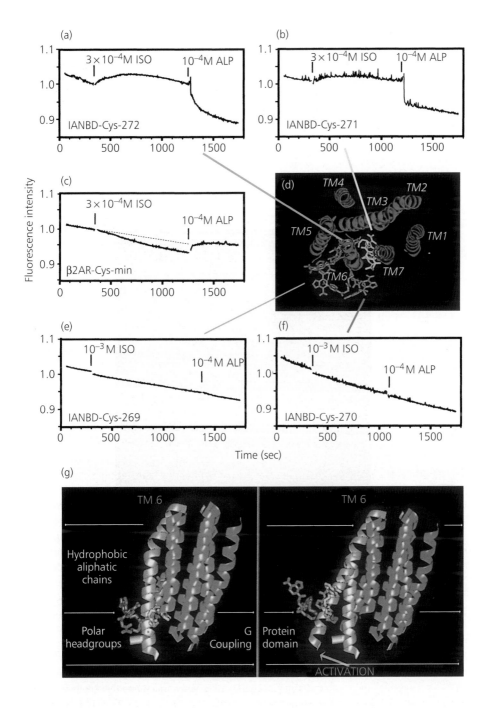

(a) 3×10^{-4}M ISO 10^{-4}M ALP
IANBD-Cys-272

(b) 3×10^{-4}M ISO 10^{-4}M ALP
IANBD-Cys-271

(c) 3×10^{-4}M ISO 10^{-4}M ALP
β2AR-Cys-min

(d) TM4 TM2 TM3 TM5 TM1 TM6 TM7

(e) 10^{-3}M ISO 10^{-4}M ALP
IANBD-Cys-269

(f) 10^{-3}M ISO 10^{-4}M ALP
IANBD-Cys-270

Fluorescence intensity

Time (sec)

(g) TM 6 TM 6
Hydrophobic aliphatic chains
Polar headgroups G Coupling
Protein domain
ACTIVATION

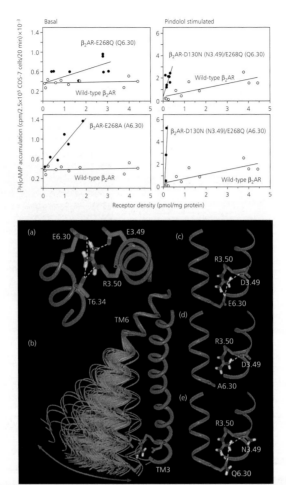

Plate 4 Activation of the β₂ adrenergic receptor involves disruption of an ionic lock between TM3 and TM6. The *upper left panels* show how mutation of Glu268[6.30] to Gln and Ala constitutively enhances basal receptor activity. The *upper right panels* show how the combined mutation of Asp130[3.49] and Glu268[6.30] dramatically enhances the efficacy of the weak partial agonist, pindolol. The *lower panel* shows molecular 3D-representations of the interaction of TM3-TM6 at their cytoplasmic ends and the effects of 6.30 mutations. The Cα traces are taken from the high-resolution structure of bovine rhodopsin. Except in panel a, the top of each panel shows the extracellular end and the bottom the intracellular end of the TMs. In panel a, an extracellular view of the high-resolution structure of rhodopsin showing the interaction between residues at the cytoplasmic ends of TM3 and TM6. In panel b, the Cα ribbons of TM3 and TM6 of rhodopsin are shown in purple. The simulated structures (light blue ribbons) are superimposed from position 1 to 17 onto the corresponding parts of TM6 (6.30 to 6.46). The blue arrow indicates the conformational space that a Pro-kink can assume relative to TM3. Panel c shows a closer view of the interactions of Glu6.30-Arg3.50-Asp3.49 in a model of the β₂ adrenergic receptor based on the rhodopsin structure. In c, the wild-type interactions are shown, with the closest contacts between Glu6.30-Arg3.50 and Arg3.50-Asp.49, which are within the distance range of an ionic interaction, shown in dashed lines. Panels d and e show the same view as in Panel c, but with alterations in putative interactions after mutation of Glu6.30 to Ala (d) or of Glu6.30 to Gln and of Asp3.49 to Asn (e).

Cys residues. Interestingly, many GPCRs including receptors for biogenic amines, peptides and many 'orphan' GPCRs contain a second conserved pair of extracellular cysteine residues linking the N terminus and third extracellular loop (see Fig. 1.3b). Mutational disruption of this disulfide bond results in a loss of high affinity binding of receptor ligands, suggesting a pivotal role of an N terminus/e3 loop-connecting disulfide bridge for proper receptor assembly (Ho *et al.* 1999; Hoffmann *et al.* 1999). In the crystal structure of rhodopsin, the N-terminal segment is located just below the e3 loop. Specific non-covalent contacts maintain the proper orientation between the rhodopsin N terminus and the extracellular loops so that an additional disulfide bridge like in other GPCRs is probably not required.

Most Family 1 GPCRs possess within their transmembrane core a number of highly conserved residues, such as an Asp residue in TMD2, a DRY motif at the TMD3/i2-transition, a Trp residue in TMD4, a Tyr residue in TMD5, a Pro residue in TMD6 and an N/DPXXY motif in TMD7 (Fig. 1.3a). For example, the DRY motif located at the boundary of TMD3 and the i2 loop is a highly conserved triplet of amino acid residues known to play an essential role in GPCR function (see Chapter 3). The crystal structure of rhodopsin proposes that the residues of the DRY motif participate in several hydrogen bonds with surrounding residues of TMD6. However, the fine structure of this region has not been finally resolved (Palczewski *et al.* 2000). It is also of note that several Family 1 GPCRs are known in which the acidic residue (Asp, Glu) within this motif is naturally substituted by His, Asn, Gln, Gly, Val, Thr, Cys, or Ser residues, questioning the general importance of a protonation event at this amino acid position for GPCR function. The fully conserved Arg residue in the DRY motive is considered to be a key residue in signal transduction of GPCRs. Replacement of the conserved Arg residue by different amino acids virtually abolished G protein coupling of many GPCR (Franke *et al.* 1992; Jones *et al.* 1995; Scheer *et al.* 1996). Therefore, the conserved Arg residue has been implicated as a central trigger of GDP release from the G protein α subunit (Acharya and Karnik 1996). Recent studies with mutants of the N-formyl peptide receptor, LHR and V2 vasopressin receptor (AVPR2) showed that, G protein coupling is only decreased, but not abolished after replacement of the Arg residue in the DRY motif (Arora *et al.* 1997; Schöneberg *et al.* 1998; Schulz *et al.* 1999) probably due to reduced receptor cell surface expression levels as response to constitutive arrestin-mediated desensitization (Barak *et al.* 2001).

The N/DPXXY motif at the cytoplasmic end of TMD7 is one of the most conserved regions among the Family 1 GPCRs (Fig. 1.3). It was shown that light activation of rhodopsin made an epitope including residues of the N/DPXXY motif accessible for an epitope-specific antibody, suggesting conformational changes of this sequence motif (Abdulaev and Ridge 1998). The Asn residue in this sequence motif is thought to play a crucial role in receptor activation and signal transduction specificity by involving small G proteins such as ARF and RhoA (Mitchell *et al.* 1998). In the cholecystokinin type B receptor, mutation of the Asn in the NPXXY motif to Ala had no effect on cell surface expression and high affinity ligand binding, but completely abolished G_q-mediated signalling (Gales *et al.* 2000). Mutational alteration of the conserved Pro residue in the N/DPXXY motif resulted in a complete loss of receptor function as demonstrated *in vivo* for the AVPR2 (Tajima *et al.* 1996) underlining the functional importance of this motif.

1.3.3 Structures involved in maintaining the inactive state

One of the truly exciting questions within the GPCR field is what structural changes a receptor undergoes during the transduction of extracellular signals. The classical 'ternary complex model' postulates that receptor activation leads to the agonist-promoted formation of an active 'ternary' complex of agonist, receptor, and G protein (de Lean *et al.* 1980). This model had to be extended in order to account for the fact that many receptors can activate G proteins in the absence of agonist (Lefkowitz *et al.* 1993). Based on these seminal observations, receptors are assumed to exist in an equilibrium between the inactive state R and the active state R* (see Chapters 3 and 8). *In vitro* mutagenesis studies with several GPCRs provided compelling evidence for the existence of intramolecular constraining determinants which stabilize the inactive receptor conformation and are described in detail within Chapter 3. Alteration of such intramolecular contact sites can lead to constitutive receptor activation.

Most of the mutations found to be responsible for constitutive receptor activity are located in the C-terminal portion of the i3 loop and within different TMDs. Interhelical salt bridges, as specific structural determinants stabilizing the inactive state, have been identified in rhodopsin (Robinson *et al.* 1992) and the α_{1B} adrenergic receptor (Porter *et al.* 1996). In the LHR, a tightly packed hydrophobic cluster and a specific H-bonding network formed between the cytoplasmic portions of TMD5 and TMD6 and the central regions of TMD6 and TMD7 is thought to maintain the inactive receptor conformation. Mutagenesis data also suggest that LHR activation is associated with the disruption of key interhelical side-chain interactions (Lin *et al.* 1997). The identification of such specific contact sites provides valuable information about the orientation of the different helices relative towards each other. Recent data with constitutive active LHR mutants implicate that in addition to interhelical interactions the inactive conformation of GPCRs is also stabilized by specific intrahelical structures (Schulz *et al.* 2000*a*; see Chapter 3).

1.3.4 **Post-translational modification**

The polypeptide chain of most GPCRs is post-translationally modified including N-glycosylation, palmitoylation, and phosphorylation. Potential N-glycosylation sites (NXS/T) and O-glycosylation sites (Sadeghi and Birnbaumer 1999; Nakagawa *et al.* 2001) are usually located within the extracellular N-terminal region, but are also found in the extracellular loops. The number and exact positions of glycosylation sites are usually not conserved among orthologs of different species. Some GPCRs, for example, the A_2 adenosine and the human α_{2B} adrenergic receptors, completely lack consensus sites for glycosylation in their N termini but are fully active in the absence of this post-translational modification. The functional relevance of post-translational modifications in GPCRs has been extensively studied in *in vitro* systems. It is well accepted that mutational disruption of potential N-glycosylation of most GPCRs has little effect on receptor function *in vitro* (Rands *et al.* 1990; Sadeghi and Birnbaumer 1999; Zeng *et al.* 1999). However, non-glycosylated receptors for parathyroid hormone and glycoprotein hormones in which glycosylation sites were mutated, are deficient in function (Zhang *et al.* 1995; Zhou *et al.* 2000). In the human calcium-sensing receptor, eight out of 11 potential N-linked glycosylation sites are actually utilized. Glycosylation of at least three sites is critical for cell surface expression of the receptor, but glycosylation does not appear to be critical for signal transduction (Ray *et al.* 1998). Interestingly, substitution mutations altering potential glycosylation sites in rhodopsin were described in families with retinitis pigmentosa (Bunge *et al.* 1993; Sullivan *et al.* 1993). These findings underline the functional importance of GPCR glycosylation *in vivo*.

Consensus acceptor phosphorylation sites for protein kinases A and C and potential receptor-specific kinase phosphorylation sites (multiple serine and threonine residues) are present in the i3 loop and the C-terminal domain. Studies with the β_2-adrenergic receptor indicated that the selectivity of receptor/G protein coupling can be regulated by receptor phosphorylation (Daaka *et al.* 1997). Most GPCRs also contain one or more conserved cysteine residues within their C-terminal tails (Fig. 1.3) which are modified by covalent attachment of palmitoyl or isoprenyl residues (Bouvier *et al.* 1995; Hayes *et al.* 1999). Depending on the specific GPCR examined, different effects on receptor phosphorylation, internalization, trafficking, and G protein-coupling profile have been described (reviewed in Bouvier *et al.* 1995; Wess 1998). However, several Family 1 GPCRs do not have Cys residues in their C-terminal tails for post-translational modification.

1.4 **GPCR assembly and oligomerization**

The successful reconstitution of adrenergic receptors from two fragments demonstrated that the integrity of the GPCR polypeptide chain is not required for proper receptor function (Kobilka *et al.* 1988). Based on these findings it has been speculated that GPCRs are composed of two or more independent folding domains. To test the hypothesis rhodopsin and the m3 muscarinic receptor were split in all three intracellular and extracellular loops. It was shown that except for a construct containing only TMD1, a significant portion of all N- and C-terminal receptor fragments studied was found to be inserted into the plasma membrane in the correct orientation even when expressed alone (Ridge *et al.* 1995; Schöneberg *et al.* 1995). Co-expression of some complementary receptor polypeptide pairs, generated by splitting GPCRs in their intra- and extracellular loops, resulted in receptors which were able to bind ligands and to mediate agonist-induced signal transduction (reviewed in Gudermann *et al.* 1997). It is noteworthy that all attempts to assemble functional receptor proteins from solubilized receptor fragments *in vitro* were unsuccessful (Schöneberg *et al.* 1997). This indicates that molecular chaperones that are likely to assist folding of the wild-type receptor protein may also play a role in facilitating complex formation. It has been shown that chaperone-dependent mechanisms are essential for proper folding of rhodopsin (Baker *et al.* 1994; Ferreira *et al.* 1996) and gonadotropin receptors (Rozell *et al.* 1998). The identity of the chaperones and the molecular mechanisms required for correct folding of other GPCRs remains to be elucidated.

The ability of functional complementation from receptor fragments is consistent with reports showing or suggesting that GPCRs can form dimers and oligomers. For several non-GPCR receptor families, such as receptor tyrosine kinases and kinase-associated cytokine receptors, agonist-induced receptor dimerization is required for initiating a signal transduction cascade. Initial evidence for GPCR dimerization came from crosslinking and photoaffinity labelling experiments with the GnRH receptor, LHR and muscarinic receptors (Conn *et al.* 1982; Avissar *et al.* 1983; Podesta *et al.* 1983). Numerous studies describing similar findings followed, but most reports of GPCR di- and oligomerization were based on co-immunoprecipitation studies. It has been argued that biochemical evidence from co-immunoprecipitation and Western blot experiments supporting the existence of GPCR oligomers is questionable, since solubilization of integral transmembrane proteins can cause artificial aggregation. However, as shown for epitope-tagged β_2 adrenergic, muscarinic and vasopressin receptors, the association is highly specific for a given receptor subtype giving rise only to homodimers (Hebert *et al.* 1998; Schulz *et al.* 2000a). In addition to investigations in transient expression systems, *in vivo* studies with D_2 and D_3 dopamine receptors (Nimchinsky *et al.* 1997; Zawarynski *et al.* 1998), somantostatin receptor type 5

($SSTR_5$)/D_2 dopamine receptors (Rocheville *et al.* 2000*a*), and rhodopsin (Colley *et al.* 1995) suggest the coexistence of receptor monomers and oligomeric complexes under physiological circumstances.

One question that arises from these studies is as to whether GPCR dimers are pre-formed or are induced in the presence of the appropriate ligand. Most co-immunoprecipitation data suggest the existence of oligomeric receptor complexes under basal conditions. Examining the biological relevance of GPCR homodimerization *in vivo*, Bouvier *et al.* 1995 used a bioluminescence resonance energy transfer (BRET) technique to study receptor–receptor interactions (Angers *et al.* 2000). It was shown that β_2-adrenergic receptors form constitutive homodimers that are expressed at the cell surface where they interact with agonists. Constitutive receptor association appears to be a general phenomenon since the yeast α-mating factor receptor forms dimers under basal conditions, as shown by a fluorescence resonance energy transfer (FRET) approach (Overton and Blumer 2000). On the other hand, there is also experimental evidence for an agonist driven oligomerization mechanism. Thus, the B_2 bradykinin receptor and the $CXCR_4$ receptor undergo receptor dimerization after ligand binding (AbdAlla *et al.* 1999; Vila-Coro *et al.* 1999).

There is growing evidence that GPCR not only exist in homodimeric structures but also in complexes formed by different GPCRs. Expression of the recombinant $GABA_{B1}$ receptor in COS cells resulted in a significantly lower agonist affinity when compared with native receptors. Interestingly, co-expression of the $GABA_{B1}$ receptor and the $GABA_{B2}$ receptor, a recently cloned novel $GABA_B$ receptor subtype, in *Xenopus* oocytes and HEK-293 cells led to efficient coupling to G protein-regulated inward rectifier K^+ channels (GIRKs) with an agonist potency in the same range as for $GABA_B$ receptors in neurons (Jones *et al.* 1998; White *et al.* 1998; Kaupmann *et al.* 1998). Encouraged by these studies, an ever growing number of heterodimeric complexes has been identified. For example, there is biochemical and pharmacological evidence that the κ and δ opioid receptors as well as μ and δ opioid receptors associate with each other. The complexes exhibit ligand binding and functional properties that are distinct from those of either receptor (Jordan and Devi 1999). Heterodimer formation was also observed for other receptor subtypes such as $5\text{-}HT_{1B}/5\text{-}HT_{1D}$ receptors and $SSTR_1/SSTR_5$ (Xie *et al.* 1999; Rocheville *et al.* 2000*b*). In a very recent study, hetero-oligomerization between the D_2 dopamine receptor and $SSTR_5$ was demonstrated, resulting in a novel receptor with enhanced functional activity (Rocheville *et al.* 2000*a*). The ability of GPCRs to heterodimerize provides a new mechanism by which a cell can fine-tune its responsiveness to a neurotransmitter via co-expression of distinct GPCR subtypes.

The molecular mechanisms and structural requirements which are responsible for GPCR oligomerization are only poorly understood. In the case of the $mGluR_5$ (Romano *et al.* 1996) and the calcium-sensing receptor (Bai *et al.* 1998; Ward *et al.* 1998), which are members of Family 3, disulfide bonds between the extracellular portions are of critical importance for receptor dimerization. In a recent crystallographic study, the homodimeric structure of the extracellular ligand-binding domain of the mGluR1 was resolved, providing direct evidence for a disulfide bond-stabilized dimer (Kunishima *et al.* 2000). In contrast, it was demonstrated that mutant calcium-sensing receptors without extracellular cysteines form dimers on the cell surface to a similar extent as observed for wild-type receptors (Zhang *et al.* 2001). Interestingly, the $GABA_{B2}$ receptor was initially discovered by a yeast two hybrid approach using the C terminus of the $GABA_{B1}$ receptor for screening a human brain cDNA library. Heterodimer formation was assumed to be mediated via a coiled-coil structure of the C termini of the two receptors (White *et al.* 1998). It was found later that a

C-terminal retention motif RXR(R) is masked by GABA$_B$ receptor dimerization allowing the plasma membrane expression of the assembled complexes (Margeta-Mitrovic *et al.* 2000). However, association of both GABA$_B$ receptors was demonstrated even in the absence of their cytoplasmic C termini (Calver *et al.* 2001; Pagano *et al.* 2001).

Most studies agree that homo- and heterodimers found for rhodopsin-like GPCRs represent non-covalent complexes. Thus, two structural models of dimer formation have been proposed for Family 1 GPCRs (Gouldson *et al.* 1998). In one dimeric structure, referred to as 'contact dimer', two tightly packed bundles of seven TMDs are positioned next to each other. The contact interface between the two monomeric receptors is assumed to be located between the lipid-orientated transmembrane receptor portions (Fig. 1.4). The so-called 'domain-swapped dimer' has been proposed to explain the reconstitution phenomenon observed with truncated and chimeric GPCRs (Maggio *et al.* 1993; Schulz *et al.* 2000*a*). In this dimer structure, the two receptor molecules fold around a hydrophilic interface by exchanging their N-terminal (TMDs1-5) and C-terminal (TMDs6-7) folding domains (Fig. 1.4). Attempting both hypothetical dimer structures data with the AVPR2 and D$_2$ dopamine receptors strongly support an oligomeric structure in which Family 1 GPCRs form contact oligomers by lateral interaction rather than by a domain-swapping mechanism (Schulz *et al.* 2000*a*; Lee *et al.* 2000). High resolution X-ray structure determinations of two heptahelical membrane proteins, the bacteriorhodopsin and the halorhodopsin, clearly show that both proteins assemble to trimers (Luecke *et al.* 1999; Kolbe *et al.* 2000). The proton pump bacteriorhodopsin shares structural similarities with the GPCR family including the assembly from multiple independent folding units. In the trimeric structure found in bacteriorhodopsin and halorhodopsin crystals, TMDs2-4 of the three molecules face each other forming an inner circle of TMDs. Structural data did not provide any support for a domain-swapping mechanism of oligomerization. Similarly, other polytopic membrane-spanning proteins which homo-oligomerize

Domain swapped dimer   Contact dimer

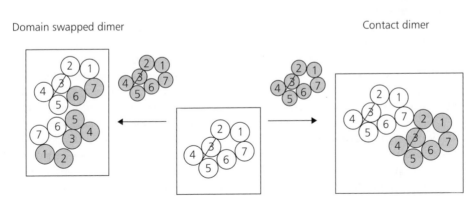

Fig. 1.4 Hypothetical structures of GPCR dimers. TMDs (numbered from 1 to 7) of GPCRs form a ring-like structure in a counter-clock wise fashion as viewed from extracellular. GPCRs are composed of at least two independent folding domains (TMDs1-5 and TMDs6-7) which are connected by the i3 loop. Accumulating evidence suggests that wild-type GPCRs can exist in dimeric complexes, and two structural models of dimer formation have been suggested (Gouldson *et al.* 1998). The contact interface of so-called 'swapped dimers' is recruited from the rearrangement of two independent folding domains of the individual receptor monomers. The ring-like TMD arrangement is still retained by the complementary exchange of the two folding domains. In contact dimers, a lateral interaction of the individual receptor molecules is assumed.

in order to build a functional complex, such as aquaporins, assemble via lateral interaction (Walz *et al.* 1997).

Taken together, current data strongly support oligomeric GPCR structures. Functional studies with mutant GPCRs provided evidence that oligomerization occurs by lateral interaction rather than by a domain-swapped mechanism. There is growing support for the idea that GPCR dimerization has consequences for physiologic receptor functions such as formation of receptor 'subtypes' with new ligand binding or signalling abilities.

1.5 Interactions between GPCRs and G proteins

Nearly two decades after the cloning of the first GPCRs, there are still many open questions relating to the mechanisms of GPCR/G protein interaction and the molecular elements determining G protein coupling specificity. Over the past few years, an increasing number of GPCRs with a broad G protein coupling profile has been identified (Gudermann *et al.* 1997). Structural elements determining signalling specificity are located in both the G protein and the receptor. Numerous *in vitro* mutagenesis studies have been performed with G protein α subunits to understand how coupling selectivity is achieved (see Chapter 4). The C terminus as well as the N terminus of the α subunit make important contributions to appropriate receptor/G protein recognition (Wess 1998).

The exact nature of G protein interaction sites within the receptor is currently unknown and may vary between the different GPCRs and G proteins. It is assumed that not only the intracellular loops but also the cytoplasmic sides of the TMDs participate in GPCR/G protein coupling. Indeed, peptides derived from the i3 loop/TMD6 junction can activate G proteins (Abell and Segaloff 1997; Varrault *et al.* 1994). Likewise, site-directed mutagenesis studies with GPCRs examined the structural elements that participate in G protein interactions and that determine the coupling profile of a given receptor. For example, the i1 loop of the formyl peptide receptor (Amatruda *et al.* 1995) and the cholecystokinin CCK_A receptor (Wu *et al.* 1997), the i2 loop of the V_{1A} vasopressin and $GABA_B$ receptors (Liu and Wess 1996; Robbins *et al.* 2001), and the i3 loop of the endothelin ET_B receptor (Takagi *et al.* 1995) have been demonstrated to participate in G protein activation. Taking advantage of chimeric GPCRs designed between structurally related receptor subtypes that are clearly distinguishable with regard to their signalling abilities, studies on muscarinic (Blüml *et al.* 1994; Blin *et al.* 1995) and vasopressin receptors (Erlenbach and Wess 1998) disclosed the importance of several distinct residues for selective G protein recognition. However, the established view of the importance of the intracellular loops in G protein interaction and specificity is challenged by recent studies with the TSHR and LHR. Large deletions or alanine replacement of most amino acid residues in the i3 loops did not abolish signal transduction, excluding a substantial participation of the i3 loop in G protein recognition at least in glycoprotein hormone receptors (Wonerow *et al.* 1998; Schulz *et al.* 1999, 2000*b*). In the m3 muscarinic receptor, a segment of 112 amino acids (central portion of the i3 loop) can be deleted without loss of receptor function (Schöneberg *et al.* 1995). Similarly, systematic reduction of the length of the i3 loop in the tachykinin NK-1 receptor revealed that most of the loop sequence can be substituted or even deleted without affecting ligand affinity or signal transduction (Nielsen *et al.* 1998). Further support for the notion that most of the i3 loop sequences are dispensable for G protein coupling comes from structural comparison of GPCR loop sequences showing that especially the i3 loop varies extremely in length and that there is no obvious sequence homology between GPCRs of similar coupling profiles.

Studies with receptor peptides that are able to activate G proteins directly indicate that the cytoplasmic extensions of the TMDs probably provide the surface for G protein interaction rather than the loops themselves (Abell and Segaloff 1997). Besides from serving simply as connectors between the TMDs, the intracellular loops may be involved in additional GPCR functions that have been discovered recently.

The ability of a GPCR to couple to more than one G protein subfamily can be conceived as a loss of specificity due to the absence of inhibitory determinants or as a gain of specific contact sites within the receptor molecule. Mutational analyses of GPCRs interacting with more than one G protein family provide evidence for both concepts. It has been demonstrated that point mutations can selectively abolish receptor coupling to one G protein subfamily (Surprenant et al. 1992; Biebermann et al. 1998). On the other hand, the coupling profile of a GPCR can be extended by mutational changes. Concomitantly with G_s activation, the LHR mediates fairly modest agonist-induced phosphoinositide breakdown via G_i recruitment. It was observed that several LHR mutations at the very N-terminal end of TMD6 profoundly enhanced agonist-induced IP accumulation, most likely via $G_{q/11}$ activation (Schulz et al. 1999). Two general mechanisms explaining multiple coupling events have been suggested—a parallel and a sequential G protein activation model. Lefkowitz and colleagues offered experimental evidence for a sequential G protein activation mechanism (Daaka et al. 1997). Agonist-induced activation of the β_2-adrenergic receptor results in cAMP formation via the G_s/adenylyl cyclase pathway followed by cAMP-dependent protein kinase-mediated receptor phosphorylation. Receptor phosphorylation represents a crucial molecular switch mechanism to allow for G_i-mediated ERK activation. These findings are of interest since several primarily G_s-coupled receptors are also capable of activating G_i. However, in the case of the LHR both signal transduction pathways, cAMP and IP formation, can be functionally separated by point mutations as shown for K583 in the e3 loop. Whereas ligand binding and agonist-induced phosphoinositide breakdown remained unaltered, cAMP formation was found to be abrogated (Gilchrist et al. 1996). These results show that ligand-independent adenylyl cyclase stimulation does not always represent a prerequisite for efficient coupling to the phospholipase C signalling pathway. Both signalling events appear to be independent, favouring a model of parallel activation of G proteins. These findings are consistent with the concept that GPCRs can exist in at least two distinct active conformations, R* and R**, which differ in their G protein coupling pattern. Based on this notion, it is also conceivable that different agonists can stabilize distinct ternary complexes. One may speculate that such pathway-selective agonists may represent the protagonists of a new class of therapeutic agents.

1.6 Future perspectives

Sequencing of the entire genomes of many higher organisms will be completed in the next few years. This information will allow the identification of new GPCRs which are more distantly related to known receptors and will help to understand the function of human GPCRs as well. Hundreds of 'orphan' GPCRs have been discovered so far. Defining the role of each GPCR and using this information to control GPCR activity therapeutically represents one of the major challenges for molecular medicine in the coming post-genomic era (see Chapters 9 and 10). Recent molecular characterization of cloned protein genes draws attention to alternative splicing and mRNA editing as a source of structural and functional diversity. Tissue-specific splicing has also been reported for several GPCRs and is likely to further increase the number of known GPCR isoforms. Undoubtedly, resolving the crystal structure of rhodopsin was a milestone in the history of GPCR research (Palczewski *et al.* 2000). This structural information provides a solid basis for further experimental structure/function relationship studies and for computed models of other GPCRs. Clearly, the success in crystallizing an integral membrane receptor will encourage projects to increase the resolution of the rhodopsin model and to analyze other GPCRs. Future attempts will aim at resolving the fine structure of different functional states in GPCR activation and the co-crystallization of a GPCR in complex with the G protein. These studies should eventually provide detailed structural information about the molecular architecture of the receptor/G protein interface. Identification of a growing number of receptor-binding proteins suggests alternative signalling mechanisms. Structural determinants within the C terminus and the intracellular loops provide potential interaction sites with other cellular proteins. The diversity of relevant interactions is likely to grow steadily, particularly since homologous and heterologous GPCR oligomerization has begun to emerge as a rather general phenomenon. The concept that monomeric GPCRs signal exclusively through heterotrimeric G proteins needs to be revised.

References

AbdAlla S, Zaki E, Lother H, and Quitterer U (1999). Involvement of the amino terminus of the B_2 receptor in agonist-induced receptor dimerization. *J Biol Chem* **274**, 26 079–84.

Abdulaev NG and Ridge KD (1998). Light-induced exposure of the cytoplasmic end of transmembrane helix seven in rhodopsin. *Proc Nat Acad Sci USA* **95**, 12 854–9.

Abell AN and Segaloff DL (1997). Evidence for the direct involvement of transmembrane region 6 of the lutropin/choriogonadotropin receptor in activating G_s. *J Biol Chem* **272**, 14 586–91.

Acharya S and Karnik SS (1996). Modulation of GDP release from transducin by the conserved Glu134-Arg135 sequence in rhodopsin. *J Biol Chem* **271**, 25 406–11.

Altenbach C, Yang K, Farrens DL, Farahbakhsh ZT, Khorana HG, and Hubbell WL (1996). Structural features and light-dependent changes in the cytoplasmic interhelical E-F loop region of rhodopsin: a site-directed spin-labeling study. *Biochemistry* **35**, 12 470–8.

Amatruda TT, Dragas-Graonic S, Holmes R, and Perez HD (1995). Signal transduction by the formyl peptide receptor. Studies using chimeric receptors and site-directed mutagenesis define a novel domain for interaction with G-proteins. *J Biol Chem* **270**, 28 010–3.

Angers S, Salahpour A, Joly E *et al.* (2000). Detection of β_2-adrenergic receptor dimerization in living cells using bioluminescence resonance energy transfer (BRET). *Proc Natl Acad Sci USA* **97**, 3684–9.

Ardati A, Goetschy V, Gottowick J *et al.* (1999). Human CRF2 alpha and beta splice variants: pharmacological characterization using radioligand binding and a luciferase gene expression assay. *Neuropharmacology* **38**, 441–8.

Arora KK, Cheng Z, and Catt KJ (1997). Mutations of the conserved DRS motif in the second intracellular loop of the gonadotropin-releasing hormone receptor affect expression, activation, and internalization. *Mol Endocrinol* **11**, 1203–12.

Arvanitakis L, Geras-Raaka E, Varma A, Gershengorn MC, and Cesarman E (1997). Human herpesvirus KSHV encodes a constitutively active G-protein-coupled receptor linked to cell proliferation. *Nature* **385**, 347–50.

Avissar S, Amitai G, and Sokolovsky M (1983). Oligomeric structure of muscarinic receptors is shown by photoaffinity labeling: subunit assembly may explain high- and low-affinity agonist states. *Proc Natl Acad Sci USA* **80**, 156–9.

Bai M, Trivedi S, and Brown EM (1998). Dimerization of the extracellular calcium-sensing receptor (CaR) on the cell surface of CaR-transfected HEK293 cells. *J Biol Chem* **273**, 23 605–10.

Baker EK, Colley NJ, and Zuker CS (1994). The cyclophilin homolog NinaA functions as a chaperone, forming a stable complex in vivo with its protein target rhodopsin. *EMBO J* **13**, 4886–96.

Baldwin JM (1994). Structure and function of receptors coupled to G proteins. *Curr Opin Cell Biol* **6**, 180–90.

Barak LS, Oakley RH, Laporte SA, and Caron MG (2001). Constitutive arrestin-mediated desensitization of a human vasopressin receptor mutant associated with nephrogenic diabetes insipidus. *Proc Natl Acad Sci USA* **98**, 93–8.

Beck M, Sakmar TP, and Siebert F (1998). Spectroscopic evidence for interaction between transmembrane helices 3 and 5 in rhodopsin. *Biochemistry* **37**, 7630–39.

Benkirane M, Jin DY, Chun RF, Koup RA, and Jeang KT (1997). Mechanism of transdominant inhibition of CCR5-mediated HIV-1 infection by ccr5Δ32. *J Biol Chem* **272**, 30 603–6.

Bermak JC, Li M, Bullock C, and Zhou QY (2001). Regulation of transport of the dopamine D1 receptor by a new membrane-associated ER protein. *Nature Cell Biol* **3**, 492–8.

Biebermann H, Schöneberg T, Schulz A *et al.* (1998). A conserved tyrosine residue (Y601) in transmembrane domain 5 determines of the human thyrotropin receptor serves as a molecular switch to determine G-protein coupling. *FASEB J* **12**, 1461–71.

Blin N, Yun J, and Wess J (1995). Mapping of single amino acid residues required for selective activation of $G_{q/11}$ by the m3 muscarinic acetylcholine receptor. *J Biol Chem* **270**, 17 741–8.

Blüml K, Mutschler E, and Wess J (1994). Identification of an intracellular tyrosine residue critical for muscarinic receptor-mediated stimulation of phosphatidylinositol hydrolysis. *J Biol Chem* **269**, 402–5.

Bond C, LaForge KS, Tian M *et al.* (1998). Single-nucleotide polymorphism in the human μ opioid receptor gene alters β-endorphin binding and activity: possible implications for opiate addiction. *Proc Natl Acad Sci USA* **95**, 9608–13.

Bouvier M, Chidiac P, Hebert TE, Loisel TP, Moffett S, and Mouillac B (1995). Dynamic palmitoylation of G-protein-coupled receptors in eukaryotic cells. *Meth Enzymol* **250**, 300–14.

Brodde OE, Buscher R, Tellkamp R, Radke J, Dhein S, and Insel PA (2001). Blunted cardiac responses to receptor activation in subjects with Thr164Ile β2-adrenoceptors. *Circulation* **103**, 1048–50.

Bunge S, Wedemann H, David D *et al.* (1993). Molecular analysis and genetic mapping of the rhodopsin gene in families with autosomal dominant retinitis pigmentosa. *Genomics* **17**, 230–3.

Burns CM, Chu H, Rueter SM *et al.* (1997). Regulation of serotonin-2C receptor G-protein coupling by RNA editing. *Nature* **387**, 303–8.

Cao TT, Deacon HW, Reczek D, Bretscher A and von Zastrow M (1999). A kinase-regulated PDZ-domain interaction controls endocytic sorting of the β2-adrenergic receptor. *Nature* **401**, 286–90.

Calver AR, Robbins MJ, Cosio C, Rice SQJ, Minton AL, Babb A *et al.* (2001). The C-terminal domain of GABA$_{B1}$ mediates intracellular trafficking, but is not required for receptor signalling. *J Neurosci* **21**, 1203–10.

Chatterjee TK, Sharma RV, and Fisher RA (1996). Molecular cloning of a novel variant of the pituitary adenylate cyclase-activating polypeptide (PACAP) receptor that stimulates calcium influx by activation of L-type calcium channels. *J Biol Chem* **271**, 32 226–32.

Claeysen S, Sebben M, Becamel C, Bockaert J, and Dumuis A (1999). A Novel brain-specific 5-HT$_4$ receptor splice variants show marked constitutive activity: role of the C-terminal intracellular domain. *Mol Pharmacol* **55**, 910–20.

Colley NJ, Cassill JA, Baker EK, and Zuker CS (1995). Defective intracellular transport is the molecular basis of rhodopsin-dependent dominant retinal degeneration. *Proc Natl Acad Sci USA* **92**, 3070–4.

Conn PM, Rogers DC, Stewart JM, Niedel J, and Sheffield T (1982). Conversion of a gonadotropin-releasing hormone antagonist to an agonist. *Nature* **296**, 653–5.

Daaka Y, Luttrell LM, and Lefkowitz RJ (1997). Switching of the coupling of the β2-adrenergic receptor to different G proteins by protein kinase A. *Nature* **390**, 88–91.

de Lean A, Stadel JM, and Lefkowitz RJ (1980). A ternary complex model explains the agonist-specific binding properties of the adenylate cyclase-coupled β-adrenergic receptor. *J Biol Chem* **255**, 7108–17.

Debouck C and Metcalf B (2000). The impact of genomics on drug discovery. *Annual Rev Pharmacol Toxicol* **40**, 193–207.

Dunham AR, Shimizu N, Roe BA *et al.* (1999). The DNA sequence of human chromosome 22. *Nature* **402**, 489–95.

Elling CE, Nielsen SM, and Schwartz TW (1995). Conversion of antagonist-binding site to metal-ion site in the tachykinin NK-1 receptor. *Nature* **374**, 74–7.

Erlenbach I and Wess J (1998). Molecular basis of V_2 vasopressin receptor/G_s coupling selectivity. *J Biol Chem* **273**, 26 549–58.

Farahbakhsh ZT, Ridge KD, Khorana HG, and Hubbell WL (1995). Mapping light-dependent structural changes in the cytoplasmic loop connecting helices C and D in rhodopsin: a site-directed spin labeling study. *Biochemistry* **34**, 8812–9.

Farrens DL, Altenbach C, Yang K, Hubbell WL, and Khorana HG (1996). Requirement of rigid-body motion of transmembrane helices for light activation of rhodopsin. *Science* **274**, 768–70.

Ferreira PA, Nakayama TA, Pak WL, and Travis GH (1996). Cyclophilin-related protein RanBP2 acts as a chaperone for red/green opsin. *Nature* **383**, 637–40.

Franke RR, Sakmar TP, Graham RM, and Khorana HG (1992). Structure and function in rhodopsin. Studies of the interaction between the rhodopsin cytoplasmic domain and transducin. *J Biol Chem* **267**, 14 767–74.

Gales C, Kowalski-Chauvel A, Dufour MN *et al.* (2000). Mutation of Asn-391 within the conserved NPXXY motif of the cholecystokinin B receptor abolishes G_q protein activation without affecting its association with the receptor. *J Biol Chem* **275**, 17 321–7.

Gilchrist RL, Ryu KS, Ji I, and Ji TH (1996). The luteinizing hormone/chorionic gonadotropin receptor has distinct transmembrane conductors for cAMP and inositol phosphate signals. *J Biol Chem* **271**, 19 283–7.

Glusman G, Yanai I, Rubin I, and Lancet D (2001). The complete human olfactory subgenome. *Genome Research* **11**, 685–702.

Gouldson PR, Snell CR, Bywater RP, Higgs C, and Reynolds CA (1998). Domain swapping in G-protein coupled receptor dimers. *Protein Engineering* **11**, 1181–93.

Grosse R, Schöneberg T, Schultz G, and Gudermann T (1997). Inhibition of gonadotropin-releasing hormone receptor signaling by expression of a splice variant of the human receptor. *Mol Endocrinol* **11**, 1305–18.

Gudermann T, Schöneberg T, and Schultz G (1997). Functional and structural complexity of signal transduction via G-protein-coupled receptors. *Ann Rev Neurosci* **20**, 399–427.

Hall RA, Premont RT, Chow CW *et al.* (1998). The β_2-adrenergic receptor interacts with the Na^+/H^+-exchanger regulatory factor to control Na^+/H^+ exchange. *Nature* **392**, 626–30.

Hamm HE (1998). The many faces of G protein signaling. *J Biol Chem* **273**, 669–72.

Hayes JS, Lawler OA, Walsh MT, and Kinsella BT (1999). The prostacyclin receptor is isoprenylated. Isoprenylation is required for efficient receptor-effector coupling. *J Biol Chem* **274**, 23 707–18.

Hebert TE, Loisel TP, Adam L, Ethier N, Onge SS, and Bouvier M (1998). Functional rescue of a constitutively desensitized β_2AR through receptor dimerization. *Biochem J* **330**, 287–93.

Ho HH, Du D, and Gershengorn MC (1999). The N terminus of Kaposi's sarcoma-associated herpesvirus G protein-coupled receptor is necessary for high affinity chemokine binding but not for constitutive activity. *J Biol Chem* **274**, 31 327–32.

Hoffmann C, Moro S, Nicholas RA, Harden TK, and Jacobson KA (1999). The role of amino acids in extracellular loops of the human P2Y1 receptor in surface expression and activation processes. *J Biol Chem* **274**, 14 639–47.

Jin J, Mao GF, and Ashby B (1997). Constitutive activity of human prostaglandin E receptor EP_3 isoforms. *Brit J Pharmacol* **121**, 317–23.

Jones KA, Borowsky B, Tamm JA *et al.* (1998). $GABA_B$ receptors function as a heteromeric assembly of the subunits $GABA_B R1$ and $GABA_B R2$. *Nature* **396**, 674–9.

Jones PG, Curtis CAM, and Hulme EC (1995). The function of a highly-conserved arginine residue in activation of the muscarinic M_1 receptor. *Eur J Pharmacol* **288**, 251–7.

Jordan BA and Devi LA (1999). G-protein-coupled receptor heterodimerization modulates receptor function. *Nature* **399**, 697–700.

Kaupmann K, Malitschek B, Schuler V *et al.* (1998). $GABA_B$-receptor subtypes assemble into functional heteromeric complexes. *Nature* **396**, 683–7.

Klein U, Ramirez MT, Kobilka BK, and von Zastrow M (1997). A novel interaction between adrenergic receptors and the α-subunit of eukaryotic initiation factor 2B. *J Biol Chem* **272**, 19 099–102.

Kobilka BK, Kobilka TS, Daniel K, Regan JW, Caron MG, and Lefkowitz RJ (1988). Chimeric α_2-, β_2-adrenergic receptors: delineation of domains involved in effector coupling and ligand binding specificity. *Science* **240**, 1310–6.

Kolbe M, Besir H, Essen LO, and Oesterhelt D (2000). Structure of the light-driven chloride pump halorhodopsin at 1.8 Å resolution. *Science* **288**, 1390–6.

Kopin AS, McBride EW, Gordon MC, Quinn SM, and Beinborn M (1997). Inter- and intraspecies polymorphisms in the cholecystokinin-B/gastrin receptor alter drug efficacy. *Proc Natl Acad Sci USA* **94**, 11 043–8.

Kudo M, Chen T, Nakabayashi K, Hsu SY, and Hsueh AJ (2000). The nematode leucine-rich repeat-containing, G protein-coupled receptor (LGR) protein homologous to vertebrate gonadotropin and thyrotropin receptors is constitutively active in mammalian cells. *Mol Endocrinol* **14**, 272–84.

Kunishima N, Shimada Y, Tsuji Y *et al.* (2000). Structural basis of glutamate recognition by a dimeric metabotropic glutamate receptor. *Nature* **407**, 971–7.

Le Gouill C, Parent JL, Rola-Pleszczynski M, and Stankova J (1997). Role of the Cys90, Cys95 and Cys173 residues in the structure and function of the human platelet-activating factor receptor. *FEBS Lett* **402**, 203–8.

Leavitt LM, Macaluso CR, Kim KS, Martin NP, and Dumont ME (1999). Dominant negative mutations in the α-factor receptor, a G protein-coupled receptor encoded by the STE2 gene of the yeast Saccharomyces cerevisiae. *Molecular General Genetics* **261**, 917–32.

Lee SP, O'Dowd BF, Ng GY *et al.* (2000). Inhibition of cell surface expression by mutant receptors demonstrates that D_2 dopamine receptors exist as oligomers in the cell. *Mol Pharmacol* **58**, 120–8.

Lefkowitz RJ, Cotecchia S, Samama P, and Costa T (1993). Constitutive activity of receptors coupled to guanine nucleotide regulatory proteins. *Trends in Pharmacol Sci* **14**, 303–7.

Lin Z, Shenker A, and Pearlstein R (1997). A model of the lutropin/choriogonadotropin receptor: insights into the structural and functional effects of constitutively activating mutations. *Protein Engineering* **10**, 501–10.

Liu F, Wan Q, Pristupa ZB, Yu XM, Wang YT, and Niznik HB (2000). Direct protein–protein coupling enables cross-talk between dopamine D5 and gamma-aminobutyric acid A receptors. *Nature* **403**, 274–80.

Liu J, Schöneberg T, van Rhee M, and Wess J (1995). Mutational analysis of the relative orientation of transmembrane helices I and VII in G protein-coupled receptors. *J Biol Chem* **270**, 19 532–9.

Liu J and Wess J (1996). Different single receptor domains determine the distinct G protein coupling profiles of members of the vasopressin receptor family. *J Biol Chem* **271**, 8772–8.

Luecke H, Schobert B, Richter HT, Cartailler JP, and Lanyi JK (1999). Structure of bacteriorhodopsin at 1.55 Å resolution. *J Mol Biol* **291**, 899–911.

Maggio R, Vogel Z, and Wess J (1993). Co-expression studies with mutant muscarinic/adrenergic receptors provide evidence for intermolecular crosstalk between G protein-linked receptors. *Proc Natl Acad Sci USA* **90**, 3103–7.

Manivet P, Mouillet-Richard S, Callebert J *et al.* (2000). PDZ-dependent activation of nitric-oxide synthases by the serotonin 2B receptor. *J Biol Chem* **275**, 9324–31.

Margeta-Mitrovic M, Jan YN, and Jan LY (2000). A trafficking checkpoint controls $GABA_B$ receptor heterodimerization. *Neuron* **27**, 97–106.

Mitchell R, McCulloch D, Lutz E *et al.* (1998). Rhodopsin-family receptors associate with small G proteins to activate phospholipase D. *Nature* 392, 411–4.

Mullikin JC, Hunt SE, Cole CG *et al.* (2000). An SNP map of human chromosome 22. *Nature* 407, 516–20.

Nakagawa M, Miyamoto T, Kusakabe R *et al.* (2001). O-Glycosylation of G-protein-coupled receptor, octopus rhodopsin. Direct analysis by FAB mass spectrometry. *FEBS Lett* 496, 19–24.

Nielsen SM, Elling CE, and Schwartz TW (1998). Split-receptors in the tachykinin neurokinin-1 system–mutational analysis of intracellular loop 3. *Eur J Biochem* 251, 217–26.

Nimchinsky EA, Hof PR, Janssen WGM, Morrison JH, and Schmauss C (1997). Expression of dopamine D_3 receptor dimers and tetramers in brain and in transfected cells. *J Biol Chem* 272, 29 229–37.

Niswender CM, Copeland SC, Herrick-Davis K, Emeson RB, and Sanders-Bush E (1999). RNA editing of the human serotonin 5-hydroxytryptamine 2C receptor silences constitutive activity. *J Biol Chem* 274, 9472–8.

Okuda-Ashitaka E, Sakamoto K, Ezashi T, Miwa K, Ito S, and Hayaishi O (1996). Suppression of prostaglandin E receptor signaling by the variant form of EP_1 subtype. *J Biol Chem* 271, 31 255–61.

Oldenhof J, Vickery R, Anafi M *et al.* (1998). SH3 binding domains in the dopamine D_4 receptor. *Biochemistry* 37, 15 726–36.

Oliveira SA and Shenk TE (2001). Murine cytomegalovirus M78 protein, a G protein-coupled receptor homologue, is a constituent of the virion and facilitates accumulation of immediate-early viral mRNA. *Proc Natl Acad Sci USA* 98, 3237–42.

Olsen MA and Schechter LE (1999). Cloning, mRNA localization and evolutionary conservation of a human 5-HT7 receptor pseudogene. *Gene* 227, 63–9.

Overton MC and Blumer KJ (2000). G-protein-coupled receptors function as oligomers in vivo. *Curr Biol* 10, 341–4.

Pagano A, Rovelli G, Mosbacher J *et al.* (2001). C-terminal interaction is essential for surface trafficking but not for heteromeric assembly of $GABA_B$ receptors. *J Neurosci* 21, 1189–202.

Palczewski K, Kumasaka T, Hori T *et al.* (2000). Crystal structure of rhodopsin: A G protein-coupled receptor. *Science* 289, 739–45.

Perlman JH, Wang W, Nussenzveig DR, and Gershengorn MC (1995). A disulfide bond between conserved extracellular cysteines in the thyrotropin-releasing hormone receptor is critical for binding. *J Biol Chem* 270, 24 682–5.

Peroutka SJ and Howell TA (1994). The molecular evolution of G protein-coupled receptors: focus on 5-hydroxytryptamine receptors. *Neuropharmacology* 33, 319–24.

Pierce KL and Regan JW (1998). Prostanoid receptor heterogeneity through alternative mRNA splicing. *Life Sci* 62, 1479–83.

Podesta EJ, Solano AR, Attar R, Sanchez ML, Molina Y, and Vedia L (1983). Receptor aggregation induced by antilutropin receptor antibody and biological response in rat testis Leydig cells. *Proc Natl Acad Sci USA* 80, 3986–90.

Porter JE, Hwa J, and Perez DM (1996). Activation of the α_{1b}-adrenergic receptor is initiated by disruption of an interhelical salt bridge constraint. *J Biol Chem* 271, 28 318–23.

Prezeau L, Gomeza J, Ahern S *et al.* (1996). Changes in the carboxyl-terminal domain of metabotropic glutamate receptor 1 by alternative splicing generate receptors with differing agonist-independent activity. *Mol Pharmacol* 49, 422–9.

Rana BK, Shiina T, and Insel PA (2001). Genetic variations and polymorphisms of g protein-coupled receptors: functional and therapeutic implications. *Annual Rev Pharmacol Toxicol* 41, 593–624.

Rands E, Candelore MR, Cheung AH, Hill WS, Strader CD, and Dixon RA (1990). Mutational analysis of β-adrenergic receptor glycosylation. *J Biol Chem* 265, 10 759–64.

Ray K, Clapp P, Goldsmith PK, and Spiegel AM (1998). Identification of the sites of N-linked glyco-sylation on the human calcium receptor and assessment of their role in cell surface expression and signal transduction. *J Biol Chem* **273**, 34 558–67.

Ridge KD, Lee SS, and Yao LL (1995). *In vivo* assembly of rhodopsin from expressed polypeptide fragments. *Proc Natl Acad Sci USA* **92**, 3204–8.

Riek RP, Rigoutsos I, Novotny J, and Graham RM (2001). Non-α-helical elements modulate polytopic membrane protein architecture. *J Mol Biol* **306**, 349–62.

Robinson PR, Cohen GB, Zhukovsky EA, and Oprian DD (1992). Constitutively active mutants of rhodopsin. *Neuron* **9**, 719–25.

Robbins MJ, Calver AR, Fillipov AK, Couve A, Moss SJ, and Pangalos MN (2001). The $GABA_{B2}$ subunit is essential for G protein coupling of the $GABA_B$ receptor heterodimer. *J Neurosci* **21**, 8043–52.

Rocheville M, Lange DC, Kumar U, Patel SC, Patel RC, and Patel YC (2000*a*). Receptors for dopamine and somatostatin: formation of hetero-oligomers with enhanced functional activity. *Science* **288**, 154–7.

Rocheville M, Lange DC, Kumar U, Sasi R, Patel RC, and Patel YC (2000*b*). Subtypes of the somatostatin receptor assemble as functional homo- and heterodimers. *J Biol Chem* **275**, 7862–9.

Romano C, Yang WL, and O'Malley KL (1996). Metabotropic glutamate receptor 5 is a disulfide-linked dimer. *J Biol Chem* **271**, 28612–6.

Rozell TG, Davis DP, Chai Y, and Segaloff DL (1998). Association of gonadotropin receptor precursors with the protein folding chaperone calnexin. *Endocrinology* **139**, 1588–93.

Sadeghi H and Birnbaumer M (1999). O-Glycosylation of the V2 vasopressin receptor. *Glycobiology* **9**, 731–7.

Samson M, Libert F, Doranz BJ *et al.* (1996). Resistance to HIV-1 infection in caucasian individuals bearing mutant alleles of the CCR-5 chemokine receptor gene. *Nature* **382**, 722–5.

Scheer A, Fanelli F, Costa T, De Benedetti PG, and Cotecchia S (1996). Constitutively active mutants of the α_{1B}-adrenergic receptor: role of highly conserved polar amino acids in receptor activation. *EMBO J* **15**, 3566–78.

Schöneberg T, Liu J, and Wess J (1995). Plasma membrane localization and functional rescue of truncated forms of a G protein-coupled receptor. *J Biol Chem* **270**, 18 000–6.

Schöneberg T, Sandig V, Wess J, Gudermann T, and Schultz G (1997). Reconstitution of mutant V_2 vasopressin receptors by adenovirus-mediated gene transfer. Molecular basis and clinical implication. *J Clin Invest* **100**, 1547–56.

Schöneberg T, Schulz A, Biebermann H *et al.* (1998). V_2 vasopressin receptor dysfunction in nephrogenic diabetes insipidus caused by different molecular mechanisms. *Human Mutation* **12**, 196–205.

Schulz A, Grosse R, Schultz G, Gudermann T, and Schöneberg T (2000*a*). Structural implication for receptor oligomerization from functional reconstitution studies of mutant V_2 vasopressin receptors. *J Biol Chem* **275**, 2381–9.

Schulz A, Bruns C, Henklein P *et al.* (2000*b*). Requirement of specific intrahelical interactions for stabilizing the inactive conformation of glycoprotein hormone receptors. *J Biol Chem* **275**, 37 860–9.

Schulz A, Schöneberg T, Paschke R, Schultz G, and Gudermann T (1999). Role of the third intracellular loop for the activation of gonadotropin receptors. *Mol Endocrinol* **13**, 181–90.

Seeman P, Nam D, Ulpian C, Liu IS, and Tallerico T (2000). New dopamine receptor, $D_{2(Longer)}$, with unique TG splice site, in human brain. *Brain Res Mol Brain Res* **76**, 132–41.

Stacey M, Lin HH, Gordon S, and McKnight AJ (2000). LNB-TM7, a group of seven-transmembrane proteins related to family-B G-protein-coupled receptors. *Trends in Biochem Sci* **25**, 284–9.

Sullivan LJ, Makris GS, Dickinson P *et al.* (1993). A new codon 15 rhodopsin gene mutation in autosomal dominant retinitis pigmentosa is associated with sectorial disease. *Arch Ophthalmol* **111**, 1512–7.

Surprenant A, Horstman DA, Akbarali H, and Limbird LE (1992). A point mutation of the β_2-adrenoceptor that blocks coupling to potassium but not calcium currents. *Science* **257**, 977–80.

Tajima T, Nakae J, Takekoshi Y *et al.* (1996). Three novel AVPR2 mutations in three Japanese families with X-linked nephrogenic diabetes insipidus. *Paedia Res* **39**, 522–6.

Takagi Y, Ninomiya H, Sakamoto A, Miwa S, and Masaki T (1995). Structural basis of G protein specificity of human endothelin receptors. A study with endothelinA/B chimeras. *J Biol Chem* **270**, 10 072–8.

Tang Y, Hu LA, Miller WE *et al.* (1999). Identification of the endophilins (SH3p4/p8/p13) as novel binding partners for the β_1-adrenergic receptor. *Proc Natl Acad Sci USA* **96**, 12 559–64.

The genome international sequencing consortium (2001). Initial sequencing and analysis of the human genome. *Nature* **409**, 860–921.

Ulrich CD II, Holtmann M, and Miller LJ (1998). Secretin and vasoactive intestinal peptide receptors: members of a unique family of G protein-coupled receptors. *Gastroenterology* **114**, 382–97.

Varrault A, Le Nguyen D, McClue S, Harris B, Jouin P, and Bockaert J (1994). 5-Hydroxytryptamine 1A receptor synthetic peptides. Mechanisms of adenylyl cyclase inhibition. *J Biol Chem* **269**, 16 720–5.

Venkatesh B, Ning Y and Brenner S (1999). Late changes in spliceosomal introns define clades in vertebrate evolution. *Proc Natl Acad Sci USA* **96**, 10 267–71.

Venter JC, Adams MD, Myers EW *et al.* (2001). The sequence of the human genome. *Science* **29**, 1304–51.

Vezza R, Habib A, and FitzGerald GA (1999). Differential signaling by the thromboxane receptor isoforms via the novel GTP-binding protein, Gh. *J Biol Chem* **274**, 12 774–9.

Vila-Coro AJ, Rodriguez-Frade JM, Martin De Ana A, Moreno-Ortiz MC, Martinez-A C, and Mellado M (1999). The chemokine SDF-1α triggers CXCR4 receptor dimerization and activates the JAK/STAT pathway. *FASEB J* **13**, 1699–710.

Walz T, Hirai T, Murata K *et al.* (1997). The three-dimensional structure of aquaporin-1. *Nature* **387**, 624–7.

Ward DT, Brown EM, and Harris HW (1998). Disulfide bonds in the extracellular calcium-polyvalent cation-sensing receptor correlate with dimer formation and its response to divalent cations *in vitro*. *J Biol Chem* **273**, 14 476–83.

Wess J (1997). G-protein-coupled receptors: molecular mechanisms involved in receptor activation and selectivity of G-protein recognition. *FASEB J* **11**, 346–54.

Wess J (1998). Molecular basis of receptor/G-protein-coupling selectivity. *Pharmacol Ther* **80**, 231–64.

West AP Jr, Llamas LL, Snow PM, Benzer S, and Bjorkman PJ (2001). Crystal structure of the ecto-domain of Methuselah, a Drosophila G protein-coupled receptor associated with extended lifespan. *Proc Natl Acad Sci USA* **98**, 3744–9.

White JH, Wise A, Main MJ *et al.* (1998). Heterodimerization is required for the formation of a functional GABA$_B$ receptor. *Nature* **396**, 679–82.

Wonerow P, Schöneberg T, Schultz G, Gudermann T, and Paschke R (1998). Deletions in the third intracellular loop of the thyrotropin receptor. A new mechanism for constitutive activation. *J Biol Chem* **273**, 7900–5.

Wu V, Yang M, McRoberts JA *et al.* (1997). First intracellular loop of the human cholecystokinin-A receptor is essential for cyclic AMP signaling in transfected HEK-293 cells. *J Biol Chem* **272**, 9037–42.

Xiao B, Tu JC, Petralia RS *et al.* (1998). Homer regulates the association of group 1 metabotropic glutamate receptors with multivalent complexes of homer-related, synaptic proteins. *Neuron* **21**, 707–16.

Xie Z, Lee SP, O'Dowd BF, and George SR (1999). Serotonin 5-HT$_{1B}$ and 5-HT$_{1D}$ receptors form homodimers when expressed alone and heterodimers when co-expressed. *FEBS Lett* **456**, 63–7.

Yeagle PL, Alderfer JL, and Albert AD (1997). Three-dimensional structure of the cytoplasmic face of the G protein receptor rhodopsin. *Biochemistry* **36**, 9649–54.

Zawarynski P, Tallerico T, Seeman P, Lee SP, O'Dowd BF, and George SR (1998). Dopamine D_2 receptor dimers in human and rat brain. *FEBS Lett* **441**, 383–6.

Zeng F-Y, Soldner A, Schöneberg T, and Wess J (1999). Putative disulfide bond in the m3 muscarinic acetylcholine receptor is essential for proper receptor trafficking but not for receptor stability and G protein coupling. *J Neurochem* **72**, 2404–14.

Zhang R, Cai H, Fatima N, Buczko E, and Dufau ML (1995). Functional glycosylation sites of the rat luteinizing hormone receptor required for ligand binding. *J Biol Chem* **270**, 21 722–8.

Zhang Z, Sun S, Quinn SJ, Brown EM, and Bai M (2001). The extracellular calcium-sensing receptor dimerizes through multiple types of intermolecular interactions. *J Biol Chem* **276**, 5316–22.

Zhou AT, Assil I, and Abou-Samra AB (2000). Role of asparagine-linked oligosaccharides in the function of the rat PTH/PTHrP receptor. *Biochemistry* **39**, 6514–20.

Zitzer H, Honck HH, Bachner D, Richter D, and Kreienkamp HJ (1999). Somatostatin receptor interacting protein defines a novel family of multidomain proteins present in human and rodent brain. *J Biol Chem* **274**, 32 997–3001.

Chapter 2

Transcriptional regulation of GPCR expression

Noel J. Buckley and Mireia Garriga-Canut

2.1 Introduction

G protein coupled receptors (GPCRs) represent the largest family of signalling molecules present in the organism and are found in all phyla from viruses through yeast to mammals. This sequence diversity is reflected in their expression patterns which invariably show that each transcript has a unique distribution. As anticipated, this diversity of gene type and distribution is most manifest in the CNS. This poses both problems and opportunities. On the one hand, such unique distributions offer potential for highly specific therapeutic targets. On the other hand, they present a problem of phenotypic regulation: How are these expression patterns established and maintained? It is this latter aspect that is the focus of this review.

Establishment of mature phenotype typically, but not always, occurs around the time of differentiation and represents activation of programmes of gene transcription. Maintenance, by contrast, occurs throughout the life of the organism and may involve both transcriptional and post-transcriptional mechanisms. Mechanisms responsible for establishment and maintenance of transcription may be quite distinct. As a primer, the following section presents a brief overview of these processes.

2.2 A transcriptional overview

Transcription is all too often graphically represented by a straight line (of DNA) broken up by boxes representing regulatory elements lying upstream of the promoter. These regulatory elements bind their cognate activator and repressor proteins which are then brought to the promoter to activate (or repress) transcription. This simple graphical, 'black-box' representation, while useful as a schematic, carries an incorrect view of the molecules and mechanisms required to carry out transcription. This view can often carry over into the types of experiments that we do and the interpretations that we draw.

Reality is somewhat different. Each cell contains about 2 m of DNA packaged into a nucleus <10 µm in diameter. This is brought about by assembling the DNA into chromatin. The chromatin is then packaged into a hierarchy of structures starting with the nucleosome, which is assembled into 30 nm fibres and finally into a metaphase chromosome. Any protein concerned with the regulation or execution of transcription must interact with and penetrate this repressive structure. Activation domains of transcription factors interact with components of the transcriptional complex and/or chromatin modifying complexes. This takes place in a two-stage process. First, chromatin remodelling complexes and/or histone modifying enzymes convert the chromatin from a 'closed' to an 'open' state. Second, as a result of this remodelling, access of proximal activators is increased. Bound activators can then interact with either the RNA polymerase holocomplex or the TFIID complex to stabilize formation of the pre-initiation complex. Activation or repression of transcription can occur by interference or modulation of any of these steps (see Lemon and Tjian 2000). Stated otherwise, the *in vivo* template for transcription is chromatin, not DNA. This perspective is underlined by the increasing number of transcription factors that are known to interact with histones via recruitment of histone modifying enzymes (see Struhl 1998) and/or chromatin remodelling activities (Flaus and Owen-Hughes 2001; Muller and Leutz 2001). If we embrace this less linear view of transcription then we must rephrase our questions and carry out different experiments if we are ever to answer the type of question: What is responsible for ensuring a hippocampal pyramidal cell expresses M1 but not M5 muscarinic cholinergic receptors? One immediate consequence of this is to examine the experimental paradigm that has, more than any other, been responsible for identifying these regulatory elements—reporter gene analysis using transient transfection.

2.3 The long and the short of reporter gene assays

A typical series of experiments aimed at understanding the basis of a particular GPCR expression pattern might go like this: (1) isolate a genomic fragment containing the GPCR gene, (2) make a series of reporter gene constructs, transiently transfect these into expressing and non-expressing cell lines and measure reporter gene activity to localize regulatory domains, (3) identify individual regulatory elements using protein/DNA interaction assays such as electromobility shift assays and DNase footprinting, and (4) use these elements to identify cognate transcription factors using gel shifts, screening expression libraries or one-hybrid assays. So what is wrong with that? There are several potential pitfalls to this approach—but they should not obscure its general usefulness. First, templates are present in transiently transfected cells as high copy episomes of many thousands (although only a small fraction are transcriptionally competent), whereas only two copies of an endogenous GPCR gene are found in a diploid cell (Smith and Hager 1997). Second, chromatin organization on episomal DNA differs from that of chromosomal genes (Archer *et al.* 1992; Smith *et al.* 1997). Third, most cell lines offer only a poor representation of their natural untransformed counterparts and this may be reflected in the molecules and mechanisms recruited to control transcription. Nevertheless, such experiments do yield valuable data and the bulk of the information that we have on regulation of GPCR transcription stems from these types of studies. Rather than seeing such experiments as definitive it is better to see them as enabling—providers of candidate regulatory elements whose role needs to be verified in a more physiological setting.

The promoters of a number of GPCRs have been examined over the last decade. There is neither space to attempt to comprehensively assess each of these studies nor justification to present such a catalogue. Instead, we have tabulated a selection of key references according to receptor type (Table 2.1). The focus presented here is upon selected examples that serve to illustrate general principles or even cautionary tales. Particular attention has been given to studies that illustrate the role of transcription in regulating endogenous GPCR expression. Many of these points can be illustrated by work carried out over the last six years by ourselves and others on the regulation of the M4 muscarinic receptor gene.

Table 2.1 Transcriptional regulation of GPCR genes: Examples have been chosen where interaction of a known transcription factor has been demonstrated by interaction either on the basis of a functional assay (using either the endogenous gene or a reporter gene) or biochemical interaction (such as electromobility shift assays, chromatin immunoprecipitation, or DNase footprinting). Studies that have demonstrated only relevant *cis* regulatory regions of a gene have not been included

Receptor	Proposed regulatory elements/transcription factors
D1A dopamine	AP-2 (Takeuchi *et al.* 1993) Sp1; Zic2 (Yang *et al.* 2000)
D2 dopamine	AP-1 (Valdenaire *et al.* 1994; Wang *et al.* 1997)
B1 bradykinin	NFκB/CRE (Yang and Polgar 1996) AP-1 (Yang *et al.* 1998*b*)
B2 bradykinin	p53; p300/CBP (Saifudeen *et al.* 2000)
β2 adrenergic	GRE/GR (Malbon and Hadcock 1988)
β1 adrenergic	RARE/TH (Bahouth *et al.* 1997) GR (Tseng and Padbury 2000)
μ opioid	PPY tract (Ko and Loh 2001)
A1 adenosine	GATA; Nkx2.5 (Rivkees *et al.* 1999) GR (Ren and Stiles 1999)
VPAC1	(Karacay *et al.* 2000)
sst2	Smad4 (Puente *et al.* 2001)
PPT-1	CRE (Qian *et al.* 2001)
GRP-R	CRE (Weber *et al.* 2000)
M1 muscarinic	PPY (Wood *et al.* 1999)
M2 muscarinic	GATA; LIF/CNTF (Rosoff and Nathanson 1998; Rosoff *et al.* 1996)
M4 muscarinic	NRSF (Wood *et al.* 1996)
NPY1	NFκB (Musso *et al.* 1997)
Thrombin	Sp1 (Wu *et al.* 1998)

2.4 GPCRs and chromatin

The muscarinic gene family was the first complete subfamily of GPCRs to be cloned (Bonner *et al.* 1987, 1988; and see chapter 18). There are five members, all of which have unique but partially overlapping expression profiles (Buckley *et al.* 1988; Vilaro *et al.* 1990). The M4 gene has a simple gene structure consisting of a single non-coding exon (containing two transcription initiation sites) and a single coding exon (Mieda *et al.* 1996, 1997; Wood *et al.* 1996, 1995). Sequence analysis of the 5′flanking region of this gene revealed the presence of a RE1/neuron restrictive silence element (NRSE) (Kraner *et al.* 1992; Mori *et al.* 1992) situated 0.5 kb upstream of the first transcription initiation site (Mieda *et al.* 1997; Wood *et al.* 1996) The RE1/NRSE is a 21 bp motif that binds a Kruppel type zinc finger transcriptional repressor REST/NRSF. Both the RE1/NRSE and REST/NRSF were originally characterized by the groups of Gail Mandel at Stony Brook and David Anderson at Caltech

(Chong *et al.* 1995; Schoenherr and Anderson 1995*a*). Based on transient transfection assays and expression studies the consensus evolved that REST/NRSF acted to silence expression of NRSE-bearing neuron-specific genes such as the SCG10 and Na type II sodium channel in non-neuronal tissue and in undifferentiated neuroepithelial cells (Schoenherr and Anderson 1995*b*; Schoenherr *et al.* 1996). The role of the NRSE^{M4} was tested within this context. As predicted, removal of the NRSE from reporter gene constructs led to activation of transcription from the M4 promoter only in NRSF-expressing cells. Likewise overexpression of NRSF led to repression of NRSE-bearing constructs while expression of dominant-negative constructs containing only the NRSF DNA binding domain, alleviated the repression. Electromobility shift assays (EMSA) also confirmed the ability of NRSF to interact with the NRSE of the M4 gene (Mieda *et al.* 1997; Wood *et al.* 1996). These data were interpreted as showing that NRSF acts to silence expression of the M4 gene in NRSF-expressing cells and were taken to re-enforce the idea that NRSF acts as a silencer of NRSE-bearing genes outside of the differentiated nervous system. However, all of these studies address the singular issue of potential—*can* NRSF interact with the NRSE^{M4} to repress transcription from the M4 promoter? None of these studies examined the interaction of NRSF with the endogenous NRSE^{M4}—*does* NRSF interact with the endogenous M4 gene? When this is done a more complicated picture emerges.

We have examined the interaction of NRSF with active and silent endogenous M4 loci. We have used rat JTC-19 fibroblasts which express endogenous NRSF and do not express M4 receptors, and PC12 pheochromocytoma cells which express M4 receptors and are one of the few cell lines that do not express NRSF (or at least very little) and into which we introduced recombinant NRSF constructs. DNase hypersensitivity, endonuclease access assays and chromatin immunoprecipitation all failed to show any evidence of interaction between NRSF and the NRSE^{M4} at the silent M4 locus in JTC-19. In contrast, these same assays showed that recombinant NRSF repressed expression of the endogenous M4 gene and did interact with the chromatin around the NRSE^{M4} of the *active* locus in PC12 cells. These data are more consistent with NRSF acting as a modulator of expression levels rather than being required for maintenance of the silenced state—quite different from the conclusions drawn from transient transfections of reporter gene constructs. Consistent with this view is the observation that relief of histone deacetylase activity by Trichostatin A (TSA) does not induce expression of the M4 gene in JTC-19 cells (Wood and Buckley, *unpublished observations*). However, this picture is further complicated when another NRSE-bearing gene, the Na type II channel gene is examined. In this case, chromatin immunoprecipitation establishes that NRSF is associated with the NRSE^{NaII} and accordingly, expression of the Na type II gene is induced by TSA (Ballas *et al.* 2001). It is important to note that transient reporter gene analysis of these two genes indicated that *both* the M4 and Na type II genes are NRSF responsive. Several other GPCR genes have NRSEs in their regulatory region including a rat CCR10 receptor (Bonini *et al.* 1997), two orphan receptors; the human GPR10 (Marchese *et al.* 1995) and the GPR6 receptor (accession U18549) and the mouse μ opioid (Pan *et al.* 1999). As yet, their functional significance has not been assessed. The lesson seems clear: the existence of a particular regulatory element in a GPCR gene does not necessarily indicate that it acts to recruit its cognate transcription factor. Only detailed analysis of the chromatin environment of individual GPCR genes will solve this issue. It is of course possible that NRSF might be required to establish but not to maintain the silenced state—this has a parallel in Drosophila where two Polycomb group complexes exist—one to establish and one to maintain silence (Farkas *et al.* 2000).

Another line of evidence pointing to a more refined role for NRSF than initially assumed comes from experiments on NRSF$^{-/-}$ mice. Although no GPCR was examined in these studies, ablation of the NRSF gene did not lead to widespread expression of NRSE-bearing genes in non-neural tissues. Only βIII-tubulin expression showed prolonged up-regulation in areas where it is normally transiently expressed (Chen *et al.* 1998). One possibility that arises from these observations is that some NRSE-bearing GPCR genes may require NRSF to establish or maintain silence while others may not.

How does this NRSF-mediated regulation of the M4 gene occur? Several studies have shown that the N-terminal repression domain of NRSF interacts with the sin3 co-repressor while the C-terminal recruits a distinct co-repressor that contains Co-REST but not sin3 (Andres *et al.* 1999; Grimes *et al.* 2000; Huang *et al.* 1999; Roopra *et al.* 2000). Both co-repressor complexes interact with histone deacteylases which can remove the N-terminal acetyl groups from histone tails of H3 and H4 causing condensation of the nucleosome and consequent repression of transcription. This remodelling is accompanied by an increase in DNase hypersensitivity, endonuclease access and formation of ordered nucleosomes around the NRSE (Wood and Buckley *unpublished observations*). There is no reason to think that these experiments are anything but typical in highlighting the apparent disparate roles attributed to particular transcription factors in regulating GPCR expression when results from transient expression of reporter genes in cultured cells are compared with those carried out on endogenous genes.

2.5 GPCR promoter architecture

Promoters can be divided into at least three classes depending upon the presence of certain core motifs. Many genes possess a TATA box 25–30 bp upstream of the transcription initiation site (TIS)—such genes normally have a singular TIS. Other genes lack a TATA box but have an initiator sequence spanning multiple start sites. Another class of genes contain neither of these motifs but are characterized by GC-richness and multiple start sites. All these architectures have been described for GPCR promoters, although the majority are of the latter category, TATA-less, GC-rich with multiple start sites.

Some GPCR genes contain multiple mixed promoters such as the rat α-1B adrenergic receptor that contains three promoters. The first lacks any consensus motif, the second is GC-rich and TATA-less (major) and the third is a TATA-containing promoter (Gao and Kunos 1994). There is little obvious contribution that the promoter architecture of a gene can tell us about its transcriptional regulation. The early notion that TATA-less GC-rich genes encode constitutively active house-keeping proteins is long overturned—not least by studies of GPCR promoters. Numerous studies have identified regulatory elements in several GPCR genes and, in a minority of studies, identification of *cis* regulatory elements or domains has been used to identify potential cognate transcription factors. In general, these sites and factors have been identified on the basis of transient reporter gene analysis and few seem to account for cell specific expression in this type of assay. Some of these details are catalogued in Table 2.1.

2.6 Cell specific transcription of GPCRs

No amount of transient transfection assays can substitute for analysing reporter gene expression in transgenic mouse models. Only then, can all the developmental constraints responsible for driving tissue and stage-specific expression be brought to bear upon the transgene. Even then, the reporter gene is integrated randomly into the chromatin and expression is subject to site-specific variegation from flanking chromatin.

Few transgenic mouse studies have been reported for GPCRs. Typically, appropriate spatiotemporal expression of a transgene is only achieved when large genomic fragments are used to drive reporter gene expression. This is partly a reflection of the need for distal regulatory elements and partly the inability of short genomic sequences to insulate themselves from the influences of the surrounding chromatin—the cause of site-specific variegation. When it comes to specificity—size matters. However, there are some exceptions. Studies of the A1 adenosine receptor showed that a 500 bp proximal promoter drove widespread expression and was sufficient to drive expression in CNS and atria (Rivkees *et al.* 1999). Interestingly, this proximal promoter sequence contains motifs for the cardiac transcription factors, GATA and Nkx2.5. Furthermore, both recombinant GATA and Nkx2.4 synergistically activated transcription from the A1 promoter, but the *in vivo* relevance of these sites was not tested in these transgenic experiments. If they recapitulate results obtained from transient reporter gene assay, then mutation of the Nkx2.4 and GATA sites should ablate cardiac expression but leave CNS expression unaffected.

When the native gene exhibits a widespread expression pattern like the A1 receptor, then it can be difficult to tell the difference between deregulated expression and correct expression. This distinction is clearer in the case of a gene that displays a very discrete pattern of expression such as the follicle-stimulating hormone (FSH) receptor that is limited to granulosa cells of the ovary and Sertoli cells of the testis. Here, over 5 kb of 5'flanking sequence was insufficient to drive reporter gene expression in Sertoli cells (Heckert *et al.* 2000). Transgenic mouse studies can also reveal differences between regulatory elements required to specify embryonic development and those required to maintain adult expression patterns. Hu *et al.* (Hu *et al.* 1999) showed that reporter genes containing 3 kb of 5'flanking sequence of the mouse μ opioid receptor gene promoter was sufficient to recapitulate embryonic expression of the endogenous gene in the forebrain, midbrain, and hindbrain. However, this same construct was insufficient to drive peripheral expression in the adult. This distinction between regulatory regions required for CNS expression and peripheral expression is also seen with the neuropeptide Y (NPY) Y1 receptor gene. In this case, a reporter gene containing 1.3 kb of 5'flanking sequence was sufficient to drive appropriate spatiotemporal expression in the embryonic and adult CNS but was not capable of driving expression in heart, liver, and kidney (Oberto *et al.* 1998). Even over 9 kb of the mGluR6 was not sufficient to repress ectopic transcription in the brain although developmentally regulated expression was seen in the retinal rod and ON-type cone bipolar cells, the major sites of expression of the endogenous mGluR1 gene (Ueda *et al.* 1997). In another study, 2.6 kb of 5'flanking sequence of the VIP1 receptor (VPAC1) was found to be sufficient to direct expression of a reporter gene to CNS and lung (Karacay *et al.* 2000). All of these studies have, to a greater or lesser extent, shown that discrete genomic fragments are capable of driving appropriate spatiotemporal expression of a reporter gene, with a fair degree of veracity. As yet, there is no report of a finer analysis that has identified specific regulatory domains that are required or sufficient to drive these expression patterns *in vivo*. Until this is achieved, the nature of the transcription factors that are responsible for these patterns of expression will remain elusive.

It seems unlikely that tissue- and stage-specific transcription of GPCRs, as a gene family, will be regulated in a co-ordinated manner. With a few notable exceptions (see below), there is no *cis*-linkage among GPCR genes and their tissue expression profiles and developmental activation are completely disparate. It seems equally likely, that, their expression will be controlled by disparate regulatory elements each operative in different tissues or at different times. This latter point appears to be borne out by the studies of Oberto et al. (Oberto et al. 1998) and Hu et al. (Hu et al. 1999) and probably reflects the fact that most GPCR expression patterns do not follow singular cell lineages. Genes expressed only in cells derived from a singular lineage might be expected to exhibit co-ordinate regulation since they are derived from the same progenitor population and consequently they have all been exposed to at least some of the same transcription factors during the cells' development. An example of such co-ordinate regulation can be seen in the noradrenergic cells of the PNS all of which are derived from the neural crest. Whereas, many of the pan-neuronal properties of sympathetic precursors are under the control of the MASH-1 bHLH protein (Sommer et al. 1995), expression of tyrosine hydroxylase and dopamine β-hydroxylase are under the control (in part) of the Phox2a and Phox2b homeobox genes (Lo et al. 1997; Pattyn et al. 2000; Yang et al. 1998a). In this way, the transmitter phenotype (in this case noradrenaline) is intimately part of the sympathoadrenal lineage and it is not surprising to see that expression of its synthetic enzymes are co-ordinately regulated at least in part, by a singular transcription factor, Phox2a. If a GPCR shared this unique expression pattern then it too, might be co-ordinately controlled by Phox2a. However, a gene such as the M1 muscarinic receptor gene, which is also expressed on all sympathetic neurons (Brown et al. 1995), is additionally expressed by many telencephalic regions (Buckley et al. 1988) where Phox2a and Phox2b are not expressed. Accordingly, M1 expression is likely controlled by different mechanisms in these two neuronal areas. Even the example given here is a gross oversimplification since there is no known metazoan gene whose expression is exquisitely regulated by only one transcription factor. Nevertheless, it is probably true that genes which are expressed across many lineages, such as is the case for many GPCRs, will probably not necessarily be co-ordinately regulated. The predicted consequence of this is that transgenic mouse studies will not reveal individual *cis* elements that are required or necessary for expression in all regions where the endogenous gene is transcribed. This concept will only be tested when sufficient GPCR genes are characterised in transgenic mice. Exceptions would be the obverse—those GPCRs whose expression is highly restricted to a particular tissue. Probably the most spatially restricted GPCRs are represented by members of the opsin gene family, expression of which are restricted to the photoreceptors of the retina. Interestingly, as little as 1.1 kb of 5′flanking sequence of the human blue opsin promoter is sufficient to restrict expression to S-cone type photoreceptors and bipolar cells (Chen et al. 1994). Similarly, a discrete 2.2 kb of 5′flanking sequence of the rhodopsin promoter was sufficient to drive expression in photoreceptor cells but expression was also seen in rods and to a lesser extent in cones (Gouras et al. 1994). Similar fragments derived from the mouse gene also drive photoreceptor expression (Lem et al. 1991)—again, a certain 'leakiness' in expression was seen with cones expressing low amounts of transgene. Earlier studies had shown that shorter constructs containing as little as 222 bp of flanking sequence of the bovine rhodopsin promoter were sufficient to confine reporter gene expression to photoreceptor cells but gave aberrant expression patterns across the retina (Zack et al. 1991). Cross species comparison of the rhodopsin promoter revealed a 102 bp domain of conserved sequence (Zack et al. 1991) approximately 1.5–2 kb (dependent upon species) upstream from the T.I.S. but deletion of this region did not result in loss of correct spatial expression, although deletion did lead to lower levels of transgene expression (Nie et al. 1996). Neither did this region confer retina-specific expression upon a heterologous promoter. Nevertheless, the possible linkage between tightly controlled expression and relatively discrete regulatory elements is seductive. Although this retina enhancer region shows homology to the locus control region (LCR) of the cone opsins, the lack of correlation between copy number and transgene expression levels clearly rules out a role of this region as a LCR in the rhodopsin gene. Although putative proximal elements and cognate transcription factors, such as Nr1 (Kumar *et al.* 1996) and the Crx1 homeobox (Chen *et al.* 1997), have been assessed *in vitro*, their *in vivo* relevance is unkown. The case for their potential *in vivo* relevance is increased by the observed synergy of their activation of the rhodopsin promoter and the demonstration of their physical interaction (Mitton *et al.* 2000). Importance of this region is further underlined by the demonstration Crx also interacts with the architectural transcription factor HMG I(Y) (Chau *et al.* 2000) and mutations in Crx have been associated with several retinal degenerative diseases including Leber congenital amaurosis and adult onset cone-rod dystrophy and retinitis pigmentosa (Sohocki *et al.* 1998).

2.7 Transcription of linked genes

Gene linkage may be a reflection of a requirement for co-ordinated *cis* transcriptional regulation, as has been shown for the globin locus and/or may be a sign of recent evolutionary divergence. The most well-known example of linkage within the GPCR family is provided by the olfactory receptor subfamily. Although there are only 47 human olfactory type GPCRs listed in the GPCR data base, initial estimates accounted for up to 906 olfactory receptor genes of which 60% were scored as pseudogenes (International Human Genome Sequencing Consortium). More recent genome mining studies (Zozulya *et al.* 2001) have revealed a total of 347 functional olfactory type GPCRs in the human genome. These can be further subdivided into subfamilies on the basis of similarity of coding regions. These subfamilies tend to exist as clusters of around 10 or so members spread over at least 25 chromosomal loci (Ben-Arie *et al.* 1994; Rouquier *et al.* 1998; Trask *et al.* 1998). Many of these clusters are located at the subteleomeric regions near the termini of the chromosomes. Subteleomeric regions of the chromosome are gene rich and show high levels of recombination resulting in increased gene duplication and it has been hypothesized that these regions act as 'olfactory receptor nurseries' (Trask *et al.* 1998). One consequence of elevated meiotic recombination would be the generation of high levels of pseudogenes—which is exactly what is seen in the olfactory gene subfamily.

In rodents, each olfactory receptor gene is expressed in one of four zones in the olfactory epithelium (Ressler 1994, 1993; Vassar *et al.* 1994, 1993) and each olfactory neuron is assumed to express only one (or very few) olfactory receptors (Malnic *et al.* 1999; Ressler, 1994, 1993; Vassar *et al.* 1994, 1993). Furthermore, allelic inactivation ensures that only one copy of each array is transcribed (Chess 1994). One further observation is that members of individual clusters tend to be expressed in the same region of the olfactory epithelium. Collectively, this would imply that within each cluster there is a mechanism to ensure co-ordinate transcriptional regulation. It was hypothesized some time ago (Chess 1994) that allelic inactivation was consistent with a stochastic *cis* mechanism of gene activation since without the allelic inactivation, stochastic activation would lead to two different genes within an array being expressed by a single olfactory neuron—a situation that is not seen. By analogy with the locus control region of the globin locus (Grosveld 1999; Grosveld *et al.* 1998), it was suggested that an equivalent region adjacent to an array of olfactory receptor genes could be responsible for the stochastic activation of a single gene within the locus. Although the mechanisms responsible for this restricted transcription are largely unknown, recent transgenic experiments have shed some light on the types of mechanisms that may be operative.

Serizawa et al. (Serizawa et al. 2000) generated a line of mice that carried a tagged endogenous MOR28 gene and a 200 kb YAC containing a differentially tagged MOR28 transgene. The salient finding was that transgene and endogenous gene were almost never co-expressed in the same olfactory neuron, thereby demonstrating that mutually exclusive expression did not require *cis* linkage. The classical picture of gene-specific transcription envisages a combination of transcription factors that uniquely specify activation of that gene locus. One of the most important and least contentious outcomes of this study is the rejection of this notion since such a model would require co-expression of transgene and endogenous gene. Although gene re-arrangement is a formal possibility, it is unlikely. The analogous system here is provided by the antigen-receptor genes in lymphocytes which undergo re-arrangement and deletion in order to bring distal promoter and enhancer elements into proximity to allow transcription. However, no evidence for DNA arrangements is apparent at the olfactory receptor locus. In a commentary by Randall Reed on the work of Serizawa (Reed 2000), it was hypothesized that two events occurred. Stage one was a 'low-probability' event in which one allele is primed for potential expression. Stage two requires transcription of the selected allele. Some 'high-probability' event or product downstream of the transcription prevents priming of other alleles. Although attractive in dispensing with the need for *cis* linkage in order to maintain stochastic selection of one gene within an array, the weakness is the absence of any direct supportive evidence.

Another subfamily of GPCRs that appears to be linked are the C-C chemokine receptors. In the human genome 6 C-C type chemokine receptor genes (CCXCR1, CCR1, CCR3, CCR2, CCR5, CCRL2) are arranged in a tandem array within a 400 kb region of chromosome 3 (3p21.3). This linkage group appears to be conserved in the mouse genome where mCCR1, mCCR2, mCCR3 and mCCR5 are all found in a cluster on chromosome 9 (9 72.0 cM). It remains to be seen if this linkage is accompanied by a co-ordinate transcriptional regulation.

2.8 Promoter polymorphisms and function

Many polymorphisms have been reported in the coding region of GPCRs and some are correlated with disease states such as the association between polymorphisms of the 5HT2A gene and psychiatric disorders (Arranz et al. 2001). Thus far relatively few have been recorded in potential regulatory regions. It is unlikely that SNPs are not present in regulatory regions of GPCR genes since, on average a SNP is present every 2 kb of the human genome. Whereas it is possible that SNPs in regulatory regions are relatively rare and probable that most are benign, their apparent absence is more likely a reflection of the fact that, in most cases, only ORFs have been searched for SNPs—if you lose your keys on a dark street, you tend to search first under the street light. Alternatively, SNPs that map to transcription factor ORFs may have an indirect effect on transcription of their target GPCR genes. An example is provided by mutations of the retina-specific homeobox gene, Crx, that regulates opsin expression and mutations of which are associated with retinal degenerative disorders including autosomal dominant cone-rod dystrophy, dominant retinitis pigmentosa and Leber congenital amaurosis (Sohocki et al. 1998).

One study of the human β_2-adrenergic receptor revealed eight polymorphisms within a 1.5 kb region upstream of the start codon. Two of these halotypes showed diminished transcription (Scott et al. 1999). However, no association has been found between β-adrenergic receptor promoter polymorphisms and hypertension (Kato *et al.* 2001). Several studies have tested the linkage between polymorphisms in the promoter of the alpha2A adrenergic receptor and mood disorders and schizophrenia, but again no association of polymorphism and behaviour has been observed (Ohara *et al.* 1998; Tsai *et al.* 2001).

2.9 Future prospects

Over a decade of work has led to the identification of domains that are responsible for modulating levels of expression and partial responsible for conferring cell-specific expression. As yet no explicit description exists that offers an explanation of the exquisite distribution of any single GPCR. This is due in part to the inherent multifactorial complexity of the problem but is probably due in equal measure to an over-reliance upon use of transient reporter gene analysis which takes little heed of the organized chromatin environment of the endogenous gene. It can be no coincidence that the GPCR gene about whose transcriptional control we know most, the rhodopsin gene, has been subject to relatively intense transgenic analysis. Wholesale application of transgenic mouse strategies and translation of piecemeal biochemical assays into array based whole-genome analyses will go some way to addressing these deficiencies. In this latter respect, following the well-worn historical precedent of borrowing technology from the yeast community will continue to provide the technical impetus. Witness, for example, the application of array-based technology to the genome-wide identification of transcription factor targets of yeast cell cycle genes, SBF and MBF (Iyer *et al.* 2001), Gal4 (Ren *et al.* 2000) and the more recent identification of mammalian targets of p53 (Wang *et al.* 2001). Obvious developmental biological constraints place a natural limitation on the pace with which transgenic mouse strains can be generated and analysed. It is likely that the rapidly advancing improvements in bioinformatic analyses of gene prediction, gene structure and diagnosis of genomic regulatory elements will provide the necessary focus to expedite *in vivo* analysis. As more genome sequences become available in the public domain, such in silico analysis of gene structure, promoter localization and enhancer/repressor identification will largely replace the tedium of RNAse protection analysis, primer extension and RACE analysis that have so often been the rate limiting step in defining potential genomic regulatory elements. Bioinformatic 'context-dependent' comparative studies (Kel *et al.* 2001) across different phyla and genes will allow explicit identification of real transcription factor binding sites, something that is not possible at present using only short motifs that occur stochastically several thousands of times per genome. Maybe then, will we be able to answer the question 'Why does a hippocampal pyramidal cell express M1 but not M5 muscarinic cholinergic receptors?'.

References

Andres ME, Burger C, Peral-Rubio MJ, Battaglioli E, Anderson ME, Grimes J *et al.* (1999). CoREST: A functional corepressor required for regulation of neural-specific gene expression. *Proc Natl Acad Sci USA* **96**, 9873–8.

Archer TK, Lefebvre P, Wolford RG, and Hager GL (1992). Transcription factor loading on the MMTV promoter: a bimodal mechanism for promoter activation [published erratum appears in Science 1992 Apr 10; 256(5054): 161]. *Science* **255**, 1573–6.

Arranz B, Rosel P, Ramirez N, and San L (2001). Genetic dysfunction of the serotonin receptor 5-HT2A in psychiatric disorders. *Actas Esp Psiquiatr* **29**, 131–8.

Bahouth SW, Cui X, Beauchamp MJ, and Park EA (1997). Thyroid hormone induces beta1-adrenergic receptor gene transcription through a direct repeat separated by five nucleotides. *J Mol Cell Cardiol* **29**, 3223–37.

Ballas N, Battaglioli E, Atouf F, Andres ME, Chenoweth J, Anderson ME *et al.* (2001). Regulation of neuronal traits by a novel transcriptional complex. *Neuron* **31**, 353–65.

Ben-Arie N, Lancet D, Taylor C, Khen M, Walker N, Ledbetter DH *et al.* (1994). Olfactory receptor gene cluster on human chromosome 17: possible duplication of an ancestral receptor repertoire. *Hum Mol Genet* **3**, 229–35.

Bonini JA, Martin SK, Dralyuk F, Roe MW, Philipson LH, and Steiner DF (1997). Cloning, expression, and chromosomal mapping of a novel human CC- chemokine receptor (CCR10) that displays high-affinity binding for MCP-1 and MCP-3. *DNA Cell Biol* **16**, 1249–56.

Bonner TI, Buckley NJ, Young AC, and Brann MR (1987). Identification of a family of muscarinic acetylcholine receptor genes. *Science* **237**, 527–32.

Bonner TI, Young AC, Brann MR, and Buckley NJ (1988). Cloning and expression of the human and rat m5 muscarinic acetylcholine receptor genes. *Neuron* **1**, 403–10.

Brown DA, Buckley NJ, Caulfield MC, Duffy SM, Jones S, Lamas JA *et al.* (1995). Coupling of muscarinic acetylcholine receptors (mAChRs) to neural ion channels: closure of some K$^+$ channels. In (ed. J. Wess) *Molecular Mechanisms of Muscarinic Acetylcholine Function*, Landes, Georgetown, TX.

Buckley NJ, Bonner TI, and Brann MR (1988). Localisation of a family of muscarinic receptor genes in rat brain. *J Neurosci* **8**, 4646–52.

Chau KY, Munshi N, Keane-Myers A, Cheung-Chau KW, Tai AK, Manfioletti G *et al.* (2000). The architectural transcription factor high mobility group I(Y) participates in photoreceptor-specific gene expression. *J Neurosci* **20**, 7317–24.

Chen J, Tucker CL, Woodford B, Szel A, Lem J, Gianella-Borradori A *et al.* (1994). The human blue opsin promoter directs transgene expression in short- wave cones and bipolar cells in the mouse retina. *Proc Natl Acad Sci USA* **91**, 2611–15.

Chen S, Wang QL, Nie Z, Sun H, Lennon G, Copeland NG *et al.* (1997). Crx, a novel Otx-like paired-homeodomain protein, binds to and transactivates photoreceptor cell-specific genes. *Neuron* **19**, 1017–30.

Chen ZF, Paquette AJ, and Anderson DJ (1998). NRSF/REST is required *in vivo* for repression of multiple neuronal target genes during embryogenesis. *Nat Genet* **20**, 136–42.

Chess A, Simon I, Cedar H, and Axel R (1994). Allelic inactivation regulates olfactory receptor gene expression. *Cell* 823–34.

Chong JA, Tapia-Ramirez J, Kim S, Toledo-Arai JJ, Zheng, Y., Boutros MC *et al.* (1995). REST: A mammalian silencer protein that restricts sodium channel gene expression to neurons. *Cell* **80**, 949–57.

Farkas G, Leibovitch BA, and Elgin SC (2000). Chromatin organization and transcriptional control of gene expression in Drosophila. *Gene* **253**, 117–36.

Flaus A and Owen-Hughes T (2001). Mechanisms for ATP-dependent chromatin remodelling. *Curr Opin Genet Dev* **11**, 148–54.

Gao B and Kunos G (1994). Transcription of the rat alpha 1B adrenergic receptor gene in liver is controlled by three promoters. *J Biol Chem* **269**, 15 762–7.

Gouras P, Kjeldbye H, and Zack DJ (1994). Reporter gene expression in cones in transgenic mice carrying bovine rhodopsin promoter/lacZ transgenes. *Vis Neurosci* **11**, 1227–31.

Grimes JA, Nielsen SJ, Battaglioli E, Miska EA, Speh JC, Berry DL *et al.* (2000). The co-repressor mSin3A is a functional component of the REST-CoREST repressor complex. *J Biol Chem* **275**, 9461–7.

Grosveld F (1999). Activation by locus control regions? *Curr Opin Genet Dev* **9**, 152–7.

Grosveld F, de Boer E, Dillon N, Gribnau J, McMorrow T, Milot E *et al.* (1998). The dynamics of globin gene expression and position effects. *Novartis Found Symp* 214, 67–79.

Heckert LL, Sawadogo M, Daggett MA, and Chen JK (2000). The USF proteins regulate transcription of the follicle-stimulating hormone receptor but are insufficient for cell-specific expression. *Mol Endocrinol* 14, 1836–48.

Hu X, Cao S, Loh HH, and Wei LN (1999). Promoter activity of mouse kappa opioid receptor gene in transgenic mouse. *Brain Res Mol Brain Res* 69, 35–43.

Huang Y, Myers SJ, and Dingledine R (1999). Transcriptional repression by REST: recruitment of Sin3A and histone deacetylase to neuronal genes. *Nat Neurosci* 2, 867–72.

Iyer VR, Horak CE, Scafe CS, Botstein D, Snyder M, and Brown PO (2001). Genomic binding sites of the yeast cell-cycle transcription factors SBF and MBF. *Nature* 409, 533–8.

Karacay B, O'Dorisio MS, Summers M, and Bruce J (2000). Regulation of vasoactive intestinal peptide receptor expression in developing nervous systems. *Ann N Y Acad Sci* 921, 165–74.

Kato N, Sugiyama T, Morita H, Kurihara H, Sato T, Yamori, Y *et al.* (2001). Association analysis of beta(2)-adrenergic receptor polymorphisms with hypertension in Japanese. *Hypertension* 37, 286–92.

Kel AE, Kel-Margoulis OV, Farnham PJ, Bartley SM, Wingender E, and Zhang MQ (2001). Computer-assisted identification of cell cycle-related genes: new targets for E2F transcription factors. *J Mol Biol* 309, 99–120.

Ko JL and Loh HH (2001). Single-stranded DNA-binding complex involved in transcriptional regulation of mouse {micro}-opioid receptor gene. *J Biol Chem* 276, 788–95.

Kraner SD, Chong JA, Tsay HJ, and Mandel G (1992). Silencing the type II sodium channel gene: a model for neural-specific gene regulation. *Neuron* 9, 37–44.

Kumar R, Chen S, Scheurer D, Wang QL, Duh E, Sung CH *et al.* (1996). The bZIP transcription factor Nrl stimulates rhodopsin promoter activity in primary retinal cell cultures. *J Biol Chem* 271, 29 612–18.

Lem J, Applebury ML, Falk JD, Flannery JG, and Simon MI (1991). Tissue-specific and developmental regulation of rod opsin chimeric genes in transgenic mice. *Neuron* 6, 201–10.

Lemon B and Tjian R (2000). Orchestrated response: a symphony of transcription factors for gene control. *Genes Dev* 14, 2551–69.

Lo L, Sommer L, and Anderson DJ (1997) MASH1 maintains competence for BMP2-induced neuronal differentiation in post-migratory neural crest cells. *Curr Biol* 7, 440–50.

Malbon CC and Hadcock JR (1988) Evidence that glucocorticoid response elements in the 5'-noncoding region of the hamster beta 2-adrenergic receptor gene are obligate for glucocorticoid regulation of receptor mRNA levels. *Biochem Biophys Res Commun* 154, 676–681.

Malnic B, Hirono J, Sato T, and Buck LB (1999) Combinatorial receptor codes for odors. *Cell* 96, 713–23.

Marchese A, Heiber M, Nguyen T, Heng HH, Saldivia VR, Cheng R *et al.* (1995). Cloning and chromosomal mapping of three novel genes, GPR9, GPR10, and GPR14, encoding receptors related to interleukin 8, neuropeptide Y, and somatostatin receptors. *Genomics* 29, 335–44.

Mieda M, Haga T, and Saffen DW (1996). Promoter region of the rat m4 muscarinic acetylcholine receptor gene contains a cell type-specific silencer element. *J Biol Chem* 271, 5177–82.

Mieda M, Haga T, and Saffen DW (1997). Expression of the rat m4 muscarinic acetylcholine receptor gene is regulated by the neuron restrictive silencer element/repressor element 1. *J Biol Chem* 272, 5854–60.

Mitton KP, Swain PK, Chen S, Xu S, Zack DJ, and Swaroop A (2000). The leucine zipper of NRL interacts with the CRX homeodomain. A possible mechanism of transcriptional synergy in rhodopsin regulation. *J Biol Chem* 275, 29 794–9.

Mori N, Schoenherr C, Vandenbergh DJ, and Anderson DJ (1992). A common silencer element in the SCG10 and type II Na^+ channel genes binds a factor present in nonneuronal cells but not in neuronal cells. *Neuron* 9, 45–54.

Muller C and Leutz A (2001). Chromatin remodeling in development and differentiation. *Curr Opin Genet Dev* 11, 167–74. ·

Musso R, Grilli M, Oberto A, Gamalero SR, and Eva C (1997). Regulation of mouse neuropeptide Y Y1 receptor gene transcription: a potential role for nuclear factor-kappa B/Rel proteins. *Mol Pharmacol* 51, 27–35.

Nie Z, Chen S, Kumar R, and Zack DJ (1996). RER, an evolutionarily conserved sequence upstream of the rhodopsin gene, has enhancer activity. *J Biol Chem* 271 2667–75.

Oberto A, Tolosano E, Brusa R, Altruda F, Panzica G, and Eva C (1998). The murine Y1 receptor 5′ upstream sequence directs cell-specific and developmentally regulated LacZ expression in transgenic mice CNS. *Eur J Neurosci* 10, 3257–68.

Ohara K, Nagai M, Tani K, Tsukamoto T, and Suzuki Y (1998). Polymorphism in the promoter region of the alpha 2A adrenergic receptor gene and mood disorders. *Neuroreport* 9, 1291–4.

Pan YX, Xu J, Bolan E, Abbadie C, Chang A, Zuckerman A *et al.* (1999). Identification and characterization of three new alternatively spliced mu-opioid receptor isoforms. *Mol Pharmacol* 56, 396–403.

Pattyn A, Goridis C, and Brunet JF (2000). Specification of the central noradrenergic phenotype by the homeobox gene Phox2b. *Mol Cell Neurosci* 15, 235–43.

Puente E, Saint-Laurent N, Torrisani J, Furet C, Schally AV, Vaysse N *et al.* (2001). Transcriptional activation of mouse sst2 somatostatin receptor promoter by transforming growth factor-beta. Involvement of Smad4. *J Biol Chem* 276, 13 461–8.

Qian J, Yehia G, Molina C, Fernandes A, Donnelly R, Anjaria D *et al.* (2001). Cloning of human preprotachykinin-I promoter and the role of cyclic adenosine 5′-monophosphate response elements in its expression by IL-1 and stem cell factor. *J Immunol* 166, 2553–61.

Reed RR (2000). Regulating olfactory receptor expression: controlling globally, acting locally. *Nat Neurosci* 3, 638–9.

Ren B, Robert F, Wyrick JJ, Aparicio O, Jennings EG, Simon I *et al.* (2000). Genome-wide location and function of DNA binding proteins. *Science* 290, 2306–9.

Ren H and Stiles GL (1999). Dexamethasone stimulates human A1 adenosine receptor (A1AR) gene expression through multiple regulatory sites in promoter B. *Mol Pharmacol* 55, 309–16.

Ressler KJ, Sullivan SL, and Buck LB (1994). Information coding in the olfactory system: Evidence for a stereotyped and highly organised epitope map in the olfactory bulb. *Cell* 79, 1245–55.

Ressler KJ, Sullivan SL, and Buck LB (1993). A zonal organization of odorant receptor gene expression in the olfactory epithelium. *Cell* 73, 597–609.

Rivkees SA, Chen M, Kulkarni J, Browne J, and Zhao Z (1999). Characterization of the murine A1 adenosine receptor promoter, potent regulation by GATA-4 and Nkx2.5. *J Biol Chem* 274, 14 204–9.

Roopra A, Sharling L, Wood IC, Briggs T, Bachfischer U, Paquette AJ *et al.* (2000). Transcriptional repression by neuron-restrictive silencer factor is mediated via the Sin3-histone deacetylase complex. *Mol Cell Biol* 20, 2147–57.

Rouquier S, Taviaux S, Trask BJ, Brand-Arpon V, van den Engh G, Demaille J *et al.* (1998). Distribution of olfactory receptor genes in the human genome. *Nat Genet* 18, 243–50.

Schoenherr CJ and Anderson DJ (1995a). The neuron-restrictive silencer factor (NRSF): A coordinate repressor of mutiple neuron-specific genes. *Science* 267, 1360–3.

Schoenherr CJ and Anderson DJ (1995b). Silencing is golden: negative regulation in the control of neuronal gene transcription. *Curr Opin Neurobiol* 5, 566–71.

Schoenherr CJ, Paquette AJ, and Anderson DJ (1996). Identification of potential target genes for the neuron-restrictive silencer factor. *Proc Natl Acad Sci USA* 93, 9881–6.

Scott MG, Swan C, Wheatley AP, and Hall IP (1999). Identification of novel polymorphisms within the promoter region of the human beta2 adrenergic receptor gene. *Br J Pharmacol* 126, 841–4.

Serizawa S, Ishii T, Nakatani H, Tsuboi A, Nagawa F, Asano M *et al.* (2000). Mutually exclusive expression of odorant receptor transgenes. *Nat Neurosci* 3, 687–93.

Smith CL and Hager GL (1997). Transcriptional regulation of mammalian genes *in vivo*. A tale of two templates. *J Biol Chem* **272**, 27 493–6.

Smith CL, Htun H, Wolford RG, and Hager GL (1997). Differential activity of progesterone and glucocorticoid receptors on mouse mammary tumor virus templates differing in chromatin structure. *J Biol Chem* **272**, 14 227–35.

Sohocki MM, Sullivan LS, Mintz-Hittner HA, Birch D, Heckenlively JR, Freund CL *et al.* (1998). A range of clinical phenotypes associated with mutations in CRX, a photoreceptor transcription-factor gene. *Am J Hum Genet* **63**, 1307–15.

Sommer L, Shah N, Rao M, and Anderson DJ (1995). The cellular function of MASH1 in autonomic neurogenesis. *Neuron* **15**, 1245–58.

Struhl K (1998). Histone acetylation and transcriptional regulatory mechanisms. *Genes Dev* **12**, 599–606.

Takeuchi K, Alexander RW, Nakamura Y, Tsujino T, and Murphy, TJ (1993). Molecular structure and transcriptional function of the rat vascular AT1a angiotensin receptor gene. *Circ Res* **73**, 612–21.

Trask BJ, Friedman C, Martin-Gallardo A, Rowen L, Akinbami C, Blankenship J *et al.* (1998). Members of the olfactory receptor gene family are contained in large blocks of DNA duplicated polymorphically near the ends of human chromosomes. *Hum Mol Genet* **7**, 13–26.

Tsai SJ, Wang YC, Yu Younger WY, Lin CH, Yang KH, and Hong CJ (2001). Association analysis of polymorphism in the promoter region of the alpha2a-adrenoceptor gene with schizophrenia and clozapine response. *Schizophr Res* **49**, 53–8.

Tseng YT and Padbury JF (2000). Transient transfection and adrenergic receptor promoter analysis. *Meth Mol Biol* **126**, 235–9.

Ueda Y, Iwakabe H, Masu M, Suzuki M, and Nakanishi S (1997). The mGluR6 5′ upstream transgene sequence directs a cell-specific and developmentally regulated expression in retinal rod and ON-type cone bipolar cells. *J Neurosci* **17**, 3014–23.

Valdenaire O, Vernier P, Maus M, Dumas-Milne-Edwards JB, and Mallet J (1994). Transcription of the rat dopamine-D2-receptor gene from two promoters. *Eur J Biochem* **220**, 577–84.

Vassar R, Chao SK, Sitcheran R, Nunez JM, Vosshall LB, and Axel R (1994). Topographic organization of sensory projections to the olfactory bulb. *Cell* **79**, 981–91.

Vassar R, Ngai J, and Axel R (1993). Spatial segregation of odorant receptor expression in the mammalian olfactory epithelium. *Cell* **74**, 309–18.

Vilaro MT, Palacios JM, and Mengod G (1990). Localisation of m5 muscarinic receptor mRNA in rat brain examined by in situ hybridisation histochemistry. *Neurosci Letts* **114**, 154–9.

Wang J, Miller JC, and Friedhoff AJ (1997). Differential regulation of D2 receptor gene expression by transcription factor AP-1 in cultured cells. *J Neurosci Res* **50**, 23–31.

Wang L, Wu Q, Qiu P, Mirza A, McGuirk M, Kirschmeier P *et al.* (2001) Analyses of p53 target genes in the human genome by bioinformatic and microarray approaches. *J Biol Chem* **276**, 43 604–10.

Weber HC, Jensen RT, and Battey JF (2000). Molecular organization of the mouse gastrin-releasing peptide receptor gene and its promoter. *Gene* **244**, 137–49.

Wood IC, Garriga-Canut M, Palmer CL, Pepitoni S, and Buckley NJ (1999). Neuronal expression of the rat M1 muscarinic acetylcholine receptor gene is regulated by elements in the first exon. *Biochem J* **340**, 475–83.

Wood IC, Roopra A, and Buckley NJ (1996). Neural specific expression of the m4 muscarinic acetylcholine receptor gene is mediated by a RE1/NRSE-type silencing element. *J Biol Chem* **271**, 14 221–5.

Wood IC, Roopra A, Harrington C, and Buckley NJ (1995). Structure of the m4 cholinergic muscarinic receptor gene and its promoter. *J Biol Chem* **270**, 30 933–40.

Wu Y, Ruef J, Rao GN, Patterson C, and Runge MS (1998). Differential transcriptional regulation of the human thrombin receptor gene by the Sp family of transcription factors in human endothelial cells. *Biochem J* **330**, 1469–74.

Yang C, Kim HS, Seo H, Kim CH, Brunet JF, and Kim KS (1998*a*). Paired-like homeodomain proteins, Phox2a and Phox2b, are responsible for noradrenergic cell-specific transcription of the dopamine beta- hydroxylase gene. *J Neurochem* 71, 1813–26.

Yang X and Polgar P (1996). Genomic structure of the human bradykinin B1 receptor gene and preliminary characterization of its regulatory regions. *Biochem Biophys Res Commun* 222, 718–25.

Yang X, Taylor L, and Polgar P (1998*b*). Mechanisms in the transcriptional regulation of bradykinin B1 receptor gene expression. Identification of a minimum cell-type specific enhancer. *J Biol Chem* 273, 10 763–70.

Yang Y, Hwang CK, Junn E, Lee G, and Mouradian MM (2000). ZIC2 and Sp3 repress Sp1-induced activation of the human D1A dopamine receptor gene. *J Biol Chem* 275, 38 863–9.

Zack DJ, Bennett J, Wang V, Davenport C, Klaunberg B, Gearhart J *et al.* (1991). Unusual topography of bovine rhodopsin promoter-*lacZ* fusion gene expression in transgenic mouse retinas. *Neuron* 6, 187–99.

Zozulya S, Echeverri F, and Nguyen T (2001), The human olfactory receptor repertoire. *Genome Biol* 2, 0018.

Chapter 3

Structural mechanics of GPCR activation

Søren G. F. Rasmussen and Ulrik Gether

3.1 Introduction

The majority of hormones, neurotransmitters, and other chemical signalling molecules in the human body exert their effects via G protein coupled receptors (GPCRs) (Strader *et al.* 1994; Ji *et al.* 1998; Gether 2000). Evidently, this extreme functional diversity raises some fundamental questions in relation to the function of GPCR at the molecular and cellular level. What are, for example, the molecular mechanisms underlying the ability of this broad variety of molecules to bind and activate receptors that presumably share a preserved overall tertiary structure? Or more specifically, what are the physical changes linking binding of an agonist to activation of intracellular signalling cascades? During the last few years our insight into these fundamental mechanisms have improved considerably. An important breakthrough has been the application of biophysical approaches that has allowed direct insight into the conformational changes that accompany activation of a GPCR (see Gether 2000; Hubbell *et al.* 2000). Another breakthrough has been the improvements in our understanding of the GPCR tertiary structure. Low-resolution structures of rhodopsin based on cryo-electron microscopy of two-dimensional crystals resolved by Schertler *et al.* (Schertler *et al.* 1993; Schertler and Hargrave 1995) became available already a few years ago. These provided important insights into the organization of the transmembrane bundle and allowed the development of tertiary structure models of GPCRs (Ballesteros and Weinstein 1995; Scheer *et al.* 1996; Baldwin *et al.* 1997). Recently, Palczewski *et al.* (Palczewski *et al.* 2000) succeeded in generating three-dimensional crystals of rhodopsin for X-ray crystallography offering for the first time a tertiary structure model of a GPCR at atomic resolution (2.8 Å) (Fig. 3.1). The structure appeared to be remarkably similar to the majority of existing receptor models that were developed based on the low-resolution structures, and accordingly a major revision of our current understanding of GPCR function has not been necessary. Nonetheless, the availability of high-resolution structural information has been able to clarify in particular previous hypothetical predictions regarding, for example, specific intramolecular interactions. Hence, it is now become possible to consider the functional roles of even individual side chains and pinpoint their specific role in receptor activation. This chapter will describe our current insight into agonist induced conformational changes in the GPCR structure leading to activation of the G protein cascade and discuss the importance of constraining intramolecular interactions that keep the receptor predominantly inactive.

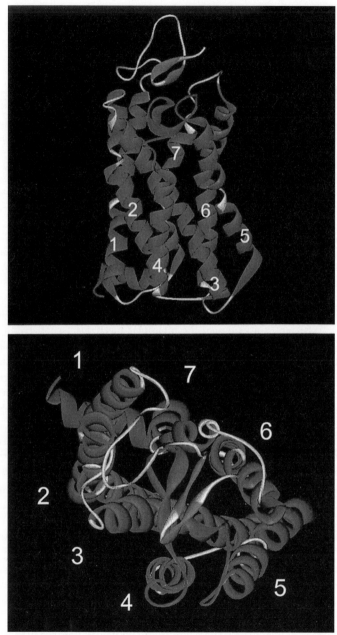

Fig. 3.1 The high-resolution structure of rhodopsin seen from the side (*upper panel*) and from the top (*lower panel*) (Palczewski *et al.* 2000). The seven transmembrane helices (indicated by numbers) are organized in a counterclockwise fashion (as seen from the extracellular side in the lower panel). A buried ligand binding crevice containing the covalently attached chromophore, retinal, is formed between the seven helices. The second extracellular loop 2 (ECL2, show in green) dives into the transmembrane domain and forms a plug in the binding crevice. (See Plate 2.)

3.2 Ligand binding in the GPCR tertiary structure

The high-resolution structure of rhodopsin unraveled at atomic resolution the seven helical-bundle characterizing GPCRs (Palczewski *et al.* 2000; Teller *et al.* 2001) (Fig. 3.1). The seven helices within the bundle are organized in a distinctive counterclockwise fashion (as seen from the extracellular side) and contain a large numbers of kinks, bends, and twists (Palczewski *et al.* 2000; Teller *et al.* 2001). Helices 1, 4, 6, and 7 are characterized by the presence of kinks at conserved proline residues while helix 3 is the longest and mostly tilted as compared to the plane of the membrane (Palczewski *et al.* 2000; Teller *et al.* 2001). In full agreement with the predictions from earlier mutagenesis studies, the chromophore 11-cis-retinal is contained within a deep binding pocket formed mainly by helices 3, 4, 5, 6, and 7 (Palczewski *et al.* 2000; Teller *et al.* 2001) (Fig. 3.1). Through formation of a Schiff base, 11-*cis*-retinal is covalently coupled to a lysine in transmembrane segment (TM) 7 (Lys296$^{7.43}$)[1] and the protonated Schiff base is paired with a glutamate (Glu113$^{3.28}$) in the outer part of TM3 (Palczewski *et al.* 2000; Teller *et al.* 2001). An interesting feature revealed by the crystal structure is that the second extracellular loop 2 (ECL2) connecting TM4 and 5 forms a lid in the binding pocket by diving down into the transmembrane region (Fig. 3.1) (Palczewski *et al.* 2000; Teller *et al.* 2001). Structurally, the loop contains two stretches of β-strands of which one is directly above retinal and contains residues forming contacts with retinal (Palczewski *et al.* 2000; Teller *et al.* 2001). Thus, retinal is completely surrounded by protein within its binding pocket (Palczewski *et al.* 2000; Teller *et al.* 2001). A key question is to what extent other GPCRs structurally resemble rhodopsin. Recently, a thorough comparison of the rhodopsin structure with data obtained from systematic application of the substituted cysteine accessibility method (SCAM) to the dopamine D$_2$ receptor was carried out (Ballesteros *et al.* 2001*b*). In the SCAM studies of the D$_2$ receptor, performed by Javitch *et al.* (Javitch *et al.* 2002) each residue in all seven TMs was one by one substituted with cysteine. The comparison indicated a remarkable structural similarity between the D$_2$ receptor and rhodopsin in the transmembrane domains (Ballesteros *et al.* 2001*b*). The amino acids residues, inferred based on the SCAM analysis to form the water accessible binding crevice in between the transmembrane helices, were almost completely consistent with the predictions from the rhodopsin structure (Ballesteros *et al.* 2001*b*).

At present it is more difficult to answer whether ECL2 is forming a plug in the binding crevice in other GPCRs as it does in rhodopsin (Fig. 3.1). If it does, it will have a major implication for our understanding of the receptor activation mechanism, hence, a key event in receptor activation would necessarily involve significant conformational changes of ECL2 allowing for rapid access of the ligand to the binding crevice. The ECL2 sequence is rather poorly conserved among the different receptors; however it does contain one of the two highly conserved cysteines known to form a disulfide bond between the top of TM3 and ECL2 (Gether 2000). It is also remarkable that several studies have suggested that the loop region may be functionally important. In the α$_{1B}$ adrenergic receptor, for example, it was found that residues in the loop are critical for the pharmacological specificity of some adrenergic ligands consistent with the notion that other small molecule ligands than 11-cis-retinal might form contacts with residues in this loop (Zhao *et al.* 1996). Furthermore, previous studies of the β$_2$ adrenergic receptor using a fluorescent ligand have demonstrated that the ligand binding

---

[1] The positions of residues are throughout the chapter indicated by their generic number followed by their number according to the Ballesteros-Weinstein nomenclature in superscript (Ballesteros and Weinstein 1995) (see legend to Fig. 3.2 for details).

site is completely inaccessible to aqueous quenchers (Tota and Strader 1990); an observation consistent with the presence of a 'plug' in the binding crevice.

Although GPCR structure might be evolutionarily highly conserved it must be emphasized that an extensive amount of evidence suggests that there is no common 'lock' for all agonist 'keys' (Schwartz and Rosenkilde 1996; Gether 2000). Accordingly, it is not required for receptor activation that all agonist molecules form a specific set of contact points in the receptor structure to promote receptor activation. Even agonists for the same receptor may not necessarily have to share the same binding site and it does not seem to be a specific requirement for receptor activation that the agonist ligand is docked in the transmembrane binding crevice (for reviews see Ji et al. 1998; Gether 2000). In other words, the mode of interaction of a given agonist depends on its individual chemical structure and whether it is an agonist or antagonist is determined by the mode of interaction and whether this mode preferentially stabilizes an inactive or an active receptor state (Ji et al. 1998; Gether 2000). In case of, for example, peptide hormones and neuropeptides, the most critical points of interactions are, in contrast to the small molecule ligands (Family A) found in the extracellular loops and in the amino terminus (Ji et al. 1998; Gether 2000). A special case is the Family 3 receptors, including among others the metabotropic glutamate and GABA receptors, which are characterized by containing their 'small molecule' ligand binding site in their large amino terminus (Ji et al. 1998; Gether 2000). Nonetheless, despite the almost extreme diversity in how structurally distinct ligands interact with their receptor, it is still highly likely that an underlying fundamental activation mechanism involving the transmembrane regions has been conserved during evolution given the common ability of the receptors to activate the same intracellular signalling pathways though the same classes of G proteins.

3.3 Conformational changes involved in receptor activation

Activation of GPCRs for 'diffusible' ligands is initiated by binding of the agonist ligand to the receptor molecule. This binding event is thought to trigger a cascade of structural changes in the receptor molecule that are capable of inducing activation of the associated G protein and subsequent stimulation of a broad variety of intracellular signal transduction pathways. Ultimate understanding of this activation mechanism requires techniques that can provide insight into the character of the physical changes which occur upon agonist binding to the receptor and accompany the transition of the receptor from the inactive to the active state. One obvious goal is to obtaining high-resolution crystal structures of GPCRs in both inactive and active states. However, due to the still prevailing difficulties in obtaining sufficient material for X-ray crystallography the only available high-resolution structure of a full-length GPCR is the structure described above of rhodopsin in its inactive dark state (Palczewski et al. 2000).

It is interesting to note that a structure of the large extracellular domain, which characterizes Family 3 receptors and contains their ligand binding domain, has been solved for the metabotrobic glutamate receptor-1 (see Chapter 4.3). The structure was solved in both the agonist-bound state and in two ligand-free conformational states of which one resembled the agonist bound state and the other 'inactive state' represented a conformation markedly different from the agonist bound active state (Kunishima et al. 2000). Although the structures provided no information about the transmembrane domains, they did show the first direct structural glimpse of what might distinguish an active state from an inactive state in at least for a Family C GPCR. Remarkably, the extracellular domain was crystallized as a dimer with a large dimeric interface (Kunishima *et al.* 2000). Comparison of the inactive state with the active state revealed major changes at the dimeric interface and it was suggested that such changes are highly critical for the receptor activation process (Kunishima *et al.* 2000). It remains unknown, however, how the signal is transmitted from the large extracellular domains to the transmembrane region of the protein.

Studies in Family 1 receptors performed during recent years have provided insight into conformational changes in the transmembrane region that characterizes transition from an inactive state to an active state. In particular the application of biophysical techniques have proven highly useful and allowed direct analysis of conformational changes in the receptor molecule (Gether 2000; Hubbell *et al.* 2000). Importantly, these techniques also allow insight into the dynamics of the receptor activation process in contrast to crystal structures that despite a high degree of structural information still only represents single pictures at fixed conformations. The majority of the initial biophysical studies were carried out in rhodopsin. There are abundant natural sources of rhodopsin and the inherent stability of the rhodopsin molecule makes it possible to produce and purify relatively large quantities of recombinant protein. Several spectroscopic techniques have been applied to rhodopsin including Fourier Transform Infrared Resonance Spectroscopy (FTIR) (Rothschild *et al.* 1983; Garcia-Quintana *et al.* 1995), Surface Plasmon Resonance (SPR) spectroscopy (Salamon *et al.* 1994), tryptophan UV-absorbance spectroscopy (Lin and Sakmar 1996) and Electron Paramagnetic Resonance Spectroscopy (EPR) (Farahbakhsh *et al.* 1995; Altenbach *et al.* 1996, 1999*a,b*). All approaches have consistently provided evidence for a significant conformational rearrangement accompanying transition of rhodopsin to metarhodopsin II. Using tryptophan UV-absorbance spectroscopy, Lin and Sakmar (1996) were able to obtain the first direct evidence that photoactivation may involve relative movements of TM3 and 6. Thus, mutation of tryptophans in TM3 and 6 eliminated the spectral differences in the UV absorbance spectra that distinguished rhodopsin from metarhodopsin II (Lin and Sakmar 1996).

In a series of elegant studies, the use of EPR spectroscopy in combination with multiple cysteine substitutions has led to further insight into the character of conformational changes accompanying photoactivation of rhodopsin (reviewed in Hubbell *et al.* 2000). Site-directed labelling of single cysteines inserted at the cytoplamic side of the transmembrane helices with sulfhydryl-specific nitroxide spin labels provided evidence for movements of particularly the cytoplasmic termination of TM6 upon light-induced activation of rhodopsin (Farahbakhsh *et al.* 1995; Altenbach *et al.* 1996, 1999*a,b*; Farrens *et al.* 1996; Langen *et al.* 1999). To investigate the character of the conformational changes, Khorana, Hubbell and coworkers have taken advantage of the magnetic dipole interaction between two nitroxide spin labels causing spectral line broadening if the two probes are less than 25 Å apart (Farrens *et al.* 1996). Pairs of sulfhydryl-reactive spin labels were incorporated into a series of double cysteine mutants enabling measurement of changes in relative distance between TM3 and TM6 (Farrens *et al.* 1996). While the movement of TM3 was interpreted as relatively small, the data pointed to a significant rigid-body movement of TM6 in a counterclockwise direction (as viewed from the extracellular side) and a movement of the cytoplasmic end of TM6 away from TM3 (Farrens *et al.* 1996). Importantly, these movements of TM6 in rhodopsin upon photoactivation have also been additionally documented by site-selective fluorescent labelling of cysteine inserted at the cytoplasmic termination of the helix (Dunham and Farrens 1999).

The first direct structural analysis of conformational changes in a GPCR activated by a diffusable ligand was performed in the β$_2$ adrenergic receptor (Chapter 3.2) using fluorescence spectroscopic techniques (Gether *et al.* 1995, 1997*a,b*; Ghanouni *et al.* 2001*b*; Jensen *et al.* 2001). The spectroscopic technique that initially was applied utilized the sensitivity of many fluorescent molecules to the polarity of their local molecular environment (Gether *et al.* 1995). The sulfhydryl reactive fluorophore IANBD (N,N′-dimethyl-N (iodoacetyl)-N′-(7-nitrobenz-2-oxa-1,3-diazol-4-yl) ethylene-diamine) was used to label free cysteine residues in purified detergent solubilized β_2 adrenergic receptor (Gether *et al.* 1995). Exposure of the IANBD-labelled receptor to agonist led to a reversible and dose-dependent decrease in emission consistent with movements of the fluorophore to a more hydrophilic environment following binding of the full agonist isoproterenol (Gether *et al.* 1995). Subsequent analysis of a series of mutant β_2 receptors with one, two, or three of the natural cysteines available for fluorescent labelling showed that IANBD bound to Cys125$^{3.44}$ in TM3 and Cys285$^{6.47}$ in TM6 (Fig. 3.2) were responsible for the observed changes in fluorescence

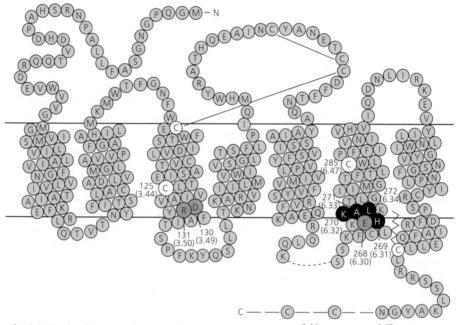

Fig. 3.2 'Snake-diagram' of the β_2 adrenergic receptor. Cys125$^{3.44}$ and Cys285$^{6.47}$ were identified as major labelling sites for the environmentally, cysteine-reactive fluorophore IANBD reporting agonist-induced conformational changes (Gether *et al.* 1997b). His269$^{6.31}$, Lys270$^{6.32}$, Ala271$^{6.33}$, and Leu272$^{6.34}$ were one by one mutated to cysteines in a background mutant with a reduced number of reactive cysteines (β2AR-Cys-min) (Jensen *et al.* 2001). Subsequent fluorescent labelling of these residues allowed identification of conformational changes at the cytoplasmic side of TM6 (Jensen *et al.* 2001). Asp130$^{3.49}$, Arg131$^{3.50}$, and Glu268$^{6.30}$ are believed to form an ionic lock that is disrupted during receptor activation (Ballesteros *et al.* 2001a). The positions of highlighted residues are indicated by their generic number followed by their number according to the Ballesteros--Weinstein nomenclature (Ballesteros and Weinstein 1995). In this scheme the most conserved residue in each helix is given the number 50, and each residue is numbered according to its position relative this conserved residue. For example, 3.49 indicates a residue in TM3, one residue aminoterminal to Arg3.50, the most conserved residue in this helix.

(Gether *et al.* 1997*b*) suggesting that movements of TM3 and 6 might occur during receptor activation (Gether *et al.* 1997*b*). It was, however, difficult to assess the precise character of the movements based on labelling of only two single sites with a rather flexible fluorescent reporter molecule. Therefore, a new set of experiments was carried out to achieve further insight into movements associated with receptor activation (Jensen *et al.* 2001). It was decided to focus on the cytoplasmic end of TM6 for two major reasons. First, the evidence for TM6 movements in response to light-activation of rhodopsin, was based on spectroscopic analysis of mutants which contained cysteine residues in this particular region of TM6, labeled with either nitroxide spin labels or fluorescent probes (Altenbach *et al.* 1996; Farrens *et al.* 1996; Dunham and Farrens 1999). Labelling of the β_2 adrenergic receptor in this region with a molecular reporter of conformational changes would thus allow a more direct comparison between rhodopsin and the β_2 adrenergic receptor. Second, many mutagenesis-based studies have indicated the importance of the cytoplasmic region of TM6 in receptor activation and G protein coupling (Strader *et al.* 1994; Wess 1998); nonetheless, its precise role is still not fully understood and conformational changes at the cytoplasmic side of TM6 had not been described for receptors activated by diffusable ligands (Gether 2000). Accordingly, four residues in the predicted cytoplasmic region of TM6 of the β_2 adrenergic receptor were substituted with cysteines in a mutant β_2 adrenergic receptor containing a reduced number of endogenous cysteines (β2AR-Cys-min) (Jensen *et al.* 2001) (Fig. 3.2). Fluorescence spectroscopy analysis of the purified and site-selectively IANBD labelled mutants suggested that the covalently attached fluorophore was exposed to a less polar environment at all four positions upon agonist binding (Jensen *et al.* 2001). Whereas evidence for only a minor change in the molecular environment was obtained for positions $269^{6.31}$ and $270^{6.32}$, the full agonist isoproterenol (ISO) caused clear dose-dependent and reversible increases in fluorescence emission at positions $271^{6.33}$ and $272^{6.34}$ (Fig. 3.3; Jensen *et al.* 2001). The magnitude of the responses correlated with the efficacy of the used agonist suggesting that the observed changes are relevant for receptor activation (Jensen *et al.* 2001). In contrast to Cys$285^{6.47}$, which is situated in a highly hydrophobic environment in the middle of the membrane, the cysteines inserted at the cytoplasmic side of TM6 reside in a very complex hydrophobic/hydrophilic environment (Figs 3.2 and 3.3). To take this into consideration when attempting to interpret the spectroscopic data a new computational method was developed. Notably, the available computational simulation methods had not incorporated the complexity of the mixed hydrophobic–hydrophilic region and only recently, a bi-phasic lipid-water solvent continuum model was developed (Ballesteros *et al.* 1998). The results of the simulations, illustrated by the most preferred conformations for each of the four IANBD-derivatized cysteine residues, are shown in context of a molecular model in Fig. 3.3D (viewed from the extracellular side).

A simple rotation of TM6 was sufficient to explain the data based on IANBD labelling of Cys$285^{6.47}$ (Gether *et al.* 1997*b*). However, the observation that IANBD at all four inserted cysteines residues most likely moves into a more hydrophobic environment upon agonist binding suggests that the helical movement is more complex and likely involves a rigid-body movement of the cytoplasmic part of TM6 away from TM3 similar to what was found for rhodopsin (Farrens *et al.* 1996). As seen from the side (Fig. 3.3G), TM6 is predicted to form a kinked α-helix due to the presence of a highly conserved proline (Pro$287^{6.50}$) (Ballesteros and Weinstein 1995). The rhodopsin structure (Palczewski *et al.* 2000), which are believed to represent the inactive state of the receptor, indicates that the cytoplasmic part of TM6 below the proline kink is almost perpendicular to the plane of the membrane, whereas the

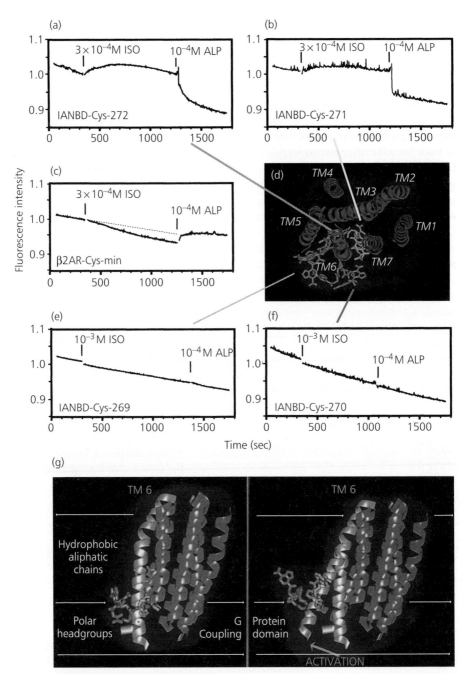

Fig. 3.3 Characterization of agonist-induced conformational changes at the cytoplasmic side of TM6 in the β_2 adrenergic receptor. His269[6.31], Lys270[6.32], Ala271[6.33], and Leu272[6.34] at the bottom of TM6 (Fig. 3.2) were one by one mutated to cysteines in a background mutant with a reduced number of reactive cysteines (β2AR-Cys-min) (Jensen *et al.* 2001). The mutants were expressed in Sf-9 insect cells, purified and labeled with the environmentally, sulfhydryl reactive fluorophore, IANBD. Panels a, b, c, e, and f show time course experiments where fluorescence intensity is measured over time in response to indicated concentrations of (-)isoproterenol (ISO)

part above the proline kink is tilted approximately 25° (Unger *et al.* 1997). A rigid body movement of the cytoplasmic part of TM6 away from TM3, and thus the receptor core, will result in large changes in the axial positioning of all four IANBD-labelled substituted cysteines. In the inactive conformation the IANBD moieties would be predicted to reside in the polar headgroup region (Fig. 3.3G). However, if the cytoplasmic part of TM6 is moved away from the receptor core all four IANBD labelled residues are brought upwards and outwards allowing them to penetrate further into the more hydrophobic region of the membrane/detergent micelles and away from the more hydrophilic polar headgroups as well as from the predicted more hydrophilic interior of the receptor protein (illustrated by the hypothetical active structure in Fig. 3.3G, right panel, where the cytoplasmic part of TM6 with the IANBD moieties attached is tilted arbitrarily away from the receptor core).

The importance of TM6 movements relative to TM3 is also supported by the possibility of inhibiting activation of rhodopsin by generation of bis-His metal ion binding sites between

---

Fig. 3.3 *(Continued)* followed by 10^{-4} M (-)alprenolol (ALP). Excitation was 481 nm and emission was measured at 530 nm. Fluorescence in the individual traces was normalized to the fluorescence observed just before addition of ligand (Jensen *et al.* 2001). In β2AR-Cys-min, ISO causes a decrease in fluorescence that can be reversed by ALP. This decrease is reported by IANBD bound to Cys125$^{3.44}$ and Cys285$^{6.47}$ (Gether *et al.* 1997b). In Cys271 and Cys272, ISO causes ALP-reversible increases in fluorescence intensity consistent with movement of the fluorophore to a more hydrophobic environment. In Cys269 and Cys270, ISO causes no apparent change in fluorescence intensity. A likely interpretation is that ISO induces an increase that counterbalances the decrease observed in the control (β2AR-Cys-min). Panel d shows an extracellular view of a receptor model with an illustrative set of the preferred conformations of the IANBD side chain covalently attached to the four substituted cysteines (color-coded). Note that IANBD attached to Cys271$^{6.33}$ (yellow) and Cys272$^{6.34}$ (purple) are facing the interior of the TM helix bundle, while IANBD attached to Cys269$^{6.31}$ (blue) and Cys270$^{6.32}$ (red) are oriented towards the lipid membrane. The preferred conformations of IANBD were determined from computational simulations as described (Jensen *et al.* 2001). Panel G shows the proposed conformations of the inactive and active states of the β2AR. The inactive conformation of the receptor (left panel) is characterized by a highly kinked TM6 helix (blue) with the cytoplasmic end in close proximity to TM3 and the helix bundle. An illustrative set of the preferred conformations of the IANBD moiety covalently attached to the four substituted cysteines at the cytoplasmic side of TM6 is shown. The hypothetical active conformation (right panel) of the receptor in which the cytoplasmic side of TM6 is moved away arbitrarily from the helix bundle and upwards towards the hydrophobic region, marked by straight lines. This putative rearrangement of TM6 moves all four IANBD labelled residues upwards and outwards allowing them to penetrate further into the more hydrophobic region of the membrane/detergent micelles and away from the more hydrophilic polar headgroups as well as from the predicted more hydrophilic interior of the receptor protein. The movement can explain the observed shift for all four IANBD labelled cysteines towards a less polar environment upon receptor activation (Jensen *et al.* 2001). Note that the movement of the cytoplasmic part of TM6 is shown to occur around the conserved proline kink but could as well involve a rigid body movement of the entire helix. However, our previous simulation of the TM6 helix indicated the possibility that the kink in the TM6 helix induced by Pro287$^{6.50}$ could behave as a flexible hinge, which can modulate the movement of the cytoplasmic side of TM6 helix relative to the extracellular region (Gether *et al.* 1997b). (Figure modified from Jensen *et al.*, 2001, *J Biol Chem* **276**, 9279–90). (See Plate 3.)

the cytoplasmic ends of these TM domains (Sheikh *et al.* 1996). The recent application of a disulfide cross-linking strategy to the M_3 muscarinic receptor, another Family 1 member, has moreover suggested that significant movements occur in the G protein coupling domain at the cytoplasmic side of TM6, although the movements predicted by the authors are somewhat distinct from those predicted in the biophysical studies (Ward *et al.* 2001). Altogether, the striking agreement between the data obtained in rhodopsin and in the β_2 adrenergic receptor strongly indicates that the activation mechanism in many aspects is similar among at least Family 1 GPCRs. It is important to note that the established importance of TM6 does not exclude that movements of other domains may contribute to receptor activation. New evidence from EPR spectroscopy studies in rhodopsin suggest that movements of the cytoplasmic portion of TM7 relative to TM1 and of TM2 relative to the so-called intracellular helix 8 (the horizontal extension of TM7) may also occur in response to photoactivation (Altenbach *et al.* 2001*a,b*). The possible importance of TM7 in receptor activation is indirectly supported by the finding that activating metal-ion binding sites can be generated between TM3 and 7 in both the β_2 adrenergic receptor and in the neurokinin-1 receptor (Elling *et al.* 1999; Holst *et al.* 2000).

3.4 Insights into the molecular mechanisms of partial agonism

A compelling question that also arises when discussing conformational changes involved in GPCR activation is the molecular mechanism underlying partial agonism, that is, the fact that some compounds elicit physiological responses lower than that of a so-called full agonist. According to the classical two-state model of receptor activation the physiological response is determined by the ability of a given ligand to alter the distribution between the inactive state (R) and the active state (R*) (Samama *et al.* 1993). While a full agonist would be predicted to move the majority of the receptor molecules into the R* state, a partial agonist would cause a less dramatic change in the distribution between R and R* (Samama *et al.* 1993). Accordingly, a lower fraction of receptor molecules would reside in R* as compared to what is seen in response to the full agonist and as a result the physiological response will be smaller. However, increasing evidence indicates that the action of partial agonists cannot be explained sufficiently within the framework of such a two-state model (see Gether 2000). Recent analysis of fusion proteins between $G\alpha_S$ and the wild-type β_2 adrenergic receptor and between $G\alpha_S$ and a constitutively activated β_2 adrenergic receptor mutant have, for example, showed an interesting discrepancy between the efficacy of ligands in stimulating GTPase activity and in their ability to stabilize the ternary complex, that is, high affinity, GTP sensitive agonist binding (Seifert *et al.* 2001). The data suggest the possibility that partial agonism at least in some cases may be explained by the ability of certain partial agonist to strongly stabilize the ternary complex resulting in a reduced rate of G protein turnover as compared to the full agonist (Seifert *et al.* 2001). Further support for a complex action of partial agonists has been obtained in recent biophysical studies. These studies include both the application of single-molecule fluorescence analysis as well as fluorescence lifetime analysis to a purified preparation of the β_2 adrenergic receptor covalently labelled with a fluorescence reporter molecule (Ghanouni *et al.* 2001*a*; Peleg *et al.* 2001). The data demonstrated clear evidence for the existence of different conformational sub-states of the β_2 adrenergic receptor that are differentially modulated by different agonists thus suggesting that different agonists stabilizes distinct conformational states of the receptor molecule (Ghanouni *et al.* 2001*a*; Peleg *et al.* 2001).

3.5 Kinetics of GPCR activation

While converging on the involvement of especially TM6 in receptor activation, the spectroscopic studies on rhodopsin and the β_2 adrenergic receptor also indicated some possible important differences between the two receptors. The EPR spectroscopic read-outs for rhodopsin demonstrated a rapid formation of the active metarhodopsin II state (within microseconds) following light-induced conversion of the prebound cis-retinal to all-trans-retinal, whereas the conversion of metarhodopsin back to the inactive metarhodopsin III state was slow with a $t_{1/2}$ of about 6 min (Farahbakhsh *et al.* 1993). In contrast to the rapid activation and the slow inactivation kinetics observed for rhodopsin, the spectroscopic analyses of the β_2 adrenergic receptor indicated slow agonist-induced conformational changes ($t_{1/2} \sim$ 2–3 min), significantly slower than the predicted association rate of the agonist (Gether *et al.* 1995, 1997*b*; Ghanouni *et al.* 2001*b*; Jensen *et al.* 2001). However, the reversal of the agonist-induced conformational change was relatively fast ($t_{1/2} \sim$ 30 sec) (Gether *et al.* 1995, 1997*b*; Ghanouni *et al.* 2001*b*; Jensen *et al.* 2001). These differences in activation kinetics could either reflect inherent differences between the two receptors or they could be a consequence of differences in how the changes were detected. Since the measurements were performed under similar conditions it is nevertheless more likely that they reflect differences between rhodopsin and a receptor activated by a 'diffusible' ligand. Of notable interest, several spectroscopic read-out have now been established either upon IANBD labelling of endogenous cysteines or cysteines at the cytoplasmic end of TM6 or upon labelling with fluoresceine maleimide in G protein coupling domain (Gether *et al.* 1995, 1997*b*; Ghanouni *et al.* 2001*b*; Jensen *et al.* 2001). This consistency for the different read-outs, and the clear correlation between change in fluorescence and biological efficacy (Gether *et al.* 1995; Ghanouni *et al.* 2001*b*; Jensen *et al.* 2001) support the contention that the slow kinetics is an intrinsic property of the receptor, at least when conformational changes are being probed on purified protein using the sulfhydryl-reactive fluorophore, IANBD.

It is important to emphasize that our experiments have been carried out in the absence of G protein and that it is very likely that the G protein can affect the kinetics of the transition between the inactive agonist bound receptor complex (the AR state) and the active agonist bound receptor complex (the AR* state). Clearly, the slow kinetics of the agonist-induced conformational change in the absence of G protein would predict a high activation energy barrier for this transition (Gether *et al.* 1997*a*). It is conceivable that the G protein is able to stabilize the agonist-receptor complex and accordingly lower the activation energy barrier substantially and cause receptor activation to occur significantly faster. This may provide an explanation for the apparent discrepancy between the slow kinetics of agonist-induced conformational changes observed for the purified β_2 adrenergic receptor with the rapid responses to agonist-stimulation of GPCRs in cells, such as, for example, activation of ion channels. Unfortunately, it has not yet been possible to test the kinetic importance of the G protein due to technical difficulties mostly due to instability of the $G\alpha_S$ protein.

3.6 GPCR activation involves disruption of stabilizing intramolecular interactions

The biophysical studies together with an array of different indirect approaches have provided evidence for movements of TM6 relative to TM3 in agonist-induced receptor activation. But what are the molecular mechanisms that control the movements of TM6 and thus govern the transition of the receptor between its inactive and active state? An important discovery in relation to this question was the now well-established fact that many GPCRs possess basal activity and accordingly can activate the G protein in the absence of agonists (Costa and Herz 1989; Samama *et al.* 1993; Chidiac *et al.* 1994). In addition, discrete mutations are able to dramatically increase this constitutive agonist-independent receptor activity (Allen *et al.* 1991; Kjelsberg *et al.* 1992; Lefkowitz *et al.* 1993; Samama *et al.* 1993; Scheer *et al.* 1996). A crucial clue about the molecular mechanisms underlying constitutive receptor activation came from a study where the naturally occurring Ala293[6.34] residue in the C-terminal part of third intracellular loop of the α_{1b} adrenergic receptor was substituted with all other possible residues. They found that the presence of all other residues than the alanine resulted in higher agonist-independent receptor activity (Kjelsberg *et al.* 1992). This led to the suggestion that constraining intramolecular interactions have been conserved during evolution to maintain the receptor preferentially in an inactive conformation in the absence of agonist. Conceivably, these inactivating constraints could be released as a part of the receptor activation mechanism, either following agonist binding or due to specific mutations, causing key sequences to be exposed to G protein. The hypothesis has been indirectly supported by the observation that constitutively activated β_2 adrenergic receptor and histamine H_2 receptor mutants are characterized by a marked structural instability (Gether *et al.* 1997a; Rasmussen *et al.* 1999; Alewijnse *et al.* 2000) and enhanced conformational flexibility (Gether *et al.* 1997a). The data imply that the mutational changes have disrupted important stabilizing intramolecular interactions in the tertiary structure, allowing the receptor more readily to undergo conversion between its inactive and active state.

If receptor activation involves disruption of stabilizing intramolecular interactions, an obvious next question is which interactions are actually broken and how is this initiated following agonist binding to the receptor molecule. Despite the availability of high-resolution structure of a GPCR, this question can still not be fully answered; however, substantial evidence suggests that at least one of the key events in the activation process among Family 1 GPCRs involves protonation of the aspartic acid or glutamic acid in the highly conserved D/E RY (Glu/Asp-Arg-Tyr) motif at the cytoplasmic side of TM3 (Fig. 3.2 and see Chapter 1). It has been suggested that this protonation event leads to release of constraining intramolecular interactions, thereby resulting in movements of TM6 and conversion of the receptor to the active state. The most direct evidence for protonation of Asp/Glu3.49 has been obtained by Sakmar and coworkers (Arnis *et al.* 1994) who compared wild-type rhodopsin and rhodopsin mutated in position Glu134[3.49] by flash photolysis allowing simultaneous measurement of photoproduct formation and rates of pH changes. Their data strongly suggested that proton uptake of Glu134[3.49] accompany formation of the metarhodopsin II state (Arnis *et al.* 1994). In the β_2 adrenergic receptor lowering of the pH has also been shown to facilitate transition of the receptor to the activated state (Ghanouni *et al.* 2000). The 'protonation hypothesis' has been further supported by the observation that charge-neutralizing mutations, which mimics the protonated state of the aspartic acid/glutamic acid, cause dramatic constitutive activation of both the adrenergic α_{1b} receptor, the β_2 adrenergic receptor and the histamin H_2 receptor (Scheer *et al.* 1996, 1997; Rasmussen *et al.* 1999, Alewijnse *et al.* 2000). Similarly, improved coupling has been observed by mutation of the aspartic acid in the GnRH receptor (Ballesteros *et al.* 1998). Mutation of the aspartic residue in the M_1 muscarinic receptor resulted in phosphoinositide turnover responses of the mutant that were quantitatively similar to the wild-type despite markedly lowered levels of expression (Lu *et al.* 1997). In parallel, constitutive activation was observed in rhodopsin following mutation of the glutamic acid found in the corresponding position of this receptor (Cohen *et al.* 1993). Finally, it was found that charge-neutralizing mutations of the aspartic acid (Asp130$^{3.49}$) in the β_2 adrenergic receptor are linked to the overall conformation of the receptor (Rasmussen *et al.* 1999). Thus, mutation of Asp130$^{3.49}$ to asparagine did not only activate the receptor but also caused a cysteine in TM6 (Cys285$^{6.47}$), which is not accessible in the wild-type receptor, to become accessible to methanethiosulfonate ethylammonium (MTSEA), a charged, sulfhydryl-reactive reagent (Rasmussen *et al.* 1999). This observation is consistent with a counter clockwise rotation (as seen from the extracellular side) and/or tilting of TM6 in the mutant receptor in agreement with the biophysical studies described above.

The network of constraining intramolecular interactions involving Asp/Glu3.49 and maintaining the receptor in its inactive state has until recently not been clear. Ballesteros *et al.* (1998) proposed that the ionic counterpart of Asp3.49 in the inactive state could be the adjacent Arg3.50 (Ballesteros *et al.* 1998) (Figs 3.2 and 3.4). In contrast, Scheer *et al.* (1996) have suggested, based on simulations in the α_{1B} adrenergic receptor, that Arg3.50 in the inactive state is constrained in a 'polar pocket' formed by residues in TM1, 2, and 7 and that it is not interacting with Asp3.49 (Scheer *et al.* 1996). The specific ionic counterpart of the arginine in the inactive state in this scheme was predicted to be the conserved aspartic acid in TM2 (Asp2.50; Asp-79 in β_2 adrenergic receptor) (Scheer *et al.* 1996). However, given that the high-resolution structure of rhodopsin indicates that Glu3.49 interacts with Arg3.50 in this receptor the most likely prediction is a similar interaction also among family A receptors containing an aspartate in position 3.49 (Fig. 3.4). In addition to the interaction with Glu3.49, the rhodopsin structure also suggested an interaction between Arg3.50 and a conserved glutamate in TM6, Glu6.30. A similar interaction was therefore proposed for the β_2 adrenergic receptor and it was suggested that the interactions formed between Asp3.49, Arg3.50, and Glu6.30 form an ionic switch that controls transition of the receptor between its inactive and active state (Ballesteros *et al.* 2001a). The hypothesis was supported by the observation that charge-neutralizing mutation of Glu6.30 alone or in combination with Asp3.49 resulted in constitutive activation and a marked increased efficacy for the partial agonist pindolol (Ballesteros *et al.* 2001a) (Fig. 3.4). The mutants showed evidence that disruption of the interaction between TM3 and TM6 not only produced constitutive receptor activation but also demonstrated a tight correlation between the extent of constitutive activation and the extent of conformational rearrangement in TM6 as determined by assessing the accessibility of MTSEA to Cys285$^{6.47}$ (Ballesteros *et al.* 2001a). Interestingly, substitution of Glu$^{6.30}$ with alanine resulted in a higher degree of constitutive activation than substitution with glutamine (Fig. 3.4). In full agreement with the hypothesis, this is presumably because glutamine can still hydrogen bond to Arg3.50 and maintain an interaction, albeit more weakly (Ballesteros *et al.* 2001a).

3.7 Is GPCR oligomerization required for receptor activation and G protein coupling?

While the crystal structure of rhodopsin suggested that this receptor exists and operates as a monomeric structure, it is apparent that many, if not all other GPCRs might form either homodimeric or even heterodimeric complexes (see Bouvier 2001 and Chapter 1). The strongest evidence for unequivocal roles of dimerization has been obtained among the Family 3 receptors (see Part 4). The crystal structure of the extracellular domain of the metabotropic glutamate receptor unraveled not only that the receptor exists as a dimeric protein but also

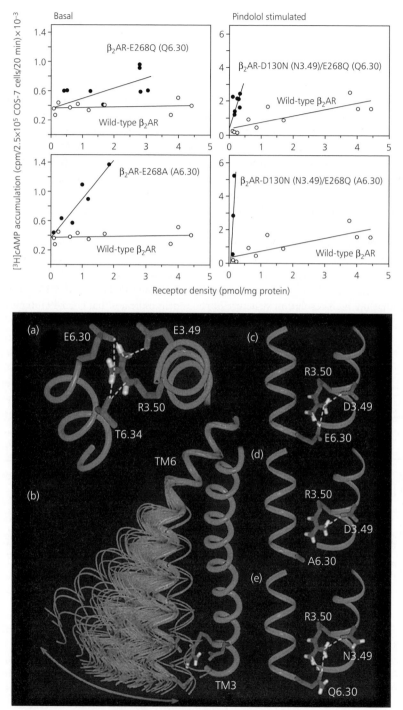

Fig. 3.4 Activation of the β$_2$ adrenergic receptor involves disruption of an ionic lock between TM3 and TM6. The *upper left panels* show how mutation of Glu268$^{6.30}$ to Gln and Ala constitutively enhances basal receptor activity. The *upper right panels* show how the combined

that dimerization of the receptor most likely plays a key role in the receptor activation mechanism (Kunishima *et al.* 2000). Moreover, studies of the metabotropic GABA$_B$ receptors, which also belong to family C, showed yet another function of dimerization. Thus, formation of a heterodimeric complex between the GABA$_{B1}$ (GBR1) and GABA$_{B2}$ (GBR2) was found to be required for surface expression of a functional receptor (see Bouvier 2001). Specifically, GBR1 requires GBR2 for proper targeting to the surface, whereas GBR2 by itself is surface expressed but does not bind GABA on its own (see Bouvier 2001). Interestingly, this heterodimerization may also be critical for receptor activation and G protein coupling since recent data have provided support for a surprising model where GABA binds to GBR1 but transmission of the signal to the G protein occurs solely via GBR2 (Calver *et al.* 2001; Robbins *et al.* 2001; Duthey *et al.* 2001; Margeta-Mitrovic *et al.* 2001). How this signal transmission takes place from one receptor to the other at the structural level is unknown as well as it remains to be clarified whether a similar scenario might be relevant for other GPCRs.

The formation of homodimers among Family 1 receptors in living cells has been supported in particular by the measurements of bioluminescence resonance energy transfer (BRET) or fluorescence resonance energy transfer (FRET) using fusion constructs between a GPCR and luciferase and/or yellow green fluorescent proteins, respectively (Bouvier 2001). In some cases, it has even been shown that agonists cause alterations in the degree of dimerization as reflected in changes in BRET or FRET. However, while the agonist-induced increase in energy transfer in the β_2 adrenergic receptor and in the thyrotropin-releasing hormone

---

Fig. 3.4 (*Continued*) mutation of Asp130$^{3.49}$ and Glu268$^{6.30}$ dramatically enhances the efficacy of the weak partial agonist, pindolol. It should be noted that mutation of Glu268$^{6.30}$ also enhances the efficacy of pindolol as well as mutation of Asp130$^{3.49}$ alone both enhances basal and pindolol-stimulated activity (data not shown, see Rasmussen *et al.* 1999; Ballesteros *et al.* 2001a). However, the combined mutations cause the most dramatic effect (Ballesteros *et al.* 2001a). The experiments were done in transiently transfected COS-7 cells (Ballesteros *et al.* 2001a). The values ([^{3}H]cAMP accumulation) are plotted as a function of receptor density, $n = 4$–10 individual transfections of wild-type, E268Q, E268A, D130N/E268Q, and D130N/E268A. The *lower panel* shows molecular 3D-representations of the interaction of TM3-TM6 at their cytoplasmic ends and the effects of 6.30 mutations. The Cα traces are taken from the high-resolution structure of bovine rhodopsin (Palczewski *et al.* 2000). Except in panel a, the top of each panel shows the extracellular end and the bottom the intracellular end of the TMs. In panel a, an extracellular view of the high-resolution structure of rhodopsin showing the interaction between residues at the cytoplasmic ends of TM3 and TM6 (Palczewski *et al.* 2000). In panel b, the Cα ribbons of TM3 and TM6 of rhodopsin are shown in purple. The simulated structures (light blue ribbons) are superimposed from position 1 to 17 onto the corresponding parts of TM6 (6.30 to 6.46) (Ballesteros *et al.* 2001a). The blue arrow indicates the conformational space that a Pro-kink can assume relative to TM3. Panel c shows a closer view of the interactions of Glu6.30-Arg3.50-Asp3.49 in a model of the β_2 adrenergic receptor based on the rhodopsin structure. In c, the wild-type interactions are shown, with the closest contacts between Glu6.30-Arg3.50 and Arg3.50-Asp.49, which are within the distance range of an ionic interaction, shown in dashed lines. Panels d and e show the same view as in Panel c, but with alterations in putative interactions after mutation of Glu6.30 to Ala (d) or of Glu6.30 to Gln and of Asp3.49 to Asn (e). (Figure modified from Ballesteros *et al.* 2001, *J Biol Chem* **276**, 29171–7). (See Plate 4.)

receptor (Angers *et al.* 2000; McVey *et al.* 2001), a decrease has recently been observed for the cholecystokinin receptor (Cheng and Miller 2001). Thus, agonist may either stabilize the dimeric complex or in some cases dissociate them. All in all, the capability of GPCRs to form dimers is now evident and in many cases a critical function has been substantiated. However, the functional implications of dimerization are surprisingly diverse and therefore there is still no general evidence among Family A receptors indicating that the formation of an oligomeric complex at the structural level is an ultimate prerequisite for receptor activation and G protein coupling. Obviously, further studies are required to clarify these aspects.

3.8 Conclusions

During the last couple of years our understanding of GPCR function at the molecular level has been considerably improved. The application of advanced biophysical techniques as well as the availability of high-resolution structural information has allowed insight both into conformational changes accompanying GPCR activation and the underlying molecular mechanims governing transition of the receptor between its active and inactive states. In particular, parallel studies in rhodopsin and the β_2 adrenergic receptor have supported a evolutionary conserved mechanism where disruption of intramolecular interactions between TM3 and TM6 leads to a major conformational change of TM6 relative to the rest of the receptor. The next key question is obviously to understand how the associated hetrotrimeric G protein senses this evolutionary conserved conformational change. High-resolution structural information of the hetrotrimeric G proteins has been available for a while (Wall *et al.* 1995; Lambright *et al.* 1996) and now a high-resolution structure of at least the inactive state of rhodopsin is available. Nevertheless, the orientation of the G protein relative to the receptor and how the signal is transmitted from the predicted G protein coupling domain to the guanine nucleotide binding domain in the G protein α-subunit still remains rather speculative. Its clarification will await further studies including hopefully at some point a high-resolution crystal structure of the entire receptor/G protein complex.

Acknowledgements

Dr. Jonathan Javitch and Lei Shi are thanked for providing the molecular model shown in Fig. 3.4 and Dr. Juan Ballesteros is thanked for providing the models shown in Fig. 3.3.

References

Alewijnse AE, Timmerman H, Jacobs EH *et al.* (2000). The effect of mutations in the DRY motif on the constitutive activity and structural instability of the histamine H(2) receptor. *Mol Pharmacol* **57**, 890–8.

Allen LF, Lefkowitz RJ, Caron MG, and Cotecchia S (1991). G-protein-coupled receptor genes as protooncogenes: constitutively activating mutation of the alpha 1B-adrenergic receptor enhances mitogenesis and tumorigenicity. *Proc Natl Acad Sci USA* **88**, 11 354–8.

Altenbach C, Cai K, Khorana HG, and Hubbell WL (1999*a*). Structural features and light-dependent changes in the sequence 306–322 extending from helix VII to the palmitoylation sites in rhodopsin: a site-directed spin-labeling study. *Biochemistry* **38**, 7931–7.

Altenbach C, Klein-Seetharaman J, Hwa J, Khorana HG, and Hubbell WL (1999*b*). Structural features and light-dependent changes in the sequence 59–75 connecting helices I and II in rhodopsin: a site-directed spin-labeling study. *Biochemistry* **38**, 7945–9.

Altenbach C, Cai K, Klein-Seetharaman J, Khorana HG, and Hubbell, WL (2001*a*). Structure and function in rhodopsin: mapping light-dependent changes in distance between residue 65 in helix TM1 and residues in the sequence 306–319 at the cytoplasmic end of helix TM7 and in helix H8. *Biochemistry* **40**, 15 483–92.

Altenbach C, Klein-Seetharaman J, Cai K, Khorana HG, and Hubbell WL (2001*b*). Structure and function in rhodopsin: mapping light-dependent changes in distance between residue 316 in helix 8 and residues in the sequence 60–75, covering the cytoplasmic end of helices TM1 and TM2 and their connection loop CL1. *Biochemistry* **40**, 15 493–500.

Altenbach C, Yang K, Farrens DL, Farahbakhsh ZT, Khorana HG, and Hubbell WL (1996). Structural features and light-dependent changes in the cytoplasmic interhelical E-F loop region of rhodopsin: a site-directed spin-labeling study. *Biochemistry* **35**, 12 470–8.

Angers S, Salahpour A, Joly E *et al.* (2000). Detection of beta 2-adrenergic receptor dimerization in living cells using bioluminescence resonance energy transfer (BRET). *Proc Natl Acad Sci USA* **97**, 3684–9.

Arnis S, Fahmy K, Hofmann KP, and Sakmar TP (1994). A conserved carboxylic acid group mediates light-dependent proton uptake and signalling by rhodopsin. *J Biol Chem* **269**, 23 879–81.

Baldwin JM, Schertler GF, and Unger VM (1997). An alpha-carbon template for the transmembrane helices in the rhodopsin family of G-protein-coupled receptors. *J Mol Biol* **272**, 144–64.

Ballesteros J, Kitanovic S, Guarnieri F *et al.* (1998). Functional microdomains in G-protein-coupled receptors. The conserved arginine-cage motif in the gonadotropin-releasing hormone receptor. *J Biol Chem* **273**, 10 445–53.

Ballesteros JA, Jensen AD, Liapakis G *et al.* (2001*a*). Activation of the beta 2-adrenergic receptor involves disruption of an ionic lock between the cytoplasmic ends of transmembrane segments 3 and 6. *J Biol Chem* **276**, 29 171–7.

Ballesteros JA, Shi, L, and Javitch JA (2001*b*). Structural mimicry in G protein-coupled receptors: implications of the high-resolution structure of rhodopsin for structure-function analysis of rhodopsin-like receptors. *Mol Pharmacol* **60**, 1–19.

Ballesteros JA and Weinstein H (1995). Integrated methods for the construction of three-dimensional models and computational probing of structure-function relations in G protein coupled receptors. *Meth Neurosci* **25**, 366–428.

Bouvier M (2001). Oligomerization of G-protein-coupled transmitter receptors. *Nat Rev Neurosci* **2**, 274–86.

Calver AR, Robbins MJ, Cosio C, Rice SQJ, Minton AL, Babb A *et al.* (2001). The C-terminal domain of GABA$_{B1}$ mediates intracellular trafficking, but is not required for receptor signalling. *J Neurosci* **21**(4), 1203–10.

Cheng ZJ and Miller LJ (2001). Agonist-dependent dissociation of oligomeric complexes of g protein-coupled cholecystokinin receptors demonstrated in living cells using bioluminescence resonance energy transfer. *J Biol Chem* **276**, 48 040–7.

Chidiac P, Hebert TE, Valiquette M, Dennis M, and Bouvier M (1994). Inverse agonist activity of beta-adrenergic antagonists. *Mol Pharmacol* **45**, 490–9.

Cohen GB, Yang T, Robinson PR, and Oprian DD (1993). Constitutive activation of opsin: influence of charge at position 134 and size at position 296. *Biochemistry* **32**, 6111–15.

Costa T and Herz A (1989). Antagonists with negative intrinsic activity at delta-opioid receptors coupled to GTP-binding proteins. *Proc Natl Acad Sci USA* **86**, 7321–5.

Dunham TD and Farrens DL (1999). Conformational changes in rhodopsin. Movement of helix f detected by site-specific chemical labeling and fluorescence spectroscopy. *J Biol Chem* **274**, 1683–90.

Duthey B, Caudron S, Perroy J *et al.* (2001). A single subunit (GB2) is required for G protein activation by the heterodimeric GABAB receptor. *J Biol Chem* **15**, 15.

Elling CE, Thirstrup K, Holst B, and Schwartz TW (1999). Exchange of agonist site with metal-ion chelator site in the β2 adrenergic receptor. *Proc Natl Acad Sci USA* **96**, 12 322–7.

Farahbakhsh ZT, Hideg K, and Hubbell WL (1993). Photoactivated conformational changes in rhodopsin: a time-resolved spin label study. *Science* **262**, 1416–9.

Farahbakhsh ZT, Ridge KD, Khorana HG, and Hubbell WL (1995). Mapping light-dependent structural changes in the cytoplasmic loop connecting helices C and D in rhodopsin: a site-directed spin labeling study. *Biochemistry* **34**, 8812–19.

Farrens DL, Altenbach C, Yang K, Hubbell WL, and Khorana HG (1996). Requirement of rigid-body motion of transmembrane helices for light activation of rhodopsin. *Science* **274**, 768–70.

Garcia-Quintana D, Francesch A, Garriga P, de Lera AR, Padros E, and Manyosa J (1995). Fourier transform infrared spectroscopy indicates a major conformational rearrangement in the activation of rhodopsin. *Biophys J* **69**, 1077–82.

Gether U (2000). Uncovering molecular mechanisms involved in activation of G protein-coupled receptors. *Endocr Rev* **21**, 90–113.

Gether U, Ballesteros JA, Seifert R, Sanders-Bush E, Weinstein H, and Kobilka BK (1997a). Structural instability of a constitutively active G protein-coupled receptor. Agonist-independent activation due to conformational flexibility. *J Biol Chem* **272**, 2587–90.

Gether U, Lin S, Ghanouni P, Ballesteros JA, Weinstein H, and Kobilka BK (1997b). Agonists induce conformational changes in transmembrane domains III and VI of the beta2 adrenoceptor. *EMBO J* **16**, 6737–47.

Gether U, Lin S, and Kobilka BK (1995). Fluorescent labeling of purified beta2-adrenergic receptor: Evidence for ligand-specific conformational changes. *J Biol Chem* **270**, 28 268–75.

Ghanouni P, Gryczynski Z, Steenhuis JJ *et al.* (2001a). Functionally different agonists induce distinct conformations in the G protein coupling domain of the beta 2 adrenergic receptor. *J Biol Chem* **276**, 24 433–6.

Ghanouni P, Schambye H, Seifert R *et al.* (2000). The effect of pH on beta(2) adrenoceptor function. Evidence for protonation-dependent activation. *J Biol Chem* **275**, 3121–7.

Ghanouni P, Steenhuis JJ, Farrens DL, and Kobilka BK (2001b). Agonist-induced conformational changes in the G-protein-coupling domain of the beta 2 adrenergic receptor. *Proc Natl Acad Sci USA* **98**, 5997–6002.

Holst B, Elling CE, and Schwartz TW (2000). Partial agonism through a zinc-Ion switch constructed between transmembrane domains III and VII in the tachykinin NK(1) receptor. *Mol Pharmacol* **58**, 263–70.

Hubbell WL, Cafiso DS, and Altenbach C (2000). Identifying conformational changes with site-directed spin labeling. *Nat Struct Biol* **7**, 735–9.

Javitch JA, Shi L, and Liapakis G (2002). Use of the substituted cysteine accessibility method to study the structure and function of G protein-coupled receptors. *Meth Enzymol* **343**, 137–56.

Jensen AD, Guarnieri F, Rasmussen SG, Asmar F, Ballesteros JA, and Gether U (2001). Agonist-induced conformational changes at the cytoplasmic side of TM6 in the {beta}2 adrenergic receptor mapped by site-selective fluorescent labeling. *J Biol Chem* **276**, 9279–90.

Ji TH, Grossmann M, and Ji I (1998). G protein-coupled receptors. I. Diversity of receptor-ligand interactions. *J Biol Chem* **273**, 17 299–302.

Kjelsberg MA, Cotecchia S, Ostrowski J, Caron MG, and Lefkowitz RJ (1992). Constitutive activation of the alpha 1B-adrenergic receptor by all amino acid substitutions at a single site. Evidence for a region which constrains receptor activation. *J Biol Chem* **267**, 1430–3.

Kunishima N, Shimada Y, Tsuji Y *et al.* (2000). Structural basis of glutamate recognition by a dimeric metabotropic glutamate receptor. *Nature* **407**, 971–77.

Lambright DG, Sondek J, Bohm A, Skiba NP, Hamm HE, and Sigler PB (1996). The 2.0 A crystal structure of a heterotrimeric G protein. *Nature* **379**, 311–19.

Langen R, Cai K, Altenbach C, Khorana HG, and Hubbell WL (1999). Structural features of the C-terminal domain of bovine rhodopsin: a site-directed spin-labeling study. *Biochemistry* **38**, 7918–24.

Lefkowitz RJ, Cotecchia S, Samama P, and Costa T (1993). Constitutive activity of receptors coupled to guanine nucelotide regulatory proteins. *Trends Pharmacol Sci* **14**, 303–7.

Lin SW and Sakmar TP (1996). Specific tryptophan UV-absorbance changes are probes of the transition of rhodopsin to its active state. *Biochemistry* **35**, 11 149–59.

Lu ZL, Curtis CA, Jones PG, Pavia J, and Hulme EC (1997). The role of the aspartate-arginine-tyrosine triad in the m1 muscarinic receptor: mutations of aspartate 122 and tyrosine 124 decrease receptor expression but do not abolish signalling. *Mol Pharmacol* **51**, 234–41.

Margeta-Mitrovic M, Jan YN, and Jan LY (2001). Function of GB1 and GB2 subunits in G protein coupling of GABAB receptors. *Proc Natl Acad Sci USA* **98**, 14 649–54.

McVey M, Ramsay D, Kellett E *et al.* (2001). Monitoring receptor oligomerization using time-resolved fluorescence resonance energy transfer and bioluminescence resonance energy transfer. The human delta-opioid receptor displays constitutive oligomerization at the cell surface, which is not regulated by receptor occupancy. *J Biol Chem* **276**, 14 092–9.

Palczewski K, Kumasaka T, Hori T *et al.* (2000). Crystal structure of rhodopsin: A G protein-coupled receptor. *Science* **289**, 739–45.

Peleg G, Ghanouni P, Kobilka BK, and Zare RN (2001). Single-molecule spectroscopy of the beta(2) adrenergic receptor: observation of conformational substates in a membrane protein. *Proc Natl Acad Sci USA* **98**, 8469–74.

Rasmussen SG, Jensen AD, Liapakis G, Ghanouni P, Javitch JA, and Gether U (1999). Mutation of a highly conserved aspartic acid in the beta2 adrenergic receptor: constitutive activation, structural instability, and conformational rearrangement of transmembrane segment 6. *Mol Pharmacol* **56**, 175–18.

Robbins MJ, Calver AR, Fillipov AK, Couve A, Moss SJ, and Pangalos MN (2001). The GABA$_{B2}$ subunit is essential for G protein coupling of the GABA$_B$ receptor heterodimer. *J Neurosci* **21**: 8043–52.

Rothschild KJ, Cantore WA, and Marrero H (1983). Fourier transform infrared difference spectra of intermediates in rhodopsin bleaching. *Science* **219**, 1333–5.

Salamon Z, Wang Y, Brown MF, Macleod HA, and Tollin G (1994). Conformational changes in rhodopsin probed by surface plasmon resonance spectroscopy. *Biochemistry* **33**, 13 706–711.

Samama P, Cotecchia S, Costa T, and Lefkowitz RJ (1993). A mutation-induced activated state of the beta2-adrenergic receptor: Extending the ternary complex model. *J Biol Chem* **268**, 4625–36.

Scheer A, Fanelli F, Costa T, De Benedetti PG, and Cotecchia S (1996). Constitutively active mutants of the alpha 1B-adrenergic receptor: role of highly conserved polar amino acids in receptor activation. *EMBO J* **15**, 3566–78.

Scheer A, Fanelli F, Costa T, De Benedetti PG, and Cotecchia S (1997). The activation process of the alpha1B-adrenergic receptor: potential role of protonation and hydrophobicity of a highly conserved aspartate. *Proc Natl Acad Sci USA* **94**, 808–13.

Schertler GF and Hargrave PA (1995). Projection structure of frog rhodopsin in two crystal forms. *Proc Natl Acad Sci USA* **92**, 11578–82.

Schertler GF, Villa C, and Henderson R (1993). Projection structure of rhodopsin. *Nature* **362**, 770–2.

Schwartz TW and Rosenkilde MM (1996). Is there a 'lock' for all agonist 'keys' in 7TM receptors? [see comments]. *Trends Pharmacol Sci* **17**, 213–16.

Seifert R, Wenzel-Seifert K, Gether U, and Kobilka BK (2001). Functional differences between full and partial agonists: evidence for ligand-specific receptor conformations. *J Pharmacol Exp Ther* **297**, 1218–26.

Sheikh SP, Zvyaga TA, Lichtarge O, Sakmar TP, and Bourne HR (1996). Rhodopsin activation blocked by metal-ion-binding sites linking transmembrane helices C and F. *Nature* **383**, 347–50.

Strader CD, Fong TM, Tota MR, Underwood D, and Dixon RAF (1994). Structure and function of G protein-coupled receptors. *Annu Rev Biochem* **63**, 101–32.

Teller DC, Okada T, Behnke CA, Palczewski K, and Stenkamp RE (2001). Advances in determination of a high-resolution three-dimensional structure of rhodopsin, a model of G-protein-coupled receptors (GPCRs). *Biochemistry* **40**, 7761–72.

Tota RT and Strader CD (1990). Characterization of the binding domain of the beta-adrenergic receptor with the fluorescent antagonist carazolol. *J Biol Chem* **265**, 16 891–7.

Unger VM, Hargrave PA, Baldwin JM, and Schertler GF (1997). Arrangement of rhodopsin transmembrane alpha-helices. *Nature* **389**, 203–6.

Wall MA, Coleman DE, Lee E *et al.* (1995). The structure of the G protein heterotrimer Gi alpha 1 beta 1 gamma 2. *Cell* **83**, 1047–58.

Ward SD, Hamdan FF, Bloodworth LM, and Wess J (2001). Conformational changes that occur during M3 muscarinic acetylcholine receptor activation probed by the use of an *in situ* disulfide cross-linking strategy. *J Biol Chem* **6**, 6.

Wess J (1998). Molecular basis of receptor/G-protein-coupling selectivity. *Pharmacol Ther* **80**, 231–64.

Zhao MM, Hwa J, and Perez DM (1996). Identification of critical extracellular loop residues involved in alpha 1-adrenergic receptor subtype-selective antagonist binding. *Mol Pharmacol* **50**, 1118–26.

Chapter 4

G protein structure diversity

Maurice K. C. Ho and Yung H. Wong

4.1 The classical GTPase cycle

Trimeric G proteins transmit signals from a huge group of heptahelical cell surface receptors to different categories of effectors such as enzymes and ion channels (Gilman 1987). The molecular diversity of trimeric G proteins signifies their pivotal role in numerous signal transduction pathways. To date, over twenty Gα, six Gβ, and thirteen Gγ subunits have been cloned (Hurowitz *et al.* 2000). The three subunits form two functional compartments—the Gα subunit and the stable Gβγ complex. Gα subunits belong to a group of enzymes known as GTP hydrolases (GTPases; Bourne *et al.* 1991). The GDP-bound form of Gα is associated with Gβγ in the resting stage. Ligand binding to an upstream receptor activates the GDP/GTP exchange of the Gα. GTP-induced conformational changes of Gα reduce its affinity for the Gβγ complex. Dissociation of GTP-bound Gα and Gβγ subunits allows their disparate interactions with effector molecules. GTP hydrolysis occurs in the Gα and returns it to the GDP-bound form with high affinity for Gβγ and receptors. This classical paradigm of trimeric G protein signalling (Fig. 4.1) has undergone substantial modifications in light of the discovery of novel signalling partners with new modes of signal modulation, such as the family of regulators of G protein signalling (RGS; see De Vries *et al.* 2000). In addition to a general account of their major properties, recent advances of G protein signalling are presented in this chapter.

4.2 Structural characteristics of Gα subunits

Gα subunits share sequence homology with p21-Ras, transcription factor EF-Tu and other GTPases in their guanine nucleotide-binding pockets (Bourne *et al.* 1990, 1991). The initial model of Gα is a Ras-like structure with a relatively large insertion between the first two guanine nucleotide-binding regions (Conklin and Bourne 1993). Resolution of crystal structures of Gα$_{t1}$ and Gα$_{i1}$ confirmed the presence of the two structural domains as the helical and GTPase domains (Sprang 1997). The helical domain encompasses seven α helices (αA to αG; Figs 3.1 and 3.2a), with the first six forming the core of the helical domain. αG is actually located on the GTPase side facing the helical domain. Similar to monomeric G proteins, the GTPase domain is comprised of five α helices (including both terminal helices) and six β strands (Figs 3.1 and 3.2a). The conformations of both GTPase and helical domains are highly similar between the GDP- and GTP-bound Gα structures. A slightly more compact structure is observed in the GTP-bound form, with the two domains getting closer to each other to bury the GTP in the cleft. Residues lying on the interface between the two portions

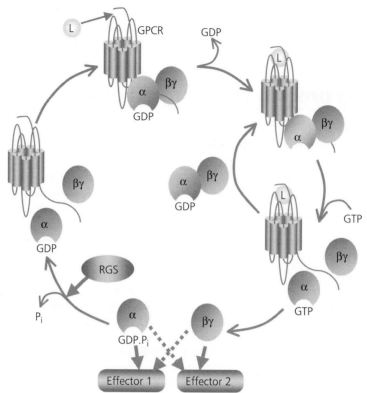

Fig. 4.1 The GTPase cycle. Ligand (L) binds to a cognate G protein-coupled receptor (GPCR)-G protein complex and triggers the release of GDP from Gα. The empty state of Gα has low affinity to ligand-bound GPCR and permits the binding of GTP. The ligand-bound GPCR can elicit another round of G protein activation by promoting nucleotide exchange of more G protein trimers. GTP-bound Gα subsequently dissociates with Gβγ and both compartments stimulate various effectors. Intrinsic GTP hydrolysis, which can be accelerated by the binding of a regulator of G protein signaling (RGS), eventually turns off the Gα and release a phosphate (Pᵢ). Finally, the GDP-bound Gα combines with Gβγ and ligand-free GPCR thereby returning to the resting stage.

of Gα form specific intramolecular interactions that are crucial for controlling the nucleotide exchange rate, basal activity, and receptor-triggered activation of G protein.

The guanine nucleotide-binding pocket of Gα subunit is comprised of five stretches of well-conserved amino acids (Bourne *et al.* 1991; Fig. 4.2), spreading along the Gα sequence but clustering in a deep cleft between the GTPase and helical domains. Mutations located in the nucleotide-binding regions produce mutants that serve as useful tools for investigating the structure–function relationships of G proteins. For example, mutations of the glutamine residue in the third nucleotide-binding region signature DVGGQR of virtually all Gα subunits produce constitutively active phenotypes, and spontaneous occurrence of these mutations are associated with a number of endocrine tumours (Lyons *et al.* 1990). Such mutations have been successfully employed to define the specific functions of various Gα subunits in regulating effectors and cell growth. The two glycines preceding the

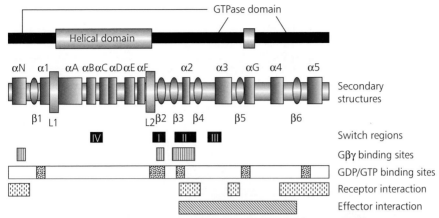

Fig. 4.2 Structure and functional domains of Gα. Various secondary and tertiary structural elements of Gα are schematically described. The nomenclature is according to the first crystal structure of GTPγS-Gα$_{t1}$ and depicted with different shapes and shadings. L1 and L2 are the two linker regions between helical and GTPase domains. Switch regions and other functional regions interacting with various molecular partners are indicated at the bottom. Most of them are discrete regions spanning along the primary structure of Gα subunit, but they are close together in the three-dimensional structures (see Fig. 3.2). Effector interacting regions are relatively diverse among various Gα subunits, but mainly within the C-terminal 40% of the primary structure.

glutamine ensure flexibility of the third nucleotide-binding region for accommodating the GTP-induced conformational changes. Mutation of either one of the glycine residues in Gα creates a dominantly negative phenotype (Osawa and Johnson 1991).

Upon activation, several regions of the Gα undergo dramatic conformational changes. The first three switches, denoted as Switch I, II, and III, cluster in the GTPase domain and a discrete Switch IV is observed in the helical domain of Gα$_{i1}$ but not Gα$_{t1}$ (see Sprang 1997; Fig. 4.2). These switch regions are involved in a series of concerted movements that facilitate nucleotide exchange and GTP hydrolysis. The first three switches overlap with two important nucleotide-binding regions, as well as a major binding surface for Gβγ and effector molecules (Fig. 4.2). The binding of GTP lifts the Switch I upward and the strictly conserved threonine (Thr-177 of Gα$_{t1}$) residue is oriented to the proximity of the γ-phosphate of GTP. Switch II covers the β3/α2/β4 structures. Extensive electrostatic and hydrophobic interactions are found between Switches II and III in the GDP-bound state. These are disrupted and several ionic bridges are re-established during activation for stabilizing the GTP-induced conformation. As compared to GTPγS-bound Gα$_{t1}$, the αB/αC loop of GTPγS-bound Gα$_{i1}$ is packed more tightly. This region is denoted as Switch IV because significant conformational changes occur during GTP hydrolysis. *Xenopus* expresses a Gα$_s$ mutant with six amino acids different from the human homologue. The mutation maps to a region corresponding to the Switch IV of Gα$_{i1}$ and this mutant exhibits a reduced guanine nucleotide exchange rate and does not activate AC (Echeverría *et al.* 2000).

The N-terminus of Gα is a single helical structure denoted as αN that stretches out from the main body and embraces the Gβγ complex (Fig. 4.3). Both the N-terminus of Gα and the C-terminus of Gγ are adjacent to each other. Although there is no observable contact

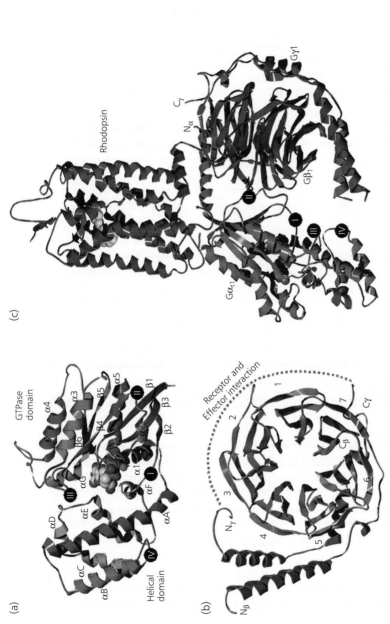

Fig. 4.3 Ribbon structural models of trimeric G protein and its receptor. The three-dimensional arrangements of the structural elements of (a) GDP-Mg^{2+}-bound Gα_{t1} subunit, (b) G$\beta_1\gamma_1$ complex, and (c) rhodopsin and trimeric G$_{t1}$ are illustrated. In (a), the N-terminal helix is omitted (but appears in (c)). The bound GDP (CPK scheme) and Mg^{2+} are displayed in space-filled models, which situates between helical and GTPase domains. Switch regions are marked with inverted letters in circles. In (b), the termini of both Gβ (N$_\beta$ and C$_\beta$) and Gγ (N$_\gamma$ and C$_\gamma$) are indicated and the seven blades are numbered as in Neer and Clapham (1997). Most of the receptor- and effector-interacting residues of Gβ subunits are on blades 1–3, as indicated. In (c), the structural models of 11-*cis*-retinal-bound rhodopsin (Palczewski *et al.* 2000) and trimeric G$_{t1}$ (Gα, Gβ and Gγ; Lambright *et al.* 1996) are assembled together according to the proposed receptor-interacting surface as described (Lichtarge *et al.* 1996; Bourne 1997; Hamm 1998). Switch regions I–IV of Gα subunit are marked similarly as in (a). Coordinates of all the structural models are obtained from the Protein Data Bank (Berman *et al.* 2000; http://www.rcsb.org/pdb) and visualized by Swiss PDB Viewer Deep View Version (Guex and Peitsch 1997; http://www.expasy.org/spdbv/) and POV-Ray 3.1 for rendering (http://www.povray.org).

between Gα and Gγ subunits, the close proximity of their two terminal regions supports the idea that the lipid moieties attached on them may strengthen the association of the trimer. The N-terminus of Gα is generally accepted as an essential G$\beta\gamma$-binding region and the large loop structure extending from blade 2 of Gβ helps to orientate the αN helix along the side of Gβ (Fig. 4.3). Another major G$\beta\gamma$-interacting surface has been mapped to Switches I and II of the Gα subunit. Within Switch II of Gα in the trimer crystals, the α2 helix adopts a loose 3_{10} helical structure resembling that observed in the GDP-bound monomer. Residues of the α2 helix swing outward and interact with Gβ. In the GTP-bound state, conformational changes in Switch I and Switch II regions of Gα disrupt most of the intermolecular interactions with G$\beta\gamma$ resulting in the release of the G$\beta\gamma$ complex.

Lipidation of Gα subunits occur at the N-terminus. Myristate is co-translationally incorporated at Gly2 of the Gα_i-subfamily members and is relatively stable (Table 4.1). The myristoylated Gα subunits have the N-terminal methionine deleted during the modification. Gα_{t1} can be exceptionally modified with other fatty acids such as stearate, laureate, and oleate which are believed to occur specifically in photoreceptor cells (Johnson *et al.* 1994). *N*-myristoylation facilitates the membrane anchorage of Gα, increasing its affinity for G$\beta\gamma$ and enhancing the regulation of AC (Wedegaertner *et al.* 1995). Reversible palmitoylation occurs on the cysteine residues at the N-terminus of almost all known Gα subunits through a thioester linkage to the side chain sulfhydryl group (Table 4.1). The same cysteine residue can be arachidonylated but its functional impact is unclear. Strikingly, Gα_{i1}, Gα_o, and Gα_z can undergo autoacylation in the presence of palmitoyl-coenzyme A (or other fatty acyl derivatives) at physiological pH and temperature (Duncan and Gilman 1996). The rate of autoacylation can be enhanced by the addition of G$\beta\gamma$ complex or increasing pH. The reversibility of palmitoylation is considered as a regulatory handle on the Gα, and a palmitoylation/depalmitoylation cycle of Gα_s has been proposed (Wedegaertner and Bourne 1994). Upon receptor activation, nucleotide exchange in Gα_s promotes its dissociation from G$\beta\gamma$. The GTP-bound Gα_s is rapidly depalmitoylated by a cytosolic palmitoylesterase and then translocates to cytosol. After GTP hydrolysis, Gα_s is recruited by G$\beta\gamma$ and returns to the plasma membrane, where the palmitoyltransferase replenishes Gα_s with palmitate moiety to form fully acylated G$_s$ protein.

Many Gα subunits are substrates for kinases (Table 4.1) and phosphorylation exerts different functional impacts on various Gα subunits. Gα_{t1} is the first Gα subunit found to be phosphorylated by both protein kinase C and insulin receptor kinase. Src kinase can phosphorylate Gα_s at Tyr37 and Tyr377. For Gα_q, phosphorylation of Tyr356 is essential in type 1a metabotropic glutamate receptor-mediated activation. Epidermal growth factor receptor (EGFR) is a membrane-bound tyrosine kinase that appears to phosphorylate and activate both Gα_s and Gα_{i1-2}. Protein kinase C phosphorylates Gα_{i2}, Gα_z, Gα_{12}, and Gα_{13}, and Gα_{16}. Uniquely, Gα_z can be phosphorylated by p21-activating kinase 1. Associations of Gα_z and Gα_{12} with G$\beta\gamma$ subunits and other molecular partners are impaired following phosphorylation of their N-terminal helices, whereas phosphorylation of Gα_{16} diminishes thyrotropin-releasing hormone receptor-induced Cl$^-$ current in *Xenopus* oocytes (reviewed in Chen and Manning 2001).

4.3 Structural characteristics of Gβ and Gγ subunits

Gβ and Gγ subunits are always associated with each other as a single functional unit. Two distinct structural domains can be identified on the Gβ. An N-terminal helix consisting of

Table 4.1 Properties of mammalian Gα subunits

Gα subunit	Human gene	Cytogenetic position	cDNA genbank accession number	Size/molecular weight[a]	% identity[b]	Lipidation[c]	Phosphorylation[c]	Toxin sensitivity
G_i Subfamily								
$G\alpha_{i1}$	GNAI1	7q21	NM_002069	354 a.a./40.3 kDa	100	Myr (G2), Palm (C3)	None	PTX
$G\alpha_{i2}$	GNAI2	3p21.3–p21.2	J03004	355 a.a./40.5 kDa	88.1	Myr (G2), Palm (C3)	PKC, EGFR (?)	PTX
$G\alpha_{i3}$	GNAI3	1p13	M27543	354 a.a./40.5 kDa	93.7	Myr (G2), Palm (C3)	None	PTX
$G\alpha_{oA}$	GNAO	16q12.1	AH002708	354 a.a./40.0 kDa	72.1	Myr (G2), Palm (C3)	None	PTX
$G\alpha_{oB}$	GNAO	16q12.1	AH002708	354 a.a./40.1 kDa	70.7	Myr (G2), Palm (C3)	None	PTX
$G\alpha_{t1}$	GNAT1	3p21.3–p21.2	X63749	350 a.a./40.0 kDa	67.5	Myr (G2), Others (G2)[d]	PKC (?)	CTX, PTX
$G\alpha_{t2}$	GNAT2	1p13	Z18859	354 a.a./40.1 kDa	69.4	Myr (G2)	None	PTX
$G\alpha_{gust}$	GNAT3	7q21	X65747 (rat)	354 a.a./40.1 kDa	67.2	Myr (G2)	None	PTX
$G\alpha_z$	GNAZ	22q11.1–q11.2	J03260	355 a.a./40.9 kDa	67.0	Ara (C3), Myr (G2), Palm (C3)	PKC (S16, 27), PAK1 (S16)	None
G_s Subfamily								
$G\alpha_{olf}$	GNAL	18p11.31	U55184	381 a.a./44.7 kDa	39.1 (75.7)	Palm (C3)	None	CTX
$G\alpha_{sL}$	GNAS	20q13.2	X04408	394 a.a./45.7 kDa	38.0 (100)	Palm (C3)	EGFR, Src (Y37, Y377?)	CTX
$G\alpha_{sS}$	GNAS	20q13.2	X04409	380 a.a./44.2 kDa	39.1 (95.9)	Palm (C3)	None	CTX
$XLG\alpha_s$	—	—	X84047 (rat)	846 a.a./94 kDa	17.6 (43.3)	Palm (C3)	None	CTX
G_q Subfamily								
$G\alpha_q$	GNAQ	9q21	U43083	359 a.a./42.2 kDa	50.5 (100)	Palm (C9, 10)	Src (Y356?)	None
$G\alpha_{11}$	GNA11	19p13.3	M69013	359 a.a./42.1 kDa	50.0 (89.4)	Palm (C9, 10)	Src	None
$G\alpha_{14}$	GNA14	9q21	NM_004297	355 a.a./41.5 kDa	50.8 (80.2)	Palm	None	None
$G\alpha_{15/16}$	GNA15	19p13.3	M63904	374 a.a./43.5 kDa	40.9 (53.7)	Palm (C9, 10)	PKC	None
G_{12} Subfamily								
$G\alpha_{12}$	GNA12	7p21-p22	L01694	379 a.a./44.1 kDa	37.1 (100)	Palm (C11)	PKC	None
$G\alpha_{13}$	GNA13	17q24	NM_006572	377 a.a./44.1 kDa	36.6 (63.0)	Palm (C11)	PKC	None

[a] The numbers of amino acids and the corresponding molecular weights are deduced from the human open reading frames. [b] The overall identities are referred to $G\alpha_{i1}$, and identities within subfamilies are referred to the first representatives are bracketed. [c] The known modification sites are shown in brackets after each type of modification. [d] Includes stearate, laureate, and oleate. Key: a.a.—amino acids, Ara—arachidonylation, CTX—cholera toxin, EGFR—epidermal growth factor receptor, Myr—myristoylation, PAK1—p21-activating kinase 1, Palm—palmitoylation, PKC—protein kinase C, PTX—pertussis toxin, Src—Src kinase, ?—to be confirmed.

20 residues sprouts from the core of Gβ and entangles with the Gγ (Fig. 4.3). The core structure of Gβ is exclusively composed of antiparallel β strands arranged in seven WD repeats (Neer *et al.* 1994). Four β strands form a twisted β sheet and the seven sheets or blades are arranged into a propeller-like structure with the narrower side making contacts largely with the Switch II region of Gα. Gγ consists of two α helices and a disordered C-terminal tail. The first helix forms a coiled-coil with the N-terminal helix of Gβ subunit, whereas the second helix and the C-terminal tail extend along the wider side of the propeller core interacting with blades 5–7 of Gβ (Fig. 4.3). The entire Gγ practically interacts with Gβ, thereby preventing their separation unless under denaturing conditions. Although direct association between $Gα_{oA}$ and $Gγ_2$ has been reported (Rahmatullah and Robishaw 1994), there is no observable physical contact between Gα and Gγ subunits in the crystal structures of trimeric G_{t1} and G_{i1}.

Accumulating evidence indicate the existence of specific combinations of Gβγ complexes (Table 4.2). Much efforts have been spent on establishing the possible combinations of functional Gβγ complexes using purified proteins, functional expression, yeast two-hybrid systems, or ribozyme approach (reviewed in Clapham and Neer 1997; Yan *et al.* 1996; Asano *et al.* 1999; Wang *et al.* 1999). The specificity of the assembly of Gβγ complex appears to be determined by a stretch of 14 residues lying within the middle of Gγ subunits. For $Gγ_1$, a motif of five amino acids, CCEEF, within the 14 residues are essential for specifying its interaction with $Gβ_1$. These residues mainly interact with blade 5 and the N-terminal helix of Gβ.

The Gγ subunit is isoprenylated at the C-terminal CAAX motif, where C is an invariant cysteine residue at −4 position of all known Gγ subunits, A is aliphatic residues and X is

Table 4.2 Specificity of the assembly of Gβ and Gγ subunits

	$Gβ_1$	$Gβ_2$	$Gβ_3$	$Gβ_4$	$Gβ_5$
$Gγ_1$	+	±	−	−	−
$Gγ_2$	++	++	+	+	+
$Gγ_3$	++++	+++	++	++	+++
$Gγ_4$	++++	+++	+	+	+++
$Gγ_5$	++++	+++	+	+	+
$Gγ_7$	++++	+++	+	+	±
$Gγ_c$	NR	NR	+	NR	NR
$Gγ_8$	NR	NR	NR	NR	NR
$Gγ_{10}$	+	+	−	NR	NR
$Gγ_{11}$	+	−	−	NR	NR
$Gγ_{12}$	NR	NR	NR	+	NR
$Gγ_{13}$	NR	NR	+	NR	NR

Key: +, dimer formation, more plus means stronger tendencies; −, no apparent association; ±, weak interactions; NR—not reported.

undefined (Lai *et al.* 1990). Types 1, 11, and cone cell-specific Gγ subunits are farnesylated and they all have a serine residue at the extreme C-terminus, whereas other Gγ subunits having a leucine at the same position are geranylgeranylated (Table 4.3). Isoprenylation of Gγ is associated with the truncation of the last three C-terminal residues and the new terminus is carboxymethylated (Casey and Seabra 1996). In addition to increasing membrane availability of Gβγ, specific prenyl groups may contribute to the G protein interactions with receptors and effectors. Interestingly, activation of neutrophil G_i by formyl peptide receptor is associated with carboxymethylation of $G\gamma_2$, suggesting that carboxymethylation of Gγ may play a role in signal transduction (Philips *et al.* 1995). $G\gamma_{12}$ is the only known Gγ subunit that is phosphorylated by protein kinase C at its N-terminal Ser1. Free $G\beta\gamma_{12}$ appears to be the preferential substrate. Phosphorylation of $G\beta\gamma_{12}$ selectively interferes its ability to interact with F-actin and phospholipase Cβ (PLCβ), but not type II AC (Ueda *et al.* 1999 and references therein).

As observed from the crystal structures of both free Gβγ and Gβγ in heterotrimer, Gβ appears as a rigid body without significant conformational changes between these two states. However, subsequent resolution of a complex between Gβγ and phosducin indicates that Gβγ

Table 4.3 Properties of mammalian Gβ/γ subunits

Gβ/γ subunit	Human gene	Cytogenetic position	cDNA genbank accession number	Size/molecular weight[a]	% identity[b]	Covalent modification[c]
Gβ						
$G\beta_1$	GNB1	1p36.2	X04526	340 a.a./37.4 kDa	100	P, R
$G\beta_2$	GNB2	7q21.1–q21.2	M16514	340 a.a./37.3 kDa	90.2	
$G\beta_3$	GNB3	12p13	M31328	340 a.a./37.2 kDa	83.2	
$G\beta_{3S}$	GNB3	12p13	—	299 a.a./32.9 kDa	74.1	
$G\beta_4$	GNB4	3q26–q27	AAG18442	340 a.a./37.6 kDa	90.8	
$G\beta_5$	GNB5	15q21	014775	353 a.a./38.6 kDa	51.2	
$G\beta_{5L}$	GNB5	15q21	AAG18444	395 a.a./43.6 kDa	45.8	
Gγ						
$G\gamma_1$	GNGT1	7q21.3	S62027	74 a.a./8.50 kDa	30.2	CM+F (C71)
$G\gamma_2$	GNG2	14q21–q22	AA868346	71 a.a./7.85 kDa	100	CM+G (C68)
$G\gamma_3$	GNG3	11p11	AF092129	75 a.a./8.30 kDa	72.0	CM+G (C72)
$G\gamma_4$	GNG4	1q43–q44	U31382	75 a.a./8.39 kDa	73.3	CM+G (C72)
$G\gamma_5$	GNG5	1p22	AF038955	68 a.a./7.32 kDa	47.8	CM+G (C65)
$G\gamma_5$-like	GNG5ps	2p12	AF188178	68 a.a./7.25 kDa	46.4	CM+G (C65)
$G\gamma_7$	GNG7	19p13.3	NM_005145	68 a.a./7.52 kDa	69.0	CM+G (C65)
$G\gamma_c$	GNGT2	17q21	014610	69 a.a./7.75 kDa	36.1	CM+G (C66)
$G\gamma_8$	GNG8	19q13.2–q13.3	AF188179	70 a.a./7.84 kDa	67.6	CM+F (C67)
$G\gamma_{10}$	GNG10	9q31–q32	U31383	68 a.a./7.21 kDa	53.5	CM+G (C65)
$G\gamma_{11}$	GNG11	7q21.3	U31384	73 a.a./8.48 kDa	30.2	CM+F (C70)
$G\gamma_{12}$	GNG12	1p31–p33	AF188181	72 a.a./8.01 kDa	61.6	CM+G (C69), P (S2)
$G\gamma_{13}$	GNG13	16p13.3	XM_012543	67 a.a./7.95 kDa	32.3	CM+G (C64)

[a] The numbers of amino acids and the corresponding molecular weights are deduced from the human open reading frames. [b]The amino acid identities are referred to $G\beta_1$ and $G\gamma_2$. [c]The known modification sites are shown in blankets after each type of modification. Key: a.a.—amino acids, CM—terminal carboxymethylation, F—farnesylation, G—geranylgeranylation, P—phosphorylation, R—ADP-ribosylation.

can exist in two distinct conformations (Loew *et al.* 1998). In the 'relaxed' state, the farnesylate of Gγ is exposed, mediating membrane association. In the 'tense' state, as observed in the phosducin-Gβγ complex, the farnesyl moiety is buried in the cavity formed between blades 6 and 7 of the Gβ subunit. Binding of phosducin induces the formation of this cavity, resulting in a switch from relaxed to tense conformation, which may sequester Gβγ from the membrane to the cytosol and turns off the signal transduction cascade.

4.4 Functional diversity of Gα subunits

The classification of G proteins is based on their amino acid identities and functional specialization of their Gα subunits (Simon *et al.* 1991). There are four major subfamilies of Gα subunits and their general properties are listed in Tables 4.1 and 4.4.

4.4.1 G$_i$ subfamily

The largest subfamily is generally known as 'inhibitory' Gα subunits, after the identification of the negative regulator of AC, Gα$_i$. All members except Gα$_z$ are subject to ADP-ribosylation mediated by the bacterial toxin from *Bordetella pertussis* (PTX), and such covalent modification prevents the G proteins from being activated by GPCRs. The three Gα$_i$ subtypes (Gα$_{i1}$, Gα$_{i2}$, and Gα$_{i3}$) can all inhibit AC to a similar extent. Little is known about the functional differences between the three subtypes. A mutation at Arg179 constitutively activates Gα$_{i2}$ and it is associated with adrenocarcinoma, ovarian cancer, and pituitary adenoma (Williamson *et al.* 1995). Mutation of Phe200 to Leu of Gα$_{i2}$ leads to a pathological status known as idiopathic ventricular tachycardia (Lerman *et al.* 1998). Gα$_{i2}$ is the most commonly used G$_i$ subtype for studying G$_i$-linked receptors. Gα$_{i2}$-deficient mice exhibit profound alterations of thymocyte maturation, elevated IgG levels in the large bowel, growth retardation and the development of lethal diffuse colitis with adenocarcinoma in the colon (Rudolph *et al.* 1995). Gα$_{i3}$ is found in both plasma membrane and Golgi apparatus indicating that Gα$_{i3}$ plays a role in membrane transport/remodelling processes. Gα$_{i3}$-deficient cardiocytes derived from embryonic stem cells show inability to activate an inwardly rectifying K$^+$ channel in response to muscarinic or adenosine receptor agonists (Sowell *et al.* 1997). The first regulator of G protein signalling identified in mammalian cells (GAIP) and nucleobindin (or Calnuc) show preferential interactions with Gα$_{i3}$ (De Vries *et al.* 1995; Lin *et al.* 2000).

Gα$_z$ is often considered as the PTX-resistant alternative of Gα$_i$, as it couples to most G$_i$-linked receptors and inhibit AC (Wong *et al.* 1992). Gα$_z$ is predominantly expressed in neuronal tissues (reviewed in Ho and Wong 2001). Specific molecular targets of Gα$_z$ include the Gα$_z$-specific GAP, RGSZ1, G protein-regulated inducers of neurite outgrowth (GRIN1 and GRIN2) that are enriched in nerve growth cones, a GTPase-activating protein of monomeric G protein Rap1 (Rap1GAP), and the transcription regulator Eya2. Genetic deletion of Gα$_z$ impairs platelet aggregation by preventing the inhibition of cAMP formation by epinephrine (Yang *et al.* 2000). Gα$_z$-deficient mice are more resistant to fatal thromboembolism, exhibit exaggerated response to cocaine, and have impaired responses to antidepressants that act as catecholamine reuptake inhibitors. The acute analgesic effect of morphine is also reduced, but another similar study indicates that deletion of Gα$_z$ facilitates the development of morphine tolerance (Hendry *et al.* 2000).

Two cDNA species of Gα$_o$ have been isolated and they are splice variants of one single gene (Murtagh *et al.* 1994). A third form, Gα$_{oC}$, is derived from Gα$_{oA}$ by deamidating the

Table 4.4 G protein-regulated effectors

Effector	Gα	Gβγ
Adenylyl cyclases		
Type I	Gα$_s$ ↑, Gα$_{i,o,z}$ ↓,	Gβ$_{1-3}$γ$_{1-2}$ (except Gβ$_2$γ$_1$) ↓
Type II	Gα$_s$ ↑	Gβ$_{1-3}$γ$_{1-2}$ (except Gβ$_2$γ$_1$) ↑*
Type III	Gα$_{s,olf}$ ↑, Gα$_i$ ↓	
Type IV	Gα$_s$ ↑	↑ *
Type V	Gα$_s$ ↑, Gα$_{i,z}$ ↓	Gβ$_1$γ$_2$ ↑
Type VI	Gα$_s$ ↑, Gα$_{i,z}$ ↓	Gβ$_1$γ$_2$ ↑
Type VII	Gα$_s$ ↑	↑ *
Type VIII	Gα$_s$ ↑, Gα$_i$ ↓	
Type IX	Gα$_s$ ↑, Gα$_i$ ↓	
Arrestin		↑
cGMP phosphodiesterases	Gα$_{t1,t2,gust}$ ↑	No effect
Dynamin I		↑
G protein receptor kinases		
GRK1		No effect
GRK2		Gβ$_{1-2}$γ$_2$ ↑
GRK3		Gβ$_{1-3}$γ$_2$ ↑
G protein-regulated inducers of neurite outgrowth		
GRIN1	Gα$_o$ ↑	
GRIN2	Gα$_{i,o,z}$ ↑	
Ion channels		
Ca^{2+}, N-type	Gα$_{i1,i2,z}$ ↓	Gβ$_{1,3}$ ↓
Ca^{2+}, L-type	Gα$_s$ ↑	
Cl$^-$, cardiac	Gα$_s$ ↑	
Cl$^-$, epithelial	Gα$_{i,o}$ ↑	
Na$^+$, cardiac	Gα$_s$ ↑	
Na$^+$, epitheial	Gα$_{i3}$ ↓	
GIRK1	Gα$_{i,z}$ ↑	↑
GIRK2	Gα$_{i,z}$ ↑	↑
GIRK4	Gα$_{i,z}$ ↑	↑
Monomeric G proteins		
CDC25^{Mm}		Gβ$_1$γ$_2$, Gβ$_1$γ$_5$, ↑
p140RasGRF		Gβ$_1$γ$_2$, Gβ$_1$γ$_5$, ↑
Raf-1		↑
Shc		↑
Rap1GAP	Gα$_{i,o,z}$ ↑	
Rap1GAPII	Gα$_{i,o,z}$ ↑	
p115RhoGEF	Gα$_{13}$ ↑	
PDZRhoGEF	Gα$_{12,13}$ ↑	
GAP1^m	Gα$_{12}$ ↑	

Table 4.4 (continued)

Effector	Gα	Gβγ
Phospholipases		
Type A_2		↑
Type C β_1	$G\alpha_{q,11,14,16}$ ↑	$G\beta_1\gamma_1$, $G\beta_1\gamma_2$, $G\beta_5\gamma_2$ ↑
Type C β_2	$G\alpha_{q,11,14,16}$ ↑	$G\beta_1\gamma_1$, $G\beta_1\gamma_2$, $G\beta_5\gamma_2$, $G\beta_3\gamma_{13}$ ↑
Type C β_3	$G\alpha_{q,11,14,16}$ ↑	$G\beta_1\gamma_1$, $G\beta_1\gamma_2$ ↑
Type C β_4	$G\alpha_{q,11,14,16}$ ↑	No effect
Type C ϵ	$G\alpha_{12}$ ↑	
Phosphoinositide 3-kinases		
Type β		↑
Type γ	$G\alpha_o$ ↑	$G\beta_1\gamma_1\pm$, $G\beta_{1-3}\gamma_2$ ↑,
Plasma membrane Ca^{2+} pump		↑
Transcription cofactor Eya2	$G\alpha_{i2,z}$ ↓	
Tyrosine kinases		
Btk	$G\alpha_{q,12}$ ↑	↑
Tsk		↑
Lck	$G\alpha_s$ ↑	
Lyn	$G\alpha_q$ ↑	
Src	$G\alpha_s$, $G\alpha_{i2}$ ↑	↑
Syk	$G\alpha_q$,$G\alpha_i$ ↑, $G\alpha_s$ ↓	

* Effective only in the presence of active $G\alpha_s$ or PKC-mediated phosphorylation.

Asn346 into Asp post-translationally (Exner *et al.* 1999). $G\alpha_o$ is neural-specific, particularly concentrated in nerve growth cones, and constitutes about 1% of the total membrane proteins in neurons. The neuronal functions of $G\alpha_o$ include inhibition of voltage-gated Ca^{2+} channels (Herlitze *et al.* 1996) and neuron-enriched type I AC (Taussig *et al.* 1994), as well as activation of mitogen-activated protein kinases (MAPKs; Moxham and Malbon, 1996). $G\alpha_o$ is implicated in the development of nerve growth cones (Igarashi *et al.* 1993), vesicle trafficking (Gasman *et al.* 1997), and survival of primary accessory olfactory neuron (Tanaka *et al.* 1999). Like $G\alpha_z$, $G\alpha_o$ interacts with Rap1GAP (Jordon *et al.* 1999) and GRIN proteins (Chen *et al.* 1999). Other neuronal-specific molecules shown to associate with $G\alpha_o$ include GAP-43 (Strittmatter *et al.* 1991), Alzheimer's disease gene presenilin-1 (Smine *et al.* 1998), Purkinje cell protein-2 (Luo and Denker 1999), amyloidogenic Aβ-(1–40) and Aβ-(25–35) (Rymer and Good 2001). Activated $G\alpha_{oA}$ can induce transformation of NIH 3T3 cells via Src and STAT3, but not MAPK, thus providing a link between G protein and STAT-related signalling pathways (Ram *et al.* 2000).

Two transducin molecules ($G\alpha_{t1}$ and $G\alpha_{t2}$) and gustducin (G_{gust}) are the major regulators of cGMP phosphodiesterases. The three related Gα subunits have very restricted expression patterns. $G\alpha_{t1}$ are $G\alpha_{t2}$ are found only in rod and cone photoreceptor cells, whereas G_{gust} is found exclusively in the taste buds of all taste papillae, brush cells in the stomach and intestine, and the vomeronasal neuroepithelium. Subsequent detection of $G\alpha_{t1}$ in taste cells links its function to bitter taste perception (see Gilbertson *et al.* 2000). Enormous efforts have been directed toward $G\alpha_{t1}$ as a prototypical model for studying receptor-G protein-effector interactions, and detailed structures of rhodopsin (the primary receptor for $G\alpha_{t1}$),

G_{t1} in various forms, and cGMP phosphodiesterase (partial structure) and other related molecules like RGS9 have been resolved (Hamm 1998; Palczewski *et al.* 2000; Slep *et al.* 2001). A mutation found in the first nucleotide binding region (Gly-38 to Asp) results in the 'Nougaret' form of congenital stationary night blindness (Dryja *et al.* 1996). G_{gust} is responsible for transducing both sweet and bitter tastes (Wong *et al.* 1996). Specific taste-sensitive GPCRs have been cloned (Hoon *et al.* 1999). Co-localization of G_{gust} with $G\beta_3$ and $G\gamma_{13}$ suggests that they form a functional heterotrimer for regulating phosphodiesterase and $PLC\beta_2$ in taste cells (Yan *et al.* 2001).

4.4.2 G_s subfamily

G_s subfamily members include two well-known ubiquitously expressed alternatively spliced variants of $G\alpha_s$, $G\alpha_{s(long)}$ and $G\alpha_{s(short)}$, the olfactory-specific $G\alpha_{olf}$ and an 'extra-large' $G\alpha_s$ or $XLG\alpha_s$ which represents the third spliced variant of $G\alpha_s$. Amino acid sequence identities among the members of G_s subfamily are highest when compared with others subfamilies and they all stimulate ACs. Except $G\alpha_{olf}$, all other $G\alpha_s$ subtypes are actually splice variants of a single gene (GNAS).

A series of $G\alpha_s$ mutations have been associated with various hormonal diseases and cancers (Weinstein and Yu 1999). Inactivating and activating mutations in the GNAS1 gene are associated with pseudohypoparathyroidism, type Ia and McCune-Albright syndrome, respectively. Regulation of the GNAS gene expression is of particular interest and it is actually an imprinted gene with the two alleles expressing different gene products. Pseudo-hypoparathyroidism type Ia is maternally inherited, which suggests that GNAS1 may be maternally imprinted. A differentially methylated locus containing an additional exon of GNAS1 has recently been identified. This exon is included within transcripts homologous to rat mRNAs encoding $XLG\alpha_s$. In contrast to $G\alpha_s$-encoding transcripts, $XLG\alpha_s$ transcripts are derived from the paternal allele exclusively. The role of $XLG\alpha_s$ in the pathogenesis of pseudohypoparathyroidism requires further exploration.

In addition to adenylyl cyclase, $G\alpha_s$ can positively regulate several types of ion channels (Table 4.4). Associations of $G\alpha_s$ with different types of tyrosine kinases permit cross talks between GPCRs and growth factor receptors. $G\alpha_s$ is activated by EGFR and Src kinase via tyrosine phosphorylation. Interestingly, Src kinase is also one of the downstream effectors of $G\alpha_s$ (Ma *et al.* 2000). $G\alpha_s$ represses adipogenesis in 3T3-L1 fibroblasts, and the process involves a direct interaction between $G\alpha_s$ and a Src-like tyrosine kinase Syk, in which the phosphorylation of Syk is repressed (Wang and Malbon 1999). Induction of adipogenesis leads to a decline of $G\alpha_s$ level and relieves Syk for phosphorylation. In contrast, $G\alpha_s$ can stimulate another tyrosine kinase Lck directly in S49 lymphoma and subsequently leading to apoptosis (Gu *et al.* 2000). These observations suggest that $G\alpha_s$ actively participates in the control of cell fate.

A family of putative odorant GPCRs exhibits restricted expressions in olfactory epithelium (Ronnett and Snyder 1992). Enhancement of both cAMP and phosphoinositide turnover has been demonstrated in response to odorants both in isolated olfactory cilia and in primary olfactory neuronal cultures. The $G\alpha$ subunit involved in olfaction is $G\alpha_{olf}$. Homozygous $G\alpha_{olf}$-knockout mice show a striking reduction in the electrophysiologic response of primary olfactory sensory neurons to a wide variety of odors, but the topographic map of primary sensory projections to the olfactory bulb remains unaltered. The homozygotes are fertile, but exhibit hyperactive behaviours and inadequate maternal behaviours in mutant females.

Recently, dopamine D_1 and adenosine A_{2a} receptors in the striatum are shown to be functionally associated with $G\alpha_{olf}$ (Corvol et al. 2001), suggesting that $G\alpha_{olf}$ is required for olfactory signal transduction and may function as a signalling molecule in the brain.

4.4.3 G_q subfamily

G_q subfamily members are regulators of phosphatidylinositol-specific PLCβ isoforms and include $G\alpha_q$, $G\alpha_{11}$, $G\alpha_{14}$, and $G\alpha_{15/16}$. $G\alpha_q$ and $G\alpha_{11}$ are found in most of the mammalian cells, while the other members show distinctive expression patterns. $G\alpha_q$ is the most extensively studied member and it activates PLCβ in response to mitogenic signals such as bombesin and lysophosphatidate. Activation of G_q by type 1a metabotropic glutamate receptor induces phosphorylation of $G\alpha_q$ at Tyr-356, whereas Burton's tyrosine kinase (Btk) is one of the direct effectors of $G\alpha_q$ (Bence et al. 1997). Mice over-expressing wild-type and constitutively activated $G\alpha_q$ in the myocardium develop hypertrophy and heart failure associated with apoptosis. Expression of $G\alpha_q$ inhibitory peptides or RGS4 attenuates development of pressure overload hypertrophy. $G\alpha_q$-deficient mice are viable but bear impaired cerebellar motor function and platelet aggregation. $G\alpha_{11}$ is closely related to $G\alpha_q$ both structurally and functionally.

$G\alpha_{14}$ is the least known member of the G_q subfamily in terms of its functional importance. $G\alpha_{14}$ shows a distinctive receptor coupling profile, with marginal overlaps with G_i- and G_s-linked receptors (Ho et al. 2001). The receptor coupling promiscuity of $G\alpha_{14}$ is significantly narrower than that of $G\alpha_{15/16}$. Most of the $G\alpha_{14}$-coupled receptors are expressed in various imflammatory cells where $G\alpha_{14}$ is mainly expressed (Nakamura et al. 1991). However, the functions of these tissues remain intact in $G\alpha_{14}$-knockout mice. An intriguing fact of $G\alpha_{14}$ is that it may mediate an inhibitory effect on phosphoinositide metabolism in *Xenopus* oocytes ectopically expressing type 1 metabotropic glutamate receptor and $G\alpha_{14}$ (Nakamura et al. 1994). The mechanism of this phenomenon is largely unknown.

$G\alpha_{15}$ and $G\alpha_{16}$ are the two orthologs cloned from mouse and human, respectively. They are promiscuous in terms of receptor coupling. Most GPCRs that normally interact with G_s or G_i can efficiently activate $G\alpha_{15/16}$ and stimulate PLCβ (Offermanns and Simon 1995). In comparison with other $G\alpha$ subunits, $G\alpha_{15/16}$ have an extraordinarily long α4/β6 loop which is one of the major receptor-interacting regions of $G\alpha$. Changes in the lengths and identities of the C-terminal region (up to α4 helix) can alter the receptor coupling specificity of $G\alpha_{16}$ (Mody et al. 2000). This region is also significantly different between $G\alpha_{15}$ and $G\alpha_{16}$, and it may contribute to the minor differences in their receptor coupling profiles. $G\alpha_{15/16}$ are expressed mainly in hematopoietic cell lineages, and T cell receptor-CD3 complex can functionally activate both $G\alpha_q$ and $G\alpha_{16}$ (Stanners et al. 1995; Zhou et al. 1998). Expression of activated mutant of $G\alpha_{16}$ markedly suppresses the growth of small cell lung carcinoma through the activation of c-Jun N-terminal kinase (JNK) and Ca^{2+} responses (Heasley et al. 1996). Control of the stimulation of $G\alpha_{16}$ in an optimal level appears to be essential for the induction of erythroid differentiation in MB-02 erythroleukemia cells (Ghose et al. 1999). Anaphylatoxin C5a receptor-mediated activation of PLC is reduced in macrophages of $G\alpha_{15}$-knockout mice. However, there is no apparent hematopoietic defect in these knockout mice and they are viable and fertile.

Functional redundancy of G_q subfamily members have been implied in various transgenic mice studies. Deletion of $G\alpha_q$ appears to produce the most severe physiological defects compared with other members. Significant overlaps of the defective consequences are observed in mice with both $G\alpha_q$ and one of the other members knocked out simultaneously.

Interestingly, other double-knockout mice with $G\alpha_{11}$, $G\alpha_{14}$, or $G\alpha_{15/16}$ deleted do not have extensive defects (for transgenic mice studies, see Offermanns 2001).

4.4.4 G_{12} subfamily

The smallest subfamily of $G\alpha$ subunits consists of only two members, $G\alpha_{12}$ and $G\alpha_{13}$. Their signalling characteristics are very diverse and distinct from the other subfamilies. Activated mutants of both $G\alpha_{12}$ and $G\alpha_{13}$ can trigger the activation of JNK in a Ras-dependent manner leading to apoptotic events (Prasad et al. 1995). The bridging molecule between $G\alpha_{12}$ and Ras has been recently identified as GAP1[m], a GTPase-activating protein of Ras, through the inspiration of the interaction between $G\alpha_{12}$ and Btk (Jiang et al. 1998). Both GAP1[m] and Btk contain a novel sequence motif known as the Btk motif, which is an essential element for interacting with $G\alpha_{12}$. $G\alpha_{12}$- and $G\alpha_{13}$-mediated activation of JNK appears to involve MEKK1, which phosphorylates and activates JNK, and apoptotic signal-regulating kinase 1 (Berestetskaya et al. 1998). Other studies also suggest that $G\alpha_{12}$-regulated JNK activity is Rac-dependent (Collins et al. 1996). The Ras- and Rac-dependent pathways seem to work cooperatively in controlling cellular transformation (Tolkacheva et al. 1997). A novel form of PLC, PLCε, has been cloned recently and it acts as another links between $G\alpha_{12}$ and Ras (Lopez et al. 2001). $G\alpha_{12}$ directly activates PLCε and it serves as a nucleotide exchange factor of Ras to trigger Ras-dependent MAPK pathways.

Activation of $G\alpha_{12}$ and $G\alpha_{13}$ leads to extensive rearrangements of cytoskeleton with the induction of tyrosine phosphorylation of focal adhesion kinase, paxillin and p130 Crk-associated substrate in a Rho-dependent manner. Two nucleotide exchange factors of Rho, namely PDZRhoGEF and p115RhoGEF, have recently been shown to functionally associate with $G\alpha_{12}$ and $G\alpha_{13}$, respectively (Fukuhara et al. 1999; Hart et al. 1998). Although $G\alpha_{12}$ can also bind p115RhoGEF, the complex is non-productive, indicating that the co-occurrence of specific types of G_{12} subfamily members and RhoGEF can determine the ultimate cell fate upon the actions of appropriate extracellular signals. Both RhoGEFs are more than bridging molecules because they also act as RGSs on their specific $G\alpha$ partners (Kozasa et al. 1998). Both $G\alpha_{12}$ and $G\alpha_{13}$ stimulate Na^+/H^+ exchange in PKC-dependent and independent manners, respectively. The $G\alpha_{12}$-mediated pathway is also dependent on Ras and PLC. For $G\alpha_{13}$, two streams of signals—one is Cdc42-MEKK1 related, another is Rho-dependent but MEKK1-independent—are responsible for the regulation of the activity of Na^+/H^+ exchanger NHE-1 (Hooley et al. 1996; Lin et al. 1996).

More recently, $G\alpha_{12}$ and $G\alpha_{13}$ have been linked to the regulation of cadherin/β-catenin signalling event. Both $G\alpha$ subunits in their activated forms directly bind to the C-terminal cytoplasmic domains of N-cadherin, E-cadherin, and cadherin-14 (Meigs et al. 2001), and thus prevent the association of β-catenin to the same region. β-catenin dissociates and translocates into the nucleus to regulate transcription activity by interacting with adenomatous polyposis coli (APC) protein (Peifer and Polakis 2000).

4.4.5 Atypical G_h

An atypical $G\alpha_h$ subunit was found to have the structure and functions resembling type II transglutaminase (Nakaoka et al. 1994). It has very low sequence identity with all other $G\alpha$ subunits ($<10\%$) and a unique 16-amino acid guanine nucleotide-binding motif is exceptionally different from the five discrete nucleotide-binding pockets of the typical $G\alpha$

subunits (Iismaa *et al.* 2000). Yet, it can directly stimulate phosphoinositide-dependent PLCδ_1 upon the activation of α_1-adrenoceptor subtypes (Baek *et al.* 2001).

4.5 Functional diversity of Gβ and Gγ subunits

The heterogeneity of the signalling properties of various G$\beta\gamma$ complexes is less apparent than that of Gα subunits. So long as a complex of G$\beta\gamma$ can be formed (as those listed in Table 4.2), its ability to regulate various effectors is almost indistinguishable from other G$\beta\gamma$ complexes. Gβ_{1-4} are highly homologous (Table 4.3) while the most diverged Gβ_5 bears unique structural and functional variance (see below). As shown in Table 4.4, the repertoire of G$\beta\gamma$-regulated effectors has significant overlaps with those of Gα and a growing number of G$\beta\gamma$-specific effectors have been identified.

Reports on defects of Gβ functions leading to genetic diseases are scarce. Polymorphism of the Gβ_3 gene has been detected with a C825T base substitution in some hypertensive patients, which results in an alternatively spliced Gβ_3 subunit losing 41 amino acids (Siffert *et al.* 1998). The deleted region corresponds to one of the seven WD repeats. The shortened Gβ_3 (named Gβ_{3S}) can still fold into a functional protein with six blades only. Gβ_{3S} promotes G$_i$-mediated Ca^{2+} signals in lymphoblasts and enhances Na$^+$/H$^+$ exchanger activity. Gβ_{3S} is subsequently correlated to lower renin and elevated diastolic blood pressure levels, obesity, type II diabetes (Siffert 2000), enhanced chemotaxis of human neutrophils in response to interleukin-8 (Virchow *et al.* 1999), elevated serum potassium and cholesterol levels (Ishikawa *et al.* 2000), and coronary artery vasoconstriction (Meirhaeghe *et al.* 2001).

Apart from its relatively low homology to other family members ($\sim$50%) and its neuronal-specific expression pattern, Gβ_5 is functionally unique (reviewed in Simonds and Zhang 2000). A splice variant of Gβ_5 with 42 additional residues at the N-terminus with unknown function has been cloned. The regulatory actions of Gβ_5-containing G$\beta\gamma$ complexes on various effectors differ from those mediated by other G$\beta\gamma$ subunits. G$\beta_5\gamma_2$ stimulates PLCβ_2 but not MAPKs, while G$\beta_1\gamma_2$ stimulates both pathways. Presumably, G$\beta_5\gamma_2$ cannot activate phosphoinositide 3-kinase that is upstream of MAPKs. G$\beta_1\gamma_2$ inhibits type I but stimulates type II ACs, whereas G$\beta_5\gamma_2$ inhibits both subtypes. G$\beta_5\gamma_2$ shows preferential association with Gα_q as well as several subtypes of RGS proteins. Gβ_5/RGS complexes defy our traditional understanding of the formation of G protein trimers. Four of the mammalian RGS proteins, including RGS6, RGS7, RGS9, and RGS11, contain a Gγ-like domain and they can replace known Gγ subunits to form tight heterodimeric complexes with Gβ_5 (Sondek and Siderovski 2001). The Gβ_5/RGS dimer appears to have a weak interaction with GDP-bound Gα subunit. A functional Gα_o/Gβ_5/RGS9 may exist and mediate muscarinic m2 receptor-induced stimulation of GIRK channel in *Xenopus* oocytes (Kovoor *et al.* 2000). Unlike other Gβ subunits, Gβ_5 is distributed equally between cytoplasmic and membrane fractions in mammalian cells. Association with RGS may help to localize Gβ_5 at the plasma membrane, since a Dishevelled/EGL-10/Pleckstrin domain is found on Gβ_5-associated RGS, and this domain is known to facilitate anchorage to membrane by an unknown mechanism.

ADP-ribosylation does not exclusively occur in Gα subunits but also Gβ subunit. Arginine-specific mono-ADP-ribosyltransferase and a cytosolic hydrolase participate in the ribosylation/deribosylation of Gβ subunit, suggesting that the function of G$\beta\gamma$ subunits is regulated by such modification. Arg129 on the blade 2 of Gβ subunit appears to be the modification site and ADP-ribosylated purified brain G$\beta\gamma$ subunits cannot inhibit type I AC (Lupi *et al.* 2000).

4.6 **Interactions between G proteins and GPCRs**

G proteins bind to their corresponding GPCRs at multiple sites on the cytoplasmic regions of the receptor. Several regions of Gα contribute toward association with GPCRs, which have been extensively reviewed (Gudermann *et al.* 1997; Bourne 1997; Hamm 1998). One of the major domains for receptor recognition is the C-terminus of Gα, where PTX-mediated ADP-ribosylation of G_i subfamily members occurs and the receptor-G protein interaction is then disrupted. Replacement of the last five amino acids of $G\alpha_q$ with those of $G\alpha_{i2}$, $G\alpha_o$, $G\alpha_z$, or $G\alpha_s$ broadens the profile of receptor coupling (Conklin *et al.* 1993, 1996). A similar strategy applied to studying the extraordinary receptor coupling property of promiscuous $G\alpha_{16}$ indicates that the C-terminus of Gα contributes to the determination of receptor coupling specificity (Mody *et al.* 2000). A peptide derived from the last 11 residues of $G\alpha_{t1}$ impeded rhodopsin-mediated activation of $G\alpha_{t1}$ (Hamm *et al.* 1988). Using a photo-activatable crosslinking reagent, the third intracellular loop of rhodopsin is crosslinked with the residues 342–345 of the extreme C-terminus of $G\alpha_{t1}$ (Cai *et al.* 2001), which clearly demonstrated the direct contact between the C-terminal tail of Gα and the cognate receptor.

$G\alpha_{oA}$ and $G\alpha_{oB}$ have identical C-terminal tails but they couple to different receptors to mediate the inhibition of calcium ion channels (Kleuss *et al.* 1991). $G\alpha_z$ couples to almost all of the G_i-linked receptors, and yet its extreme C-terminus diverges from those of $G\alpha_i$ subtypes (Ho and Wong 1998). Several loop regions adjacent to both termini of Gα subunit are also important for defining the receptor coupling property of G protein. Multiple regions of $G\alpha_{16}$ are required for interacting with C5a receptor (Lee *et al.* 1995) and one of them are the residues 220–240 (including the α2/β4 loop) of $G\alpha_{16}$. However, mutations of the α2/β4 loop of $G\alpha_s$ do not impede receptor coupling (Grishina and Berlot, 2000). Instead, the same study points to the importance of the α3/β5 loop in receptor coupling, where homologous mutations of $G\alpha_s$ residues to those of $G\alpha_{i2}$ increase the affinity of the mutant to β-adrenoceptor. The α4 helix and α4/β6 loop has been demonstrated to be a major receptor contact site in two successive chimera studies (Bae *et al.* 1997). In addition to the C-terminal tail, the α4/β6 loop can also be crosslinked with the third intracellular loop of rhodopsin, suggesting a direct contact between the two structural domains (Cai *et al.* 2001). Sequence alignment of all Gα subunits showed that $G\alpha_{16}$ has particularly long α4/β6 loop, and its promiscuous receptor coupling property is greatly impaired when it is trimmed down to a length comparable to other Gα subunits (authors' unpublished results). It is obvious that different modes of actions may occur between specific receptors and G proteins.

The N-terminus of Gα is also involved in receptor recognition. Deletion of first six amino acids of $G\alpha_q$ can also allow the mutant to couple to G_i-linked receptors (Kostenis *et al.* 1997). Substitution of the N-terminal 36 residues of $G\alpha_{t1}$ with the corresponding region of $G\alpha_z$ enables the chimera to couple to δ-opioid receptor (Ho and Wong 2000). The portion of the N-terminal helix closer to the core of Gα is in close proximity with the C-terminal tail, as revealed in the crystal structures of various Gα subunits, and the third intracellular loop of receptor (Itoh *et al.* 2001).

Gβγ complex may also be important for determining the efficiency and specificity of receptor coupling (see Gudermann *et al.* 1997). A peptide derived from a cytoplasmic portion of the α_2-adrenergic receptor binds to a defined site of the Gβ subunit, and the identity of Gγ subunit may be involved in specifying receptor-G protein interaction. The N-terminus of Gα subunit and the C-terminus of Gγ subunit are in close proximity and both are lying on the plasma membrane-facing surface of the heterotrimer (Wall *et al.* 1995, Fig. 3.2c). A model for the receptor-Gβγ complex predicts a contact between the receptor's C-terminal tail and the Gβγ dimer (Lichtarge *et al.* 1996, Fig. 3.2c). Another detailed mapping of the receptor- and effector-interacting surfaces of Gβγ subunits suggests that blades 1-3 contribute most of the protein-protein interactions (Ford *et al.* 1998).

4.7 **Interactions between G proteins and downstream effectors**

Most of the information about the effector binding domains of Gα subunits is obtained from the construction of chimeric proteins and subsequent site-directed mutagenesis targeted to identify specific effector interacting residues. The C-terminal portion of Gα$_s$ (residues 235–356) contained the AC-activating domain (Osawa *et al.* 1990). Mutations within this region of Gα$_s$ decreased its ability to stimulate AC (Itoh and Gilman 1991). The AC activating domain is comprised of three discrete stretches of residues of Gα$_s$ (Berlot and Bourne 1992), corresponding to three of the loops (α2/β4, α3/β5, and α4/β6 loops) observed on Gα crystals. The Switch IV region (αB/αC junction) may also be a crucial but indirect element for AC stimulation (Echeverría *et al.* 2000).

Similarly, other Gα-effector interaction surfaces have also mapped to comparable regions. The AC inhibiting residues of Gα$_{i2}$ are located at the Switch II region and the α4/β6 loop. Similar regions are also identified as the AC-interacting domains of Gα$_z$, suggesting that both Gα$_{i2}$ and Gα$_z$ may act on the same contact sites of AC (Ho and Wong 2000). For Gα$_q$, discrete regions between α3 helix and β6 strand and the N-terminal cysteine residues are essential for PLC activation. For Gα$_{t1}$, the α2, α3 and α4 helices and the loops followed form the interacting surfaces for the γ subunit of cGMP phosphodiesterase (PDEγ) (for details, see Hamm 1998).

It has been shown that tubulin associated with the synaptic membrane can interact with G$_s$ and G$_{i2}$ (Wang *et al.* 1990). For Gα$_s$, residues 54–212 are important for the GTP transfer from tubulin, while the first 54 amino acids are required for the ability of tubulin to stimulate AC (Popova *et al.* 1994). It has been postulated that the active Gα$_s$ conformation provoked by nucleotide transfer from tubulin is stabilized by Gα$_s$-tubulin interaction leading to extended stimulation of AC. Gα$_s$ suppresses the commitment of 3T3-L1 cells to adipocytes in response to dexamethasone, but Gα$_{i2}$ enhances the adipogenesis. The process is independent to AC regulation (Wang and Malbon 1996) and the amino acids 146–220 of Gα$_s$ covering Switches I and II are likely the suppressor domain for adipogenesis (Liu *et al.* 1998).

4.8 **Unravelling the missing links and new paradigms**

In the world of monomeric G proteins, the identities and functional roles of the correspond- ing GTPase-activating proteins (GAPs) have been extensively investigated. A number of GAPs for monomeric G proteins actually serve as their immediately downstream effectors, and it is believed that the same principle applies to trimeric G proteins. Supporting evidences for this analogy are the interactions between Gα$_q$/PLCβ (Berstein *et al.* 1992) and Gα$_{t1}$/cGMP PDEγ (Arshavsky and Bownd 1992). However, the GAP role of the G protein effectors could not fit perfectly into some of the fast signalling events involving G proteins, such as the regulation of the opening time of ion channels and the turnover of the photo-activation of rhodopsin-transducin signalling pathway. Recent identification of RGS (regulators of G protein signalling) proteins provides the missing link on the regulatory mechanisms of the GTPase activity of Gα (see De Vries *et al.* 2000). RGS proteins which appear to accelerate the GTP hydrolysis of Gα are covered in detail in Chapter 7. There are indications of specific regulatory roles of RGS proteins on individual GPCR signalling pathway, which may be dependent on a combination of cell type-specific expression, tissue distribution, intracellular localization, post-translational modifications, and other functional domains on the RGS. Based on a functional screen of the pheromone response pathway in yeast, a group of three diverse proteins known as the activators of G protein signalling (AGS1-3) have been identified and they activate trimeric G protein signalling pathways in the absence of a typical receptor (reviewed in Cismowski *et al.* 2001). AGS1 defines a distinct member of Ras-related proteins. AGS2 is identical to mouse Tctex1, a protein that exists as a light chain component of dynein with unknown signalling function. AGS3 possesses a novel functional motif termed G protein regulatory/GoLoco motifs and it can be found in a number of distinct proteins including Rap1GAP, human mosaic protein LGN, Purkinje cell protein-2. Each AGS protein activates G protein signalling, they do so by different mechanisms within the context of the G protein activation/deactivation cycle. AGS proteins provide unexpected mechanisms for regulating trimeric G protein signalling pathways and additional surprises are bound to lie ahead of us.

Acknowledgements

Relevant works generated from the authors' laboratory were supported by the Research Grants Council of Hong Kong (HKUST 6096/98M and 2/99C).

References

Arshavsky VY and Bownds MD (1992). Regulation of deactivation of photoreceptor G protein by its target enzyme and cGMP. *Nature* 357, 416–7.

Asano T, Morishita R, Ueda H, and Kato K (1999). Selective association of G protein β$_4$ with γ$_5$ and γ$_{12}$ subunits in bovine tissues. *J Biol Chem* 274, 21 425–9.

Bae H, Anderson K, Flood LA, Skiba NP, Hamm HE, and Graber SG (1997). Molecular determinants of selectivity in 5-hydroxytryptamine$_{1B}$ receptor-G protein interactions. *J Biol Chem* 272, 32 071–7.

Baek KJ, Kang S, Damron D, and Im M (2001). Phospholipase Cδ$_1$ is a guanine nucleotide exchanging factor for transglutaminase II (Gα$_h$). and promotes α$_{1B}$-adrenoreceptor-mediated GTP binding and intracellular calcium release. *J Biol Chem* 276, 5591–7.

Bence K, Ma W, Kozasa T, and Huang XY (1997). Direct stimulation of Bruton's tyrosine kinase by G$_q$-protein α-subunit. *Nature* 389, 296–9.

Berestetskaya YV, Faure MP, Ichijo H, and Voyno-Yasenetskaya TA (1998). Regulation of apoptosis by α-subunits of G$_{12}$ and G proteins via apoptosis signal-regulating kinase-1. *J Biol Chem* 273, 27 816–23.

Berlot CH and Bourne HR (1992). Identification of effector-activating residues of G$_s$α. *Cell* 68, 911–22.

Berman HM, Bhat TN, Bourne PE *et al.* (2000). The protein data bank and the challenge of structural genomics. *Nature Structural Biol* 7, 957–9.

Berstein G, Blank JL, Jhon DY, Exton JH, Rhee SG, and Ross EM (1992). Phospholipase C-β$_1$ is a GTPase-activating protein for G$_{q/11}$, its physiologic regulator. *Cell* 70, 411–8.

Bourne HR (1997). How receptors talk to trimeric G proteins. *Curr Opin Cell Biol* 9, 134–42.

Bourne HR, Sanders DA, and McCormick F (1990). The GTPase superfamily: a conserved switch for diverse cell functions. *Nature* 348, 125–32.

Bourne HR, Sanders DA, and McCormick F (1991). The GTPase superfamily: conserved structure and molecular mechanism. *Nature* 349, 117–27.

Cai K, Itoh Y, and Khorana HG (2001). Mapping of contact sites in complex formation between transducin and light-activated rhodopsin by covalent crosslinking: use of a photoactivatable reagent. *Proc Natl Acad Sci USA* **98**, 4877–82.

Casey PJ and Seabra MC (1996). Protein prenyltransferases. *J Biol Chem* **271**, 5289–92.

Chen CA and Manning DR (2001). Regulation of G proteins by covalent modification. *Oncogene* **20**, 1643–52.

Chen LT, Gilman AG, and Kozasa T (1999). A candidate target for G protein action in brain. *J Biol Chem* **274**, 26 931–8.

Cismowski MJ, Takesono A, Bernard ML, Duzic E, and Lanier SM (2001). Receptor-independent activators of heterotrimeric G-proteins. *Life Sci* **68**, 2301–8.

Clapham DE and Neer EJ (1997). G protein βγ subunits. *Ann Rev Pharmacol Toxicol* **37**, 167–203.

Collins LR, Minden A, Karin M, and Brown JH (1996). $G\alpha_{12}$ stimulates c-Jun NH_2-terminal kinase through the small G proteins Ras and Rac. *J Biol Chem* **271**, 17 349–53.

Conklin BR and Bourne HR (1993). Structural elements of Gα subunits that interact with Gβγ, receptors, and effectors. *Cell* **73**, 631–41.

Conklin BR, Farfel Z, Lustig KD, Julius D, and Bourne HR (1993). Substitution of three amino acids switches receptor specificity of $G_{q\alpha}$ to that of $G_{i\alpha}$. *Nature* **363**, 274–6.

Conklin BR, Herzmark P, Ishida S *et al.* (1996). Carboxyl-terminal mutations of G_{qa} and G_{sa} that alter the fidelity of receptor activation. *Mol Pharmacol* **50**, 885–90.

Corvol JC, Studler JM, Schonn JS, Girault JA, and Herve D (2001). $G\alpha_{olf}$ is necessary for coupling D_1 and A_{2a} receptors to adenylyl cyclase in the striatum. *J Neurochem* **76**, 1585–8.

De Vries L, Mousli M, Wurmser A, and Farquhar MG (1995). GAIP, a protein that specifically interacts with the trimeric G protein $G\alpha_{i3}$, is a member of a protein family with a highly conserved core domain. *Proc Natl Acad Sci USA* **92**, 11 916–20.

De Vries L, Zheng B, Fischer T, Elenko E, and Farquhar MG (2000). The regulator of G protein signaling family. *Ann Rev Pharmacol Toxicol* **40**, 235–71.

Dryja TP, Hahn LB, Reboul T, and Arnaud B (1996). Missense mutation in the gene encoding the α subunit of rod transducin in the Nougaret form of congenital stationary night blindness. *Nature Genet* **13**, 358–60.

Duncan JA and Gilman AG (1996). Autoacylation of G protein alpha subunits. *J Biol Chem* **271**, 23 594–600.

Echeverria V, Hinrichs MV, Torrejon M *et al.* (2000). Mutagenesis in the switch IV of the helical domain of the human $G_{s\alpha}$ reduces its GDP/GTP exchange rate. *J Cell Biochem* **76**, 368–75.

Exner T, Jensen ON, Mann M, Kleuss C, and Nurnberg B (1999). Posttranslational modification of $G\alpha_{o1}$ generates $G\alpha_{o3}$, an abundant G protein in brain. *Proc Natl Acad Sci USA* **96**, 1327–32.

Ford CE, Skiba NP, Bae H *et al.* (1998). Molecular basis for interactions of G protein βγ subunits with effectors. *Science* **280**, 1271–4.

Fukuhara S, Murga C, Zohar M, Igishi T, and Gutkind JS (1999). A novel PDZ domain containing guanine nucleotide exchange factor links heterotrimeric G proteins to Rho. *J Biol Chem* **274**, 5868–79.

Gasman S, Chasserot-Golaz S, Popoff MR, Aunis D, and Bader MF (1997). Trimeric G proteins control exocytosis in chromaffin cells. G_o regulates the peripheral actin network and catecholamine secretion by a mechanism involving the small GTP-binding protein Rho. *J Biol Chem* **272**, 20 564–71.

Ghose S, Porzig H, and Baltensperger K (1999). Induction of erythroid differentiation by altered $G\alpha_{16}$ activity as detected by a reporter gene assay in MB-02 cells. *J Biol Chem* **274**, 12 848–54.

Gilbertson TA, Damak S, and Margolskee RF (2000). The molecular s physiology of taste transduction. *Curr Opin Neurobiol* **10**, 519–27.

Gilman AG (1987). G proteins: transducers of receptor-generated signals. *Ann Rev Biochem* **56**, 615–49.

Grishina G and Berlot CH (2000). A surface-exposed region of G_{sa} in which substitutions decrease receptor-mediated activation and increase receptor affinity. *Mole Pharmacol* **57**, 1081–92.

Gu C, Ma YC, Benjamin J, Littman D, Chao MV, and Huang XY (2000). Apoptotic signaling through the β-adrenergic receptor. A new G_s effector pathway. *J Biol Chem* **275**, 20 726–33.

Gudermann T, Schoneberg T, and Schultz G (1997). Functional and structural complexity of signal transduction via G-protein-coupled receptors. *Ann Rev Neurosci* **20**, 399–427.

Guex N and Peitsch MC (1997). SWISS-MODEL and the Swiss-PdbViewer: an environment for comparative protein modeling. *Electrophoresis* **18**, 2714–23.

Hamm HE (1998). The many faces of G protein signaling. *J Biol Chem* **273**, 669–72.

Hamm HE, Deretic D, Arendt A, Hargrave PA, Koenig B, and Hofmann KP (1988). Site of G protein binding to rhodopsin mapped with synthetic peptides from the α subunit. *Science* **241**, 832–5.

Hart MJ, Jiang X, Kozasa T *et al.* (1998). Direct stimulation of the guanine nucleotide exchange activity of p115 RhoGEF by $Gα_{13}$. *Science* **280**, 2112–4.

Heasley LE, Zamarripa J, Storey B *et al.* (1996). Discordant signal transduction and growth inhibition of small cell lung carcinomas induced by expression of GTPase-deficient $Gα_{16}$. *J Biol Chem* **271**, 349–54.

Hendry IA, Kelleher KL, Bartlett SE *et al.* (2000). Hypertolerance to morphine in G_{za}-deficient mice. *Brain Res* **870**, 10–19.

Herlitze S, Garcia DE, Mackie K, Hille B, Scheuer T, and Catterall WA (1996). Modulation of Ca2+ channels by G-protein beta gamma subunits. *Nature* **380**, 258–62.

Ho MK and Wong YH (1998). Structure and function of the pertussis-toxin-insensitive G_z protein. *Biol Signals and Receptors* **7**, 80–9.

Ho MK and Wong YH (2000). The amino terminus of $Gα_z$ is required for receptor recognition, whereas its α4/β6 loop is essential for inhibition of adenylyl cyclase. *Mol Pharmacol* **58**, 993–1000.

Ho MK and Wong YH (2001). G_z signaling: emerging divergence from G_i signaling. *Oncogene* **20**, 1615–25.

Ho MK, Yung LY, Chan JS, Chan JH, Wong CS, and Wong YH (2001). $Gα_{14}$ links a variety of G_i- and G_s-coupled receptors to the stimulation of phospholipase C. *Brit J Pharmacol* **132**, 1431–40.

Hooley R, Yu CY, Symons M, and Barber DL (1996). $Gα_{13}$ stimulates Na^+-H^+ exchange through distinct Cdc42-dependent and RhoA-dependent pathways. *J Biol Chem* **271**, 6152–8.

Hoon MA, Adler E, Lindemeier J, Battey JF, Ryba NJ, and Zuker CS (1999). Putative mammalian taste receptors: a class of taste-specific GPCRs with distinct topographic selectivity. *Cell* **96**, 541–51.

Hurowitz EH, Melnyk JM, Chen YJ, Kouros-Mehr H, Simon MI, and Shizuya H (2000). Genomic characterization of the human heterotrimeric G protein α, β, and γ subunit genes. *DNA Res* **7**, 111–20.

Igarashi M, Strittmatter SM, Vartanian T, and Fishman MC (1993). Mediation by G proteins of signals that cause collapse of growth cones. *Science* **259**, 77–9.

Iismaa SE, Wu MJ, Nanda N, Church WB, and Graham RM (2000). GTP binding and signaling by G_h/transglutaminase II involves distinct residues in a unique GTP-binding pocket. *J Biol Chem* **275**, 18 259–65.

Ishikawa K, Imai Y, Katsuya T *et al.* (2000). Human G-protein β3 subunit variant is associated with serum potassium and total cholesterol levels but not with blood pressure. *Am J Hypertension* **13**, 140–5.

Itoh H and Gilman AG (1991). Expression and analysis of G_{sa} mutants with decreased ability to activate adenylylcyclase. *J Biol Chem* **266**, 16 226–31.

Itoh Y, Cai K, and Khorana HG (2001). Mapping of contact sites in complex formation between light-activated rhodopsin and transducin by covalent crosslinking: use of a chemically preactivated reagent. *Proc Natl Acad Sci USA* **98**, 4883–7.

Jiang Y, Ma W, Wan Y, Kozasa T, Hattori S, and Huang XY (1998). The G protein $Gα_{12}$ stimulates Bruton's tyrosine kinase and a rasGAP through a conserved PH/BM domain. *Nature* **395**, 808–13.

Johnson RS, Ohguro H, Palczewski K, Hurley JB, Walsh KA, and Neubert TA (1994). Hetero-geneous N-acylation is a tissue-and species-specific posttranslational modification. *J Biol Chem* **269**, 21 067–71.

Jordan JD, Carey KD, Stork PJ, and Iyengar R (1999). Modulation of rap activity by direct interaction of $G\alpha_o$ with Rap1 GTPase-activating protein. *J Biol Chem* **274**, 21 507–10.

Kleuss C, Hescheler J, Ewel C, Rosenthal W, Schultz G, and Wittig B (1991). Assignment of G-protein subtypes to specific receptors inducing inhibition of calcium currents. *Nature* **353**, 43–8.

Kostenis E, Degtyarev MY, Conklin BR, and Wess J (1997). The N-terminal extension of $G\alpha_q$ is critical for constraining the selectivity of receptor coupling. *J Biol Chem* **272**, 19 107–10.

Kovoor A, Chen CK, He W, Wensel TG, Simon MI, and Lester HA (2000). Co-expression of $G\beta_5$ enhances the function of two $G\gamma$ subunit-like domain-containing regulators of G protein signaling proteins. *J Biol Chem* **275**, 3397–402.

Kozasa T, Jiang X, Hart MJ *et al.* (1998). p115 RhoGEF, a GTPase activating protein for $G\alpha_{12}$ and $G\alpha_{13}$. *Science* **280**, 2109–11.

Lai RK, Perez-Sala D, Canada FJ, and Rando RR (1990). The γ subunit of transducin is farnesylated. *Proc Natl Acad Sci USA* **87**, 7673–7.

Lambright DG, Sondek J, Bohm A, Skiba NP, Hamm HE, and Sigler PB (1996). The 2.0 Å crystal structure of a heterotrimeric G protein. *Nature* **379**, 311–9.

Lee CH, Katz A, and Simon MI (1995). Multiple regions of $G\alpha_{16}$ contribute to the specificity of activation by the C5a receptor. *Mol Pharmacol* **47**, 218–23.

Lerman BB, Dong B, Stein KM, Markowitz SM, Linden J, and Catanzaro DF (1998). Right ventricular outflow tract tachycardia due to a somatic cell mutation in G protein subunit α_{i2}. *J Clini Invest* **101**, 2862–8.

Lichtarge O, Bourne HR, and Cohen FE (1996). Evolutionarily conserved $G\alpha\beta\gamma$ binding surfaces support a model of the G protein-receptor complex. *Proc Natl Acad Sci USA* **93**, 7507–11.

Lin P, Fischer T, Weiss T, and Farquhar MG (2000). Calnuc, an EF-hand Ca(2+). binding protein, specifically interacts with the C-terminal α5-helix of $G\alpha_{i3}$. *Proc Natl Acad Sci USA* **97**, 674–9.

Lin X, Voyno-Yasenetskaya TA, Hooley R, Lin CY, Orlowski J, and Barber DL (1996). $G\alpha_{12}$ differentially regulates Na^+-H^+ exchanger isoforms. *J Biol Chem* **271**, 22 604–10.

Liu X, Malbon CC, and Wang HY (1998). Identification of amino acid residues of $G_{s\alpha}$ critical to repression of adipogenesis. *J Biol Chem* **273**, 11 685–94.

Loew A, Ho YK, Blundell T, and Bax B (1998). Phosducin induces a structural change in transducin βγ. *Structure* **6**, 1007–19.

Lopez I, Mak EC, Ding J, Hamm HE, and Lomasney JW (2001). A novel bifunctional phospholipase C that is regulated by $G\alpha_{12}$ and stimulates the Ras/mitogen-activated protein kinase pathway. *J Biol Chem* **276**, 2758–65.

Luo Y and Denker BM (1999). Interaction of heterotrimeric G protein $G\alpha_o$ with Purkinje cell protein-2. Evidence for a novel nucleotide exchange factor. *J Biol Chem* **274**, 10 685–8.

Lupi R, Corda D, and Di Girolamo M (2000). Endogenous ADP-ribosylation of the G protein β subunit prevents the inhibition of type 1 adenylyl cyclase. *J Biol Chem* **275**, 9418–24.

Lyons J, Landis CA, Harsh G *et al.* (1990). Two G protein oncogenes in human endocrine tumors. *Science* **249**, 655–9.

Ma YC, Huang J, Ali S, Lowry W, and Huang XY (2000). Src tyrosine kinase is a novel direct effector of G proteins. *Cell* **102**, 635–46.

Meigs TE, Fields TA, McKee DD, and Casey PJ (2001). Interaction of $G\alpha_{12}$ and $G\alpha_{13}$ with the cyto-plasmic domain of cadherin provides a mechanism for β-catenin release. *Proc Natl Acad Sci USA* **98**, 519–24.

Meirhaeghe A, Bauters C, Helbecque N *et al.* (2001). The human G-protein β_3 subunit C825T polymorphism is associated with coronary artery vasoconstriction. *Eur Heart J* **22**, 845–8.

Mody SM, Ho MK, Joshi SA, and Wong YH (2000). Incorporation of $G\alpha_z$-specific sequence at the carboxyl terminus increases the promiscuity of $G\alpha_{16}$ toward G_i-coupled receptors. *Mol Pharmacol* 57, 13–23.

Moxham CM and Malbon CC (1996). Insulin action impaired by deficiency of the G-protein subunit $G_{i\alpha2}$. *Nature* 379, 840–4.

Murtagh JJ Jr, Moss J, and Vaughan M (1994). Alternative splicing of the guanine nucleotide-binding regulatory protein $G_{o\alpha}$ generates four distinct mRNAs. *Nucleic Acids Res* 22, 842–9.

Nakamura F, Ogata K, Shiozaki K *et al.* (1991). Identification of two novel GTP-binding protein α-subunits that lack apparent ADP-ribosylation sites for pertussis toxin. *J Biol Chem* 266, 12 676–81.

Nakamura K, Nukada T, Haga T, and Sugiyama H (1994). G protein-mediated inhibition of phosphoinositide metabolism evoked by metabotropic glutamate receptors in frog oocytes. *J Physiol* 474, 35–41.

Nakaoka H, Perez DM, Baek KJ *et al.* (1994). G_h: a GTP-binding protein with transglutaminase activity and receptor signaling function. *Science* 264, 1593–6.

Neer EJ, Schmidt CJ, Nambudripad R, and Smith TF (1994). The ancient regulatory-protein family of WD-repeat proteins. *Nature* 371, 297–300.

Offermanns S (2001). In vivo functions of heterotrimeric G-proteins: studies in Gα-deficient mice. *Oncogene* 20, 1635–42.

Offermanns S and Simon MI (1995). $G\alpha_{15}$ and $G\alpha_{16}$ couple a wide variety of receptors to phospholipase C. *J Biol Chem* 270, 15 175–80.

Osawa S, Dhanasekaran N, Woon CW, and Johnson GL (1990). $G\alpha_i$-$G\alpha_s$ chimeras define the function of α chain domains in control of G protein activation and βγ subunit complex interactions. *Cell* 63, 697–706.

Osawa S and Johnson GL (1991). A dominant negative $G\alpha_s$ mutant is rescued by secondary mutation of the α chain amino terminus. *J Biol Chem* 266, 4673–6.

Palczewski K, Kumasaka T, Hori T *et al.* (2000). Crystal structure of rhodopsin: A G protein-coupled receptor. *Science* 289, 739–45.

Peifer M and Polakis P (2000). Wnt signaling in oncogenesis and embryogenesis—a look outside the nucleus. *Science* 287, 1606–9.

Philips MR, Staud R, Pillinger M *et al.* (1995). Activation-dependent carboxyl methylation of neutrophil G-protein γ subunit. *Proc Natl Acad Sci USA* 92, 2283–7.

Popova JS, Johnson GL, and Rasenick MM (1994). Chimeric $G\alpha_s/G\alpha_{i2}$ proteins define domains on $G\alpha_s$ that interact with tubulin for β-adrenergic activation of adenylyl cyclase. *J Biol Chem* 269, 21 748–54.

Prasad MV, Dermott JM, Heasley LE, Johnson GL, and Dhanasekaran N (1995). Activation of Jun kinase/stress-activated protein kinase by GTPase-deficient mutants of $G\alpha_{12}$ and $G\alpha_{13}$. *J Biol Chem* 270, 18 655–9.

Rahmatullah M and Robishaw JD (1994). Direct interaction of the α and γ subunits of the G proteins. Purification and analysis by limited proteolysis. *J Biol Chem* 269, 3574–80.

Ram PT, Horvath CM, and Iyengar R (2000). Stat3-mediated transformation of NIH-3T3 cells by the constitutively active Q205L $G\alpha_o$ protein. *Science* 287, 142–4.

Ronnett GV and Snyder SH (1992). Molecular messengers of olfaction. *Trends in Neurosci* 15, 508–13.

Rudolph U, Finegold MJ, Rich SS *et al.* (1995). $G_{i2}\alpha$ protein deficiency: a model of inflammatory bowel disease. *J Clin Immunol* 15, 101–5.

Rymer DL and Good TA (2001). The role of G protein activation in the toxicity of amyloidogenic Aβ-(1-40), Aβ-(25-35), and bovine calcitonin. *J Biol Chem* 276, 2523–30.

Siffert W (2000). G protein β_3 subunit 825T allele, hypertension, obesity, and diabetic nephropathy. *Nephrology, Dialysis, Transplantation* 15, 1298–306.

Siffert W, Rosskopf D, Siffert G *et al.* (1998). Association of a human G-protein β_3 subunit variant with hypertension. *Nature Genetics* 18, 45–8.

Simon MI, Strathmann MP, and Gautam N (1991). Diversity of G proteins in signal transduction. *Science* 252, 802–8.

Simonds WF and Zhang JH (2000). New dimensions in G protein signalling: Gβ_5 and the RGS proteins. *Pharmaceutica Acta Helvetiae* 74, 333–6.

Smine A, Xu X, Nishiyama K *et al.* (1998). Regulation of brain G-protein G$_o$ by Alzheimer's disease gene presenilin-1. *J Biol Chem* 273, 16 281–8.

Sondek J and Siderovski DP (2001). Gγ-like (GGL) domains: new frontiers in G-protein signaling and beta-propeller scaffolding. *Biochem Pharmacol* 61, 1329–37.

Sowell MO, Ye C, Ricupero DA *et al.* (1997). Targeted inactivation of α_{i2} or α_{i3} disrupts activation of the cardiac muscarinic K$^+$ channel, IK$^+$ Ach, in intact cells. *Proc Natl Acad Sci USA* 94, 7921–6.

Sprang SR (1997). G proteins, effectors and GAPs: structure and mechanism. *Curr Opin Struct Biol* 7, 849–56.

Stanners J, Kabouridis PS, McGuire KL, and Tsoukas CD (1995). Interaction between G proteins and tyrosine kinases upon T cell receptor.CD3-mediated signaling. *J Biol Chem* 270, 30 635–42.

Strittmatter SM, Valenzuela D, Sudo Y, Linder ME, and Fishman MC (1991). An intracellular guanine nucleotide release protein for G$_o$. GAP-43 stimulates isolated α subunits by a novel mechanism. *J Biol Chem* 266, 22 465–71.

Tanaka M, Treloar H, Kalb RG, Greer CA, and Strittmatter SM (1999). G$_o$ protein-dependent survival of primary accessory olfactory neurons. *Proc Natl Acad Sci USA* 96, 14 106–11.

Taussig R, Tang WJ, and Gilman AG (1994). Expression and purification of recombinant adenylyl cyclases in Sf9 cells. *Meth Enzymol* 238, 95–108.

Tolkacheva T, Feuer B, Lorenzi MV, Saez R, and Chan AM (1997). Cooperative transformation of NIH3T3 cells by Gα_{12} and Rac1. *Oncogene* 15, 727–35.

Ueda H, Yamauchi J, Itoh H *et al.* (1999). Phosphorylation of F-actin-associating G protein γ_{12} subunit enhances fibroblast motility. *J Biol Chem* 274, 12 124–8.

Virchow S, Ansorge N, Rosskopf D, Rubben H, and Siffert W (1999). The G protein β_3 subunit splice variant Gβ_3-s causes enhanced chemotaxis of human neutrophils in response to interleukin-8. *Naunyn Schmiedebergs Archives in Pharmacol* 360, 27–32.

Wall MA, Coleman DE, Lee E *et al.* (1995). The structure of the G protein heterotrimer G$_i\alpha_1\beta_1\gamma_2$. *Cell* 83, 1047–58.

Wang HY and Malbon CC (1996). The G$_{s\alpha}$/G$_{i\alpha2}$ axis controls adipogenesis independently of adenylylcyclase. *Inter J Obes Relat Metab Disord* 20, S26–31.

Wang HY and Malbon CC (1999). G$_{s\alpha}$ repression of adipogenesis via Syk. *J Biol Chem* 274, 32159–66.

Wang N, Yan K, and Rasenick MM (1990). Tubulin binds specifically to the signal-transducing proteins, G$_{s\alpha}$ and G$_{i\alpha1}$. *J Biol Chem* 265, 1239–42.

Wang Q, Mullah BK, and Robishaw JD (1999). Ribozyme approach identifies a functional association between the G protein $\beta_1\gamma_7$ subunits in the β-adrenergic receptor signaling pathway. *J Biol Chem* 274, 17365–71.

Wedegaertner PB and Bourne HR (1994). Activation and depalmitoylation of G$_{s\alpha}$. *Cell* 77, 1063–70.

Wedegaretner PB, Wison PT, and Bourne HR (1995). Lipid modifications of trimeric G proteins. *J Biol Chem* 270, 503–6.

Weinstein LS and Yu S (1999). The Role of Genomic imprinting of Gα in the pathogenesis of Albright hereditary osteodystrophy. *Trends Endocrinol Metab* 10, 81–5.

Williamson EA, Johnson SJ, Foster S, Kendall-Taylor P, and Harris PE (1995). G protein gene mutations in patients with multiple endocrinopathies. *J Clin Endocrinol Metab* 80, 1702–5.

Wong GT, Gannon KS, and Margolskee RF (1996). Transduction of bitter and sweet taste by gustducin. *Nature* 381, 796–800.

Wong YH, Conklin BR, and Bourne HR (1992). G$_z$-mediated hormonal inhibition of cyclic AMP accumulation. *Science* 255, 339–42.

Yan K, Kalyanaraman V, and Gautam N (1996). Differential ability to form the G protein βγ complex among members of the β and γ subunit families. *J Biol Chem* **271**, 7141–6.

Yan W, Sunavala G, Rosenzweig S, Dasso M, Brand JG, and Spielman AI (2001). Bitter taste transduced by PLC-β_2-dependent rise in IP_3 and α-gustducin-dependent fall in cyclic nucleotides. *American J Physiol Cellular Physiol* **280**, C742–51.

Yang J, Wu J, Kowalska MA *et al.* (2000). Loss of signaling through the G protein, G_z, results in abnormal platelet activation and altered responses to psychoactive drugs. *Proc Natl Acad Sci USA* **97**, 9984–9.

Zhou J, Stanners J, Kabouridis P, Han H, and Tsoukas CD (1998). Inhibition of TCR/CD3-mediated signaling by a mutant of the hematopoietically expressed G_{16} GTP-binding protein. *Eur J Immunol* **28**, 1645–55.

Chapter 5

GPCR signalling through enzymatic cascades

R. A. John Challiss and Stefan R. Nahorski

5.1 Introduction

It is now appreciated that the G protein-coupled receptors (GPCRs) constitute a superfamily of proteins that allow cells within multicellular organisms to respond to their environment and to exchange information. Of the 600-plus GPCRs already identified in the human genome (Venter *et al.* 2001), many are expressed exclusively, or at their highest levels, within the central nervous system (CNS). Despite this abundance of GPCRs in the CNS, it is only recently that we have begun to appreciate how these receptors contribute to the control of specific aspects of neuronal function. In this Chapter we will discuss the intracellular pathways regulated by GPCRs with a view to providing an insight into the diversity of mechanisms by which this receptor superfamily can mediate and modulate neuronal signalling.

5.2 Effectors and the regulation of second messengers

Early research into the signal transduction pathways utilized by GPCRs focussed upon pathways regulating the cellular concentration of second messenger molecules, notably cyclic AMP. Cyclic AMP regulates different cell activities through its ability to bind to specific protein kinases (cyclic AMP-dependent protein kinase (PKA)) leading to phosphorylation, and activity modification, of key protein substrates (Schramm and Selinger 1984; Francis and Corbin 1999). Such research also revealed that GPCRs regulate enzymes, which either synthesize (adenylyl cyclase/cyclic AMP) or breakdown (cyclic GMP phosphodiesterase/cyclic GMP) second messengers, via a family of heterotrimeric guanine nucleotide-binding proteins, termed G proteins (Rodbell 1980; Gilman 1987; Simon *et al.* 1991). Thus, these early studies led to the establishment of the 'universal' tripartite receptor-G protein-effector sequence by which GPCRs were thought to exert all of their cellular actions (Casey and Gilman 1988; Birnbaumer 1992). Although this tripartite view of GPCR signalling has pertained for many years, very recent experimental data have begun to uncover GPCR signalling pathways that appear to operate independently of G proteins (Table 5.1). A number of excellent reviews dealing with G protein-independent GPCR signalling have appeared recently (Hall *et al.* 1999; Heuss and Gerber 2000; Milligan and White 2001).

Many GPCRs can influence the activity of adenylyl cyclases through activation of stimulatory G_s, or inhibitory G_i proteins. The cloning of nine mammalian adenylyl cyclase isoforms,

Table 5.1 Non-ion channel effectors for GPCRs

Effector	Second messenger/mediator	G protein involvement
Second messenger-linked		
Adenylyl cyclase[1]	$\updownarrow$ cyclic AMP	$G_s/G_{olf}(+)/G_{i/o}/G_z$ $(-)$ $G_{\beta\gamma}$-subunits
Cyclic nucleotide	$\downarrow$ cyclic GMP	G_t
phosphodiesterase	$\downarrow$ cyclic AMP	G_{gust}
Phospholipase C[2]	$\uparrow$ Ins(1,4,5)P_3 $\uparrow$ DAG $\downarrow$ PtdIns(4,5)P_2	$G_{q/11}$, G_h $G_{\beta\gamma}$-subunits
Phosphoinositide	$\uparrow$ PtdIns(3,4,5)P_3	$G_{\beta\gamma}$-subunits
3-kinase (PI3K)	$\uparrow$ PtdIns(3,4)P_2	
Phospholipase A_2	$\uparrow$ arachidonate $\uparrow$ lyso-PtdCho $\uparrow$ lyso-PtdEth	G_i $G_{\beta\gamma}$-subunits
Phospholipase D	$\uparrow$ phosphatidic acid $\uparrow$ lyso-PtdOH $\uparrow$ DAG	$G_{q/11}$, G_i
Sphingosine kinase	$\uparrow$ sphingosine 1-phosphate	$G_{q/11}$, $G_{12/13}$
Adaptor-linked		
Guanine nucleotide exchange factors		
e.g. p115[Rho-GEF]	Rho	$G_{12/13}$
Non-receptor tyrosine kinases		
e.g. Src/Pyk2	Shc/Grb2/SOS	G_s, $G_{i/o}$, $G_{q/11}$, $G_{12/13}$
Btk	PI3K, PLC-γ	$G_{\beta\gamma}$-subunits
Other adaptor molecules		
e.g. β-arrestins	Src/JNK3	⌐ G protein-
Homers	Shank/IP$_3$R/RyR	>independent
NHERF	Na$^+$/H$^+$-exchange	⌐

[1] The mammalian adenylyl cyclase isoenzymes (AC1-AC9) are differentially regulated by $G_s/G_{i/o}$ and $G_{\beta\gamma}$-subunits. [2] PLC-β isoenzymes (PLC-β1-PLCβ4) exhibit different sensitivities to $G_{q/11}$ and $G_{\beta\gamma}$-subunits. PLC-δ isoenzymes are reported to be regulated by transglutaminase II (G_h). PLC-ϵ may also be regulated by $G_{\beta\gamma}$-subunits.

all of which are expressed in the CNS, has allowed detailed characterizations to be undertaken (see Sunahara *et al.* 1996). From such work it is now clear that adenylyl cyclases are regulated not only by G_s/G_i-coupled GPCRs, but also that G_q-coupled GPCRs can affect the activities of some isoenzymes through changes in the intracellular concentration of Ca^{2+} ($[Ca^{2+}]_i$) and protein kinase C activity. The profound stimulatory or inhibitory effects of changes in $[Ca^{2+}]$ also suggest that specific adenylyl cyclase isoenzymes represent a point of integration between GPCR and non-GPCR-mediated signalling pathways (Cooper *et al.* 1995; Hanoune and Defer 2001). This view is supported by the differential localization of adenylyl cyclase isoenzymes within neurons and evidence is accumulating to implicate specific isoenzymes in a variety of neuronal functions including long-term potentiation (LTP) and long-term depression (LTD) (Cooper *et al.* 1995; Hanoune and Defer 2001). In addition to advances in our understanding of how cyclic AMP levels are regulated within neurons, parallel advances have been made with respect to events downstream of cyclic AMP. Thus, PKA can be localized within cells through reversible interactions with anchoring proteins (Mochly-Rosen 1995), and cyclic AMP has been recognized as an important nuclear signal affecting the cell at a transcriptional level. In addition, it is now known that the actions of cyclic AMP are not universally mediated by PKA; additional cyclic AMP binding proteins (de Rooij *et al.* 1998; Kawasaki *et al.* 1998) contribute to the diversity of action of this second messenger.

Work over the past 20 years has considerably expanded the cellular metabolites considered to have second messenger (or putative second messenger) function beyond the cyclic nucleotides (Table 5.1). Thus, phospholipase C isoenzymes (β, γ, δ, and ε) generate both inositol 1,4,5-trisphosphate (IP$_3$) and *sn*-1,2-diacylglycerol (DAG) (Berridge 1993; Rhee 2001). The former inositol polyphosphate can mobilize Ca^{2+} from endoplasmic reticular stores through interaction with a receptor-operated Ca^{2+} channel gated by IP$_3$ (Wilcox *et al.* 1998), while DAG acts through a subset of the protein kinase C isoenzymes (the cPKC and nPKCs (Tanaka and Nishizuka 1994)). The substrate for phospholipase Cs, the minor membrane phospholipid phosphatidylinositol 4,5-bisphosphate (PIP$_2$), is also considered to share characteristics with classical second messengers and to regulate a number of key cell modalities (Toker 1998) including synaptic function (Osborne *et al.* 2001). Furthermore, PIP$_2$ can be 3-phosphorylated by phosphoinositide 3-kinases (PI3Ks) to generate phosphatidylinositol 3,4,5-trisphosphate (PIP$_3$), which together with its immediate metabolite, phosphatidylinositol 3,4-bisphosphate, have demonstrable second messenger activities (Rameh and Cantley 1999; Lemmon and Ferguson 2000). In addition, a number of lipid metabolites, including long chain polyunsaturated fatty acids (e.g. arachidonic acid, generated by phospholipase A$_2$ (Leslie 1997; Six and Dennis 2000)), phosphatidic acid (generated by phospholipase D (Liscovitch *et al.* 2000; Fang *et al.* 2001), sphingosine 1-phosphate (generated by sphingosine kinase (Pyne and Pyne 2000)) and other sphingolipids, have all been proposed to be intracellular messenger molecules.

5.3 Effectors and the regulation of intracellular Ca²⁺ concentration

Ca^{2+} is a ubiquitous intracellular messenger that controls a multitude of activities within cells (Berridge 1998; Berridge *et al.* 2000). Increases in $[Ca^{2+}]_i$ can exert direct effects on Ca^{2+}-sensitive proteins or, more commonly, exert effects via Ca^{2+}-binding proteins such as calmodulin (Berridge *et al.* 2000; Chin and Means 2000). Ca^{2+}/calmodulin can, in turn, regulate a variety of proteins, either by direct binding or via regulation of phosphorylation status through Ca^{2+}/calmodulin-dependent protein kinases (Corcoran and Means 2001; Soderling *et al.* 2001) or phosphatases (calcineurin) (Rusnak and Mertz 2000; Crabtree 2001). GPCRs can regulate $[Ca^{2+}]_i$ through a variety of mechanisms. At the level of the plasma membrane GPCRs can directly modulate voltage-operated Ca^{2+} channels (Dolphin 1998; Miller 1998), or act through less direct pathways to reduce Ca^{2+}-influx (e.g. by membrane hyperpolarization through K^+-channel activation), or to promote Ca^{2+}-efflux (e.g. by increasing plasma membrane Ca^{2+}-ATPase activity) (Wickman and Clapham 1995; Penniston and Enyedi 1998).

Probably the most studied mechanism linking GPCRs to increases in $[Ca^{2+}]_i$ involves the phospholipase C/inositol 1,4,5-trisphosphate (IP$_3$) signalling pathway, where IP$_3$ gates Ca^{2+}-mobilization through activation of IP$_3$ receptors in the endoplasmic reticulum (ER)

(Berridge 1993). The mobilization of Ca^{2+} from ER stores can, in turn, trigger another process termed store-operated (or 'capacitative') Ca^{2+} entry, where ER Ca^{2+}-store depletion in some way gates Ca^{2+} entry via a plasma membrane cation channel (Putney 1986; Parekh and Penner 1997). This Ca^{2+}-signalling pathway has been most studied in non-excitable cells; however, there is increasing evidence that the GPCR-regulated phospholipase C/IP$_3$ signalling pathway may also play an important role in neuronal cell function.

In addition to IP$_3$ receptors, the ER may also express other proteins involved with Ca^{2+}-mobilization. Ryanodine receptors are the best-characterized receptor-operated Ca^{2+}-channels that are responsible for Ca^{2+}-induced Ca^{2+}-release. Thus, global or localized changes in cytoplasmic Ca^{2+} can activate ryanodine receptors to cause a further rise through ER Ca^{2+}-mobilization (Verkhratsky and Shmigol 1996; Berridge et al. 2000). IP$_3$ receptors and ryanodine receptors possess structural similarities and are regulated in analogous ways. Thus, the ability of changes in IP$_3$ concentration to activate types 1-3 IP$_3$ receptors can be profoundly affected by cytoplasmic Ca^{2+}. Conversely, Ca^{2+} activation of type 2 ryanodine receptors is modulated by the level of cyclic ADP-ribose in the cell, which, in turn, can be regulated by cell-surface receptors (Higashida et al. 2001). Other metabolites (e.g. NAADP, sphingosine 1-phosphate) are also proposed as regulators of ER Ca^{2+}-mobilization, however, for these putative messengers information on the Ca^{2+}-release mechanisms is lacking at present (Berridge et al. 2000; Rizzuto 2001).

In the CNS, Ca^{2+} entry, via receptor-mediated (e.g. NMDA receptor) and voltage-gated mechanisms, has been the focus of much research, and extracellular Ca^{2+} has long been considered to be the major source for changes in $[Ca^{2+}]_i$. However, recent studies have suggested that intracellular Ca^{2+} stores may also fulfil specific neuronal functions, in particular, to play a role in activity-dependent synaptic plasticity (Emptage 1999; Rose and Konnerth 2001). Anatomical studies have shown that the ER ramifies into almost all neuronal cellular compartments, including dendritic spines. Furthermore, a number of studies have demonstrated key roles for GPCR-mediated Ca^{2+} mobilization in bringing about changes in synaptic efficacy. Perhaps the most complete picture has emerged for the role of group I metabotropic glutamate (mGlu) receptors (Hermans and Challiss 2001) in LTD at parallel fibre-Purkinje cell synapses in the cerebellum. Parallel fibre firing causes glutamate release and a highly localized increase in $[Ca^{2+}]_i$ in single spines or spinodendritic domains of the Purkinje cell. The involvement of mGlu1 receptors in cerebellar LTD has been demonstrated by the knockout (Aiba et al. 1994; Conquet et al. 1994) and selective rescue (Ichise et al. 2000) of mGlu1 receptor function in transgenic mice. Further, the pathway linking mGlu1 receptor activation to PLC and IP$_3$-mediated Ca^{2+} release has been shown to be essential in establishing LTD (Inoue et al. 1998; Rose and Konnerth 2001), with very recent studies highlighting the roles of specific PLC isoenzymes and classical PKCs in the pathway (Hirono et al. 2001). The crucial role of the ER as a Ca^{2+} source in LTD has also been elegantly demonstrated using mutant rats and mice where the ER does not ramify into dendritic spines (Miyata et al. 2000). Collectively, these studies demonstrate clear roles for ER Ca^{2+}-stores and GPCR-mediated Ca^{2+} mobilization in modifying synaptic function, and highlight the potential of this source of Ca^{2+} to provide highly compartmentalized and temporally resolved Ca^{2+} signals (Rose and Konnerth 2001). These and other studies are beginning to provide an insight into the likely roles of IP$_3$ and ryanodine receptors in the CNS.

Until recently, neurobiologists have, in the main, viewed the vast array of GPCRs expressed in the CNS and their diverse signalling with respect to playing a modulatory role upon ionotropic-mediated synaptic transmission. Thus, GPCR activation, by regulating directly or indirectly ion channel function, can modulate both pre- and post-synaptic events to alter the temporal and spatial integration of synaptic activity that in turn alters the frequency and timing of action potentials in central neurons.

It seems possible that this view will change as our understanding of GPCR-mediated signalling in neurons evolves, particularly in relation to its association with signalling complexes in discrete neuronal compartments. For example, computational analysis of GPCR signalling, both in simple neuronal systems, as well as in some mammalian CNS pathways, suggests that synaptically activated GPCRs can play a role in information transfer independent of voltage-mediated signals (Bourne and Nicoll 1993; Katz and Clemens 2001). Thus, perhaps second messenger signalling may eventually be considered to play a more direct, integral part in short-term information transfer in the CNS. However, it is also becoming increasingly clear that GPCR signalling cascades in the CNS do play a crucial role in long-term changes brought about through changes in gene transcription and protein synthesis. We emphasize this aspect of signalling in the remainder of this Chapter.

5.4 Effectors and signalling to the nucleus

A number of major breakthroughs in the 1980s and 1990s substantially increased our appreciation of the true versatility of signalling events initiated by GPCRs. The fact that different ion channels can act as direct GPCR effectors (acting immediately downstream of $G\alpha$-GTP or $G\beta\gamma$ subunits) will be discussed in the next Chapter. Another key advance came with the appreciation that at least some members of the GPCR superfamily can affect cellular activity at the level of gene transcription (Collins *et al.* 1992). Here, the mitogen-activated protein (MAP) kinases will be considered as an example of a nuclear signalling pathway responsive to GPCR activation, and the regulation of the transcription factor cyclic AMP response element-binding protein (CREB) by cell-surface receptors (including GPCRs) will also be discussed in a neuronal context.

5.4.1 GPCRs and mitogen-activated protein (MAP) kinases

Many papers published over the past 10–15 years have brought together a compelling case for GPCRs possessing the potential to bring about phenotypic changes in the cell, including oncogenic transformation (Dhanasekaran *et al.* 1995; Gutkind 1998). Thus, activation of an array of GPCRs, which couple preferentially to different G proteins, has been shown to impact upon cell fate in a variety of cell-types (Gutkind 1998) and these effects can often be mimicked by expression of constitutively active $G_s\alpha$, $G_i\alpha$, $G_q\alpha$, or $G_{12/13}\alpha$ mutant proteins (Dhanasekaran *et al.* 1995). An important, highly conserved, family of protein kinases, collectively termed the MAP kinases, have been established as important relays linking signals arriving at the cell-surface to cellular changes at the level of transcription and translation (Davis 1993; Schaeffer and Weber 1999). At the core of MAP kinase pathways is a cassette of three kinases acting in series (MAPKKK $\rightarrow$ MAPKK $\rightarrow$ MAPK; see Fig. 5.1). The most studied of these involves dual phosphorylation of extracellular signal-regulated protein kinases (ERK1/2; also known as p44^{MAPK} and p42^{MAPK}, respectively) at threonine and tyrosine residues (in the TEY-motif) by the MAPKKs, MEK1/2, which in turn are dually phosphorylated at serine/threonine residues by the MAPKKK, Raf-1 (Kolch 2000). At least three additional MAP kinase pathways have been recognized in

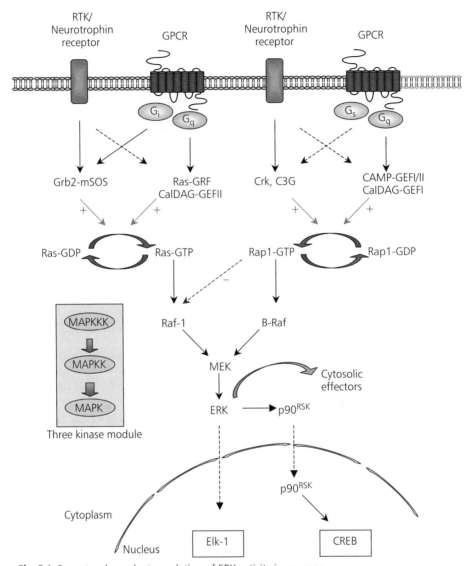

Fig. 5.1 Receptor-dependent regulation of ERK activity in neurons.

mammalian systems, the c-Jun N-terminal protein kinases (JNK1/2/3, also known as stress-activated protein kinases, SAPK1α/β/γ), the p38 MAP kinases (p38α/β/γ/δ) and ERK5 (Schaeffer and Weber 1999; Kolch 2000). Here, we will focus on the pathways linking GPCRs to ERK1/2, firstly in a general context and then with specific reference to neuronal function.

ERK1/2 were initially shown to be downstream targets of growth factor receptor tyrosine kinases (RTKs) (Schlessinger 1993), capable of phosphorylating specific cell proteins, including a variety of transcription factors (Davis 1993; Kolch 2000). RTK activation leads to dimerization and autophosphorylation of specific tyrosine residues within the intracellular domains of the receptor to create docking sites for src-homology-2 (SH2) domain-containing proteins. One such protein, Grb2 (growth factor receptor binding protein-2) binds to specific intracellular tyrosine-phosphorylated sites (e.g. ^{740}Y in the PDGF-β receptor). The tethered Grb2 acts as an adaptor for the guanine nucleotide exchange factor (GEF), mSOS (mammalian 'Son-of-Sevenless' protein), with which it is constitutively associated. mSOS in turn recruits p21^{ras} and facilitates GTP-loading of Ras leading to its activation. Ras-GTP binds to a number of proteins including Raf-1 leading to the translocation of this MAPKKK to the plasma membrane and activation of the Raf-1-MEK1/2-ERK1/2 cascade (Fig. 5.1). It is likely that endocytosis of the complex, in whole or in part, is a prerequisite for efficient ERK activation by at least some RTKs (Vieira *et al.* 1996). ERK1/2 activation leads to the proline-directed serine/threonine phosphorylation of specific cytoplasmic substrate proteins, and often to nuclear translocation, where ERK1/2 can directly affect transcription through direct phosphorylation of ternary complex factors (e.g. Elk-1), or indirectly through activation of other protein kinases (e.g. p90^{RSK}—see next section) that, in turn, regulate transcription factor activity (Davis 1993; Treisman 1996).

Initial investigations of how GPCRs link to MAPK/ERK activation, which often involved transient expression of GPCRs, MAPKs and putative intermediary components in non-neuronal cell-lines, suggested that traditional G protein-effector-second messenger pathways might directly regulate components of the MAPK cascade in either a Ras-dependent or -independent manner (Alblas *et al.* 1993; Howe and Marshall 1993; Kolch *et al.* 1993; Offermanns *et al.* 1993; van Corven *et al.* 1993), with both Gα- and/or $\beta\gamma$-subunits being implicated as transducers of MAPK signalling (Gutkind 1998). For many cell systems it could be demonstrated that $\beta\gamma$-subunits play the major role in determining Ras-dependent ERK activity downstream of G$_{i/o}$-coupled GPCRs, whereas both G$_{q/11}\alpha$- and $\beta\gamma$-subunits have been implicated downstream of G$_{q/11}$-coupled GPCRs in mediating both Ras-dependent and -independent ERK activation (Crespo *et al.* 1994; Faure *et al.* 1994; Koch *et al.* 1994; Hawes *et al.* 1995).

Attempts to delineate the sequence of proteins linking GPCR/G protein activation to the MAPK/ERK pathway have yielded a wide range of potential signalling intermediates. A common feature of many of these proteins is their ability to act as adaptors linking GPCRs to the Ras activation sequence utilized by RTKs. Thus, a number of GPCRs have been shown to promote the phosphorylation of the adaptor protein Shc leading to Grb2-mSOS recruitment (Ohmichi *et al.* 1994; van Biesen *et al.* 1995; Chen *et al.* 1996). GPCRs have also been shown to 'transactivate' some RTKs (Daub *et al.* 1996; Luttrell *et al.* 1997; Herrlich *et al.* 1998). This process was initially considered to occur independently of ligand activation of the RTK, with the receptor undergoing ligand-independent phosphorylation (by non-receptor tyrosine kinases (PTKs)) to allow it to act as a signalling scaffold. However, more recent data have now shown that GPCR activation may stimulate metalloproteinases capable of releasing RTK ligands (Prenzel *et al.* 1999). In addition to RTKs acting as signalling 'scaffolds' for the GPCR-ERK pathway a number of other protein scaffolds have been identified. These include integrins (Schlaepfer *et al.* 1994; Slack 1998; Litvak *et al.* 2000; Short *et al.* 2000) and the GPCRs themselves.

An involvement of PTKs, such as Src (and other Src family proteins such as Yes, Fyn, and Lyn), Pyk2 and focal adhesion kinase (FAK) has also been shown for a number of G$_{i/o}$- and G$_{q/11}$-coupled GPCRs (Dikic *et al.* 1996; Luttrell *et al.* 1996; Della Rocca *et al.* 1997). The linkage between G proteins and PTKs in some cases can be clearly defined; for example, Pyk2 is a Ca^{2+}/calmodulin- and PKC-regulated enzyme (Lev *et al.* 1995; Dikic *et al.* 1996). A further (indirect) link between G$\beta\gamma$ and PTKs has been shown for at least some members of the class 1 PI3Ks (Hawes *et al.* 1996; Lopez-Ilasaca *et al.* 1997), suggesting a Gβγ-PI3K-Src sequence. Finally, a direct Gβγ interaction has been proposed for some PTKs (Gutkind 1998). Another fascinating development has been the finding that some GPCRs may link to ERK signalling via a different G protein subpopulation to that utilized in their coupling to classical effectors. Thus, Daaka and colleagues have shown that β_2-adrenoceptors, expressed in HEK293 cells, couple to a Src- and Ras-dependent ERK pathway via βγ-subunits released by G_i protein activation. These data led to the proposal of a model where the β_2-adrenoceptor-G_s protein interaction becomes rapidly uncoupled and the receptor, instead of desensitizing with respect to all G protein interactions, 'switches' to activate G_i proteins (Daaka *et al.* 1997). It is noteworthy that there are a number of examples of GPCRs, considered to be $G_{q/11}$ or G_s-coupled that link to ERK via pertussis toxin-sensitive $G_{i/o}$ proteins (Ferraguti *et al.* 1999; Soeder *et al.* 1999), and therefore it is possible that ERK activation arising through GPCR switching may be a widespread phenomenon.

The finding that covalent modification of GPCRs, normally associated with receptor desensitization, could instead switch the GDP/GTP exchange activity of the receptor from one subpopulation of G proteins to another led to investigations of whether proteins previously considered only to be associated with the uncoupling of GPCRs (i.e. G protein-coupled receptor kinases and arrestins) might fulfil other functions (Table 5.1). Using the HEK293-β_2-adrenoceptor model system it was shown that the 'desensitized' β_2-adrenoceptor/β-arrestin1 complex recruits Src (Luttrell *et al.* 1999). In addition, the clathrin-adaptor function of β-arrestin was shown to be essential, as internalization of the β_2-adrenoceptor/β-arrestin1-Src complex was necessary for ERK activation (Luttrell *et al.* 1999). A comparable GPCR/β-arrestin2 scaffold has been reported to recruit the MAP-KKK/MAPKK/MAPK components ASK1/MKK4/JNK3 (McDonald *et al.* 2000). Endocytosis of the GPCR as a crucial aspect of ERK signalling has subsequently been shown for a subset of GPCRs that stimulate ERK activity (Pierce *et al.* 2001) and there appears to be receptor subtype variation in whether endocytosis of the entire signalling complex (DeFea *et al.* 2000), or only a terminal (MEK/ERK) portion (Kranenburg *et al.* 1999) is essential. As a variant on this theme, some GPCRs are reported to be able directly to recruit Src providing a scaffold independent of β-arrestin for building ERK activation complexes, or to activate other signalling pathways (Cao *et al.* 2000; Heuss and Gerber 2000). Direct, G protein-independent GPCR recruitment of Src family PTKs is believed to depend on proline-rich motifs present in the 3rd intracellular loop or C-terminal tails of only a subset of GPCRs (e.g. D_4-dopamine, $mGlu_{1a}$ and $mGlu_{5a}$ receptors (Pierce *et al.* 2001; Hermans and Challiss 2001)).

The work outlined above has highlighted the fact that GPCR phosphorylation may switch G protein-effector signalling, rather than simply arresting it. Furthermore, the demonstration that β-arrestin isoforms may act as scaffolds to assemble different proteins that regulate distinct MAPK pathways, together with the fact that GPCR/β-arrestin isoform specificity has been reported (Penn *et al.* 2001), leads to the intriguing possibility that receptor phosphorylation and selective β-arrestin recruitment may contribute to defining the (MAPK) pathways activated downstream of a particular GPCR. The ability of β-arrestins to act as scaffolds for assembling specific MAPK signalling complexes is reminiscent of the role of Ste5 in *S. cerevisiae* (Schaeffer and Weber 1999), and further potential scaffolding proteins are being identified in mammalian systems, including JNK-interacting (JIP) and MEK partner (MP) proteins and MKK4, suggesting a further mechanism of imposing specificity and control to MAPK signalling (Hagemann and Blank 2001).

Despite the considerable progress that has been made in mapping the biochemical pathways GPCRs can utilize to couple to ERK activation, it has proved to be very difficult to draw up general rules. There are a number of reasons for this. First, the pathway linking a GPCR to ERK activation appears to be highly dependent on cell background and it has been shown on many occasions that it is possible for a specific GPCR to couple to ERK1/2 activation via different pathways in different cell-types (Duckworth and Cantley 1997; Della Rocca *et al.* 1999). A further difficulty arises from the fact that a single GPCR might activate multiple parallel or overlapping pathways capable of linking to ERK activation (Duckworth and Cantley 1997; Della Rocca *et al.* 1999; Blaukat *et al.* 2000) and these pathways may show an interdependency or redundancy with respect to both the magnitude and duration of the consequent ERK activation. *In vivo* it is likely that such apparent anarchy does not reign and instead specificity is achieved through a variety of mechanisms. These might include the expression of only a subset of intermediary components in a particular subtype and compartmentation (e.g. through signalling complex assembly on component-defining scaffolds) within the cell.

5.4.2 GPCR and ERK signalling in the CNS

The biochemical elucidation of the RTK-Ras-ERK pathway (Schlessinger 1993) was quickly followed by demonstrations that other signalling events could also trigger its activation to regulate a wide range of cell outcomes in a many tissues, including the CNS (Ghosh and Greenberg 1995; Grewal *et al.* 1999). In parallel with studies showing ERK activation by GPCRs, studies in neuronal cell-lines (notably PC12 rat pheochromocytoma cells) demonstrated Ras/ERK activation by Ca^{2+}-influx (through voltage-dependent or receptor-dependent mechanisms), suggesting that changes in neuronal excitability might also regulate this pathway (Rosen *et al.* 1994; Rusanescu *et al.* 1995; Finkbeiner and Greenberg 1996). Alongside this initial functional demonstration it was also shown that many of the components of the Ras/ERK pathway are enriched in the mammalian CNS, where they exhibit an unequal regional and sub-cellular distribution, and are expressed at highest levels in brain regions capable of undergoing long-term changes in synaptic efficacy (Finkbeiner and Greenberg 1996; Impey *et al.* 1999).

Although it has come to be generally accepted that MAPK/ERK signalling is crucial for an array of neuronal functions, and considerable progress has been made with respect to implicating ERK regulation by neurotrophins/growth factors, and ionotropic receptor- and voltage-dependent mechanisms, relatively few studies have been performed with respect to GPCR inputs to the ERK pathway in neurons (see for example Roberson *et al.* 1999; Watanabe *et al.* 2000). This is surprising as the progress that has been made strongly suggests that metabotropic mechanisms can directly regulate MAPK/ERK activation and therefore influence neuronal survival, differentiation and plasticity (Finkbeiner and Greenberg 1996; Impey *et al.* 1999; Sweatt 2001). For example, cyclic AMP has been shown to regulate the transcription factor Elk-1 by an ERK-dependent mechanism in PC12 cells (Vossler *et al.* 1997). The pathway leading to ERK activation involves the small GTPase Rap1 that preferentially activates the neuronal MAPKKK, B-Raf (Fig. 5.1). Subsequent work showed that PC12 cell differentiation, mediated by the neurotrophin, nerve growth factor (NGF), also requires ERK activation downstream of Rap1/B-Raf (York *et al.* 1998). Furthermore, the sustained nature of the ERK activation mediated by Rap1/B-Raf may 'programme' the cell to differentiate, rather than to proliferate (Marshall 1995).

Table 5.2 Guanine nucleotide exchange factors: links between neuronal GPCR-second messenger signalling and the ERK pathway

Guanine nucleotide exchange factor (GEF)	Upstream regulators	Small GTPase target
mSOS	Pyk2/Src/Shc/Grb2	Ras
Ras-GRF (CDC25^{Mm})	Gβγ, Ca^{2+}/calmodulin	Ras
CalDAG-GEF-II (Ras-GRP)	DAG, Ca^{2+}	Ras (Rap1)
CalDAG-GEF-III	DAG, Ca^{2+}	Ras, Rap1, Rap2
C3G	FRS2, Crk cyclic AMP/PKA	Rap1
cAMP-GEF-I (Epac1)	cyclic AMP	Rap1
cAMP-GEF-II	cyclic AMP	Rap1
CalDAG-GEF-I	Ca^{2+}, DAG	Rap1

Ras/Rap1 guanine nucleotide exchange factors (GEFs) reported to be expressed in the CNS. mSOS and C3G are regulated by growth factor/neurotrophin-regulated tyrosine kinase/adaptor activities. However, many GEFs are directly or indirectly regulated by changes in second messenger and/or Ca^{2+} concentration, providing a mechanism by which GPCR regulation of adenylyl cyclase/phospholipase C activities can influence ERK pathways downstream of Raf-1/B-Raf.

Ras and Rap1 require GEFs to facilitate GTP for GDP exchange, and GTPase activating proteins (GAPs) to stimulate GTP hydrolysis (Fig. 5.1). Efforts to identify GEF and GAP activities in the CNS (and other tissues) have yielded a number of novel proteins (Grewal *et al.* 1999; Bos *et al.* 2001), many of which can be directly or indirectly regulated by mediators downstream of GPCRs (see Table 5.2). Thus, the neuronal Ras-GEF, Ras-GRF (also known as CDC25^{Mm}) is directly regulated by Ca^{2+}/calmodulin and Gβγ-subunits (Farnsworth *et al.* 1995; Mattingly and Macara 1996) and transgenic knockout experiments have implicated it in different forms of synaptic plasticity, perhaps those especially requiring metabotropic receptor inputs (Brambilla *et al.* 1997). Another Ras-GEF, CalDAG-GEFII (also known as RasGRP) possesses, as its name implies, binding domains for both Ca^{2+} and DAG (Ebinu *et al.* 1998). In addition to the likely modulation by GPCRs of Ras-GTP loading, there is also the potential for regulation of Ras-GTPase activity. Thus, the brain-specific RasGAP, SynGAP is associated with post-synaptic density (PSD) complexes and is negatively regulated by Ca^{2+}/calmodulin-dependent kinase II activity (Chen *et al.* 1998).

A number of Rap-GEFs and Rap-GAPs have also been reported, with many again showing a neuron-specific expression, and binding motifs that allow direct or indirect regulation by second messengers (Fig. 5.1; Table 5.2). The first Rap-GEF to be identified was C3G (Crk, SH3-domain-binding guanine-nucleotide releasing factor), which appears to play a similar role to mSOS in linking neurotrophin receptor activation to GTP-loading of Rap1, and is regulated upstream by FRS2-Crk adaptors (Kao *et al.* 2001). In contrast, cyclic AMP binding directly regulates the Rap1-GEFs, Epac1/cAMP-GEFI and cAMP-GEFII (de Rooij *et al.* 1998; Kawasaki *et al.* 1998) and these proteins are likely to account for cyclic AMP dependent, but PKA-independent activation of the Rap1/B-Raf/ERK pathway (Grewal *et al.* 1999; Bos *et al.* 2001). Finally, CalDAG-GEFI and CalDAG-GEFIII are directly regulated by Ca^{2+} and DAG and can facilitate guanine nucleotide exchange on Rap1 (and 2, and some other small GTPases; Bos *et al.* 2001). Although little is presently know about the regulation of neuronal Rap-GAP activities, there is evidence for G protein regulation of both Rap1-GAPI and Rap-GAPII (Jordan *et al.* 1999; Bos *et al.* 2001).

The identification of families of GEF/GAPs that regulate Ras/Rap1 GTP-loading/GTP hydrolysis through changes in Ca^{2+}, cyclic AMP or DAG levels (Table 5.2) strongly suggests a signalling pathway through which G_s- and G_q-coupled GPCRs can regulate ERK activity in neurons. Emerging experimental data suggest that GPCRs are likely to utilize these pathways (Guo *et al.* 2001) and RTK/PTK-dependent pathways (Rosenblum *et al.* 2000; Peavy *et al.* 2001) to regulate ERK activity in the CNS.

5.5 Transcriptional regulation by GPCRs

An example of GPCR regulation of transcription emerged from studies of the hormonal regulation of the synthesis and release of somatostatin and other neuropeptides. It was shown that hormones (such as glucagon and adrenaline) that elevate cyclic AMP and activate PKA lead to the phosphorylation of cyclic AMP response element (CRE)-binding protein (CREB), which binds to regulatory CRE regions (5'-**TGACGTCA**-3') upstream of genes encoding some neuropeptides (and other proteins such as the immediate early gene *c-fos* (Bonni *et al.* 1995)), as a first step in the transcriptional process (Montminy and Bilezikjian 1987; Gonzalez and Montminy 1988; Sheng *et al.* 1990). Serine-133 phosphorylated, dimeric CREB recruits CREB-binding protein (CBP) leading to the assembly of the transcriptional complex (Nakajima *et al.* 1997; Shaywitz and Greenberg 1999). The CREB basal transcriptional machinery alone, or in cooperation with other transcriptional complexes (Roesler *et al.* 1995), can then initiate gene transcription. To date approx. 150 genes encoding neuro-peptides and growth factors, metabolic enzymes, transcriptional components, and cell cycle regulatory proteins have been reported to contain CRE motifs within their promoter regions (Mayr and Montminy 2001).

Although it was first believed that CREB was exclusively regulated by cyclic AMP/PKA, subsequent studies have shown that a variety of other protein kinases have the potential to phosphorylate CREB, and some of these have been shown to do so *in vivo*, in partic-ular, Ca^{2+}/calmodulin-dependent protein kinase IV (Ho *et al.* 2000) and members of the ribosomal S6 kinase (RSK) family (e.g. p90^{RSK}; MSK-1, Shaywitz and Greenberg 1999; Mayr and Montminy 2001). CREB can be phosphorylated at sites other than serine-133 (notably Serine-142; Sun *et al.* 1994), however, the regulatory significance of such phosphorylation is presently not known, whereas the ability of each protein kinase to activate transcription has been shown in the vast majority of instances to be causally linked to the phosphorylation of this specific serine-133 residue (Shaywitz and Greenberg 1999). In neurons a crucial role for CREB has been demonstrated in growth factor-dependent survival (Riccio *et al.* 1999) and CREB activation has been implicated in neuronal synaptic plasticity (Bartsch *et al.* 1998). Although cyclic AMP/PKA is likely to play a role in CREB regulation in neurons, it is generally believed that Ca^{2+} may be the pre-eminent regulator of CREB activation and much attention has been focused on the abilities of Ca^{2+} influx, via receptor-operated and voltage-operated channels, to activate this transcription factor (Shaywitz and Greenberg 1999; West *et al.* 2001). Although it has been reported that CREB synthesis/activity can be regulated outside of the nucleus (e.g. within dendrites; Crino *et al.* 1998), most studies have found CREB to be present in the nucleus, and it is thought that CREB forms an inactive complex at CRE regions (West *et al.* 2001). Irrespective of the kinase responsible for CREB phosphorylation, a coincident rise in nuclear $[Ca^{2+}]$ appears to be required to assemble the transcriptional complex (Chawla *et al.* 1998; Hu *et al.* 1999).

Like other mechanisms of transcriptional control, the duration of CREB phosphorylation is a crucial determinant of whether certain genes are transcribed (Bito *et al.* 1996). Thus, an increasing number of studies have shown that different mechanisms of elevating neuronal $[Ca^{2+}]_i$ can result in very different patterns of CREB-dependent gene transcription (Shaywitz and Greenberg 1999; West *et al.* 2001). Comparisons between receptor-operated (e.g. NMDA receptor-dependent) and voltage-operated (e.g. L-type VOCC dependent) mechanisms have begun to reveal the complexity of such regulation (Sala *et al.* 2000; Dolmetsch *et al.* 2001). In addition, the ability of different extracellular stimuli to cause sequential activation, or differential longevity, of protein kinase activities in cells can also impart specificity to target gene expression in a variety of cell-types, including neurons (Mayr *et al.* 2001; Wu *et al.* 2001; Zanassi *et al.* 2001). In non-neuronal cells it has been demonstrated that increases in $[Ca^{2+}]_i$, mediated by GPCRs, can influence transcriptional activity and that frequency, as well as amplitude encoding of the GPCR-evoked Ca^{2+}-signal may determine the pattern of such activity (Dolmetsch *et al.* 1998; Crabtree 1999). However, in neurons it is possible that the GPCR-PLC-Ca^{2+} signalling pathway plays a more subtle role, shaping the Ca^{2+} signal initiated by Ca^{2+}-influx (Bading 2000; Hardingham *et al.* 2001).

A further mechanism by which GPCR-mediated signalling can regulate CREB activation is by determining the duration of CREB phosphorylation, as it is known that the regulation of the phosphorylation status of CREB by protein phosphatases can profoundly influence the repertoire of genes transcribed (Shaywitz and Greenberg 1999). For example, in the developing striatum sustained activation of CREB is required for *c-fos* expression and this is achieved by regulation of proteins responsible for both the phosphorylation and dephosphorylation of CREB (Liu and Graybiel 1996). Thus, it is possible for G_s-linked neurotransmitter receptors (e.g. D_1/D_5 dopamine receptors) to promote the nuclear translocation of PKA (Hagiwara *et al.* 1993*a*), to cause both direct CREB phosphorylation *and* attenuation of dephosphorylation by PKA phosphorylation of dopamine- and cyclic AMP-regulated 32 kDa phosphoprotein (DARPP-32), a potent inhibitor of protein phosphatase-1 (Greengard *et al.* 1999), which may be the key CREB phosphatase in some tissues (Hagiwara *et al.* 1993*b*). This raises the possibility that activation of the cyclic AMP/PKA/DARPP-32 pathway (Fig. 5.2) by G_s-linked GPCRs can play an important role by consolidating CREB phosphorylation catalysed by other protein kinase activities.

Finally, experimental data arising from studies of CREB-dependent changes in transcriptional activity in learning and memory paradigms, and work to establish the downstream consequences of ERK activation (Impey *et al.* 1998; Grewal *et al.* 1999; Roberson *et al.* 1999; Sweatt 2001) provide a basis for the proposal that ERK activation of CREB, through p90[RSK], may be a crucial mechanism for CREB regulation. Furthermore, cyclic AMP/Ca^{2+} inputs to CREB activation may also occur additionally, or perhaps predominantly, through ERK regulation at the level Ras/Rap1 activation.

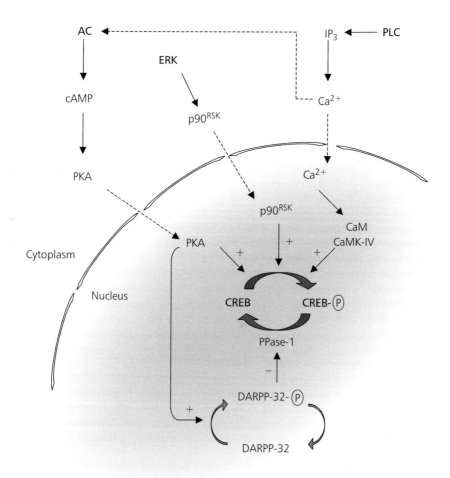

Fig. 5.2 Regulation of CREB activity downstream of cyclic AMP, Ca^{2+} and ERK: coordinate regulation of protein kinases and protein phosphatases.

5.6 **Conclusions**

In the CNS, GPCRs are at last emerging from the shadow cast by voltage-dependent/ionotropic paradigms for neuronal information transfer. In addition, to gaining a better understanding of the traditional 'neuromodulatory' roles of GPCRs, we are beginning to see the first clear examples of GPCR regulation of independent, dynamic biochemical signalling networks involving not only short-term information transfer, but also bringing about adaptative changes involving regulation at the level of gene transcription. Many important challenges lie ahead if we are to gain a comprehensive understanding of how GPCR signalling specificity might be encoded in neurons with respect to the coordinate regulation multiple downstream effectors.

Acknowledgements

We thank Sukhwinder Thandi for assistance with the Figures and her helpful comments on earlier drafts of the manuscript. Cited and ongoing studies from the authors' own laboratories are supported by the Wellcome Trust and the Medical Research Council.

References

Aiba A, Kano M, Chen C, Stanton ME, Fox GD, Herrup K *et al.* (1994). Deficient cerebellar long-term depression and impaired motor learning in mGluR1 mutant mice. *Cell* **79**, 377–88.

Alblas J, van Corven EJ, Hordijk PL, Milligan G, and Moolenaar WH (1993). G_i-mediated activation of the $p21^{ras}$-mitogen-activated protein kinase pathway by α_2-adrenergic receptors expressed in fibroblasts. *J Biol Chem* **268**, 22 235–8.

Bading H (2000). Transcription-dependent neuronal plasticity. The nuclear calcium hypothesis. *Euro J Biochem* **267**, 5280–3.

Bartsch D, Casadio A, Karl KA, Serodio P, and Kandel ER (1998). CREB1 encodes a nuclear activator, a repressor, and a cytoplasmic modulator that form a regulatory unit crucial for long-term facilitation. *Cell* **95**, 211–23.

Berridge MJ (1993). Inositol trisphosphate and calcium signalling. *Nature* **361**, 315–25.

Berridge MJ (1998). Neuronal calcium signalling. *Neuron* **21**, 13–26.

Berridge MJ, Lipp P, and Bootman MD (2000). The versatility and universality of calcium signalling. *Nature Rev Mol Cell Biol* **1**, 11–21.

Birnbaumer L (1992). Receptor-to-effector signalling through G proteins: Roles for $\beta\gamma$ dimers as well as α subunits. *Cell* **71**, 1069–72.

Bito H, Deisseroth K, and Tsien RW (1996). CREB phosphorylation and dephosphorylation: a Ca^{2+}- and stimulus duration-dependent switch for hippocampal gene expression. *Cell* **87**, 1203–14.

Blaukat A, Barac A, Cross MJ, Offermanns S, and Dikic I (2000). G protein-coupled receptor-mediated mitogen-activated protein kinase activation through cooperation of G_q and G_i signals. *Mol Cell Biol* **20**, 6837–48.

Bonni A, Ginty DD, Dudek H, and Greenberg ME(1995). Serine-133-phosphorylated CREB induces transcription via a cooperative mechanism that may confer specificity to neurotrophin signals. *Mole Cell Neurosci* **6**, 168–83.

Bos JL, de Rooij J, and Reedquist KA (2001). Rap1 signalling: adhering to new models. *Nature Rev Mol Cell Biol* **2**, 369–77.

Bourne HR and Nicoll R (1993). Molecular machines integrate coincident synaptic signals. *Cell* **72** (Suppl.), 65–75.

Brambilla R, Gnesutta N, Minichiello L, White G *et al.* (1997). A role for the Ras signalling pathway in synaptic transmission and long-term memory. *Nature* **390**, 281–6.

Cao WH, Luttrell LM, Medvedev AV, Pierce KL, Daniel KW, Dixon TM *et al.* (2000). Direct binding of activated c-Src to the β_3-adrenergic receptor is required for MAP kinase activation. *J Biol Chem* **275**, 38 131–4.

Casey PJ and Gilman AG (1988). G protein involvement in receptor-effector coupling. *J Biol Chem*, **263**, 2577–80.

Chawla S, Hardingham GE, Quinn DR, and Bading H (1998). CBP: a signal-regulated transcriptional coactivator controlled by nuclear calcium and CaM kinase IV. *Science*, **281**, 1505–9.

Chen HJ, Rojas-Soto M, Oguni A, and Kennedy MB (1998). A synaptic Ras GTPase activating protein ($p135^{SynGAP}$) inhibited by CaM Kinase II. *Neuron* **20**, 895–904.

Chen Y-H, Grall D, Salcini AE, Pelicci PG, Pouysségur J, and van Obberghen-Schilling E (1996). Shc adaptor proteins are key transducers of mitogen signalling mediated by the G protein-coupled thrombin receptor. *EMBO J* **15**, 1037–44.

Chin D and Means AR (2000). Calmodulin: a prototypical calcium sensor. *Trends in Cell Biol* 10, 322–8.

Collins S, Caron MG, and Lefkowitz RJ (1992). From ligand-binding to gene-expression—new insights into the regulation of G protein-coupled receptors. *Trends in Biochem Sci* 17, 37–9.

Conquet F, Bashir ZI, Davies CH, Daniel H, Ferraguti F, Bordi F *et al.* (1994). Motor deficit and impairment of synaptic plasticity in mice lacking mGluR1. *Nature* 372, 237–43.

Cooper DMF, Mons N, and Karpen JW (1995). Adenylyl cyclases and the interaction between calcium and cAMP signalling. *Nature* 374, 421–4.

Corcoran EE and Means AR (2001). Defining Ca^{2+}/calmodulin-dependent protein kinase cascades in transcriptional regulation. *J Biol Chem* 276, 2975–8.

Crabtree GR (1999). Generic signals and specific outcomes: signalling through Ca^{2+}, calcineurin, and NF-AT. *Cell* 96, 611–14.

Crabtree GR (2001). Calcium, calcineurin, and the control of transcription. *J Biol Chem* 276, 2313–16.

Crespo P, Xu N, Simonds WF, and Gutkind JS (1994). Ras-dependent activation of MAP kinase pathways mediated by G-protein $\beta\gamma$ subunits. *Nature* 369, 418–20.

Crino P, Khodakhah K, Becker K, Ginsberg S, Hemby S, and Eberwine J (1998). Presence and phosphorylation of transcription factors in developing dendrites. *Proc Natl Acad Sci USA* 95, 2313–18.

Daaka Y, Luttrell LM and Lefkowitz RJ (1997). Switching of the capacity of the β_2-adrenergic receptor to different G proteins by protein kinase A. *Nature* 390, 88–91.

Daub H, Ulrich-Weiss F, Wallasch C, and Ullrich A (1996). Role of transactivation of the EGF receptor in signalling by G-protein-coupled receptors. *Nature* 379, 557–60.

Davis RJ (1993). The mitogen-activated protein kinase signal transduction pathway. *J Biol Chem* 268, 14 553–6.

DeFea KA, Zalevsky J, Thoma MS, Dery O, Mullins RD, and Bunnett NW (2000). β-arrestin-dependent endocytosis of PAR2 is required for intracellular targeting of activated ERK1/2. *J Cell Biol* 148, 1267–81.

Della Rocca GJ, van Biesen T, Daaka Y, Luttrell DK, Luttrell LM, and Lefkowitz RJ (1997). Ras-dependent mitogen-activated protein kinase activation by G protein-coupled receptors. *J Biol Chem* 272, 19 125–32.

Della Rocca GJ, Maudsley S, Daaka Y, Lefkowitz RJ, and Luttrell LM (1999). Pleiotropic coupling of G protein-coupled receptors to the mitogen-activated protein kinase cascade. *J Biol Chem* 274, 13 978–84.

de Rooij J, Zwartkruis FJT, Verheijen MHG, Cool RH, Nijman SMB, Wittinghofer A *et al.* (1998). Epac is a Rap1 guanine-nucleotide-exchange factor directly activated by cyclic AMP. *Nature* 396, 474–7.

Dharnasekaran N, Heasley LE, and Johnson GL (1995). G protein-coupled receptor systems involved in cell growth and oncogenesis. *Endocrine Reviews* 16, 259–70.

Dikic I, Tokiwa G, Lev S, Courtneidge SA, and Schlessinger J (1996). A role for Pyk2 and Src in linking G-protein-coupled receptors with MAP kinase activation. *Nature* 383, 547–50.

Dolmetsch RE, Pajvani U, Fife K, Spotts JM, and Greenberg ME (2001). Signalling to the nucleus by an L-type calcium channel–calmodulin complex through the MAP kinase pathway. *Science* 294, 333–9.

Dolmetsch RE, Xu K, and Lewis RS (1998). Calcium oscillations increase the efficiency and specificity of gene expression. *Nature* 392, 933–6.

Dolphin AC (1998). Mechanisms of modulation of voltage-dependent calcium channels by G proteins. *J Physiol* 506, 3–11.

Duckworth BC and Cantley LC (1997). Conditional inhibition of the mitogen-activated protein kinase cascade by wortmannin. *J Biol Chem* 272, 27 665–70.

Ebinu JO, Bottorff DA, Chan EY, Stang SL, Dunn RJ, and Stone JC (1998). RasGRP, a Ras guanyl nucleotide-releasing protein with calcium and diacylglycerol-binding motifs. *Science* 280, 1082–6.

Emptage NJ (1999). Calcium on the up: supralinear calcium signalling in central neurons. *Neuron* 24, 495–7.

Fang Y, Vilella-Bach M, Bachmann R, Flanigan A, and Chen J (2001). Phosphatidic acid-mediated mitogenic activation of mToR signalling. *Science* 294, 1942–5.

Farnsworth CL, Freshney NW, Rosen LB, Ghosh A, Greenberg ME, and Feig LA (1995). Calcium activation of Ras mediated by neuronal exchange factor Ras-GRF. *Nature* 376, 524–7.

Faure M, Voyno-Yasenetskaya TA, and Bourne HR (1994). Cyclic AMP and βγ-subunits of heterotrimeric G proteins stimulate the mitogen-activated protein kinase pathway in COS-7 cells. *J Biol Chem* 269, 7851–4.

Ferraguti F, Baldani GB, Corsi M, Nakanishi S, and Corti C (1999). Activation of the extracellular signal-regulated kinase 2 by metabotropic glutamate receptors. *Euro J Neurosci* 11, 2073–82.

Finkbeiner S and Greenberg ME (1996). Ca^{2+}-dependent routes to Ras: mechanisms for neuronal survival, differentiation and plasticity? *Neuron* 16, 233–6.

Francis SH and Corbin JD (1999). Cyclic nucleotide-dependent protein kinases: Intracellular receptors for cAMP and cGMP action. *Critical Reviews in Clinical Laboratory Sciences* 36, 275–328.

Ghosh A and Greenberg ME (1995). Calcium signalling in neurons: molecular mechanisms and cellular consequences. *Science* 268, 239–47.

Gilman AG (1987). G proteins: transducers of receptor-generated signals. *Annual Rev Biochem*, 56 615–49.

Gonzalez GA and Montminy MR (1989). Cyclic AMP stimulates somatostatin gene transcription by phosphorylation of CREB at serine-133. *Cell* 59, 675–80.

Greengard P, Allen PB, and Nairn AC (1999). Beyond the dopamine receptor: the DARPP-32/protein phosphatase-1 cascade. *Neuron* 23, 435–47.

Grewal SS, York RD, and Stork PJS (1999). Extracellular-signal-regulated kinase signalling in neurons. *Curr Opin Neurobiol* 9, 544–53.

Guo FF, Kumahara E, and Saffen D (2001). A CalDAG-GEFI/Rap1/B-raf cassette couples M_1 muscarinic acetylcholine receptors to the activation of ERK1/2. *J Biol Chem* 276, 25 568–81.

Gutkind JS (1998). The pathways connecting G protein-coupled receptors to the nucleus through divergent mitogen-activated protein kinase cascades. *J Biol Chem* 273, 1839–42.

Hagemann K and Blank JL (2001). The ups and downs of MEK kinase interactions. *Cell Signal* 13, 863–75.

Hagiwara M, Brindle P, Harootunian A, Armstrong R, Rivier J, Vale W *et al.* (1993*a*). Coupling of hormonal stimulation and transcription via cyclic AMP-responsive factor CREB is rate limited by nuclear entry of protein kinase A. *Mol Cell Biol* 13, 4852–9.

Hagiwara M, Alberts A, Brindle P, Meinkoth J, Feramisco J, Deng T *et al.* (1993*b*). Transcriptional attenuation following cAMP induction requires PP-1-mediated dephosphorylation of CREB. *Cell* 70, 105–13.

Hall RA, Premont RT, and Lefkowitz RJ (1999). Heptahelical receptor signalling: Beyond the G protein paradigm. *J Cell Biol* 145, 927–32.

Hanoune J and Defer N (2001). Regulation and role of adenylyl cyclase isoforms. *Annual Rev Pharmacol Toxicol* 41, 145–74.

Hardingham GE, Arnold FJL, and Bading H (2001). Nuclear calcium signalling controls CREB-mediated gene expression triggered by synaptic activity. *Nature Neurosci* 4, 261–7.

Hawes BE, van Biesen T, Koch WJ, Luttrell LM, and Lefkowitz RJ (1995). Distinct pathways of G_i- and G_q-mediated mitogen-activated protein kinase activation. *J Biol Chem* 270, 17 148–53.

Hawes BE, Luttrell LM, van Biesen T, and Lefkowitz RJ (1996). Phosphatidylinositol 3-kinase is an early intermediate in the Gβγ-mediated mitogen-activated protein kinase signalling pathway. *J Biol Chem* 271, 12 133–6.

Hermans E and Challiss RAJ (2001). Structural, signalling and regulatory properties of the group I metabotropic glutamate receptors: prototypic family C G-protein-coupled receptors. *Biochem J* 359, 465–84.

Herrlich A, Daub H, Knebel A, Herrlich P, Ullrich A, Schultz G, and Gudermann T (1998). Ligand-independent activation of PDGF receptor is a necessary intermediate in lysophosphatidic acid-stimulated mitogenic activity in L cells. *Proc Natl Acad Sci USA* **95**, 8985–90.

Heuss C and Gerber U (2000). G-protein-independent signalling by G-protein-coupled receptors. *Trends Neurosci* **23**, 469–75.

Higashida H, Hashii M, Yokoyama S, Hoshi N, Asai K, and Kato T (2001). Cyclic ADP-ribose as a potential second messenger for neuronal Ca^{2+} signalling. *J Neurochem* **76**, 321–31.

Hirono M, Sugiyama T, Kishimoto Y, Sakai I *et al.* (2001). Phospholipase Cβ4 and protein kinase Cα and/or CβI are involved in the induction of long term depression in cerebellar Purkinje cells. *J Biol Chem* **276**, 45 236–42.

Ho N, Liauw JA, Blaeser F, Wei F, Hanissian S *et al.* (2000). Impaired synaptic plasticity and cAMP response element-binding protein activation in Ca^{2+}/calmodulim-dependent protein kinase type IV/Gr-deficient mice. *J Neurosci* **20**, 6459–72.

Howe LR and Marshall CJ (1993). Lysophosphatidic acid stimulates mitogen-activated protein kinase activation via a G-protein-coupled pathway requiring p21^{ras} and p74^{raf-1}. *J Biol Chem* **268**, 20 717–20.

Hu SC, Chrivia J, and Ghosh A (1999). Regulation of CBP-mediated transcription by neuronal calcium signalling. *Neuron* **22**, 799–808.

Ichise T, Kano M, Hashimoto K, Yanagihara D, Nakao K, Shigemoto R *et al.* (2000). mGluR1 in cerebellar Purkinje cells essential for long-term depression, synapse elimination, and motor coordination. *Science* **288**, 1832–5.

Impey S, Obrietan K, and Storm DR (1999). Making new connections: role of ERK/MAP kinase signalling in neuronal plasticity. *Neuron* **23**, 11–14.

Impey S, Obrietan K, Wong ST, Poser S, Yano S, Wayman G, Deloulme JC, Chan G, and Storm DR (1997). Cross talk between ERK and PKA is required for Ca^{2+} stimulation of CREB-dependent transcription and ERK nuclear translocation. *Neuron* **21**, 869–83.

Inoue T, Kato K, Kohda K, and Mikoshiba K (1998). Type 1 inositol 1,4,5-trisphosphate receptor is required for induction of long-term depression in cerebellar Purkinje neurons. *J Neurosci* **18**, 5366–73.

Jordan JD, Carey KD, Stork PJS, and Iyengar R (1999). Modulation of Rap activity by direct interaction of Gα_o with Rap1 GTPase-activating protein. *J Biol Chem* **274**, 21 507–10.

Kao SC, Jaiswal RK, Kolch W, and Landreth GE (2001). Identification of the mechanisms regulating the differential activation of MAPK cascade by EGF and NGF in PC12 cells. *J Biol Chem* **276**, 18 169–77.

Katz PS and Clemens S (2001). Biochemical networks in nervous systems: expanding neuronal information capacity beyond voltage signals. *Trends Neurosci* **24**, 18–25.

Kawasaki H, Springett GM, Mochizuki N, Toki S, Nakaya M, Matsuda M *et al.* (1998). A family of cAMP-binding proteins that directly activate Rap1. *Science* **282**, 2275–9.

Koch WJ, Hawes BE, Allen LF, and Lefkowitz RJ (1994). Direct evidence that G$_i$-coupled receptor stimulation of mitogen-activated protein kinase is mediated by G$_{\beta\gamma}$ activation of p21^{ras}. *Proc Nat Acad Sci USA* **91**, 12706–10.

Kolch W, Heidecker G, Kochs G, Hummel R, Vahidi H, Mischak H *et al.* (1993). Protein kinase Cα activates Raf-1 by direct phosphorylation. *Nature* **364**, 249–52.

Kolch W (2000). Meaningful relationships: the regulation of the Ras/Raf/MEK/ERK pathway by protein interactions. *Biochem J* **351**, 289–305.

Kranenburg O, Verlaan I, and Moolenaar WH (1999). Dynamin is required for the activation of mitogen-activated protein (MAP) kinases by MAP kinase kinase. *J Biol Chem* **274**, 35 301–04.

Lemmon MA and Ferguson KM (2000). Signal-dependent membrane targeting by pleckstrin homology (PH) domains. *Biochem J* **350**, 1–18.

Leslie CC (1997). Properties and regulation of cytosolic phospholipase A$_2$. *J Biol Chem* **272**, 16 709–712.

Lev S, Moreno H, Martinez R, Canoll O, Peles E, Musacchio JM *et al.* (1995). Protein-tyrosine kinase Pyk2 involved in Ca^{2+}-induced regulation of ion channel and MAP kinase functions. *Nature* 376, 737–45.

Liscovitch M, Czarny M, Fiucci G, and Tang X (2000). Phospholipase D: molecular and cell biology of a novel gene family. *Biochem J* 345, 401–15.

Litvak V, Tian D, Shaul YD, and Lev S (2000). Targeting of Pyk2 to focal adhesions as a cellular mechanism for convergence between integrins and G protein-coupled receptor signalling cascades. *J Biol Chem* 275, 32 736–46.

Liu F-C and Graybiel AM (1996). Spatiotemporal dynamics of CREB phosphorylation: transient versus sustained phosphorylation in the developing striatum. *Neuron* 17, 1133–44.

Lopez-Ilasaca M, Crespo P, Pellici PG, Gutkind JS, and Wetzker R (1997). Linkage of G-protein-coupled receptors to the MAPK signalling pathway through PI 3-kinase γ. *Science* 275, 394–7.

Luttrell LM, Hawes BE, van Biesen T, Luttrell DK, Lansing TJ, and Lefkowitz RJ (1996). Role of c-Src tyrosine kinase in G protein-coupled receptor- and Gβγ subunit-mediated activation of mitogen-activated protein kinases. *J Biol Chem* 271, 19 443–50.

Luttrell LM, Della Rocca GJ, van Biesen T, Luttrell DK, and Lefkowitz RJ (1997). Gβγ subunits mediate Src-dependent phosphorylation of the epidermal growth factor receptor. *J Biol Chem* 272, 4637–44.

Luttrell LM, Ferguson SSG, Daaka Y, Miller WE, Maudsley S, Della Rocca GJ, *et al.* (1999). β-arrestin-dependent formation of β2 adrenergic receptor-Src protein kinase complexes. *Science* 283, 655–61.

Marshall CJ (1995). Specificity of receptor tyrosine kinase signalling: transient versus sustained extracellular signal-regulated kinase activation. *Cell* 80, 179–85.

Mattingly RR and Macara IG (1996). Phosphorylation-dependent activation of the Ras-GRF/CDC25^{Mm} exchange factor by muscarinic receptors and G-protein βγ subunits. *Nature* 382, 268–72.

Mayr B and Montminy M (2001). Transcriptional regulation by the phosphorylation-dependent factor CREB. *Nature Rev Mol Cell Biol* 2, 599–609.

Mayr BM, Canetteri G, and Montminy MR (2001). Distinct effects of cAMP and mitogenic signals on CREB-binding protein recruitment impart specificity to target gene activation via CREB. *Proc Natl Acad Sci USA* 98, 10 936–41.

McDonald PH, Chow CW, Miller WE, Laporte SA, Field ME, Lin FT *et al.* (2000). β-arrestin 2: a receptor-regulated MAPK scaffold for the activation of JNK3. *Science* 290, 1574–7.

Miller RJ (1998). Presynaptic receptors. *Ann Rev Pharmacol Toxicol* 38, 201–27.

Milligan G and White JH (2001). Protein–protein interactions at G-protein-coupled receptors. *Trends Pharmacol Sci* 22, 513–18.

Miyata M, Finch EA, Khrioug L, Hashimoto K, Hayasaka S, Oda SI *et al.* (2000). Local calcium release in dendritic spines required for long-term synaptic depression. *Neuron* 28, 233–44.

Mochly-Rosen D (1995). Localization of protein kinases by anchoring proteins: A theme in signal transduction. *Science* 268, 247–51.

Montminy MR and Bilezikjian LM (1987). Binding of a nuclear-protein to the cyclic AMP response element of the somatostatin gene. *Nature* 328, 175–8.

Nakajima T, Uchida C, Anderson SF, Lee CG, Hurwitz J, Parvin JD *et al.* (1997). RNA helicase A mediates association of CBP with RNA polymerase II. *Cell* 90, 1107–12.

Offermanns S, Bombien E, and Schultz G (1993). Stimulation of tyrosine phosphorylation and mitogen-activated-protein (MAP) kinase activity in human SH-SY5Y neuroblastoma cells by carbachol. *Biochem J* 294, 545–50.

Ohmichi M, Sawada T, Kanda Y, Koike K, Hirota K, Miyake A, and Saltiel AR (1994). Thyrotropin-releasing hormone stimulates MAP kinase activity in GH3 cells by divergent pathways. *J Biol Chem* 269, 3783–8.

Osborne SL, Meunier FA, and Schiavo G (2001). Phosphoinositides as key regulators of synaptic function. *Neuron* 32, 9–12.

Parekh AB and Penner R (1997). Store depletion and calcium influx. *Physiol Rev* 77, 901–30.

Peavy RD, Chang MSS, Sanders-Bush E, and Conn PJ (2001). Metabotropic glutamate receptor5-induced phosphorylation of extracellular signal-regulated kinase in astrocytes depends on trans-activation of the epidermal growth factor receptor. *J Neurosci* 21, 9619–28.

Penn RB, Pascual RM, Kim YM, Mundell SJ, Krymskaya VP, Panettieri RA *et al.* (2001). Arrestin specificity for G-protein-coupled receptors in human airway smooth muscle. *J Biol Chem* 276, 32 648–56.

Penniston JT and Enyedi A (1998). Modulation of the plasma membrane Ca^{2+} pump. *J Membr Biol* 165, 101–9.

Pierce KL, Luttrell LM, and Lefkowitz RJ (2001). New mechanisms in heptahelical receptor signalling to mitogen-activated protein kinase cascades. *Oncogene* 20, 1532–9.

Prenzel N, Zwick E, Daub H, Leserer M, Abraham R, Wallasch C *et al.* (1999). EGF receptor transactivation by G-protein-coupled receptors requires metalloproteinase cleavage of proHB-EGF. *Nature* 402, 884–8.

Putney JW (1986). A model for receptor-regulated calcium entry. *Cell Calcium* 7, 1–12.

Pyne S and Pyne NJ (2000). Sphingosine 1-phosphate signalling in mammalian cells. *Biochem J* 349, 385–402.

Rameh LE and Cantley LC (1999). The role of phosphoinositide 3-kinase lipid products in cell function. *J Biol Chem* 274, 8347–50.

Rhee SG (2001). Regulation of phosphoinositide-specific phospholipase C. *Ann Rev Biochem* 70, 281–312.

Riccio A, Ahn S, Davenport CM, Blendy JA, and Ginty DD (1999). Mediation by a CREB family transcription factor of NGF-dependent survival of sympathetic neurons. *Science* 286, 2358–61.

Rizzuto R (2001). Intracellular Ca^{2+} pools in neuronal signalling. *Curr Opin Neurobiol* 11, 306–11.

Roberson ED, English JD, Adams JP, Selcher JC, Kondratick C, and Sweatt JD (1999). The mitogen-activated protein kinase cascade couples PKA and PKC to cAMP response element binding protein phosphorylation in area CA1 of hippocampus. *J Neurosci* 19, 4337–48.

Rodbell M (1980). The role of hormone receptors and GTP-regulatory proteins in membrane transduction. *Nature* 284, 17–22.

Roesler WJ, Graham JG, Kolen R, Klemm DJ, and McFie PJ (1995). The cAMP response element-binding protein synergizes with other transcription factors to mediate cAMP responsiveness. *J Biol Chem* 270, 8225–32.

Rose CR and Konnerth A (2001). Stores not just for storage: Intracellular calcium release and synaptic plasticity. *Neuron* 31, 519–22.

Rosen LB, Ginty DD, Weber MJ, and Greenberg ME (1994). Membrane depolarisation and calcium influx stimulate MEK and ERK kinase via activation of Ras. *Neuron* 12, 1207–21.

Rosenblum K, Futter M, Jones M, Hulme EC, and Bliss TVP (2000). ERKI/II regulation by the muscarinic acetylcholine receptors in neurons. *J Neurosci* 20, 977–85.

Rusanescu G, Qi H, Thomas SM, Brugge JS, and Halegoua S (1995). Calcium influx induces neurite growth through a Src-Ras signalling cassette. *Neuron* 15, 1415–25.

Rusnak F and Mertz P (2000). Calcineurin: form and function. *Physiol Rev* 80, 1483–521.

Sala C, Rudolph-Correia S, and Sheng M (2000). Developmentally regulated NMDA receptor-dependent dephosphorylation of cAMP response element-binding protein (CREB) in hippocampal neurons. *J Neurosci* 20, 3529–36.

Schaeffer HJ and Weber MJ (1999). Mitogen-activated protein kinases: specific messages from ubiquitous messengers. *Mol Cell Biol* 19, 2435–44.

Schlaepfer DD, Hanks SK, Hunter T, and van der Geer P (1994). Integrin-mediated signal transduction linked to Ras pathway by Grb2 binding to focal adhesion kinase. *Nature* 372, 786–91.

Schlessinger J (1993). How receptor tyrosine kinases activate Ras. *Trends Biochem Sci* 18, 273–5.

Schramm M and Selinger Z (1984). Message transmission: receptor-controlled adenylate cyclase system. *Science* **225**, 1350–6.

Shaywitz AJ and Greenberg ME (1999). CREB: a stimulus-induced transcription factor activated by a diverse array of extracellular signals. *Ann Rev Biochem* **68**, 821–61.

Sheng M, McFadden G, and Greenberg ME (1990). Membrane depolarization and calcium induce c-fos transcription via phosphorylation of transcription factor CREB. *Neuron* **4**, 571–82.

Short SM, Boyer JL, and Juliano RL (2000). Integrins regulate the linkage between upstream and downstream events in G-protein-coupled receptor signalling to mitogen-activated protein kinase. *J Biol Chem* **275**, 12 970–7.

Simon MI, Strathmann MP and Gautam N (1991). Diversity of G proteins in signal transduction. *Science* **252**, 802–8.

Six DA and Dennis EA (2000). The expanding superfamily of phospholipase A_2 enzymes: classification and characterization. *Biochim Biophys Acta* **1488**, 1–19.

Slack BE (1998). Tyrosine phosphorylation of paxillin and focal adhesion kinase by activation of muscarinic m3 receptors is dependent on integrin engagement by the extracellular matrix. *Proc Natl Acad Sci USA* **95**, 7281–6.

Soderling TR, Chang B, and Brickey D (2001). Cellular signalling through multifunctional Ca^{2+}/calmodulin-dependent protein kinase II. *J Biol Chem* **276**, 3719–22.

Soeder KJ, Snedden SK, Cao WH, Della Rocca GJ, Daniel KW, Luttrell LM *et al.* (1999). The β_3-adrenergic receptor activates mitogen-activated protein kinase in adipocytes through a G_i-dependent mechanism. *J Biol Chem* **274**, 12 017–22.

Sun PQ, Enslen H, Myung PS, and Maurer RA (1994). Differential activation of CREB by Ca^{2+}/calmodulin-dependent protein kinases type II and type IV involves phosphorylation of a site that negatively regulates activity. *Genes Dev* **8**, 2527–39.

Sunahara RK, Dessauer CW, and Gilman AG (1996). Complexity and diversity of mammalian adenylyl cyclases. *Ann Rev Pharmacol Toxicol* **36**, 461–80.

Sweatt JD (2001). The neuronal MAP kinase cascade: a biochemical signal transduction system subserving synaptic plasticity and memory. *J Neurochem* **76**, 1–10.

Tanaka C and Nishizuka Y (1994). The protein kinase C family for neuronal signalling. *Ann Rev Neurosci* **17**, 551–67.

Toker A (1998). The synthesis and cellular roles of phosphatidylinositol 4,5-bisphosphate. *Curr Opin Cell Biol* **10**, 254–61.

Treisman R (1996). Regulation of transcription by MAP kinase cascades. *Curr Opin Cell Biol* **8**, 205–15.

van Biesen T, Hawes BE, Luttrell DK, Krueger KM, Touhara K, Porfiri E *et al.* (1995). Receptor-tyrosine-kinase- and $G_{\beta\gamma}$-mediated MAP kinase activation by a common signalling pathway. *Nature* **376**, 781–4.

van Corven EJ, Hordijk PL, Medema RH, Bos JL, and Moolenaar WH (1993). Pertussis toxin-sensitive activation of p21[ras] by G protein-coupled receptor agonists in fibroblasts. *Proc Natl Acad Sci USA* **90**, 1257–61.

Venter JC, Adams MD, Myers EW, Li PW *et al.* (2001). The sequence of the human genome. *Science* **291**, 1304–49.

Verkhratsky A and Shmigol A (1996). Calcium-induced calcium-release in neurons. *Cell Calcium* **19**, 1–14.

Vieira AV, Lamaze C, and Scmid SL (1996). Control of EGF receptor signalling by clathrin-mediated endocytosis. *Science* **274**, 2086–89.

Vossler MR, Yao H, York RD, Pan MG, Rim CS, and Stork PJS (1997). Cyclic AMP activates MAP kinase and Elk-1 through a B-Raf- and Rap-1-dependent pathway. *Cell* **89**, 73–82.

Watanabe AM, Zaki PA, and O'Dell TJ (2000). Coactivation of β-adrenergic and cholinergic receptors enhances the induction of long-term potentiation and synergistically activates mitogen-activated protein kinase in the hippocampal CA1 region. *J Neurosci* **20**, 5924–31.

West AE, Chen WG, Dalva MB, Dolmetsch RE, Kornhauser JM, Shaywitz AJ *et al.* (2001). Calcium regulation of neuronal gene expression. *Proc Natl Acad Sci USA* **98**, 11 024–31.

Wickman K and Clapham DE (1995). Ion channel regulation by G proteins. *Physiol Rev* **75**, 865–85.

Wilcox RA, Primrose WU, Nahorski SR, and Challiss RAJ (1998). New developments in the molecular pharmacology of the *myo*-inositol 1,4,5-trisphosphate receptor. *Trends Pharmacol Sci* **19**, 467–75.

Wu GY, Deisseroth K, and Tsien RW (2001). Activity-dependent CREB phosphorylation: convergence of a fast, sensitive calmodulin kinase pathway and a slow, less sensitive mitogen-activated protein kinase pathway. *Proc Natl Acad Sci USA* **98**, 2808–13.

York RD, Yao H, Dillon T, Ellig CL, Eckert SP, McCleskey EW, and Stork PJS (1998). Rap1 mediates sustained MAP kinase activation induced by nerve growth factor. *Nature* **392**, 622–6.

Zanassi P, Paolillo M, Feliciello A, Avvedimento EV, Gallo V, and Schinelli S (2001). cAMP-dependent protein kinase induces cAMP-response element-binding protein phosphorylation via an intracellular calcium release.ERK-dependent pathway in striatal neurons. *J Biol Chem* **276**, 11 487–95.

Chapter 6

GPCR signalling through ion channels

Christov Roberson and David E. Clapham

6.1 Introduction

G proteins act as relay stations which sense structural changes on the receptor protein in response to ligand binding and translate this action into the onset of a cellular response, as illustrated in the previous chapter. This is often manifest as alterations in the activity of a number of different intracellular biochemical cascades. However, many GPCRs have also been shown to interact with ion channels in complex ways. When considering the coupling mechanisms that link seven-transmembrane receptors to the activation or inhibition of ion channels a number of questions arise. First, one must establish which G protein subclass is involved in the regulatory step. Thus far, at least 20 different G-alpha subtypes have been identified and classified into four major subfamilies based on sequence homology (Simon *et al.* 1991; Hepler and Gilman 1992; Chapter 1.4). The sensitivity of different $G\alpha$ subunits to certain bacterial toxins such as pertussis toxin (PTX) or cholera toxin (CTX) is generally used as an initial aid in this identification. Furthermore, it must be shown which of the two functionally relevant G protein subunits, $G\alpha$ or $G\beta\gamma$, is the active participant in the regulatory step. In certain instances both subunits may be required to elicit an expected response. Subsequently, one must determine whether the ion channel is regulated directly by the active G protein subunits themselves, or indirectly through the actions of other G protein-stimulated effectors. Several criteria can serve to validate the involvement of G proteins in the activation or inhibition of a particular ion channel under study. Channel activity must always be conditionally and reversibly modified by the activation of a relevant G protein-linked receptor. In addition, the use of poorly hydrolyzable GTP analogs such as GTP-γ-S or GMP-PNP should allow for channel modulation even in the absence of receptor stimulation. Satisfying both of these minimal criteria supports the assumption that G proteins carry signals from upstream receptors to downstream ion channels. Further empirical evidence must then seek to identify the pertinent steps involved in the transduction cascade.

It is important to point out that virtually all ion channels are susceptible to some form of regulation by G protein-mediated pathways. This is because the generation of large pools of cytosolic second messengers will activate enzymes that routinely modify portions of the channel protein exposed on the inside of a cell. Such global adaptations are particularly relevant when receptors are stimulated for prolonged periods of time to initiate long-term physiological and metabolic changes. In these instances, changes in channel behaviour often outlast the duration of the original stimulus due to covalent alterations of the protein. However, many receptors in the heart and nervous system are intricately involved in matching ongoing electrical activity to the constantly changing demands of the extracellular environment.

Various neurotransmitters modulate membrane excitability by activating receptors that uniquely affect the properties of particular channels while not influencing the activities of others. In contrast to global modulation, this form of control is generally quite specific and tends to occur on a much more rapid timescale. Many examples of ion channel regulation have been discovered over the past several years which can all be attributed to the activation of a unique G protein-coupled receptor. In each case the available repertoire of proteins involved in the regulatory pathway combines to generate distinct responses that are specifically suited to the needs of individual cells.

Two main modes of ion channel regulation can be distinguished based on the number of steps required to elicit a response. (1) Direct modes, in which G protein subunits themselves physically interact with discrete regions of the channel protein to control gating. (2) Indirect modes, in which the functions of various channels are modified by the actions of specific enzymes, second messengers, regulatory proteins and lipids, all of which act downstream of the G protein activation step. In the following sections, we present selected examples for each type of modulation and review the experimental data used to support their validity.

6.2 **Direct regulation of ion channels**

Direct G protein modulation of ion channels refers to instances in which G protein subunits physically interact with discrete regions of a channel complex to cause changes in the probabilities of opening or closing. Accordingly, changes in channel activity will generally occur in the absence of cytosolic second messengers and are mostly independent of other receptor-activated enzymes. Before considering in detail the events involved in direct G protein gating of ion channels, it is worthwhile to review certain criteria that should be met in order to suggest a direct interaction between these two proteins.

1. The activity of the channel is reversibly altered by the activation of a suitable G protein-coupled receptor, and this regulation is dependent on GTP.

2. Poorly-hydrolyzable GTP analogues such as GTP-γ-S or guanylyl-imidodiphosphate (GMP-PNP) promote similar changes in channel activity that are independent of receptor activation.

3. The addition of purified and active G protein subunits to the channel is sufficient to trigger changes in channel activity.

4. A physical association between G proteins and channel subunits can be demonstrated within the confines of an intact cell.

Although the fulfillment of all the above criteria is strongly suggestive of a direct interaction between G proteins and ion channels, they do not constitute ultimate proof that this reciprocal interaction alone is what underlies the observed changes in channel behaviour. The third point illustrates the delicate nature of this issue. Due to the heterogeneous composition of the plasma membrane surrounding the channel, the question remains as to whether purified G protein subunits directly bind to distinct regions of the ion channel to control gating, or whether these proteins simply facilitate the interaction of the channel with surrounding lipids or other membrane-anchored co-factors. In such instances, the direct binding of G proteins to the channel might still be necessary to support and sustain these interactions. Recent evidence suggests that G protein subunits may very well act in concert with other regulatory factors to modify channel activity (see Huang *et al.* 1998;

Sui *et al.* 1998). Nonetheless, the continuous presence of the relevant G protein subunit, in its active form, must remain an absolute requirement for maintaining the observed changes in channel behaviour. Furthermore, direct G protein regulation implies that any changes in channel activity cannot proceed in the presence of other co-factors alone.

Direct modulation represents the simplest mechanism by which ion channel activity can be influenced through activated receptors, yet it offers several distinct advantages that are well suited to the needs of many different cells. In contrast to second messenger pathways which seem to favour prolonged, amplified responses, direct modes provide effective ways of narrowing down the pool of potential effector targets. Furthermore, direct gating interactions can greatly help to minimize temporal constraints that are imposed by indirect processes involving several intermediate steps. This form of control is thus advantageous in cases where incoming signals are received at high frequencies and must be rapidly processed.

6.2.1 Inwardly rectifying potassium channels

Perhaps the best-studied example of an ion channel that undergoes direct modulation by G proteins is the G protein-gated inwardly rectifying potassium channel (GIRK) expressed in brain, heart, and pancreatic tissues. GIRK channels comprise a subfamily of inwardly rectifying K^+ channels that currently contains four distinct members (GIRK 1–4) along with several alternatively-spliced isoforms (Doupnik *et al.* 1995). In cardiac atrial myocytes and atrial pacemaker cells, the prototype member of this potassium channel family, termed $I_{K(ACh)}$, mediates the parasympathetic slowing of heart rate. The binding of acetylcholine (ACh) released by vagal nerve terminals to muscarinic type 2 (M_2) receptors (Chapter 18) on the surfaces of pacemaker cells activates a PTX-sensitive G protein to mediate a rapid membrane hyperpolarization which decreases excitability (Hartzell 1988). It is now widely recognized that the opening of this inward rectifier K^+ channel is triggered by direct contacts between activated G protein subunits and distinct regions of the channel protein.

A series of classical observations were important in revealing the direct nature of this interaction. Early electrophysiological studies of cardiac atrial fibers described an increase in potassium flux across the outer membrane following parasympathetic nerve stimulation (Del Castillo and Katz 1955; Hutter and Trautwein 1956, 1957). The channel responsible for this increase in K^+ permeability was shown to have unique biophysical properties that differed from the background potassium conductance (I_{K1}) present in the same cells (Sakmann *et al.* 1983). Evidence for the involvement of G proteins in the activation of $I_{K(ACh)}$ was provided by several findings. Pfaffinger *et al.* (1985) showed that receptor-induced activation of the current in atrial myocytes was abolished by pretreatment of cells with pertussis toxin, a bacterial toxin which ADP-ribosylates and inactivates G proteins of the G_i subfamily. Moreover, intracellular perfusion with a non-hydrolyzable GTP-analogue (GTP-γ-S) resulted in the irreversible and receptor-independent activation of the same potassium current (Breitwieser and Szabo 1985). Further studies by Kurachi *et al.* (1986) determined that $I_{K(ACh)}$ channel activity recorded in cell-attached patches of atrial myocytes declined upon excision of the patch into GTP-free solution, but was recovered when GTP was subsequently added back to the bath.

Although the above findings allowed only for speculation regarding the steps that link G proteins to the activation of GIRK channels, a provocative discovery by Soejima and Noma (1984) shed further light on the nature of the regulatory cascade. Their studies had found that when channels were measured in cell-attached patches on atrial cells, $I_{K(ACh)}$

activity increased only when acetylcholine was present in the recording pipette, but not when applied to the bath solution surrounding the cells. These results argued against the involvement of diffusible second messengers such as cAMP or cGMP in the activation of GIRK currents, and left unanswered the question of how channel openings were coupled to receptor activation.

The first straightforward evidence implicating a direct role for G proteins in the activation of $I_{K(ACh)}$ was provided by Logothetis *et al.* (1987), who showed that cardiac GIRK channels were strongly activated after the addition of purified Gβγ subunits to the cytoplasmic surface of a membrane patch containing the channel. By measuring channel activity in the inside-out patch configuration (Hamill *et al.* 1981), this study convincingly demonstrated that $I_{K(ACh)}$ open-probability was increased more than 500-fold in response to Gβγ application, and that this rise in activity could be sustained even in the absence of native cytosolic constituents. This was surprising, given that the large body of literature had implicated Gα subunits as the functional arm in the G protein complex. A number of independent studies have since confirmed the notion that Gβγ is the primary activator of GIRK channels expressed within the heart and throughout the brain (Logothetis *et al.* 1988; Ito *et al.* 1992; Reuveny *et al.* 1994; Wickman *et al.* 1994). The cloning of various GIRK genes has provided a means of expressing these channels in cellular environments different from those of their native tissues in order to examine their modes of activation (illustrated in Fig. 6.1). The expression of recombinant GIRK1/GIRK4 channel complexes in various cell lines gives rise to potassium currents with single-channel properties that are indistinguishable from native cardiac $I_{K(ACh)}$ and which are activated by purified Gβγ dimers (Wickman *et al.* 1994; Krapivinsky *et al.* 1995a). GIRK channels found in the brain also show specific activation by Gβγ subunits when studied in a variety of heterologous expression systems (Kofuji *et al.* 1995; Velimirovic *et al.* 1996; Jelacic *et al.* 1999, 2000).

A number of residues on the surface of the Gβγ molecule appear to be important for this interaction. Thus far, site-directed mutagenesis strategies have identified distinct regions on the Gβ subunit that are thought to directly interact with GIRK channels. Ford *et al.* (1998) found that mutations of specific residues in the distal amino terminus of Gβ (blade 1;

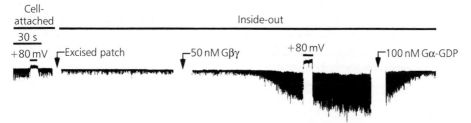

Fig. 6.1 Gβγ subunits directly activate GIRK channels. In cells transiently expressing recombinant GIRK1 and GIRK4 channel subunits, inwardly-rectifying potassium currents can be readily measured in the cell-attached configuration. Patch excision into GTP-free bath solution causes a decline in channel activity, which can be strongly recovered upon addition of 50 nM purified Gβγ to the inside surface of the patch. The subsequent addition of 100 nM Gα-GDP reverses the effect by forming inactive Gαβγ heterotrimers. The membrane potential was held at −80 mV with brief steps to +80 mV to verify inward rectification. Reproduced from Krapivinsky *et al.* (1995) *Nature*, **374**, 135–41.

Chapter 4) caused a measurable decline in GIRK current activation when the subunits were transiently expressed in *Xenopus* oocytes. The majority of these sites are masked by the amino-terminal helix of Gα under conditions where the two G protein subunits do not dissociate (Lambright *et al.* 1996). Further mutations of residues on the top surface of Gβ that face the switch regions of Gα-GDP have also caused these subunits to become less effective at stimulating channel activity. Albsoul-Younes *et al.* (2001) purified a series of mutant βγ subunits and tested their abilities to activate neuronal GIRK channels in excised patches from rat brainstem neurons. Their results revealed additional residues exposed on the outer surfaces of the blades forming the Gβ-propeller (Chapter 4) that may be important for enhancing channel activity. Though less attention has been directed towards identifying similar regions on the Gγ protein, it remains possible that this subunit serves other functions than to simply anchor the βγ dimer to the membrane (see Kawano *et al.* 1999). The sites on Gβγ that are involved in effector regulation have been more carefully defined for adenylyl cyclase and phospholipase C β, and it seems that there is at least a partial overlap with those regions found to be important for GIRK activation (Li *et al.* 1998).

The previous results strongly support the idea that the stimulation of M_2 muscarinic receptors in the heart activates $I_{K(ACh)}$ by causing the release of free Gβγ dimers which then go on to gate the channel. Proof of a direct interaction between these two participants came from studies showing that purified Gβγ binds specifically to immunoprecipitated GIRK1 and GIRK4 channel subunits (Krapivinsky *et al.* 1995*b*). The interaction was found with both native and recombinant channels, while no detectable binding was revealed for either inactive Gα-GDP or activated Gα-GTPγS. Further protein-binding studies using affinity-purified recombinant GIRK1 and Gβγ subunits have produced similar results.

Considerable effort has been directed towards identifying regions on GIRK proteins that interact with Gβγ subunits. Numerous sites located within the intracellular amino- and carboxyterminal tails of both GIRK1 and GIRK4 subunits have been implicated in Gβγ binding based on mutagenesis and peptide-competition results (Inanobe *et al.* 1995; Kunkel and Peralta 1995; Huang *et al.* 1997). Native cardiac $I_{K(ACh)}$ exists as a heterotetrameric complex formed by two GIRK1 and two GIRK4 channel subunits (Corey *et al.* 1998). Purified Gβγ interacts specifically with fusion proteins containing the hydrophilic amino- and carboxyterminal domains of GIRK1 (Huang *et al.* 1995). Studies by Krapivinsky *et al.* (1998) showed that a C-terminal region of the GIRK4 protein close to the second transmembrane domain is critically involved in Gβγ-dependent activation of the GIRK1/GIRK4 channel complex. It seems likely that the intracellular stretches of both channel subunits combine with one another to form a high-affinity binding site for Gβγ inside the cell. The precise mechanisms by which the binding of G proteins to the cytoplasmic tail regions translates into the opening of the channel pore remain to be revealed.

6.2.2 Calcium channels

The release of neurotransmitters from synaptic nerve terminals is triggered by the influx of calcium ions (Ca^{2+}) through voltage-gated Ca^{2+} channels (VDCC) located near vesicle docking sites. Calcium ions act in concert with distinct components of the presynaptic docking machinery to facilitate the fusion of synaptic vesicles with the plasma membrane. Modulating the entry of Ca^{2+} into the nerve terminal thus represents a major means by which neurotransmitter secretion can be controlled. There is substantial evidence that most VDCC are influenced by metabotropic pathways in many neurons and secretory cells (Hille 1994;

Dolphin 1995). These channels are placed into discrete subfamilies based on their voltage-dependence of activation as well as their sensitivity to different pharmacological toxins. In both neurons and cardiomyocytes, high-threshold activated channels (L,N,P,Q,R-type) open in response to large depolarizations relative to the resting membrane potential. Multiple neurotransmitters including dopamine, GABA and glutamate act through $G_{i/o}$-coupled receptors to inhibit the opening of N- and P/Q-type Ca^{2+} channels in the central nervous system (reviewed by Hille 1994). G proteins were first implicated in this inhibition by studies showing that PTX blocks the response to some transmitters while GTPγS causes channel inhibition in the absence of receptor activation (Holz et al. 1986; Dolphin and Scott 1987). Macroscopic measurements of Ca^{2+} currents have revealed that the receptor-induced inhibition is due to a shift in the speed and voltage-dependence of channel opening. Receptor stimulation shifts a fraction of the channels into a 'reluctant' gating mode which decreases their voltage-sensitivity and makes them less likely to open at moderate depolarizing potentials (Bean 1989). The agonist-induced inhibition can be transiently overcome by large depolarizing (facilitating) prepulses that counteract the voltage-dependent block of the channels (Hille 1994).

Results from many studies revealed that these changes in N-type currents are largely mediated by a membrane-delimited pathway that does not involve second messengers (Forscher et al. 1986; Lipscombe et al. 1989; Shapiro and Hille 1993). Instead, channel inhibition can be mimicked in rat sympathetic ganglion neurons by the transient overexpression of exogenous Gβγ subunits even in the absence of receptor activation (Ikeda 1996; Herlitze et al. 1996). Similar results have been obtained in cell lines expressing recombinant P/Q-type Ca^{2+} channel subunits as well as Gβγ dimers (Herlitze et al. 1996). Purified Gβγ protein is also able to reproduce the receptor-evoked inhibition of endogenous Ca^{2+} channels when directly introduced into chick dorsal root ganglion (DRG) cells as well as rat superior cervical ganglion cells (Diversé-Pierluissi et al. 1995; Herlitze et al. 1996). Gβγ subunits bind to a cytoplasmic linker region found between transmembrane domains I and II in the α_1 subunit of the N-type Ca^{2+} channel. This region contains a βγ-binding motif (QXXER) that is found in adenylyl cyclase and is required for G protein regulation of the enzyme (Chen et al. 1995). De Waard et al. (1997) found that fusion proteins encompassing this region of the α_{1A} channel subunit bound specifically to in vitro translated as well as rat brain-solubilized Gβγ subunits, whereas no binding was detected with activated Gα-GTPγS. Mutations of certain amino acids in a segment containing the QXXER motif together with several flanking residues prevented the inhibitory effects of exogenous Gβγ subunits on Ca^{2+} channels expressed in Xenopus oocytes. In addition, peptides containing the QXXER sequence are able to interfere with Gβγ-induced inhibition of Ca^{2+} currents in HEK 293 cells presumably by acting as binding competitors (Zamponi et al. 1997). An interesting observation derives from the finding that the same cytoplasmic linker region that binds to Gβγ has also been shown to interact with the Ca^{2+} channel β-subunit (Pragnell et al. 1994; Witcher et al. 1995). This protein influences the voltage-sensitivity of the channel when co-expressed together with α_1 subunits. The binding of G proteins to the channel may therefore alter the voltage-dependence of activation by controlling various aspects of this interaction.

Additional domains within the amino terminus, the cytoplasmic linker regions, as well as the C-terminus have all been purported to interact with G Protein βγ subunits (Zhang et al. 1996; De Waard et al. 1997; Furukawa et al. 1998; Canti et al. 1999). Taken together, these results suggest that neurotransmitters bind to their cognate receptors to activate a PTX-sensitive G protein which goes on to depress both N- and P/Q-type Ca^{2+} currents *via* a membrane-delimited pathway (i.e. one that does not involve cytosolic transduction elements) that most likely results from direct interactions with the channel. Gβγ subunits presumably bind to several different regions of the channel protein to promote a shift in the voltage-dependence of activation, which causes fewer channels to open with moderate depolarizations (-50 to $0\,mV$). The consequences of such regulation can have enormous impacts on the frequency with which nerve impulses are transmitted between cells, and they are likely to play important roles in the modulation of synaptic strength.

6.3 Indirect regulation of ion channels

Aside from direct interactions that were described in the previous sections, ion channels also undergo extensive modulation by processes that link them indirectly to G proteins. In this respect, a hallmark feature of signalling pathways involving GPCRs is the production of cytosolic second messengers. The enormous power of these receptors to influence so many diverse reactions inside any given cell derives in large part from their ability to stimulate enzymes that respond to changes in the concentrations of compounds such as cAMP or cGMP (Chapter 5). Few proteins are resistant to the widespread consequences of phosphory-lation or dephosphorylation that occur in response to many G protein-linked signals, and the series of events that originate in one area of the cell can often influence the activities of proteins found great distances away in separate locations. This allows channels to be connected to a much larger network of regulatory machinery and thereby greatly enhances their ability to be precisely tuned by external signals. Responses involving the indirect actions of second messengers generally tend to be longer lasting and also somewhat slower than direct routes. Ion channels that are targeted by these pathways will therefore often alter membrane excitability over more prolonged periods of time.

In addition to the minimal criteria that are required to establish a link between G proteins and ion channels, indirect pathways predict that any observed changes in channel activity should be replicated by the addition of the relevant enzyme or second messenger itself in the absence of G proteins. Though many tools are now available for bypassing the G protein activation step to cause changes in the levels of various second messengers, progress is at times hampered by the technical difficulties associated with such experiments. Furthermore, the use of pharmacological drugs to activate or inhibit the actions of specific enzymes is often confounded by the nonselective nature of their effects. Nonetheless, these compounds continue to serve as invaluable tools in trying to reconstruct the transduction pathways that connect G protein-linked receptors with certain ion channels. It is beyond the scope of this chapter to provide an extensive review of all second messenger pathways involved in channel modulation. Instead, we will describe selected examples of well-characterized enzymes and second messengers frequently implicated in the regulation of channel activity. Table 6.1 lists some further examples of ion channels that are affected by such pathways.

6.3.1 cAMP and protein kinase A

The receptor-mediated activation of adenylyl cyclase (AC) to generate cyclic AMP (cAMP) represents the classical example of a soluble intracellular mediator that regulates a great number of events in nearly all mammalian cells studied to date. The binding of cAMP to the regulatory subunit of protein kinase A (PKA) stimulates the enzyme to phosphorylate a diverse group of substrates involved in the initiation, maintenance and inhibition of the

Table 6.1 Common examples of currents modulated by neurotransmitters

Transmitter	Target receptors	Currents modified	Type of regulation	Selected references
Acetylcholine	muscarinic m2, m4	I_{GIRK}	direct ↑	(1)
		$I_{K,ATP}$	direct ↑	(2)
	muscarinic m1, m3	I_M	indirect ↓	(3, 4)
		$I_{K,Ca}$	indirect ↓	(5)
		$I_{Ca,N}$	indirect ↓	(6)
Dopamine	D2,D3	I_{GIRK}	direct ↑	(7, 8)
	D2,D3	$I_{Ca,N}$	direct ↓	(9)
	D1	I_{Na}	indirect ↓	(10)
GABA	$GABA_B$	I_{GIRK}	direct ↑	(11)
Norepinephrine	α_1-AR	$I_{K,leak}$	indirect ↓	(12)
	β-AR	$I_{K,Ca}$	indirect ↓	(13)
		$I_{Ca,L}$	indirect ↑	(14)
Serotonin	5-HTs	I_{GIRK}	direct ↑	(15, 16)
		$I_{Ca,N}$	direct ↓	(17)
		$I_{K,Ca}$	indirect ↓	(13)
		$I_{K,leak}$	indirect ↓	(12)
		$I_{Na,K}$	indirect ↑	(18)

1. Logothetis DE et al. (1987). Nature 325 (6102), 321–6.
2. Ito H et al. (1992). J Gen Physiol 99(6), 961–83.
3. Brown DA et al. (1989). J Physiol 413, 469–88.
4. Marrion NV (1993). Neuron 11(1), 77–84.
5. Madison DV et al. (1987). J Neurosci 7(3), 733–41.
6. Beech DJ et al. (1992). Neuron 8(1), 97–106.
7. Pillai G et al. (1998). Neuropharmacology 37(8), 983–7.
8. Kuzhikandathil EV and Oxford GS (2000). J Gen Physiol 115(6), 697–706.
9. Marchetti C et al. (1986). Pflugers Arch 406(2), 104–11.
10. Cantrell AA et al. (1997). Neurosci 17(19), 7330–8.
11. Sodickson DL and Bean BP (1996). J Neurosci 16(20), 6374–85.
12. Talley EM et al. (2000). Neuron 25(2), 399–410.
13. Pedarzani P and Storm JF (1993). Neuron 11(6), 1023–35.
14. Osterrieder W et al. (1982). Nature 298(5874), 576–8.
15. Andrade R et al. (1986). Science 234(4781), 1261–5.
16. Andrade R and Nicoll RA (1987). J Physiol 394, 99–124.
17. Bayliss, DA et al. (1997). J Neurophysiol 77(3), 1362–74.
18. Cardenas CG et al. (1999). J Physiol 518(2): 507–23.

signalling response. The effects of cAMP augmentation on channel function have been extensively studied in the heart. Sympathetic nerve stimulation increases heart rate and contractive force by releasing norepinephrine onto β-adrenergic receptors expressed within cardiac tissue (Hartzell 1988; Trautwein and Hescheler 1990, Chapter 12). The activation of these receptors leads to a significant potentiation of the high-threshold activated (L-type) Ca^{2+} current which causes heart cells to depolarize (Bean 1985). Pharmacological agents that directly activate AC activity such as forskolin are able to mimic the effects of receptor stimulation, and the current increase is also observed upon intracellular perfusion with cAMP in the absence of G protein activation (Tsien et al. 1972; Morad et al. 1981; Fischmeister and Hartzell 1986). In addition, direct injection of the purified catalytic subunit of PKA into isolated cardiomyocytes produces similar increases in L-type currents (Osterrieder et al. 1982). The enhancement of I_{Ca} results from the phosphorylation of L-type Ca^{2+} channels by PKA on specific sites contained within the α_1 and β channel subunits (see Catterall 2000). PKA phosphorylation appears to cause increases in both channel open-probability as well as single-channel mean open time in response to physiological depolarizations (Reuter *et al.* 1982; Cachelin *et al.* 1983; Yue *et al.* 1990).

PKA also phosphorylates the rat brain voltage-gated Na$^+$ channel on a number of residues in the α subunit to cause a suppression of the net current in response to membrane depolarization (Costa *et al.* 1982; Cantrell *et al.* 1999; Smith and Goldin 1997). The decrease in sodium current can be reproduced in *Xenopus* oocytes expressing the cloned channel and injected with cAMP or the regulatory subunit of PKA, thereby obviating the need to activate G proteins (Gershon *et al.* 1992). In cultures of dissociated hippocampal neurons, Na$^+$ current inhibition caused by receptor activation is mimicked by the isolated addition of specific PKA activators to the cells (Cantrell *et al.* 1999). Interestingly, PKA-dependent reductions in peak Na$^+$ currents can depend upon convergent activation of PKC and consequent phosphorylation of different sites along the cytoplasmic linkers found between domains I and IV of the sodium channel. Even more fascinating is the finding that such convergence is observed only at relatively hyperpolarized membrane potentials. At more depolarized potentials ($-70\,\text{mV}$ and above), significant PKA modulation of the channel can occur even in the absence of PKC activation (Cantrell *et al.* 1999). Such complex regulation enables the sodium channel to act as a coincidence detector which integrates changes in membrane potential with G protein-coupled alterations in PKA and PKC activity (Cantrell and Catterall 2001).

6.3.2 Calcium and inositol trisphosphate (IP$_3$)

Calcium ions represent one of the most common and versatile second messengers found inside a cell. Changes in the cytosolic free Ca^{2+} concentration affect numerous intracellular events ranging from the short-term modulation of accessory proteins to long-lasting changes in gene expression. Besides crossing the cell membrane through voltage-gated channels from the outside, Ca^{2+} can enter the cytosol from internal stores in response to many environmental signals. Several G protein-linked receptors activate a PTX-insensitive G protein to stimulate the activity of membrane-bound phospholipase Cβ (PLCβ). This enzyme specifically cleaves phosphatidylinositol(4,5)bisphosphate (PIP$_2$) in the membrane to generate two important second messengers, diacylglycerol (DAG) and inositol 1,4,5-trisphosphate (IP$_3$) (Berridge 1993; Clapham 1995). IP$_3$ triggers Ca^{2+} release from internal stores by binding to and opening specific channels found along the surface of the endoplasmic reticulum (ER). Signals are transduced throughout the cell by stimulating the activity of various enzymes such as calcium-sensitive or calcium/calmodulin-dependent kinases and phosphatases, several of which modify the functions of ion channels. Protein kinase C (PKC) represents a well-known example of a kinase whose activity is partially dependent on the binding of Ca^{2+} to its regulatory domain. Another Ca^{2+}-sensitive kinase, calcium/calmodulin-dependent kinase II (CaM kinase II), is expressed at high levels throughout different regions of the central nervous system and has been linked to the phosphorylation of various channels and receptors (Schulman *et al.* 1995; Ghosh and Greenberg 1995; Soderling 1996). There is considerable evidence that activated CaMK II enhances AMPA receptor currents in dissociated cultures of hippocampal as well as spinal cord neurons (McGlade-McCulloh *et al.* 1993; Kolaj *et al.* 1994). The intracellular perfusion of dorsal horn neurons with an activated form of CaMK II leads to significant increases in AMPA-induced macroscopic currents which are not observed with denatured forms of the enzyme. Similar findings have been obtained with CA1 pyramidal neurons recorded in brain slice preparations (Lledo *et al.* 1995). This effect is thought to occur at least partially through the direct phosphorylation of glutamate receptors by CaMK II in response to Ca^{2+} elevations in the cell (McGlade-McCulloh *et al.* 1993; Tan *et al.* 1994).

6.3.3 Protein kinase C (PKC)

Like PKA and CaMK II, PKC is a serine/threonine kinase that is a direct target of G protein regulatory pathways. Four classical PKC isoforms (α, βI, βII, and γ) are activated by the binding of diacylglycerol (DAG) and Ca^{2+} to two stretches of conserved residues located within the kinase regulatory domain (Tanaka and Nishizuka 1994). Numerous channels have been implicated as targets of PKC phosphorylation, but only rarely have the specific enzyme isoforms involved in their modulation been identified. This is largely due to the current lack of isoform-specific pharmacological activators or blockers of these kinases. In rat cortical pyramidal neurons as well as hippocampal CA3 cells, the stimulation of PKC by phorbol esters depresses the inhibitory effects of glutamate on N-type Ca^{2+} channels, thereby enhancing synaptic transmitter release (Swartz *et al.* 1993). The effect can be blocked by the inclusion of a PKC-specific inhibitor (PKC 19–36) in the pipette. As outlined earlier, glutamate acts through metabotropic receptors to reduce Ca^{2+} currents *via* a $\beta\gamma$-dependent channel inhibition. The role of PKC in reversing the G$\beta\gamma$-mediated effect was examined more closely by Zamponi *et al.* (1997), who found that certain peptides comprising the cytoplasmic I–II linker region of the α_{1B} Ca^{2+} channel subunit acted as substrates for PKC phosphorylation *in vitro*. G$\beta\gamma$ subunits normally bind to this linker region on the native channel to inhibit its activity. The perfusion of non-phosphorylated peptides into cells that express recombinant N-type Ca^{2+} channel subunits was shown to interfere with G$\beta\gamma$-mediated channel inhibition (Zamponi *et al.* 1997). However, prior phosphorylation of these peptides by PKC decreased their abilities to interrupt this process, suggesting that kinase modulation antagonizes the binding of $\beta\gamma$ subunits to the peptides and to the corresponding regions of the channel.

L-type Ca^{2+} channels found in the heart are both positively and negatively regulated by PKC and often exhibit a biphasic response consisting of an initial current increase followed by a subsequent delayed decrease (Lacerda *et al.* 1988; Tseng and Boyden 1991; Satoh 1992). When the cardiac α_{1C} Ca^{2+} channel is expressed in *Xenopus* oocytes, currents are inhibited by phorbol ester-induced PKC stimulation after a brief initial increase (Bourinet *et al.* 1992). The kinase recognition motif on this channel has been mapped to a short stretch of amino acids within the amino terminus of the protein which is unique among members of the $Ca_V 1.2$ subfamily. Mutations of two critical serine residues in this region completely abrogate the regulatory effects of the kinase, providing clear evidence that the observed decline in Ca^{2+} current involves a direct action of PKC on the channel (McHugh *et al.* 2000).

6.4 G protein independent regulation of ion channels

Many G protein coupled receptors, including $\beta 2$ adrenergic receptors, exhibit G protein independent signal transduction (Chapter 7). This mechanism of effector activation/inhibition adds a further dimension to the mechanisms by which GPCRs regulate ion channel function. Thus, for example, the metabotropic glutamate receptor mGluR1 (Chapter 29) generates an excitatory postsynaptic current in hippocampal CA3 pyramidal neurones that arises because of a G protein independent activation of a src-family tyrosine kinase associated signalling pathway that, in turn, activates a non-specific cationic conductance (Heuss *et al.* 1999).

It is believed that src associates with mGluR1 either directly or *via* an adaptor protein similar to the situation in β2 adrenergic receptors where stimulated receptors form a complex with activated c-src bound to β-arrestin (a deactivator of G protein mediated signalling; Chapter 7). One possibility is that Homer proteins fulfil this function, another is that an SH3 domain on mGluR1 directly interacts with the tyrosine kinase. Other direct interactions of mGluRs with ion channels, such as the ligand gated ionotropic NMDA receptor, have also been suggested (Chapter 7).

6.5 Physiological integration of GPCR signalling cascades

This chapter has reviewed some basic concepts regarding the interactions that occur between seven-transmembrane receptors and ion-selective channels. Yet how do these pathways combine to control the patterns of excitability that underlie neuronal processing in the brain or the cardiac output of the heart? In this respect, a particularly fascinating aspect of receptor-mediated activation of ion channels lies in the range of combinatorial outputs that can be generated by the simultaneous stimulation of different receptors within the same cell. Numerous cells in the central nervous system are responsive to a large assortment of neurotransmitters that act upon G protein-linked receptors, many of which go on to modulate ion channel function. In several cases, these receptors converge upon common G proteins to initiate overlapping pathways, whereas in other instances the pathways diverge to generate independent signals. The large array of intracellular proteins that are affected by these signals allows for modulation at nearly all stages of the response (Chapters 5 and 7). Receptor subtype diversity adds yet another level of control (Chapters in Parts 2, 3, and 4)—cells respond differently to external signals depending not only on which agonists are present, but also on the types of receptors that the agonists can bind. Perhaps it is this enormous capacity for intervention at multiple steps of the signalling pathway that has made G protein-coupled signalling cascades the systems of choice for most neuroactive transmitters. In turn, understanding the crucial molecular mechanisms involved in physiological and pathophysiological functions remains an elusive challenge despite many years of ongoing investigation.

Taking the brain as a test case the metabotropic effects of several neurotransmitters have been most extensively studied in cells of the cerebral cortex and thalamus. These structures are densely innervated by projection fibres that originate in various brainstem nuclei and project diffusely across different areas of the brain. The addition of acetylcholine to many neocortical pyramidal cells causes a considerable reduction in spike-frequency adaptation, the gradual decrease in firing rate that normally follows a train of action potentials (Madison *et al.* 1987; McCormick and Williamson 1989). This effect is due in part to the inhibition of a Ca^{2+}-activated potassium current (I_{AHP}) which activates in response to calcium influx and prevents these cells from reaching their firing thresholds (Andrade and Nicoll 1987). The reduction in I_{AHP} thus enhances the effects of subsequent excitatory inputs and causes these cells to become more active. Acetylcholine also inhibits at least two other types of potassium currents in a variety of neuronal cells including the voltage-activated M-current (I_M) as well as a resting (leak) potassium conductance (Halliwell 1986; McCormick 1992). In all cases, decreases in these potassium currents gradually shift the transmembrane potential towards more positive, depolarized values and thereby facilitate the activation of voltage-sensitive channels that are open in this range.

Many thalamic neurons receive similar modulatory inputs from cells that originate in the brainstem. Projection fibres that emanate from cell bodies in the locus coeruleus provide noradrenergic innervation to various thalamic nuclei where they increase the excitability of cells in areas such as the lateral geniculate nucleus (LGN) and the nucleus reticularis (Rogawski and Aghajanian 1980; McCormick and Prince 1988). In thalamocortical relay cells, such changes in electrical responsiveness are correlated with different states of wakefulness and attention, and they also influence many aspects of sensory processing in the cortex (McCormick 1989; Swadlow and Gusev 2001). The modulation of membrane excitability in the brain is particularly important at excitatory synapses, where changes in the patterns of synaptic activity combine with changes in synaptic strength to affect the storage of new information during learning, or the reinforcement of synaptic connections during development (Bear and Malenka 1994). Future research should seek to find out more about how individual components of GPCR cascades such as those illustrated above collectively interact to modify membrane excitability and physiological/pathophysiological processes.

References

Albsoul-Younes AM, Sternweis PM, Zhao P, Nakata H, Nakajima S, Nakajima Y, and Kozasa T (2001). Interaction sites of the G protein beta subunit with brain G protein-coupled inward rectifier K+ channel. *J Biol Chem* **276**, 12 712–7.

Andrade R and Nicoll RA (1987). Pharmacologically distinct actions of serotonin on single pyramidal neurones of the rat hippocampus recorded *in vitro*. *J Physiol* **394**, 99–124.

Bean BP (1985). Two kinds of calcium channels in canine atrial cells. Differences in kinetics, selectivity, and pharmacology. *J Gen Physiol* **86**, 1–30.

Bean BP (1989). Neurotransmitter inhibition of neuronal calcium currents by changes in channel voltage dependence. *Nature* **340**, 153–6.

Bear MF and Malenka RC (1994). Synaptic plasticity: LTP and LTD. *Curr Opin Neurobiol* **4**, 389–99.

Berridge MJ (1993). Inositol trisphosphate and calcium signalling. *Nature* **361**, 315–25.

Bourinet E, Fournier F, Lory P, Charnet P, and Nargeot J (1992). Protein kinase C regulation of cardiac calcium channels expressed in Xenopus oocytes. *Pflugers Arch* **421**, 247–55.

Breitwieser GE and Szabo G (1985). Uncoupling of cardiac muscarinic and beta-adrenergic receptors from ion channels by a guanine nucleotide analogue. *Nature* **317**, 538–40.

Cachelin AB, de Peyer JE, Kokubun S, and Reuter H (1983). Ca2+ channel modulation by 8-bromocyclic AMP in cultured heart cells. *Nature* **304**, 462–4.

Canti C, Page KM, Stephens GJ, and Dolphin AC (1999). Identification of residues in the N terminus of alpha1B critical for inhibition of the voltage-dependent calcium channel by Gbeta gamma. *J Neurosci* **19**, 6855–64.

Cantrell *et al.* (1999). Voltage dependent neuromodulation of Na+ channels by D1-like dopamine receptors in rat hippocampal neurons. *J Neurosci* **19**, 5301–10.

Cantrell AR and Catterall WA (2001). Neuromodulation of Na+ channels: an unexpected form of cellular plasticity. *Nature Neurosci* **2**, 397–407.

Catterall WA (2000). Structure and regulation of voltage-gated Ca2+ channels. *Annu Rev Cell Dev Biol* **16**, 521–55.

Chen J, DeVivo M, Dingus J, Harry A, Li J, Sui J *et al.* (1995). A region of adenylyl cyclase 2 critical for regulation by G protein beta gamma subunits. *Science* **268**, 1166–9.

Clapham DE (1995). Calcium signaling. *Cell* **80**, 259–68.

Corey S, Krapivinsky G, Krapivinsky L, and Clapham DE (1998). Number and stoichiometry of subunits in the native atrial G-protein-gated K+ channel, IKACh. *J Biol Chem* **273**, 5271–8.

Costa MR, Casnellie JE, and Catterall WA (1982). Selective phosphorylation of the alpha subunit of the sodium channel by cAMP-dependent protein kinase. *J Biol Chem* **257**, 7918–21.

De Waard M, Liu H, Walker D, Scott VE, Gurnett CA, and Campbell KP (1997). Direct binding of G-protein betagamma complex to voltage-dependent calcium channels. *Nature* 385, 446–50.

Del Castillo J and Katz B (1955). Production of membrane potential changes in the frog's heart by inhibitory nerve impulses. *Nature* 175, 1035.

Diverse-Pierluissi M, Goldsmith PK, and Dunlap K (1995). Transmitter-mediated inhibition of N-type calcium channels in sensory neurons involves multiple GTP-binding proteins and subunits. *Neuron* 14, 191–200.

Dolphin AC (1995). The G.L. Brown Prize Lecture. Voltage-dependent calcium channels and their modulation by neurotransmitters and G proteins. *Exp Physiol* 80, 1–36.

Dolphin AC and Scott RH (1987). Calcium channel currents and their inhibition by (−)-baclofen in rat sensory neurones: modulation by guanine nucleotides. *J Physiol* 386, 1–17.

Doupnik CA, Davidson N, and Lester HA (1995). The inward rectifier potassium channel family. *Curr Opin Neurobiol* 5, 268–77.

Fischmeister R and Hartzell HC (1986). Mechanism of action of acetylcholine on calcium current in single cells from frog ventricle. *J Physiol* 376, 183–202.

Ford CE, Skiba NP, Bae H, Daaka Y, Reuveny E, Shekter LR *et al.* (1998). Molecular basis for interactions of G protein betagamma subunits with effectors. *Science* 280, 1271–4.

Forscher P, Oxford GS, and Schulz D (1986). Noradrenaline modulates calcium channels in avian dorsal root ganglion cells through tight receptor-channel coupling. *J Physiol* 379, 131–44.

Furukawa T, Miura R, Mori Y, Strobeck M, Suzuki K, Ogihara Y *et al.* (1998). Differential interactions of the C terminus and the cytoplasmic I–II loop of neuronal Ca2+ channels with G-protein alpha and beta gamma subunits. II. Evidence for direct binding. *J Biol Chem* 273, 17 595–603.

Gershon E, Weigl L, Lotan I, Schreibmayer W, and Dascal N (1992). Protein kinase A reduces voltage-dependent Na+ current in Xenopus oocytes. *J Neurosci* 12, 3743–52.

Ghosh A and Greenberg ME (1995). Calcium signaling in neurons: molecular mechanisms and cellular consequences. *Science* 268, 239–47.

Halliwell JV (1986). M-current in human neocortical neurones. *Neurosci Lett* 67, 1–6.

Hamill OP, Marty A, Neher E, Sakmann B, and Sigworth FJ (1981). Improved patch-clamp techniques for high-resolution current recording from cells and cell-free membrane patches. *Pflugers Arch* 391, 85–100.

Hartzell HC (1988). Regulation of cardiac ion channels by catecholamines, acetylcholine and second messenger systems. *Prog Biophys Mol Biol* 52, 165–247.

Hepler JR and Gilman AG (1992). G proteins. *Trends Biochem Sci* 17, 383–7.

Herlitze S, Garcia DE, Mackie K, Hille B, Scheuer T, and Catterall WA (1996). Modulation of Ca2+ channels by G-protein beta gamma subunits. *Nature* 380, 258–62.

Heuss C, Scanziani M, Gahwiler B, and Gerber U (1999). G-protein independent signalling mediated by metabotropic glutamate receptors. *Nature Neurosci* 2, 1070–7.

Hille B (1994). Modulation of ion-channel function by G-protein-coupled receptors. *Trends Neurosci* 17, 531–6.

Holz GGT, Rane SG, and Dunlap K (1986). GTP-binding proteins mediate transmitter inhibition of voltage-dependent calcium channels. *Nature* 319, 670–2.

Huang CL, Feng S, and Hilgemann DW (1998). Direct activation of inward rectifier potassium channels by PIP2 and its stabilization by Gbetagamma. *Nature* 391, 803–6.

Huang CL, Jan YN, and Jan LY (1997). Binding of the G protein betagamma subunit to multiple regions of G protein-gated inward-rectifying K+ channels. *FEBS Lett* 405, 291–8.

Huang CL, Slesinger PA, Casey PJ, Jan YN, and Jan LY (1995). Evidence that direct binding of G beta gamma to the GIRK1 G protein-gated inwardly rectifying K+ channel is important for channel activation. *Neuron* 15, 1133–43.

Hutter OF and Trautwein W (1956). Vagal and sympathetic effects on the pacemaker fibers in the sinus venosus of the heart. *J Gen Physiol* 39, 715–33.

Hutter OF and Trautwein W (1957). Effect of vagal stimulation on the sinus venosus of the frog heart. *Nature* **176**, 512–13.

Ikeda SR (1996). Voltage-dependent modulation of N-type calcium channels by G-protein beta gamma subunits. *Nature* **380**, 255–8.

Inanobe A, Morishige KI, Takahashi N, Ito H, Yamada M, Takumi T *et al.* (1995). G beta gamma directly binds to the carboxyl terminus of the G protein-gated muscarinic K+ channel, GIRK1. *Biochem Biophys Res Commun* **212**, 1022–8.

Ito H, Tung RT, Sugimoto T, Kobayashi I, Takahashi K, and Katada T (1992). On the mechanism of G protein beta gamma subunit activation of the muscarinic K+ channel in guinea pig atrial cell membrane. Comparison with the ATP-sensitive K+ channel. *J Gen Physiol* **99**, 961–83.

Jelacic TM, Kennedy ME, Wickman K, and Clapham DE (2000). Functional and Biochemical Evidence for G-protein-gated Inwardly Rectifying K+ (GIRK) Channels Composed of GIRK2 and GIRK3. *J Biol Chem* **275**, 36 211–16.

Jelacic TM, Sims SM, and Clapham DE (1999). Functional expression and characterization of G-protein-gated inwardly rectifying K+ channels containing GIRK3. *J Membr Biol* **169**, 123–9.

Kawano T, Chen L, Watanabe SY, Yamauchi J, Kaziro Y, Nakajima Y *et al.* (1999). Importance of the G protein gamma subunit in activating G protein-coupled inward rectifier K(+) channels. *FEBS Lett* **463**, 355–9.

Kofuji P, Davidson N, and Lester HA (1995). Evidence that neuronal G-protein-gated inwardly rectifying K+ channels are activated by G beta gamma subunits and function as heteromultimers. *Proc Natl Acad Sci USA* **92**, 6542–6.

Kolaj M, Cerne R, Cheng G, Brickey DA, and Randic M (1994). Alpha subunit of calcium/calmodulin-dependent protein kinase enhances excitatory amino acid and synaptic responses of rat spinal dorsal horn neurons. *J Neurophysiol* **72**, 2525–31.

Krapivinsky G, Gordon EA, Wickman K, Velimirovic B, Krapivinsky L, and Clapham DE (1995*a*). The G-protein-gated atrial K+ channel IKACh is a heteromultimer of two inwardly rectifying K(+)-channel proteins. *Nature* **374**, 135–41.

Krapivinsky G, Kennedy ME, Nemec J, Medina I, Krapivinsky L, and Clapham DE (1998). Gbeta binding to GIRK4 subunit is critical for G protein-gated K+ channel activation. *J Biol Chem* **273**, 16 946–52.

Krapivinsky G, Krapivinsky L, Wickman K, and Clapham DE (1995*b*). G beta gamma binds directly to the G protein-gated K+ channel, IKACh. *J Biol Chem* **270**, 29 059–62.

Kunkel MT and Peralta EG (1995). Identification of domains conferring G protein regulation on inward rectifier potassium channels. *Cell* **83**, 443–9.

Kurachi Y, Nakajima T, and Sugimoto T (1986). Acetylcholine activation of K+ channels in cell-free membrane of atrial cells. *Am J Physiol* **251**, H681–4.

Lacerda AE, Rampe D, and Brown AM (1988). Effects of protein kinase C activators on cardiac Ca2+ channels. *Nature* **335**, 249–51.

Lambright DG, Sondek J, Bohm A, Skiba NP, Hamm HE, and Sigler PB (1996). The 2.0 A crystal structure of a heterotrimeric G protein. *Nature* **379**, 311–9.

Li Y, Sternweis PM, Charnecki S, Smith TF, Gilman AG, Neer EJ *et al.* (1998). Sites for Galpha binding on the G protein beta subunit overlap with sites for regulation of phospholipase Cbeta and adenylyl cyclase. *J Biol Chem* **273**, 16 265–72.

Lipscombe D, Kongsamut S, and Tsien RW (1989). Alpha-adrenergic inhibition of sympathetic neurotransmitter release mediated by modulation of N-type calcium-channel gating. *Nature* **340**, 639–42.

Lledo PM, Hjelmstad GO, Mukherji S, Soderling TR, Malenka RC, and Nicoll RA (1995). Calcium/calmodulin-dependent kinase II and long-term potentiation enhance synaptic transmission by the same mechanism. *Proc Natl Acad Sci USA* **92**, 11 175–9.

Logothetis DE, Kim DH, Northup JK, Neer EJ, and Clapham DE (1988). Specificity of action of guanine nucleotide-binding regulatory protein subunits on the cardiac muscarinic K+ channel. *Proc Natl Acad Sci USA* **85**, 5814–8.

Logothetis DE, Kurachi Y, Galper J, Neer EJ, and Clapham DE (1987). The beta gamma subunits of GTP-binding proteins activate the muscarinic K+ channel in heart. *Nature* **325**, 321–6.

Madison DV, Lancaster B, and Nicoll RA (1987). Voltage clamp analysis of cholinergic action in the hippocampus. *J Neurosci* **7**, 733–41.

McCormick DA (1989). Cholinergic and noradrenergic modulation of thalamocortical processing. *Trends Neurosci* **12**, 215–21.

McCormick DA (1992). Neurotransmitter actions in the thalamus and cerebral cortex and their role in neuromodulation of thalamocortical activity. *Prog Neurobiol* **39**, 337–88.

McCormick DA and Prince DA (1988). Noradrenergic modulation of firing pattern in guinea pig and cat thalamic neurons, in vitro. *J Neurophysiol* **59**, 978–96.

McCormick DA and Williamson A (1989). Convergence and divergence of neurotransmitter action in human cerebral cortex. *Proc Natl Acad Sci USA* **86**, 8098–102.

McGlade-McCulloh E, Yamamoto H, Tan SE, Brickey DA, and Soderling TR (1993). Phosphorylation and regulation of glutamate receptors by calcium/calmodulin-dependent protein kinase II. *Nature* **362**, 640–2.

McHugh D, Sharp EM, Scheuer T, and Catterall WA (2000). Inhibition of cardiac L-type calcium channels by protein kinase C phosphorylation of two sites in the N-terminal domain. *Proc Natl Acad Sci USA* **97**, 12334–8.

Morad M, Sanders C, and Weiss J (1981). The inotropic actions of adrenaline on frog ventricular muscle: relaxing versus potentiating effects. *J Physiol* **311**, 585–604.

Osterrieder W, Brum G, Hescheler J, Trautwein W, Flockerzi V, and Hofmann F (1982). Injection of subunits of cyclic AMP-dependent protein kinase into cardiac myocytes modulates Ca2+ current. *Nature* **298**, 576–8.

Pfaffinger PJ, Martin JM, Hunter DD, Nathanson, NM, and Hille B (1985). GTP-binding proteins couple cardiac muscarinic receptors to a K channel. *Nature* **317**, 536–8.

Pragnell M, De Waard M, Mori Y, Tanabe T, Snutch TP, and Campbell KP (1994). Calcium channel beta-subunit binds to a conserved motif in the I–II cytoplasmic linker of the alpha 1-subunit. *Nature* **368**, 67–70.

Reuter H, Stevens CF, Tsien RW, and Yellen G (1982). Properties of single calcium channels in cardiac cell culture. *Nature* **297**, 501–4.

Reuveny E, Slesinger PA, Inglese J, Morales, JM, Iniguez-Lluhi JA, Lefkowitz RJ et al. (1994). Activation of the cloned muscarinic potassium channel by G protein beta gamma subunits. *Nature* **370**, 143–6.

Rogawski MA and Aghajanian GK (1980). Modulation of lateral geniculate neurone excitability by noradrenaline microiontophoresis or locus coeruleus stimulation. *Nature* **287**, 731–4.

Sakmann B, Noma A, and Trautwein W (1983). Acetylcholine activation of single muscarinic K+ channels in isolated pacemaker cells of the mammalian heart. *Nature* **303**, 250–3.

Satoh H (1992). Inhibition in L-type Ca2+ channel by stimulation of protein kinase C in isolated guinea pig ventricular cardiomyocytes. *Gen Pharmacol* **23**, 1097–102.

Schulman H, Heist K, and Srinivasan M (1995). Decoding Ca2+ signals to the nucleus by multifunctional CaM kinase. *Prog Brain Res* **105**, 95–104.

Shapiro MS and Hille B (1993). Substance P and somatostatin inhibit calcium channels in rat sympathetic neurons via different G protein pathways. *Neuron* **10**, 11–20.

Simon MI, Strathmann MP, and Gautam N (1991). Diversity of G proteins in signal transduction. *Science* **252**, 802–8.

Smith RD and Goldin AL (1997). Phosphorylation at a single site in the rat brain sodium channel is necessary and sufficient for current reduction by protein kinase A. *J Neurosci* **17**, 6086–93.

Soderling TR (1996). Modulation of glutamate receptors by calcium/calmodulin-dependent protein kinase II. *Neurochem Int* **28**, 359–61.

Soejima M. and Noma A (1984). Mode of regulation of the ACh-sensitive K-channel by the muscarinic receptor in rabbit atrial cells. *Pflugers Arch* **400**, 424–31.

Sui JL, Petit-Jacques J, and Logothetis DE (1998). Activation of the atrial KACh channel by the betagamma subunits of G proteins or intracellular Na+ ions depends on the presence of phosphatidylinositol phosphates. *Proc Natl Acad Sci USA* **95**, 1307–12.

Swadlow HA and Gusev AG (2001). The impact of 'bursting' thalamic impulses at a neocortical synapse. *Nat Neurosci* **4**, 402–8.

Swartz KJ, Merritt A, Bean BP, and Lovinger DM (1993). Protein kinase C modulates glutamate receptor inhibition of Ca2+ channels and synaptic transmission. *Nature* **361**, 165–8.

Tan SE, Wenthold RJ, and Soderling TR (1994). Phosphorylation of AMPA-type glutamate receptors by calcium/calmodulin-dependent protein kinase II and protein kinase C in cultured hippocampal neurons. *J Neurosci* **14**, 1123–9.

Tanaka C and Nishizuka Y (1994). The protein kinase C family for neuronal signaling. *Annu Rev Neurosci* **17**, 551–67.

Trautwein W and Hescheler J (1990). Regulation of cardiac L-type calcium current by phosphorylation and G proteins. *Annu Rev Physiol* **52**, 257–74.

Tseng GN and Boyden PA (1991). Different effects of intracellular Ca and protein kinase C on cardiac T and L Ca currents. *Am J Physiol* **261**, H364–79.

Tsien RW, Giles W, and Greengard P (1972). Cyclic AMP mediates the effects of adrenaline on cardiac purkinje fibres. *Nat New Biol* **240**, 181–3.

Velimirovic BM, Gordon EA, Lim NF, Navarro B, and Clapham DE (1996). The K+ channel inward rectifier subunits form a channel similar to neuronal G protein-gated K+ channel. *FEBS Lett* **379**, 31–7.

Wickman KD, Iniguez-Lluhl JA, Davenport PA, Taussig R, Krapivinsky GB, Linder ME *et al.* (1994). Recombinant G-protein beta gamma-subunits activate the muscarinic-gated atrial potassium channel. *Nature* **368**, 255–7.

Witcher DR, De Waard M, Liu H, Pragnell M, and Campbell KP (1995). Association of native Ca2+ channel beta subunits with the alpha 1 subunit interaction domain. *J Biol Chem* **270**, 18 088–93.

Yue DT, Herzig S, and Marban E (1990). Beta-adrenergic stimulation of calcium channels occurs by potentiation of high-activity gating modes. *Proc Natl Acad Sci USA* **87**, 753–7.

Zamponi GW, Bourinet E, Nelson D, Nargeot J, and Snutch TP (1997). Crosstalk between G proteins and protein kinase C mediated by the calcium channel alpha1 subunit. *Nature* **385**, 442–6.

Zhang JF, Ellinor PT, Aldrich RW, and Tsien RW (1996). Multiple structural elements in voltage-dependent Ca2+ channels support their inhibition by G proteins. *Neuron* **17**, 991–1003.

Chapter 7

Protein regulators of GPCR function

John R. Hepler and Julie A. Saugstad

7.1 Introduction

As described in previous chapters many extracellular ligands such as neurotransmitters, hormones, cytokines, and sensory input rely upon G protein-coupled receptors (GPCRs) and linked signalling pathways to exert their actions at target cells. The established dogma is that these extracellular stimuli utilize a receptor, a heterotrimeric guanine nucleotide binding protein (G protein αβγ subunit complex), and downstream effector proteins to transmit their signals across the plasma membrane (Hepler and Gilman 1992; Clapham and Neer 1997; Hamm 1998). Specifically, agonist occupancy of GPCRs causes exchange of GTP for GDP on Gα that activates Gα and stimulates its dissociation from the Gβγ complex. Free Gα-GTP and Gβγ then directly regulate the activity of well-defined effector proteins (e.g. adenylyl cyclases, phospholipases, ion channels) that, in the case of enzyme targets, generate second messenger molecules. In this model, second messengers activate signalling cascades that account for the many cellular actions initiated by the activated receptor. However, a wealth of new information highlights the limitations of this model and suggests that GPCRs utilize a growing list of additional protein binding partners to carry out their cellular actions (Bockaert and Pin 1999; Hall *et al.* 1999*a*). In some cases, these proteins regulate GPCR oligomerization, subcellular targeting, and pharmacology as well as the strength and duration of receptor-mediated signalling events. In other cases, these binding partners serve as previously unrecognized effectors or adaptors/scaffolds for other signalling proteins that propagate ligand-directed cellular signalling events independent of G proteins. This chapter focuses on the properties and cellular roles for these newly appreciated non-G protein binding partners of GPCRs.

7.2 Proteins regulating receptor desensitization

Agonist occupancy of GPCRs initiates activation of downstream signalling cascades that can often be followed by rapid attenuation of receptor responsiveness. This process of desensitization can result from a complex series of cellular protein interactions that mediate both acute and chronic receptor down-regulation. Desensitization can be either homologous where only the response of the activated receptor is desensitized or heterologous where activation of one receptor causes desensitization of others which stimulate/inhibit similar effector cascades. While the former involves decreased receptor G protein coupling the latter can also involve changes in the activity of effector molecules downstream of the G protein. Desensitization can occur rapidly within seconds to minutes where phosphorylation of the receptor and/or internalization of the receptor to an intracellular compartment occurs. Alternatively, desensitization can be long-term (down-regulation) involving changes in expression and stability of mRNA for GPCR cascades and changes in receptor/G protein levels. Indeed such long-term down-regulation can account for tolerance as a consequence of chronic drug treatment or progressive pathological state.

Classically, the β_2 adrenergic receptor has been used to study both homologous and heterologous desensitization. While the latter occurs through PKA phosphorylation of serine residues in intracellular loop 3 and the C-terminal tail, homologous desensitization occurs through a separate mechanism in which G protein-coupled receptor serine/threonine kinases (GRKs) phosphorylate the receptor. Depending on the receptor studied GRKs generally phosphorylate one or more serine and/or threonine residues in either the third intracellular loop or the C-terminal tail domains of the GPCR (Bunemann and Hosey 1999). No obvious consensus sequences are known to dictate which specific residues are preferred by GRK as phosphorylation sites *in vivo*. However, GRKs bind directly to GPCRs and following GRK phosphorylation of the GPCR, hydrophilic soluble protein arrestins are recruited from the cytosol to bind the phosphorylated sites within the third intracellular loop or the C-terminal tail. These arrestins form a stable complex with the receptor that prevents further G protein binding and promotes targeting of the GPCR/arrestin complex to specialized microdomains of the plasma membrane including clathrin coated pits and lipid rafts/caveolae thereby aiding internalization (endocytosis). Thus, GRKs work coordinately with arrestins to regulate the strength and duration of GPCR signalling events (Ferguson 2001). This regulation is complex in that many isoforms of GRK and arrestins exist. Thus, seven distinct isoforms of GRK (GRK1–7) have been identified ranging in size from 62–80 kDa. Whereas GRK1, 4, and 7 are limited in their tissue expression patterns to retinal rods and cones (GRK1 and 7, respectively) and testis (GRK4), GRK2, 3, 5, and 6 are found in many tissues (Pitcher *et al.* 1998*a*; Bunemann and Hosey 1999). GRKs share a conserved structure over the first 450 amino acids and within this region all family members contain an RGS-like domain and a catalytic domain. Outside of these domains, GRKs display highly divergent carboxy (C) termini that confer the capacity for different mechanisms of regulated membrane association (Penn *et al.* 2000). These include consensus sites for lipid acylation including farnesylation (GRK1), gerenyl gerenylation (GRK7), and palmitoylation (GRK4 and 6); a pleckstrin homology (PH) domain that promotes G$\beta\gamma$ binding (GRK2 and 3); phospholipid phosphatidyl inositol bisphosphate (PIP$_2$) binding (GRK2, 3, and 5); and a polycationic binding domain (GRK1, 5, and 6). Cellular mechanisms that regulate GRK recruitment from the cytosol to the plasma membrane to interact directly with GPCRs is described in detail elsewhere (Penn *et al.* 2000; Ferguson 2001). With respect to arrestins, four mammalian isoforms exist including arrestin-1, arrestin-2, β-arrestin-1, and β-arrestin-2 (Miller and Lefkowitz 2001). The visual arrestins are limited in their expression to rods and cones and block signalling by rhodopsin. In contrast, the β-arrestins are expressed in many cells and tissues and modulate the signalling and trafficking of numerous GPCRs (Miller *et al.* 2000). Given the many isoforms of arrestins and GRKs it is possible to encode specific interactions with different receptors. Thus, visual arrestin exhibits greatest affinity for phosphorylated rhodopsin with little affinity for β-adrenergic or muscarinic receptors. Furthermore, the level and rate of desensitization of a given receptor type can vary from cell type to cell type possibly because of differential expression of GRK and arrestin isoforms.

While the role of GRKs and arrestins in GPCR desensitization have been extensively studied and reviewed in greater detail elsewhere (Penn *et al.* 2000; Ferguson 2001; Miller and Lefkowitz 2001) it should also be pointed out that these proteins are multifunctional engaging additional binding partners in many more unsuspected GPCR signalling functions far beyond just receptor desensitization. These newly appreciated roles for GRKs and arrestins as adaptors or scaffolds that link GPCRs to signalling proteins and pathways are described in more detail below.

7.3 Proteins regulating receptor trafficking and clustering

The vast majority of GPCRs can undergo agonist-induced internalization into an endosomal compartment with or without bound ligand. This can occur very rapidly (e.g. 12% of muscarinic acetylcholine receptors in neuroblastoma cells are removed from the cell surface within a minute of agonist activation) or over a more prolonged time period. Once the agonist is removed from the GPCR the receptor can be recycled back to the plasma membrane. Classically, when the rate of internalization and recycling are at equilibrium 5% of receptors are turned over every minute. These dynamic changes in receptor expression at the cell surface provide a mechanism for processes such as: (i) short-term desensitization, (ii) initiation to long-term desensitization, (iii) retrograde signalling whereby internalized agonist receptor complex acts as a signalling mechanism transported down axons, (iv) dephosphorylation, and (v) morphological change. The mechanism by which receptors are internalized is via clathrin-coated pits and often involves cytoskeletal elements. Smooth pits have also been implicated for certain receptors, such as CCK (Chapter 14), although the relative importance of this mechanism versus clathrin coated pits is unclear. The structural requirements for internalization vary from receptor to receptor. Thus for β2 adrenoreceptors Tyr326 at the junction of the seventh TM domain and the C-terminal sequence is thought to be important but the analogous aromatic amino acid forming part of an 'NPXY' domain in GRP and angiotensin receptors is not implicated in this process. In general mutations in the second and third intracellular loops and C-terminal tail have been implicated in internalization with many producing subtle changes in the rates of internalization versus recycling rather than a clear cut on off modulation of one or both processes. Certainly, it is quite likely that cytoskeletal interactions, agonist binding kinetics and differential kinase phosphorylation can markedly affect internalization. Below is a description of some of the protein interactions that have a bearing on receptor clustering and trafficking:

7.3.1 GRKs

While established models suggest that GRKs interact only transiently with GPCRs to phosphorylate sites that promote arrestin binding, evidence suggests that GRKs may form a stable complex with GPCRs and other proteins. Recent studies indicate that GRKs engage a growing list of binding partners besides GPCRs, supporting the idea that GRKs are multifunctional proteins. Besides GPCRs, GRK isoforms have been shown to directly bind to the Ca^{++} binding proteins recoverin (GRK1) and calmodulin (GRKs2–6); caveolin (GRK2, others); actin (GRK5), tubulin (GRK2 and GRK5); synucleins (GRK5); and Gqα (GRK2) and Gβγ (GRK2 and GRK3) (for review see Penn et al. 2000 and references therein).

   In many cases, binding of these proteins to GRKs either directly stimulates or inhibits GRK catalytic activity. However, in other cases GRKs may provide a functional link between GPCRs and these signalling proteins. Caveolin is a scaffolding protein that is the principle structural protein of caveolae, that is, subsets of the plasma membrane specializations known as lipid rafts (Brown and London 1998). Caveolin can bind a remarkable variety of signalling proteins that concentrate in caveolae, including certain GPCRs (Okamoto *et al.* 1998). It is possible that GRKs may contribute to GPCR localization to caveolae by interacting with caveolin (Carman *et al.* 1999*a*). The cytoskeletal proteins tubulin, actin, and synucleins also directly interact with GRKs. Both α- and β-synucleins are phosphorylated by GRK5 (Pronin *et al.* 2000), whereas tubulin is phosphorylated by both GRK2 and GRK5 (Carman *et al.* 1998; Pitcher *et al.* 1998*b*). Although actin is not a substrate for GRK phosphorylation, it binds to the N terminus of GRK5 to inhibit its kinase activity (Freeman *et al.* 1998). Taken together, these observations raise the intriguing possibility that GRKs may physically link GPCRs to proteins of the cytoskeleton and consequently influence GPCR subcellular localization, either at the plasma membrane or during intracellular trafficking following receptor internalization.

7.3.2 Arrestins

New findings demonstrate that β-arrestins also act as adaptors that directly link GPCRs to proteins involved with endocytosis (Fig. 7.1a). Consistent with this idea, β-arrestin binding

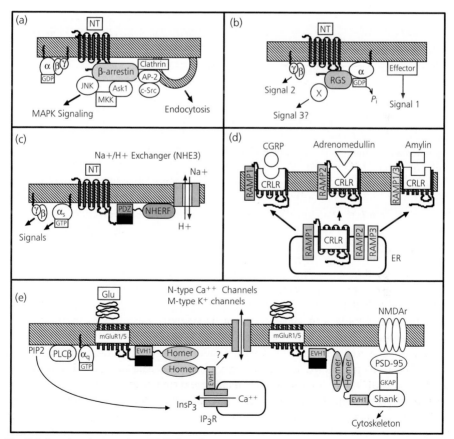

Fig. 7.1 Proteins that regulate GPCR functions.

to phosphorylated receptors is required for receptors to be targeted to internalized vesicles (Ferguson *et al.* 1996). Other studies demonstrate that β-arrestins bind the heavy chain of clathrin by way of a specific C-terminal binding motif (L-I-E-L/F) (Goodman *et al.* 1996, 1997) depending on the phosphorylation state of β-arrestin (Lin *et al.* 1997). β-arrestins also directly link GPCRs to other proteins involved in formation of endocytotic vesicles. Adapter protein-2 (AP-2) binds clathrin and is essential for GPCR targeting to coated pits and endocytic vesicles (Laporte *et al.* 1999). β-arrestins bind directly to AP-2, and complex formation between GPCRs, β-arrestin, AP-2 and clathrin are essential for targeting GPCRs to clathrin-coated pits (Laporte *et al.* 2000). Recent studies also indicate that certain GPCRs interact directly with the N-ethyl maleimide fusion protein (NSF), that is, an ATPase involved with intracellular membrane trafficking (McDonald *et al.* 1999). Taken together, these observations indicate that β-arrestins act as adaptors that directly link GPCRs to proteins important for endocytosis. Emerging models propose that, in certain cases, receptor/β-arrestin complex formation and interaction with clathrin and AP-2 is essential for targeted GPCR localization and intracellular trafficking (Miller *et al.* 2000).

7.4 **Protein regulation of receptor signalling**

Besides roles in regulating desensitization and receptor clustering/trafficking, many regulatory proteins modify the signal transduction pathways initiated/inhibited by GPCRs. Below we consider the roles of GRKs, arrestins and RGS proteins as exemplars of such regulators.

7.4.1 **GRKs**

In addition to their roles in GPCR desensitization and receptor clustering/trafficking, GRKs also bind directly to G protein subunits. GRK2 and GRK3 each contain C-terminal PH domains, which bind Gβγ subunits (Pitcher *et al.* 1992; Carman *et al.* 2000). While Gβγ association is an established mechanism for recruiting GRKs to the plasma membrane to regulate GPCR function, it is possible that GRK2 and GRK3 may also form a stable complex with GPCRs and Gβγ to modulate receptor-directed G protein signalling. GRK2 also binds to Gqα through its RGS-like domain (Carman *et al.* 1999*b*). RGS domains are present in many proteins and bind directly with activated Gα subunits, and all GRKs contain an RGS-like domain. While the functional consequences of Gα interactions with GRK are unclear at present, it is possible that selective GRK/Gα complex formation could either dictate which receptors GRKs modulate, contribute to desensitization of G protein signalling, and/or propagate as yet unrecognized ligand and GPCR-directed signalling events.

7.4.2 **Arrestins**

Activation of many GPCRs stimulates cell growth and differentiation in a variety of physiological and experimental systems (Gutkind 1998; Pierce *et al.* 2001). In many cases, these actions are correlated with receptor capacity to stimulate mitogen-activated protein kinase (MAPK) signalling pathways. Although mechanisms responsible for GPCR-directed stimulation of MAPK pathways are poorly understood, recent studies demonstrate that β-arrestins serve as adaptor/scaffolding proteins which form stable complexes with certain GPCRs and various intracellular non-receptor kinases (Miller and Lefkowitz 2001; Pierce *et al.* 2001). β-arrestins recruit members of the c-Src tyrosine kinases to β-adrenergic receptors (Luttrell *et al.* 1999; Miller *et al.* 2000) and neurokinin-1 receptors (DeFea *et al.* 2000*a*). Similarly, β-arrestins recruit the Src family kinases c-Fgr and Hck to the interleukin CXCR-1 receptors (Barlic *et al.* 2000). In each of these cases, β-arrestin-mediated complex formation confers kinase activity, albeit indirectly, to associated receptors. Thus, GPCRs can signal similarly to other non-G protein-coupled growth factor receptors with intrinsic tyrosine kinase catalytic activity.

β-arrestins also complex directly with kinases that comprise MAPK signalling cascades, in particular those of the extracellular regulated kinases (ERKs) and c-jun N terminal kinases (JNKs) (Fig. 7.1a). Diverse upstream signals activate raf-1 kinase that phosphorylates and activates MEK1/2 kinases. Activated MEK1/2 then phosphorylate and activate ERK1/2 kinases leading to transcriptional regulation of target genes involved in cell growth and differentiation (Pearson *et al.* 2001). In a separate but parallel MAPK cascade, upstream activation of the serine/threonine kinase ASK1 leads to phosphorylation and activation of MKK4/7 kinases which, in turn, phosphorylate and activate JNK kinases (JNK1–3). Recent studies indicate that certain GPCR/β-arrestin complexes bind cognate proteins of both the ERK and JNK pathways. Agonist-occupied protease-activated receptors (PAR-2) and angiotensin II type 1a receptors (AT1aR) each form stable complexes with β-arrestins, raf-1 and Erk2 (DeFea *et al.* 2000*b*). In the case of AT1aR, MEK1 also complexes with β-arrestin. Similarly, AT1a1 receptors can complex with β-arrestin, ASK1 and JNK3, and stimulation of AT1aR leads to activation of JNK3 (McDonald *et al.* 2000). These findings demonstrate that β-arrestins are hormone and GPCR-regulated scaffolding proteins for certain branches of the ERK and JNK signalling cascades.

Collectively, these observations indicate that β-arrestins serve as adaptor proteins that physically and functionally link GPCRs to components of both the endocytotic pathways (clathrin, AP-2, NSF) and MAPK signalling cascades. Emerging models propose that following agonist activation GRKs are recruited to the plasma membrane to phosphorylate receptors. In turn, β-arrestins are recruited to bind GRK-phosphorylated GPCRs to form a stable complex that is recognized by clathrin and AP-2 for targeted internalization. This stable GPCR/β-arrestin complex can also serve as a platform to assemble raf-1, MEK1, and ERK2 or, alternatively, ASK1, MKK4, and JNK3 (Fig. 7.1a). As such, β-arrestins apparently act in a manner analogous to the cytosolic scaffolding proteins MP-1 and JIP that assemble kinases involved with ERK and JNK signalling, respectively (Schaeffer *et al.* 1998; Yasuda *et al.* 1999). Such assemblies provide a cellular mechanism for promoting and localizing highly efficient MAPK signalling events. GPCR/β-arrestin complexes allow a novel mechanism, whereby MAPK may phosphorylate substrates at cellular sites such as the plasma membrane or internalized vesicles at various temporal and spatial stages of sorting.

7.4.3 **RGS**

Other proteins besides β-arrestins and GRKs contribute to GPCR desensitization and couple receptors directly to novel signalling proteins and pathways. The regulators of G protein signalling (RGS proteins) are a family (>30 mammalian isoforms) of multifunctional signalling proteins that bind Gα subunits to modulate linked signalling events (De Vries *et al.* 2000; Ross and Wilkie 2000; Zhong and Neubig 2001). RGS proteins contain a conserved 130 amino acid core domain (RGS domain) that directly binds activated Gα-GTP subunits and serves as a GTPase-activating protein (GAP) and/or effector antagonists for Gα. In reconstitution assays using purified proteins, most RGS proteins block the function of Giα and Gqα family members whereas certain family members are selective regulators of G12/13α; no

RGS proteins have yet been reported that directly regulate Gsα functions. As a class, proteins with RGS domains negatively regulate receptor-directed G protein signalling.

Aside from their shared capacity to block G protein signalling, RGS proteins differ widely in their overall size and amino acid identity, and possess a remarkable variety of structural domains and motifs (Hepler 1999; Siderovski *et al.* 1999). Based on RGS domain amino acid identities and comparative overall structural and functional similarities, RGS proteins have been classified into six distinct subfamilies (RZ, R4, R7, R12, RA, and RL) (Ross and Wilkie 2000). Members of the RZ and R4 subfamilies, with a few exceptions, are small 20–30 kDa proteins that contain short amino (N) and C-terminal regions flanking the RGS domain. In contrast, the R7, R12, RA, and RL subfamily members, with a few exceptions, are much larger proteins (up to 160 kDa) that possess longer N- and C termini flanking the RGS domain that contain various binding domains and motifs for other proteins. Among the reported non-Gα binding partners for various RGS proteins are Gβ5 subunits, activated forms of the small G proteins rho and rap1/2, 14 : 3 : 3, adenomous polyposis coli, β-catenin, glycogen synthase kinase-3β, coatomer proteins, as well as others (for review see De Vries *et al.* 2000; Ross and Wilkie 2000). Thus RGS proteins serve both as inhibitors of Gα-directed signalling events and as scaffold/effector proteins that link activated Gα subunits to diverse signalling proteins and pathways.

Although RGS proteins bind activated Gα-GTP subunits, they may interact directly with GPCRs to modulate their functions (Fig. 7.1b). RGS12 is alternatively spliced (Chatterjee and Fisher 2000) and longer variant forms contain an N terminal PDZ domain that recognizes specific binding motifs on target proteins. Many GPCRs contain PDZ binding motifs at their C termini, and a screen of receptor C termini revealed that the PDZ domain of RGS12 binds to a specific PDZ binding motif of the interleukin CXCR2 receptor, but not other receptors (Snow *et al.* 1998*a*). This observation suggests that certain GPCRs serve active roles in recruiting RGS proteins to regulate function of linked G proteins.

RGS protein binding to both GPCRs and G proteins may contribute to signalling selectivity. Accumulated evidence indicates that most, but not all, RGS proteins exhibit a surprising lack of selectivity for the target Gα *in vitro*. With few exceptions, most RGS proteins are GAPs for most Giα family members and/or Gq/11α. How then does an RGS protein decide which Gα to selectively regulate in a cellular context? Recent studies provide compelling evidence that RGS/G protein interactions are dictated by linked GPCR in cells. Wilkie and co-workers demonstrated that purified RGS1, RGS4, and RGS16, when introduced directly into cells, selectively inhibited inositol lipid/Ca^{++} signalling by the m3 muscarinic cholinergic receptor (m3ChoR) compared with the CCK receptor (Xu *et al.* 1999); in stark contrast, RGS2 displayed no preference. Receptor selectivity of RGS4 is conferred by its N-terminal domain since truncated RGS4 lacking this domain exhibited no receptor-selectivity. However, RGS4 selectivity for m3ChoR was restored by combined addition of the N terminus and the RGS core domain (Zeng *et al.* 1998).

Additional evidence for direct RGS/GPCR interactions has come from genetic studies in *Caenorhabditis elegans*. The worm RGS proteins Eat-16 and Egl-10 and members of the mammalian R7 family of RGS proteins (RGS7, RGS6, RGS9, RGS11) each contain an RGS domain, a poorly understood DEP (dishevelled, egl-10, pleckstrin) domain, and a highly unusual G protein gamma-like (GGL) domain. GGL domains specifically bind Gβ5 subunits in mammals or the equivalent (GPB-2) in worms (Cabrera *et al.* 1998; Snow *et al.* 1998*b*; Hajdu-Cronin *et al.* 1999). Gβ5 is unusual among identified mammalian Gβ subunits in that it is the only isoform localized to the cytosol and it shares only 53% amino acid identity with the other homologous Gβ subunits (β1–4) (Watson *et al.* 1996). By forming a complex, RGS/Gβ5 resemble conventional G protein Gβγ subunits with an RGS domain attached. In *Caenorhabditis elegans*, serotonin receptors linked to worm Goα at the neuromuscular junction stimulate egg-laying behaviour, and muscarinic receptors linked to Gq oppose these actions. Muscarinic agonists and Egl-10/GPB-2 selectively block Goα actions, whereas serotonin and Eat-16/GPB-2 selectively block Gq actions (Hajdu-Cronin *et al.* 1999). Speculative models (Sierra *et al.* 2000) propose that serotonin receptors and Goα complex with Eat-16/GPB-2 and that serotonin-directed receptor activation releases Eat-16 to block Gq actions. In reciprocal fashion, muscarinic-directed dissociation of Egl-10/GPB-2 from Gqα blocks Goα actions. Consistent with this idea, mammalian RGS9/Gβ5 complex substitutes for conventional Gβγ in supporting Gα/β5/RGS coupling to receptors in reconstitution systems using purified proteins (T.K. Harden and A.G. Gilman, personal communication).

Collectively, these findings predict that certain RGS proteins serve as multifunctional modulators of GPCR signalling. Direct RGS coupling to GPCRs may dictate which G proteins and linked signalling pathways are regulated. Formation of a stable GPCR/Gα/RGS complex could, in fact, facilitate more efficient and finely tuned signalling since the GAP activity of RGS would allow rapid exchange and rebinding of GTP on Gα to sustain, rather than inhibit the signalling event (Ross and Wilkie 2000). Finally, RGS proteins may serve as scaffolding proteins that assemble multiple signalling proteins in a complex to switch GPCR and G protein signalling between classical second messenger pathways and novel signalling pathways.

7.5 Regulatory roles of novel GPCR binding proteins

7.5.1 Homer/Vesl

In recent years, a family of proteins has emerged that facilitate protein interactions via their PDZ (*PSD-95/Discs large/ZO-1*) domains and that function to colocalize receptors and ion channels with their downstream signalling components. Homer proteins contain a PDZ-like EVH1 domain that interacts with metabotropic glutamate receptors (mGluRs) (Fig. 7.1e). Homer1a (H1a), also known as *Vesl* (*VASP/Ena-related gene up-regulated during seizure and LTP*) was initially identified as an immediate early gene that is dynamically regulated by synaptic activity and that specifically binds to the C-terminal tail of the Gq-coupled mGluRs 1 and 5 (Brakeman *et al.* 1997; Kato *et al.* 1998). Homer proteins have been isolated from *Drosophila*, rat, mouse, and human brain, (Kato *et al.* 1998; Xiao *et al.* 1998) and contribute to a number of synaptic functions (for review see Xiao *et al.* 2000). With the exception of H1a, a rapidly inducible form of Homer, all other Homers are constitutively expressed. They contain a leucine zipper motif and a coiled-coil (CC) region in the C-terminal domain, while the N-terminal domain contains a single PDZ motif that is located within a region with homology to the EVH1/WH1 (Ena/VASP homology 1/WASP homology 1). Constitutively expressed Homers multimerize via their CC domain, while the N-terminal domains bind simultaneously to mGluR C-terminal domains and inositol trisphosphate (IP3) receptors (Tu *et al.* 1998). Induction of H1a, a short form of Homer lacking the CC region leads to the uncoupling of mGluR/Homer complexes from IP3 receptors and an apparent loss of receptor activity.

The CC domains of Homer can also bind to Shank, a family of scaffold proteins that interact with a variety of membrane and cytoplasmic proteins at postsynaptic sites of excitatory synapses (see Sheng and Kim 2000). The three members of the Shank family (1, 2, and 3) contain a PDZ domain, a Src homology 3 (SH3) domain, ankyrin repeats, a proline-rich region and a sterile alpha motif (SAM) domain and thus are designed for maximizing complex formations between unrelated proteins. For example, the proline-rich region of Shank binds directly to Homer, the PDZ domain of Shank interacts with GKAP, and the SAM domain of Shank interacts with the actin cytoskeleton (Fig. 7.1e). GKAP binds to PSD95 that in turn binds to the NMDA receptor, thus Shank serves as central scaffolding proteins that indirectly links mGluRs to NMDA receptors and the cytoskeletal matrix. However, the calcium-independent α-latrotoxin receptor CL1, a seven transmembrane domain GPCR, has been shown to directly interact with Shank1 (Tobaben *et al.* 2000) and suggests that other GPCRs can directly bind to Shank as well.

Homer also couples mGluRs to other intracellular effectors; the constitutive and inducible forms of Homer can compete to regulate the coupling of group I mGluRs to N-type calcium and M-type potassium channel currents in neurons (Kammermeier *et al.* 2000). Moreover, stimulation of the NMDA receptor or application of brain-derived neurotrophic factor selectively up-regulate H1a mRNA via the MAPK cascade (Sato *et al.* 2001). In addition, the direct binding of Homer to the C-terminal tail of the mGluRs controls the constitutive activity observed for mGluRs 1a and 5 when overexpressed in heterologous cells, and expression of endogenous H1a in neurons induces the constitutive activity of mGluR1a or 5 indicating that they can directly activate the receptors independent of agonist activation (Ango *et al.* 2001). Thus, these studies indicate that Homer proteins can directly couple to mGluRs to regulate receptor responses in neurons, and that either synaptic activity, extracellular stimuli or intracellular proteins can all regulate mGluR responses by altering their interactions with Homer and ultimately their interaction with downstream effectors. Homer 1b (H1b) is a constitutively expressed long form of Homer that has been shown to regulate the trafficking and surface expression of group I mGluRs (Roche *et al.* 1999). In these experiments either mGluR5 alone or mGluR5 coexpressed with H1a resulted in the insertion of mGluR5 into the plasma membrane. However, coexpression of mGluR5 with H1b inhibited the surface expression of mGluR5 and caused it to be retained in the endoplasmic reticulum. This inhibition required a direct interaction between mGluR5 and H1b and revealed a novel mechanism for regulating synapse-specific receptor expression (Roche *et al.* 1999). Further studies revealed that mGluR5 is exclusively localized in cell bodies when transfected alone in cultured cerebellar granule cells, but is localized to dendrites when coexpressed with H1b or 1c, and is found in both dendrites and axons when cotransfected with H1a (Ango *et al.* 2000). In addition, depolarization of neurons induced a transient expression of H1a that resulted in persistent neurite localization of mGluR5. These studies show that Homer proteins can target mGluR5 to axons or dendrites in response to neuronal activity and thus may play important roles in defining the molecular makeup of neurons and/or glia in response to synaptic plasticity.

7.5.2 **Cupidin**

Cupidin is an isoform of Homer/Vesl that links Gq-coupled mGluRs to the actin cytoskeleton in mouse cerebellar granule cells (Shiraishi *et al.* 1999). The N-terminal domain of Cupidin binds to the C-terminal tail of mGluRs and also interacts with actin. The C-terminal portion of Cupidin binds Cdc42, a member of the Rho family of small GTPases, in a GTP-dependent manner. Thus Cupidin serves as a post-synaptic scaffold protein linking mGluRs to both the actin cytoskeleton and Rho proteins, and also suggests that Cupidin may be involved in the dynamic morphological changes that occur during synapse formation or plasticity in neuron.

7.5.3 **PICK1**

PICK1 (*protein interacting with C kinase*) is a PDZ-containing protein that binds to mGluR7a, a presynaptic autoreceptor (Boudin *et al.* 2000). Coexpression of PICK1 and mGluR7a in heterologous expression systems results in the aggregation of mGluR7a. In addition, PICK1 causes a reduction in PKCα-evoked phosphorylation of mGluR7a in *in vitro* phosphorylation assays (Dev *et al.* 2000) and thus modulates the desensitization of mGluR7a responses. Interestingly, calmodulin (CaM) and Gβγ subunits bind directly to a sudomain of the C-terminal tail of mGluR7 in a mutually exclusive manner (Nakajima *et al.* 1999; O'Connor *et al.* 1999). This binding is calcium-dependent and occurs on the mGluR7 site that is phosphorylated by PKC. Thus CaM binding is prevented by PKC phosphorylation and conversely, phosphorylation of the receptor is inhibited by CaM binding (Nakajima *et al.* 1999). In addition, CaM binding promotes dissociation of Gβγ from the receptor making Gβγ available to inhibit voltage-dependent calcium channels (O'Connor *et al.* 1999) and resulting in the inibition of glutamate release from the terminal.

7.5.4 **NHERF**

The C-terminal tail of β2-Adrenergic receptors (β2-ARs) are directly bound by the Na^+/H^+ exchanger regulatory factor (NHERF), a protein that regulates the activity of the Na^+/H^+ exchanger type 3 (NHE3) and subsequently regulates cellular pH (Hall *et al.* 1998b) (Fig. 7.1c). While NHERF does not directly regulate β2-ARs *per se*, it does directly link these receptors to the G protein-independent modulation (inhibition) of NHE3. Binding of NHERF to β2-ARs and NHE3 is agonist-dependent but is independent of receptor-mediated activation of adenylyl cyclase. NHERF contains two PDZ domains, the first of which is required for binding to the amino acid motif D-S/T-x-L located at the C-terminal tail of β2-ARs. This amino acid binding motif is well conserved in other GPCRs and ion channels, and NHERF was subsequently shown to bind to the purinergic P2Y1 receptor (D-T-S-L) and the cystic fibrosis transmembrane conductance regulator (D-T-R-L) as well (Hall *et al.* 1998a). NHERF also associates and is constitutively phosphorylated by GRK6A on Ser289, but does not associate with GRK6B or GRK6C, alternatively spliced variants that differ from GRK6A at their extreme C termini (Hall *et al.* 1999b).

7.5.5 **InaD**

Phototransduction is the fastest known GPCR cascade taking just tens of milliseconds to go from light activation of rhodopsin to the generation of a receptor potential, and less than 100 ms to shut off after the stimulus is terminated (Ranganathan and Stevens 1995). In the *Drosophila* photoreceptor system, the *inaD* gene encodes a five PDZ domain protein that assembles different components of the phototransduction system including the transient receptor potential channel (TRP), eye-protein kinase C (eye-PKC), and phospholipase C-β (PLC-β). Null *inaD* mutations that occur in the PDZ domains of InaD result in a dramatic reorganization of signalling molecules and a total loss of transduction complexes (Tsunoda *et al.* 1997), suggesting that InaD is essential for coordination of the components involved in the activation (PLC-β and TRP) and the deactivation (eye-PKC) of phototransduction.

7.5.6 **Polyproline-containing proteins**

Several GPCRs contain polyproline motifs within their intracellular domains that mediate protein interactions with SH3 domains. For example, the third intracellular loop of β1-adrenergic receptors (β1-ARs) and β2-ARs are highly conserved except that β1-ARs contain a 24 amino acid polyproline motif in the middle of the loop. Thus this motif may in part be responsible for the differences in signalling exhibited by β-ARs. Protein interaction studies revealed that SH3p4/p8/p13, also known as endophilin 1/2/3, binds to the third intracellular loop of the β1-AR, but not the β2-AR (Tang *et al.* 1999). This interaction is mediated by binding of the C-terminal SH3 domain of SH3p4 to the β1-AR, and overexpression of SH3p4 promotes agonist-induced internalization and decreases the Gs coupling efficacy of β1-ARs in HEK293 cells, while having no effect on β2-ARs. Polyproline motifs are found in several other GPCRs including the β3-AR, the β2A-AR, and dopamine D4 receptors suggesting a previously unrecognized role for SH3-containing proteins in the regulation of GPCR signalling.

7.5.7 **CortBP1**

Yeast two-hybrid studies show that the C-terminal tail of the somatostatin2A (sst2A) receptor binds to cortactin-binding protein 1 (CortBP1; Zitzer *et al.* 1999*b*) also termed proline rich synapse associated protein, ProSAP1 (Boeckers *et al.* 1999) or Shank2. CortBP1 is a PDZ-containing protein that binds to the SH3 domain of cortactin, an actin-binding protein involved in the restructuring of the cortical actin cytoskeleton. The C-terminal tail of the sst2A receptor binds to the PDZ domain of CortBP1 that in turn binds to cortactin and forms a scaffold between the receptors and the cytoskeleton. Coexpression of both proteins results in a redistribution of CortBP1 to the cell membrane that is enhanced when cells are treated with somatostatin receptor inducing factor (SRIF) suggesting that agonist activation of the receptor increases the accessibility of the C-terminal domain for the CortBP1 PDZ domain. Using the same approach as above a human somatostatin receptor interacting protein (SSTRIP), a protein that has strong homology with the rat protein Shank1/Synamon was also shown to interact with sst2A (Zitzer *et al.* 1999*a*). In addition, the sst1 receptor was shown to interact with the human homologue of Skb1 (Skb1Hs; Schwarzler *et al.* 2000) and while the function of Skb1Hs is not fully understood the interaction required the full C-terminal sequence of sst1. In addition, cotransfection of both proteins in HEK cells resulted in a higher number of binding sites for SRIF suggesting that Skb1Hs may aid in targeting the sst1 receptor to the cell surface (Schwarzler *et al.* 2000). Thus CortBP1, SSTRIP, and Skb1Hs define a novel family of multidomain proteins containing a PDZ domain, ankyrin repeats, SH3 binding regions and a SAM domain, and are involved in the targeting and localization of somatostatin receptors to the plasma membrane.

7.6 **Accessory proteins that regulate GPCR expression, pharmacology, and localization**

7.6.1 **RAMPs**

GPCRs are monomeric, seven transmembrane spanning proteins, the majority of which are easily expressed in heterologous expression systems. However, a small subset of GPCRs exist that are resistant to expression and/or activation. The reason for this resistance was unclear until recently when new studies identified a family of accessory proteins that are necessary for the expression of some GPCRs, RAMPS (receptor activity modifying proteins) (Fig. 7.1d). RAMPs are predicted to exert their effects on receptor expression in three ways: they transport receptors to the surface of the plasma membrane, they define the pharmacology of receptors, and they determine the glycosylation state of receptors (for review, see Foord SM 1999). For example, the calcitonin receptor-like receptors (CRLRs) require the coexpression of RAMP1, 2, or 3 to yield CGRP-, adrenomedullin-, or amylin-activated receptors, respectively (Buhlmann et al. 1999; McLatchie LM 1998; Muff et al. 1999). Structural studies reveal that the N-terminal domain of RAMPs is the critical determinant for both the glycosylation state and the ligand-binding site of CRLRs (Fraser et al. 1999). In addition, RAMP1 directly interacts with the receptor and defines the ligand recognition sites (Leuthäuser et al. 2000). While studies on the role of RAMPs in regulating receptor expression have mainly been directed at the calcitonin family of receptors, RAMP mRNAs are very widely expressed thus their role in regulating the expression of other GPCRs remains to be determined. For example, mGluR7 is a GPCR curiously resistant to expression and/or activation in heterologous expression systems (Okamoto et al. 1994; Saugstad et al. 1994) and thus may require the coexpression of a RAMP for optimal activation.

7.6.2 Other accessory proteins

Accessory proteins are cell-specific biological factors that are necessary for the correct localization of GPCRs to the plasma membrane and are distinct from RAMPS in that they are not involved in defining the pharmacological profile of GPCRs. The first description of accessory proteins came about from studies on opsins in which ninaA, a *Drosophila melanogaster* photoreceptor cyclophilin, was shown to be required for the localization of a subset of opsins to the rhabdomeres (Colley et al. 1991). NinaA and its mammalian homologue RanBP2 bind to specific opsins and act as chaperone proteins involved in the correct folding or transport of the receptor to the plasma membrane.

Another family of GPCRs that require cell-specific factors for their correct localization and expression is olfactory receptors. These receptors are extremely difficult to express in heterologous systems and only traffic to the plasma membrane in mature olfactory receptor neurons. For example, in *Caenorhabditis elegans*, the *odr-10* gene encodes a seven transmembrane domain olfactory receptor that is inefficiently transported to the plasma membrane when expressed in heterologous cells (Zhang et al. 1997). This and other similar observations of inefficient expression of mammalian olfactory receptors led to the identification of the *odr-4* and *odr-8* genes in *Caenorhabditis elegans* (Dwyer et al. 1998). ODR-4 encodes a novel 445 amino acid protein that contains one C-terminal transmembrane domain, is expressed in chemosensory neurons and facilitates odorant receptor folding and/or trafficking to the plasma membrane. The molecular mechanism(s) involved in this facilitation are not known but several models have been proposed based on experimental observations. ODR-4 could serve as a chaperone protein to aid in protein folding, it could be necessary for sorting odorant receptors into the correct secretory vesicles, or it could be involved in targeting receptor-containing vesicles to the cilia (Dwyer et al. 1998). A more recent model has emerged in which olfactory receptor trafficking to the plasma membrane involves a two-step process in that one accessory protein may be necessary to release olfactory receptors from the endoplasmic reticulum to the golgi apparatus and/or endosomes, and a second accessory protein may be required for olfactory receptor release from the golgi/endosomes to the plasma membrane (Gimelbrant *et al.* 2001).

7.7 Conclusion

Emerging models describing how GPCRs function to transduce signals from the outside to the inside of the cell have greatly expanded in the last decade as new information high-lights the limitations of established models. While the tripartite model including the GPCRs, G proteins, and effectors remains intact, it is apparent that GPCRs utilize a growing list of additional protein binding partners to carry out their cellular actions. Newly recognized binding partners serve to regulate GPCR oligomerization, pharmacology, and subcellular targeting, as well as determine the strength and duration of receptor-mediated signalling events. In other cases, these binding partners serve as previously unrecognized effectors or adaptors/scaffolds for other signalling proteins that propagate hormone-directed cellular sig-nalling events independent of G proteins. The interacting and regulatory proteins included in this chapter appear to be the first examples of many yet to be discovered. A full under-standing of how these proteins interact with one another to modulate GPCR signalling in the CNS remains a formidable and exciting challenge for future research.

References

Ango F, Pin JP, Tu JC, Xiao B, Worley PF, Bockaert J, and Fagni L (2000). Dendritic and axonal targeting of type 5 metabotropic glutamate receptor is regulated by homer1 proteins and neuronal excitation. *J Neurosci* 20, 8710–6.

Ango F, Prezeau L, Muller T, Tu JC, Xiao B, Worley PF *et al.* (2001). Agonist-independent activation of metabotropic glutamate receptors by the intracellular protein Homer. *Nature* 411, 962–5.

Barlic J, Andrews JD, Kelvin AA, Bosinger SE, DeVries ME, Xu L *et al.* (2000). Regulation of tyrosine kinase activation and granule release through beta-arrestin by CXCRI. *Nat Immunol* 1, 227–33.

Bockaert J and Pin JP (1999). Molecular tinkering of G protein-coupled receptors: an evolutionary success. *EMBO J* 18, 1723–9.

Boeckers TM, Kreutz MR, Winter C, Zuschratter W, Smalla KH, Sanmarti-Vila L *et al.* (1999). Proline-rich synapse-associated protein-1/cortactin binding protein 1 (ProSAP1/CortBP1) is a PDZ-domain protein highly enriched in the postsynaptic density. *Ann Anat* 183, 101.

Boudin H, Doan A, Xia J, Shigemoto R, Huganir RL, Worley P *et al.* (2000). Presynaptic clustering of mGluR7a requires the PICK1 PDZ domain binding site. *Neuron* 28, 485–97.

Brakeman PR, Lanahan AA, O'Brien R, Roche K, Barnes CA, Huganir RL *et al.* (1997). Homer: a protein that selectively binds metabotropic glutamate receptors. *Nature* 386, 284–8.

Brown DA, and London E (1998). Functions of lipid rafts in biological membranes. *Annu Rev Cell Dev Biol* 14, 111–36.

Buhlmann N, Leuthauser K, Muff R, Fischer JA, and Born W (1999). A receptor activity modifying protein (RAMP)2-dependent adrenomedullin receptor is a calcitonin gene-gelated peptide receptor when coexpressed with human RAMP1. *Endocrinology* 140, 2883–90.

Bunemann M and Hosey MM (1999). G-protein coupled receptor kinases as modulators of G-protein signalling. *J Physiol* 517, 5–23.

Cabrera JL, de Freitas F, Satpaev DK, and Slepak VZ (1998). Identification of the Gbeta5-RGS7 protein complex in the retina. *Biochem Biophys Res Commun* 249, 898–902.

Carman CV, Barak LS, Chen C, Liu-Chen LY, Onorato JJ, Kennedy SP *et al.* (2000). Mutational analysis of Gbetagamma and phospholipid interaction with G protein-coupled receptor kinase 2. *J Biol Chem* 275, 10 443–52.

Carman CV, Lisanti MP, and Benovic JL (1999*a*). Regulation of G protein-coupled receptor kinases by caveolin. *J Biol Chem* **274**, 8858–64.

Carman CV, Parent JL, Day PW, Pronin AN, Sternweis PM, Wedegaertner PB *et al.* (1999*b*). Selective regulation of Galpha(q/11) by an RGS domain in the G protein-coupled receptor kinase, GRK2. *J Biol Chem* **274**, 34 483–92.

Carman CV, Som T, Kim CM, and Benovic JL (1998). Binding and phosphorylation of tubulin by G protein-coupled receptor kinases. *J Biol Chem* **273**, 20 308–16.

Chatterjee TK and Fisher RA (2000). Novel alternative splicing and nuclear localization of human RGS12 gene products. *J Biol Chem* **275**, 29 660–71.

Clapham DE and Neer EJ (1997). G protein beta gamma subunits. *Annu Rev Pharmacol Toxicol* **37**, 167–203.

Colley NJ, Baker EK, Stamnes MA, and Zuker CS (1991). The cyclophilin homolog ninaA is required in the secretory pathway. *Cell* **67**, 255–63.

De Vries L, Zheng B, Fischer T, Elenko E, and Farquhar MG (2000). The regulator of G protein signalling family. *Annu Rev Pharmacol Toxicol* **40**, 235–71.

DeFea KA, Vaughn ZD, O'Bryan EM, Nishijima D, Dery O, and Bunnett NW (2000*a*). The proliferative and antiapoptotic effects of substance P are facilitated by formation of a beta-arrestin-dependent scaffolding complex. *Proc Natl Acad Sci USA* **97**, 11 086–91.

DeFea KA, Zalevsky J, Thoma MS, Dery O, Mullins RD, and Bunnett NW (2000*b*). Beta-arrestin-dependent endocytosis of proteinase-activated receptor 2 is required for intracellular targeting of activated ERK1/2. *J Cell Biol* **148**, 1267–81.

Dev KK, Nakajima Y, Kitano J, Braithwaite SP, Henley JM, and Nakanishi S (2000). PICK1 interacts with and regulates PKC phosphorylation of mGLUR7. *J Neurosci* **20**, 7252–7.

Dwyer ND TE, Sengupta P, and Bargmann CI (1998). Odorant receptor localization to olfactory cilia is mediated by ODR-4, a novel membrane-associated protein. *Cell* **93**, 455–66.

Ferguson SS (2001). Evolving concepts in G protein-coupled receptor endocytosis: the role in receptor desensitization and signalling. *Pharmacol Rev* **53**, 1–24.

Ferguson SS, Downey WE III, Colapietro AM, Barak LS, Menard L, and Caron MG (1996). Role of beta-arrestin in mediating agonist-promoted G protein-coupled receptor internalization. *Science* **271**, 363–6.

Foord SM and Marshall FH (1999). RAMPs: accessory proteins for seven transmembrane domain receptors. *Trends Pharmacol Sci* **20**, 184–7.

Fraser NJ, Wise A, Brown J, McLatchie LM, Main MJ, and Foord SM (1999). The amino terminus of receptor activity modifying proteins is a critical determinant of glycosylation state and ligand binding of calcitonin receptor-like receptor. *Mol Pharmacol* **55**, 1054–9.

Freeman JL, De La Cruz EM, Pollard TD, Lefkowitz RJ, and Pitcher JA (1998). Regulation of G protein-coupled receptor kinase 5 (GRK5) by actin. *J Biol Chem* **273**, 20 653–7.

Gimelbrant AA, Haley SL, and McClintock TS (2001). Olfactory receptor trafficking involves conserved regulatory steps. *J Biol Chem* **276**, 7285–90.

Goodman OB Jr, Krupnick JG, Gurevich VV, Benovic JL, and Keen JH (1997). Arrestin/clathrin interaction. Localization of the arrestin binding locus to the clathrin terminal domain. *J Biol Chem* **272**, 15 017–22.

Goodman OB Jr, Krupnick JG, Santini F, Gurevich VV, Penn RB, Gagnon AW *et al.* (1996). Beta-arrestin acts as a clathrin adaptor in endocytosis of the beta2-adrenergic receptor. *Nature* **383**, 447–50.

Gutkind JS (1998). The pathways connecting G protein-coupled receptors to the nucleus through divergent mitogen-activated protein kinase cascades. *J Biol Chem* **273**, 1839–42.

Hajdu-Cronin YM, Chen WJ, Patikoglou G, Koelle MR, and Sternberg PW (1999). Antagonism between G(o)alpha and G(q)alpha in Caenorhabditis elegans: the RGS protein EAT-16 is necessary for G(o)alpha signalling and regulates G(q)alpha activity. *Genes Dev* **13**, 1780–93.

Hall RA, Ostedgaard LS, Premont RT, Blitzer JT, Rahman N, Welsh MJ *et al.* (1998*a*). A C-terminal motif found in the beta2-adrenergic receptor, P2Y1 receptor and cystic fibrosis transmembrane conductance regulator determines binding to the Na+/H+ exchanger regulatory factor family of PDZ proteins. *Proc Natl Acad Sci USA* **95**, 8496–501.

Hall RA, Premont RT, Chow CW, Blitzer JT, Pitcher JA, Claing A *et al.* (1998*b*). The beta2-adrenergic receptor interacts with the Na+/H+-exchanger regulatory factor to control Na+/H+ exchange. *Nature* **392**, 626–30.

Hall RA, Premont RT, and Lefkowitz RJ (1999*a*). Heptahelical receptor signaling: beyond the G protein paradigm. *J Cell Biol* **145**, 927–32.

Hall RA, Spurney RF, Premont RT, Rahman N, Blitzer JT, Pitcher JA *et al.* (1999*b*). G protein-coupled receptor kinase 6A phosphorylates the Na(+)/H(+) exchanger regulatory factor via a PDZ domain-mediated interaction. *J Biol Chem* **274**, 24328–34.

Hamm HE (1998). The many faces of G protein signaling. *J Biol Chem* **273**, 669–72.

Hepler JR (1999). Emerging roles for RGS proteins in cell signalling. *Trends Pharmacol Sci* **20**, 376–82.

Hepler JR, and Gilman AG (1992). G proteins. *Trends Biochem Sci* **17**, 383–7.

Kammermeier PJ, Xiao B, Tu JC, Worley PF, and Ikeda SR (2000). Homer proteins regulate coupling of group I metabotropic glutamate receptors to N-type calcium and M-type potassium channels. *J Neurosci* **20**, 7238–45.

Kato A, Ozawa F, Saitoh Y, Fukazawa Y, Sugiyama H, and Inokuchi, K (1998). Novel members of the Vesl/Homer family of PDZ proteins that bind metabotropic glutamate receptors. *J Biol Chem* **273**, 23969–75.

Laporte SA, Oakley RH, Holt JA, Barak LS, and Caron MG (2000). The interaction of beta-arrestin with the AP-2 adaptor is required for the clustering of beta 2-adrenergic receptor into clathrin-coated pits. *J Biol Chem* **275**, 23 120–6.

Laporte SA, Oakley RH, Zhang J, Holt JA, Ferguson SS, Caron MG *et al.* (1999). The beta2-adrenergic receptor/betaarrestin complex recruits the clathrin adaptor AP-2 during endocytosis. *Proc Natl Acad Sci USA* **96**, 3712–7.

Leuthäuser K, Gujer R, Aldecoa A, Mkinney RA, Muff R, Fischer JA *et al.* (2000). Receptor-activity-modifying protein 1 forms heterodimers with two G-protein-coupled receptors to define ligand recognition. *Biochem J* **351**, 347–51.

Lin FT, Krueger KM, Kendall HE, Daaka Y, Fredericks ZL, Pitcher JA *et al.* (1997). Clathrin-mediated endocytosis of the beta-adrenergic receptor is regulated by phosphorylation/dephosphorylation of beta-arrestin1. *J Biol Chem* **272**, 31051–7.

Luttrell LM, Ferguson SS, Daaka Y, Miller WE, Maudsley S, Della Rocca GJ *et al.* (1999). Beta-arrestin-dependent formation of beta2 adrenergic receptor-Src protein kinase complexes. *Science* **283**, 655–61.

McDonald PH, Chow C-W, Miller WE, Laporte SA, Field ME, Lin F-T *et al.* (2000). Beta-arrestin 2: a receptor-regulated MAPK scaffold for the activation of JNK3. *Science* **290**, 1574–7.

McDonald PH, Cote NL, Lin FT, Premont RT, Pitcher JA, and Lefkowitz RJ (1999). Identification of NSF as a beta-arrestin1-binding protein. Implications for beta2-adrenergic receptor regulation. *J Biol Chem* **274**, 10677–80.

McLatchie LM, FN, Main MJ, Wise A, Brown J, Thompson N, Solari R, Lee MG, and Foord SM (1998). RAMPs regulate the transport and ligand specificity of the calcitonin-receptor-like receptor. *Nature* **393**, 333–9.

Miller WE and Lefkowitz RJ (2001). Expanding roles for beta-arrestins as scaffolds and adapters in GPCR signalling and trafficking. *Curr Opin Cell Biol* **13**, 139–45.

Miller WE, Maudsley S, Ahn S, Khan KD, Luttrell LM, and Lefkowitz RJ (2000). beta-arrestin1 interacts with the catalytic domain of the tyrosine kinase c-SRC. Role of beta-arrestin1-dependent targeting of c-SRC in receptor endocytosis. *J Biol Chem* **275**, 11 312–9.

Muff R, Buhlmann N, Fischer JA, and Born W (1999). An amylin receptor is revealed following co-transfection of a calcitonin receptor with receptor activity modifying proteins-1 or -3. *Endocrinology* **140**, 2924–7.

Nakajima Y, Yamamoto T, Nakayama T, and Nakanishi S (1999). A relationship between protein kinase C phosphorylation and calmodulin binding to the metabotropic glutamate receptor subtype 7. *J Biol Chem* **274**, 27 573–7.

O'Connor V, El Far O, Bofill-Cardona E, Nanoff C, Freissmuth M, Karschin A *et al.* (1999). Calmodulin dependence of presynaptic metabotropic glutamate receptor signaling. *Science* **286**, 1180–4.

Okamoto N, Hori S, Akazawa C, Hayashi Y, Shigemoto R, Mizuno N *et al.* (1994). Molecular characterization of a new metabotropic glutamate receptor mGluR7 coupled to inhibitory cyclic AMP signal transduction. *J Biol Chem* **269**, 1231–6.

Okamoto T, Schlegel A, Scherer PE, and Lisanti MP (1998). Caveolins, a family of scaffolding proteins for organizing "preassembled signalling complexes" at the plasma membrane. *J Biol Chem* **273**, 5419–22.

Pearson G, Robinson F, Beers Gibson T, Xu BE, Karandikar M, Berman T *et al.* (2001) Mitogen activated protein (MAP) kinase pathways: regulation and physiological functions. *Endocrine Rev* **22**, 153–83.

Penn RB, Pronin AN, and Benovic JL (2000). Regulation of G protein-coupled receptor kinases. *Trends Cardiovasc Med* **10**, 81–9.

Pierce KL, Luttrell LM, and Lefkowitz RJ (2001). New mechanisms in heptahelical receptor signaling to mitogen activated protein kinase cascades, *Oncogene* **20**,1532–9.

Pitcher JA, Freedman NJ, and Lefkowitz RJ (1998*a*). G protein-coupled receptor kinases. *Annu Rev Biochem* **67**, 653–92.

Pitcher JA, Hall RA, Daaka Y, Zhang J, Ferguson SS, Hester S *et al.* (1998*b*). The G protein-coupled receptor kinase 2 is a microtubule-associated protein kinase that phosphorylates tubulin. *J Biol Chem* **273**, 12 316–24.

Pitcher JA, Inglese J, Higgins JB, Arriza JL, Casey PJ, Kim C (1992). Role of beta gamma subunits of G proteins in targeting the beta-adrenergic receptor kinase to membrane-bound receptors. *Science* **257**, 1264–7.

Pronin AN, Morris AJ, Surguchov A, and Benovic JL (2000). Synucleins are a novel class of substrates for G protein-coupled receptor kinases. J Biol Chem **275**, 26 515–22.

Ranganathan R and Stevens CF (1995). Arrestin binding determines the rate of inactivation of the G protein-coupled receptor rhodopsin *in vivo*. *Cell* **81**, 841–8.

Roche KW, Tu JC, Petralia RS, Xiao B, Wenthold RJ, and Worley PF (1999). Homer 1b regulates the trafficking of group I metabotropic glutamate receptors. *J Biol Chem* **274**, 25 953–7.

Ross EM and Wilkie TM (2000). GTPase-activating proteins for heterotrimeric G proteins: regulators of G protein signaling (RGS) and RGS-like proteins. *Annu Rev Biochem* **69**, 795–827.

Sato M, Suzuki K, and Nakanishi S (2001). NMDA receptor stimulation and brain-derived neurotrophic factor upregulate Homer 1a mRNA via the mitogen-activated protein kinase cascade in cultured cerebellar granule cells. *J Neurosci* **21** 3797–805.

Saugstad JA, Kinzie JM, Mulvihill ER, Segerson TP, and Westbrook GL (1994). Cloning and expression of a new member of the L-2-amino-4-phosphonobutyric acid-sensitive class of metabotropic glutamate receptors. *Mol Pharmacol* **45**, 367–72.

Schaeffer HJ, Catling AD, Eblen ST, Collier LS, Krauss A, and Weber MJ (1998). MP1: a MEK binding partner that enhances enzymatic activation of the MAP kinase cascade. *Science* **281**, 1668–71.

Schwarzler A, Kreienkamp HJ, and Richter D (2000). Interaction of the somatostatin receptor subtype 1 with the human homolog of the Shk1 kinase-binding protein from yeast. *J Biol Chem* **275**, 9557–62.

Sheng M and Kim E (2000). The Shank family of scaffold proteins. *J Cell Sci* **113**, 1851–6.

Shiraishi Y, Mizutani A, Bito H, Fujisawa K, Narumiya S, Mikoshiba K *et al.* (1999). Cupidin, an isoform of Homer/Vesl, interacts with the actin cytoskeleton and activated rho family small GTPases and is expressed in developing mouse cerebellar granule cells. *J Neurosci* **19**, 8389–400.

Siderovski DP, Strockbine B, and Behe CI (1999). Whither goest the RGS proteins?, *Crit Rev Biochem Mol Biol* **34**, 215–51.

Sierra DA, Popov S, and Wilkie TM (2000). Regulators of G-protein signaling in receptor complexes. *Trends Cardiovasc Med* **10**, 263–8.

Snow BE, Hall RA, Krumins AM, Brothers GM, Bouchard D, Brothers CA *et al.* (1998*a*). GTPase activating specificity of RGS12 and binding specificity of an alternatively spliced PDZ (PSD-95/Dlg/ZO-1) domain. *J Biol Chem* **273**, 17 749–55.

Snow BE, Krumins AM, Brothers GM, Lee SF, Wall MA, Chung S *et al.* (1998*b*). A G protein gamma subunit-like domain shared between RGS11 and other RGS proteins specifies binding to Gbeta5 subunits. *Proc Natl Acad Sci USA* **95**, 13 307–12.

Tang Y, Hu LA, Miller WE, Ringstad N, Hall, RA, Pitcher JA *et al.* (1999). Identification of the endophilins (SH3p4/p8/p13) as novel binding partners for the beta1-adrenergic receptor. *Proc Natl Acad Sci USA* **96**, 12 559–64.

Tobaben S, Sudhof TC, and Stahl B (2000). The G protein-coupled receptor CL1 interacts directly with proteins of the Shank family. *J Biol Chem* **275**, 36204–10.

Tsunoda S, Sierralta J, Sun Y, Bodner R, Suzuki E, Becker A *et al.* (1997). A multivalent PDZ-domain protein assembles signalling complexes in a G-protein-coupled cascade. *Nature* **388**, 243–9.

Tu JC, Xiao B, Yuan JP, Lanahan AA, Leoffert K, Li M *et al.* (1998). Homer binds a novel proline-rich motif and links group 1 metabotropic glutamate receptors with IP3 receptors. *Neuron* **21**, 717–26.

Watson AJ, Aragay AM, Slepak VZ, and Simon MI (1996). A novel form of the G protein beta subunit Gbeta5 is specifically expressed in the vertebrate retina. *J Biol Chem* **271**, 28154–60.

Xiao B, Tu JC, Petralia RS, Yuan JP, Doan A, Breder CD *et al.*(1998). Homer regulates the association of group 1 metabotropic glutamate receptors with multivalent complexes of homer-related, synaptic proteins. *Neuron* **21**, 707–16.

Xiao B, Tu JC, and Worley PF (2000). Homer: a link between neural activity and glutamate receptor function. *Curr Opin Neurobiol* **10**, 370–4.

Xu X, Zeng W, Popov S, Berman DM, Davignon I, Yu K *et al.* (1999). RGS proteins determine signaling specificity of Gq-coupled receptors. *J Biol Chem* **274**, 3549–56.

Yasuda J, Whitmarsh AJ, Cavanagh J, Sharma M, and Davis RJ (1999). The JIP group of mitogen-activated protein kinase scaffold proteins. *Mol Cell Biol* **19**, 7245–54.

Zeng W, Xu X, Popov S, Mukhopadhyay S, Chidiac P, Swistok J *et al.* (1998). The N-terminal domain of RGS4 confers receptor-selective inhibition of G protein signaling. *J Biol Chem* **273**, 34687–90.

Zhang Y, Chou JH, Bradley J, Bargmann CI, and Zinn K (1997). The Caenorhabditis elegans seven-transmembrane protein ODR-10 functions as an odorant receptor in mammalian cells. *Proc Natl Acad Sci USA* **94**, 12 162–7.

Zhong H and Neubig RR (2001) Regulators of G protein signalling proteins: Novel multifunctional drug targets. *J Pharmacol Exp Therapeutics* **297**, 837–45.

Zitzer H, Honck HH, Bachner D, Richter D, and Kreienkamp HJ (1999*a*). Somatostatin receptor interacting protein defines a novel family of multidomain proteins present in human and rodent brain. *J Biol Chem* **274**, 32 997–3001.

Zitzer H, Richter D, and Kreienkamp HJ (1999*b*). Agonist-dependent interaction of the rat somatostatin receptor subtype 2 with cortactin-binding protein 1. *J Biol Chem* **274**, 18 153–6.

Chapter 8

Quantitative pharmacology of GPCRs

Terry Kenakin

8.1 **Introduction**

Historically, the science of pharmacology originated through the observation and quantification of drug effect on physiological systems. This was necessary because the testing of experimental chemicals on physiological systems, for reasons of safety and ethics, necessarily had to be conducted in non-human systems and the results of those studies extrapolated to humans. Quantitative drug response analysis is the common currency of this process since it defines the difference between therapeutic and toxic effects. For drugs to be useful, their effects have to be quantified and this requires the development of tools which come directly from the discipline of drug receptor theory. This chapter outlines the tools used to quantify the activity of drugs on GPCRs.

Ligands have two overt properties. These are affinity, the property that causes the ligand and receptor to remain in close proximity, and efficacy, the property of the ligand that causes the receptor to change its behaviour towards the host cell. These properties historically have been considered as separate. There often appear to be two classes of ligands, those with only affinity and those with both affinity and efficacy. Therefore, the latter group appears to be a subset of the former. In fact, there are examples of separate structure-activity relationships for changing affinity and efficacy. In these cases it appears that separate properties of molecules contribute differently to a ligand's total affinity and efficacy. However, this historical distinction is based on empirical observations and is not rooted in the molecular dynamics of drug-receptor interaction. In fact, the same molecular forces that cause a ligand to bind to the receptor are also involved in causing that binding to change receptor behaviour and thus initiate a response. Therefore, affinity and efficacy are linked on a molecular and thermodynamic level.

8.2 **The relationship between affinity and efficacy**

There are numerous lines of evidence to indicate that proteins do not possess one single tertiary conformation, but rather exist in an ensemble of micro-conformations (Hilser and Freire 1996, 1997; Hilser *et al.* 1998; Woodward *et al.* 1982; Woodward 1993; Karplus and Petsk 1990). The driving force to change these micro-conformational states is thermal energy (Hilser *et al.* 1997; Hvidt and Nielsen 1966; Bai *et al.* 1993, 1995). Thus, an ensemble of receptor conformations exists because of a characteristic low Gibbs free energy; a change in the free energy will herald a change in the makeup of the conformational ensemble. The equation for defining the change in the free energy of a given receptor state i, in comparison with a reference state 0, as a function of the binding of a ligand [A] to both states is given by (Freire 2000):

$$\Delta G_i = \Delta G_i^0 - RT \ln \left(\frac{1 + K_{a,i} \, [A]}{1 + K_{a,0} \, [A]} \right), \tag{1}$$

where $\Sigma \Delta G_i^0$ refers to the free energy of state i in the absence of ligand and $K_{a,0}$ and $K_{a,i}$ refer to the respective affinities of ligand [A] for both states. Correspondingly, the free energy changes for a complete ensemble of i micro-states is given by:

$$\sum \Delta G_i = \sum \Delta G_i^0 - RT \sum \ln \left(\frac{1 + K_{a,i} \, [A]}{1 + K_{a,0} \, [A]} \right). \tag{2}$$

It can be seen from equation (2) that if the affinity of the ligand for *any* state i ($K_{a,i}$) differs from the reference state ($K_{a,0}$) then $\sum \Delta G_i \neq \sum \Delta G_i^0$. Under these circumstances there will be a change in free energy of the ensemble and this would cause the formation of a new ensemble according to that free energy change. Therefore, if the ligand has differential affinity for *any* micro-conformation, it will enrich that conformation at the expense of others and thus alter the makeup of the original unbound ensemble. In this way a ligand with macro-affinity will create a new ensemble of receptor conformations. What this may mean pharmacologically depends on how many micro-species of the ligand-bound ensemble correspond to micro-conformations with biological activity, that is, the extent of the *intersection* of the ligand-bound ensemble with biologically active receptor ensembles.

At this point it is relevant to define what is meant by 'biologically active' receptor ensembles. These are collections of receptor species that are able to interact with the host system to induce a change. This could be in the form of activation of various G proteins, response production through mechanisms not utilizing G proteins (Hall *et al.* 1999), propensity to desensitize, be phosphorylated, form oligmers with other receptors or each other, internalize and, in general, perform any other biochemical activity that would change the presentation of the receptor to the host. These various occurrences should all be viewed as efficacy (Kenakin 2001*a*). Therefore, in view of the fact that ligand binding almost certainly changes the distribution of receptor conformations, it should not be expected *a priori* that a ligand which possesses macro-affinity (i.e. binds to the receptor) will not have some form of corresponding 'efficacy' in terms of producing a receptor species that reacts differently towards the host system. The challenge then is to have the appropriate measuring system to observe the full range of activities of ligands. This is because our vantage points define activity. The *non-observation* of certain activities should not be taken as evidence of the *absence* of those activities. In the human body, a great many vantage points are available to detect drug activity. Therefore, activities may become apparent *in vivo* that were not apparent in *in vitro* test systems.

8.3 System dependent and independent measurements

Quantification of ligand activity at GPCRs is unique in that GPCRs are societal proteins residing in synoptic systems. Therefore, the activity of ligands on GPCRs can vary with respect to the type of system in which the receptor resides. This immediately differentiates the types of measurements that can be made on GPCRs, namely system dependent and system independent ligand activity. The aim of quantitation of ligand-GPCR activity is to obtain system independent measurements since these can be used to predict effects in other systems, and eventually, in therapeutic settings.

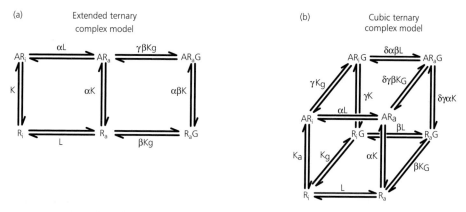

Fig. 8.1 Two widely applied models of GPCR systems; the extended ternary complex model (ETC model; Samama *et al.* 1993) and the cubic ternary complex model (CTC model, Wiess *et al.* 1996*a,b,c*).

There are two generally used models, based upon linkage theory (Wyman 1965, 1967; Weber 1972, 1975; De Lean *et al.* 1980) for GPCR behaviour. These are the extended ternary complex (ETC) model (Samama *et al.* 1993) and the cubic ternary complex (CTC) model (Weiss *et al.* 1996*a,b,c*); these are shown schematically in Fig. 8.1. The latter is thermodynamically more complete but also more cumbersome version of the former. There is evidence to show that some receptor types require the more thermodynamic completeness and are thus more appropriately modelled with the CTC model (Kenakin *et al.* 2000). Expressions using both models will be given in this chapter. Figure 8.1 indicates the meaning of the various terms. This chapter will discuss methods attempting to measure system independent estimates of affinity (K_A) and efficacy (α, γ, δ).

8.4 Measurements of affinity with binding studies

The interaction of ligands with receptors can be quantified as the strength of attraction between the ligand and the receptor (affinity) and also as the observation of the change in the receptor system as a result of the binding. Binding studies quantify the proportion of receptor occupied by ligand for any given concentration of ligand. This method requires a tracer molecule, the disposition of which, whether bound to the receptor or not, can be discerned. If the ligand is a tracer itself (i.e. radioligand), then the binding can be measured directly (saturation binding). However, since most ligands are not radioactive, an indirect method, namely the effect of a non-radioactive ligand on a bound tracer molecule, is used to determine the affinity of the non-radioactive ligand.

Binding experiments primarily yield information about affinity although there are techniques which also yield estimates of efficacy (*vide infra*). The first step is to characterize the saturation binding of a tracer radioligand. Once this is done, the effects of other ligands on the binding of the tracer ligand can furnish information regarding the molecular interaction of the ligand with the receptor and the concentrations at which these interactions occur. Detailed descriptions of methods to optimize and quantify binding are beyond the scope of this chapter (see Limbird 1996; Klotz 1997; Winzor and Sawyer 1995). Simultaneous comparison of total and non-specific (not receptor related) binding yields a sigmoid saturation curve (preferable to linear transformations—see Klotz 1982, 1997; Klotz and Hunston 1984). In general, the effect of ligand-binding on a single bound concentration of radioligand yields a sigmoid displacement curve.

For competitive antagonism with the tracer ligand, whereby both the radioactive and non-radioactive ligand compete for the same site on the receptor, parallel displacement of the tracer curve is produced (see Fig. 8.2a). The displacement of different concentrations of radioligand is shown in Fig. 8.2b; a characteristic pattern is observed, whereby the ligand reduces the radioactive ligand binding completely to non-specific (nsb) levels and the IC_{50} (concentration producing 50% inhibition of the binding) is linearly related to the concentration of the radioligand (Cheng and Prusoff 1973).

$$IC_{50} = K_I \left(1 + \frac{[A]}{K_d} \right) \tag{3}$$

K_I and K_d refer to the equilibrium dissociation constants of the receptor-antagonist and receptor-radiolabel complex, respectively and [A] the concentration of radiolabel. Under ideal conditions, a system independent measure of affinity (K_I) can be derived.

In the case where the antagonism precludes binding of the radioligand (either by non-competitive or pseudo-irreversible interaction), a depression of the saturation binding curve is obtained with no dextral displacement (see Fig. 8.2c). The displacement curves are still depressed to non-specific levels but there is no increase in the IC_{50} with increasing radioligand concentration (see Fig. 8.2d).

Another type of ligand interaction, whereby the tracer and non-radioactive ligands bind to distinctly different sites on the receptor, is termed allosteric. The effects on saturation binding can be very similar to competitive ligands except that they are saturable, that is, when the 'allosteric' site for the ligand is saturated, the effect is maximal. The maximal effect on the affinity of the tracer ligand is given by a cooperativity factor (denoted α) which can be inhibitory (i.e. a ligand with $\alpha = 0.1$ produces a maximal 10-fold decrease in the affinity of the tracer ligand) or potentiating (i.e. $\alpha = 10$ leads to a 10-fold enhancement of tracer affinity). Figure 8.3a shows the effect of an allosteric modulator with $\alpha = 0.1$ on saturation binding. Figure 8.3b shows the resulting displacement curve. A characteristic of allosteric inhibition is the potential to *not* decrease radioligand completely to non-specific levels. Also, the relationship between radioligand concentration and allosteric ligand IC_{50} is hyperbolic

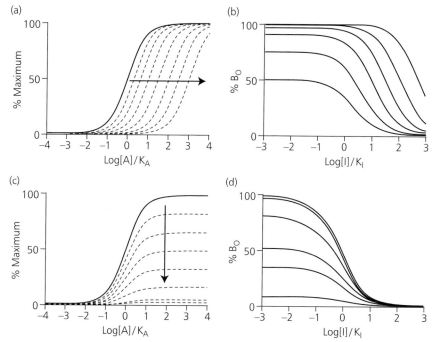

Fig. 8.2 Effects of antagonists on tracer ligand binding. (a) Effect of a simple competitive antagonist on a saturation binding curve of a tracer ligand. Parallel shifts to the right with no diminution of maxima result. (b) Effects of a range of concentrations of a simple competitive antagonist on a various single concentrations of tracer ligand. Binding is diminished to nsb levels with shifts to the right of the IC_{50} for inhibition. (c) Effect of a non-competitive antagonist (binding of antagonist precludes binding of tracer ligand) on a saturation binding curve for a tracer ligand. (d) Effects of a range of concentrations of a non-competitive antagonist on various single concentrations of tracer ligand. Binding is diminished to nsb levels with no shift to the right of the IC_{50} for inhibition.

and not linear:

$$IC_{50} = \frac{K_1((\text{[A]}/K_d)(1-\alpha) + \alpha + 1)}{\alpha(\text{[A]}/K_d(1-\alpha) + \alpha + 1)} \tag{4}$$

An allosteric enhancer shifts the saturation binding curve to the left (Fig. 8.3c) with a corresponding increase in ordinate values for tracer binding in the displacement curve (Fig. 8.3d).

The most sensitive method of detecting allosteric effects through ligand binding is to examine the *rate* of tracer ligand association and dissociation in the presence of the suspected allosteric ligand. The rationale for this idea is that an allosteric ligand will affect a conformational change in the receptor which will, in turn, alter the orthosteric association and/or dissociation constant of the tracer ligand. In theory, allosteric ligands can either decrease both the association and dissociation constant of the tracer ligand. Therefore, kinetic experiments may detect allosteric effects that may not otherwise be detected in equilibrium binding experiments. For example, the allosteric modulator Tetra-W84 decreases *both* the association

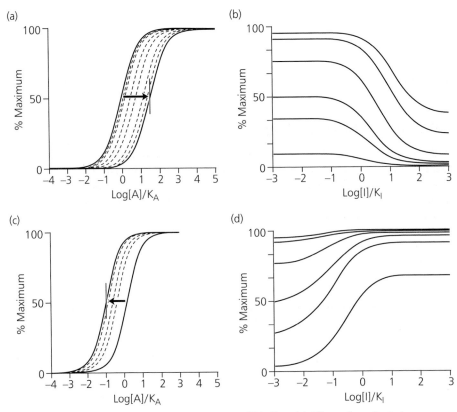

Fig. 8.3 Effects of allosteric ligands on tracer ligand binding. (a) Effect of an allosteric modulator with $\alpha = 0.1$ on a saturation binding curve of a tracer ligand. Parallel shifts to the right with diminution of maxima result until a 10-fold maximum shift is obtained. (b) Effects of a range of concentrations of the same allosteric modulator on a various single concentrations of tracer ligand. Binding may not be diminished to nsb levels with higher concentrations of tracer ligand and limited shifts to the right of the IC_{50} for inhibition are observed up to a maximum value. (c) Effect of an allosteric enhancer (increased affinity) on a saturation binding curve for a tracer ligand. (d) Effects of a range of concentrations of allosteric enhancer on various single concentrations of tracer ligand. Binding of tracer is increased by the allosteric enhancer up to the maximal asymptote for tracer binding.

and dissociation rate constants of $[^3H]$-N-Methylscopolamine at muscarinic acetylcholine receptors (Chapter 18), an allosteric effect clearly detected in kinetic experiments (Kostenis and Mohr 1996). However, since both rate constants are affected, there is no obvious change in the equilibrium affinity of the tracer in the presence of Tet-W84, thus equilibrium binding does not detect the allosteric effect of this ligand.

Assuming the kinetics of the allosteric ligand are faster than the tracer ligand (as is almost always the case), the observed rate of dissociation of the tracer ligand ([A]), in the presence of the allosteric ligand ([C]) is given by (Christopoulos 2000):

$$k_{\text{off-obs}} = \frac{\alpha[C]\, k_{\text{offAC}}/K_C + k_{\text{offA}}}{1 + \alpha[C]/K_C} \tag{5}$$

where α is the allosteric constant (maximal change in affinity imparted by the allosteric ligand to the the tracer ligand) and K_C is the equilibrium dissociation constant of the allosteric ligand–receptor complex. The rate of dissociation of the tracer ligand from the receptor is given by k_{offA} in the absence of C and k_{offAC} in the presence of C. It can be seen that, with increasing concentrations of allosteric ligand, there will be a linear change in the observed dissociation rate constant.

8.5 Complex binding phenomena

With synoptic GPCR systems, G protein binding can greatly modify the interaction of ligands with receptors, that is, the observed affinities may be modified by the influence of the G protein. For example, if the ligand alters the affinity of the receptor for G proteins, then the observed overall affinity will be an amalgam of ligand affinity for the receptor and the degree of modification of G protein affinity (referred to as receptor isomerization). Thus the binding of ligand [A] to receptor [R] and subsequent isomerization to a receptor state R* is shown in the scheme below:

$$A + R \xrightarrow{K_A} AR \xrightarrow{\varsigma} AR^* \tag{6}$$

where K_A is the equilibrium dissociation constant of the ligand–receptor complex (1/affinity) and ς is a term quantifying the ability of the ligand to induce receptor isomerization (conversion from R to R*). In such a system, the observed affinity of A for the receptor (denoted K_{obs}) is augmented by the ability of the ligand to isomerize the receptor to a different species (Colquhoun 1985):

$$K_{obs} = \frac{K_A}{(1 + \varsigma)} \tag{7}$$

Thus, if the ligand isomerizes the receptor, the observed affinity of the ligand for the receptor will be greater than the affinity of [A] for R (the observed equilibrium dissociation constant of the agonist–receptor complex will be $<K_A$). This system modification is affected by the relative stoichiometry between receptor and G protein. For example, the observed affinity of a ligand for a receptor, in terms of the ETC model, is given as:

$$K_{obs} = \frac{K_A(1 + L(+ \beta[G]/K_G))}{1 + \alpha L(1 + \gamma\beta[G]/K_G)} \tag{8}$$

As can be seen from equation (8), there are efficacy terms (namely α and γ) which dictate the observed potency of the ligand, that is, the change of affinity of the receptor upon ligand binding and the propensity of the ligand to cause formation of the active state receptor Ra.

There are two practical ramifications of these effects. The first is that, observed affinity may be a result of ligand affinity *and* efficacy and thus may change in different systems due to differences in receptor/G protein relative stoichiometry (Kenakin 1997a). This is a problem when system independent measures of affinity are required for prediction of therapeutic effect. There is also the possibility, especially in engineered recombinant receptor systems, that system relative stoichiometry (receptor/G protein ratios) can confound observation of characteristic binding patterns and lead to complex binding curves (Kenakin 1997a). For example, Fig. 8.4a shows competitive displacement curves of an antagonist radioligand by a non-radioactive agonist ligand systems with varying but limiting amounts of G protein.

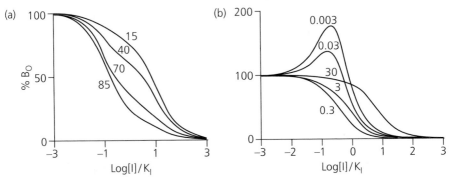

Fig. 8.4 Complex binding kinetics for stoichiometrically deficient (a) or oligomeric binding (b) systems. a. Displacement of an antagonist radioligand by an agonist radioligand in four different systems of varying amount of G protein available for high affinity binding. Numbers next to the curves correspond to the amount of G protein expressed as a percentage of the amount of receptor. (b) Displacement of a submaximal amount of tracer radioligand by concentrations of the same non-radioactive ligand in a system where the receptor forms a dimer and the binding of the ligand favours dimerization and thus enhances the affinity of the receptor for the ligand. Numbers refer to various concentrations of radioligand (ordinate values normalized).

Finally, GPCRs are known to form homodimers (Mukhopadhyay *et al.* 2000; Cvejic and Devi 1997, 2000; Jordan *et al.* 2000; Hebert *et al.* 1996; Angers 2000; Zeng *et al.* 2000), heterodimers (Pauwells *et al.* 1998; Benkirane *et al.* 1997; White *et al.* 1998; Jones *et al.* 1998; Marshall *et al.* 1999; Rocheville *et al.* 2000) and higher order oligomers (Kuhman *et al.* 2000) (Chapter 1) and this also can lead to complex binding phenomena. Figure 8.4b shows displacement of a radioligand which interacts with a GPCR dimer by ligands with differing propensity to affect the dimerization process (Kenakin 2001*b*). This figure shows a system whereby the receptor forms a dimer and the binding of the ligand to that dimer is enhanced. An interesting effect is created whereby the addition of non-radioligand actually *increases* radioligand binding before it decreases it within a certain concentration range (see Fig. 8.4b).

The measurement of the ability of ligands to isomerize the receptor to a form that binds to G proteins can be a measure of ligand efficacy. Thus, the more a ligand facilitates the conversion from R to R* (see scheme 6) then the greater is the ligand efficacy; in essence the magnitude of the term ς (equation (7)) corresponds to the magnitude of the ligand efficacy. Experimentally, in binding assays, the affinity of the ligand is measured in the binding assay as a displacement curve under conditions where G protein coupling can occur (observed affinity according to equation (7)). To measure efficacy, the affinity then is measured again under conditions whereby G protein binding is cancelled. This latter condition can be achieved with an excess amount of unhydolyzable analogue of GTP in the assay to prevent accumulation of the ARaG complex (see Fig. 8.1). Under these conditions, a measure of K_A (affinity with no efficacy modification) is obtained. The ratio of these affinities (referred to as the 'GTP-shift') represents the ability of the ligand to induce G protein coupling (i.e. efficacy). Within the ETC and CTC model, the relative GTP-shift for two agonists designated A and B is represented by:

ETC Model:

$$\text{GTP-shift} = \frac{[1 + \alpha_A L(1 + \gamma_A \beta[G]/K_G)](1 + \alpha_B L)}{[1 + \alpha_B L(1 + \gamma_B \beta[G]/K_G)](1 + \alpha_A L)} \tag{9}$$

CTC Model:

$$\text{GTP-shift} = \frac{[1 + \alpha_A L + \gamma_A[G]/K_G(1 + \delta_A \alpha_A \beta L)](1 + \alpha_B L + \gamma_B[G]/K_G)}{[1 + \alpha_B L + \gamma_B[G]/K_G(1 + \delta_B \alpha_B \beta L)](1 + \alpha_A L + \gamma_A[G]/K_G)} \tag{10}$$

Note that under the conditions of non-limiting G protein, the magnitude of the GTP-shift depends upon efficacy terms α, γ, and δ. It should be pointed out that GTP-shift experiments are kinetic in nature and depend upon the rate of GDP–GTP exchange. For some systems this can be quite slow and thus incomplete cancellation of G protein coupling can occur.

8.6 Functional studies

Functional studies either directly observe the effects of ligands on receptor systems when those ligands possess efficacy (agonists, inverse agonists) or measure the effects of ligands on tracer agonists by observation of the modification of tracer response. It should be noted that ligand-mediated agonist response results from the activation of a receptor to produce a stimulus and the subsequent processing of that stimulus through complicated biochemical cascade reactions to yield an observed response. Therefore, it is not possible to directly equate the amount of response with the degree of receptor occupancy to gauge the 'power' of a ligand to induce that response. Instead, indirect approaches, which compare equieffective agonist concentrations based on the null method, are used to estimate relative affinity and efficacy in functional studies.

With the ability to transfect surrogate cells with GPCRs has come the ability to create functional assays in reporter, melanophore, second messenger, and yeast formats (Chapter 10). This technology therefore allows the measurement of agonism for virtually any receptor known to couple to G proteins. The tools to quantify this agonism (and antagonism of agonist response) historically have been developed in isolated tissue systems but can directly be applied to recombinant functional assays.

8.6.1 Affinity measurements in functional studies

As with binding, antagonists produce distinctive effects on dose-response curves for agonists. In all cases but non-competitive and pseudo-irrevesible antagonism, the effects are the same as those observed on tracer saturation binding curves. Thus, a competitive antagonist produces parallel dextral displacement of dose-response curves with no diminution of maximal response (see Fig. 8.2a). In this case, the affinity of the antagonist can be estimated by Schild analysis, whereby the dose-ratio (DR) for agonism (quantified as the ratio of EC_{50} [molar concentration producing 50% maximal response] obtained in the presence and absence of antagonist) is compared to the concentration of the antagonist according to the

Schild equation (Arunlakshana and Schild 1959):

$$Log(DR - 1) = Log\,[B] - LogK_B \qquad (11)$$

The concentration of the antagonist is denoted as [B] and K_B is the equilibrium dissociation constant of the antagonist–receptor complex (reciprocal of affinity). Thus a regression of Log(DR−1) values upon Log[B] values should be linear with a slope of unity and an intercept of Log K_B (Arunlakshana and Schild 1959).

While Schild analysis is historically the most often utilized method to measure competitive antagonist affinity, it is not the optimal method. Another approach, namely a non-linear fit of the data (with visualization of the data with a 'Clark plot'—Stone and Angus 1976; Stone 1980; Lew and Angus 1996), does not over-emphasize control pEC_{50} (as does Schild analysis) thereby giving a more balanced estimate of antagonist affinity. While superior to the Schild method it should be noted that Schild analysis is rapid, intuitive, and can be used to detect non-equilibrium steady-states in the system that can corrupt estimates of affinity. Also non-linear regression requires matrix algebra to estimate the error of the pK_B. While error estimates are given with many commercially available software packages for curve fitting, they are difficult to obtain without these (from first principles). In contrast, manual calculation with Schild analysis furnishes an estimate of the error for the pK_B from the linear regression using all of the data.

For the non-linear procedure, the pEC_{50}s ($-$ log of the EC_{50} values) of the agonist dose-response curves are fit to the equation:

$$pEC_{50} = -Log([B] + 10^{-pK_B}) - Log\,c \qquad (12)$$

where [B] is the concentration of the antagonist and pK_B and c are fitting constants. Note that the control pEC_{50} is used with [B] $= 0$. The relationship between the pEC_{50} and increments of antagonist concentration can be shown in a Clark plot of pEC_{50} versus $-Log([B] + 10^{-pK_B})$. Constructing such a plot is useful because, although it is not used in any calculation of the pK_B, it allows visualization of the data to ensure that the plot is linear and has a slope of unity (Lew and Angus 1996).

Competitive antagonists can also possess efficacy and thereby produce low levels of tissue response. If the maximal level of response is lower than the system maximum, these compounds are referred to as partial agonists. Providing a significantly greater response can be obtained with a full agonist in the presence of the partial agonist, the affinity of the partial agonist can be measured with Schild analysis. As a first approximation it should be noted that the EC_{50} of the partial agonist (molar concentration producing 50 per cent of the maximal response *to the partial agonist*) is equal to the K_B of the partial agonist. The positive efficacy of the partial agonist will introduce a minor error into the estimation, the magnitude of which is proportional to the efficacy of the partial agonist. Thus, the relationship between the EC_{50} of a partial agonist and its affinity is given by:

ETC Model

$$\frac{EC_{50}}{K_B} = \frac{[1 + L(1 + \beta[G]/K_G)]}{1 + \alpha L(1 + \gamma\beta[G]/K_G)} \qquad (13)$$

CTC Model

$$\frac{EC_{50}}{K_B} = \frac{[1 + L + [G]/K_G(1 + \beta L)]}{1 + \alpha L + \gamma[G]/K_G(1 + \alpha\beta\delta L)} \qquad (14)$$

In the case of non-competitive and/or pseudo-irreversible antagonism, the receptor occupancy by the agonist is precluded by the antagonist causing a direct depression of the maximal receptor occupancy curve (Fig. 8.1c). However, due to the amplication factors in stimulus-response mechanisms, maximal tissue responses for high efficacy agonists can be attained, in some systems, with sub-maximal receptor occupancy, that is, a powerful agonist may need only 10% of the existing receptor population to produce maximal tissue response. Thus, until 90% of receptors are inactivated by antagonist, the maximal response for the agonist will be attained and the observed antagonism will resemble simple competitive antagonism until that point. Therefore, the characteristic pattern of non-competitive antagonism on agonist response is a mixture of dextral displacement and eventual depression of maximal response (see Fig. 8.5a).

There is another mechanism whereby the pattern shown in Fig. 8.5a can evolve, namely in cases of hemi-equilibria. Under equilibrium conditions, the relative receptor occupancies of an agonist and competitive antagonist adjust according to their respective affinities and relative concentrations. However, if insufficient time is allowed for this to occur, then a truncation of the response will ensue. This is especially the case with slowly dissociating antagonists since the antagonist occupancy must re-adjust to the presence of the agonist during the measurement of response. If this does not occur then the maximal asymptote of response is effectively truncated and a series of depressions of maximal responses with increasing concentrations of a competitive antagonist are observed. Thus, the curves are shifted to the right according to the Schild equation but with depressed maxima. While this apparently resembles non-competitive antagonism, it actually is due to competititve antagonism with insufficient time to observe equilibrium response.

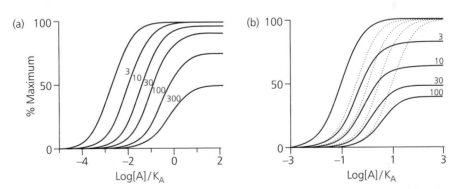

Fig. 8.5 Depression of functional maximal responses by non-competitive antagonism (a) and hemi-equilibrium kinetics (b). (a) Inhibition of agonist response by a non-competitive antagonist in a system where receptor stimulus is highly amplified (i.e. 10% receptor occupancy by the agonist generates the system maximum response). Dextral displacement of the curve with little diminution of maximum can be observed before depression occurs. Numbers refer to values of $[B]/K_B$. (b) Effect of a slowly dissociating simple competitive antagonist on agonist response measured within a time frame insufficient to allow equilibration of the receptors with agonist and antagonist. If kinetic conditions would allow equilibrium to be achieved, the curves would be the dotted lines. Under hemi-equilibrum conditions, depressions of the maximal response are observed. Numbers refer to $[B]/K_B$.

Non-competitive antagonism can be quantified by the method of Gaddum (Gaddum *et al.* 1955) in functional studies. Thus, equiactive concentrations of agonist in the absence (denoted [A]) and presence (denoted [A']) of a given concentration of non-competitive antagonist (the concentration must be sufficient to depress the maximal response to the agonist) are compared according to a double reciprocal equation of the form (Gaddum *et al.* 1955):

$$1/[A] = 1/[A'](1 + [B]/K_B) + \alpha[B]/K_B K_A \qquad (15)$$

where K_A and K_B refer to the equilibrium dissociation constants of the agonist and antagonist receptor complexes, respectively and α the allosteric factor for the non-competitive antagonist (change in affinity of the agonist for the receptor imparted by the antagonist). Note that there may not be a change in the affinity of the receptor for the agonist (α may be unity) but that the receptor may be made inoperative when occupied by the non-competitive antagonist. Equation (15) describes a straight line the slope of which can be used to estimate the K_B of the antagonist with the equation:

$$K_B = [B]/(\text{slope} - 1) \qquad (16)$$

Best results with this method are obtained when the maximum response is depressed below 50 per cent control levels. There are other transformations which circumvent the obvious weighting disadvantages of double reciprocal plots that can be used with this method (see Kenakin 1997*b*). Equiactive responses are most readily obtained by comparing the control and depressed concentration-response curves at various levels of response.

8.6.2 Measurement of efficacy in functional studies

Theoretically, efficacy is the power of an agonist to impart stimulus to a system for a given receptor occupancy. Thus, if one agonist produces 50 per cent maximal system response through occupation of 10 per cent of the existing receptor population while another requires 40 per cent receptor occupation, then the former can be thought to have four times the 'efficacy' of the latter in terms of producing tissue response. The method of Furchgott (1966) does this through comparison of equiactive responses on log occupancy-response curves (response expressed as a function of the logarithm of receptor occupancy). However, a pre-requisite to the correct use of this method is an unbiased estimate of agonist affinity (needed to calculate receptor occupancy). As noted previously, the very fact that receptor isomerization of the receptor by agonists corrupts estimates of affinity (see equation (7)) makes the use of this method circular and flawed.

An alternative method is to compare the relative maximal response of agonists as an estimate of relative efficacy. Thus, comparisons are made at saturating concentrations of agonist (with respect to receptor occupancy) and affinity ceases to be an issue. The relative maxima of two agonists can be a useful surrogate measurement of relative efficacy providing that the system maximal response is not obtained. Often because of the amplifying effects of cellular stimulus–response mechanisms, receptor stimulus exceeds the system capability to register response and a 'tissue' maximal response for a number of agonists is observed, that is, all are full agonists. This does not mean that the agonists are of identical efficacy, but rather that the efficacy of the agonists exceeds the capability of the system to differentiate them. Generally, the further away from the receptor the response is measured, in terms of biochemical reactions that take place after receptor activation, the more amplification occurs. Thus, the best opportunity to measure relative maxima as a reflection of true agonist efficacy is at the earliest step in the stimulus–response chain. This is activation of the G protein. Thus, differences in the capability of agonists to stimulate exchange of GDP to GTP on the G protein can be a useful measure of relative efficacy (i.e. Jasper *et al.* 1998; Umland *et al.* 2001).

In terms of the models of GPCR systems, the relative maxima of two agonists denoted A and B are given as:

ETC Model

$$\text{Relative Maxima}_{A/B} = \frac{\gamma_A \alpha_A [1 + \alpha_B L (1 + \gamma_B \beta\, [G]/K_G)]}{\gamma_B \alpha_B [1 + \alpha_A L (1 + \gamma_A \beta [G]/K_G)]} \tag{17}$$

CTC model

$$\text{Relative Maxima}_{A/B} = \frac{\delta_A \gamma_A \alpha_A [1 + \alpha_B L + \gamma_B [G]/K_G (1 + \alpha_B \delta_B \beta)]}{\delta_B \gamma_B \alpha_B [1 + \alpha_A L + \gamma_A [G]/K_G (1 + \alpha_A \delta_A \beta)]} \tag{18}$$

It can be seen that this experimentally derived parameter depends upon efficacy terms and does not involve affinity.

8.7 Constitutively active GPCR systems

It can be seen intuitively from both the ETC and CTC models that active state receptors can form spontaneously according to the magnitude of the allosteric constant L. When this occurs to the extent that a response is observed, the system is referred to as being constitutively active. Thus, the system itself emanates an elevated basal activity resulting from spontaneous receptor activation. Experimentally, constitutive GPCR systems can be engineered through over-expression of receptors (Whaley *et al.* 1994; Barker *et al.* 1994; Van Sande *et al.* 1995; Chen *et al.* 2000), G proteins (Senogles *et al.* 1990), point mutation (Kjelsberg *et al.* 1992; Scheer *et al.* 1996, 1997; Porter *et al.* 1996) and changes in biochemical milieu (i.e. removal of Na^+, Costa and Herz 1989; Tian *et al.* 1994). Such systems then can detect positive agonism (providing the system maximum is not attained) and, in fact, are more sensitive to positive agonists than are quiescent (non-constitutively active) systems. The effects of a positive agonist on systems of increasing amounts of constitutive activity are shown in Fig. 8.6a. Constitutively active GPCR systems can also be used to detect inverse agonism caused by ligands that preferentially stabilize the inactive state of the receptor. Figure 8.6b shows the effects of an inverse agonist on GPCR systems of various levels of constitutive activity. Providing that the inverse agonist does not completely eliminate constitutive activity (i.e. reduce the observed basal response to zero), the EC_{50} is a reasonable estimate of the reciprocal of affinity (K_B). As with partial agonists, the antagonism of agonist responses by an inverse agonist can be used to measure the affinity of the inverse agonist. Equiactive concentrations of the agonist can be used for analysis in Schild regressions (see Fig. 8.6c). However, the effects of the inverse agonist on basal response cause a slight over-estimation of the inverse agonist affinity with this method.

Theoretically, constitutive receptor activity can be studied either in binding or functional studies. However, in practice, functional studies are much more practical. This is because, the amount of constitutively active receptor species producing the signal, namely RaG, is usually relatively low due to the low magnitude of most receptor allosteric constants. Changes in

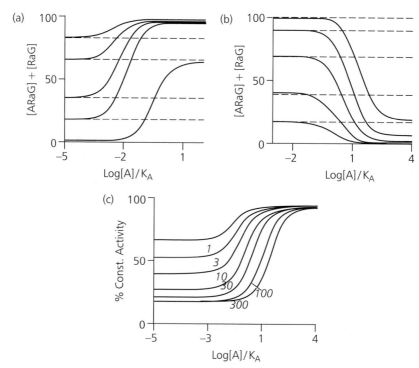

Fig. 8.6 Effects of agonists (a) and inverse agonists (b and c) in constitutively active GPCR systems. (a) Under conditions of increasing constitutive activity, baseline response increases as well as sensitivity to agonists. (b) Inverse agonists depress constitutive activity in a concentration-dependent manner. (c) Antagonism of agonist-induced response by an inverse agonist. Dextral displacement of the agonist dose--response curve is observed concomitantly with depressed basal constitutive activity.

very low levels of RaG are difficult to quantify with binding. In contrast, the amplification of receptor signalling produced by cellular stimulus response mechanisms allow very small levels of RaG to produce measurable and stable levels of tissue response.

8.8 Conclusions

This chapter briefly discusses techniques used to measure drug affinity and effciacy in GPCR systems. The key objective of this exercise is to quantify drug activity in a system independent manner so that predictions of drug activity can be made for therapeutic advantage. It should be noted that GPCRs are unique in the sense that ligand activity can be modified by the proteins with which the receptor interacts. From this viewpoint, the study of GPCRs in recombinant systems of varying makeup should yield valuable information about how receptors translate chemical information into biological activity.

References

Angers S (2000). Detection of β_2-adrenergic receptor dimerization in living cells using bioluminescence resonance energy transfer (BRET). *Proc Natl Acad Sci USA* **97**, 3684–9.

Bai Y, Milne JS, Mayne L, and Englander SW (1993). Primary structure effects on peptide group hydrogen exchange. *Proteins* **17**, 75–86.

Bai Y, Sosnick TR, Mayne L, and Englander SW (1995). Protein folding intermediates: native-state hydrogen exchange. *Science* **269**, 192–7.

Barker EL, Westphal RS, Schmidt D, and Sanders-Bush E (1994). Constitutively active 5-hydroxytryptamine2C receptors reveal novel inverse agonist activity of receptor ligands. *J Biol Chem* **269**, 11 687–90.

Benkirane M, Jin D-Y, Chun RF, Koup RA, and Jeang K-T (1997). Mechanism of transdominant inhibition of CCR-5 mediated HIV-1 infection by ccr5Δ32. *J Biol Chem* **272**, 30 603–6.

Chen G, Way J, Armour S *et al.* (2000). Use of constitutive G-protein coupled receptor activity for drug discovery. *Mol Pharmacol* **57**, 125–34.

Cheng YC and Prusoff WH (1973). Relationship between the inhibition constant (Ki) and the concentration of inhibitor which causes 50 percent inhibition (I50) of an enzymatic reaction. *Biochem Pharmacol* **22**, 3099–108.

Christopoulos A (2000). Overview of receptor allosterism. In (eds, SJ Enna, MW Williams, JW Ferkany, RD Posolt, JP Sullivan, and TP Kenakin) *Current Protocols in Pharmacology*, John Wiley and Sons, New York, pp. 1.21.1–1.21.45.

Colquhoun D (1985). Imprecision in presentation of binding studies. *Trends Pharmacol Sci* **6**, 197.

Costa T and Herz A (1989). Antagonists with negative intrinsic activity at δ-opioid receptors coupled to GTP-binding proteins. *Proc Natl Acad Sci USA* **86**, 7321–25.

Cvejic S and Devi LA (1997). Dimerization of the δ opioid receptor. *J Biol Chem* **272**, 26959–64.

De Lean A, Stadel JM, and Lefkowitz RJ (1980). A ternary complex explains the agonist-specific binding properties of the adenylate cyclase-coupled β-adrenergic receptor. *J Biol Chem* **255**, 7108–17.

Freire E (2000). Can allosteric regulation be preditcted from structure? *Proc Nat Acad Sci USA* **97**, 11 680–82.

Furchgott RF (1966). The use of β-haloalkylamines in the differentiation of receptors and in the determination of dissociation constants of receptor-agoinist complexes. In (eds, NJ Harper and AB Simmonds) *Advances in Drug Research*, vol. 3 ed. Academic Press, London, New York, pp. 21–55.

Gaddum JH, Hameed KA, Hathway DE, and Stephens FF (1955). Quantitative studies of antagonists for 5-hydroxytryptamine. *Q J Exp Physiol* **40**, 49–74.

Hall RA, Premont RT, and Lefkowitz RJ (1999). Heptahelical receptor signaling: Beyond the G-protein paradigm. *J Cell Biol* **145**, 927–32.

Hebert TE, Moffett S, Morello J-P, Loisel TP, Bichet DG, Barret C, and Bouvier M (1996). A peptide derived from a β_2-adrenergic receptor transmembrane domain inhibits both receptor dimerization and activation *J Biol Chem* **272**, 16 384–92.

Hilser VJ and Freire E (1996). Structure-based calculation of the equilibrium folding pathway of proteins: correlation with hydrogen exchange protection factors. *J Mol Biol* **262**, 756–72.

Hilser VJ and Freire E (1997). Predicting the equilibrium protein folding pathway: structure-based analysis of staphylococcal nuclease. *Protein Struct Funct Genet* **27**, 171–83.

Hilser VJ, Townsend BD, and Freire E (1997). Structure-based statistical thermodynamic analysis of T4 lysozyme mutants: structural mapping of cooperative interactions. *Biophys Chem* **64**, 69–79.

Hilser VJ, Dowdy D, Oas TG, and Freire E (1998). The structural distribution of cooperative interactions in proteins: analysis of the native state ensemble. *Proc Natl Acad Sci USA* **95**, 9903–8.

Hvidt A and Nielsen S (1966). Hydrogen exchange in proteins. *Adv Prot Chem* **21**, 287–386.

Jasper JR, Lesnick JD, Chang LK *et al.* (1998). Ligand efficacy and potency at recombinant α_2-adrenergic receptors. *Biochem Pharmacol* **55**, 1035–43.

Jones KE, Borowsky B, Tamm JA *et al.* (1998). GABAB receptors function as a heteromeric assembly of the subunits GAGABR1 and GABABR2. *Nature* **396**, 674–77.

Jordan BA, Cvejic S and Devi LA (2000). Opioids and their complicated receptor complexes. *Neuropsychopharmacology* **23**, S5–S18.

Kenakin TP (1997a). Differences between natural and recombinant G protein-coupled receptor systems with varying receptor/G protein stoichiometry. *Trends Pharmacol Sci* **18**, 456–64.

Kenakin TP (1997b). Competitive antagonism. In *The Pharmacologic Analysis of Drug-Receptor Interaction.* (3rd edition), Lippincott-Raven, New York, pp. 331–74.

Kenakin TP (2000). Efficacy: Molecular mechanisms and operational methods of measurement. A new algorithm for the prediction of side effects. In (eds, TP Kenakin and JA Angus) *The Pharmacology of Functional, Biochemical, and Recombinant Systems Handbook of Experimental Pharmacology*, Vol. 148, Springer, Heidelberg, Germany, pp. 183–216.

Kenakin TP, Morgan P, Lutz M, and Weiss J (2000). The evolution of drug-receptor models: the cubic ternary complex model for G-protein coupled receptors. In (eds, TP Kenakin and JA Angus) *The Pharmacology of Functional, Biochemical, and Recombinant Systems Handbook of Experimental Pharmacology*, Vol. 148, Springer, Heidelberg, Germany, pp. 147–66.

Kenakin TP (2001a). Efficacy at G-Protein Coupled Receptors. *Annu Rev Pharmacol Toxicol* **42**, 349–79.

Kenakin TP (2001b). The pharmacologic consequences of modeling synaptic receptor systems. In (ed. A. Christopoulos) *Biomedical Applications of Computer Modeling*, CRC Press, Boca Raton.

Kjelsberg MA, Cottechia S, Pstrowski J, Caron MG, and Lefkowitz RJ (1992). Constitutive activation of the α1B-adrenergic receptor by all amino acid substitutions at a single site. *J Biol Chem* **267**, 1430–33.

Klotz IM (1982). Numbers of receptor sites from Scatchard graphs: Facts and fantasies. *Science* **217**, 1247–9.

Klotz IM and Hunston DH (1984) Mathematical models for ligand-receptor binding *J Biol Chem* **259**, 10 060–2.

Klotz IM (1997). *Ligand-receptor energetics: A guide for the perplexed.* John Wiley and Sons, New York.

Kostenis E and Mohr K (1996). Composite action of allosteric modulators on ligand binding *Trends Pharmacol Sci* **17**, 443–4.

Kuhmann SE, Platt EJ, Kozak SL, and Kabat D (2000). Cooperation of multiple CCR5 co-receptors is required for infections by human immunodeficiency virus type 1. *J Virol* **74**, 7005–15.

Lew MJ and Angus JA (1996). Analysis of competitive agonist-antagonist interactions by nonlinear regression *Trends Pharmacol Sci* **16**, 328–37.

Limbird LE (1996). *Cell Surface Receptors: A Short Course on Theory and Methods.* (2nd edition) Kluwer Academic Publishers, Boston.

Mukhopadhyay S, McIntosh HH, Houston DB, and Howlett AC (2000). The CB1 cannabinoid receptor juxtamembrane C-terminal peptide confers activation to specific G-proteins in brain. *Mol Pharmacol* **57**, 162–70.

Marshall FH, Jones KA, Kaupmann K, and Bettler B (1999). GABA$_B$ receptors—the first 7TM heterodimers. *Trends Pharmacol Sci* **20**, 396–9.

Pauwells PJ, Dupuis DS, Perez M, and Halazy S (1998). Dimerization of 8-OH-DPAT increases activity at serotonin 5-HT1A receptors. *Naunyn Schmiedeberg's Arch Pharmacol* **358**, 404–10.

Onaran HO and Costa T (1997). Agonist efficacy and allosteric models of receptor action. *Ann N Y Acad Sci* **812**, 98–110.

Porter JE, Hwa J, and Perez DM (1996). Activation of the α1b-adrenergic receptor is initiated by disruption of an interhelical salt bridge constraint. *J Biol Chem* **271**, 28 318–23.

Rocheville M, Lange DC, Kumar U, Patel SC, Patel RC, and Patel YC (2000). Receptors for dopamine and somatostatin: formation of hetero-oligomers with enhanced functional activity. *Science* **288**, 154–57.

Romano C, Yang W-L, and O'Malley KL (1996). Metabotropic glutamate receptor 5 is a disulfide-linked dimer. *J Biol Chem* **271**, 28 612–20.

Samama P, Cotecchia S, Costa T, and Lefkowitz RJ (1993). A mutation-induced activated state of the β_2-adrenergic receptor: extending the ternary complex model *J Biol Chem* **268**, 4625–36.

Scheer A, Fanelli F, Costa T, De Benedetti PG, and Cotecchia S (1996). Constitutively active mutants of the α1B-adrenergic receptor: role of highly conserved polar amino acids in receptor activation. *EMBO J* 15, 3566–78.

Scheer A, Fanelli F, Costa T, De Benedetti PG, and Cotecchia S (1997). The activation process of the α_{1B}-adrenergic receptor: potential role of protonation and hydrophobicity of a highly conserved aspartate. *Proc Natl Acad Sci USA* 94, 808–13.

Senogles SE, Spiegel AM, Pardrell E, Iyengar R, and Caron M (1990). Specificity of rceptor G-protein interactions *J Biol Chem* 265, 4507–14.

Stone M (1980). The Clark plot: A semi-historical study. *J Pharm Pharmacol* 32, 81–86.

Stone M and Angus JA (1978). Developments of computer-based estimation of pA2 values and associated analysis. *J Pharmacol Exp Ther* 207, 705–18.

Tian W-N, Duzic E, Lanier SM, and Deth RC (1994). Determinants of α_2-adrenergic receptor activation of G-proteins: evidence for a precoupled receptor/G-Protein state. *Mol Pharmacol* 45, 524–31.

Umland SP, Wan Y, Shah H, Billah M, Egan RW, and Hey JA (2001). Receptor reserve analysis of the human α_{2C}-adrenoceptor using $[^{35}S]$GTPγS and cAMP functional assays. *Eur J Pharmacol* 411, 211–21.

Van Sande J, Swillens S, Gerard C, Allgeier A, Massart C, Vassart G, and Dumont JE (1995). In Chinese hamster ovary K1 cells dog and human thyrotropin receptors activate both cyclic AMP and the phosphoinositol 4,5-bisphosphates cascades in the presence of thyrotropin and the cyclic AMP cascade in its absence. *Eur J Biochem* 229, 338–43.

Whaley BS, Yuan N, Birnbaumer L, Clark RB, and Barber R (1994). Differential expression of the β_2-adrenergic receptor modifies agonist stimulation of adenylyl cyclase: a quantitative evaluation. *Mol Pharmacol* 45, 481–9.

White JH, Wise A, Main MJ *et al.* (1998). Heterodimerization is required for the formation of a functional GABAB receptor. *Nature* 396, 679–82.

Weber G (1972). Ligand binding and internal equilibria in proteins. *Biochemistry* 11, 864–78.

Weber G (1975). Energetics of ligand binding to proteins. *Adv Protein Chem* 29, 1–83.

Weiss JM, Morgan PH, Lutz MW, and Kenakin TP (1996*a*). The cubic ternary complex receptor-occupancy model. I. Model description. *J Theoret Biol* 178, 151–67.

Weiss JM, Morgan PH, Lutz MW, and Kenakin TP (1996*b*). The cubic ternary complex receptor-occupancy model. II. Understanding apparent affinity. *J Theoret Biol* 178, 169–82.

Weiss JM, Morgan PH, Lutz MW, and Kenakin TP (1996*c*). The cubic ternary complex receptor-occupancy model. III. Resurrecting efficacy. *J Theoret Biol* 181, 381–97.

Winzor, DJ and Sawyer WH (1995). *Quantitative Characterization of Ligand Binding*. Wiley-Liss, New York.

Woodward C, Simon I, and Tuchsen E (1982). Hydrogen exchange and the dynamic structure of proteins. *Mol Cell Biochem* 48, 135–60.

Woodward C (1993). Is the slow-exchange core the protein folding core? *Trends Biol Sci* 18, 359–6031.

Wyman J (1965). The binding potential, a neglected linkage concept *J Mol Biol* 11, 631–67.

Wyman J (1967). Allosteric linkage *J Amer Chem Soc* 89, 2202–32.

Zeng F-Y and Wess J (2000). Molecular aspects of muscarinic receptor dimerization. *Neuropsychopharmacology* 23, S19–32.

Chapter 9

Bioinformatic mining for GPCRs

D. Malcolm Duckworth and David Michalovich

9.1 Introduction

The advent of large-scale cDNA and genome sequencing efforts has revolutionized how we identify new genes and has also helped established the discipline of bioinformatics as a major component of both academic and industrial research programs. The mining for novel GPCRs is a good case in point with both industrial and academic researchers utilizing expressed sequence tag (EST) and genomic sequence data (see later) as their starting point to hunt for new receptors. In recent years bioinformatic search strategies have yielded the second GABA$_B$ receptor subunit (Jones *et al.* 1998; Kaupmann *et al.* 1998; see chapter 28), the human orexin receptors (Sakurai *et al.* 1998) and the third and fourth histamine receptors [reviewed by (Hough 2001); see chapter 17] to name just a few examples.

In the following chapter we intend to describe how bioinformatics tools can be applied to the identification, characterization, and disease association of GPCRs. The chapter is intended to provide an overview of the applications, tools, and methodologies required to identify and characterize GPCR sequences at both the DNA and protein level. Although focused on GPCRs the techniques are also applicable to other gene families.

Many of the tools described are available for free public use on the internet. However, access to a UNIX based computer system greatly enhances the depth of bioinformatic analysis that can be achieved as many of the tools have been developed for implementation on this system. A good starting guide both to bioinformatic tools and setting up a bioinformatic UNIX work station can be found in 'Developing Bioinformatics Computer Skills' (Gibas and Jambeck 2001).

9.2 Identification of mammalian GPCRs via bioinformatics approaches

9.2.1 Data sources

Three centres provide the main repositories for sequence data. The European Molecular Biology Laboratory (EMBL) data library located at the European Bioinformatics Institute (EBI) (http://www.ebi.ac.uk/Databases/index.html). GenBank® the National Institute for Health (NIH) database (http://www.ncbi.nlm.nih.gov/Genbank/index.html) located at the National Centre for Biotechnology Information (NCBI) and thirdly the DNA Data Bank of Japan (DDBJ) (http://www.ddbj.nig.ac.jp/). The three centres make up the International Nucleotide Sequence Database Collaboration.

Each centre provides the facilities for researchers to submit new sequences to the databases. Once submitted the sequence is provided with a unique accession number and placed into a subdivision of the database based on taxonomy or sequencing project, that is, Primate (gb_pr), EST (gb_est), high throughput genomic (HTG) (gb_htg). Beyond the sequence, other information such as the sequence description, organism, author references, biological sequence features and cross-references to other databases are captured. The information is shared freely between each centre, negating the need to query each database. Further services are provided by the database centres allowing users to query the databases via text searches using user interfaces such as NCBI's Entrez system (http://www.ncbi.nlm.nih.gov/Entrez/) or the EBI's Sequence Retrieval System (http://srs6.ebi.ac.uk/).

Parallel to the public sequencing efforts a number of companies, such Human Genome Sciences, Incyte and Celera, have established proprietary EST and genomic sequence databases.

9.2.2 DNA databanks

The nucleotide sequence component of the public databases can be divided into two main categories, namely complimentary DNA (cDNA) derived sequence and genomic sequence. Historically, the cDNA sequences present in the databases would be derived from focused gene cloning experiments. These would therefore represent well-characterized sequence for a defined gene. More recently the advent of improved DNA sequencing methodologies has allowed large-scale genome projects to be undertaken generating raw sequence data for which the function is unknown. Uncharacterized sequence therefore provides a good starting point for novel gene identification and annotation. Figure 9.1 provides an overview of the sequences present in the public databases.

The first of the high throughput sequencing projects to make its mark was the public and private EST sequencing efforts (Boguski et al. 1993). EST sequences are approximately 500

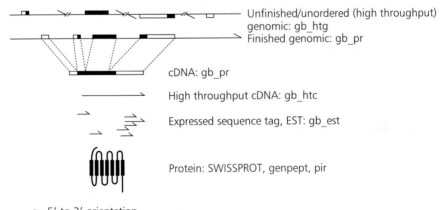

Unfinished/unordered (high throughput) genomic: gb_htg
Finished genomic: gb_pr
cDNA: gb_pr
High throughput cDNA: gb_htc
Expressed sequence tag, EST: gb_est
Protein: SWISSPROT, genpept, pir

→ 5′ to 3′ orientation
▱ Exon (transcribed region), filled box represents coding region

Fig. 9.1 Describes the relationship of unfinished genomic sequence through to protein sequence. The GENBANK subdivision for each sequence source is given next to the text description.

nucleotide long sequence reads from the 5′ and 3′ end of cloned cDNAs. The sequences are single pass reads and therefore the quality can be poor. cDNA cloning artifacts are also represented in EST databases. These artifacts include miss-priming events and aberrant transcripts resulting in ESTs representing truncated cDNAs, the 5′ and 3′ annotation of the EST being in reverse orientation with respect to the genuine transcript, and contamination of un-spliced genomic sequence. Depending on the quality of the cDNA library which is sequenced the ESTs may represent the true 5′ and 3′ ends of a cDNA insert; however, for larger mRNAs, only partial cDNA sequence may be provided. In this case the derived 5′ EST sequences may fall in the coding region of the true transcript or may solely represent 3′ untranslated region sequence.

Greater value can be obtained from EST sequences by assembling overlapping EST reads. This can be further enriched by using the 'clone identifier' information, allowing discreet EST clusters (contigs) to be linked by 5′–3′ information independently of sequence overlap. This type of analysis can allow researchers to walk through an unknown cDNA sequence rapidly. For an example of this method see the Human Gene Indices and TIGR Human Consensus (THC) resource (http://www.tigr.org/tdb/tgi.shtml). A related approach is taken by the NCBI's UNIGENE, a system which automatically partitions GenBank sequences into a non-redundant set of gene-oriented clusters. This captures both cDNA and EST sequence as well as comparisons to model organism sequences such as mouse, *Caenorhabditis elegans*, *Drosophila melanogaster* and *Saccharomyces cerevisiae*. Unlike TIGR the related EST sequences are collected into sequence 'bins' but the assembled EST contigs are not available. The largest representative sequence can, however, be obtained.

High throughput cDNA (HTcDNA or HTC) represents unfinished cDNA sequence and may include 5′ UTR and 3′ UTR regions and coding region (http://www.ebi.ac.uk/embl/Documentation/Release_notes/relnotes66/relnotes.html#htc). Example of these cDNAs are those produced by the RIKEN Genomic Sciences Center (Kawai *et al.* 2001). These can represent full length or partial transcripts. Once the sequences are finished they are moved to the appropriate taxonomic division of the database.

The majority of human nucleotide sequence in the public database is now derived from the human genome sequencing project. This was initiated during the late 1980s and launched in 1990, for a background see (Lander *et al.* 2001). Similarly to cDNA sequence, the quantity of genomic derived sequence in the public database has seen a rapid expansion. Previously to the high throughput sequencing projects, genomic sequence would be derived through specific gene focused projects. Routinely these would be cosmid vector cloned inserts of around 30 kb or may only be the sequence around exons.

The advent of high throughput genome sequence has added three further representations of genomic sequence to the database. The public human genome sequencing effort has focused on sequencing bacterial artificial chromosome (BAC) clones, which can hold around 200 kb of DNA. Each BAC clone is 'shotgun' cloned into smaller vectors, these are sequenced and contiguous reads which represent a sub-sequence of the BAC are built up, at this point the BAC sequence is given an accession number and submitted to the database as unfinished high throughput genomic sequence (HTGS). The order and orientation of the contigs are unknown until the sequence of the BAC is completed. This makes gene hunting in unfinished sequence problematic. Once the genomic sequence is completed, the entry is moved from the HTGS division of the database into its taxonomic division. Depending on the sequencing centre, biological sequence features such as predicted genes, repeat elements etc are annotated in the entry.

Genome Survey Sequence (GSS) represent a third division of database which contains raw genomic sequence, GSSs are end reads of BAC clones, approximately 500 bp in length, the genomic equivalent of ESTs. Their primary function is to aid the assembly of BAC clones into overlapping assemblies. Exonic sequence can be found in these small genomic reads and therefore GSSs can be used for novel gene hunting. This is exemplified in the identification of a second melanin-concentrating hormone receptor, MCH2, cloned via the identification of a putative coding exon in a GSS sequence (Hill *et al.* 2001; Sailer *et al.* 2001).

9.2.3 **Protein databanks**

Protein sequences derived from cDNA and genomic sequence can be found in two main databases, SWISSPROT/TrEMBL (Bairoch and Apweiler 2000) and PIR (Barker *et al.* 2001). SWISSPROT is a highly curated database of protein sequence, derived from the EMBL nucleotide database. There is a minimal level of redundancy in the data. Each entry is highly integrated with other bioinformatic databases. TrEMBL (translated EMBL) is a supplement of SWISSPROT and represents the translations of EMBL nucleotide sequence not yet integrated into SWISSPROT (http://www.ebi.ac.uk/swissprot/ and http://www.expasy.ch/sprot/sprot-top.html). Protein Information Resource-Protein Sequence Database (PIR-PSD) provides a similar resource to SWISSPROT (http://pir.georgetown.edu/).

9.2.4 **Database searching using BLAST and FASTA**

The aim of this section is to give an overview of the tools available for database searching and sequence comparison, it is not intended to give an in-depth description of sequence alignment algorithms. For the theory and methodology behind tools such as The basic local alignment search tool (BLAST) and FASTA, refer to the original research papers or a bioinformatic text book such as an 'Introduction to Bioinformatics' by (Attwood and Parry-Smith 1999).

The BLAST program is based on a heuristic sequence comparison algorithm used to search sequence databases for optimal local alignments to a query (Altschul *et al.* 1990, 1997).

The BLAST suite of program's support a number of different query options. DNA sequence query to DNA database (BLASTN), protein sequence to protein database (BLASTP), six-frame translated DNA sequence query to protein database (BLASTX), protein to six-frame translated DNA database (TBLASTN) and translated DNA sequence against a translated DNA sequence database (TBLASTX). BLAST can be downloaded for local use or can be run at a number of public web sites such as the NCBI (http://www.ncbi.nlm.nih.gov/BLAST/) or DDBJ (http://www.ddbj.nig.ac.jp/E-mail/homology.html).

Position Specific Iterated BLAST (PSI-BLAST) (Altschul *et al.* 1997) is a hybrid version of BLAST used for searching protein sequence against a protein database. This combines the speed of the BLAST pairwise alignment algorithm with the advantages of searching with a sequence profile. On the first run of the program a normal BLASTP search is carried out. Sequences, which are found within a pre-determined threshold are used to build a profile. On subsequent searches, or iterations, the profile is used to search the database and the profile refined based on the new matches identified. This method has the advantage that more distant sequence relationships can be found. However, care must be taken to mask out low complexity regions, as these can cause the profile to degenerate and align false positive matches. Applying this method to GPCR searching can be problematic due to the relative low complexity of transmembrane regions. A public PSI-BLAST server can be found via http://www.ncbi.nlm.nih.gov/BLAST/.

FastA, described by Lipman and Pearson in 1985 (Lipman and Pearson 1985) is based around the idea of identifying short regions of sequence common to both sequences. These are then linked together in a heuristic manner similar to BLAST.

9.2.5 Identification of GPCRs from high throughput sequence sources

One of the main applications of bioinformatics tools to GPCR research has been the identification of new family members. This section will discuss the techniques and tools used in mining both EST and genomic sequence sources.

Searching EST data

As discussed earlier ESTs represent partial cDNA sequence and are therefore less problematic to search than genomic sequence, although in most cases only partial sequence can be assembled. Other factors to consider are that some GPCRs are expressed at low levels in a tissue and will therefore be underrepresented in cDNA libraries and hence EST databases.

A gene mining session would commonly start with a BLAST search against an EST database. To begin a search a query sequence needs to be selected. One point to consider is the domain structure of the query sequence. If other domains are present results may be returned which are not pertinent to the intended query. For instance the rhodopsin family glycoprotein hormone receptors contain a 7 transmembrane domain (TMD) spanning region but also multiple leucine rich repeat (LRR) domains (Jiang et al. 1995). A search with this type of receptor will identify other rhodopsin GPCRs but will also return a large number of LRR containing non-GPCR sequences.

Once the query protein sequence has been selected the nucleotide database should be searched with TBLASTN. We could search the database using a nucleotide sequence and BLASTN, however, far better results will be obtained by using TBLASTN and a protein query, since a gene family will show greater conservation at the amino acid level than at the nucleotide level. Once the results are returned, the alignments (high scoring pairs or HSPs) can be viewed. Knowledge of the conserved residues that describe the GPCR family is of great advantage in looking through the matches. The E-value, a statistical score based on the probability of finding an exact match in the database by chance, can also be used to judge the significance of the match. For globular proteins an E-value of 10^{-3} or lower is deemed significant, however, the structural constraints for related globular proteins are far more rigid than for GPCRs. This is due to the seven transmembrane spanning regions, which can vary more widely in their amino acid composition of hydrophobic residues, but still retain the ability to form a hydrophobic membrane spanning helix. E-values of 9.0 or higher can be obtained and still represent family members.

When an EST match has been identified, the clone identifier should be used to check for $5'$ or $3'$ partner ESTs. As described above, consulting UNIGENE and TIGR HGI will add value to the match by finding overlapping and $5'$–$3'$ linked ESTs. The sequence quality of an EST can be poor, but using the tool ESTWISE (http://www.sanger.ac.uk/Software/Wise2/) allows frame shifts to be tracked through the sequence and extend the region of identity further than that found with BLAST. ESTWISE can also be run as the primary search tool on an EST database. A successful EST mining approach that led to the identification of five novel rhodopsin family GPCRs was recently reported (Wittenberger *et al.* 2001).

Searching genomic sequence

Mammalian genomic sequence can be queried in two ways, first directly by searching the DNA sequence using tools such as BLAST, or second by searching predicted open reading frames from genomic sequence. The latter will be dealt with in genome annotation projects section below. Genomic sequence can be approached initially in a similar way to ESTs. However, genomic sequence contains intron breaks between coding regions, which can result in the matching HSPs being short in length and numerous for a multi-exonic gene. Once the original HSP is found, progression of the potential novel gene can be greatly enhanced by examining the genomic DNA and building a gene model. The aim of building a gene model is to try and identify the coding exons and build a virtual cDNA. Unlike EST sequence sources, it is possible to assemble the complete coding region from genomic sequence if the whole gene is there; this can speed up time in the lab significantly. One difficult challenge to overcome, however, is the fact that much of the genomic BAC sequences contain unordered sequence fragments or contigs, which could split a gene of interest over a number of contigs. To overcome this problem, each contig should be treated as a separate sequence; these can be turned into a BLAST database and searched with the target sequence. Contigs that contain consecutive HSPs can be reassembled in order and the prediction programs can then be run. As genomic sequencing comes to a close, genome assembly projects such as Golden Path, will negate this need for contig reassembly (Lander *et al.* 2001). Figure 9.2 describes a gene model building project.

Although web tools are available to help gene model building, access to a UNIX system and sequence manipulation tools such as EMBOSS (Rice *et al.* 2000) suite of programs allows greater working flexibility. Furthermore many of the advanced gene prediction tools are faster to use at the UNIX command line level and can handle larger sequences for analysis.

First we should discuss some of the tools available to aid gene prediction. Gene prediction programs fall into two groups, *ab initio* and homology driven, and both have their place in building the gene model. Greatest success is achieved when the results from both methodologies are combined and compared. Prior to running the prediction programs the genomic sequence to be studied should be masked for genomic repeats using programs such as REPEATMASKER (http://ftp.genome.washington.edu/cgi-bin/RepeatMasker).

Ab initio, prediction programs look for gene signals in the raw genomic DNA and build an exon model. Table 9.1 provides a list of gene prediction software. The sequence signals used by prediction programs include the GC content of coding versus non-coding DNA, identification of open reading frames, splice site prediction and in some cases promoter assignment and polyadenylation signal prediction. One of the most commonly used of these tools is GENSCAN (Burge and Karlin 1997). GENSCAN uses a Hidden Markov Model of a gene to predict the virtual sequence. It should be noted that false exons can also be predicted, genes spliced together, exons missed and genes split apart. Care therefore needs to be taken when using output from these programs. Additionally in the case of rhodopsin family of GPCRs many members have been found to be single coding exon genes (Gentles and Karlin 1999) therefore these can be missed by exon prediction programs.

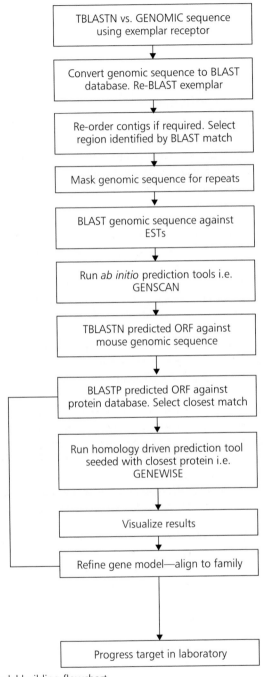

Fig. 9.2 A gene model building flowchart.

Table 9.1 Sample of gene prediction programs and their relevant web sites

Ab intio	Web site	Reference
GENSCAN	http://genes.mit.edu/GENSCAN.html	Burge and Karlin 1997
FGENESH	http://genomic.sanger.ac.uk/gf/gf.shtml	Solovyev and Salamov 1997
MZEF	http://argon.cshl.org/genefinder/	Zhang 1997
GRAIL II	http://compbio.ornl.gov/public/tools/	Xu et al. 1994
HMMgene	http://www.cbs.dtu.dk/services/HMMgene/	Krogh 1997
Homology driven		
GENEWISE	http://www.sanger.ac.uk/Software/Wise2/	Birney and Durbin 2000
FGENESH+	http://genomic.sanger.ac.uk/gf/gf.shtml	See web site
PROCRUSTES	http://www-hto.usc.edu/software/procrustes/	Gelfand et al. 1996

Homology driven tools compare a protein sequence, or profile of a gene family against the genomic DNA, these incorporate a splicing model, such that the identity can be split across exons, but the phase of the exon boundary must be maintained. GENEWISE (Birney and Durbin 2000) is a good example of this type of application, although slower to run than the *ab initio* programs, excellent results can be achieved. It is worth comparing the local alignment option with the global alignment, as this can help identify the N-terminal exons. A list of gene prediction programs is provided in Table 9.1.

Further to using prediction tools, the region of masked genomic sequence should also be compared to EST databases, this can support the exons predicted by the gene finding tools. Use of ESTs to support predictions is exemplified in the identification of the Family 3 receptor GPRC5B (Robbins *et al.* 2000). It is important to remember that ESTs can also be found in untranslated regions, matches which do not lie in a coding exon but say close to a predicted stop codon add further evidence to a gene being present. Furthermore, comparison to other syntenic vertebrate genomic sequence can highlight phylogenic footprints supporting evidence for coding exons, a useful tool for such analysis is PIPMAKER (Schwartz *et al.* 2000) (http://bio.cse.psu.edu/pipmaker/).

Once the prediction tools have been run and comparisons to ESTs and other genomes made, the exons need to be assembled and compared back to the family of interest. It is at this point that the model is handcrafted such that the prediction makes the most biological sense, that is, a predicted exon which inserted a large sequence in the middle of a transmembrane helix would not be likely to occur. This requires assembling the high confidence exons, adhering to splice junction rules. Visualization tools such as GENOTATOR can help the researcher compare the output from the prediction tools and EST searches (Harris 1997) (http://www.fruitfly.org/~nomi/genotator/). The process should be iterated until the optimal prediction is achieved.

Human genome annotation projects

Building gene models can be a difficult procedure. Public automated genome annotation projects are now underway, and are improving constantly. These include the ENSEMBL annotation effort and NCBIs genome view. These combine prediction tools, EST data, and model organism data to provide exon predictions and genome annotation. The resulting predicted open reading frames can be searched by chromosomal location and also by sequence query directly via the following web sites (http://www.ensembl.org/) (http://www.ncbi.nlm.nih.gov/genome/guide/human/). Once a novel gene has been identified the next task is to characterize the protein sequence in greater depth. The following section describes the bioinformatic tools which can help us achieve this. Many of these tools are useful in enhancing the gene prediction, for instance checking the transmembrane organization of a predicted ORF. Again it is the combination of tools which gives the best results.

9.3 Characterization of GPCR sequences

In the process of identifying a new GPCR, the sequence analysis will have clearly placed the receptor into one of the three families, and provided strong clues as to further classification. To identify the activating ligand and/or discover the biological function of the receptor, requires in-depth analysis of the sequence at the protein and nucleotide level. The aim of this section is to describe bioinformatic methods, highlight resources, and point to pertinent literature to help determine key features of a receptor sequence. Some protein motifs appear to be specific for small sub-families based on current knowledge, yet, this type of information will increase as the complete set of GPCRs becomes known and, collectively, should become a valuable resource if the data can be captured in searchable form.

The analysis of sequences is broken down into specific sections which, more or less, form a workflow process for an analyst. It is impossible to be exhaustive and cover all methods. Thus, some sections cover the analysis process generally, some are more specific, but where this occurs, it is because in the authors' opinion they represent the future trend for protein family analysis where the emphasis is shifting to specific details crucial to biological understanding and relevance.

9.3.1 The initiating methionine

An incorrectly predicted initiator methionine can lead to failure to obtain cell-surface expression, or failure to be activated by ligands. Rarely is the initiating methionine defined by experimental N-terminal sequencing studies; thus, there is a clear need to be able to confidently predict the correct one. There are several lines of evidence to suggest the correct one may have been chosen, such as an ATG codon in a favourable context for initiation, or the presence of an upstream in-frame terminator codon, or the prediction of a signal peptide-like sequence at the amino terminus, all of which have some validity (Kozak 1996). The availability of genomic sequence makes the task of identifying an upstream in-frame terminator codon relatively straightforward and any extra sequence evidence from ESTs is also helpful. Comparison with species orthologues, where known or predicted, will also add weight to a correct prediction, since the nucleotide sequence identity decreases going from coding region into the 5' UTR (Makalowski et al. 1996).

9.3.2 Signal peptides

Not all GPCRs are predicted to have signal peptides (e.g. many mammalian Family A GPCRs) but it is important to check, especially if N-terminal peptide-tagging experiments are to be performed. There are several reported methods in the literature and on the web, but SignalP is the probably the most used publicly available program (Nielsen et al. 1999), accessible over the internet at the Center for Biological Sequence Analysis (http://www.cds.dtu.dk/services).

Excellent instructions for using the program are given on the web site, and it is important to use only the N-terminal part of the sequence. This will avoid predictions of abnormally long signal peptides that are likely to be false predictions within transmembrane domains.

9.3.3 Transmembrane regions

To date there is only one three-dimensional crystal structure of a GPCR, that of bovine rhodopsin at 2.8A (Palczewski *et al.* 2000). Although this may be used as a yardstick for assessing the reliability of transmembrane domain (TMD) prediction methods for type I receptors, predictions will have to be relied upon for the foreseeable future. The are many available methods for defining TMD regions based on the primary amino acid sequence (see the ExPASy web site for a list: http://www/expasy.ch/tools). It is also instructive to look at the predictions in tandem with simple hydrophobicity plots (e.g. pepplot, in GCG) (Kyte and Doolittle 1982) as part of the prediction process.

Good results have been obtained when multiple aligned sequences can be analysed; however, in many cases, predictions on single-sequences is important. The Dense Alignment Surface (DAS) method (http://www.sbc.su.se/~miklos/DAS/) was introduced in an attempt to improve sequence alignments in the G protein coupled receptor family of transmembrane proteins (Cserzo *et al.* 1994) and as now been generalized this method to predict transmembrane segments in any integral membrane protein. DAS is based on low-stringency dot-plots of the query sequence against a collection of non-homologous membrane proteins using a special scoring matrix. TMPRED (http://www.ch.embnet.org/software/TMPRED_form.html), uses an algorithm based on the statistical analysis of data in TMbase, which are data largely based on SwissProt data on transmembrane proteins. Other tools use machine learning methods, such as TMHMM (http://www.cds.dtu.dk/services/TMHMM/), HMMTOP (http://www.enzim.hu/hmmtop/index.html). These are based on training sets of transmembrane proteins and negative sets of non-membrane proteins. With all methods, improvement should be seen when actual crystal structures of membrane proteins become more common. A modified version of TMHMM for use with GPCRs, 7TMHMM, will soon be published.

In terms of which program performs the best, it is important to remember that all methods are predictions. Furthermore, a high degree of the underlying data are common to all and comes from the expertly analysed sources. All programs have user-friendly web interfaces but it is the display of results that becomes the factor in deciding user preference. Most produce graphic output but SOSUI (http://sosui.proteome.bio.tuat.ac.jp/sosuiframe0.html) (Hirokawa *et al.* 1998) produces the most extensive (table, hydropathy plot, helical wheel, and a cartoon).

9.3.4 Motifs and domains

The literature on interacting proteins with GPCRs and the extensive role of GPCRs in numerous signalling pathways is ever growing [for a recent review see (Marinissen and Gutkind 2001) and chapters 5, 6, and 7]. This section will overview some resources that allow: (i) ways to identify motifs that characterize a sequence as a member of a specific GPCR sub-family, (ii) identification of motifs that are not specific to the GPCR superfamily but are additional features of the extra and intracellular domains of GPCRs, and (iii) identification of motifs that lead to post-translational events or impact on cell signalling and cell trafficking.

With respect to the last category, the interactions with G protein-receptor kinases or arrestins will not be covered here but this is an active research area (Oakley *et al.* 2001).

(i) While it is clear that a new receptor can be classified into a subfamily with some confidence based on the initial pairwise sequence searches, using BLAST, for example, it is necessary to analyse sequences at a more specific level. Thus, GPCR fingerprints (searchable in PRINTS: http://www.bioinf.man.ac.uk/dbbrowser/PRINTS/) were developed, which allow characterization of regions of sequence in a more fine-grained way, for example, a comparison at a level of the individual TM Ds (Attwood *et al.* 2000). Contrast this with other profile databases, for example, PROSITE (http://www.expasy.ch/prosite/) and the more sophisticated PFAM (http://sanger.ac.uk/Pfam/) (Bateman *et al.* 1999), which provide sequence information at the higher GPCR family level. The program InterPro (http://www.ebi.ac.uk/interpro/scan.html) (Apweiler *et al.* 2001) is an integrated documentation resource of functional descriptions and literature for protein families, domains and functional sites, devloped as a means of rationalizing many complimentary efforts, such as those quoted above. It allows a user to visually compare the results of all these methods. iProclass (http://pir.georgetown.edu/iproclass/) is another integrated resource based around PIR and SWISSPROT that provides comprehensive family relationships and structural/functional and features of proteins (Wu *et al.* 2001).

(ii) In addition to the domain search databases mentioned (PRINTs, PFAM, PROSITE, InterPro) there are other databases that allow the identification of domains in the N and C termini of GPCRs, which are not specific to the GPCR superfamily. SMART (http://smart.embl-heidelberg.de/)—a simple modular architecture research tool (Schultz *et al.* 1998) contains extensively annotated data on signalling domains classified by extra and intracellular location, as alignments, profiles and hidden Markov models. This is a coordinated source of literature and sequence information. ProDom (http://www.toulouse.inra.fr/prodom.html) (Corpet *et al.* 2000) derives alignments based on PSI-BLAST. Each resource has its own style and the inter-relatedness is evident from the way they mostly link to one another. A feature of some members of Type III GPCRs is the coiled coil motif found in the cytoplasmic C termini. These are known to be important for the GABAB receptors and are one of the reasons why the known subtypes, GABABR1 and GABABR2, heterodimerize. Indeed, this feature was key to the yeast two hybrid experiments that led to the discovery of the second subtype (Kammerer *et al.* 1999; Kuner *et al.* 1999; White *et al.* 1998). Predicting these is straightforward using programs such as COILS (http://www.ch.embnet.org/software/COILS_form.html) or Multicoil (http://nightingale.lcs.mit.edu/cgi-bin/multicoil) (Wolf *et al.* 1997).

(iii) Sequence motifs that lead to post-translational processing, N-glycosylation sites, for example, (found in the extracellular domains of GPCRs) can be readily predicted using PROSITE. Sequence motifs that influence both cellular trafficking events (e.g. manifest in cloning studies by lack of cell surface expression) and cell signalling events, are less well understood. These cannot be predicted reliably. Just as the knowledge of amino acid residues involved in the interactions of receptors with heterotrimeric G proteins (discussed separately below) has been generated by years of biochemical study, the same will be true for understanding the sequence basis of trafficking and signalling pathways. The PSORT program (http://psort.nibb.ac.jp/) (Nakai and Horton 1999) is a good general place to start for a wide range of predicted features, including some of those covered in earlier sections. More specifically, it is becoming increasingly recognized through work on ion channels that there are more ER retention signals than first thought, and these affect the trafficking of membrane proteins to the cell surface (Teasdale and Jackson 1996; Zerangue *et al.* 1999; Ma *et al.* 2001). Some of the known motifs are RXR(R), KKXX, and RKR, generally found in the C-terminus of sequences. These can be easily searched for textually, or by using a pattern searching programs, such as in GCG. It is the RXRR motif present in the GABA$_{B1}$ receptor that results in the receptor being retained in the ER. Coexpression with the GABA$_{B2}$ homologue results in masking of the motif by the coiled coil interacting domains, and allows the heterodimer to reach the cell surface (Margeta-Mitrovic *et al.* 2000; Calver *et al.* 2001, see chapter 28).

There are other examples of motifs that currently appear subfamily specific, which over time may form a part of a more integrated picture and allow predictions to be made. A specific example concerns the interaction of some members of the metabotropic glutamate receptor family with Homer adapter proteins. These interactions have been explained through identification of a proline-rich sequence (PPSPF) in the C terminus of the receptors (Xiao *et al.* 2000). These adaptor proteins are important again, not only in the intracellular signalling process of the receptors, but also for getting the receptors to the cell surface (Roche *et al.* 1999). Several receptors (dopamine D4, muscarinic M4) have been shown to have polyproline based motifs (PXXP) in their third intracellular loops which interact with SH3 binding domains of adapter proteins, for example, Grb2 (Ren *et al.* 1993; Oldenhof *et al.* 1998; Tang *et al.* 1999). Recent experiments on mutant dopamine D3 receptors indicate that the picture for some receptors may be more complicated than originally thought (Oldenhof *et al.* 2001).

PDZ domains (constituents of scaffold proteins) are easy to predict, but the interacting proteins less so. For one PDZ domain protein, PSD-95 (Cho *et al.* 1992), originally identified as part of the post-synaptic density, the interacting proteins are known to interact through their C termini. This is true for the beta-adrenergic receptors (Hu *et al.* 2000 and refs therein) where the signature ESKV from the beta 1 subtype is important, and is consistent with the generalized pattern S/TXV(L/I). Recently, the multi-PDZ domain protein MUPP1 has been shown to interact with the extreme C-terminal SSV sequence of the 5-HT2C receptor (Becamel *et al.* 2001).

These specific examples are important because they indicate that while some general domain features of receptor sequences are easily identified, teasing out more specific features to understand the biology, will only emerge from very specific experimentation.

(iv) Knowledge of the receptor interactions with heterotrimeric G proteins has important implications for work on all receptors, particularly orphans. Being able to predict the correct heterotrimeric G protein family-receptor association will help assay development and functional studies. For any receptor, possible functional effects of genetic variations will be important in the context of disease associations. The current knowledge on key amino acid residues has been gleaned from years of extensive biochemical studies, typically by performimg site-directed mutagenesis experiments and generating chimeric receptors. The results of these studies demonstrate that most of the intracellular regions (loops and C terminus) of a receptor are implicated (Wess 1998).

Computational approaches to predict the G protein coupling profiles, using primary sequence data, have been published for known receptors. Vriend and colleagues applied a sequence analysis method (corelated mutation analysis—CMA) on aminergic and adenosine receptors (Horn *et al.* 2000). The conclusion was that a weak signal could be detected in receptor families where specialization for coupling to a given G protein occurred during a recent divergent evolutionary event, for example, with the muscarinic receptors.

An alternative data mining method for predicting coupling specificity of G protein coupled receptors to their G proteins, has combined pattern discovery and membrane topology prediction. The patterns were derived by analysing a set of receptor sequences for which the literature indicated clear non-promiscuous coupling to G proteins. The method discovered patterns of amino acid residues in the intracellular domains that are specific for coupling to particular functional classes of G proteins and hence can be used to predict coupling specificity of orphans (Moller *et al.* 2001).

9.3.5 Alignments

For orphan receptors (see chapter 10), it is important to know how the sequences align with other members of the GPCR family primarily to provide clues to possible ligands [for a recent example, (Chambers *et al.* 2000)]. Multiple alignments, however, have many uses, such as the identification of important conserved residues, structure, and function prediction, demonstration of homology and inference of evolutionary relationships. In the case of GPCRs, they have evolved in more than one direction, for example, towards G proteins (as noted in the previous section for muscarinic receptors) and towards ligands, as in the case of histamine receptors (Zhu *et al.* 2001).

Phylogeny is a specialist subject and accurate analyses are time consuming and best left to specialists. The best alignment will be the one that represents the most likely evolutionary picture but because of the complexity in trying to achieve these, fast approximate methods have been developed (Duret and Abdeddaim 2000). There are many multiple sequence alignment programs and ClustalW (Thompson *et al.* 1994) is one of the more widely used. It is a progressive global alignment method that is most applicable where the sequences to be aligned are related over their entire length. Alignments can be viewed and annotated using packages such as JalView Clamp (http://ww.ebi.ac.uk/~michele/jalview/), belvu (http://www.sanger.ac.ukPfam/help/belvu_setup.shtml) and Genedoc (Nicholas, http://www.cris.com/~Ketchup/genedoc.shtml). Phylogenetic trees can be generated by a multitude of methods (see http://evolution.genetics.washington.edu/phylip/software.html). The methods PAUP* (http://www.lms.si.edu/PAUP/) (Swofford 1999) and Phylip (http://evolution.genetics.washington.edu/phylip.html) (Felsenstein 2000) are commonly used to calculate consensus trees as can be seen in the examples of the receptor for UDP-glucose (Chambers *et al.* 2000) and discovery of new receptors from ESTs (Wittenberger *et al.* 2001). TreeView (Page 1996) is a popular package used to view generated trees on a PC.

9.4 In silico analysis of biological function and disease association

Finding associations to link a receptor with biological function and disease is a complex process that needs integration of information from sequence databases and literature. In addition to searching the obvious human data sources, one must not forget that information from other species may be important in some cases. It is impossible to be comprehensive and include all available resources so the process described below is given from the perspective of starting with a sequence and adding information to that knowledge (Fig. 9.3).

In order to be able to interrogate the web resources the chromosomal localization or genetic map position of the receptor are required, together with any information on close genetic markers (single nucleotide polymorphisms—SNPs, or microsatellites). Mapping

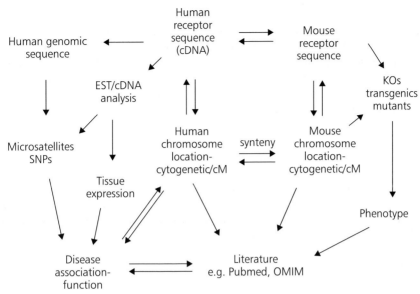

Fig. 9.3 Relationships of gene sequence and genetic data sources that can be used to find disease associations.

information for a gene can be obtained in a variety of ways. Given that most of the human genome sequence is now available it is easy to locate a gene on a chromosome by pairwise sequence comparison methods. There are many web sites that allow this to be done. Ensembl (http://www.ensembl.org), the NCBI site (http://www.ncbi.nlm.nih.gov/BLAST/) and the Human Genome Browser (http://genome.ucsc.edu/goldenPath/hgTracks.html) are just three of the excellent resources available. The OMIM database—Online Mendelian Inheritance in Man—(http://www.ncbi.nlm.nih.gov/Omim/) is a catalogue of human genes and genetic disorders based on the published literature. It is now incorporated in the NCBI's Entrez system and can thus be queried in the same way as GenBank and PubMed (http://www.ncbi.nlm.nih.gov/entrez/query.fcgi?db=OMIM). This source will provide information on mapping and known diseases linking to that gene or region. Another useful resource for linking information is the NCBI's map viewer (http://www.ncbi.nlm.nih.gov/PMGIFs/Genomes/MapViewerHelp.html) which is a comprehensive viewer that provides searching opportunities at four levels: (1) the Organism home page summarizing the resources available for that organism; (2) a Genome view displaying the complete genome as a set of chromosome ideograms; (3) a Map view presenting one or more maps of a selected chromosome; and (4) a Sequence view of a chromosome region graphically depicting the biological features annotated to that region. With extensive links, a user can quickly gather information related to genes of interest.

With regard to identification of genetic variation of the receptor sequence, there are other web sources in addition to the ones cited above that may provide further information. HGBASE (http://hgbase.cgr.ki.se/) is an attempt to summarize all known sequence variations in the human genome and includes highly curated information on SNPs, indels, simple tandem repeats and other sequence alternatives (Brookes 2001). The CGAP Genetic Annotation Initiative (http://lpg.nci.nih.gov/) is a research program to explore and apply technology for the identification of genetic variation of genes involved in cancer, but is in fact a source for examining variation in all genes. This is a data mining exercise, and differences observed in ESTs are a key source of data. The SNPs covered are classed as candidate (predicted), validated (observed in an experiment) and confirmed (tested in a minimum number of five CEPH families for Mendelian transmission and placed in genetic reference maps). This site also links to HGBASE and dbSNP, the NCBI database of SNPs (http://www.ncbi.nlm.nih.gov/SNP/).

It is important to remember that studies on mouse receptors, knockouts, for example, or existence of mouse mutants, also increase our understanding of the human receptors. There are several web sources for finding the equivalent mouse receptors and the syntenic regions of mouse chromosomes. A comprehensive source is the mouse genome informatics (MGI) site (http://www.informatics.jax.org/) where one can view mammalian homology maps, and link even further to lower organisms. If the mouse chromosomal region is known it is possible to see if any phenotypes have been linked to the region, for example, the data provided by the mouse ENU Mutagenesis Programme (http://www/mgu/har/mrc/ac/uk/mutabase/) can suggest links to phenotypes (Nolan *et al.* 2000). This web resource is linked back to the MGI.

Sequence variation can also be analysed in terms of the significance of any nucleotide change to protein sequence and receptor function. It is useful to know if any predicted change has been observed before. A useful reference source for GPCRs is the database tGRAP (originally tinyGRAP) (http://tinygrap.uit.no/), which is a mutation database of GPCR mutation data containing over 10,000 mutations from close to 1400 papers (Beukers *et al.* 1999). This data source has multiple sequence alignments of family members, with TMDs highlighted, and also links to SWISSPROT entries.

9.5 Concluding remarks

We are rapidly approaching the stage when the human genome sequence will be complete. The set of human GPCRs is now almost complete but the challenges ahead are no less awesome than when we first set out to determine the full complement. There are still orphan receptors waiting to be paired with ligands. We need to understand the sequence variation of individual receptors and how this variation is translated into variant and functional forms of proteins, and what the biological relevance is. There are favoured sub-families of receptors where the pharmaceutical industry has had tremendous success in designing selective and safe drugs, take the aminergics for example. Soon we will see an increasing number of effective medicines against peptide-liganded and prostaglandin-related receptors. Continued detailed biological work on the newer members of these families, in order to understand their biological significance and thus allowing the continuation of the successful drug discovery trend, will generate much new data on the part these receptor play in signalling processes and disease. The next steps for the bioinformaticians will be to bring order to this new information.

Acknowledgements

The authors would like to thank their colleagues at GlaxoSmithKline and Inpharmatica for helpful discussions and comments.

References

Altschul SF, Gish W, Miller W, Myers EW, and Lipman DJ (1990). Basic local alignment search tool. *J Mol Biol* 215, 403–10.

Altschul SF, Madden TL, Schaffer AA *et al.* (1997). Gapped BLAST and PSI-BLAST: a new generation of protein database search programs. *Nucleic Acids Res* 25, 3389–402.

Apweiler R, Attwood TK, Bairoch A *et al.* (2001). The InterPro database, an integrated documentation resource for protein families, domains and functional sites. *Nucleic Acids Res* 29, 37–40.

Attwood TK, Croning MD, Flower DR *et al.* (2000). PRINTS-S: the database formerly known as PRINTS. *Nucleic Acids Res* 28, 225–7.

Attwood TK and Parry-Smith DJ (1999). *Introduction to Bioinformatics*, (1st edition) *Cell and Molecular Biology in Action Series*: Longman Limited, Harlow, UK.

Bairoch A and Apweiler R (2000). The SWISS-PROT protein sequence database and its supplement TrEMBL in 2000. *Nucleic Acids Res* 28, 45–48.

Barker WC, Garavelli JS, Hou Z *et al.* (2001). Protein information resource: a community resource for expert annotation of protein data. *Nucleic Acids Res* 29, 29–32.

Bateman A, Birney E, Durbin R, Eddy SR, Finn RD, and Sonnhammer EL (1999). Pfam 3.1: 1313 multiple alignments and profile HMMs match the majority of proteins. *Nucleic Acids Res* 27, 260–2.

Becamel C, Figge A, Poliak S *et al.* (2001). Interaction of serotonin 5-hydroxytryptamine type 2C receptors with PDZ10 of the multi-PDZ domain protein MUPP1. *J Biol Chem* 276, 12974–82.

Beukers MW, Kristiansen I, IJzerman AP, and Edvardsen I (1999). TinyGRAP database: a bioinformatics tool to mine G-protein-coupled receptor mutant data. *Trends Pharmacol Sci* 20, 475–7.

Birney E and Durbin R (2000). Using GeneWise in the Drosophila annotation experiment. *Genome Res* 10, 547–8.

Boguski MS, Lowe TM, and Tolstoshev CM (1993). dbEST—database for "expressed sequence tags". *Nat Genet* 4, 332–3.

Calver AR, Robbins MJ, Cosio C *et al.* (2001). The C-terminal domains of the GABA(b) receptor subunits mediate intracellular trafficking but are not required for receptor signaling. *J Neurosci* 21, 1203–10.

Chambers JK, Macdonald LE, Sarau HM *et al.* (2000). A G protein-coupled receptor for UDP-glucose. *J Biol Chem* 275, 10767–71.

Cho KO, Hunt CA, and Kennedy MB (1992). The rat brain postsynaptic density fraction contains a homolog of the Drosophila discs-large tumor suppressor protein. *Neuron* 9, 929–42.

Corpet F, Servant F, Gouzy J, and Kahn D (2000). ProDom and ProDom-CG: tools for protein domain analysis and whole genome comparisons. *Nucleic Acids Res* 28, 267–9.

Cserzo M, Bernassau JM, Simon I, and Maigret B (1994). New alignment strategy for transmembrane proteins. *J Mol Biol* 243, 388–96.

Duret L and Abdeddaim S (2000). Multiple alignments for structural, functional, or phylogenetic analyses of homologous sequences, In (eds, D. Higgins and W. Taylor) *Bioinformatics: Sequence, Structure and Databanks*, Oxford University Press, Oxford UK, pp. 51–76.

Felsenstein J (2000). PHYLIP (Phylogeny Inference Package), Version 3.6, Department of Genetics, University of Washington, Seattle.

Gentles AJ and Karlin S (1999). Why are human G-protein-coupled receptors predominantly intronless? *Trends Genet* 15, 47–9.

Gibas C and Jambeck P (2001). *Developing Bioinformatic Computer Skills*, First edition, O'Reilly & Associates Inc., Sebastopol, California.

Harris NL (1997). Genotator: a workbench for sequence annotation. *Genome Res* 7, 754–62.

Hill J, Duckworth M, Murdock P *et al.* (2001). Molecular cloning and functional characterization of MCH2, a novel human MCH receptor. *J Biol Chem* 276, 20125–9.

Hirokawa T, Boon-Chieng S, and Mitaku S (1998). SOSUI: classification and secondary structure prediction system for membrane proteins. *Bioinformatics* **14**, 378–9.

Horn F, van der Wenden EM, Oliveira L, IJzerman AP, and Vriend G (2000). Receptors coupling to G proteins: is there a signal behind the sequence? *Proteins* **41**, 448–59.

Hough LB (2001). Genomics meets histamine receptors: new subtypes, new receptors. *Mol Pharmacol* **59**, 415–19.

Hu LA, Tang Y, Miller WE *et al.* (2000). Beta 1-adrenergic receptor association with PSD-95. Inhibition of receptor internalization and facilitation of beta 1-adrenergic receptor interaction with N-methyl-D-aspartate receptors. *J Biol Chem* **275**, 38 659–66.

Jiang X, Dreano M, Buckler DR *et al.* (1995). Structural predictions for the ligand-binding region of glycoprotein hormone receptors and the nature of hormone-receptor interactions. *Structure* **3**, 1341–53.

Jones KA, Borowsky B, Tamm JA *et al.* (1998). GABA(B) receptors function as a heteromeric assembly of the subunits GABA(B)R1 and GABA(B)R2. *Nature* **396**, 674–9.

Kammerer RA, Frank S, Schulthess T, Landwehr R, Lustig A, and Engel J (1999). Heterodimerization of a functional GABAB receptor is mediated by parallel coiled-coil alpha-helices. *Biochemistry* **38**, 13 263–9.

Kaupmann K, Malitschek B, Schuler V *et al.* (1998). GABA(B)-receptor subtypes assemble into functional heteromeric complexes. *Nature* **396**, 683–7.

Kawai J, Shinagawa A, Shibata K *et al.* (2001). Functional annotation of a full-length mouse cDNA collection. *Nature* **409**, 685–90.

Kozak M (1996). Interpreting cDNA sequences: some insights from studies on translation. *Mamm Genome* **7**, 563–74.

Kuner R, Kohr G, Grunewald S, Eisenhardt G, Bach A, and Kornau HC (1999). Role of heteromer formation in GABAB receptor function. *Science* **283**, 74–7.

Kyte J and Doolittle RF (1982). A simple method for displaying the hydropathic character of a protein. *J Mol Biol* **157**, 105–32.

Lander ES, Linton LM, Birren B *et al.* (2001). Initial sequencing and analysis of the human genome. *Nature* **409**, 860–921.

Lipman DJ and Pearson WR (1985). Rapid and sensitive protein similarity searches. *Science* **227**, 1435–41.

Ma D, Zerangue N, Lin YF *et al.* (2001). Role of ER export signals in controlling surface potassium channel numbers. *Science* **291**, 316–19.

Makalowski W, Zhang J, and Boguski MS (1996). Comparative analysis of 1196 orthologous mouse and human full-length mRNA and protein sequences. *Genome Res* **6**, 846–57.

Margeta-Mitrovic M, Jan YN, and Jan LY (2000). A trafficking checkpoint controls GABA(B) receptor heterodimerization. *Neuron* **27**, 97–106.

Marinissen MJ and Gutkind JS (2001). G-protein-coupled receptors and signaling networks: emerging paradigms. *Trends Pharmacol Sci* **22**, 368–76.

Moller S, Vilo J, and Croning MDR (2001). Prediction of the coupling specificity of G protein-coupled receptors to their G proteins. *Bioinformatics* **17**, Suppl. 1, S174–S181.

Nakai K and Horton P (1999). PSORT: a program for detecting sorting signals in proteins and predicting their subcellular localization. *Trends Biochem Sci* **24**, 34–6.

Nielsen H, Brunak S, and von Heijne G (1999). Machine learning approaches for the prediction of signal peptides and other protein sorting signals. *Protein Eng* **12**, 3–9.

Nolan PM, Peters J, Strivens M *et al.* (2000). A systematic, genome-wide, phenotype-driven mutagenesis programme for gene function studies in the mouse. *Nat Genet* **25**, 440–3.

Oakley RH, Laporte SA, Holt JA, Barak LS, and Caron MG (2001). Molecular determinants underlying the formation of stable intracellular G protein-coupled receptor-beta-arrestin complexes after receptor endocytosis*. *J Biol Chem* **276**, 19 452–60.

Oldenhof J, Ray A, Vickery R, and Van Tol HH (2001). SH3 ligands in the dopamine D3 receptor. *Cell Signal* 13, 411–16.

Oldenhof J, Vickery R, Anafi M *et al.* (1998). SH3 binding domains in the dopamine D4 receptor. *Biochemistry* 37, 15 726–36.

Page RD (1996). TreeView: an application to display phylogenetic trees on personal computers. *Comput Appl Biosci* 12, 357–8.

Palczewski K, Kumasaka T, Hori T *et al.* (2000). Crystal structure of rhodopsin: A G protein-coupled receptor. *Science* 289, 739–45.

Ren R, Mayer BJ, Cicchetti P, and Baltimore D (1993). Identification of a ten-amino acid proline-rich SH3 binding site. *Science* 259, 1157–61.

Rice P, Longden I, and Bleasby A (2000). EMBOSS: the european molecular biology open software suite. *Trends Genet* 16, 276–7.

Robbins MJ, Michalovich D, Hill J *et al.* (2000). Molecular cloning and characterization of two novel retinoic acid- inducible orphan G-protein-coupled receptors (GPRC5B and GPRC5C). *Genomics* 67, 8–18.

Roche KW, Tu JC, Petralia RS, Xiao B, Wenthold RJ, and Worley PF (1999). Homer 1b regulates the trafficking of group I metabotropic glutamate receptors. *J Biol Chem* 274, 25 953–7.

Sailer AW, Sano H, Zeng Z *et al.* (2001). Identification and characterization of a second melanin-concentrating hormone receptor, MCH-2R. *Proc Natl Acad Sci USA* 98, 7564–9.

Sakurai T, Amemiya A, Ishii M *et al.* (1998). Orexins and orexin receptors: a family of hypothalamic neuropeptides and G protein-coupled receptors that regulate feeding behavior. *Cell* 92, 573–85.

Schultz J, Milpetz F, Bork P, and Ponting CP (1998). SMART, a simple modular architecture research tool: identification of signaling domains. *Proc Natl Acad Sci USA*, 95, 5857–64.

Schwartz S, Zhang Z, Frazer KA *et al.* (2000). PipMaker—a web server for aligning two genomic DNA sequences. *Genome Res* 10, 577–86.

Swofford DL (1999). PAUP* (Phylogenetic Analysis Using Parsimony (*and other methods), Version 4, Sinauer Associates, Sunderland, MA.

Tang Y, Hu LA, Miller WE *et al.* (1999). Identification of the endophilins (SH3p4/p8/p13) as novel binding partners for the beta1-adrenergic receptor. *Proc Natl Acad Sci USA* 96, 12 559–64.

Teasdale RD and Jackson MR (1996). Signal-mediated sorting of membrane proteins between the endoplasmic reticulum and the golgi apparatus. *Annu Rev Cell Dev Biol* 12, 27–54.

Thompson JD, Higgins DG, and Gibson TJ (1994). CLUSTAL W: improving the sensitivity of progressive multiple sequence alignment through sequence weighting, position-specific gap penalties and weight matrix choice. *Nucleic Acids Res* 22, 4673–80.

Wess J (1998). Molecular basis of receptor/G-protein-coupling selectivity. *Pharmacol Ther* 80, 231–64.

White JH, Wise A, Main MJ *et al.* (1998). Heterodimerization is required for the formation of a functional GABA(B) receptor. *Nature* 396, 679–82.

Wittenberger T, Schaller HC, and Hellebrand S (2001). An expressed sequence tag (EST) data mining strategy succeeding in the discovery of new G-protein coupled receptors. *J Mol Biol* 307, 799–813.

Wolf E, Kim PS, and Berger B (1997). MultiCoil: a program for predicting two- and three-stranded coiled coils. *Protein Sci* 6, 1179–89.

Wu CH, Xiao C, Hou Z, Huang H, and Barker WC (2001). iProClass: an integrated, comprehensive and annotated protein classification database. *Nucleic Acids Res* 29, 52–4.

Xiao B, Tu JC, and Worley PF (2000). Homer: a link between neural activity and glutamate receptor function. *Curr Opin Neurobiol* 10, 370–4.

Zerangue N, Schwappach B, Jan YN, and Jan LY (1999). A new ER trafficking signal regulates the subunit stoichiometry of plasma membrane K(ATP) channels. *Neuron* 22, 537–48.

Zhu Y, Michalovich D, Wu H *et al.* (2001). Cloning, expression, and pharmacological characterization of a novel human histamine receptor. *Mol Pharmacol* 59, 434–41.

Chapter 10

Characterization of orphan GPCRs

Mark Fidock

10.1 Introduction

Bioinformatic analysis of sequence databases to identify new GPCR family members based on homology and conserved 'motifs' to known GPCRs suggests that the total number of GPCRs is likely to be around 350 (this does not include olfactory, taste or opsin receptors, see chapter 1). Of these 350, the number which are currently classified as 'orphans' (that is receptors without a defined physiologically relevant ligand) is in the region of 175. These 'orphans' typically show low sequence identity (<40%) to known receptors and are distributed throughout the entire GPCR phylogenetic tree. A large number show greater sequence identity to each other than to any known GPCR suggesting they are new subfamilies with distinct, possibly novel ligands.

The pharmaceutical industry and academic community continues to invest considerable resources and efforts in the GPCR family, as it is justifiably perceived as a source of chemically attractive therapeutic targets. There is an ever-increasing effort to evaluate orphan receptors and determine their physiological role. Indeed over the past two years there has been a linear growth in the number of publications describing the pairing of orphan GPCRs with their cognate ligand (Fig. 10.1). The substantial growth of sequence data, bioinformatic advances in sequence analysis, and the development of sophisticated high-throughput screening technologies may explain this increase in success. This chapter will focus on the advances made in 'enabling' technologies for pairing orphan GPCRs with their cognate or surrogate ligands, and methods to determine which receptors will be of likely therapeutic potential.

10.2 Characterizing orphan receptors using a reverse pharmacology approach

Although complex and technically challenging, the strategy that has become standard for the identification of ligands for orphan GPCRs is known as the 'reverse pharmacology' approach, which is to clearly distinguish it from the more classical approach to drug discovery (Stadel *et al.* 1997). Historically, the classical approach was initiated by the discovery of a biological or physiological activity that could be attributed to a specific ligand, this was then used to pharmacologically characterize the tissue to determine its therapeutic potential. Subsequently, the ligand was used to isolate its corresponding receptor for use as a drug target in high-throughput screening. The reverse approach begins with an orphan receptor of unknown function that is used as a 'hook' to fish for its ligand. The ligand is then used to characterize the function of the receptor in normal cell signalling and to establish its

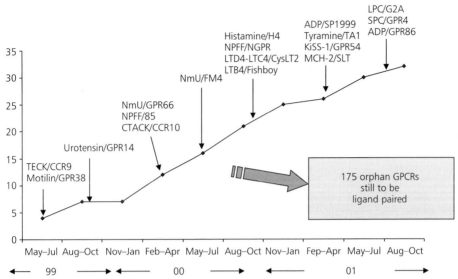

Fig. 10.1 Demonstrates the linear rise in the number of publications describing the pairing of orphan GPCRs with their cognate ligand. A number of examples have been highlighted illustrating the important new biological mediators that have been discovered.

pathophysiological role or therapeutic potential. In parallel, high-throughput screening is initiated in order to develop selective chemical tools that will realize the therapeutic value of the receptor.

A possible strategy to characterize orphan GPCRs is described in Fig. 10.2. Using bio-informatic analyses such as basic local alignment search tool (BLAST) (Altschul *et al.* 1990) searching with the known members of the GPCR superfamily as a probe set, plus the use of Hidden Markov Models (Grundy *et al.* 1997) and motif searches, we believe we have identified almost all the 'orphan' GPCRs contained within the human genome (see chapter 9). This totals, at the time of writing, to around 140 family 1: rhodopsin like, 30 family 2: calcitonin like and 5 family 3 receptors: metabotropic like. Full-length cDNAs have been isolated using standard molecular biology techniques or are available as off-the-shelf reagents. In order to prioritize those receptors for screening, tissue expression patterns are determined. In the first instance, information from expressed sequence tag (EST) databases such as that produced by Incyte Pharmaceuticals provides a limited view of expression. This is because GPCRs are generally expressed at low levels, therefore frequently below the sensitivity of their database, so a negative result is not particularly informative, but the results can be obtained instantly and direct a more focused study. More sensitive techniques using either quantitative PCR, for example, Taqman against a panel of cDNA's from important therapeutic tissues, or immunohistochemistry studies to identify expression in important cell types of therapeutic interest (e.g. Lifespan Inc, Seattle) provides data of the highest quality. In addition, Deltagen (Redwood City) produces a database containing information on numerous mouse knockouts. This is used increasingly to highlight important orphan receptors based on observed phenotypes. Once prioritized, the receptors are expressed in mammalian cells for functional analysis. The expression system chosen is of critical importance for success; cell lines with a good history of GPCR expression, which contain a wide

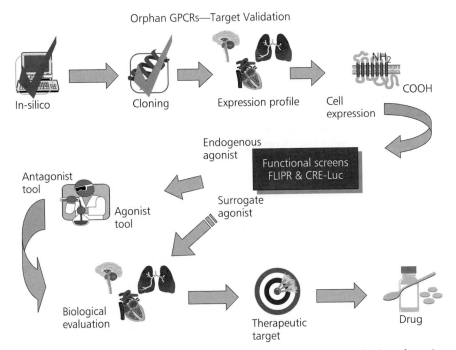

Fig. 10.2 Reverse pharmacology strategy for the identification and characterization of novel receptor:ligand pairs. The identification and cloning of the GPCR gene family is now complete. The remaining challenge is to identify ligands for orphan receptors and to define their pathophysiological role.

variety of G proteins to allow functional downstream coupling to effector systems are key. Human embryonic kidney 293 (HEK293) and Chinese hamster ovary (CHO) are often the cell lines of choice. The expressed receptor is then screened in a variety of functional assays to identify an activating ligand and since it is very difficult to predict the coupling system of a receptor, it is essential to configure the assays so as to detect the broadest array of coupling mechanisms (Kostenis 2001). However, such systems are still unlikely to work universally. Challenges to overcome include, for example, the presence of endogenous receptors in such mammalian host cell lines, and especially clonal variations in the patterns of endogenous receptor expression in cells derived from the same parental cell line.

10.2.1 Applications of the 'reverse pharmacology' strategy

From the discovery that the ligand for the orphan receptor G-21 was serotonin (Fargin *et al.* 1988), which represented the first example of 'ligand pairing', the methodology has not really evolved appreciably *per se*. The greatest advances have been in the 'technology', for example, sensitivity of the assays to increase signal-to-noise, and the miniaturization of assay formats to allow high-throughput. Nowadays, it is very much the standard to be screening in 384 well format, but in general, we continue to use changes in cAMP levels or mobilization of intracellular calcium as endpoints. Other, more generic assays have been applied, making use of natural phenomena such as the measurement of colour changes using frog melanophores

(Birgul *et al.* 1999; Lynch *et al.* 1999; Lenz *et al.* 2001) or changes in extracellular pH using the microphysiometer (Habata *et al.* 1999; Hinuma *et al.* 2000).

The advantage of the melanophore technology is that the pigment bearing cells contain all the major G proteins (G_q, G_s, and G_i) allowing a functional response to be elicited without detailed knowledge of the receptor's coupling mechanism. In melanophores, G_s or G_q activation results in pigment dispersion, while G_i mediated signalling leads to pigment aggregation. The technology is ideally suited for use in multi-well screening and pharmacological profiling of lead compounds. However, the technology is not considered to be high-throughput by today's standards. The attractiveness of this technology for many of the major pharma groups is reduced due to the difficulty in obtaining melanophores, the current need for high-throughput assay formats, and the ease-of-use and generic nature of today's mammalian expression systems.

The microphysiometer, which measures extracellular media acidification in response to perturbations of energy consumption in mammalian cells, provides a low-throughput capability. Since activation of any signal-transduction pathway results in energy usage, this approach also obviates the need to define intracellular signalling mechanistically. This assay has been used to identify the natural ligand for the orphan GPCR APJ, termed apelin (Habata *et al.* 1999), and the receptor for mammalian RF-amide peptides (Hinuma *et al.* 2000). However, in the absence of improvements for a higher throughput capacity assay, broader application of the above technologies will be limited.

The current orphan strategy makes use of promiscuous and chimeric G proteins which allow orphan receptors to be functionally coupled through a common pathway, resulting in calcium mobilization via phospholipase C (PLC) in standard mammalian cell-lines. While unlikely to work for all GPCRs, it has considerably broadened the range of receptors that will give measurable calcium mobilization responses. The approach makes use of the naturally occurring promiscuous G proteins ($G_{\alpha 15}$ and $G_{\alpha 16}$) (Offermanns and Simon 1995), and chimeric G_{α} subunits (G_{qi5}), in which the C terminal five amino acids of the G_q subunit are replaced by corresponding amino acids from the adenylyl-cyclase linked G_i subunit (Milligan and Rees 1999). The G_q family subunit, G_{q11}, was used in the discovery of the ghrelin receptor (Howard *et al.* 2000) and more recently, several groups identified the melanin-concentrating hormone 1 and 2 (Chambers *et al.* 1999; Sailer *et al.* 2001), and the NPFF1 (Elshourbagy *et al.* 2000) receptors using Ca^{2+}-based assays in which chimeric G proteins were used.

To date, the most successful strategy to identify ligands for orphan receptors has been accomplished using assays that report changes in intracellular Ca^{2+} as a result of PLC activation. Changes in calcium mobilization can be easily detected using standard fluorescent based methods, via a high-throughput imaging system such as FLIPR™ (Fluorescent Imaging Plate Reader, Molecular Devices). Several important biological mediators defined by this means are summarized in Table 10.1. The advantages of these assays are their high-throughput capability (automated and 96 or 384-well plate capacity), robustness (high signal-to-noise ratio arising from large intracellular amplification) and flexibility (mode of detection and varied instrumentation). This technology has proved successful in our labs, with a number of ligand pairs identified. For example, we and others, have found the endogenous ligand for FM-4 to be Neuromedin U (Howard *et al.* 2000) and for HG57 to be LTC_4 (Heise *et al.* 2000). These being the NUM2R and CysLT2 receptors, respectively.

As already stated, the use of promiscuous G proteins has allowed the effective coupling of G_q and G_i receptors to calcium mobilization, but the ability to couple those receptors that stimulate adenylyl-cyclase via the G protein G_s to calcium mobilization has proven much

Table 10.1 Identification of ligands for orphan GPCRs. The ligand pairings are grouped on the basis of the mechanism by which they were discovered. The citation is only illustrative and not comprehensive as in a number of cases several groups have identified the same ligand pair. An attempt to cite the first publication or back-to-back publications for each of the receptors has been made.

Measured activity	Orphan GPCR paired	Ligand	Putative function	References
Ca^{2+}	HFGAN72	Orexin	Feeding/sleep-wakefulness	Sakurai et al. 1998
	GHS-R	Ghrelin	GH secretion/Adiposity	Kojima et al. 1999
	GPR-9–6	TECK	Gastric inflammation	Zaballos et al. 1999
	GPR24	Melanin-concentrating hormone	Feeding	Chambers et al. 1999; Saito et al. 1999
	HG55	Leukotriene D4	Bronchial constriction	Lynch et al. 1999
	GPR38	Motilin	Gastric motility	Feighner et al. 1999
	GPR14	Urotensin II	Vasoconstriction	Ames et al. 1999
	OGR-1	Spingosylphosphorylcholine	Cell proliferation	Xu et al. 2000
	HG57	Leukotriene C4	Bronchial constriction	Heise et al. 2000
	444-J19	Leukotriene B4	Bronchial constriction	Wang et al. 2000
	GPR2	Eskine	Leukocyte traffiking	Jarmin et al. 2000
	FM-3	Neuromedin U	Feeding	Fujii et al. 2000
	FM-4	Neuromedin U	Feeding	Howard et al. 2000
	HLWAR77	NPFF	Pain modulation	Elshourbagy et al. 2000
	SP1999	ADP	Thrombosis	Zhang et al. 2001
	SLT	Melanin-concentrating hormone	Obesity	Sailer et al. 2001
	CRTH2	Prostaglandin D2	Induces TH2 chemotaxis	Hirai et al. 2001
	GPR54	KiSS-1	Trophoblast Invasion	Ohtaki et al. 2001
	G2A	Lysophosphatidylcholine		Kabarowski et al. 2001
	GPR4	Spingosylphosphorylcholine	Atherosclerosis/ inflammation	Zhu et al. 2001
cAMP	ORL-1	Nociceptin	Anxiety/memory	Reinscheid et al. 1995
	EDG-1	Spingosine-1-phosphate	Cell differentiation/ growth	Lee et al. 1998
	GPCR97	Histamine	CNS-obesity, psychiatry	Lovenberg et al. 1999
	H4	Histamine	Inflammation	Oda et al. 2000
	TDAG8	Psychosine	Cytokinesis	Im et al. 2001
	GPR86	ADP	Hematopoiesis/ inflammation	Communi et al. 2001
	TA1	Tyramine	Depression/electrolyte homeostatis	Borowsky et al. 2001
Energy utilization	APJ	Apelin	Immune response	Habata et al. 1999
	OT7TO22	RF-amides 1–3	Pain	Hinuma et al. 2000
GIRK channels	DAR-1	Allostatin	?	Birgul et al. 1999
	DAR-2	Allostatin	?	Lenz et al. 2000
Arach. acid	GPR10	Prolactin releasing peptide	Prolactin secretion	Hinuma et al. 1998
Yeast pheromone	EDG-2	Lysophosphatidic acid	Cell differentiation/ growth	Erickson et al. 1998
	KIAA0001	UDP-glucose	?	Chambers et al. 2000

more difficult (Milligan and Rees 1999). However, the availability of assays that measure the cytoplasmic level of cAMP via a reporter gene construct, for example, making use of the cAMP response elements (CRE) upstream of either β-lactamase or Luciferase reporter genes, has provided a sensitive and reasonably straightforward assay to allow high-throughput screening. We have made extensive use of this assay in parallel to the FLIPR™ screen as our primary screening cascade and it has delivered a number of successes. For example, we and others have identified an orphan GPCR that is highly expressed in the kidney and is activated by tyramine and β-phenylethylamine; this being the first member of the Trace Amine Receptor family (Borowsky *et al.* 2001). Other alternatives to the reporter-based methods for measuring cAMP levels include radioimmunoasay and scintillation proximity assay (Amersham Pharmacia Biotech). These assays have been used to identify the ligands for the orphan receptor GPR97 as the Histamine H_3 receptor (Lovenberg *et al.* 1999) and GPR86 as the ADP receptor (Communi *et al.* 2001) (Table 10.1).

Functional expression of GPCRs in yeast offers an alternative high-throughput option for screening. The pheromone pathway has been engineered to permit functional coupling of human GPCRs to growth on a selective media or reporter gene expression (Price *et al.* 1995). In comparison to mammalian cells, yeast cells are robust, cheap and straightforward to handle. This methodology has had limited success (Erickson *et al.* 1998; Chambers *et al.* 2000) compared to the other systems described above. This is probably attributable either to the difficulty in demonstrating functional expression of the GPCR in yeast or to ligand binding to components of the yeast cell wall.

In industry, the ability to screen in a high-throughput manner is key as it is essential to screen an orphan receptor against a wide range of putative ligands. We have amassed a collection of over 2000 naturally occurring putative GPCR ligands (bioactive peptides, small molecules, lipids, and carbohydrates) which these constitute the first phase in the screening process. This 'ligands file' contains all known or suspected GPCR ligands, together with a large number of molecules for which the receptor is unknown. It contains a number of bioactive peptides from lower species and novel peptides derived through *in-silico* analysis of the human genome. Screening against this ligand file has provided us with a number of successes. If no hits are identified we initiate a second round of screening for novel ligands using biological extracts of tissues, fluids, and cell supernatants. In addition we may also screen against a file of small organic molecules, selected computationally based on known GPCR pharmacophores, in order to identify surrogate ligands. These can be used as tool compounds to explore receptor function in more physiological systems. Once a ligand has been identified a high-throughput screen is initiated to identify antagonists of the receptor.

10.3 Realizing the potential therapeutic value of de-orphanization

This second phase is by far the most challenging, this involves using the chemical tools (agonist/antagonist) for the orphan receptor in disease or mechanistically relevant physiological assays to evaluate the role of the receptor and its potential as a therapeutic target for drug discovery. It is essential to have a detailed knowledge of the receptor's expression at the cell level, in both normal and diseased tissue to aid in directing functional studies most efficiently. We and others discovered a novel histamine receptor H_4 (H_4R) (Oda *et al.* 2000), which demonstrated a restricted expression pattern, predominately in peripheral blood leukocytes. A paper was identified describing a novel histamine receptor on eosinophils (Raible *et al.* 1995) whose pharmacology could not be rationalized in terms of H_1, H_2 or H_3 receptor characteristics, that is, an example of 'missing pharmacology'. Continuation of our studies with relevant *in-vitro* assays demonstrated a role for the histamine H_4 receptor in histamine driven eosinophil chemotaxis, as such it is a potential target in inflammatory diseases (Fig. 10.3). Briefly, chemotaxis studies on isolated human eosinophils confirmed that histamine is chemotactic, and that selective agonists of the known histamine receptors (H_1, H_2, and H_3) do not induce such a response. Furthermore, studies employing histamine-receptor antagonists showed only H_3 antagonists clobenpropit and thioperamide inhibited chemotaxis due to their mixed H_3/H_4 pharmacology. Since these compounds are also antagonists of the H_4 receptor we postulated that the receptor mediating histaminergic chemotaxis is the histamine H_4 (O'Reilly et al. 2002).

Alternative target validation tools include transgenic mouse knockout lines. These can be useful in evaluating receptor function, although care is needed when evaluating these results as species difference in receptor expression may be unrepresentative of man. Using the histamine H_4 receptor as an example, it has been reported that the murine histamine H_4 receptor demonstrates CNS expression specifically in the hippocampus (Zhu et al. 2001). However, in the reports concerning the human histamine H_4, we and others (Oda et al. 2000;

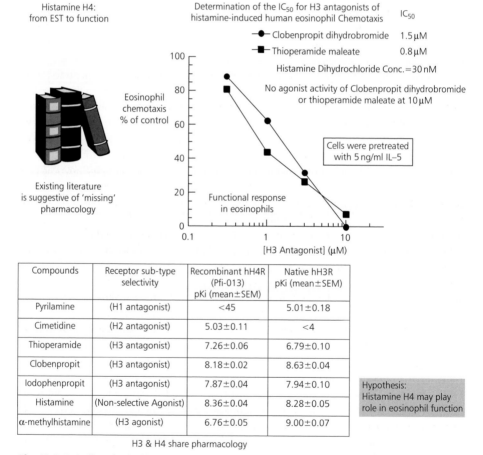

Compounds	Receptor sub-type selectivity	Recombinant hH4R (Pfi-013) pKi (mean±SEM)	Native hH3R pKi (mean±SEM)
Pyrilamine	(H1 antagonist)	<45	5.01±0.18
Cimetidine	(H2 antagonist)	5.03±0.11	<4
Thioperamide	(H3 antagonist)	7.26±0.06	6.79±0.10
Clobenpropit	(H3 antagonist)	8.18±0.02	8.63±0.04
Iodophenpropit	(H3 antagonist)	7.87±0.04	7.94±0.10
Histamine	(Non-selective Agonist)	8.36±0.04	8.28±0.05
α-methylhistamine	(H3 agonist)	6.76±0.05	9.00±0.07

H3 & H4 share pharmacology

Fig. 10.3 Describes the isolation and characterization of the human histamine H_4 receptor. The pharmacological profile of H_4 and the 'missing pharmacology' as described by Raible et al. (1994) led us to test the hypothesis and demonstrate that H_4 may play a role in eosinophil chemotaxis.

O'Reilly *et al.* 2002) have not been able to demonstrate any expression of this receptor in the brain.

10.4 Alternative technologies to identify ligands for orphan receptors

Further refinements to the generic-screening cascade described above can and have been made. For example, introducing signal peptides into the N terminal of GPCRs as a chimera to direct the orphan receptor to the cell surface. This approach may be considered controversial since the exact nature of ligand binding is unknown and there may be instances where ligand binding would be altered or inhibited and therefore not identified. The class of receptors most at risk may be the protease activated and chemokine receptors where it is known that the N-terminal domain is important. Synaptic Pharmaceutical Corporation is an example of a company who have been reported to screen orphan receptors using signal peptides in their expression strategy. Their 'Universal Functional Assay' still employs the 'reverse pharmacology' approach. In essence, this FLIPR based assay is based on; (1) expression engineered to ensure cell surface localization (2) novel promiscuous G proteins (its ability to couple a greater diversity of receptors than those G proteins more generally available), and (3) an extensive library of novel bioactive molecules for screening. Using this technology platform, Synaptic identified ligands for NPFF1, NPFF2 (Bonini *et al.* 2000), NMU1, NMU2 (Raddatz *et al.* 2000), and the Trace amine receptor TA1 (Borowsky *et al.* 2001).

An alternative approach to delineating function is to design a strategy which allows surrogate ligands to be identified either as agonists or inverse agonists. This can be achieved through screening constitutively active GPCRs against a chemical file of synthetic organic compounds. The principal of constitutive activation is making use of the phenomenon that receptor's can be stabilized in an active confirmation in the absence of their natural ligand (Costa and Hertz 1989; Samama *et al.* 1993, see chapter 8). Briefly, GPCRs are hypothesized to exist in a dynamic equilibrium between an inactive (R) and active (R*) state, in the absence of ligand the vast majority of the receptor population are in the inactive state (Kenakin *et al.* 1996). Upon agonist binding, the equilibrium favours R* by stabilizing the active confirmation. There are two established methods to produce constitutively active GPCRs. The first is through over expression in recombinant systems to a point of observing constitutive activity, achieved by raising the number of R* in absolute and relative terms (Kenakin *et al.* 1996). This population of R* can be further isolated using for instance fluorescent activated cell sorting (FACs) methodology, where ligand independent activation of the receptor is measured via reporter gene activity, thus providing a homogenous population of activated receptors for screening (Hong *et al.* 2000). Alternatively, changing a single amino acid in the third intracellular loop, of the β-2-adrenergic receptor, has been shown to induce constitutive activity (Kjelsberg *et al.* 1992). These constitutively active mutants (CAMs) are locked in the activated state, allowing standard high-throughput screening which detects compounds either enhancing CAM activity (i.e. agonists), or decreasing the activity (i.e. inverse agonsits). These tools can then be used to define the pathophysiological role of the receptor. The advantage of this approach is that one is likely to identify 'hits' at a greater frequency than screening against known or suspected biologically active molecules. The greatest disadvantage is that the identity of the endogenous ligand remains unknown and so any additional biological information and 'missing pharmacology' that may aid the quest in demonstrating the therapeutic value of the discovery is lacking, thereby making this a much more challenging task. Arena Pharmaceuticals have been using this approach exclusively to identify lead compounds for orphan receptors (Behan and Chalmers 2001). They have, as expected, identified chemical leads for a number of receptors and have examples in a number of different therapeutic areas. Their most advanced discoveries are small molecule modulators of the human 18F orphan GPCR, for the potential treatment of metabolic diseases and obesity. Preliminary experiments of the 18F GPCR modulators in animal models have shown that acute administration of these compounds leads to a reduction in food intake, a decrease in body weight and an increase in fat metabolism (Investigational Drugs Database (IDdb)).

A further approach developed by Acadia Pharmaceuticals makes use of the functional assay system called receptor selection and amplification technology (R-SAT). R-SAT is a cell-based assay system that has been used to detect agonist activity across a diverse range of GPCRs (Messier *et al.* 1995; Burstein *et al.* 1997). In the R-SAT assay, a series of potential target GPCRs can be mixed together and transfected into cells in culture. In the absence of an added test compound, the cells which express the GPCRs continue to show normal growth kinetics until they encounter another cell, at which point all cellular growth ceases. If a test compound activates a GPCR, the cells that express that receptor are able to grow in preference to all other cells in the culture. Cells, which express a GPCR activated by a compound, are therefore selected and amplified in the culture relative to those that are transfected with other target GPCRs. In short, the technology uses the principle of genetic selection as a method to evaluate compound/target interactions. A number of features of the R-SAT system make it an efficient and productive method. First, a strong correlation has been demonstrated between receptor pharmacology, as determined by the R-SAT system and response in humans (Weiner *et al.* 2001). Second, the technology allows a group of genes or a whole gene family to be tested simultaneously, that is, multiplexed. This feature, combined with the simplicity of the assay format, allows the studies to be automated and performed at ultra high-throughput.

The final methodology exploits the observation that GPCR desensitization is a well-conserved mechanism which is reasonably common amongst GPCRs and involves the arrestin family of proteins (Krupnick and Benovic 1998). Briefly, agonist binding to the GPCR activates G proteins and ultimately results in second messenger cascades to elicit a physiological response. In most cases receptors occupied by agonists undergo rapid phosphorylation by GPCR Kinases (GRKs). This in turn allows β-arrestin to bind to the GPCR preventing any further interaction between the GPCR and G protein. Internalization of the GPCR via clathrin coated pits follows as a result of arrestin–clathrin interactions (Krupnick and Benovic 1998). By employing fusion constructs of green fluorescent protein (GFP)-β arrestin, this receptor desensitization can be visualized using confocal microscopy. In the absence of receptor activation in a transfected cell line, fluorescence associated with the β-arrestin-GFP is evenly distributed throughout the cytoplasm. Upon receptor activation by an agonist, cytosolic β-arrestin-GFP translocates to the cell surface reflecting the interaction with the activated receptor (Evans *et al.* 2001). The use of this technology has been reported for ligand confirmation at individual receptors (Evans *et al.* 2001), and Norak Biosciences have developed the system for high-throughput use in collaboration with Amersham Pharmacia Biotech. Irrespective of the throughput, both applications are based around the same methodology.

10.5 **Summary**

Clearly, scientists now have access to high-throughput technologies to identify ligands for orphan receptors. A number of these are able to produce a functional readout, directed through a single pathway, allowing ligands to be found without knowledge of the receptors endogenous coupling mechanism. Based on the diversity of ligands which have been paired, it is likely that various assay formats and a much larger array of ligands will be needed to pair the remaining orphans. But, this may still not yield the complete picture of receptor : ligand pairs as we need to bear in mind the lessons learned from the calcitonin receptor-like receptor (CRLR) which requires receptor activity modifying proteins (RAMPs) to express the full array of pharmacological phenotypes (McLatchie *et al.* 1998, see chapter 7). This highlights the importance of additional variables in assay design to characterize orphan GPCRs and discover new ligands. Although it is unclear to what extent GPCRs employ the RAMP proteins but recently another example has been reported, functional amylin receptors are dependent on RAMP1 or RAMP3 coexpression (Christopoulos *et al.* 1999). Another variable is the recent observations that GPCRs oligomerise. Both homo- and hetero-examples have been described and the importance clearly demonstrated in the discovery of the GABA$_B$ receptor (Jones *et al.* 1998; White *et al.* 1998; Kaupman *et al.* 1998, see chapter 28). However, the importance, in the broadest context, of these observations pharmacologically is the subject of considerable debate (Bouvier 2001). Clearly, if oligomerization is involved in trafficking or receptor function this could have an effect on the ability to identify ligands for orphan receptors. However, the best measure of success for any initiative to identify ligands for orphan receptors will be the number of new, validated targets, and ultimately those that lead to novel therapies in the clinic. Time will be the judge and jury.

References

Altschul SF, Gish W, Miller W, Myers EW, and Lipman DJ (1990). Basic local alignment search tool. *J Mol Biol* 215, 403–10.

Ames RS, Sarau HM, Chambers JK *et al.* (1999). Human urotensin-II is a potent vasoconstrictor and agonist for the orphan receptor GPR14. *Nature* 401, 282–6.

Behan DP and Chalmers DT (2001). The use of constitutively active receptors for drug discovery at the G protein-coupled receptor gene pool. *Curr Opin Drug Discovery and Dev* 4, 548–60.

Birgul N, Weise C, Kreienkamp HJ, and Richter D (1999). Reverse physiology in drosophila: identification of a novel allatostatin-like neuropeptide and its cognate receptor structurally related to the mammalian somatostatin/galanin/opioid receptor family. *Eur Mol Biol Organisation J* 18, 5892–900.

Bonini JA, Jones KA, Adham N *et al.* (2000). Identification and characterisation of two G protein-coupled receptors for neuropeptide FF. *J Biol Chem* 275, 39 324–31.

Borowsky B, Adham N, Jones KA *et al.* (2001). Trace amines: Identification of a family of mammalian G protein-coupled receptors. *Proc Natl Acad Sci USA* 98, 8966–71.

Bouvier M (2001). Oligomerization of G protein-coupled transmitter receptors. *Nature Rev Neurosci* 2, 274–86.

Burstein ES, Brauner-Osborne H, Spalding TA, Conklin BR, and Brann MR (1997). Interactions of muscarinic receptors with the heterotrimeric G proteins Gq and G12: transduction of proliferative signals. *J Neurochem* 68, 525–33.

Chambers JK, Ames RS, Bergsma D *et al.* (1999). Melanin-concentrating hormone is the cognate ligand for the orphan G-protein-coupled receptor SLC-1. *Nature* 400, 261–5.

Chambers JK, MacDonald LE, Sarau HM *et al.* (2000). A G protein-coupled receptor for UDP-glucose. *J Biol Chem* 275, 10 767–71.

Christopoulos G, Perry KJ, Morfis M *et al.* (1999). Multiple amylin receptors arise from receptor activity-modifying protein interaction with the calcitonin receptor gene product. *Mol Pharmacol* **56**, 235–42.

Communi D, Gonzalez NS, Detheux M *et al.* (2001). Identification of a novel human ADP receptor coupled to G(i). *J Biol Chem* **276**, 41 479–85.

Costa T and Hertz A (1989). Antagonists with negative intrinsic activity at the delta opioid receptors coupled to GTP-binding proteins. *Proc Natl Acad Sci USA* **86**, 7321–25.

Elshourbagy NA Ames RS Fitzgerald LR *et al.* (2000). Receptor for the pain modulatory neuropeptides FF and AF is an orphan G protein-coupled receptor. *J Biol Chem* **275**, 25 965–71.

Erickson JR, Wu JJ, Goddard JG *et al.* (1998). Edg-2/Vzg-1 couples to the yeast pheromone response pathway selectively in response to lysophosphatidic acid. *J Biol Chem* **273**, 1506–10.

Evans NA, Groarke DA, Warrack J *et al.* (2001). Visualising differences in ligand-induced β-arrestin-GFP interactions and trafficking between three recently characterised G protein-coupled receptors. *J Neurochem* **77**, 476–85.

Fargin A, Raymond JR, Lohse MJ *et al.* (1988). The genomic clone G-21 which resembles a beta-adrenergic receptor sequence encodes the 5HT1A receptor. *Nature* **335**, 358–60.

Feighner SD, Tan CP, McKee KK *et al.* (1999). Receptor for motilin identified in the human gastrointestinal system. *Science* **284**, 2148–8.

Fujii R, Hosoya M, Fukusumi S *et al.* (2000). Identification of neuromedin U as the cognate ligand of the orphan receptor FM-3. *J Biol Chem* **275**, 21 068–74.

Grundy WN, Bailey TL, Elkan CP *et al.* (1997). Meta MEME: Motfi-based hidden Markov models of protein familes. *Comp Appl Biosci* **13**, 397–406.

Habata Y, Fujii R, Hosoya M *et al.* (1999). Apelin, the natural ligand of the orphan receptor APJ, is abundantly secreted in the colostrum. *Biochem Biophy Acta* **1452**, 25–35.

Heise CE, O'Dowd BF, Figueroa DJ *et al.* (2000). Characterisation of the human cysteinyl leukotriene 2 receptor. *J Biol Chem* **275**, 30 531–36.

Hinuma S, Habata Y, Fujii R *et al.* (1998). A prolactin-releasing peptide in the brain. *Nature* **394**, 211–2.

Hinuma S, Shintani Y, Fukusumi S *et al.* (2000). New neuropeptides containing carboxy-terminal Rfamide and their receptors in mammals. *Nature Cell Biol* **2**, 703–8.

Hirai H, Tanaka K, Yoshie O *et al.* (2001). Prostaglandin D2 selectively induces chemotaxis in T-helper type 2 cells, eosinophils, and basophils via seven-transmembrane receptor CRTH2. *J Experimental Medicine* **193**, 255–61.

Howard AD, Wang R, Pong SS *et al.* (2000). Identification of receptors for neuromedin U and its role in feeding. *Nature* **406**, 70–4.

Im DS, Heise CE, Nguyen T *et al.* (2001). Identification of a molecular target of psychosine and its role in globoid cell formation. *J Cell Biol* **153**, 429–34.

Investigational Drugs Database (IDdb) (1999). Current Drugs Limited.

Jarmin DI, Rits M, Bota D *et al.* (2000). Cutting edge: identification of the orphan receptor G-protein-coupled receptor 2 as CCR10, a specific receptor for the chemokine Eskine. *J Immunol* **164**, 3460–4.

Jones KA, Borowsky B, Tamm JA *et al.* (1998). GABA(B) receptors function as a heteromeric assembly of the subunits GABA(B)R1 and GABA(B)R2. *Nature* **396**, 674–9.

Jordan BA and Devi LA (1999). G-protein-coupled receptor heterodimerisation modulates receptor function. *Nature* **399**, 697–700.

Kabarowski JH, Zhu K, Le LQ *et al.* (2001). Lysophosphatidylcholine as a ligand for the immunoregulatory receptor G2A. *Science* **293**, 702–5.

Kaupmann K, Malitschek B, Schuler V *et al.* (1998). GABA(B)-receptor subtypes assemble into functional heteromeric complexes. *Nature* **396**, 683–7.

Kenakin TP (1996). The classification of seven transmembrane receptors in recombinant expression systems. *Pharmacol Rev* **48**, 413–63.

Kjelsberg MA, Cotecchia S, Ostrowski J *et al.* (1992). Constitutive activation of the alpha 1B adrenergic receptor by all amino acid substitutions at a single site. Evidence for a region which constrains receptor activation. *J Biol Chem* **267**, 1430–33.

Kojima M, Hosoda H, Date Y *et al.* (1999). Ghrelin is a growth-hormone-releasing acylated peptide from stomach. *Nature* **402**, 656–60.

Kostenis E (2001). Is Gα16 the optimal tool for fishing ligands of orphan G-protein coupled receptors? *Trends in Pharmacol Sci* **22**, 560–64.

Krupnick JG and Benovic JL (1998). The role of receptor kinases and arrestins in G protein-coupled receptor regulation. *Ann Rev Pharmacol Toxicol* **38**, 289–319.

Lee MJ, Van Brocklyn JR, Thangada S *et al.* (1998). Spingosine-1-phosphate as a ligand for the G protein-coupled receptor EDG-1. *Science* **279**, 1552–5.

Lenz C, Williamson M, Hansen GN, and Grimmelikhuijzen CJ (2001). Identification of four Drosophila allatostatins as the cognate ligands for the Drosophila orphan receptor DAR-2. *Biochem Biophy Res Communications*, 286, 1117–22.

Lovenberg TW, Roland BL, Wilson SJ *et al.* (1999). Cloning and functional expression of the human histamine H3 receptor. *Mol Pharmacol* **55**, 1101–7.

Lynch KR, O'Neill GP, Liu Q *et al.* (1999). Characterisation of the human cysteinyl leukotriene CysLT1 receptor. *Nature* **399**, 789–93.

McLatchie LM, Fraser NJ, Main MJ *et al.* (1998). RAMPs regulate the transport and ligand specificity of the calcitonin-receptor-like receptor. *Nature* **393**, 333–39.

McVey M, Ramsay D, Kellett E *et al.* (2001). Monitoring receptor oligomerisation using time-resolved fluorescence resonance energy transfer and bioluminescence resonance energy transfer. The human delta opioid receptor displays constitutive oligomerisation at the cell surface, which is not regulated by receptor occupancy. *J Biol Chem* **276**, 14 092–99.

Messier TL, Dorman CM, Brauner-Osborne H, Eubanks D, and Brann MR (1995). High-throughput assays of cloned adrenergic, muscarinic, neurokinin, and neurotrophin receptors in living mammalian cells. *Pharmacol Toxicol* **76**, 308–11.

Milligan G and Rees S (1999). Chimeric Gα proteins : their potential use in drug discovery. *Trends in Pharmacol Sci* **20**, 118–24.

Oda T, Morikawa N, Saito Y *et al.* (2000). Molecular cloning and characterisation of a novel member of the histamine receptor family. *J Biol Chem* **275**, 36 781–86.

Offermanns S and Simon MI (1995). G alpha 15 and G alpha 16 couple a wide variety of receptors to phospholipase C. *J Biol Chem* **270**, 15 175–80.

Ohtaki T, Shintani Y, Honda S *et al.* (2001). Metastasis suppressor gene KiSS-1 encodes peptide ligand of a G protein-coupled receptor. *Nature* **411**, 613–17.

O'Reilly M, Alpert R, Jenkinson S *et al.* (2002). Identification of a histamine H4 receptor on human eosinophils—role in eosinophil chemotaxis. *J Receptor and Signal Transduction Research* (Accepted).

Price LA, Kajkowski EM, Hadcock JR *et al.* (1995). Functional coupling of a mammalian somatostatin receptor to the yeast pheromone response pathway. *Mol Cell Biol* **15**, 6188–95.

Raible DG, Lenahan T, Fayvilevich Y *et al.* (1994). Pharmacological characterisation of a novel histamine receptor on eosinophils. *American Journal Respiratory Critical Care Medicine* 149, 1506–11.

Raddatz R, Wilson AE, Artymyshyn R *et al.* (2000). Identification and characterisation of two neuromedin U receptors differentially expressed in peripheral tissues and the central nervous system. *J Biol Chem* **275**, 32 452–9.

Reinscheid RK, Nothacker HP, Bourson A *et al.* (1995). Orphanin FQ: a neuropeptide that activates an opioidlike G protein-coupled receptor. *Science* **270**, 792–4.

Sailer AW, Sano H, Zeng Z *et al.* (2001). Identification and characterisation of a second melanin-concentrating hormone receptor, MCH-2R. *Proc Natl Acad Sci USA* **98**, 7564–69.

Saito Y, Nothacker HP, Wang *et al.* (1999). Molecular characterisation of the melanin-concentrating-hormone receptor. *Nature* **400**, 265–9.

Sakurai T, Amemiya A, Ishii M *et al.* (1998). Orexins and orexin receptors: a family of hypothalamic neuropeptides and G protein-coupled receptors that regulate feeding behaviour. *Cell* **92**, 573–85.

Samama P, Cotecchia S, Costa T, and Lefkowitz RJ (1993). A mutation-induced activated state of the β2-adrenergic receptor : Extending the ternary complex model. *J Biol Chem* **268**, 4625–36.

Stadel JM, Wilson S, and Bergsma DJ (1997). Orphan G protein-coupled receptors: a neglected opportunity for pioneer drug discovery. *Trends in Pharmacol Sci* **18**, 430–37.

Venter JC, Adams MD, Myers EW *et al.* (2001). The sequence of the human genome. *Science* **291**, 1304–51.

Wang S, Gustafson E, Pang L *et al.* (2000). A novel hepatointestinal leukotriene B4 receptor. Cloning and functional characterisation. *J Biol Chem* **275**, 27 000–4.

Weiner DM, Burstein ES, Nash N *et al.* (2001). 5-hydroxytryptamine2A receptor inverse agonists as antipsychotics. *J Pharmacol Experimental Therapeutics* **299**, 268–76.

White JH, Wise A, Main MJ *et al.* (1998). Heterodimerization is required for the formation of a functional GABA(B) receptor. *Nature* **396**, 679–82.

Xing H, Tran HC, Knapp TE *et al.* (2000). A Fluorescent reporter assay for the detection of ligands activating G protein-coupled receptors. *J Receptor and Signal Transduction* **20**, 189–210.

Xu Y, Zhu K, Hong G *et al.* (2000). Spingosylphosphorylcholine is a ligand for the ovarian cancer G protein-coupled receptor 1. *Nature Cell Biol* **2**, 261–7.

Zaballos A, Gutierrez J, Varona R *et al.* (1999). Cutting Edge: identification of the orphan chemokine receptor GPR-9–6 as CCR9, the receptor for the chemokine TECK. *J Immunol* **162**, 5671–5.

Zhang FL, Luo L, Gustafson E *et al.* (2001). ADP is the cognate ligand for the orphan G protein-coupled receptor SP1999. *J Biol Chem* **276**, 8608–15.

Zhu K, Baudhuin LM, Hong G *et al.* (2001). Spingosylphosphorylcholine and lysophosphatidylcholine are ligands for the G protein-coupled receptor GPR4. *J Biol Chem* **276**, 41325–35.

Zhu Y, Michalovich D, Wu HL *et al.* (2001). Cloning and expression and pharmacological characterisation of a novel human histamine receptor. *Mol Pharmacol* **59**, 434–41.

Part 2

Family 1 GPCRs

Chapter 11

Adenosine receptors

Bertil B. Fredholm

11.1 Introduction

Adenosine is not a neurotransmitter but rather a paracrine neuromodulator. This obviously does not mean that adenosine is not biologically important—it just means that we must think differently about its role compared to, for example, dopamine or noradrenaline.

There are two main ways in which adenosine levels can increase: it can be released from cells primarily by means of bi-directional transporters (Thorn and Jarvis 1996; Griffiths *et al.* 1997; Dunwiddie and Diao 2000), or it can be formed from breakdown of extracellular adenine nucleotides by means of a multitude of ecto-nucleotidases (Zimmermann 2000). Judging by the large number of studies using techniques such as microdialysis adenosine levels are likely to be in the range of 20 to 300 nM (Ballarín *et al.* 1991; Baranowski and Westenfelder 1994; Pazzagli *et al.* 1995; Huston *et al.* 1996; Porkka-Heiskanen *et al.* 1997). Furthermore, these studies show that adenosine levels can be altered by hypoxia, by drugs and by behavioural states. Thus, cells in the nervous system will always be exposed to adenosine at concentrations close to 100 nM. Locally, levels can increase in response to synaptic activity, to increased metabolic demand and to several other stimuli. Globally, levels may change as a consequence of decreased supply of oxygen or glucose. However, compared to many transmitters, changes in adenosine are likely to occur over wider domains and with a slower time course.

11.2 Molecular characterization

Despite the fact that adenosine receptors were well characterized and partially purified, the two first adenosine receptors cloned, A_1 and A_{2A}, came from a library of orphan receptors from the dog thyroid (Maenhaut *et al.* 1990; Libert *et al.* 1991). Soon the same receptors were cloned from rat and humans (Mahan *et al.* 1991; Furlong *et al.* 1992), and a related receptor, the A_{2B} receptor, was cloned from rat brain (Stehle *et al.* 1992). Whereas these receptors had all been predicted from extensive pharmacological studies, the fourth receptor, A_3 was more unexpected (Zhou *et al.* 1992). These four adenosine receptors have now been cloned from a variety of mammalian and non-mammalian species (Table 11.1). A_1, A_{2A}, and A_{2B} receptors are well conserved among mammals, whereas A_3 receptors show considerable structural variability.

For all four adenosine receptors the coding region is split up by an intron in a region corresponding to the second intracellular loop (Fredholm *et al.* 2000). When the structure of the A_1 receptor was first reported the presence of two major transcripts were noted. The

Table 11.1 Adenosine receptors

	Adenosine A$_1$	Adenosine A$_{2A}$	Adenosine A$_{2B}$	Adenosine A$_3$
Alternative names	R$_i$; A1R	A$_{2a}$, R$_s$; A2AR	A$_{2b}$, R$_s$; A2BR	A3R
Structural information (Accession no.)	h 326 aa (P30542) r 326 aa (P25099) m 326 aa (Q60612)	h 410 aa (P29274) r 409 aa (P30543) m 409 aa (UO5672)	h 328 aa (P29275) r 332 aa (P29276) m 332 aa (UO5673)	h 318 aa (P33765) r 320 aa (P28647) m 320 aa (AF069778)
Chromosomal location	1q32.1	22q11.2	17p11.2–12	1p13.3
Selective agonists	CPA, CCPA, CHA	CGS21680, HE-NECA, CV1808, CV1674, ATL146e	—	Cl-IB-MECA
Selective antagonists	DPCPX, 8-cyclopentyltheophylline, WRC0571	SCH 58261, (moderately selective) ZM241385, KF17387, CSC	MRS1754, enprofylline, alloxazine	MRS1220, MRE3008-F20, MRS1191, MRS1523
Radioligands	[^{3}H]-DPCPX, [^{3}H]-CHA	[^{3}H]-CGS21680, [^{3}H]-SCH58261, [^{3}H]-ZM241385	([^{3}H]-ZM241385, [^{3}H]-DPCPX)	[^{3}H]-MRE3008-F20
G protein coupling	G$_i$/G$_o$	G$_s$/G$_{olf}$	G$_s$/G$_q$	G$_i$/G$_o$
Expression profile	brain including cerebral cortex, cerebellum, hippocampus, dorsal horn of spinal cord, eye, adrenal gland	striatopallidal GABAergic neurons, caudate-putamen, nucleus accumbens, tuberculum olfactorium, olfactory bulb, low in other brain regions	blood vessels, eye, median eminence, mast cells, low levels in adrenal and pituitary glands	brain but particularly cerebellum, hippocampus

Physiological function	bradycardia, inhibition of lipolysis, reduced glomerular filtration, tubulo-glomerular feedback, antinociception, reduction of sympathetic and parasympathetic activity, presynaptic inhibition, neuronal hyperpolarization, ischemic preconditioning	regulation of sensorimotor integration in basal ganglia, inhibition of platelet aggregation and polymorphonuclear leukocytes, vasodilatation, protection against ischemic damage, stimulation of sensory nerve activity	relaxation of smooth muscle in vasculature and intestine, inhibition of monocyte and macrophage function, stimulation of mast cell mediator release (some species)	enhancement of mediator release from mast cells (some species), preconditioning (some species)
Knockout phenotype	anxiety, hyperalgesia, decreased tolerance to hypoxia, loss of tubulo-glomerular feedback	anxiety, hypoalgesia, hypertension, increased tolerance to ischemia, altered sensitivity to motor stimulant drugs, decreased platelet aggregation	—	altered inflammatory reactions, altered release of inflammatory mediators, decreased edema
Disease relevance	acute and chronic pain, renal failure, sleep disorders, epilepsy, obesity, brain and cardiac ischemia	Parkinson's disease, schizophrenia, asthma, inflammation	asthma, inflammation	asthma, inflammation, cardiac ischemia

functional consequences of this have since been elucidated (Ren and Stiles 1994, 1995). Transcripts containing exons 4, 5, and 6 were found in all tissues expressing the receptor, whereas transcripts containing exons 3, 5, and 6 were in addition found in tissues such as brain, testis, and kidney that express high levels of the receptor. There are two promoters, a proximal one denoted promoter A, and a distal one denoted promoter B, which are about 600 bp apart. The A_{2A} receptor shows one hybridizing transcript in most tissues examined (Maenhaut *et al.* 1990; Stehle *et al.* 1992). It should also be mentioned that the human adenosine A_{2A} receptor is polymorphic. In particular, a (silent) T1083C mutation occurs in various populations, more frequently in Caucasians than in Asians (Deckert *et al.* 1996). The rat A_{2B} receptor shows two hybridizing transcripts of 1.8 and 2.2 kb, where the latter is predominant (Stehle *et al.* 1992). This could, in analogy with the above, suggest the presence of multiple promoters. The human A_3 receptor shows two transcripts: the most abundant being approximately 2 kb in size, and the less abundant about 5 kb (Atkinson *et al.* 1997), perhaps indicating similarities with the A_1 receptor gene.

The four adenosine receptor subtypes are asparagine-linked glycoproteins and all but the A_{2A} have sites for palmitoylation near the carboxyl terminus (Linden 2001). Depalmitoylation of A_3 (but not A_1) receptors renders them susceptible to phosphorylation by G protein-coupled receptor kinases (GRKs), which in turn results in rapid phosphorylation and desensitization (Palmer and Stiles 2000).

11.3 Cellular and subcellular localization

The distribution of receptors tells us where agonists and antagonists given to the intact organism may act. Furthermore, the rather low levels of endogenous adenosine present under basal physiological conditions have the potential of activating receptors where they are abundant, but not where they are sparse (Kenakin 1993, 1995; Svenningsson *et al.* 1999*c*; Fredholm *et al.* 2001).

There is much information on the distribution of the A_1 and A_{2A} receptors because good pharmacological tools including radioligands are available. Several studies have used antibodies to demonstrate the widespread distribution of adenosine A_1 and A_{2A} receptors in the brain. This includes regions such as the cerebral cortex, cerebellum hippocampus, with A_{2A} receptors being particularly enriched in the striatum (Swanson *et al.* 1995; Rosin *et al.* 1998; Hettinger *et al.* 2001). In the case of the A_{2B} and A_3 receptors the data are less impressive and one needs to primarily rely mRNA expression data. Some of this information is summarized in Table 11.1.

Although receptor protein and the corresponding mRNA message are in general co-localized, there are some important differences. For example, in several regions of the central nervous system, receptor binding and expression of mRNA transcripts do not exactly match (Johansson *et al.* 1993*a*), and the two are differently regulated by long-term antagonist treatment (Johansson *et al.* 1993*a*) and during development (Ådén *et al.* 2000). Much of the differential distribution can probably be explained by the fact that a substantial number of adenosine A_1 receptors are present at nerve terminals. A similar explanation probably underlies the observations that A_{2A} receptors are present in globus pallidus, despite the fact that A_{2A} receptor mRNA cannot be detected there (Svenningsson *et al.* 1997, 1999*c*). These receptors are probably located at the terminals of the striatopallidal GABAergic neurons (Rosin *et al.* 1998; Svenningsson *et al.* 1999*b*; Linden 2001).

11.4 **Pharmacology**

Ideally, agonists and antagonists should differ in potency by at least two orders of magnitude at different receptor subtypes in order to be really useful in receptor classification. This is rarely the case for the compounds often used in classifying adenosine receptors. Nevertheless, with a judicious use of agonists and antagonists at A_1, A_{2A}, and A_3 receptors in *in vitro* experiments, strong conclusions can be drawn (see Table 11.1). The situation is less fortunate *in vivo* as the pharmacokinetics of these compounds has not been studied extensively, and because differences in potency at the different receptors may be offset by huge differences in receptor abundance.

11.4.1 **Agonists**

In the case of human, rat, and mouse, CCPA (and to a somewhat lesser extent CPA) represents a selective full agonist at the adenosine A_1 receptors. Partial agonists are also available, including compounds substituted in the C8-position (Roelen *et al.* 1996) or the ribose $5'$-substituent (van der Wenden *et al.* 1998). These compounds show tissue selectivity *in vivo*. NECA was long considered to be a selective adenosine A_2 receptor agonist, but it is now realized that it is nonselective. Based on evidence that 2-substitution of NECA increased selectivity, CGS21680 was developed as an A_{2A} receptor selective agonist (Hutchison *et al.* 1989). However, in humans it is less potent and less selective than in rats (Kull *et al.* 1999). There is an additional problem with CGS21680 as a tool: it also binds to sites unrelated to A_{2A} receptors (Johansson *et al.* 1993*b*; Cunha *et al.* 1996; Lindström *et al.* 1996). This means that at least in organs or cells with few A_{2A} receptors, effects of CGS21680 must be viewed with scepticism. In the case of A_{2B} receptors the most potent agonists have affinities only marginally below 1 μM. Furthermore, selectivity is negligible. The most potent A_3 receptor agonist is Cl-IB-MECA displaying 2500 and 1400 fold selectivity over adenosine A_1 and adenosine A_{2A} receptors, respectively.

11.4.2 **Antagonists**

The adenosine A_1 receptor antagonist DPCPX also interacts with appreciable affinity with A_{2B} receptors, but binding of DPCPX is virtually eliminated in animals lacking the A_1 receptor (Johansson *et al.* 2001), suggesting that this presents a minor problem. There are several useful A_{2A} receptor antagonists. The most selective so far is SCH58261. The structurally related ZM241385 is more readily available (Poucher *et al.* 1995), but shows appreciable affinity to A_{2B} receptors (Ongini *et al.* 1999). The situation is somewhat more favourable in the case of A_{2B} antagonists, where some potent and relatively selective antagonists have been found (Kim *et al.* 2000). The A_3 receptor is—in contrast to the other adenosine receptors—notably insensitive to several xanthines. Hence, most A_3 antagonists have a non-xanthine structure—including dihydropyridines, pyridines and flavonoids (Baraldi *et al.* 2000). One of the most selective compounds (for human, but not rat A_3 receptors) is MRE-3008-F20, which is also a useful antagonist radioligand at human A_3 receptors (Varani *et al.* 2000). MRS1523 and MRS1191 are two compounds that also show marked species difference and are selective for only the human A_3 receptor (Jacobson *et al.* 1997).

11.4.3 **Allosteric modulators**

Another interesting class of compounds acting on adenosine A_1 receptors are the so-called allosteric enhancers. The prototype here is PD81723 (Bruns and Fergus 1990), which has been shown by various research groups to (allosterically) increase agonist binding and effect (e.g. Linden 1997).

11.5 **Signal transduction and receptor modulation**

The adenosine A_1 and A_2 receptors were initially subdivided on the basis of their inhibiting and stimulating adenylyl cyclase, respectively. Indeed, A_1 and A_2 receptors are coupled to members of the G_i group and G_s group of G proteins, respectively. The A_3 receptor is also G_i coupled. In addition, there is some evidence from transfection experiments that the adenosine receptors may signal via other G proteins, but it is not known if such coupling is physiologically important. Recently, evidence was presented that whereas the A_{2A} receptor is coupled to G_s in most peripheral tissues it is coupled to G_{olf} in striatum (Kull *et al.* 2000). Endogenous A_{2B} receptors of HEK 293 cells, human HMC-1 mast cells and canine BR mast cells are dually coupled to G_s and G_q (Auchampach *et al.* 1997; Linden *et al.* 1999).

After activation of the G proteins, enzyme and ion channel activity is modulated. A_1 receptors mediate inhibition of adenylyl cyclase, activation of several types of K^+-channels (probably via β,γ-subunits), inactivation of N, P, and Q-type Ca^{2+} channels, activation of phospholipase $C\beta$ etc. The same appears to be true for A_3 receptors. In CHO cells transfected with the human A_3 adenosine receptor both adenylyl cyclase inhibition and a Ca^{2+} signal are mediated via a $G_{i/o}$-dependent pathway (Klotz *et al.* 2000). Given that many of the steps in the signalling cascade involve signal amplification it is not surprising that the position of the dose-response curve for agonists will depend on which particular effect is measured (Baker *et al.* 2000). Both A_{2A} and A_{2B} receptors stimulate the formation of cyclic AMP, but other actions, including mobilization of intracellular calcium have also been described (Mirabet *et al.* 1997). Actions of adenosine A_{2A} receptors on neutrophil leukocytes are due in part to cyclic AMP (Fredholm *et al.* 1996; Sullivan *et al.* 2001) but cyclic AMP-independent effects of A_{2A} receptor activation in these cells have also been suggested (Cronstein 1994).

Activation of A_1 receptors can dose- and time-dependently activate ERK1/2 via β,γ-subunits released from pertussis toxin-sensitive $G_{i/o}$ proteins and phosphoinositol-3-kinase (Faure *et al.* 1994; Dickenson *et al.* 1998; Schulte and Fredholm 2000). Activation of A_{2A} receptors also increases MAPK activity (Sexl *et al.* 1997) but the signalling pathways used by the A_{2A} receptor seem to vary with the cellular background and the signalling machinery that the cell possesses (Seidel *et al.* 1999). A_{2A} receptor activation may not only stimulate, but also inhibit ERK phosphorylation (Hirano *et al.* 1996; Arslan and Fredholm 2000), probably via PKA-dependent phosphorylation of Raf-1. The adenosine A_{2B} receptor is the only subtype which so far has been shown to activate not only ERK1/2, but also JNK and p38 (Feoktistov *et al.* 1999), perhaps via activation of $G_{q/11}$, PLC, genistein-insensitive tyrosine kinases, ras, B-raf and MEK1/2 (Gao *et al.* 1999). Studies in transfected cells (Schulte and Fredholm 2000) show a nearly 100-fold higher potency of both NECA and adenosine in inducing ERK1/2 phosphorylation than in inducing cyclic AMP production. The EC_{50} value for ERK1/2 phosphorylation in transfected CHO cells lies in the nanomolar range, whereas cyclic AMP production is half-maximally activated around 1–5 μM NECA. This emphasizes that G protein-coupled adenosine receptors can have substantially different potencies on different signalling pathways in the same cellular system. The adenosine A_3 receptor activates ERK1/2 in human foetal astrocytes (Neary *et al.* 1998) and in CHO cells (Schulte and Fredholm 2000).

11.6 Physiology and disease relevance

Recently, genetically modified mice have been generated, that provide insights into the physiology and pathophysiology of the different adenosine receptors. The adenosine A_{2A} receptor was first to be knocked out (Ledent *et al.* 1997; Chen *et al.* 1999). Using these mice it has been shown that A_{2A} receptors play a role in mediating pain via peripheral sites, inhibiting platelet aggregation and regulating blood pressure, and are critically important for the motor stimulant effects of caffeine (Ledent *et al.* 1997; El Yacoubi *et al.* 2000). Mice with a targeted disruption of the A_3 receptor show a decreased effect of adenosine analogues on mast cell degranulation (Salvatore *et al.* 2000) and a consequent decrease in vascular permeability (Tilley *et al.* 2000). These animals also, surprisingly, show increased cardiovascular effects of administered adenosine (Zhao *et al.* 2000). Given the species differences in A_3 receptor distribution and pharmacology it is, however, not clear that the roles of this receptor are similar in humans. Mice with a targeted disruption of A_1 receptors have also been generated (Johansson *et al.* 2001). These mice show increased anxiety and are hyperalgesic, indicating a role for A_1 receptors in mediating endogenous antinociception. In these animals, the effect of adenosine on excitatory neurotransmission is totally eliminated and the neuronal response to hypoxia is markedly altered (Dunwiddie *et al.* 2000). Interestingly, all these mice show essentially normal viability and fertility.

11.6.1 Schizophrenia

Adenosine A_{2A} receptors are highly enriched in the basal ganglia, where they are present in highest abundance on a subset of the GABAergic output neurons, namely those that project to the globus pallidus (Schiffmann *et al.* 1991; Fink *et al.* 1992; Svenningsson *et al.* 1997; Rosin *et al.* 1998). These neurons also express dopamine D_2 receptors and it is abundantly clear that the two receptors interact in binding assays (Ferré *et al.* 1991), on signal transduction (Kull *et al.* 1999), and behaviourally (Fink *et al.* 1992; Fredholm *et al.* 1999; Svenningsson *et al.* 1999*b*). It appears that one major role of dopamine in the striatum is to suppress signaling via A_{2A} receptors (Svenningsson *et al.* 1999*a*; Chen *et al.* 2001). Given that antagonists of D_2 receptors are used in the treatment of schizophrenia, there are interesting implications for the pathophysiology of that disease and/or its treatment. When tonic adenosine A_{2A} receptor activation is unopposed by dopamine for a long period of time, the response to adenosine receptor stimulation decreases (Zahniser *et al.* 2000), even though significant effects remain (Chen *et al.* 2001).

11.6.2 Parkinson's disease

Of even greater interest is the use of A_{2A} antagonists in treatment of Parkinson's disease (Fredholm *et al.* 1976; Ongini and Fredholm 1996; Ferré *et al.* 1997; Svenningsson *et al.* 1999*b*; Impagnatiello *et al.* 2000). The effect of an A_{2A} receptor antagonist is synergistic with dopamine receptor agonists in primate models of Parkinson's disease (Kanda *et al.* 2000), and several A_{2A} antagonists are being developed as adjuncts to L-DOPA treatment.

11.6.3 **Ischemia**

Adenosine protects tissues from ischemic damage (e.g. brain and heart) through multiple receptor subtypes (Marangos 1990; Rudolphi *et al.* 1992; Fredholm 1996). A_1 and possibly A_3 receptor activation produces preconditioning to protect the heart and other tissues from subsequent ischemic injury. Rapid preconditioning is mediated by a pathway including PKC and increased mitochondrial K_{ATP} channel activation (Sato *et al.* 2000). In addition, A_1 and A_{2A} receptor knockout mice studies suggest that these receptors regulate ischemic brain damage in adult mice (Chen *et al.* 1999; Dunwiddie *et al.* 2000).

11.6.4 **Inflammation**

There is evidence that IgE antibodies and mast cells play a central role in the symptoms and pathology of asthma. Aerosolized adenosine causes mast-cell dependent bronchoconstriction in asthmatic subjects, but causes bronchodilation in non-asthmatics (Cushley *et al.* 1983; Vilsvik *et al.* 1990). Moreover, the nonselective adenosine receptor antagonist theophylline is widely used as an anti-asthmatic drug although its mechanism of action is uncertain. A related xanthine, enprofylline (3-propylxanthine), is also therapeutically efficacious in the treatment of asthma, and was thought to act through a non-adenosine receptor-mediated mechanism due to its low affinity at A_1 and A_{2A} receptors. Recently, attention has shifted to the A_{2B} and A_3 receptor subtypes found on mast cells, that, when activated, facilitate antigen-mediated mast cell degranulation. The adenosine A_3 receptor was initially implicated as the receptor subtype that triggers the degranulation of rat RBL 2H3 mast-like cells (Ramkumar *et al.* 1993). There is also evidence of mast cell degranulation when agonists of A_3 receptors are administered to rats or mice (Fozard *et al.* 1996; Tilley *et al.* 2000). In contrast, the A_{2B} receptor has been implicated as the receptor subtype that facilitates the release of allergic mediators from canine and human HMC-1 mastocytoma cells (Feoktistov and Biaggioni 1995; Auchampach *et al.* 1997). A role for A_{2B} receptors in human asthma is also suggested by the efficacy of enprofylline, which at therapeutic concentrations of 20–50 μM, only blocks the A_{2B} receptor subtype (Fredholm and Persson 1982; Linden *et al.* 1999). In sum, the literature indicates that the release of allergic mediators from mast cells is mediated by A_3 receptors and/or A_{2B} receptors. This may result from tissue difference and/or species differences, with rodent mast cells (rat, guinea-pig, and mouse) responding primarily to A_3, and canine or human mast cells mainly to A_{2B} adenosine receptor stimulation. Finally, there is also good evidence that under *in vivo* conditions A_{2A} receptors on leukocytes can be tonically activated by endogenous adenosine (Cronstein 1994; Fredholm 1997). Nevertheless, drugs that enhance endogenous adenosine and/or directly stimulate A_{2A} receptors produce additional anti-allergic effects.

11.7 **Concluding remarks**

Interest in modulation of this receptor system exists for numerous disease indications and it is remarkable that despite intensive efforts, relatively few adenosine receptor ligands have made it into clinical trials. Adenosine itself restores normal heart rhythm in patients with abnormally rapid heartbeats originating in the upper chambers of the heart, so-called paroxysmal supraventricular tachycardia. Adenosine is also used as an adjunct to thallium cardiac imaging in the evaluation of coronary artery disease in patients unable to exercise adequately. With respect to antagonists it can be argued that caffeine, the most widely used psychotropic substance, has its main action via adenosine receptors. The same might hold true for theophylline and enprofylline, both used in the treatment of asthma. Currently, several adenosine agonists and antagonists are under development for a variety of indications. Thus, adenosine receptors remain an attractive target for drug development.

References

Arslan G and Fredholm BB (2000). Stimulatory and inhibitory effects of adenosine A_{2A} receptors on nerve growth factor-induced phosphorylation of extracellular regulated kinases 1/2 in PC12 cells. *Neurosci Lett* 292, 183–6.

Atkinson MR, Townsend-Nicholson A, Nicholl JK, Sutherland GR, and Schofield PR (1997). Cloning, characterisation and chromosomal assignment of the human adenosine A3 receptor (ADORA3) gene. *Neurosci Res* 29(1), 73–9.

Auchampach JA, Jin X, Wan TC, Caughey GH, and Linden J (1997). Canine mast cell adenosine receptors: cloning and expression of the A3 receptor and evidence that degranulation is mediated by the A2B receptor. *Mol Pharmacol* 52(5), 846–60.

Baker SP, Scammells PJ, and Belardinelli L (2000). Differential A(1)-adenosine receptor reserve for inhibition of cyclic AMP accumulation and G-protein activation in DDT(1) MF-2 cells. *Br J Pharmacol* 130(5), 1156–64.

Ballarín M, Fredholm BB, Ambrosio S, and Mahy N (1991). Extracellular levels of adenosine and its metabolites in the striatum of awake rats: inhibition of uptake and metabolism. *Acta Physiol Scand* 142, 97–103.

Baraldi PG, Cacciari B, Romagnoli R, Merighi S, Varani K, Borea PA *et al.* (2000). A(3) adenosine receptor ligands: history and perspectives. *Med Res Rev* 20(2), 103–28.

Baranowski RL and Westenfelder C (1994). Estimation of renal interstitial adenosine and purine metabolites by microdialysis. *Am J Physiol* 267, F174–82.

Bruns RF and Fergus JH (1990). Allosteric enhancement of adenosine A1 receptor binding and function by 2-amino-3-benzoylthiophenes. *Mol Pharmacol* 38, 939–49.

Chen JF, Huang Z, Ma J, Zhu J, Moratalla R, Standaert D *et al.* (1999). A(2A) adenosine receptor deficiency attenuates brain injury induced by transient focal ischemia in mice. *J Neurosci* 19(21), 9192–200.

Chen JF, Moratalla R, Impagnatiello F, Grandy DK, Cuellar B, Rubinstein M *et al.* (2001). The role of the D(2) dopamine receptor (D(2)R) in A(2A) adenosine receptor (A(2A)R)-mediated behavioral and cellular responses as revealed by A(2A) and D(2) receptor knockout mice. *Proc Natl Acad Sci USA* 98(4), 1970–5.

Cronstein BN (1994). Adenosine, an endogenous anti-inflammatory agent. *J Appl Physiol* 76, 5–13.

Cunha RA, Johansson B, Constantino MD, Sebastião AM, and Fredholm BB (1996). Evidence for high-affinity binding sites for the adenosine A_{2A} receptor agonist [^{3}H] CGS 21680 in the rat hippocampus and cerebral cortex that are different from striatal A_{2A} receptors. *Naunyn Schmiedebergs Arch Pharmacol* 353, 261–71.

Cushley MJ, Tattersfield AE, and Holgate ST (1983). Inhaled adenosine and guanosine on airway resistance in normal and asthmatic subjects. *Br J Clin Pharmacol* 15(2), 161–5.

Deckert J, Nothen MM, Rietschel M, Wildenauer D, Bondy B, Ertl MA *et al.* (1996). Human adenosine A2a receptor (A2aAR) gene: systematic mutation screening in patients with schizophrenia. *J Neural Transm* 103, 1447–55.

Dickenson JM, Blank JL, and Hill SJ (1998). Human adenosine A1 receptor and P2Y2-purinoceptor-mediated activation of the mitogen-activated protein kinase cascade in transfected CHO cells. *Br J Pharmacol* 124, 1491–9.

Dunwiddie TV and Diao L (2000). Regulation of extracellular adenosine in rat hippocampal slices is temperature dependent: role of adenosine transporters. *Neuroscience* 95(1), 81–8.

Dunwiddie TV, Masino SA, Poelchen W, Diao L, Johansson B, and Fredholm B (2000). Altered electrophysiological sensitivity to A_1 but not GABA$_B$ agonists in the hippocampal CA1 region in adenosine A_1 receptor knockout mice. *Soc Neurosci Symposia* **26**, 816.18.

El Yacoubi M, Ledent C, Menard JF, Parmentier M, Costentin J, and Vaugeois JM (2000). The stimulant effects of caffeine on locomotor behaviour in mice are mediated through its blockade of adenosine A(2A) receptors. *Br J Pharmacol* **129**(7), 1465–73.

Faure M, Voyno-Yasenetskaya TA, and Bourne HR (1994). cAMP and beta gamma subunits of heterotrimeric G proteins stimulate the mitogen-activated protein kinase pathway in COS-7 cells. *J Biol Chem* **269**, 7851–4.

Feoktistov I and Biaggioni I (1995). Adenosine A2b receptors evoke interleukin-8 secretion in human mast cells. An enprofylline-sensitive mechanism with implications for asthma. *J Clin Invest* **96**, 1979–86.

Feoktistov I, Goldstein AE, and Biaggioni I (1999). Role of p38 mitogen-activated protein kinase and extracellular signal-regulated protein kinase kinase in adenosine A2B receptor-mediated interleukin-8 production in human mast cells. *Mol Pharmacol* **55**, 726–34.

Ferré S, Fredholm BB, Morelli M, Popoli P, and Fuxe K (1997). Adenosine-dopamine receptor-receptor interactions as an integrative mechanism in the basal ganglia. *Trends Neurosci* **20**, 482–7.

Ferré S, von Euler G, Johansson B, Fredholm BB, and Fuxe K (1991). Stimulation of high-affinity adenosine A_2 receptors decreases the affinity of dopamine D_2 receptors in rat striatal membranes. *Proc Natl Acad Sci USA* **88**, 7238–41.

Fink JS, Weaver DR, Rivkees SA, Peterfreund RA, Pollack AE, Adler EM *et al.* (1992). Molecular cloning of the rat A_2 adenosine receptor: selective co-expression with D_2 dopamine receptors in rat striatum. *Brain Res Mol Brain Res* **14**, 186–95.

Fozard JR, Pfannkuche HJ, and Schuurman HJ (1996). Mast cell degranulation following adenosine A3 receptor activation in rats. *Eur J Pharmacol* **298**(3), 293–7.

Fredholm BB (1996). Adenosine and neuroprotection. In *Neuroprotective Agents and Cerebral Ischemia*, Green AR, Cross AJ. London: Academic Press, pp. 259–80.

Fredholm BB (1997). Purines and neutrophil leukocytes. *Gen Pharmacol* **28**, 345–50.

Fredholm BB, Arslan G, Halldner L, Kull B, Schulte G, and Wasserman W (2000). Structure and function of adenosine receptors and their genes. *Naunyn-Schmiedeberg's Arch Pharmacol* **362**, 364–74.

Fredholm BB, Bättig K, Holmén J, Nehlig A, and Zvartau E (1999). Actions of caffeine in the brain with special reference to factors that contribute to its widespread use. *Pharmacol Rev* **51**, 83–153.

Fredholm BB, Fuxe K, and Agnati L (1976). Effect of some phosphodiesterase inhibitors on central dopamine mechanisms. *Eur J Pharmacol* **38**, 31–8.

Fredholm BB, Irenius E, Kull B, and Schulte G (2001). Comparison of the potency of adenosine as an agonist at human adenosine receptors expressed in Chinese hamster ovary cells. *Biochem Pharmacol* **61**, 443–8.

Fredholm BB and Persson CGA (1982). Xanthine derivatives as adenosine receptor antagonists. *Eur J Pharmacol* **81**, 673–6.

Fredholm BB, Zhang Y, and van der Ploeg I (1996). Adenosine A$_{2A}$ receptors mediate the inhibitory effect of adenosine on formyl-Met-Leu-Phe-stimulated respiratory burst in neutrophil leukocytes. *Naunyn Schmiedebergs Arch Pharmacol* **354**, 262–7.

Furlong TJ, Pierce KD, Selbie LA, and Shine J (1992). Molecular characterization of a human brain adenosine A_2 receptor. *Brain Res Mol Brain Res* **15**, 62–6.

Gao Z, Chen T, Weber MJ, and Linden J (1999). A2B adenosine and P2Y2 receptors stimulate mitogen-activated protein kinase in human embryonic kidney-293 cells. Cross-talk between cyclic AMP and protein kinase C pathways. *J Biol Chem* **274**, 5972–80.

Griffiths M, Beaumont N, Yao SY, Sundaram M, Boumah CE, Davies A *et al.* (1997). Cloning of a human nucleoside transporter implicated in the cellular uptake of adenosine and chemotherapeutic drugs [see comments]. *Nat Med* **3**, 89–93.

Hettinger BD, Lee A, Linden J, and Rosin DL (2001). Ultrastructural localization of adenosine A2A receptors suggests multiple cellular sites for modulation of GABAergic neurons in rat striatum. *J Comp Neurol* **431**(3), 331–46.

Hirano D, Aoki Y, Ogasawara H, Kodama H, Waga I, Sakanaka C *et al.* (1996). Functional coupling of adenosine A2a receptor to inhibition of the mitogen-activated protein kinase cascade in Chinese hamster ovary cells. *Biochem J* **316**, 81–6.

Huston JP, Haas HL, Boix F, Pfister M, Decking U, Schrader J *et al.* (1996). Extracellular adenosine levels in neostriatum and hippocampus during rest and activity periods of rats. *Neuroscience* **73**, 99–107.

Hutchison AJ, Webb RL, Oei HH, Ghai GR, Zimmerman MB, and Williams M (1989). CGS 21680C, an A2 selective adenosine receptor agonist with preferential hypotensive activity. *J Pharmacol Exp Ther* **251**, 47–55.

Impagnatiello F, Bastia E, Ongini E, and Monopoli A (2000). Adenosine receptors in neurological disorders. *Emerging Therapeutic Targets* **4**(5), 635–64.

Jacobson KA *et al.* (1997). Pharmacological characterisation of novel A3 adenosine receptor selective antagonists. *Neuropharm* **36**, 1157–65.

Johansson B, Ahlberg S, van der Ploeg I, Brené S, Lindefors N, Persson H *et al.* (1993*a*). Effect of long term caffeine treatment on A_1 and A_2 adenosine receptor binding and on mRNA levels in rat brain. *Naunyn Schmiedebergs Arch Pharmacol* **347**, 407–14.

Johansson B, Georgiev V, Parkinson FE, and Fredholm BB (1993*b*). The binding of the adenosine A_2 receptor selective agonist [^{3}H]CGS 21680 to rat cortex differs from its binding to rat striatum. *Eur J Pharmacol Mol Pharmacol Sect* **247**, 103–10.

Johansson B, Halldner L, Dunwiddie TV, Masino SA, Poelchen W, Giménez-Llort L *et al.* (2001). Hyperalgesia, anxiety, and decreased hypoxic neuroprotection in mice lacking the adenosine A_1 receptor. *Proc Natl Acad Sci USA* **98**, 9407–12.

Kanda T, Jackson MJ, Smith LA, Pearce RK, Nakamura J, Kase H *et al.* (2000). Combined use of the adenosine A(2A) antagonist KW-6002 with L-DOPA or with selective D1 or D2 dopamine agonists increases antiparkinsonian activity but not dyskinesia in MPTP-treated monkeys. *Exp Neurol* **162**(2), 321–7.

Kenakin T (1993). Pharmacologic analysis of drug-receptor interaction, 2nd Edition, Raven Press, New York.

Kenakin T (1995). Agonist-receptor efficacy I: mechanisms of efficacy and receptor promiscuity. *Trends Pharmacol Sci* **16**, 188–92.

Kim YC, Ji X, Melman N, Linden J, and Jacobson KA (2000). Anilide derivatives of an 8-phenylxanthine carboxylic congener are highly potent and selective antagonists at human A(2B) adenosine receptors. *J Med Chem* **43**(6), 1165–72.

Klotz KN, Quitterer U, and Englert M (2000). Effector coupling of human A3 adenosine receptors (Abstract). *Drug Dev Res* **50**, 80.

Kull B, Arslan G, Nilsson C, Owman C, Lorenzen A, Schwabe U *et al.* (1999). Differences in the order of potency for agonists, but not antagonists, at human and rat adenosine A_{2A} receptors. *Biochem Pharmacol* **57**, 65–75.

Kull B, Svenningsson P, and Fredholm BB (2000). Adenosine A_{2A} receptors are co-localized with and activate G_{olf} in rat striatum. *Mol Pharmacol* **58**, 771–7.

Ledent C, Vaugeois JM, Schiffmann SN, Pedrazzini T, El Yacoubi M, Vanderhaeghen JJ *et al.* (1997). Aggressiveness, hypoalgesia and high blood pressure in mice lacking the adenosine A2A receptor. *Nature* **388**, 674–8.

Libert F, Schiffmann SN, Lefort A, Parmentier M, Gerard C, Dumont JE *et al.* (1991). The orphan receptor cDNA RDC7 encodes an A1 adenosine receptor. *EMBO J* **10**, 1677–82.

Linden J (1997). Allosteric enhancement of adenosine receptors. In (eds, KA Jacobson and MF Jarvis) *Purinergic Approaches in Experimental Therapeutics*, Wiley-Liss, New York, pp. 85–97.

Linden J (2001). Molecular approach to adenosine receptors: receptor-mediated mechanisms of tissue protection. *Annu Rev Pharmacol Toxicol* 41, 775–87.

Linden J, Thai T, Figler H, Jin X, and Robeva AS (1999). Characterization of human A(2B) adenosine receptors: radioligand binding, western blotting, and coupling to G(q) in human embryonic kidney 293 cells and HMC-1 mast cells. *Mol Pharmacol* 56(4), 705–13.

Lindström K, Ongini E, and Fredholm BB (1996). The selective adenosine A_{2A} receptor antagonist SCH 58261 discriminates between two different binding sites for [^{3}H]-CGS 21680 in the rat brain. *Naunyn Schmiedebergs Arch Pharmacol* 354, 539–41.

Maenhaut C, Van Sande J, Libert F, Abramowicz M, Parmentier M, Vanderhaegen JJ *et al.* (1990). RDC8 codes for an adenosine A2 receptor with physiological constitutive activity. *Biochem Biophys Res Commun* 173, 1169–78.

Mahan LC, McVittie LD, Smyk-Randall EM, Nakata H, Monsma Jr FJ, Gerfen CR *et al.* (1991). Cloning and expression of an A1 adenosine receptor from rat brain. *Mol Pharmacol* 40, 1–7.

Marangos PJ (1990). Adenosinergic approaches to stroke therapeutics. *Med Hypotheses* 32, 45–9.

Mirabet M, Mallol J, Lluis C, and Franco R (1997). Calcium mobilization in Jurkat cells via A2b adenosine receptors. *Br J Pharmacol* 122, 1075–82.

Neary JT, McCarthy M, Kang Y, and Zuniga S (1998). Mitogenic signaling from P1 and P2 purinergic receptors to mitogen-activated protein kinase in human fetal astrocyte cultures. *Neurosci Lett* 242, 159–62.

Ongini E, Dionisotti S, Gessi S, Irenius E, and Fredholm BB (1999). Comparison of CGS 15943, ZM 241385 and SCH 58261 as antagonists at human adenosine receptors. *Naunyn-Schmiedeberg's Arch Pharmacol* 359, 7–10.

Ongini E and Fredholm BB (1996). Pharmacology of adenosine A2A receptors. *Trends Pharmacol Sci* 17, 364–72.

Palmer TM and Stiles GL (2000). Identification of threonine residues controlling the agonist-dependent phosphorylation and desensitization of the rat A(3) adenosine receptor. *Mol Pharmacol* 57(3), 539–45.

Pazzagli M, Corsi C, Fratti S, Pedata F, and Pepeu G (1995). Regulation of extracellular adenosine levels in the striatum of aging rats. *Brain Res* 684, 103–6.

Porkka-Heiskanen T, Strecker RE, Thakkar M, Bjorkum AA, Greene RW, and McCarley RW (1997). Adenosine: a mediator of the sleep-inducing effects of prolonged wakefulness. *Science* 276, 1265–8.

Poucher SM, Keddie JR, Singh P, Stoggall SM, Caulkett PW, and Jones G *et al.* (1995). The *in vitro* pharmacology of ZM 241385, a potent, non-xanthine A_{2a} selective adenosine receptor antagonist. *Br J Pharmacol* 115, 1096–102.

Ramkumar V, Stiles GL, Beaven MA, and Ali H (1993). The A3 adenosine receptor is the unique adenosine receptor which facilitates release of allergic mediators in mast cells. *J Biol Chem* 268, 16 887–90.

Ren H and Stiles GL (1995). Separate promoters in the human A1 adenosine receptor gene direct the synthesis of distinct messenger RNAs that regulate receptor abundance. *Mol Pharmacol* 48, 975–80.

Ren J and Stiles G (1994). Characterization of the human A_1 adenosine receptor gene. *J Biol Chem* 269, 3104–10.

Roelen H, Veldman N, Spek AL, von Frijtag Drabbe Kunzel J, Mathot RA, and IJzerman AP (1996). N6,C8-distributed adenosine derivatives as partial agonists for adenosine A1 receptors. *J Med Chem* 39, 1463–71.

Rosin DL, Robeva A, Woodard RL, Guyenet PG, and Linden J (1998). Immunohistochemical localization of adenosine A2A receptors in the rat central nervous system. *J Comp Neurol* 401(2), 163–86.

Rudolphi KA, Schubert P, Parkinson FE, and Fredholm BB (1992). Neuroprotective role of adenosine in cerebral ischaemia. *Trends Pharmacol Sci* 13, 439–45.

Salvatore CA, Tilley SL, Latour AM, Fletcher DS, Koller BH, and Jacobson MA (2000). Disruption of the A(3) adenosine receptor gene in mice and its effect on stimulated inflammatory cells. *J Biol Chem* 275(6), 4429–34.

Sato T, Sasaki N, O'Rourke B, and Marban E (2000). Adenosine primes the opening of mitochondrial ATP-sensitive potassium channels: a key step in ischemic preconditioning? *Circulation* 102(7), 800–5.

Schiffmann SN, Jacobs O, and Vanderhaeghen JJ (1991). Striatal restricted adenosine A2 receptor (RDC8) is expressed by enkephalin but not by substance P neurons: an in situ hybridization histochemistry study. *J Neurochem* 57, 1062–7.

Schulte G and Fredholm BB (2000). Human adenosine A_1, A_{2A}, A_{2B}, and A_3 receptors expressed in Chinese hamster ovary cells all mediate the phosphorylation of extracellular-regulated kinase 1/2. *Mol Pharmacol* 58(3), 477–82.

Seidel MG, Klinger M, Freissmuth M, and Holler C (1999). Activation of mitogen-activated protein kinase by the A(2A)-adenosine receptor via a rap1-dependent and via a p21(ras)-dependent pathway. *J Biol Chem* 274(36), 25 833–41.

Sexl V, Mancusi G, Holler C, Gloria-Maercker E, Schutz W, and Freissmuth M (1997). Stimulation of the mitogen-activated protein kinase via the A2A-adenosine receptor in primary human endothelial cells. *J Biol Chem* 272, 5792–9.

Stehle JH, Rivkees SA, Lee JJ, Weaver DR, Deeds JD, and Reppert SM (1992). Molecular cloning and expression of the cDNA for a novel A2-adenosine receptor subtype. *Mol Endocrinol* 6, 384–93.

Sullivan GW, Rieger JM, Scheld WM, MacDonald TL, and Linden J (2001). Cyclic AMP-dependent inhibition of human neutrophil oxidative activity by substituted 2-propynylcyclohexyl adenosine A2A receptor agonists. *Br J Pharmacol* 132(5), 1017–26.

Svenningsson P, Fourreau L, Bloch B, Fredholm BB, Gonon F, and Le Moine C (1999*a*). Opposite tonic modulation of dopamine and adenosine on c-*fos* mRNA expression in striatopallidal neurons. *Neuroscience* 89, 827–37.

Svenningsson P, Le Moine C, Fisone G, and Fredholm BB (1999*b*). Distribution, biochemistry and function of striatal adenosine A2A receptors. *Prog Neurobiol* 59, 355–96.

Svenningsson P, Le Moine C, Kull B, Sunahara R, Bloch B, and Fredholm BB (1997). Cellular expression of adenosine A_{2A} receptor messenger RNA in the rat central nervous system with special reference to dopamine innervated areas. *Neuroscience* 80, 1171–85.

Svenningsson P, Nomikos GG, and Fredholm BB (1999*c*). The stimulatory action and the development of tolerance to caffeine is associated with alterations in gene expression in specific brain regions. *J Neurosci* 19, 4011–22.

Swanson TH, Drazba JA, and Rivkees SA (1995). Adenosine A1 receptors are located predominantly on axons in the rat hippocampal formation. *J Comp Neurol* 363, 517–31.

Thorn JA and Jarvis SM (1996). Adenosine transporters. [Review] [75 refs]. *Gen Pharmacol* 27, 613–20.

Tilley SL, Wagoner VA, Salvatore CA, Jacobson MA, and Koller BH (2000). Adenosine and inosine increase cutaneous vasopermeability by activating A(3) receptors on mast cells. *J Clin Invest* 105(3), 361–7.

van der Wenden EM, Carnielli M, Roelen HC, Lorenzen A, von Frijtag Drabbe Kunzel JK, and IJzerman AP (1998). 5'-substituted adenosine analogs as new high-affinity partial agonists for the adenosine A1 receptor. *J Med Chem* 41(1), 102–8.

Varani K, Merighi S, Gessi S, Klotz KN, Leung E, Baraldi PG *et al.* (2000). [(3)H]MRE 3008F20: a novel antagonist radioligand for the pharmacological and biochemical characterization of human A(3) adenosine receptors. *Mol Pharmacol* 57(5), 968–75.

Vilsvik JS, Persson CG, Amundsen T, Brenna E, Naustdal T, Syvertsen U *et al.* (1990). Comparison between theophylline and an adenosine non-blocking xanthine in acute asthma. *Eur Respir J* 3, 27–32.

Zahniser NR, Simosky JK, Mayfield RD, Negri CA, Hanania T, Larson GA *et al.* (2000). Functional uncoupling of adenosine A_{2A} receptors and reduced response to caffeine in mice lacking dopamine D_2 receptors. *J Neurosci* 20(16), 5949–57.

Zhao Z, Makaritsis K, Francis CE, Gavras H, and Ravid K (2000). A role for the A3 adenosine receptor in determining tissue levels of cAMP and blood pressure: studies in knock-out mice. *Biochim Biophys Acta* **1500**(3), 280–90.

Zhou QY, Li C, Olah ME, Johnson RA, Stiles GL, and Civelli O (1992). Molecular cloning and characterization of an adenosine receptor: the A3 adenosine receptor. *Proc Natl Acad Sci USA* **89**, 7432–6.

Zimmermann H (2000). Extracellular metabolism of ATP and other nucleotides. *Naunyn Schmiedebergs Arch Pharmacol* **362**(4–5), 299–309.

Ådén U, Herlenius E, Tang L-Q, and Fredholm BB (2000). Maternal caffeine intake has minor effects on adenosine receptor ontogeny in the rat brain. *Pediat Res* **48**, 177–83.

Chapter 12

Adrenergic receptors

J. Paul Hieble and Robert R. Ruffolo, Jr.

12.1 Introduction

The adrenoceptors (adrenergic receptors) mediate the diverse effects of the neurotransmitters of the sympathetic nervous system, norepinephrine, and epinephrine, at virtually all sites throughout the body. During the last century, the adrenoceptors have been extensively studied by a variety of functional and molecular techniques, and have been progressively subdivided (Hieble *et al.* 1995*a*). Currently, nine adrenoceptors have been cloned, and they have been divided into three major categories, the α_1-adrenoceptors, α_2-adrenoceptors, and β-adrenoceptors. Although there are pharmacological data that suggest the existence of additional adrenoceptor subtypes, it now seems likely that this results from multiple 'affinity states' of a particular adrenoceptor subtype, rather than from a new adrenoceptor protein yet to be cloned.

12.2 Molecular characterization

The adrenoceptors belong to the large family of 7-transmembrane-spanning, G protein coupled receptors, and there is 32–54% amino acid identity between the individual adrenoceptor proteins (Fig. 12.1). This receptor family was originally subdivided on the basis of pharmacological differentiation years before individual subtypes were cloned. In this respect, the demonstration that prejunctional effects had a different pharmacological profile from actions mediated by postjunctional α-adrenoceptors, such as vascular contraction, originally led to the division of α-adrenoceptors into the α_1- and α_2-subtypes (Langer 1974; Starke *et al.* 1974). The most likely sites for interaction of the catecholamines with conserved structural elements of the adrenoceptors have been identified (Strader *et al.* 1994) and are referred to in Chapter 1.

Three distinct α_1-adrenoceptor proteins have been cloned; after some confusion in nomenclature, it has now been established that these three recombinant α_1-adrenoceptors, which are designated as α_{1a}, α_{1b}, and α_{1d}, correspond to the pharmacologically defined α_{1A}, α_{1B}, and α_{1D} adrenoceptors in native tissues (Hieble *et al.* 1995*b*). Multiple splice variants of the α_{1a}-adrenoceptor have been identified; however they appear to have identical pharmacological characteristics (Chang *et al.* 1998). An additional α_1-adrenoceptor, designated as the α_{1L}-adrenoceptor, appears to mediate the contraction of several vascular and urogenital tissues. This receptor has not been cloned, and recent evidence suggests that the α_{1a}-adrenoceptor may have either α_{1A}- or α_{1L}-pharmacology, depending on the particular assay, tissue, or experimental conditions employed (Ford *et al.* 1997). Three α_2-adrenoceptor

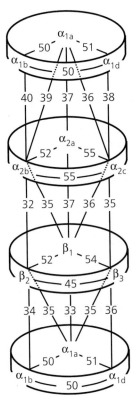

Fig. 12.1 Structural similarity, based on overall amino acid identity, between the nine adrenoceptors. Based on this similarity, the adrenoceptors can be divided into three groups (α_1, α_2, β), with each group further divided into three subtypes. Structural similarity is higher (45–55 per cent identity) within either of the three groups than between individual members of different groups (33–40 per cent identity).

proteins have been cloned. These recombinant receptors, designated as α_{2a}, α_{2b}, and α_{2c}, result in four discrete pharmacological profiles, since the α_{2a}-adrenoceptor appears to exist as species orthologs, with those of human, pig, and rabbit having a profile designated as α_{2A}, while those of rat, mouse, guinea pig, and cow exhibit pharmacologic profiles designated as α_{2D}. Three β-adrenoceptor proteins have been cloned, and the characteristics of these recombinant receptors correspond with those of the three well characterized β-adrenoceptors on native tissues, designated as β_1, β_2, and β_3.

12.3 Cellular and subcellular localization

α_1-Adrenoceptor subtypes can be differentially distributed in certain tissues. This differential distribution can reflect either expression of receptor protein/message or pharmacological characteristics in functional assays. For example, the predominant α_1-adrenoceptor found in rat spleen is the α_{1B} subtype, and the functional pharmacological profile of α_1-adrenoceptor agonists (i.e. contraction) in this tissue has α_{1B}-adrenoceptor characteristics, as determined by the use of subtype selective antagonists (Aboud et al. 1993; Stam et al. 1998). Although all of the α_1-adrenocetpor subtypes are present in most blood vessels, the pharmacological characteristics of agonist-induced contraction can vary, such that the responses may be α_{1A} (human microvessels, rat caudal artery), α_{1B} (rabbit or canine aorta) or α_{1D} (rat aorta).

The α_1-adrenoceptor subtypes are differentially distributed within the central nervous system. Highest levels of α_{1a}-adrenoceptor mRNA are found in regions of the olfactory system, hypothalamic nuclei, and in regions of the brainstem and spinal cord related to motor function (Day et $al.$ 1997). Studies in rat brain have shown α_{1b}-adrenoceptor mRNA to be concentrated in cerebral cortex, thalamus, raphe nuclei, cranial, and spinal motorneurons, as well as in the pineal gland. mRNA for the α_{1d}-adrenoceptor is localized to the olfactory bulb, cerebral cortex, hippocampus, dentate gyrus, reticular thalamic nucleus, motor neurons, and the inferior olivary complex. In the thalamus, the α_{1b}-and α_{1d}-adrenoceptors have a complimentary distribution (Nicholas et $al.$ 1996). Much less information is available on the distribution of α_1-adrenoceptors in the human brain. However, α_1-adrenoceptors are found in neocortex and dentate gyrus, although their distribution differs substantially from that observed in the rat (Zilles et $al.$ 1993)

α_2-Adrenoceptors are also widely distributed throughout the body. Studies in the rat demonstrate the existence of mRNA for all three α_2-adrenoceptor subtypes in the central nervous system. α_{2a}-Adrenoceptor mRNA is most widely distributed, being found in the cerebral cortex, locus coeruleus, amygdala, hypothalamic paraventricular nucleus, nucleus tractus solitarii, ventrolateral reticular formation, spinal cord, and dorsal root ganglia. Message for the α_{2b}-adrenoceptor is found almost exclusively in the thalamus. The α_{2c}-adrenoceptor is found in olfactory bulb, islands of Calleja, cerebral and cerebellar cortex, hippocampal formation and dorsal root ganglia (Nicholas et $al.$ 1996). In human brain, radioligand binding assays demonstrate the presence of α_{2A}-adrenoceptor protein in frontal cortex, cerebellum, and hippocampal formation, with another subtype (α_{2B} or α_{2C}) predominant in neostriatum (Grijabla et $al.$ 1996).

In rat brain, mRNA for the β_1-adrenoceptor is widely distributed with radioligand assays using dissected rat brain showing that β_1-adrenoceptors are typically associated with forebrain structures (cerebral cortex, striatum, and hippocampus), In contrast, message for the β_2-adrenoceptor is concentrated in olfactory bulb, hippocampal formation, piriform cortex, and cerebellar cortex (Nicholas et $al.$ 1996) with densest binding in the cerebellum (Nicholas et $al.$ 1996). Interestingly, it has been suggested that the β-adrenoceptors of rat cerebral cortex are localized primarily on glial cells (Strong et $al.$ 1991). The β_3-adrenoceptor does not appear to be present in the central nervous system.

12.4 Pharmacology

12.4.1 Agonists

The most selective α_{1A}-adrenoceptor subtype agonists that are currently available include SKF-89748 and A61603. No selective agonists are available for the α_{1B}- or α_{1D}-adrenoceptor subtype. A similar situation applies to α_2-adrenoceptors. Thus, for the α_{2A}-adrenoceptor subtype guanfacine and oxymetazoline are the agonists of choice although the latter exhibits only partial agonist activity at this receptor subtype and is a full agonist at α_{1A}-adrenoceptors. Many more useful pharmacological tools are available for β-adrenoceptor characterization. Thus, agonists capable of selectively activating β_1-, β_2-, or β_3-adrenoceptors such as xamoterol, salbutamol, and BRL37344 are frequently used to characterize the functional effects of these receptor subtypes, respectively. Species differences appear to be important for the β_3-adrenoceptor, since several selective β_3-adrenoceptor agonists can activate rodent, but not human, β_3-adrenoceptors.

12.4.2 **Antagonists**

Selective antagonists are available for most of the α_1-adrenoceptor subtypes (Table 12.1); however, the degree of selectivity varies, and consistent results between functional and radioligand binding assays is not always observed (Stam *et al.* 1998). Classically, though SNAP-5089, L765314, and SKF105854 are used to block α_{1A}, α_{1B}, and α_{1D} adrenoceptors, respectively. α_2 adrenoceptor mediated effects are classically differentiated from α_1 in that the former are sensitive to antagonists such as yohimbine and idazoxan. Within the α_2 adrenoceptor family α_{2A}- and α_{2D}-adrenoceptor mediated responses can be differentiated by the low sensitivity of the α_{2D}-adrenoceptor to blockade by the commonly used antagonists, yohimbine, and rauwolscine (Bylund *et al.* 1995; Hieble *et al.* 1996).

Many useful pharmacological tools are available for β-adrenoceptor characterization. These include antagonists selective for β_1-, β_2-, or β_3-adrenoceptors subtypes such as atenolol, ICI118551 and SR58894, respectively (Hieble 1991).

12.5 **Signal transduction and receptor modulation**

α_1-Adrenoceptors are widely distributed within the CNS, where their activation generally results in depolarization and increased neuronal firing. Most of the peripheral actions of α_1-adrenoceptors are mediated through $G_{q/11}$ coupling to the inositol phosphate pathway, while there is evidence for activation of adenylyl cyclase within the CNS. Perhaps the most extensively characterized action of α_2-adrenoceptors is the prejunctionally mediated inhibition of neurotransmitter release from many peripheral and central neurons. In line with other receptors with similar actions α_2-adrenoceptor mediated effects are mediated through the inhibition of adenylyl cyclase as a consequence of interaction of the agonist-receptor complex with $G_{i/o}$, although other second messengers remain to be characterized. Most β-adrenoceptor mediated actions involve stimulation of adenylyl cyclase through the interaction of the agonist-receptor complex with G_s.

12.6 **Physiology and disease relevance**

α_1-Adrenoceptors are widely distributed in both central and peripheral sites, and are activated either by norepinephrine released from sympathetic nerve terminals, or by epinephrine released from the adrenal medulla and some central adrenergic neurons. Peripheral α_1-adrenoceptors mediate a variety of functions, including contraction of smooth muscle, cardiac stimulation, cellular proliferation/apoptosis and activation of hepatic gluconeogenesis and glycogenolysis. The presence of α_1-adrenoceptors in layers of human hippocampus having a concentration of glutaminergic synapses suggests the possibility for noradrenergic modulation of glutaminergic transmission (Zilles *et al.* 1993). Furthermore, activation of central α_1-adrenoceptors, presumably in the hypothalamus, stimulates the secretion of several pituitary hormones in human volunteers (Al-Damlugji and Francis 1993).

Activation of prejunctional α_2-autoreceptors on sympathetic neurons results in a sympatholytic action. α_2-Adrenoceptors are also present at postjunctional sites, where they mediate actions such as smooth muscle contraction, platelet aggregation, and inhibition of insulin secretion (Ruffolo *et al.* 1993; Hieble and Ruffolo 1996). The α_{2A}- or α_{2D}-adrenoceptor subtype (depending on species) appears to be responsible for many other

α_2-adrenoceptor mediated responses, including the major component of prejunctional modulation of sympathetic neurotransmission (Trendelenburg *et al.* 1997). However, while knockout or mutation of the α_{2a}-adrenoceptor subtype produces only partial attenuation of α_2-adrenoceptor mediated inhibition of transmitter release, elimination of both α_{2a}- and α_{2c}-adrenoceptors results in complete loss of prejunctional modulation of adrenergic neurotransmission, and induces pathologic effects related to excess adrenergic tone (Hein *et al.* 1999). Hence, the α_{2c}-adrenoceptor can also participate in prejunctional modulation of neurotransmission.

Important physiological consequences of β-adrenoceptor activation include stimulation of cardiac rate and force, relaxation of vascular, urogenital, and bronchial smooth muscle, stimulation of renin secretion from the juxta-glomerular apparatus, stimulation of insulin and glucagon secretion from the endocrine pancreas, stimulation of glycogenolysis in liver and skeletal muscle, and stimulation of lipolysis in the adipocyte. Prejunctional β-adrenoceptors are present on some central and peripheral nerve terminals, where their activation results in facilitation of stimulation-evoked neurotransmitter release (Majewski and Rand 1984). However, in contrast to the prejunctional α_2-adrenoceptors, these prejunctional receptors do not appear to have major physiological significance.

Compounds interacting with several of the adrenoceptors have proven to be important drugs, the primary examples being nonselective β-adrenoceptor antagonists for hypertension and heart failure, β_2-adrenoceptor agonists for asthma and α_1-adrenoceptor antagonists for benign prostatic hyperplasia and hypertension (Ruffolo *et al.* 1995).

12.6.1 Arousal, attention, and sleep

Centrally acting α_1-adrenoceptor agonists often potentiate motor stimulation or arousal produced by other pharmacological (Holz *et al.* 1982) or behavioural (Astrachan *et al.* 1983a,b) challenges, and will increase vigilance when administered alone (Nozulak *et al.* 1992). α_1-Adrenoceptor agonists can reverse sedation induced by norepinephrine depletion produced by a dopamine β-hydroxylase inhibitor (Nozulak *et al.* 1992). Likewise, they can reverse cataplexy in a genetic canine model of narcolepsy (Renaud *et al.* 1991). Based on studies in rats and primates, modafinil, now marketed for the treatment of narcolepsy, may act through the activation of central α_1-adrenoceptors (Duteil *et al.* 1990), although based on its structure and peripheral pharmacological effects, this drug would not be expected to be an α-adrenoceptor agonist. The precise functional roles of each individual α_1-adrenoceptor subtype within the CNS are not yet known. It has been proposed that both α_{1B}- and α_{2A}-adrenoceptors are present on cell bodies of presympathetic ganglionic neurons in the intermediolateral cell column of the thoracic spinal cord, mediating an excitatory and inhibitory action, respectively (Nicholas *et al.* 1995). This reciprocal action is consistent with the observation that central α_1-adrenoceptor stimulation will attenuate the sedative actions of an α_2-adrenoceptor agonist in the rat (Guo *et al.* 1991). Mice lacking the α_{1b}-adrenoceptor show different behavioural effects from wild-type mice, in which enhanced reactivity to new situations is observed. Correlation of affinity for recombinant α_1-adrenoceptors with their activity in the canine narcolepsy model suggests that the ability of α_1-adrenoceptor agonists to reverse cataplexy in this model results from α_{1B}-adrenoceptor activation (Nishino *et al.* 1993).

Table 12.1 α_1-, α_2-, and β-adrenoceptors

α_1-Adrenoceptors	α_{1A}	α_{1B}	α_{1D}
Alternative names	α_{1a}, α_{1c}	α_{1b}	α_{1d}, $\alpha_{1a/d}$
Structural information (Accession no.)	h 466 aa(P35348)[AS], r 466 aa(P43140)[AS], m 466 aa (P97718)	h 519 aa (P35368), r 515 aa (P15823), m 514 aa (P97717)	h 572 aa (P25100), r 561 aa (P23944), m 562 aa (P97714)
Chromosomal location	8	5q33	20p13
Selective agonists	oxymetazoline, SKF89748, A61603	—	—
Selective antagonists	SNAP5089, SNAP5272, Ro70004, RS17053, KMD3213	(+)-Cyclazosin, L765314	BMY7378, SKF105854
Radioligands	[^{3}H]-Prazosin, [^{125}I]-HEAT, [^{125}I]-L762459	[^{3}H]-Prazosin, [^{125}I]-HEAT	[^{3}H]-Prazosin, [^{125}I]-HEAT
G protein coupling	G_q/G_{11}	G_q/G_{11}	G_q
Expression profile	olfactory system, hypothalamus, brainstem, spinal cord, prostate, vas deferens, heart, blood vessels	cerebral cortex, thalamus, raphe, motor neurons, pineal gland, spleen, kidney, heart, blood vessels	cerebral cortex, hippocampus, dentate gyrus, thalamus, motor neurons, inferior olivary complex, olfactory bulb, aorta, large blood vessels
Physiological function	smooth muscle and myocardial contraction	smooth muscle contraction	smooth muscle contraction
Knockout phenotype	—	decreased vasocontrictor response, selective learning deficits	—
Disease relevance	attention disorders, narcolepsy, feeding disorders, Parkinson's disease, cognitive decline	neurodegenerative diseases, epilepsy, attention disorders, narcolepsy, feeding disorders, Parkinson's disease, cognitive decline	attention disorders, narcolepsy, feeding disorders, Parkinson's disease, cognitive decline

α_2-Adrenoceptors	α_{2A}	α_{2B}	α_{2C}
Alternative names	α_2-C10, RG20, α_{2D} (rat)	α_2-C2, RNG	α_2-C4
Structural information (Accession no.)	h 450 aa (P08913) r 450 aa (P22909) m 450 aa (Q01338)	h 451 aa (P18089) r 453 aa (P19328) m 455 aa (P30545)	h 461 aa (P18825) r 458 aa (P22086) m 458 aa (Q01337)
Chromosomal location	10q23–25	2	4
Selective agonists	oxymetazoline (partial), guanfacine	—	—
Selective antagonists	BRL44408, BRL48962	imiloxan, rauwolscine	MK912
Radioligands	[^{3}H]-Rauwolscine, [^{3}H]-RX821002	[^{3}H]-Rauwolscine, [^{3}H]-RX821002	[^{3}H]-Rauwolscine, [^{3}H]-RX821002
G protein coupling	G_i/G_o	G_i/G_o	G_i/G_o
Expression profile	cerebral cortex, hippocampus, locus coeruleus, amygdala, hypothalamus, spinal cord, platelets, adipocytes, aorta, kidney, spleen	thalamus, cerebellum, septum, striatum, olfactory tubercle, liver, spleen, heart, kidney, neonatal lung	cerbral cortex, cerebellum, hipocampus, striatum, olfactory bulb, islands of Cajella, heart, lung, aorta, kidney
Physiological function	major presynaptic receptor subtype regulating norepinephrine release from sympathetic nerves, hypotension, analgesia, sedation, anaesthesia	regulation of the sympathetic nervous system, vasoconstriction	regulation of the sympathetic nervous system
Knockout phenotype	increase in sympathetic activity with resting tachycardia, depletion of cardiac tissue norepinephrine concentration, down-regulation of cardiac β receptors	decreased vascular response to agonists	decreased hypothermic response, reduced behavioural despair, increased startle response
Disease relevance	depression, pain, addiction, attention deficit hyperactivity disorder, hypertension, intra-ocular pressure in glaucoma		

Table 12.1 (*Continued*)

β-Adrenoceptors	β1	β2	β3
Alternative names	—	—	—
Structural information (Accession no.)	h 477 aa (P08588), r 466 aa (P18090), m 466 aa (P34971)	h 413 aa (P07550), r 418 aa (P10608), m 418 aa (P18762)	h 408 aa (P13945)[AS], r 400 aa (P26255)[AS], m 400 aa (25962)[AS]
Chromosomal location	10q24–26	5q31–32	8p11–12
Selective agonists	norepinephrine, T0509, denopamine, xamoterol (partial)	procaterol, salbutamol, fenoterol	BRL37344, CL316243, SB226552
Selective antagonists	CGP20712A, betaxolol, atenolol, bisoprolol, practolol	ICI118551, butoxamine, α-methylpropranolol	SR58894, SR59230A
Radioligands	[^{125}I]-iodoncyanopindolol, [^{3}H]-CGP12177, [^{3}H]-dihydroalprenolol	[^{125}I]-iodoncyanopindolol, [^{3}H]-CGP12177, [^{3}H]-dihydroalprenolol	[^{125}I]-iodoncyanopindolol, [^{3}H]-CGP12177, [^{3}H]-dihydroalprenolol
G protein coupling	G_s	G_s	G_s
Expression profile	widely expressed in brain inculding cerebral cortex, hippocampus, diencephalon, cerebellum, caudate	cerebral cortex, piriform cortex, hippocampus, cerebellum, caudate, diencephalon, olfactory bulb	adipose, gall bladder, small intestine, stomach, prostate, atrium, bladder
Physiological function	increased heart rate and force of contraction	smooth muscle relaxation	lipolysis, cardioinhibition
Knockout phenotype	decreased cardiac response to agonists	increased exercise capacity	increased fat accumulation
Disease relevance	hypertension, congestive heart failure, myocardial infarction, angina, cardial arrhythmia	same as β1 but may also be useful for the treatment of asthma	same as β1 but may also be useful for the treatment of diabetes and obesity

12.6.2 Thermoregulation

Adrenoceptors have also been implicated in thermoregulation. Thus, rodents exhibit a hypothermic response to an α_2-adrenoceptor agonist that was attenuated in α_{2c}-adrenoceptor knockout mice and accentuated by α_{2c}-adrenoceptor overexpression (Sallinen et al. 1997). The β_3-adrenoceptor is responsible for lipolysis in white adipose tissue and thermogenesis in the brown adipose tissue found in rodents.

12.6.3 Anxiety

α_{2C}-adrenoceptors appear to be involved in the startle reflex and isolation induced aggression, since knockout of α_{2c}-adrenoceptors in mice results in an enhanced startle response and shortened attack latency, whereas overexpression of this receptor produced the opposite effect on both parameters (Sallinen et al. 1998). Behavioural despair in the forced swimming test was enhanced or inhibited by α_{2c}-adrenoceptor overexpression or knockout, respectively (Sallinen et al. 1999).

12.6.4 Alzheimer's disease and cognition

α_1-Adrenoceptor agonists can reverse the cognitive impairment produced by a neurotoxin selective for noradrenergic neurons (Nozulak et al. 1992). Learning behaviour in a water-maze was impaired, in mice lacking the α_{1b}-adrenoceptor although the ability to escape onto a visible platform was unaffected (Spreng et al. 2001). Overexpression of the α_{2c}-adrenoceptor impairs the performance of mice in a water maze (Bjorklund et al. 2000).

12.6.5 Parkinson's disease

A centrally active α_1-adrenoceptor agonist, SDZ NVI-085 (Naphtoxazine), improved the performance of patients with Parkinson's Disease in some, but not all, tests for attentional deficits (Bedard et al. 1998). SDZ NVI-085 has shown activity in many of the animal assays for central α_1-adrenoceptor activation. However, this compound has potent antagonist activity at 5-HT$_2$ receptors (Lachnit et al. 1996) which could contribute to some of its CNS effects. The human striatum exhibits dense β_1-adrenoceptor binding and in MPTP treated monkeys it is clear that β-adrenoceptor antagonists reduce tremor. Overexpression of either wild-type or constitutively active α_{1b}-adrenoceptors results in granulovacular neurodegeneration, beginning in the areas of the brain expressing highest levels of the α_{1b}-adrenoceptor (cortex, cerebellum, and hypothalamus), and progressing to other brain areas with age (Zujik et al. 2000). These mice exhibit symptoms consistent with neuronal damage, including parkinsonian hindlimb symptoms associated with paralysis and tremor.

12.6.6 Epilepsy

Overexpression of either wild-type or constitutively active α_{1b}-adrenoceptors results in grand mal seizures by 12 months of age. The severity of the seizure disorder was proportional to the level of α_{1B}-adrenoceptor activation, with mice expressing constitutively active receptors being most affected. α_{1B}-Adrenoceptor blockade with terazosin partially reversed the behavioural pathology.

12.6.7 **Drug dependence**

The ability of α_2-adrenoceptor antagonists to alleviate many of the symptoms of opiate withdrawal most likely results from the ability of both opiate agonists and α_2-adrenoceptor agonists to inhibit locus coeruleus firing (Aghajanian 1978), although an action at spinal centres may also contribute (Buccafusco 1990).

12.6.8 **Other CNS indications**

Despite clinical trials for several indications, no molecule specifically designed to block the α_2-adrenoceptor has been successfully developed as a drug, although α_2-adrenoceptor antagonist activity may play an important role in antidepressants such as mirtazepine (Langer 1997). In addition to their use as antihypertensives, selective centrally-acting α_2-adrenoceptor agonists are used as adjuncts to general anesthesia, in the treatment of opiate withdrawal, attention-deficit hyperactivity disorder and Tourette's syndrome; the anaesthetic and analgesic actions of α_2-adrenoceptor agonists appearing to be mediated by the α_{2A}-adrenoceptor.

Central β-adrenoceptors appear to play a role in movement disorders and in the human striatum the dense β-adrenoceptor binding that is normally seen decreases in the late stages of Huntington's chorea (Waeber *et al.* 1991). It is well established that chronic treatment with tricyclic antidepressants produces downregulation of central β-adrenoceptors (Sulser *et al.* 1984), presumably as a consequence of increased synaptic levels of the catecholamine neurotransmitters as a result of blockade of neuronal re-uptake. β-Adrenoceptors may be involved in the antinociceptive activity of the tricyclic antidepressants, since β-adrenoceptor antagonists can reverse this activity (Mico *et al.* 1997). Other functions attributed to central β-adrenoceptors include adaptation to stress, memory, and learning, control of respiration and glial proliferation (Doze *et al.* 1998). Several β-adrenoceptor antagonists have also been evaluated for a variety of CNS applications; perhaps the most convincing results are the use of these agents in treating anxiety of varying etiology (Turner 1991), although it is not clear whether a central or peripheral site of action is involved. Despite these CNS indications knockouts of β_1-(Rohrer *et al.* 1996), β_2-(Chruschinski *et al.* 1999) and β_3-adrenoceptors (Susulic *et al.* 1995) as well as mice lacking both β_1- and β_2-adrenoceptors (Rohrer *et al.* 1999) do not overtly exhibit phenotypic changes reflecting deficits in CNS function, although detailed behavioural studies have not yet been reported.

12.6.9 **Peripheral indications**

Knockout and overexpression experiments have been reported only for the α_{1b}-adrenoceptor subtype. Knockout of the α_{1b}-adrenoceptor results in impaired vascular responsiveness to phenylephrine, demonstrated both in isolated blood vessels and through measurement of blood pressure responses in anesthetized mice (Cavalli *et al.* 1997). Cardiac specific overexpression of the α_{1b}-adrenoceptor results in an impairment of cardiac function without cardiomyocyte hypertrophy (Grupp *et al.* 1998). Increased α_{1B}-adrenoceptor density in the heart impairs the responsiveness to β-adrenoceptor agonists, perhaps as a result of heterologous desensitization (Lemire *et al.* 1998). In many cases, the particular subtype involved in an α_1-adrenoceptor mediated response has not yet been identified. This is due in part to the lack of subtype selective antagonists suitable for *in vivo* evaluation. Depending on the species and/or vascular bed, each α_1-adrenoceptor subtype can contribute to vascular contraction.

For example, contraction of the rat caudal artery is mediated by the α_{1A}-adrenoceptor, the rat aorta by the α_{1D}-adrenoceptor, and many canine and human vessels by the α_{1L}-adrenoceptor (Muramatsu *et al.* 1998). Knockout of the α_{1B}-adrenoceptor significantly attenuates the pressor response to α_1-adrenoceptor agonists in the mouse. α_2-Adrenoceptors within the CNS have long been known to be involved in the antihypertensive action of clonidine and other α_2-adrenoceptor agonists (Timmermans *et al.* 1982). Although the involvement of a non-adrenergic imidazoline receptor in this action has been postulated (Ernsberger *et al.* 1990), the failure of α_2-adrenoceptor agonists to produce a sympatholytic action in mice where the α_{2a}-adrenoceptor has been knocked out or mutated (Hein *et al.* 1999) supports functional experiments in rats and rabbits suggesting an α_2-adrenoceptor mediated action, at least for systemically administered agonists (Hieble and Kolpak 1993; Szabo *et al.* 1993; Urban *et al.* 1995). The hypertensive action of α_2-adrenoceptor agonists likely results from activation of vascular α_{2B}-adrenoceptors (Hein *et al.* 1999). Consistent with the apparent role of the α_{2B}-adrenoceptor in the maintenance of vascular tone, mice lacking this receptor subtype failed to develop salt-induced hypertension (Makaritsis *et al.* 1999). Hence, a selective α_{2A}-adrenoceptor agonist may be preferable as a centrally active drug, although sedative and antihypertensive actions can apparently not be dissociated by subtype selectivity. Knockout of the α_{2c}-adrenoceptor has no apparent cardiovascular effect. Recent data also indicate that the α_{2C}-adrenoceptor participates in cold-induced augmentation of α-adrenoceptor mediated vasoconstriction (Chotani *et al.* 2000). Although multiple β-adrenoceptor subtypes can participate in the cardiac stimulation produced by the catecholamines, gene knockout experiments demonstrate that the β_1-adrenoceptor subtype plays the major role. Experimental evidence has been presented supporting the presence of an additional β-adrenoceptor, tentatively designated as the β_4-adrenoceptor, modulating cardiac contractility (Kaumann and Molenaar 1997). However, it now seems likely that these data may be explained by multiple affinity states of the β_1-adrenoceptor (Oostendorp *et al.* 2000).

Mice lacking the β_1 adrenoceptor fail to respond to the inotropic action of β-adrenoceptor agonists, confirming the importance of this subtype in the control of cardiac contractility. However, maximal exercise capacity is not reduced (Rohrer *et al.* 1998). Mice lacking the β_2-adrenoceptor have a normal response to exogenous β-adrenoceptor agonists, and have even greater exercise capacity than wild-type mice. However, these animals become hypertensive during exercise and have a lower respiratory exchange ratio, suggesting influences of the β_2-adrenoceptor on energy metabolism (Chruscinski *et al.* 1999). Mice lacking both β_1- and β_2-adrenoceptors have normal basal cardiovascular parametres and normal exercise capacity, although the ability of exercise or the administration of exogenous agonists to increase heart rate is blunted (Rohrer *et al.* 1999). Mice lacking the β_3-adrenoceptor show mild increases in body fat stores, and do not show metabolic responses to a selective β_3-adrenoceptor agonists (Susulic *et al.* 1995). In several cases, an increased responsiveness to one of the remaining β-adrenoceptors is observed when one of the subtypes is knocked out. Thus, there is increased β_1-adrenoceptor responsiveness in β_3-adrenoceptor knockout animals, and enhanced β_3-adrenoceptor responsiveness in β_1/β_2 knockout animals. These observations support the concept that many physiological functions can be mediated by multiple β-adrenoceptor subtypes. Cardiac-directed overexpression of the human β_1-adrenoceptor results in the accumulation of fibrous tissue between cardiac myocytes, myocyte hypertrophy, and myofibrilar disarray (Bisognano *et al.* 2000). These changes result in cardiac dysfunction in older animals (Bisognano *et al.* 2000; Engelhardt *et al.* 1999). Overexpression of human β-adrenoceptors in mouse heart enhances basal cardiac function, due to constitutive activity of the expressed receptors (Liggett *et al.* 2000). Fibrotic cardiomyopathy is observed in mice expressing high levels of the β_2-adrenoceptor, with the severity and rate of onset being dependent on the level of receptor expression. Overexpression of cardiac β_2-adrenoceptors exacerbates functional deterioration following pressure overload as a result of aortic stenosis (Du *et al.* 2000).

Many important drugs target the β-adrenoceptor. β-Adrenoceptor antagonists, either selective or nonselective for the β_1-adrenoceptor, are widely used as antihypertensives. The mechanism for this action is still not clearly understood, but may involve, at least in part, an action on central β-adrenoceptors or inhibition of renin release via the β_1-adrenoceptor. Several β-adrenoceptor antagonists, including carvedilol, which combines nonselective β-adrenoceptor blockade with α_1-adrenoceptor blockade (Hieble *et al.* 1998), have been shown to produce a dramatic reductions in the mortality and morbidity associated with congestive heart failure (Nuttall *et al.* 2000). β-Adrenoceptor antagonists have other cardiovascular applications, including the prevention and treatment of myocardial infarction, angina pectoris, and cardiac arrhythmia.

Bronchodilation is mediated primarily by the β_2-adrenoceptor. Selective overexpression of β_2-adrenoceptors in airway epithelium or smooth muscle results in decreased sensitivity to methacholine-induced bronchoconstriction (McGraw *et al.* 2000), and an increased sensitivity to a β-adrenoceptor agonists. Selective β_2-adrenoceptor agonists are commonly used as bronchodilators in the treatment of asthma and other conditions associated with inappropriate bronchoconstriction. These agents are synergistic with inhaled steroids, and may have other beneficial actions, such as reduction of neutrophil infiltration (Howarth *et al.* 2000).

Contraction of prostatic and urethral smooth muscle appears to be mediated by the α_{1L}-adrenoceptor. α_1-adrenoceptor antagonists having selective affinity for α_{1A}- and α_{1L}-adrenoceptors, as well as antagonists having affinity for both α_{1A}- and α_{1D}-adrenoceptors, are currently being developed for the treatment of benign prostatic hyperplasia, although it has not yet been established that these drugs are clinically superior to the non-subtype selective α_1-adrenoceptor antagonists which have been proven to be effective for this indication.

Intra-ocular administration of an α_2-adrenoceptor agonist will reduce intra-ocular pressure in glaucoma as will intra-ocular administration of nonselective β-adrenoceptor antagonists. None of the agonists employed clinically have pharmacologically significant selectivity between any of the α_2-adrenoceptor subtypes. Selective β_3-adrenoceptor agonists are being developed for the treatment of type II diabetes and obesity. Although these agonists are clearly effective in animal models, and can produce metabolic effects in man (Weyer *et al.* 1998) clinical efficacy has been elusive. This is partially a result of pharmacological differences between rodent and human β_3-adrenoceptors (Sennitt *et al.* 1998) causing many compounds to have much lower efficacy at the human receptor. Knockout of the β_1-adrenoceptor has a marked effect on embryonic viability, although the few homozygous animals surviving appear normal (Rohrer *et al.* 1996). This effect on viability is not observed with knockout of the β_2- or β_3-adrenoceptor subtypes.

References

Aboud R, Shafii M, and Docherty JR (1993). Investigation of the subtypes of alpha 1-adrenoceptor mediating contractions of rat aorta, vas deferens and spleen. *Br J Pharmacol* 109, 80–7.

Aghajanian GK (1978). Tolerance of locus coeruleus neurons to morphine and suppression to withdrawal response by clonidine. *Nature* **176**, 186–88.

Al-Damlughi S and Francis D (1993). Activation of central alpha-1 adrenoceptors in humans stimulates secretion of prolactin and TSH, as well as ACTH. *Am J Physiol* **264**, E208–14.

Astrachan DI, Gallager DW, and Davis M (1983*a*). Behavior and binding: desensitization to alpha 1-adrenergic stimulation of acoustic startle is associated with a decrease in alpha 1-adrenoceptor binding sites. *Brain Res* **276**(1), 183–7.

Astrachan DI, Davis M, and Gallager DW (1983*b*). Behavior and binding: correlations between alpha 1-adrenergic stimulation of acoustic startle and alpha 1-adrenoceptor occupancy and number in rat lumbar spinal cord. *Brain Res* **260**, 81–90.

Bedard MA, el Massioui F, Malapani C *et al.* (1998). Attentional deficits in Parkinson's disease: partial reversibility with naphtoxazine (SDZ NVI-085), a selective noradrenergic alpha-1 agonist. *Clin Neuropharmacol* **21**, 108–17.

Bisognano JD, Weinberger HD, Bohlmeyer TJ *et al.* (2000). Myocardial directed overexpression of the human beta(1) adrenergic receptor in transgenic mice. *J Mol Cell Cardiol* **32**, 817–30.

Bjorklund M, Sirvio J, Riekkinen M *et al.* (2000). Overexpression of alpha2C-adrenoceptors impairs water maze navigation. *Neuroscience* **95**, 481–7.

Buccafusco JJ (1990). Participation of different brain regions in the anti-narcotic withdrawal action of clonidine in the dependent rat. *Brain Res* **513**, 8–14.

Bylund DB, Regan JW, Faber JE *et al.* (1995). Vascular α-adrenoceptors: From the gene to the human. *Canadian J Physiol Pharmacol* **73**, 533–53.

Cavalli A, Lattion AL, Hummler E *et al.* (1997). Decreased blood pressure response in mice deficient of the alpha1b-adrenergic receptor. *Proc Nat Acad Sci USA* **94**, 11 589–94.

Chang DJ, Chang TK, Yamanishi SS *et al.* (1998). Molecular cloning, genomic characterization and expression of novel human alpha1A-adrenoceptor isoforms. *FEBS Lett* **422**, 279–83.

Chotani MA, Flavahan SA, Mitra S, Daunt D, and Flavahan NA (2000). Silent α_{2C} adrenergic receptors enable cold-induced vasoconstriction in cutaneous arteries. *Am J Physiol* **278**, H1075–83.

Chruscinski AJ, Rohrer DK, Schauble E, Desai KH, Bernstein D, and Kobilka BK (1999). Targeted disruption of the β2 adrenergic receptor gene. *J Biol Chem* **274**, 16694–700.

Day HE, Campeau S, Watson SJ Jr, and Akil H (1997). Distribution of alpha 1a-, alpha 1b-, and alpha 1d-adrenergic receptor mRNA in the rat brain and spinal cord. *J Chem Neuroanat* **13**, 115–39.

Doze P, van Waarde A, Elsinga PH, van-Loenen Weemaes AM, and Vaalburg W (1998). Validation of S-1'-[18F] fluorocarazolol for *in vivo* imaging and quantification of cerebral beta-adrenoceptors. *Eur J Pharmacol* **353**, 215–26.

Du XJ, Autelitano DJ, Dilley RJ, Wang B, Dart AM, and Woodcock EA (2000). Beta (2)-adrenergic receptor overexpression exacerbates development of heart failure after aortic stenosis. *Circulation* **101**, 71–7.

Duteil J, Rambert FA, Pessonnier J, Hermant JF, Gombert R, and Assous E (1990). Central alpha-1 adrenergic stimulation in relation to the behavior stimulating effect of modafinil: studies with experimental animals. *Eur J Pharmacol* **180**, 49–58.

Engelhardt S, Hein L, Wiesmann F, and Lohse MJ (1999). Progressive hypertrophy and heart failure in beta-1 adrenergic receptor transgenic mice. *Proc Nat Acad Sci USA* **96**, 7059–64.

Ernsberger P, Giuliano R, Willette RN, and Reis DJ (1990). Role of imidazole receptors in the vasodepressor response to clonidine analogs in the rostral ventrolateral medulla. *J Pharmacol Experimental Therapeutics* **253**, 403–18.

Ford AP, Daniels DV, Chang DJ *et al.* (1997). Pharmacological pleiotropism of the human recombinant α_{1A}-adrenoceptor: Implications for α1-adrenoceptor classification. *Br J Pharmacol* **121**, 1127–35.

Grijalba B, Callado LF, Javier-Meana J, Garcia-Sevilla JA, and Pazos A (1996). Alpha-2 adrenoceptor subtypes in human brain. a pharmacological delineation of [3H] RX-821002 binding to membranes and tissue sections. *Eur J Pharmacol* **310**, 83–93.

Grupp IL, Lorenz JN, Walsh RA, Boivin GP, and Rindt H (1998). Overexpression of alpha-1B adrenergic receptor induces left ventricular dysfunction in the absence of hypertrophy. *Am J Physiol* **275**, H1338–50.

Guo TZ, Tinklenberg J, Oliker R, and Maze M (1991). Central alpha-1 adrenoceptor stimulation functionally antagonizes the hypnotic response to dexmedetomidine, an alpha-2 adrenoceptor agonist. *Anesthesiology* **75**, 252–6.

Hein L, Altman JD, and Kobilka BK (1999). Two functionally distinct α_2-adrenergic receptors regulate sympathetic neurotransmission. *Nature* **402**, 181–4.

Hieble JP (1991). Structure activity relationships of beta-adrenoceptor agonists and antagonists. In (eds, RR Ruffolo, Jr, S Karger, AG, Basel), *Beta-Adrenoceptors: Molecular Biology, Biochemistry and Pharmacology, Progress in Basic and Clinical Pharmacology*, Vol. 7, pp. 105–72.

Hieble JP, Bondinell WE, and Ruffolo RR Jr (1995*a*). Alpha and beta adrenoceptors: From the gene to the clinic. Part 1. *J Medicinal Chem* **38**, 3415–44.

Hieble JP, Bylund DB, Clarke DE *et al.* (1995*b*). International union of pharmacology recommendation for nomenclature of α_1-adrenoceptors: consensus update. *Pharmacol Rev* **47**, 267–70.

Hieble JP and Kolpak DC (1993). Mediation of the hypotensive action of systemic clonidine in the rat by α_2-adrenoceptors. *Br J Pharmacol* **110**, 1635–9.

Hieble JP, Naselsky DP, Arch JRS *et al.* (1998). Affinity of carvedilol for recombinant human adrenoceptors. *Pharmacol Communications* **10**, 43–9.

Hieble JP and Ruffolo RR Jr (1996). Subclassification and nomenclature of α_1- and α_2-adrenoceptors. In ed. E. Jucker, *Progress in Drug Research* pp. 81–130, Birkhauser-Verlag.

Hieble JP, Ruffolo RR, Jr, and Starke K (1996). Subclassification of α_2-adrenoceptors: An overview. In eds, SM Lanier, LE Limbird, α_2-*Adrenergic Receptors: Structure, Function and Therapeutic Applications* Harwood, pp. 1–18.

Holz WC, Hieble JP, Gill CA, DeMarinis RM, and Pendleton RG (1982). Alpha adrenergic agents. 3. Behavioral Effects of 2-Aminotetralins. *Psychopharmacol* **77**, 259.

Howarth PH, Beckett P, and Dahl R. (2000). The effect of long-acting beta2-agonists on airway inflammation in asthmatic patients. *Respiratory Med* **94** (Suppl F), S22–5.

Kaumann AJ and Molenaar P (1997). Modulation of human cardiac function through 4 β-adrenoceptor populations. *Naunyn-Schmiedeberg's Arch Pharmacol* **355**, 667–81.

Lachnit WG, Ford AP, and Clarke DE (1996). SDZ NVI 085, an alpha 1A-adrenoceptor agonist with 5-HT2A receptor antagonist properties. *Eur J Pharmacol* **297**, 83–6.

Langer SZ (1997). 25 years since the discovery of presynaptic receptors: present knowledge and future perspectives. *Trends Pharmacol Sci* **18**, 95–9.

Langer SZ (1974). Presynaptic regulation of catecholamine release. *Br J Pharmacol* **60**, 481–97.

Lemire I, Allen BG, Rindt H, and Hebert TE (1998). Cardiac specific overexpression of alpha1B AR regulated beta AR activity via molecular crosstalk. *J Mol Cell Cardiol* **30**, 1827–39.

Liggett SB, Tepe NM, and Lorenz JN (2000). Early and delayed consequences of beta(2) adrenergic receptor overexpression in mouse hearts: critical role for expression level. *Circulation* **101**, 1707–14.

Majewski H and Rand MJ (1984). Prejunctional β-adrenoceptors and hypertension: a hypothesis revisited. *Trends Pharmacol Sci* **5**, 500–2.

Makaritsis KP, Handy DE, Johns C, Kobilka B, Gavras I, and Gavras H (1999). Role of the alpha$_{2B}$-adrenergic receptor in the development of salt-induced hypertension. *Hypertension* **33**, 14–7.

McGraw DW, Forbes SL, Mak JC *et al.* (2000). Transgenic overexpression of beta(2) adrenergic receptors in airway epithelial cells decreases bronchoconstriction. *Am J Physiol* **279**, L379–89.

Mico JA, Gibert-Rahola J, Casas J, Rojas O, Serrano MI, and Serrano JS (1997). Implication of beta-1 and beta-2 adrenergic receptors in the antinociceptive effect of trycyclic antidepressants. *Eur J Neuropsychopharmacol* **7**, 139–45.

Muramatsu I, Murata S, Isaka M *et al.* (1998). Alpha 1-adrenoceptor subtypes and two receptor systems in vascular tissues. *Life Sci* **62**, 1461–5.

Nicholas AP, Hokfelt T, and Pieribone VA (1996). The distribution and significance of CNS adrenoceptors examined with *in situ* hybridization. *Trends Pharmacol Sci* 17, 245–55.

Nicholas AP, Peribone V, Dagerlind Å, Meister B, Elde R, and Hökfelt T (1995). *In Situ* hybridization. A complementary method to radioligand mediated autoradiography for localizing adrenergic, alpha-2 receptor producing cells. *Annals of the New York Acad Sci* 763, 222–42.

Nishino S, Fruhstorfer B, Arrigoni J, Guilleminault C, Dement WC, and Mignot E (1993). Further characterization of the alpha-1 receptor subtype involved in the control of cataplexy in canine narcolepsy. *J Pharmacol and Experimental Therapeutics* 264, 1079–84.

Nozulak J, Vigouret JM, Jaton AL *et al.* (1992). Centrally acting α1-adrenoceptor agonists based on hexahydronaphth[2,3-b]oxazines and octahydrobenzo[g]quinolines. *J Med Chem* 35, 480–9.

Nuttall SL, Langford NJ, and Kendall MJ (2000). Beta-blockers in heart failure. 1. Clinical evidence. *J Clin Pharmacy & Therapeutics* 25, 395–8.

Oostendorp J, Preitner F, Moffatt J, Jimenez M, Giacobino JP, Molenaar P *et al.* (2000). Contribution of beta-adrenoceptor subtypes to relaxation of colon and esophagus and pacemaker activity of ureter in wildtype and beta(3)-adrenoceptor knockout mice. *Brit J Pharmacol* 130, 747–58.

Renaud A, Nishino S, Dement WC, Guilleminault C, and Mignot E (1991). Effects of SDZ NVI-085, a putative subtype-selective alpha-1 agonist, on canine cataplexy, a disorder of rapid eye movement sleep. *Eur J Pharmacol* 205, 11–6.

Rohrer DK, Chruscinski A, Schauble EH, Bernstein D, and Kobilka BK (1999). Cardiovascular and metabolic alterations in mice lacking both β1 and β2 adrenergic receptors. *J Biol Chem* 274, 16 701–8.

Rohrer DK, Desai KH, Jasper JR *et al.* (1996). Targeted disruption of the mouse β1-adrenergic receptor gene: Developmental and cardiovascular effects. *Proc Nat Acad Sci USA* 93, 7375–80.

Rohrer DK, Schauble EH, Desai KH, Kobilka BK, and Bernstein D (1998). Alterations in dynamic heart rate control in the beta-1 adrenergic receptor knockout mouse. *Am J Physiol* 274, H1184–93.

Ruffolo RR Jr, Bondinell WE, and Hieble JP (1995). Alpha- and beta-adrenoceptors: From the gene to the clinic. Part 2. *J Med Chem* 38, 3681–716.

Ruffolo RR Jr, Nichols AJ, and Stadel JM *et al.* (1993). Pharmacologic and therapeutic applications of a2-adrenoceptor subtypes. *Ann Rev Pharmacol Toxicol* 33, 243–79.

Sallinen J, Haapalinna A, Kobilka BK, and Scheinin M. (1998). Adrenergic alpha 2C receptors modulate the acoustic startle reflex, prepulse inhibition and aggression in mice. *J Neurosci* 18, 3035–42.

Sallinen J, Haapalinna A, MacDonald E *et al.* (1999). Genetic alteration of the alpha2-adrenoceptor subtype c in mice affects the development of behavioral despair and stress-induced changes in plasma corticosterone levels. *Mol Psychiatry* 4, 443–52.

Sallinen J, Link RE, Haapalinna A *et al.* (1997). Genetic alteration of alpha 2C-adrenoceptor expression in mice: influence on locomotor, hypothermic, and neurochemical effects of dexmedetomidine, a subtype-nonselective alpha 2-adrenoceptor agonist. *Mol Pharmacol* 51, 36–46.

Sennitt MV, Kaumann AJ, Molenaar P *et al.* (1998). The contribution of classical $\beta_{1/2}$ and atypical β-adrenoceptors to the stimulation of human white adipocyte lipolysis and right atrial appendage contraction by novel β_3-adrenoceptor agonists of differing selectivities. *J Pharmacol Experimental Therapeutics* 285, 1084–95.

Spreng M, Cotecchia S, and Schenk F (2001). A behavioral study of alpha-1b adrenergic receptor knockout mice: Increased reaction to novelty and selectively reduced learning capacities. *Neurobiol Learning & Memory* 75, 214–29.

Stam WB, Van der Graaf PH, and Saxena PR (1998). Functional characterisation of the pharmacological profile of the putative alpha1B-adrenoceptor antagonist, (+)-cyclazosin. *Eur J Pharmacol* 361, 79–83.

Starke K, Montel H, Gayk W, and Merker R(1974). Comparison of the effects of clonidine on pre- and postsynaptic adrenoceptors in the rabbit pulmonary artery. *Naunyn-Schmiedeberg's Arch Pharmacol* 285, 133–50.

Strader CD, Fong TM, and Tota MR (1994). Underwood D. Dixon RA. Structure and function of G protein-coupled receptors. *Ann Rev of Biochem* 63, 101–32.

Strong R, Huang JS, Huang SS, Chung HD, Hale C, and Burke WJ (1991). Degeneration of the cholinergic innervation of the locus ceruleus in Alzheimer's disease. *Brain Res* 542, 23–8.

Sulser F, Gillespie DD, Mishra R, and Manier DH (1984). Desensitization by antidepressants of central norepinephrine receptor systems coupled to adenylate cyclase. *Annals of the New York Acad Sci* 430, 91–101.

Susulic VS, Frederich RC, Lawitt J *et al*. (1995). Targeted disruption of the β3-adrenergic receptor gene. *J Biol Chem* 270, 29 483–92.

Szabo B, Urban R, and Starke K (1993). Sympathoinhibition by rilmenidine in conscious rabbits: involvement of alpha 2-adrenoceptors. *Naunyn-Schmiedeberg's Arch Pharmacol* 348, 593–600.

Timmermans PBMWM and van Zwieten PA (1982). α2-Adrenoceptor: classification, localization, mechanisms and targets for drugs. *J Medicinal Chem* 25, 1389–401.

Trendelenburg AU, Sutej I, Wahl CA, Molderings GJ, Rump LC, and Starke K (1997). A re-investigation of questionable subclassifications of presynaptic alpha2-adrenoceptors: rat vena cava, rat atria, human kidney and guinea pig urethra. *Naunyn-Schmiedeberg's Arch Pharmacol* 356, 721–37.

Turner P (1991). Clinical psychopharmacology of beta-adrenoceptor antagonism in treatment of anxiety. *Annals of the Acad Medicine, Singapore* 20, 43–5.

Urban R, Szabo B, and Starke K (1995). Involvement of alpha 2-adrenoceptors in the cardiovascular effects of moxonidine. *Eur J Pharmacol* 282, 19–28.

Waeber C, Rigo M, Chinaglia G, Probst A, and Palacios JM (1991). Beta-adrenergic receptor subtypes in the basal ganglia of patients with Huntington's chorea and Parkinson's disease. *Synapse* 8, 270–80.

Weyer C, Tataranni PA, Snitker S, Danforth E Jr, and Ravussin E (1998). Increase in insulin action and fat oxidation after treatment with CL-316,243, a highly selective β 3-adrenoceptor agonist in humans. *Diabetes* 47, 1555–61.

Zilles K, Qu M, and Schleicher A (1993). Regional distribution and heterogeneity of alpha-adrenoceptors in the rat and human central nervous system. *J Hirnforschung* 34, 123–32.

Zujic MJ, Sands S, Ros SA *et al*. (2000). Overexpression of the α 1B- adrenergic receptor causes apoptotic neurodegeneration: Multiple system atrophy. *Nature Med* 6, 1388–94.

Chapter 13

Cannabinoid receptors

D. A. Kendall and S. P. H. Alexander

13.1 Introduction

Cannabinoid receptors are seven transmembrane receptors that mediate the central and peripheral actions of extracts from the cannabis plant. Although many endogenous agonists of cannabinoid receptors have been described, a categorical role of these (mainly fatty acid derived) molecules as *the* endogenous ligand for cannabinoid receptors is currently lacking. Putative endogenous ligands of the cannabinoid receptors, which have been termed endocannabinoids, include anandamide, 2-arachidonylethanolamide, homo-γ-linolenylethanolamide, docosatetra-7,10,13,16-enylethanolamide and 2-arachidonyl glyceryl ether. It should be pointed out, however, that the putative endocannabinoid anandamide is also a full agonist at the VR1 vanilloid receptor (Smart *et al.* 2000). This is a Ca^{2+}-gating ion channel receptor that is located on the central and peripheral terminals of sensory nerves as well as in a number of different brain regions (Guo *et al.* 1999). Indeed, it could be argued that the VR1 represents an ionotropic class of cannabinoid receptor analogous to the nicotinic acetylcholine receptor. Since many of these endogenous ligands are labile and subject to metabolism through various routes, notably the hydrolytic enzyme anandamide amidohydrolase (Ueda *et al.* 2000), assays of cannabinoid receptor binding and function may be complicated. In *in vitro* assays, serine protease inhibitors such as phenylmethylsulphonylfluoride (PMSF) may be added to the medium to increase endocannabinoid lifetime. Furthermore, the life-cycle of endocannabinoids is distinct from the classical view of a transmitter/hormone, in that synthetic pathways appear to generate ligands 'on demand' (Di Marzo *et al.* 1994). Thus, *N*-acylphosphatidylethanolamine (NAPE), a putative precursor for anandamide has been suggested to be generated through the action of a calcium-stimulated *N*-acyltransferase (Sugiura *et al.* 1996). The resulting NAPE is then proposed to be hydrolysed by a specific phospholipase D activity to produce anandamide (Sugiura *et al.* 1996). A by-product of NAPE formation via the calcium-sensitive acyltransferase has been suggested to be 1-lyso-2-arachidonoyl-phosphatidylcholine (from a 1,2-diarachidonoylphosphatidylcholine acyl donor) (Di Marzo *et al.* 1996). Upon activation of phospholipase C, this lysophospholipid would be hydrolysed to form choline phosphate and 2-arachidonoylglycerol. Both anandamide and 2-arachidonoylglycerol have been reported to accumulate in brain *post mortem* (Sugiura *et al.* 2001). Termination of endocannabinoid signalling appears to be via uptake from the extracellular medium, through a specific transporter, and metabolism by intracellular hydrolases, such as anandamide amidohydrolase (Di Marzo and Deutsch 1998; Piomelli *et al.* 1998).

Two confirmed cannabinoid receptor subtypes coupled to the G_i/o family of G proteins mediate the effects of endocannabinoids. Interestingly, one of these, the CB_2 receptor, was thought to be the specific target for palmitoylethanolamide (Facci *et al.* 1995) although this suggestion has recently been challenged (Ross *et al.* 2000). Despite this the wide-spread distribution of these receptors in the central and peripheral nervous systems, as well as non-nervous tissue implicates this receptor class in diverse physiological and pathophysiological roles in the body, including regulation of mood, appetite, pain sensation, vascular, and non-vascular smooth muscle tone and immune function. Here we review the complex molecular mechanisms mediating these responses.

13.2 **Molecular characterization**

Currently, there appear to be only two subtypes of cannabinoid receptor, CB_1 and CB_2. The *cnr1* gene coding for the CB_1 cannabinoid receptor has been localized to chromosome 6 in man (6q14–15) (Hoehe *et al.* 1991), while the *cnr1* gene is found on mouse chromosome 4. Splice variants of the human (but not mouse or rat) CB_1 receptor have been reported (Rinaldi-Carmona *et al.* 1996a) as well as a human 'silent' polymorphism. The CB_2 cannabinoid receptor is coded for by the *cnr2* gene (1p35–36) in man but as yet, there is no evidence for alternative splicing of CB_2 cannabinoid receptors from man, mouse, or rat.

The cannabinoid receptors themselves are structurally unexceptional in comparison with other members of the rhodopsin-like G protein-coupled receptors (Table 13.1), with a length within the middle range of such proteins. Although CB_1 and CB_2 receptors are relatively well-conserved between species, the intra-species comparison of receptors suggests little similarity (the human CB_2 receptor, for example, exhibits only 44% amino acid homology with the CB_1 receptor (Munro *et al.* 1993)). Each cannabinoid receptor subtype consists of a single polypeptide chain that spans the membrane seven times, with the amino terminal being extracellular and the carboxy terminal intracellular. The amino termini contain consensus sites for N-linked glycosylation. Site-directed mutagenesis and chimeric studies on cannabinoid receptors have been relatively limited. The fourth and fifth transmembrane domains of the CB_1 receptor appear to be intimately involved in high affinity binding of CB_1-selective antagonists (Shire *et al.* 1996a). Chimeric receptors in which the second extracellular loops of the CB_1 and CB_2 receptors were swopped failed to alter binding of antagonist to the CB_1 receptor, but effectively abolished agonist binding to both mutants. A Ser $\rightarrow$ Ala mutation in the CB_2 receptor second extracellular loop also eliminated agonist binding, but, in the CB_1 receptor resulted in the sequestration of the mutated receptors intracellularly (Shire *et al.* 1996a).

In the *Xenopus* oocyte expression system, desensitization of the CB_1 receptor in the continuous presence of agonist is reported to depend upon the presence of GRK3 and β-arrestin 2 (Jin *et al.* 1999). The C-terminal tail is essential to the process since desensitization was abolished by truncation at residue 418 without affecting agonist activation. A deletion mutant lacking residues 418–439 did not desensitize, indicating that residues within this region are crucial. Desensitization appears to involve a phosphorylation event because mutation of either of two putative phosphorylation sites (Ser426Ala or Ser430Ala) significantly attenuated desensitization. In contrast, receptor activation in CB_1-transfected CHO cells results in rapid internalization that is independent of C-terminal phosphorylation. The CB_1 receptor inverse agonist/antagonist SR141716A caused an increase in cell surface receptor density

Table 13.1 Cannabinoid receptors

Cannabinoid receptors	Cannabinoid CB$_1$	Cannabinoid CB$_2$
Alternative names	THC receptor cannabinol receptor central cannabinoid receptor	peripheral cannabinoid receptor
Structural information (Accession no.)	h 472 aa (P21554)[AS] r 473 aa (P20272) m 473 aa (P47746)	h 360 aa (P34972) r 360 aa (Q9QZN9) m 347 aa (P47936)
Chromosomal location	6q14–q15	1p35–p36
Selective agonists	ACEA, ACPA, methanandamide, moderately selective: anandamide, cannabinol	L759633, HU308, L759656, JWH015, WIN55212-2
Selective antagonists	LY320135, SR141716A, AM281	SR144528, AM630, JTE907
Radioligands	[^{3}H]-CP55940, [^{3}H]-HU243, [^{3}H]-WIN552122, [^{3}H]-SR141716A	[^{3}H]-CP55940, [^{3}H]-HU243, [^{3}H]-WIN55212-2
G protein coupling	G$_i$/G$_o$	G$_i$/G$_o$
Expression profile	widespread in the CNS, including basal ganglia, hippocampus, cerebellum, cerebral cortex, also in innervation by peripheral nervous system, testes, spleen	spleen, lymph nodes, bone marrow, tonsils, peripheral blood, retina, leukocytes, macrophages
Physiological function	modulation of neurotransmitter release	not clearly established
Knockout phenotype	increased aggression, enhanced memory and LTP, reduced addictive effects of opiates, hypoalgesia, hypoactivity, increased mortality, retrograde synaptic inhibition in the hippocampus	—
Disease relevance	pain, neurodegenerative diseases, schizophrenia, addiction, spasticity associated with multiple sclerosis or spinal cord injury, haemorrhagic shock, cancer, feeding disorders, emesis, glaucoma	not clear but possibly pain and cancer

(Rinaldi-Carmona *et al.* 1998*b*). In experimental animals, repeated administration of cannabinoid agonists results in loss of CB$_1$ receptors and/or reduced coupling to G proteins in a variety of brain regions (Breivogel *et al.* 1999; Rubino *et al.* 2000*a,b*). The CB$_2$ receptor is also reported to undergo desensitization upon agonist exposure (Derocq *et al.* 2000) although the molecular mechanisms controlling the process have yet to be as fully investigated as for the CB$_1$.

The existence of a third CB receptor has recently been suggested on the basis of several reports of unique pharmacology. These include: (1) in the rat mesenteric artery, the endogenous CB receptor agonist anandamide, but not the synthetic agonist HU210, causes relaxation blocked by the CB_1 receptor antagonist SR141716A, (2) an analogue of cannabidiol ('abnormal cannabidiol') causes mesenteric vasodilation in CB_1/CB_2 knockout mice and this is antagonized by cannabidiol (Jarai et al. 1999), (3) in CB_1 knockout mice, the behavioural effects of THC, but not anandamide, are decreased, (4) in the same knockout mice, anandamide (but not THC) is able to enhance ^{35}S-GTPγS binding to brain membranes but this effect is insensitive to CB_1 and CB_2 antagonists (Di Marzo et al. 2000), (5) in rat astrocytes in vitro, the highly potent CB agonists WIN 55212 and CP55940 inhibit β-adrenoceptor-mediated cAMP formation, in an SR141716-insensitive fashion, in the absence of any evidence for CB_2 receptors.

There is also some evidence that the antinociceptive actions of anandamide might be mediated by a novel receptor. Thus, antinociceptive effects of anandamide in the mouse tail flick test have been reported to be insensitive to SR141716A (Adams et al. 1998). Similarly, Smith et al. (1998) reported that SR141716A antagonized THC but not anandamide-evoked antinociception in the rat paw pressure test (Smith et al. 1998).

13.3 Cellular and subcellular localization

High affinity agonist radioligands are available for both CB_1 and CB_2 receptors, although there is little selectivity apparent to distinguish the two receptors (Table 13.1). An antagonist radioligand, [^{3}H]-SR141716A, is available with good selectivity for the CB_1 receptor. Due to the hydrophobic nature of these radioligands, however, binding is usually conducted in the presence of bovine serum albumin (BSA, 0.1–5 mg/ml) with filtration taking place over vacuum and washes employing solutions with high concentrations of BSA also (Pertwee 1997). As would be expected for a G protein-coupled receptor, it has been observed that Mg^{2+} ions enhance, while GTP analogues inhibit, agonist radioligand binding to rat brain preparations (Devane et al. 1988). Sulfhydryl reagents appear to reduce binding of agonist radioligand to rat brain membranes (Lu et al. 1993), such that the presence of both free sulphydryl groups and disulphide bridges appear necessary for ligand recognition. Using these approaches in conjunction with in situ hybridization and functional analyses, cannabinoid receptor localization has been defined throughout the brain and spinal cord, as well as on peripheral nerves and immune cells. CB_1 and CB_2 receptors are located predominantly on cells of the nervous and immune systems, respectively. More specific localization details are described below.

CB_1 receptor distribution—CB_1 receptors are among the most abundant G protein coupled receptors in the central nervous system. In early binding experiments with an agonist radioligand, a density of almost 2 pmol/mg protein in a P2 fraction from rat cerebral cortex was observed (Devane et al. 1988). The CB_1 receptor is distributed differentially throughout the central nervous system with highest levels being found in the cerebral cortex, hippocampus, cerebellum, and striatum. CB_1 cannabinoid receptor mRNA in the rat brain appears to show a distinct developmental profile in relation to radioligand binding, in that mRNA levels are at adult levels at postnatal day 3. Receptor number, on the other hand, doubles with increasing age, with some lag after mRNA levels have achieved a plateau (Mclaughlin et al. 1994). From E11, CB_1 receptor mRNA is apparent in the embryonic rat neuronal tube, and later in the CNS (Buckley et al. 1998). Message was also visible in sympathetic and parasympathetic ganglia, in the retina and in enteric ganglia of the alimentary system. Indeed, the CB_1 receptor can be found prejunctionally at the autonomic neuromuscular junction and classical assays of CB_1 receptor function include inhibition of electrically-evoked contractions in the guinea-pig ileum (parasympathetic nervous system) and mouse vas deferens (sympathetic nervous system). High levels of CB_1 message have also been noted in the thyroid and adrenal glands with receptor protein being detected in the testes (Gerard et al. 1991), spleen (Shire et al. 1996b), B cells and natural killer cells (Galiegue et al. 1995).

CB_2 receptor distribution—Northern blot analysis of CB_2 receptor distribution in man suggested initially a localization outside the CNS, particularly in macrophage populations in the spleen (Munro et al. 1993). RT-PCR techniques suggest the presence of CB_2 mRNA in the spleen (Galiegue et al. 1995), lymphocytes and tonsilar B cells (Galiegue et al. 1995; Marchand et al. 1999) and retina (Lu et al. 2000). It has also been reported that adult rat sensory nerves synthesize both CB_1 and CB_2 receptor protein (Ross et al. 2001), although others (Hohmann et al. 1999) have not been able to locate CB_2 receptor mRNA in rat cultured dorsal root ganglion cells. In ontogenic studies, CB_2 receptor mRNA appeared to be expressed exclusively in the liver of embryonic rats (from E13 onwards), but was undetectable in the adult liver (Buckley et al. 1998).

13.4 Pharmacology

13.4.1 Agonists

Since the dawn of recorded medicine, cannabis extracts have been used as medication or for recreational purposes. The major active ingredient in extracts from the cannabis plant is Δ^9-tetrahydrocannabinol (THC, Fig. 13.1), although many related molecules (such as cannabinol and cannabidiol) are also present. There are also a number of synthetic subtype selective cannabinoid receptor agonists (Table 13.2; Fig. 13.1) and antagonists (Fig. 13.2) available. Agonists tend to be one of four major groups: 'classical' cannabinoids, structurally related to THC; non-classical cannabinoids, such as CP55940; aminoalkylindoles, such as WIN 55,212-2 and fatty acid derivatives such as anandamide. Agonists selective for either CB_1 or CB_2 receptors have been described and are listed in Table 13.1.

13.4.2 Antagonists

Of the antagonists that have been developed, the best characterized and most widely-employed is the CB_1 receptor-selective ligand SR141716A (Fig. 13.2, Table 13.3), which was discovered as a result of drug library screening (Rinaldi-Carmona et al. 1994). The antagonist is effective both in vitro and in vivo and displays very little activity at other receptor classes. It is apparent, however, that at relatively high concentrations ($>1\,\mu M$), SR141716A is able to block gap junctions (Chaytor et al. 1999) and to inhibit vanilloid VR1 receptors (De Petrocellis et al. 2001), but whether these effects contribute to its overall pharmacological profile in vivo is unclear. AM281, a structurally-related compound, shows relatively high CB_1 receptor affinity (Table 13.3), with greater aqueous solubility, and has been labelled with iodine-131 for use as a SPECT ligand in vivo (Gifford et al. 1997b). AM630 (Ross et al. 1999), with reduced efficacy at CB_1 receptors, and SR144528 (Rinaldi-Carmona et al. 1998a) show good selectivity as CB_2-selective antagonists (Fig. 13.2, Table 13.3).

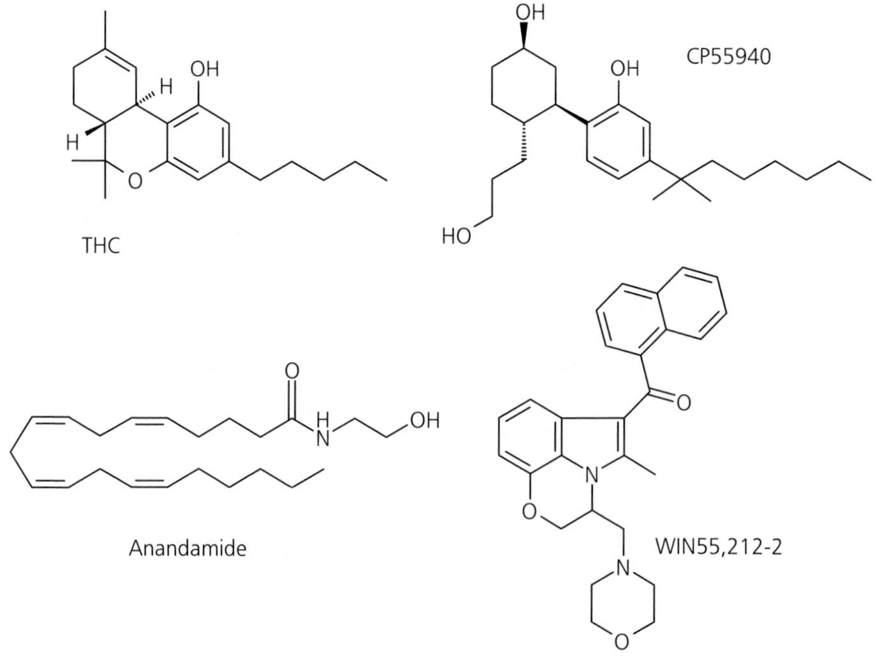

Fig. 13.1 Cannabinoid receptor agonists.

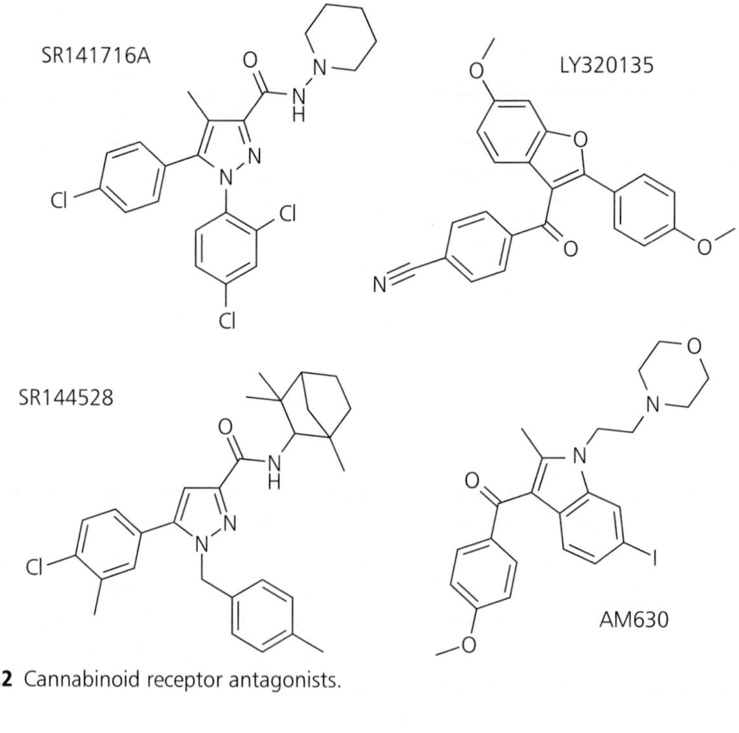

Fig. 13.2 Cannabinoid receptor antagonists.

Table 13.2 Agonist affinity at cannabinoid receptors

Compound	CB$_1$ cannabinoid receptors		CB$_2$ cannabinoid receptors		Selectivity
	CB$_1$ affinity (nM)	Source	CB$_2$ affinity (nM)	Source	CB$_1$: CB$_2$
ACEA	1.4	Human receptor*[§a]	>2000	Rat spleen*[a]	<0.0007
ACPA	2.2	Human receptor*[§a]	700	Rat spleen*[a]	0.0031
Methanandamide	20	Rat brain*[b]	815	Mouse spleen*[b]	0.025
Anandamide	89	Human receptor*[§c]	371	Human receptor*[§c]	0.24
2AG	472	Human receptor (COS-7 cells)*[d]	1400	Human receptor (COS-7 cells)*[d]	0.34
CP55940	0.58	Human receptor*[§c]	0.69	Human receptor*[§c]	0.84
THC	41	Human receptor*[§c]	36	Human receptor*[§c]	1.1
Cannabidiol	4350	Rat receptor*[§c]	2860	Human receptor*[§c]	1.5
Cannabinol	1130	Human receptor (L cells)*[e]	301	Human receptor (AtT-20 cells)*[e]	3.8
WIN55212-2	1.9	Human receptor*[§c]	0.28	Human receptor*[§c]	6.8
JWH015	383	Human receptor*[§c]	13.8	Human receptor*[§c]	27.8
L759656	4900	Human receptor*[§f]	11.8	Human receptor*[§f]	415
HU308	>10000	Rat brain[†g]	23	Rat receptor[†§g]	>435
L759633	1000	Human receptor*[§e]	0.4	Human receptor*[§e]	2500

* Competition for [^{3}H]-CP55,940 binding; [†] Competition for [^{3}H]-HU243 binding; [§] Expressed in Chinese Hamster ovary cells; [a](Hillard et al. 1999); [b](Khanolkar et al. 1996); [c](Showalter et al. 1996); [d](Mechoulam et al. 1995); [e](Felder et al. 1995); [f](Ross et al. 1999); [g](Hanus et al. 1999).

Key: 2AG: 2-arachidonoylglycerol; ACEA: arachidonoyl-2-chloroethylamide; ACPA arachidonoyl-cyclopropylamide; CP55940: (-)-3-[2-hydroxy-4-(1,1-dimethylheptyl)phenyl]-4-(3-hydroxypropyl)cyclohexan-1-ol; JWH133: 3-(1',1'-dimethylbutyl)-1-deoxy-Δ^8-tetrahydrocannabinol; L759633: (6aR,10aR)-3-(1,1-dimethylheptyl)-1-methoxy-6,6,9-trimethyl-6a,7,10,10a-tetrahydro-6H-benzo[c]chromene; L759656: (6aR,10aR)-3-(1,1-dimethylheptyl)-1-methoxy-6,6-dimethyl-9-methylene-6a,7,8,9,10,10a-hexahydro-6H-benzo[c]chromene; methanandamide: (R)-(+)-arachidonoyl-1'-hydroxy-2'-propylamide; THC: Δ^9-tetrahydrocannabinol; WIN55212-2: (R)-(+)-[2,3-dihydro-5-methyl-3-[(4-morpholino)methyl]pyrrolo-[1,2,3-de]-1,4-benzoxazin-6-yl](1-naphthyl)methanone.

There are reports of SR141716A having CB$_1$ receptor inverse agonist properties in cell lines, peripheral organs and brain (Landsman et al. 1997; MacLennan et al. 1998; Pan et al. 1998; Meschler et al. 2000). Other less well characterized CB$_1$ receptor ligands also appear to exhibit inverse agonist properties (Landsman et al. 1998; Meschler et al. 2000). Recent studies measuring [^{35}S]-GTPγS binding in brain sections, however, indicate that SR141716 is a competitive antagonist in the nanomolar concentration range but an inverse agonist at micromolar levels (Sim-Selley et al. 2001). Furthermore, it appears that the apparent inverse

Table 13.3 Antagonist affinity at cannabinoid receptors

Compound	CB$_1$ cannabinoid receptors		CB$_2$ cannabinoid receptors		Selectivity
	CB$_1$ affinity (nM)	Source	CB$_2$ affinity (nM)	Source	CB$_1$: CB$_2$
LY320135	224	Human receptor*§a	>10000	Human receptor*§a	<0.022
SR141716A	11	Human receptor*§b	702	Human receptor*§b	0.016
	2	Rat brain*c			
	2	Mouse vas deferens§d			
AM630	5152	Human receptor*§e	31	Human receptor*§e	166
SR144528	305	Rat brain*f	0.3	Rat spleen*f	508
	437	Human receptor*§f	0.6	Human receptor*§f	728
AM281	14	Rat hippocampus*g	—	—	—

* Competition for [3H]-CP55940 binding; †Competition for [^{3}H]-HU243 binding; §Expressed in Chinese hamster ovary cells; ¶Antagonism of the cannabinoid inhibition of electrically-evoked contractions; a(Felder *et al.* 1998); b(Showalter *et al.* 1996); c(Rinaldi-Carmona *et al.* 1994); d(Petwee *et al.* 1995); e(Ross *et al.* 1999); f(Rinaldi-Carmona *et al.* 1998); g(Gifford *et al.* 1997)

Key: AM281 N-(morpholin-4-yl)-5-(4-iodophenyl)-1-(2,4-dichlorophenyl)-4-methyl-1H-pyrazole-3-carboxamide; AM630: 6-iodopravadoline; LY320135: [6-methoxy-2-(4-methoxyphenyl)benzo[b]thien-3-yl][4-cyanophenyl]methanone; SR141716A: N-(piperidin-1-yl)-5-(4-chlorophenyl)-1-(2,4-dichlorophenyl)-4-methyl-1H-pyrazole-3carboxamide hydrochloride; SR144528: N-([1S]-endo-1,3,3-trimethylbicyclo[2.2.1]heptan-2-yl)-5-(4-chloro-3-methylphenyl)-1-(4-methylbenzyl)-pyrazole-3-carboxamide.

agonist effect is either not cannabinoid CB$_1$ receptor-specific or that different sites on the cannabinoid CB$_1$ receptor mediate antagonist and inverse agonist effects.

SR144528 and JTE-907, a novel selective ligand for CB$_2$ cannabinoid receptors, have also been reported to have inverse agonist properties at human overexpressed CB$_2$ receptors (Shire *et al.* 1999; Iwamura *et al.* 2001).

13.5 Signal transduction and receptor modulation

Binding of agonist at cannabinoid receptors leads to activation of predominantly pertussis toxin-sensitive G proteins. These, in turn, cause inhibition of adenylyl cyclase and voltage-activated calcium channels or activation of potassium channels. The most common molecular/cellular assays of CB$_1$ receptor function, therefore, involve enhancement of [^{35}S]-GTPγS binding to preparations from brain tissue or cells and the inhibition of cyclic AMP accumulation in intact brain slice or cell preparations. In isolated tissues, CB$_1$ receptor function can be studied by examining inhibition of electrically-evoked neurotransmitter release in the guinea-pig ileum or mouse vas deferens through measurement of alterations in contractile responses (see Table 13.4). From such studies in a variety of preparations, it is apparent that CB$_1$ receptor inhibition of transmitter release is associated with a direct inhibition of calcium currents, while the enhancement of potassium fluxes may be secondary to inhibition of adenylyl cyclase activity (Pertwee 1997).

Turning to enzymatic pathways, heterologous expression of CB$_1$ receptors in CHO or L cells revealed a lack of coupling to the phospholipase C pathway (Felder *et al.* 1992).

Table 13.4 Cannabinoid receptor regulation of transmitter/hormone release *in vitro*

Neurotransmitter	Effect of agonist	Effect of CB$_1$ antagonist	Source
5HT	↓	Reversal	Mouse cerebral cortical slices[a]
ACh	↓ (THC)		Human platelets (from migraineurs)[b]
	↓	Enhancement	Rat hippocampal slices[c]
	↓	Enhancement	Mouse urinary bladder[d]
	↓	Enhancement	Guinea-pig ileum[e]
	↓	No enhancement	Rat striatal slices[f]
	↓		Human ileum[g]
	↓	Enhancement	Rat cortical synaptosomes[h]
	↓	Enhancement	Rat hippocampal synaptosomes[h]
	=		Rat striatal synaptosomes[h]
	↓	No effect	Guinea-pig trachea[i]
	↓	Enhancement	Mouse hippocampal slices[j]
	=		Mouse striatal slices[k]
ACTH	↑	Not reversed	Rat anterior pituitary cells[l]
CCK	↓		Rat hippocampal slices[m]
	=		Rat cortical slices[n]
CGRP	↑ (AEA)	Not reversed	Rat mesenteric artery[o]
CRF	↑	Reversal	Rat median eminence fragments[p]
DA	↓	Enhancement	Guinea-pig retina[q]
	↓	Reversal	Rat striatal slices[r]
	↓	Reversal	Rat striatal slices[s]
	=	No effect	Rat striatal slices[t]
	↑ (THC)		Rat nucleus accumbens slices[u]
GABA	↓	Reversal	Cultured rat hippocampal neurones[v]
	↓	Reversal	Rat nucleus accumbens[w]
	=		Rat substantia nigra[x]
	↓		Rat hippocampal slices[y]
	↓	Reversal	Rat hippocampal slices[z]
	↓	Reversal	Rat striatal slices[aa]
	↓	Reversal	Human hippocampal slices[ab]
	↓	Reversal	Rat nucleus accumbens (shell)[ac]
GH	↑	Not reversed	Rat anterior pituitary cells[ah]
Glu	↓	Reversal	Rat substantia nigra pars reticulata[ad]
	↓	Reversal	Rat striatal slices[ae]
	↓	Reversal	Mouse nucleus accumbens[af]
GnRH	↑	Reversal	Rat median eminence fragments[ag]

Table 13.4 (Contd.)

Neurotransmitter	Effect of Agonist	Effect of CB$_1$ Antagonist	Source
LH	↓	Reversal	Rat anterior pituitary cells[ai]
	↓	Reversal	Mouse vas deferens[aj]
	=	No effect	Rat hippocampal slices[ak]
	↓	Reversal	Guinea-pig hippocampal slices[al]
NA	↓	Reversal	Human hippocampal slices[am]
	↓	Reversal	Human atrial segments[an]
	=	No effect	Rat hippocampal slices[ao]
	=	No effect	Mouse hippocampal slices[ap]
PRL	↓	Reversal	Rat anterior pituitary cells[aq]
Test	↓		Mouse testes[ar]

[a] (Nakazi et al. 2000); [b](Volfe et al. 1985); [c](Gifford and Ashby 1996); [d](Pertwee et al. 1996); [e](Pertwee and Fernando 1996); [f](Gifford et al. 1997b); [g](Croci et al. 1998); [h](Gifford et al. 2000); [i](Spicuzza et al. 2000); [j](Kathmann et al. 2001); [k](Kathmann et al. 2001); [l](Wenger et al. 2000); [m](Beinfeld and Connolly 2001); [n](Beinfeld and Connolly 2001); [o](Zygmunt et al. 1999); [p](Prevot et al. 1998); [q](Schlicker et al. 1996); [r](Cadogan et al. 1997); [s](Kathmann et al. 1999); [t](Szabo et al. 1999); [u](Szabo et al. 1999); [v](Ohno-Shosaku et al. 2001; Irving et al. 2000); [w](Hoffman and Lupica 2001); [x](Romero et al. 1997); [y](Wilson and Nicoll 2001); [z](Katona et al. 1999; Hajos et al. 2000); [aa](Szabo et al. 1998); [ab](Katona et al. 2000); [ac](Hoffman and Lupica 2001); [ad](Szabo et al. 2000); [ae](Gerdeman and Lovinger 2001; Huang et al. 2001); [af](Robbe et al. 2001); [ag](Prevot et al. 1998); [ah](Wenger et al. 2000); [ai](Wenger et al. 2000); [aj](Ishac et al. 1996); [ak](Gifford et al. 1997a); [al](Schlicker et al. 1997); [am](Schlicker et al. 1997); [an](Molderings et al. 1999); [ao](Schlicker et al. 1997); [ap](Schlicker et al. 1997); [aq](Wenger et al. 2000); [ar](Dalterio et al. 1977).

Key: 5HT serotonin; ACh acetylcholine; AEA anandamide; ACTH adrenocorticotrophic hormone; CCK cholecystokinin; CGRP calcitonin gene-related peptide; CRF corticotropin-releasing factor; DA dopamine; GABA γ-aminobutyric acid; GH growth hormone; Glu glutamate; GnRH gonadotropin-releasing hormone; LH luteinizing hormone; NA noradrenaline; PRL prolactin; Test testosterone; THC tetrahydrocannabinol; ↑ increase/stimulation; ↓ decrease/inhibition; = no effect.

Indeed, in rat hippocampal slices, THC appears to inhibit phosphoinositide turnover, albeit through a pertussis toxin-insensitive mechanism (Nah et al. 1993).

Given that arachidonic acid derivatives are putative endogenous ligands of the cannabinoid receptors, it is of note that cannabinoid receptor activation has been reported to couple to an increase in arachidonic acid release (Burstein et al. 1994) and eicosanoid biosynthesis, as well as synthesis of the endocannabinoid anandamide (Hunter and Burstein 1997). Furthermore, it is apparent that activation of cannabinoid receptors by either anandamide or THC in WI-38 fibroblasts leads to stimulation of the ERK/MAP kinase signal transduction pathway which, in turn, leads to increased phosphorylation and activation of the arachidonate-specific cytoplasmic phospholipase A$_2$ (Wartmann et al. 1995). ERK/MAP kinase activation has also been reported to couple cannabinoid receptor activation to other effectors, including the NHE-1 isoform of the Na$^+$/H$^+$ exchanger (Bouaboula et al. 1999) and glucose metabolism in primary astrocytes (Sanchez et al. 1998). Other studies suggest, however, that cannabinoids can inhibit the activation of ERK/MAP kinases via CB$_2$ receptor activation (Faubert and Kaminski 2000).

In cultured macrophage-like cells, it is apparent that CB$_2$ receptors couple to the inhibition of nitric oxide production stimulated by bacterial lipopolysaccharide (Ross et al. 2000). However, whether this phenomenon is involved in the putative immunosuppressive actions of cannabinoids in vivo remains to be elucidated.

13.6 **Physiology and disease relevance**

CB_1 receptors are thought to mediate the psychoactive responses to administration of extracts from the cannabis plant, whereas CB_2 receptors mediate the non-psychoactive immune responses. It has been appreciated for decades that application of cannabis extracts and synthetic cannabinoids *in vivo* elicits a classical 'tetrad' of symptoms in experimental animals: catalepsy, hypothermia, antinociception, and hypokinesia. *In vivo*, these responses depend upon CB_1 receptor activation because they are reversed by the CB_1-selective antagonist, SR141716 and are absent in CB_1 knockout mice (Ledent *et al.* 1999; Mascia *et al.* 1999). Many of the behavioural effects of cannabinoids are consistent with effects on hormone/transmitter release observed *in vivo* (Table 13.5). Anandamide, however, exhibits several properties at variance with other agonists. In particular, hypothermia, antinociception, and catalepsy induced by the endocannabinoid are not reversed by SR141716. Cannabinoid-related processes seem also to be involved in cognition, memory, anxiety, control of appetite, emesis, inflammatory, and immune responses (Chaperon and Thiebot 1999).

Table 13.5 Cannabinoid receptor regulation of hormone/transmitter release *in vivo*

Neurotransmitter	Effect of agonist	Effect of CB₁ antagonist	Source
Ach	↑	Enhancement	Rat prefrontal cortex and hippocampus[a]
	↑		Rat prefrontal cortex and hippocampus[b]
ACTH	↑	Reversal	Rat plasma[c]
Cort	↑	Reversal	Rat plasma[d]
CRF		Stimulation after repeated agonist application	Rat amygdala[e]
DA	= (THC)		Rat striatum[f]
	↑ (THC)		Rat striatum[h]
	↑ (THC)	Reversed by opioid antagonist	Rat nucleus accumbens (shell)[g]
FSH	↓ (THC)		Rat plasma[i]
	= (THC)		Rat plasma[j]
LH	↓ (THC)		Rat plasma[k]
PRL	↓ (THC)		Rat plasma[l]
Test	↑ (THC)		Mouse plasma[m]

[a] (Gessa *et al.* 1998); [b](Acquas *et al.* 2000); [c](Weidenfeld *et al.* 1994; De Fonseca *et al.* 1995; Manzanares *et al.* 1999); [d](Weidenfeld *et al.* 1994; De Fonseca *et al.* 1995; Manzanares *et al.* 1999); [e](De Fonseca *et al.* 1997; [f](Castaneda *et al.* 1991); [g](Tanda *et al.* 1997); [h](Malone and Taylor 1999); [i](Fernández-Ruiz *et al.* 1992); [j](de Miguel *et al.* 1998); [k](Fernández-Ruiz *et al.* 1992; de Miguel *et al.* 1998); [l](Fernández-Ruiz *et al.* 1992; de Miguel *et al.* 1998); [m](Dalterio *et al.* 1981).

Key: ACh: acetylcholine; ACTH: adrenocorticotrophic hormone; Cort: corticosterone; CRF: corticotropin-releasing factor; DA: dopamine; FSH: follicle-stimulating hormone; LH: luteinizing hormone; PRL: prolactin; Test: testosterone; THC: tetrahydrocannabinol; ↑: increase/stimulation; ↓: decrease/inhibition; = no effect.

13.6.1 Neurodegeneration

Exogenous and endogenous cannabinoids reduce hypoxic neuronal damage in cell culture and are neuroprotective in animal models of global and focal ischemia. These effects appear to be mediated via CB_1 receptors and an induction of CB_1 protein has been reported in an experimental stroke model (Jin *et al.* 2000). There is a major loss of cannabinoid receptor binding in post mortem brains from Huntington's patients. In particular, there is drop out of binding in the substantia nigra (Glass *et al.* 1993) and globus pallidus (Richfield and Herkenham 1994).

13.6.2 Drug dependence

Targeted disruption of the *cnr1* gene (Ledent *et al.* 1999; Mascia *et al.* 1999) has been described. The marked similarities between opioid and cannabinoid receptor responses led to an investigation of the effects of morphine in CB_1 receptor knockout mice. The results of these studies were that CB_1 receptors appeared to regulate mesolimbic dopaminergic transmission in the nucleus accumbens which is believed to be involved in the reinforcing effects of morphine. Furthermore, while *in vivo* the acute effects of opiates in CB_1 deficient mice were unaffected, the reinforcing properties and severity of the withdrawal syndrome were strongly reduced.

13.6.3 Pain and analgesia

One of the most likely clinical applications for cannabinoid receptor ligands is in the treatment of pain. Interestingly, different receptors might be involved in different sorts of pain (see Pertwee 2001, for a comprehensive review) and it is clear that the cannabinoids can act at a variety of anatomical locations (peripheral, spinal, and at higher brain centres).

13.6.4 Locomotor dysfunction

There is objective evidence in animal models of multiple sclerosis (MS) for the control of spasticity by cannabinoids (Baker *et al.* 2000, 2001) and clinical trials of THC preparations in MS are proceeding. There is pharmacological evidence for the tonic activity of an endocannabinoid system involving both CB_1 and CB_2 receptors in the control of spasticity and tremor.

13.6.5 Schizophrenia

An association between schizophrenia and elevated CNS levels of anandamide and palmitoylethanolamide has been described (Leweke *et al.* 1999). This provides an objective basis for the on-going clinical trials of SR141716A as a potential neuroleptic agent.

13.6.6 Feeding and satiety

Plant cannabinoids are well known, anecdotally, to increase food consumption in recreational users and preparations of THC are used therapeutically to increase appetite, for example, in AIDS patients (Mechoulam *et al.* 1998). In addition, anandamide, acting through the CB_1 receptor, has been shown to increase feeding in laboratory animals (Hao *et al.* 2000). Food-deprived knockout mice have been reported to eat less than wild-type littermates and the nutritional regulator leptin reduces the levels of the endocannabinoids anandamide and

2-AG in the rat hypothalamus (Di Marzo *et al.* 2001). The CB_1 antagonist SR141716 reduces food intake in wild-type but not CB_1 knockout mice and the accumulated evidence therefore supports a physiological role for the CB_1 receptor in the control of feeding. These findings indicate that selective CB_1 agonists and antagonists might have medicinal uses as orexigenic and anorectic agents, respectively.

13.6.7 Emesis

THC and its synthetic analogues (nabilone and levonantradol) are used to prevent emesis in cancer patients receiving chemotherapy. Moreover, the anti-emetic potency of CP55940 is 45 times greater than that of THC. Conversely, the CB_1 antagonist SR141716 causes emesis in an animal model (the least shrew, *Cryptotis parva*; (Darmani 2001)). It remains to be seen whether emetic effects of CB_1 antagonists might limit their use in humans.

13.6.8 Peripheral indications

CB_2 receptors are thought to mediate the non-psychoactive, immune responses to adminis-tration of extracts from the cannabis plant a suggestion supported by studies on mice carrying a targeted deletion of the *cnr2* gene (like *cnr1* on chromosome 4, Buckley *et al.* 2000). Spe-cifically, CB_2 receptor stimulation appears to cause immunosuppresssion, although, as yet, there are no reports of additional physiological/pathophysiological roles for CB_2 receptors.

It has been reported that activation of peripheral CB_1 receptors in rats contributes to haemorrhagic hypotension, and that anandamide, produced by macrophages, may be a mediator of this effect (Wagner *et al.* 1997). This indicates a possible role for antagonists of CB_1 receptors in the treatment of shock.

Cannabinoids have anti-tumoral actions. THC induces apoptosis of transformed neural cells in culture and WIN 55212-2 is reported to cause regression of malignant gliomas in rats via an action involving sustained ceramide accumulation and both CB_1 and CB_2 receptors (Galve-Roperh *et al.* 2000).

The CB_1 receptor is present in human ciliary body and cannabinoid receptor agonists such as WIN 55212-2 are able to reduce intraocular pressure in glaucoma patients resistant to conventional therapies (Porcella *et al.* 2001).

13.7 Concluding remarks

Despite thousands of years of proximity to cannabinoids, we have only comparatively recently been in a position to exploit cannabinoid receptors rationally. It is evident that CB_1 receptors have a number of possible roles in disease-related conditions and it is likely that a number of novel therapeutic avenues will be explored based on these receptors. Future therapies targeting CB_2 receptors appear much more distant.

References

Abood ME, Ditto KE, Noel MA, Showalter VM, and Tao Q (1997). Isolation and expression of a mouse CB1 cannabinoid receptor gene. Comparison of binding properties with those of native CB1 receptors in mouse brain and N18TG2 neuroblastoma cells. *Biochem Pharmacol* **53**, 207–14.

Acquas E, Pisanu A, Marrocu P, and Di Chiara G (2000). Cannabinoid CB_1 receptor agonists increase rat cortical and hippocampal acetylcholine release *in vivo*. *Eur J Pharmacol* **401**, 179–85.

Adams IB, Compton DR, and Martin BR (1998). Assessment of anandamide interaction with the cannabinoid brain receptor: SR 141716A antagonism studies in mice and autoradiographic analysis of receptor binding in rat brain. *J Pharmacol Experimental Therapeutics* **284**, 1209–17.

Baker D, Pryce G, Croxford JL *et al.* (2000). Cannabinoids control spasticity and tremor in a multiple sclerosis model. *Nature* **404**, 84–7.

Baker D, Pryce G, Croxford JL *et al.* (2001). Endocannabinoids control spasticity in a multiple sclerosis model. *FASEB J* **15**, 300–2.

Bayewitch M, Avidorreiss T, Levy R, Barg J, Mechoulam R, and Vogel Z (1995). The peripheral cannabinoid receptor: adenylate cyclase inhibition and G-protein coupling. *FEBS Lett* **375**, 143–7.

Beinfeld MC and Connolly K (2001). Activation of CB1 cannabinoid receptors in rat hippocampal slices inhibits potassium-evoked cholecystokinin release, a possible mechanism contributing to the spatial memory defects produced by cannabinoids. *Neurosci Lett* **301**, 69–71.

Bouaboula M, Desnoyer N, Carayon P, Combes T, and Casellas P (1999). G_i protein modulation induced by a selective inverse agonist for the peripheral cannabinoid receptor CB2: implication for intracellular signalization cross-regulation. *Mol Pharmacol* **55**, 473–80.

Breivogel CS, Childers SR, Deadwyler SA, Hampson RE, Vogt LJ, and Sim-Selley LJ (1999). Chronic Δ^9-tetrahydrocannabinol treatment produces a time-dependent loss of cannabinoid receptors and cannabinoid receptor-activated G proteins in rat brain. *J Neurochem* **73**, 2447–59.

Buckley NE, Hansson S, Harta G, and Mezey E (1998). Expression of the CB1 and CB2 receptor messenger RNAs during embryonic development in the rat. *Neuroscience* **82**, 1131–49.

Buckley NE, Mccoy KL, Mezey E *et al.* (2000). Immunomodulation by cannabinoids is absent in mice deficient for the cannabinoid CB2 receptor. *Eur J Pharmacol* **396**, 141–9.

Burstein SH, Budrow J, Debatis M, Hunter SA, and Subramanian A (1994). Phospholipase participation in cannabinoid-induced release of free arachidonic acid. *Biochem Pharmacol* **48**, 1253–64.

Cadogan A-K, Alexander SPH, Boyd EA, and Kendall DA (1997). Influence of cannabinoids on electrically-evoked dopamine release and cyclic AMP generation in the rat striatum. *J Neurochem* **69**, 1131–7.

Castaneda E, Moss DE, Oddie SD, and Whishaw IQ (1991). THC does not affect striatal dopamine release: microdialysis in freely moving rats. *Pharmacol Biochem Behavior* **40**, 587–91.

Chaperon F and Thiebot MH (1999). Behavioral effects of cannabinoid agents in animals. *Critical Rev Neurobiol* **13**, 243–81.

Chaytor AT, Martin PEM, Evans WH, Randall MD, and Griffith TM (1999). The endothelial component of cannabinoid-induced relaxation in rabbit mesenteric artery depends on gap junctional communication. *J Physiol (London)* **520**, 539–50.

Croci T, Manara L, Aureggi G *et al.* (1998). *In vitro* functional evidence of neuronal cannabinoid CB1 receptors in human ileum. *Br J Pharmacol* **125**, 1393–5.

Dalterio S, Bartke A, and Burstein S (1977). Cannabinoids inhibit testosterone secretion by mouse testes *in vitro*. *Science* **196**, 1472–3.

Dalterio S, Bartke A, and Mayfield D (1981). Delta9-tetrahydrocannabinol increase plasma testosterone concentrations in mice. *Science* **213**, 581–3.

Darmani NA (2001). Δ^9-tetrahydrocannabinol and synthetic cannabinoids prevent emesis produced by the cannabinoid CB1 receptor antagonist/inverse agonist SR 141716A. *Neuropsychopharmacology* **24**, 198–203.

De Fonseca FR, Carrera MRA, Navarro M, Koob GF, and Weiss F (1997). Activation of corticotropin-releasing factor in the limbic system during cannabinoid withdrawal. *Science* **276**, 2050–4.

De Fonseca FR, Villanua MA, Munoz RM, Sanmartinclark O, and, Navarro M (1995). Differential effects of chronic treatment with either dopamine D-1 or D-2 receptor agonists on the acute neuroendocrine actions of the highly potent synthetic cannabinoid HU-210 in male rats. *Neuroendocrinology* **61**, 714–21.

de Miguel R, Romero J, Munoz RM *et al.* (1998). Effects of cannabinoids on prolactin and gonadotrophin secretion: involvement of changes in hypothalamic gamma-aminobutyric acid (GABA) inputs. *Biochem Pharmacol* **56**, 1331–8.

De Petrocellis L, Bisogno T, Maccarrone M, Davis JB, Finazzi-Agro A, and Di Marzo V (2001). The activity of anandamide at vanilloid VR1 receptors requires facilitated transport across the cell membrane and is limited by intracellular metabolism. *J Biol Chem* **276**, 12 856–63.

Derocq JM, Jbilo O, Bouaboula M, Segui M, Clere C, and Casellas P (2000). Genomic and functional changes induced by the activation of the peripheral cannabinoid receptor CB2 in the promyelocytic cells HL-60. Possible involvement of the CB2 receptor in cell differentiation. *J Biol Chem* **275**, 15621–8.

Devane WA, Breuer A, Sheskin T, Jarbe TUC, Eisen MS, and Mechoulam R (1992). A novel probe for the cannabinoid receptor. *J Med Chem* **35**, 2065–9.

Devane WA, Dysarz FA, Johnson MR, Melvin LS, and Howlett AC (1988). Determination and characterization of a cannabinoid receptor in rat brain. *Mol Pharmacol* **34**, 605–13.

Di Marzo V, Breivogel CS, Tao Q *et al.* (2000). Levels, metabolism, and pharmacological activity of anandamide in CB1 cannabinoid receptor knockout mice: Evidence for non-CB1, non-CB2 receptor-mediated actions of anandamide in mouse brain. *J Neurochem* **75**, 2434–44.

Di Marzo V, De Petrocellis L, Sugiura T, and Waku K (1996). Potential biosynthetic connections between the two cannabimimetic eicosanoids, anandamide and 2-arachidonoyl-glycerol, in mouse neuroblastoma cells. *Biochem Biophys Res Communications* **227**, 281–8.

Di Marzo V and Deutsch DG (1998). Biochemistry of the endogenous ligands of cannabinoid receptors. *Neurobiol Disease* **5**, 386–404.

Di Marzo V, Fontana A, Cadas H *et al.* (1994). Formation and inactivation of endogenous cannabinoid anandamide in central neurons. *Nature* **372**, 686–91.

Di Marzo V, Goparaju SK, Wang L *et al.* (2001). Leptin-regulated endocannabinoids are involved in maintaining food intake. *Nature* **410**, 822–5.

Facci L, Daltoso R, Romanello S, Buriani A, Skaper SD, and Leon A (1995). Mast cells express a peripheral cannabinoid receptor with differential sensitivity to anandamide and palmitoylethanolamide. *Proc Natl Acad Sci USA* **92**, 3376–80.

Faubert BL and Kaminski NE (2000). AP-1 activity is negatively regulated by cannabinol through inhibition of its protein components, c-fos and c-jun. *J Leukocyte Biol* **67**, 259–66.

Felder CC, Joyce KE, Briley EM *et al.* (1998). LY320135, a novel cannabinoid CB1 receptor antagonist, unmasks coupling of the CB1 receptor to stimulation of cAMP accumulation. *J Pharmacol Experimental Therapeutics* **284**, 291–7.

Felder CC, Joyce KE, Briley EM *et al.* (1995). Comparison of the pharmacology and signal transduction of the human cannabinoid CB1 and CB2 receptors. *Mol Pharmacol* **48**, 443–50.

Felder CC, Veluz JS, Williams HL, Briley EM, and Matsuda LA (1992). Cannabinoid agonists stimulate both receptor- and non-receptor-mediated signal transduction pathways in cells transfected with and expressing cannabinoid receptor clones. *Mol Pharmacol* **42**, 838–45.

Fernández-Ruiz JJ, Navarro M, Hernandez ML, Vaticon D, and Ramos JA (1992). Neuroendocrine effects of an acute dose of $\Delta 9$-tetrahydrocannabinol-changes in hypothalamic biogenic-amines and anterior-pituitary hormone-secretion. *Neuroendocrinol Lett* **14**, 349–55.

Gadzicki D, Muller-Vahl K, and Stuhrmann M (1999). A frequent polymorphism in the coding exon of the human cannabinoid receptor (CNR1) gene. *Mol Cell Probes* **13**, 321–3.

Galiegue S, Mary S, Marchand J *et al.* (1995). Expression of central and peripheral cannabinoid receptors in human immune tissues and leukocyte subpopulations. *Eur J Biochem* **232**, 54–61.

Galve-Roperh I, Sanchez C, Cortes ML, del Pulgar TG, Izquierdo M, and Guzman M (2000). Antitumoral action of cannabinoids: Involvement of sustained ceramide accumulation and extracellular signal-regulated kinase activation. *Nature Medicine* **6**, 313–19.

Gebremedhin D, Lange AR, Campbell WB, Hillard CJ, and Harder DR (1999). Cannabinoid CB1 receptor of cat cerebral arterial muscle functions to inhibit L-type Ca2+ channel current. *Am J Physiol Heart Circ Physiol* **276**, H2085–93.

Gerard CM, Mollereau C, Vassart G, and Parmentier M (1991). Molecular cloning of a human cannabinoid receptor which is also expressed in testis. *Biochem J* **279**, 129–34.

Gerdeman G and Lovinger DM (2001). CB1 cannabinoid receptor inhibits synaptic release of glutamate in rat dorsolateral striatum. *J Neurophysiol* **85**, 468–71.

Gessa GL, Casu MA, Carta G, and Mascia MS (1998). Cannabinoids decrease acetylcholine release in the medial-prefrontal cortex and hippocampus, reversal by SR 141716A. *Eur J Pharmacol* **355**, 119–24.

Gifford AN and Ashby CR (1996). Electrically evoked acetylcholine release from hippocampal slices is inhibited by the cannabinoid receptor agonist, WIN 55212-2, and is potentiated by the cannabinoid antagonist, SR 141716A. *J Pharmacol Experimental Therapeutics* **277**, 1431–6.

Gifford AN, Bruneus M, Gatley SJ, and Volkow ND (2000). Cannabinoid receptor-mediated inhibition of acetylcholine release from hippocampal and cortical synaptosomes. *Br J Pharmacol* **131**, 645–50.

Gifford AN, Samiian L, Gatley SJ, and Ashby CR (1997a). Examination of the effect of the cannabinoid receptor agonist, CP 55,940, on electrically evoked transmitter release from rat brain slices. *Eur J Pharmacol* **324**, 187–92.

Gifford AN, Tang YJ, Gatley SJ, Volkow ND, Lan RX, and Makriyannis A (1997b). Effect of the cannabinoid receptor SPECT agent, AM 281, on hippocampal acetylcholine release from rat brain slices. *Neurosci Lett* **238**, 84–6.

Glass M, Faull RL, and Dragunow M (1993). Loss of cannabinoid receptors in the substantia nigra in Huntington's disease. *Neuroscience* **56**, 523–7.

Griffin G, Tao Q, and Abood ME (2000). Cloning and pharmacological characterization of the rat CB2 cannabinoid receptor. *J Pharmacol Experimental Therapeutics* **292**, 886–94.

Guo A, Vulchanova L, Wang J, Li X, and Elde R (1999). Immunocytochemical localization of the vanilloid receptor 1 (VR1): relationship to neuropeptides, the $P2X_3$ purinoceptor and IB4 binding sites. *Eur J Neurosci* **11**, 946–58.

Hajos N, Katona I, Naiem SS *et al.* (2000). Cannabinoids inhibit hippocampal GABAergic transmission and network oscillations. *Eur J Neurosci* **12**, 3239–49.

Hanus L, Breuer A, Tchilibon S *et al.* (1999). HU-308: A specific agonist for CB_2, a peripheral cannabinoid receptor. *Proc Natl Acad Sci USA* **96**, 14 228–33.

Hao SZ, Avraham Y, Mechoulam R, and Berry EM (2000). Low dose anandamide affects food intake, cognitive function, neurotransmitter and corticosterone levels in diet-restricted mice. *Eur J Pharmacol* **392**, 147–56.

Hillard CJ, Manna S, Greenberg MJ *et al.* (1999). Synthesis and characterization of potent and selective agonists of the neuronal cannabinoid receptor (CB1). *J Pharmacol Experimental Therapeutics* **289**, 1427–33.

Hoehe MR, Caenazzo L, Martinez MM *et al.* (1991). Genetic and physical mapping of the human cannabinoid receptor gene to chromosome 6q14-q15. *New Biol* **3**, 880–5.

Hohmann AG and Herkenham M (1999). Cannabinoid receptors undergo axonal flow in sensory nerves. *Neuroscience* **92**, 1171–5.

Hoffman AF and Lupica CR (2001). Direct actions of cannabinoids on synaptic transmission in the nucleus accumbens: A comparison with opioids. *J Neurophysiol* **85**, 72–83.

Huang CC, Lo SW, and Hsu KS (2001). Presynaptic mechanisms underlying cannabinoid inhibition of excitatory synaptic transmission in rat striatal neurons. *J Physiol London* **532**, 731–48.

Hunter SA and Burstein SH (1997). Receptor mediation in cannabinoid stimulated arachidonic acid mobilization and anandamide synthesis. *Life Sci* **60**, 1563–73.

Irving AJ, Coutts AA, Harvey J *et al.* (2000). Functional expression of cell surface cannabinoid CB1 receptors on presynaptic inhibitory terminals in cultured rat hippocampal neurons. *Neuroscience* **98**, 253–62.

Ishac EJN, Jiang L, Lake KD, Varga K, Abood ME, and Kunos G (1996). Inhibition of exocytotic noradrenaline release by presynaptic cannabinoid CB_1 receptors on peripheral sympathetic nerves. *Br J Pharmacol* **118**, 2023–8.

Iwamura H, Suzuki H, Ueda Y, Kaya T, and Inaba T (2001). *In vitro* and *in vivo* pharmacological characterization of JTE907, a novel selective ligand for cannabinoid CB2 receptor. *J Pharmacol Experimental Therapeutics* **296**, 420–5.

Jarai Z, Wagner JA, Varga K *et al.* (1999). Cannabinoid-induced mesenteric vasodilation through an endothelial site distinct from CB1 or CB2 receptors. *Proc Natl Acad Sci USA* **96**, 14 136–41.

Jin KL, Mao XO, Goldsmith PC, and Greenberg DA (2000). CB2 cannabinoid receptor induction in experimental stroke. *Annals Of Neurol* **48**, 257–61.

Jin WZ, Brown S, Roche JP *et al.* (1999). Distinct domains of the CB1 cannabinoid receptor mediate desensitization and internalization. *J Neurosci* **19**, 3773–80.

Kathmann M, Bauer U, Schlicker E, and Gothert M (1999). Cannabinoid CB1 receptor-mediated inhibition of NMDA- and kainate-stimulated noradrenaline and dopamine release in the brain. *Naunyn-Schmiedebergs Arch Pharmacol* **359**, 466–70.

Kathmann M, Weber B, Zimmer A, and Schlicker E (2001). Enhanced acetylcholine release in the hippocampus of cannabinoid CB1 receptor-deficient mice. *Br J Pharmacol* **132**, 1169–73.

Katona I, Sperlagh B, Magloczky Z *et al.* (2000). Gabaergic interneurons are the targets of cannabinoid actions in the human hippocampus. *Neuroscience* **100**, 797–804.

Katona I, Sperlagh B, Sik A *et al.* (1999). Presynaptically located CB1 cannabinoid receptors regulate GABA release from axon terminals of specific hippocampal interneurons. *J Neurosci* **19**, 4544–58.

Khanolkar AD, Abadji V, Lin S *et al.* (1996). Head group analogs of arachidonylethanolamide, the endogenous cannabinoid ligand. *J Medicinal Chem* **39**, 4515–19.

Landsman RS, Burkey TH, Consroe P, Roeske WR, and Yamamura HI (1997). SR141716a is an inverse agonist at the human cannabinoid CB1 receptor. *Eur J Pharmacol* **334**, R1–2.

Landsman RS, Makriyannis A, Deng HF, Consroe P, Roeske WR, and Yamamura HI (1998). AM630 is an inverse agonist at the human cannabinoid CB1 receptor. *Life Sci* **62**, PL109–3.

Ledent C, Valverde O, Cossu C *et al.* (1999). Unresponsiveness to cannabinoids and reduced addictive effects of opiates in CB1 receptor knockout mice. *Science* **283**, 401–4.

Leweke FM, Giuffrida A, Wurster U, Emrich HM, and Piomelli D (1999). Elevated endogenous cannabinoids in schizophrenia. *Neuroreport* **10**, 1665–9.

Lu QJ, Straiker A, Lu QX, and Maguire G (2000). Expression of CB2 cannabinoid receptor mRNA in adult rat retina. *Visual Neurosci* **17**, 91–5.

Lu RS, Hubbard JR, Martin BR, and Kalimi MY (1993). Roles of sulfhydryl and disulfide groups in the binding of CP-55, 940 to rat brain cannabinoid receptor. *Mol Cell Biochem* **121**, 119–26.

MacLennan SJ, Reynen PH, Kwan J, and Bonhaus DW (1998). Evidence for inverse agonism of SR141716A at human recombinant cannabinoid CB_1 and CB_2 receptors. *Br J Pharmacol* **124**, 619–22.

Malone DT and Taylor DA (1999). Modulation by fluoxetine of striatal dopamine release following Δ^9-tetrahydrocannabinol: a microdialysis study in conscious rats. *Br J Pharmacol* **128**, 21–6.

Manzanares J, Corchero J, and Fuentes JA (1999). Opioid and cannabinoid receptor-mediated regulation of the increase in adrenocorticotropin hormone and corticosterone plasma concentrations induced by central administration of Δ^9-tetrahydrocannabinol in rats. *Brain Res* **839**, 173–9.

Marchand J, Bord A, Penarier G, Laure F, Carayon P, and Casellas P (1999). Quantitative method to determine mRNA levels by reverse transcriptase polymerase chain reaction from leukocyte subsets purified by fluorescence-activated cell sorting: application to peripheral cannabinoid receptors. *Cytometry* **35**, 227–34.

Mascia MS, Obinu MC, Ledent C *et al.* (1999). Lack of morphine-induced dopamine release in the nucleus accumbens of cannabinoid CB1 receptor knockout mice. *Eur J Pharmacol* **383**, R1–2.

Matsuda LA, Lolait SJ, Brownstein MJ, Young AC, and Bonner TI (1990). Structure of a cannabinoid receptor and functional expression of the cloned cDNA. *Nature* **346**, 561–4.

Mclaughlin CR, Martin BR, Compton DR, and Abood ME (1994). Cannabinoid receptors in developing rats: detection of mRNA and receptor binding. *Drug and Alcohol Dependence* **36**, 27–31.

Mechoulam R, Benshabat S, Hanus L *et al.* (1995). Identification of an endogenous 2-monoglyceride, present in canine gut, that binds to cannabinoid receptors. *Biochem Pharmacol* **50**, 83–90.

Mechoulam R, Hanus L, and Fride E (1998). Towards cannabinoid drugs—revisited. *Prog Med Chem* **35**, 199–243.

Meschler JP, Kraichely DM, Wilken GH, and Howlett AC (2000). Inverse agonist properties of N-(piperidin-1-yl)-5-(4-chlorophenyl)-1-(2,4-dichlorophenyl)-4-methyl-1H-pyrazole-3-carboxamide HCl (SR141716A) and 1-(2-chlorophenyl)-4-cyano-5-(4-methoxyphenyl)-1H-pyrazole-3-carboxylic acid phenylamide (CP-272871) for the CB1 cannabinoid receptor. *Biochem Pharmacol* **60**, 1315–23.

Molderings GJ, Likungu J, and Gothert M (1999). Presynaptic cannabinoid and imidazoline receptors in the human heart and their potential relationship. *Naunyn-Schmiedebergs Arch Pharmacol* **360**, 157–64.

Munro S, Thomas KL, and Abu-Shaar M (1993). Molecular characterization of a peripheral receptor for cannabinoids. *Nature* **365**, 61–5.

Nah S-Y, Saya D, and Vogel Z (1993). Cannabinoids inhibit agonist-stimulated formation of inositol phosphates in rat hippocampal cultures. *Eur J Pharmacol* **246**, 19–24.

Nakazi M, Bauer U, Nickel T, Kathmann M, and Schlicker E (2000). Inhibition of serotonin release in the mouse brain via presynaptic cannabinoid CB1 receptors. *Naunyn-Schmiedebergs Arch Pharmacol* **361**, 19–24.

Ohno-Shosaku T, Maejima T, and Kano M (2001). Endogenous cannabinoids mediate retrograde signals from depolarized postsynaptic neurons to presynaptic terminals. *Neuron* **29**, 729–38.

Pan XH, Ikeda SR, and Lewis DL (1998). SR 141716A acts as an inverse agonist to increase neuronal voltage-dependent Ca2+ currents by reversal of tonic CB1 cannabinoid receptor activity. *Mol Pharmacol* **54**, 1064–72.

Pertwee RG (1997). Pharmacology of cannabinoid CB_1 and CB_2 receptors. *Pharmacol Therapeutics* **74**, 129–80.

Pertwee RG (2001). Cannabinoid receptors and pain. *Prog Neurobiol* **63**, 569–611.

Pertwee RG and Fernando SR (1996). Evidence for the presence of cannabinoid CB_1 receptors in mouse urinary bladder. *Br J Pharmacol* **118**, 2053–8.

Pertwee RG, Fernando SR, Nash JE, and Coutts AA (1996). Further evidence for the presence of cannabinoid CB_1 receptors in guinea-pig small intestine. *Br J Pharmacol* **118**, 2199–05.

Pertwee RG, Griffin G, Lainton JAH, and Huffman JW (1995). Pharmacological characterization of three novel cannabinoid receptor agonists in the mouse isolated vas deferens. *Eur J Pharmacol* **284**, 241–7.

Petitet F, Marin L, and Doble A (1996). Biochemical and pharmacological characterization of cannabinoid binding sites using [^{3}H]SR141716A. *Neuroreport* **7**, 789–92.

Piomelli D, Beltramo M, Giuffrida A, and Stella N (1998). Endogenous cannabinoid signaling. *Neurobiol Disease* **5**, 462–73.

Porcella A, Maxia C, Gessa GL, and Pani L (2001). The synthetic cannabinoid WIN55212-2 decreases the intraocular pressure in human glaucoma resistant to conventional therapies. *Eur J Neurosci* **13**, 409–12.

Prevot V, Rialas CM, Croix D *et al.* (1998). Morphine and anandamide coupling to nitric oxide stimulates GnRH and CRF release from rat median eminence: neurovascular regulation. *Brain Res* **790**, 236–44.

Richfield EK and Herkenham M (1994). Selective vulnerability in Huntingtons disease. Preferential loss of cannabinoid receptors in lateral globus pallidus. *Annals Neurology* 36, 577–84.

Rinaldi-Carmona F, Casellas P, Ferrara P, and Le Fur G (1996a). Characterization of two cloned human CB1 cannabinoid receptor isoforms. *J Pharmacol Experimental Therapeutics* 278, 871–8.

Rinaldi-Carmona M, Barth F, Heaulme M *et al.* (1994). SR141716A, a potent and selective antagonist of the brain cannabinoid receptor. *FEBS Lett* 350, 240–4.

Rinaldi-Carmona M, Barth F, Millan J *et al.* (1998a). SR 144528, the first potent and selective antagonist of the CB2 cannabinoid receptor. *J Pharmacol Experimental Therapeutics* 284, 644–50.

Rinaldi-Carmona M, Le Duigou A, Oustric D *et al.* (1998b). Modulation of CB1 cannabinoid receptor functions after a long-term exposure to agonist or inverse agonist in the Chinese hamster ovary cell expression system. *J Pharmacol Experimental Therapeutics* 287, 1038–47.

Rinaldi-Carmona M, Pialot F, Congy C *et al.* (1996b). Characterization and distribution of binding sites for [^{3}H]-SR 141716A, a selective brain (CB1) cannabinoid receptor antagonist, in rodent brain. *Life Sci* 58, 1239–47.

Robbe D, Alonso G, Duchamp F, Bockaert J, and Manzoni OJ (2001). Localization and mechanisms of action of cannabinoid receptors at the glutamatergic synapses of the mouse nucleus accumbens. *J Neurosci* 21, 109–16.

Romero J, De Miguel R, Ramos JA, and Fernández-Ruiz JJ (1997). The activation of cannabinoid receptors in striatonigral GABAergic neurons inhibited GABA uptake. *Life Sci* 62, 351–63.

Ross RA, Brockie HC, and Pertwee RG (2000). Inhibition of nitric oxide production in RAW264.7 macrophages by cannabinoids and palmitoylethanolamide. *Eur J Pharmacol* 401, 121–30.

Ross RA, Brockie HC, Stevenson LA *et al.* (1999). Agonist-inverse agonist characterization at CB1 and CB2 cannabinoid receptors of L759633, L759656 and AM630. *Br J Pharmacol* 126, 665–72.

Ross RA, Coutts AA, McFarlane SM, Anavi-Goffer S, Irving AJ, Pertwee RG *et al.* (2001). Actions of cannabinoid receptor ligands on rat cultured sensory neurones: implications for antinociception *Neuropharmacology* 40, 221–32.

Rubino T, Vigano D, Costa B, Colleoni M, and Parolaro D (2000a). Loss of cannabinoid-stimulated guanosine 5′-O-(3-[^{35}S]thiotriphosphate) binding without receptor down-regulation in brain regions of anandamide-tolerant rats. *J Neurochem* 75, 2478–84.

Rubino T, Vigano D, Massi P, and Parolaro D (2000b). Changes in the cannabinoid receptor binding, G protein coupling, and cyclic AMP cascade in the CNS of rats tolerant to and dependent on the synthetic cannabinoid compound CP55,940. *J Neurochem* 75, 2080–6.

Sanchez C, Galve-Roperh I, Rueda D, and Guzman M (1998). Involvement of sphingomyelin hydrolysis and the mitogen-activated protein kinase cascade in the Δ^9-tetrahydrocannabinol-induced stimulation of glucose metabolism in primary astrocytes. *Mol Pharmacol* 54, 834–43.

Schatz AR, Lee M, Condie RB, Pulaski JT, and Kaminski NE (1997). Cannabinoid receptors CB1 and CB2: a characterization of expression and adenylate cyclase modulation within the immune system. *Toxicol Appl Pharmacol* 142, 278–87.

Schlicker E, Timm J, and Gothert M (1996). Cannabinoid receptor-mediated inhibition of dopamine release in the retina. *Naunyn-Schmiedebergs Arch Pharmacol* 354, 791–5.

Schlicker E, Timm J, Zentner J, and Gothert M (1997). Cannabinoid CB$_1$ receptor-mediated inhibition of noradrenaline release in the human and guinea-pig hippocampus. *Naunyn-Schmiedebergs Arch Pharmacol* 356, 583–9.

Shire D, Calandra B, Bouaboula M *et al.* (1999). Cannabinoid receptor interactions with the antagonists SR 141716A and SR 144528. *Life Sci* 65, 627–35.

Shire D, Calandra B, Delpech M *et al.* (1996a). Structural features of the central cannabinoid CB1 receptor involved in the binding of the specific CB1 antagonist SR 141716A. *J Biol Chem* 271, 6941–6.

Shire D, Calandra B, Rinaldi-Carmona M *et al.* (1996b). Molecular cloning, expression and function of murine CB2 peripheral cannabinoid receptor. *Biochim Biophys Acta-Lipids And Lipid Metab* 1307, 132–6.

Showalter VM, Compton DR, Martin BR, and Abood ME (1996). Evaluation of binding in a transfected cell line expressing a peripheral cannabinoid receptor (CB2): identification of cannabinoid receptor subtype selective ligands. *J Pharmacol Experimental Therapeutics* **278**, 989–99.

Sim-Selley LJ, Brunk LK, and Selley DE (2001). Inhibitory effects of SR141716A on G-protein activation in rat brain. *Eur J Pharmacol* **414**, 135–43.

Slipetz DM, O'Neill GP, Favreau L *et al.* (1995). Activation of the human peripheral cannabinoid receptor results in inhibition of adenylyl cyclase. *Mol Pharmacol* **48**, 352–61.

Smart D, Gunthorpe MJ, Jerman JC *et al.* (2000). The endogenous lipid anandamide is a full agonist at the human vanilloid receptor (hVR1). *Br J Pharmacol* **129**, 227–30.

Smith FL, Fujimori K, Lowe J, and Welch SP (1998). Characterization of Δ^9-tetrahydrocannabinol and anandamide antinociception in nonarthritic and arthritic rats. *Pharmacol Biochem Behavior* **60**, 183–91.

Soderstrom K and Johnson F (2001). Zebra finch cb1 cannabinoid receptor: pharmacology and *in vivo* and *in vitro* effects of activation. *J Pharmacol Experimental Therapeutics* **297**, 189–97.

Soderstrom K, Leid M, Moore FL, and Murray TF (2000). Behavioral, pharmacological, and molecular characterization of an amphibian cannabinoid receptor. *J Neurochem* **75**, 413–23.

Song ZH and Bonner TI (1996). A lysine residue of the cannabinoid receptor is critical for receptor recognition by several agonists but not WIN55212-2. *Mol Pharmacol* **49**, 891–6.

Spicuzza L, Haddad EB, Birrell M *et al.* (2000). Characterization of the effects of cannabinoids on guinea-pig tracheal smooth muscle tone: role in the modulation of acetylcholine release from parasympathetic nerves. *Br J Pharmacol* **130**, 1720–6.

Sugiura T, Kondo S, Sukagawa A *et al.* (1996). Transacylase-mediated and phosphodiesterase-mediated synthesis of *N*-arachidonoylethanolamine, an endogenous cannabinoid receptor ligand, in rat brain microsomes: comparison with synthesis from free arachidonic acid and ethanolamine. *Eur J Biochem* **240**, 53–62.

Sugiura T, Yoshinaga N, and Waku K (2001). Rapid generation of 2-arachidonoylglycerol, an endogenous cannabinoid receptor ligand, in rat brain after decapitation. *Neurosci Lett* **297**, 175–8.

Szabo B, Dorner L, Pfreundtner C, Norenberg W, and Starke K (1998). Inhibition of GABAergic inhibitory postsynaptic currents by cannabinoids in rat corpus striatum. *Neuroscience* **85**, 395–403.

Szabo B, Muller T, and Koch H (1999). Effects of cannabinoids on dopamine release in the corpus striatum and the nucleus accumbens *in vitro*. *J Neurochem* **73**, 1084–9.

Szabo B, Wallmichrath I, Mathonia P, and Pfreundtner C (2000). Cannabinoids inhibit excitatory neurotransmission in the substantia nigra pars reticulata. *Neuroscience* **97**, 89–97.

Tanda G, Pontieri FE, and Di Chiara G (1997). Cannabinoid and heroin activation of mesolimbic dopamine transmission by a common μ 1 opioid receptor mechanism. *Science* **276**, 2048–50.

Ueda N, Puffenbarger RA, Yamamoto S, and Deutsch DG (2000). The fatty acid amide hydrolase (FAAH). *Chem Phys Lipids* **108**, 107–21.

Volfe Z, Dvilansky A, and Nathan I (1985). Cannabinoids block release of serotonin from platelets induced by plasma from migraine patients. *Int J Clin Pharmacol Res* **5**, 243–6.

Wagner JA, Varga K, Ellis EF, Rzigalinski BA, Martin BR, and Kunos G (1997). Activation of peripheral CB1 cannabinoid receptors in haemorrhagic shock. *Nature* **390**, 518–21.

Wartmann M, Campbell D, Subramanian A, Burstein SH, and Davis RJ (1995). The MAP kinase signal-transduction pathway is activated by the endogenous cannabinoid anandamide. *FEBS Lett* **359**, 133–6.

Weidenfeld J, Feldman S, and Mechoulam R (1994). Effect of the brain constituent anandamide, a cannabinoid receptor agonist, on the hypothalamo-pituitary-adrenal axis in the rat. *Neuroendocrinology* **59**, 110–12.

Wenger T, Jamali KA, Juaneda C, Bacsy E, and Tramu G (2000). The endogenous cannabinoid, anandamide regulates anterior pituitary secretion *in vitro*. *Addiction Biol* **5**, 59–64.

Wilson RI and Nicoll RA (2001). Endogenous cannabinoids mediate retrograde signalling at hippocampal synapses. *Nature* 410, 588–92.

Yamaguchi F, Macrae AD, and Brenner S (1996). Molecular cloning of two cannabinoid type 1-like receptor genes from the puffer fish fugu rubripes. *Genomics* 35, 603–5.

Zygmunt PM, Petersson J, Andersson DA *et al.* (1999). Vanilloid receptors on sensory nerves mediate the vasodilator action of anandamide. *Nature* 400, 452–7.

Chapter 14

Cholecystokinin receptors

Florence Noble and Bernard P. Roques

14.1 Introduction

Cholecystokinin (CCK) is a peptide originally discovered in the gastrointestinal tract (Ivy and Oldberg 1928), but also found in high density in the mammalian CNS (Vanderhaeghen *et al.* 1975). This peptide, initially characterized as a 33 amino acid sequence, is present in a variety of biologically active molecular forms (Rehfeld and Nielsen 1995) derived from a 115 amino acid precursor molecule. The CCK fragments are involved in numerous physiological functions such as feeding behaviour, central respiratory control and cardiovascular tonus, vigilance states, memory processes, nociception, emotional and motivational responses. The C-terminal octapeptide fragment CCK_8, constitutes one of the major neuropeptides in the brain, and binds with nanomolar affinities with two different receptors, designated CCK_1 (CCK_A) and CCK_2 (CCK_B/gastrin) receptors. The N-terminus of these heptahelical receptors is located extracellularly and the C-terminal domain is often anchored in the plasma membrane by means of a cysteine bearing a fatty acid.

14.2 Molecular characterization

Receptors for CCK were first characterized on pancreatic acinar cells and identified as CCK_1 receptors (Sankaran *et al.* 1980). Subsequently, a second receptor, the CCK_2 receptor, was identified in brain which exhibits a distinct pharmacology (Innis and Snyder 1980).

The human genes for the CCK_1 and CCK_2 receptors are organized in a similar manner consisting of five exons and four introns. Exon one encodes the putative extracellular amino terminus of the receptor. Exon 2 and 3 encode TM regions I–IV and exon 4 encodes the fifth TM region and an initial portion of the third intracellular loop. Exon 5 encodes the C-terminal part of this intracellular loop, the remaining TM VI and VII and the intracellular C-terminal domain (Song *et al.* 1993).

A splice variant of the CCK_2 receptor was isolated from human stomach (Miyake 1995). This cDNA differed from initially cloned cDNA in the 5′-end region and encoded a truncated isoform in which the putative N-terminal extracellular domain was completely lost. Isolation of genomic CCK_2 receptor DNA revealed that the gene structure was similar to that previously reported except that the first intron contained the sequence for an alternative first exon. The alternative usage of this exon causes no change in translated receptor protein. Another splice variant was characterized in exon 4, resulting in the presence of two CCK_2 receptor transcripts differing by a block of five amino acids (GGAGP) within the third intracellular loop. No significant difference in agonist affinity or signal transduction was measured between the shorter and longer isoforms (Wank 1995). The shorter transcript is predominant in the stomach. Several CCK_2 receptor mRNA isoforms from rat brain tissues and fundus glands have also been isolated, including truncated mRNA species (Jagerschmidt *et al.* 1994). Unspliced precursor mRNA and the mature form were identified in the cerebral cortex, hypothalamus, and hippocampus in apparently differing proportions according to the region examined, suggesting that the expression of the CCK_2 receptor could be modulated at a post-transcriptional level. In the cerebellum, only a completely unspliced mRNA form was found, which is in agreement with previous studies showing that CCK_2 receptor-binding sites are not expressed in this structure in rat (Pélaprat *et al.* 1987).

CCK_1 and CCK_2 receptors have been cloned from several species and have approximately 50 per cent homology to each other (Wank *et al.* 1992; see Noble *et al.* 1999; Fig. 14.1). At the amino acid level, the CCK_1 receptor is highly conserved, with an overall sequence homology of 80 per cent and a pairwise amino acid sequence identity of 87–92 per cent in humans, guinea pig, rat, and rabbit. Similarly, the CCK_2 receptor is highly conserved in humans, canine, guinea pig, calf, rabbit, and rat, with an overall identity of 72 per cent and pairwise amino acid sequence identities of 84–93 per cent (see Wank 1995). The deduced sequences of the rat CCK_1 and CCK_2 receptors correspond to 429 and 452 amino acid proteins, respectively. Hydropathy analysis of the primary sequence of CCK_1 and CCK_2 receptors predicts seven transmembrane-spanning domains (TM) as expected for a member of GPCR superfamily (Dohlman *et al.* 1991). In agreement with the heavy and variable degrees of glycosylation reported using ligand affinity cross linking techniques (de Weerth *et al.* 1993), at least three consensus sequence sites for N-linked glycosylation (Asn-X-Ser/Thr) have been identified in the CCK_1 and CCK_2 receptors sequences. There are multiple potential serine and threonine phosphorylation sites in the CCK_2 receptor, one for protein kinase C (PKC) (serine 82, in the first intracellular loop), and two for protein kinase A (PKA) (serine 154 in the second intracellular loop and serine 442 in the cytoplasmic tail). Similar to the CCK_2 receptor, CCK_1 receptor has three consensus sequences for PKC phosphorylation in the third intracellular loop, and one site in the cytoplasmic tail of the rat pancreatic CCK_1 receptor (Ozcelebi and Miller 1995). Moreover, in both receptors there are two cysteines in the first and second extracellular loops, which may form a disulfide bridge required for stabilization of the tertiary structure as demonstrated for other receptors belonging to the GPCR superfamily (Silvente-Poirot *et al.* 1998), and a cysteine in the C-terminus of the receptor which may serve as a membrane-anchoring palmitoylation site as demonstrated for rhodopsin and the β2-adrenergic receptors (O'Dowd *et al.* 1988; Ovchinikov *et al.* 1988).

Finally, based on pharmacological and biochemical studies, the existence of further subtypes of CCK_1 and CCK_2 receptors has been postulated (Durieux *et al.* 1986; Knapp *et al.* 1990; Talkad *et al.* 1994). Nevertheless, at this time only two genes have been cloned. Gastrin receptors in the stomach and CCK_2 receptors in the brain were initially viewed as distinct CCK receptors on the basis of their difference in affinity for CCK- and gastrin-like peptides (Menozzi *et al.* 1989). Endogenous peptide agonists CCK_8 [Asp-Tyr(SO_3H)-Met-Gly-Trp-Met-Asp-Phe-NH_2] and gastrin [H_2N-Gln-Gly-Pro-Trp-Met-Glu-Glu-Glu-Glu-Glu-Ala-Tyr(SO_3H)-Gly-Trp-Met-Asp-Phe-NH_2] share the same COOH-terminal pentapeptide amide sequence but differ in sulfation at the sixth (gastrin) or seventh (CCK) tyrosyl residue. Agonist binding studies on brain membranes and parietal cells show a six to ten fold and one to two fold higher affinity for CCK than for gastrin, respectively (Jensen *et al.* 1990). These small differences in agonist binding have generated controversy regarding the existence of subtypes within this receptor class. The identification of a single CCK_2 receptor encoding

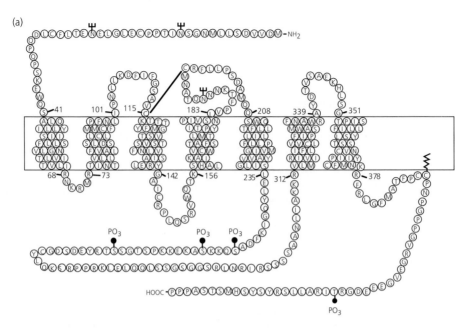

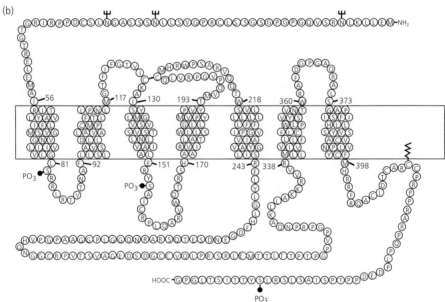

Fig. 14.1 Schematic representation of the rat CCK$_1$ (a) and CCK$_2$ (b) receptor showing the postulated transmembrane topology, sites for putative NH$_2$-linked glycosylation (tridents), serine and threonine phosphorylation by PKC and protein kinase A (PO$_3$) and conserved cysteines in the first and second extracellular loops, possibly forming a disulfide bridge, and a possible palmitoylated conserved cysteine in the cytoplasmic tail.

gene through low- and high-stringency hybridization of cDNA and genomic libraries and Northern and Southern blot analysis in numerous species indicates that gastrin receptors do indeed correspond to the CCK_2 receptors located in the gastrointestinal tract and do not constitute a third type of CCK receptor (Wank 1995). However, a third receptor subtype has been pharmacologically identified, although not yet cloned. This receptor is a gastrin preferring receptor, described for the first time in an immortalized fibroblast cell line (Swiss 3T3 cells), and which discriminates between ionidated and glycine-extended gastrins (Singh et al. 1995). Based on the guidelines defined by the International Union of Pharmacology (IUPHAR), Committee on receptor Nomenclature and Drug Classification, this receptor cannot be registered, as the formal demonstration of its existence by cloning and sequencing has not been done.

14.3 Cellular and subcellular localization

CCK_1 receptors are found principally in the gastrointestinal tract and some brain structures, while CCK_2 receptors are widely distributed in the CNS, particular regions of the gastrointestinal tract and on pancreatic acinar- and parietal-cells (Moran et al. 1986; Pélaprat et al. 1987; Jensen et al. 1994). Specific CCK binding sites were characterized in membranes from brain homogenates two decades ago (Innis and Snyder 1980; Saito et al. 1980). Autoradiographic studies using CCK related peptides that do not distinguish between the two CCK receptors in several species (e.g. rat, guinea pig, monkey, humans) showed high densities of CCK binding sites in several areas, including the cerebral cortex, striatum, olfactory bulb, and tubercle and certain amygdaloid nuclei. Moderate levels were observed in the hippocampus, claustrum, substantia nigra, superior colliculus, periaqueductal gray matter, and pontine nuclei. Low densities were reported in thalamic and hypothalamic nuclei and in the spinal cord. Nevertheless, it is important to note that species specific heterogeneity in tissue expression is apparent in different structures and indicates that the results of studies performed in one species may not necessarily be generalized to other species (Niehoff 1989). For example, in the cerebellum, high densities of CCK binding sites were present in guinea pig, human, and mouse, whereas only low levels were detected in rat (Zarbin et al. 1983; Gaudreau et al. 1985).

With the advent of specific radioligands both CCK_1 and CCK_2 receptors were found in brain structures. However, the vast majority of CCK receptors in the CNS are of the CCK_2 type, with CCK_1 receptors restricted to discrete regions (Moran et al. 1986). Nevertheless, recent reports on the immunohistochemical distribution of CCK_1 receptors in the rat CNS, described numerous brain regions displaying CCK_1 receptor-like immunoreactivity (Mercer and Beart 1997). Thus, CCK_1 receptors were found in the interpeduncular nucleus, area postrema, medial nucleus tractus solitarius and with additional areas of binding found in the habenular nuclei, dorsomedial nucleus of the hypothalamus, central amygdala, nucleus accumbens, superior colliculus, periaqueductal gray matter, olivary nuclei, anteroventral thalamic nuclei (see Noble et al. 1999). The precise anatomical localization of the two CCK receptor types serves to provide morphological substrates for many of the diverse functions attributed to neural CCK (see Section 14.6 and Crawley et al. 1994).

In the rat, CCK_2 mRNA was shown to be widely distributed in areas such as the cerebral cortex, the olfactory regions, the hippocampal formation, the septum, the amygdala, the basal ganglia, and related regions (nucleus accumbens, caudate putamen, substantia nigra), the interpeduncular nucleus and the cerebellum (Honda et al. 1993). This mRNA localization is largely consistent with previously reported histochemical binding studies (Pélaprat *et al.* 1987; Niehoff 1989), except for regions such as the cerebellum, where neither CCK_2 receptors nor prepro-CCK mRNA were previously detected (Pélaprat *et al.* 1987; Lanaud *et al.* 1989). This is consistent with the absence of mature mRNA encoding the sequence of the CCK_2 receptor (Jagerschmidt *et al.* 1994).

14.4 Pharmacology

14.4.1 Agonists

At CCK_1 receptors, sulphated CCK_8 was the minimal sequence for high affinity binding, whereas CCK_5, CCK_4, gastrin, and unsulphated CCK_8 interact with CCK_2 receptors albeit with affinities that are lower than sulphated CCK_8. Only a few compounds have been reported to be CCK_1 selective agonists. Most of them are peptides, such as A71378 [des-NH_2-Tyr(SO_3H)-Nle-Gly-Trp-Nle-(NMe)Asp-Phe-NH_2], or A71623 and A70874. Recently, a series of 1,5-benzodiazepines acting as CCK_1 receptor agonists *in vitro* and *in vivo* were discovered. Potency within this series was modulated by substituents on the N1-anilinoacetamide moiety (Aquino *et al.* 1996) and by substitution and/or replacement of the phenylurea moiety in the C3-position (GW5823, GW7854; Henke *et al.* 1997; Table 14.1).

Different strategies have been followed to design potent and selective agonists of CCK_2 receptors. In spite of its intrinsic flexibility, CCK_8 was found by NMR to exist preferentially in a folded form in aqueous solution with a proximity between Asp1 and Gly4 (Fournié-Zaluski *et al.* 1986). This property was used to synthesize cyclic peptides such as BC254 and BC197, which were found to be highly potent and selective CCK_2 agonists (Charpentier *et al.* 1988*a*, 1989; Table 14.2).

Another approach towards CCK_2 receptor agonists was to protect CCK_8 from degrading enzymes such as aminopeptidase A (Migaud *et al.* 1996) and a thiol/serine protease cleaving this peptide at the Met-Gly bond (Rose *et al.* 1996). The biologically active Boc[$Nle^{28,31}$]CCK_{27-33} (BDNL; Ruiz-Gayo *et al.* 1985) was used as the parent compound to design enzyme resistant analogs. In this compound, the major sites of cleavage are at the Trp^{30}/Nle^{31} and Nle^{28}/Gly^{29} bonds. Consequently, several enzyme resistant BDNL analogs containing either a retro-inverso 28–29 amide bond, or a (NMe)Nle^{31} residue, or a combination of these two modifications have been synthesized (Charpentier *et al.* 1988*b*). This led to BC264, a highly potent CCK_2 receptor agonist which exhibits similar affinities ($K_D = 0.1$–0.5 nM) in all species (Charpentier *et al.* 1988*b*; Durieux *et al.* 1991). The behavioural results obtained with BC264 such as increase in memory processes, increase in dopamine release in mesolimbic and nigrostrial pathways (see Daugé and Roques 1995), suggest that the development of non-peptide CCK_2 selective agonists endowed with good stability and bioavailability should provide useful pharmacological tools and possibly interesting therapeutic agents. In order to design such derivatives, the C-terminal tetrapeptide, showing significant CCK_2 affinity and selectivity was used as scaffold, although it has been shown to trigger panic attacks in humans (de Montigny 1989; Bradwejn *et al.* 1991). Several modifications of CCK_4 increased its CCK_2 receptor selectivity, such as N-terminal protection of the tetrapeptide in Boc-CCK_4 (Harhammer *et al.* 1991) and modifications of the different constituting amino acids with replacement of Met by Nle or (NMe)Nle (Corringer *et al.* 1993). NMR and molecular dynamics studies indicated that the CCK_2

Table 14.1 CCK$_1$-receptor agonists

des-NH$_2$-Tyr(SO$_3$H)-Nle-Gly-Trp-Nle-(NMe)Asp-Phe-NH$_2$	A-71378
Boc-Trp-Lys(o-tolylaminocarbonyl) -Asp-MePhe-NH$_2$	A-71623
Boc-Trp-Lys(p-hydroxycinnamoyl)-Asp-(NMe)Phe-NH$_2$	A-70874

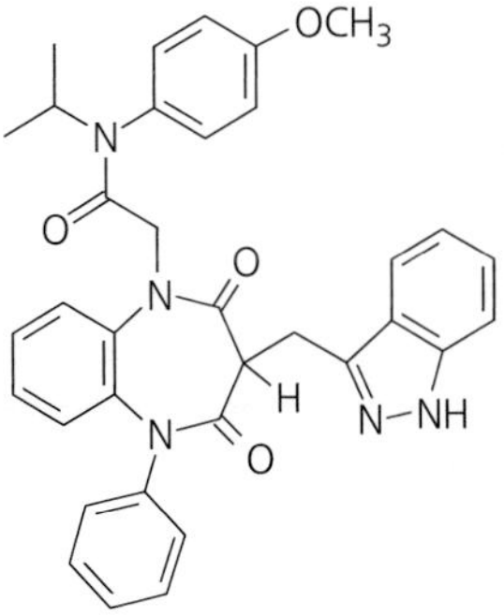

GW-5823

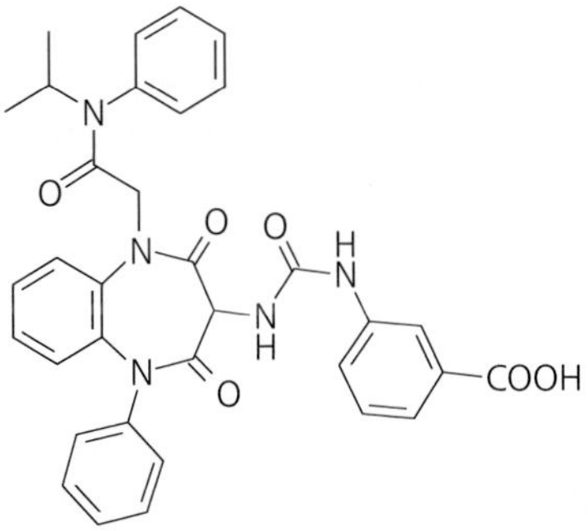

GW-7854

Table 14.2 CCK$_2$-receptor agonists

Boc-D.Asp-Tyr(SO$_3$H)-Nle-D.Lys-Trp-Nle-Asp-Phe-NH$_2$	BC 197
Boc-Tyr(SO$_3$H)-gNle-mGly-Trp-NMe(Nle)-Asp-Phe-NH$_2$	BC 264
Boc-γD-Glu-Tyr(SO$_3$H)-Nle-D.Lys-Trp-Nle-Asp-Phe-NH$_2$	BC 254

receptor-selective CCK$_4$ analogs adopt on S-shaped conformation with a relatively well-defined orientation of the side chains (Goudreau *et al.* 1994). The same type of folded structure has been reported for several potent agonists derived from CCK$_4$ and containing a [trans-3-propyl-L-proline] (Nadzan *et al.* 1991), a diketopiperazine skeleton (Shiosaki *et al.* 1990), or a [(alkylthio)proline] residue (Kolodziej *et al.* 1995). Moreover, to stabilize the bioactive conformation of CCK$_2$ receptor agonists with the aim of designing non-peptide ligands, macrocyclic constrained CCK_4 analogs have been developed, endowed with CCK_2 agonist properties and the ability to cross the blood–brain barrier (Blommaert *et al.* 1997). Based on these results, compounds combining the modifications introduced in BC264 and CCK_4 such as RB400 were found to be highly potent (Million *et al.* 1997).

14.4.2 Antagonists

Much of the early research regarding the physiological effects of CCK was hindered by the lack of selective antagonists. The first CCK antagonists were derived from a naturally occurring benzodiazepine, asperlicin, which has been isolated from the fungus *Aspergillus alliaceus* (Chang *et al.* 1985; Table 14.3). The high *in vitro* and *in vivo* potency of asperlicin at CCK_1 receptors conferred clear advantages over previously reported CCK antagonists as a tool for investigating the physiological and pharmacological actions of CCK. The 5-phenyl-1,4-benzodiazepine ring was used as a model to design improved CCK receptor antagonists (Evans *et al.* 1986), leading to several compounds such as L364718 (MK329, devazepide) which remained for several years the most potent CCK antagonist with a good selectivity for CCK_1 receptors. Various benzodiazepine derivatives were developed, such as FK480 which is a highly selective and potent CCK_1 receptor antagonist (Ito *et al.* 1994). Several other potent and selective antagonists of the CCK_1 receptor have been described, including glutamic acid derivatives, such as loxiglumide (CR1505) or lorglumide (CR1409; Makovec *et al.* 1995) and other dipeptoids such as 2-NAP [2-naphtalenesulfonyl 1-aspartyl-(2-phenethyl)amide] (Hull *et al.* 1993), or PD140548 (Boden *et al.* 1993). Interest in nonpeptide CCK receptor selective ligands has directed efforts towards the incorporation of conformationally restricted structures as spacers between Trp and Phe residues in the sequence of the CCK receptor endogenous ligands CCK_4. Thus, recently, a new series of CCK_4 restricted analogs with a 3-oxoindolizidine ring were synthesized, for example, IQM95333 (Martin-Martinez *et al.* 1997). Another CCK_1 receptor antagonist, SR27897, which is chemically unrelated to peptoids, benzodiazepine, or glutamic acid derivatives has also been developed (Table 14.3).

Turning to the CCK_2 receptor the moderate affinity of L364718 for CCK_2 receptors suggested that the benzodiazepine nucleus might also hold a key to selective ligands for these receptors. The first compound of interest developed using this strategy was L365260 (Bock *et al.* 1989). Undoubtedly, the benzodiazepine template which is present in asperlicin, L364718 and L365260 has been the most exploited structure in the development of CCK receptor antagonists. This is particularly true for the CCK_2 receptor where subtantial investment by the pharmaceutical industry has resulted in the generation of more potent and selective antagonists incorporating the benzodiazepine moiety, such as L740093 (Patel *et al.* 1994), YM022 (Nishida *et al.* 1994) and YF476 (Takinami *et al.* 1997; Table 14.4).

A second approach in the development of non-peptide antagonists of the CCK2 receptor has been the synthesis of 'dipeptoids' (Horwell *et al.* 1991). This led to tryptophan dipeptoid derivatives such as PD134308 (CI988) with nanomolar affinity for CCK_2 receptors. A direct comparison of the structure of these compounds showed that their sizes could be reduced to increase their lipophilicity. Indeed, the clinical development of CI988 was limited due to its poor bioavailability, which was attributed to poor absorption and efficient hepatic extraction. Several modifications have been performed, leading to compounds such as CI1015 (Trivedi *et al.* 1998) or RB211 (Blommaert *et al.* 1993).

Three other series have been described, leading to the synthesis of derivatives that have both excellent selectivity and high affinity for CCK_2 receptor. The 4-benzamido-5-oxopentanoic

Table 14.3 CCK$_1$-receptor antagonists

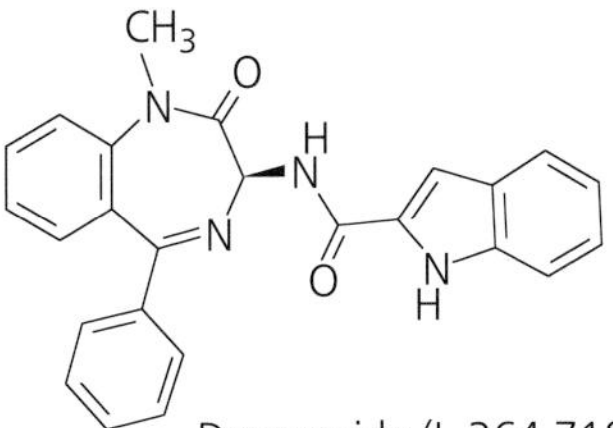

Devazepide (L-364,718)

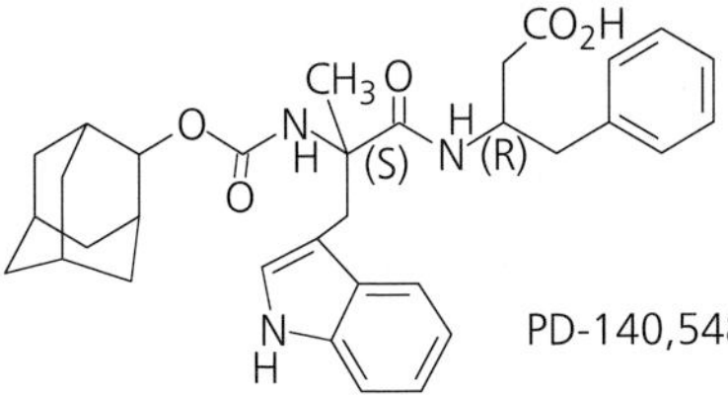

PD-140,548

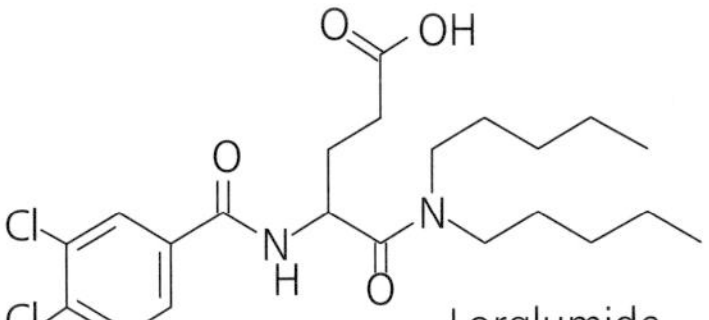

Lorglumide

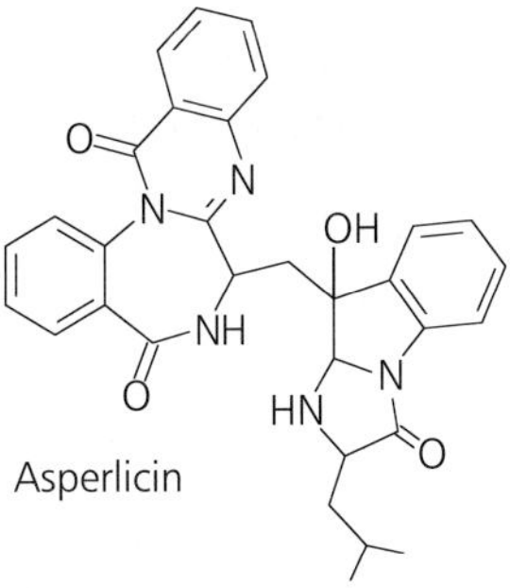

Asperlicin

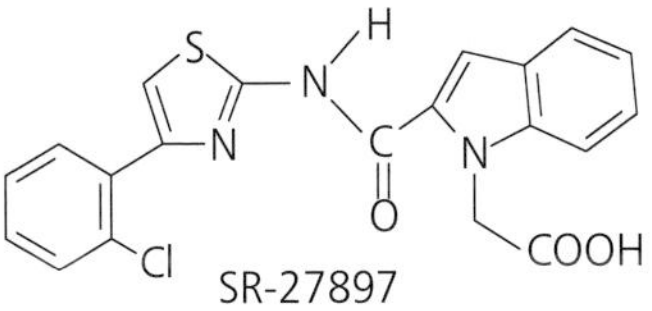

SR-27897

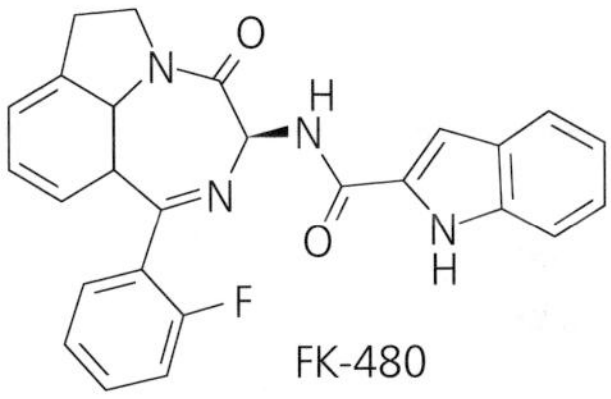

FK-480

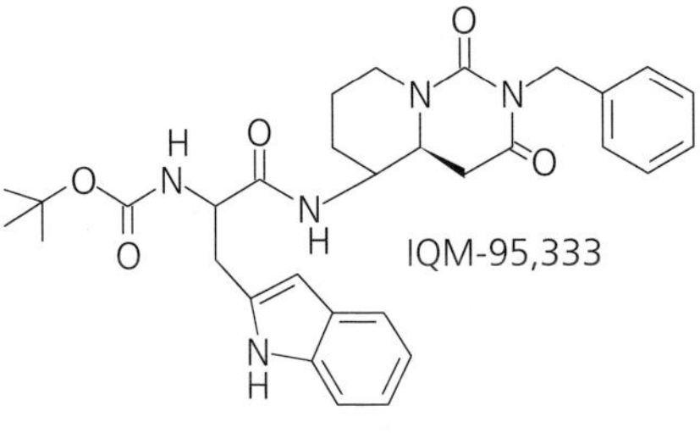

IQM-95,333

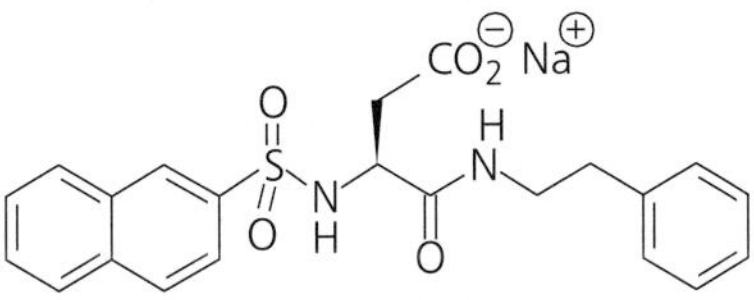

2-NAP

derivatives, the diphenylpyrazolidinone series and the ureidoacetamides of which CR2194 (Revel *et al.* 1992), LY262691 (Howbert *et al.* 1992) and RP73870 (Pendley *et al.* 1995), respectively, are representative examples.

With the use of site-directed mutagenesis, it has been shown that CCK peptide agonists interact with multiple amino acids in the extracellular domain of CCK receptors and that CCK$_1$ and CCK$_2$ receptors have distinct binding sites despite their shared high affinity for CCK$_8$ (see Noble *et al.* 1999). Moreover, the reversal of the relative affinity for L364718 and L365260 between canine gastrin receptor and both rat and human CCK$_2$ receptors noted earlier has been explained by an interspecies variation of a single amino acid in TMVI (Leu355 in dog versus the corresponding Val349 in humans; Marino *et al.* 1993). The lack

Table 14.4 CCK$_2$-receptor antagonists

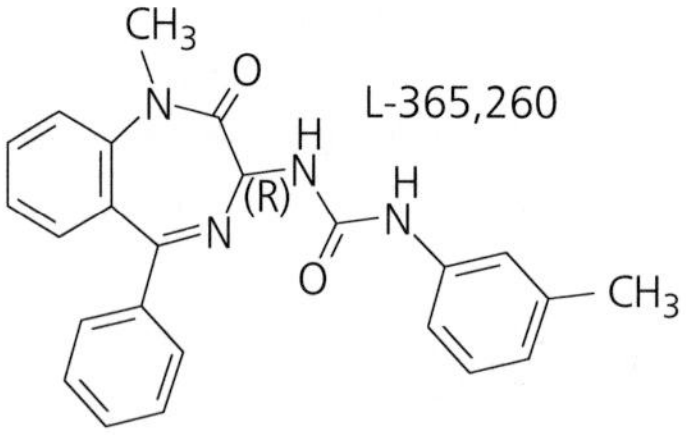

L-365,260

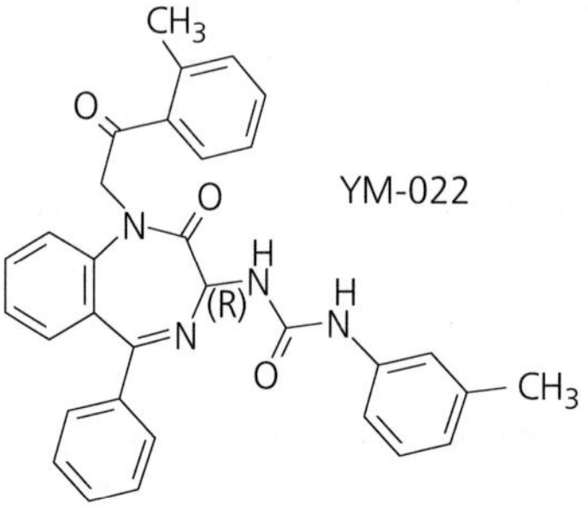

CR 2194

YM-022

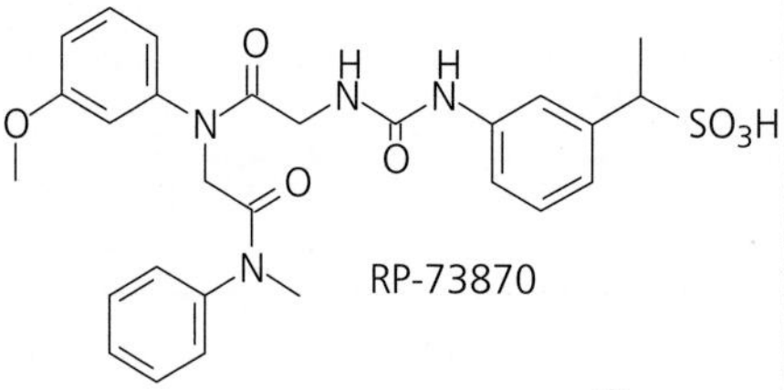

L-740,093

RP-73870

YF-476

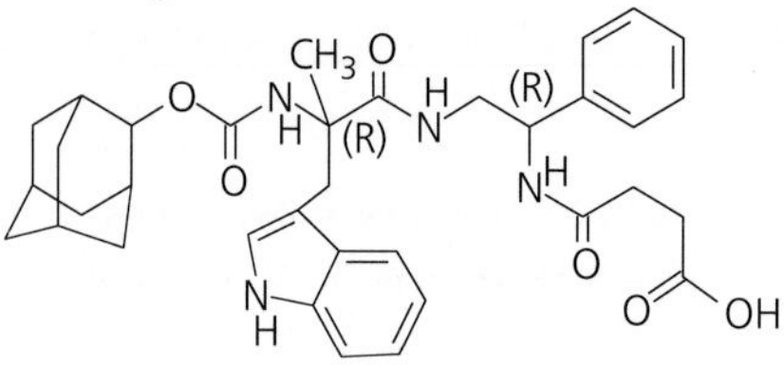

PD-134,308

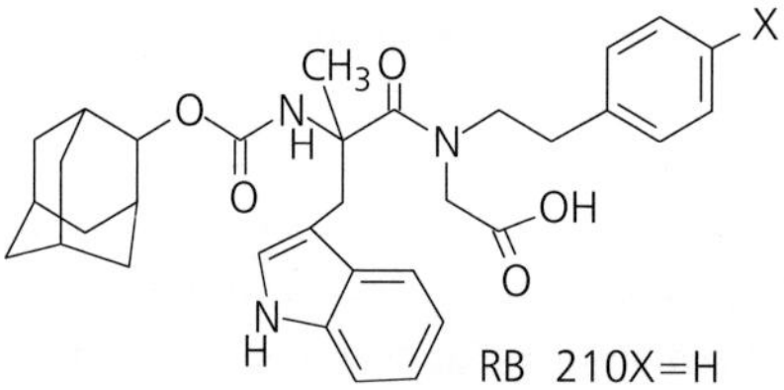

RB 210 X=H
RB 211 X=Cl

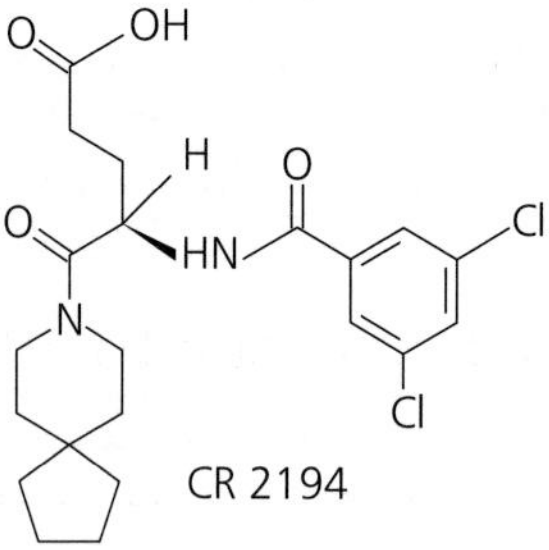

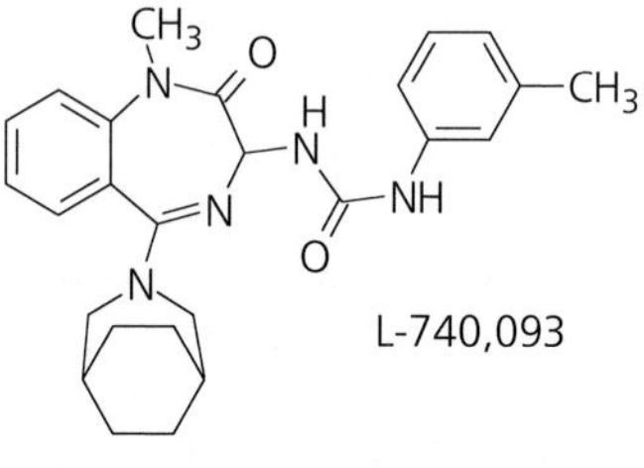

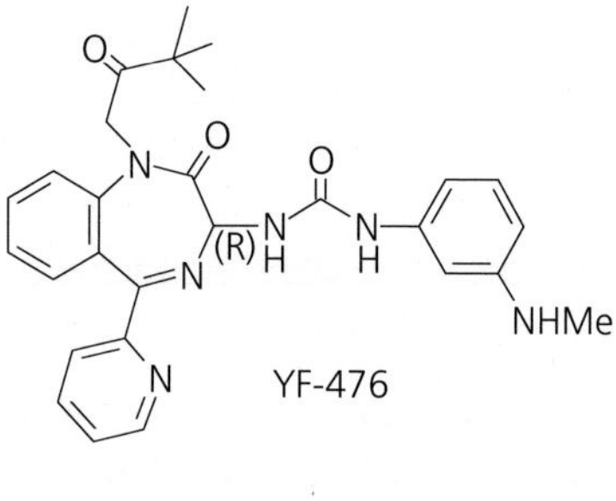

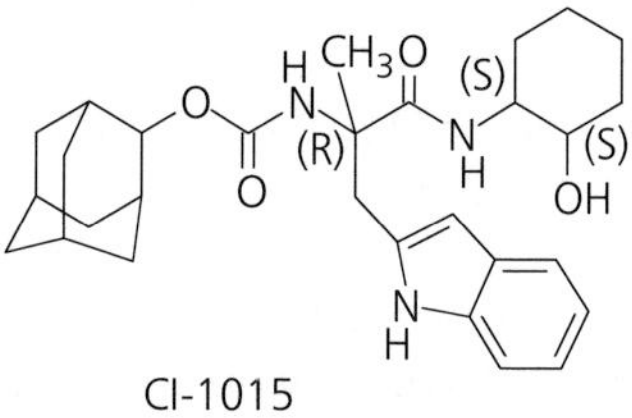

CI-1015

of effect of mutations in TMVI and TMVII on agonist affinity suggests that agonist- and antagonist binding sites are, at best, only partially overlapping. Other studies have also shown that antagonists highly selective for CCK_1 receptors had increased affinities for H381L and H381F CCK_2 receptor mutants. Thus, His381 residue is essential for CCK_2 versus CCK_1 antagonist selectivity (Jagerschmidt et al. 1996). This result is in good agreement with an independent study (Kopin et al. 1995), in which the authors predicted the position of residues within the TM helices of the human CCK receptors involved in CCK_1/CCK_2 selectivity.

14.5 Signal transduction and receptor modulation

The signal transduction cascade for CCK_1 receptors has been essentially characterized in rat pancreatic acini. In this system it has been well established that CCK_1 receptor is capable of coupling through G_q and G_s to both phospholipase C and adenylyl cyclase, respectively, at physiological concentrations (Piiper et al. 1997; Wu et al. 1997). On the other hand, it has been demonstrated that CCK_1 receptors are coupled to the phospholipase A2 (PLA2). Moreover, it was recently shown that mitogen-activated protein kinase (MAPK) and c-Jun-NH_2-terminal kinases (JNK, which phosphorylate serine residues of c-Jun) are rapidly activated by the octapeptide CCK_8 in rat pancreas both in vitro and in vivo (Dabrowski et al. 1996). All these signalling pathways have not yet been confirmed in the CNS, although it is tempting to speculate that the intracellular effectors coupled to CCK_1 receptors are identical. This lack of characterization in the brain is largely due to the difficulty of working with isolated neurones. Thus, for a long time, central CCK_2 receptors have not been proved to be linked to a well characterized second-messenger system in the brain, including the phosphoinositide system, although phosphoinositide metabolism was shown to be affected by CCK in neuroblastoma (Barrett et al. 1989) and in an embryonic pituitary cell line (Lo and Hughes 1988). More recently, Zhang et al. (1992) showed that CCK_8 increased the turnover of phosphoinositides and IP_3 in dissociated neonatal rat brain cells, in which both CCK_1 and CCK_2 receptors were expressed. Despite this, it was not possible, using synaptoneurosomes from guinea pig cortex, to support their possible coupling with adenylyl cyclase or PLC, although Ca^{2+} release from intracellular stores, possibly via a G protein-independent mechanisms was triggered by a CCK analog (Galas et al. 1992). However, in a mammalian expression system it has been shown that CCK_2 receptors are coupled to a phospholipase pathway leading to the release of arachidonic acid via a PTX-sensitive G protein (Pommier et al. 1999) and to a MAPK pathway (Taniguchi et al. 1994).

The region of the CCK_2 receptor interacting with G_q leading to activation of PLC was determined using Lys333Met, Lys334Thr, and Arg335Leu mutations transiently expressed in COS-7 cells and Xenopus oocytes. These mutations resulted in the loss of G_q activation without affecting receptor affinity (Wang 1997). Site directed mutagenic replacement of Asp100 in the rat CCK_2 receptor, a highly conserved residue in TMII of most GPCRs, results in a 50 per cent reduction in CCK_8 stimulated phophoinositide turnover with no change in CCK_8 affinity and only a small decrease in antagonist affinity (Jagerschmidt et al. 1995). These data led to the hypothesis that Asp100 points in the direction of the cluster of basic amino acids (Lys333/Lys334/Arg335) located in the third intracellular loop of the receptor at the bottom of the TMVI. Another residue, Phe347, which belongs to the TMVI domain was also identified as essential for the signal transduction process. Thus, exchange of Phe347 for alanine, disrupts phosphatidylinositol signalling pathway, without affecting the binding of CCK receptor agonists (Jagerschmidt et al. 1998). This amino acid could be implicated in transduction processes, through its role in agonist induced changes in receptor conformation and subsequent triggering of G protein activation. Indeed, the exchange of Phe347 for alanine could produce a conformational change in the sequence containing the basic triplet, located just beneath TMVI.

14.6 **Physiology and disease relevance**

In line with its distribution in brain, CCK is involved in the modulation/control of multiple central functions. In particular numerous experimental and clinical studies have shown that CCK, through its action at CCK_1 and CCK_2 receptors, (Table 14.5) participates in the neurobiology of feeding, satiety, cardiovascular regulation, anxiety, pain, depression, psychosis, analgesia, memory, neuroendocrine control, and osmotic stress (see Crawley and Corwin 1994).

14.6.1 **Schizophrenia**

In the brain, it has been shown that the CCK_1 receptor modulates CCK-stimulated dopamine release in the posterior nucleus accumbens (Josselyn and Vaccarino 1996). Thus, it was speculated that alterations in CCK_1 receptor lead to an increase in dopamine release, which may in turn constitute a predisposition to schizophrenia. This was confirmed by an association analysis conducted between unrelated schizophrenic patients and healthy volunters, which confirmed that the 201A allele frequency was higher in the schizophrenic group, especially in the paranoid type, than in the control group (Tachikawa *et al.* 2000). The same type of study has been performed on the CCK_2 receptors. However, the results suggest that the CCK_2 receptor gene polymorphisms have no association with schizophrenia (Tachikawa *et al.* 1999).

The precise role of CCK in schizophrenia remains incompletely understood. Methodological problems, study groups of patients that were too small and patient heterogeneity might have contributed to inconsistent results. Nevertheless, overall the available data suggest that schizophrenia may be associated with reduced CCK anxiety (Ferrier *et al.* 1983; Carruthers *et al.* 1984). Several open studies reported that administration of nonselective CCK receptor agonists improved psychotic symptoms in schizophrenic patients when added to ongoing neuroleptic treatment (Payeur *et al.* 1993).

14.6.2 **Anxiety**

The initial suggestion that the CCK system might be involved in anxiety came from experiments of Bradwejn and de Montigny (1984) showing that benzodiazepine receptor agonists could attenuate CCK induced excitation of rat hippocampal neurones. Subsequent clinical studies demonstrated that bolus injections of the CCK_2 receptor agonist CCK_4 or pentagastrin provoke panic attacks in patients with panic disorders (Bradwejn *et al.* 1991, 1992). Recent investigations have revealed that the panicogenic effects of CCK_2 receptor agonists are not limited to patients with panic disorder, because individuals with social phobia, generalized anxiety disorder, obsessive compulsive disorder, and premenstrual dysphoric disorder also exhibit an augmented behavioural response to these ligands (Le Melledo *et al.* 1995; De Leeuw *et al.* 1996; van Vliet *et al.* 1997). Interestingly, a significant association exists between panic disorder and polymorphism of the CCK_2 receptor gene (Kennedy *et al.* 1999). The neurobiological mechanisms by which CCK_2 receptor agonists

Table 14.5 Cholecystokinin receptors

Cholecystokinin receptors	CCK$_1$	CCK$_2$
Alternative names	CCK$_A$, CCK-A	CCK$_B$, CCK-B, CCK-B/gastrin
Structural information (Accession no.)	h 428 aa (P32238) r 444 aa (P30551) m 436 aa (O08786)	h 447 aa (P32239) AS r 452 aa (P30553) AS m 453 aa (P56481)
Chromosomal location	4p16.2–15.1	11p15.4
Selective agonists	A71378, A71623, A70874, GW5823, GW7854	BC254, BC197, BC264, CCK-8ns, CCK-4, Gastrin, RB400
Selective antagonists	Devazepide (L364718), FK480, Lorglumide (CR1409), PD140548, IQM95333, SR27897	L365260, YM022, PD134308, LY262691, CI1015, RB211, RP73870
Radioligands	[^{3}H]-devazepide, [^{3}H] or [^{125}I]-CCK$_8$	[^{3}H]-L365,260, [^{3}H]-PD140,376, [^{125}I]-PD142,308, [^{3}H] or [^{125}I]-gastrin, [^{3}H] or [^{125}I]-CCK$_8$
G protein coupling	G$_q$/G$_s$	G$_q$/G$_i$/G$_o$
Expression profile	found in discrete brain regions, nucleus tractus solitarius, area postrema, interpenduncular nucleus, high in the periphery, gall bladder, pancreas, pyloric sphincter and vagal afferent fibres	highly expressed in brain, cerebral cortex, caudate nucleus, limbic system, also expressed in the gastrointestinal tract and pancreatic acinar and parietal cells
Physiological function	mediates action of CCK on cell proliferation, pancreatic enzyme secretion, gallbladder contraction, dopamine related behaviours and release; satiety, nociception	mediates action of CCK on increased neuronal firing rates, nociception, anxiety, attention, memory, respiration, dopamine related behaviours and dopamine release
Knock-out phenotype	learning and memory functions are impaired in OLETF rats which lack the CCK1 receptor (a model of obese, non insulin dependent diabetes)	impaired performance in memory tests, up-regulation of the opioid system, sensitisation of the dopamine system
Disease association	acute pancreatitis, schizophrenia, cognitive decline	panic attacks, depression, cognitive decline

provoke panic and concomitant biological changes (robust increase in heart rate, blood pressure, hypothalamic-pituitary-adrenal axis activity, elevated blood levels of dopamine, epinephrine, norepinephrine, and neuropeptide Y) have been the subject of considerable research activity. Animal studies suggest that anxious behaviour induced by various CCK fragments is associated with selective CCK_2 receptor stimulation (Harro and Oreland 1993). However, the anxiogenic effects of CCK peptides in animals have not been observed by all investigators, and the relevant negative findings should not be ignored (Shlik et al. 1997). The effect of CCK compounds could vary considerably because of existing differences in the distribution and binding characteristics of CCK receptor types and/or affinity states among species. Recently, the effects of the selective CCK_2 receptor agonist BC264 and BC197 and of the nonselective CCK receptor agonist BDNL were investigated in rats subjected to the elevated plus maze (Fig. 14.1). Surprisingly, BDNL and BC197 did induce anxiogenic like effects although BC264 was devoid of any effect (Derrien et al. 1994c; Fig. 14.2).

14.6.3 Depression

One of the physiological actions of the neuropeptide CCK seems to involve modulation of the nigrostriatal and mesolimbic dopaminergic pathways, suggesting a role of CCK in mood disorders. Several studies have shown that selective CCK_2 receptor agonists accentuated the suppression of motility test in mice, an animal model used to select antidepressant drugs. Moreover, this effect was inhibited by L365260, demonstrating the selective involvement of the CCK_2 receptors (Derrien et al. 1994a). However, the most interesting results were obtained with the CCK_2 antagonist, which alone decreased motor inhibition in shocked mice and induced antidepressant-like effect in the forced swim test in mice (Smadja et al. 1995). However, relatively little is known about the role of CCK in clinical depression. Several laboratories have demonstrated that patients with major depression display cerebrospinal fluid CCK concentrations comparable to those of control subjects (Geracioti et al. 1993). However, there is some evidence that an increase in cerebrospinal fluid CCK levels can occur in particularly severe depression (Löfberg et al. 1998). On the other hand, postmortem studies have revealed that compared with healthy controls and patients with schizophrenia, suicide victims have elevated prepro-CCK mRNA levels and an increased density of CCK-containing neurones in the dorsolateral prefrontal cortex and a high density of CCK-receptors in the frontal cortex (Ferrier et al. 1983).

14.6.4 Alzheimer's disease and cognition

On the basis of anatomical data, studies of CCK on memory processes have constituted an important field of investigation, as this neuropeptide is present in regions such as limbic structure and cortical areas, which are implicated in the control of cognitive processes, motivational and emotional behaviours. It has been suggested that CCK_1 and CCK_2 receptors have different roles in learning and memory functions (Harro and Oreland 1993). In particular, a balance between CCK_1 receptor mediated facilitatory effects and CCK_2 receptor mediated inhibitory effects on memory retention has been postulated (Lemaire et al. 1992). However, there are conflicting reports on the effects of CCK_2 receptor agonists in animal models of memory. For instance, although some groups have reported that selective CCK_2 receptor agonists (CCK_4, BC264) impair memory (Lemaire et al. 1992; Derrien et al. 1994b), others have found that these peptides enhance memory (Gerhardt et al. 1994). Treatment

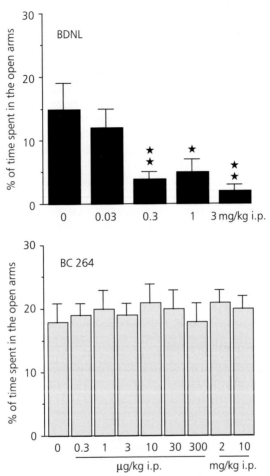

Fig. 14.2 Effects of i.p. injection of BDNL and BC 264 administered 30 min before the experiment in the elevated plus maze. The behavioural responses of rats were measured in the elevated plus-maze for 5 min and are expressed as the percentage of time spent in open arms. * $p < 0.05$ and ** $p < 0.01$ as compared to control group.

with BC264 has also been described to elicit prominent hypervigilance in monkeys and to increase behavioural arousal in rats (review in Daugé and Roques 1995). The latter findings suggest a possible role for CCK_2 receptor in attentional activation that can facilitate learning. The apparent discrepancies observed with CCK agonists indicate that CCK_2 receptors could have different functions involving alternate neuronal pathways, according to the task carried out by the animal. To date only few studies have been devoted to the effects of CCK receptor agonists on human memory. Recently, Shlik *et al.* (1998) found that the continuous administration of the selective CCK_2 receptor agonists, CCK_4, had no effect on psychomotor performance, although it produced impairment in cognitive tests of free recall and recognition. The results of this study suggest that CCK_4 may exert a negative influence on memory consolidation and retrieval.

14.6.5 **Pain and analgesia**

A large body of evidence has now been accumulated supporting physiological interaction between CCK and the endogenous opioid peptides, enkephalins (see Chapter 20). The opioid and CCK systems are critically involved, generally in an opposite manner in various physiological processes, including respiratory functions and cardiovascular tonus. It has been demonstrated that activation of CCK_2 receptors can negatively modulate the opioid system. This is supported by data obtained with selective CCK_2 receptor antagonists. Indeed, these ligands strongly potentiate the antinociceptive and antiallodynic effects of morphine or endogenous enkephalins protected from their enzymatic degradation (Fournié-Zaluski *et al.* 1992; Valverde *et al.* 1994; Coudoré-Civiale *et al.* 2001; Fig. 14.3). In addition, CCK_2 receptor antagonists suppressed the development of autonomy behaviour in a model of neuropathic pain in rat and efficiently relieved the allodynia-like symptoms in spinally injured rats (see Roques and Noble 1996).

Behavioural studies showed that blockade of CCK_1 and CCK_2 receptors produced opposite effects on the opioid induced reduction of conditioned suppression motility due to endogenous enkephalins protected from peptidase inactivation by the dual inhibitor RB101 (Fournié-Zaluski *et al.* 1992). Thus, the antidepressant-like effects of RB101 were suppressed by the CCK_1 receptor antagonist L364718 and enhanced by the CCK_2 receptor antagonist L365,260 (Smadja *et al.* 1995).

14.6.6 **Peripheral indications**

In the periphery, CCK_1 receptors have been shown to be involved in several disease states. For instance, it is well established that activation of both MAPKs and JNKs may be of importance in the early pathogenesis of acute pancreatitis (Dabrowski *et al.* 1996). Moreover, a case report

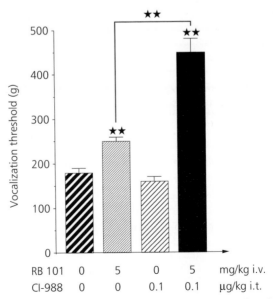

Fig. 14.3 Antinociceptive effects of RB 101 i.v. on the paw pressure-induced vocalization threshold in diabetic rats. * $p < 0.01$ as compared to control group.

of a woman with gallstones and obesity was ascribed to abnormal processing of transcripts from a normal CCK_1 receptor gene that resulted in the predominance of mRNA with a 262 bp deletion corresponding to the third exon. Although this mutation could negatively affect expression or coupling to G proteins, neither *in vivo* nor *in vitro* data were obtained in support of such inferences (Miller *et al.* 1995).

14.7 Concluding remarks

The actions of CCK are mediated by at least two distinct receptors whose existence has been confirmed by molecular cloning. Nevertheless, the large variety of functions mediated by CCK receptors, as well as pharmacological studies, cleary suggests the existence of heterogeneity for these receptors.

The physiological and pathophysiological implications of these receptors can now be further investigated in CCK_2 receptor deficient mice obtained through gene targeting, and in OLETF rats, which have no functional CCK_1 receptors. It is believed that several potential clinical applications with selective agonists and/or antagonists could be developed in a near future.

References

Aquino CJ, Armour DR, Berman JM, Birkemo LS, Carr RAE, Croom DK *et al.* (1996). Discovery of 1,5-benzodiazepines with peripheral cholecystokinin (CCK-A) receptor agonist activity. 1. Optimization of the agonist "trigger". *J Med Chem* **39**, 562–9.

Barrett RW, Steffey ME, and Wolfram CAW (1989). Type-A CCK receptors in CHP 212 neuroblastoma cells: evidence for association with G protein and activation of phosphoinositide hydrolysis. *Mol Pharmacol* **35**, 394–400.

Blommaert AG, Weng JH, Dorville A, McCort I, Ducos B, Durieux C *et al.* (1993). Cholecystokinin peptidomimetics as selective CCK-B antagonists: design, synthesis, and *in vitro* and *in vivo* biochemical properties. *J Med Chem* **36**, 2868–77.

Blommaert AGS, Dhôtel H, Ducos B, Durieux C, Goudreau N, Bado A *et al.* (1997). Structure-based design of new constrained cyclic agonists of the cholecystokinin CCK-B receptor. *J Med Chem* **40**, 647–58.

Bock MG, DiPardo RM, Evans BE, Rittle KE, Whitter WL, Veber DF *et al.* (1989). Benzodiazepine gastrin and brain cholecystokinin receptor ligands: L-365,260. *J Med Chem* **32**, 13–16.

Boden PR, Higginbottom M, Hill DR, Horwell DC, Hughes J, Rees DC *et al.* (1993). Cholecystokinin dipeptoid antagonists: design, synthesis, and anxiolytic profile of some novel CCK-A and CCK-B selective and "mixed" CCK-A/CCK-B antagonists. *J Med Chem* **36**, 552–65.

Bradwejn J and de Montigny C (1984). Benzodiazepines antagonize cholecystokinin-induced activation of rat hippocampal neurons. *Nature* **312**, 363–4.

Bradwejn J, Koszycki D, and Shriqui C (1991). Enhanced sensitivity to cholecystokinin tetrapeptide in panic disorder. *Arch Gen Psychiatry* **48**, 603–10.

Bradwejn J, Koszycki D, Payeur R, Bourin M, and Borthwick (1992). Replication of action of cholecystokinin tetrapeptide in panic disorder: clinical and behavioral findings. *Am J Psychiatry* **149**, 962–4.

Carruthers B, Dawbarn D, De Quidt M, Emson PC, Hunter J, and Reynolds GP (1984). Changes in neuropeptide content of amygdala in schizophrenia. *Br J Pharmacol* **81** (**Suppl**), 190P.

Chang RSL, Lotti VJ, Monaghan RL, Birnbaum J, Stapley EO, Goetz MA (1985). A potent nonpeptide cholecystokinin antagonist selective for peripheral tissues isolated from *Aspergillus alliaceus*. *Science* **230**, 177–9.

Charpentier B, Pélaprat D, Durieux C, Dor A, Reibaud M, Blanchard JC *et al.* (1988*a*). Cyclic cholecystokinin analogues with high selectivity for central receptors. *Proc Natl Acad Sci USA* **85**, 1968–72.

Charpentier B, Durieux C, Pélaprat D, Dor A, Reibaud M, Blanchard JC *et al.* (1988*b*). Enzyme-resistant CCK analogs with high affinities for central receptors. *Peptides* **9**, 835–41.

Charpentier B, Dor A, Roy P, England P, Pham H, Durieux C *et al.* (1989). Synthesis and binding affinities of cyclic and related linear analogues of CCK$_8$ selective for central receptors. *J Med Chem* **31**, 1184–90.

Corringer PJ, Weng JH, Ducos B, Durieux C, Boudeau P, Böhme A *et al.* (1993). CCK-B agonist or antagonist activities of structurally hindered and peptidase-resistant Boc-CCK$_4$ derivatives. *J Med Chem* **36**, 166–72.

Coudoré-Civiale MA, Meen M, Fournié-Zaluski MC, Boucher M, Roques BP, and Eschalier A (2001). Enhancement of the effects of a complete inhibitor of enkephalin-catabolizing enzymes, RB 101, by a cholecystokinin-B receptor antagonist in diabetic rats. *Br J Pharmacol* **133**, 179–85.

Crawley JN and Corwin RL (1994). Biological actions of cholecystokinin. *Peptides* **15**, 731–55.

Dabrowski A, VanderKuur JA, Carter-Su C, and Williams JA (1996). Cholecystokinin stimulates formation of Shc-Grb2 complex in rat pancreatic acinar cells through a protein kinase C-dependent mechanism. *J Biol Chem* **271**, 27 125–9.

Daugé V and Roques BP (1995). Opioid and CCK systems in anxiety and reward. In (eds, J Bradwejn and E Vasar), *Cholecystokinin and Anxiety: From Neuron to Behavior*, R.G. Landes Company, Springer, pp. 151–171.

De Leeuw AS, den Boer JA, Slaap BR, and Westenberg HG (1996). Pentagastrin has panic-inducing properties in obsessive compulsive disorder. *Psychopharmacology* **126**, 339–44.

De Montigny C (1989). Cholecystokinin tetrapeptide induces panic-like attacks in healthy volunteers. *Arch Gen Psychiatry* **46**, 511–17.

De Weerth A, Pisegna JR, Huppil K, and Wank SA (1993). Molecular cloning, functional expression and chromosomal localization of the human cholecystokinin type A receptor. *Biochem Biophys Res Commun* **194**, 811–18.

Derrien M, Durieux M, and Roques BP (1994*a*). Antidepressant-like effects of CCK-B antagonists in mice: antagonism by naltrindole. *Br J Pharmacol* **111**, 956–60.

Derrien M, Daugé V, Blommaert A, and Roques BP (1994*b*). The selective CCK-B agonist, BC 264, impairs socially reinforced memory in the three-panel runway test in rats. *Behav Brain Res* **65**, 139–46.

Derrien M, McCort-Tranchepain I, Ducos B, Roques BP, and Durieux C (1994*c*). Heterogeneity of CCK-B receptors involved in animal models of anxiety. *Pharmacol Biochem Behav* **49**, 133–41.

Dohlman HG, Thorner J, Caron MG, and Lefkowitz RJ (1991). Model systems for the study of seven-transmembrane-segment receptors. *Ann Rev Biochem* **60**, 653–80.

Durieux C, Coppey M, Zajac JM, and Roques BP (1986). Occurrence of two cholecystokinin binding sites in guinea pig brain cortex. *Biochem Biophys Res Commun* **137**, 1167–73.

Durieux C, Ruiz-Gayo M, and Roques BP (1991). *In vivo* binding affinities of cholecystokinin agonists and antagonists determined using the selective CCK-B agonist [^{3}H]pBC 264. *Eur J Pharmacol* **209**, 185–93.

Evans BE, Bock MG, Rittle KE, de Pardo RM, Whitter WL, Veber DF *et al.* (1986). Design of potent, orally effective, nonpeptidal antagonists of the peptide hormone cholecystokinin. *Proc Natl Acad Sci USA* **83**, 4948–22.

Faris PL, Komisaruk BR, Watkins LR, and Mayer DJ (1983). Evidence for the neuropeptide cholecystokinin as an antagonist of opiate analgesia. *Science* **219**, 310–12.

Ferrier IN, Roberts GW, Crow TJ, Johnstone EC, Owens DG, Lee YC *et al.* (1983). Reduced cholecystokinin-like and somatostatin-like immunoreactivity in limbic lobe is associated with negative symptoms in schizophrenia. *Life Sci* **33**, 475–82.

Fournié-Zaluski MC, Belleney J, Lux B, Durieux C, Gérard G *et al.* (1986). Conformational analysis of neuronal cholecystokinin CCK_{26-33} and related fragments by ^1H NMR spectroscopy, fluorescence tranfer measurements and calculations. *Biochem* 25, 3778–87.

Fournié-Zaluski MC, Coric P, Turcaud S, Lucas E, Noble F, Maldonado R *et al.* (1992). Mixed-inhibitor-prodrug as a new approach towards systemically active inhibitors of enkephalin degrading enzymes. *J Med Chem* 35, 2474–81.

Galas MC, Bernard N, and Martinez J (1992). Pharmacological studies on CCK-B receptors in guinea pig synaptoneurosomes. *Eur J Pharmacol* 226, 35–41.

Gaudreau P, St. Pierre S, Pert CB, and Quirion R (1985). Cholecystokinin receptors in mammalian brain: a comparative characterization and visualization. *Ann NY Acad Sci* 448, 198–219.

Geracioti TD, Nicholson WE, Orth DN, Ekhator NN, and Loosen PT (1993). Cholecystokinin in human cerebrospinal fluid: concentrations, dynamics, molecular forms and relationship to fasting and feeding in health, depression and alcoholism. *Brain Res* 629, 260–8.

Gerhardt P, Voits M, Fink H, and Huston JP (1994). Evidence for mnemotropic action of cholecystokinin fragments Boc-CCK-4 and CCK-8S. *Peptides* 15, 689–97.

Goudreau N, Weng JH, and Roques BP (1994). Conformational analysis of CCK-B agonists using ^1H-NMR and restrained molecular dynamics: comparison of biologically active Boc-Trp-(NMe)Nle-Asp-Phe-NH_2 and inactive Boc-Trp-(NMe)Phe-Asp-Phe-NH_2. *Biopolymers* 34, 155–9.

Harhammer R, Schafer U, Henklein P, Ott T, and Repke H (1991). CCK-8-related C-terminal tetra-peptides: affinities for central CCKB and peripheral CCKA receptors. *Eur J Pharmacol* 209, 263–6.

Harro J and Oreland L (1993). Cholecystokinin receptors and memory: a radial maze study. *Pharmacol Biochem Behav* 44, 509–17.

Henke BR, Aquino CJ, Birkemo LS, Croom DK, Dougherty RW, Ervin GN (1997). Optimization of 3-(1H-indazol-3-ylmethyl)-1,5-benzodiazepines as potent, orally active CCK-A agonists. *J Med Chem* 40, 2706–25.

Honda T, Wada E, Battey JF, and Wank SA (1993). Differential gene expression of CCK_A and CCK_B receptors in the rat brain. *Mol Cell Neurosci* 4, 143–54.

Horwell DC, Hughes J, Hunter JC, Pritchard MC, Richardson RS, Roberts E *et al.* (1991). Rationally designed "dipeptoid" analogues of CCKα-Methyltryptophan derivatives as highly selective and orally active gastrin and CCK-B antagonists with potent anxiolytic properties. *J Med Chem* 34, 404–14.

Howbert JJ, Lobb KL, Brown RF, Reel JK, Neel DA, Mason NR *et al.* (1992). A novel series of non-peptide CCK and gastrin antagonists: medicinal chemistry and electrophysiological demonstration of antagonism. In (eds, CT Dourish and SJ Cooper), *Multiple Cholecystokinin Receptors Progress Toward CNS Therapeutic Targets*, Oxford University Press, London, pp. 28–37.

Hull RAD, Shankley NP, Harper EA, Gerskowitch VP, and Black JW (1993). 2-Naphthalenesulphonyl L-aspartyl-(2-phenethyl)amide (2-NAP)- a selective cholecystokinin CCK-A receptor antagonist. *Br J Pharmacol* 108, 734–40.

Innis RB and Snyder SH (1980). Distinct cholecystokinin receptors in brain and pancreas. *Proc Natl Acad Sci USA* 77, 6917–21.

Ito H, Sogabe H, Nakarai T, Sato Y, Tomoi M, Kadowaki M *et al.* (1994). Pharmacological profile of FK-480, a novel cholecystokinin type-A receptor antagonist: comparison to loxiglumide. *J Pharmacol Exp Ther* 268, 571–5.

Ivy AC and Oldberg E (1928). A hormone mechanism for gallbladder contraction and evacuation. *Am J Physiol* 86, 599–613.

Jagerschmidt A, Popovici T, O'Donohue M, and Roques BP (1994). Identification and characterization of various cholecystokinin B receptor mRNA forms in rat brain tissue and partial determination of the cholecystokinin B receptor gene structure. *J Neurochem* 63, 1199–206.

Jagerschmidt A, Guillaume N, Goudreau N, Maigret B, and Roques BP (1995). Mutation of Asp^{100} in the second transmembrane domain of the cholecystokinin B receptor increases antagonist binding and reduces signal transduction. *Mol Pharmacol* 48, 783–9.

Jagerschmidt A, Guillaume-Rousselet N, Vickland ML, Goudreau N, Maigret B, and Roques BP (1996). His^{381} of the rat CCK$_B$ receptor is essential for CCK$_B$ versus CCK$_A$ receptor antagonist selectivity. *Eur J Pharmacol* **296**, 97–106.

Jagerschmidt A, Guillaume N, Roques BP, and Noble F (1998). Binding sites and transduction process of the cholecystokinin B receptor: involvement of highly conserved aromatic residues of the transmembrane domains evidenced by site-directed mutagenesis. *Mol Pharmacol* **53**, 878–85.

Jensen RT, Huang SC, von Schrenck T, Wank SA, and Gardner JD (1990). Cholecystokinin receptor antagonists: ability to distinguish various classes of cholecystokinin receptors. In (eds, JT Thompson, CM Townsend, GA Greely, PL Rayford, CW Wooper, PO Singh and N Rubin), *Gastrointestinal Endocrinology: Receptors and Post-receptor Mechanisms*, Academic, New York, pp. 95–113.

Jensen RT, Qian JM, Lin JT, Mantey SA, Pisegna JR, and Wank SA (1994). Distinguishing multiple CCK receptor subtypes: Studies with guinea pig chief cells and transfected human CCK receptors. *Ann NY Acad Sci* **713**, 88–106.

Josselyn SA and Vaccarino FJ (1996). Acquisition of conditioned reward blocked by intra-accumbens infusion of PD-140548, a CCK-A receptor antagonist. *Pharmacol Biochem Behav* **55**, 439–44.

Kennedy JL, Bradwejn J, Koszycki D, King N, Crowe R, Vincent J *et al.* (1999). Investigation of cholecystokinin system genes in panic disorder. *Mol Psychiatry* **4**, 284–5.

Knapp RJ, Vaughn LK, Fang SN, Bogert CL, Yamamura MS, Hruby VJ *et al.* (1990). A new, highly selective CCK-B receptor radioligand ([^{3}H][N-methyl-Nle^{28,31}]CCK$_{26-33}$): evidence for CCK-B receptor heterogeneity. *J Pharmacol Exp Ther* **255**, 1278–86.

Kolodziej SA, Nikiforovich GV, Skeean R, Lignon MF, Martinez J, and Marshall GR (1995). Ac-[3- and 4-alkylthioproline^{31}]-CCK$_4$ analogs: synthesis, and implication for the CCK-B receptor-bound conformation. *J Med Chem* **38**, 137–49.

Kopin AS, McBride EW, Quinn SM, Kolakowski LF, and Beinborn M (1995). The role of the cholecystokinin-B/gastrin receptor transmembrane domains in determining affinity for subtype-selective ligands. *J Biol Chem* **270**, 5019–23.

Lanaud P, Popovici T, Normand E, Lemoine C, Bloch B, and Roques BP (1989). Distribution of CCK mRNA in particular regions (hippocampus, periaqueductal grey and thalamus) of the rat by in situ hybridization. *Neurosci Lett* **104**, 38–42.

Le Melledo JM, Bradwejn J, Koszycki D, and Bichet D (1995). Premenstrual dysphoric disorder and response to cholecystokinin-tetrapeptide. *Arch Gen Psychiatry* **52**, 605–6.

Lemaire M, Piot O, Roques BP, Böhme AG, and Blanchard JC (1992). Evidence for an endogenous cholecystokininergic balance in social memory. *NeuroReport* **3**, 925–32.

Lo WWY and Hughes J (1988). Differential regulation of cholecystokinin- and muscarinic-receptor-mediated phosphoinositide turnover in flow 900 cells. *Biochem J* **251**, 625–30.

Löfberg C, Agren H, Harro J, and Oreland L (1998). Cholecystokinin in CSF from depressed patients: possible relations to severity of depression and suicidal behaviour. *Eur Neuropsychopharmacol* **8**, 153–7.

Makovec F, Chiste R, Bani M, Pacini MA, Setnikar I, and Rovati LA (1995). New glutaramic acid derivatives with potent competitive and specific cholecystokinin-antagonistic activity. *Arzneimittelforschung* **35**, 1048–51.

Marino CR, Leach SD, Schaefer JF, Miller L, and Gorelick FS (1993). Characterization of cAMP-dependent protein kinase activation by CCK in rat pancreas. *FEBS Lett* **316**.

Martin-Martinez M, Bartolomé-Nebreda JM, Gomez-Monterrey I, Gonzalez-Muniz R, Garcia-Lopez MT, Ballaz S, Barber A *et al.* (1997). Synthesis and stereochemical structure-activity relationships of 1,3-dioxoperhydropyrido[1,2-*c*]pyrimidine derivatives: potent and selective cholecystokinin-A receptor antagonists. *J Med Chem* **40**, 3402–07.

Menozzi D, Gardner JD, and Maton PN (1989). Properties of receptors for gastrin and CCK on gastric smooth muscle cells. *Am J Physiol* **257**, G73–G79.

Mercer LD and Beart PM (1997). Histochemistry in rat brain and spinal cord with an antibody directed at the cholecystokinin A receptor. *Neurosci Lett* **225**, 97–100.

Migaud M, Durieux C, Viereck J, Soroca-Lucas E, Fournié-Zaluski MC, and Roques BP (1996). The *in vivo* metabolism of cholecystokinin (CCK-8) is essentially ensured by aminopeptidase A. *Peptides* **17**, 601–07.

Miller LJ, Holicky E, Ulrich CD, and Wieben ED (1995). Abnormal processing of the human cholecystokinin receptor gene in association with gallstones and obesity. *Gastroenterology* **109**, 1375–80.

Million ME, Léna I, Da Nascimento S, Noble F, Daugé V, Garbay C *et al.* (1997). Development of new potent agonists able to interact with two postulated subsites of the cholecystokinin CCK-B receptor. *Lett Peptide Sci* **4**, 407–10.

Miyake A (1995). A truncated isoform of human CCK-B/gastrin receptor generated by alternative usage of a novel exon. *Biochem Biophys Res Commun* **208**, 230–7.

Moran TH, Robinson PH, Goldrich MS, and McHugh PR (1986). Two brain cholecystokinin receptors: implications for behavioral actions. *Brain Res* **362**, 175–9.

Nadzan AM, Garvey DS, Holladay MW, Shiosaki K, Tufano MD, Shue YK *et al.* (1991). Design of cholecystokinin analogs with high affinity and selectivity for brain receptors. In (eds, JA Smith and JE Rivier), *Peptides, Chemistry and Biology, Proc. 12th American Peptide Symposium*, ESCOM, Leiden, pp. 101–102.

Niehoff DL (1989). Quantitative autoradiographic localization of cholecystokinin receptors in rat and guinea pig brain using ^{125}I-Bolton-Hunter-CCK$_8$. *Peptides* **10**, 265–74.

Nishida A, Miyata K, Tsutsumi R, Yuki H, Akuzawa S, Kobayashi A *et al.* (1994). Pharmacological profile of (R)-1-[2,3-dihydro-1-(2'-methyl-phenacyl)-2-oxo-5-phenyl-1H-1,4-benzodiazepin-3-yl]-3-(3-methylphenyl)urea (YM022), a new potent and selective gastrin/cholecystokinin-B receptor antagonist, *in vitro* and *in vivo*. *J Pharmacol Exp Ther* **269**, 725–31.

Noble F, Derrien M, and Roques BP (1993). Modulation of opioid analgesia by CCK at the supraspinal level: evidence of regulatory mechanisms between CCK and enkephalin systems in the control of pain. *Br J Pharmacol* **109**, 1064–70.

Noble F, Wank S, Crawley J, Bradwejn J, Seroogy K, Hamon M *et al.* (1999). International Union of Pharmacology. XXI. Structure, distribution, and functions of cholecystokinin receptors. *Pharmacol Rev* **51**, 745–81.

O'Dowd B, Hnatowich M, Caron MG, Lefkowitz RJ, and Bouvier M (1988). Site-directed mutagenesis of the cytoplasmatic domains of the human β-2 adrenergic receptor. Localization of regions involved in G-protein-receptor coupling. *J Biol Chem* **264**, 7964–4569.

Ovchinikov YA, Ablulajew NG, and Bogachuck AS (1988). Two adjacent cysteine residues in the C-terminal cytoplasmatic fragment of bovine rhodopsin are palmitoylated. *FEBS Lett* **230**, 1–5.

Ozcelebi F and Miller LJ (1995). Phosphopeptide mapping of cholecystokinin receptors on agonist-stimulated native pancreatic acinar cells. *J Biol Chem* **270**, 3435–41.

Patel S, Smith AJ, Chapman KL, Fletcher AE, Kemp JA, Marshall GR *et al.* (1994). Biological properties of the benzodiazepine amide derivative L-740,093, a cholecystokinin-B/gastrin receptor antagonist with high affinity *in vitro* and high potency *in vivo*. *Mol Pharmacol* **46**, 943–8.

Payeur R, Nixon MK, Bourin M, Bradwejn J, and Legrand JM (1993). The potential role of cholecystokinin in schizophrenia: review and update. *Eur Psychiatry* **8**, 67–78.

Pendley CE, Fitzpatrick LR, Capolino AJ, Davis MA, Esterline NJ, Jakubowska A *et al.* (1995). RP 73870, a gastrin/cholecystokinin-B receptor antagonist with potent anti-ulcer activity in the rat. *J Pharmacol Exp Ther* **273**, 1015–22.

Pélaprat D, Broer Y, Studler JM, Peschanski M, Tassin JP, Glowinski J *et al.* (1987). Autoradiography of CCK receptors in the rat brain using [^{3}H]Boc[Nle$_{28,31}$]CCK$_{27-33}$ and [^{125}I]Bolton-Hunter CCK$_8$. *Neurochem Int* **10**, 495–508.

Piiper A, Stryjek-Kaminska D, Klengel R, and Zeuzem S (1997). CCK, carbachol, and bombesin activate distinct PLC-beta isoenzymes via Gq/11 in rat pancreatic acinar membranes. *Am J Physiol* **272**, G135–G140.

Pommier B, Da Nascimento S, Dumont S, Bellier B, Million E, Garbay C *et al.* (1999). The CCK-B receptor is coupled to two effector pathways through pertussis toxin sensitive and insensitive G proteins. *J Neurochem* **73**, 281–8.

Rehfeld JH and Nielsen FC (1995). Molecular forms and regional distribution of cholecystokinin in the central nervous system. In (eds, J Bradwejn and E Vasar), *Cholecystokinin and Anxiety*, RG landes Company, Austin, pp. 33–56.

Revel L, Ferrari F, Makovec F, Rovati LC, and Impicciatore M (1992). Characterization of antigastrin activity in vivo of CR 2194, a new R-4-benzamido-5-oxo-pentanoic acid derivative. *Eur J Pharmacol* **216**, 217–24.

Roques BP and Noble F (1996). Association of enkephalin catabolism inhibitors and CCK-B antagonists: a potential use in the management of pain and opioid addiction. *Neurochem Res* **21**, 1395–409.

Rose C, Vargas F, Facchinetti P, Bourgeat P, Bambal RB, Bishop PB *et al.* (1996). Characterization and inhibition of a cholecystokinin-inactivating serine peptidase. *Nature (Lond.)* **380**, 403–9.

Ruiz-Gayo M, Daugé V, Menant I, Bégué D, Gacel G, and Roques BP (1985). Synthesis and biological activity of Boc(Nle28, Nle31)CCK27-33 a highly potent CCK8 analogue. *Peptides* **6**, 415–20.

Saito AH, Sankaran H, Goldfine ID, and Williams JA (1980). Cholecystokinin receptors in the brain: characterization and distribution. *Science* **208**, 1155–6.

Sankaran H, Goldfine ID, Deveney CW, Wong KY, and Williams JA (1980). Binding of cholecystokinin to high affinity receptors on isolated rat pancreatic acini. *J Biol Chem* **255**, 1849–53.

Shiosaki K, Graig R, Lin CW, Barrett R, Miller T, Witte D *et al.* (1990). Toward development of peptidomimetics: diketopiperazine templates for the Trp-Met segment of CCK₄. In (eds, JE Rivier and GR Marshall), *Peptides: Chemistry, Structure and Biology, Proc. 11th American Peptide Symposium*, ESCOM, Leiden, pp 978–80.

Shlik J, Aluoja A, Vasar V, Vasar E, Podar T, and Bradwejn J (1997). Effects of citalopram treatment on behavioural, cardiovascular, and neuroendocrine response to cholecystokinin tetrapeptide challenge in patients with panic disorder. *J Psychiatry Neurosci* **22**, 332–40.

Shlik J, Koszycki D, and Bradwejn J (1998). Decrease in short-term memory function induced by CCK-4 in healthy volunteers. *Peptides* **19**, 969–75.

Silvente-Poirot S, Escrieut C, and Wank SA (1998). Role of the extracellular domains of the cholecystokinin receptor in agonist binding. *Mol Pharmacol* **54**, 364–71.

Singh P, Owlia A, Espeijo R, and Dai B (1995). Novel gastrin receptors mediate mitogenic effects of gastrin and processing intermediates of gastrin on Swiss 3T3 fibroblasts. *J Biol Chem* **270**, 8429–38.

Smadja C, Maldonado R, Turcaud S, Fournié-Zaluski MC, and Roques BP (1995). Opposite role of CCK-A and CCK-B receptors in the modulation of endogenous enkephalins antidepressant-like effects. *Psychopharmacology* **128**, 400–8.

Song I, Brown DR, Wiltshire RN, Gantz I, Trent JM, and Yamada T (1993). The human gastrin/cholecystokinin type B receptor gene: alternative splice donor site in exon 4 generates two variant mRNAs. *Proc Natl Acad Sci USA* **90**, 9085–9.

Tachikawa H, Harada S, Kawanishi Y, Okubo T, and Shiraishi H (1999). Novel polymorphism in the promoter and coding regions of the human cholecystokinin B receptor gene: an association analysis with schizophrenia. *Am J Med Genet* **88**, 700–4.

Tachikawa H, Harada S, Kawanishi Y, Okubo T, and Shiraishi H (2000). Novel polymorphisms of the human cholecystokinin A receptor gene: an association analysis with schizophrenia. *Am J Med Genet* **96**, 141–5.

Takinami Y, Yuki H, Nishida A, Akuzawa S, Uchida A, Takemoto Y *et al.* (1997). YF476 is a new potent and selective gastrin/cholecystokinin-B receptor antagonist *in vitro* and *in vivo*. *Aliment Pharmacol Ther* **11**, 113–20.

Talkad VD, Fortune KP, Pollo DA, Shah GN, Wank SA, and Gardner JD (1994). Direct demonstration of three different states of the pancreatic cholecystokinin receptor. *Proc Natl Acad Sci USA* **91**, 1868–72.

Taniguchi T, Matsui T, Ito M, Murayama T, Tsukamoto T, Katakami Y *et al.* (1994). Cholecystokinin-B/gastrin receptor signaling pathway involves tyrosine phosphorylations of p125^{FAK} and p42^{MAP}. *Oncogene* **9**, 861–7.

Trivedi BK, Padia JK, Holmes A, Rose S, Wright DS, Hinton JP *et al.* (1998). Second generation "peptoid" CCK-B receptor antagonists: identification and development of N-(adamantyloxycarbonyl)-α-methyl-(R)-tryptophan derivative (CI-1015) with an improved pharmacokinetic profile. *J Med Chem* **41**, 38–45.

Valverde O, Maldonado R, Fournié-Zaluski MC, and Roques BP (1994). Cholecystokinin B antagonists strongly potentiate antinociception mediated by endogenous enkephalins. *J Pharmacol Exp Ther* **270**, 77–88.

van Vliet IM, Westenberg HG, Slaap BR, den Boer JA, and Ho Pian KL (1997). Anxiogenic effects of pentagastrin in patients with social phobia and healthy controls. *Biol Psychiatry* **42**, 76–8.

Vanderhaeghen JJ, Signeau JC, and Gepts W (1975). New peptide in the vertebrate CNS reacting with antigastrin antibodies. *Nature (Lond.)* **257**, 604–05.

Wang H-L (1997). Basic amino acids at the C-Terminus of the third intracellular loop are required for the activation of phospholipase C by cholecystokinin-B receptors. *J Neurochem* **68**, 1728–35.

Wank SA, Harkins RT, Jensen RT, Shapira H, de Weerth A, and Slattery T (1992). Purification, molecular cloning, and functional expression of the cholecystokinin receptor from rat pancreas. *Proc Natl Acad Sci USA* **89**, 3125–9.

Wank SA (1995). Cholecystokinin receptors. *Am J Physiol* **269**, G628–G646.

Wu V, Yang M, McRoberts JA, Ren J, Seensalu R, Zeng N *et al.* (1997). First intracellular loop of the human cholecystokinin-A receptor is essential for cyclic AMP signaling in transfected HEK-293 cells. *J Biol Chem* **272**, 9037–42.

Zarbin MA, Innis RB, Wamsley JK, Snyder SH, and Kuhar MJ (1983). Autoradiographic localization of cholecystokinin receptors in rodent brain. *J Neurosci* **3**, 877–906.

Zhang LJ, Lu XY, and Han JS (1992). Influences of cholecystokinin octapeptide on phosphoinositide turnover in neonatal-rat brain cells. *Biochem J* **285**, 847–50.

Chapter 15

Dopamine receptors

Lucien Gazi and Philip G. Strange

15.1 Introduction

Dopamine is one of the major neurotransmitters in the central nervous system. Central dopaminergic systems participate in a number of important physiological functions. These systems control voluntary as well as involuntary motor movements, execution of learned motor programs, and regulate the secretion of prolactin and corticotrophin. In addition, central dopaminergic systems mediate reward, enhance mood, and are important for functions like working memory and goal-oriented behaviour. The effects of dopamine tend to be relatively slow so that it may be considered to be a modulator of fast synaptic transmission. The physiological functions of dopamine are mediated by several receptors, which belong to the superfamily of G protein-coupled receptors (GPCRs). Here, we present the main features of these receptors, from their molecular and pharmacological properties to the clinical perspective.

15.2 Molecular characterization

Originally, on the basis of biochemical and pharmacological criteria, Kebabian and Calne (1979) classified dopamine receptors into D_1 and D_2 types. D_1 receptors were defined as those stimulating adenylate cyclase activity while D_2 receptors inhibited adenylate cyclase activity. The application of molecular biology techniques, however, revealed the existence of more than two subtypes of dopamine receptor. Hence, there are currently five dopamine receptor subtypes known. Based on structural and pharmacological properties, these receptors were subdivided into two families termed D_1-like (D_1 and D_5) and D_2-like (D_2, D_3, and D_4) (Hartman and Civelli 1997).

15.2.1 D_1-like receptors

The dopamine D_1 receptor The gene encoding the dopamine D_1 receptor was cloned in the early 1990s by four independent laboratories (Dearry *et al.* 1990; Monsma *et al.* 1990; Sunahara *et al.* 1990; Zhou *et al.* 1990). It was identified in several species, including opossum, goldfish, tilapia, frog, chicken, and *Drosophila* (see Neve and Neve 1997). The mammalian D_1 receptor consists of a 446-amino acid protein with a Mr $\sim$49,000, and strong sequence homology among different species. For example, the human and rhesus D_1 receptors differ by only two amino acid residues (Machida *et al.* 1992).

The rat dopamine D_1 receptor gene is approximately 4 kb in size, consisting of two exons separated by a 115-bp intron (Zhou *et al.* 1992). However, since the intron is in the 5'-untranslated region, the rat dopamine D_1 receptor is virtually intronless. The human dopamine D_1 receptor gene is located on the long arm of chromosome 5, at q35.1 (Grandy *et al.* 1990) and has a similar organization, with very minor differences (Minowa *et al.* 1992). All the mammalian D_1 receptors have several potential sites for N-linked glycosylation and phosphorylation. For the human D_1 receptor, these include Asn-5 at the amino terminus and Asn-175 in the third extracellular domain for glycosylation. Potential phosphorylation by PKA may occur on Thr-136 in the second cytoplasmic loop and Thr-268 at the C-terminal end of the third cytoplasmic loop, while multiple serine and threonine residues found in the long cytoplasmic tail of mammalian D_1 receptors represent potential sites of phosphorylation by G protein-coupled receptor kinases.

The dopamine D_5 receptor　　The first description of the cloning of the human dopamine D_5 receptor gene was published in 1991 (Sunahara *et al.* 1991). The human D_5 receptor was isolated by using sequences derived from the D_1 dopamine receptor (Sunahara *et al.* 1991; Grandy *et al.* 1991) to probe genomic libraries. Hence the D_5 receptor sequence was found to be more closely related to the D_1 receptor than to the D_2-like dopamine receptors. In addition to the human dopamine D_5 receptor, the rat (Tiberi *et al.* 1991), and the chicken (Demchyshyn *et al.* 1995) D_5 receptor genes have also been cloned. The human dopamine D_5 receptor gene was localized to the short arm of chromosome 4 (Polymeropoulos *et al.* 1991) and, like the dopamine D_1 gene, contains no introns within its coding region (Tiberi *et al.* 1991; Sunahara *et al.* 1991).

The human dopamine D_5 receptor is 477-amino acids in length, sharing $\sim$60% amino acid identity with the D_1 receptor, while the rat D_5 receptor has $\sim$83% amino acid sequence homology with the human D_5 receptor. At least two potential sites for N-linked glycosylation (Asn-7 and Asn-199) and two potential sites for phosphorylation by PKA (Thr-153 and Ser-260) are found in the human D_5 receptor.

15.2.2 D_2-like receptors

The dopamine D_2 receptor　　The gene encoding the D_2 subtype was the first dopamine receptor gene to be identified. The rat cDNA was cloned by homology with the hamster β_2 adrenergic receptor (Bunzow *et al.* 1988). Soon after the publication of the rat dopamine D_2 receptor sequence, several studies reported the existence of a longer form of the receptor, generated by alternative splicing of the same gene (Chio *et al.* 1990; Dal Toso *et al.* 1989; Giros *et al.* 1989; Grandy *et al.* 1989; Miller *et al.* 1989; Monsma *et al.* 1989; O'Malley *et al.* 1990; Rao *et al.* 1990; Selbie *et al.* 1989). The two isoforms of the dopamine D_2 receptors were termed D_{2S} (short) and D_{2L} (long). The longer isoform contains a 29 amino acid insert in the third intracellular loop. *In vivo*, D_{2L} is thought to be the predominant form, although some variability in the proportions of the two isoforms occurs among different brain regions (Neve *et al.* 1991; Snyder *et al.* 1991).

The human dopamine D_2 receptor gene has been localized to chromosome 11q22–q23 (Grandy *et al.* 1989) and contains eight exons (Gandelman *et al.* 1992), with alternative splicing occuring at exon 6, yielding mRNAs that differ by 87 nucleotides (generating D_{2S} and D_{2L}). The sequence of that exon is well-conserved across species, indicating an important function for the region (Fryxell 1994).

The rat dopamine D_{2S} and D_{2L} receptors are 415 and 444 residue proteins (Bunzow *et al.* 1988; Monsma *et al.* 1989), with a Mr $\sim$47,000 and 50,900, respectively. The human counterparts contain only 414 and 443 amino acids, respectively, due to the absence of an isoleucine within the C-terminal half of the third intracellular loop of the human receptor. Dopamine D_2 receptors identified in different species (e.g. murine, *Xenopus* and bovine (Chio *et al.* 1990; Montmayeur *et al.* 1991; Martens *et al.* 1991)) show a high degree of sequence homology with the rat and human D_2 receptors.

Like the D_1 receptors, the mammalian D_2 receptors have consensus sites for N-linked glycosylation, including Asn-5, Asn-17, and Asn-23. Several potential sites of phosphorylation by PKA have been identified, such as Ser-147 and 148 in the second cytoplasmic loop, Ser-229, -296, -354, and -364 in the third cytoplasmic loop.

The dopamine D_3 receptor   The cloning of the cDNA encoding the dopamine D_3 receptor was reported in 1990. Sokoloff *et al.* (1990) used a protocol based on sequence homology with the existing dopamine D_2 receptor to identify this new dopamine receptor. The sequence of the receptor encoded by rat D_3 cDNA was found to share an overall 52 per cent amino acid identity with the D_2 receptor, with >70 per cent identity within the transmembrane regions. The cDNA's encoding human receptors have been also cloned and characterized (MacKenzie *et al.* 1994; Pilon *et al.* 1994) and the human dopamine D_3 receptor gene mapped to chromosome 3q13.3 (Le Corniat *et al.* 1991). The dopamine D_3 receptor genes from various species possess several exons and introns in their sequences. The murine and human D_3 receptor genes have seven exons and six introns (Park *et al.* 1995), whereas the rat gene possesses only five introns (Giros *et al.* 1991).

The rat D_3 receptor is a 446 amino acid protein, with $\sim$90 per cent and 97 per cent overall amino acid sequence homology with the human and murine forms of the receptor, respectively. However, the human D_3 receptor is 46 amino acid shorter than the rat D_3 receptor (Giros *et al.* 1991). Two murine isoforms of D_3 receptors resulting from alternative splicing of the gene have been identified. The long form (D_{3L}) corresponds to the rat D_3 receptor (446 amino acids), whereas the murine D_{3S} lacks 21 amino acids compared to rat and human D_3 receptors (Giros *et al.* 1991; Park *et al.* 1995).

A truncated form of dopamine D_3 receptor, which lacks the sixth and the seventh transmembrane segments found in the full-length D_3 receptor was identified and named D_3nf (Schmauss *et al.* 1993; Liu *et al.* 1994; Nimchinsky *et al.* 1997). The function of the D_3nf is not entirely clear, but recent evidence shows that it may act as a dominant-negative regulator of D_3 receptor activity (Karpa *et al.* 2000).

Potential sites for N-linked glycosylation on the dopamine D_3 receptors from different species include the Asn-12, -19, and -97 residues, while a potential site for phosphorylation by PKA, corresponding to the Ser-348 of the rat receptor has been identified. In addition, several further phosphorylation sites within the third cytoplasmic loop have been postulated.

The dopamine D_4 receptor   The cloning of the DNA encoding the dopamine D_4 receptor was first reported by Van Tol *et al.* in early 1991 (Van Tol *et al.* 1991). As with the discovery of other dopamine receptor subtypes, the dopamine D_4 receptor was identified by homology sequence with the dopamine D_2 receptor. However, initial cloning of the full-length cDNA appeared unsuccessful, and this first description of dopamine D_4 receptor was based on a gene/cDNA hybrid. The dopamine D_4 receptor was found to share 41 per cent sequence homology with the D_2 receptor and 39 per cent with the D_3 receptor, with higher percentage of homologies observed within the transmembrane regions. The rat and the murine dopamine D_4 genes were cloned independently of the human receptor (O'Malley et al. 1992; Fishburn et al. 1995). The gene encoding the human dopamine D_4 receptor was found to be located on chromosome 11p15.5 (Gelernter et al. 1992; Petronis et al. 1993). This gene has five exons and is contained within 5 kb of DNA, whereas the rat D_4 gene is a 3.5 kb, with four exons (Van Tol et al. 1991; O'Malley et al. 1992).

Shortly after the description of the D_4 receptor it was found that in some species, the receptor has a striking structural feature, consisting of the presence of a 16-amino acid direct repeat in the third cytoplasmic loop. In humans, up to 19 different repeat units present in 2–10 copies have been described (the 9 repeat variant is not seen). This polymorphism has also been described in nonhuman primates (Livak et al. 1995; Matsumoto et al. 1995), but does not exist in the rat D_4 receptor. The most common allele found in humans has 4 direct repeats ($D_{4.4}$) (Lichter et al. 1993), which is a 449 amino acid residues protein, sharing 73 per cent amino acid identity with the rat D_4 receptor. Potential sites for N-linked glysosylation and phosphorylation by PKA of the human dopamine D_4 receptor are located on Asn-3 and Ser-234, respectively.

15.3 Cellular and subcellular localization

The high degree of homology between the dopamine receptors has complicated the development of selective radioligands and antibodies capable of defining the localization of dopamine receptor subtypes. In particular for the receptors belonging to the same family (D_1- or D_2-like), the radioligands available have poor selectivity. Hence for D_1 and D_5 receptors the two currently known radioligands are [^{3}H]SCH 23390 and [^{125}I]SCH 23982. These ligands can be used for the labelling of both receptors. The D_2-like receptors (D_2, D_3, and D_4) also share common radioligands. These include [^{3}H]nemonapride and [^{3}H]spiperone. However, some radioligands like [^{3}H]raclopride and [^{3}H]PD128907 can distinguish between D_2/D_3 and D_4 receptors.

Several specific antibodies that recognize the different subtypes of dopamine receptors are now available. Antibodies selective for D_1 (MoraFerrer et al. 1997; Luedtke et al. 1999), D_2 (Fishburn et al. 1994; Khan et al. 1998), D_3 (Ariano and Sibley 1994), D_4 (Ariano et al. 1997; Defagot et al. 1997; Lanau et al. 1997) and D_5 (Bergson et al. 1995; Khan et al. 2000) receptors have been described. These antibodies have been primarily used to histochemically localise the different subtypes of dopamine receptors, but also to identify regions of the receptors involved in the interaction of dopamine receptor with G proteins (Neve et al. 1997). Below is a summary of the localization data currently available on dopamine receptor subtypes.

15.3.1 D_1-like receptors

Dopamine D_1 receptors are widely distributed within the brain, with high densities found in the striatum, nucleus accumbens, olfactory tubercle, and frontal cortex (Tiberi et al. 1991). Overall, in the central nervous system (CNS) the D_1 (and D_2) receptors show the highest densities when compared to the other dopamine receptor subtypes.

The D_5 receptor mRNA is most abundant in hippocampus, hypothalamus, and midbrain (Tiberi et al. 1991; Sunahara et al. 1991; Laurier et al. 1994), with lower levels detected in the striatum but at $\sim 1/10$ the abundance of the D_1 receptor mRNA.

15.3.2 D₂-like receptors

The dopamine D_2 receptors are the most abundant of the five dopamine receptor subtypes. They are widely distributed in the brain, the highest density of receptors being localized in the brain regions involved in the control of motor activity such as striatum.

The dopamine D_3 receptor is less abundant and its distribution is much more restricted compared to that of the D_1 and D_2 receptors. Dopamine D_3 receptors are present in greatest abundance in the ventral forebrain, the islands of Calleja, and the nucleus accumbens, whereas lower levels have been detected within the hypothalamus, hippocampus, and substantia nigra (Bouthenet *et al.* 1991).

The dopamine D_4 receptor in humans is found in greatest abundance within the retina. In the brain, high concentrations of the dopamine D_4 receptor have been identified in the hippocampus, the thalamus, the entorhinal cortex, and prefrontal cortex, the hypothalamus, with lower level being found in the striatum (Van Tol *et al.* 1991). Overall, D_4 receptor mRNA is estimated to be one to two orders of magnitude less abundant than that for the D_2 receptor.

In addition to post-synaptic actions, dopamine receptors can function as autoreceptors and therefore control the homeostasis of neurons by regulating cellular functions such as neurotransmitter release, synthesis, and impulse flow. Indeed, dopamine autoreceptors have been located on soma, dendrites, and terminals of most dopaminergic neurons where they fulfill different roles. Thus, stimulation of dopamine autoreceptors in the somatodendritic region slows the firing rate of dopamine neurons, whereas stimulation of autoreceptors located on dopamine nerve terminals inhibits dopamine synthesis and release (see Neve *et al.* 1997). Most evidence suggests that the dopamine autoreceptors are from the D_2-like family. In this respect, data from transgenic mice lacking the D_{2L} receptor (Usiello *et al.* 2000) suggests that the D_{2L} receptors act mainly at post-synaptic sites, whereas D_{2S} serve presynaptic autoreceptor functions. In addition, the D_{2S} receptors appeared to inhibit D_1 receptor-mediated functions.

15.4 Pharmacology

15.4.1 Agonists

The D_1-like receptors show moderate affinities for typical dopamine receptor agonists such as SKF38393, SKF82526, and dihydrexidine. While slight differences in affinities for D_1 and D_5 receptors are exhibited by these compounds, no single compound can be described as a truly subtype-selective agonist. With respect to D_2-like receptors, most typical dopamine agonists exhibit slightly higher affinity for the D_3 subtype. However, again no single compound can be viewed as a subtype selective agonist. Thus, in addition to the compounds listed in Table 15.1 such as PD128907 which exhibits the receptor preference $D_3 \geq D_2 > D_4$ and PD168077 which exhibits greatest receptor preference for D_4 receptors, other D_2-like selective agonists of choice include quinpirole and NO437.

15.4.2 Antagonists

The original definition of the D_1 and D_2 subtypes based on biochemical and pharmacological criteria noted the ability of certain drugs such as sulpiride and metoclopramide to act as antagonists of the D_2 actions of dopamine but not the D_1 actions. Subsequent work lead to the synthesis of selective antagonists SCH 23390 and raclopride for the D_1 and D_2 subtypes, respectively. Following the identification of the different dopamine receptor subtypes using molecular biological techniques, there have been active programmes to synthesize compounds more selective for D_2-like receptor subtypes. Thus, a number of improvements on the prototypical agent raclopride's spectrum of higher affinity for D_2 and D_3 receptors than D_4 receptors have been made. In particular, L741626, SB-277011-A, and L745870 are D_2 selective (~40 fold), D_3 selective (~100 fold), and D_4 selective (~2000 fold) antagonists, respectively (Bowery et al. 1996; Stemp et al. 2000; Patel et al. 1997). The aminotetralins UH232 and AJ76 have been reported to be selective D_2-like autoreceptor antagonists but these compounds also exhibit activity at D_3 receptors where UH232 acts as a partial agonist.

15.5 Signal transduction and receptor modulation

After the molecular cloning of the five subtypes of dopamine receptors, extensive work has been done aiming to determine the structural requirements for their interaction with G proteins. The approaches used involved mutagenesis, use of chimeras and synthetic peptides, as well as anti-receptor antibodies. The results of these studies have suggested an important role of the putative third intracellular loop for all the GPCRs in general in the interaction of the receptor with G proteins. For example, several studies using chimeric receptors have found that the third intracellular loop of α- and β-adrenergic receptors can determine the second messenger pathway to which the receptor is coupled (Kobilka et al. 1988; Strader et al. 1987). In the case of dopamine D_1 and D_2 receptors, peptides derived from different regions of the third intracellular loop have been used as probes for regions of the receptors that interact with G proteins (Konig et al. 1994; Voss et al. 1993; Luttrel et al. 1993; Hawes et al. 1994; Malek et al. 1993). Another approach has examined chimeras between D_1 and D_2 or between D_2 and D_3 receptors. These studies have shown that regions from the second intracellular loop of the receptor are also important for G protein interaction. For example, one group constructed a series of chimeras consisting of macaque D_1 receptor containing increasing amount of rat D_{2S} sequence (Kozell et al. 1994). Chimeras that contain the third intracellular loop of the D_1 receptor stimulate cAMP formation. A chimera containing the third intracellular loop from D_2 receptor, but the second intracellular loop from the D_1 receptor has no effect on cAMP formation, whereas a chimera containing both the second intracellular loop and the third intracellular loop of D_2 receptor was able to inhibit cAMP formation (Kozell et al. 1994). Other studies using D_3/D_2 chimeras showed that the third intracellular loop of the D_2 receptor was not sufficient for the coupling to certain second messenger pathways (McAllister et al. 1993; Robinson et al. 1994, 1996). Finally, approaches using anti-receptor antibodies and site-directed mutagenesis identified some regions internal to the third intracellular loop of D_2 receptor (Plug et al. 1992; Boundy et al. 1993), as well as residues in the transmembrane domains of the D_1 (Pollok et al. 1992) and D_2 (Neve et al. 1991) receptors, important for receptor interaction with G proteins. Overall the interaction of dopamine receptors with G proteins involves different domains of the receptors.

Having identified five dopamine receptor subtypes, one of the main challenges was to identify the second messenger pathways specific to each receptor. Expression of individual dopamine receptor subtypes in recombinant cell lines has enabled the characterization of the second messenger pathways to which the five receptors are coupled.

Table 15.1 Dopamine receptors

	D₁-like		D₂-like		
	D_1	D_5	D_2	D_3	D_4
Alternative names	D_{1A}	D_{1B}	—	—	—
Structural information (Accession no.)	h 446 aa (P21728) r 446 aa (P18901)	h 477 aa (P21918) r 475 aa (P25115)	h 443 aa (P14416)[AS] r 444 aa (P13953)[AS] m 444 aa (X55674)	h 400 aa (P35462)[AS] r 446 aa (P19020) m 446 aa (P30728)	h 387 aa (P21917) r 387 aa (P30729) m 387 aa (P51436)
Chromosomal location	5q35.1	4p15.1–16.1	11q22–23	3q13.3	11p15.5
Selective agonists	SKF38393, SKF81297, dihydrexine		bromocriptine, lisuride	quinelorane, PD128907, pramipexole	PD168077
Selective antagonists	SCH23390, SCH39166, SKF83566		L741626, raclopride	SB277011, S33084, S14297	L745870, U101958, L741742
Radioligands	[³H]-SCH23390 [¹²⁵I]-SCH23982	[³H]-SCH23390 [¹²⁵I]-SCH23982	[³H]-spiperone [³H]-raclopride [³H]-nemonapride	[³H]-spiperone [³H]-raclopride [³H]-nemonapride	[³H]-spiperone [³H]-nemonapride
G protein coupling	G_s/G_q	G_s/G_q	$G_i/G_o/G_z$	G_i/G_o	G_i/G_o
Expression profile	striatum, nucleus accumbens, olfactory tubercle, frontal cortex, hypothalamus, thalamus	hippocampus, thalamus, lateral mamillary nucleus, striatum, cerebral cortex	striatum, nucleus accumbens, olfactory tubercle, cerebral cortex	nucleus accumbens, olfactory tubercle, islands of Calleja, cerebral cortex	frontal cortex, midbrain, amygdala, hippocampus, medulla, retina

Physiological function	control of motor function and cardiovascular function, regulation of immediate early gene expression		control of motor function, behaviour, control of prolactin and MSH secretion, cardiovascular function, regulation of immediate early gene expression		
Knockout phenotype	retarded postnatal development, increased basal locomotor activity, decreased locomotor response to a novel environment, altered cognitive behaviour	increased horizontal and rearing activity, reduced anxiety, superior rotarod performance	retarded postnatal development, postural abnormalities, decreased reward response to morphine, bradykinesia, prolonged immobility disrupted pre-pulse inhibition, absence of haloperidol induced catalepsy	increased locomotor activation in a novel environment	decreased spontaneous locomotor and rearing activity, reduced novelty related exploration, superior rotarod performance
Disease relevance	Parkinson's disease, cognitive decline, neurodegenerative disease	not clear	Parkinson's disease, schizophrenia, neurodegenerative disease, addiction	Parkinson's disease, schizophrenia, neurodegenerative disease	schizophrenia, attention deficit hyperactivity disorder

15.5.1 **The D$_1$-like receptors**

The two members of the D$_1$-like receptor family (D$_1$ and D$_5$) share a high degree of structural homology and pharmacological similarity. In all the recombinant cell lines expressing these subtypes tested to date, the D$_1$ and D$_5$ receptors stimulate cAMP formation in response to an agonist, an effect which involves the activation of the G proteins of the Gs family. The D$_5$ receptor has been shown to activate adenylate cyclase activity independent of agonist (constitutive activation) and the D$_1$-like receptors cloned from *Xenopus laevi* and from chicken also mediate dopamine stimulation of adenylate cyclase (Tiberi *et al.* 1994; Sugamori *et al.* 1994; Liu *et al.* 1992). The D$_1$-like receptors were also found to couple to several other pathways in different cell lines. These include changes in the intracellular Ca^{2+} concentrations, modulation of Na$^+$/H$^+$ exchanger activity, and modulation of K$^+$ currents (Frail *et al.* 1993; Undie *et al.* 1994; Felder *et al.* 1993; Laitinen 1993; Pedersen *et al.* 1994). However, the direct involvement of Gs proteins in these effects is not clear. They may rather involve activation of protein kinases such as protein kinase A or activation of phospholipase C (Frail *et al.* 1993). Some of the results observed on the intracellular Ca^{2+} and phospholipase C activation were controversial. For example, when expressed in HEK 293 cells, the human as well as the goldfish D$_1$ receptor increased intracellular Ca^{2+} (Undie *et al.* 1994). However, in baby hamster kidney cells, neither D$_1$, nor D$_5$ receptor causes any change in intracellular Ca^{2+} levels (Hayes *et al.* 1992). These discrepancies may arise from the difference between the systems in which the receptors were expressed.

15.5.2 **The D$_2$-like receptors**

The D$_2$-like family of dopamine receptors (D$_2$, D$_3$, and D$_4$) is much more complex, when compared to the D$_1$-like receptor family. The three receptors are believed to play an important role in diseases such as Parkinson's disease and schizophrenia; however, it is not entirely clear which of these receptors may be important as a potential target for drugs used in the treatment of these diseases (see section below). Analysing the G protein coupling and second messenger pathways of each of the three receptors may help elucidate the physiological function of these receptors. In this respect, numerous α, β, and γ subunits of G proteins have been identified (Chapter 10; Liu *et al.* 1994; Neer 1995; Gharhremani *et al.* 1999) which may differentially couple specific dopamine receptor subtypes to distinct second messenger pathways. Extensive work has been performed on the D$_2$-like family of receptors to identify the specific G protein (s) to which each dopamine receptor may be coupling. The techniques used include determination of the G protein content of the cell lines in which a receptor was expressed, the use of mutated or chimeric G proteins, 'knock-out' of a particular G protein subunit and '*in vivo* reconstitution' of receptor/G protein interaction. Below we review what is currently known about the signal transduction pathways activated by the D$_2$-like receptors.

The dopamine D$_2$ receptor   The two isoforms of dopamine D$_2$ receptor (D$_{2S}$ and D$_{2L}$) have been shown to inhibit adenylate cyclase activity. This effect is triggered by activation of G proteins of the G$_{i/o}$ family as demonstrated by the sensitivity of the response to pertussistoxin. The D$_{2S}$ seems to cause a greater maximal inhibition of cAMP formation than the D$_{2L}$ (Montmayeur *et al.* 1991, 1993; Felder *et al.* 1991; Hayes *et al.* 1992). These observations were attributed to differential coupling of the two isoforms to Gα G proteins. The D$_{2S}$ appeared to interact more efficiently with Gαi1 and Gαi3 whereas the D$_{2L}$ has a better coupling to Gαi2 over Gαi1 and Gαi3 (Montmayeur *et al.* 1993). In another study using mutant

G proteins, it was found that D_{2S} inhibits cAMP accumulation exclusively via $G\alpha i2$, whereas D_{2L} appeared to affect this pathway through $G\alpha i3$ (Senogles 1994). These two studies are in direct disagreement and the difference between them may reflect the cell lines used in these studies (JEG-3 cells versus GH_4C_1 cells). It could also be possible that the $\beta\gamma$ subunits of G proteins play an important role in the specificity of interaction between the receptor and a given G protein (see Clapham *et al.* 1993), since, for example, the $\beta\gamma$ subunits can trigger the coupling of D_2 receptor to adenylate cyclase type II stimulation in HEK 293 cells (Lustig *et al.* 1993) as well as MAP-kinase activation (Faure *et al.* 1994).

Our laboratory recently used an approach consisting of the reconstitution of D_{2S} and D_{2L} interaction with defined G protein subunits using the baculovirus expression system. We have observed that when the rat or human D_{2L} receptor was reconstituted with $G\alpha i1$, $G\alpha i2$, $G\alpha i3$, or $G\alpha o$ along with $G\beta 1$ and $G\gamma 2$ G protein subunits, the D_{2L} receptor appeared to couple preferentially to $G\alpha o$ G protein. In fact, assessment of the signalling using $[^{35}S]GTP\gamma S$ binding showed higher relative efficacies and potencies for most of the dopamine receptor ligands at D_2-$G\alpha o$ than at D_2-$G\alpha i2$. Interestingly, some compounds like S-(-)-3PPP and p-tyramine could stimulate $[^{35}S]GTP\gamma S$ binding at D_2-Go, but not at D_2-Gi2 (Cordeaux *et al.* 2001; Gazi *et al.* 2001; L Gazi and PG Strange unpublished). Similar observations were made with D_{2S} reconstituted with $G\alpha i2$ and $G\alpha o$ (Nickolls *et al.* 2001). Our results show a clear difference in the coupling of each isoform of D_2 receptor with different subtypes of G proteins, but they do not reveal a clear difference between D_{2S} and D_{2L} regarding their interaction and signalling via the G protein subunits used.

The D_2 receptor can also potentiate the release of arachidonic acid in many cell lines. This effect was found to be PTX-sensitive by some groups (Kanterman *et al.* 1991; Piomelli *et al.* 1991), but not by others (Liu *et al.* 1992). The D_2 receptor was also found to either increase or decrease the intracellular Ca^{2+} levels (Vallar *et al.* 1990; Nussinovitch *et al.* 1992; Liu *et al.* 1994; Seabrook *et al.* 1994). These observations show that the same receptor may produce an opposite effect on the same pathway. One possible explanation is a difference in the cell line used for expression, related to factors such as receptor and/or G protein expression levels. The dopamine D_2 receptor can also modulate Ca^{2+} currents and voltage-activated K^+ currents through activation of $G\alpha o$ and $G\alpha i3$, respectively (Lledo *et al.* 1992). Furthermore, modulation of K^+ currents has been shown to be both PTX-sensitive or insensitive (Einhorn *et al.* 1991; Baertschi *et al.* 1992; Lledo *et al.* 1992; Memo *et al.* 1992; Castellano *et al.* 1993) suggesting further complexities in the coupling of the receptor to the effector. Inhibition of dopamine release (Nussinovitch *et al.* 1992), as well as modulation of the activity of Na^+/H^+ exchanger by dopamine D_2 receptors was demonstrated in several systems (Ganz *et al.* 1990; Neve *et al.* 1992; Chio *et al.* 1994; Tang *et al.* 1994). Similar to its effect on intracellular Ca^{2+} levels, the dopamine D_2 receptor was found to either inhibit (Florio *et al.* 1992; Senogles 1994) or stimulate (Lajiness *et al.* 1993) cell growth.

The dopamine D_3 receptor   Despite the high degree of homology between the dopamine D_3 and D_2 receptors, the coupling of the dopamine D_3 receptor to activation of second messenger pathways was shown to be different from those observed for the D_2 receptors. In fact, demonstration of the coupling of D_3 receptor to second messenger pathways has been the most problematic among all the different subtypes of dopamine receptors. Early work on the rat D_3 receptor showed that the binding of dopamine to that receptor in CHO cells was unaffected by GppNHp (Sokoloff *et al.* 1990). Similar results were later observed with GTP in other expression system (Boundy *et al.* 1993; Freedman *et al.* 1994). These findings suggested that the D_3 receptor did not couple to G proteins or that the binding characteristics of agonists to D_3 receptor were different from those of D_2 or D_4 receptors. However, several studies have subsequently reported a reduction of agonist affinity at D_3 receptors by guanine nucleotides in different cell lines (MacKenzie *et al.* 1994; Castro *et al.* 1993; Seabrook *et al.* 1992; Pilon *et al.* 1994; McAllister *et al.* 1995). Indeed, the D_3 receptor has been shown to inhibit adenylate cyclase activity in several cell lines (Potenza *et al.* 1994; Tang *et al.* 1994; Hall *et al.* 1999), but not in others (Pilon *et al.* 1994; McAllister *et al.* 1995; Seabrook *et al.* 1994). However, in all cases, the magnitude of the inhibition of cAMP accumulation by D_3 receptor was less than that caused by D_2 receptors. Freedman *et al.* (Freedman *et al.* 1994) found a small increase in arachidonic acid release in CHO cells expressing the D_3 receptor, but others failed to show an effect of D_3 receptors on arachidonic acid release (Pilon *et al.* 1994; Seabrook *et al.* 1994). The D_3 receptor was able to inhibit high threshold Ca^{2+} currents in differentiated NG 108-15 cells (Seabrook *et al.* 1994) in a PTX-sensitive fashion, an effect absent in GH_4C_1 cells. In MN9D cells the D_3 receptor inhibited dopamine release, apparently by a mechanism involving potassium channel activation (Tang *et al.* 1994). The D_3 receptor was also reported to modulate the activity of Na^+/H^+ exchanger (Coldwell *et al.* 1999) and stimulate [^3H]thymidine uptake (Pilon *et al.* 1994; McAllister *et al.* 1995).

The dopamine D_4 receptor The D_4 receptor can couple to most of the second messenger pathways modulated by the D_2 receptor. Hence, the D_4 receptor can inhibit the activity of adenylate cyclase in different cell lines (Chio *et al.* 1994; Gazi *et al.* 1998, 2000), an effect which is PTX-sensitive. Kazmi *et al.* (2000) have extended these studies using an *in vitro* reconstitution approach to study possible differences in the interaction of the isoforms of the D_4 receptor with G proteins. Surprisingly, no differences could be observed between the isoforms of D_4 receptor regarding their interaction with G proteins. As such, the exact role of the repeat sequences found in the third intracellular loop of D_4 receptors remains mysterious. This aside, the D_4 receptor can also potentiate arachidonic acid release (Castro *et al.* 1993) and modulate the activity of Na^+/H^+ exchanger in many systems (Chio *et al.* 1994; Coldwell *et al.* 1999; Gazi *et al.* 2000). The D_4 receptor was also reported to inhibit an L-type calcium current as well as potassium current (Mei *et al.* 1995; Wilke *et al.* 1998), and modulate the activity of N-acetyl-transferase, an enzyme involved in the biosynthesis of melatonin (Zawilska *et al.* 1997). Thus, it seems that the dopamine D_4 receptor coupling to second messenger pathways is more comparable to that of the dopamine D_2 receptor than that of the dopamine D_3 receptor.

15.6 Physiology and disease relevance

Dopamine was considered for a long time to be a precursor of noradrenaline, but eventually it was recognized as a neurotransmitter in its own right, fulfilling specific physiological roles including the control of different aspects of voluntary and involuntary motor movements and execution of learned motor programs. The central dopaminergic system also regulates the secretion of prolactin and corticotrophin, and is involved in the mediation of reward, control of mood, and working memory as well as goal-oriented behaviour. It is believed that several physiological effects of dopamine are linked to specific brain regions. Thus, the striatum which possesses the highest density of dopaminergic terminals is thought to control motor programs essential for the execution of complex motor acts. This concept is being continually redefined since current evidence suggests that it is the dorsolateral part of the striatum that is important for these motor functions, whereas the ventral or limbic part of the striatum, including nucleus accumbens and olfactory tubercle, are the sites involved in the mediation of rewarding and motivational process related to emotions and goal-directed behaviour (Di Chiara *et al.* 1992). The dopaminergic neurons involved in the regulation of neuroendocrine functions (i.e. prolactin secretion) are mainly localized in the hypothalamus. Despite this segregation of physiological functions by brain region, it is generally difficult to ascribe specific physiological functions to the different dopamine receptor subtypes. This is because in most brain regions examined there are mixtures of subtypes. Also drugs with high selectivity for the different subtypes do not yet exist. In the motor parts of the striatum the predominant receptor subtypes are D_1 and D_2 so control of movement by dopamine may depend on both of these receptor subtypes. The D_3 and D_4 subtypes are found mainly in limbic and cortical brain regions so are probably involved in functions of dopamine such as reward and control of mood and this has made them important targets for drug design. The anterior pituitary contains only the D_2 subtype so the control of prolactin release by dopamine is mediated via this subtype.

The dopaminergic system has been the focus of much research over the past 40 years because several pathological conditions such as Parkinson's disease, schizophrenia, bipolar disorder, manic depression, Tourette's syndrome, and hyperprolactinaemia are believed to be associated with either dopaminergic system dysfunction or side effect profiles of drugs used to treat these disorders. Extensive work has also been performed to analyse the genetic relationship between dopamine receptors and the diseases mentioned above (reviewed in Missale *et al.* 1998). However, the results that have been produced are equivocal. Thus no linkage of D_2 and D_4 receptors to bipolar disorder has been found (Missale *et al.* 1998) and the association between the dopamine D_2 receptor gene and susceptibility to alcoholism is controversial (Missale *et al.* 1998). More recently, some evidence has been reported for the genetic association of dopamine D_4 receptor polymorphisms with behavioural disorders such as attention deficit hyperactivity disorder (ADHD) (see review in Todd and O'Malley 2001).

15.6.1 Schizophrenia

Dopamine receptor antagonists have been used over several decades to alleviate hallucinations and delusions that occur in schizophrenia. The relationship between schizophrenia and dopaminergic systems was postulated in the early 1960s, with further confirmation coming from findings in the 1970s that there was a strong correlation between the affinity of antipsychotic drugs for the pharmacologically defined dopamine D_2 receptors and their clinically effective concentrations (for review see Strange 2001). The major drawback of most of these drugs, however, was the induction of severe extrapyramidal symptoms. In addition, the majority of these drugs were effective against the positive symptoms of schizophrenia but did not have any effect on the negative symptoms of the disease (Strange 2001).

After the cloning of the five dopamine receptor subtypes, it was apparent that the antipsychotic drugs could have been acting at D_2, D_3, or D_4 receptors and this opened up the possibility of developing more effective antipsychotic agents with reduced side effects.

To establish the roles of each subtype, knockout mice lacking dopamine receptors of all D_2-like family members have been generated and extensively characterized (Ralph *et al.* 1999; Boulay *et al.* 2000; Rubinstein *et al.* 1997; Dulawa *et al.* 1999). Notably, mice lacking the D_2 receptor but not the D_3 or the D_4 receptor show disruptions of prepulse inhibition in response to amphetamine administration (Ralph *et al.* 1999). This paradigm is characteristic of schizophrenia and the latter results suggest that the model reflects only a D_2 receptor involvement and not a D_3 or D_4 receptor. Interestingly, a dopamine D_4 receptor selective ligand was found to be ineffective on most of the available animal models of schizophrenia although this may be because these models are 'specific' to the dopamine D_2 receptors. More recently, Usiello *et al.* (2000) reported a selective knock-out of the dopamine D_{2L} receptor isoform. The corresponding mice did not exhibit catalepsy induced by the typical neurolpetic haloperidol, suggesting that drugs such as this may act selectively on specific isoforms of dopamine receptor. Although interesting, these data need more work to confirm the selective targeting of a subtype of receptor by a given compound. Additionally, the study reported evidence that suggested that D_{2S} and D_{2L} were located presynaptically and postsynaptically, respectively.

Complementing these studies extensive efforts have been undertaken to find a possible genetic linkage of the five dopamine receptors to the aetiology of schizophrenia. They all systematically excluded any association in many populations, including Japanese, Swedish, Italian, Irish Californian, and Amish (Missale *et al.* 1998). However, in a report by Crocq *et al.* (1992) on two French and British populations, a small but significant risk of schizophrenia associated with homozygosity at D_3 receptor was found. These findings were later confirmed by a study conducted on a large population from different European centres (Spurlock *et al.* 1998). Other studies on D_4 receptors have excluded a clear linkage between the different alleles of the D_4 receptor and the risk of schizophrenia (Missale *et al.* 1998). There have been several papers in the literature claiming that the D_4 receptors may be increased in the brain of schizophrenic patients. However, these results have not been reproduced by other groups (Strange 2001).

These findings aside there has been great interest in D_3 and D_4 receptors as potential targets for the development of novel antipsychotics. This was partly because these subtypes were found largely in limbic/cortical regions of brain with low levels in striatal regions. Hence a D_3 or D_4 selective drug might be antipsychotic without the motor side effects, the latter being thought to be elicited via actions at striatal D_2 receptors. For the D_4 receptor, interest was raised by the first study describing its cloning which reported a 15-fold higher affinity of clozapine (an atypical antipsychotic that produces limited extrapyramidal side effects and causes some improvement of the negative symptoms of schizophrenia) for D_4 over D_2 receptor (Van Tol *et al.* 1991) although this difference in affinity is not maintained in more controlled studies. Most antipsychotics have high affinities for the D_3 receptor, but some clinically effective antipsychotics have a low affinity for the D_4 receptor so that D_4 receptor occupancy may not be mandatory for antipsychotic action. Selective ligands have been developed for dopamine D_3 and D_4 receptors, with many of them being used in animal models of behavioural disorders. In the case of dopamine D_4 receptors, one ligand (L-745,870) was tested in schizophrenic patients but was found to be devoid of antipsychotic properties (Kramer *et al.* 1997). Some other groups have developed ligands with high affinity for D_4 receptor and 5-HT_{2A} receptor, but these were also not effective in clinical trials (Truffinet *et al.* 1999). These studies reinforce the idea that actions at D_4 receptors are not a prerequisite for antipsychotic effect. At present, therefore, it seems that antipsychotic effect may involve actions at D_2 or D_3 receptors and we await the results of clinical trials with selective compounds for these two receptors to define their relative importance.

15.6.2 **Parkinson's disease**

A number of dopamine agonists have been shown to be effective monotherapies in the early stages of Parkinson's disease and also as adjuncts to levodopa treatment in advanced PD. These include pergolide, pramipexole, bromocriptine, and ropinirole, which have high affinity for both D_2 and D_3 receptors (Tan and Jankovic 2001). Selective D_1 agonists have shown convincing antiparkinsonian efficacy in experimental models of PD, including CY 208-43, which produces mild clinical benefit in PD patients. D_1 agonists may also be useful against L-DOPA-induced dyskinesia as demonstrated in MPTP marmosets, and at low doses can produce synergistic responses with D_2/D_3 agonists (Pearce 1996). Recent evidence suggests that dopamine agonists may also be neuroprotective, although the exact mechanisms involved remain controversial, due partly to the fact that most compounds studied are mixed D_2/D_3 agonists. Several reports suggest that these neuroprotective effects are independent of dopamine receptor activation and may be due to antioxidant activity via the direct scavenging of free radicals or increased activity of radical-scavenging enzymes (Le and Jankovic 2001). However, in contrast, there are some instances where the neuroprotective effects of dopamine agonists have been blocked by selective dopamine receptor antagonists. Further studies are required to investigate potential neuroprotective mechanisms of dopamine agonists, especially given the emerging clinical data showing a potential slowing of the rate of disease progression with such agents.

15.7 **Concluding remarks**

Despite the huge amount of work done on dopaminergic systems, and in particular the progress made during the last decade, there is still much to be learnt about dopamine receptors. Additional dopamine receptors may yet be identified and this may shed further light on the ambiguous results found with some dopamine receptor ligands. One of the most interesting features of dopamine receptors, as well as all other GPCRs, is the recently discovered phenomenon of oligomerization of these receptors. Previously, it had been assumed that these receptors functioned as monomers, but evidence for oligomerization of dopamine receptors has now appeared in several systems. Indeed, our laboratory has recently reported the dimerization of dopamine D_2 receptor in CHO cells (Armstrong *et al.* 2001), an observation which confirms previous finding from other groups (Lee *et al.* 2000; Zawarynski *et al.* 1998). Evidence of heterodimerization between dopamine D_5 receptors and $GABA_A$ receptors (Liu *et al.* 2000) or between dopamine D_2 receptors and somatostatin (sst5) receptors (Rocheville *et al.* 2000) has also been forthcoming. In respect to the former interaction direct binding of the D_5 carboxy terminal domain with the second intracellular loop of the $GABA_A$ γ_{2S} receptor subunit enables mutual inhibitory functional interactions between the two receptor systems. Thus the D_5 receptor can dynamically regulate inhibitory synaptic strength through an interaction independent of classical G protein signal transduction cascades. In the case of the sst5/D_2 receptor interaction agonists of both receptors promote receptor heterooligomerization to form a receptor complex that possesses enhanced functional activity when it is occupied by both agonists. The issue now will be to understand the exact function of these homo- and hetero-oligomers and their possible role(s) in the physiological and pathophysiological conditions. From a strictly mechanistic point of view, it will be of great interest to analyse how different G protein subtypes can affect receptor oligomerization.

Acknowledgements

Work in the authors' lab was supported by the BBSRC and the Wellcome Trust.

References

Ariano MA and Sibley DR (1994). Dopamine receptor distribution in the CNS: elucidation using anti-peptide antisera directed against D_{1A} and D_3 subtypes. Brain Res 649, 95–110.

Ariano MA, Wang J, Noblett KL, Larson ER, and Sibley DR (1997). Cellular distribution of the rat D4 dopamine receptor protein in the CNS using anti-receptor antisera. Brain Res 752, 26–34.

Armstrong D and Strange PG (2001). Dopamine D2 receptor dimer formation: evidence from ligand binding. J Biol Chem 276, 22 621–9.

Baertschi AJ, Audigier Y, Lledo P-M, Israel J-M, Bockaert J, and Vincent J-D (1992). Dialysis of lactotropes with antisense oligonucleotides assigns guanine nucleotide binding protein subtypes to their channel effectors. Mol Endocrinol 6, 2257–65.

Bergson C, Mrzljak L, Lidow MS, Goldman-Rakic PS, and Levenson R (1995). Characterization of subtype-specific antibodies to the human D5 dopamine receptor: studies in primate brain and transfected mammalian cells. Proc Natl Acad Sci USA 92, 3468–72.

Boulay D, Depoortere R, Oblin A, Sanger DJ, Schoemacher H, and Perrault G (2000). Haloperidol-induced catalepsy is absent in dopamine D(2), but maintained in dopamine D(3) receptor knock-out mice. Eur J Pharmacol 391, 63–73.

Boundy VA, Luedtke RR, Artymyshyn RP, Filtz TM, and Molinoff PB (1993). Development of poly-clonal anti-D_2 dopamine receptor antibodies using sequence-specific peptides. Mol Pharmacol 43, 666–76.

Boundy VA, Luedtke RR, Gallitano AL, Smith JE, Filtz TM, Kallen RG, and Molinoff PB (1993). Expression and characterization of the rat D_3 dopamine receptor: pharmacological properties and development of antibodies. J Pharmacol Exp Ther 264, 1002–11.

Bouthenet M-L, Souil E, Martres MP, Sokoloff P, Giros B, and Schwartz J-C (1991). Localisation of dopamine D_3 receptor mRNA in the rat brain using in situ hybridization hystochemistry: comparison with dopamine D_2 receptor mRNA. Brain Res 564, 203–19.

Bowery BJ, Razzaque Z, Emms F, Patel S, Freedman S, Bristow L et al. (1996). Antagonism of the effects of (+)-PD 128907 on midbrain dopamine neurones in rat brain slices by a selective D2 receptor antagonist L-714,626. Br J Pharmacol 119, 1491–7.

Bunzow JR, Van Tol HHM, Grandy DK, Albert P, Salon J, Christie M et al. (1988). Cloning and expression of a rat D_2 dopamine receptor. Nature 336, 783–7.

Castellano MA, Liu L-X, Monsma FJ, Sibley DR, Kapatos G, and Chiodo LA (1993). Transfected D_2 short dopamine receptors inhibit voltage-dependent potassium current in neuroblastoma x glioma hybrid (NG108-15) cells. Mol Pharmacol 44, 649–56.

Castro SW and Strange PG (1993). Coupling of D_2 and D_3 dopamine receptors to G-proteins. FEBS Lett 315, 223–6.

Chio CL, Hess GF, Graham RS, and Huff RM (1990). A second molecular form of D_2 dopamine receptor in rat brain and bovine caudate nucleus. Nature 343, 266–9.

Chio CL, Drong RF, Riley DT, Gill GS, Slightom JL, and Huff RM (1994). D_4 receptor-mediated signaling events determined in transfeceted chinese hamster ovary cells. J Biol Chem 269, 11 813–19.

Chio CL, Lajiness ME, and Huff RM (1994). Activation of heterologously expressed D_3 dopamine receptors: comparison with D_2 dopamine receptors. Mol Pharmacol 45, 51–60.

Clapham DE and Neer EJ (1993). New roles for G-protein $\beta\gamma$-dimers in transmembrane signalling. Nature 365, 403–6.

Coldwell MC, Boyfield I, Brown AM, Stemp G, and Middlemiss DN (1999). Pharmacological char-acterization of extracellular acidification rate responses in human $D_{2(long)}$, D_3 and $D_{4.4}$ receptors expressed in Chinese hamster ovary cells. Br J Pharmacol 127, 1135–44.

Cordeaux Y, Nickolls SA, Flood LA, Graber SG, and Strange PG (2001). Agonist regulation of D_2 dopamine receptor/G protein interaction: evidence for agonist selection of G protein subtype. *J Biol Chem* **276** 28 667–75.

Crocq MA, Mant R, Asherson J, Williams J, Hode Y, Mayerova A *et al.* (1992). Association between schizophrenia and homozygosity at the dopamine D_3 receptor gene. *J Med Genet* **29**, 858–60.

Dal Toso R, Sommer B, Ewert M, Herb A, Pritchett DB, Bach A *et al.* (1989). The dopamine D_2 receptor: Two molecular forms generated by alternative splicing. *EMBO J* **8**, 4025–34.

Dearry A, Gringrich JA, Falardeau P, Fremeau RTJr, Bates MD, and Caron MG (1990). Molecular cloning and expression of the gene for a human D1 dopamine receptor. *Biochem* **29**, 2335–42.

Defagot MC, Malchiodi EL, Villar MJ, and Antonelli MC (1997). Distribution of D4 dopamine receptor in rat brain with sequence-specific antibodies. *Brain Res Mol Brain Res* **45**, 1–12.

Demchyshyn LL, Sugamori KS, Lee FJS, Hamadanizadeh SA, and Niznik HB (1995). The dopamine D1D receptor. Cloning and characterization of three pharmacologically distinct D1-like receptor from *Gallus domesticus J Biol Chem* **270**, 4005–12.

Di Chiara G, Morelli M, Aquas E, and Carboni E (1992). Funtions of dopamine in the extrapyramidal and limbic systems. Clues for the mechanism of drug action. *Arzneimittelforschung* **42**, 231–7.

Dulawa SC, Grandy DK, Low MJ, Paupus MP, and Geyer MA (1999). Dopamine D4 receptor-knock-out mice exhibit reduced exploration of novel stimuli. *J Neurosci* **19**, 9550–6.

Einhorn LC, Gregerson KA, and Oxford GS (1991). D_2 dopamine receptor activation of potassium channels in identified rat lactotrophs: whole-cell and single-channel recording. *J Neurosci* **11**, 3727–37.

Faure M, Voyno-Yasenetskaya TA, and Bourne HR (1994). cAMP and $\beta\gamma$ subunits of heterotrimeric G proteins stimulate the mitogen-activated protein kinase pathway in COS-7 cells *J Biol Chem* **269**, 7851–4.

Felder CC, Albrecht FE, Campbell T, Eisner GM, and Jose PA (1993). cAMP-dependent, G protein-linked inhibition of Na^+/H^+ exchange in renal brush border by D_1 dopamine agonists. *Am J Physiol* **264**, F1032–7.

Felder CC, Williams HL, and Axelrod J (1991). A transduction pathway associated with receptors coupled to the inhibitory guanine nucleotide binding protein G_i that amplifies ATP-mediated arachidonic acid release. *Proc Natl Acad Sci USA* **88**, 6477–80.

Fishburn CS, Carmon S, and Fuchs S (1995). Molecular cloning and characterisation of the gene encoding the murine D_4 dopamine receptor. *FEBS Lett* **361**, 215–19.

Fishburn CS, David C, Carmon S, Wein C, and Fuchs S (1994). *In vitro* translation of D2 dopamine receptors and their chimaeras: analysis of subtype-specific antibodies. *Biochem Biophys Res Commun* **205**, 1460–6.

Florio T, Pan M-G, Newman B, Hershberger RE, Civelli O, and Stork PJS (1992). Dopaminergic inhibition of DNA synthesis in pituitary tumor cells is associated with phosphotyrosine phosphatase activity. *J Biol Chem* **267**, 24 169–72.

Frail DE, Manelli AM, Witte DG, Lin CW, Steffey ME, and MacKenzie RG (1993). Cloning and characterization of a truncated dopamine D_1 receptor from goldfish retina: stimulation of cyclic AMP production and calcium mobilization. *Mol Pharmacol* **44**, 1113–18.

Freedman SB, Patel S, Marwood R, Emms F, Seabrook GR, Knowles MR *et al.* (1994). Expression and pharmacological characterization of the human D_3 dopamine receptor. *J Pharmacol Exp Ther* **268**, 417–26.

Fryxell KJ (1994). The evolution of the dopamine receptor family. In (ed. HB Niznik), *Dopamine Receptor and Transporters* Dekker, New York, pp. 237–60.

Gandelman K-Y, Harmon S, Todd RD, and O'Malley KL (1992). Analysis of the structure and expression of the human dopamine D2A receptor gene. *J Neurochem* **56**, 1024–9.

Ganz MB, Pachter JA, and Barber DL (1990). Multiple receptor coupled to adenylate cyclase regulate Na-H exchange independent of cAMP. *J Biol Chem* **265**, 8989–92.

Gazi L, Bobirnac I, Danzeisen M, Schupbach E, Bruinvels AT, Geisse S *et al.* (1998). The agonist activities of the putative antipsychotic agents, L-745,870 and U-101958 in HEK293 cells expressing the human dopamine $D_{4.4}$ receptor. *Br J Pharmacol* **124**, 889–96.

Gazi L, Nickolls SA, and Strange PG (2001). Human dopamine $D_{2L(long)}$ receptor coupling to $G_{\alpha i2}$ and $G_{\alpha o}$ G proteins in Sf9 insect cells. *Abstract at British Pharmacological Society Dublin Meeting* P106.

Gazi L, Schoeffter P, Nunn C, Croskery K, Hoyer D, and Feuerbach D (2000). Cloning, expression, functional coupling and pharmacological characterization of the rat dopamine D_4 receptor. *Naunyn Schmiedeberg's Arch Pharmacol* **361**, 555–64.

Gelernter J, Kennedy JL, Van Tol HHM, Civelli O, and Kidd KK (1992). The D4 dopamine receptor (DRD4) maps to distal 11p close to HRAS. *Genomics* **13**, 208–10.

Gharhremani MH, Cheng PH, Lembo PMC, and Albert PR (1999). Distinct roles for G alpha(i)2, Galpha(i)3, and G beta gamma in modulation of forskolin- or G(s)-mediated cAMP accumulation and calcium mobilization by dopamine D2S receptors. *J Biol Chem* **274**, 9238–45.

Giros B, Martres MP, Pilon C, Sokoloff P, and Schwartz J-C (1991). Shorter variant of the dopamine D3 receptor produced through various patterns of alternative splicing. *Biochem Biophys Res Commun* **176**, 1584–92.

Giros B, Sokoloff P, Martres M-P, Riou J-F, Emorine LJ, and Schwartz J-C (1989). Alternative splicing directs the expression of two D_2 dopamine receptor isoforms. *Nature* **342**, 923–6.

Grandy DK, Litt M, Allen L, Bunzow JR, Machioni M, Makam H *et al.* (1989). The human dopamine D2 receptor is located on chromosome 11 at q22-q23 and identifies a TaqI RFLP. *Am J Hum Genet* **45**, 778–85.

Grandy DK, Machioni MA, Makam H, Stofko RE, Alfano M, Frothingham L *et al.* (1989). Cloning of the cDNA and a gene for a human D_2 dopamine receptor. *Proc Natl Acad Sci USA* **86**, 9762–6.

Grandy DK, Zhang Y, Bouvier C, Zhou Q-Y, Johnson RA, Allen L *et al.* (1991). Multiple human D_5 receptor genes: a functional receptor and two pseudogenes. *Proc Natl Acad Sci USA* **88**, 9175–9.

Grandy DK, Zhou Q-Y, Allen L, Litt R, Magenis RE, Civelli O *et al.* (1990). A human dopamine D1 receptor gene is located on chromosome 5 at q35.1 and identifies an ExoRI RFLP. *Am J Hum Genet* **47**, 828–34.

Hall DA and Strange PG (1999). Comparison of the ability of dopamine receptor agonists to inhibit forskolin-stimulated adenosine $3'5'$-cyclic monophosphate (cAMP) accumulation via D2L (long isoform) and D3 receptor expressed in Chinese hamster ovary (CHO) cells. *Biochem Pharmacol* **58**, 285–9.

Hartman DS and Civelli O (1997). Dopamine receptor diversity: molecular and pharmacological perspectives. In (ed. E Jucker) *Progress in Drug Research*, Vol. 4, Basel: Birkhäuser, 173–94.

Hawes BE, Luttrel LM, Exum ST, and Lefkowitz RJ (1994). Inhibition of G protein-coupled receptor signaling by expression of cytoplasmic domains of the receptor. *J Biol Chem* **269**, 15 776–85.

Hayes G, Biden TJ, Selbie LA, and Shine J (1992). Structural subtypes of the dopamine D_2 receptor are functionally distinct: expression of the cloned D_{2A} and D_{2B} subtypes in a heterologous subtypes. *Mol Endocrinol* **6**, 920–6.

Kanterman RY, Mahan LC, Briley EM, Monsma FJ, Sibley DR, Axelrod J *et al.* (1991). Transfected D_2 dopamine receptors mediate the potentiation of arachidonic acid release in Chinese hamster ovary cells. *Mol Pharmacol* **39**, 364–9.

Karpa KD, Lin R, Kabbani N, and Levenson R (2000). The dopamine D3 receptor interacts with itself and the truncated D3 splice variant D3nf: D3-D3nf interaction causes mislocalization of D3 receptors. *Mol Pharmacol* **58**, 677–83.

Kazmi MA, Snyder LA, Cypess AM, Graber SG, and Sakmar TP (2000). Selective reconstitution of human D4 dopamine receptor variants with Gi alpha subtypes. *Biochemistry* **39**, 3734–44.

Kebabian JW and Calne DB (1979). Multiple receptors for dopamine. *Nature* **277**, 93–6.

Khan ZU, Gutierrez A, Martin R, Penafiel A, Rivera A, and De La Calle A (1998). Differential regional and cellular distribution of dopamine D2-like receptor: immunocytochemical study of subtype-specific antibodies in rat and human. *J Comp Neurol* **402**, 353–71.

Khan ZU, Gutierrez A, Martin R, Penafiel A, Rivera A, and De La Calle A (2000). Dopamine D5 receptor of rat and human brain. *Neuroscience* **100**, 689–99.

Kobilka BK, Kobilka TS, Daniel K, Regan JW, Caron MG, and Lefkowitz RJ (1988). Chimeric α_2-, β_2-adrenergic receptor: delineation of domains involved in effector coupling and ligand binding specificity. *Science* **240**, 1310–16.

Konig B and Gratzel M (1994). Site of dopamine D_1 receptor binding to G_s protein mapped with synthetic peptides. *Biochim Biophys Acta* **1223**, 261–6.

Kozell LB, Machida CA, Neve RL, and Neve KA (1994). Chimeric D_1/D_2 dopamine receptors. Distinct determinant of selective efficacy, potency, and signal transduction. *J Biol Chem* **269**, 30 299–306.

Kramer MS, Last B, Getson A, and Reines S (1997). The effects of a selective D_4 dopamine receptor antagonist (L-745,870) in acutely psychotic in patients with schizophrenia. *Arch Gen Psychiatry* **54**, 567–72.

Laitinen JT (1993). Dopamine stimulates K^+ efflux in the chick retina via D_1 receptors independently of adenylyl cyclase activation. *J Neurochem* **61**, 1461–9.

Lajiness ME, Chio CL, and Huff RM (1993). D_2 dopamine receptor stimulation of mitogenesis in transfected Chinese hamster ovary cells: relationship to dopamine stimulation of tyrosine phosphorylation. *J Pharmacol Exp Ther* **267**, 1573–81.

Lanau F, Brockhaus M, Pink JR, Franchet C, Wildt-Perinic D, Goepfert C *et al.* (1997). Development and characterization of antibodies against the N terminus of the human dopamine D4 receptor. *J Neurochem* **69**, 2169–78.

Laurier LG, O'Dowd BF, and George SR (1994). Heterogeneous tissue-specific transcription of dopamine receptor subtype messenger RNA in rat brain. *Mol Brain Res* **25**, 344–50.

Le W-D and Jankovic J (2001). Are dopamine receptor agonists neuroprotective in Parkinson's disease? *Drugs and Aging* **18**(6), 389–96.

Le Corniat M, Hillion J, Martres MP, Giros B, Pilon C, Schwartz J-C *et al.* (1991). Chromosomal localisation of the human D_3 dopamine receptor gene. *Hum Genet* **87**, 618–20.

Lee SP, O'Dowd BF, Ng GYK, Varghese G, Akil H, Mansour A *et al.* (2000). Inhibition of cell surface expression by mutant receptors demonstrates that D2 dopamine receptors exist as oligomers in cell. *Mol Pharmacol* **58**, 120–8.

Lichter JB, Barr CL, Kennedy JL, Van Tol HHM, Kidd KK, and Livak KJ (1993). A hyper-variable segment in the human dopamine receptor D4 (DRD4) gene. *Hum Mol Genet* **2**, 757–63.

Liu F, Wan Q, Pristupa ZB, Yu X-M, Wang YT, and Niznik HB (2000). Direct protein-protein coupling enables cross-talk between dopamine D5 and gamma-aminobutyric acid A receptors. *Nature* **403**, 274–80.

Liu K, Bergson C, Levenson R, and Schmauss C (1994). On the origin of mRNA encoding the truncated dopamine D3-type receptor D3nf and detection of D3nf-like immunoreactivity in human brain. *J Biol Chem* **269**, 29 220–6.

Liu YF, Civelli O, Grandy DK, and Albert PR (1992). Differential sensitivity of the short and long human dopamine D_2 receptor subtypes to protein kinase C. *J Neurochem* **59**, 2311–17.

Liu YF, Civelli O, Zhou Q-Y, and Albert PR (1992). Cholera toxin-sensitive $3'5'$-cyclic adenosine monophosphate and calcium signals of the human dopamine D_1 receptor: selective potentiation by proteine kinase A. *Mol Endocrinol* **6**, 1815–24.

Liu YF, Jakobs KH, Rasenick MM, and Albert PR (1994). G rotein specificity in receptor-effector coupling. Analysis of the roles of G_o and G_{i2} in GH_4C_1 pituitary cells. *J Biol Chem* **269**, 13 880–6.

Livak KJ, Rogers J, and Lichter JB (1995). Variability of dopamine D4 receptor (DRD4) gene sequence within and among nonhuman primate species. *Proc Natl Acad Sci USA* **92**, 427–31.

Lledo P-M, Homburger V, Bockaert J, and Vincent J-D (1992). Differential G protein-mediated coupling of D_2 dopamine receptors to K^+ and Ca^{2+} currents in rat pituitary cells. *Neuron* 8, 455–63.

Luedtke RR, Griffin SA, Conroy SS, Jin X, Pinto A, and Sesack SR (1999). Immunoblot and immunohistochemical comparison of murine monoclonal antibodies specific for the rat D_{1a} and D_{1b} dopamine receptor subtypes. *J Neuroimmunol* 101, 170–87.

Lustig KD, Conklin BR, Herzmark P, Taussig R, and Bourne HR (1993). Type II adenylyl cyclase integrates coincident signals from G_s, G_i, and G_q. *J Biol Chem* 268, 13 900–05.

Luttrel LM, Ostrowsky J, Cotecchia S, Kendall H, and Lefkowitz RJ (1993). Antagonism of catecholamine receptor signaling by expression of cytoplasmic domain of the receptor. *Science* 259, 1453–7.

Machida CA, Searles RP, Nipper V, Brown JA, Kozel LB, and Neve KA (1992). Molecular cloning and expression of the rhesus macaque D1 dopamine receptor gene. *Mol Pharmacol* 41, 652–9.

MacKenzie RG, Vanleeuwen D, Pugsley TA, Shih Y-H, Demattos S, Tang L *et al.* (1994). Characterization of the human dopamine D_3 receptor expressed in transfected cell lines. *Eur J Pharmacol-Mol Pharmacol* 266, 79–85.

MacKenzie RG, Vanleeuwen D, Pugsley TA, Shih Y-H, Demttos S, Tang L *et al.* (1994). Characterization of the human dopamine D_3 receptor expressed in transfected cell lines. *Eur J Pharmacol* 266, 79–85.

Malek D, Munch G, and Palm D (1993). Two sites in the third inner loop of the dopamine D_2 receptor are involved in functional G protein-mediated coupling to adenylate cyclase. *FEBS Lett* 325, 215–19.

Martens GJM, Molhuizen HOF, Groneveld D, and Roubos EW (1991). Cloning and sequence analysis of brain cDNA encoding a Xenopus D2 dopamine receptor. *FEBS Lett* 281, 85–9.

Matsumoto M, Hidaka K, Tada S, Tasaki Y, and Yamaguchi T (1995). Polymorphic tandem repeats in dopamine D4 receptor are spread over primate species. *Biochem Biophys Res Commun* 207, 467–75.

McAllister G, Knowles MR, Patel S, Marwood R, Emms F, Seabrook GR *et al.* (1993). Characterisation of a chimeric hD_3/D_2 dopamine receptor expressed in CHO cells. *FEBS Lett* 324, 81–6.

McAllister G, Knowles MR, Ward-Booth SM, Sinclair HA, Patel S, Marwood R *et al.* (1995). Functional coupling of human D_2, D_3, and D_4 dopamine receptors in HEK293 cells. *J Receptor Signal Trans Res* 15, 267–81.

Mei YA, Griffon N, Buquet C, Martres MP, Vaudry H, Schwartz J-C *et al.* (1995). Activation of dopamine D4 receptor inhibits an L-type calcium current in cerebellar granule cells. *Neuroscience* 68, 107–16.

Memo M, Pizzi M, Belloni M, Benarese M, and Spano PF (1992). Activation of dopamine D_2 receptors linked to voltage-sensitive potassium channels reduces forskolin-induced cyclic AMP formation in rat pituitary cells. *J Neurochem* 59, 1829–35.

Miller JC, Wang Y, and Filer D (1990). Identification by sequence analysis of a second rat brain cDNA encoding the dopamine (D2) receptor. *Biochem Biophys Res Commun* 166, 109–12.

Minowa MT, Minowa T, Monsma FJ, Sibley DR, and Mouradian MM (1992). Characterization of the $5'$ flanking region of the human D_{1A} dopamine receptor gene. *Proc Natl Acad Sci USA* 89, 3045–9.

Missale C, Nash R, Robinson SW, Jaber M, and Caron MG (1998). Dopamine receptors: from structure to function. *Physiolgical Rev* 78, 189–224.

Monsma FJ, McVittie LD, Gerfen CR, Mahan LC, and Sibley DR (1989). Multiple dopamine D2 receptors produced by alternative RNA splicing. *Nature* 342, 926–9.

Monsma FJJr, Shen Y, Gerfen CR, Mahan LC, and Sibley DR (1990). Molecular cloning and expression of a D_1 dopamine receptor linked to adenylyl cyclase activation. *Proc Natl Acad Sci USA* 87, 6723–7.

Montmayeur JP and Borelli E (1991). Transcription mediated by cAMP-responsive promoter element is reduced upon activation of dopamine D_2 receptors. *Proc Natl Acad Sci USA* 88, 3135–9.

Montmayeur JP, Bausero P, Amlaiky N, Moroteaux L, Hen R, and Borelli E (1991). Differential expression of the mouse D_2 dopamine receptor isoforms. *FEBS Lett* 278, 239–43.

Montmayeur JP, Guiramand J, and Borelli E (1993). Preferential coupling between dopamine D_2 receptors and G proteins. *Mol Endocrinol* 7, 161–70.

MoraFerrer C, Yazulla S, and Studholme KM (1997). A novel D1-dopamine receptor antibody labels D1-receptors in lower and higher vertebrate retinas. *Invest Ophth Vis Sci* 38, 217–17.

Neer EJ (1995). Heterotrimeric G proteins: organisers of transmembrane signals. *Cell* 80, 249–57.

Neve KA and Neve RL (1997). *The Dopamine Receptors*, Humana Press, Totowa, NJ.

Neve KA, Cox BA, Henningsen RA, Spanoyannis A, and Neve RL (1991). Pivotal role of aspartate-80 in the regulation of dopamine D_2 receptor affinity for drugs and inhibition of adenylyl cyclase. *Mol Pharmacol* 39, 733–9.

Neve KA, Kozlowski MR, and Rosser MP (1992). Dopamine D_2 receptor stimulation of Na^+/H^+ exchange assessed by quantification of extracellular acidification. *J Biol Chem* 267, 25 748–53.

Neve KA, Neve RL, Fidel S, Janowsky A, and Higgins GA (1991). Increased abundance of alternatively spliced forms of D-2 receptor mRNA after denervation. *Proc Natl Acad Sci USA* 88, 2802–06.

Nickolls SA and Strange PG (2001). Preferential coupling of D_{2S} dopamine receptors to alphao G proteins over alphai2 G proteins. *Br J Pharmacol* 133, p. 129.

Nimchinsky EA, Hof PR, Janssen WGM, Morrison JH, and Schmauss C (1997). Expression of dopamine receptor dimers and tetramers in brain and in transfected cells. *J Biol Chem* 272, 29229–37.

Nussinovitch I and Kleinhaus AL (1992). Dopamine inhibits voltage-activated calcium channel currents in rat pars intermedia pituitary cells. *Brain Res* 574, 49–55.

O'Malley KL, Mack KJ, Gandelman K-Y, and Todd RD (1990). Organisation and expression of the rat D2A receptor gene: identification of alternative transcripts and a variant donor splice site. *Biochemistry* 29, 1367–71.

O'Malley KL, Harmon S, Tang L, and Todd RD (1992). The rat dopamine D_4 receptor: sequence, gene structure, and demonstration of expression in the cardiovascular system. *New Biologist* 4, 137–46.

Park B-H, Fishburn CS, Carmon S, Accili D, and Fuchs S (1995). Structural organization of the murine D_3 dopamine receptor gene. *J Neurochem* 64, 482–6.

Patel S, Freedman S, Chapman KL, Emms F, Fletcher AE, Knowles M *et al.* (1997). Biological profile of L-745,870, a selective antagonist with high affinity for the dopamine D4 receptor. *J Pharmacol Exp Ther* 283, 636–47.

Pearce RKB (1996) Dopamine D1 receptor agonists and dopamine reuptake blockers: new treatment statagems fo Parkinson's disease. *Exp Opin Ther Patents* 6(10), 949–53.

Pedersen UB, Norby B, Jensen AA, Schiodt M, Hansen A, Suhr-Jessen P *et al.* (1994). Characteristics of stably expressed human D_{1a} and D_{1b} receptors: atypical behavior of the dopamine D_{1b} receptor. *Eur J Pharmacol* 267, 85–93.

Petronis A, Van Tol HHM, Lichter JB, Livak KJ, and Kennedy JL (1993). The D4 dopamine receptor gene maps on 11p proximal to HRAS *Genomics* 18, 161–3.

Pilon C, Levesque D, Dimitriadou V, Griffon N, Martres MP, Schwartz J-C *et al.* (1994). Functional coupling of the human dopamine D_3 receptor in a transfected NG 108-15 neuroblastoma-glioma hybrid cell line. *Eur J Pharmacol-Mol Pharmacol* 268, 129–39.

Pilon C, Levesque D, Dimitriadou V, Griffon N, Martres MP, Schwartz J-C *et al.* (1994). Functional coupling of the human dopamine D_3 receptor in a transfected NG108-15 neuroblastoma-glioma hybrid cell line. *Eur J Pharmacol* 268, 129–39.

Piomelli D, Pilon C, Giros B, Sokoloff P, Martres M-P, and Schwartz J-C (1991). Dopamine activation of the arachidonic acid cascade as a basis for D_1/D_2 synergism. *Nature* 353, 164–367.

Plug MJ, Dijk J, Maassen A, and Moller W (1992). An anti-peptide antibody that recognizes the dopamine D_2 receptor from bovine striatum. *Eur J Biochem* 206, 123–30.

Pollok NJ, Manelli AM, Hutchins CW, Steffey ME, MacKenzie RG, and Frail DE (1992). Serine mutations in transmembrane V of the dopamine D_1 receptor affect ligand interactions and receptor activation. *J Biol Chem* 267, 17 780–6.

Polymeropoulos MH, Xiao H, and Merril CR (1991). The human D5 dopamine receptor (DRD5) maps on chromosome 4. *Genomics* 11, 777–8.

Potenza MN, Graminski GF, Schmauss C, and Lerner MR (1994). Functional expression and characterization of human D_2 and D_3 dopamine receptors. *J Neurosci* 14, 1463–76.

Ralph RJ, Varty GB, Kelly MA, Wang YM, Caron MG, Rubinstein M *et al.* (1999). The dopamine D2, but not D3 or D4, receptor subtype is essential for the disruption of prepulse inhibition produced by amphetamine in mice. *J Neurosci* 19, 4627–33.

Rao DD, McKelvy KJ, Gandelman K-Y, and MacKenzie RG (1990). Two forms of the rat D_2 dopamine receptor as revealed by the polymerase chain reaction. *FEBS Lett* 263, 18–22.

Robinson SW and Caron MG (1996). Chimeric D_2/D_3 dopamine receptor efficiently inhibit adenylyl cyclase in HEK293 cells. *J Neurochem* 67, 212–19.

Robinson SW, Jarvie KR, and Caron MG. (1994). High affinity agonist binding to the dopamine D_3 receptor: chimeric receptor delineate a role for intracellular domain. *Mol Pharmacol* 46, 352–6.

Rocheville M, Lange DC, Kumar U, Patel SC, Patel RC, and Patel YC (2000). Receptors for dopamine and somatostatin: formation of hetero-oligomers with enhanced functional activity. *Science* 288, 154–8.

Rubinstein M, Phillips TJ, Bunzow JR, Falzone TL, Dziewczapolski G, Zhang G *et al.* (1997). Mice lacking dopamine D4 receptors are supersensitive to ethanol, cocaine, and metamphetamine. *Cell* 90, 991–1001.

Schmauss C, Haroutunian V, Davis KL, and Davidson M (1993). Selective loss of dopamine D3-type receptor mRNA expression in partial and motor cortices of patients with chronic schizophrenia. *Proc Natl Acad Sci USA* 19, 8942–6.

Seabrook GR, Kemp JA, Freedman SB Patel S, Sinclair HA, and McAllister G (1994). Functional expression of human D3 dopamine receptors in differentiated neuroblastoma x glioma NG108-15 cells. *Br J Pharmacol* 111, 391–3.

Seabrook GR, Knowles MR, Brown N, Myers J, Sinclair H, Patel S *et al.* (1994). Pharmacology of high-threshold calcium currents in GH_4C_1 pituitary cells and their regulation by activation of dopamine D_2 and D_4 dopamine receptors. *Br J Pharmacol* 112, 728–34.

Seabrook GR, Patel S, Marwood R, Emms F, Knowles MR Freedman SB *et al.* (1992). Stable expression of human D_3 dopamine receptors in GH_4C_1 pituitary cells. *FEBS Lett* 312, 123–6.

Selbie LA, Hayes G, and Shine J (1989). The major dopamine D2 receptor: molecular analysis of the human D2A subtype. *DNA* 8, 683–9.

Senogles I (1994). The D_2 dopamine receptor mediates inhibition of growth inGH4ZR7 cells: involvement of protein kinase-$C\varepsilon$. *Endocrinology* 134, 783–9.

Senogles SE (1994). The D_2 dopamine receptor isoforms signal through distinct $Gi\alpha$ proteins to inhibit adenylyl cyclase. A study with site-directed mutant $Gi\alpha$ proteins. *J Biol Chem* 269, 23 120–7.

Snyder LA, Roberts JL, and Sealfon SC (1991). Distribution of dopamine receptor mRNA splice variants in the rat by solution hybridisation/protection assay. *Neurosci Lett* 122, 37–40.

Sokofoff P, Giros B, Martres MP, Bouthenet ML, and Schwartz J-C (1990). Molecular cloning and characterization of a novel dopamine receptor (D_3) as target for neuroleptics. *Nature* 347, 146–51.

Spurlock G, Williams J, McGuffin P, Aschauer HN, Lenzinger E, Fuchs K *et al.* (1998). European multicentre association study of schizophrenia: a study of the DRD2 Ser311Cys and DRD3 Ser9Gly polymorphisms. *Am J Med Genet* 81, 24–8.

Stemp G, Ashmeade T, Branch CL, Hadley MS, Hunter AJ, Johnson DJ *et al.* (2000). Design and synthesis of trans-N-[4-[2-(6-cyano-1,2,3,4-tetraisoquinolin-2-yl)ethyl]-4-quinolinec (SB-277011): a potent and selective dopamine D(3) receptor antagonist with high oral bioavailability and CNS penetration in rat. *J Med Chem* 43, 1878–85.

Strader CD, Dixon RAF, Cheung AH, Candelore MR, Blake AD, and Sigal IL (1987). Mutations that uncouple the β-adrenergic receptor from G_s and increase agonist affinity. *J Biol Chem* 262, 16 439–43.

Strange PG (2001). Antipsychotic drugs: importance of dopamine receptors for mechanisms of therapeutic actions and side effects. *Pharmacol Rev* 53, 1–15.

Sugamori KS, Demchyshyn LL, Chung M, and Niznik HB (1994). D_{1A}, D_{1B}, and D_{1C} dopamine receptor from Xenepus laevis. *Proc Natl Acad Sci USA* **91**, 10 536–40.

Sunahara RK, Guan HC, O'Dowd BF, Seeman P, Laurier LG, Ng G *et al.* (1991). Cloning of the gene for a human dopamine D_5 receptor with higher affinity for dopamine than D_1. *Nature* **350**, 614–19.

Sunahara RK, Niznik HB, Weiner DM, Stormann TM, Brann MR, Kennedy JL *et al.* (1990). Human dopamine D_1 receptor encoded by an intronless gene on chromosome 5. *Nature* **347**, 80–3.

Tan E-K and Jankovic J (2001). Choosing dopamine agonists in Parkinson's disease. *Clin Neuropharmacol* **24**(5), 247–53.

Tang L, Todd RD, and O'Malley KL (1994) Dopamine D_2 and D_3 receptors inhibit dopamine release. *J Pharmacol Exp Ther* **270**, 475–9.

Tang L, Todd RD, Heller A, and O'Malley KL (1994). Pharmacological and functional characterization of D_2, D_3 and D_4 dopamine receptors in fibroblast and dopaminergic cell lines. *J Pharmacol Exp Ther* **268**, 495–502.

Tiberi M and Caron MG (1994). High agonist-independant activity is a distinguishing feature of the dopamine D_{1B} receptor subtype. *J Biol Chem* **269**, 27 925–31.

Tiberi M, Jarvie KR, Silvia C, Faladeau P, Gringrich JA, Godinot N *et al.* (1991). Cloning, molecular characterization, and chromosomal assignment of a gene encoding a second D_1 dopamine receptor subtype: differential expression pattern in rat brain compared with the D_{1A} receptor. *Proc Natl Acad Sci USA* **88**, 7491–5.

Todd RD and O'Malley KL (2001). The dopamine receptor DRD4 gene: are duplications distracting? *TiPS* **22**, 55–6.

Truffinet P, Tamminga CA, Fabre LF, Meltzer HY, Riviere M-E, and Papillon-Downy C (1999). Placebo-controlled study of the D_4/5-HT_{2A} agonist Fanaserin in the treatment of schizophrenia. *Am J Psychiatry* **156**, 419–25.

Undie AS and Friedman E (1994). Stimulation of dopamine D_1 receptor enhances inositol phosphates formation in rat brain. *J Pharmacol Exp Ther* **253**, 987–92.

Usiello A, Baik J-H, Rouge-Pont F, Picetti R, Dierich A, LeMeur M *et al.* (2000). Distinct functions of the two isoforms of dopamine D2 receptors. *Nature* **408**, 199–202.

Vallar L, Muca C, Magni M, Albert P, Bunzow J *et al.* (1990). Differential coupling of dopaminergic D_2 receptors expressed in different cell types. Stimulation of phosphsinositol 4,5-bisphosphate hydrolysis in Ltk^- fibroblasts, hyperpolazisation, and cytosolic-free Ca^{2+} concentration decrease in GH_4C_1 cells. *J Biol Chem* **265**, 10 320–6.

Van Tol HHM, Bunzow JR, Guan H-C, Sunahara RK, Seeman P, Niznik HB, and Civelli O (1991). Cloning of the gene for a human dopamine D4 receptor with high affinity for the antipsychotic clozapine. *Nature* **350**, 610–4.

Voss T, Wallner E, Czernilofsky AP, and Freissmuth M (1993). Amphipathic α-helical structure does not predict the ability of receptor-derived synthetic peptides to interact with guanine nucleotide-binding regulatory proteins. *J Biol Chem* **268**, 4637–42.

Wilke RA, Hsu S-F, and Meyer BJ (1998). Dopamine D_4 receptor mediated inhibition of potassium current in Neurophysial Nerve Terminals. *J Pharmacol Exp Ther* **284**, 542–8.

Zawarynski P, Tallerico T, Seeman P, Lee SP, O'Dowd BF, and George SR (1998). Dopamine D2 receptor dimers in human and rat brain. *FEBS Lett* **456**, 63–7.

Zawilska JB and Nowak JZ (1997). Dopamine D_4-like receptors in vertebrate retina: does the retina offer a model for the D_4-receptor analysis? *Pol J Pharmacol* **49**, 201–11.

Zhou Q-Y, Grandy DK, Thambi L, Kushner JA, Van Tol HHM, Cone R *et al.* (1990). Cloning and expression of human and rat D_1 dopamine receptors. *Nature* **347**, 76–9.

Zhou Q-Y, Li C, and Civelli O (1992). Characterization of gene organisation and promoter region of the rat dopamine D_1 receptor gene. *J Neurochem* **59**, 1875–83.

Chapter 16

Galanin receptors

U. Kahl, Ü. Langel, and T. Bartfai

16.1 Introduction

Galanin was discovered in 1983 by Tatemoto *et al.* who isolated the 29 amino acids long neuropeptide from porcine intestine, by using a chemical method for detection of amidated C-terminals of peptides (Tatemoto *et al.* 1983). Since then, the galanin sequence from a large number of species has been determined (Table 16.1). In all species so far, apart from man, galanin has 29 amino acids and is C-terminally amidated. In humans the galanin sequence is 30 amino acids long, without C-terminal amidation, and instead has a free carboxyl group in its C-terminal end. The N-terminal 14 amino acids of the molecule are 100 per cent homologous in all species, with the exception of galanin from tuna fish, where the ordinary serine in position 6 is exchanged for a proline. The first 13 amino acids of galanin also constitute the part of the molecule that has been shown to be crucial for binding of galanin to galanin binding sites in the central nervous system (CNS) (Table 16.2) (Land *et al.* 1991). The primary sequence of the C-terminal domain of galanin varies between species, and the biological function of this part of the peptide is not completely understood. Studies of the binding of galanin(1–16) versus galanin(1–29) suggest that the C-terminal portion accounts for only 1/7 of the binding energy of the full-length peptide to its receptors in the rat brain (Land *et al.* 1991; Kask *et al.* 1996). It has been proposed that the C-terminal sequence serves as a protector to proteolytic attacks, and that it has a structure-stabilizing effect in solution. Proton NMR and CD dichroism spectroscopy studies show that the galanin structure has a different degree of rigidity depending on the environment. In water, the galanin conformation is fairly flexible and not very stable, whereas in interaction with SDS micelles, which serve as models of biological membranes, there is a much higher level of stability (Ohman *et al.* 1995, 1998). The proposed model is that galanin adopts a hairpin-like structure with an N-terminal fragment of galanin (galanin(1–10)) adopting an α-helical secondary structure, followed by a β-turn around the proline in position 13 (Fig. 16.1).

The galanin cDNA encodes a precursor peptide (preprogalanin) consisting of the 29 amino acids long galanin and then a glycine, which is the amide donor for N-terminal amidation, followed by a dibasic cleavage site, and then the 60 amino acids long galanin message associated peptide (GMAP). The function of GMAP remains unknown, although it has been shown that stoichiometric quantities of GMAP are produced by galanin-producing cells.

Until recently, galanin had not been shown to belong to any known family of neuropeptides. In 1999 however, a novel galanin-related peptide was found in porcine hypo-thalamus. The peptide was shown to be 60 amino acids long, and to have a non-amidated

Table 16.1 Galanin sequences from different species

	1	6	11	16	21	26	
Human	GWTLN	SAGYL	LGPHA	VGNHR	SFSDK	NGLTS	
Porcine	GWTLN	SAGYL	LGPHA	IDNHR	SFHDK	YGLA	amide
Bovine	GWTLN	SAGYL	LGPHA	LDSHR	SFQDK	HGLA	amide
Rat	GWTLN	SAGYL	LGPHA	IDNHR	SHSDK	HGLT	amide
Mouse	GWTLN	SAGYL	LGPHA	IDNHR	SFSDK	HGLT	amide
Sheep	GWTLN	SAGYL	LGPHA	IDNHR	SFHDK	HGLA	amide
Chicken	GWTLN	SAGYL	LGPHA	VDNHR	SFNDK	HGFT	amide
Bowfin	GWTLN	SAGYL	LGPHA	VDNHR	SLNDK	HGLA	amide
Alligator	GWTLN	SAGYL	LGPHA	IDNHR	SFSDK	HGIA	amide
Trout	GWTLN	SAGYL	LGPHA	IDGHR	TLSDK	HGLA	amide
Frog	GWTLN	SAGYL	LGPHA	IDNHR	SFNDK	HGLA	amide

C-terminal, and was given the name GALP, galanin-like peptide (Ohtaki *et al.* 1999). GALP binds to galanin receptors but GALP-specific receptors that do not bind galanin have not yet been found, despite limited expression cloning efforts using labelled GALP as ligand. The structures for rat and human GALP were also deduced from the cDNAs and show great degree of homology between the three species (porcine, human, and rat), with full conservation of the 9–21 sequence that is identical to galanin 1–13.

Both galanin and galanin receptors show a widespread anatomical distribution throughout the central and peripheral nervous systems (Fig. 16.2). In many brain regions galanin coexists with other neuropeptides and/or small molecule classical transmitters (Chan-Palay 1988; Crawley and Wenk 1989; Merchenthaler 1991; Melander *et al.* 1985, 1986*b*; Melander and Staines 1986; Skofitsch and Jacobowitz 1985; Skofitsch *et al.* 1989). Like many other neuropeptides, galanin often serves as a modulator of other transmitters' action. In most cases of coexistence with classical neurotransmitters the co-released galanin acts as a pre-synaptic inhibitor of the release of the classical neurotransmitters (e.g. ACh, NE, and glutamate).

16.2 **Molecular characterization**

In 1987, Servin *et al.* managed to identify and biochemically characterize a galanin–galanin receptor binding complex in the rat brain, which on SDS-PAGE was revealed as a 56 kD protein (Amiranoff *et al.* 1989; Fisone *et al.* 1989; Servin *et al.* 1987), suggesting that it may be a GPCR, as it indeed also turned out to be, as shown in cloning studies later (see below).

So far, three subtypes of the galanin receptor, a typical 7-transmembrane domain G protein-coupled receptor, have been cloned and named GalR1, GalR2, and GalR3, respectively (Table 16.3) (Branchek *et al.* 2000). The three receptor subtypes differ in terms of their primary sequence, genomic organization, pharmacological profile, functional coupling, and

Table 16.2 Displacement binding of galanin and galanin fragments in the rat CNS

	1	16	29	IC_{50} (nM)
(1–29)	GWTLNSAGYLLGPHAIDNHRSFHDKYGLA			0.8
(1–16)	----------------			7.0
(17–29)			----------------------------	>10 000
(1–13)	-----------			150
[AC-GLY^1]-(1–16)	---------------			500
[Ala^1](1–16)	A---------------			90
[Ala^2](1–16)	-A--------------			>10 000
[Ala^3](1–16)	--A-------------			230
[Ala^4](1–16)	---A------------			90
[Ala^5](1–16)	----A-----------			1900
[Ala^6](1–16)	-----A----------			60
[Ala^8](1–16)	--------A-------			200
[Ala^9](1–16)	---------A------			1900
[Ala^{10}](1–16)	----------A-----			2400
[Ala^{11}](1–16)	-----------A----			1200
[Ala^{12}](1–16)	-----------A----			1000
[Ala^{13}](1–16)	------------A---			80
[Ala^{14}](1–16)	-------------A--			20
[Ala^{16}](1–16)	---------------A			50

distribution. These variations all contribute to the ligand selectivity, signalling divergence, and specific importance of each respective subtype in various systems.

The first galanin receptor (GalR1) was cloned by Habert-Ortoli *et al.* from Bowes human melanoma cell line (Habert-Ortoli *et al.* 1994), a little more than a decade after the discovery of the endogenous ligand galanin. This discovery was followed by the cloning of the rat GalR1 homologue in 1995, and two more subtypes (GalR2 and GalR3) have since then also been cloned from several species (Ahmad *et al.* 1998; Bloomquist *et al.* 1998; Borowsky *et al.* 1998; Burgevin *et al.* 1995; Fathi *et al.* 1997, 1998*a*; Howard *et al.* 1997; Iismaa *et al.* 1998; Jacoby *et al.* 1997; Kolakowski *et al.* 1998; Lorimer and Benya 1996; Lorimer *et al.* 1997; Pang *et al.* 1998; Parker *et al.* 1995; Smith *et al.* 1997; Sullivan *et al.* 1997; Wang *et al.* 1997*a–c*).

For each galanin receptor subtype there is a high sequence homology between different species. Between receptor subtypes however, the sequence similarity is lower; the three receptors reveal only around 35–50 per cent amino acid identity. In addition, the GalR1 gene

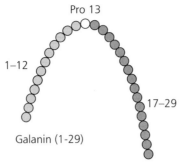

Fig. 16.1 Schematic drawing of the proposed 'hairpin' structure of galanin, with β turn at Pro[13].

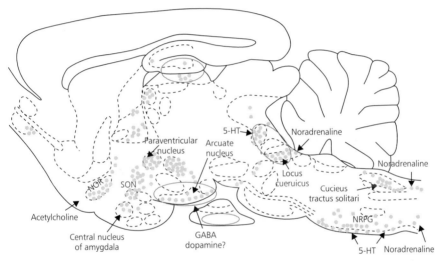

Fig. 16.2 Distribution of galanin-positive cell groups and galanin-binding sites in the rat brain.

has a different intron/exon organization, as compared to the genes for GalR2 and GalR3 (Iismaa *et al.* 1998; Kolakowski *et al.* 1998), pointing at, at least two different paths for galanin receptors throughout evolution. The highest homology between the three different galanin receptor subtypes is found in the transmembrane domains, while the extracellular and intracellular regions display a lower degree of similarity. This is in good agreement with the suggestion that the latter are responsible for selective binding of ligands, and for differential G protein coupling and signalling pathways (Berthold and Bartfai 1997).

16.2.1 **The GalR1 receptor**

In both rats and humans, GalR1 is composed of 349 amino acids, in mouse 348, and in some species and cell lines even fewer (Branchek *et al.* 1998; Jacoby *et al.* 1997). The human and rat receptors share the same consensus sites for N-linked glycosylation and intracellular phosphorylation, with the exception of two additional phosphorylation sites in the human receptor.

Table 16.3 Galanin receptors

	GalR1	GalR2	GalR3
Alternative names	GN-1		
Preferred endogenous ligand	Galanin	GALP	—
Structural information (Accession no.)	h 349 aa (P42711) r 346 aa (Q62805) m 348 aa (P56479)	h 387 aa (O43603) r 372 aa (O88854) m 371 aa (O08726)	h 368 aa (O60755) r 370 aa (O88626) m 370 aa (O88853)
Chromosomal location	18q23	17q25.3	22q13.1
Selective agonists	Galanin 1–29	(partially selective) galanin 2–29, [D-Trp]-galanin 1–29, galanin 2–11	(Partially selective) galanin 2–29
Selective antagonists	M15 = M35	M35 > M15	—
Radioligands	[^{125}I]-galanin	[^{125}I]-galanin	[^{125}I]-galanin
G protein coupling	G_i/G_o	$G_i/G_o/G_q/G_{11}$	G_i/G_o
Expression profile	Hypothalamus, amygdala, hippocampus, thalamus, brainstem, spinal cord, DRG	Cerebral cortex, hypothalamus, hippocampus, dentate gyrus, amygdala, mammiliary nuclei, cerebellum, DRG, pituitary, vas deferens, prostate, uterus, ovary, stomach, intestine, pancreas	Cerebral cortex, hypothalamus, medulla oblongata, caudate putamen, cerebellum, spinal cord, pituitary, heart, spleen, testis, liver, kidney, stomach, adrenal cortex, lung, uterus, vas deferens, choroid plexus
Physiological function	Regulation of horomone and neurotransmitter release	Regulation of the release of prolactin and growth hormone	Unclear
Knockout phenotype	Altered parasympathetic tone, enhanced hippocampal excitability, seizures	—	—
Disease relevance	Neurodegeneration, epilepsy, stroke, eating disorders, cognitive decline, pain, depression, gut motility, diabetes	Neurodegeneration, epilepsy, stroke, eating disorders, cognitive decline, pain, depression, diabetes	Eating disorders, pain, depression, diabetes

In humans, GalR1 has been mapped to chromosome 18q23, and in mouse to 18E4, which is homologous with the human chromosomal position (Jacoby *et al.* 1997; Nicholl *et al.* 1995). The human, as well as the mouse, GalR1 receptor gene contains three exons, where exon 1 encodes the NH_2-terminal end and the first five transmembrane (TM) domains, exon 2 the third extracellular loop, and exon 3 the remainder of the receptor, from TM 6 to the COOH-terminus (Iismaa *et al.* 1998).

The interactions between galanin and the GalR1 receptor were studied by mutagenesis studies of the GalR1 receptor, and by use of galanin analogues in which the side chains of the amino acids that were assumed to interact with the receptor were substituted (Fig. 16.3) (Berthold *et al.* 1997; Kask *et al.* 1996). Using computer modelling and the rhodopsin scaffold for the GalR1 and a mutagenesis approach, galanin was docked into the GalR1 receptor, with interactions between Trp^2 of the ligand and His264 or His267 of the receptor, with Phe282 of the receptor and Tyr^9 in galanin, and through a hydrogen bond between Glu271 of the receptor and the N-terminus of galanin (Fig. 16.4). Subsequent mutagenesis studies indicated that the N-terminus of galanin might also interact with Phe115 in TMIII (Berthold *et al.* 1997). The assumed interactions accounted for about 80 per cent of the observed binding free energy of the galanin-GalR1 interactions, although the model suggested that the C-terminal half of the galanin molecule is not in contact with the receptor. This again suggests that in the case of the GalR1 subtype it is the N-terminal portion of galanin that is important for binding to the receptor, and the extracellular domains of the receptor are most important for binding of the peptide ligand, as expected. It is noteworthy that in this model the N-terminus

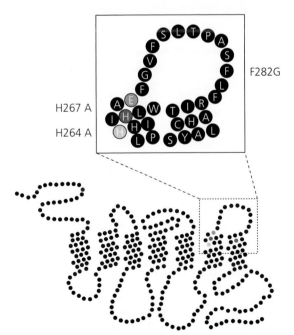

Fig. 16.3 A large number of residues in the human GalR1 were mutated into alanines, or in some cases other amino acids. This initial mutagenesis study showed that His264, His267, and Glu271 at the top of TMVI, and Phe282 in TMVII were important for binding of galanin to the receptor.

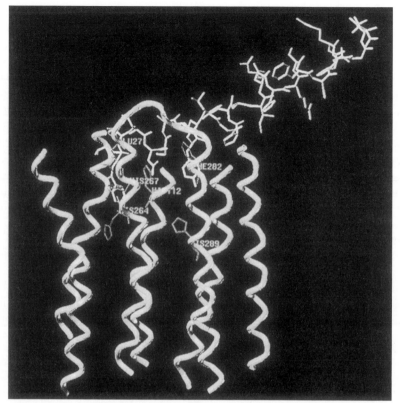

Fig. 16.4 Computer model based on the coordinates for rhodopsin, with the galanin molecule docked into its binding site at the human GalR1. In the proposed model, the N-terminus of galanin hydrogen bonds with Glu271 of the receptor, Trp^2 of galanin interacts with the Zn^{2+}-sensitive pair of His264 and His267 of TMVI, and Tyr^9 of galanin interacts with Phe282 of TMVII, while the C-terminus of galanin is pointing towards the N-terminus of the receptor. Subsequent mutations of residues in the receptor indicated that Phe115 in TMIII might also interact with the N-terminus of galanin.

of the receptor does not interact with galanin, and the interactions are concentrated to the top of transmembrane domains III, VI, and VII (Figs 16.3 and 16.4). Glycosylation of the receptor does not seem to be important for binding of galanin.

16.2.2 **The GalR2 receptor**

The GalR2 galanin receptor was originally cloned from the rat and consists of 372 amino acids, including three consensus sites for extracellular glycosylation, and several intracellular phosphorylation sites that differ from the GalR1 receptor (Smith *et al.* 1997). The similarity between rat GalR2 and rat or human GalR3 is around 55–60 per cent, while the homology with the rat or human GalR1 receptor is slightly lower, about 40 per cent. The human GalR2 receptor contains 387 amino acids, with a 15 amino acids shorter C-terminus (Bloomquist *et al.* 1998; Borowsky *et al.* 1998; Fathi *et al.* 1997).

The human GalR2 receptor gene is located on chromosome 17q25.3, and in mouse GalR2 has been mapped to chromosome 11 (Fathi *et al.* 1998*a*; Kolakowski *et al.* 1998; Pang *et al.* 1998). The exon/intron organization in the human GalR2 gene is different from the one observed in the GalR1 gene. The coding sequence in human GalR2 is encoded on two exons, separated by a 1.4 kb intron (Fathi *et al.* 1998*a*; Iismaa *et al.* 1998). Exon 1 encodes the NH_2-terminal end and the first three TM domains, exon 2 encodes the remainder of the receptor, from intracellular loop 2 to the COOH-terminus. The rat GalR2 gene has been shown to have the same structure (Howard *et al.* 1997).

16.2.3 The GalR3 receptor

Like GalR2, the GalR3 receptor was first cloned from rat (Smith *et al.* 1998; Wang *et al.* 1997*a*). Rat GalR3 is a 370 amino acid protein, 36 per cent similar to rat GalR1, and with 55 per cent homology to rat GalR2. Human GalR3 was later cloned from a human genomic library, and was shown to encode a 368 amino acids long protein, sharing 90 per cent sequence identity with rat GalR3. Both rat and human GalR3 have a single consensus site for N-linked glycosylation, and several phosphate acceptor sites for intracellular protein kinase C phosporylation.

In humans, GalR3 is mapped to 22q12.2–13.1, and the mouse homologue is located on mouse chromosome 15 (Kolakowski *et al.* 1998). In similarity with GalR2, the gene for human GalR3 is interrupted by an intron. This intron is in GalR3 somewhat smaller than in the GalR2 gene, around 1 kb (Iismaa *et al.* 1998). The NH_2-terminal and the first three TM domains are encoded by exon 1, and exon 2 encodes the remaining part of the receptor, that is, TM IV–VII and the COOH-terminal.

16.2.4 Novel Galanin receptor subtypes?

Interestingly, none of the three so far cloned receptor subtypes have been found to bind galanin(3–29) with any significant affinity. This N-terminally truncated galanin analogue was reported by Wynick *et al.* to replace ^{125}I-Bolton-Hunter-labelled galanin with high affinity in membranes of rat pituitary, and to act as a galanin agonist on prolactin secretion from the pituitary (Wynick *et al.* 1993). The same peptide was also reported to have affinity for galanin receptors in rat jejunal preparations (Rossowski *et al.* 1990). This points at the presence of a galanin N-terminus-independent receptor, which is yet to be cloned.

Recently, another potential galanin receptor subtype was suggested in a paper by Xu *et al.* (Xu *et al.* 1999). This receptor was found in neurones in the CA3 area of the hippocampus, as shown with electrophysiological recordings. These neurones responded with hyperpolarization to galanin(1–15), but not to full-length galanin, galanin(1–16), or D-Trp^2-galanin, which indicates the possible existence of a galanin(1–15)-selective hippocampal galanin receptor.

16.3 Cellular and subcellular localization

Both galanin receptors show a widespread anatomical distribution throughout the central and peripheral nervous systems (Fig. 16.2). In 1986, both Melander *et al.* and Skofitsch *et al.* used receptor autoradiography to determine the distribution of [^{125}I]-galanin binding sites in the rat CNS (Fig. 16.2) (Melander *et al.* 1986*a*; Skofitsch *et al.* 1986). Dense galanin binding was observed in the prefrontal cortex, the anterior nuclei of the olfactory bulb, several nuclei of the amygdaloid complex, the dorsal septal area, dorsal bed nucleus of the stria terminalis, the ventral pallidum, the internal medullary laminae of the thalamus, medial pretectal nucleus, nucleus of the medial optic tract, borderline area of the caudal spinal trigeminal nucleus adjacent to the spinal trigeminal tract, the substantia gelatinosa, and the superficial layers of the dorsal spinal cord. Moderate binding was observed in the piriform, periamygdaloid, entorhinal, insular cortex and the subiculum, the nucleus accumbens, medial forebrain bundle, anterior hypothalamic, ventromedial, dorsal premamillary, lateral and periventricular thalamic nuclei, the subzona incerta, Forel's field H1 and H2, periventricular gray matter, medial and superficial gray strata of the superior colliculus, dorsal parts of the central gray, peripeduncular area, the interpeduncular nucleus, substantia nigra zona compacta, ventral tegmental area, the dorsal and ventral parabrachial and parvocellular reticular nuclei. Mapping of galanin receptors was later performed in human post-mortem brain, and showed that the distribution is similar to that in the rat brain (Kohler and Chan-Palay 1990). However it is important to note that the primate and human brain has a higher density of galanin receptors than the rodent brain. Galanin receptors were also detected by autoradiography in the spinal cord and several peripheral tissues, such as the kidneys, the pancreas, the gastrointestinal tract, and the respiratory tract (Arvidsson et al. 1991; Waters and Krause 2000; Wiesenfeld-Hallin et al. 1992).

In the rat, GalR1 mRNA has been detected by Northern blot analysis in the brain, spinal cord and RIN14b cells. The GalR1 mRNA distribution in the rat CNS has been determined by in situ hybridization, which shows that rGalR1 expression occurs in areas where galanin and [^{125}I]-galanin binding sites have been found previously, that is, in the hypothalamus (supraoptic nucleus), amygdala, ventral hippocampus, thalamus, brainstem (medulla oblongata, locus coeruleus and lateral parabranchial nucleus), and spinal cord (dorsal horn), showing a good agreement between mRNA and gene product data (Burgevin et al. 1995; Gustafson et al. 1996; Parker et al. 1995). The expression of GalR1 mRNA further shows plasticity. Hypothalamic GalR1 mRNA is elevated more in females than males, varies throughout the estrous cycle, and in the dorsal root ganglia of the spinal cord, GalR1 mRNA has been shown to increase after inflammation or peripheral nerve injury. Lactation and hypophysectomy, on the other hand, decrease GalR1 mRNA expression (Faure-Virelizier et al. 1998; Landry et al. 1998; Sten Shi et al. 1997). Similar binding profiles for GalR1 have been observed in [^{125}I]-galanin bindning assays in humans and rats.

In comparison with rat GalR1, rat GalR2 mRNA is more widely distributed, with expression in many areas of the brain, where the highest levels are found in the hypothalamus, hippocampus, amygdala, piriform cortex, dentate gyrus, mammilary nuclei, and cerebellar cortex (Fathi et al. 1997; O'Donnell et al. 1999). Peripherally, rat GalR2 is found in the vas deferens, prostate, uterus, ovary, stomach, large intestine, dorsal root ganglia, and cells derived form the pancreas (Howard et al. 1997; O'Donnell et al. 1999; Sten Shi et al. 1997; Wang et al. 1997b). GalR2 has also been detected in the pituitary (Fathi et al. 1998a).

Northern blot studies of GalR3 mRNA initially showed that GalR3 mRNA is present in the heart, spleen, and testis (Wang et al. 1997a). Later, with the help of more sensitive methods, GalR3 transcripts were also detected in the CNS, with the highest levels in the hypothalamus and pituitary. Other regions, like the olfactory bulb, cerebral cortex, medulla oblongata, caudate putamen, cerebellum, and spinal cord, but not the hippocampus and substantia nigra, were also found to contain GalR3 mRNA (Smith et al. 1998). In the periphery, GalR3 mRNA is widespread, with existence in the liver, kidney, stomach, testicles, adrenal cortex, lung, adrenal medulla, uterus, vas deferens, choroid plexus, and dorsal root ganglia (Smith *et al.* 1998).

16.4 **Pharmacology**

16.4.1 **Agonists**

A limiting factor in the pharmacological characterization of galanin receptors is that there have not been any really potent and selective non-peptidergic agonists or antagonists available. This despite large random screening efforts from many pharmaceutical companies, who have used millions of compounds and all the three cloned galanin receptor subtypes. The only tools available for pharmacological differentiation of the three receptor subtypes have been peptide galanin analogues with various amino acid changes in one or more positions, or chimeric peptide ligands consisting of the N-terminal part of galanin (galanin(1–13)) joined to fragments from other peptides (Table 16.4).

Based on L-Ala substitutions of amino acids in the galanin sequence, where it was deduced that Trp^2, Asn^5, and Tyr^9 are the most important pharmacophores in the interaction between galanin and GalR1 (Land *et al.* 1991; Kask *et al.* 1996), together with the assumption that galanin (1–10) forms an alpha-helix in presence of the receptor, a tripeptide library with Trp, Asn, and Tyr analogues was synthesized and deconvoluted. Galnon (Fmoc-Cha-Lys-amidomethylcoumarin), a member of this three amino acid/peptidomimetic library, has a molecular weight below 600 Da, and was shown to act as a galanin receptor agonist with central effects (Fig. 16.5). The problem with galnon is that it is still a peptidomimetic with sensitivity for peptidase-catalyzed degradation, has low affinity, and does not posses strong receptor subtype selectivity. Hence, galnon is only an interesting lead for the development of centrally and peripherally acting galanin receptor agonists.

Rat GalR2 and GalR1 share similar pharmacological profiles, and both possess high affinity for full-length and N-terminal fragments of galanin. The affinity of galanin (2–11) at galanin receptors in the spinal cord, and this fragment's activity as an agonist, has been reported recently (Liu *et al.* 2001). GalR2 displays low affinity for galanin (3–29) (Fathi *et al.* 1997; Howard *et al.* 1997; Smith *et al.* 1997, 1998; Wang *et al.* 1997*a*). Rat GalR2 can be distinguished from GalR1 in that it has higher relative affinity for D-Trp^2-galanin and galanin (2–29) (Wang *et al.* 1997*a*). Human GalR2 has basically the same pharmacological profile as rat GalR2 (Bloomquist *et al.* 1998).

Another recent finding was the identification and cloning of GALP from porcine hypothalamus (Ohtaki *et al.* 1999). Interestingly, amino acids 9–21 in GALP were shown to be identical to amino acids 1–13 in galanin, that is the region known to be important

Table 16.4 Chimeric galanin peptides

Peptide	Sequence
M15	galanin(1–13)-substance P(5–11)-amide
M35	galanin(1–13)-bradykinin(2–9)-amido
M40	galanin(1–13)-Pro-Pro-Ala-Leu-Ala-Leu-Ala-amide

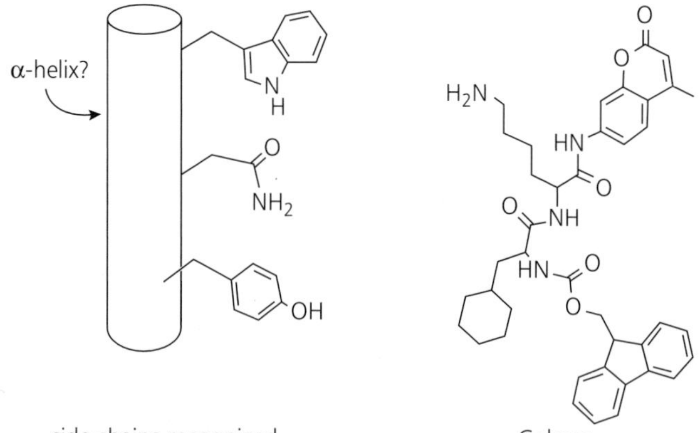

side chains recognized
by the galanin receptor

Galnon

Fig. 16.5 The non-peptide galanin agonist, galnon, was generated through a library approach, which was based on earlier findings that Trp^2, Asn^5 and Tyr^9 are the major pharmacophores of galanin.

for galanin receptor binding. Binding studies showed that GALP bound to heterogenously expressed galanin type 1 and type 2 galanin receptors (GalR1 and GalR2) and also increased $[^{35}S]$-GTPγS binding, suggesting that it can act as an alternative agonist at these receptors. GALP further seems to recognize the GalR2 subtype with 15–20-fold higher affinity for GalR2 than for GalR1. Hence, GALP seems to be an endogenous ligand that preferentially binds the GalR2 receptor, with even higher affinity than galanin itself.

Pharmacologically, GalR3 and GalR2 can be separated from GalR1 in that they bind porcine galanin (1–29) with 10-fold higher affinity than galanin (2–29), while GalR1 has a 100-fold preference for galanin (1–29), compared to galanin (2–29) (Wang *et al.* 1998). In addition, galanin (1–16) has low affinity for GalR3, but high affinity for GalR1 and GalR2 (Wang *et al.* 1997*a*, 1998). Human and rat GalR3 share similar pharmacological profiles in binding assays.

16.4.2 **Antagonists**

Although recently, two small, non-peptidergic compounds were discovered (Fig. 16.6) by random screening of large chemical libraries and libraries of natural products, and both these compounds behave as antagonists at galanin receptors (Puar *et al.* 2000; Scott *et al.* 2000). One of these compunds, spirocoumaranon (Sch 202596), is a fungal metabolite, which binds to human GalR1 with an IC$_{50}$ of 1.4 μM. The other one, dithiepine-1,1,4,4-tetroxide (from Johnson and Johnson), has an IC$_{50}$ of 0.17 μM at the human GalR1. These compounds however, have low affinity, no galanin receptor subtype specificity, and their chemical structure is not ideal for future medicinal chemistry development. In addition, since galanin is not a member of a family of peptides, few endogenous cross-reactive ligands are expected to be found at this point. The number of receptor subtypes might although increase. Hence, the compounds that are now beginning to evolve will have to be refined in terms of subtype specificity and affinity, for these to be useful as ligands to distinguish different subtypes of the galanin receptor.

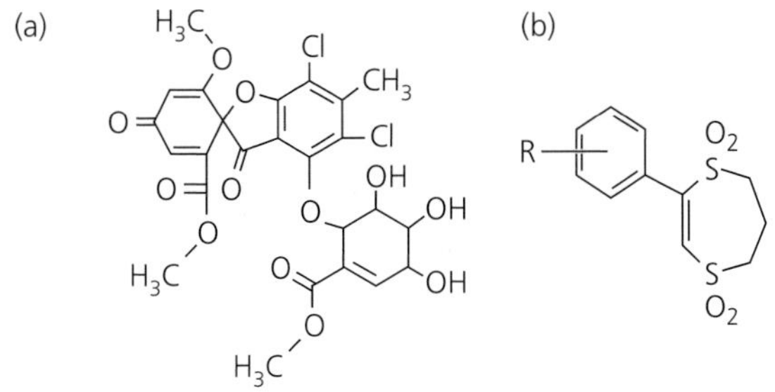

Fig. 16.6 Non-peptide antagonists, Sch202596 (A) and dithiepine-1,1,4,4-tetroxide (B).

Table 16.5 Pharmacological application of galanin receptor subtype-specific ligands

Subtype specificity	Site of action	Indication	Neurochemical effects to be measured
Type I	Hippocampus	Anticonvulsant Neuroprotection, Ischaemia	Glu ↓, GABA −
	LC, DRN	Antidepressant	NE ↓, 5-HT ↓
	Spinal cord	Analgesia, Allodynia	SP release ↓ Synergy with opates ↓
Type I/II	Nucleus basalis 'Trophic factor' Hippocampus/LC	Disease modifier Alzheimer's disease Cognitive enhancer	Cholinergic Neuron survival ↑ Ach ↑, NE ↑
Type II/III	Pituitary Hypothalamus	Dwarfism Feeding disorder	GH ↑ Modification of leptin, NPY, DA response

GalR2 also binds chimeric peptides consisting of the N-terminal part of galanin (galanin (1–13)) covalently connected to sequences derived from other known peptides. The chimeric peptides M15, M35, and M40 (Table 16.5) proved to be high affinity antagonists in most *in vivo* experiments, and were helpful in defining the roles of galanin. However, when used in very high concentrations *in vitro*, these chimeric peptides act as partial agonists because they all present the galanin (1–13) sequence.

16.5 Signal transduction and receptor modulation

Several actions of galanin on effector systems have been reported and include inhibition of adenylyl cyclase, activation of ATP-sensitive K^+ channels, inhibition of L- and N-type Ca^{2+} channels, stimulation and/or inhibition of inositol phospholipid turnover, stimulation of phospholipases C and A_2, activation of MAP kinase, and mitogenesis (Branchek *et al.* 1998). In most pharmacological studies, galanin appears as a strongly hyperpolarizing peptide with actions similar to that of NPY and the enkephalins. In many brain areas and in the spinal cord, galanin, similarly to NPY and the enkephalins, hyperpolarizes presynaptic terminals. It also suppresses the stimulus-evoked release of different neurotransmitters, like glutamate, acetylcholine, and dopamine. Galanin also hyperpolarizes cell bodies of locus coeruleus noradrenergic neurones.

Surprisingly little is known about which G proteins mediate the signal between the three galanin receptor subtypes and the different second messengers, although most studies point at pertussis toxin ADP-ribosylable G_i/G_o proteins. Furthermore, very few studies of G protein signalling by galanin have been carried out directly on brain tissue. In 1987, Fisone *et al.* reported that galanin inhibits acetylcholine release from the rat hippocampus (Fisone *et al.* 1987), and recently, Wang *et al.* reported that galanin inhibits acetylcholine release from cerebral cortex via a PTX-sensitive G_i protein (Wang *et al.* 1999). The authors suggested that, in the cerebral cortex of the rat, $G\alpha_{i3}$ mediated galanin-induced inhibition of spontaneous acetylcholine release, while $G\alpha_{i1}$ was responsible for galanin-mediated inhibition of induced acetylcholine release (Fisone *et al.* 1987; Wang *et al.* 1999). Consistent with its coupling to $G\alpha_{i1}$, galanin in another study was shown to inhibit forskolin-induced cAMP production in the rat hypothalamus and the entorhinal cortex (Billecocq *et al.* 1994). This study did however not determine exactly which G protein(s) coupled to which galanin receptor. Hence, much work still has to be performed before we have a clear picture of the signalling pathways that the different galanin receptors are coupled to in different cell types.

Regarding the GalR1-mediated second messenger signalling, the cloned and endogenously expressed GalR1 has been shown to inhibit cAMP production, open G protein-coupled, inwardly rectifying K^+ channels, and stimulate MAP kinase activity (Burgevin *et al.* 1995; Habert-Ortoli *et al.* 1994; Parker *et al.* 1995; Smith *et al.* 1997; Wang *et al.* 1998). These effects are all sensitive to pertussis toxin-catalyzed ADP ribosylation, which indicates coupling of GalR1 via G proteins of the $G_{i/o}$ subtypes.

GalR2 has been shown to couple to several intracellular pathways. One involves activation of phospholipase C, and binding of galanin to GalR2 affects pertussis toxin-sensitive inositol phosphate hydrolysis, mobilizes intracellular Ca^{2+}, and mediates Ca^{2+}-dependent Cl^- channel activation (Borowsky *et al.* 1998; Fathi *et al.* 1997, 1998*b*; Wang *et al.* 1998). GalR2 can also inhibit cAMP production (Fathi *et al.* 1998*a*, Wang *et al.* 1998).

It should be pointed out that many second messenger coupling studies with cloned galanin receptors have been carried out in cells with transiently or stably transfected receptors. This approach may sometimes result in interpretations that pertain more to the *in vitro* system than to the real *in vivo* conditions, since the overexpression of receptors often found in such cells may lead to shifted receptor/G protein stoichiometry and interactions. It may be so, that the G protein type that the receptor normally prefers and requires under endogenous conditions required to transmit signalling is missing in a particular type of cell used for expression of the GalR, and as a result there may instead be receptor-G protein interactions occurring that never normally would occur *in vivo*.

Wang *et al.* performed signalling experiments with GalR1 and GalR2 expressed in both CHO cells and COS-7 cells (Wang *et al.* 1998). The IP_3 accumulation, MAP kinase activity, PKC inhibition and cAMP production were measured in the two cell lines. Based on their findings, the authors suggested that GalR1 preferentially couples to G_i, while GalR2 couples to G_i as well as G_o and $G_{q/11}$. In another report by Kolakowski *et al.* the effect of galanin on adenylyl cyclase, phospholipase C (PLC), and intracellular accumulation of Ca^{2+} via the cloned human GalR2 and GalR3 was investigated, as assessed by aequorin luminescence in HEK293 fibroblast cells and pigment aggregation in *Xenopus* melanophores (Kolakowski *et al.* 1998). In this study, the authors suggested that GalR2 couples to PLC and elevates calcium levels, by coupling to $G_{q/11}$ and/or G_s, but not to G_i. At GalR3, in contrast, galanin inhibited adenylyl cyclase via the activation of G_i. Wittau *et al.* recently published a paper, where the signalling pathways for galanin in small lung cancer cells were investigated (Wittau *et al.* 2000). In this study, the GalR2, when expressed in various cells, was shown to activate multiple signalling pathways by coupling to G_q, G_o and G_{12} after interaction with galanin. However, because of the problems that come along with 'heterologous' expression systems, the suggestions made in the reports by Wang, Kolakowski, and Wittau ought to be taken with some caution.

GalR3 has been proven to stimulate pertussis toxin-sensitive activation of inwardly recti-fying K^+ currents when expressed in *Xenopus* oocytes, and has therefore been proposed to act through $G_{i/o}$ proteins.

16.6 Physiology and disease relevance

Galanin has been found to influence several physiological processes, such as cognition and memory, the release of various neurotransmitters and hormones, feeding, motility of the digestive tract, nociception, nerve regeneration, and sexual behaviour and reproduction (Bartfai *et al.* 1993; Crawley 1996, 1999). In most cases, galanin acts as a hyperpolarizing peptide, like in pancreatic β-cells, where it suppresses the release of insulin. The only reported stimulatory action of galanin on the release of a signal substance is on the release of growth hormone in the anterior pituitary, where growth hormone release is increased after i.v. galanin injection in humans.

The distribution and observed pharmacology at the GalR1 receptor centrally and peripher-ally indicates that galanin may act through GalR1 receptor to inhibit the release of hormones (Bartfai *et al.* 1993). GalR1 may also mediate the effects of galanin on the regulation of insulin and glucose homeostasis, being the primary pancreatic receptor for galanin that mediates the stress-induced, sympathetic (noradrenergic, galaninergic) inhibition of insulin release.

Judging from the intracellular coupling and widespread distribution observed for rat GalR2 mRNA, there are many suggested roles for this receptor. GalR2 may mediate the effects of galanin on the release of prolactin and growth hormone, and regulate lactation, feeding, memory, emotion, nociception, nerve regeneration, cardiovascular and pancreatic function, and reproduction. The localization of GalR3 in the regions mentioned above suggests that it may be involved in mechanisms of emotion, feeding, pituitary hormone release, pancreatic function, nociception, and various metabolic pathways.

Table 16.3 indicates the potential pharmacological applications that galanin receptor agon-ists and antagonists could have in neurology and psychiatry. These potential therapeutic applications have been assembled from a very large body of pharmacological experiments *in vivo*, using intracerebroventricularly or intrathecally applied galanin (0.1–10 nmol) and chi-meric type galanin receptor antagonists in different behavioural and endocrine paradigms relevant to feeding, emotion, memory, nociception, gut secretion and motility (Bartfai *et al.* 1993).

16.6.1 Epilepsy

The discovery of galanin's effects on epilepsy, where it acts as an anticonvulsant, is a good example of the pre-synaptic inhibitory effects of galanin on glutamate but not on GABA release (Ben-Ari 1990), and shows how it acts as a postsynaptic hyperpolarizing substance to reduce run-away discharges alike (Mazarati *et al.* 1992, 1998, 2000, 2001). The galanin agonist, galnon, suppresses phenyltetrazole-induced (40 mg/kg) seizures when injected i.p. (2 mg/kg) (Saar, 2002).

Additional information on the pharmacological potential of galanin receptor ligands is now forthcoming from transgenic studies on the role of galanin (Table 16.6). The galanin-deficient mice generated by Wynick *et al.* (1998) has lowered seizure threshold, and some impairment of the development of sensory neurones. Mice strains that over-express galanin under the DβH promoter in noradrenergic cells (Steiner *et al.* 2001), or under the PDGF β-chain promoter in most forebrain neurones (Blakeman *et al.* 2001), have elevated seizure thresholds, owing to the hyperpolarizing effects of galanin in the hippocampus at both CA1/CA3 and the dentate gyrus area (Kokaia *et al.* 2001).

The most instructive for the role of galanin receptors are the GalR1($-/-$) mice generated by Iisma *et al.* which have been studied in several contexts (Iismaa, unpublished material). The most striking phenotype so far found (see above) is the occurrence of spontaneous seizures in some founders, strongly suggesting that GalR1 agonists will be potent anticonvulsant drugs. GalR1($-/-$) transgenic mice have been generated by truncated disruption of the GalR1 gene and these animals are being characterized in several laboratories including our own. In some founders spontaneous seizures were observed (Iismaa, Bartfai, unpublished observations). These data are in congruence with the preliminary data from Mazarati, Wasterlain and Langel, who used antisense oligonucleotides to the GalR1 to knock down the expression of GalR1 *in vivo*, and then found increased seizure susceptibility when GalR1 function was impaired (Mazarati *et al.* 2001).

Table 16.6 Galanin transgenic mice strains suggest that galanin controls hippocampal excitability in seizure models via the galanin type 1 receptor (Ga1R1)

Strain	Expression pattern	Major phenotype
Galanin ($-/-$)[1]	Galanin deficient	Sensory neurones ↓ enhanced hippocampal excitability
GALOE—DβH[2]	3–5 fold overexpression in NE neurones	Suppressed hippocampal excitability Cognitive deficit
GALOE—PDGFβ[3]	5–20 fold overexpression in all CNS and most peripheral neurones	Suppressed hippocampal excitability Altered parasympathetic tone
Ga1R1 ($-/-$)[4]	Ga1R1 deficient	Enhanced hippocampal excitability Spontaneous seizures in some founder lines

[1] Holmes *et al.* 2000.
[2] Steiner *et al.* 2001 (Galanin gene proceeded by the dopamine beta hydroxylase promoter).
[3] Holmgren, Kahl *et al.* 2000 (Galanin gene proceeded by the dopamine beta PDGFb promoter).
[4] Iismaa *et al.* 2001.

16.6.2 Alzheimer's disease and cognition

At GalR1, galanin inhibits the release of hormones or neurotransmitters, affecting memory (Bartfai *et al.* 1993). Galanin acts as a postsynaptic hyperpolarizing substance to reduce LTP (Mazarati *et al.* 1992, 1998, 2000, 2001). One of the most intriguing features of galanin neurobiology is the overexpression of galaninergic fibres around the few surviving cholinergic cell bodies in autopsy material from patients with Alzheimer's disease, primarily in cholinergic basal forebrain neurones (Chan-Palay 1988; Kohler *et al.* 1989; Mufson *et al.* 1998, 2000). One explanation for this phenomenon may be that galanin acts as a trophic factor in this area, promoting growth and survival of some ACh-containing neurones. Since there is a very large loss of cholinergic cells in the Alzheimer's disease brain, factors like galanin, which may influence the survival and function of those cells that project to the hippocampus and play a major role in cognitive function, is of great importance.

16.6.3 Pain and analgesia

GalR1 has also been implicated in pain sensation (Bartfai *et al.* 1993), and galnon acts as a galanin receptor agonist in the spinal cord (Wiesenfeld-Hallin, preliminary results). GalR2 (but not GalR1) mRNA was found to be significantly increased after injury of the facial nerve in an experimental model, and the levels GalR2 mRNA are further elevated in dorsal root ganglia following inflammation (Burazin and Gundlach 1998; Sten Shi *et al.* 1997). Surprisingly, both GalR2 and GalR1 mRNA is downregulated in dorsal root ganglia after axotomy, where the levels of the endogenous ligand galanin is significantly (10–20-fold) upregulated (Sten Shi *et al.* 1997). One explanation for this finding could be the presence of additional, yet unidentified, galanin receptor subtype(s) in these neurones.

16.7 Concluding remarks

The wealth of the pharmacological effects and potency of galanin hence make the galanin receptor subtypes promising drug targets (Table 16.4). Continuing efforts in galanin research will likely reveal additional receptor subtypes, and hopefully there will also soon be more specific and potent ligands to the different subtypes. Such ligands could then serve as leads in the development of specific drugs targeted at central as well as peripheral binding sites for galanin. Considering its observed function in various important biological systems, its widespread distribution, and its modulatory effects on other transmitters' actions, these potential drugs could have applications in the treatment of disorders related to obesity, pain, epilepsy, depression, memory impairment, and neurodegeneration.

References

Ahmad S, O'Donnell D, Payza K, Ducharme J, Menard D, Brown W *et al.* (1998). Cloning and evaluation of the role of rat GALR-2, a novel subtype of galanin receptor, in the control of pain perception. *Ann N Y Acad Sci*, **863**, 108–19.

Amiranoff B, Lorinet AM, and Laburthe M (1989). Galanin receptor in the rat pancreatic beta cell line Rin m 5F. Molecular characterization by chemical cross-linking. *J Biol Chem* **264**, 20 714–7.

Arvidsson U, Ulfhake B, Cullheim S, Bergstrand A, Theodorson E, and Hökfelt T (1991). Distribution of 125I-galanin binding sites, immunoreactive galanin, and its coexistence with 5-hydroxytryptamine in the cat spinal cord: biochemical, histochemical, and experimental studies at the light and electron microscopic level. *J Comp Neurol* **308**, 115–38.

Bartfai T, Hökfelt T, and Langel Ü (1993). Galanin—a neuroendocrine peptide. *Crit Rev Neurobiol* **7**, 229–74.

Ben-Ari Y (1990). Modulation of ATP sensitive K+ channels: a novel strategy to reduce the deleterious effects of anoxia. *Adv Exp Med Biol* **268**, 481–9.

Berthold M and Bartfai T (1997). Modes of peptide binding in G protein-coupled receptors. *Neurochem Res* **22**, 1023–31.

Berthold M, Kahl U, Jureus A, Kask K, Nordvall G, Langel, Ü, and Bartfai T (1997). Mutagenesis and ligand modification studies on galanin binding to its GTP-binding-protein-coupled receptor GalR1. *Eur J Biochem* **249**, 601–6.

Billecocq A, Hedlund PB, Bolanos-Jimenez F, and Fillion G (1994). Characterization of galanin and 5-HT1A receptor coupling to adenylyl cyclase in discrete regions of the rat brain. *Eur J Pharmacol* **269**, 209–17.

Blakeman KH, Holmberg K, Hao JX, Xu XJ, Kahl U, Lendahl U *et al.* (2001). Mice over-expressing galanin have elevated heat nociceptive threshold. *Neuroreport* **12**, 423–5.

Bloomquist BT, Beauchamp MR, Zhelnin L, Brown SE, Gore-Willse AR, Gregor P, and Cornfield LJ (1998). Cloning and expression of the human galanin receptor GalR2. *Biochem Biophys Res Commun* **243**, 474–9.

Borowsky B, Walker MW, Huang LY, Jones KA, Smith KE, Bard J *et al* (1998). Cloning and characterization of the human galanin GALR2 receptor. *Peptides* **19**, 1771–81.

Branchek T, Smith KE, and Walker MW (1998). Molecular biology and pharmacology of galanin receptors. *Ann N Y Acad Sci* **863**, 94–107.

Branchek TA, Smith KE, Gerald C, and Walker MW (2000). Galanin receptor subtypes. *Trends Pharmacol Sci* **21**, 109–17.

Burazin TC and Gundlach AL (1998). Inducible galanin and GalR2 receptor system in motor neuron injury and regeneration. *J Neurochem* **71**, 879–82.

Burgevin MC, Loquet I, Quarteronet D, and Habert-Ortoli E (1995). Cloning, pharmacological characterization, and anatomical distribution of a rat cDNA encoding for a galanin receptor. *J Mol Neurosci* **6**, 33–41.

Chan-Palay V (1988). Neurons with galanin innervate cholinergic cells in the human basal forebrain and galanin and acetylcholine coexist. *Brain Res Bull* **21**, 465–72.

Crawley JN (1996). Minireview. Galanin-acetylcholine interactions: relevance to memory and Alzheimer's disease. *Life Sci* **58**, 2185–99.

Crawley, JN (1999). The role of galanin in feeding behavior. *Neuropeptides* **33**, 369–75.

Crawley JN and Wenk GL (1989). Co-existence of galanin and acetylcholine: is galanin involved in memory processes and dementia? *Trends Neurosci* **12**, 278–82.

Fathi Z, Battaglino PM, Iben LG, Li H, Baker E, Zhang D *et al.* (1998a). Molecular characterization, pharmacological properties and chromosomal localization of the human GALR2 galanin receptor. *Brain Res Mol Brain Res* **58**, 156–69.

Fathi Z, Cunningham AM, Iben LG, Battaglino PB, Ward SA, Nichol KA *et al.* (1997). Cloning, pharmacological characterization and distribution of a novel galanin receptor. *Brain Res Mol Brain Res* **51**, 49–59.

Fathi Z, Cunningham AM, Iben LG, Battaglino PB, Ward SA, Nichol KA *et al.* (1998b). Cloning, pharmacological characterization and distribution of a novel galanin receptor. *Brain Res Mol Brain Res* **53**, 348.

Faure-Virelizier C, Croix D, Bouret S, Prevot V, Reig S, Beauvillain JC, and Mitchell V (1998). Effects of estrous cyclicity on the expression of the galanin receptor Gal-R1 in the rat preoptic area: a comparison with the male. *Endocrinology* **139**, 4127–39.

Fisone G, Berthold M, Bedecs K, Unden A, Bartfai T, Bertorelli R et al. (1989). N-terminal galanin-(1–16) fragment is an agonist at the hippocampal galanin receptor. *Proc Natl Acad Sci USA* **86**, 9588–91.

Fisone G, Wu CF, Consolo S, Nordstrom O, Brynne N, Bartfai T et al. (1987). Galanin inhibits acetylcholine release in the ventral hippocampus of the rat: histochemical, autoradiographic, *in vivo*, and *in vitro* studies. *Proc Natl Acad Sci USA* **84**, 7339–43.

Gustafson EL, Smith KE, Durkin MM, Gerald C, and Branchek TA (1996). Distribution of a rat galanin receptor mRNA in rat brain. *Neuroreport* **7**, 953–7.

Habert-Ortoli E, Amiranoff B, Loquet I, Laburthe M, and Mayaux JF (1994). Molecular cloning of a functional human galanin receptor. *Proc Natl Acad Sci USA* **91**, 9780–3.

Howard AD, Tan C, Shiao LL, Palyha OC, McKee KK, Weinberg DH et al. (1997). Molecular cloning and characterization of a new receptor for galanin. *FEBS Lett* **405**, 285–90.

Iismaa TP, Fathi Z, Hort YJ, Iben LG, Dutton JL, Baker E et al. (1998). Structural organization and chromosomal localization of three human galanin receptor genes. *Ann N Y Acad Sci* **863**, 56–63.

Jacoby AS, Webb GC, Liu ML, Kofler B, Hort YJ, Fathi Z et al. (1997). Structural organization of the mouse and human GALR1 galanin receptor genes (Galnr and GALNR) and chromosomal localization of the mouse gene. *Genomics* **45**, 496–508.

Kask K, Berthold M, Kahl U, Nordvall G, and Bartfai T (1996). Delineation of the peptide binding site of the human galanin receptor. *Embo J* **15**, 236–44.

Kohler C and Chan-Palay V (1990). Galanin receptors in the post-mortem human brain. Regional distribution of 125I-galanin binding sites using the method of *in vitro* receptor autoradiography. *Neurosci Lett* **120**, 179–82.

Kohler C, Persson A, Melander T, Theodorsson E, Sedvall G, and Hokfelt T (1989). Distribution of galanin-binding sites in the monkey and human telencephalon: preliminary observations. *Exp Brain Res* **75**, 375–80.

Kokaia M, Holmberg K, Nanobashvili A, Xu ZQ, Kokaia Z, Lendahl U et al. (2001). Suppressed kindling epileptogenesis in mice with ectopic overexpression of galanin. *Proc Natl Acad Sci USA* **98**, 14006–11.

Kolakowski LF, Jr, O'Neill GP, Howard AD, Broussard SR, Sullivan KA, Feighner SD et al. (1998). Molecular characterization and expression of cloned human galanin receptors GALR2 and GALR3. *J Neurochem* **71**, 2239–51.

Land T, Langel U, Low M, Berthold M, Unden A, and Bartfai T (1991). Linear and cyclic N-terminal galanin fragments and analogs as ligands at the hypothalamic galanin receptor. *Int J Pept Protein Res* **38**, 267–72.

Landry M, Aman K, and Hokfelt T (1998). Galanin-R1 receptor in anterior and mid-hypothalamus: distribution and regulation. *J Comp Neurol* **399**, 321–40.

Liu HX, Brumovsky P, Schmidt R, Brown W, Payza K, Hodzic L et al. (2001). Receptor subtype-specific pronociceptive and analgesic actions of galanin in the spinal cord: selective actions via GalR1 and GalR2 receptors. *Proc Natl Acad Sci USA* **98**, 9960–4.

Lorimer DD and Benya RV (1996). Cloning and quantification of galanin-1 receptor expression by mucosal cells lining the human gastrointestinal tract. *Biochem Biophys Res Commun* **222**, 379–85.

Lorimer DD, Matkowskj K, and Benya RV (1997). Cloning, chromosomal location, and transcriptional regulation of the human galanin-1 receptor gene (GALN1R). *Biochem Biophys Res Commun* **241**, 558–64.

Mazarati A, Langel U, and Bartfai T (2001). Galanin: an endogenous anticonvulsant? *Neuroscientist* **7**, 506–17.

Mazarati AM, Halaszi E, and Telegdy G (1992). Anticonvulsive effects of galanin administered into the central nervous system upon the picrotoxin-kindled seizure syndrome in rats. *Brain Res* **589**, 164–6.

Mazarati AM, Hohmann JG, Bacon A, Liu H, Sankar R, Steiner RA *et al.* (2000). Modulation of hippocampal excitability and seizures by galanin. *J Neurosci* **20**, 6276–81.

Mazarati AM, Liu H, Soomets U, Sankar, R, Shin D, Katsumori H *et al.* (1998). Galanin modulation of seizures and seizure modulation of hippocampal galanin in animal models of status epilepticus. *J Neurosci* **18**, 10 070–7.

Melander T, Hokfelt T, and Rokaeus A (1986*a*). Distribution of galaninlike immunoreactivity in the rat central nervous system. *J Comp Neurol* **248**, 475–517.

Melander T, Hokfelt T, Rokaeus A, Cuello AC, Oertel WH, Verhofstad A, and Goldstein M (1986*b*). Coexistence of galanin-like immunoreactivity with catecholamines, 5-hydroxytryptamine, GABA and neuropeptides in the rat CNS. *J Neurosci* **6**, 3640–54.

Melander T and Staines WA (1986). A galanin-like peptide coexists in putative cholinergic somata of the septum-basal forebrain complex and in acetylcholinesterase-containing fibers and varicosities within the hippocampus in the owl monkey (Aotus trivirgatus). *Neurosci Lett* **68**, 17–22.

Melander T, Staines WA, Hokfelt T, Rokaeus A, Eckenstein F, Salvaterra PM, and Wainer BH (1985). Galanin-like immunoreactivity in cholinergic neurons of the septum-basal forebrain complex projecting to the hippocampus of the rat. *Brain Res* **360**, 130–8.

Merchenthaler I (1991). The hypophysiotropic galanin system of the rat brain. *Neuroscience* **44**, 643–54.

Mufson EJ, Deecher DC, Basile M, Izenwasse S, and Mash DC (2000). Galanin receptor plasticity within the nucleus basalis in early and late Alzheimer's disease: an *in vitro* autoradiographic analysis. *Neuropharmacology* **39**, 1404–12.

Mufson EJ, Kahl U, Bowser R, Mash DC, Kordower JH, and Deecher DC (1998). Galanin expression within the basal forebrain in Alzheimer's disease. Comments on therapeutic potential. *Ann N Y Acad Sci* **863**, 291–304.

Nicholl J, Kofler B, Sutherland GR, Shine J, and Iismaa TP (1995). Assignment of the gene encoding human galanin receptor (GALNR) to 18q23 by in situ hybridization. *Genomics* **30**, 629–30.

O'Donnell D, Ahmad S, Wahlestedt C, and Walker P (1999). Expression of the novel galanin receptor subtype GALR2 in the adult rat CNS: distinct distribution from GALR1. *J Comp Neurol* **409**, 469–81.

Ohman A, Lycksell PO, Andell, S, Langel U, Bartfai T, and Graslund A (1995). Solvent stabilized solution structures of galanin and galanin analogs, studied by circular dichroism spectroscopy. *Biochim Biophys Acta* **1236**, 259–65.

Ohman A, Lycksell PO, Jureus A, Langel, U, Bartfai T, and Graslund A (1998). NMR study of the conformation and localization of porcine galanin in SDS micelles. Comparison with an inactive analog and a galanin receptor antagonist. *Biochemistry* **37**, 9169–78.

Ohtaki T, Kumano S, Ishibashi Y, Ogi K, Matsui H, Harada M *et al.* (1999). Isolation and cDNA cloning of a novel galanin-like peptide (GALP) from porcine hypothalamus. *J Biol Chem* **274**, 37 041–5.

Pang L, Hashemi T, Lee HJ, Maguire M, Graziano MP, Bayne M *et al.* (1998). The mouse GalR2 galanin receptor: genomic organization, cDNA cloning, and functional characterization. *J Neurochem* **71**, 2252–9.

Parker EM, Izzarelli DG, Nowak HP, Mahle CD, Iben LG, Wang J, and Goldstein ME (1995). Cloning and characterization of the rat GALR1 galanin receptor from Rin14B insulinoma cells. *Brain Res Mol Brain Res* **34**, 179–89.

Puar MS, Barrabee E, Hallade M, and Patel M (2000). Sch 420789: a novel fungal metabolite with phospholipase D inhibitory activity. *J Antibiot (Tokyo)* **53**, 837–8.

Rossowski WJ, Rossowski TM, Zacharia S, Ertan A, and Coy DH (1990). Galanin binding sites in rat gastric and jejunal smooth muscle membrane preparations. *Peptides* **11**, 333–8.

Saar K, Mazarati AM, Mahlapuu R, Hallnemo G, Soomets U, Kilk K, Hellberg S, Pooga M, Tolf B-R, Shi TS, Hökfelt T, Wasterlain C, Bartfai T, and Langel U (2002). Anticonvulsant activity of a nonpeptide galanin receptor agonist. *Proc Natl Acad Sci USA* **99**, 7136–41.

Scott MK, Ross TM, Lee DH, Wang HY, Shank RP, Wild KD *et al.* (2000). 2,3-Dihydro-dithiin and-dithiepine-1,1,4,4-tetroxides: small molecule non-peptide antagonists of the human galanin hGAL-1 receptor. *Bioorg Med Chem* 8, 1383–91.

Servin AL, Amiranoff B, Rouyer-Fessard C, Tatemoto K, and Laburthe M (1987). Identification and molecular characterization of galanin receptor sites in rat brain. *Biochem Biophys Res Commun* 144, 298–306.

Skofitsch G and Jacobowitz DM (1985). Galanin-like immunoreactivity in capsaicin sensitive sensory neurons and ganglia. *Brain Res Bull* 15, 191–5.

Skofitsch G, Jacobowitz DM, Amann R, and Lembeck F (1989). Galanin and vasopressin coexist in the rat hypothalamo-neurohypophyseal system. *Neuroendocrinology* 49, 419–27.

Skofitsch G, Sills MA, and Jacobowitz DM (1986). Autoradiographic distribution of 125I-galanin binding sites in the rat central nervous system. *Peptides* 7, 1029–42.

Smith KE, Forray C, Walker MW, Jones KA, Tamm JA, Bard J *et al.* (1997). Expression cloning of a rat hypothalamic galanin receptor coupled to phosphoinositide turnover. *J Biol Chem* 272, 24 612–6.

Smith KE, Walker MW, Artymyshyn R, Bard J, Borowsky B, Tamm JA *et al.* (1998). Cloned human and rat galanin GALR3 receptors. Pharmacology and activation of G-protein inwardly rectifying K+ channels. *J Biol Chem* 273, 23 321–6.

Steiner RA, Hohmann JG, Holmes A, Wrenn CC, Cadd G, Jureus A *et al.* (2001). Galanin transgenic mice display cognitive and neurochemical deficits characteristic of Alzheimer's disease. *Proc Natl Acad Sci USA* 98, 4184–9.

Sten Shi TJ, Zhang X, Holmberg K, Xu ZQ and Hokfelt T (1997). Expression and regulation of galanin-R2 receptors in rat primary sensory neurons: effect of axotomy and inflammation. *Neurosci Lett* 237, 57–60.

Sullivan KA, Shiao LL and Cascieri MA (1997). Pharmacological characterization and tissue distribution of the human and rat GALR1 receptors. *Biochem Biophys Res Commun* 233, 823–8.

Tatemoto K, Rokaeus A, Jornvall H, McDonald TJ, and Mutt V (1983). Galanin—a novel biologically active peptide from porcine intestine. *FEBS Lett* 164, 124–8.

Wang HY, Wild KD, Shank RP, and Lee DH (1999). Galanin inhibits acetylcholine release from rat cerebral cortex via a pertussis toxin-sensitive G(i)protein. *Neuropeptides* 33, 197–205.

Wang S, Hashemi T, Fried S, Clemmons AL, and Hawes BE (1998). Differential intracellular signaling of the GalR1 and GalR2 galanin receptor subtypes. *Biochemistry* 37, 6711–7.

Wang S, Hashemi T, He C, Strader C, and Bayne M (1997a). Molecular cloning and pharmacological characterization of a new galanin receptor subtype. *Mol Pharmacol* 52, 337–43.

Wang S, He C, Hashemi T, and Bayne M (1997b). Cloning and expressional characterization of a novel galanin receptor. Identification of different pharmacophores within galanin for the three galanin receptor subtypes. *J Biol Chem* 272, 31 949–52.

Wang S, He C, Maguire MT, Clemmons AL, Burrier RE, Guzzi MF *et al.* (1997c). Genomic organization and functional characterization of the mouse GalR1 galanin receptor. *FEBS Lett* 411, 225–30.

Waters SM, and Krause JE (2000). Distribution of galanin-1, -2 and -3 receptor messenger RNAs in central and peripheral rat tissues. *Neuroscience* 95, 265–71.

Wiesenfeld-Hallin Z, Bartfai T, and Hokfelt T (1992). Galanin in sensory neurons in the spinal cord. *Front Neuroendocrinol* 13, 319–43.

Wittau N, Grosse R, Kalkbrenner F, Gohla A, Schultz G, and Gudermann T (2000). The galanin receptor type 2 initiates multiple signaling pathways in small cell lung cancer cells by coupling to G(q), G(i) and G(12) proteins. *Oncogene* 19, 4199–209.

Wynick D, Small CJ, Bloom SR, and Pachnis V (1998). Targeted disruption of the murine galanin gene. *Ann N Y Acad Sci* 863, 22–47.

Wynick D, Smith DM, Ghatei M, Akinsanya K, Bhogal R, Purkiss P *et al.* (1993). Characterization of a high-affinity galanin receptor in the rat anterior pituitary: absence of biological effect and reduced membrane binding of the antagonist M15 differentiate it from the brain/gut receptor. *Proc Natl Acad Sci USA* **90**, 4231–5.

Xu Z. Q, Ma X, Soomets U, Langel U, and Hokfelt T (1999). Electrophysiological evidence for a hyperpolarizing, galanin (1–15)-selective receptor on hippocampal CA3 pyramidal neurons. *Proc Natl Acad Sci USA* **96**, 14 583–7.

Chapter 17

Histamine receptors

Lindsay B. Hough and Rob Leurs

17.1 Introduction

Histamine [2-(4-imidazolyl)ethylamine, HA] is an endogenous biogenic amine synthesized in and released from several kinds of cells, and involved in many kinds of biological signalling (Uvnas 1991). Mast cells (a family of bone marrow-derived secretory cells) store much of the body's HA in very high concentrations. These cells respond with HA release to a myriad of diverse signals (including tissue damage, antigen exposure, and some kinds of drugs), yet the significance of mast cell HA under strictly physiological conditions remains unknown. In contrast, HA in the gastric mucosa is stored in and released from specialized cells resembling enterochromaffin cells, where it plays a vital physiological role as a mediator of normal gastric acid secretion. HA is also a brain neurotransmitter (Brown *et al.* 2001; Hough 1999). Located in the tuberomammillary nucleus of the hypothalamus, histaminergic neurones project widely throughout the brain and, as discussed in this chapter, are thought to participate in several brain functions.

This chapter reviews information about the protein biosensors which recognize HA and initiate the transduction of extracellular HA levels into an incredible repertoire of biological responses. Related topics which are also discussed include the drugs known to act on these receptors, and the physiological and pathological significance of the receptors. Many earlier reviews on HA receptors have been published (Hill *et al.* 1997; Leurs *et al.* 1998; Hough 2001; Arrang *et al.* 1995; Leurs *et al.* 1995*b*; Hill 1992) and the reader is also referred to broader reviews of HA (Brown *et al.* 2001; Hough 1999; Onodera *et al.* 1994; Schwartz *et al.* 1991; Hough 1988; Silver *et al.* 1996; Barke and Hough 1993). Topics such as the synthesis, storage, release, and metabolism of HA (and drugs acting on these mechanisms) are not discussed in the present chapter, but are covered in some of these other reviews.

17.2 Molecular characterization

For many years the molecular characteristics of the targets for HA remained unknown. Until the early nineteen nineties, the classification of the HA receptors was based solely on pharmacological findings. Although biochemical data suggested that HA receptors were G protein-coupled receptors (GPCRs) (e.g. signal transduction pathways, GTP-shifts of agonist binding), unequivocal proof of this was not provided for any HA receptor until 1991. Now, four different HA receptor subtypes (all GPCRs) have been identified. Except for the H_3 and H_4 receptor, the homology between the various HA receptors is very low (Fig. 17.1), even in the seven transmembrane domains. Nevertheless, all four of these receptor proteins contain a negatively charged aspartate residue in transmembrane (TM) 3, which is considered to be the anchor point for the protonated amine function of HA. The receptor proteins all contain the various conserved amino acids of the rhodopsin-like family (e.g. Asp in TM2, 2 conserved Cys residues forming a disulphide bridge, Asp-Arg-Tyr/Phe at the end of TM3, Asn-Pro-X-X-Tyr motif at the end of TM6), including an eighth alpha helical structure in the C-terminal tail. In this section, we will review the present biochemical and molecular biological data on the characteristics of the HA receptors.

17.2.1 H_1 receptor

The identification of the H_1 receptor protein was initiated with the use of [^{3}H]azidobenzamide by Yamashita et al. (1991b) to irreversibly label high amounts of receptor peptides (53–58 kDa) in bovine adrenal medulla membranes. Following these findings, this group cloned the gene encoding the bovine H_1 receptor (Yamashita et al. 1991a) by an elegant approach of expression cloning. Injection of bovine adrenal medulla mRNA into Xenopus oocytes resulted in the induction of HA-induced chloride current. By fractionation, a single cDNA was finally isolated. This H_1-receptor cDNA encodes for a 491 amino acid receptor protein (apparent molecular weight 56 kDa) with all of the structural features of a GPCR (see Fig. 17.1) (Yamashita et al. 1991a). Since biochemical studies indicate that the H_1-receptor protein is glycosylated (Garbarg et al. 1985), the predicted molecular weight of 56 kDa is certainly underestimated. Currently, genes for the rat (Fujimoto et al. 1993), guinea pig (Traiffort et al. 1994), mouse (Inoue et al. 1996) and human (De Backer et al. 1993; Fukui et al. 1994; Moguilevsky et al. 1994) H_1 receptor have been cloned. The H_1 receptor protein is encoded by a single exon and contains 486 (rat), 488 (guinea pig, mouse), 491 (bovine), or 487 (human) amino acids. De Backer et al. (1998) recently identified a previously unknown intron of approximately 5.8 kB in the 5' untranslated region immediately upstream of the start codon. The importance of this intron has not been clarified. The homology between the H_1-receptor proteins across species is very high in the transmembrane domains (ca. 90 per cent), but is significantly lower in the intracellular and extracellular parts. The human H_1 receptor is located on chromosome 3 (Table 17.1).

Site-directed mutagenesis studies indicate that binding of agonists and antagonists mainly occurs within the TM domain. The protonated amine function of HA interacts with the Asp[107] residue in TM3 (Ohta et al. 1994), whereas the binding of the imidazole ring is suggested to occur via hydrogen bonding contacts with Asp[198] and Lys[191] residues in TM5 (Leurs et al. 1994a, 1995a). Like HA, the antagonists also use the conserved aspartate residue in TM3 as a counter ion for their protonated amine function (Ohta et al. 1994). Mutagenesis studies revealed that Phe[433] and Phe[435] are likely to be involved in the binding of the aromatic rings of the H_1 antagonists. Moreover, an additional interaction point (Lys[191] in TM5) was suggested for the acidic side-chain in the non-sedative, zwitterionic H_1 antagonists acrivastine and cetirizine (Gillard et al. 2002; Wieland et al. 1999), whereas Thr[194] is suggested to regulate the known H_1 antagonist stereospecificity (Gillard et al. 2002).

17.2.2 H_2 receptor

The gene and cDNA encoding the H_2 receptor have been identified in several species including man (Gantz et al. 1991a,b; Ruat et al. 1991; Traiffort et al. 1995). Using the polymerase chain reaction (PCR) with degenerate oligonucleotides based on the homology between various GPCRs, and canine gastric parietal cDNA, Gantz et al. (1991b) obtained the first gene

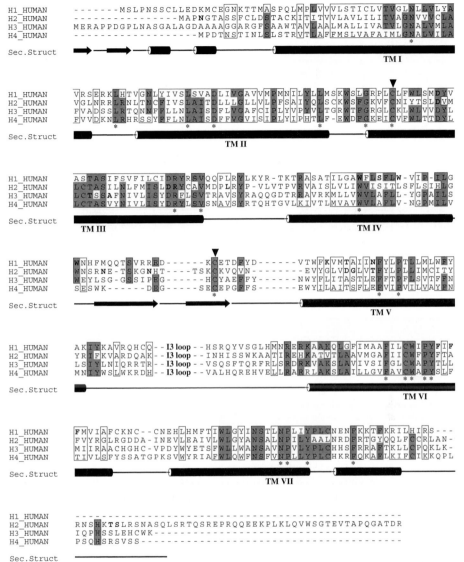

Fig. 17.1 Amino acid sequence alignment of the four human HA receptors using ClustalX (Thompson *et al.* 1997) (see also ftp://ftp-igbmc.u-strasbg.fr/pub/ClustalX/). For ease of presentation, portions of the third intracellular loops (I3) have been omitted. The boxes indicate relative homologous protein domains, whereas the shaded areas indicate complete amino acid conservation between the subtypes. The asterisks indicate amino acid conservation with the rhodopsin sequence, whereas the filled triangles indicate the two Cys residues that are considered to form a disulphide bridge. Residues in bold have been mutated and studied for their functional role in receptor function (ligand binding, signalling, glycosylation, phosphorylation), as indicated in the tinyGRAP/GPCRDB database (Kristiansen *et al.* 1996) (see also http://tinyGRAP.uit.no/). Below the alignment the secondary structure prediction based on the homology with the rhodopsin structure is indicated (arrow = beta-sheet, cylinder = alpha-helix). The putative TM are indicated as well.

Table 17.1 Synopsis of histamine receptors

	H₁	H₂	H₃	H₄
Alternative names	none	none	none	none
Structural information (Accession no.) * = splice variants rep.	Human: 487 aa (P35367) Rat: 486 aa (NP_058714) Mouse: 488 aa (NP_032311)	Human: 359 aa (NP_071640) Rat: 358 aa (NP_037097) Mouse: 358 aa (NP_032312)	Human: 445 aa (Q9Y5N1)* Rat: 445 aa (Q9QYN8)* Mouse: 445 aa (P58406)	Human: 390 aa (Q9H3N8) Rat: 391 aa (AAK97381) Mouse: 391 aa (AAK97380)
Chromosomal location (Human)	3p25	5q35.3	20q13.33	18q11.2
Selective agonists	2-Thiazolylethylamine Histaprodifen	Dimaprit Amthamine Impromidine	R-α-Methylhistamine[a] Imetit[a] Immepip[a]	None (Histamine[a] Immepip[a])
Selective antagonists	Pyrilamine (i.e. Mepyramine) d-Chlorpheniramine	Ranitidine Famotidine Iodoaminopotentidine Zolantidine	Thioperamide[a] Clobenpropit Ciproxyfan	None (Thioperamide[a])
Radioligands	³H-Pyrilamine ¹¹C-Doxepin	³H-Tiotidine ¹²⁵I-Iodoaminopotentidine	³H-R-α-methylhistamine ³H-Nα-methylhistamine ¹²⁵I-Iodophenpropit ¹²⁵I-Iodoproxyfan	³H-Histamine ³H-Nα-methylhistamina

	Gαq/11	Gαs	Gαi/o	Gαi/o
G protein coupling	Gαq/11	Gαs	Gαi/o	Gαi/o
Expression profile	Heart, smooth muscle, CNS, endothelium, adrenal medulla, sensory neurones	Gastric mucosa, T-lymphocytes, mast cells, basophils, CNS	CNS	Eosinophils, basophils
Physiological function	Endothelium-derived vasodilation; increased vascular permeability; intestinal and smooth muscle contraction; CNS wakefulness; decreased appetite; CNS depolarization	Gastric secretion; increased chronotropic and inotropic effects; CNS inhib of voltage- and calcium-activated K channels	Inhibition of release of several transmitters; increased sleep; decreased food and water intake	Possible immunomodulation
Knockout phenotype	Impaired locomotor and exploratory behaviour; defective leptin (feeding) and orexin A (arousal) responses; decreased nociception, increased IgE antibody responses	Gastric hypertropy and hypergastrinemia; increased T-helper cell cytokine release	Unknown	Unknown
Disease association	Alzheimer's, epilepsy, narcolepsy, obesity	Gastric acid disorders, migraine	ADHD, Alzheimer's, epilepsy, obesity	Unknown

A summary of the main features of the four known histamine GPCRs is shown. [a]Many 'selective' H_3 agonists and H_3 antagonists also have appreciable activity at H_4 receptors.

encoding of the H_2 receptor, which is intronless (Gantz *et al.* 1991*b*). The sequences from various species show considerable homology (80–90 per cent) and encode for a receptor protein of 359 (dog, man, and guinea-pig) or 358 (rat and mouse) amino acids, with an apparent molecular weight of approximately 40 kDa. The human receptor is located on chromosome 5. Since a consensus sequence for N-linked glycosylation is present in the N-terminal extracellular tail, the actual molecular weight of the receptor is significantly higher. Currently, a growing body of evidence suggests that GPCR dimers can be formed and may play an important role in receptor function. Interestingly, H_2-receptor oligomers have been observed in Sf9 cells (Beukers *et al.* 1997; Fukushima *et al.* 1997), although the significance of this finding remains uncertain.

With respect to receptor recognition, the H_2 receptor (like the H_1) uses the highly conserved aspartate residue in TM3 (Fig. 17.1) for the binding of HA and H_2-receptor antagonists (Gantz *et al.* 1992). Furthermore, mutagenesis studies indicate that the Thr and Asp residues in TM5 of the H_2 receptor are likely to interact with HA and the antagonist [^{3}H]tiotidine. More details on the ligand binding site of the human H_2 receptor are currently not available.

17.2.3 H_3 receptor

Whereas the genes encoding the H_1 and H_2 receptors were cloned in 1991, the molecular architecture of the H_3 receptor was not known until 1999 when Lovenberg *et al.* (1999) finally showed that the H_3 receptor also belongs to the GPCR super-family. The cloning of the H_3-receptor gene also shows the impact of genome projects on the life sciences. In a search for orphan GPCRs, a GPCR-related EST-sequence was identified *in silico* and used to clone a full-length human cDNA. The cDNA contained an open reading frame of 445 amino acids with an Asp residue in TM3 (Fig. 17.1), which suggested the protein to be a biogenic amine receptor. After expression of the cDNA, functional responses to HA were observed, and the protein was shown to be the H_3 receptor. This protein shows very low homology with other GPCRs. Overall homology between the H_3 receptor and the H_1 and H_2 receptor amounts only to 22 and 20 per cent, respectively. Within the TM domains the homology is still only 27 and 33 per cent, respectively. This remarkably low sequence homology explains why the H_3-receptor gene was not cloned by homology screening with H_1- or H_2-receptor specific probes. With the information on the human H_3-receptor cDNA, the rat, guinea-pig, and mouse cDNAs were cloned; all of these proteins have 445 amino acids (Drutel *et al.* 2001; Lovenberg *et al.* 2000; Tardivel-Lacombe *et al.* 2000).

The H_3-receptor gene is located on chromosome 20 and contains at least three introns (Fig. 17.2). Consequently, various isoforms have been identified for the human, rat, and guinea-pig H_3 receptor as a result of alternative splicing (Coge *et al.* 2001*a*; Drutel *et al.* 2001; Tardivel-Lacombe *et al.* 2000). For the rat H_3 receptor, splicing of the first intron results in the introduction of a stop codon after TM2 and a truncated, non-functional receptor (not shown, see Drutel *et al.* 2001). Splicing of the third intron results in H_3-receptor proteins, which lack 32 (H_{3B}) or 48 amino acids (H_{3C}) in the third intracellular loop (I3) (Drutel *et al.* 2001). One isoform with a deletion in I3 was also found for the guinea-pig H_3 receptor (Tardivel-Lacombe *et al.* 2000). The I3 loop is known to be important for the GPCR signalling and indeed important differences have in this respect been reported for the three various rat isoforms. Finally, alternative splicing of a putative fourth intron eliminates the part encoding TM7, the C-terminal tail and the stop codon and introduces a new open reading frame with a putative new seventh TM domain. Combination of the splicing of this

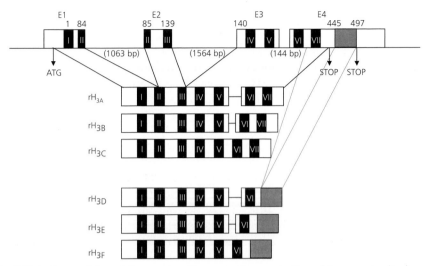

Fig. 17.2 Isoforms of the rat H₃ receptor. The genomic organization of the receptor (top) shows four exons (boxes, E1–E4) and three introns (sizes given in parentheses). Sequences corresponding to translated TM regions are identified as roman numerals (I–VII). An alternatively spliced region (grey box) is depicted between the stop sequences. Six isoforms of the receptor (labelled on the left) have been shown to form as a result of alternative splicing.

fourth intron and alternative splicing of the third intron results in three additional isoforms. Preliminary data (Leurs *et al.* unpublished) indicate that these new proteins are not able to bind ligands. For the human H₃-receptor, a similar complexity has been observed. Currently, at least nine different human H₃-receptor isoforms have been identified as a result of various alternative splicing events (Coge *et al.* 2001*a*; Wellendorf *et al.* 2000). Again, splicing of the first intron at the C-terminal end of TM2 occurs, resulting in an inframe deletion of 14 amino acids and a non-functional receptor. Splicing within the I3 region results in the generation of H₃-receptor proteins with 30, 80, and 116 amino acid deletions within I3 or a 144 amino acid deletion within I3 and TM6 and TM7. Another splicing event results in the deletion of 119 amino acids within TM5 and I3, whereas splicing within TM4 results in a frameshift and the loss of TM4, TM5, and a part of I3. Finally, double splicing events create H₃-receptor proteins with either the 80 or 144 amino acid deletion and a novel eight amino acid C-terminus. Currently, only binding data and functional data have been reported for the various 80 and the 144 amino acid deletion variants and the H₃-receptor protein that is generated after the frame shift. Only the two 80 amino acid deletion variants act as functional receptor proteins, although some controversy exists regarding their signalling (Coge *et al.* 2001*a*; Wellendorf *et al.* 2000). As found for the rat H₃ₐ and H₃ᵦ isoforms, one study reported that the 80 amino acid deletion variant signals more efficiently compared with the 445 amino acid receptor protein and the former also shows a higher affinity for H₃ agonists in binding studies (Wellendorf *et al.* 2000).

Based on the amino acid alignment (Fig. 17.1) and an intial 3D homology model, a role for Asp[114] has been suggested to be important in ligand binding to the H₃ receptor (Ligneau *et al.* 2000). Very recently, mutagenesis of the H₃ receptor has been used to explore agonist binding and receptor activation (Uveges *et al.* 2002). Mutation of Asp[114] indeed abolishes radioligand binding and agonist action. Moreover, an alanine scan of TM5 has indicated that Glu^{206} plays a key role in the interaction of agonist with the H_3 receptor (Uveges *et al.* 2002). With respect to antagonist binding to the receptor, Thr^{119} and Ala^{122} are essential (probably due to indirect interactions) (Ligneau *et al.* 2000), explaining the marked species differences for some H_3 antagonists (e.g. thioperamide, ciproxyfan), but not for others (clobenpropit) (Ligneau *et al.* 2000).

17.2.4 H_4 receptor

A new HA receptor on chromosome 18 was rapidly identified by various groups by the use of the sequence information on the H_3 receptor (Liu *et al.* 2001*a*; Morse *et al.* 2001; Nakamura *et al.* 2000; Nguyen *et al.* 2001; Oda *et al.* 2000; Zhu *et al.* 2001). Within the transmembrane domains, the new H_4 receptor shows approximately 60 per cent homology with the H_3 receptor, whereas little overall sequence homology with the other HA receptors can be noticed (Fig. 17.1). The high homology with the H_3 receptor also explains the overlapping pharmacology; many H_3 agonists and antagonists also act on the H_4 receptor, although differences have been observed. The human H_4 receptor gene exhibits an exon/intron arrangement identical to the H_3 receptor gene (Oda *et al.* 2000). In view of the discovery of multiple H_3-receptor isoforms, H_4 isoforms may be identified soon. No actual data on the ligand binding site are currently available, but a remarkable species specificity has been reported (Liu *et al.* 2001*b*). The amino acid sequences of rat, mouse, guinea-pig and human H_4 receptors are overall 65–70 per cent homologous and show substantial differences in their pharmacological profile; the affinity of $[^3H]$-HA varies considerably, from 4 nM (human) to 136 nM (rat) (Liu *et al.* 2001*b*).

17.2.5 Additional sites of HA action

There is ample evidence indicating that HA can induce responses which are not mediated by any of the four GPCRs discussed presently, although the significance of these remains uncertain. HA can modulate NMDA receptor activity by an action at the polyamine binding site (Brown *et al.* 2001). HA can also act on ion channels directly in invertebrates (Brown *et al.* 2001; Gisselmann *et al.* 2002; Zheng *et al.* 2002). In mammals, there is evidence for histaminergic ionotropic transmission which may involve both excitatory and inhibitory receptors (Hatton and Yang 1996, 2001), but these receptors have not yet been isolated or characterized. There may also be additional HA-activated GPCRs yet to be discovered and characterized.

17.3 Cellular and subcellular localization

Radioligands for the H_1 receptor have been available for twenty-five years. Studies with $[^3H]$-mepyramine (i.e. pyrilamine) have shown a diverse distribution of H_1 binding sites, including CNS, retina, adrenal medulla, cerebral and non-cerebral endothelial cells, astrocytes, lymphocytes as well as vascular, intestinal, and genitourinary smooth muscle (Hill 1990; Hill *et al.* 1997). Within the CNS, both brain and spinal cord binding has been observed. Marked species differences among mammals have been noted in the regional distrbution of H_1 binding (Palacios *et al.* 1981; Yanai *et al.* 1990*b*), unlike the distribution of histaminergic neurones, which is highly conserved within this group (Panula and Airaksinen 1991). H_1 binding has also been observed in the superficial dorsal horn and within dorsal root ganglia, suggesting the presence of these receptors on primary afferent fibres (Taylor *et al.* 1982; Ninkovic and Hunt 1985). Although useful as an H_1 ligand, labelled mepyramine can also bind to non-H_1 sites in liver and in some cell lines (Liu *et al.* 1994; Leurs *et al.* 1989; Hill *et al.* 1997). Quinine reduces some (but not all) of this binding (Hill *et al.* 1997), and recent knock-out studies confirm the caution which must be used with this ligand in studying H_1 receptor binding outside the CNS (Yanai *et al.* 1998) Often overlooked is the very high affinity H_1 binding exhibited by tricyclic antidepressants, which permits $[^{11}C]$-doxepin to label human brain receptors in living subjects (Yanai *et al.* 1992*b*). The technique yields activity in good agreement with results from binding studies in animals (Yanai *et al.* 1990*a*), and it has been used to assess the brain penetration of systemically-administered H_1 antagonists in man (Yanai *et al.* 1995). Other radioligands for the H_1 receptor have also been used successively (see Hill *et al.* 1997). In brain, H_1-receptor mRNA distribution generally follows the distribution of H_1 ligand binding. Exceptions to this are in the cerebellum and hippocampal complex in guinea pig (Traiffort *et al.* 1994). In the rat, message levels are lower than binding signals in hypothalamus (Lintunen *et al.* 1998). In the guinea-pig dorsal root ganglion, *in situ* hybridization experiments have confirmed the existence of H_1 receptors on sensory neurones (Kashiba *et al.* 2001).

In contrast to the H_1-receptor methods, radioligand binding methods for the H_2 receptor have been difficult to develop. Radiolabelled cimetidine yielded specific binding, but not to the H_2 receptor (Smith *et al.* 1980), whereas $[^3H]$-tiotidine gave an accurate, but low specific binding signal in brain (Gatjkowski *et al.* 1983). Development of $[^{125}I]$-iodopotentidine permitted identification and characterization of H_2 binding sites in guinea pig (Ruat *et al.* 1990) and primate brain (Martinez-Mir *et al.* 1990). In both, highest H_2 densities were found in caudate, putamen, accumbens, and in some layers of the cerebral cortex. Although a neuronal localization is clear, brain H_2 receptors are also found in astrocytes, choroid plexus, and brain microvessel fractions. The H_2-receptor activation of adenylate cyclase has been of great value is discerning H_2-receptor localization (Johnson 1982). In addition to the well-established existence of these receptors in heart and gastric muscosa, the H_2 receptor-linked cyclase has permitted identification of the receptors in lymphocytes, basophils, neutrophils, adipocytes, vascular and other smooth muscle, kidney and thyroid (Johnson 1982). *In situ* hybridization studies of H_2-receptor mRNA distribution in brain have confirmed high receptor levels in striatal, limbic, and hippocampal areas (Vizuete *et al.* 1997; Karlstedt *et al.* 2001). In other areas, mismatches between receptor binding and receptor message were observed, providing important clues to the cellular location of binding sites. In the developing rat brain, H_2 receptor message followed closely the ontogeny of histaminergic neurones, consistent with a suggested trophic role for the receptor (Karlstedt *et al.* 2001). A variety of techniques have confirmed that in the gastric mucosa, the H_2-receptor mRNA and protein are found within parietal cells (Diaz *et al.* 1994).

The CNS contains the great majority of mammalian H_3 receptors. The high affinity and selectivity of R-α-methylhistamine for H_3 receptors has permitted detailed examination of the distribution of these receptors in brain. In rats, high densities of H_3 receptors were seen in cerebral cortex, hippocampal formation, amygdala, nucleus accumbens, globus pallidus, and striatum (Pollard *et al.* 1993). This distribution is similar to the histaminergic innervation, but mismatches between receptor levels and fibre densities have also been noted. Specifically, H_3 receptor density was only moderate in hypothalamus, which contains dense terminals and histaminergic perikarya (Pollard *et al.* 1993). Most striking is the high density of H_3 receptors in several thalamic nuclei, where histaminergic fibres are rare (Pollard *et al.* 1993), but brain mast cells are numerous (Hough *et al.* 1985). H_3-receptor distribution within primate brain resembles that found in rat brain, with particularly dense labelling detected in globus pallidus, molecular layer of the dentate gyrus, and limbic and frontal cortex (Martinez-Mir *et al.* 1990). Brain autoradiography with radiolabelled N^{α}-methylhistamine, (which also possesses high H_3 affinity) has shown distributions of H_3 receptors in rat (Cumming *et al.* 1991) and guinea-pig (Korte *et al.* 1990) brain comparable to those seen with R-α-methylhistamine.

Although H_3 receptors were discovered as autoreceptors which directly modulate histaminergic tone (Arrang *et al.* 1987), lesion studies imply that the majority of H_3 binding sites are localized post-synaptically, where they would function as heteroreceptors. Thus, lesions of the histaminergic system lead to increases (not decreases) in H_3 binding in the target areas (Pollard *et al.* 1993). Similarly, quinolinic acid lesions of the striatum decreased H_3 binding in the striatum and elsewhere, suggesting that the receptors are localized on striatonigral neurones (Cumming *et al.* 1991; Hoon Ryu *et al.* 1994). Radioligand binding studies have also detected H_3 receptors in several peripheral guinea-pig tissues, including intestine, pancreas, and lung (Hill *et al.* 1997). Although morphological evidence is lacking, functional evidence suggests that sympathetic autonomic fibres and vascular endothelial cells possess H_3 receptors (discussed below). Embryonic liver and adipose tissues show H_3-receptor expression, consistent with an earlier hypothesis of HA as a mediator of tissue growth (Heron *et al.* 2001).

The distribution of H_3 receptor mRNA has confirmed a nearly-exclusive CNS localization for this receptor. Strong hybridization was observed in cerebral cortex, thalamus, striatum, and portions of hippocampus in rat (Lovenberg *et al.* 1999) and guinea pig (Tardivel-Lacombe *et al.* 2000). Also consistent with other literature, H_3 message was detected in tuberomammillary nuclei and in locus coeruleus (Lovenberg *et al.* 1999). Probes designed to detect three of the H_3 isoforms in rat brain have been used to estimate the distribution of the H_{3A}, H_{3B}, and H_{3C} receptors (Drutel *et al.* 2001). Strongest signals were found with H_{3C} probes, suggesting that the smallest of the proteins is most prevalent. H_{3A} and H_{3B} signals were weaker but detectable in certain brain areas (Drutel *et al.* 2001). As mentioned, there is PCR evidence for the existence of multiple H_3 mRNA isoforms in human brain (Wellendorf *et al.* 2000; Coge *et al.* 2001*a*), but this has not been clarified with Northern blots. Human H_3-receptor mRNA was detected in several peripheral tissues by PCR, but could not be confirmed by Northern blot analysis (Lovenberg *et al.* 1999). Recently, an antibody raised against a decapeptide portion I3 of the human H_3 receptor was shown to cross-react with rat and mouse brain proteins (Chazot *et al.* 2001). Two distinct bands of cross-reactivity were discerned with molecular weights of 62 and 93 kDa; these await further identification. Regional distribution of this immunoreactivity closely paralleled the characteristics of H_3 receptors discussed above (Chazot *et al.* 2001).

The distribution of the H_4 receptor is quite different from that of the H_3 receptor. Abundant expression of H_4 mRNA was found in bone marrow, monocytes, neutrophils, and lung (Liu *et al.* 2001*a*; Zhu *et al.* 2001), with moderate levels observed in spleen, leucocytes (particularly eosinophils), intestine, T-lymphocytes, mast cells, heart, and kidney (Oda *et al.* 2000; Morse *et al.* 2001; Nguyen *et al.* 2001). Although several labs found no evidence for the existence of H_4-receptor message in the human brain by Northern analysis or RT-PCR (Oda *et al.* 2000; Nakamura *et al.* 2000; Morse *et al.* 2001; Nguyen *et al.* 2001), others found low levels of this message in human brain by RNase protection (Liu *et al.* 2001*a*) and by RT-PCR (Coge *et al.* 2001*b*). *In situ* hybridization experiments found H_4-receptor expression in the mouse hippocampus, but not in other brain areas (Zhu *et al.* 2001).

17.4 **Pharmacology**

17.4.1 **Agonists**

In contrast to the plethora of drugs available to selectively *block* H_1 receptors, a relatively small number of drugs have been found which are able to selectively *activate* these receptors. The 2-pyridinyl and 2-thiazolyl isosteres of HA show some selectivity toward H_1 versus other HA receptors (Hill *et al.* 1997), but these drugs are less potent than HA at H_1 receptors, and can have non-HA actions *in vivo*. Discovery that 2-substituted HA derivatives showed H_1-selectivity and retained potency led to development of 2-(3-(trifluoromethyl) phenyl)-HA as an experimental H_1 agonist (Leschke *et al.* 1995).

It is interesting to note that other 2-substituted phenylhistamine derivatives with H_1 agonist activity were found to activate G proteins by an H_1-receptor independent mechanism (Seifert *et al.* 1994*b*). Subsequent carefully-performed experiments showed the mechanism to be a mastoparan-like direct activation of G proteins (Detert *et al.* 1995), an action which depends on the detergent-like cationic-amphiphilic properties of some of these agents. This mechanism is thought to account for the mast cell HA-releasing properties of many compounds, including local anesthetics, contrast media, opioids, neuromuscular blockers, even some H_1 antagonists (Burde *et al.* 1996; Klinker and Seifert 1997).

More recent exploration of 2-substituted HA derivatives has led to the discovery of histaprodifen and methylhistaprodifen (Fig. 17.3), which are highly selective and potent H_1 agonists (Elz *et al.* 2000). These compounds show at least ten-fold selectivity for the H_1 receptor as compared with several other receptors (Elz *et al.* 2000) and have excellent H_1 activity *in vivo* (Malinowska *et al.* 1999). Although further characterization of the selectivity of these compounds is needed, they show great promise as emerging tools for exploring drug action and physiology of the H_1 receptor.

Dimaprit, the first selective H_2 agonist (Parsons *et al.* 1977), has been widely used *in vivo* and *in vitro*. Although not an imidazole, the compound closely mimics imidazole tautomerism, a characteristic thought to be essential for H_2 receptor activation (Leurs *et al.* 1995*b*). Dimaprit has an H_2 potency approximately equal to that of HA, but demonstrates extraordinary selectivity as compared with the other HA receptors (Hill *et al.* 1997). Combinations of modelling and testing of dimaprit derivatives led to development of amthamine, another highly selective and slightly more potent H_2 agonist which shows good *in vivo* activity (Leurs *et al.* 1995*b*). An even more potent H_2 agonist is impromidine (Leurs *et al.* 1995*b*; Hill *et al.* 1997), which is 48 times more potent than HA on the guinea-pig atrial H_2 receptor. However, impromidine has much lower potency on other H_2-receptor responses, suggesting a limited intrinsic efficacy for this compound. It must be kept in mind that impromidine and dimaprit can also act as H_3 antagonists (Arrang *et al.* 1983; Hill *et al.* 1997). Impromidine can also block H_1 receptors at higher (i.e. micromolar) concentrations (Hill *et al.* 1997). Administered *in vivo*, dimaprit and amthamine have been shown to produce CNS and cardiovascular effects which are distinct from H_2 receptor actions (Coruzzi *et al.* 1996; Swaab *et al.* 1992).

The discovery that R-α-methylhistamine is a potent and selective agonist which can inhibit the neuronal release of HA was a cornerstone in establishing the existence of the H_3 receptor (Arrang *et al.* 1983). This compound has not only served as a critical radioligand for characterizing the receptor, but also as a powerful *in vivo* tool for revealing H_3 functions. It should be noted that although R-α-methylhistamine has very high selectivity for the H_3 receptor, it

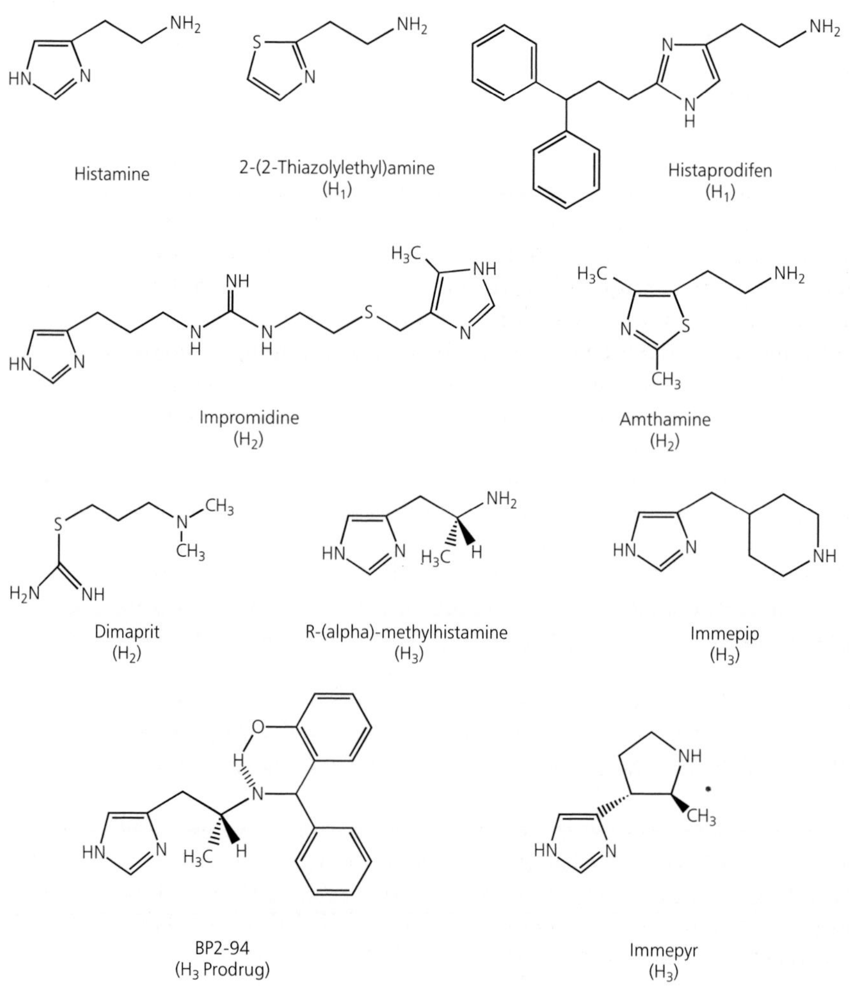

Fig. 17.3 Structures of HA agonists.

also has significant H_1 activity which can be demonstrated *in vivo* (Hey *et al.* 1992; Endou *et al.* 1994; Jennings *et al.* 1996; Hegde *et al.* 1994). The compound can also act at the adrenergic α_2 and H_4 receptors.

Like HA, R-α-methylhistamine is highly polar and shows minimal (but measurable (Taylor *et al.* 1992)) brain penetration after systemic dosing. In addition, R-α-methylhistamine in plasma was rapidly inactivated in man by histamine N-methyltransferase following systemic administration (Rouleau *et al.* 1997). Although it was known that this compound was a substrate for histamine N-methyltransferase (Hough *et al.* 1981), the short half-life seen in man was not predicted in animal studies due to a species difference in enzyme activity. To circumvent the rapid metabolism and poor brain penetration of R-α-methylhistamine, pro-drugs of this H_3 agonist (e.g. BP2-94) have been synthesized and tested (Krause *et al.* 1995). This pro-drug approach succeeded in achieving more persistent levels of the active metabolite

R-α-methylhistamine, although neither BP2-94 nor its active metabolite was detected in the brain (Rouleau *et al.* 1997).

Several other H_3 agonists have been described which are closely related to the HA/R-α-methylhistamine structure. These include imetit (Garbarg *et al.* 1992; van der Goot *et al.* 1992), immepip (Vollinga *et al.* 1994), and immepyr (Shih *et al.* 1995). Unlike R-α-methylhistamine, these newer H_3 agonists lack significant H_1 activity and are also not substrates for histamine N-methyltransferase. It is interesting to note that although immepip and imetit are both imidazoles with ammonium-containing side chains at physiological pH (and thus might not be expected to penetrate the brain (Garbarg *et al.* 1992)), there is evidence that these drugs act at brain H_3 receptors after systemic administration (Garbarg *et al.* 1992; Blandina *et al.* 1996; Jansen *et al.* 1998). Another HA analogue, impentamine, has very high affinity for H_3 receptors (Vollinga *et al.* 1995). This drug behaved as a partial agonist in brain H_3 assays, but as an antagonist on the peripheral H_3 receptor bioassay (Leurs *et al.* 1996). In addition, impentamine and the other HA homologues show different potencies on the brain and peripheral H_3 receptors (Leurs *et al.* 1996; Harper *et al.* 1999). Impentamine showed nearly a ten-fold difference in agonist potencies among the cloned H_3 subtypes found in the rat brain (Drutel *et al.* 2001), which may be relevant to these observations. More recently, ether and carbamate derivatives have been discovered which behave *in vivo* as full H_3 agonists, yet these drugs completely lack side chain amines, and thus cannot exist as monocations (Sasse *et al.* 1999). Since all known agonists acting at HA receptors were previously thought to require this cation (Hough 1999; Leurs *et al.* 2000; Lovenberg *et al.* 1999), this is of considerable interest.

Because the discovery of H_4 receptors is so recent, potent drugs capable of selectively activating or inhibiting these receptors have not yet been found (Hough 2001). The high affinity H_3 agonists all activate the H_4 receptor, but with a reduced potency. The H_3 antagonist thioperamide has a 5–10-fold lower activity at the H_4 as compared with the H_3 receptor. Other H_3 antagonists such as clobenpropit show partial agonist activity at the H_4 receptor. It is clear that H_3 and H_4 activities can be separated, since a recently-described non-imidazole H_3 antagonist was shown to lack activity at H_4 (Liu *et al.* 2001a). The emerging pharmacology of the H_4 receptor was recently reviewed (Hough 2001).

17.4.2 **Antagonists**

Drugs which act at the H_1 receptor have been known for many years. Discovery of the first HA antagonist (over fifty years ago) led to an increased understanding of anaphylaxis and allergy, set the stage for demonstrating the likely existence of more than one type of HA receptor, and indirectly led to the subsequent development of phenothiazines and tricyclic antidepressants (Green 1983). Hundreds of the so-called 'classical' or first-generation H_1-antagonists have been developed and many are still in clinical use. These compounds all share two aromatic rings and a basic amine moiety, and have been further divided into five chemical classes. Members of this group include diphenhydramine, *d*-chlorpheniramine, and pyrilamine. All of these drugs have good brain-penetrating characteristics. Pyrilamine (also named mepyramine) remains a prototype H_1 antagonist with very high affinity and excellent selectivity for the H_1 receptor. Although all of the first-generation compounds have some local anesthetic and anti-muscarinic properties, for pyrilamine these occur at concentrations substantially greater than those acting at the H_1 receptor. In cases where H_1 antagonists have activity but the role of the H_1 receptor is uncertain, *d*- and *l*-chlorpheniramine can be used to classify HA responses, since these compounds share many activities but differ by several hundred-fold in their anti-H_1 activity (Hill *et al.* 1997). The classical H_1 antagonists have very good anti-allergic properties, and are also used widely as over-the-counter sleep aids.

The sedative properties of the classical H_1 blockers limit their day-time use and have fostered the development of the second-generation or 'non-classical' H_1 antagonists. These compounds include astemizole, cetirizine, terfenadine, fexofenadine, and loratidine (Fig. 17.4). Several of the second-generation H_1 antagonists enjoy wide popularity and have good anti-allergic activity. As compared with classical H_1 antagonists, these newer drugs show much lower brain penetration, commensurate with greatly reduced sedative side effects. Although differences in brain versus peripheral H_1 receptor affinity for these drugs had been suggested to account for the peripheral selectivity of these compounds, this has not been validated (Ter Laak *et al.* 1993), and there is no convincing evidence for the existence of more than one type of H_1 receptor. Some of the early second-generation compounds (e.g. terfenadine) were found to act at cardiac potassium channels, leading to rare but dangerous toxicity (Woosley 1996). Many of the second-generation H_1 antagonists are metabolites of other H_1 antagonists. For example, cetirizine, fexofenadine, and desloratidine are active metabolites of hydroxyzine, terfenadine, and loratidine, respectively. Detailed models of the interactions between H_1 antagonists and the H_1 receptor have been developed (Ter Laak *et al.* 1995).

Sir James Black and colleagues synthesized and characterized the first H_2 receptor antagonists, unequivocally demonstrating the existence of more than one type of HA receptor (Black *et al.* 1972). The work showed clearly a physiological role for gastric HA and revolutionized the treatment of peptic ulcer disease. Presently, there are four H_2 antagonists in wide clinical use: cimetidine, ranitidine, famotidine, and nizatidine. Surprisingly, cimetidine's Kd on the H_2 receptor is only 0.8 μM (Hill *et al.* 1997), yet the drug has been used successfully by millions of patients with only a few side effects. Cimetidine is known to inhibit some forms of cytochrome P450, which can lead to clinically significant drug–drug interactions (Furuta *et al.* 2001). Many other non-H_2 actions of cimetidine have been found at higher drug concentrations (see table 3 in Hough *et al.* 1997). Because of its low potency and non-H_2 actions, cimetidine is not a good experimental agent to study H_2 receptor actions. Ranitidine (Kd = 63 nM (Ganellin 1982) or famotidine (Kd = 17 nM (Cooper *et al.* 1990)) are good alternatives with much higher H_2 potency and lacking many of cimetidine's non-H_2 actions. Among the most potent known H_2 antagonist is iodoaminopotentidine (Kd = 30 nM, (Hirschfeld *et al.* 1992)), which, as mentioned, is a very useful tool for receptor studies, but has also been used to study H_2 receptor function. A very large number of H_2 antagonists have been developed over the two decades which followed the commercial success of this class of drugs, and a diverse structure–activity relationship (SAR) has evolved (Cooper *et al.* 1990). Newer H_2 antagonists continue to be developed (Anglada *et al.* 1997; Kijima *et al.* 1998).

Because many H_2 antagonists retain HA's polar imidazole group (or another isosteric polar group), most H_2 antagonists do not reach the CNS after systemic dosing. Thus, intracerebral or intraventricular injections of H_2 antagonists were needed to study H_2 receptor functions in the CNS. To circumvent this, Young *et al.* (1988) described the development and characterization of zolantidine, a brain-penetrating H_2 receptor antagonist (Calcutt *et al.* 1988). Although never developed for human use, zolantidine is active *in vivo* and has been used to investigate many brain roles for H_2 receptors.

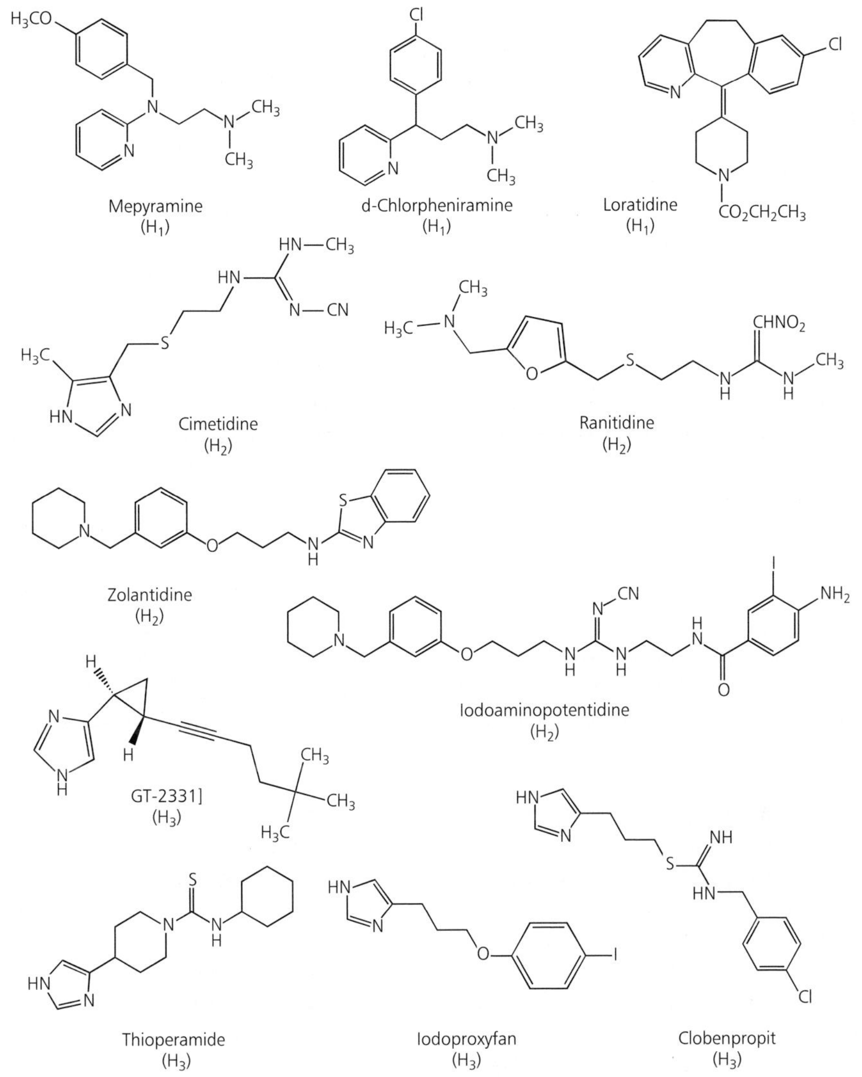

Fig. 17.4 Structures of HA antagonists.

Discovery of the H_3 receptor followed the observation that extra-synaptosomal HA was capable of downregulating HA synthesis and release. Initial characterization of this response showed it to be distinct from both H_1 and H_2 receptors (Arrang *et al.* 1983). The unequivocal existence of the new receptor was established by development of thioperamide, a potent and selective H_3 antagonist (Arrang *et al.* 1987). The activity of thioperamide in a many different biochemical, physiological, and behavioural assays has documented the significance of HA and the H_3 receptor in the nervous system, and thioperamide remains a prototype H_3 antagonist. However, at concentrations higher than the H_3 Kd, thioperamide can act at other receptors, including $5HT_3$, and H_4 (Hough, 2001; Leurs *et al.* 1995*c*). Other findings suggest that thioperamide may also act directly to inhibit GABA transport (Yamamoto *et al.*

1997) and drug metabolizing enzymes (Alves-Rodrigues *et al.* 1996). Binding experiments with the cloned H_3 receptors have confirmed that thioperamide has a high affinity for the rat receptor (Ki = 4 nM), but shows a ten-fold lower affinity for the human receptor (Lovenberg *et al.* 2000), even though these receptors differ by only a few amino acids (Ligneau *et al.* 2000). As mentioned above, there is evidence that H_3 receptors are constitutively active *in vivo*, and thioperamide (and many other H_3 antagonists) may behave *in vivo* as an inverse agonist (Morisset *et al.* 2000). Systematic variation in structures of H_3 agonists led to the development of the H_3 antagonist clobenpropit, which can be viewed as an isothiourea congener of thioperamide (van der Goot *et al.* 1992; Leurs *et al.* 1995*b*). Clobenpropit has very high affinity for both the rat and the human H_3 receptor (Ki = 0.4–0.6 nM, Lovenberg *et al.* 2000).

The SARs derived from early studies of thioperamide-like drugs suggested the existence of fairly stringent structural requirements for blocking H_3 receptors, including an imidazole nucleus, a conformationally restricted side chain, a thiourea polar group, and a hydrophobic tail (Leurs and Timmerman 1992; Leurs *et al.* 1995*b*). However, more recent work has shown the existence of a much broader SAR for H_3 antagonists, which has led to a large expansion of known active structures. Development of GT-2016 (Tedford *et al.* 1995) eliminated the thiourea group of thioperamide which was thought to be the cause of liver toxicity after chronic administration. Discovery of ciproxifan (Ligneau *et al.* 1998) demonstrated a potent H_3 antagonist lacking the polar group. Remarkably, potent H_3 antagonists have been more recently discovered which have no heteroatom whatsoever in their side chain. These include GT-2331 (Tedford *et al.* 1999) and several others (Stark *et al.* 1998; De Esch *et al.* 1999). Newer H_3 antagonists continue to be developed by several academic and corporate laboratories. Most interesting among these compounds is the discovery that potent H_3 antagonists can be synthesized which lack imidazole (Ganellin *et al.* 1998; Meier *et al.* 2001; Bennani *et al.* 2001; Carruthers *et al.* 2002). Removal of imidazole seems to be desirable in order to improve brain penetration and also minimize H_4 antagonist activity, which is a feature of many imidzole-containing H_3 antagonists (Liu *et al.* 2001*a*). Molecular models of H_3 antagonist action have been developed and tested recently (De Esch *et al.* 2001).

17.5 Signal transduction and receptor modulation

17.5.1 H_1-receptor signalling

The H_1 receptor belongs to the family of Ca^{2+}—mobilizing GPCRs. Activation of this receptor leads to the hydrolysis of phosphatidyl 4,5-biphosphate ($PI(4,5)P_2$) resulting in the formation of the second messengers inositol (1,4,5) trisphosphate ($InsP_3$) and 1,2-diacylglycerol (DAG). HA induces the production of inositol phosphates in many tissues including airway smooth muscle, intestinal smooth muscle, vascular smooth muscle, and heart (Barnes 1991; Claro *et al.* 1989; Donaldson and Hill 1986; Orellana *et al.* 1987; Sakuma *et al.* 1988). In guinea-pig brain, the magnitude of this response correlates well with the density of H_1 receptors (Carswell and Young 1986; Daum *et al.* 1983). Studies with bacterial toxins, $G\alpha$-subunit antibodies and $[\alpha\text{-}^{32}P]$GTP azidoanilide incoporation into $G\alpha$-subunits have shown that the stimulation of phospholipase C (PLC) occurs via $G\alpha_{q/11}$ proteins (Gutowski *et al.* 1991; Kühn *et al.* 1996; Leopoldt *et al.* 1997; Orellana *et al.* 1987).

The main physiological consequence of the production of inositol phosphates is the elevation of intracellular Ca^{2+}, characterized by a rapid transient rise of the intracellular Ca^{2+}

concentration, which is followed by a sustained elevation of the Ca^{2+} concentration. Experiments in Ca^{2+}—free medium suggest that the sustained response is highly dependent on the influx of extracellular calcium, whereas the transient increase is caused by the release of Ca^{2+} from intracellular Ca^{2+} stores (Leurs *et al.* 1994*b*). The generated InsP$_3$ is responsible for the rapid, transient release component; InsP$_3$ receptors in the endoplasmic reticulum bind InsP$_3$ and release Ca^{2+} into the cytosol. Recent evidence suggests that the formation of DAG is probably involved in the Ca^{2+} influx. DAG activates protein kinase C (PKC) and protein kinase D. Activated PKC has many cellular functions, including desensitization of the H$_1$ receptor via receptor phosphorylation (Fujimoto *et al.* 1999; Smit *et al.* 1992) and transcriptional downregulation of H$_1$-receptor expression (Pype *et al.* 1998). Moreover, the H$_1$ receptor—mediated production of DAG has been shown to stimulate Ca^{2+} influx by activating transient receptor potential (TRP) channels (homologs of *Drosophila* TRP proteins (Hofmann *et al.* 1999)), which are thought to mediate capacitative Ca^{2+} entry (Birnbaumer *et al.* 1996).

Since Ca^{2+} and PKC are involved in the regulation of many cellular functions, the stimulation of PLC activity can explain a wide variety of secondary signalling events after stimulation of the H$_1$ receptor. These include the Ca^{2+}/calmodulin-dependent activation of the nitric oxide synthetase/guanylate cyclase pathway (Casale *et al.* 1985; Duncan *et al.* 1980; Hattori *et al.* 1988; Leurs *et al.* 1991*a*; Schmidt *et al.* 1990; Sertl *et al.* 1987; Yuan *et al.* 1993), the release of arachidonic acid via phoshoplipase A$_2$ activation along with the subsequent generation of arachidonic acid metabolites such as prostacyclin and thromboxane A$_2$ (Baenziger *et al.* 1980; Leurs *et al.* 1994*b*; Resink *et al.* 1987), the modulation of cAMP via crosstalk with other receptors (e.g. H$_2$-, adenosine A$_2$-, and vasoactive intestinal polypeptide receptors (Al-Gadi and Hill 1987; Donaldson *et al.* 1989; Magistretti and Schorderet 1985; Marley *et al.* 1991; Palacios *et al.* 1978)) and the activation of the transcription factors (e.g. nuclear factor of activated T cells [NF-AT] (Boss *et al.* 1998) and nuclear factor kappa B [NF-κB]) (Aoki *et al.* 1998; Bakker *et al.* 2001; Hu *et al.* 1999).

Besides the activation of second messenger pathways via the $G\alpha_{q/11}$ subunits, several H$_1$ receptor responses are mediated via other G protein signalling modules. In specific cell types the H$_1$ receptor has been shown to activate pertussis toxin (PTX)—sensitive $G_{i/o}$ proteins linked to the elevation of intracellular $[Ca^{2+}]$ (Seifert *et al.* 1994*a*), to the modulation of a non-selective cation current (Wang and Kotlikoff 2000) or to the activation of phospholipase A$_2$ (Leurs *et al.* 1994*b*; Murayama *et al.* 1990). Moreover, activation of phospholipase D by the H$_1$ receptor is secondary to the activation of PLC (Natarajan and Garcia 1993). However, in 1321N1 astrocytoma cells, part of the HA-stimulated phospholipase D activity occurs in the absence of PKC activation (Dawson *et al.* 1993; Natarajan and Garcia 1993) and has been suggested to be mediated via the small G protein ARF (Mitchell *et al.* 1998).

In the nervous system, H$_1$ receptor stimulation nearly always leads to depolarization and/or an increase in firing frequency (Brown *et al.* 2001). Stimulatory mechanisms which have been identified include blockade of a leak potassium conductance, activation of a calcium-independent, tetrodotoxin-insensitive sodium current, enhancement of NMDA currents, and activation of other cation channels. H$_1$-mediated hyperpolarizations also occur (Brown *et al.* 2001).

H$_1$-receptor responsiveness is well regulated by desensitization, internalization, and downregulation. H$_1$-receptor desensitization can be both homologous and heterologous; the latter is often the result of excessive PKC activation (Leurs *et al.* 1991*b*; Smit *et al.* 1992; Zamani *et al.* 1995). Recent studies of Fukui and coworkers showed PKC-mediated phosphorylation of Ser[396] and, especially, Ser[398]; seem involved in the PKC-induced desensitization (Fujimoto *et al.* 1999). In bovine airway smooth muscle cells PKC-mediated blunting of HA-induced contractions was accompanied by a transcriptional downregulation of H_1-receptor mRNA (Pype *et al.* 1998). Young and coworkers were able to show H_1-receptor internalization upon short-term HA exposure of U373 MG astrocytoma cells (Hishinuma and Young 1995), whereas in transfected CHO cells a PKC-independent H_1-receptor downregulation was observed (Smit *et al.* 1996*c*). Detailed information on the molecular aspects of these processes is not presently available.

17.5.2 H_2-receptor signalling

It is generally accepted that the H_2 receptor is positively coupled to the adenylyl cyclase system. The H_2 receptor activates adenylyl cyclase in membrane fractions in a guanyl nucleotide—sensitive manner (Johnson 1992). Direct evidence for the involvement a G_s-protein was shown by an enhanced GTP azidoanilide-labelling of G_s-like proteins in H_2 receptor expressing Sf9 cells (Leopoldt *et al.* 1997). A large number of studies have shown that HA increases the levels of cAMP in brain, stomach, heart and adipose tissue in several species, including man (Hill *et al.* 1997; Schwartz *et al.* 1991).

In addition to elevation of cAMP levels via G_s-proteins, the H_2 receptor may activate PLC via different G proteins. In differentiated HL-60 cells (Gespach *et al.* 1982; Seifert *et al.* 1992), parietal cells (Negulescu and Machen 1988), human keratinocytes (Koizumi and Ohkawara 1999), and HEPA, COS-7, and HEK-293 cells transfected with the canine H_2 receptor (DelValle *et al.* 1992; Kühn *et al.* 1996; Wang *et al.* 1997), an H_2-receptor mediated increase of the intracellular Ca^{2+} concentration was observed. HA was also found to increase the levels of InsP$_3$ in various cell types (DelValle *et al.* 1992; Fitzsimons *et al.* 1999; Kühn *et al.* 1996; Legnazzi *et al.* 1998; Wang *et al.* 1996), and activation of both PKCα and PKCβ has been observed (Wang *et al.* 1997). In COS-7 cells, co-expression of $G_{q/11}$-family members of G proteins, but not G_i or G_s, leads to a further H_2-receptor mediated increase in InsP$_3$ formation. Moreover, in guinea-pig membranes, HA stimulates via both H_1- and H_2-receptor incorporation of [α-^{32}P]GTP azidoanilide into $G_{q/11}$-like proteins (Kühn *et al.* 1996).

In the nervous system, H_2 receptors promote excitation by inhibition of a calcium-activated potassium conductance (Brown *et al.* 2001). This action is thought to occur through the cAMP-protein kinase A pathway. H_2-mediated depolarizations in pyramidal cells and thalamus can also occur through cAMP modulation of cation channels (Brown *et al.* 2001). However, recent work has demonstrated that H_2 receptor activation can paradoxically dampen high frequency firing by blocking Kv3.2, a voltage-gated potassium channel (Rudy and McBain 2001).

The H_2 receptor is known to be rapidly desensitized in a homologous or heterologous manner in a variety of cell lines (Holden *et al.* 1987; Sawutz *et al.* 1984; Schreurs *et al.* 1984; Smit *et al.* 1994). Several studies have pointed to phosphorylation of Ser/Thr residues in I3 by GRK2 and/or GRK3 in the process of homologous desensitization (Nakata *et al.* 1996; Rodriguez-Pena *et al.* 2000; Shayo *et al.* 2001), whereas the induction of phosphodiesterase activity was often suggested to explain heterologous desensitization (Holden *et al.* 1987). The H_2 receptor will internalize within 10–60 min of agonist exposure (Smit *et al.* 1995). Following exposure for a few hours, expression of the H_2 receptor is downregulated via both cAMP-dependent and cAMP-independent mechanisms (Smit *et al.* 1996*b*).

17.5.3 **H$_3$-receptor signalling**

Although the H$_3$ receptor was first suggested to exist in 1983, detailed information on this receptor's transduction mechanisms was only obtained recently. Until the cloning of the receptor cDNA in 1999 (Lovenberg *et al.* 1999), the most convincing evidence had been obtained by Clark and Hill (1996). In rat brain membranes, a PTX-sensitive stimulation of [^{35}S]GTPγS binding was observed. In agreement with these observations, a PTX-sensitive inhibition of cAMP accumulation in response to H$_3$ agonists has been observed in a variety of transfected cells expressing the human or rat H$_3$ receptor (Drutel *et al.* 2001; Lovenberg *et al.* 1999; Morisset *et al.* 2000). Prior to the H$_3$ cloning, inhibitory effects of PTX were known on H$_3$ signalling (Clark and Hill 1996; Endou *et al.* 1994; Nozaki and Sperelakis 1989; Oike *et al.* 1992; Poli *et al.* 1993; Takeshita *et al.* 1998), and these had already supported the involvement of a G$_{i/o}$-protein in the H$_3$-receptor mediated responses, but the modulation of cAMP levels in brain tissue had not been previously observed (Garbarg *et al.* 1989). Recently, it was reported that the H$_3$-mediated regulation of histidine decarboxylase activity can be controlled via the cAMP-adenylate cyclase-PKA pathway, but direct modulation of brain cAMP levels were not investigated (Gomez-Ramirez *et al.* 2002). In transfected cell lines, the H$_3$ receptor has also been reported to activate [^{35}S]GTPγS binding (Morisset *et al.* 2000), arachidonic acid release (Morisset *et al.* 2000), the MAPK pathway (Drutel *et al.* 2001), the Na$^+$/H$^+$ exchanger (Silver *et al.* 2001) and to inhibit Ca^{2+} influx and exocytosis of [^{3}H]noradrenaline from transfected SH-SY5Y-H$_3$ cells (Silver *et al.* 2002).

Although the mechanisms of H$_3$ receptor signalling in physiological systems are not clear, it is known that H$_3$ activation leads to inhibition of transmitter release. The list of transmitters modulated by H$_3$ receptors includes not only HA (the basis for the autoreceptor function), but many others: glutamate, norepinephrine, dopamine, acetylcholine, serotonin, and several peptides (Brown *et al.* 2001; Hough 1999). An inhibition of high threshold Ca^{2+} currents has been shown to mediate some of these effects. Activation of inwardly-rectifying potassium channels, another mechanism commonly used by PTX-sensitive autoreceptors, has not been demonstrated for the H$_3$ receptor (Brown *et al.* 2001).

With respect to the regulation of H$_3$-receptor responses, no detailed information is currently available. In the guinea-pig intestine, H$_3$-receptor responses are desensitized rapidly (Perez-Garcia *et al.* 1998), but the actual biochemical mechanisms have so far not been investigated.

17.5.4 **H$_4$-receptor signalling**

Like the H$_3$ receptor, the H$_4$ receptor belongs to the family of G$_{i/o}$ coupled receptors. Studies with transfected cells show that the H$_4$ receptor inhibits forskolin-induced cAMP accumulation (Liu *et al.* 2001*a*; Nakamura *et al.* 2000; Oda *et al.* 2000), stimulates [^{35}S]GTPγS binding (Morse *et al.* 2001), and activates MAP kinase (Morse *et al.* 2001) in a PTX-sensitive manner. Co-expression studies using either Gα$_{15}$ or Gα$_{16}$ (Morse *et al.* 2001; Oda *et al.* 2000) have shown that the H$_4$ receptor can also activate these G proteins, leading to a mobilization of intracellular Ca^{2+}]. As the receptor is strongly expressed in leucocytes, coupling of the receptor to $G_{15/16}$ seems to represent a natural second messenger pathway of the H_4 receptor (Oda et al. 2000). Indeed, Raible et al. (1994) reported on a new HA receptor on human eosinophils linked to mobilization of intracellular $[Ca^{2+}]$; this response had a pharmacological profile resembling that of the H_4 receptor. At this moment, no data on the regulation of H_4-receptor responses are known.

17.5.5 Constitutive HA receptor signalling

In the field of GPCRs it is currently well accepted that receptors can induce signal transduction without the presence of the agonist. The notion of constitutive receptor activity has also resulted in the introduction of the class of inverse agonists, ligands able to reduce agonist-independent signalling. Recently, all human HA receptors have been shown to have constitutive activity (Alewijnse et al. 1998; Bakker et al. 2000; Morisset et al. 2000; Morse et al. 2001; Wieland et al. 2001). Overexpression of the H_1 receptor results in an agonist-independent production of InsP$_3$ and NF-κB activation (Bakker et al. 2000, 2001), which can be inhibited by H_1 antagonists. Among these are well-known therapeutics, such as cetirizine (Zyrtec®), loratadine (Claritin®), and epinastine (Flurinol®). Similarly, all well-known therapeutics acting at the H_2 receptor (e.g. cimetidine, Tagamet®, rantidine, Zantac®) can act as inverse agonists at constitutively active H_2 receptors, thereby reducing basal cellular cAMP levels in transfected cell lines (Alewijnse et al. 1998; Smit et al. 1996a). Also, the $G_{i/o}$ coupled H_3 and H_4 receptor show constitutive activity in transfected cells as determined by modulation of cAMP levels (Morisset et al. 2000; Wieland et al. 2001), arachidonic acid release or [^{35}S]GTPγS binding (Morisset et al. 2000; Morse et al. 2001). Interestingly, the H_3 receptor is one of the few examples of a G protein coupled receptor showing a high level of constitutive activity in a physiological system (Morisset et al. 2000; Wieland et al. 2001). These findings strongly suggest that constitutively active H_3 receptors are relevant for the regulation of histaminergic (and possibly other) neuronal activity.

17.6 Physiology and disease relevance

As discussed in the introduction, HA has significant actions within and outside of the CNS. HA actions that occur outside of the brain are given in the next section, and these are organized by receptor. HA actions in the CNS, organized by function, are in the subsequent sections.

17.6.1 Peripheral indications

HA can produce a wide variety of powerful physiological changes as a result of activation of H_1 receptors. Diseases involving mast cells such as urticaria pigmentosa or systemic mastocytosis also show many of these responses. Patients with allergies also experience many of these symptoms. It has been known for many years that HA is potent spasmogen on intestinal and bronchiolar smooth muscle, actions which are mediated predominantly by H_1 receptors (Ash and Schild 1966). Asthmatics show exaggerated H_1 bronchocontriction, but H_1 antagonists are of little benefit for treating most patients. However, H_1 receptors can induce coronary vasospasm, and may contribute to myocardial infarction (Gupta et al. 2001; Nakamura 2000). H_1 effects on blood vessels consist of direct contractile actions on some vascular smooth muscle, but indirect vasodilation of arterioles by endothelium-dependent processes. Thus, H_1 receptors on endothelial cells activate the release of endothelium-derived relaxant factor, now known to be nitric oxide. Other H_1 actions on endothelial cells include changes in morphology (accounting for increased vascular permeability), and release of many other substances, including prostacyclins, platelet-aggregating factor, and interleukins (Hill 1990; Hill et al. 1997). HA actions on the H_1 receptor may contribute to the development of atherosclerosis. For example, HA released from platelets and mast cells following vascular injury stimulates smooth muscle cell migration and proliferation, and H_1 antagonists prevent the intimal hyperplasia following vascular injury (Miyazawa et al. 1998). In addition, HA reduces plasma HDL cholesterol through an H_1-receptor mechanism (Liao et al. 1997). HA also activates H_1 receptors on adrenal chromaffin cells to promote the release of catecholamines (Borges 1994). H_1 receptors can also modify cardiac contractility and conduction (Hill et al. 1997). Recent knock-out studies have helped to clarify HA's complex actions on the immune system. Thus, HA is released from mast cells following expsoure to antigens, but this released HA acts as a feedback suppressor of humoral immunity through H_1 receptors located on TH_1 helper cells (Jutel et al. 2001).

The liberation of HA in the skin following injury is known to result in activation of afferent fibres through the H_1 receptor. This effect results in the 'flare' response (vasodilation in nearby areas which is mediated by a combination of nerves and mast cells) and also stimulates itching. Recent studies suggest that specific subsets of primary afferents (Schmelz 2002) and spinothalamic lamina I neurones (Andrew and Craig 2001) subserve itching. Several findings also suggest that H_1 receptors on peripheral fibres can contribute nociceptive influences (further discussed under CNS analgesia).

As mentioned, H_2 receptors were discovered as mediators of gastric acid secretion. All organisms with true stomachs have high concentrations of gastric HA, and demonstrate H_2 receptor mediated increases in gastric acid. Pharmacological evidence that the gastric release of HA and its subsequent activation of H_2 receptors occurs physiologically includes the fact that H_2 blockers antagonize the actions of all known secretogogues (Black 1979). Recent studies with H_2 receptor–deficient mutant mice confirm that this receptor is needed for normal gastric development, and for gastrin- and HA-stimulated acid secretion (Kobayashi et al. 2000). Interestingly, muscarinic cholinergic agents still activate acid secretion in these mutant mice.

When stimulated, H_2 receptors increase cardiac rate and force of contraction (Levi et al. 1982). These effects are most pronounced in guinea pig and man. H_2 agonists have been developed as experimental agents for treating heart failure. Although cardiac H_2 receptors may be activated as a result of immunological challenge (Del Balzo et al. 1988), H_2 receptors in the heart are not normally activated, and H_2 antagonists have no effects on the heart in healthy subjects. It is interesting to point out that if H_2 receptors in the heart and gastric mucosa were both constitutively active in man, and if cimetidine acts in vivo as an inverse agonist (as discussed above), then this drug should reduce heart rate as well as gastric acid secretion in humans, but it does not. Further work is needed to address this issue.

H_2 receptors also contribute to HA-mediated vasodilation (Levi et al. 1982), such that blood pressure disturbances due to exaggerated HA release are best treated with a combination of H_1 and H_2 blockers. Other H_2 receptors lead to relaxation of non-vascular smooth muscle, and also inhibition of acitivity in neutrophils, basophils, and subsets of T-lymphocytes (Hill 1990). Recent results with H_2 receptor–deficient mice indicate that the H_2 receptor can inhibit both T_H1 and T_H2 classes of T helper cells (Jutel et al. 2001).

Although H_3 receptors have their highest density in the brain, peripheral H_3 receptors seem capable of reducing the activity of sympathetic, sensory, and enteric nerves. Postganglionic adrenergic transmission is suppressed by H_3 activation (Hutchison and Hey 1994; Silver *et al.* 2002). In the heart, activation of H_3 receptors reduces both normal exocytotic release of norepinephrine, but also the excessive non-exocytotic norepinephrine release following myocardial ischaemia (Levi and Smith 2000). Sensory neurones (possibly C fibres) are thought to have inhibitory H_3 receptors, suggested to explain the anti-inflammatory activity of the H_3 agonist prodrug BP2-94 (Rouleau *et al.* 1997). Although most studies of peripheral H_3 receptors have focused on the neuronal H_3 components, there is also evidence that H_3 receptor agonists can act directly on smooth muscle (Cardell and Edvinsson 1994). H_3 receptors are also in the stomach, where they may protect against muscosal damage when activated (Morini *et al.* 1996). Results of some of these studies may require re-evaluation in light of the H_4 activity of many H_3 agonists and antagonists.

Because the discovery of the H_4 receptor is so recent, and since no selective drugs have yet been discovered, very little is known about the physiological functions for this receptor. However, it seems likely that H_4 agonists and/or antagonists could be used to modulate inflammatory or immunological responses.

17.7 CNS functions for HA

17.7.1 Arousal, attention, and sleep

Although HA was suggested to be a waking substance in the brain many years ago (Monnier *et al.* 1970), it is now clear that histaminergic neurones and brain H_1 receptors are essential for normal regulation of sleep–wake cycles. Findings in support of this conclusion have been summarized (Leurs *et al.* 1998; Brown *et al.* 2001): (1) CNS HA injections induce electrographic arousal, effects which are blocked by H_1 antagonists (2) pharmacological or surgical manipulation of the histaminergic tuberomammillary system induces hypersomnolence, and (3) histaminergic neurone activity and brain HA turnover rates follow sleep–wake cycles, with highest activity during wakeful periods. Current thought is that histaminergic neurones function within hypothalamic sleep–wake circuits, with inputs from orexin A—containing (wake-promoting) neurones and preoptic sleep regulatory centres, and outputs to several critical areas (Brown *et al.* 2001; Eriksson *et al.* 2001; Saper *et al.* 2001). Thus, the sedative profile of brain-penetrating H_1 antagonists (which accounts for their clinical utility as over-the-counter sleep aids) is believed to be due to blockade of brain H_1 receptors. In addition, the wake-promoting drug modafinil, whose mechanism remains elusive, was recently shown to activate histaminergic neurones (Scammell *et al.* 2000). H_3 receptor modulation of the histaminergic system shows the expected effects on wakefulness. Thus, H_3 antagonists, which act on pre-synaptic H_3 receptors to increase the release of neuronal HA, promote wakefulness and decrease rapid eye movement (REM) and slow-wave sleep, whereas H_3 agonists reduce neuronal HA release and increase slow-wave sleep (Lin *et al.* 1990; Leurs *et al.* 1998). Recent studies suggest that sleep disorders may have a histaminergic component; most interesting are results showing: (1) animal models of narcolepsy (in which the orexin [hypocretin] Hcrtr-2 receptor is mutated) have defective histaminergic transmission (Nishino *et al.* 2001), and (2) orexin A—induced arousal is abolished in H_1 receptor-deficient mice (Huang *et al.* 2001).

17.7.2 Alzheimer's disease and cognition

Extensive literature suggests that the brain histaminergic system can modulate learning and reinforcement. However, many details are lacking, and some findings seem contradictory. For example, several workers have found that microinjections of HA into certain brain areas facilitate cognitive performance and task recall, and pharmacological depletion of brain HA has the opposite effect (Kamei *et al.* 1993; Leurs *et al.* 1998). Furthermore, both H_1 receptor agonists, and an inhibitor of brain HA metabolism reversed scopolamine amnesia in a passive avoidance test (Malmberg-Aiello *et al.* 2000). In contrast, H_1 antagonists have been found to reduce some kinds of cognitive performance and to impair recall (Kamei and Tasaka 1991). However, a number of studies have also shown that tuberomammillary lesions (which were thought to deplete neuronal HA) actually *improve* performance on several kinds of learning tasks (Klapdor *et al.* 1994; Frisch *et al.* 1998). Moreover, in some studies H_1 antagonists have been shown to produce reinforcing and memory-promoting effects (Frisch *et al.* 1997). Thus, HA, H_1 antagonists, and tuberomammillary lesions have all been reported to enhance memory. While this paradox has not yet been resolved, all of the key findings have recently been observed by the same laboratory and discussed (Hasenohrl *et al.* 2001). The 'direction' of cognitive modulation by brain H_3 receptors seems somewhat more clear. Thus, H_3 antagonists reverse scopolamine amnesia in several tests; this effect is antagonized and enhanced, respectively, by blockers of H_1 and H_2 receptors, respectively (Miyazaki *et al.* 1995, 1997). Similarly, thioperamide enhances passive avoidance responding in senescence-accelerated mice, but not in normal mice (Meguro *et al.* 1995). GT-2227, an experimental H_3 antagonist, improved passive avoidance responding in 3 week-old rat pups (Yates *et al.* 1999). H_3 agonists reduce cortical acetylcholine release, and impair object recognition and passive avoidance responding (Blandina *et al.* 1996). A case can therefore be made that H_3 antagonists show cognition-enhancing properties, and may be useful in several clinical settings (Leurs *et al.* 1998), even though the mechanisms by which these drugs act remain to be completely clarified. H_2 agonists seem to impair cognitive performance, an effect reversed by H_2 blockers, but the H_2 blocker alone had no effect (Onodera *et al.* 1998). There are extensive interactions between the histaminergic and cholinergic systems which seem consistent with a cognitive role for the former as well as the latter (Passani *et al.* 2000; Passani and Blandina 1998).

Several studies suggest that components of the histaminergic system may be altered in aging or dementia. Compared with age-matched controls, HA content was reduced by 42–53 per cent in hypothalamus, hippocampus, and temporal cortex of post-mortem Alzheimer brains (Panula *et al.* 1998). Alzheimer's patients also show reduced H_1 receptor binding in frontal and temporal cortical areas when compared with age-matched controls (Higuchi *et al.* 2000). Older humans show decreases in brain H_1 receptors as compared with younger controls (Yanai *et al.* 1992*a*).

17.7.3 Satiety/Feeding

Activation of brain H_1 receptors is known to supress food intake (Brown *et al.* 2001; Hough 1988; Lecklin *et al.* 1998). The ventromedial nucleus of the hypothalamus, thought to be important in satiety signalling, contains H_1 receptors which can contribute to the suppression of eating. H_1 agonists and HA-releasing drugs reduce feeding behaviour, whereas blockade of H_1 receptors or depletion of neuronal HA have the opposite effect (Sakata *et al.* 1988;

Fukagawa *et al.* 1989). However, the latter only occurs during certain portions of the diurnal cycle, consistent with known circadian variations in histaminergic activity. Consistent with this, H_3 antagonists reduce food intake, and this effect is inhibited by H_1 antagonists (Leurs *et al.* 1998). Defective histaminergic transmission seems to be a feature of some animal models of obesity. For example, genetically obese Zucker rats and db/db obese mice (both of which have defective leptin receptors) have reduced hypothalamic HA levels (Machidori *et al.* 1992; Yoshimatsu *et al.* 1999). Similarly, in mice, leptin activates hypothalamic HA turnover and suppresses feeding; the latter effect was abolished by inhibition of HA synthesis (Yoshimatsu *et al.* 1999) and was not observed in H_1 receptor-deficient mice (Morimoto *et al.* 1999). However, centrally-administered leptin did not alter histaminergic dynamics in rats (Lecklin *et al.* 2000). The histaminergic system can regulate other aspects of brain energy balance in addition to feeding behaviour. For example, H_1 receptors also activate glycogenolysis and reduce thermoregulatory set point (Brown *et al.* 2001; Hough 1988; Sakata and Yoshimatsu 1995).

17.7.4 Neuroendocrine

There is good evidence that histaminergic neurones can activate H_1 receptors in the supra-optic nucleus of the hypothalamus to increase the release of arginine vasopressin. Early key findings (Hough 1988) include high densities of histaminergic fibres and H_1 receptors in this area, increases in plasma vasopressin levels following intracerebral HA injections, and antagonism of these effects by H_1 blockers. More recent *in vitro* experiments have confirmed monosynaptic inputs from tuberomammillary neurones to vasopressinergic sur-praoptic cells; activation of the former induces increased dye coupling in the latter, and this effect is inhibited by H_1 antagonists (Hatton and Yang 1996). However, an *in vivo* experiment attempting to demonstrate vasopressin release following tuberomammillary stimulation did not reach the same conclusion (Akins and Bealer 1993). Other *in vivo* studies in rats have shown that dehydration activates the histaminergic system, and the dehydration-induced stimulation of vasopressin release is reduced by inhibitors of HA synthesis and by H_1 ant-agonists (Kjaer *et al.* 1994). Interestingly, when the latter experiment was performed in humans, H_2 antagonists, but not H_1 antagonists, inhibited the vasopressin release. (Kjaer *et al.* 2000). Histaminergic neurones are also able to activate the synthesis and release of other hypothalamic-hypophyseal hormones, including oxytocin, CRH, ACTH and beta-endorphin; Both H_1 and H_2 receptors appear to contribute to these effects (Kjaer *et al.* 1998).

17.7.5 Drinking

HA acts at hypothalamic H_1 receptors to reduce drinking behaviour and to regulate overall water balance, but it is less clear the extent to which these findings apply to physiological drinking (Lecklin *et al.* 1998). Exceptions may be the regulation of vasopressin release (discussed above) and food-associated drinking behaviour; the latter seems to have an H_1 and H_3 component (Kraly *et al.* 1996).

17.7.6 Pain and analgesia

Although HA has an H_1-mediated pronociceptive (i.e. pain-*enhancing*) effect on afferent fibres (and possibly in the spinal cord), HA is able to activate antinociceptive (pain-*relieving*) responses in the brain stem through an action on both H_1 and H_2 receptors (Glick and Crane 1978; Thoburn *et al.* 1994). Inhibitors of brain HA metabolism also have analgesic activity (Malmberg-Aiello *et al.* 1997). Based on these findings, one would expect a HA-releasing drug like thioperamide to induce analgesia, but it does not (Li *et al.* 1996). H_1 antagonists block HA analgesia, but these drugs also have their own analgesic properties, depending on the test and the species. It is not clear to what extent these effects represent actions at peripheral versus CNS H_1 receptors (Ghelardini *et al.* 1998; Mobarakeh *et al.* 2000; Rumore and Schlichting 1986; Olsen *et al.* 2002). It does seem clear, however, that brain H_2 receptor activation is an important component of both opioid and stress-induced non-opioid pain-relieving responses (Barke and Hough 1994; Hough and Nalwalk 1992; Hough *et al.* 1990; Gogas and Hough 1989). The findings that H_2 agonists do not penetrate the blood–brain barrier and that some of these drugs produce brain damage (Swaab *et al.* 1992) has limited development of new analgesics based on the H_2 receptor.

In doses larger than those needed to block HA receptors, several (but not all) H_2 and H_3 antagonists induce powerful morphine-like pain relief when injected directly into the brain (Hough *et al.* 1997, 2000). The prototype analgesic discovered in these studies, named improgan, is derived from the H_2 antagonist cimetidine. In contrast with cimetidine, however, improgan lacks activity at over 50 receptors, including all known HA and opioid receptors (Hough *et al.* 2000; Hough 2001). Impentamine, a congener of HA, also induces improgan-like analgesia, suggesting the possibility that these drugs act at an unknown HA receptor, but this has not been established (Hough *et al.* 1999).

17.7.7 Epilepsy

HA increases seizure threshold through an action on the brain H_1 receptor, and high doses of H_1 antagonists have the opposite effect (Yokoyama *et al.* 1994, 1996). H_3 antagonists have an anticonvulsant effect, possibly due to the increased release of neuronal HA (Murakami *et al.* 1995; Kakinoki *et al.* 1998). Metoprine, the inhibitor of brain HA metabolism, has a similar effect (Kakinoki *et al.* 1998). High doses of H_2 antagonists (which do not readily penetrate the brain) cause seizures, and some data suggest that the effects are due to H_2 receptor actions (Shimokawa *et al.* 1996). This is not likely to be the case, however, since large doses of the brain-penetrating H_2 antagonist zolantidine do not induce seizures (Hough unpublished).

17.7.8 Other CNS actions

There is evidence that histaminergic mechanisms can contribute to vestibular disturbances, findings which may explain the efficacy of H_1 antagonists in motion sickness (Takeda *et al.* 1986). Several studies have shown that HA in the brain is increased as a result of thiamine deficiency (Panula *et al.* 1992; Langlais *et al.* 1994; McRee *et al.* 2000), and both brain damage (Langlais *et al.* 1994) and muricidal behaviour (Onodera *et al.* 1993) induced by this condition were reversed by inhibition of HA biosynthesis, implicating HA as an important mediator of Wernicke's encephalopathy. Centrally-administered HA increases blood pressure by mechanisms which may include H_1-mediated stimulation of sympathetic outflow, but this is not clear (Brown *et al.* 2001).

17.8 Concluding remarks

It is remarkable that biological organisms can develop so many ways to use the same molecule to accomplish so many tasks. It is even more remarkable that HA, a molecule with such diverse activities, accomplishes this with such a small number of receptors. As mentioned, additional HA receptors are likely to be discovered, and the H_3 isoforms being presently characterized also may be shown to account for some of this diversification.

Acknowledgements

The authors thank members of the Hough Lab for proof-reading and assistance in preparing Fig 17.3. Dr. Aldo Jongejan helped in preparation of Fig. 17.1. We also thank the National Institute on Drug Abuse (DA-03814) for support (LBH).

References

Akins VF and Bealer SL (1993). Hypothalamic histamine release, neuroendocrine and cardiovascular responses during tuberomammillary nucleus stimulation in the conscious rat. *Neuroendocrinology* 57, 849–55.

Al-Gadi M and Hill SJ (1987). The role of calcium in the cyclic AMP response to histamine in rabbit cerebral cortical slices. *Br J Pharmacol* 91, 213–22.

Alewijnse AE, Smit MJ, Hoffmann M, Verzijl D, Timmerman H, and Leurs R (1998). Constitutive activity and structural instability of the wild-type human H_2 receptor. *J Neurochem* 71, 799–807.

Alves-Rodrigues A, Leurs R, Wu TS, Prell GD, Foged C, and Timmerman H (1996). [^{3}H]-thioperamide as a radioligand for the histamine H3 receptor in rat cerebral cortex. *Br J Pharmacol* 118, 2045–52.

Andrew D and Craig AD (2001). Spinothalamic lamina I neurones selectively responsive to cutaneous warming in cats. *J Physiol* 537, 489–95.

Anglada L, Raga M, Márquez M, Sacristán A, Castelló JM, and Ortiz JA (1997). Synthesis and assessment of formamidines as new histamine H_2-receptor antagonists. *Arzneimittel-Forschung Drug Res* 47, 431–4.

Aoki Y, Qiu D, Zhao GH, and Kao PN (1998). Leukotriene B_4 mediates histamine induction of NF-κB and IL-8 in human bronchial epithelial cells. *Am J Physiol* 274, 1030–9.

Arrang JM, Garbarg M, Lancelot JC, Lecomte J-M, Pollard H, Robba M *et al.* (1987). Highly potent and selective ligands for histamine H3-receptors. *Nature* 327, 117–123.

Arrang JM, Garbarg M, and Schwartz JC (1983). Auto-inhibition of brain histamine release mediated by a novel class (H3) of histamine receptors. *Nature* 302, 832–7.

Arrang JM, Drutel G, Garbarg M, Ruat M, Traiffort E, and Schwartz J-C (1995). Molecular and functional diversity of histamine receptor subtypes. *Ann NY Acad Sci* 757, 314–23.

Ash ASF and Schild HO (1966). Receptors mediating some actions of histamine. *Br J Pharmacol* 27, 427–39.

Baenziger NL, Force LE, and Becherer PR (1980). Histamine stimulate prostacyclin synthesis in cultured human umbilical vein endothelial cells. *Biochem Biophys Res Commun* 92, 1435–40.

Bakker RA, Schoonus S, Smit MJ, Timmerman H, and Leurs R (2001). Histamine H_1-receptor activation of NF-κB: roles for Gβγ and $Gα_{q/11}$-subunits in constitutive and agonist-mediated signaling. *Mol Pharmacol* 60, 1133–42.

Bakker RA, Wieland K, Timmerman H, and Leurs R (2000). Constitutive activity of the histamine H(1) receptor reveals inverse agonism of histamine H(1) receptor antagonists. *Eur J Pharmacol* 387, R5–7.

Barke KE and Hough LB (1993). Opiates, mast cells and histamine release. *Life Sci* 53, 1391–9.

Barke KE and Hough LB (1994). Characterization of basal and morphine-induced histamine release in the rat periaqueductal gray. *J Neurochem* 63, 238–44.

Barnes PJ (1991). Histamine receptors in the lung. In (eds. H Timmerman and H van der Goot) *New Perspectives in Histamine Research*, Birkhauser Verlag, Basel, pp. 103–22.

Bennani, Youssef L, Faghih, and Ramin. (2001). Aminoalkoxybiphenylcarboxamides as histamine-3 receptor ligands and their therapeutic applications. Abbott Laboratories (Abbott Park, IL.) 715601(*6316475*), 1–18.

Beukers MW, Klaassen CH, De Grip WJ, Verzijl D, Timmerman H, and Leurs R (1997). Heterologous expression of rat epitope-tagged histamine H_2 receptors in insect Sf9 cells. *Br J Pharmacol* 122, 867–74.

Birnbaumer L, Zhu X, Jiang M, Boulay G, Peyton M, Vannier B *et al.* (1996). On the molecular basis and regulation of cellular capacitative calcium entry: roles for Trp proteins. *Proc Natl Acad Sci USA* 93, 15 195–202.

Black JW (1979). The Riddle of Gastric Histamine. In (ed. TO Yellin) *Histamine Receptors*, SP Medical & Scientific Books, New York, p. 23.

Black JW, Duncan WAM, Durant GJ, Ganellin CR, and Parsons ME (1972). Definition and antagonism of histamine H2-receptors. *Nature* 236, 385–90.

Blandina P, Giorgetti M, Bartolini L, Cecchi M, Timmerman H, Leurs R *et al.* (1996). Inhibition of cortical acetylcholine release and cognitive performance by histamine H_3 receptor activation in rats. *Br J Pharmacol* 119, 1656–64.

Borges R (1994). Histamine H_1 receptor activation mediates the preferential release of adrenaline in the rat adrenal gland. *Life Sci* 54, 631–40.

Boss V, Wang X, Koppelman LF, Xu K, and Murphy TJ (1998). Histamine induces nuclear factor of activated T cell-mediated transcription and cyclosporin A-sensitive interleukin-8 mRNA expression in human umbilical vein endothelial cells. *Mol Pharmacol* 54, 264–72.

Brown RE, Stevens DR, and Haas HL (2001). The physiology of brain histamine. *Prog Neurobiol* 63, 637–72.

Burde R, Dippel E, and Seifert R (1996). Receptor-independent G protein activation may account for the stimulatory effects of first-generation H1-receptor antagonists in HL- 60 cells, basophils, and mast cells. *Biochem Pharmacol* 51, 125–31.

Calcutt CR, Ganellin CR, Griffiths R, Leigh BK, Maguire JP, Mitchell RC *et al.* (1988). Zolantidine (SK&F 95282) is a potent selective brain-penetrating histamine H_2-receptor antagonist. *Br J Pharmacol* 93, 69–78.

Cardell LO and Edvinsson L (1994). Characterization of the histamine receptors in the guinea-pig lung: evidence for relaxant histamine H_3 receptors in the trachea. *Br J Pharmacol* 111, 445–54.

Carruthers, NI, Li, Xiaobing, and Lovenberg, TW (2002). Method for using 2-aryloxyalkyl-aminobenzoxazoles and 2-aryloxyalkylaminobenzothiazoles as H3 antagonists. 20010044439, 1-14-2002. 3-29-0001.

Carswell H and Young JM (1986). Regional variation in the characteristics of histamine H_1-agonist mediated breakdown of inositol phospholipids in guinea-pig brain. *Br J Pharmacol* 89, 809–17.

Casale TB, Rodbard D, and Kaliner M (1985). Characterization of histamine H_1 receptors on human peripheral lung. *Biochem Pharmacol* 34, 3285–92.

Chazot PL, Hann V, Wilson C, Lees G, and Thompson CL (2001). Immunological identification of the mammalian H3 histamine receptor in the mouse brain. *Neuroreport* 12, 259–62.

Clark EA and Hill SJ (1996). Sensitivity of histamine H_3 receptor agonist-stimulated $[^{35}S]GTPgamma[S]$ binding to pertussis toxin. *Eur J Pharmacol* 296, 223–5.

Claro E, Garcia A, and Picatoste F (1989). Carbachol and histamine stimulation of guanine-nucleotide-dependent phosphoinositide hydrolysis in rat brain cortical membranes. *Biochem J* 261, 29–35.

Coge F, Guenin SP, Audinot V, Renouard-Try A, Beauverger P, Macia C *et al.* (2001*a*). Genomic organization and characterization of splice variants of the human histamine H3 receptor. *Biochem J* 355, 279–88.

Coge F, Guenin SP, Rique H, Boutin JA, and Galizzi JP (2001*b*). Structure and expression of the human histamine H4-receptor gene. *Biochem Biophys Res Commun* 284, 301–9.

Cooper DG, Young RC, Durant GJ, and Ganellin CR (1990). Histamine receptors. *Comp Med Chem* 3, 323–421.

Coruzzi G, Gambarelli E, Bertaccini G, and Timmerman H (1996). Cardiovascular effects of the novel histamine H_2 receptor agonist amthamine: interaction with the adrenergic system. *Naunyn Schmiedebergs Arch Pharmacol* 353, 417–22.

Cumming P, Shaw C, and Vincent SR (1991). High affinity histamine binding site is the H_3 receptor: characterization and autoradiographic localization in rat brain. *Synapse* 8, 144–51.

Daum PR, Downes CP, and Young JM (1983). Histamine-induced inositol phospholipid breakdown mirrors H_1-receptor density in brain. *Eur J Pharmacol* 87, 497–8.

Dawson G, Dawson SA, and Post GR (1993). Regulation of phospholipase D activity in a human oligodendroglioma cell line (HOG). *J Neurosci Res* 34, 324–30.

De Backer MD, Gommeren W, Moereels H, Nobels G, Van Gompel P, Leysen JE, and Luyten WHML (1993). Genomic cloning, heterologous expression and pharmacological characterization of a human histamine H1 receptor. *Biochem Biophys Res Commun* 197, 1601–8.

De Backer MD, Loonen I, Verhasselt P, Neefs JM, and Luyten WH (1998). Structure of the human histamine H1 receptor gene. *Biochem J* 335, 663–70.

De Esch IJ, Gaffar A, Menge WM, and Timmerman H (1999). Synthesis and histamine H3 receptor activity of 4-(n-alkyl)-1H-imidazoles and 4-(omega-phenylalkyl)-1H-imidazoles. *Bioorg Med Chem* 7, 3003–9.

De Esch IJ, Mills JE, Perkins TD, Romeo G, Hoffmann M, Wieland K *et al.* (2001). Development of a pharmacophore model for histamine H3 receptor antagonists, using the newly developed molecular modeling program SLATE. *J Med Chem* 44, 1666–74.

Del Balzo U, Polley MJ, and Levi R (1988). Activation of the third complement component (C3) and C3a generation in cardiac anaphylaxis: histamine release and associated inotropic and chronotropic effects. *J Pharmacol Exp Ther* 246, 911–16.

DelValle J, Wang L, Gantz I, and Yamada T (1992). Characterization of H_2 histamine receptor: linkage to both adenylate cyclase and $[Ca^{2+}]_i$ signaling systems. *Am J Physiol Gastrointest Liver Physiol* 263, G967–72.

Detert H, Hagelüken A, Seifert R, and Schunack W (1995). 2-substituted histamines with G-protein-stimulatory activity. *Eur J Med Chem* 30, 271–6.

Diaz J, Vizuete M-L, Traiffort E, Arrang J-M, Ruat M, and Schwartz J-C (1994). Localization of the histamine H_2 receptor and gene transcripts in rat stomach: back to parietal cells. *Biochem Biophys Res Commun* 198, 1195–202.

Donaldson J, Brown AM, and Hill SJ (1989). Temporal changes in the calcium-dependence of the histamine H_1-receptor-stimulation of cyclic AMP accumulation in guinea-pig cerebral cortex. *Br J Pharmacol* 98, 1365–75.

Donaldson J and Hill SJ (1986). Histamine-induced hydrolysis of polyphosphoinostides in guinea-pig ileum and brain. *Eur J Pharmacol* 124, 255–65.

Drutel G, Peitsaro N, Karlstedt K, Wieland K, Smit MJ, Timmerman H *et al.* (2001). Identification of rat H(3) receptor isoforms with different brain expression and signaling properties. *Mol Pharmacol* 59, 1–8.

Duncan PG, Brink C, Adolphson RL, and Douglas JS (1980). Cyclic nucleotides and contraction/relaxation in airway muscle: H_1 and H_2 agonists and antagonists. *J Pharmacol Exp Ther* 215, 434–42.

Elz S, Kramer K, Pertz HH, Detert H, Ter Laak AM, Kuhne R, and Schunack W (2000). Histaprodifens: synthesis, pharmacological *in vitro* evaluation, and molecular modeling of a new class of highly active and selective histamine H(1)-receptor agonists. *J Med Chem* **43**, 1071–84.

Endou M, Poli E, and Levi R (1994). Histamine H_3-receptor signaling in the heart: possible involvement of G_i/G_o proteins and N-type Ca^{++} channels. *J Pharmacol Exp Ther* **269**, 221–9.

Eriksson KS, Sergeeva O, Brown RE, and Haas HL (2001). Orexin/hypocretin excites the histaminergic neurons of the tuberomammillary nucleus. *J Neurosci* **21**, 9273–9.

Fitzsimons C, Durán H, Engel N, Molinari B, and Rivera E (1999). Changes in H_2 receptor expression and coupling during Ca^{2+}-induced differentiation in mouse epidermal keratinocytes. *Inflamm Res* **48**(Suppl 1), 73–4.

Frisch C, Hasenöhrl RU, Haas HL, Weiler HT, Steinbusch HW, and Huston JP (1998). Facilitation of learning after lesions of the tuberomammillary nucleus region in adult and aged rats. *Exp Brain Res* **118**, 447–56.

Frisch C, Hasenöhrl RU, and Huston JP (1997). The histamine H_1-antagonist chlorpheniramine facilitates learning in aged rats. *Neurosci Lett* **229**, 89–92.

Fujimoto K, Horio Y, Sugama K, Ito S, Qi Liu Y, and Fukui H (1993). Genomic cloning of the rat histamine H1 receptor. *Biochem Biophys Res Commun* **190**, 294–301.

Fujimoto K, Ohta K, Kangawa K, Kikkawa U, Ogino S, and Fukui H (1999). Identification of protein kinase C phosphorylation sites involved in phorbol ester-induced desensitization of the histamine H_1 receptor. *Mol Pharmacol* **55**, 735–42.

Fukagawa K, Sakata T, Shiraishi T, Yoshimatsu H, Fujimoto K, Ookuma K, and Wada H (1989). Neuronal histamine modulates feeding behavior through H_1-receptor in rat hypothalamus. *Am J Physiol* **256**, R605–11.

Fukui H, Fujimoto K, Mizuguchi H, Sakamoto K, Horio Y, Takai S *et al.* (1994). Molecular cloning of the human histamine H_1 receptor gene. *Biochem Biophys Res Commun* **201**, 894–901.

Fukushima Y, Asano T, Saitoh T, Anai M, Funaki M, Ogihara T *et al.* (1997). Oligomer formation of histamine H_2 receptors expressed in Sf9 and COS7 cells. *FEBS Lett* **409**, 283–6.

Furuta S, Kamada E, Suzuki T, Sugimoto T, Kawabata Y, Shinozaki Y, and Sano H (2001). Inhibition of drug metabolism in human liver microsomes by nizatidine, cimetidine and omeprazole. *Xenobiotica* **31**, 1–10.

Ganellin CR (1982). Chemistry and structure-activity relationships of drugs acting at histamine receptors. In (eds. CR Ganellin and ME Parsons), *Pharmacology of Histamine Receptors*, John Wright & Sons, Ltd., Bristol, pp. 10–102.

Ganellin CR, Leurquin F, Piripitsi A, Arrang J-M, Garbarg M, Ligneau X *et al.* (1998). Synthesis of potent, non-imidazole histamine H_3 antagonists. *Arch Pharm Pharm Med Chem* **331**, 395–404.

Gantz I, DelValle J, Wang L, Tashiro T, Munzert G, Guo Y-J *et al.* (1992). Molecular basis for the interaction of histamine with the histamine H2 receptor. *J Biol Chem* **267**, 20 840–3.

Gantz I, Munzert G, Tashiro T, Schaffer M, Wang L, DelValle J, and Yamada T (1991*a*). Molecular cloning of the human histamine H_2 receptor. *Biochem Biophys Res Commun* **178**, 1386–92.

Gantz I, Schaffer M, DelVelle J, Logsdon C, Campbell V, Uhler M, and Yamada T (1991*b*). Molecular cloning of a gene encoding the histamine H_2 receptor. *Proc Natl Acad Sci USA* **88**, 429–33.

Garbarg M, Arrang J-M, Rouleau A, Ligneau X, Dam Trung Tuong M, Schwartz JC, and Ganellin CR (1992). S-[2-(4-imidazolyl)ethyl]isothiourea, a highly specific and potent histamine H_3 receptor agonist. *J Pharmacol Exp Ther* **263**, 304–10.

Garbarg M, Tuong MDT, Gros C, and Schwartz JC (1989). Effects of histamine H_3-receptor ligands on various biochemical indices of histaminergic neuron activity in rat brain. *Eur J Pharmacol* **164**, 1–11.

Garbarg M, Yeramian E, Körner M, and Schawartz JC (1985). Biochemical studies of cerebral H_1 receptors. In (eds CR Ganellin and JC Schwartz), *Frontiers in Histamine Research*, Pergamon Press, Oxford, pp. 9–25.

Gatjkowski GA, Norris DB, Rising TJ, and Wood TP (1983). Specific binding of 3H-tiotidine to histamine H2 receptors in guinea pig cerebral cortex. *Nature* **304**, 65–7.

Gespach C, Saal F, Cost H, and Abita JP (1982). Identification and characterization of surface receptors for histamine in the human promyelocytic leukemia cell line HL-60. Comparison with human peripheral neutrophils. *Mol Pharmacol* **22**, 547–53.

Ghelardini C, Galeotti N, and Bartolini A (1998). No development of tolerance to analgesia by repeated administration of H_1 antagonists. *Life Sci* **63**, L317–22.

Gillard M, Van Der Perren C, Moguilevsky N, Massingham R, and Chatelain P (2002). Binding characteristics of cetirizine and levocetirizine to human H_1 histamine receptors: Contribution of Lys^{191} and Thr^{194}. *Mol Pharmacol* **61**, 391–9.

Gisselmann G, Pusch H, Hovemann BT, and Hatt H (2002). Two cDNAs coding for histamine-gated ion channels in D. melanogaster. *Nat Neurosci* **5**, 11–12.

Glick SD and Crane LA (1978). Opiate-like and abstinence-like effects of intracerebral histamine administration in rats. *Nature* **273**, 547–9.

Gogas KR and Hough LB (1989). Inhibition of naloxone—resistant antinociception by centrally—administered H_2 antagonists. *J Pharmacol Exp Ther* **248**, 262–7.

Gomez-Ramirez J, Ortiz J, and Blanco I (2002). Presynaptic H_3 autoreceptors modulate histamine synthesis through cAMP pathway. *Mol Pharmacol* **61**, 239–45.

Green JP (1983). Histamine receptors in brain. *Handbook Psychopharmacol* **17**, 385–420.

Gupta MK, Gupta P, and Rezai F (2001). Histamine—can it cause an acute coronary event? *Clin Cardiol* **24**, 258–9.

Gutowski S, Smrcka A, Nowak L, Wu DG, Simon M, and Sternweis PC (1991). Antibodies to the α_q subfamily of guanine nucleotide-binding regulatory protein α subunits attenuate activation of phosphatidylinositol 4,5-bisphosphate hydrolysis by hormones. *J Biol Chem* **266**, 20 519–24.

Hasenohrl RU, Kuhlen A, Frisch C, Galosi R, Brandao ML, and Huston JP (2001). Comparison of intra-accumbens injection of histamine with histamine H1-receptor antagonist chlorpheniramine in effects on reinforcement and memory parameters. *Behav Brain Res* **124**, 203–11.

Harper EA, Shankley NP, and Black JW (1999). Evidence that histamine homologues discriminate between H3-receptors in guinea-pig cerebral cortex and ileum longitudinal muscle myenteric plexus. *Br J Pharmacol* **128**, 751–9.

Hatton GI and Yang QZ (1996). Synaptically released histamine increases dye coupling among vaso-pressinergic neurons of the supraoptic nucleus: Mediation by H_1 receptors and cyclic nucleotides. *J Neurosci* **16**, 123–9.

Hatton GI and Yang QZ (2001). Ionotropic histamine receptors and H2 receptors modulate supraoptic oxytocin neuronal excitability and dye coupling. *J Neurosci* **21**, 2974–82.

Hattori Y, Sakuma I, and Kanno M (1988). Differential effects of histamine mediated by histamine H_1- and H_2-receptors on contractility, spontaneous rate and cyclic nucleotides in the rabbit heart. *Eur J Pharmacol* **153**, 221–9.

Hegde SS, Chan P, and Eglen RM (1994). Cardiovascular effects of R-α-methylhistamine, a selective histamine H_3 receptor agonist, in rats: Lack of involvement of histamine H_3 receptors. *Eur J Pharmacol* **251**, 43–51.

Heron A, Rouleau A, Cochois V, Pillot C, Schwartz JC, and Arrang JM (2001). Expression analysis of the histamine H(3) receptor in developing rat tissues. *Mech Dev* **105**, 167–73.

Hey JA, Del Prado M, Egan RW, Kreutner W, and Chapman RW (1992). (R)-alpha-methylhistamine augments neural, cholinergic bronchospasm in guinea pigs by histamine H1 receptor activation. *Eur J Pharmacol* **211**, 421–6.

Higuchi M, Yanai K, Okamura N, Meguro K, Arai H, Itoh M *et al.* (2000). Histamine H(1) receptors in patients with Alzheimer's disease assessed by positron emission tomography [In Process Citation]. *Neurosci* **99**, 721–9.

Hill SJ (1990). Distribution, properties, and functional characteristics of three classes of histamine receptor. *Pharmacol Rev* 42, 45–83.

Hill SJ (1992). Histamine receptor agonists and antagonists. *Neurotrans* 8, 1–5.

Hill SJ, Ganellin CR, Timmerman H, Schwartz JC, Shankley NP, Young, JM *et al.* (1997). International Union of Pharmacology. XIII. Classification of histamine receptors. [Review] [385 refs]. *Pharmacol Rev* 49, 253–78.

Hirschfeld J, Buschauer A, Elz S, Schunack W, Ruat M, Traiffort E, and Schwartz JC (1992). Iodoaminopotentidine and related compounds: A new class of ligands with high affinity and selectivity for the histamine H_2 receptor. *J Med Chem* 35, 2231–8.

Hishinuma S and Young JM (1995). Characteristics of the binding of [3H]-mepyramine to intact human U373 MG astrocytoma cells: evidence for histamine-induced H1-receptor internalisation. *Br J Pharmacol* 116, 2715–23.

Hofmann T, Obukhov AG, Schaefer M, Harteneck C, Gudermann T, and Schultz G (1999). Direct activation of human TRPC6 and TRPC3 channels by diacylglycerol. *Nature* 397, 259–63.

Holden CA, Chan SC, Norris S, and Hanifin JM (1987). Histamine induced elevation of cyclic AMP phosphodiesterase activity in human monocytes. *Agents Actions* 22, 36–42.

Hoon Ryu J, Yanai K, Iwata R, Ido T, and Watanabe T (1994). Heterogeneous distributions of histamine H3, dopamine D1 and D2 receptors in rat brain. *Neuroreport* 5, 621–4.

Hough LB (1988). Cellular localization and possible functions for brain histamine: recent progress. In (eds GA Kerkut and JW Phillis) *Progress in Neurobiology*, Vol. 30 Pergamon Press, Oxford, pp. 469–505.

Hough LB (1999). Histamine. In (eds GJ Siegel, BW Agranoff, RW Albers, SK Fisher, and S Uhlén) *Basic Neurochemistry: Molecular, Cellular and Medical Aspects*, Lippincott-Raven, Philadelphia, pp. 293–313.

Hough LB (2001). Genomics meets histamine receptors: new subtypes, new receptors. *Mol Pharmacol* 59, 1–5.

Hough LB, Goldschmidt RC, Glick SD, and Padawer J (1985). Mast cells in rat brain: Characterization, localization, and histamine content. In (eds. CR Ganellin and JC Schwartz) *Frontiers in histamine research: A tribute to Heinz Schild Advances in the biosciences*, Pergamon Press, New York, pp. 131–40.

Hough LB, Khandelwal JK, and Mittag TW (1981). Alphamethylhistamine methylation by histamine methyltransferase. *Agents Actions* 11, 427–30.

Hough LB and Nalwalk JW (1992). Modulation of morphine antinociception by antagonism of H_2 receptors in the periaqueductal gray. *Brain Res* 588, 58–66.

Hough LB, Nalwalk JW, Barnes WG, Warner LM, Leurs R, Menge WMPB *et al.* (2000). A third life for burimamide: discovery and characterization of a novel class of non-opioid analgesics derived from histamine antagonists. In (eds. SD Glick and IM Maisonneuve) *New Medications for Drug Abuse*, New York Acad. Sci., New York, pp. 25–40.

Hough LB, Nalwalk JW, and Battles AM (1990). Zolantidine-induced attenuation of morphine antinociception in rhesus monkeys. *Brain Res* 526, 153–5.

Hough LB, Nalwalk JW, Leurs R, Menge WMPB, and Timmerman H (1999). Antinociceptive activity of impentamine, a histamine congener, after CNS administration. *Life Sci* 64, PL79–86.

Hough LB, Nalwalk JW, Li BY, Leurs R, Menge WMPB, Timmerman H *et al.* (1997). Novel qualitative structure-activity relationships for the antinociceptive actions of H2 antagonists, H3 antagonists and derivatives. *J Pharmacol Exp Ther* 283, 1534–43.

Hu Q, Deshpande S, Irani K, and Ziegelstein RC (1999). $[Ca^{2+}]_i$ Oscillation frequency regulates agonist-stimulated NF-κB transcriptional activity. *J Biol Chem* 274, 33 995–8.

Huang ZL, Qu WM, Li WD, Mochizuki T, Eguchi N, Watanabe T *et al.* (2001). Arousal effect of orexin A depends on activation of the histaminergic system. *Proc Natl Acad Sci USA* 98, 9965–70.

Hutchison RW and Hey JA (1994). Pharmacological characterization of the inhibitory effect of (R)-α-methylhistamine on sympathetic cardiopressor responses in the pithed guinea-pig. *J Auton Pharmacol* **14**, 393–402.

Inoue I, Taniuchi I, Kitamura D, Jenkins NA, Gilbert DJ, Copeland NG, and Watanabe T (1996). Characteristics of the mouse genomic histamine H1 receptor gene. *Genomics* **36**, 178–81.

Jansen FP, Mochizuki T, Yamamoto Y, Timmerman H, and Yamatodani A (1998). *In vivo* modulation of rat hypothalamic histamine release by the histamine H3 receptor ligands, immepip and clobenpropit. Effects of intrahypothalamic and peripheral application. *Eur J Pharmacol* **362**, 149–55.

Jennings LJ, Salido GM, Pariente JA, Davison JS, Singh J, and Sharkey KA (1996). Control of exocrine secretion in the guinea-pig pancreas by histamine H₃ receptors. *Can J Physiol Pharmacol* **74**, 744–52.

Johnson CL (1982). Histamine receptors and cyclic nucleotides. In (eds CR Ganellin and ME Parsons) *Pharamcology of Histamine Receptors*, John Wright & Son, Bristol, pp. 146–216.

Johnson CL (1992). Histamine receptors and cyclic nucleotides. In (eds JC Schwartz and H Haas) *The Histamine Receptor*, Wiley-Liss, New York, pp. 129–43.

Jutel M, Watanabe T, Klunker S, Akdis M, Thomet OA, Malolepszy J *et al.* (2001). Histamine regulates T-cell and antibody responses by differential expression of H1 and H2 receptors. *Nature* **413**, 420–5.

Kakinoki H, Ishizawa K, Fukunaga M, Fujii Y, and Kamei C (1998). The effects of histamine H₃-receptor antagonists on amygdaloid kindled seizures in rats. *Brain Res Bull* **46**, 461–5.

Kamei C, Okumura Y, and Tasaka K (1993). Influence of histamine depletion on learning and memory recollection in rats. *Psychopharmacol* **111**, 376–82.

Kamei C and Tasaka K (1991). Participation of histamine in the step-through active avoidance response and its inhibition by H₁-blockers. *Jpn J Pharmacol* **57**, 473–82.

Karlstedt K, Senkas A, Ahman M, and Panula P (2001). Regional expression of the histamine H(2) receptor in adult and developing rat brain. *Neurosci* **102**, 201–8.

Kashiba H, Fukui H, and Senba E (2001). Histamine H1 receptor mRNA is expressed in capsaicin-insensitive sensory neurons with neuropeptide Y-immunoreactivity in guinea pigs. *Brain Res* **901**, 85–93.

Kijima H, Isobe Y, Muramatsu M, Yokomori S, Suzuki M, and Higuchi S (1998). Structure-activity characterization of an H₂-receptor antagonist, 3-amino-4-[4-[4-(1-piperidinomethyl)-2-pyridyloxy]-*cis*-2-butenylamino]-3-cyclobutene-1,2-dione hydrochloride (IT-066), involved in the insurmountable antagonism against histamine-induced positive chronotropic action in guinea pig atria. *Biochem Pharmacol* **55**, 151–7.

Kjær A, Knigge U, Jorgensen H, and Warberg J (2000). Dehydration-induced vasopressin secretion in humans: involvement of the histaminergic system. *Am J Physiol Endocrinol Metab* **279**, E1305–10.

Kjær A, Knigge U, Rouleau A, Garbarg M, and Warberg J (1994). Dehydration-induced release of vasopressin involves activation of hypothalamic histaminergic neurons. *Endocrinology* **135**, 675–81.

Kjær A, Larsen PJ, Knigge U, Jorgensen H, and Warberg J (1998). Neuronal histamine and expression of corticotropin-releasing hormone, vasopressin and oxytocin in the hypothalamus: relative importance of H₁ and H₂ receptors. *Eur J Endocrinol* **139**, 238–43.

Klapdor K, Hasenöhrl RU, and Huston JP (1994). Facilitation of learning in adult and aged rats following bilateral lesions of the tuberomammillary nucleus region. *Behav Brain Res* **61**, 113–16.

Klinker JF and Seifert R (1997). Morphine and muscle relaxants are receptor-independent G-protein activators and cromolyn is an inhibitor of stimulated G-protein activity. *Inflamm Res* **46**, 46–50.

Kobayashi T, Tonai S, Ishihara Y, Koga R, Okabe S, and Watanabe T (2000). Abnormal functional and morphological regulation of the gastric mucosa in histamine H2 receptor-deficient mice. *J Clin Invest* **105**, 1741–9.

Koizumi H and Ohkawara A (1999). H₂ histamine receptor-mediated increase in intracellular Ca^{2+} in cultured human keratinocytes. *J Dermatol Sci* **21**, 127–32.

Korte A, Myers J, Shih N-Y, Egan RW, and Clark MA (1990). Characterization and tissue distribution of H_3 histamine receptors in guinea pigs by N^α-methylhistamine. *Biochem Biophys Res Commun* **168**, 979–86.

Kraly FS, Keefe ME, Tribuzio RA, Kim YM, Finkell J, and Braun CJ (1996). H_1, H_2, and H_3 receptors contribute to drinking elicited by exogenous histamine and eating in rats. *Pharmacol Biochem Behav* **53**, 347–54.

Krause M, Rouleau A, Stark H, Luger P, Lipp R, Garbarg M *et al.* (1995). Synthesis, X-ray crystallography, and pharmacokinetics of novel azomethine prodrugs of (R)-α-methylhistamine: highly potent and selective histamine H_3 receptor agonists. *J Med Chem* **38**, 4070–9.

Kristiansen K, Dahl SG, and Edvardsen O (1996). A database of mutants and effects of site-directed mutagenesis experiments on G protein-coupled receptors. *Proteins* **26**, 81–94.

Kühn B, Schmid A, Harteneck C, Gudermann T, and Schultz G (1996). G proteins of the G_q family couple the H_2 histamine receptor to phospholipase C. *Mol Endocrinol* **10**, 1697–707.

Langlais PJ, Zhang SX, Weilersbacher G, Hough LB, and Barke KE (1994). Histamine-mediated neuronal death in a rat model of wernicke's encephalopathy. *J Neurosci Res* **38**, 565–74.

Lecklin A, Etu-Seppälä P, Stark H, and Tuomisto L (1998). Effects of intracerebroventricularly infused histamine and selective H_1, H_2 and H_3 agonists on food and water intake and urine flow in Wistar rats. *Brain Res* **793**, 279–88.

Lecklin A, Hermonen P, Tarhanen J, and Mannisto PT (2000). An acute i.c.v. infusion of leptin has no effect on hypothalamic histamine and tele-methylhistamine contents in Wistar rats. *Eur J Pharmacol* **395**, 113–19.

Legnazzi BL, Monczor F, Rivera E, Bergoc R, and Davio C (1998). Histamine receptors in human epithelial cells—characterization of the receptor G-protein-effector system. *Inflamm Res* **47**, 40–1.

Leopoldt D, Harteneck C, and Nürnberg B (1997). G proteins endogenously expressed in *Sf9* cells: interactions with mammalian histamine receptors. *Naunyn Schmiedeberg's Arch Pharmacol* **356**, 216–24.

Leschke C, Elz S, Garbarg M, and Schunack W (1995). Synthesis and histamine H_1 receptor agonist activity of a series of 2-phenylhistamines, 2-heteroarylhistamines, and analogues. *J Med Chem* **38**, 1287–94.

Leurs R, Bast A, and Timmerman H (1989). High affinity, saturable [^{3}H]mepyramine binding sites on rat liver plasma membrane do not represent histamine H_1-receptors. A warning. *Biochem Pharmacol* **38**, 2175–80.

Leurs R, Blandina P, Tedford CE, and Timmerman H (1998). Therapeutic potential of histamine H_3 receptor agonists and antagonists. *Trends Pharmacol Sci* **19**, 177–83.

Leurs R, Brozius MM, Jansen W, Bast A, and Timmerman H (1991*a*). Histamine H_1-receptor-mediated cyclic GMP production in guinea-pig lung tissue is an L-arginine-dependent process. *Biochem Pharmacol* **42**, 271–7.

Leurs R, Hoffmann M, Wieland K, and Timmerman H (2000). H3 receptor gene is cloned at last. *Trends Pharmacol Sci* **21**, 11–12.

Leurs R, Kathmann M, Vollinga RC, Menge WMPB, Schlicker E, and Timmerman H (1996). Histamine homologues discriminating between two functional H_3 receptor assays. Evidence for H_3 receptor heterogeneity? *J Pharmacol Exp Ther* **276**, 1009–15.

Leurs R, Smit MJ, Brozius MM, Jansen W, Bast A, and Timmerman H (1991*b*). Is protein kinase C involved in histamine H1-receptor desensitization? *Agents Actions Suppl* **33**, 393–402.

Leurs R, Smit MJ, Meeder R, Ter Laak AM, and Timmerman H (1995*a*). Lysine^{200} located in the fifth transmembrane domain of the histamine H_1 receptor interacts with histamine but not with all H_1 agonists. *Biochem Biophys Res Commun* **214**, 110–17.

Leurs R, Smit MJ, Tensen CP, Ter Laak AM, and Timmerman H (1994a). Site-directed mutagenesis of the histamine H_1-receptor reveals a selective interaction of asparagines[207] with subclasses of H_1-receptor agonists. *Biochem Biophys Res Commun* **201**, 295–301.

Leurs R, Smit MJ, and Timmerman H (1995b). Molecular pharmacological aspects of histamine receptors. *Pharmacol Ther* **66**, 413–63.

Leurs R and Timmerman H (1992). The histamine H_3 receptor: a target for developing new drugs. *Progr Drug Res* **39**, 127–65.

Leurs R, Traiffort E, Arrang JM, Tardivel-Lacombe J, Ruat M, and Schwartz J-C (1994b). Guinea pig histamine H_1 receptor. II. Stable expression in Chinese hamster ovary cells reveals the interaction with three major signal transduction pathways. *J Neurochem* **62**, 519–27.

Leurs R, Tulp MTM, Menge WMPB, Adolfs MJP, Zuiderveld OP, and Timmerman H (1995c). Evaluation of the receptor selectivity of the H_3 receptor antagonists, iodophenpropit and thioperamide: an interaction with the 5-HT_3 receptor revealed. *Br J Pharmacol* **116**, 2315–21.

Levi R, Owen DAA, and Trzeciakowski JP (1982). Actions of histamine on the heart and vasculature. In (eds CR Ganellin and ME Parsons) *Pharmacology of Histamine Receptors*, Wright PSG, Bristol.

Levi R and Smith NC (2000). Histamine H(3)-receptors: a new frontier in myocardial ischemia. *J Pharmacol Exp Ther* **292**, 825–30.

Li BY, Nalwalk JW, Barker LA, Cumming P, Parsons ME, and Hough LB (1996). Characterization of the antinociceptive properties of cimetidine and a structural analog. *J Pharmacol Exp Ther* **276**, 500–8.

Liao W, Rudling M, and Angelin B (1997). Novel effects of histamine on lipoprotein metabolism: suppression of hepatic low density lipoprotein receptor expression and reduction of plasma high density lipoprotein cholesterol in the rat. *Endocrinology* **138**, 1863–70.

Ligneau X, Lin JS, Vanni-Mercier G, Jouvet M, Muir JL, Ganellin CR *et al.* (1998) Neurochemical and behavioral effects of ciproxifan, a potent histamine H_3-receptor antagonist. *J Pharmacol Exp Ther* **287**, 658–66.

Ligneau X, Morisset S, Tardivel-Lacombe J, Gbahou F, Ganellin CR, Stark H *et al.* (2000). Distinct pharmacology of rat and human histamine H(3) receptors: role of two amino acids in the third transmembrane domain. *Br J Pharmacol* **131**, 1247–50.

Lin J-S, Sakai K, Vanni-Mercier G, Arrang JM, Garbarg M, Schwartz JC, and Jouvet M (1990). Involvement of histaminergic neurons in arousal mechanisms demonstrated with H_3-receptor ligands in the cat. *Brain Res* **523**, 325–30.

Lintunen M, Sallmen T, Karlstedt K, Fukui H, Eriksson KS, and Panula P (1998). Postnatal expression of H_1-receptor mRNA in the rat brain: correlation to L-histidine decarboxylase expression and local upregulation in limbic seizures. *Eur J Neurosci* **10**, 2287–301.

Liu C, Ma X-J, Jiang X, Wilson SJ, Hofstra C, Blevitt J *et al.* (2001a). Cloning and pharmacological characterization of a fourth histamine receptor (H_4) expressed in bone marrow. *Mol Pharmacol* **59**, 420–6.

Liu C, Wilson SJ, Kuei C, and Lovenberg TW (2001b). Comparison of human, mouse, rat, and guinea pig histamine H4 receptors reveals substantial pharmacological species variation. *J Pharmacol Exp Ther* **299**, 121–30.

Liu YQ, Horio Y, Fujimoto K, and Fukui H (1994). Does the [^{3}H]mepyramine binding site represent the histamine H_1 receptor? Re-examination of the histamine H_1 receptor with quinine. *J Pharmacol Exp Ther* **268**, 959–64.

Lovenberg TW, Pyati J, Chang H, Wilson SJ, and Erlander MG (2000). Cloning of rat histamine H(3) receptor reveals distinct species pharmacological profiles. *J Pharmacol Exp Ther* **293**, 771–8.

Lovenberg TW, Roland BL, Wilson SJ, Jiang X, Pyati J, Huvar A *et al.* (1999). Cloning and functional expression of the human histamine H_3 receptor. *Mol Pharmacol* **55**, 1101–7.

Machidori H, Sakata T, Yoshimatsu H, Ookuma K, Fujimoto K, Kurokawa M *et al.* (1992). Zucker obese rats: defect in brain histamine control of feeding. *Brain Res* **590**, 180–6.

Magistretti PJ and Schorderet M (1985). Norepinephrine and histamine potentiate the increases in cyclic adenosine $3':5'$-monophosphate elicited by vasoactive intestinal polypeptide in mouse cerebral cortical slices: mediation by α_1-adrenergic and H_1-histaminergic receptors. *J Neurosci* 5, 362–8.

Malinowska B, Piszcz J, Schlicker E, Kramer K, Elz S, and Schunack W (1999). Histaprodifen, methyl-histaprodifen, and dimethylhistaprodifen are potent H1-receptor agonists in the pithed and in the anaesthetized rat. *Naunyn Schmiedeberg's Arch Pharmacol* 359, 11–16.

Malmberg-Aiello P, Ipponi A, Bartolini A, and Schunack W (2000). Antiamnesic effect of metoprine and of selective histamine H(1) receptor agonists in a modified mouse passive avoidance test. *Neurosci Lett* 288, 1–4.

Malmberg-Aiello P, Lamberti C, Ipponi A, Hänninen J, Ghelardini C, and Bartolini A (1997). Effects of two histamine-N-methyltransferase inhibitors, SKF 91488 and BW 301U, in rodent antinociception. *Naunyn Schmiedeberg's Arch Pharmacol* 355, 354–60.

Marley PD, Thomson KA, Jachno K, and Johnston MJ (1991). Histamine-induced increases in cyclic AMP levels in bovine adrenal medullary cells. *Br J Pharmacol* 104, 839–46.

Martinez-Mir MI, Pollard H, Moreau J, Arrang JM, Ruat M, and Traiffort E (1990). Three histamine receptors (H1, H2 and H3) visualized in the brain of human and non-human primates. *Brain Res* 526, 322–7.

McRee RC, Terry-Ferguson M, Langlais PJ, Chen Y, Nalwalk JW, Blumenstock FA, and Hough LB (2000). Increased histamine release and granulocytes within the thalamus of a rat model of Wernicke's encephalopathy. *Brain Res* 858, 227–36.

Meguro K-I, Yanai K, Sakai N, Sakurai E, Maeyama K, Sasaki H, and Watanabe T (1995). Effects of thioperamide, a histamine H_3 antagonist, on the step-through passive avoidance response and histidine decarboxylase activity in senescence-accelerated mice. *Pharmacol Biochem Behav* 50, 321–5.

Meier G, Apelt J, Reichert U, Grassmann S, Ligneau X, Elz S, (2001). Influence of imidazole replacement in different structural classes of histamine H(3)-receptor antagonists. *Eur J Pharm Sci* 13, 249–59.

Mitchell R, McCulloch D, Lutz E, Johnson M, MacKenzie C, and Fennell M (1998). Rhodopsin-family receptors associate with small G proteins to activate phospholipase D. *Nature* 392, 411–4.

Miyazaki S, Imaizumi M, and Onodera K (1995). Effects of thioperamide on the cholinergic system and the step-through passive avoidance test in mice. *Methods Find Exp Clin Pharmacol* 17, 653–8.

Miyazaki S, Onodera K, Imaizumi M, and Timmerman H (1997). Effects of clobenpropit (VUF-9153), a histamine H3-receptor antagonist, on learning and memory, and on cholinergic and monoaminergic systems in mice. *Life Sci* 61, 355–61.

Miyazawa N, Watanabe S, Matsuda A, Kondo K, Hashimoto H, Umemura K, and Nakashima M (1998). Role of histamine H1 and H2 receptor antagonists in the prevention of intimal thickening. *Eur J Pharmacol* 362, 53–9.

Mobarakeh JI, Sakurada S, Katsuyama S, Kutsuwa M, Kuramasu A, Lin ZY et al. (2000). Role of histamine H(1) receptor in pain perception: a study of the receptor gene knockout mice [In Process Citation]. *Eur J Pharmacol* 391, 81–9.

Moguilevsky N, Varsalona F, Noyer M, Gillard M, Guillaume JP, Garcia L et al. (1994). Stable expression of human H_1-histamine-receptor cDNA in Chinese hamster ovary cells. Pharmacological characterisation of the protein, tissue distribution of messenger RNA and chromosomal localisation of the gene. *Eur J Biochem* 224, 489–95.

Monnier M, Sauer R, and Hatt AM (1970). The activating effect of histamine on the central nervous system. *Int Rev Neurobiol* 12, 265–305.

Morimoto T, Yamamoto Y, Mobarakeh JI, Yanai K, Watanabe T, Watanabe T, and Yamatodani A (1999). Involvement of the histaminergic system in leptin-induced suppression of food intake. *Physiol Behav* 67, 679–83.

Morini G, Grandi D, and Bertaccini G (1996). Comparison of the protective effect of (*R*)-α-methylhistamine and its prodrugs on gastric mucosal damage induced by ethanol in the rat. *Gen Pharmacol* 27, 1391–4.

Morisset S, Rouleau A, Ligneau X, Gbahou F, Tardivel-Lacombe J, Stark H *et al.* (2000). High constitutive activity of native H3 receptors regulates histamine neurons in brain. *Nature* 408, 860–4.

Morse KL, Behan J, Laz TM, West RE, Jr., Greenfeder SA, Anthes JC *et al.* (2001). Cloning and characterization of a novel human histamine receptor. *J Pharmacol Exp Ther* 296, 1058–66.

Murakami K, Yokoyama H, Onodera K, Iinuma K, and Watanabe T (1995). AQ-0145, a newly developed histamine H_3 antagonist, decreased seizure susceptibility of electrically induced convulsions in mice. *Methods Find Exp Clin Pharmacol* 17, 70–3.

Murayama T, Kajiyama Y, and Nomura Y (1990). Histamine-stimulated and GTP-binding proteins-mediated phopholipase A_2 activation in rabbit platelets. *J Biol Chem* 265, 4290–5.

Nakamura M (2000). Our animal model of coronary spasm—my personal view. *J Atheroscler Thromb* 6, 1–12.

Nakamura T, Itadani H, Hidaka Y, Ohta M, and Tanaka K (2000). Molecular cloning and characterization of a new human histamine receptor, HH4R. *Biochem Biophys Res Commun* 279, 615–20.

Nakata H, Kinoshita Y, Kishi K, Fukuda H, Kawanami C, Matsushima Y *et al.* (1996). Involvement of β-adrenergic receptor kinase-1 in homologous desensitization of histamine H_2 receptors in human gastric carcinoma cell line MKN-45. *Digestion* 57, 406–10.

Natarajan V and Garcia JG (1993). Agonist-induced activation of phospholipase D in bovine pulmonary artery endothelial cells: regulation by protein kinase C and calcium. *J Lab Clin Med* 121, 337–47.

Negulescu PA and Machen TE (1988). Intracellular Ca regulation during secretagogue stimulation of the parietal cell. *Am J Physiol* 254, 130–40.

Nguyen T, Shapiro DA, George SR, Setola V, Lee DK, Cheng R *et al.* (2001). Discovery of a novel member of the histamine family. *Mol Pharmacol* 59, 427–33.

Ninkovic M and Hunt SP (1985). Opiate and histamine H1 receptors are present on some substance p-containing dorsal root ganglion cells. *Neurosci Lett* 53, 133–7.

Nishino S, Fujiki N, Ripley B, Sakurai E, Kato M, Watanabe T *et al.* (2001). Decreased brain histamine content in hypocretin/orexin receptor-2 mutated narcoleptic dogs. *Neurosci Lett* 313, 125–8.

Nozaki M and Sperelakis N (1989). Pertussis toxin effects on transmitter release from perivascular nerve terminals. *Am J Physiol* 256, 455–9.

Oda T, Morikawa N, Saito Y, Masuho Y, and Matsumoto S (2000). Molecular cloning and characterization of novel type of histamine receptor preferentially expressed in leukocytes. *J Biol Chem* 275, 36 781–6.

Ohta K, Hayashi H, Mizuguchi H, Kagamiyama H, Fujimoto K, and Fukui H (1994). Site-directed mutagenesis of the histamine H_1 receptor: roles of aspartic acid[107], asparagine[198] and threonine[194]. *Biochem Biophys Res Commun* 203, 1096–101.

Oike M, Kitamura K, and Kuriyama H (1992). Histamine H_3-receptor activation augments voltage-dependent Ca^{2+} current via GTP hydrolysis in rabbit saphenous artery. *J Physiol (Lond)* 448, 133–52.

Olsen UB, Eltorp CT, Ingvardsen BK, Jorgensen TK, Lundbaek JA, Thomsen C, and Hansen AJ (2002). ReN 1869, a novel tricyclic antihistamine, is active against neurogenic pain and inflammation. *Eur J Pharmacol* 435, 43–57.

Onodera K, Miyazaki S, and Imaizumi M (1998). Cognitive involvement by negative modulation of histamine H_2 receptors in passive avoidance task in mice. *Methods Find Exp Clin Pharmacol* 20, 307–10.

Onodera K, Yamatodani A, and Watanabe T (1993). Effect of alpha-fluoromethylhistidine on brain histamine and noradrenaline in muricidal rats. *Methods Find Exp Clin Pharmacol* 15, 423–7.

Onodera K, Yamatodani A, Watanabe T, and Wada H (1994). Neuropharmacology of the histaminergic neuron system in the brain and its relationship with behavioral disorders. *Prog Neurobiol* 42, 685–702.

Orellana S, Solski PA, and Brown JH (1987). Guanosine 5'-O-(thiotriphosphate)-dependent inositol trisphosphate formation in membranes is inhibited by phorbol ester and protein kinase C. *J Biol Chem* **262**, 1638–43.

Palacios JM, Garbarg M, Barbin G, and Schwartz JC (1978). Pharmacological characterization of histamine receptors mediating the stimulation of cyclic AMP accumulation in slices from guinea-pig hippocampus. *Mol Pharmacol* **14**, 971–82.

Palacios JM, Wamsley JK, and Kuhar MJ (1981). The distribution of histamine H1-receptors in the rat brain: An autoradiographic study. *Neurosci* **6**, 15–37.

Panula P and Airaksinen MS (1991). The histaminergic neuronal system as revealed with antisera against histamine, In (eds T Watanabe and H Wada) *Histaminergic Neurons: Morphology and Function*, CRC Press, Boca Raton, pp. 127–43.

Panula P, Rinne J, Kuokkanen K, Eriksson KS, Sallmen T, Kalimo H, and Relja M (1998). Neuronal histamine deficit in Alzheimer's disease. *Neurosci* **82**, 993–7.

Panula P, Tuomisto L, Karhunen T, Sarviharju M, and Korpi ER (1992). Increased neuronal histamine in thiamine-deficient rats. *Agents Actions* **36** (Suppl. C), C354–7.

Parsons ME, Owen DAA, Ganellin CR, and Durant GJ (1977). Dimaprit-[S-[3-(N,N-dimethylamino)propyl]isothiourea]—a highly specific histamine H2-receptor agonist. Part 1. Pharmacology. *Agents Actions* **7**, 31–7.

Passani MB, Bacciottini L, Mannaioni PF, and Blandina P (2000). Central histaminergic system and cognition. *Neurosci Biobehav Rev* **24**, 107–13.

Passani MB and Blandina P (1998). Cognitive implications for H3 and 5-HT3 receptor modulation of cortical cholinergic function: a parallel story. *Methods Find Exp Clin Pharmacol* **20**, 725–33.

Perez-Garcia C, Morales L, and Alguacil LF (1998). Histamine H3 receptor desensitization in the guinea-pig ileum. *Eur J Pharmacol* **341**, 253–6.

Poli E, Pozzoli C, Coruzzi G, and Bertaccini G (1993). Histamine H3-receptor-induced inhibition of duodenal cholinergic transmission is independent of intracellular cyclic AMP and GMP. *Gen Pharmacol* **24**, 1273–8.

Pollard H, Moreau J, Arrang JM, and Schwartz JC (1993). A detailed autoradiographic mapping of histamine H_3 receptors in rat brain areas. *Neurosci* **52**, 169–89.

Pype JL, Mak JC, Dupont LJ, Verleden GM, and Barnes PJ (1998). Desensitization of the histamine H_1-receptor and transcriptional down-regulation of histamine H_1-receptor gene expression in bovine tracheal smooth muscle. *Br J Pharmacol* **125**, 1477–84.

Raible DG, Lenahan T, Fayvilevich Y, Kosinski R, and Schulman ES (1994). Pharmacologic characterization of a novel histamine receptor on human eosinophils. *Am J Resp Crit Care Med* **149**, 1506–11.

Resink TJ, Grigorian G, Moldabaeva AK, Danilov SM, and Buhler FR (1987). Histamine-induced phosphoinositide metabolism in cultured human umbilical vein endothelial cells. Association with thromboxane and prostacyclin release. *Biochem Biophys Res Commun* **144**, 438–46.

Rodriguez-Pena MS, Timmerman H, and Leurs R (2000). Modulation of histamine H(2) receptor signalling by G-protein-coupled receptor kinase 2 and 3. *Br J Pharmacol* **131**, 1707–15.

Rouleau A, Garbarg M, Ligneau X, Mantion C, Lavie P, Advenier C, Lecomte J-M, Krause M, Stark H, Schunack W, and Schwartz J-C (1997). Bioavailability, antinociceptive and antiinflammatory properties of BP 2-94, a histamine H_3 receptor agonist prodrug. *J Pharmacol Exp Ther* **281**, 1085–94.

Ruat M, Traiffort E, Arrang JM, Leurs R, and Schwartz JC (1991). Cloning and tissue expression of a rat histamine H_2-receptor gene. *Biochem Biophys Res Commun* **179**, 1470–78.

Ruat M, Traiffort E, Bouthenet ML, Schwartz JC, Hirschfeld J, Buschauer A, and Schunack W (1990). Reversible and irreversible labeling and autoradiographic localization of the cerebral histamine H_2 receptor using [^{125}I]iodinated probes. *Proc Natl Acad Sci USA* **87**, 1658–62.

Rudy B and McBain CJ (2001). Kv3 channels: voltage-gated K+ channels designed for high-frequency repetitive firing. *Trends Neurosci* **24**, 517–26.

Rumore MM and Schlichting DA (1986). Clinical efficacy of antihistaminics as analgesics. *Pain* **25**, 7–22.

Sakata T, Ookuma K, Fukagawa K, Fujimoto K, Yoshimatsu H, Shiraishi T, and Wada H (1988). Blockade of the histamine H1-receptor in the rat ventromedial hypothalmus and feeding elicitation. *Brain Res* **441**, 403–7.

Sakata T and Yoshimatsu H (1995). Homeostatic maintenance regulated by hypothalamic neuronal histamine. *Methods Find Exp Clin Pharmacol* **17** (Suppl C), 51–6.

Sakuma I, Gross SS, and Levi R (1988). Positive inotropic effect of histamine on guinea pig left atrium: H_1-receptor-induced stimulation of phosphoinositide turnover. *J Pharmacol Exp Ther* **247**, 466–72.

Saper CB, Chou TC, and Scammell TE (2001). The sleep switch: hypothalamic control of sleep and wakefulness. *Trends Neurosci* **24**, 726–31.

Sasse A, Stark H, Reidemeister S, Hüls A, Elz S, Ligneau X *et al.* (1999). Novel partial agonists for the histamine H3 receptor with high *in vitro* and *in vivo* activity. *J Med Chem* **42**, 4269–74.

Sawutz DG, Kalinyak K, Whitsett JA, and Johnson CL (1984). Histamine H2 receptor desensitization in HL-60 human promyelocytic leukemia cells. *J Pharmacol Exp Ther* **231**, 1–7.

Scammell TE, Estabrooke IV, McCarthy MT, Chemelli RM, Yanagisawa M, Miller MS, and Saper CB (2000). Hypothalamic arousal regions are activated during modafinil-induced wakefulness. *J Neurosci* **20**, 8620–28.

Schmelz M (2002). Itch—mediators and mechanisms. *J Dermatol Sci* **28**, 91–6.

Schmidt HH, Zernikow B, Baeblich S, and Bohme E (1990). Basal and stimulated formation and release of L-arginine-derived nitrogen oxides from cultured endothelial cells. *J Pharmacol Exp Ther* **254**, 591–7.

Schreurs J, Dailey MO, and Schulman H (1984). Pharmacological characterization of histamine H2 receptors on clonal cytolytic T lymphocytes. Evidence for histamine-induced desensitization. *Biochem Pharmacol* **33**, 3375–82.

Schwartz JC, Arrang JM, Garbarg M, Pollard H, and Ruat M (1991). Histaminergic transmission in the mammalian brain. *Physiol Rev* **71**, 1–51.

Seifert R, Grünbaum L, and Schultz G (1994a). Histamine H_1-receptors in HL-60 monocytes are coupled to G_i-proteins and pertussis toxin-insensitive G-proteins and mediate activation of Ca^{2+} influx without concomitant Ca^{2+} mobilization from intracellular stores. *Naunyn Schmiedeberg's Arch Pharmacol* **349**, 355–61.

Seifert R, Hagelüken A, Höer A, Höer D, Grünbaum L, Offermanns S *et al.* (1994b). The H_1 receptor agonist 2-(3-chlorophenyl)histamine activates G_i proteins in HL-60 cells through a mechanism that is independent of known histamine receptor subtypes. *Mol Pharmacol* **45**, 578–86.

Seifert R, Höer A, Schwaner I, and Buschauer A (1992). Histamine increases cytosolic Ca^{2+} in HL-60 promyelocytes predominantly via H_2 receptors with an unique agonist/antagonist profile and induces functional differentiation. *Mol Pharmacol* **42**, 235–41.

Sertl K, Casale TB, Wescott SL, and Kaliner MA (1987). Immunohistochemical localization of histamine-stimulated increases in cyclic GMP in guinea pig lung. *Am Rev Respir Dis* **135**, 456–62.

Shayo C, Fernandez N, Legnazzi BL, Monczor F, Mladovan A, Baldi A, and Davio C (2001). Histamine H2 receptor desensitization: involvement of a select array of G protein-coupled receptor kinases. *Mol Pharmacol* **60**, 1049–56.

Shih N-Y, Lupo AT, Jr, Aslanian R, Orlando S, Piwinski JJ, Green MJ *et al.* (1995). A novel pyrrolidine analog of histamine as a potent, highly selective histamine H_3 receptor agonist. *J Med Chem* **38**, 1593–9.

Shimokawa M, Yamamoto K, Kawakami J, Sawada Y, and Iga T (1996). Neurotoxic convulsions induced by histamine H_2 receptor antagonists in mice. *Toxicol Appl Pharmacol* **136**, 317–323.

Silver R, Silverman A, Vitkovic L, and Lederhendler II (1996). Mast cells in the brain: evidence and functional significance. *Trends Neurosci* **19**, 25–31.

Silver RB, Mackins CJ, Smith NC, Koritchneva IL, Lefkowitz K, Lovenberg TW, and Levi R (2001). Coupling of histamine H_3 receptors to neuronal Na^+/H^+ exchange: a novel protective mechanism in myocardial ischemia. *Proc Natl Acad Sci USA* **98**, 2855–9.

Silver RB, Poonwasi KS, Seyedi N, Wilson SJ, Lovenberg TW, and Levi R (2002). Decreased intracellular calcium mediates the histamine H_3-receptor-induced attenuation of norepinephrine exocytosis from cardiac sympathetic nerve endings. *PNAS* **99**, 501–6.

Smit MJ, Bloemers SM, Leurs R, Tertoolen LG, Bast A, de Laat SW, and Timmerman H (1992). Short-term desensitization of the histamine H_1 receptor in human HeLa cells: involvement of protein kinase C dependent and independent pathways. *Br J Pharmacol* **107**, 448–55.

Smit MJ, Leurs R, Alewijnse AE, Blauw J, Amerongen GPV, Van de Vrede Y, Roovers E, and Timmerman H (1996*a*). Inverse agonism of histamine H_2 antagonists accounts for upregulation of spontaneously active histamine H_2 receptors. *Proc Natl Acad Sci USA* **93**, 6802–7.

Smit MJ, Leurs R, Shukrula SR, Bast A, and Timmerman H (1994). Rapid desensitization of the histamine H_2 receptor on the human monocytic cell line U937. *Eur J Pharmacol Mol Pharmacol* **288**, 17–25.

Smit MJ, Roovers E, Timmerman H, Van de Vrede Y, Alewijnse AE, and Leurs R (1996*b*). Two distinct pathways for histamine H_2 receptor down-regulation. *J Biol Chem* **271**, 7574–82.

Smit MJ, Timmerman H, Alewijnse AE, Punin M, Van den Nieuwenhof I, Blauw J *et al.* (1995). Visualization of agonist-induced internalization of histamine H2 receptors. *Biochem Biophys Res Commun* **214**, 1138–45.

Smit MJ, Timmerman H, Hijzelendoorn JC, Fukui H, and Leurs R (1996*c*). Regulation of the human histamine H_1 receptor stably expressed in Chinese hamster ovary cells. *Br J Pharmacol* **117**, 1071–80.

Smith IR, Cleverley MT, Ganellin CR, and Metters KM (1980). Binding of [^{3}H]cimetidine to rat brain tissue. *Agents Actions* **10**, 422–6.

Stark H, Ligneau X, Arrang JM, Schwartz JC, and Schunack W (1998). General construction pattern of histamine H_3-receptor antagonists: Change of a paradigm. *Bioorg Med Chem Lett* **8**, 2011–16.

Swaab R, Battles AM, Hough LB, and Bruner CA (1992). Dimaprit-induced neurotoxicity. *Agents Actions* **35**, 179–84.

Takeda N, Morita M, Kubo T, Yamatodani A, Watanabe T, Wada H, and Matsunaga T (1986). Histaminergic mechanism of motion sickness. *Acta Otolaryngol* **101**, 416–21.

Takeshita Y, Watanabe T, Sakata T, Munakata M, Ishibashi H, and Akaike N (1998). Histamine modulates high-voltage-activated calcium channels in neurons dissociated from the rat tuberomammillary nucleus. *Neurosci* **87**, 797–805.

Tardivel-Lacombe J, Rouleau A, Heron A, Morisset S, Pillot C, Cochois V *et al.* (2000). Cloning and cerebral expression of the guinea pig histamine H3 receptor: evidence for two isoforms. *Neuro Report* **11**, 755–9.

Taylor JE, Yaksh TL, and Richelson E (1982). Histamine H_1 receptors in the brain and spinal cord of the cat. *Brain Res* **243**, 391–4.

Taylor SJ, Michel AD, and Kilpatrick GJ (1992). *In vivo* occupancy of histamine H_3 receptors by thioperamide and (R)-α-methylhistamine measured using histamine turnover and an *ex vivo* labeling technique. *Biochem Pharmacol* **44**, 1261–7.

Tedford CE, Phillips JG, Gregory R, Pawlowski GP, Fadnis L, Khan MA *et al.* (1999). Development of trans-2-[1H-imidazol-4-yl] cyclopropane derivatives as new high-affinity histamine H3 receptor ligands. *J Pharmacol Exp Ther* **289**, 1160–8.

Tedford CE, Yates SL, Pawlowski GP, Nalwalk JW, Hough LB, Khan MA *et al.* (1995). Pharmacological characterization of GT-2016, a non-thiourea-containing histamine

H$_3$ receptor antagonist, *in vitro* and *in vivo* studies. *J Pharmacol Exp Ther* **275**, 598–604.

Ter Laak AM, Donne-Op den Kelder GM, Bast A, and Timmerman H (1993). Is there a difference in the affinity of histamine H1 receptor antagonists for CNS and peripheral receptors? An *in vitro* study. *Eur J Pharmacol* **232**, 199–205.

Ter Laak AM, Venhorst J, Donne-Op den Kelder GM, and Timmerman H (1995). The histamine H1-receptor antagonist binding site. A stereoselective pharmacophoric model based upon (semi-)rigid H1-antagonists and including a known interaction site on the receptor. *J Med Chem* **38**, 3351–60.

Thoburn KK, Hough LB, Nalwalk JW, and Mischler SA (1994). Histamine-induced modulation of nociceptive responses. *Pain* **58**, 29–37.

Thompson JD, Gibson TJ, Plewniak F, Jeanmougin F, and Higgins DG (1997). The CLUSTAL_X windows interface: flexible strategies for multiple sequence alignment aided by quality analysis tools. *Nucleic Acids Res* **25**, 4876–82.

Traiffort E, Leurs R, Arrang JM, Tardivel-Lacombe J, Diaz J, Schwartz J-C, and Ruat M (1994). Guinea pig histamine H$_1$ receptor. I. Gene cloning, characterization, and tissue expression revealed by in situ hybridization. *J Neurochem* **62**, 507–18.

Traiffort E, Vizuete M-L, Tardivel-Lacombe J, Souil E, Schwartz JC, and Ruat M (1995). The guinea pig histamine H2 receptor, gene cloning, tissue expression and chromosomal localization of its human counterpart. *Biochem Biophys Res Commun* **211**, 570–7.

Uvnas, B (1991). *Histamine and Histamine Antagonists*. Springer-Verlag, Berlin.

Uveges AJ, Kowal D, Zhang Y, Spangler TB, Dunlop J, Semus S, and Jones PG (2002). The role of transmembrane helix 5 in agonist binding to the human H3 receptor. *J Pharmacol Exp Ther* **301**, 451–8.

van der Goot H, Schepers MJP, Sterk GJ, and Timmerman H (1992). Isothiourea analogues of histamine as potent agonists or antagonists of the histamine H$_3$-receptor. *Eur J Med Chem* **27**, 511–17.

Vizuete M-L, Traiffort E, Bouthenet ML, Ruat M, Souil E, Tardivel-Lacombe J, and Schwartz JC (1997). Detailed mapping of the histamine H$_2$ receptor and its gene transcripts in guinea-pig brain. *Neurosci* **80**, 321–43.

Vollinga RC, De Koning JP, Jansen FP, Leurs R, Menge WMPB, and Timmerman H (1994). A new potent and selective histamine H$_3$ receptor agonist, 4-(1H-imidazol-4-ylmethyl)piperidine. *J Med Chem* **37**, 332–3.

Vollinga RC, Menge WMPB, Leurs R, and Timmerman H (1995). Homologs of histamine as histamine H$_3$ receptor antagonists: a new potent and selective H$_3$ antagonist, 4(5)-(5-aminopentyl)-1H-imidazole. *J Med Chem* **38**, 266–71.

Wang L, Gantz I, and Del Valle J (1996). Histamine H$_2$ receptor activates adenylate cyclase and PLC via separate GTP-dependent pathways. *Am J Physiol* **271**, 613–20.

Wang LD, Hoeltzel M, Butler K, Hare B, Todisco A, Wang M, and Del Valle J (1997). Activation of the human histamine H$_2$ receptor is linked to cell proliferation and c-*fos* gene transcription. *Am J Physiol Cell Physiol* **273**, C2037–45.

Wang YX and Kotlikoff MI (2000). Signalling pathway for histamine activation of non-selective cation channels in equine tracheal myocytes. *J Physiol* **523** (Pt 1), 131–8.

Wellendorf P, Goodman MW, Nash NR, and Weiner DM (2000). Molecular cloning and expression of functionally distinct isoforms of the human H3 receptor. International Sendai Histamine Symposium, Sendai, Japan, 50. 11-22-2000.

Wieland K, Laak AM, Smit MJ, Kühne R, Timmerman H, and Leurs R (1999). Mutational analysis of the antagonist-binding site of the histamine H$_1$ receptor. *J Biol Chem* **274**, 29 994–30 000.

Wieland K, Bongers G, Yamamoto Y, Hashimoto T, Yamatodani A, Menge WMBP *et al.* (2001). Constitutive activity of histamine H_3 receptors stably expressed in SK-N-MC cells: display of agonism and inverse agonism by H_3 antagonists. *J Pharmacol Exp Ther* **299**, 908–14.

Woosley RL (1996). Cardiac actions of antihistamines. *Annu Rev Pharmacol Toxicol* **36**, 233–52.

Yamamoto Y, Mochizuki T, Okakura-Mochizuki K, Uno A, and Yamatodani A (1997). Thioperamide, a histamine H_3 receptor antagonist, increases GABA release from the rat hypothalamus. *Methods Find Exp Clin Pharmacol* **19**, 289–98.

Yamashita M, Fukui H, Sugama K, Horio Y, Ito S, Mizuguchi H, and Wada H (1991*a*). Expression cloning of a cDNA encoding the bovine histamine H_1 receptor. *Proc Natl Acad Sci USA* **88**, 11 515–19.

Yamashita M, Ito S, Sugama K, Fukui H, Smith B, Nakanishi K, and Wada H (1991*b*). Biochemical characterization of histamine H_1 receptors in bovine adrenal medulla. *Biochem Biophys Res Commun* **177**, 1233–9.

Yanai K, Ryu JH, Watanabe T, Iwata R, Ido T, Sawai Y *et al.* (1995). Histamine H_1 receptor occupancy in human brains after single oral doses of histamine H_1 antagonists measured by positron emission tomography. *Br J Pharmacol* **116**, 1649–55.

Yanai K, Son LZ, Endou M, Sakurai E, and Watanabe T (1998). Targeting disruption of histamine H_1 receptors in mice: Behavioral and neurochemical characterization. *Life Sciences* **62**, 1607–10.

Yanai K, Watanabe T, Hatazawa J, Itoh M, Nunoki K, Hatano K *et al.* (1990*a*). Visualization of histamine H_1 receptors in dog brain by positron emission tomography. *Neuorsci Lett* **118**, 41–4.

Yanai K, Watanabe T, Meguro K, Yokoyama H, Sato I, Sasano H *et al.* (1992*a*). Age-dependent decrease in histamine H_1 receptor in human brains revealed by PET. *Neuroreport* **3**, 433–6.

Yanai K, Watanabe T, Yokoyama H, Meguro K, Hatazawa J, Itoh M *et al.* (1992*b*). Histamine H_1 receptors in human brain visualized *in vivo* by [^{11}C]doxepin and positron emission tomography. *Neurosci Lett* **137**, 145–8.

Yanai K, Yagi N, Watanabe T, Itoh M, Ishiwata K, Ido T, and Matsuzawa T (1990*b*). Specific binding of [^{3}H]pyrilamine to histamine H_1 receptors in guinea pig brain *in vivo*: determination of binding parameters by a kinetic four-compartment model. *J Neurochem* **55**, 409–20.

Yates SL, Phillips JG, Gregory R, Pawlowski GP, Fadnis L, Khan MA *et al.* (1999). Identification and pharmacological characterization of a series of new 1H-4-substituted-imidazoyl histamine H3 receptor ligands. *J Pharmacol Exp Ther* **289**, 1151–9.

Yokoyama H, Onodera K, Iinuma K, and Watanabe T (1994). 2-thiazolylethylamine, a selective histamine H_1 agonist, decreases seizure susceptibility in mice. *Pharmacol Biochem Behav* **47**, 503–7.

Yokoyama H, Sato M, Iinuma K, Onodera K, and Watanabe T (1996). Centrally acting histamine H_1 antagonists promote the development of amygdala kindling in rats. *Neurosci Lett* **217**, 194–6.

Yoshimatsu H, Itateyama E, Kondou S, Tajima D, Himeno K, Hidaka S *et al.* (1999). Hypothalamic neuronal histamine as a target of leptin in feeding behavior. *Diabetes* **48**, 2286–91.

Young RC, Mitchell RC, Brown TH, Ganellin CR, Griffiths R, Jones M *et al.* (1988). Development of a new physicochemical model of brain penetration and its application to the design of centrally acting H_2 receptor histamine antagonists. *J Med Chem* **31**, 656–71.

Yuan Y, Granger HJ, Zawieja DC, DeFily DV, and Chilian WM (1993). Histamine increases venular permeability via a phospholipase C-NO synthase-guanylate cyclase cascade. *Am J Physiol* **264**, 1734–9.

Zamani MR, Dupere JR, and Bristow DR (1995). Receptor-mediated desensitisation of histamine H1 receptor-stimulated inositol phosphate production and calcium mobilisation in GT1–7 neuronal cells is independent of protein kinase C. *J Neurochem* **65**, 160–9.

Zheng Y, Hirschberg B, Yuan J, Wang AP, Hunt DC, Ludmerer SW, Schmatz DM, and Cully DF (2002). Identification of two novel Drosophila melanogaster histamine-gated chloride channel subunits expressed in the eye. *J Biol Chem* 277, 2000–5.

Zhu Y, Michalovich D, Wu H-L, Tan KB, Dytko GM, Mannan IJ *et al.* (2001). Cloning, expression and pharmacological characterization of a novel human histamine receptor. *Mol Pharmacol* 59, 434–41.

Chapter 18

Muscarinic receptors

John Ellis

18.1 Introduction

The pharmacology of acetylcholine is in many ways a very old subject. The use of antimuscarinic plant extracts as both poisons and therapies can be traced back to ancient times. Thus, Linnaeus gave the systematic name Atropa belladonna to the deadly nightshade, in homage to Atropos of Greek mythology, who cuts the thread of life, and to Italian women, who reportedly used the same plant as a topical cosmetic to dilate the pupils (Brown and Taylor 2001). Indeed, Socrates may have been poisoned with henbane (Hyoscyamus niger), rather than with hemlock (Burgen 1995). Many cholinergic compounds were isolated to purity or near-purity in the nineteenth century, including atropine and acetylcholine, and it was the muscarinic action of acetylcholine on the heart that established the chemical basis of neurotransmission in Otto Loewi's famous experiment (Smith 1996). Just prior to that, Dale had differentiated muscarinic and nicotinic types of responses to acetylcholine, a definition that served well for much of the rest of the century. Muscarinic actions are typically distinguishable from nicotinic by their slower onset and longer duration, anatomical localization, and antagonism by highly selective antagonists. In the peripheral nervous system, muscarinic receptors are most strikingly associated with the parasympathetic end-organs of the autonomic system, but in this review these autonomic effects will only be discussed if they illustrate a basic physiological action of the receptors or if they have generated therapeutic interest; comprehensive coverage can be found in pharmacology texts (Hoffman and Taylor 2001).

18.2 Molecular characterization

Until about 25 years ago, it seemed that the postulation of a single type of receptor was sufficient to accommodate virtually all of the existing pharmacological data (Beld *et al.* 1975; Inch and Brimblecombe 1974). However, the development of antagonists that proved to be moderately selective, such as 4DAMP (Barlow *et al.* 1976) and pirenzepine (Hammer *et al.* 1980), suggested that different muscarinic subtypes predominated in different tissues (e.g. forebrain vs. heart vs. smooth muscle). These initial pharmacological data implied most strongly that the muscarinic receptors in the heart differed from those in the brain, so the first sequence data for muscarinic receptors were partial amino acid sequences obtained from porcine receptors purified from these two tissues (Haga and Haga 1983; Peterson *et al.* 1984); these sequences were used to clone the respective cDNAs (Kubo *et al.* 1986*a,b*; Peralta *et al.* 1987*b*). Subsequently, complete families of five muscarinic receptor subtypes were cloned from human and rat libraries (Bonner *et al.* 1987, 1988). The existence of these subtypes has been confirmed in these species and the sequences are highly conserved among mammals (Liao *et al.* 1989; Peralta *et al.* 1987*a*). These receptors are all predicted to possess seven transmembrane (TM) domains and are named M_1–M_5, (Table 18.1) according to the order of their cloning (Caulfield and Birdsall 1998). No new members of the mammalian family have been discovered in more than a decade and no splice variants have been reported. Readers of the literature of the late 1980s and early 1990s should be aware that there was a conflicting nomenclature at that time that agreed on M_1, M_2, and M_5, but transposed M_3 and M_4. A simplified way of distinguishing these subtypes is that the M_3 receptor is the longest subtype and couples to phosphoinositide metabolism via the G_q family, whereas M_4 couples to G_i mediated effects. Although coupling-specific sequences do exist, the muscarinic receptors comprise an unusually closely related GPCR family (Schwartz *et al.* 1997; Vernier *et al.* 1995).

The various subtypes possess between two and five consensus sequences for N-linked glycosylation on their extracellular N-terminals. The importance of these sites varies among the subtypes. Elimination of the glycosylation sites by site-directed mutagenesis in the M_2 receptor does not affect expression, binding affinity, or function (van Koppen and Nathanson 1990). However, related studies on the M_3 receptor in Sf9 cells found that glycosylation is important for full expression; treatment with tunicamycin also reduced expression (Behr *et al.* 1998). Earlier studies found that tunicamycin reduced the expression and altered the subcellular distribution of muscarinic receptors in N1E-115 cells (Liles and Nathanson 1986), which express the M_1 and M_4 subtypes (McKinney *et al.* 1991).

Like other GPCRs, the muscarinic receptors all possess a cysteine in the C-terminal cytoplasmic tail that may be palmitoylated to provide a membrane anchor. Elimination of this residue in the M_2 receptor does not affect expression, ligand affinities, or regulation of cyclic AMP levels in CHO cells (van Koppen and Nathanson 1991). However, in studies with Sf9 cells, the elimination of the cysteine residue reduces receptor-G protein interactions somewhat (Hayashi and Haga 1997).

Two other important cysteine residues are present in all of the muscarinic subtypes and in most of the GPCRs of the rhodopsin family. They are located in the second outer loop, just above TM3, and in the middle of the third outer loop of the receptors. These residues are believed to form a disulfide bond that links these two loops; they are also implicated in the formation of covalent receptor dimers (Zeng and Wess 1999). Mechanistic studies have also supported the existence of cooperative interactions between muscarinic binding sites (Chidiac *et al.* 1997; Hirschberg and Schimerlik 1994; Mattera *et al.* 1985). A number of studies using chimeric and split M_3 receptors, in combination with immunoprecipitation and binding assays, have demonstrated that this subtype forms oligomers with itself, but not with M_1 or M_2 receptors, that some of these oligomers are expressed on the cell surface, that the oligomers can bind muscarinic ligands, and that oligomerization is not stimulated by agonist stimulation (Zeng and Wess 2000). Solubilized M_2 receptors have also been found to exist as dimers or larger oligomers; in agreement with the conclusions of the M_3 studies, oligomers were not observed when solubilized receptors from singly transfected cells were mixed and the presence of muscarinic ligands did not alter the distribution of oligomers (Park *et al.* 2001).

Chimeric studies have shown that the ability of muscarinic receptors to activate G proteins, and the specificity of that activation, is a function of the sequence of the second and third intracellular loops (Kubo *et al.* 1988; Wess *et al.* 1990, 1989; Wong *et al.* 1990; Wong and Ross 1994). Within the third loop, it is the membrane-proximal regions that dictate coupling and specificity. Studies of the N-terminal end of this loop indicate that the helical nature of TM5 continues into the cytoplasm and that a series of hydrophobic residues on one face of the helix are essential (Bluml *et al.* 1994*b*; Hill-Eubanks *et al.* 1996). A second hydrophobic helical surface is implicated at the C-terminal end of the loop (Kostenis *et al.* 1998). As long as these membrane-proximal regions of the loop remain, the great majority of the third intracellular loop can be deleted without much effect on the binding or the G protein signalling properties of the receptor (Hulme *et al.* 2001; Lee and Fraser 1993; Pals-Rylaarsdam *et al.* 1995; Shapiro and Nathanson 1989).

The ability of muscarinic receptors to be activated by agonists can be reduced by the activation of the receptors themselves (homologous desensitization) or by the actions of other receptors (heterologous desensitization). Heterologous desensitization may be mediated by phosphorylation of the receptors by second-messenger stimulated kinases, especially PKA and PKC. There is no doubt that the receptors are substrates for phosphorylation by these enzymes, but receptor activity is not always affected (reviewed by Wess 1996). Homologous desensitization is also believed to be a result of receptor phosphorylation, but in this case the receptor appears to be required to be in the active conformation (i.e. agonist-bound) to be phosphorylated. The enzymes that carry out this phosphorylation comprise a family of six G protein-coupled receptor kinases (GRKs); the muscarinic receptors are phosphorylated by GRK2 and GRK3, which are recruited to the membrane and activated by the presence of free $\beta\gamma$ subunits that are released upon G protein activation (Premont *et al.* 1995; Wess 1996). Serine and threonine residues in acidic regions in the middle of the i3 loop are the targets of these enzymes (a part of the loop that is not necessary for G protein activation, as noted above). Recently, the M_3 receptor has been found to be phosphorylated in an agonist-dependent manner in this same region of the third inner loop by an unrelated enzyme, casein kinase 1α (CK1α); introduction of a catalytically inactive form of CK1α into intact cells inhibited the incorporation of phosphate into the receptor (Budd *et al.* 2000).

Beyond desensitization, prolonged activation of muscarinic receptors leads to receptor sequestration (in which the receptors are not at the cell surface but are still accessible to hydrophobic ligands and can be recycled to the cell membrane) and eventually to downregulation (a permanent loss of the receptor that requires protein synthesis for recovery). The first stage, desensitization, is closely related to the phosphorylation described above, along with the binding of arrestin molecules to the phosphorylated receptor. However, the processes of desensitization, sequestration, and downregulation may not always be sequentially related and may involve somewhat different receptor domains (Bunemann *et al.* 1999; Hosey *et al.* 1999; Wess 1996).

Many studies have indicated that the binding site for acetylcholine and competitive antagonists lies within a pocket formed by the transmembrane regions of the receptor (Hulme *et al.* 1990). Peptide labelling studies and mutational studies both point to an aspartate residue in TM3 (conserved among all of the biogenic amine receptors) as a crucial component for the binding of classical muscarinic ligands (Curtis *et al.* 1989; Fraser *et al.* 1989; Spalding *et al.* 1994). The helical bundles are predicted to lie in a counter-clockwise arrangement, as viewed from the outside of the cell. This is consistent with the apparently sterically unfavourable packing of $M_2{}^{423}$Thr (TM7) and $M_5{}^{37}$Thr (TM1) in certain chimeric receptors, in which these residues would be pointing toward each other; reversion of either residue to alanine restored favourable structure and antagonist binding (Wess 1996). However, somewhat surprisingly, the expected steric hindrance is not observed in the TM7 point mutation of M_5, or in some other receptors and constructs (Buller *et al.* 2002).

Table 18.1 Muscarinic receptors

	M_1	M_2	M_3	M_4	M_5
Alternative names	—	—	—	—	—
Structural information (accession no.)	h 460 aa (P11229) r 460 aa (P08482) m 460 aa (P12657)	h 466 aa (P08172) r 466 aa (P10980) m 466 aa (Q9ERZ4)	h 590 aa (P20309) r 589 aa (P08483) m 589 aa (Q9ERZ3)	h 479 aa (P08173) r 478 aa (P08485) m 479 aa (P32211)	h 532 aa (P08912) r 531 aa (P08911)
Chromosomal location	11q12–13	7q31–36	1q43–44	11p12–11.2	15q26
Selective agonists	—	—	—	—	—
Selective antagonists	Pirenzepine, MT7, guanylpirenzepine	Tripitramine, gallamine, m2-toxin	Darifenacin	PD102807, MT3	—
Radioligands	[^{3}H]-NMS, [^{3}H]-QNB, [^{3}H]-pirenzepine	[^{3}H]-NMS, [^{3}H]-QNB, [^{3}H]-acetylcholine	[^{3}H]-NMS, [^{3}H]-QNB, [^{3}H]-darifenacin	[^{3}H]-NMS, [^{3}H]-QNB, [^{3}H]-acetylcholine	[^{3}H]-NMS, [^{3}H]-QNB
G protein coupling	G_q/G_{11}	G_i/G_o	G_q/G_{11}	G_i/G_o	G_q/G_{11}
Expression profile	Cerebral cortex, hippocampus, striatum, ganglia, thalamus, glandular tissues	Brainstem, thalamus, cerebral cortex, hippocampus, striatum smooth muscle, heart, lung	Cerebral cortex, hippocampus, smooth muscle, glandular tissues	Striatum, cerebral cortex, hippocampus	Substantia nigra

Physiological function	Slow EPSP by inhibition of M-current in SCG and CNS, LTP in hippocampus	Vagal inhibition of cardiac rate and force of contraction, inhibition of Ca^{2+} channels and activation of K^+ channels, inhibition of acetylcholine release, antagonism of sympathetic relaxation of smooth muscle	Smooth muscle contraction, glandular secretion	Regulation of locomotor activity	Stimulation of dopamine release
Knockout phenotype	Loss of: muscarinic inhibition of M-current in SCG, slow inhibition of Ca^{2+} currents in SCG, pilocarpine-induced seizures, muscarinic-mediated phase advance in circadian rhythm in SCN, reduction in: LTP in hippocampus, performance in spatial memory tasks, PLC and MAPK activation by muscarinic agonists in forebrain regions	Loss of: muscarinic inhibition of acetylcholine release, muscarinic bradycardia, muscarinic fast inhibition of Ca^{2+} channels in SCG, oxotremorine-induced tremor, reduction in: muscarinic-induced analgesia and hypothermia	Bladder distension, urine retention, decrease in body weight, size of fat pads, and plasma leptin concentration large reduction in: muscarinic-induced salivation, pupillary constriction, salivation	Increased basal and dopamine-stimulated locomotor activity loss of residual muscarinic analgesia from M_2 KO	Large decrease in muscarinic stimulation of dopamine release, minor decrease in muscarinic-induced salivation, increased water consumption following deprivation
Disease relevance	Alzheimer's disease, cognitive decline, schizophrenia	Alzheimer's disease, cognitive decline, acute pain	COPD, urinary incontinence, irritable bowel syndrome	Parkinson's disease, schizophrenia, neuropathic pain	Parkinson's disease, schizophrenia, addiction

Early mutational studies focused on hydroxyl groups and found that agonist binding was strongly influenced by conserved threonine and tyrosine residues in TM5 and TM6 (Wess *et al.* 1991). Since then, alanine-scanning mutagenesis has located about a dozen residues in transmembrane regions that are moderately to strongly important in the binding of both NMS and acetylcholine; several others, including threonines in TM5, selectively influence acetylcholine affinity (Hulme *et al.* 2001). On the other hand, the replacement of an asparagine residue in TM6 reduces the affinity of many antagonists (those related to atropine or pirenzepine) by up to 30 000-fold (Bluml *et al.* 1994*a*; Hulme *et al.* 2001). Atropine may interact with this residue by forming two hydrogen bonds, in much the same way that propranolol interacts with the TM7 asparagine of mutant $5HT_{1D\beta}$ receptors (Glennon *et al.* 1996; Spalding *et al.* 1998).

The agonist-binding pocket seems to be so highly conserved that there is very little binding selectivity among agonists, yet most agonists have the highest affinity for the M_2 subtype; this selectivity between M_2 and M_3 can be reversed by exchanging the third intracellular loops of the receptors (Wess *et al.* 1990). However, there are many moderately selective antagonists (see Table 18.2). For some of these antagonists, the molecular basis of these selectivities has

Table 18.2 Typical affinities of antagonists for mammalian muscarinic receptor subtypes[a]

Antagonist	Muscarinic receptor subtype				
	M_1	M_2	M_3	M_4	M_5
NMS	10.3	10.1	10.3	10.6	10.0
Atropine	9.5	9.2	9.6	9.4	9.5
Pirenzepine	8.2	6.6	6.9	7.5	6.8
Guanylpirenzepine	7.7	5.5	6.5	6.5	6.8
UH-AH 37	8.7	7.4	8.2	8.3	8.3
4-DAMP	9.1	8.3	9.2	8.9	9.0
Methoctramine	7.4	8.0	6.7	7.6	7.0
Himbacine	7.1	8.1	7.1	8.2	6.3
AF-DX 116	5.9	6.9	6.1	6.5	5.6
AF-DX 384	7.5	8.4	7.4	8.2	6.3
Tripitramine	8.6	9.5	7.3	8.0	7.4
Darifenacin	7.7	7.2	8.7	7.9	8.1
PD 102807	5.3	5.7	6.2	7.3	5.2

[a] Data are expressed as pK_D or equivalent (pK_i, pK_B) and were complied from previous reviews (Jones *et al.* 1992; Caulfield and Birdsall 1998).

been addressed by the use of chimeric receptors composed of sequences drawn from high affinity and low affinity subtypes (Brann *et al.* 1993). In general, it appears that multiple epitopes play a role in these selectivities. However, UH-AH 37 has lower affinity for the M_2 subtype than for M_5 (or any of the other subtypes) and chimeric studies suggested that this differential affinity was determined by a limited structural region (Wess *et al.* 1992). Subsequent studies found that an alanine residue in TM6 (three residues below the asparagine described above) is responsible for the low affinity at M_2 (Ellis and Seidenberg 2000*b*); the other subtypes all possess a threonine at this location. The closely related compound pirenzepine showed a similar sensitivity to these ala⇔thr mutations and it is likely that the unusually high affinity of the chick M_2 receptor for pirenzepine (Tietje and Nathanson 1991) is due to the presence of the threonine at this site. A growing number of compounds interact with an allosteric site on muscarinic receptors (see below) and the chimeric approach has been used here also. The subtype selectivity of some of these compounds is dictated by the second outer loop of the receptor and by specific residues near the junction of the third outer loop and TM7 (Buller *et al.* 2002; Gnagey *et al.* 1999). Interestingly, UH-AH 37 interacts with NMS both competitively and allosterically at muscarinic receptors; its allosteric subtype-selectivity depends on an epitope outside of TM6, probably within TM7 (Ellis and Seidenberg 1999).

18.3 Cellular and subcellular localization

Early studies of the detailed distribution of muscarinic receptors in nervous tissue employed autoradiographic techniques with highly specific muscarinic radioligands (Kuhar and Yamamura 1975, 1976). These studies were extended as the ability to discern receptor sub-types improved, based initially on the selectivity of pirenzepine, but the fact remains that a full set of subtype-specific muscarinic radioligands does not exist. Some ingenious approaches have been developed to circumvent this problem, based on the use of non-selective and somewhat selective radioligands in combination with somewhat selective unlabelled competitors. Additionally, some studies have taken advantage of the subtype-selective *kinetics* of muscarinic ligands, which can be quite pronounced even for ligands that do not discriminate subtypes on the basis of affinity (Ellis *et al.* 1991; Waelbroeck *et al.* 1986, 1990). A combined approach has been used to selectively label all five subtypes of muscarinic receptors in binding and autoradiographic studies (Flynn *et al.* 1997). Selective antibodies have also been used to quantitate the levels of expression of individual receptor subtypes in brain regions and in the periphery, through immunoprecipitation assays (Dorje *et al.* 1991; Levey *et al.* 1991; Li *et al.* 1991; Wall *et al.* 1991*a,b*). Overall, these approaches have led to the conclusion that all five receptor subtypes are found in brain, with M_1 predominating in the forebrain and M_2 predominating in the thalamus, brainstem, and cerebellum. The M_4 subtype is present in significant quantities in many regions of the brain, but is especially prominent in the striatum. The M_3 and M_5 subtypes have accounted for only a small proportion of the muscarinic receptors in all brain regions studied. In the periphery, the M_2 receptor is the major subtype expressed in smooth muscle, although the M_3 receptor appears to be functionally dominant in spite of its lower expression. The M_2 subtype has traditionally been considered to be *the* mammalian cardiac receptor, and expression studies support this. However, recent studies have suggested that there are small but functionally important components of other subtypes, in the rat (Sharma *et al.* 1996), dog (Shi *et al.* 1999), and human (Wang *et al.* 2001). In contrast, there is no M_1 expression in mouse heart (Hamilton *et al.* 2001). The M_2

subtype also accounts for most of the expression in sympathetic ganglia, although there is a significant M_1 component. M_1 and M_3 are the most abundant subtypes in glandular tissues. The M_4 subtype is abundant in rabbit lung (Dorje *et al.* 1991; Lazareno *et al.* 1990), but apparently not in the lung tissue of other species (Zaagsma *et al.* 1997).

The expression patterns described above are in good agreement with the distribution of receptor subtype mRNA, when the potential differentiation of mRNA and protein in neurones is taken into account (Buckley *et al.* 1988). For example, the M_2 receptor protein is relatively abundant in lung even though its mRNA is not, because the receptors lie on terminals of axons whose cell bodies lie in the parasympathetic ganglion (Jacoby and Fryer 2001). (As described above for smooth muscle generally, the less abundant M_3 receptors appear to play the major role in airway constriction.) Also, within the substantia nigra, pars compacta, the only mRNA found for muscarinic receptors is that of M_5, and it is largely co-localized with D_2 dopamine receptors (Weiner *et al.* 1990). Thus, the M_1, M_4, and M_5 receptor subtypes are potentially well positioned to interact with the dopamine system in the control of motor function.

Electron microscopic immunocytochemistry has identified M_2 receptors in cholinergic synaptic terminals, by co-localization with the synthetic enzyme choline acetyltransferase or with the vesicular acetylcholine transporter (Rouse *et al.* 1997). This provides the structural background for functional studies that have found the release of acetylcholine to be regulated by the M_2 subtype (Raiteri *et al.* 1990). Muscarinic inhibition of acetylcholine release is also dramatically reduced in M_2 KO mice (Birdsall *et al.* 2001). In the hippocampus, M_2 receptors are found as presynaptic autoreceptors on the septo-hippocampal afferents. Lesioning studies were able to identify muscarinic receptors on the other well known input paths to hippocampal granule cell dendrites; M_1 was the predominant post-synaptic receptor, while M_2, M_3 and M_4 were commonly found to be presynaptic heteroceptors (Rouse *et al.* 1999).

18.4 **Pharmacology**

18.4.1 **Agonists**

It was noted at the beginning of this chapter that it has historically been very difficult to discern muscarinic receptor subtypes by pharmacological criteria. This difficulty persists and is especially troublesome for muscarinic agonists. In terms of affinities, most muscarinic agonists are somewhat selective for the M_2 and/or M_4 subtypes; indeed, receptor-binding studies with tritiated acetylcholine itself have been carried out successfully at these two subtypes (Gnagey and Ellis 1996; Lazareno *et al.* 1998). This is particularly problematic because it is believed that M_1-selective agonists may be beneficial in the treatment of Alzheimer's disease (see below). The reason for the difficulties in generating selective agonists and competitive antagonists is probably that the amino acid sequences of the muscarinic subtypes are very highly conserved in the transmembrane regions, which form the ligand-binding pocket (Hulme *et al.* 1990; Jones *et al.* 1992). As noted above, acetylcholine and competitive ligands bind to an aspartate residue in TM3 and to other supplementary sites that also lie well within the transmembrane region of the receptor (TM5 and TM6). In the absence of affinity-based selectivity, the design of new agonists has increasingly been driven by *functional* selectivity. However, the potencies and efficacies of agonists in functional assays are usually driven by tissue-specific properties that include receptor density (e.g. receptor reserve) and the intracellular environment (e.g. downstream coupling factors). This strategy has not been dramatically effective so far, perhaps because an unfavourable receptor density or intra-cellular environment is likely to be the very problem that agonist therapy aims to correct. Furthermore, where these factors are very favourable, most agonists should be functionally selective. Nonetheless, it is possible that some drugs could be highly efficacious full agonists at one subtype and very weak partial agonists (or antagonists) at the others and thereby be functionally selective except when the tissue characteristics are extremely unfavourable. There may even be useful functional differences among muscarinic agonists that are not related to tissue-specific properties at all. As indicated below, muscarinic receptors activate multiple intracellular responses and some of these seem to progress in parallel rather than in sequence (e.g. perhaps by activation of multiple G proteins). As has been demonstrated for other GPCRs (Watson *et al.* 2000), specific agonists may be found to differentially activate these responses in ways that are therapeutically relevant.

18.4.2 Antagonists

Considerable efforts over the last thirty years have developed an array of selective muscarinic antagonists. However, the subtype-selectivities of even the most selective muscarinic antagonists are relative rather than absolute (see Table 18.2). Functional and binding studies on cloned receptor subtypes expressed in cell lines that lack endogenous muscarinic receptors have permitted clear and accurate determinations of antagonist affinities. These affinities have proven to be independent of the cell type used for expression and correlate well with the results of careful pharmacological studies carried out in many intact tissues, (Caulfield and Birdsall 1998; Jones *et al.* 1992). Nonetheless, multiple subtypes may influence some responses, which places further strain on an already limited pharmacology.

It is simply not possible to assign a response to a particular subtype, based solely on blockade by a particular antagonist. The most common error in this regard is probably the use of 4-DAMP as an 'M$_3$-selective' antagonist. This antagonist is moderately (about 10-fold) selective for the M$_3$ subtype, compared to M$_2$, and this is useful in smooth muscle, in which only these two subtypes are expressed. However, it has virtually no selectivity among M$_1$, M$_3$, M$_4$, and M$_5$ (see Table 18.2).

18.4.3 Allosteric modulators

Another approach to the development of subtype selective agonists/antagonists would be to target less conserved regions of the receptor, assuming that this is possible. Such com-pounds would presumably not have much structural similarity to acetylcholine or classical agonists and antagonists. One M$_1$-selective agonist, dubbed AC-42, has recently been repor-ted to have been discovered by high-throughput functional screening of chemical libraries (Spalding *et al.* 2001) and, unlike acetylcholine, its potency is sensitive to changes in epi-topes in the third outer loop and the N-terminus. Interestingly, the third outer loop and adjacent portions of TM7 seem to be part of an increasingly well-defined allosteric site on muscarinic receptors (Buller *et al.* 2002; Ellis *et al.* 1993). The characterization of this site grew out of investigations into anomalous actions of the neuromuscular blocking agent gal-lamine (Clark and Mitchelson 1976; Ellis and Hoss 1982; Stockton *et al.* 1983). In kinetic studies of the binding of labelled classical ligands, gallamine has been shown to interact with an allosteric site on all five muscarinic receptor subtypes (Ellis *et al.* 1991). The use of kinetic assays simplifies the interpretation of the interaction, because regulation of the rate of dissociation of one ligand by another requires the presence of two separate binding sites (reviewed by Christopoulos *et al.* 1998; Ellis 1997). The analysis of equilibrium (or pseudo-equilibrium) data can be more problematic, but allows the determination of the degree of cooperativity between the two ligands that are simultaneously bound. Some of these interactions are positively cooperative. For example, alcuronium enhances the binding of the classical muscarinic antagonist N-methylscopolamine (Proska and Tucek 1994; Tucek *et al.* 1990), while brucine and related compounds enhance the affinity of acetylcholine in binding and functional assays (Birdsall *et al.* 1997; Lazareno *et al.* 1998). Appropriate allosteric interactions may provide significant therapeutic advantages; the degree of cooperativity can define a maximal, concentration-independent, effect (Ehlert 1988) and, especially important in neuronal signalling, the patterning of signal input may be preserved (Ellis and Seidenberg 2000*a*). Allosteric ligands offer a unique twist on functional selectivity that has been called absolute selectivity (Birdsall *et al.* 1997). That is, if an allosteric enhancing agent is positively cooperative with the endogenous agonist at one subtype, but neutral at all others, it will be selective in a concentration-independent manner. The advantages of allosteric drugs can be appreciated in the safety and efficacy of benzodiazepines that enhance the affinity of GABA for the $GABA_A$ receptor; there are no therapeutic applications for directly acting GABA agonists. Recent reports suggest that there are multiple allosteric sites on muscarinic receptors (Lazareno *et al.* 2000).

18.4.4 Muscarinic snake venom toxins

A family of snake venom toxins includes some members that do exhibit extreme subtype-selectivity. These peptides are approximately 65 residues in length and contain four disulfide bonds that are common to the larger family of 'three-fingered' toxins (named after the configuration of the loops formed by the disulfides). The most selective of these is MT-7 (also known as m_1-toxin or m1-toxin1), which is exceptionally specific (and essentially irreversible) for the M1 subtype over all other muscarinic receptors (Adem and Karlsson 1997; Potter 2001). Two other toxins, MT-1 and MT-2, are selective for M_1 and M_4 and act as agonists at the M_1 subtype (Jerusalinsky and Harvey 1994). MT-3 (also called m4-toxin) binds to M_4 receptors with 100-fold higher affinity than M_1, and has negligible affinity for the other subtypes (Adem and Karlsson 1997). Another recently isolated and partially sequenced toxin is highly selective for the M_2 subtype (Potter 2001).

18.5 Signal transduction and receptor modulation

A simplified but still useful statement of muscarinic effector coupling is that the 'odd-numbered' receptors preferentially regulate phosphoinositide metabolism and intracellular calcium levels via the G_q family of G proteins, while the 'even-numbered' receptors couple preferentially to the inhibition of adenylate cyclase via the G_i family. However, this situation has grown ever more complicated over time, due to cross-talk between the two pathways, activity of G protein $\beta\gamma$ subunits as well as α subunits, activation of other G protein families, differential regulation of multiple effector isoforms, and even activation of G protein-independent pathways.

The regulation of cyclic AMP levels illustrates these complexities. The G_i-coupled subtypes inhibit several types of adenylate cyclase by activating the (pertussis toxin sensitive)

α_i subunit. However, $\beta\gamma$ subunits are released simultaneously and can increase cyclic AMP levels in tissues that express other isoforms of AC, for example, Types II and IV in the olfactory bulb (Olianas *et al.* 1998). Muscarinic receptor subtypes can also couple weakly to G_s, so that under the right conditions (e.g. high levels of receptor expression) adenylate cyclase can be significantly activated by this path (Michal *et al.* 2001; Migeon and Nathanson 1994). The G_q-coupled receptors activate PLC to raise intracellular calcium levels and activate PKC (see below). The elevation of calcium can activate both calcium-calmodulin sensitive phosphodiesterase (lowering cyclic AMP levels) and calcium-calmodulin sensitive AC isoforms (raising cyclic AMP levels); the result will depend on tissue-specific expression patterns of these enzymes (Choi *et al.* 1992; Simonds 1999; Tanner *et al.* 1986). Calmodulin-dependent protein kinases may further modulate the activity of some isoforms of AC, as can PKA and PKC (Hurley 1999).

The regulation of phosphoinositide metabolism is somewhat less complicated, but shows similar patterns to the regulation of cyclic AMP levels. It has been known that M_1, M_3, and M_5 receptors robustly stimulate the production of inositol phosphates and diacyglycerol by a pertussis toxin-insensitive pathway, whereas M_2 and M_4 produce a much weaker effect that is sensitive to pertussis toxin (Jones *et al.* 1991; Peralta *et al.* 1988; Wess *et al.* 1990). The robust effect is mediated by α subunits, which stimulate PLCβ1 better than PLCβ3 or PLCβ2, while the pertussis toxin-sensitive effect is mediated by $\beta\gamma$ subunits, which stimulate PLCβ3 better than PLCβ2 or PLCβ1 (Rhee and Bae 1997). The M_1, M_3, and M_5 receptors stimulate the release of arachidonic acid with similar potencies to those seen in phosphoinositide assays (Conklin *et al.* 1988); however, pretreatment with phorbol esters enhanced the release of arachidonic, while inhibiting the PLC response. The arachidonic acid is probably derived mostly from phosphatidylcholine by the action of PLA$_2$, secondary to calcium influx and activation of PKC. PKC does not activate PLA$_2$ directly, but can trigger activation of the MAPK cascade; phosphorylation by MAPK does activate PLA$_2$ (Leslie 1997). When the M_2 and M_4 subtypes are activated, they do not stimulate PLA$_2$ by themselves, but the response to ATP and ionophore is augmented in CHO cells (Felder *et al.* 1991), possibly through MAPK activation (see below).

The M_1, M_3, and M_5 receptors also mediate activation of another phospholipase, PLD. This enzyme acts on phosphatidylcholine to produce phosphatidic acid and choline, though its activity is usually monitored by the formation of phosphatidylalcohols via a transphosphatidylation assay, because choline itself can be generated through other phospholipases (Klein *et al.* 1996). Although the same subtypes activate the two responses, PLD is not a downstream effect of PLC/Ca^{2+}/PKA activation. Rather, the PLD response is sensitive to inhibition of the small G proteins ARF and Rho by brefeldin and C3 botulinum toxin, respectively (Mitchell *et al.* 1998). It is an open question whether ARF and Rho couple directly with the M_3 receptor to activate PLD (Hall *et al.* 1999).

Muscarinic receptors modulate a variety of ion channels in the CNS and in the periphery, including many potassium conductances and several types of calcium conductances (Jones 1993; Smith 1996). One class of potassium channels, the G protein-coupled inwardly rectifying K^+ (GIRK) channels, is stimulated by G_i-coupled receptors (Clapham and Neer 1997). In the heart, M_2 receptors activate G_i, releasing $\beta\gamma$ subunits that in turn activate the GIRK channels directly; activation of these channels decreases excitability and slows the heart rate. RGS proteins may play a dual role in this system, elevating the basal level of $\beta\gamma$ subunits, but also accelerating the rate of deactivation of the receptor-stimulated $\beta\gamma$ subunits (Hosey *et al.* 1999). Another potassium current has been named the M-current (I_M), because it was initially discovered as a novel action of muscarinic agonists (Brown and Adams 1980). This voltage-dependent, non-inactivating, current has been found in many neurones, but has been most intensely studied in the superior cervical ganglion (SCG). In transfected cells, G_q-coupled receptors inactivate I_M in a manner that suggests a diffusible (second messenger) mediator, although it is not clear which messenger is responsible (Robbins et al. 1993). Pharmacological data has identified the M_1 subtype as the cause of cholinergic inhibition of I_M in rat sympathetic neurones (Bernheim et al. 1992) and muscarinic inhibition is absent in the SCG of M_1 knockout mice (Hamilton et al. 1997). The M_1 receptor also inhibits N- and L-type calcium currents in the SCG and, as with I_M, the inhibition has a slow time course consistent with second messenger diffusion and is not sensitive to pertussis toxin. Again in agreement with data from the rat, muscarinic inhibition of these calcium currents was absent in M_1 KO mice (Shapiro et al. 2001). There is also a fast muscarinic inhibition of N- and P/Q-type calcium channels that is pertussis toxin sensitive. This muscarinic effect has pharmacological characteristics of the M_4 receptor in the rat SCG, but was fully present in M_4 KO mice and absent in M_2 KO mice (Shapiro et al. 2001). Thus, although the structures of the muscarinic receptors are highly conserved among mammals (see above), it should be kept in mind that the expression and/or tissue-specific coupling patterns may differ.

Stimulation of cell proliferation through activation of extracellular signal-regulated kinases (ERKs) was initially associated with tyrosine kinase receptors, but GPCRs are now known to activate these pathways as well (Marinissen and Gutkind 2001). The route from membrane receptor to the nucleus involves many intermediate enzymes and coupling proteins, many of which are expressed in a cell-specific manner. Thus, the M_1, M_3, and M_5 subtypes stimulate growth and induce focus formation in NIH-3T3 cells, but the M_2 and M_4 subtypes do not (Gutkind et al. 1991). On the other hand, the M_2 receptor activates the ERK pathway in rat 1a cells, while the M_1 receptor does not; rather, the M_1 receptor can inhibit part of the activation produced through tyrosine kinase receptor activation (Russell et al. 1994). In an analogous manner to the ion channel regulation described above, the G_i-linked activation is mediated through the actions of freed $\beta\gamma$ subunits, acting through PI3 kinase and (many) other intermediates (Gutkind et al. 1997). The M_1 receptor may activate the ERK pathway through $G_{\alpha q}$ in a PKC-dependent manner in COS-7 cells (Hawes et al. 1995) or in a partially PKC-dependent manner in NIH-3T3 cells (Crespo et al. 1994).

The stimulation of proliferation of NIH-3T3 cells has proven to be a sensitive indicator of G_q-coupled muscarinic receptor activation, whether by agonist stimulation or by manipulation of the receptor or its environment; the use of this assay to screen M_5 receptor mutants that were generated by random saturation mutagenesis has identified a series of constitutively activating mutations in TM6, suggesting that TM6 acts as a ligand-dependent switch (Spalding et al. 1998). Recently, all five muscarinic receptor subtypes have been shown to be strongly activated by homologous mutations at the top of TM6 (Ford et al. 2002). However, the same mutation at the top of TM6 in the M_1 receptor does not seem to produce robust constitutive activation of PLC, perhaps reflecting the additional pathways that may be involved in the proliferative response in NIH-3T3 cells, as noted above (Crespo et al. 1994). In this regard, deletion of a serine-threonine rich portion of the third inner loop of the M_3 receptor greatly reduces both the phosphorylation of the receptor and the ability of the receptor to activate the ERK pathway, without inhibiting coupling to the phosphoinositide response (Budd et al. 2001); however, ERK activation is also dependent on PKC activity, suggesting that *both* PLC-initiated PKC activity and another mechanism that is dependent on the phosphorylated portion of the third loop are necessary for the ERK response.

18.6 Physiology and disease relevance

The existence of five subtypes of muscarinic receptors does entail a certain pharmacological complexity, but it also provides the opportunity for finer pharmacological control. This seems to be quite necessary, considering the number of functions and/or illnesses that are modulated by muscarinic activities in the autonomic and central nervous systems. A partial list includes cardiac, smooth muscle, and gland function, affective illness (Janowsky *et al.* 1994), learning and memory (Segal and Auerbach 1997), REM sleep (Capece and Lydic 1997), thermoregulation (Dilsaver *et al.* 1991), cell proliferation (see above), and also analgesia, obstructive airway disease, schizophrenia, Parkinson's disease, and Alzheimer's disease. Clearly, *more* subtypes are needed! And worse yet, the available agonists and even antagonists are not very selective (see above), so it is difficult to be sure what receptor subtype is involved in each case. Nonetheless, strides are being made in many areas in the development of muscarinic therapeutic approaches. The comments below are amplified in recent reviews (Birdsall *et al.* 2001; Eglen *et al.* 2001; Felder *et al.* 2000).

18.6.1 Parkinson's disease

The symptoms of Parkinson's disease include a resting tremor and difficulty in initiating and executing movement. These functions seem to be under the control of a balanced interplay between cholinergic and dopaminergic activity, as symptoms are exacerbated by dopamine antagonists or muscarinic agonists and improved by dopamine agonists (e.g. the precursor levodopa) or muscarinic antagonists. The relatively selective expression of the M_4 receptor in the striatum and the pharmacological characterization of animal models of Parkinson's disease (Mayorga *et al.* 1999) have suggested that an M_4-selective antagonist might be beneficial and avoid significant side-effects. Studies with knockout mice have reinforced the importance of the M_4 subtype in both basal and dopamine-induced locomotor behaviour (Gomeza *et al.* 2001).

18.6.2 Schizophrenia

At a simplistic level, there is a reverse relationship of dopamine and muscarinic activity in schizophrenia, compared to Parkinson's disease. Once again, there is a balance between the two systems, but in schizophrenia dopamine agonists and muscarinic antagonists exacerbate symptoms. At high doses, muscarinic antagonists are known to induce positive, negative, and cognitive symptoms that are similar to those of schizophrenia (Felder *et al.* 2000). Indeed, a class of muscarinic antagonists has been labelled 'the psychotomimetic glycolate esters' (Abood 1968). Pharmacological studies of animal models in intact and knockout animals suggest that the M_4 subtype may again be involved in a functional antagonism with dopamine receptors (Bymaster *et al.* 1999; Felder *et al.* 2001), while M_1 activation may be helpful on cognitive deficits. Finally, the selective co-localization of M_5 receptors on dopamine neurones, coupled with evidence that M_5 activation promotes dopamine release, suggests that a selective M_5 antagonist could be beneficial, though none are known at present.

18.6.3 Alzheimer's disease and cognition

One of the most prominent and therapeutically challenging conditions is Alzheimer's disease. The muscarinic involvement is evidenced by a pronounced and selective loss of cholinergic neurones and the similarity of the cognitive deficits to those that can be induced by muscarinic antagonists (Bartus *et al.* 1982; Drachman and Leavitt 1974; Whitehouse *et al.* 1982). However, various forms of cholinergic replacement therapy have met with little success, in sharp contrast to the benefits of dopamine therapy in Parkinson's disease. The limited success to date may indicate that muscarinic receptors are not in fact intimately involved and, indeed, other avenues (amyloid and tau proteins, neurotrophic factors) are being pursued by many laboratories (Hardy *et al.* 1998). It might be argued, then, that when amyloid deposits form, target cells in the cortex and hippocampus die and the loss of trophic factors leads secondarily to the loss of presynaptic cholinergic cells in the basal forebrain. However, studies suggest that loss of the target cells does *not* in fact lead to cholinergic cell death (Kordower *et al.* 1992). Rather, muscarinic agonists have been shown to regulate processing of the amyloid precursor protein (Growdon 1997; Nitsch *et al.* 1992) and to reduce the hyperphosphorylated state of tau protein (Genis *et al.* 1999; Sadot *et al.* 1996); these actions are expected to reduce the formation of the plaques and tangles that are characteristic of Alzheimer's disease. The bottom line is that cholinesterase inhibitors (donepezil, rivastigmine, galantamine, tacrine) have been the only approved drugs for the treatment of Alzheimer's disease. The lack of dramatic success of these cholinergic replacement therapies may be due to the non-specific activation of many receptor subtypes. Inappropriate activation of receptors may lead to side effects or may even negate the positive effects of appropriate activation. A variety of evidence suggests that loss of activation of the M_1 subtype is most responsible for AD symptoms and pharmaceutical companies have been developing and testing agonists that are M_1-selective by various criteria, though once again with little success so far (Eglen *et al.* 2001). Yet another approach to raising acetylcholine levels is to inhibit presynaptic autoreceptors and the selective M_2 antagonist SCH72788 has recently been shown to be successful in microdialysis and behavioural studies in rats (Lachowicz *et al.* 2001).

Each of these approaches has advantages and disadvantages (assuming that some sort of replacement therapy is called for). A truly selective agonist would be subtype specific, but would completely disrupt the spatial and temporal patterning of receptor activation, by activating all of the receptors all of the time; further, such a treatment might desensitize the receptors. Inhibiting acetylcholinesterase is probably less disruptive to spatial and temporal patterning, but is not subtype-specific at all. A specific M_2 receptor antagonist would probably be even more faithful to spatial and temporal patterning than a cholinesterase inhibitor, but would not be selective among the remaining subtypes. An allosteric enhancing agent with absolute selectivity might be able to provide subtype-selectivity while preserving spatio-temporal patterning, but the current prototype ligands have likely toxicity problems (e.g. strychnine-like or curare-like). It is not even clear at this time whether a neurotransmitter-like or hormone-like activation of the receptor is the most important. The current treatment with cholinesterase inhibitors is considered to stabilize cognitive decline, rather than to reverse the existing deficits (Giacobini 2000), and this could be the result of a hormone-like activation of muscarinic receptors that may slow the formation of plaques and tangles (see above). If so, then selective M_1 agonists, which would have better side-effect profiles, may be the best choice for treatment. If spatial and temporal patterning of activation is as important, or more important, then directly acting agonists may not be such a good choice. As more selective and well-characterized agents are developed, it should become possible to determine which strategy will be best.

18.6.4 **Peripheral indications**

In peripheral systems, the therapeutic use of muscarinic agents revolves primarily around the wanted and unwanted effects of the M_3 subtype in glands and smooth muscle. Although M_2 receptors are expressed at much higher levels and do contribute indirectly to the enhanced contractility of smooth muscle by antagonizing relaxation induced by sympathetic activity, M_3 receptors are of primary importance in the direct contraction of smooth muscle (Ehlert et al. 1999). Thus, M_3 antagonists are sought to reduce the stimulation of contraction of the bladder in urinary incontinence (Nilvebrant 2001), of the ileum in irritable bowel syndrome (Eglen et al. 2001), and of airway smooth muscle in chronic obstructive pulmonary disease (Jacoby and Fryer 2001). A common side effect is the inhibition of salivation (i.e. dry mouth), which is also mediated largely by the M_3 subtype. The involvement of the same subtype in the therapeutic and the side effect makes it especially difficult to attenuate the unwanted action. On the other hand, activation of M_3 receptors is a problem in the muscarinic treatment of glaucoma; muscarinic agonists beneficially reduce intraocular pressure via unknown receptor subtypes, but activation of M_3 receptors in the iris sphincter and ciliary muscles can lead to night blindness and blurred vision. Recently, a compound that activates other muscarinic receptors, but not M_3, has been found to significantly reduce intraocular pressure (Gil et al. 2001). All of the above actions of the M_3 receptor are consistent with the phenotype of knockout mice lacking this subtype (Matsui et al. 2000; Yamada et al. 2001), although the in vivo function of the gastrointestinal tract appeared to be normal. Additionally, the knockout animals exhibited low body weight, decreased fat pads, and decreased serum leptin and insulin concentrations, suggesting a central effect of M_3 receptors in food intake regulation (Yamada et al. 2001).

18.7 **Concluding remarks**

While there is still much to be learned, many advances have been made in understanding the properties of the individual subtypes of muscarinic receptors since their cloning in the late 1980s. Much of this progress stems from the ability to examine the receptors in isolation by expressing them in recombinant systems. The tissue distributions of the subtypes have been established, even down to cellular and subcellular levels in some cases, thanks to the specificity of antibodies and in situ hybridization. The subtype dependence of many physiological responses has been determined, through the use of knockout mice and panels of antagonists. However, the most dramatic deficit continues to be the lack of highly selective agonists and antagonists, with which to turn responses on or off selectively in vivo. Advances in this area are likely to develop from a better understanding of the structural features that comprise the binding sites for the most selective agents currently available and may depend on the synthesis or discovery of agonists and antagonists that bind to less conserved regions of the receptors.

References

Abood LG (1968). The psychotomimetic glycolate esters and related drugs. In *Psychopharmacology—a review of progress 1957–1967* (ed. DH Efron) U. S. Goverment Printing Office, Washington, pp. 683–92.

Adem A and Karlsson E (1997). Muscarinic receptor subtype selective toxins. *Life Sci* **60**, 1069–76.

Barlow RB, Berry KJ, Glenton PA, Nilolaou NM, and Soh KS (1976). A comparison of affinity constants for muscarine-sensitive acetylcholine receptors in guinea-pig atrial pacemaker cells at 29 degrees C and in ileum at 29 degrees C and 37 degrees C. *Br J Pharmacol* **58**, 613–20.

Bartus RT, Dean RL, 3rd, Beer B, and Lippa AS (1982). The cholinergic hypothesis of geriatric memory dysfunction. *Science* **217**, 408–14.

Behr J, Haase W, Maul G, Vasudevan S, and Reilander H (1998). Effect of N-glycosylation on production of the rat m3 muscarinic acetylcholine receptor in baculovirus-infected insect cells. *Biochem Soc Trans* **26**, 704–9.

Beld AJ, Van Den Hoven S, Wouterse AC, and Zegers MA (1975). Are muscarinic receptors in the central and peripheral nervous system different? *Eur J Pharmacol* **30**, 360–3.

Bernheim L, Mathie A, and Hille B (1992). Characterization of muscarinic receptor subtypes inhibiting Ca2+ current and M current in rat sympathetic neurons. *Proc Natl Acad Sci USA* **89**, 9544–8.

Birdsall NJ, Farries T, Gharagozloo P, Kobayashi S, Kuonen D, Lazareno S *et al.* (1997). Selective allosteric enhancement of the binding and actions of acetylcholine at muscarinic receptor subtypes. *Life Sci* **60**, 1047–52.

Birdsall NMJ, Nathanson NM, and Schwartz RD (2001). Muscarinic receptors: it's a knockout. *Trends Pharmacol Sci* **22**, 215–9.

Bluml K, Mutschler E, and Wess J (1994a). Functional role in ligand binding and receptor activation of an asparagine residue present in the sixth transmembrane domain of all muscarinic acetylcholine receptors. *J Biol Chem* **269**, 18 870–6.

Bluml K, Mutschler E, and Wess J (1994b). Insertion mutagenesis as a tool to predict the secondary structure of a muscarinic receptor domain determining specificity of G-protein coupling. *Proc Natl Acad Sci USA* **91**, 7980–4.

Bonner TI, Buckley NJ, Young AC, and Brann MR (1987). Identification of a family of muscarinic acetylcholine receptor genes. *Science* **237**, 527–32.

Bonner TI, Young AC, Brann MR, and Buckley NJ (1988). Cloning and expression of the human and rat m5 muscarinic acetylcholine receptor genes. *Neuron* **1**, 403–10.

Brann MR, Klimkowski VJ, and Ellis J (1993). Structure/function relationships of muscarinic acetylcholine receptors. *Life Sci* **52**, 405–12.

Brown DA and Adams PR (1980). Muscarinic suppression of a novel voltage-sensitive K+ current in a vertebrate neurone. *Nature* **283**, 673–6.

Brown JH and Taylor P (2001). Muscarinic receptor agonists and antagonists. In (eds. JG Hardman and LE Limbird) *Goodman and Gilman's The pharmacological basis of therapeutics* McGraw-Hill, New York, pp. 155–173.

Buckley NJ, Bonner TI, and Brann MR (1988). Localization of a family of muscarinic receptor mRNAs in rat brain. *J Neurosci* **8**, 4646–52.

Budd DC, McDonald JE, and Tobin AB (2000). Phosphorylation and regulation of a Gq/11-coupled receptor by casein kinase 1alpha. *J Biol Chem* **275**, 19 667–75.

Budd DC, Willars GB, McDonald JE, and Tobin AB (2001). Phosphorylation of the Gq/11-coupled m3-muscarinic receptor is involved in receptor activation of the ERK-1/2 mitogen-activated protein kinase pathway. *J Biol Chem* **276**, 4581–7.

Buller S, Zlotos DP, Mohr K, and Ellis J (2002). Allosteric site on muscarinic acetylcholine receptors: a single amino acid in TM7 is critical to the subtype selectivities of caracurine V derivatives and alkane-bisammonium ligands. *Mol Pharmacol* **61**, 160–8.

Bunemann M, Lee KB, Pals-Rylaarsdam R, Roseberry AG, and Hosey MM (1999). Desensitization of G-protein-coupled receptors in the cardiovascular system. *Annu Rev Physiol* **61**, 169–92.

Burgen AS (1995). The background of the muscarinic system. *Life Sci* **56**, 801–6.

Bymaster FP, Shannon HE, Rasmussen K, DeLapp NW, Ward JS, Calligaro DO *et al.* (1999). Potential role of muscarinic receptors in schizophrenia. *Life Sci* **64**, 527–34.

Capece ML and Lydic R (1997). cAMP and protein kinase A modulate cholinergic rapid eye movement sleep generation. *Am J Physiol* **273**, R1430–40.

Caulfield MP and Birdsall NJ (1998). International Union of Pharmacology. XVII. Classification of muscarinic acetylcholine receptors. *Pharmacol Rev* **50**, 279–90.

Chidiac P, Green MA, Pawagi AB, and Wells JW (1997). Cardiac muscarinic receptors. Cooperativity as the basis for multiple states of affinity. *Biochemistry* **36**, 7361–79.

Choi EJ, Wong ST, Hinds TR, and Storm DR (1992). Calcium and muscarinic agonist stimulation of type I adenylcyclase in whole cells. *J Biol Chem* **267**, 12 440–2.

Christopoulos A, Lanzafame A, and Mitchelson F (1998). Allosteric interactions at muscarinic cholinoceptors. *Clin Exp Pharmacol Physiol* **25**, 185–94.

Clapham DE and Neer EJ (1997). G protein beta gamma subunits. *Annu Rev Pharmacol Toxicol* **37**, 167–203.

Clark AL and Mitchelson F (1976). The inhibitory effect of gallamine on muscarinic receptors. *Br J Pharmacol* **58**, 323–31.

Conklin BR, Brann MR, Buckley NJ, Ma AL, Bonner TI, and Axelrod J (1988). Stimulation of arachidonic acid release and inhibition of mitogenesis by cloned genes for muscarinic receptor subtypes stably expressed in A9 L cells. *Proc Natl Acad Sci USA* **85**, 8698–702.

Crespo P, Xu N, Daniotti JL, Troppmair J, Rapp UR, and Gutkind JS (1994). Signaling through transforming G protein-coupled receptors in NIH 3T3 cells involves c-Raf activation. Evidence for a protein kinase C- independent pathway. *J Biol Chem* **269**, 21 103–9.

Curtis CA, Wheatley M, Bansal S, Birdsall NJ, Eveleigh P, Pedder EK *et al.* (1989). Propylbenzilylcholine mustard labels an acidic residue in transmembrane helix 3 of the muscarinic receptor. *J Biol Chem* **264**, 489–95.

Dilsaver SC, Majchrzak MJ, Snider RM, and Davidson RK (1991). A nicotinic receptor antagonist enhances the hypothermic response to a muscarinic agonist. *Prog Neuropsychopharmacol Biol Psychiatry* **15**, 539–49.

Dorje F, Levey AI, and Brann MR (1991). Immunological detection of muscarinic receptor subtype proteins (m1–m5) in rabbit peripheral tissues. *Mol Pharmacol* **40**, 459–62.

Drachman DA and Leavitt J (1974). Human memory and the cholinergic system. A relationship to aging? *Arch Neurol* **30**, 113–21.

Eglen RM, Choppin A, and Watson N (2001). Therapeutic opportunities from muscarinic receptor research. *Trends Pharmacol Sci* **22**, 409–14.

Ehlert FJ (1988). Estimation of the affinities of allosteric ligands using radioligand binding and pharmacological null methods. *Mol Pharmacol* **33**, 187–94.

Ehlert FJ, Sawyer GW, and Esqueda EE (1999). Contractile role of M2 and M3 muscarinic receptors in gastrointestinal smooth muscle. *Life Sci* **64**, 387–94.

Ellis J (1997). Allosteric binding sites on muscarinic receptors. *Drug Dev Res* **40**, 193–204.

Ellis J and Hoss W (1982). Competitive interaction of gallamine with multiple muscarinic receptors. *Biochem Pharmacol* **31**, 873–6.

Ellis J, Huyler J, and Brann MR (1991). Allosteric regulation of cloned m1-m5 muscarinic receptor subtypes. *Biochem Pharmacol* **42**, 1927–32.

Ellis J and Seidenberg M (1999). Competitive and allosteric interactions of 6-chloro-5,10-dihydro-5-[(1- methyl-4-piperidinyl)acetyl]-11H-di benzo[b,e][1, 4]diazepine-11-one hydrochloride (UH-AH 37) at muscarinic receptors, via distinct epitopes. *Biochem Pharmacol* **57**, 181–6.

Ellis J and Seidenberg M (2000a). Interactions of alcuronium, TMB-8, and other allosteric ligands with muscarinic acetylcholine receptors: studies with chimeric receptors. *Mol Pharmacol* **58**, 1451–60.

Ellis J and Seidenberg M (2000b). Site-directed mutagenesis implicates a threonine residue in TM6 in the subtype selectivities of UH-AH 37 and pirenzepine at muscarinic receptors. *Pharmacology* **61**, 62–9.

Ellis J, Seidenberg M, and Brann MR (1993). Use of chimeric muscarinic receptors to investigate epitopes involved in allosteric interactions. *Mol Pharmacol* **44**, 583–8.

Felder CC, Bymaster FP, Ward J, and DeLapp N (2000). Therapeutic opportunities for muscarinic receptors in the central nervous system. *J Med Chem* **43**, 4333–53.

Felder CC, Porter AC, Skillman TL, Zhang L, Bymaster FP, Nathanson NM *et al.* (2001). Elucidating the role of muscarinic receptors in psychosis. *Life Sci* **68**, 2605–13.

Felder CC, Williams HL, and Axelrod J (1991). A transduction pathway associated with receptors coupled to the inhibitory guanine nucleotide binding protein Gi that amplifies ATP-mediated arachidonic acid release. *Proc Natl Acad Sci USA* **88**, 6477–80.

Flynn DD, Reever CM, and Ferrari-Dileo G (1997). Pharmacological strategies to selectively label and localize muscarinic receptor subtypes. *Drug Dev Res* **40**, 104–16.

Ford DJ, Essex A, Spalding TA, Burstein ES, and Ellis J (2002). Homologous mutations near the junction of the sixth transmembrane domain and the third extracellular loop lead to constitutive activity and enhanced agonist affinity at all muscarinic receptor subtypes. *J Pharmacol Exp Ther* **300**, 810–7.

Fraser CM, Wang CD, Robinson DA, Gocayne JD, and Venter JC (1989). Site-directed mutagenesis of m1 muscarinic acetylcholine receptors: conserved aspartic acids play important roles in receptor function. *Mol Pharmacol* **36**, 840–7.

Genis I, Fisher A, and Michaelson DM (1999). Site-specific dephosphorylation of tau of apolipoprotein E-deficient and control mice by M1 muscarinic agonist treatment. *J Neurochem* **72**, 206–13.

Giacobini E (2000). Cholinesterase inhibitors stabilize Alzheimer disease. *Neurochem Res* **25**, 1185–90.

Gil D, Spalding T, Kharlamb A, Skjaerbaek N, Uldam A, Trotter C *et al.* (2001). Exploring the potential for subtype-selective muscarinic agonists in glaucoma. *Life Sci* **68**, 2601–4.

Glennon RA, Dukat M, Westkaemper RB, Ismaiel AM, Izzarelli DG, and Parker EM (1996). The binding of propranolol at 5-hydroxytryptamine1D beta T355N mutant receptors may involve formation of two hydrogen bonds to asparagine. *Mol Pharmacol* **49**, 198–206.

Gnagey A and Ellis J (1996). Allosteric regulation of the binding of [3H]acetylcholine to m2 muscarinic receptors. *Biochem Pharmacol* **52**, 1767–75.

Gnagey AL, Seidenberg M, and Ellis J (1999). Site-directed mutagenesis reveals two epitopes involved in the subtype selectivity of the allosteric interactions of gallamine at muscarinic acetylcholine receptors. *Mol Pharmacol* **56**, 1245–53.

Gomeza J, Zhang L, Kostenis E, Felder CC, Bymaster FP, Brodkin J *et al.* (2001). Generation and pharmacological analysis of M2 and M4 muscarinic receptor knockout mice. *Life Sci* **68**, 2457–66.

Growdon JH (1997). Muscarinic agonists in Alzheimer's disease. *Life Sci* **60**, 993–8.

Gutkind JS, Crespo P, Xu N, Teramoto H, and Coso OA (1997). The pathway connecting m2 receptors to the nucleus involves small GTP-binding proteins acting on divergent MAP kinase cascades. *Life Sci* **60**, 999–1006.

Gutkind JS, Novotny EA, Brann MR, and Robbins KC (1991). Muscarinic acetylcholine receptor subtypes as agonist-dependent oncogenes. *Proc Natl Acad Sci USA* **88**, 4703–7.

Haga K and Haga T (1983). Affinity chromatography of the muscarinic acetylcholine receptor. *J Biol Chem* **258**, 13 575–9.

Hall RA, Premont RT, and Lefkowitz RJ (1999). Heptahelical receptor signaling: beyond the G protein paradigm. *J Cell Biol* **145**, 927–32.

Hamilton SE, Hardouin SN, Anagnostaras SG, Murphy GG, Richmond KN, Silva AJ *et al.* (2001). Alteration of cardiovascular and neuronal function in M1 knockout mice. *Life Sci* **68**, 2489–93.

Hamilton SE, Loose MD, Qi M, Levey AI, Hille B, McKnight GS *et al.* (1997). Disruption of the m1 receptor gene ablates muscarinic receptor-dependent M current regulation and seizure activity in mice. *Proc Natl Acad Sci USA* **94**, 13 311–6.

Hammer R, Berrie CP, Birdsall NJ, Burgen AS, and Hulme EC (1980). Pirenzepine distinguishes between different subclasses of muscarinic receptors. *Nature* **283**, 90–2.

Hardy J, Duff K, Hardy KG, Perez-Tur J, and Hutton M (1998). Genetic dissection of Alzheimer's disease and related dementias: amyloid and its relationship to tau. *Nat Neurosci* **1**, 355–8.

Hawes BE, van Biesen T, Koch WJ, Luttrell LM, and Lefkowitz RJ (1995). Distinct pathways of Gi- and Gq-mediated mitogen-activated protein kinase activation. *J Biol Chem* **270**, 17 148–53.

Hayashi MK and Haga T (1997). Palmitoylation of muscarinic acetylcholine receptor m2 subtypes: reduction in their ability to activate G proteins by mutation of a putative palmitoylation site, cysteine 457, in the carboxyl-terminal tail. *Arch Biochem Biophys* **340**, 376–82.

Hill-Eubanks D, Burstein ES, Spalding TA, Brauner-Osborne H, and Brann MR (1996). Structure of a G-protein-coupling domain of a muscarinic receptor predicted by random saturation mutagenesis. *J Biol Chem* **271**, 3058–65.

Hirschberg BT and Schimerlik MI (1994). A kinetic model for oxotremorine M binding to recombinant porcine m2 muscarinic receptors expressed in Chinese hamster ovary cells. *J Biol Chem* **269**, 26 127–35.

Hoffman BB and Taylor P (2001). Neurotransmission. In *Goodman and Gilman's The pharmacological basis of therapeutics* (eds. JG Hardman and LE Limbird) McGraw-Hill, New York, pp. 115–53.

Hosey MM, Pals-Rylaarsdam R, Lee KB, Roseberry AG, Benovic JL, Gurevich VV, and Bunemann M (1999). Molecular events associated with the regulation of signaling by M2 muscarinic receptors. *Life Sci* **64**, 363–8.

Hulme EC, Birdsall NJ, and Buckley NJ (1990). Muscarinic receptor subtypes. *Annu Rev Pharmacol Toxicol* **30**, 633–73.

Hulme EC, Lu ZL, Bee M, Curtis CA, and Saldanha J (2001). The conformational switch in muscarinic acetylcholine receptors. *Life Sci* **68**, 2495–500.

Hurley JH (1999). Structure, mechanism, and regulation of mammalian adenylyl cyclase. *J Biol Chem* **274**, 7599–602.

Inch TD and Brimblecombe RW (1974). Antiacetylcholine drugs: chemistry, stereochemistry, and pharmacology. *Int Rev Neurobiol* **16**, 67–144.

Jacoby DB and Fryer AD (2001). Anticholinergic therapy for airway diseases. *Life Sci* **68**, 2565–72.

Janowsky DS, Overstreet DH, and Nurnberger JI, Jr. (1994). Is cholinergic sensitivity a genetic marker for the affective disorders? *Am J Med Genet* **54**, 335–44.

Jerusalinsky D and Harvey AL (1994). Toxins from mamba venoms: small proteins with selectivities for different subtypes of muscarinic acetylcholine receptors. *Trends Pharmacol Sci* **15**, 424–30.

Jones SV (1993). Muscarinic receptor subtypes: modulation of ion channels. *Life Sci* **52**, 457–64.

Jones SV, Heilman CJ, and Brann MR (1991). Functional responses of cloned muscarinic receptors expressed in CHO-K1 cells. *Mol Pharmacol* **40**, 242–7.

Jones SVP, Levey AI, Weiner DM, Ellis J, Novotny E, Yu S-H, Dorje F *et al.* (1992). Muscarinic acetylcholine receptors. In *Molecular biology of G-protein-coupled receptors* (ed. MR Brann) Birkhauser, Boston, pp. 170–97.

Klein J, Lindmar R, and Loffelholz K (1996). Muscarinic activation of phosphatidylcholine hydrolysis. *Prog Brain Res* **109**, 201–8.

Kordower JH, Burke-Watson M, Roback JD, and Wainer BH (1992). Stability of septohippocampal neurons following excitotoxic lesions of the rat hippocampus. *Exp Neurol* **117**, 1–16.

Kostenis E, Zeng FY, and Wess J (1998). Structure-function analysis of muscarinic acetylcholine receptors. *J Physiol Paris* **92**, 265–8.

Kubo T, Bujo H, Akiba I, Nakai J, Mishina M, and Numa S (1988). Location of a region of the muscarinic acetylcholine receptor involved in selective effector coupling. *FEBS Lett* **241**, 119–25.

Kubo T, Fukuda K, Mikami A, Maeda A, Takahashi H, Mishina M *et al.* (1986*a*). Cloning, sequencing and expression of complementary DNA encoding the muscarinic acetylcholine receptor. *Nature* **323**, 411–6.

Kubo T, Maeda A, Sugimoto K, Akiba I, Mikami A, Takahashi H *et al.* (1986*b*). Primary structure of porcine cardiac muscarinic acetylcholine receptor deduced from the cDNA sequence. *FEBS Lett* **209**, 367–72.

Kuhar M and Yamamura HI (1976). Localization of cholinergic muscarinic receptors in rat brain by light microscopic radioautography. *Brain Res* **110**, 229–43.

Kuhar MJ and Yamamura HI (1975). Light autoradiographic localisation of cholinergic muscarinic receptors in rat brain by specific binding of a potent antagonist. *Nature* **253**, 560–1.

Lachowicz JE, Duffy RA, Ruperto V, Kozlowski J, Zhou G, Clader J *et al.* (2001). Facilitation of acetylcholine release and improvement in cognition by a selective M2 muscarinic antagonist, SCH 72788. *Life Sci* **68**, 2585–92.

Lazareno S, Buckley NJ, and Roberts FF (1990). Characterization of muscarinic M4 binding sites in rabbit lung, chicken heart, and NG108–15 cells. *Mol Pharmacol* **38**, 805–15.

Lazareno S, Gharagozloo P, Kuonen D, Popham A, and Birdsall NJ (1998). Subtype-selective positive cooperative interactions between brucine analogues and acetylcholine at muscarinic receptors: radioligand binding studies. *Mol Pharmacol* **53**, 573–89.

Lazareno S, Popham A, and Birdsall NJ (2000). Allosteric interactions of staurosporine and other indolocarbazoles with N-[methyl-(3)H]scopolamine and acetylcholine at muscarinic receptor subtypes: identification of a second allosteric site. *Mol Pharmacol* **58**, 194–207.

Lee NH and Fraser CM (1993). Cross-talk between m1 muscarinic acetylcholine and beta 2-adrenergic receptors. cAMP and the third intracellular loop of m1 muscarinic receptors confer heterologous regulation. *J Biol Chem* **268**, 7949–57.

Leslie CC (1997). Properties and regulation of cytosolic phospholipase A2. *J Biol Chem* **272**, 16 709–12.

Levey AI, Kitt CA, Simonds WF, Price DL, and Brann MR (1991). Identification and localization of muscarinic acetylcholine receptor proteins in brain with subtype-specific antibodies. *J Neurosci* **11**, 3218–26.

Li M, Yasuda RP, Wall SJ, Wellstein A, and Wolfe BB (1991). Distribution of m2 muscarinic receptors in rat brain using antisera selective for m2 receptors. *Mol Pharmacol* **40**, 28–35.

Liao CF, Themmen AP, Joho R, Barberis C, Birnbaumer M, and Birnbaumer L (1989). Molecular cloning and expression of a fifth muscarinic acetylcholine receptor. *J Biol Chem* **264**, 7328–37.

Liles WC and Nathanson NM (1986). Regulation of neuronal muscarinic acetylcholine receptor number by protein glycosylation. *J Neurochem* **46**, 89–95.

Marinissen MJ and Gutkind JS (2001). G-protein-coupled receptors and signaling networks: emerging paradigms. *Trends Pharmacol Sci* **22**, 368–76.

Matsui M, Motomura D, Karasawa H, Fujikawa T, Jiang J, Komiya Y *et al.* (2000). Multiple functional defects in peripheral autonomic organs in mice lacking muscarinic acetylcholine receptor gene for the M3 subtype. *Proc Natl Acad Sci USA* **97**, 9579–84.

Mattera R, Pitts BJ, Entman ML, and Birnbaumer L (1985). Guanine nucleotide regulation of a mammalian myocardial muscarinic receptor system. Evidence for homo- and heterotropic cooperativity in ligand binding analyzed by computer-assisted curve fitting. *J Biol Chem* **260**, 7410–21.

Mayorga AJ, Cousins MS, Trevitt JT, Conlan A, Gianutsos G, and Salamone JD (1999). Characterization of the muscarinic receptor subtype mediating pilocarpine-induced tremulous jaw movements in rats. *Eur J Pharmacol* **364**, 7–11.

McKinney M, Anderson DJ, Vella-Rountree L, Connolly T, and Miller JH (1991). Pharmacological profiles for rat cortical M1 and M2 muscarinic receptors using selective antagonists: comparison with N1E-115 muscarinic receptors. *J Pharmacol Exp Ther* **257**, 1121–9.

Michal P, Lysikova M, and Tucek S (2001). Dual effects of muscarinic M(2) acetylcholine receptors on the synthesis of cyclic AMP in CHO cells: dependence on time, receptor density and receptor agonists. *Br J Pharmacol* **132**, 1217–28.

Migeon JC and Nathanson NM (1994). Differential regulation of cAMP-mediated gene transcription by m1 and m4 muscarinic acetylcholine receptors. Preferential coupling of m4 receptors to Gi alpha-2. *J Biol Chem* 269, 9767–73.

Mitchell R, McCulloch D, Lutz E, Johnson M, MacKenzie C, Fennell M *et al.* (1998). Rhodopsin-family receptors associate with small G proteins to activate phospholipase D. *Nature* 392, 411–4.

Nilvebrant L (2001). Clinical experiences with tolterodine. *Life Sci* 68, 2549–56.

Nitsch RM, Slack BE, Wurtman RJ, and Growdon JH (1992). Release of Alzheimer amyloid precursor derivatives stimulated by activation of muscarinic acetylcholine receptors. *Science* 258, 304–7.

Olianas MC, Ingianni A, and Onali P (1998). Role of G protein betagamma subunits in muscarinic receptor-induced stimulation and inhibition of adenylyl cyclase activity in rat olfactory bulb. *J Neurochem* 70, 2620–7.

Pals-Rylaarsdam R, Xu Y, Witt-Enderby P, Benovic JL, and Hosey MM (1995). Desensitization and internalization of the m2 muscarinic acetylcholine receptor are directed by independent mechanisms. *J Biol Chem* 270, 29 004–11.

Park P, Sum CS, Hampson DR, Van Tol HH, and Wells JW (2001). Nature of the oligomers formed by muscarinic m2 acetylcholine receptors in Sf9 cells. *Eur J Pharmacol* 421, 11–22.

Peralta EG, Ashkenazi A, Winslow JW, Ramachandran J, and Capon DJ (1988). Differential regulation of PI hydrolysis and adenylyl cyclase by muscarinic receptor subtypes. *Nature* 334, 434–7.

Peralta EG, Ashkenazi A, Winslow JW, Smith DH, Ramachandran J, and Capon DJ (1987*a*). Distinct primary structures, ligand-binding properties and tissue-specific expression of four human muscarinic acetylcholine receptors. *Embo J* 6, 3923–9.

Peralta EG, Winslow JW, Peterson GL, Smith DH, Ashkenazi A, Ramachandran J *et al.* (1987*b*). Primary structure and biochemical properties of an M2 muscarinic receptor. *Science* 236, 600–5.

Peterson GL, Herron GS, Yamaki M, Fullerton DS, and Schimerlik MI (1984). Purification of the muscarinic acetylcholine receptor from porcine atria. *Proc Natl Acad Sci USA* 81, 4993–7.

Potter LT (2001). Snake toxins that bind specifically to individual subtypes of muscarinic receptors. *Life Sci* 68, 2541–7.

Premont RT, Inglese J, and Lefkowitz RJ (1995). Protein kinases that phosphorylate activated G protein-coupled receptors. *FASEB J* 9, 175–82.

Proska J and Tucek S (1994). Mechanisms of steric and cooperative actions of alcuronium on cardiac muscarinic acetylcholine receptors. *Mol Pharmacol* 45, 709–17.

Raiteri M, Marchi M, and Paudice P (1990). Presynaptic muscarinic receptors in the central nervous system. *Ann N Y Acad Sci* 604, 113–29.

Rhee SG and Bae YS (1997). Regulation of phosphoinositide-specific phospholipase C isozymes. *J Biol Chem* 272, 15 045–8.

Robbins J, Marsh SJ, and Brown DA (1993). On the mechanism of M-current inhibition by muscarinic m1 receptors in DNA-transfected rodent neuroblastoma x glioma cells. *J Physiol* 469, 153–78.

Rouse ST, Marino MJ, Potter LT, Conn PJ, and Levey AI (1999). Muscarinic receptor subtypes involved in hippocampal circuits. *Life Sci* 64, 501–9.

Rouse ST, Thomas TM, and Levey AI (1997). Muscarinic acetylcholine receptor subtype, m2: diverse functional implications of differential synaptic localization. *Life Sci* 60, 1031–8.

Russell M, Winitz S, and Johnson GL (1994). Acetylcholine muscarinic m1 receptor regulation of cyclic AMP synthesis controls growth factor stimulation of Raf activity. *Mol Cell Biol* 14, 2343–51.

Sadot E, Gurwitz D, Barg J, Behar L, Ginzburg I, and Fisher A (1996). Activation of m1 muscarinic acetylcholine receptor regulates tau phosphorylation in transfected PC12 cells. *J Neurochem* 66, 877–80.

Schwartz TW, Perlman S, Rosenkilde MM, and Hjorth SA (1997). How receptor mutagenesis may confirm or confuse receptor classification. *Ann N Y Acad Sci* 812, 71–84.

Segal M and Auerbach JM (1997). Muscarinic receptors involved in hippocampal plasticity. *Life Sci* 60, 1085–91.

Shapiro MS, Gomeza J, Hamilton SE, Hille B, Loose MD, Nathanson NM *et al.* (2001). Identification of subtypes of muscarinic receptors that regulate Ca2+ and K+ channel activity in sympathetic neurons. *Life Sci* 68, 2481–7.

Shapiro RA and Nathanson NM (1989). Deletion analysis of the mouse m1 muscarinic acetylcholine receptor: effects on phosphoinositide metabolism and down-regulation. *Biochemistry* 28, 8946–50.

Sharma VK, Colecraft HM, Wang DX, Levey AI, Grigorenko EV, Yeh HH, and Sheu SS (1996). Molecular and functional identification of m1 muscarinic acetylcholine receptors in rat ventricular myocytes. *Circ Res* 79, 86–93.

Shi H, Wang H, and Wang Z (1999). Identification and characterization of multiple subtypes of muscarinic acetylcholine receptors and their physiological functions in canine hearts. *Mol Pharmacol* 55, 497–507.

Simonds WF (1999). G protein regulation of adenylate cyclase. *Trends Pharmacol Sci* 20, 66–73.

Smith CUM (1996). *Elements of molecular neurobiology.* Wiley, New York.

Spalding TA, Birdsall NJ, Curtis CA, and Hulme EC (1994). Acetylcholine mustard labels the binding site aspartate in muscarinic acetylcholine receptors. *J Biol Chem* 269, 4092–7.

Spalding TA, Burstein ES, Henderson SC, Ducote KR, and Brann MR (1998). Identification of a ligand-dependent switch within a muscarinic receptor. *J Biol Chem* 273, 21 563–8.

Spalding TA, Trotter C, LaPaglia AK, Skjaerbaek N, Uldam AL, Hacksell U, and Brann MR (2001). Characterization of a novel series of M1-selective muscarinic agonists that act through an ectopic site. *Soc Neurosci Abstr* 27, 549.13.

Stockton JM, Birdsall NJ, Burgen AS, and Hulme EC (1983). Modification of the binding properties of muscarinic receptors by gallamine. *Mol Pharmacol* 23, 551–7.

Tanner LI, Harden TK, Wells JN, and Martin MW (1986). Identification of the phosphodiesterase regulated by muscarinic cholinergic receptors of 1321N1 human astrocytoma cells. *Mol Pharmacol* 29, 455–60.

Tietje KM and Nathanson NM (1991). Embryonic chick heart expresses multiple muscarinic acetylcholine receptor subtypes. Isolation and characterization of a gene encoding a novel m2 muscarinic acetylcholine receptor with high affinity for pirenzepine. *J Biol Chem* 266, 17 382–7.

Tucek S, Musilkova J, Nedoma J, Proska J, Shelkovnikov S, and Vorlicek J (1990). Positive cooperativity in the binding of alcuronium and N- methylscopolamine to muscarinic acetylcholine receptors. *Mol Pharmacol* 38, 674–80.

van Koppen CJ and Nathanson NM (1990). Site-directed mutagenesis of the m2 muscarinic acetylcholine receptor. Analysis of the role of N-glycosylation in receptor expression and function. *J Biol Chem* 265, 20 887–92.

van Koppen CJ and Nathanson NM (1991). The cysteine residue in the carboxyl-terminal domain of the m2 muscarinic acetylcholine receptor is not required for receptor-mediated inhibition of adenylate cyclase. *J Neurochem* 57, 1873–7.

Vernier P, Cardinaud B, Valdenaire O, Philippe H, and Vincent JD (1995). An evolutionary view of drug-receptor interaction: the bioamine receptor family. *Trends Pharmacol Sci* 16, 375–81.

Waelbroeck M, Gillard M, Robberecht P, and Christophe J (1986). Kinetic studies of [3H]-N-methylscopolamine binding to muscarinic receptors in the rat central nervous system: evidence for the existence of three classes of binding sites. *Mol Pharmacol* 30, 305–14.

Waelbroeck M, Tastenoy M, Camus J, and Christophe J (1990). Binding of selective antagonists to four muscarinic receptors (M1 to M4) in rat forebrain. *Mol Pharmacol* 38, 267–73.

Wall SJ, Yasuda RP, Hory F, Flagg S, Martin BM, Ginns EI, and Wolfe BB (1991*a*). Production of antisera selective for m1 muscarinic receptors using fusion proteins: distribution of m1 receptors in rat brain. *Mol Pharmacol* 39, 643–9.

Wall SJ, Yasuda RP, Li M, and Wolfe BB (1991*b*). Development of an antiserum against m3 muscarinic receptors: distribution of m3 receptors in rat tissues and clonal cell lines. *Mol Pharmacol* 40, 783–9.

Wang H, Han H, Zhang L, Shi H, Schram G, Nattel S, and Wang Z (2001). Expression of multiple subtypes of muscarinic receptors and cellular distribution in the human heart. *Mol Pharmacol* **59**, 1029–36.

Watson C, Chen G, Irving P, Way J, Chen WJ, and Kenakin T (2000). The use of stimulus-biased assay systems to detect agonist-specific receptor active states: implications for the trafficking of receptor stimulus by agonists. *Mol Pharmacol* **58**, 1230–8.

Weiner DM, Levey AI, and Brann MR (1990). Expression of muscarinic acetylcholine and dopamine receptor mRNAs in rat basal ganglia. *Proc Natl Acad Sci USA* **87**, 7050–4.

Wess J (1996). Molecular biology of muscarinic acetylcholine receptors. *Crit Rev Neurobiol* **10**, 69–99.

Wess J, Bonner TI, Dorje F, and Brann MR (1990). Delineation of muscarinic receptor domains conferring selectivity of coupling to guanine nucleotide-binding proteins and second messengers. *Mol Pharmacol* **38**, 517–23.

Wess J, Brann MR, and Bonner TI (1989). Identification of a small intracellular region of the muscarinic m3 receptor as a determinant of selective coupling to PI turnover. *FEBS Lett* **258**, 133–6.

Wess J, Gdula D, and Brann MR (1991). Site-directed mutagenesis of the m3 muscarinic receptor: identification of a series of threonine and tyrosine residues involved in agonist but not antagonist binding. *Embo J* **10**, 3729–34.

Wess J, Gdula D, and Brann MR (1992). Structural basis of the subtype selectivity of muscarinic antagonists: a study with chimeric m2/m5 muscarinic receptors. *Mol Pharmacol* **41**, 369–74.

Whitehouse PJ, Price DL, Struble RG, Clark AW, Coyle JT, and Delon MR (1982). Alzheimer's disease and senile dementia: loss of neurons in the basal forebrain. *Science* **215**, 1237–9.

Wong SK, Parker EM, and Ross EM (1990). Chimeric muscarinic cholinergic: beta-adrenergic receptors that activate Gs in response to muscarinic agonists. *J Biol Chem* **265**, 6219–24.

Wong SK and Ross EM (1994). Chimeric muscarinic cholinergic:beta-adrenergic receptors that are functionally promiscuous among G proteins. *J Biol Chem* **269**, 18 968–76.

Yamada M, Miyakawa T, Duttaroy A, Yamanaka A, Moriguchi T, Makita R *et al.* (2001). Mice lacking the M3 muscarinic acetylcholine receptor are hypophagic and lean. *Nature* **410**, 207–12.

Zaagsma J, Roffel AF, and Meurs H (1997). Muscarinic control of airway function. *Life Sci* **60**, 1061–8.

Zeng F and Wess J (2000). Molecular aspects of muscarinic receptor dimerization. *Neuropsychopharmacology* **23**, S19–31.

Zeng FY and Wess J (1999). Identification and molecular characterization of m3 muscarinic receptor dimers. *J Biol Chem* **274**, 19 487–97.

Chapter 19

Neuropeptide Y receptors

Yvan Dumont, John Paul Redrobe, and
Rémi Quirion

19.1 Introduction

Neuropeptide Y (NPY) was isolated from porcine brain almost two decades ago (Tatemoto and Mutt 1980). This 36 amino acid residues shares high sequence homology and structural identity with two other peptides, namely peptide YY (PYY) and pancreatic polypeptide (PP) (Tatemoto *et al.* 1982). All three peptides have thus been included in the same peptide family called the NPY family (Table 19.1). NPY is one of the most abundant peptide found in the central nervous system (CNS) of all mammals, including human (Chan-Palay *et al.* 1985; Chronwall *et al.* 1985) while PYY and PP are mostly found in endocrine cells of the intestine (Solomon 1985). Additionally PYY is present in the brainstem and various hypothalamic nuclei (Ekman *et al.* 1986). These peptides, especially NPY and PYY, are among the most conserved peptides during evolution (Larhammar 1996*a*; Larhammar *et al.* 2001).

The human gene coding for NPY has been located on chromosome HSA7q15.1 in proximity to the HOXA cluster (HSA7q15–q14) (Baker *et al.* 1995) whereas PYY and PP genes are located only 10 kb apart from each other on chromosome HSA17q21.1 close to the HOXB cluster (HSA17q21–22) (Hort *et al.* 1995). The human NPY gene is divided into four exons and three introns coding for a 98 amino acids precursor protein (Allen *et al.* 1987; Minth *et al.* 1984). The first exon contains the 5′-untranslated domain. The second exon codes for the initiating codon and the main part of the mature NPY sequence up to glutamine 34. The third exon codes for the last two amino acid (arginine and tyrosine), the glycine amine donor site, the dibasic cleavage site, and the main portion of CPON. The fourth exon includes the end of CPON and the 3′-untranslated region (Cerda-Reverter and Larhammar 2000). The genomic organization of PYY and PP are highly similar to that of NPY. The main difference is found in intron sizes of PYY and PP genes, which are smaller than in the NPY gene (Cerda-Reverter and Larhammar 2000). The structural organization of the precursor protein is almost identical for all three peptides with a N-terminal sequence of 28–29 amino acid signal, a peptide coding region which contain a Gly-Lys-Arg amidation cleavage sequence and a C-terminal peptide of about 30 amino acids (Leiter *et al.* 1987). The mature peptides contain 36 amino acid residues with an amidated tyrosine at the C-terminal portion, several other tyrosine residues in positions 1, 20, 21, and 27 and proline residues in positions 2, 5, and 8 (Larhammar 1996*a*).

Few investigators have evaluated the secondary and tertiary structure of the NPY peptide family. Using a combination of molecular modelling and structural dynamics, comparison of the possible structure of NPY with that of aPP has shown that structural elements found in aPP were maintained in pNPY and pPYY (Allen *et al.* 1987; MacKerell Jr. 1988); the tertiary structure being determined using X-ray crystallography (Blundell *et al.* 1981; Glover *et al.* 1983). The theoretical secondary and tertiary structures of NPY and PYY may therefore be similar to that of aPP. In the proposed model, N- and C-termini are stabilized by the intramolecular association of the hydrophobic moieties of a polyproline type II helix (residues 1–9) and the amphiphilic α-helix (residues 14–30); these two structures being connected by a type II β-turn (Allen *et al.* 1987; MacKerell Jr. 1988; MacKerell Jr. *et al.* 1989; McLean *et al.* 1990; Minakata *et al.* 1989). This confers a hairpin-like structure to these molecules where the tyrosine residue in position 1 and amino acid residues 30–36 are located in close proximity to each other (Keire *et al.* 2000*a*). Structure-activity studies have also demonstrated that the central segment of NPY, amino acid residues 19–23, is important for confering the helical conformation and high affinities for the Y_1, Y_2, Y_4, and Y_5 receptor subtypes (Cabrele *et al.* 2001). In addition, it has been proposed that in solution, NPY may exist as a dimer via the interactions of hydrophobic residues in the α-helical region (Saudek and Pelton 1990). However, it has recently been shown that at physiological concentrations (nanomolar), PYY and NPY mostly exist as monomers (Keire *et al.* 2000*a,b*).

19.2 **Molecular characterization**

At present, five NPY receptor subtypes have been cloned and designated as Y_1, Y_2, Y_4, Y_5, and Y_6 (Table 19.1). All belong to the seven transmembrane G protein-coupled receptor of the rhodopsin family (Michel *et al.* 1998). Although NPY and PYY have very high affinities for the cloned Y_1, Y_2, Y_5, and Y_6 subtypes (Table 19.1) (Michel *et al.* 1998), these receptors display relatively low sequence identities between each other (about 30–50 per cent; Fig. 19.1). In fact, NPY receptors appear to be the most divergent receptors among a given receptor family (Larhammar 1996*b*, 1997) and some NPY receptor subtypes even have higher homology for other families of G protein-coupled receptors (Bonini *et al.* 2000; Elshourbagy *et al.* 2000; Parker *et al.* 2000).

19.2.1 **Y_1 receptor**

The first NPY receptor to be cloned was initially reported as an orphan receptor isolated by screening a rat forebrain cDNA library (Eva *et al.* 1990). Upon transfection into a cell line, this clone demonstrated a ligand selectivity profile that was typical of the Y_1 receptor: PYY ≥ NPY ≥ [Leu^{31}, Pro^{34}]NPY > NPY$_{2-36}$ ≫ hPP > NPY$_{13-36}$ (Krause *et al.* 1992). Subsequently, the human (Herzog *et al.* 1992; Larhammar *et al.* 1992), mouse (Eva *et al.* 1992; Nakamura *et al.* 1995), guinea pig (Berglund *et al.* 1999) porcine, dog (Malmstrom *et al.* 1998) and monkey (Gehlert *et al.* 2001) Y_1 receptor cDNAs were isolated.

The human Y_1 receptor gene is located on chromosome HSA4q31.3–q32 and consists of three exons (Herzog *et al.* 1993*a*). The first exon contains the 5′ untranslated region while exon 2 contains both the 5′ untranslated domain and the coding region up to just after the fifth seven transmembrane domain while exon 3 encodes for the remaining sequence and the 3′ untranslated domain. The first intron contains at least three distinct promoters: possible response elements for the glucocorticoid receptor and potential binding sites for the AP-1, AP-2 as well as NF-kappa B transcription factors. In addition, analysis of Y_1 transcripts in

Table 19.1 Neuropeptide Y receptors

	Y_1	Y_2	Y_4	Y_5	Y_6
Alternative names	—	PYY preferring	PP_1	Atypical Y_1 feeding receptor	Y_5, Y_{2b}, PP_2
Preferred endogenous ligand	Neuropeptide Y (NPY), peptide YY (PYY)	NPY, PYY, NPY_{3-36} PYY_{3-36}	Pancreatic polypeptide (PP)	NPY, PYY, NPY_{3-36} PYY_{3-36}	NPY, PYY, NPY_{3-36} PYY_{3-36}, PP
Structural information (Accession no.)	h 384 aa (P25929) r 382 aa (P21555) m 382 aa (Q04573)^{AS}	h 381 aa (P40146) r 385 aa (NP_076458) m 385 aa (P97295)	h 375 aa (P50391) r 375 aa (Q63447) m 375 aa (Q61041)	h 445 aa (Q15761) r 456 aa (Q63634) m 466 aa (O70342)	h 290 aa (Y59431) m 371 aa (Q61212)
Chromosomal location	4q31.3–32	4q31	10q11.2	4q31.3–q32	5q31
Selective agonists	[Pro^{34}]NPY, [Pro^{34}]PYY [Leu^{31}, Pro^{34}]NPY, [Leu^{31}, Pro^{34}]PYY,	NPY_{13-36}, PYY_{13-36} [Ahx^{5-24}, γ-Glu^2-ϵ-Lys^{20}]NPY, C2-NPY	[Leu^{31}, Pro^{34}]NPY, [Leu^{31}, Pro^{34}]PYY, GR231118	[Leu^{31}, Pro^{34}]NPY, [Leu^{31}, Pro^{34}]PYY, h PP, [Ala^{31}, Aib^{32}]NPY, hPP_{1-17}, [Ala^{31}, Aib^{32}]NPY	Not well defined
Selective antagonists	BIBP3226, BIBO3304 GR231118, J104870 J115814, GI264879A	BIIE0246	—	L152804, CGP71683A	—
Radioligands	[^{125}I]-PYY, [^{125}I]-[Leu^{31}, Pro^{34}]PYY, [^{125}I]-GR231118	[^{125}I]-PYY, [^{125}I]-PYY_{3-36}	[^{125}I]-PPs, [^{125}I]-PYY, [^{125}I]-[Leu^{31}, Pro^{34}]PYY, [^{125}I]-GR231118	[^{125}I]-PYY, [^{125}I]-[Leu^{31}, Pro^{34}]PYY	[^{125}I]-PYY
G protein coupling	G_i/G_o	G_i/G_o	G_i/G_o	G_i/G_o	G_i/G_o

Expression profile	Cerebral cortex, thalamus, brain stem, smooth muscle of blood vessels	Hippocampus, brain stem nuclei, hypothalamus, gastrointestinal tract, smooth muscle of blood vessels	Colon, small intestine, prostate, very low in brain, paraventricular hypothalamus, interpeduncular nucleus	Hippocampus, plexiform cortex of the olfactor bulb, suprachiasmatic and accruate nuclei	Not fully characterised
Physiological function	Vasoconstriction, regulation of food intake, anxiety-related behaviours, regulation of transmitter release	Inhibition of glutamate release, inhibition of noradrenaline release, learning and memory	Possible in regulation of LH secretion	Possibly regulation of food intake	Unclear
Knockout phenotypes	Hyperalgesia, altered feeding in response to NPY, increased body temperure with antisense knockdown	Increased body weight and food intake		Susceptible to seizures	No obvious phenotype
Disease relevance	Anxiety, feeding disorders, pain	Epilepsy, cognitive decline	Unclear	Feeding disorders	Unclear

(a) Human Y$_1$

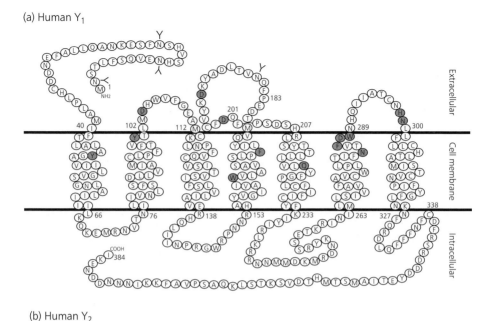

(b) Human Y$_2$

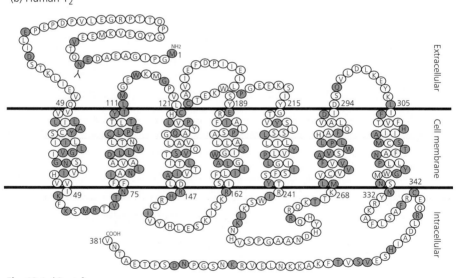

Fig. 19.1 (*Contd*)

various cell types demonstrates that tissue-specific activation of the three promoters occurs (Ball *et al.* 1995).

The human Y$_1$ receptor gene codes for a 384 amino acids protein that has all the characteristics of the GPCR family including potential glycosylation sites in the N-terminal portion and in the second extracellular loop, four extracellular cysteines in positions 33, 113, 198, and 296 which may form two disulfide bridges (Cys 33 and 296 and Cys 113 and 198), the presence of an intracellular cysteine in the C-terminal portion at position 338 that may be used

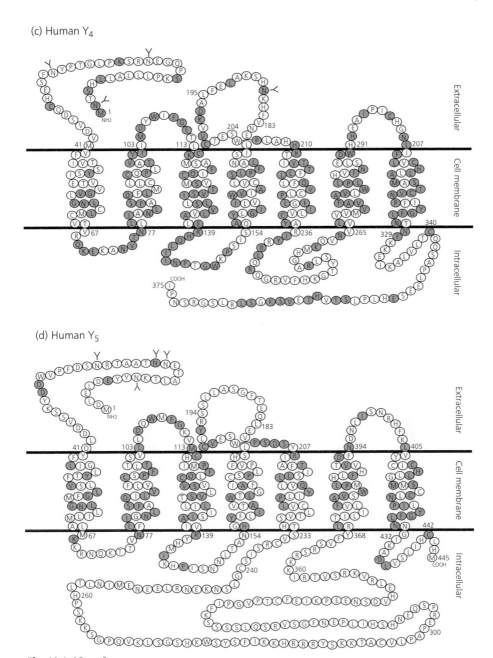

Fig. 19.1 (Contd)

for the attachment of palmitate residues into the cell membrane and possible phosphorylation sites in the intracellular domain (Fig. 19.1a). All mammalian Y_1 receptors cloned thus far display 90–95 per cent homology with the human Y_1 receptor (Larhammar *et al.* 2001). Additionally, two variants of the mouse Y_1 receptors have been identified with a different

(e) Human Y$_6$

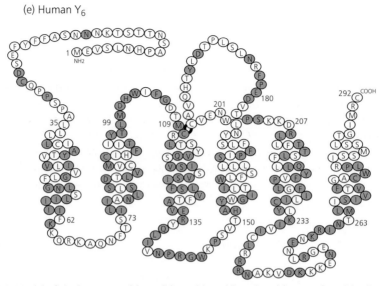

Fig. 19.1 Model of the human Y$_1$ (a), Y$_2$ (b), Y$_4$ (c) Y$_5$ (d) and Y$_6$ (e) receptors with alignment of the potential extracellular, cytoplasmic and transmembrane domains. Each circle represents a single amino acid using the one letter symbol. Possible glycosylation sites are represented by Y. In Fig. 19.1a, gray circles represent amino acid residues that are implicated in the interaction with either NPY or PYY. In Fig. 19.1b–e, gray circles represent amino acids that are identical to the human Y$_1$ receptor.

sequence from amino acid residue 303 (Nakamura *et al.* 1995). This generates a truncated form with an incomplete seven transmembrane domain. Both isoforms bind NPY but the shorter one does not initiate second messenger responses. It is unknown at this time if this alternative splicing also occurs in human. However, the presence of termination codons in intron 2 of human Y$_1$ cDNA clones (Herzog *et al.* 1993*a*) could eventually result in a protein lacking the last two seven transmembrane domains and the C-terminal portion of the Y$_1$ receptor protein.

19.2.2 **Y$_2$ receptor**

Wahlestedt and co-workers (1986) were the first to propose the existence of the Y$_2$ receptor based on the differential potencies of NPY and C-terminal fragments such as NPY$_{13–36}$ or PYY$_{13–36}$ in various bioassays. Expression screening from cDNA libraries of neuroblastoma cells known as SMS-KAN (Rose *et al.* 1995), human hippocampus (Gerald *et al.* 1995) or human brain (Gehlert *et al.* 1996*a*) eventually lead to the isolation of a human cDNA receptor clone which possesses a pharmacological binding profile similar to the Y$_2$ receptor. This receptor has now been cloned from various species including the rat (St Pierre *et al.* 1998), bovine (Ammar *et al.* 1996), mouse (Nakamura *et al.* 1996), guinea pig, porcine, dog (Malmstrom *et al.* 1998) and monkey (Gehlert *et al.* 2001). High homology (90 and 95 per cent in the 7TM) is observed between species (Larhammar *et al.* 2001). However, rather surprisingly considering that NPY and PYY possess very high affinities for both the Y$_1$ and Y$_2$ receptor subtypes, the overall homology between the Y$_1$ and Y$_2$ receptors is only 31 per cent, which explains the failure of homology screening approaches to clone the Y_2 receptor.

The human Y_2 receptor gene is located in close proximity to the Y_1 and Y_5 genes on chromosome HSA4q31 and has a single intron of approximately 4.5 kb located in the 5'-untranslated region (Ammar et al. 1996). The human Y_2 receptor gene codes for a 381 amino acid protein and possesses the typical seven transmembrane helix receptor structure which includes a single glycosylation site in the N-terminal region, two extracellular cysteines in positions 122 and 203 that may form a disulphide bridge, a single cysteine in the C-terminal region at position 342 that could serve as an attachment site for palmitate and phosphorylation sites in the intracellular domain (Fig. 19.1b). The amino acid sequence of the Y_2 receptor protein reported by Rose et al. (1995) differs by one amino acid from that reported by Gerald et al. (1995) and Gehlert et al. (1996a), differences that are most likely to be allelic.

19.2.3 Y_3 receptor

Various groups have proposed the existence of a receptor that possesses high affinity for NPY but not PYY in several assays including the rat brain (Grundemar et al. 1991), rat colon (Dumont et al. 1994), rat lung (Hirabayashi et al. 1996) and rat and bovine adrenals (Wahlestedt et al. 1992). However, evidence for the existence of such a subtype is still circumstantial and the clone that has been reported as a Y_3 receptor (Rimland et al. 1991) does not bind NPY (Herzog et al. 1993b; Jazin et al. 1993) and actually belongs to the cytokine receptor family (Feng et al. 1996). Early on, we suggested that the rat colon was a prototypical Y_3 in vitro bioassay (Dumont et al. 1994). However, a more extensive study of the contractile effect induced by NPY and related peptides including PYY_{3-36}, hPP and [Leu^{31}, Pro^{34}]NPY suggests that the contraction induced by NPY and related agonists in this bioassay was due to the activation of both Y_2 and Y_4 receptors (Pheng et al. 1999); a situation supported by recent molecular studies confirming the presence of Y_2 and Y_4 mRNA in this tissue (Feletou et al. 1998). The cloning of a genuine Y_3 receptor is still awaited. It may be that the Y_3 receptor protein is a G protein-coupled receptor for which the expression at the cell surface is dependent on the presence of regulating activity modifying protein (RAMP) as seen for CRLR in generating a functional CGRP receptor (for review, Juaneda et al. 2000). Alternatively, the Y_3 receptor could exist as a dimer of any of the cloned NPY receptors, as recently reported for opioid receptors (see Chapter 21). Further studies will be required to verify these hypotheses.

19.2.4 Y_4 receptor

The use of sequence homology screening with the Y_1 receptor probe lead to the isolation of a new human NPY receptor cDNA. This receptor was originally designated as either PP_1 (Gehlert et al. 1996c; Lundell et al. 1995) or Y_4 (Bard et al. 1995). After transfection in cell lines, the expressed protein demonstrated very high affinity for PP-related peptides such as human (h) PP, porcine (p) PP and bovine (b) PP (sub nM range). Homologues of the Y_4 receptor have now been cloned from mouse (Gregor et al. 1996b), rat (Lundell et al. 1996; Yan et al. 1996) and guinea pig (Eriksson et al. 1998). Sequence homology between human and other species Y_4 receptors is one of the lowest (less than 75 per cent) reported for orthologous G protein coupled receptors between different mammalian species (Larhammar *et al.* 2001). Moreover, the human Y_4 receptor protein has higher homology with the human Y_1 (43 per cent) than the human Y_2 (34 per cent) receptor (Larhammar 1996*b*).

The human Y_4 receptor gene is located on chromosome HSA10q11.2–q21 (Wraith *et al.* 2000) and codes for a 375 amino acid residue protein. No information is currently available regarding the genomic organization of the Y_4 receptor gene. As for all members of this peptide family, the Y_4 receptor protein has all the characteristics of a GPCR including four glycosylation sites (three at the N-terminal segment and one in the second extra-cellular loop), a seven transmembrane helix structure, four extracellular cysteines in positions 34, 114, 201, and 298, an intracellular cysteine in position 340 and the presence of several serines and threonines in the C-terminal and intracellular loops which may serve as phosphorylation sites by various protein kinases (Fig 19.1c). The extracellular cysteine can form a disulphide bridge between Cys 34 and 298 and a second disulphide bond between Cys 114 and 210. The intracellular cysteine located at the C-terminal tail likely enables the attachment of palmitate residues.

19.2.5 Y_5 receptor

In the nineteen nineties, the existence of an atypical receptor subtype was proposed on the basis of the effects of NPY and long C-terminal fragments, such as NPY$_{2-36}$, on food intake. This receptor was referred to as the atypical Y_1 or 'feeding' receptor (Quirion *et al.* 1990). More recently, the profile of another NPY receptor cloned from human and rat tissues was classified as the Y_5 subtype and was found to have a pharmacological profile similar to that of the atypical feeding receptor (Gerald *et al.* 1996; Hu *et al.* 1996). This Y_5 receptor has now been cloned from various species including mouse (Nakamura *et al.* 1997), dog (Borowsky *et al.* 1998), guinea pig (Lundell *et al.* 2001) and monkey (Gehlert *et al.* 2001).

The human Y_5 receptor gene is transcribed in opposite directions to the Y_1 gene from a common promoter region on chromosome HSA4q31–q32 (Herzog *et al.* 1997). The transcription of both genes from opposite strands of the same DNA sequence suggests that transcriptional activation of one will have an effect on the regulation of the other. As both Y_1 and Y_5 receptors are thought to play important roles in the regulation of food intake, the coordinated expression of their specific genes may be important in the modulation of NPY-induced feeding behaviours. The human Y_5 gene codes for a protein of 456 amino acids and has the usual characteristics of a GPCR including glycosylation sites in the N-terminal region and possible cysteine disulphide bridge in the extracellular domain (Cys 114 and 198), a cysteine residue at the C-terminal segment (which could serve as attachment for palmitate) and phosphorylation sites in the intracellular domain. In comparison to other NPY receptor subtypes, the Y_5 receptor protein has a very long third intracellular loop and an unusual short C-terminal tail (Fig. 19.1d).

19.2.6 Y_6 receptor

Three groups have reported the cloning of another NPY receptor in mice, rabbit, monkey, and human tissues (Gregor *et al.* 1996*a*; Matsumoto *et al.* 1996; Weinberg *et al.* 1996) now known as the Y_6 receptor (Michel *et al.* 1998). Upon transfection of the mouse and rabbit receptor clone into cell lines, distinct pharmacological profiles have been reported with similarities to the Y_2 (Matsumoto *et al.* 1996) or Y_4 (Gregor *et al.* 1996*a*) or Y_5 (Weinberg *et al.* 1996) receptors. Rather surprisingly, transfection of the human Y_6 receptor cDNA failed to be fully translated and to generate a functional receptor. It is now known that the Y_6 receptor is not expressed in the rat (Burkhoff *et al.* 1998), while in human and primates, the cDNA contains a single base deletion resulting in the expression of a non-functional NPY receptor protein which is truncated from the sixth transmembrane domain (Gregor *et al.* 1996*a*; Matsumoto *et al.* 1996). Pseudogenes have also been reported in the guinea pig (Starback *et al.* 2000) and pig (Wraith *et al.* 2000) while in the dog, it is expressed as a functional receptor (Borowsky *et al.* 1998).

In the human, the Y_6 gene is located on chromosome HSA5q31 (Wraith *et al.* 2000). However, despite high levels of expression of Y_6 mRNA in primates and human tissues, the physiological function of this truncated protein is unknown. Nevertheless, it is important to remember that the Y_6 receptor is a functional receptor in mouse, rabbit, and dog which is particularly important in mice knockout models in which the Y_6 receptor could compensate for other deleted NPY receptor proteins.

19.2.7 Peptide–receptor interactions

To understand the interaction of peptides with their receptors, two approaches are usually used. The first, known as structure-activity studies, aims to identify peptide sequence and amino acid residues that are important for receptor recognition and activation (e.g. see, Beck-Sickinger 1997). The second, more recently developed is known as site-directed mutagenesis of the cloned receptor protein (Schwartz 1994). These two approaches provide complementary valuable information that is indispensable to our understanding of interactions between GPCR and their respective ligands. As expected, each method has its limitations and must be performed using proper control to avoid misinterpretation of the obtained data (Ingenhoven and Beck-Sickinger 1999; Schwartz 1994).

Most structure-activity studies have focused on Y_1 and Y_2 receptor subtypes. The full length of NPY-like peptides and a close proximity between the N and C-termini are crucial for the recognition of the Y_1 receptor subtype while the C-terminal portion is important for interaction with the Y_2 subtype (Wahlestedt and Reis 1993). In addition, alanine scan revealed that positively charged arginine residues in positions 33 and 35 play a very important role for binding to the Y_1 and Y_2 receptor subtypes (Beck-Sickinger *et al.* 1994). Interestingly, the C-terminal moiety (Tyr-Arg-XXX-Arg-Tyr-NH2) is identical for NPY, PYY, and PP in all mammals (Larhammar 1996*a*). This suggests that positively charged residues located in the C-terminal region of the peptide interact with negatively charged amino acid residues of either the Y_1, Y_2, Y_4, Y_5, or Y_6 receptor subtypes. Additionally, deleting the middle portion of NPY and joining the N-terminal and C-termini by a linker is well tolerated by the Y_2 but not by the Y_1 (Beck-Sickinger *et al.* 1992, 1993; Rist *et al.* 1996) and Y_5 (Beck-Sickinger 1997) receptor subtype. This result indicates that negatively charged amino acid residues (Asp or D and Glu or E) located in the β-turn are involved either to maintain a proper hairpin-like structure of the peptide which is important for interaction with the Y_1 and possibly Y_5 receptor subtypes, or to interact with positively charged amino acid residues located on the extracellular domain of NPY receptors. This is in agreement with Ala scan data which revealed that replacing Asp or Glu in positions 10, 11, 15, and 16 resulted in a significant decrease in affinity of the substituted NPY analogues for the Y_1, but not Y_2 receptor subtypes (Beck-Sickinger *et al.* 1994).

Using site-directed mutagenesis, it has been shown that replacing Asp residues in positions 104, 194, 200, and 287 of the Y_1 receptor protein resulted in a complete loss of detectable [^{125}I]NPY binding (Walker et al. 1994), suggesting the formation of bridges between positively and negatively charged amino acid residues of NPY and the Y_1 receptor, respectively. Computer-assisted modelling suggested that the C-terminal tyrosine amide moiety of NPY might dock at a pocket formed by hydrophobic amino acids of transmembrane domains 1, 2, 6, and 7 of the Y_1 receptor (Walker et al. 1994). Accordingly, further studies revealed that substitution of the Tyr residue in position 100 by Phe yielded a receptor having no detectable affinity for NPY (Sautel et al. 1995). Additionally, the Phe residue in position 286 and the His in position 298 are critical for the interaction of NPY with the Y_1 receptor protein (Sautel et al. 1995). Computer-assisted modelling suggested that the carboxyl tyrosine binds to an hydrophobic pocket formed by residue Tyr100, Phe286 and His298 located at the surface of the Y_1 receptor (Sautel et al. 1995). In these studies, the absence of specific NPY binding was not due to a lack of expression or to the inability of the receptor to reach the cell surface (Sautel et al. 1995; Walker et al. 1994). However, using site-directed mutagenesis, it is not possible to distinguish between direct and indirect (conformational changes that will affect the three-dimensional structure of the receptor) effects. In order to potentially distinguish between direct and indirect effects of mutants, several ligands must be used. For example, using Y_1 receptor antagonists, it was shown that alanine mutants in position Tyr100, Asp104, Trp288, and His298 of the human Y_1 receptor had not effect on specific antagonist binding ([^{3}H]BIBP3226 (Sautel et al. 1996), [^{125}I]GR231118 and [^{3}H]J-104870 (Kanno et al. 2001)) while no or low levels of specific [^{125}I]NPY (Sautel et al. 1996) and [^{125}I]PYY (Kanno et al. 2001) binding were detected. In contrast, other mutations affected the binding of antagonists but not of agonists. For example, substitution of Tyr211 by an alanine residue markedly reduced specific [^{3}H]BIBP3226 binding (Sautel et al. 1996) but had not effect on specific [^{125}I]NPY (Sautel et al. 1996), [^{125}I]PYY, [^{125}I]GR231118, and [^{3}H]J-104770 (Kanno et al. 2001) binding. In addition, mutation of the Phe173 residue by Ala resulted in the loss of specific [^{3}H]J-104870 binding whereas an Ala mutant of Lys303 decreased specific [^{125}I]GR231118 binding while not affecting [^{3}H]BIBP3226, [^{125}I]PYY, and [125I]NPY binding (Kanno et al. 2001; Sautel et al. 1996).

Overlaps between different radioligands have also been reported, revealing that different radioligands interact with the same amino acid residues of the Y_1 receptor. Substitution of Tyr47 or Asn299 resulted in the loss of both [^{125}I]PYY and [^{125}I]GR23111 binding while mutation of Asn283 resulted in [^{125}I]NPY, [^{125}I]PYY, [^{3}H]BIBP3226, and [^{3}H]J104870 binding (Kanno et al. 2001; Sautel et al. 1996) while not affecting [^{125}I]GR231118 binding (Kanno et al. 2001). Mutation of amino acid residues Trp163 and Asp287 by an alanine residue resulted in a dramatic decrease of specific binding for all radioligands (Kanno et al. 2001; Sautel et al. 1996). Interestingly, the Asp287 residue has been suggested to interact with the guanidine group of NPY and BIBP3226 (Sautel et al. 1996), which is also present in PYY and GR231118. Furthermore, the importance of Asp287 is supported by the fact that this amino acid residue is well conserved in all receptor subtypes (Fig. 19.1). Finally, since it has been suggested that NPY preferentially binds at the surface of the transmembrane segment (Sautel et al. 1995), it is rather surprising that substitution of Trp in position 163, which is located deep in the fourth transmembrane domain almost abolished binding of all radioligands. It may be that this substitution induces a change in the three-dimensional conformation of the Y_1 receptor.

Site-directed mutagenesis information is not currently available regarding other NPY receptor subtypes. However, considering that the Y_2 antagonist BIIE0246 has been found to be inactive on the cloned chicken Y_2 receptor (Salaneck *et al.* 2000), comparative studies between the chicken and human Y_2 receptor could be a starting approach to evaluate key amino acid residues that are implicated in the interaction between NPY and the Y_2 receptor subtype. In addition as specific and selective antagonists are discovered, they will be very useful tools to investigate differential interaction between agonists, antagonists, and the various NPY receptor subtypes. Complementing these studies will be those using antibodies and fluorescent peptide probes (Gaudreau *et al.* 1999; Dumont *et al.* 2001). Already antibodies against selected NPY receptors have been used by several laboratories (Eckard *et al.* 1999; Grove *et al.* 2000; Ji *et al.* 1994; St Pierre *et al.* 2000b; Wieland *et al.* 1998a; Yang *et al.* 1996; Zhang *et al.* 1994). Fluorescent probes and the use of NPY receptors linked to green fluorescent protein (GFP) will allow receptor dynamics such as internalization and recycling to be investigated.

19.3 Cellular and subcellular localization

In peripheral tissues, the Y_1 subtype is abundantly expressed in blood vessels where it mediates vasoconstriction (Lundberg *et al.* 1996; Wahlestedt and Reis 1993). In the rat brain, the localization of the Y_1 receptor mRNA (Eva *et al.* 1990; Larsen *et al.* 1993; Parker and Herzog 1999; Tong *et al.* 1997) closely matches to that of the Y_1 receptor protein (Dumont *et al.* 1990, 1993, 1996; Dumont and Quirion 2000; Gehlert *et al.* 1992; Gehlert and Gackenheimer 1996; Schober *et al.* 1998) with predominant expression in the cerebral cortex, thalamus, and brainstem nuclei. Species differences have been noted on the distribution of the Y_1 receptor subtype (Dumont *et al.* 1997, 1998b, 2000c; Jacques *et al.* 1997) but it is thought that among its many functions the Y_1 receptor subtype may also act as an autoreceptor in the CNS (Pickel *et al.* 1998; St Pierre *et al.* 2000a).

The Y_2 receptor subtype is also expressed peripherally where it is predominantly found at the pre-synaptic nerve terminal involved in the inhibition of neurotransmitter release, especially noradrenaline (Smith-White *et al.* 2001; Wahlestedt *et al.* 1986). The existence of post-synaptic Y_2 receptors has also been reported in blood vessels (Pheng *et al.* 1997). The presence of Y_2 receptors has also been described in the gastrointestinal tract (Cox *et al.* 1998; Cox and Cuthbert 1990; Pheng *et al.* 1999; Playford and Cox 1996) where it is often referred to as the PYY preferring receptor (Goumain *et al.* 1998a; Voisin *et al.* 1990). In the rat CNS, Y_2 receptor mRNA and protein are abundantly expressed in the hippocampal formation and brainstem nuclei while moderate levels of receptors are detected in various hypothalamic nuclei (Dumont *et al.* 1990, 1993, 1996, 2000b; Gustafson *et al.* 1997; Larsen and Kristensen 1998; Parker and Herzog 1999). In fact, Y_2 receptor mRNA is discretely distributed in the rat brain and its localization is largely similar to that of the Y_2 receptor protein (Dumont *et al.* 2000c). However, as for the Y_1 subtype, species differences exist in the level of expression of the Y_2 receptor in various brain regions (Dumont *et al.* 1997, 1998b, 2000c).

Little is known about the precise distribution of the Y_3 subtype although functional studies suggest both central and peripheral localization. Y_4 receptor mRNA, on the other hand, has been detected in the colon, small intestine, and prostate in humans (Lundell *et al.* 1995) as well as in the rat intestine (Goumain *et al.* 1998b). In the rat brain, only low levels of expression of the Y_4 receptor mRNA have been detected thus far (Parker and Herzog 1999). In addition, a few studies using $[^{125}I]hPP$ (Trinh *et al.* 1996) or $[^{125}I]bPP$ (Gehlert *et al.* 1997;

Whitcomb *et al.* 1997) have confirmed the restricted distribution of PP (Y_4-like) binding sites in the rat brain.

In situ hybridization signals of Y_5 receptor mRNA are evident in the external plexiform layer of the olfactory bulb, anterior olfactory nuclei, hippocampus, suprachiasmatic and arcuate nuclei (Larsen and Kristensen 1998; Parker and Herzog 1998, 1999); a distribution pattern that is relatively similar to that observed for Y_5 receptor protein except for very low levels of Y_5 receptor protein in the arcuate nucleus (Dumont *et al.* 1998a; Dumont *et al.* 2000c). In humans strong *in situ* hybridization signals of Y_5 receptor mRNA are also present in the human brain at the levels of the arcuate nucleus which is a key structure implicated in feeding behaviours (Jacques *et al.* 1998). To date little is known about the precise localization of Y_6 receptors.

19.4 **Pharmacology**

19.4.1 **Agonists**

Endogenous ligands

Neuropeptide Y and PYY have low nanomolar affinities for Y_1, Y_2, Y_5 and Y_6 receptors while PPs are much less potent (over 100 nM; except human PP on Y_5) (Michel *et al.* 1998). In contrast, PPs are more potent on the Y_4 subtype than PYY and NPY (Bard *et al.* 1995; Gehlert *et al.* 1996c; Lundell *et al.* 1995, 1996). Furthermore, these endogenous peptides are processed at their N-termini by a dipeptidyl peptidase IV enzyme which remove the first two amino acid residues to generate C-terminal fragments 3–36 (Grandt *et al.* 1992; Medeiros and Turner 1994b, 1996; Mentlein *et al.* 1993). These fragments, especially NPY_{3-36} and PYY_{3-36} demonstrate marked decreases in affinity for the Y_1 subtype while being as potent as the native peptides on the Y_2 receptor (Dumont *et al.* 1994, 1995; Gerald *et al.* 1995; Michel *et al.* 1998; Rose *et al.* 1995). As observed for the native peptides, these endogenous C-terminal fragments are potent agonists on the Y_5 (Gerald *et al.* 1996; Hu *et al.* 1996) and Y_6 (Matsumoto *et al.* 1996; Weinberg *et al.* 1996) receptors. In addition, the NPY fragment NPY_{1-30} has been isolated from the rat brain (Wahlestedt *et al.* 1990) and it has been shown that both PYY (Medeiros and Turner 1994a) and NPY (Medeiros and Turner 1996) can be hydrolyzed by neutral endopeptidase-24.11 to generate this fragment. The physiological relevance of this fragment is unknown but could be related to the termination of NPY and PYY effects by generating an inactive fragment since N-terminal fragments have very poor affinities for a variety of NPY receptors (Martel *et al.* 1990). On the other hand, it has been reported that NPY_{1-30} could decrease spontaneous locomotor activity and induce hypothermia, suggesting possible limited activities in some paradigms (Bouali *et al.* 1994).

Synthetic agonists

Over the last twelve years, several synthetic peptide analogues of NPY, PYY and PP have been developed. The first replacing the amino acid residue (glutamine) in position 34 of NPY by a proline resulted in an analogue, [Pro^{34}]NPY which possesses higher affinity for the Y_1 over the Y_2 subtype (Schwartz *et al.* 1990) and enabled identification of the differential localization of the Y_1-like and Y_2-like receptors in the rat brain (Dumont *et al.* 1990). Similar results have been reported using analogues such as [Leu^{31}, Pro^{34}]NPY and [Leu^{31}, Pro^{34}]PYY (Dumont *et al.* 1994; Fuhlendorff *et al.* 1990a,b; Jorgensen *et al.* 1990; Schwartz *et al.* 1990). However, following the discovery of the Y_4, Y_5 and Y_6 subtypes it was demonstrated that these substituted NPY and PYY analogues behaved as potent agonists of these receptor subtypes (Bard *et al.* 1995; Gehlert *et al.* 1996*b*; Gerald *et al.* 1996; Hu *et al.* 1996; Lundell *et al.* 1995; Matsumoto *et al.* 1996; Weinberg *et al.* 1996).

Pharmacological approaches, and specifically the differential potencies of NPY and C-terminal fragments such as NPY_{13-36} or PYY_{13-36} in various bioassays initiated the suggestion that Y_1 and Y_2 receptors existed (Wahlestedt *et al.* 1986). It was subsequently found that deletion of the tyrosine residue in position 1 markedly decreased the affinity of NPY for the cloned Y_1 (Herzog *et al.* 1992; Larhammar *et al.* 1992) but not Y_2 (Gehlert *et al.* 1996*a*; Gerald *et al.* 1995), Y_5 (Gerald *et al.* 1996; Hu *et al.* 1996) or Y_6 (Matsumoto *et al.* 1996; Weinberg *et al.* 1996) receptors transfected in cell lines. Furthermore, deletion in the N-terminal region to generate C-terminal fragments 13–36 (Gerald *et al.* 1995; Martel *et al.* 1990; Wahlestedt *et al.* 1986), 16–36 (Colmers *et al.* 1991) and 18–36 (Boublik *et al.* 1989*a,b*; Feinstein *et al.* 1992) demonstrated agonistic activity in various Y_2 assays while being almost inactive on the Y_1, Y_4 and Y_5 subtypes (Dumont *et al.* 1994, 1995, 1998*a*; Gehlert *et al.* 1996*a,c*; Gerald *et al.* 1996). It thus appears that the C-terminal portions of NPY and PYY are sufficient for Y_2 receptor recognition. Additionally, the PP-like hairpin loop does not seem to be essential but may help to stabilize the C-terminal hexapeptide amide, improving its affinity for the Y_2 receptor. Accordingly, several shortened and truncated modified C-terminal fragments of NPY produced by cyclization of the N- and C-terminal segments of the discontinuous peptide have been developed. They usually displayed an affinity 40 times greater than that seen for NPY_{13-36} on the Y_2 receptors. Among them, two of these peptide analogues have demonstrated good selectivity for Y_2 (1–10 nM) versus Y_1 and Y_5 (300–700 nM) receptors, namely $[Cys^2, aminooctanoic\ acid^{5-24}, D\text{-}Cys^{27}]NPY$ also known as C2-NPY (Beck-Sickinger 1997; Gerald *et al.* 1996; McLean *et al.* 1990) and $[Ahx^{5-24}, \gamma\text{-}Glu^2\text{-}\varepsilon\text{-}Lys^{20}]NPY$ (Rist *et al.* 1997). Additionally, glutamine in position 34 plays an essential role to maintain a proper C-terminal orientation of the peptide in order to facilitate the interaction between the peptide and the Y_2 receptor protein (Cabrele *et al.* 2001; Fuhlendorff *et al.* 1990*a,b*; Schwartz *et al.* 1990).

As with Y_1 receptors $[Leu^{31}, Pro^{34}]NPY$ and $[Leu^{31}, Pro^{34}]PYY$ demonstrate agonist activity at the Y_4 receptor. GR231118 represents a small molecule Y_4 receptor agonist although this compound also has activity at other receptor subtypes, in particular antagonism at the Y_1 receptor. Rather surprisingly, despite the fact that GR231118 and $[Leu^{31}, Pro^{34}]PYY$ have very high to high affinity for the Y_4 receptor subtype, their radiolabelled counterparts do not label all sites recognized by $[^{125}I]hPP$ (Dumont *et al.* 1996, 1998*b*, 2000*c*; Dumont and Quirion 2000; Gehlert *et al.* 1997; Schober *et al.* 2000). These data may indicate that labelling seen with $[^{125}I]hPP$, especially in the paraventricular hypothalamic nucleus and the interpenduncular nuclei may represent a yet unknown NPY-like receptor. In this respect it is interesting that recently it was demonstrated that in Y_4-transfected cells, $[^{125}I]hPP$ recognized two binding while $[^{125}I]PYY$ bound only to the high affinity site/state (Berglund *et al.* 2001). These data could also explain why different apparent affinities were obtained for NPY and PYY when using $[^{125}I]PP$ or $[^{125}I]PYY$ as radioligands (Bard *et al.* 1995; Gehlert *et al.* 1996*c*; Gregor *et al.* 1996*b*; Lundell *et al.* 1995, 1996; Yan *et al.* 1996). Accordingly, depending upon the ratio between high and low affinity states and the affinity of the radioligand for each component, different radioligands may reveal either high or low levels of binding sites. Further studies using highly selective agonist and antagonist radioligands as well as molecular approaches will be required to resolve this issue.

One of the key features of the Y_5 receptor subtype is the high affinity of this protein for NPY, PYY, long C-terminal fragments of NPY and PYY (such as NPY_{2-36}, NPY_{3-36} and PYY_{3-36}), the proline substituted analogues in position 34 of NPY and PYY ([Leu^{31}, Pro^{34}]NPY, [Leu^{31}, Pro^{34}]PYY and [Pro^{34}]NPY) and human PP, while shorter C-terminal fragments (NPY$_{13-36}$ and PYY$_{13-36}$) and rat PP are much less potent (Gerald *et al.* 1996; Hu *et al.* 1996; Michel *et al.* 1998). Most recently, highly potent and rather selective Y$_5$ agonists have been characterized. At least four of these peptides, [Ala^{31}, Aib^{32}]NPY, hPP$_{1-17}$[Ala^{31}, Aib^{32}]NPY and [cPP$_{1-7}$, NPY$_{19-29}$, Ala^{31}, Aib^{32}, Gln^{34}]hPP (Cabrele *et al.* 2000) as well as [K^4, NPY$_{19-23}$]hPP$_{2-36}$ (McCrea *et al.* 2000) demonstrate affinities in the nM range (0.3–2 nM) for the human Y$_5$ receptor transfected in HEK293 cells. Their respective affinities for the Y$_1$, Y$_2$ and Y$_4$ receptor subtypes are markedly lower. Hence, these selective Y$_5$ agonists even if of peptidergic nature, should prove useful to study further the role of this receptor on the effects induced by NPY and homologues. To date, no specific agonists have been produced which target the Y$_6$ receptor.

19.4.2 **Antagonists**

One of the first non-peptide Y$_1$ antagonists to be reported was R-N^2-(Diphenylacetyl)-N-(4-hydroxyphenyl)-methyl argininamide, known as BIBP3226 (Rudolf *et al.* 1994). This antagonist behaves as a competitive, selective and specific Y$_1$ receptor antagonist in various binding assays and *in vitro* and *in vivo* bioassays (e.g. see, Doods *et al.* 1995, 1996; Jacques *et al.* 1995; Rudolf *et al.* 1994; Wieland *et al.* 1995) without any significant activity for the cloned Y$_2$, Y$_4$, and Y$_5$ (Gehlert *et al.* 1996*a*,*c*; Gerald *et al.* 1996) receptors.

However, most recently it has been reported that BIBP3226 was able to compete for specific NPFF receptors with affinities of 100 nM or less (Bonini *et al.* 2000; Mollereau *et al.* 2001) acting as an antagonist on the NPFF2 receptor subtype (Mollereau *et al.* 2001). Hence, the full specificity of BIBP3226 for the Y$_1$ versus NPFF receptors needs to be considered especially in *in vivo* studies. Another highly potent and selective non-peptide receptor antagonist, ((R)-N-[[4-(aminocarbonylaminomethyl)-phenyl]methyl]-N2-(diphenylacetyl)-argininamide tri-fluoroacetate), known as BIBO3304, possesses low nanomolar affinity for both human and rat Y$_1$ receptors, a 10-fold greater affinity than BIBP3226 and is inactive at Y$_2$, Y$_4$, and Y$_5$ receptors (Dumont *et al.* 2000*a*; Wieland *et al.* 1998*b*). A number of other non-peptide antagonists for the Y$_1$ receptor have also been reported including SR120819A (Serradeil-Le Gal *et al.* 1995), 1-(1-[3-((3s)(3-piperidyl))-propyl]-2-[(4-chlorophenoxyl)-methyl]indol-3-yl]-2-(4-piperidylpiperidyl)ethan-1-one, code name LY357897 (Hipskind *et al.* 1997), 2-[(4-chlorophenoxy)methyl]benzimidazoles (Zarrinmayeh *et al.* 1998), diaminoalkyl substituted benzimidazole (Zarrinmayeh *et al.* 1999), 1-substituted-4- methylbenzimidazole (Zimmerman *et al.* 1998), GI264879A (Daniels *et al.* 2001), imidazolylpropylguanidine (Dove *et al.* 2000), J-104870 (Kanatani *et al.* 1999), J-115814 (Kanatani *et al.* 2001), H409/22 (Malmstrom *et al.* 2000), and benzazepines derivatives (Murakami *et al.* 1999; Shigeri *et al.* 1998). However, the selectivity of these compounds for NPY and NPFF receptors has not been fully demonstrated and they are usually less potent than BIBO3304. Finally, a Y$_1$ peptide antagonist homodimeric Ile-Glu-Pro-Dpr-Tyr-Arg-Leu-Arg-Tyr-CONH$_2$, known as either 1229U91, GW1229, or GR231118 (Bitran *et al.* 1997; Daniels *et al.* 1995; Parker *et al.* 1998) has been rather extensively studied. However, this Y$_1$ receptor antagonist also possesses potent agonist effects on the Y$_4$ (Kanatani *et al.* 1998; Parker *et al.* 1998; Schober *et al.* 1998) and Y$_6$ (Parker *et al.* 1998) receptors and is a weak agonist at the Y$_2$, Y$_5$ (Kanatani *et al.* 1998; Parker *et al.* 1998) and NPFF2 receptors (Mollereau *et al.* 2001).

Recently, Doods *et al.* (1999) have reported that BIIE246 behaves as a potent and selective Y_2 receptor antagonist while no agonist or antagonist activities were observed at the Y_1, Y_4, and Y_5 receptors. These results have been confirmed using various *in vitro* bioassays and receptor binding assays (Dumont *et al.* 2000*b*; King *et al.* 2000). Thus, BIIE246 represents the first potent antagonist to investigate further the role of Y_2 receptor subtype in the organism.

Y_4 receptor antagonists have yet to be developed. However, a few non-peptide Y_5 receptor antagonists with affinity in the nM range have recently been reported including CGP71683A (Criscione *et al.* 1998), L-152804 (Kanatani *et al.* 2000*a*), Pyrrolo[3,2-d] pyrimidine derivatives (Norman *et al.* 2000), 2-substituted 4-amino-quinazolin derivatives (Rueeger *et al.* 2000) and alpha-substituted N-(sulfonamino)alkyl-beta aminotetralins (McNally *et al.* 2000*a,b*; Youngman *et al.* 2000). These molecules have yet to be fully characterized in various *in vitro* and *in vivo* assays. For example, while CGP71683A (Dumont *et al.* 2000*a*) and L-152804 (Kanatani *et al.* 2000*a*) demonstrated selectivity for Y_5 versus Y_1, Y_2, and Y_4, further studies revealed that CGP71683A possesses high affinity for muscarinic, adrenergic and serotonergic uptake receptors (Zuana *et al.* 2001). Hence, further studies are required to establish the full selectivity of these molecules for the Y_5 receptor not only versus other NPY receptors but also for other GPCRs. No antagonists for the Y_6 receptor exist.

19.5 Signal transduction and receptor modulation

While generally NPY receptors are believed to couple to $G_{i/o}$ and inhibition of adenylate cyclase it is clear that promiscuous coupling to other signal transduction pathways can occur. Thus, in HEK293 cells transfected with the human Y_1 receptor cDNA, the receptor is coupled to a pertussis toxin-sensitive G protein that mediates the inhibition of cyclic AMP accumulation, while in CHO cells the Y_1 receptor is coupled to elevation of intracellular Ca^{++} (Ball *et al.* 1995; Herzog *et al.* 1992). Likewise, transfection of different cell types with human Y_2 receptor cDNA, can generate a receptor that could coupled to a pertussis toxin-sensitive G protein that mediates the inhibition of cAMP accumulation and the elevation of intracellular Ca^{++} (Rose *et al.* 1995). A similar situation arises using transfection of the human or rat Y_5 receptor cDNA (Gerald *et al.* 1996; Hu *et al.* 1996).

19.6 Physiology and disease relevance

Numerous studies have addressed the physiological functions of NPY and its congeners in the central and peripheral nervous systems (Table 19.1; for recent reviews see for example, Dumont *et al.* 2000*c*; Inui 1999; Vezzani *et al.* 1999). For example, both NPY and PYY have been shown to produce decreases in sexual behaviour (Clark *et al.* 1985) and shifts in circadian rhythms (Albers *et al.* 1984; Albers and Ferris 1984). Below are more extensive details of some of the most striking physiological functions of NPY receptors.

19.6.1 Feeding and satiety

Intracerebroventricular (icv) injections of NPY or PYY consistently demonstrate that these peptides are among the most potent substances known thus far to stimulate feeding behaviours (Inui 1999; Stanley and Leibowitz 1984). Thus, reducing the synthesis of NPY by icv administration of NPY antisense oligonucleotides causes a marked reduction in food intake (Akabayashi *et al.* 1994). Most evidence points towards an involvement of the Y_1 receptor in feeding behaviours given the effectiveness of a number of Y_1 receptor antagonists such as

BIBO3304 (Wieland *et al.* 1998*b*), J104870 (Kanatani *et al.* 1999), J115814 (Kanatani *et al.* 2001), BIBP3226 (Kask *et al.* 1998), GR231118 (Ishihara *et al.* 1998; Kanatani *et al.* 1996, 1998) or GI264879A (Daniels *et al.* 2001) to inhibit food intake. On the other hand, mice lacking Y_1 receptors do not display major abnormalities in feeding behaviour (Pedrazzini *et al.* 1998). However, in these animals, food intake induced by NPY is profoundly altered (Kanatani *et al.* 2000*b*). Altogether, these data suggest that the Y_1 subtype plays an important role in feeding behaviours. However, it has also been reported that Y_2 receptors may be involved in feeding in that Y_2 knockout mice exhibit an increase in body weight and food intake (Naveilhan *et al.* 1999) suggesting that the Y_2 receptor subtype may negatively regulate feeding behaviours. Interestingly, BIIE0246, a Y_2 receptor antagonist, inhibits NPY release in the hypothalamus confirming its role as autoreceptor (King *et al.* 2000). The precise physiological role of the Y_5 receptor subtypes is still debated despite its initial labelling as the receptor involved in NPY-induced feeding (Gerald *et al.* 1996). In this respect, most of the recent studies have failed to provide clear evidence for such a role. A notion exemplified by the Y_5 receptor knockout mice which exhibit completely normal feeding behaviour (Marsh *et al.* 1998) and respond in a similar manner to wild types in NPY-induced food intake studies (Kanatani *et al.* 2000*b*). Furthermore, the effect of NPY on food intake in wild-type animals was unaltered by L-152804, a Y_5 antagonist while that of bovine PP was reduced (Kanatani *et al.* 2000*a*). Interestingly however, rather selective Y_5 agonists (see Section 19.4), while being mostly inactive on the Y_1, Y_2 and Y_4 subtypes were able to stimulate food intake (Cabrele *et al.* 2000, 2001; McCrea *et al.* 2000). Hence, more than one NPY receptor subtypes may regulate NPY-induced food intake.

19.6.2 Drug dependence

Interestingly, NPY can also modulate alcohol consumption. Thus, NPY knockout mice exhibit increased alcohol consumption (Thiele *et al.* 1998). whereas transgenic mice overexpressing NPY were shown to be low alcohol drinkers (Thiele *et al.* 1998).

19.6.3 Alzheimer's disease and cognition

NPY and its congeners are reported to affect cognition although in a complex way. Thus, NPY and PYY can facilitate learning and memory processes (Flood *et al.* 1987; Redrobe *et al.* 1999) whereas, paradoxically, transgenic NPY-overexpressing rats display memory impairments (Thorsell *et al.* 2000).

19.6.4 Pain and analgesia

Y_1-knockout mice develop hyperalgesia to acute thermal, cutaneous, and visceral chemical pain, and exhibit mechanical hypersensitivity (Naveilhan *et al.* 2001), suggesting a role of the Y_1 receptor subtype in nociception.

19.6.5 Anxiety

It has also been reported that NPY and PYY induce anxiolytic-like effects (Heilig *et al.* 1993); an effect overtly evident in transgenic NPY-overexpressing rats (Thorsell *et al.* 2000). Icv injection of an NPY Y_1 receptor antisense oligonucleotide (Wahlestedt *et al.* 1993) and NPY gene knockout strategies produce anxiogenic-like behaviours (Bannon *et al.* 2000) and currently it is the Y_1 receptor subtype that is the primary focus for the anxiolytic-like effects of NPY (for review see Kask *et al.* 2001).

19.6.6 Epilepsy

NPY inhibits glutamatergic excitatory synaptic transmission and inhibits epileptiform activity in hippocampal slices (Colmers and Bleakman 1994; Vezzani *et al.* 1999). Furthermore, NPY-knockout mice exhibit mild spontaneous seizures (Erickson *et al.* 1996). The general consensus of opinion is that the Y_2 receptor subtype is involved in the inhibition of glutamate release (Colmers *et al.* 1987; Colmers and Bleakman 1994; Kombian and Colmers 1992) leading to the anti-epileptic effects of NPY (Klapstein and Colmers 1997; Vezzani *et al.* 1999).

19.6.7 Neuroendocrine

Both NPY and PYY modulate various neuroendocrine secretions (Kalra and Crowley 1992). Thus, reducing the synthesis of NPY by icv administration of NPY antisense oligonucleotides results in an attenuation of progesterone-induced LH surge (Kalra *et al.* 1995). It is possible that these effects may, at least in part, reflect Y_4 receptor activity since two recent reports have suggested its possible involvement in LH secretion (Jain *et al.* 1999; Raposinho *et al.* 2000).

19.6.8 Thermoregulation

Both NPY and PYY produce hypothermia (Esteban *et al.* 1989). Interestingly, this effect can also be produced through icv injection of NPY Y1 antisense oligonucleotides (Lopez-Valpuesta *et al.* 1996).

19.7 Concluding remarks

Major species differences exist in the expression and distribution of a given NPY receptor in the CNS. Hence data obtained in rodents cannot be fully extrapolated to human. Furthemore, recently develop selective NPY receptor agonists and antagonists should facilitate the identification of the respective role(s) of each NPY receptor in the broad spectrum of effects induced by NPY and its congeners.

References

Akabayashi A, Wahlestedt C, Alexander JT, and Leibowitz SF (1994). Specific inhibition of endogenous neuropeptide Y synthesis in arcuate nucleus by antisense oligonucleotides suppresses feeding behavior and insulin secretion. *Brain Res Mol Brain Res* **21**, 55–61.

Albers HE and Ferris CF (1984). Neuropeptide Y: role in light-dark cycle entrainment of hamster circadian rhythms. *Neurosci Lett* **50**, 163–8.

Albers HE, Ferris CF, Leeman SE, and Goldman BD (1984). Avian pancreatic polypeptide phase shifts hamster circadian rhythms when microinjected into the suprachiasmatic region. *Science* **223**, 833–5.

Allen J, Novotny J, Martin J, and Heinrich G (1987). Molecular structure of mammalian neuropeptide Y: analysis by molecular cloning and computer-aided comparison with crystal structure of avian homologue. *Proc Natl Acad Sci USA* **84**, 2532–6.

Ammar DA, Eadie DM, Wong DJ, Ma YY, Kolakowski LF, Jr., Yang-Feng TL, and Thompson DA (1996). Characterization of the human type 2 neuropeptide Y receptor gene (NPY2R) and localization to the chromosome 4q region containing the type 1 neuropeptide Y receptor gene. *Genomics* **38**, 392–8.

Baker E, Hort YJ, Ball H, Sutherland GR, Shine J, and Herzog H (1995). Assignment of the human neuropeptide Y gene to chromosome 7p15.1 by nonisotopic in situ hybridization. *Genomics* **26**, 163–4.

Ball HJ, Shine J, and Herzog H (1995). Multiple promoters regulate tissue-specific expression of the human NPY-Y1 receptor gene. *J Biol Chem* **270**, 27 272–6.

Bannon AW, Seda J, Carmouche M, Francis JM, Norman MH, Karbon B, and McCaleb ML (2000). Behavioral characterization of neuropeptide Y knockout mice. *Brain Res* **868**, 79–87.

Bard JA, Walker MW, Branchek TA, and Weinshank RL (1995). Cloning and functional expression of a human Y4 subtype receptor for pancreatic polypeptide, neuropeptide Y, and peptide YY. *J Biol Chem* **270**, 26 762–5.

Beck-Sickinger AG (1997). The importance of various parts of the NPY molecule for receptor recognition. In (eds., L Grundemar and SR Bloom) *Neuropeptide Y and Drug Development*, Academic Press, London, UK, pp. 107–26.

Beck-Sickinger AG, Grouzmann E, Hoffmann E, Gaida W, van Meir EG, Waeber B, and Jung G (1992). A novel cyclic analog of neuropeptide Y specific for the Y2 receptor. *Eur J Biochem* **206**, 957–64.

Beck-Sickinger AG, Koppen H, Hoffmann E, Gaida W, and Jung G (1993). Cyclopeptide analogs for characterization of the neuropeptide Y Y2-receptor. *J Recept Res* **13**, 215–28.

Beck-Sickinger AG, Wieland HA, Wittneben H, Willim KD, Rudolf K, and Jung G (1994). Complete L-alanine scan of neuropeptide Y reveals ligands binding to Y1 and Y2 receptors with distinguished conformations. *Eur J Biochem* **225**, 947–58.

Berglund MM, Holmberg SK, Eriksson H, Gedda K, Maffrand JP, Serradeil-Le Gal C *et al.* (1999). The cloned guinea pig neuropeptide Y receptor Y1 conforms to other mammalian Y1 receptors. *Peptides* **20**, 1043–53.

Berglund MM, Lundell I, Eriksson H, Soll R, Beck-Sickinger AG, and Larhammar D (2001). Studies of the human, rat, and guinea pig Y4 receptors using neuropeptide Y analogues and two distinct radioligands. *Peptides* **22**, 351–6.

Bitran M, Daniels AJ, and Boric MP (1997). GW1229, a novel neuropeptide Y Y1 receptor antagonist, inhibits the vasoconstrictor effect on neuropeptide Y in the hamster microcirculation. *Eur J Pharmacol* **319**, 43–7.

Blundell TL, Pitts JE, Tickle IJ, and Wood SP (1981). The conformation and receptor binding of pancreatic hormones. *Biochem Soc Trans* **9**, 31–2.

Bonini JA, Jones KA, Adham N, Forray C, Artymyshyn R, Durkin MM *et al.* (2000). Identification and characterization of two G protein-coupled receptors for neuropeptide FF. *J Biol Chem* **275**, 39 324–31.

Borowsky B, Walker MW, Bard J, Weinshank RL, Laz TM, Vaysse P *et al.* (1998). Molecular biology and pharmacology of multiple NPY Y5 receptor species homologs. *Regul Pept* **75–6**, 45–53.

Bouali SM, Fournier A, St Pierre S, and Jolicoeur FB (1994). *In vivo* central actions of NPY(1–30), an N-terminal fragment of neuropeptide Y. *Peptides* **15**, 799–802.

Boublik J, Scott N, Taulane J, Goodman M, Brown M, and Rivier J (1989*a*). Neuropeptide Y and neuropeptide Y18–36. Structural and biological characterization. *Int J Pept Protein Res* **33**, 11–15.

Boublik JH, Scott NA, Brown MR, and Rivier JE (1989*b*). Synthesis and hypertensive activity of neuropeptide Y fragments and analogues with modified N- or C-termini or D-substitutions. *J Med Chem* **32**, 597–601.

Burkhoff A, Linemeyer DL, and Salon JA (1998). Distribution of a novel hypothalamic neuropeptide Y receptor gene and it's absence in rat. *Brain Res Mol Brain Res* **53**, 311–16.

Cabrele C, Langer M, Bader R, Wieland HA, Doods HN, Zerbe O, and Beck-Sickinger AG (2000). The first selective agonist for the neuropeptide YY_5 receptor increases food intake in rats. *J Biol Chem* **275**, 36 043–8.

Cabrele C, Wieland HA, Langer M, Stidsen CE, and Beck-Sickinger AG (2001). Y-receptor affinity modulation by the design of pancreatic polypeptide/neuropeptide Y chimera led to Y5-receptor ligands with picomolar affinity. *Peptides* 22, 365–78.

Cerda-Reverter JM and Larhammar D (2000). Neuropeptide Y family of peptides: structure, anatomical expression, function, and molecular evolution. *Biochem Cell Biol* 78, 371–92.

Chan-Palay V, Allen YS, Lang W, Haesler U, and Polak JM (1985). Cytology and distribution in normal human cerebral cortex of neurons immunoreactive with antisera against neuropeptide Y. *J Comp Neurol* 238, 382–9.

Chronwall BM, DiMaggio DA, Massari VJ, Pickel VM, Ruggiero DA, and O'Donohue TL (1985). The anatomy of neuropeptide-Y-containing neurons in rat brain. *Neuroscience* 15, 1159–81.

Clark JT, Kalra PS, and Kalra SP (1985). Neuropeptide Y stimulates feeding but inhibits sexual behavior in rats. *Endocrinology* 117, 2435–42.

Colmers WF and Bleakman D (1994). Effects of neuropeptide Y on the electrical properties of neurons. *Trends Neurosci* 17, 373–9.

Colmers WF, Klapstein GJ, Fournier A, St Pierre S, and Treherne KA (1991). Presynaptic inhibition by neuropeptide Y in rat hippocampal slice in vitro is mediated by a Y2 receptor. *Br J Pharmacol* 102, 41–4.

Colmers WF, Lukowiak K, and Pittman QJ (1987). Presynaptic action of neuropeptide Y in area CA1 of the rat hippocampal slice. *J Physiol (Lond)* 383, 285–99.

Cox HM and Cuthbert AW (1990). The effects of neuropeptide Y and its fragments upon basal and electrically stimulated ion secretion in rat jejunum mucosa. *Br J Pharmacol* 101, 247–52.

Cox HM, Tough IR, Ingenhoven N, and Beck-Sickinger AG (1998). Structure-activity relationships with neuropeptide Y analogues: a comparison of human Y1-, Y2- and rat Y2-like systems. *Regul Pept* 75–6, 3–8.

Criscione L, Rigollier P, Batzl-Hartmann C, Rueger H, Stricker-Krongrad A, Wyss P *et al.* (1998). Food intake in free-feeding and energy-deprived lean rats is mediated by the neuropeptide Y5 receptor. *J Clin Invest* 102, 2136–45.

Daniels AJ, Chance WT, Grizzle MK, Heyer D, and Matthews JE (2001). Food intake inhibition and reduction in body weight gain in rats treated with GI264879A, a non-selective NPY-Y1 receptor antagonist. *Peptides* 22, 483–91.

Daniels AJ, Matthews JE, Slepetis RJ, Jansen M, Viveros OH, Tadepalli A *et al.* (1995). High-affinity neuropeptide Y receptor antagonists. *Proc Natl Acad Sci USA* 92, 9067–71.

Doods H, Gaida W, Wieland HA, Dollinger H, Schnorrenberg G, Esser F *et al.* (1999). BIIE0246: a selective and high affinity neuropeptide Y Y2 receptor antagonist. *Eur J Pharmacol* 384, R3–R5.

Doods HN, Wieland HA, Engel W, Eberlein W, Willim KD, Entzeroth M *et al.* (1996). BIBP 3226, the first selective neuropeptide Y1 receptor antagonist: a review of its pharmacological properties. *Regul Pept* 65, 71–7.

Doods HN, Wienen W, Entzeroth M, Rudolf K, Eberlein W, Engel W, and Wieland HA (1995). Pharmacological characterization of the selective nonpeptide neuropeptide Y Y1 receptor antagonist BIBP 3226. *J Pharmacol Exp Ther* 275, 136–42.

Dove S, Michel MC, Knieps S, and Buschauer A (2000). Pharmacology and quantitative structure-activity relationships of imidazolylpropylguanidines with mepyramine-like substructures as non-peptide neuropeptide Y Y1 receptor antagonists. *Can J Physiol Pharmacol* 78, 108–15.

Dumont Y, Cadieux A, Doods H, Fournier A, and Quirion R (2000a). Potent and selective tools to investigate neuropeptide Y receptors in the central and peripheral nervous systems: BIB03304 (Y1) and CGP71683A (Y5). *Can J Physiol Pharmacol* 78, 116–25.

Dumont Y, Cadieux A, Doods H, Pheng LH, Abounader R, Hamel E *et al.* (2000b). BIIE0246, a potent and highly selective non-peptide neuropeptide Y Y2 receptor antagonist. *Br J Pharmacol* 129, 1075–88.

Dumont Y, Cadieux A, Pheng LH, Fournier A, St Pierre S, and Quirion R (1994). Peptide YY derivatives as selective neuropeptide Y/peptide YY Y1 and Y2 agonists devoided of activity for the Y3 receptor sub-type. *Brain Res Mol Brain Res* **26**, 320–4.

Dumont Y, Fournier A, and Quirion R (1998*a*). Expression and characterization of the neuropeptide Y Y5 receptor subtype in the rat brain. *J Neurosci* **18**, 5565–74.

Dumont Y, Fournier A, St Pierre S, and Quirion R (1993). Comparative characterization and auto-radiographic distribution of neuropeptide Y receptor subtypes in the rat brain. *J Neurosci* **13**, 73–86.

Dumont Y, Fournier A, St Pierre S, and Quirion R (1995). Characterization of neuropeptide Y binding sites in rat brain membrane preparations using [125I][Leu31, Pro34]peptide YY and [125I]peptide YY3-36 as selective Y1 and Y2 radioligands. *J Pharmacol Exp Ther* **272**, 673–80.

Dumont Y, Fournier A, St Pierre S, and Quirion R (1996). Autoradiographic distribution of[125I][Leu31, Pro34]PYY and [125I]PYY3-36 binding sites in the rat brain evaluated with two newly developed Y1 and Y2 receptor radioligands. *Synapse* **22**, 139–58.

Dumont Y, Fournier A, St Pierre S, Schwartz TW, and Quirion R (1990). Differential distribution of neuropeptide Y1 and Y2 receptors in the rat brain. *Eur J Pharmacol* **191**, 501–3.

Dumont Y, Jacques D, Bouchard P, and Quirion R (1998*b*). Species differences in the expression and distribution of the neuropeptide Y Y1, Y2, Y4 and Y5 receptors in rodents, guinea pig and primates brains. *J Comp Neurol* **402**, 372–84.

Dumont Y, Jacques D, St Pierre JA, and Quirion, R (1997). Neuropeptide Y receptor types in the mammalian brain: Species differences and status in the human central nervous system. In (eds. L Grundemar and SR Bloom) *Neuropeptide Y and Drug Development*, Academic Press, London, UK, pp. 57–86.

Dumont Y, Jacques D, St Pierre JA, Tong Y, Parker R, Herzog H, and Quirion R (2000*c*). Neuropeptide Y, peptide YY and pancreatic polypeptide receptor proteins and mRNAs in mammalian brains. In (eds., R Quirion, A Bjorklund and T Hokfelt) *Handbook of Chemical Neuroanatomy, Vol 16 Peptide Receptor, Part 1*, Elsevier, London, UK, pp. 375–475.

Dumont Y and Quirion R (2000). [125I]-GR231118: a high affinity radioligand to investigate neuropeptide Y Y1 and Y4 receptors. *Br J Pharmacol* **129**, 37–46.

Dumont Y, St-Pierre JA, Thakur M, Pheng LH, Langlois D, Slon-Usakiewicz J *et al.* (2001). Charac-terization of BODIPY-TMR-conjugated fluo-neuropeptide Y analogs in Y1, Y2, Y4 and Y5 receptor binding assays. *Soc Neurosci Abst* **30**, Program No 30.15.

Eckard CP, Beck-Sickinger AG, and Wieland HA (1999). Comparison of antibodies directed against receptor segments of NPY-receptors. *J Recept Signal Transduct Res* **19**, 379–94.

Ekman R, Wahlestedt C, Bottcher G, Sundler F, Hakanson R, and Panula P (1986). Peptide YY-like immunoreactivity in the central nervous system of the rat. *Regul Pept* **16**, 157–68.

Elshourbagy NA, Ames RS, Fitzgerald LR, Foley JJ, Chambers JK, and Szekeres PG (2000). Receptor for the pain modulatory neuropeptides FF and AF is an orphan G protein-coupled receptor. *J Biol Chem* **275**, 25 965–71.

Erickson JC, Clegg KE, and Palmiter RD (1996). Sensitivity to leptin and susceptibility to seizures of mice lacking neuropeptide Y. *Nature* **381**, 415–21.

Eriksson H, Berglund MM, Holmberg SK, Kahl U, Gehlert DR, and Larhammar D (1998). The cloned guinea pig pancreatic polypeptide receptor Y4 resembles more the human Y4 than does the rat Y4. *Regul Pept* **75–6**, 29–37.

Esteban J, Chover AJ, Sanchez PA, Mico JA, and Gibert-Rahola J (1989). Central administration of neuropeptide Y induces hypothermia in mice. Possible interaction with central noradrenergic systems. *Life Sci* **45**, 2395–400.

Eva C, Keinanen K, Monyer H, Seeburg P, and Sprengel R (1990). Molecular cloning of a novel G protein-coupled receptor that may belong to the neuropeptide receptor family. *FEBS Lett* **271**, 81–4.

Eva C, Oberto A, Sprengel R, and Genazzani E (1992). The murine NPY-1 receptor gene. Structure and delineation of tissue-specific expression. *FEBS Lett* **314**, 285–8.

Feinstein RD, Boublik JH, Kirby D, Spicer MA, Craig AG, Malewicz K *et al.* (1992). Structural requirements for neuropeptide Y18–36-evoked hypotension: a systematic study. *J Med Chem* **35**, 2836–43.

Feletou M, Rodriguez M, Beauverger P, Germain M, Imbert J, Dromaint S *et al.* (1998). NPY receptor subtypes involved in the contraction of the proximal colon of the rat. *Regul Pept* **75–6**, 221–9.

Feng Y, Broder CC, Kennedy PE, and Berger EA (1996). HIV-1 entry cofactor: functional cDNA cloning of a seven-transmembrane, G protein-coupled receptor. *Science* **272**, 872–7.

Flood JF, Hernandez EN, and Morley JE (1987). Modulation of memory processing by neuropeptide Y. *Brain Res* **421**, 280–90.

Fuhlendorff J, Gether U, Aakerlund L, Langeland-Johansen N, Thogersen H, Melberg SG *et al.* (1990*a*). [Leu31, Pro34]neuropeptide Y: a specific Y1 receptor agonist. *Proc Natl Acad Sci USA* **87**, 182–6.

Fuhlendorff J, Johansen NL, Melberg SG, Thogersen H, and Schwartz TW (1990*b*). The antiparallel pancreatic polypeptide fold in the binding of neuropeptide Y to Y1 and Y2 receptors. *J Biol Chem* **265**, 11 706–12.

Gaudreau P, Langlois D, Dumont Y, Fournier A, and Quirion R (1999). Development of fluorescent analogs of neuropeptide Y. *Conf Proc 5th Int NPY Meet, Caymans Island*, April 17–22, 1999.

Gehlert DR, Beavers LS, Johnson D, Gackenheimer SL, Schober DA, and Gadski RA (1996*a*). Expression cloning of a human brain neuropeptide Y Y2 receptor. *Mol Pharmacol* **49**, 224–8.

Gehlert DR and Gackenheimer SL (1996). Unexpected high density of neuropeptide Y Y1 receptors in the guinea pig spleen. *Peptides* **17**, 1345–8.

Gehlert DR, Gackenheimer SL, and Schober DA (1992). [Leu31-Pro34] neuropeptide Y identifies a subtype of 125I-labeled peptide YY binding sites in the rat brain. *Neurochem Int* **21**, 45–67.

Gehlert DR, Gackenheimer SL, Schober DA, Beavers L, Gadski R, Burnett JP *et al.* (1996*b*). The neuropeptide Y Y1 receptor selective radioligand, [125I][Leu31, Pro34]peptide YY, is also a high affinity radioligand for human pancreatic polypeptide 1 receptors. *Eur J Pharmacol* **318**, 485–90.

Gehlert DR, Schober DA, Beavers L, Gadski R, Hoffman JA, Smiley DL *et al.* (1996*c*). Characterization of the peptide binding requirements for the cloned human pancreatic polypeptide-preferring receptor. *Mol Pharmacol* **50**, 112–18.

Gehlert DR, Schober DA, Gackenheimer SL, Beavers L, Gadski R, Lundell I, and Larhammar D (1997). [125I]Leu31, Pro34-PYY is a high affinity radioligand for rat PP1/Y4 and Y1 receptors: evidence for heterogeneity in pancreatic polypeptide receptors. *Peptides* **18**, 397–401.

Gehlert DR, Yang P, George C, Wang Y, Schober D, Gackenheimer S *et al.* (2001). Cloning and characterization of Rhesus monkey neuropeptide Y receptor subtypes. *Peptides* **22**, 343–50.

Gerald C, Walker MW, Criscione L, Gustafson EL, Batzl-Hartmann C, Smith KE *et al.* (1996). A receptor subtype involved in neuropeptide-Y-induced food intake. *Nature* **382**, 168–71.

Gerald C, Walker MW, Vaysse PJ, He C, Branchek TA, and Weinshank RL (1995). Expression cloning and pharmacological characterization of a human hippocampal neuropeptide Y/peptide YY Y2 receptor subtype. *J Biol Chem* **270**, 26 758–61.

Glover I, Haneef I, Pitts J, Wood S, Moss D, Tickle I, and Blundell T (1983). Conformational flexibility in a small globular hormone: x-ray analysis of avian pancreatic polypeptide at 0.98-A resolution. *Biopolymers* **22**, 293–304.

Goumain M, Voisin T, Lorinet AM, Balasubramaniam A, and Laburthe M (1998*a*). Pharmacological profile of the rat intestinal crypt peptide YY receptor vs. the recombinant rat Y5 receptor. *Eur J Pharmacol* **362**, 245–9.

Goumain M, Voisin T, Lorinet AM, and Laburthe M (1998*b*). Identification and distribution of mRNA encoding the Y1, Y2, Y4, and Y5 receptors for peptides of the PP-fold family in the rat intestine and colon. *Biochem Biophys Res Commun* **247**, 52–6.

Grandt D, Teyssen S, Schimiczek M, Reeve JR, Jr., Feth F, Rascher W *et al.* (1992). Novel generation of hormone receptor specificity by amino terminal processing of peptide YY. *Biochem Biophys Res Commun* **186**, 1299–306.

Gregor P, Feng Y, DeCarr LB, Cornfield LJ, and McCaleb ML (1996*a*). Molecular characterization of a second mouse pancreatic polypeptide receptor and its inactivated human homologue. *J Biol Chem* **271**, 27 776–81.

Gregor P, Millham ML, Feng Y, DeCarr LB, McCaleb ML, and Cornfield LJ (1996*b*). Cloning and characterization of a novel receptor to pancreatic polypeptide, a member of the neuropeptide Y receptor family. *FEBS Lett* **381**, 58–62.

Grove KL, Campbell RE, Ffrench-Mullen JM, Cowley MA, and Smith MS (2000). Neuropeptide Y Y5 receptor protein in the cortical/limbic system and brainstem of the rat: expression on gamma-aminobutyric acid and corticotropin-releasing hormone neurons. *Neuroscience* **100**, 731–40.

Grundemar L, Wahlestedt C, and Reis DJ (1991). Neuropeptide Y acts at an atypical receptor to evoke cardiovascular depression and to inhibit glutamate responsiveness in the brainstem. *J Pharmacol Exp Ther* **258**, 633–8.

Gustafson EL, Smith KE, Durkin MM, Walker MW, Gerald C, Weinshank R, and Branchek TA (1997). Distribution of the neuropeptide Y Y2 receptor mRNA in rat central nervous system. *Brain Res Mol Brain Res* **46**, 223–35.

Heilig M, McLeod S, Brot M, Heinrichs SC, Menzaghi F, Koob GF, and Britton KT (1993). Anxiolytic-like action of neuropeptide Y: mediation by Y1 receptors in amygdala, and dissociation from food intake effects. *Neuropsychopharmacology* **8**, 357–63.

Herzog H, Baumgartner M, Vivero C, Selbie LA, Auer B, and Shine J (1993*a*). Genomic organization, localization, and allelic differences in the gene for the human neuropeptide Y Y1 receptor. *J Biol Chem* **268**, 6703–7.

Herzog H, Darby K, Ball H, Hort Y, Beck-Sickinger A, and Shine J (1997). Overlapping gene structure of the human neuropeptide Y receptor subtypes Y1 and Y5 suggests coordinate transcriptional regulation. *Genomics* **41**, 315–19.

Herzog H, Hort YJ, Ball HJ, Hayes G, Shine J, and Selbie LA (1992). Cloned human neuropeptide Y receptor couples to two different second messenger systems. *Proc Natl Acad Sci USA* **89**, 5794–8.

Herzog H, Hort YJ, Shine J, and Selbie LA (1993*b*). Molecular cloning, characterization, and localization of the human homolog to the reported bovine NPY Y3 receptor: lack of NPY binding and activation. *DNA Cell Biol* **12**, 465–71.

Hipskind PA, Lobb KL, Nixon JA, Britton TC, Bruns RF, Catlow J *et al.* (1997). Potent and selective 1,2,3-trisubstituted indole NPY Y-1 antagonists. *J Med Chem* **40**, 3712–14.

Hirabayashi A, Nishiwaki K, Shimada Y, and Ishikawa N (1996). Role of neuropeptide Y and its receptor subtypes in neurogenic pulmonary edema. *Eur J Pharmacol* **296**, 297–305.

Holliday ND, Pollock EL, Tough IR, and Cox HM (2000). PYY preference is a common characteristic of neuropeptide Y receptors expressed in human, rat, and mouse gastrointestinal epithelia. *Can J Physiol Pharmacol* **78**, 126–33.

Hort Y, Baker E, Sutherland GR, Shine J, and Herzog H (1995). Gene duplication of the human peptide YY gene (PYY) generated the pancreatic polypeptide gene (PPY) on chromosome 17q21.1. *Genomics* **26**, 77–83.

Hu Y, Bloomquist BT, Cornfield LJ, DeCarr LB, Flores-Riveros JR, Friedman L *et al.* (1996). Identification of a novel hypothalamic neuropeptide Y receptor associated with feeding behavior. *J Biol Chem* **271**, 26 315–19.

Ingenhoven N and Beck-Sickinger AG (1999). Molecular characterization of the ligand-receptor interaction of neuropeptide Y. *Curr Med Chem* **6**, 1055–66.

Inui A (1999). Neuropeptide Y feeding receptors: are multiple subtypes involved?. *Trends Pharmacol Sci* **20**, 43–6.

Ishihara A, Tanaka T, Kanatani A, Fukami T, Ihara M, and Fukuroda T (1998). A potent neuropeptide Y antagonist, 1229U91, suppressed spontaneous food intake in Zucker fatty rats. *Am J Physiol* 274, R1500–R1504.

Jacques D, Cadieux A, Dumont Y, and Quirion R (1995). Apparent affinity and potency of BIBP3226, a non-peptide neuropeptide Y receptor antagonist, on purported neuropeptide Y Y1, Y2 and Y3 receptors. *Eur J Pharmacol* 278, R3–R5.

Jacques D, Dumont Y, Fournier A, and Quirion R (1997). Characterization of neuropeptide Y receptor subtypes in the normal human brain, including the hypothalamus. *Neuroscience* 79, 129–48.

Jacques D, Tong Y, Shen SH, and Quirion R (1998). Discrete distribution of the neuropeptide Y Y5 receptor gene in the human brain: an in situ hybridization study. *Brain Res Mol Brain Res* 61, 100–7.

Jain MR, Pu S, Kalra PS, and Kalra SP (1999). Evidence that stimulation of two modalities of pituitary luteinizing hormone release in ovarian steroid-primed ovariectomized rats may involve neuropeptide Y Y1 and Y4 receptors. *Endocrinology* 140, 5171–7.

Jazin EE, Yoo H, Blomqvist AG, Yee F, Weng G, Walker MW *et al.* (1993). A proposed bovine neuropeptide Y (NPY) receptor cDNA clone, or its human homologue, confers neither NPY binding sites nor NPY responsiveness on transfected cells. *Regul Pept* 47, 247–58.

Ji RR, Zhang X, Wiesenfeld-Hallin Z, and Hokfelt T (1994). Expression of neuropeptide Y and neuropeptide Y Y1 receptor mRNA in rat spinal cord and dorsal root ganglia following peripheral tissue inflammation. *J Neurosci* 14, 6423–34.

Jorgensen JC, Fuhlendorff J, and Schwartz TW (1990). Structure-function studies on neuropeptide Y and pancreatic polypeptide—evidence for two PP-fold receptors in vas deferens. *Eur J Pharmacol* 186, 105–14.

Juaneda C, Dumont Y, and Quirion R (2000). The molecular pharmacology of CGRP and related peptide receptor subtypes. *Trends Pharmacol Sci* 21, 432–8.

Kalra PS, Bonavera JJ, and Kalra SP (1995). Central administration of antisense oligodeoxynucleotides to neuropeptide Y (NPY) mRNA reveals the critical role of newly synthesized NPY in regulation of LHRH release. *Regul Pept* 59, 215–20.

Kalra SP and Crowley WR (1992). Neuropeptide Y: a novel neuroendocrine peptide in the control of pituitary hormone secretion, and its relation to luteinizing hormone. *Front Neuroendocrinol* 13, 1–46.

Kanatani A, Hata M, Mashiko S, Ishihara A, Okamoto O, Haga Y *et al.* (2001). A typical Y1 receptor regulates feeding behaviors: effects of a potent and selective Y1 antagonist, J-115814. *Mol Pharmacol* 59, 501–5.

Kanatani A, Ishihara A, Asahi S, Tanaka T, Ozaki S, and Ihara M (1996). Potent neuropeptide Y Y1 receptor antagonist, 1229U91: blockade of neuropeptide Y-induced and physiological food intake. *Endocrinology* 137, 3177–82.

Kanatani A, Ishihara A, Iwaasa H, Nakamura K, Okamoto O, Hidaka M *et al.* (2000a). L-152,804: orally active and selective neuropeptide Y Y5 receptor antagonist. *Biochem Biophys Res Commun* 272, 169–73.

Kanatani A, Ito J, Ishihara A, Iwaasa H, Fukuroda T, Fukami T *et al.* (1998). NPY-induced feeding involves the action of a Y1-like receptor in rodents. *Regul Pept* 75–6, 409–15.

Kanatani A, Kanno T, Ishihara A, Hata M, Sakuraba A, Tanaka T *et al.* (1999). The novel neuropeptide Y Y1 receptor antagonist J-104870: a potent feeding suppressant with oral bioavailability. *Biochem Biophys Res Commun* 266, 88–91.

Kanatani A, Mashiko S, Murai N, Sugimoto N, Ito J, Fukuroda T *et al.* (2000b). Role of the Y1 receptor in the regulation of neuropeptide Y-mediated feeding: comparison of wild-type, Y1 receptor-deficient, and Y5 receptor-deficient mice. *Endocrinology* 141, 1011–16.

Kanno T, Kanatani A, Keen SL, Arai-Otsuki S, Haga Y, Iwama T *et al.* (2001). Different binding sites for the neuropeptide Y Y1 antagonists 1229U91 and J-104870 on human Y1 receptors. *Peptides* 22, 405–13.

Kask A, Harro J, von Horsten S, Redrobe JP, Dumont Y, and Quirion R. (2002). Receptor subtypes and brain sites mediating anxiolytic-like activity of neuropeptide Y. *Neurosci Biobehav Rev*, 26, 259–83.

Kask A, Rago L, and Harro J (1998). Evidence for involvement of neuropeptide Y receptors in the regulation of food intake: studies with Y1-selective antagonist BIBP3226. *Br J Pharmacol* 124, 1507–15.

Keire DA, Kobayashi M, Solomon TE, and Reeve JR, Jr. (2000*a*). Solution structure of monomeric peptide YY supports the functional significance of the PP-fold. *Biochemistry* 39, 9935–42.

Keire DA, Mannon P, Kobayashi M, Walsh JH, Solomon TE, and Reeve JR, Jr. (2000*b*). Primary structures of PYY, [Pro34]PYY, and PYY3-36 confer different conformations and receptor selectivity. *Am J Physiol* 279, G126–G131.

King PJ, Williams G, Doods H, and Widdowson PS (2000). Effect of a selective neuropeptide Y Y2 receptor antagonist, BIIE0246 on neuropeptide Y release. *Eur J Pharmacol* 396, R1–R3.

Klapstein GJ and Colmers WF (1997). Neuropeptide Y suppresses epileptiform activity in rat hippocampus. *in vitro. J Neurophysiol* 78, 1651–61.

Kombian SB and Colmers WF (1992). Neuropeptide Y selectively inhibits slow synaptic potentials in rat dorsal raphe nucleus in vitro by a presynaptic action. *J Neurosci* 12, 1086–93.

Krause J, Eva C, Seeburg PH, and Sprengel R (1992). Neuropeptide Y1 subtype pharmacology of a recombinantly expressed neuropeptide receptor. *Mol Pharmacol* 41, 817–21.

Larhammar D (1996*a*). Evolution of neuropeptide Y, peptide YY and pancreatic polypeptide. *Regul Pept* 62, 1–11.

Larhammar D (1996*b*). Structural diversity of receptors for neuropeptide Y, peptide YY and pancreatic polypeptide. *Regul Pept* 65, 165–74.

Larhammar D (1997). Extraordinary structural diversity of NPY family receptors. In (eds L Grundemar and SR Bloom) *Neuropeptide Y and Drug Development*, Academic Press, London, UK, pp. 87–105.

Larhammar D, Blomqvist AG, Yee F, Jazin E, Yoo H, and Wahlested C (1992). Cloning and functional expression of a human neuropeptide Y/peptide YY receptor of the Y1 type. *J Biol Chem* 267, 10 935–8.

Larhammar D, Wraith A, Berglund MM, Holmberg SK, and Lundell I (2001). Origins of the many NPY-family receptors in mammals. *Peptides* 22, 295–307.

Larsen PJ and Kristensen P (1998). Distribution of neuropeptide Y receptor expression in the rat suprachiasmatic nucleus. *Brain Res Mol Brain Res* 60, 69–76.

Larsen PJ, Sheikh SP, Jakobsen CR, Schwartz TW, and Mikkelsen JD (1993). Regional distribution of putative NPY Y1 receptors and neurons expressing Y1 mRNA in forebrain areas of the rat central nervous system. *Eur J Neurosci* 5, 1622–37.

Leiter AB, Toder A, Wolfe HJ, Taylor IL, Cooperman S, Mandel G, and Goodman RH (1987). Peptide YY. Structure of the precursor and expression in exocrine pancreas. *J Biol Chem* 262, 12 984–8.

Lopez-Valpuesta FJ, Nyce JW, and Myers RD (1996). NPY-Y1 receptor antisense injected centrally in rats causes hyperthermia and feeding. *Neuroreport* 7, 2781–4.

Lundberg JM, Modin A, and Malmstrom RE (1996). Recent developments with neuropeptide Y receptor antagonists. *Trends Pharmacol Sci* 17, 301–4.

Lundell I, Blomqvist AG, Berglund MM, Schober DA, Johnson D, Statnick MA *et al.* (1995). Cloning of a human receptor of the NPY receptor family with high affinity for pancreatic polypeptide and peptide YY. *J Biol Chem* 270, 29 123–8.

Lundell I, Eriksson H, Marklund U, and Larhammar D (2001). Cloning and characterization of the guinea pig neuropeptide Y receptor Y5. *Peptides* 22, 357–63.

Lundell I, Statnick MA, Johnson D, Schober DA, Starback P, Gehlert DR, and Larhammar D (1996). The cloned rat pancreatic polypeptide receptor exhibits profound differences to the orthologous receptor. *Proc Natl Acad Sci USA* 93, 5111–15.

MacKerell AD, Jr. (1988). Molecular modeling and dynamics of neuropeptide Y. *J Comput Aided Mol Des* 2, 55–63.

MacKerell AD, Jr., Hemsen A, Lacroix JS, and Lundberg JM (1989). Analysis of structure-function relationships of neuropeptide Y using molecular dynamics simulations and pharmacological activity and binding measurements. *Regul Pept* 25, 295–313.

Malmstrom RE, Alexandersson A, Balmer KC, and Weilitz J (2000). *In vivo* characterization of the novel neuropeptide Y Y1 receptor antagonist H 409/22. *J Cardiovasc Pharmacol* 36, 516–25.

Malmstrom RE, Hokfelt T, Bjorkman JA, Nihlen C, Bystrom M, Ekstrand AJ, and Lundberg JM (1998). Characterization and molecular cloning of vascular neuropeptide Y receptor subtypes in pig and dog. *Regul Pept* 75–6, 55–70.

Marsh DJ, Hollopeter G, Kafer KE, and Palmiter RD (1998). Role of the Y5 neuropeptide Y receptor in feeding and obesity. *Nat Med* 4, 718–21.

Martel JC, Fournier A, St Pierre S, Dumont Y, Forest M, and Quirion R (1990). Comparative structural requirements of brain neuropeptide Y binding sites and vas deferens neuropeptide Y receptors. *Mol Pharmacol* 38, 494–502.

Matsumoto M, Nomura T, Momose K, Ikeda Y, Kondou Y, Akiho H *et al.* (1996). Inactivation of a novel neuropeptide Y/peptide YY receptor gene in primate species. *J Biol Chem* 271, 27 217–20.

McCrea K, Wisialowski T, Cabrele C, Church B, Beck-Sickinger A, Kraegen E, and Herzog H (2000). 2–36[K4,RYYSA(19–23)]PP a novel Y5-receptor preferring ligand with strong stimulatory effect on food intake. *Regul Pept* 87, 47–58.

McLean LR, Buck SH, and Krstenansky JL (1990). Examination of the role of the amphipathic alpha-helix in the interaction of neuropeptide Y and active cyclic analogues with cell membrane receptors and dimyristoylphosphatidylcholine. *Biochemistry* 29, 2016–22.

McNally JJ, Youngman MA, Lovenberg TW, Nepomuceno D, Wilson S, and Dax SL (2000a). N-acylated alpha-(3-pyridylmethyl)-beta-aminotetralin antagoinsts of the human neuropeptide Y Y5 receptor. *Bioorg Med Chem Lett* 10, 1641–3.

McNally JJ, Youngman MA, Lovenberg TW, Nepomuceno DH, Wilson SJ, and Dax SL (2000b). N-(sulfonamido)alkyl[tetrahydro-1H-benzo[e]indol-2-yl]amines: potent antagonists of human neuropeptide Y Y5 receptor. *Bioorg Med Chem Lett* 10, 213–16.

Medeiros MD and Turner AJ (1994a). Processing and metabolism of peptide-YY: pivotal roles of dipeptidylpeptidase-IV, aminopeptidase-P, and endopeptidase-24.11. *Endocrinology* 134, 2088–94.

Medeiros Md and Turner AJ (1996). Metabolism and functions of neuropeptide Y. *Neurochem Res* 21, 1125–32.

Medeiros MS and Turner AJ (1994b). Post-secretory processing of regulatory peptides: the pancreatic polypeptide family as a model example. *Biochimie* 76, 283–7.

Mentlein R, Dahms P, Grandt D, and Kruger R (1993). Proteolytic processing of neuropeptide Y and peptide YY by dipeptidyl peptidase IV. *Regul Pept* 49, 133–44.

Michel MC, Beck-Sickinger A, Cox H, Doods HN, Herzog H, Larhammar D *et al.* (1998). XVI. International Union of Pharmacology recommendations for the nomenclature of neuropeptide Y, peptide YY, and pancreatic polypeptide receptors. *Pharmacol Rev* 50, 143–50.

Minakata H, Taylor JW, Walker MW, Miller RJ, and Kaiser ET (1989). Characterization of amphiphilic secondary structures in neuropeptide Y through the design, synthesis, and study of model peptides. *J Biol Chem* 264, 7907–13.

Minth CD, Bloom SR, Polak JM, and Dixon JE (1984). Cloning, characterization, and DNA sequence of a human cDNA encoding neuropeptide tyrosine. *Proc Natl Acad Sci USA* 81, 4577–81.

Mollereau C, Gouarderes C, Dumont Y, Kotani M, Detheux M, Doods H *et al.* (2001). Agonist and antagonist activities on human NPFF2 receptors of the NPY ligands GR231118 and BIBP3226. *Br J Pharmacol* 133, 1–4.

Murakami Y, Hara H, Okada T, Hashizume H, Kii M, Ishihara Y *et al.* (1999). 1,3-Disubstituted benzazepines as novel, potent, selective neuropeptide Y Y1 receptor antagonists. *J Med Chem* 42, 2621–32.

Nakamura M, Aoki Y, and Hirano D (1996). Cloning and functional expression of a cDNA encoding a mouse type 2 neuropeptide Y receptor. *Biochim Biophys Acta* **1284**, 134–7.

Nakamura M, Sakanaka C, Aoki Y, Ogasawara H, Tsuji T, Kodama H *et al*. (1995). Identification of two isoforms of mouse neuropeptide Y-Y1 receptor generated by alternative splicing. Isolation, genomic structure, and functional expression of the receptors. *J Biol Chem* **270**, 30 102–10.

Nakamura M, Yokoyama M, Watanabe H, and Matsumoto T (1997). Molecular cloning, organization and localization of the gene for the mouse neuropeptide Y-Y5 receptor. *Biochim Biophys Acta* **1328**, 83–9.

Naveilhan P, Hassani H, Canals JM, Ekstrand AJ, Larefalk A, Chhajlani V *et al*. (1999). Normal feeding behavior, body weight and leptin response require the neuropeptide Y Y2 receptor. *Nat Med* **5**, 1188–93.

Naveilhan P, Hassani H, Lucas G, Blakeman KH, Hao JX, Xu XJ *et al*. (2001). Reduced antinociception and plasma extravasation in mice lacking a neuropeptide Y receptor. *Nature* **409**, 513–17.

Norman MH, Chen N, Chen Z, Fotsch C, Hale C, Han N *et al*. (2000). Structure-activity relationships of a series of Pyrrolo[3, 2-d]pyrimidine derivatives and related compounds as neuropeptide Y5 receptor antagonists. *J Med Chem* **43**, 4288–312.

Parker EM, Babij CK, Balasubramaniam A, Burrier RE, Guzzi M, Hamud F *et al*. (1998). GR231118 (1229U91) and other analogues of the C-terminus of neuropeptide Y are potent neuropeptide Y Y1 receptor antagonists and neuropeptide Y Y4 receptor agonists. *Eur J Pharmacol* **349**, 97–105.

Parker R, Liu M, Eyre HJ, Copeland NG, Gilbert DJ, Crawford J *et al*. (2000). Y-receptor-like genes GPR72 and GPR73: molecular cloning, genomic organisation and assignment to human chromosome 11q21.1 and 2p14 and mouse chromosome 9 and 6. *Biochim Biophys Acta* **1491**, 369–75.

Parker RM and Herzog H (1998). Comparison of Y-receptor subtype expression in the rat hippocampus. *Regul Pept* **75–6**, 109–15.

Parker RM and Herzog H (1999). Regional distribution of Y-receptor subtype mRNAs in rat brain. *Eur J Neurosci* **11**, 1431–48.

Pedrazzini T, Seydoux J, Kunstner P, Aubert JF, Grouzmann E, Beermann F, and Brunner HR (1998). Cardiovascular response, feeding behavior and locomotor activity in mice lacking the NPY Y1 receptor. *Nat Med* **4**, 722–6.

Pheng LH, Fournier A, Dumont Y, Quirion R, and Regoli D (1997). The dog saphenous vein: a sensitive and selective preparation for the Y2 receptor of neuropeptide Y. *Eur J Pharmacol* **327**, 163–7.

Pheng LH, Perron A, Quirion R, Cadieux A, Fauchere JL, Dumont Y, and Regoli D (1999). Neuropeptide Y-induced contraction is mediated by neuropeptide Y Y2 and Y4 receptors in the rat colon. *Eur J Pharmacol* **374**, 85–91.

Pickel VM, Beck-Sickinger AG, Chan J, and Weiland HA (1998). Y1 receptors in the nucleus accumbens: ultrastructural localization and association with neuropeptide Y. *J Neurosci Res* **52**, 54–68.

Playford RJ and Cox HM (1996). Peptide YY and neuropeptide Y: two peptides intimately involved in electrolyte homeostasis. *Trends Pharmacol Sci* **17**, 436–8.

Quirion R, Martel JC, Dumont Y, Cadieux A, Jolicoeur F, St Pierre S, and Fournier A (1990). Neuropeptide Y receptors: autoradiographic distribution in the brain and structure-activity relationships. *Ann N Y Acad Sci* **611**, 58–72.

Raposinho PD, Broqua P, Hayward A, Akinsanya K, Galyean R, Schteingart C *et al*. (2000). Stimulation of the gonadotropic axis by the neuropeptide Y receptor Y1 antagonist/Y4 agonist 1229U91 in the male rat. *Neuroendocrinology* **71**, 2–7.

Redrobe JP, Dumont Y, St Pierre JA, and Quirion R (1999). Multiple receptors for neuropeptide Y in the hippocampus: putative roles in seizures and cognition. *Brain Res* **848**, 153–66.

Rimland J, Xin W, Sweetnam P, Saijoh K, Nestler EJ, and Duman RS (1991). Sequence and expression of a neuropeptide Y receptor cDNA. *Mol Pharmacol* **40**, 869–75.

Rist B, Ingenhoven N, Scapozza L, Schnorrenberg G, Gaida W, Wieland HA, and Beck-Sickinger AG (1997). The bioactive conformation of neuropeptide Y analogues at the human Y2-receptor. *Eur J Biochem* **247**, 1019–28.

Rist B, Zerbe O, Ingenhoven N, Scapozza L, Peers C, Vaughan PF *et al.* (1996). Modified, cyclic dodecapeptide analog of neuropeptide Y is the smallest full agonist at the human Y2 receptor. *FEBS Lett* **394**, 169–73.

Rose PM, Fernandes P, Lynch JS, Frazier ST, Fisher SM, Kodukula K *et al.* (1995). Cloning and functional expression of a cDNA encoding a human type 2 neuropeptide Y receptor [published erratum appears in *J Biol Chem* 1995 Dec 1;270(48):29038], *J Biol Chem* **270**, 22 661–4.

Rudolf K, Eberlein W, Engel W, Wieland HA, Willim KD, Entzeroth M *et al.* (1994). The first highly potent and selective non-peptide neuropeptide Y Y1 receptor antagonist: BIBP3226. *Eur J Pharmacol* **271**, R11–R13.

Rueeger H, Rigollier P, Yamaguchi Y, Schmidlin T, Schilling W, Criscione L *et al.* (2000). Design, synthesis and SAR of a series of 2-substituted 4-amino-quinazoline neuropeptide Y Y5 receptor antagonists. *Bioorg Med Chem Lett* **10**, 1175–9.

Salaneck E, Holmberg SK, Berglund MM, Boswell T, and Larhammar D (2000). Chicken neuropeptide Y receptor Y2: structural and pharmacological differences to mammalian Y2. *FEBS Lett* **484**, 229–34.

Saudek V and Pelton JT (1990). Sequence-specific 1H NMR assignment and secondary structure of neuropeptide Y in aqueous solution. *Biochemistry* **29**, 4509–15.

Sautel M, Martinez R, Munoz M, Peitsch MC, Beck-Sickinger AG, and Walker P (1995). Role of a hydrophobic pocket of the human Y1 neuropeptide Y receptor in ligand binding. *Mol Cell Endocrinol* **112**, 215–22.

Sautel M, Rudolf K, Wittneben H, Herzog H, Martinez R, Munoz M *et al.* (1996). Neuropeptide Y and the nonpeptide antagonist BIBP 3226 share an overlapping binding site at the human Y1 receptor. *Mol Pharmacol* **50**, 285–92.

Schober DA, Gackenheimer SL, Heiman ML, and Gehlert DR (2000). Pharmacological characterization of (125)I-1229U91 binding to Y1 and Y4 neuropeptide Y/Peptide YY receptors. *J Pharmacol Exp Ther* **293**, 275–80.

Schober DA, Van Abbema AM, Smiley DL, Bruns RF, and Gehlert DR (1998). The neuropeptide Y Y1 antagonist, 1229U91, a potent agonist for the human pancreatic polypeptide-preferring (NPY Y4) receptor. *Peptides* **19**, 537–42.

Schwartz TW (1994). Locating ligand-binding sites in 7 TM receptors by protein engineering. *Curr Opin Biotechnol* **5**, 434–44.

Schwartz TW, Fuhlendorff J, Kjems LL, Kristensen MS, Vervelde M, O'Hare M *et al.* (1990). Signal epitopes in the three-dimensional structure of neuropeptide Y. Interaction with Y1, Y2, and pancreatic polypeptide receptors. *Ann N Y Acad Sci* **611**, 35–47.

Serradeil-Le Gal C, Valette G, Rouby PE, Pellet A, Oury-Donat F, Brossard G *et al.* (1995). SR 120819A, an orally-active and selective neuropeptide Y Y1 receptor antagonist. *FEBS Lett* **362**, 192–6.

Shigeri Y, Ishikawa M, Ishihara Y, and Fujimoto M (1998). A potent nonpeptide neuropeptide Y Y1 receptor antagonist, a benzodiazepine derivative. *Life Sci* **63**, L-60.

Smith-White MA, Hardy TA, Brock JA, and Potter EK (2001). Effects of a selective neuropeptide Y Y2 receptor antagonist, BIIE0246, on Y2 receptors at peripheral neuroeffector junctions. *Br J Pharmacol* **132**, 861–8.

Solomon TE (1985). Pancreatic polypeptide, peptide YY, and neuropeptide Y family of regulatory peptides. *Gastroenterology* **88**, 838–41.

St Pierre JA, Dumont Y, Nouel D, Herzog H, Hamel E, and Quirion R (1998). Preferential expression of the neuropeptide Y Y1 over the Y2 receptor subtype in cultured hippocampal neurons and cloning of the rat Y2 receptor. *Br J Pharmacol* **123**, 183–94.

St Pierre JA, Nouel D, Dumont Y, Beaudet A, and Quirion R (2000*a*). Association of neuropeptide Y Y1 receptors with glutamate-positive and NPY-positive neurons in rat hippocampal cultures. *Eur J Neurosci* 12, 1319–30.

St Pierre JA, Nouel D, Dumont Y, Beaudet A, and Quirion R (2000*b*). Sub-population of cultured hippocampal astrocytes expresses neuropeptide Y Y1 receptors. *Glia* 30, 82–91.

Stanley BG and Leibowitz SF (1984). Neuropeptide Y: stimulation of feeding and drinking by injection into the paraventricular nucleus. *Life Sci* 35, 2635–42.

Starback P, Wraith A, Eriksson H, and Larhammar D (2000). Neuropeptide Y receptor gene y6: multiple deaths or resurrections? *Biochem Biophys Res Commun* 277, 264–9.

Tatemoto K, Carlquist M, and Mutt V (1982). Neuropeptide Y–a novel brain peptide with structural similarities to peptide YY and pancreatic polypeptide. *Nature* 296, 659–60.

Tatemoto K and Mutt V (1980). Isolation of two novel candidate hormones using a chemical method for finding naturally occurring polypeptides. *Nature* 285, 417–18.

Thiele TE, Marsh DJ, Marie L, Bernstein IL, and Palmiter RD (1998). Ethanol consumption and resistance are inversely related to neuropeptide Y levels. *Nature* 396, 366–9.

Thorsell A, Michalkiewicz M, Dumont Y, Quirion R, Caberlotto L, Rimondini R *et al.* (2000). Behavioral insensitivity to restraint stress, absent fear suppression of behavior and impaired spatial learning in transgenic rats with hippocampal neuropeptide Y overexpression. *Proc Natl Acad Sci USA* 97, 12 852–7.

Tong Y, Dumont Y, Shen SH, and Quirion R (1997). Comparative developmental profile of the neuropeptide Y Y1 receptor gene and protein in the rat brain. *Brain Res Mol Brain Res* 48, 323–32.

Trinh T, van Dumont Y, and Quirion R (1996). High levels of specific neuropeptide Y/pancreatic polypeptide receptors in the rat hypothalamus and brainstem. *Eur J Pharmacol* 318, R1–R3.

Vezzani A, Sperk G, and Colmers WF (1999). Neuropeptide Y: emerging evidence for a functional role in seizure modulation. *Trends Neurosci* 22, 25–30.

Voisin T, Rouyer-Fessard C, and Laburthe M (1990). Distribution of common peptide YY-neuropeptide Y receptor along rat intestinal villus-crypt axis. *Am J Physiol* 258, G753–G759.

Wahlestedt C, Grundemar L, Hakanson R, Heilig M, Shen GH, Zukowska-Grojec Z, and Reis DJ (1990). Neuropeptide Y receptor subtypes, Y1 and Y2. *Ann N Y Acad Sci* 611, 7–26.

Wahlestedt C, Pich EM, Koob GF, Yee F, and Heilig M (1993). Modulation of anxiety and neuropeptide Y-Y1 receptors by antisense oligodeoxynucleotides. *Science* 259, 528–31.

Wahlestedt C, Regunathan S, and Reis DJ (1992). Identification of cultured cells selectively expressing Y1-, Y2-, or Y3-type receptors for neuropeptide Y/peptide YY. *Life Sci* 50, L7–12.

Wahlestedt C and Reis DJ (1993). Neuropeptide Y-related peptides and their receptors–are the receptors potential therapeutic drug targets?. *Annu Rev Pharmacol Toxicol* 33, 309–52.

Wahlestedt C, Yanaihara N, and Hakanson R (1986). Evidence for different pre- and post-junctional receptors for neuropeptide Y and related peptides. *Regul Pept* 13, 307–18.

Walker P, Munoz M, Martinez R, and Peitsch MC (1994). Acidic residues in extracellular loops of the human Y1 neuropeptide Y receptor are essential for ligand binding. *J Biol Chem* 269, 2863–9.

Weinberg DH, Sirinathsinghji DJ, Tan CP, Shiao LL, Morin N, Rigby MR *et al.* (1996). Cloning and expression of a novel neuropeptide Y receptor. *J Biol Chem* 271, 16 435–8.

Whitcomb DC, Puccio AM, Vigna SR, Taylor IL, and Hoffman GE (1997). Distribution of pancreatic polypeptide receptors in the rat brain. *Brain Res* 760, 137–49.

Wieland HA, Eckard CP, Doods HN, and Beck-Sickinger AG (1998*a*). Probing of the neuropeptide Y-Y1-receptors interaction with anti-receptor antibodies. *Eur J Biochem* 255, 595–603.

Wieland HA, Engel W, Eberlein W, Rudolf K, and Doods HN (1998*b*). Subtype selectivity of the novel nonpeptide neuropeptide Y Y1 receptor antagonist BIBO 3304 and its effect on feeding in rodents. *Br J Pharmacol* 125, 549–55.

Wieland HA, Willim KD, Entzeroth M, Wienen W, Rudolf K, Eberlein W *et al.* (1995). Subtype selectivity and antagonistic profile of the nonpeptide Y1 receptor antagonist BIBP 3226. *J Pharmacol Exp Ther* 275, 143–9.

Wraith A, Tornsten A, Chardon P, Harbitz I, Chowdhary BP, Andersson L *et al.* (2000). Evolution of the neuropeptide Y receptor family: gene and chromosome duplications deduced from the cloning and mapping of the five receptor subtype genes in pig. *Genome Res* 10, 302–10.

Yan H, Yang J, Marasco J, Yamaguchi K, Brenner S, Collins F, and Karbon W (1996). Cloning and functional expression of cDNAs encoding human and rat pancreatic polypeptide receptors. *Proc Natl Acad Sci USA* 93, 4661–5.

Yang SN, Bunnemann B, Cintra A, and Fuxe K (1996). Localization of neuropeptide Y Y1 receptor-like immunoreactivity in catecholaminergic neurons of the rat medulla oblongata. *Neuroscience* 73, 519–30.

Youngman MA, McNally JJ, Lovenberg TW, Reitz AB, Willard NM, Nepomuceno DH *et al.* (2000). Alpha-Substituted N-(sulfonamido)alkyl-beta-aminotetralins: potent and selective neuropeptide Y Y5 receptor antagonists. *J Med Chem* 43, 346–50.

Zarrinmayeh H, Nunes AM, Ornstein PL, Zimmerman DM, Arnold MB, Schober DA *et al.* (1998). Synthesis and evaluation of a series of novel 2-[(4-chlorophenoxy)methyl]benzimidazoles as selective neuropeptide Y Y1 receptor antagonists. *J Med Chem* 41, 2709–19.

Zarrinmayeh H, Zimmerman DM, Cantrell BE, Schober DA, Bruns RF, Gackenheimer SL *et al.* (1999). Structure-activity relationship of a series of diaminoalkyl substituted benzimidazole as neuropeptide Y Y1 receptor antagonists. *Bioorg Med Chem Lett* 9, 647–52.

Zhang X, Bao L, Xu ZQ, Kopp J, Arvidsson U, Elde R, and Hokfelt T (1994). Localization of neuropeptide Y Y1 receptors in the rat nervous system with special reference to somatic receptors on small dorsal root ganglion neurons. *Proc Natl Acad Sci USA* 91, 11 738–42.

Zimmerman DM, Cantrell BE, Smith EC, Nixon JA, Bruns RF, Gitter B *et al.* (1998). Structure-activity relationships of a series of 1-substituted-4-methylbenzimidazole neuropeptide Y-1 receptor antagonists. *Bioorg Med Chem Lett* 8, 473–6.

Zuana OD, Sadlo M, Germain M, Feletou M, Chamorro S, Tisserand F *et al.* (2001). Reduced food intake in response to CGP 71683A may be due to mechanisms other than NPY Y5 receptor blockade. *Int J Obes Relat Metab Disord* 25, 84–94.

Chapter 20

Opioid receptors

François Cesselin and Michel Hamon

20.1 Introduction

Opioid receptors are the targets of analgesic compounds, of which the alkaloid from opium, morphine, remains the most widely used for relieving severe pain. The existence of opioid receptors was first postulated in the 1950s, and established in 1973 by three groups independently (Pert and Snyder 1973; Simon *et al.* 1973; Terenius 1973). Early binding studies with various ligands suggested the existence of multiple types of opioid receptors. This concept was then actually demonstrated by molecular biology approaches. Three major receptor types were cloned and sequenced, and their genomic structures determined in the 1990s (Evans *et al.* 1992; Kieffer *et al.* 1992; Chen *et al.* 1993*a,b*; Fukuda *et al.* 1993; Meng *et al.* 1993; Wang *et al.* 1993; Yasuda *et al.* 1993). The nomenclature of opioid receptors is controversial. They were originally designed using a Greek letter terminology (μ, ∂, and κ). A 1996 proposal from the IUPHAR Receptor Nomenclature Committee to use a non-Greek letter nomenclature in order to follow general guidelines that would be applicable to all receptors led to rename the ∂, κ, and μ types, OP_1, OP_2, and OP_3, respectively (Dhawan *et al.* 1996). However, this nomenclature was not widely accepted, and has now been replaced by a revised recommendation from IUPHAR (Cox 2000). To date, the common name for this receptor family continues to be OP (for opioid peptides), and either a Greek μ, ∂, or κ, or a capitalized M, D, or K letter is added in front to nominate each type, leading to μOP or MOP, ∂OP or DOP, and κOP or KOP, for the main three types (see Table 20.1).

The endogenous ligands of OP receptors are opioid peptides that derive from precursor proteins known as proopiomelanocortin (from which originate β-endorphin and related compounds), proenkephalin (that gives enkephalins and other peptides) and prodynorphin (source of dynorphins, neoendorphins and related molecules). In addition, the more recently isolated endomorphin-1 and -2 may represent a class of highly selective MOP receptor endogenous ligands (Zadina *et al.* 1997). OP and OP receptors play key roles in the control of behaviours that are essential for self and species survival, including responses to noxious stimulation and stress, reward and motivation. Opioidergic systems also control autonomic functions, including respiration, thermoregulation, gastrointestinal motility, and modulate immune responses (see Dhawan *et al.* 1996). It is thus not surprising that opioid drugs produce a multitude of effects in human. These effects include analgesia, euphoria or dysphoria, respiratory depression, myosis, constipation, and changes in the endocrine and immune systems, with differences depending on the opioid receptor involved. Tolerance to these effects develops to a varying degree, and physical dependence is a consequence of chronic opioid use.

Although considerable progress in the knowledge of OP receptor functioning at the molecular level emerged following their cloning, numerous aspects of OP receptor pharmacology and physiology remain unclear. Reviewing these data within the limited format of this chapter is a challenging goal that imposes citation of mainly review articles from which, however, the reader can retrieve the original references.

20.2 Molecular characterization

Comparison of OP receptor sequences with those of other GPCRs shows that they display some similarity with somatostatin, angiotensin, and chemoattractant receptors. The amino acid sequences of the three human OP receptors are 85–90 per cent identical to their rodent counterparts and about 60 per cent identical to each other, as well as to another receptor called opioid receptor-like 1 (ORL-1; Bunzow et al. 1994; Chen et al. 1994; Mollereau et al. 1994; Nishi et al. 1994; Wang JB et al. 1994). Because ORL-1 appears to have a very low affinity for most opioid ligands (but see Hawkinson et al. 2000), including naloxone, it is considered to belong to a non-opioid branch of the OP receptor family, which led to name it OP4 (Dhawan et al. 1996) and now NOP, in the recently revised nomenclature (Cox 2000). An endogenous agonist at the NOP receptor was identified as a 17 amino acid peptide called either nociceptin (Meunier et al. 1995) or orphanin FQ (Reinscheid et al. 1995). Nociceptin/orphanin FQ (N/OFQ) is the product of a precursor protein encoded by a gene having clear-cut sequence homology with the genes encoding the precursors of OP (Mollereau et al. 1996). These data suggest that the system utilizing N/OFQ as an endogenous ligand has probably emerged from gene duplication followed by subsequent evolutionary modification (Danielson and Dores 1999). Originally characterized as a nociception-producing peptide ('nociceptin'), N/OFQ has now been shown to exert diverse effects on nociception (hyperalgesia, allodynia, analgesia) as well as on other functions and behaviours such as locomotion, feeding, anxiety, spatial attention, reproductive behaviour, and opiate tolerance. The exact nature of the relationships between N/OFQ and authentic opioids remains controversial. Like other putative 'anti-opioid' peptides, N/OFQ should in fact be considered as an 'opioid-modulating' peptide (Cesselin 1995).

In contrast to ORL-1/NOP, the receptor that was initially called zeta and is now named OGFr, for 'Opioid Growth Factor' receptor, has no homology with the three main types of OP receptors (Zagon et al. 2000). It could be the target of met-enkephalin, that one can therefore consider as an endogenous OGF. By acting at OGFr, met-enkephalin appears to tonically modulate cell proliferation and tissue organization during development, cancer, cellular renewal, wound healing, and angiogenesis (Zagon et al. 2000).

Among the three main OP receptors, the most divergent regions between species are the N- and C-terminal domains, where variations from one OP receptor type to another are also the largest in the same species. The importance of extracellular loops for the recognition of both alkaloid and peptide ligands has been well documented (Fukuda et al. 1995; Meng et al. 1995; Wang et al. 1995). Several pieces of evidence support the concept that determinants of the OP receptor binding pocket differ, at least partly, from ligand to ligand, notably from peptide to non-peptide ligand. The emerging picture is that the receptor is capable of considerable plasticity regarding the recognition of ligands and subsequent events leading to receptor activation. Indeed, competition binding assays with MOP receptor mutants revealed that ligand binding affinity can be influenced by more than 20 different amino acid residues

Table 20.1 Opioid receptors

	MOP	DOP	KOP	NOP
Alternative names	μ, MOR, OP3	δ, DOR, OP1	κ, KOR, OP2	ORL1, LY322, N/OFQ receptor, OP4
Preferred endogenous ligand	Endomorphin 1 Endomorphin 2 β-endorphin Enkephalins	Enkephalins	Dynorphins	N/OFQ
Structural information (Accession no.)	h 400 aa (P35372)[AS] r 398 aa (P33535)[AS] m 398 aa (P42866)	h 372 aa (P41143) r 372 aa (P33533) m 372 aa (P32300)	h 380 aa (P41145) r 380 aa (P34975) m 380 aa (P33534)	h 370 aa (P41146)[AS] r 367 aa (P35370) m 367 aa (P35377)[AS]
Chromosomal location	6q24–25	1p34.3–36.1	8q11.12	20q13.2–13.3
Selective agonists	Endomorphin 1 & 2, DAMGO, PL017, TAPS, DALDA, dermorphin, Sufentanil	BUBUC, deltorphin I, deltorphin II, DPDPE, DSTLET, DTLET, SNC80, TAN67	EMD61753, CI977, fedotozine, PD117302, TRK820, U50488, U69593	N/OFQ, Ro646198,
Inverse agonists	β-chlornaltrexamine (βCNA), 7-Benzylidenenaltrexone (BNTX), clocinnamox (C-CAM)	BNTX, clocinnamox, ICI174864, naltriben (NTB)	—	—
Selective antagonists	CTOP, CTAP naloxonazine (irreversible)	DALCE, naltrindole, TIPP, TIPPψ	DIPPA, norbinaltorphimine (nor-BNI), UPHIT, 5′-guanidinonaltrindole (GNTI)	J113397, naloxone benzoylhydrazone (NalBzoH), $[Nphe^1]$N/OFQ(1–13)NH_2
Radioligands	$[^3H]$-DAMGO	$[^3H]$-deltorphin 1, $[^3H]$deltorphin 2, $[^3H]$-DPDPE, $[^3H]$-pcl-DPDPE, $[^3H]$-TIPP, $[^3H]$-TIPPψ	$[^3H]$-CI977, $[^3H]$-PD117,302, $[^3H]$-U69,593, $[^3H]$-nor-BNI	$[^3H]$-N/OFQNH_2, $[^{125}I]$-$[Tyr14]$N/OFQ
G protein coupling	$G_o/G_i/G_z/G_{15-16}$	$G_o/G_i/G_z/G_{16}$	$G_o/G_i/G_z/G_{16}$	$G_o/G_i/G_z/G_{16}$

Expression profile	Caudate putamen, neocortex, nucleus accumbens, thalamus, nucleus accumbens, hippocampus, amygdala, gastrointestinal tract, immune cells	Olfactory bulb, cerebral cortex, caudate putamen, nucleus accumbens, thalamus, hypothalamus, gastrointestinal tract, immune cells	Nucleus accumbens, claustrum, dorsal endopiriform nucleus, interpedoncular nucleus, gastrointestinal tract, immune cells	Cerebral cortex, anterior olfactory nucleus, lateral septum, ventral forebrain, hypothalamus, hippocampus, amygdala, gastrointestinal tract, immune cells
Main physiological function	Modulation of nociception, locomotor activity, sexual activity and immune system function	Modulation of nociception and mood	Modulation of visceral nociception	Acts as an 'opioid-modulating' peptide
KO phenotypes	Reduction in nociceptive thresholds, locomotor activity, and sexual activity; lack of morphine induced behaviour; increase in immune cell activity	Abolition of spinal, but not supraspinal, induction of analgesia by selective DOP receptor agonists; reduction of tolerance to the analgesic effects of morphine; enhancement of anxiogenic- and depression-like responses in relevant tests	No major changes; attenuation of morphine withdrawal signs	Improved spatial attention and memory
Disease relevance	Drug addiction, pain, idiopathic absence epilepsy	Drug addiction, pain, depression, schizophrenia, Parkinson's disease, inflammation, lung cancer	Addiction	Addiction, pain, cognitive decline

(Surrat *et al.* 1994; Xu *et al.* 1999). However, MOP selectivity seems to depend mainly on four amino acids only (Surrat *et al.* 1994; Mansour *et al.* 1997; Larson *et al.* 2000).

At least two MOP receptor subtypes have been defined on the basis of pharmacological criteria. The MOP1 subtype, that mediates only part of the effects of morphine, is selectively antagonized by naloxazone and naloxonazine. Morphine effects still occurring in the presence of these antagonists, and that can therefore be ascribed to the MOP2 subtype, include respiratory depression and inhibition of gastrointestinal transit, which raises the possibility that selective MOP1 agonists (which are not yet available) would not trigger these two major side effects of morphine (Dhawan *et al.* 1996). In respect to DOP receptors, several groups distinguish the DOP1 subtype, at which D-[Pen^{2,5}]-enkephalin (Tyr-D-Pen-Gly-Phe-D-Pen) (DPDPE) acts as an agonist and 7-benzylidenenaltrexone (BNTX) and Tyr-D-Ala^2-Gly-Phe-Leu-Cys (DALCE) as antagonists, and the DOP2 subtype which is activated by deltorphin II, an amphibian peptide, and Tyr-D-Ser-Gly-Phe-Leu-Thr (DSLET), and blocked by naltriben (NTB) and a naltrindole derivative, naltrindole 5'-isothiocyanate (Jordan *et al.* 2000). Pharmacological studies also suggested the existence of multiple KOP receptors. The best characterized subtype, referred to as KOP1, binds arylacetamide compounds such as U50488H, U69593, PD117302, and CI977 (Dhawan *et al.* 1996). The existence of other KOP receptor subtypes has been postulated to account for the fact that benzomorphan opioid ligands, such as bremazocine, label a greater number of KOP receptor sites than arylacetamide ligands under conditions where binding onto MOP and DOP receptors is suppressed (see Dhawan *et al.* 1996). These other subtypes, however, are still to be characterized with really selective ligands. Recently, the existence of two pharmacologically distinct NOP receptor subtypes has also been suggested (Letchworth *et al.* 2000).

The relationships between these pharmacologically defined OP receptor subtypes and the cloned OP receptors have still to be established. However, the high affinity of the cloned MOP receptor for naloxonazine may be consistent with the idea that it actually corresponds to the MOP1 subtype. On the other hand, the higher binding potency and stronger efficacy of NTB compared to BNTX at the cloned DOP receptor have led to the hypothesis that this receptor is in fact the DOP2 subtype. In the case of the cloned KOP receptor, its high affinity for U50488H suggests that it corresponds to the pharmacologically defined KOP1 subtype (see Dhawan *et al.* 1996). To date, however, no other DOP, MOP, or KOP receptor-encoding cDNA that would correspond to other receptor subtypes have been cloned, and these putative assignments remain elusive. Accordingly, the rather complex pharmacology of OP receptor subtypes can probably not be explained by the existence of multiple encoding genes (although it cannot be ruled out that other OP receptor genes have still to be discovered). A more probable explanation is that the pharmacologically defined receptor subtypes represent either (i) alternatively spliced variants of the cloned MOP, DOP, and KOP receptors (ii) protein products from the same mRNA that have undergone different post-translational modifications or even (iii) identical receptor proteins associated with different proteins (such as G proteins or other OP receptors in heterodimer complexes, see below) in the cell membrane.

Although OP receptor genes are located on distinct chromosomes, they have very similar genomic structures, with multiple introns. Following the initial description of the MOP receptor, two splice variants were described that differ at the intracellular C-terminus (Bare *et al.* 1994). The gene encoding the first variant lacks exon 4, whereas this exon is replaced by exon 5 in the gene encoding the second variant. Subsequently, other variants were identified in which exon 4 is replaced by combinations of additional exons. These variants are actually expressed in brain where they exhibit different regional distributions (Abbadie *et al.* 2000). In line with these data, antisense mapping studies showed that several MOP receptor variants apparently contribute to morphine-induced analgesia (Pasternak 2001). In the NG108–15 cell line, Northern blot analysis revealed the existence of four distinct mRNA coding for the DOP receptor even though only one protein species was found (Kim *et al.* 1995). In addition, alternative splicing of DOP primary transcript has been demonstrated in the mouse brain (Gavériaux-Ruff *et al.* 1997). In line with the existence of DOP receptor subtypes issued from alternative splicing, antisense oligonucleotides directed against three exons of the DOP receptor gene displayed differential ability to prevent the effects of DOP receptor agonists at spinal and supraspinal levels (Rossi *et al.* 1997). The mouse KOP receptor gene is known to contain four exons and use two promoters. Promoter 1 directs the expression of two mRNA isoforms produced by alternative splicing at intron 1. Therefore, a total of three KOP receptor isoforms can be generated (Wei *et al.* 2000). Alternative splicing of KOP primary transcript has actually been observed in human immune cells (Gavériaux-Ruff *et al.* 1997). In case of the NOP receptor, a number of splice variants have also been identified in the mouse brain (Peluso *et al.* 1998; Xie *et al.* 1999). Although their regional distributions are clearly different, they apparently exhibit the same pharmacological properties.

Evidence that DOP and KOP receptors can form homodimers has been reported (Jordan *et al.* 2000). In addition, OP receptor heterodimers have been described and some authors have proposed that the pharmacologically defined receptor subtypes may in fact correspond to distinct opioid receptor complexes (Zhu *et al.* 1999). In line with this concept, studies in knock-out mice lacking MOP receptors suggested that the presence of MOP receptors is required to obtain full DOP receptor-mediated responses (Kieffer 1999). Indeed, MOP and DOP receptors are able to form heterodimers having signalling properties different from those of each individual receptor component (George *et al.* 2000). On the other hand, KOP receptors can make heterodimers with DOP receptors, but not MOP receptors (Jordan and Devi 1999; Wessendorf and Dooyema 2001). KOP/DOP receptor complexes were found to bind benzomorphans with high affinity, but did not recognize the highly selective KOP1 agonist U69593, thereby providing support to the actual existence of other KOP receptor subtypes. Recently, hetero-oligomers implicating not only DOP or KOP receptors but also other GPCRs (notably β2-adrenoreceptors) have been described in HEK293S and CHO cells (Jordan *et al.* 2001; McVey *et al.* 2001). Such OP receptor oligomerization has clear-cut consequences on the cell trafficking and signalling properties of the receptors (Jordan and Devi 1999; Jordan *et al.* 2001).

20.3 Cellular and subcellular localization

In situ hybridization, immunohistochemistry, and autoradiography of radioligands bound onto brain sections (see Dhawan *et al.* 1996) showed that the highest density of MOP receptors is found in the caudate putamen. MOP receptor density then diminishes in the following order: neocortex, thalamus, nucleus accumbens, hippocampus, and amygdala. In case of DOP receptors, the same approaches demonstrated that they are especially abundant in olfactory bulb, neocortex, caudate putamen, and nucleus accumbens. Thalamus, hypo-thalamus, and brainstem have only moderate to poor DOP receptor density. On the other hand, brain areas with the highest densities of KOP receptors are the nucleus accumbens, claustrum, dorsal endopiriform nucleus, and interpedoncular nucleus; in contrast, caudal brain areas are relatively poor in KOP receptors. With regard to NOP receptors, the cerebral cortex, anterior olfactory nucleus, lateral septum, ventral forebrain, several hypothalamic nuclei, hippocampal formation, and amygdala are the regions where they are expressed at high density. In contrast, the caudate-putamen and the cerebellum are almost devoid of NOP receptors (Bunzow *et al.* 1994; Mollereau and Mouledous 2000).

Immunocytochemical labelling at the ultrastructural level showed that DOP receptors are targeted primarily to axons where they most likely function as pre-synaptic receptors controlling neurotransmitter release (Arvidsson *et al.* 1995*a*). In contrast, MOP receptors appear to be preferentially targeted to the somatodendritic domain where they participate in a post-synaptic control of the neuronal discharge (Arvidsson *et al.* 1995*c*). A more complex situation applies to KOP receptors because immunocytochemical studies showed that these receptors are addressed to both axonal and somatodendritic compartments in neurones (although the majority of immunostaining is seen in the latter compartment, see Arvidsson *et al.* 1995*b*). A similar subcellular distribution has been described for the NOP receptors, with both nerve fibres and neuronal cell bodies (notably in the hippocampus) being labelled by specific antibodies; however, the proportion of NOP receptors associated with axonal profiles seems to be much larger than that associated with the somatodendritic domain of neurones (Anton *et al.* 1996).

It is of interest that cells and tissues can express two or more OP receptors, congruent with the idea that these receptors can form heterodimers. This is notably the case for several neuroblastoma cell lines which coexpress MOP and DOP receptors, or even all three OP receptors. In patches of the rat striatum, DOP receptor immunoreactivity is commonly seen within the cytoplasm of spiny and aspiny neurones, many of which also express the MOP receptor. Similarly, co-localization of both MOP and DOP receptors has been demonstrated in neurones of the dorsal root ganglia and dorsal spinal cord (Cheng *et al.* 1997). Evidence for the co-localization of both DOP and KOP receptors has also been reported in dorsal root ganglia, spinal cord, hippocampus, and posterior pituitary, where these receptors are associated with structures resembling large dense-core vesicles (see Zhang X *et al.* 1998). Finally, it should be noted that in addition to the reported OP receptor expression in the CNS, a variety of peripheral tissues, notably the gastrointestinal tract and immune cells, have been shown to express OP receptors (see Dhawan *et al.* 1996).

20.4 **Pharmacology**

It has to be emphasized that relevant data in the literature are often heterogeneous with marked variations in dissociation constant values for a given opioid ligands reported from one study to another. These discrepancies reflect differences in cell types and assay conditions used for the pharmacological characterization of OP receptors. The data in Table 20.1 summarizes the main ligands available to date and only a few additional comments are made in the following sections.

20.4.1 **Endogenous ligands**

Among the endogenous opioid peptides, dynorphin A (1–13) preferentially interacts with the KOP receptor whereas enkephalins preferentially interact with both DOP and MOP receptors, as does β-endorphin. However, these preferences are not exclusive. For instance, the heptadecapeptide dynorphin A is not only a high affinity KOP receptor agonist but can also activate (in a higher concentrations range) human MOP, DOP, and NOP receptors

(Zhang S *et al.* 1998). Inverse agonist action of leu-enkephalin at DOP receptors, that probably explains its spinal antianalgesic effects, has been recently reported by Rady *et al.* (2001).

The two tetrapeptides, endomorphin-1 (Tyr-Pro-Trp-Phe-NH$_2$) and endomorphin-2 (Tyr-Pro-Phe-Phe-NH$_2$), are the likely endogenous ligands for the MOP receptors. Indeed, their affinity and specificity for these receptors are the highest when compared with any other endogenous compound found so far in the mammalian nervous system (Zadina *et al.* 1997). As expected, both endomorphins are potent antinociceptive molecules through MOP receptor activation. According to Sakurada *et al.* (1999), endomorphin-1 would act predominantly at MOP2 subtype whereas endomorphin-2 would be a preferential MOP1 agonist. In addition to the endogenous ligand for NOP receptor, N/OFQ, other biologically active peptides can derive from its precursor, notably a heptadecapeptide called orphanin FQ2 or nociceptin II, and nocistatin (Lee *et al.* 1999). Although none of them recognize NOP receptors, nocistatin acts as a functional antagonist of N/OFQ (Calo *et al.* 2000).

20.4.2 Agonists

Most of opioid analgesics currently used in the clinic are agonists at MOP receptors. The chemical structures of these analgesics vary markedly from minor changes compared to the morphine alkaloid structure (e.g. codeine) to completely different structures (e.g. fentanyl and methadone). Studies with MOP receptors stably expressed in CHO cells confirmed that the prototypical MOP ligands, DAMGO (Tyr-D-Ala-Gly-MePhe-Met(O)-ol) and PL017 (Tyr-Pro-τ-MePhe-D-Pro-NH$_2$), are recognized with high affinity by these receptors. Ligands really selective of the MOP receptor are the agonist PL017 and the antagonist CTOP (D-Phe-Cys-Tyr-D-Trp-Orn-Thr-Pen-Thr-NH$_2$) (Dhawan *et al.* 1996). Originally isolated from frog skin, [Lys^7]dermorphin displays a very high affinity (K$_i$ = 0.12 nM) for MOP receptors (Erspamer 1992). The dermorphin tetrapeptide analogues, TAPS (Tyr-D-Arg^2-Phe-sarcosine^4), DALDA ([D-Arg^2, Lys^4]dermorphin-(1-4)-amide, or H-Tyr-D-Arg-Phe-Lys-NH$_2$) and [Dmt^1]DALDA (i.e. H-Dmt-D-Arg-Phe-Lys-NH$_2$, with Dmt = 2′,6′-dimethylTyr) show even higher MOP receptor agonistic potency and selectivity (K$_i$DOP/K$_i$MOP > 10 000 for the last compound), and metabolic stability (Schiller *et al.* 2000).

Prototypical DOP receptor agonists, such as DPDPE and deltorphin II, display high affinity for this receptor, but are not always selective because, for instance, DPDPE also has a high affinity for MOP receptors. Deltorphin II is a selective DOP agonist which seems to interact preferentially with the DOP2 subtype whereas another amphibian peptide, deltorphin I (or C), as well as DPDPE, are rather selective DOP1 agonists.

Fedotozine[(+)-(1R)-1-phenyl-1-[(3,4,5-trimethoxy) benzyloxymethyl]-N,N-dimethyl-n-propylamine] is an arylacetamide agonist that interacts preferentially with the KOP1 receptor subtype at the periphery (Delvaux 2001), whereas TRK820 [(-)-17-cyclopropylmethyl-3-14b-dihydroxy-4,5a-epoxy-6b-[N-methyl-trans-3-(3-furyl) acrylamide] morphinan hydrochloride] (Endoh *et al.* 2001) is a selective, centrally active, KOP receptor agonist.

Extensive binding studies have clearly demonstrated that NOP receptors are not recognized by known opioids, although its high affinity endogenous ligand, N/OFQ, does share part of its amino acid sequence with that of dynorphin A (Meunier *et al.* 1995, Reinscheid *et al.* 1995). Recently, a series of non-peptide mimetics of N/OFQ were reported mainly in patent literature (Barlocco *et al.* 2000). In particular, the compound Ro646198 ([1S,3aS]-8-(2,3,3a,4,5,6-hexahydro-1H-phenalen-1-yl)-1-phenyl-1,3,8-triaza-spiro[4.5]decan-4one) is a high affinity agonist at NOP receptors, which has a 100-fold lower potency at other OP receptors (Jenck *et al.* 2000).

20.4.3 Antagonists

There have been relatively few truly selective antagonists reported for OP receptors. Nevertheless CTOP and CTAP can be considered as satifactory pharmacological tools to efficiently block MOP receptors both *in vitro* and *in vivo*. With regard to DOP receptors, naloxone, buprenorphine and naltrindole clearly act as silent antagonists (Neilan *et al.* 1999), but their action is obviously not confined to these receptors. In contrast, the antagonist TippΨ [H-Tyr-Tic[Ψ, CH_2NH]Phe-Phe-OH (Tic = 1, 2, 3, 4-tetrahydroisoquinoline-3-carboxylic acid)], which displays subnanomolar affinity for DOP receptors, was initially considered as being highly selective of these receptors, and indeed, it is actually much more (by 500-fold) selective than naltrindole (Schiller *et al.* 1999). However, it was subsequently shown to exhibit some agonistic property, causing inhibition of DOP receptor-coupled adenylyl cyclase activity in various preparations (Martin *et al.* 2001). Similarly, WIN44413 [quadazocine, (-)-1-cyclopentyl-5-(1,2,3,4,5,-hexahydro-8-hydroxy-3,6,11-trimethyl-2,6-methano-3-benzazocin-11-yl)-3-pentanone methanesulfonate], initially designed as a selective KOP receptor antagonist, acts in fact as a non-selective OP receptor antagonist. In contrast, 5′-guanidinonaltrindole (GNTI), which is about two-fold more potent than norBNI, can still be considered as a really selective KOP receptor antagonist (Metzger *et al.* 2001). With regard to NOP receptors, J113397 (1-[(3R,4R)-1-cyclooctylmethyl-3-hydroxymethyl-4-piperidyl]-3-ethyl-1,3-dihydro-2H-benzimidazol-2-one) has been shown to act as a potent and rather selective antagonist (Kawamoto *et al.* 1999). In contrast, the analogue [Phe1Ψ(CH_2-NH)Gly^2]N/OFQ-(1–13)-NH_2, which was initially reported to behave as an antagonist, is in fact an agonist at NOP receptors (see Mollereau and Mouledous 2000). Recently, [Nphe^1]N/OFQ(1–13)NH_2 has been described as a pure NOP receptor antagonist, but with a relatively low potency (Calo *et al.* 2000).

20.4.4 Inverse agonists

β-chlornaltrexamine (β-CNA), clocinnamox (C-CAM), and BNTX are inverse agonists at constitutively active MOP receptors in some transfected cell lines (Burford *et al.* 2000). Naloxone and naltrexone are silent antagonists for untreated cells, but inverse agonists at MOP receptors that become constitutively active after morphine pretreatment (Wang *et al.* 2001). Constitutive activity has also been described for DOP receptor in some cell lines (Neilan *et al.* 1999), which allows the demonstration that ICI174864, the DOP receptor-selective antagonists BNTX and NTB, and the irreversible MOP receptor-antagonist C-CAM, are in fact DOP receptor inverse agonists.

20.5 Signal transduction and receptor modulation

In a variety of cell types, OP receptors are negatively coupled with adenylyl cyclase and modulate a number of membrane conductances through activation of $G_{i/o}$ proteins (Dhawan *et al.* 1996). In particular, OP receptor agonists activate inwardly rectifying K^+ currents and inhibit N- and P/Q-type Ca^{2+} channels, most likely via G protein βγ-subunits. The increase in K^+ conductance and decrease in Ca^{2+} conductance underlie the negative effect of OP receptor stimulation on plasma membrane excitability. On the other hand, OP receptor-mediated inhibition of adenylyl cyclase has various intracellular consequences on the transcription of genes, the activity of phosphatases and kinases, and other effectors (see Law and Loh 1999). Recently, a novel set of second messenger systems has been shown to be activated by OP receptors, through the $\beta\gamma$-subunits of G proteins coupled with the receptors (Ford et al. 1998). These include the activation of phospholipase C, protein kinase C (PKC), G protein receptor kinases (GRKs), dynamin 1, and the mitogen-activated protein kinase (MAPK) cascade (Zhang et al. 1999).

The opioid signalling cascades very probably involve cellular proteins other than the receptors, G proteins, and the effectors. Indeed, even if one considers the possible homo- or heterodimerization of OP receptors, the molecular size of G proteins ($\sim$80 kDa) and that of the OP receptors (58–65 kDa), the apparent molecular size of opioid receptor binding complexes is much higher (200–350 kDa; see Law et al. 2000) than that expected from the association of these components alone. Among other proteins which also contribute to these complexes, Law et al. (2000) recently identified a 22 kDa protein which is tightly associated with the MOP receptor; interestingly, calmodulin also shares this property (Wang et al. 1999).

Morphine, through the stimulation of DOP receptors, has been shown to promote the phosphorylation of the transcription factor 'cAMP responsive element binding protein' (CREB) in NG108–15 cells. This effect neither depends on protein synthesis nor involves the cAMP pathway, but requires Ca^{2+}/calmodulin and activation of PKC. Accordingly, in these cells, opioids affect two different intracellular cascades, an inhibitory one which involves the cAMP pathway, and a stimulatory one involving Ca^{2+} and the PKC pathway (Bilecki et al. 2000). In fact, already in the nineteen seventies, evidence has been reported that opioids can exert excitatory effects which are sensitive to either pertussis toxin (PTX) or cholera toxin (CTX) (Dhawan et al. 1996). Interestingly, stimulation of NOP receptors can also trigger adenylyl cyclase activation in some cell types (Onali et al. 2001).

Following the cloning of OP receptors, as well as that of G proteins, it has become possible to reconstitute combinations of receptor and G proteins in reasonably well-defined systems. These studies confirmed that OP receptors can couple to the PTX-sensitive G_{i1-3} and G_{o1-2}. In addition, they showed that PTX insensitive G proteins such as G_Z and G_{16} can also interact with the receptors. In line with these data, native OP receptors seem to actually interact with different G proteins. Thus, morphine, heroin, methadone, and buprenorphine clearly show different patterns of G protein activation associated with MOP receptor-mediated supraspinal antinociception. Studies of ligand interactions with MOP receptor mutants suggested that receptor occupancy by opioid drugs belonging to different classes might alter MOP receptor conformation in various, specific, fashions, thereby leading to the activation of different G proteins (Yu et al. 1997).

Phosphorylation of OP receptors in response to agonist stimulation has been demonstrated by several groups (see Dhawan et al. 1996). Phosphorylation takes place on the carboxyl tail (Afify et al. 1998) and/or the third intracellular loop (Koch et al. 1997). This process is receptor type-specific and may differ as a function of the agonist tested. For example, exposure to potent full agonists, such as sufentanil, etorphine or DAMGO led to strong MOP receptor phosphorylation, whereas methadone, morphine, β-endorphin and enkephalins produced an intermediate effect, and the partial agonist buprenorphine only minimally promoted receptor phosphorylation (Yu et al. 1997). PKC, Ca^{2+}/calmodulin-dependent protein kinases, GRKs, as well as MAPKs can phosphorylate OP receptors, thereby contributing to their agonist-mediated functional desensitization. In addition, agonist-independent mechanisms, such as heterologous desensitization, can also regulate OP receptor functioning. It has notably been shown that phosphorylation by PKC accounts for heterologous desensitization of DOP receptors, whereas homologous desensitization of these receptors involves phosphorylation (of other amino acid residues) by GRK (Xiang *et al.* 2001). Although phosphorylation of OP receptors is thought to be the initial step in OP receptor desensitization, there may be no direct correlation between the loss of agonist-induced inhibition of adenylyl cyclase activity on the one hand, and agonist-induced OP receptor phosphorylation on the other hand. This is notably the case for the MOP receptor, which led El Kouhen *et al.* (1999) to conclude that receptor phosphorylation is not an obligatory event for the agonist (DAMGO)-induced loss of adenylyl cyclase regulation by the receptor.

In addition to preventing receptor/G protein interactions, arrestins initiate OP receptor endocytosis in a dynamin-dependent process via clathrin-coated pits. Thus, overexpression of arrestin or GRK leads to enhanced agonist-induced internalization (Zhang J *et al.* 1998). Internalization of MOP or DOP receptors occurs within a few minutes of the continued presence of agonist and has been observed in both transfected cell lines (Keith *et al.* 1998) and myenteric (Sternini *et al.* 1996) and hippocampal neurones (Whistler *et al.* 1999). Phosphorylation of receptors has been suggested as the trigger for endocytosis to occur; however, this does not appear to be uniform among OP receptors. For some authors, agonist-induced phosphorylation of the MOP receptor even appears of subordinate significance for receptor internalization (Schulz *et al.* 1999). Furthermore, phosphorylation does not appear to be required for DOP receptor internalization. In that case, functional DOP receptors would exist as homodimers, and, upon prolonged stimulation, they would dissociate to monomers to be internalized (Cvejic and Devi 1997). Once receptors are internalized, they can be resensitized and recycled back to the plasma membrane. The requirement of internalization for resensitization has been demonstrated for both MOP and DOP receptors (Hasbi *et al.* 2000). Rapid MOP receptor endocytosis in neurones and transfected cells can be triggered by opioids, including enkephalins and endomorphins, and by opiates, like etorphine. In contrast, morphine, which binds to MOP receptors with high affinity and activates the same intracellular pathways as other opiates, does not appear to induce rapid receptor endocytosis in neurones *in vivo* or in transfected cells *in vitro* (Harrison *et al.* 2000). This has led to the suggestion that opioid agonists may differ in their ability to induce endocytosis of MOP receptors independently of their efficacy to trigger the receptor signalling cascades (Whistler *et al.* 1999). According to Whistler *et al.* (1999), the inability of morphine to promote receptor internalization would be related to the high capacity of this alkaloid to induce tolerance and dependence phenomena.

Prolonged and repeated exposure to opioid agonists reduces the responsiveness of OP receptors. This reduction in receptor function is hypothesized to contribute to opioid tolerance, dependence, and addiction in humans. Substantial experimental evidence allowed the identification of several events that are related to the agonist-induced loss of receptor function: (1) receptor desensitization (2) receptor internalization, and (3) receptor downregulation, as detailed above for MOP receptors. A prominent role for the ubiquitin/proteasome pathway in agonist-induced downregulation and basal turnover of MOP and DOP receptors has been proposed (Chaturvedi *et al.* 2001). In HEK293S cells expressing FLAG-tagged OP receptors, and Neuro2A cells, but not C6 glioma cells, functional PTX-sensitive G proteins are required for agonist-induced downregulation of the MOP but not the DOP receptor (Chaturvedi *et al.* 2001). The insensitivity of DOP receptor downregulation to PTX may indicate that receptor proteolysis is independent of G protein coupling, or that the agonist-stimulated receptor couples to PTX-insensitive G proteins. It is established that chronic administration of morphine, albeit with less efficacy than DAMGO, completely desensitizes and downregulates MOP receptors. However, as emphasized above, numerous reports indicate that morphine does not induce receptor phosphorylation and internalization (Keith *et al.* 1998). MAPK activation occurs when OP receptors are internalized (Polakiewicz *et al.* 1998), and may result in stimulation of transcription factors, such as CREB. Thus, the agonists etorphine, DAMGO, DPDPE and U69593, which cause receptor internalization, also stimulate the MAPK pathway (Belcheva *et al.* 1998). In contrast, morphine does not affect this pathway, although it provokes an upregulation of adenylyl cyclase activity, a cellular marker of dependence. In mice lacking β-arrestin-2, desensitization of the MOP receptor does not occur after chronic morphine treatment, and these animals fail to develop antinociceptive tolerance. However, the deletion of β-arrestin-2 does not prevent the chronic morphine-induced upregulation of adenylyl cyclase activity, and the mutant mice still become physically dependent on the drug (Bohn *et al.* 2000).

20.6 Physiology and Disease Relevance

20.6.1 Drug dependence

The cloning of the human MOP receptor gene has prompted studies of DNA sequence variations within the gene, and their possible association with substance (heroin/cocaine) dependence. More than 40 variants were identified within the human MOP receptor gene (Hoehe *et al.* 2000). Six mutations are in the coding region, and five of them modify the encoded protein sequence. Thus, the A118G polymorphism, that results in an Asn/Asp change at amino acid residue 40, is associated with differences in receptor binding and signal transduction between the two alleles (Bond *et al.* 1998). However, neither this polymorphism, nor the Ala6Val polymorphism, appear to be associated with the risk of developing drug addiction (Gelernter *et al.* 1999), and contradictory results have been published concerning the association of the A118G polymorphism with alcohol dependence (Gscheidel *et al.* 2000). A mutation found in the third intracellular loop, which replaces a serine by a proline (S268P), represents a loss-of-function mutation for the human MOP receptor, which may have an incidence on opioid-regulated behaviours or drug addiction (Befort *et al.* 2001).

Studies of natural sequence variations in human DOP receptor gene showed that they have little influence on ligand binding or receptor downregulation but can otherwise modify receptor density and signalling. Although the amino acid sequence remains unchanged in the product of the allelic variant T307C of the gene encoding the DOP receptor, it is interesting to note that the biallelic CC genotype was found more frequently, and the TT genotype less frequently, in heroin addicts than in control subjects. This would suggest that the C allele predisposes to heroin abuse (Mayer *et al.* 1997). However, this conclusion has not been supported by a more recent study that failed to detect any significant association between genetic variation of the DOP receptor and the risk for heroin or alcohol abuse (Franke *et al.* 1999).

20.6.2 Pain and analgesia

The molecular cloning of OP receptors and the availability of gene targeting technology provide the opportunity to characterize unambiguously the physiological role of each receptor type and to re-evaluate its possible contribution to the effects of opioid drugs. MOP receptor knockout mice exhibit reduced locomotor activity compared to the wild type, as well as decreased sexual activity, increased immune cell activity (Tian *et al.* 1997), and lower nociceptive thresholds (Uhl *et al.* 1999). As expected, disruption of the MOP receptor gene allowed the demonstration that the encoded receptor is a mandatory component of the opioid system for the main actions of morphine, including analgesia and lethality (Loh *et al.* 1998). In particular, homozygous MOP knock-out mice dot not exhibit morphine-induced locomotion, tolerance, physical dependence, and reward, as well as the effects of the alkaloid on gastrointestinal transit, macrophage phagocytosis, and secretion of tumor necrosis factor-α. However, other morphine effects on immune cell functions, such as the inhibition of IL-1 and IL-6 secretion by macrophages (Roy *et al.* 1998), are not altered by elimination of the MOP receptors, thereby suggesting their mediation through other receptors. *In vitro* studies using various cell types have shown that some morphine responses persist after inactivation of as many as 90 per cent of the initial MOP receptor content, while others are attenuated after inactivating only a few percentage of these receptors. Varying levels of MOP receptor reserve could thus exist in different MOP-expressing cells. In this respect, experiments using heterozygous MOP mice which express half of wild-type MOP receptor levels are especially relevant. Indeed, studies of morphine actions in these mice supported the idea that different MOP receptor reserves exist in brain circuits that mediate distinct opioid effects. Heterozygous animals display attenuated locomotion, reduced morphine self-administration, rightward shifts in morphine lethality dose/effect relationships, and variable behaviours in place preference tests compared to wild-type mice. In contrast, they demonstrate full physical dependence, as measured by naloxone-precipitated abstinence following five days of morphine administration. Although neuroadaptive changes at sites other than MOP receptors might also be involved, these data support the idea that interindividual differences in levels of MOP receptor expression can differentially affect one versus another action of opioid drugs (Sora *et al.* 2001).

DOP receptor signalling appears to be unaltered in MOP knock-out mutants. However, in different strains of MOP receptor deficient mice, DOP receptor-mediated analgesia is either abolished, impaired, or maintained in models of acute or transient pain (Hosohata *et al.* 2000), whereas it is maintained in stress-induced analgesia (LaBuda *et al.* 2000) and enhanced in a model of persistent inflammatory pain (Qiu *et al.* 2000). These data suggest that the genetic background may play a critical role in opioid-mediated physiological control of nociceptive mechanisms, possibly through strain differences in MOP/DOP receptor interactions.

Studies in DOP knock-out mutants have shown that DOP receptors are absolutely needed for spinal, but not supraspinal, induction of analgesia by selective DOP receptor agonists. Interestingly, these mutants fail to develop tolerance to the analgesic effects of morphine (Zhu *et al.* 1999). This finding is congruent with previous data showing that the pharmacological blockade of DOP receptors attenuates the development of morphine tolerance and physical dependence. Accordingly, compounds that possess a MOP agonist/DOP antagonist profile would be especially interesting therapeutic agents (see Wells *et al.* 2001). However other pharmacological combinations should also be considered because morphine withdrawal is attenuated in KOP knock-out mice (Simonin *et al.* 1998), and NOP knock-out mutants hardly develop tolerance to morphine (Ueda *et al.* 1997).

20.6.3 Parkinson's disease

DOP receptors are abundantly expressed in rat and human caudate putamen. Interestingly, the non-peptide DOP receptor agonist SNC80 has been demonstrated to induce anti-Parkinsonian-like behaviours in rats resembling those elicited by dopamine receptor agonists acting at D_1 and/or D_2 receptors (Spina *et al.* 1998). Such agonists have also been shown to facilitate D_1 and D_2 receptor agonist turning behaviour in unilaterally 6-OHDA lesioned rats, although they have no effect when administered alone (Pinna and Di Chiara 1998). Further evidence for an interaction between DOP receptors and the dopaminergic system include the observation that D_1 receptors are upregulated in young rats following SNC80 treatment during preweaning (Shieh *et al.* 1997). Finally, it should be investigated whether DOP receptor agonists might have some neuroprotective action in Parkinson's disease as a recent report demonstrated that activation of neuronal DOP receptors (but not MOP or KOP receptors) can protect cortical neurones from glutamate excitotoxicity (Zhang *et al.* 2000).

20.6.4 Alzheimer's disease, cognition, and anxiety

NOP knock-out mice were reported to display improved spatial attention and memory (Manabe *et al.* 1998). In addition, consistent anxiogenic- and depressive-like responses in DOP knock-out mice indicate that DOP receptor stimulation might normally exert a positive influence on mood (Filliol *et al.* 2000). These observations should lead to further investigations regarding the potential interest of OP receptor ligands not only for alleviating pain but also for improving cognitive functions and mood in some psychopathological disorders.

20.7 Concluding remarks

There is obviously considerable interest in opioid receptors because they are well recognized as targets of the most potent analgesic drugs available since the early age of human life on earth. Historically, conclusions from pharmacological studies rely on the use of MOP, DOP, and KOP receptor ligands whose selectivity may be questioned, particularly under *in vivo* conditions where the route of administration and the pharmacokinetic properties of these ligands are critical. However, with the recent advent of the cloning of these receptors, the generation of multiple transgenic lines and the discovery of opioid receptor heterodimerization, futher major advances in the understanding and development of new and selective opioid receptor ligands can be imagined in the coming decades. It is hoped that these studies will help us in the development of not only better, less addictive, analgesics, but also in molecules useful in a number of other disorders such as inflammatory conditions, anxiety and depression, Parkinson's disease, and cognitive impairment.

References

Abbadie C, Pan Y, Drake CT, and Pasternak GW (2000). Comparative immunohistochemical distributions of carboxy terminus epitopes from the mu-opioid receptor splice variants MOR-1D, MOR-1 and MOR-1C in the mouse and rat CNS. *Neuroscience* **100**, 141–53.

Afify EA, Law PY, Riedl M, Elde R, and Loh HH (1998). Role of carboxyl terminus of μ- and ∂-opioid receptor in agonist-induced down-regulation. *Mol Brain Res* **54**, 24–34.

Anton B, Fein J, To T, Li X, Silberstein L, and Evans CJ (1996). Immunohistochemical localization of ORL-1 in the central nervous system of the rat. *J Com Neurol* **368**, 248–68.

Arvidsson U, Dado RJ, Riedl M *et al.* (1995a) ∂-opioid receptor immunoreactivity: distribution in brain stem and spinal cord, and relationship to biogenic amines and enkephalin. *J Neurosci* **15**, 1215–35.

Arvidsson U, Riedl M, Chakrabarti S *et al.* (1995b). The κ-opioid receptor is primarily postsynaptic: combined immunohistochemical localization of the receptor and endogenous opioids. *Proc Natl Acad Sci USA* **92**, 5062–6.

Arvidsson U, Riedl M, Chakrabarti S *et al.* (1995c). Distribution and targeting of a mu-opioid receptor (MOR1) in brain and spinal cord. *J Neurosci* **15**, 3328–41.

Bare LA, Mansson E, and Yang D (1994). Expression of two variants of the human mu opioid receptor mRNA in SK-N-SH cells and human brain. *FEBS Lett* **354**, 213–6.

Barlocco D, Cignarella G, Giardina GAM, and Toma L (2000). The opioid-receptor-like 1 (ORL-1) as a potential target for new analgesics. *Eur J Med Chem* **35**, 275–82.

Befort K, Filliol D, Décaillot FM, Gavériaux-Ruff C, Hoehe MR, and Kieffer BL (2001). A single nucleotide polymorphic mutation in the human μ-opioid receptor severely impairs receptor signaling. *J Biol Chem* **276**, 3130–7.

Belcheva MM, Vogel Z, Ignatova E *et al.* (1998). Opioid modulation of extracellular signal-regulated protein kinase activity is Ras-dependent and involves Gβγ subunits. *J Neurochem* **70**, 635–45.

Bilecki W, Höllt V, and Przewlocki R (2000). Acute ∂-opioid receptor activation induces CREB phosphorylation in NG108-15 cells. *Eur J Pharmacol* **390**, 1–6.

Bohn LM, Gainetdinov RR, Lin FT, Lefkowitz RJ, and Caron MG (2000). μ-opioid receptor desensitization by β-arrestin-2 determines morphine tolerance but not dependence. *Nature* **408**, 720–3.

Bond C, LaForge KS, Tian M *et al.* (1998). Single-nucleotide polymorphism in the human mu opioid receptor gene alters beta-endorphin binding and activity: possible implications for opiate addiction. *Proc Natl Acad Sci USA* **95**, 9608–13.

Bunzow JR, Saez C, Mortrud M *et al.* (1994). Molecular cloning and tissue distribution of a putative member of the rat opioid receptor gene family that is not a μ, ∂ and κ opioid receptor type. *FEBS Lett* **347**, 284–8.

Burford NT, Wang D, and Sadée W (2000). G-protein coupling of μ-opioid receptors (OP3): elevated basal signalling activity. *Biochem J* **348**, 531–7.

Calo G, Guerrini R, Rizzi A, Salvadori S, and Regoli D (2000). Pharmacology of nociceptin and its receptor: a novel therapeutic target. *Br J Pharmacol* **129**, 1261–83.

Cesselin F (1995). Opioid and anti-opioid peptides. *Fundam Clin Pharmacol* **9**, 409–33.

Chaturvedi K, Bandari P, Chinen N, and Howells RD (2001). Proteasome involvement in agonist-induced down-regulation of μ and ∂ opioid receptors. *J Biol Chem* **276**, 12 345–55.

Chen Y, Fan Y, Liu J *et al.* (1994). Molecular cloning, tissue distribution and chromosomal localization of a novel member of the opioid receptor gene family. *FEBS Lett* **347**, 279–83.

Chen Y, Mestek A, Liu J, Hurley JA, and Yu L (1993a). Molecular cloning and expression of a mu-opioid receptor from rat brain. *Mol Pharmacol* **44**, 8–12.

Chen Y, Mestek A, Liu J, and Yu L (1993b). Molecular cloning of a rat κ opioid receptor reveals sequence similarities to the μ and ∂ opioid receptors. *Biochem J* **295**, 625–8.

Cheng P, Liu-Chen L, and Pickel V (1997). Dual ultrastructural immunocytochemical labeling of mu and delta opioid receptors in the superficial layers of the rat cervical spinal cord. *Brain Res* **778**, 367–80.

Cox BM (2000). *Opioid Receptors. The IUPHAR receptor Compendium*, IUPHAR Media Publ., London, pp. 321–33.

Cvejic S and Devi LA (1997). Dimerization of the ∂ opioid receptor: implication for a role in receptor internalization. *J Biol Chem* **272**, 26 959–64.

Danielson PB and Dores RM (1999). Molecular evolution of the opioid/orphanin FQ family. *Gen Comp Endocrinol* 113, 168–86.

Delvaux M (2001). Pharmacology and clinical experience with fedotozine. *Expert Opinion on Investigational Drugs*, 10, 97–110

Dhawan BN, Cesselin F, Raghubir R *et al.* (1996). International Union of Pharmacology. XII. Classification of opioid receptors. *Pharmacol Rev* 48, 567–92.

El Kouhen R, Maestri-El Kouhen O, Law PY, and Loh HH (1999). The absence of correlation between the loss of [D-Ala^2, MePhe^4, Gly^5-ol]enkephalin inhibition of adenylyl cyclase activity and agonist-induced μ-opioid receptor phosphorylation. *J Biol Chem* 274, 9207–15.

Endoh T, Tajima A, Izumoto N *et al.* (2001). TRK-820, a selective κ-opioid agonist, produces potent antinociception in cynomolgus monkeys. *J J Pharmacol* 85, 282–90.

Erspamer V (1992). The opioid peptides of the amphibian skin. *Int J Dev Neurosci* 10, 3–30.

Evans CJ, Keith DE, Morrison H, Magendzo K, and Edwards RH (1992). Cloning of a delta opioid receptor by functional expression. *Science* 258, 1952–5.

Filliol D, Ghozland S, Chluba J, *et al.* (2000). Mice deficient for ∂- and μ-opioid receptors exhibit opposing alterations of emotional responses. *Nat Genet* 25, 195–200.

Ford CE, Skiba NP, Bae H *et al.* (1998). Molecular basis for interactions of G protein beta/gamma subunits with effectors. *Science* 280, 1271–4.

Franke P, Nöthen MM, Wang T *et al.* (1999). Human ∂-opioid receptor gene and susceptibility to heroin and alcohol dependence. *Am J Med Genet (Neuropsychiatr Genet)* 88, 462–4.

Fukuda K, Kato S, and Mori K (1995). Location of regions of the opioid receptor involved in selective agonist binding. *J Biol Chem* 270, 6702–9.

Fukuda K, Kato S, Mori K, Nishi M, and Takeshima H (1993). Primary structures and expression from cDNAs of rat opioid receptor ∂- and μ-subtypes. *FEBS Lett* 327, 311–4.

Gavériaux-Ruff C, Peluso J, Befort K, Simonin F, Zilliox C, and Kieffer BL (1997). Detection of opioid receptor mRNA by RT-PCR reveals alternative splicing for the ∂- and κ-opioid receptors. *Mol Brain Res* 48, 298–304.

Gelernter J, Kranzler H, and Cubells J (1999). Genetics of two μ opioid receptor gene (OPRM1) exon I polymorphisms: population studies, and allele frequencies in alcohol- and drug-dependent subjects. *Mol Psychiatry* 4, 476–83.

George SR, Fan T, Xie Z *et al.* (2000). Oligomerization of μ- and ∂-opioid receptors. Generation of novel functional properties. *J Biol Chem* 275, 26 128–35.

Gscheidel N, Sander T, Wendel B, *et al.* (2000). Five exon 1 variants of mu opioid receptor and vulnerability to alcohol dependence. *Polish J Pharmacol* 52, 27–31.

Harrison C, Rowbotham DJ, Grandy DK, and Lambert DG (2000). Endomorphin-1 induced desensitization and down-regulation of the recombinant mu-opioid receptor. *Br J Pharmacol* 131, 1220–6.

Hasbi A, Allouche S, Sichel F *et al.* (2000). Internalization and recycling of delta-opioid receptor are dependent on a phosphorylation-dephosphorylation mechanism. *J Pharmacol Exp Ther* 293, 237–47.

Hawkinson JE, Acosta-Burruel M, and Espitia SA (2000). Opioid activity profiles indicate similarities between the nociceptin/orphanin FQ and opioid receptors. *Eur J Pharmacol* 389, 107–14.

Hoehe MR, Kopke K, Wendel B *et al.* (2000). Sequence variability and candidate gene analysis in complex disease: association of mu opioid receptor gene variation with substance dependence. *Hum Mol Genet* 9, 2895–908.

Hosohata Y, Vanderah TW, Burkey TH *et al.* (2000). Delta-opioid receptor agonists produce anti-nociception and [^{35}S]GTPgammaS binding in mu receptor knockout mice. *Eur J Pharmacol* 388, 241–8.

Jenck F, Wichmann J, Dautzenberg FM *et al.* (2000). A synthetic agonist at the orphanin FQ: nociceptin receptor ORL1: anxiolytic profile in the rat. *Proc Natl Acad Sci USA* 97, 4938–43.

Jordan BA, Cvejic S, and Devi LA (2000). Opioids and their complicated receptor complexes. *Neuropsychopharmacology* 23, S5–18.

Jordan BA and Devi LA (1999). G-protein-coupled receptor heterodimerization modulates receptor function. *Nature* 399, 697–700.

Jordan BA, Trapaidze N, Gomes I, Nivarthi R, and Devi LA (2001). Oligomerization of opioid receptors with β2-adrenergic receptors: a role in trafficking and mitogen-activated protein kinase activation. *Proc Natl Acad Sci USA* 98, 343–8.

Kawamoto H, Ozaki S, Itoh Y *et al.* (1999). Discovery of the first potent and selective small molecule opioid receptor-like (ORL1) antagonist: 1-[(3R,4R)-1-cyclooctylmethyl-3-hydroxymethyl-4-piperidyl]-3-ethyl-1,3- dihydro-2H-benzimidazol-2-one (J-113397). *J Med Chem* 42, 5061–3.

Keith DE, Anton B, Murray SR *et al.* (1998). μ-Opioid receptor internalization: opiate drugs have differential effects on a conserved endocytic mechanism in vitro and in the mammalian brain. *Mol Pharmacol* 53, 377–84.

Kieffer BL (1999). Opioids: first lessons from knockout mice. *Trends Pharmacol Sci* 20, 19–26.

Kieffer B, Befort K, Gaveriaux-Ruff C, and Hirth CG (1992). The delta opioid receptor: isolation of a cDNA by expression cloning and pharmacological characterization. *Proc Natl Acad Sci USA* 89, 12 048–52.

Kim DS, Chin H, and Klee WA (1995). Agonist regulation of the expression of the delta opioid receptor in NG108-15 cells. *FEBS Lett* 376, 11–4.

Koch T, Kroslak T, Mayer P, Raulf E, and Höllt V (1997). Site mutation in the rat μ-opioid receptor demonstrates the involvement of calcium/calmodulin-dependent protein kinase II in agonist-mediated desensitization. *J Neurochem* 69, 1767–70.

LaBuda CJ, Sora I, Uhl GR, and Fuchs PN (2000). Stress-induced analgesia in μ-opioid receptor knockout mice reveals normal function of the ∂-opioid receptor system. *Brain Res* 869, 1–5.

Larson DL, Jones RM, Hjorth SA, Schwartz TA, and Portoghese PS (2000). Binding of norbinaltorphimine (norBNI) congeners to wild-type and mutant mu and kappa opioid receptors: molecular recognition loci for the pharmacophore and address components of kappa antagonists. *J Med Chem* 43, 1573–6.

Law PY and Loh HH (1999). Regulation of opioid receptor activities. *J Pharmacol Exp Ther* 289, 607–24.

Law PY, Tine SJ, McLeod LA, and Loh HH (2000). Association of a lower molecular weight protein to the μ-opioid receptor demonstrated by ^{125}I-β-endorphin cross-linking studies. *J Neurochem* 75, 164–73.

Lee TL, Fung FM, Chen FG *et al.* (1999). Identification of human, rat and mouse nocistatin in brain and human cerebrospinal fluid. *Neuroreport* 10, 1537–41.

Letchworth SR, Mathis JP, Rossi GC, Bodnar RJ, and Pasternak GW (2000). Autoradiographic localization of ^{125}I[Tyr^{14}]orphanin FQ/nociceptin and ^{125}I[Tyr^{10}]orphanin FQ/nociceptin binding sites in rat brain. *J Comp Neurol* 423, 319–29.

Loh HH, Liu HC, Cavalli A, Yang W, Chen YF, and Wei LN (1998). Mu opioid receptor knockout in mice: effects on ligand-induced analgesia and morphine lethality. *Brain Res Mol Brain Res* 54, 321–6.

Manabe T, Noda Y, Mamiya T *et al.* (1998). Facilitation of long-term potentiation and memory in mice lacking nociceptin receptors. *Nature* 394, 577–81.

Mansour A, Taylor LP, Fine JL *et al.* (1997) Key residues defining the μ-opioid receptor binding pocket: a site-directed mutagenesis study. *J Neurochem* 68, 344–53.

Martin NA, Terruso MT, and Prather PL (2001). Agonist activity of the ∂-antagonists TIPP and TIPP-Ψ in cellular models expressing endogenous or transfected ∂-opioid receptors. *J Pharmacol Exp Ther* 298, 240–8.

Mayer P, Rochlitz H, Rauch E *et al.* (1997). Association between a delta opioid receptor gene polymorphism and heroin dependence in man. *Neuroreport* 8, 2547–50.

McVey M, Ramsay D, Kellett E *et al.* (2001). Monitoring receptor oligomerization using time-resolved fluorescence resonance energy transfer and bioluminescence resonance energy transfer. The human

∂-opioid receptor displays constitutive oligomerization at the cell surface which is not regulated by receptor occupancy. *J Biol Chem* **276**, 14 092–9.

Meng F, Hoversten MT, Thompson RC, Taylor L, Watson SJ, and Akil H (1995). A chimeric study of the molecular basis of affinity and selectivity of the kappa and the delta opioid receptors. Potential role of extracellular domains. *J Biol Chem* **270**, 12730–6.

Meng F, Xie GX, Thompson RC *et al.* (1993). Cloning and pharmacological characterization of a rat κ opioid receptor. *Proc Natl Acad Sci USA* **90**, 9954–8.

Metzger TG, Paterlini MG, Ferguson DM, and Portoghese PS (2001). Investigation of the selectivity of oxymorphone- and naltrexone-derived ligands via site-directed mutagenesis of opioid receptors: exploring the 'address' recognition locus. *J Med Chem* **44**, 857–62.

Meunier JC, Mollereau C, Toll L *et al.* (1995). Isolation and structure of the endogenous agonist of opioid receptor-like ORL1 receptor. *Nature* **377**, 532–5.

Mollereau C and Mouledous L (2000). Tissue distribution of the opioid receptor-like (ORL1) receptor. *Peptides* **21**, 907–17.

Mollereau C, Parmentier M, Mailleux P *et al.* (1994). ORL1, a novel member of the opioid receptor family: cloning, functional expression and localization. *FEBS Lett* **341**, 33–8.

Mollereau C, Simons MJ, Soularue P *et al.* (1996). Structure, tissue distribution, and chromosomal localization of the prepronociceptin gene. *Proc Natl Acad Sci USA* **93**, 8666–70.

Neilan CL, Akil H, Woods JH, and Traynor JR (1999). Constitutive activity of the ∂-opioid receptor expressed in C6 glioma cells: identification of non-peptide ∂-inverse agonists. *Br J Pharmacol* **128**, 556–62.

Nishi M, Takeshima H, Mori M, Nakagawara K, and Takeuchi T (1994). Structure and chromosomal mapping of genes for the mouse κ-opioid receptor and an opioid receptor homologue (MORC). *Biochem Biophys Res Commun* **205**, 1353–7.

Onali P, Ingianni A, and Olianas MC (2001). Dual coupling of opioid receptor-like (ORL1) receptors to adenylyl cyclase in the different layers of the rat main olfactory bulb. *J Neurochem* **77**, 1520–30.

Pasternak GW (2001). Incomplete cross tolerance and multiple mu opioid peptide receptors. *Trends Pharmacol Sci* **22**, 67–70.

Peluso J, Laforge KS, Matthes HW, Kreek MJ, Kieffer BL, and Gaveriaux-Ruff C (1998). Distribution of nociceptin/orphanin FQ receptor transcript in human central nervous system and immune cells. *J Neuroimmunol* **81**, 184–92.

Pert C and Snyder S (1973). Opiate receptor demonstration in nervous tissue. *Science* **179**, 1011–4.

Pinna A and Di Chiara G (1998). Dopamine-dependent behavioural stimulation by non-peptide delta opioids BW373U86 and SNC80: 3. Facilitation of D1 and D2 responses in unilaterally 6-hydroxydopamine-lesioned rats. *Behav Pharmacol* **9**, 15–21.

Polakiewicz RD, Schieferl SM, Dorner LF, Kansra V, and Comb MJ (1998). A mitogen-activated protein kinase pathway is required for μ-opioid receptor desensitization. *J Biol Chem* **273**, 12402–6.

Qiu C, Sora I, Ren K, Uhl G, and Dubner R (2000). Enhanced ∂-opioid receptor-mediated antinociception in μ-opioid receptor-deficient mice. *Eur J Pharmacol* **387**, 163–9.

Rady JJ, Holmes BB, Tseng LF, and Fujimoto JM (2001). Inverse agonist action of leu-enkephalin at ∂2-opioid receptors mediates spinal antianalgesia. *J Pharmacol Exp Ther* **297**, 582–9.

Reinscheid RK, Nothacker HP, Bourson A *et al.* (1995). A neuropeptide that activates an opioid-like G protein-coupled receptor. *Science* **270**, 792–4.

Rossi GC, Su W, Lewenthal L, Su H, and Pasternak GW (1997). Antisense mapping DOR-1 in mice: further support for ∂ receptor subtypes. *Brain Res* **753**, 176–9.

Roy S, Barke RA, and Loh HH (1998). Mu-opioid receptor-knockout mice: role of μ-opioid receptor in morphine mediated immune functions. *Brain Research Mol Brain Res* **61**, 190–4.

Sakurada S, Zadina JE, Kastin AJ *et al.* (1999). Differential involvement of μ-opioid receptor subtypes in endomorphin-1- and -2-induced antinociception. *Eur J Pharmacol* **372**, 25–30.

Schiller PW, Nguyen TM, Berezowska I *et al.* (2000). Synthesis and in vitro opioid activity profiles of DALDA analogues. *Eur J Med Chem* **35**, 895–901.

Schiller PW, Weltrowska G, Berezowska I *et al.* (1999). The TIPP opioid peptide family: development of delta antagonists, delta agonists, and mixed mu agonist/delta antagonists. *Biopolymers* **51**, 411–25.

Schulz R, Wehmeyer A, Murphy J, and Schulz K (1999). Phosducin, β-arrestin and opioid receptor migration. *Eur J Pharmacol* **375**, 349–57.

Shieh G-J, Ravis WR, and Walters DE (1997) Up-regulation of dopamine D1-receptors in the brain of 28-day-old rats exposed to the delta (δ) opioid agonist SNC80 during the preweaning period. *Brain Res Dev Brain Res* **103**, 209–11.

Simon E, Hiller J, and Edelman I (1973). Stereospecific binding of the potent narcotic analgesic ^{3}H-etorphine to rat brain homogenate. *Proc Natl Acad Sci USA* **70**, 1947–51.

Simonin F, Valverde O, Smadja C *et al.* (1998). Disruption of the κ-opioid receptor gene in mice enhances sensitivity to chemical visceral pain, impairs pharmacological actions of the selective κ-agonist U-50, 488H and attenuates morphine withdrawal. *EMBO J* **17**, 886–97.

Sora I, Elmer G, Funada M *et al.* (2001). Mu opiate receptor gene dose effects on different morphine actions: evidence for differential in vivo mu receptor reserve. *Neuropsychopharmacology* **25**, 41–54.

Spina L, Longoni R, Mulas A, Chang KJ, and Di Chiara G (1998). Dopamine-dependent behavioural stimulation by non-peptide delta opioids BW373U86 and SNC 80 : 1. Locomotion, rearing and stereotypies in intact rats. *Behav Pharmacol* **9**, 1–8.

Sternini C, Spann M, Anton B *et al.* (1996). Agonist-selective endocytosis of mu opioid receptor by neurons in vivo. *Proc Natl Acad Sci USA* **93**, 9241–6.

Surrat CK, Johnson PS, Moriwaki A *et al.* (1994). Mu opiate receptor. Charged transmembrane domain amino acids are critical for agonist recognition and intrinsic activity. *J Biol Chem* **269**, 20 548–53.

Terenius L (1973). Characteristics of the receptor for narcotic analgesics in synaptic plasma membrane fraction from rat brain. *Acta Pharmacol Toxicol* **32**, 317–29.

Tian M, Broxmeyer HE, Fan Y *et al.* (1997). Altered hematopoiesis, behavior, and sexual function in μ opioid receptor-deficient mice. *J Exp Med* **185**, 1517–22.

Ueda H, Yamaguchi T, Tokuyama S, Inoue M, Nishi M, and Takeshima H (1997). Partial loss of tolerance liability to morphine in mice lacking the nociceptin receptor gene. *Neurosci Lett,* **237**, 136–8.

Uhl GR, Sora I, and Wang Z (1999). The μ opiate receptor as a candidate gene for pain: polymorphisms, variations in expression, nociception, and opiate responses. *Proc Natl Acad Sci USA* **96**, 7752–5.

Wang D, Raehal KM, Bilsky EJ, and Sadée W (2001). Inverse agonists and neutral antagonists at μ opioid receptor (MOR): possible role of basal receptor signaling in narcotic dependence. *J Neurochem,* **77**, 1590–600.

Wang D, Sadée W, and Quillian JM (1999). Calmodulin binding to G protein-coupling domain of opioid receptors. *J Biol Chem* **274**, 22 081–8.

Wang JB, Imai Y, Eppler CM, Gregor P, Spivak CE and Uhl GR (1993). μ Opiate receptor: cDNA cloning and expression. *Proc Natl Acad Sci USA* **90**, 10 230–4.

Wang JB, Johnson PS, Imai Y *et al.* (1994). cDNA cloning of an orphan opiate receptor gene family member and its splice variant. *FEBS Lett* **348**, 75–9.

Wang WW, Shahrestanifar M, Jin J, and Howells R (1995). Studies on μ and ∂ opioid receptor selectivity utilizing chimeric and site-mutagenized receptors. *Proc Natl Acad Sci USA* **92**, 12 436–40.

Wei LN, Hu X, Bi J, and Loh HH (2000). Post-transcriptional regulation of mouse κ opioid receptor expression. *Mol Pharmacol* **57**, 401–8.

Wells JL, Barttlett JL, Ananthan S, and Bilsky EJ (2001). *In vivo* pharmacological characterization of SoRI 9409, a nonpeptidic opioid μ-agonist/∂-antagonist that produces limited antinociceptive tolerance and attenuates morphine physical dependence. *J Pharmacol Exp Ther* **297**, 597–605.

Wessendorf MW and Dooyema J (2001). Coexistence of kappa- and delta-opioid receptors in rat spinal cord axons. *Neurosci Lett* **298**, 151–4.

Whistler J, Chuang HH, Chu P, Jan LY, and von Zastrow M (1999). Functional dissociation of μ opioid signaling and endocytosis: implications for the biology of opiate tolerance and dependence. *Neuron* 23, 737–46.

Xiang B, Yu GH, Chen L *et al.* (2001). Heterologous activation of protein kinase C stimulates phosphorylation of ∂-opioid receptor at serine 344, resulting in β-arrestin- and clathrin-mediated receptor internalization. *J Biol Chem* 276, 4709–16.

Xie GX, Meuser T, Pietruck C, Sharma M, and Palmer PP (1999). Presence of opioid receptor-like (ORL1) receptor mRNA splice variants in peripheral sensory and sympathetic neuronal ganglia. *Life Sci* 64, 2029–37.

Xu W, Ozdener F, Li JG *et al.* (1999). Functional role of the spatial proximity of Asp[114](2.50) in TMH 2 and Asn[332](7.49) in TMH 7 of the mu opioid receptor. *FEBS Lett* 447, 318–24.

Yasuda K, Raynor K, Kong H *et al.* (1993). Cloning and functional comparison of kappa and delta opioid receptors from mouse brain. *Proc Natl Acad Sci USA* 90, 6736–40.

Yu Y, Zhang L, Yin X, Sun H, Uhl GR, and Wang JB (1997). μ Opioid receptor phosphorylation, desensitization and ligand efficacy. *J Biol Chem* 272, 28869–74.

Zadina JE, Hackler L, Ge LJ, and Kastin AJ (1997). A potent and selective endogenous agonist for the μ-opiate receptor. *Nature* 386, 499–502.

Zagon IS, Verderame MF, Zimmer WE, and McLaughlin PJ (2000). Molecular characterization and distribution of the opioid growth factor receptor (OGFr) in mouse. *Mol Brain Res* 84, 106–14.

Zhang J, Ferguson SS, Barak LS *et al.* (1998). Role for G protein-coupled receptor kinase in agonist-specific regulation of μ-opioid receptor responsiveness. *Proc Natl Acad Sci USA* 95, 7157–62.

Zhang J, Haddad GG and Xia Y (2000). Delta-, but not mu- and kappa-, opioid receptor activation protects neocortical neurons from glutamate-induced excitotoxic injury. *Brain Res* 885, 143–53.

Zhang S, Tong Y, Tian M *et al.* (1998). Dynorphin A as a potential endogenous ligand for four members of the opioid receptor gene family. *J Pharmacol Exp Ther* 286, 136–41.

Zhang X, Bao L, Arvidsson U, Elde R, and Hökfelt T (1998). Localization and regulation of the ∂-opioid receptor in dorsal root ganglia and spinal cord of the rat and monkey: evidence for association with the membrane of large dense-core vesicles. *Neuroscience* 82, 1225–42.

Zhang Z, Xin SM, Wu GX, Zhang WB, Ma L, and Pei G (1999). Endogenous ∂-opioid and ORL1 receptors couple to phosphorylation and activation of p38 MAPK in NG108-15 cells and this is regulated by protein kinase A and protein kinase C. *J Neurochem* 73, 1502–9.

Zhu Y, King M, Schuller A *et al.* (1999). Retention of supraspinal delta-like analgesia and loss of morphine tolerance in delta opioid receptor knockout mice. *Neuron* 24, 243–52.

Chapter 21

Purinergic receptors

Brian F. King and Geoffrey Burnstock

21.1 Introduction

The term *purinergic receptor* (or *purinoceptor*) was first introduced to describe classes of membrane receptors that, when activated by either neurally released ATP (*P2 purinoceptor*) or its breakdown product adenosine (*P1 purinoceptor*), mediated relaxation of gut smooth muscle (Burnstock 1972, 1978). P2 purinoceptors were further divided into five broad phenotypes (P2X, P2Y, P2Z, P2U, and P2T) according to pharmacological profile and tissue distribution (Burnstock and Kennedy 1985; Gordon 1986; O'Connor *et al.* 1991; Dubyak 1991). Thereafter, they were reorganized into families of metabotropic ATP receptors (P2Y, P2U, and P2T) and ionotropic ATP receptors (P2X and P2Z) (Dubyak and El-Moatassim 1993), later redefined as extended P2Y and P2X families (Abbracchio and Burnstock 1994).

In the early 1990s, cDNAs were isolated for three heptahelical proteins—called $P2Y_1$, $P2Y_2$, and $P2Y_3$—with structural similarities to the rhodopsin GPCR template. At first, these three GPCRs were believed to correspond to the P2Y, P2U, and P2T receptors. However, the complexity of the P2Y receptor family was underestimated. At least 15, possibly 16, heptahelical proteins have been associated with the P2Y receptor family (King *et al.* 2001, see Table 21.1). Multiple expression of P2Y receptors is considered the norm in all tissues (Ralevic and Burnstock 1998) and mixtures of P2 purinoceptors have been reported in central neurones (Chessell *et al.* 1997) and glia (King *et al.* 1996). The situation is compounded by P2Y protein dimerization to generate receptor assemblies with subtly distinct pharmacological properties from their constituent components (Filippov *et al.* 2000). Also, the range of naturally occurring nucleotides capable of stimulating P2Y receptors has extended beyond ATP and its immediate breakdown products (Jacobson *et al.* 2000).

21.2 Molecular characterization

Cloning and heterologous expression of each of the three first P2Y receptors in oocytes, resulted in increased Ca^{2+}-mobilization following activation by ATP (Lustig *et al.* 1993; Webb *et al.* 1993; Barnard *et al.* 1994). These three GPCRs were assimilated into the $P2Y_{1-n}$ receptor family, as opposed to the cloned ionotropic ATP receptors that now form the $P2X_{1-n}$ receptor family (Abbracchio and Burnstock 1994). The $P2Y_1$ receptor was cloned from chick brain and its agonist potency profile (2-MeSATP > ATP ≥ ADP) approximated the native P2Y phenotype (Webb *et al.* 1993). The $P2Y_2$ receptor was cloned from murine neuroblastoma (NG108-15) cells and its agonist profile (ATP ≈ UTP > ATPγS) approximated the native P2U phenotype (Lustig *et al.* 1993). The $P2Y_3$ receptor was cloned from chick brain and its preference for ADP over ATP appeared to approximate the P2T phenotype (Barnard *et al.* 1994). Thus, a seemingly perfect correspondence was established between these recombinant P2Y receptors and subtypes of metabotropic ATP receptors in mammalian tissues. However, since that time, many other GPCRs associated with the P2Y receptor family have been identified and now include $P2Y_{1-12}$, turkey p2y, skate p2y, and the human UDP-glucose receptor. A dinucleotide receptor, $P2Y_{Ap4A}$ or its older name P2D, is anticipated but has not yet been cloned (Fredholm *et al.* 1997). The fifteen cloned receptor proteins are 328–532 amino acids in length and represent some of the shortest GPCRs found in mammalian cells. They possess seven hydrophobic regions, forming the transmembrane spanning regions TM1-VII, which lie between an extracellular N-terminus (21–51 residues in length) and a cytosolic C-terminus (16–217 residues in length) possessing multiple consensus motifs for phosphorylation by intracellular kinases. Alignment of the protein sequences for the TM1-VII region reveals 17–62 per cent identity (35–80 per cent similarity).

The human $P2Y_1$ receptor is found on chromosome 3q25 (Ayyanathan *et al.* 1996). $P2Y_1$ orthologues (89–98 per cent identical over TMI-VII) have been cloned from brain tissue of cow, chick, and turkey. Transcripts for mouse and rat orthologues are also present in nervous tissue. Recently, a $P2Y_1$-deficient mouse model has been generated with a phenotype showing decreased platelet aggregation and increased bleeding (Fabre *et al.* 1999; Leon *et al.* 1999). $P2Y_2$ orthologues (94 per cent identical) have been cloned from canine, mouse, and rat tissues. Rat $P2Y_2$ was cloned from a pituitary cDNA library (Chen *et al.* 1996), although $P2Y_2$ transcripts are more commonly associated with epithelial cell lines than brain derived cell lines. It should be borne in mind that the anterior pituitary is derived from epithelial tissue, and this may explain why $P2Y_2$ was found in the rat pituitary. The human $P2Y_2$ receptor gene located on chromosome 11q13.5–14.1, lies adjacent to the $P2Y_6$ gene at 11q13.3–13.5 (Pidlaoan *et al.* 1997). In addition, a $P2Y_2$-deficient mouse model has been generated which suggests this receptor is critical in regulating airway epithelial ion transport but not ion transport in non-respiratory epithelia (Homolya *et al.* 1999). The human $P2Y_2$ receptor protein shows polymorphism at position 334 (an arginine–cysteine transition), due to a replication error at nucleotide 1000 (thymine for cytosine), although this mutation does not significantly alter functionality (Janssens *et al.* 1999).

The human $P2Y_4$ gene encodes a receptor stimulated by UTP and antagonized by ATP (Kennedy *et al.* 2000), and is located in region q13 of chromosome X (Nguyen *et al.* 1995). By contrast, both ATP and UTP stimulate the mouse and rat $P2Y_4$ receptors (Bogdanov *et al.* 1998; Webb *et al.* 1998; Lazarowski *et al.* 2001). Rat $P2Y_4$ was cloned from brain (Webb *et al.* 1998) and heart (Bogdanov *et al.* 1998). The human $P2Y_4$ protein is 89 per cent identical to mouse and rat orthologues, whilst the two rodent $P2Y_4$ receptors are 95 per cent identical (over TMI-VII). The structural basis for the divergent pharmacology between human and rodent $P2Y_4$ receptors has not been elucidated. Agonist-activated human $P2Y_4$ receptors are rapidly internalized by a phosphorylation process involving serine-333 and serine-334 on the C-terminus (Brinson and Harden 2001).

The open reading frame for the $P2Y_5$ receptor is contained in intron 17 of the human retinoblastoma susceptibility gene, a tumour-suppressing gene located on chromosome 13q14.12–14.2 (Herzog *et al.* 1996). Opinions vary on whether the $P2Y_5$ receptor is a functional or orphan GPCR. The chick orthologue (83 per cent identical over TM1-VII) avidly binds $[^{35}S]$-dATPαS (Webb *et al.* 1996*b*), but turkey $P2Y_5$ does not respond functionally to either dATPαS or other nucleotides (Li *et al.* 1997). The human $P2Y_5$ receptor, when expressed in oocytes, is weakly stimulated by ATP and slowly activates the PLC_β/Ca^{2+}

Table 21.1 Purinergic receptors

	P2Y$_1$	P2Y$_2$	P2Y$_4$	P2Y$_6$	P2Y$_{11}$	P2Y$_{12}$
Alternative names	P$_{2Y}$	P$_{2U}$ or nucleotide	Uridine nucleotide or pyrimidinoceptor	Uridine nucleotide	—	P2Y$_{ADP}$, P$_{2T}$
Structural Information (Accession no.)	h 373 aa (U42029) r 373 aa (U22830) m 373 aa (U22829)	h 377 aa (U07225) r 374 aa (U09402) m 373 aa (L14751)	h 365 aa (U40223) r 361 aa (Y14705) m 361aa (AF277752)	h 328 aa (X97058) r 328 aa (Q63371) m 328aa (AF298899)	h 371 aa (AF030335)	h 342 aa (AF313449) r 341 aa (AF313450)
Gene location (human)	3q25	11q13.5–14.1	Xq13	11q13.3–13.5	19p31	3q24–25
Selective agonists	2-MeSADP, 2-MeSATP, ADPβS	Up$_4$U or UTPγS	Up$_3$U or UTP	UDPβS or UDP	AR-C67085	2-MeSADP,
Selective antagonist	MRS 2179 MRS 2279	—	—	—	—	C1330-7 AR-C67085
Radioligands	[^{35}S]-dATPαS	—	—	—	—	[^{3}H]-ADP
G protein coupling	G$_q$/G$_{11}$	G$_q$/G$_{11}$G$_i$	G$_q$/G$_{11}$G$_i$	G$_q$/G$_{11}$G$_i$	G$_q$/G$_s$	G$_i$/G$_o$

Expression profile	Cerebral cortex, cerebellum, hippocampus, caudate nucleus, putamen, corpus callosum, midbrain, subthalamic nuclei, DRG, astrocytes, placenta, heart, muscle, prostate, intestine, platelets, bone, pancreas	Pituitary, heart, blood vessels, lung, kidney, placenta, skeletal muscle, endocrine tissue, bone, astrocytes	Brain, placenta, heart, epithelium, pancreas, smooth muscle, kidney, intestine, liver	Kidney, lung, spleen, thymus, placenta, heart, bone, smooth muscle, epithelium, intestine	Brain, spleen, placenta, intestine, smooth muscle, granulocytes	Brain, platelets
Physiological function	Astrogliosis, transmitter release, bone resorption	Astrogliosis, transmitter release, mucus escalation, airway hydration	Glial growth, epithelial growth, SM growth	Lymphocytic maturation	Granulocyte differentiation	Haemostasis
Knockout phenotype	Resistance to thromboembolism, defective platelet aggregation	Reduced nucleotide induced airway epithelial ion transport	—	—	—	Bleeding disorder, defective platelet aggregation
Disease relevance	Stroke, epilepsy, neurodegeneration, thromboembolism, cardiovascular disease, osteoporosis	Cystic fibrosis, chronic bronchitis, dry eye	Proliferative disorders of the kidney epithelia	Inflammatory bowel disease	Neutropenia, leukaemia	Thromboembolism, cardiovascular disease

pathway (King and Townsend-Nicholson 2000). Human P2Y$_5$ seems to couple inefficiently to G$_q$ isoforms and its role in nucleotide signalling is unclear.

The chick P2Y$_3$ receptor is proposed to be an orthologue of mammalian P2Y$_6$ receptors—on the basis of similarity in protein sequence (78 per cent over TMI-VII) and a shared pharmacological profile (Li *et al.* 1998). The latter receptor gene is located on human chromosome 11q13.3–13.5, lying adjacent to the P2Y$_2$ gene (q13.5–14.1; Pidlaoan *et al.* 1997; Somers *et al.* 1997). Three forms of human P2Y$_6$ cDNA were found by RT-PCR amplification (Maier *et al.* 1997). Two forms contain the coding region for P2Y$_6$ but possess different 5$'$-untranslated regions, probably the consequence of alternative gene splicing. The third cDNA appears to be a pseudogene and shows a frame shift in the coding region that cannot be translated into protein. P2Y$_6$ transcripts for coding cDNAs were found in a series of brain-derived cell lines (Maier *et al.* 1997). P2Y$_6$ orthologues (88 per cent identical over TMI-VII) have been cloned from rat and mouse.

The human P2Y$_{11}$ gene is located on chromosome 19p31 (Suarez-Heurta *et al.* 2000), where an intron interrupts the coding sequence at the 5$'$-end and separates the first 3 codons from the remainder of the coding region (Communi *et al.* 1997). The P2Y$_{11}$ and SSF1 genes on chromosome 19 can undergo intergenic splicing to create a fusion protein (Communi *et al.* 2001). This SSF1-P2Y$_{11}$ protein is functionally indistinguishable from human P2Y$_{11}$ itself. A canine P2Y$_{11}$ receptor has been cloned from a kidney epithelial cell line (Zambon *et al.* 2001).

P2Y$_{12}$ is structurally related to the UDP-glucose receptor (49 per cent identical) but distinct from P2Y$_1$ (22 per cent identical) and is located on chromosome 3q24–25, adjacent to genes for the UDP-glucose and P2Y$_1$ receptors (Hollopeter *et al.* 2001). Both P2Y$_1$ and P2Y$_{12}$ are present in human blood platelets where they play a key role in haemostasis. A mutated P2Y$_{12}$ receptor arises through a two base pair deletion at the coding region for residue 240, a transcriptional frame shift and premature truncation of the protein (Hollopeter *et al.* 2001). The truncated P2Y$_{12}$ protein is non-functional and, in one patient, was associated with a mild bleeding disorder. P2Y$_{12}$ transcripts were also found in rat glioma cells and rat brain.

The skate p2y receptor was cloned from liver, but also found in brain, and is 69 per cent similar to human P2Y$_1$ (Dranoff *et al.* 2000). This chordate p2y receptor was claimed to be the most primitive form of the P2Y$_1$ receptor. In contrast, the turkey p2y receptor was cloned from a cDNA library from whole blood, but the physiological role of this receptor is unknown. This avian p2y receptor inhibits cAMP levels via the G$_i$/AC pathway, as well as stimulating IP$_3$ production via the G$_q$/PLC$_\beta$ pathway (Boyer *et al.* 2000). The receptor is structurally related to the *Xenopus* p2y$_8$ receptor (67 per cent identical, 79 per cent similar over TMI-VII). Both tp2y and xp2y$_8$ receptors are stimulated by most naturally occurring nucleoside triphosphates (ATP, CTP, GTP, ITP, UTP) and couple strongly to G$_q$/PLC$_\beta$ (Bogdanov *et al.* 1997; Boyer *et al.* 2000). Expression of amphibian xp2y$_8$ receptor is confined to the neural plate of developing *Xenopus* embryos and closely related to periods of neurogenesis (Bogdanov *et al.* 1997). P2Y$_9$ (or P2Y$_5$-like) and P2Y$_{10}$ receptors are considered to be orphan GPCRs (Janssens *et al.* 1997; Ralevic and Burnstock 1998). Human P2Y$_7$ receptor was wrongly identified as a nucleotide receptor (Akbar *et al.* 1996). This P2Y-like GPCR, located on chromosome 14q11.2–q12 (Owman *et al.* 1996), has been reclassified as a chemoattractant leucotriene B$_4$ (LTB$_4$) receptor (Yokomizo *et al.* 1997). The P2Y-like protein (fb1) cloned from human foetal hippocampus is located on chromosome 2q21 (Blasius *et al.* 1998) and joins an extended family of orphan GPCRs with low sequence homology to functional

P2Y receptors (Marchese *et al.* 1999). Finally, the human UDP-glucose receptor which is probably the most unusual of the known P2Y-like receptors recognizes UDP-glucose and UDP-galactose as its natural ligands, but not UTP or UDP and related mononucleotides (Chambers *et al.* 2000). This receptor is 49 per cent identical to $P2Y_{12}$ and, like the latter, can couple to pertussis toxin sensitive G_i protein (Chambers *et al.* 2000). The physiological role of the UDP-glucose receptor is unknown, but transcripts are spread widely throughout the neuraxis. The UDP-glucose receptor is also structurally related to $P2Y_5$, $P2Y_9$, and $P2Y_{10}$ receptors.

21.3 Cellular and subcellular localization

Presently, only six P2Y receptors $P2Y_1$, $P2Y_2$, $P2Y_4$, $P2Y_6$, $P2Y_{11}$, and $P2Y_{12}$ are accepted as clearly defined, distinct, nucleotide receptors in the P2Y receptor family. The UDP-glucose receptor fulfils many criteria for acceptance, but the physiological role for UDP-glucose signalling is mainly unexplored. The remainder of this review will concentrate on the six, accepted, P2Y receptors.

The $P2Y_1$ receptor was isolated first from chick brain, in which exceedingly high levels of $P2Y_1$ receptor expression are observed (37 pmol mg^{-1} protein for [^{35}S]-dATPαS binding (K_d, 9 nM); *cf.* 1–2 pmol mg^{-1} for muscimol binding at GABA$_A$) (Webb *et al.* 1993, 1994). Mammalian orthologues were later cloned from corpus callosum, as well as from endothelial cells, insulinoma cells, and placenta. $P2Y_1$-like immunoreactivity is located throughout the human, rat, and bovine neuraxis and concentrated in neuronal cells in cerebral and cerebellar cortex, hippocampus, caudate nucleus, putamen, subthalamic nucleus, and midbrain (Moore *et al.* 2000; Moran-Jimenez and Matute 2000). $P2Y_1$ receptor transcripts are present in rat brain cortical astrocytes (Webb *et al.* 1996a) and $P2Y_1$-like immunoreactivity also observed in rat and bovine brain astrocytes (Moran-Jimenez and Matute 2000). [^{35}S]-dATPαS binding was observed throughout the rat neuraxis at putative $P2Y_1$ receptors (Simon *et al.* 1997). The related compound, [^{35}S]-ATPαS, binds with high affinity (K_d, 10.5 nM) to $P2Y_1$-like receptors in rat cortex synaptosomes (Schäfer and Reiser 1999). Outside the brain, $P2Y_1$ transcripts have been found in mammalian placenta, heart, blood vessels, skeletal muscle, pancreas, blood platelets, and leucocytes, prostate, ovary, small and large intestine, and in some large DRG neurones (Ralevic and Burnstock 1998; King *et al.* 2001). $P2Y_1$ transcripts are abundant in developing limb buds, mesonephros, brain, somites, and facial primordia in the chick embryo (Meyer *et al.* 1999). At the subcellular level, a $P2Y_1$-GFP construct expressed in HEK 293 cells was identified in plasmalemma, endoplasmic reticulum, Golgi and microsomal fractions but was absent from nuclear and mitochondrial fractions (Vöhringer *et al.* 2000).

$P2Y_2$ receptors were first isolated from a murine neuroblastoma x glioma hybrid cell line (Lustig *et al.* 1993) and later cloned from alveolar cells, bone, epithelial, endothelial, and pituitary cells. $P2Y_2$ receptor transcripts are found in mammalian heart and vasculature, lung, kidney, osteoblasts, placenta, skeletal muscle, and endocrine tissues (King *et al.* 2001). On a functional level, $P2Y_2$-like receptors are found in astrocytes, chromaffin cells, epithelia, endothelia, fibroblasts, glia, hepatocytes, keratinocytes, leucocytes, myocytes, pituitary cells, and tumour cells (Ralevic and Burnstock 1998). $P2Y_2$ receptors tagged with Haemagglutinin A (HA) showed a punctate distribution on the surface of 1321N1 cells and, after agonist activation, $P2Y_2$-HA immunoreactivity was randomly internalized into the cytosolic pool (Sromek and Harden 1998).

Less information on the distribution of the remainder of the P2Y receptors exists. $P2Y_4$ receptors were cloned from human placenta (Communi *et al.* 1995) and later isolated from mammalian brain, epithelial cells, heart, and pancreas. $P2Y_4$ transcripts are also found in astrocytes, epithelial lining of hollow organs, kidney, leucocytes, and vascular smooth muscle (King *et al.* 2001). $P2Y_4$-like receptors were identified functionally in the jejunal lining in $P2Y_2$R-deficient mice (Cressman *et al.* 1999). The $P2Y_6$ receptor was cloned first from rat aortic smooth muscle (Chang *et al.* 1995) and later from human placenta and T-lymphocytes. $P2Y_6$ transcripts are found in mammalian bone, epithelia, heart, kidney, leucocytes, lung, spleen, and thymus (King *et al.* 2001). A $P2Y_6$-like receptor was identified functionally in the epithelial lining of gallbladder and trachea in $P2Y_2$ receptor-deficient mice (Cressman *et al.* 1999) and a $P2Y_6$-like receptor has also been reported in human nasal epithelial cells (Lazarowski *et al.* 1997). The $P2Y_{11}$ receptor was cloned from human placenta and its transcripts are found in human spleen, intestine, and granulocytes (HL-60 cells) (Communi *et al.* 1997). The $P2Y_{11}$ receptor approximates the native P2Y receptor in HL-60 cells (Conigrave *et al.* 1998; Communi *et al.* 1999), and appears to be directly involved in the differentiation of human granulocytes into neutrophils (Communi *et al.* 2000). Transcripts for the canine $P2Y_{11}$ receptor are more widespread and also found in brain (Zambon *et al.* 2001). $P2Y_{12}$ receptors were cloned from rat and human blood platelets and its transcripts are abundant in platelets but also present in brain (Hollopeter *et al.* 2001). $P2Y_{12}$ receptor transcripts were found in rat C6-2B glioma cells. The $P2Y_{12}$ receptor is pharmacologically similar to the human platelet P2T receptor and the endogenous P2Y receptor of C6-2B glioma cells (Boyer *et al.* 1993; Hollopeter *et al.* 2001).

21.4 Pharmacology

The pharmacology of the P2Y receptors is complex and involves a wide range of purine- and pyrimidine-based, mononucleotidic and dinucleotidic, compounds (King *et al.* 2001). Many ligands are naturally occurring, but synthetic nucleotides are now available to test against P2Y receptor subtypes.

21.4.1 Agonists

Human $P2Y_1$ receptors are activated fully by ADP and the naturally occurring dinucleotide, Ap_4A. ATP can act either as a full agonist, partial agonist, or antagonist depending on receptor reserve (King *et al.* 2001). Other mononucleotides (e.g. CTP, GTP, ITP, UTP and their immediate breakdown products) are inactive. The synthetic alkylthio-ATP derivatives are potent agonists (e.g. 2-MeSATP, 2-MeSADP, 2-HT-ATP, PAPET-ATP), as are phosphorothioate ATP derivatives (e.g. ATPγS and ADPβS), while methylene phosphonate-ATP derivatives (α, β-meATP and β, γ-meATP) are inert (Jacobson *et al.* 2000). At recombinant $P2Y_1$ receptors, dATPαS is either a weak agonist (Simon *et al.* 1995) or an antagonist (King and Townsend-Nicholson 2000), whereas at $P2Y_1$-like receptors in rat brain, ATPαS and ATP are equipotent agonists (Schäfer and Reiser 1999). Human $P2Y_2$ receptors are activated equally by ATP and UTP, as well as by the dinucleotides Ap_4A and Up_4U (King *et al.* 2001; Pendergast *et al.* 2001). The phosphorothioate derivatives ATPγS and UTPγS are potent stimulants, but other major classes of synthetic nucleotides are not. Human $P2Y_4$ receptors are activated by UTP and Up_3U (Pendergast *et al.* 2001), while CTP, GTP, ITP, and Ap_4A are considered to be weak agonists (Jacobson *et al.* 2000; King *et al.* 2001). In contrast, rat and mouse $P2Y_4$ receptors are activated equally by UTP and ATP. Human P2Y$_6$ receptors are activated by UDP and, to a lesser extent, ADP with all other nucleoside triphosphates being very weak agonists (Communi *et al.* 1996*b*). Of the synthetic compounds, UDPβS and Up$_3$U are both potent agonists at P2Y$_6$ receptors (Malmsjo *et al.* 2000; Pendergast *et al.* 2001). Rat and mouse orthologues show a similar pharmacological profile to human P2Y$_6$ (Filippov *et al.* 1999; Lazarowski *et al.* 2001). The human P2Y$_{11}$ receptor is activated by ATP and ADP (Communi *et al.* 1999) and the synthetic nucleotides BzATP, deoxyATP, 2-MeSATP, and AR-C67085 (2-propyl-D-β, γ-dichloromethylene-ATP) are also all potent agonists (Communi *et al.* 1999; Qi *et al.* 2001). In contrast, some synthetic nucleotides—ADPβS, ATPγS, and A3P5PS— are partial agonists, although, under some circumstances, they can also act as antagonists (Communi *et al.* 1999). Human P2Y$_{12}$ receptors are activated by ADP, 2-MeSADP and to a lesser extent by ATPγS (Hollopeter *et al.* 2001).

21.4.2 Antagonists

Some synthetic adenosine 3′, 5′-bisphosphate derivatives are potent antagonists (e.g. MRS 2179 and MRS 2279) (Nandanan *et al.* 2000) at P2Y$_1$ receptors and the classical P2 receptor antagonists (PPADS, Reactive blue-2 and suramin) also inhibit P2Y$_1$ receptor activity (King *et al.* 2001). Suramin is also the one and only weak antagonist of P2Y$_2$ receptors. For human P2Y$_4$, ATP is reported to be the most potent competitive antagonist (K_b, 0.7 μM; Kennedy *et al.* 2000). Weak antagonist activity has also been reported for PPADS, suramin and Reactive blue-2 at P2Y$_4$, P2Y$_6$, and P2Y$_{11}$ receptors (Robaye *et al.* 1997; Bogdanov *et al.* 1998; Lazarowski *et al.* 2001; King *et al.* 2001;). Human P2Y$_{12}$ is antagonized by 2-MeSAMP (Hollopeter *et al.* 2001) and C13307 (Scarborough *et al.* 2001), while the native form of the receptor is potently blocked by ARC67085 (Humphries *et al.* 1995). P2Y$_1$ antagonists, in the form of adenosine 3′, 5′-bisphosphate derivatives are inert at human P2Y$_{12}$.

21.5 Signal transduction and receptor modulation

Most of the recombinant P2Y receptors (P2Y$_1$, P2Y$_2$, P2Y$_4$, P2Y$_6$, P2Y$_{11}$) couple via the G$_q$/PLCβ pathway to cause IP$_3$ production, Ca^{2+}-mobilization and activation of Ca^{2+}-dependent reporter currents in heterologous expression systems (King *et al.* 2001). When expressed in cultured sympathetic neurones, some P2Y receptors inhibit native Ca^{2+} and K$^+$ currents by a direct action on ion channels by G protein catalytic and regulatory subunits (Filippov *et al.* 1997, 1998, 1999, 2000). Endogenous metabotropic P2 receptors affect a much wider range of intracellular signalling pathways and utilize PLCβ, PLD, PLA$_2$, AC, MEP/MAP kinases and Rho-dependent kinase, as well as coupling directly to some ion channels. The narrow selectivity of recombinant P2Y subtypes may only reflect the limited availability of signalling pathways in expression systems used so far.

A native P2Y$_1$-like receptor, in the clonal line (B10) of rat brain capillary endothelial cells, appeared to couple negatively to adenylate cyclase and inhibit cAMP levels through a PTX-sensitive G protein (Webb *et al.* 1996). The possibility that recombinant rP2Y$_1$ receptors might affect cAMP production was investigated by expression into 1321N1 and C6 rat glioma cells that, respectively, utilize G$_q$/PLCβ and G$_i$/AC signalling mechanisms (Schachter *et al.* 1997). Experiments showed that rat P2Y$_1$ receptors selected only the G$_q$/PLCβ pathway in 1321N1 cells, and not the G$_i$/AC pathway in C6 glioma cells. Although B10 cells possess P2Y$_1$ transcripts, it was later shown that B10 cells also possessed a P2Y$_{12}$-like receptor that could activate the G_i/AC pathway and help explain earlier results (Simon *et al.* 2001). The available evidence suggests that known species orthologues of P2Y$_1$ couple primarily to the G_q/PLC$_\beta$ pathway. The skate p2y receptor, considered to be the most primitive form of P2Y$_1$, is the only GPCR in skate liver to signal via the G_q/PLC$_\beta$ pathway (Dranoff *et al.* 2000). Apart from the G_q/PLC$_\beta$ signalling, P2Y$_1$ receptors directly inhibit N-type Ca^{2+}-currents in rat sympathetic neurones (Filippov *et al.* 2000).

The P2Y$_2$ receptor appears to couple mainly to the G_q/PLC$_\beta$ pathway, although 35 per cent of the evoked Ca^{2+}-signal is inhibited by PTX (Erb *et al.* 1993; Parr *et al.* 1994). PLC$_\beta$ activation, via G_α of PTX-insensitive G_q and $G_{\beta,\gamma}$ complex of PTX-sensitive G_i, could account for the P2Y$_2$-induced Ca^{2+}-signal (Lustig *et al.* 1996). In *Xenopus* oocytes, P2Y$_2$ receptors couple directly to co-expressed K^+ channels of the Kir 3.0 subfamily via PTX-sensitive $G_{i/o}$ proteins (Mosbacher *et al.* 1998). In sympathetic neurons, P2Y$_2$ receptors inhibit N-type Ca^{2+}-currents via a PTX-sensitive mechanism (Filippov *et al.* 1997, 1998). A native P2Y$_2$-like receptor, in canine MDCK-D1 epithelial cells, was reported to couple indirectly to G_s/AC through an indomethacin-sensitive pathway (Zambon *et al.* 2000). Dual signalling also occurs with the human P2Y$_4$ receptor, since PTX limits the Ca^{2+} signal by 60 per cent in the first 30 s of agonist activation but fails to inhibit Ca^{2+}_i levels following prolonged (>300 s) agonist activation (Communi *et al.* 1996a). By contrast, P2Y$_6$ receptor signalling via Ca^{2+}-mobilisation was reported to be PTX-insensitive (Chang *et al.* 1995; Robaye *et al.* 1997). However, P2Y$_6$ receptors inhibit N-type Ca^{2+}-currents via a PTX-sensitive mechanism in sympathetic neurons (Filippov *et al.* 1999).

The P2Y$_{11}$ receptor couples strongly to the G_q/PLC$_\beta$ pathway, but also activates the G_s/AC pathway (Communi *et al.* 1997, 1999; Qi *et al.* 2001). It was reported that inositol hydrolysis and Ca^{2+}-mobilisation, via the G_q/PLC$_\beta$ pathway, could potentiate cAMP production via the G_s/AC pathway in 1321N1 and CHO-K1 cells (Qi *et al.* 2001). This potentiating effect may help explain differences in agonist potencies when a range of nucleotides was tested against the two signalling pathways (Communi *et al.* 1999; Qi *et al.* 2001). The signalling and pharmacological properties of P2Y$_{11}$ mirror the endogenous P2Y receptor in HL-60 cells (Conigrave *et al.* 1998; Suh *et al.* 2000).

P2Y$_1$ receptors couple to the G_i/AC pathway to inhibit cAMP production, an effect blunted by PTX (Hollopeter *et al.* 2001). While human P2Y$_{12}$ receptors directly inhibit cAMP production in CHO cells, receptor activation was otherwise assessed in oocytes by $G_{\beta,\gamma}$ subunit stimulation of Kir 3.1 and 3.4 ion channels co-expressed with P2Y$_{12}$ (Hollopeter *et al.* 2001).

21.6 **Physiology and disease relevance**

The multiplicity of P2Y receptor subtypes and ubiquitous presence in all human tissues indicates nucleotidic signalling is important in the major physiological systems. The role of purines and pyrimidines in the pathophysiology of disease, has been broadly described in a number of reviews over the last five years (Burnstock 1997, 2002; Abbracchio and Burnstock 1998; Burnstock and Williams 2000; Williams and Jarvis 2000; Boeynaems *et al.* 2001).

21.6.1 **Diseases of the central nervous system**

The widespread distribution and density of P2Y$_1$ receptors in the neuraxis indicates a role in central transmission. P2Y$_1$-like receptors can inhibit (von Kügelgen *et al.* 1994, 1997;

Bennett and Boarder 2000; Mendoza-Fernandez et al. 2000) or facilitate (Zhang et al. 1995) transmitter release from central neurones. These pre-synaptic effects on transmitter release could be explained by the known effector systems for $P2Y_1$ receptors—for example, a direct inhibitory action on N-type Ca^{2+}-channels to limit exocytosis (Filippov et al. 2000) or large Ca^{2+}-transients via the G_q/PLC_β pathway to stimulate exocytosis (Mirinov 1994; Schäfer and Reiser 1999). With regard to disease states, the presence of $P2Y_1$ and P2U-like receptor receptors on astrocytes (King et al. 1996) coupled to their signalling through the ERK pathway (Neary et al. 1998, 1999, 2000), has implicated these receptors in the mitogenic action of ATP in reactive astrogliosis. This is a hyperplastic condition associated with CNS injury in a number of conditions including trauma, stroke, epilepsy and Alzheimer's disease and multiple sclerosis. Astrocytes support the viability of neurones and $P2Y_1$ and P2U-like receptor signalling is implicated in the reparative processes following such CNS injury (Ciccarelli et al. 2001). It has been difficult to define a clear role for $P2Y_2$ receptors in the nervous system due to a lack of selective agonists and antagonists able to distinguish this receptor from the other P2Y subtypes showing a P2U-like phenotype (i.e. $P2Y_4$ and $P2Y_6$). The recent discovery of the di-uridine polyphosphate series, particularly Up_4U as a stable agonist of $P2Y_2$ receptors, may alter these circumstances. A P2U-like receptor has been implicated in the regulation of transmitter release from hypothalamic vasopressin neurones (Hiruma and Bourque 1995) and paravertebral sympathetic neurones (Boehm 1998).

21.6.2 Peripheral indications

$P2Y_1$-like receptors in rodent osteoclasts show an interesting chemosensitivity to extracellular pH (Hoebertz et al. 2001) and this modulatory effect of H^+ ions may have some bearing on the purinergic (P2Y- and P2X-based) control of central respiratory drive during acidosis (Ralevic et al. 1999). In addition, $P2Y_2$ receptors are known to stimulate the mucus escalator in lung and directly affect airway hydration (Yerxa 2001). Consequently, the utility of uridine-based nucleotides (e.g. INS365 (Up_4U)) is under investigation for chronic bronchitis and cystic fibrosis (Shaffer et al. 1998).

P2Y receptors are being pursued for a number of potential peripheral disease indications. For example, $P2Y_2$ receptors may have an ameliorating influence in dry eye by stimulating tear production (Pintor and Peral 2001; Yerxa 2001) and $P2Y_4$ receptors have been implicated in proliferative disorders of the kidney (Harada et al. 2000). Reports also suggest that modulation of $P2Y_6$ receptor activity may be implicated in nucleotide-promoted colonic damage in inflammatory bowel disease (Somers et al. 1998). The $P2Y_{11}$ receptor has been strongly implicated in the differentiation of HL60 cells from granulocytes to neutrophils (Communi et al. 2000) and it has therefore been postulated that $P2Y_{11}$ receptor-directed ligands may be important in neutropenia (a loss of neutrophils) and some forms of leukaemia (Boeynaems et al. 2001). Finally, the $P2Y_{12}$ receptor corresponds to the G_i-linked ADP receptor in human blood platelets (Hollopeter et al. 2001) and, like the $P2Y_1$ receptor, is intimately involved in the clotting response (Hourani 2001). Use of a highly selective and competitive antagonist (C13307) has been developed for the $P2Y_{12}$ receptor and should help to clarify the role of this receptor in disorders where the regulation of blood clotting may be of importance (Scarborough et al. 2001).

21.7 **Concluding remarks**

The understanding of purinergic GPCRs has advanced significantly in the last ten years, with the cloning of multiple P2Y receptor subtypes and isolated study of their pharmacological and signalling properties. Advances have also been made in the medical chemistry of P2Y receptors, with the beginnings of agonist and antagonist selectivity for various P2Y subtypes. Purinergic signalling in the nervous system is less than clear, but an improved armament of selective ligands is now showing benefit. The role of P2Y receptors in diseases states is slowly being revealed and clinical trials have begun in key therapeutic areas, although CNS investigations are in their infancy. The role of purinergic signalling in neural development is now being considered, as are conditions for upregulation of P2Y receptors at various stages in the life cycle. The complexity of purinergic signalling represents a major challenge, which has now been taken up by many laboratories in academia and industry.

Note: A G_i-coupled $P2Y_{13}$ receptor—activated by ADP and inhibited by PTX—has been found in brain and spleen (Communi *et al. J Biol Chem* **276**, 41 479–85, December 2001).

References

Abbracchio MP and Burnstock G (1994). Purinoceptors: are there families of P2X and P2Y purinoceptors? *Pharmacol Ther* **64**, 445–75.

Abbracchio MP and Burnstock G (1998). Purinergic signalling: pathophysiological roles. *Jap J Pharmacol* **78**, 113–45.

Akbar GKM, Dasari VR, Webb TE, Ayyanathan K, Pillarisetti K, Sandhu AK, *et al.* (1996). Molecular cloning of a novel P2 purinoceptor from human erythroleukemia cells. *J Biol Chem* **271**, 18 363–7.

Ayyanathan K, Naylor SL, and Kunapuli SP (1996). Structural characterization and fine chromosomal mapping of the human $P2Y_1$ purinergic receptor gene ($P2RY_1$). *Somat Cell Mol Genet* **22**, 419–24.

Barnard EA, Burnstock G, and Webb TE (1994). G protein-coupled receptors for ATP and other nucleotides: a new receptor family. *Trends Pharmacol Sci* **15**, 67–70.

Bennett GC and Boarder MR (2000). The effect of nucleotides and adenosine on stimulus-evoked glutamate release from rat brain cortical slices. *Br J Pharmacol* **131**, 617–23.

Blasius R, Weber RG, Lichter P, and Ogilvie A (1998). A novel orphan G protein-coupled receptor primarily expressed in the brain is localized on human chromosomal band 2q21. *J Neurochem* **70**, 1357–65.

Boehm S (1998). Selective inhibition of M-type potassium channels in rat sympathetic neurons by uridine nucleotide preferring receptors. *Br J Pharmacol* **124**, 1261–9.

Boeynaems JM, Robaye B, Janssens R, Suarez-Huerta N, and Communi D (2001). Overview of P2Y receptors as therapeutic targets. *Drug Dev Res* **52**, 187–9.

Bogdanov YD, Dale L, King BF, Whittock N, and Burnstock G (1997). Early expression of a novel nucleotide receptor in the neural plate of *Xenopus* embryos. *J Biol Chem* **272**, 12 583–90.

Bogdanov YD, Wildman SS, Clements MP, King BF, and Burnstock G (1998). Molecular cloning and characterization of rat $P2Y_4$ nucleotide receptor. *Br J Pharmacol* **124**, 428–30.

Boyer JL, Delaney SM, Villanueva D, and Harden TK (2000). A molecularly identified P2Y receptor simultaneously activates phospholipase C and inhibits adenylyl cyclase and is nonselectively activated by all nucleoside triphosphates. *Mol Pharmacol* **57**, 805–10.

Boyer JL, Lazarowski ER, Chen XH, and Harden TK (1993). Identification of a P2Y-purinergic receptor that inhibits adenylyl cyclase. *J Pharmacol Exp Ther* **267**, 1140–6.

Brinson AE and Harden TK (2001). Differential regulation of the uridine nucleotide-activated P2Y$_4$ and P2Y$_6$ receptors. Ser-333 and Ser-334 in the carboxyl terminus are involved in agonist-dependent phosphorylation desensitization and internalization of the P2Y$_4$ receptor. *J Biol Chem* **276**, 11 939–48.

Burnstock G (1972). Purinergic nerves. *Pharmacol Rev* **24**, 509–81.

Burnstock G (1978). A basis for distinguishing two types of purinergic receptor. In (ed. RW Straub and L Bolis) *Cell Membrane Receptors for Drugs and Hormones: A Multidisciplinary Approach*, Raven Press, New York, pp. 107–18.

Burnstock G (1997). The past, present and future of purine nucleotides as signalling molecules. *Neuropharmacol* **36**, 1127–39.

Burnstock G (2002). Potential therapeutic targets in the rapidly expanding field of purinergic signalling. *Clin Med* **2**, 45–53.

Burnstock G and Kennedy C (1985). Is there a basis for distinguishing two types of P2 purinoceptor? *Gen Pharmacol* **16**, 433–40.

Burnstock G and Williams M (2000). P2 purinergic receptors: modulation of cell function and therapeutic potential. *J Pharmacol Exp Ther* **295**, 862–9.

Chambers JK, Macdonald LE, Sarau HM, Ames RS, Freeman K, Foley JJ *et al.* (2000). A G protein-coupled receptor for UDP-glucose. *J Biol Chem* **275**, 10 767–71.

Chang K, Hanaoka K, Kumada M, and Takuwa Y (1995). Molecular cloning and functional analysis of a novel P2 nucleotide receptor. *J Biol Chem* **270**, 26 152–8.

Chen ZP, Krull N, Levy A, and Lightman SL (1996). Molecular cloning and functional characterization of a rat pituitary G-protein coupled adenosine triphosphate (ATP) receptor. *Endocrinol* **137**, 1833–40.

Chessell IP, Michel AD, and Humphrey PPA (1997). Functional evidence for multiple purinoceptor subtypes in the rat medial vestibular nucleus. *Neurosci* **77**, 783–91.

Ciccarelli R, Ballerini O, Sabatino G, Rathbone MP, D'Onofrio M, Caciagli F, and Iorio P (2001). Involvement of astrocytes in purine-based reparative processes in the brain. *Int J Dev Neurosci* **19**, 395–414.

Communi D, Govaerts C, Parmentier M, and Boeynaems JM (1997). Cloning of a human purinergic P2Y receptor coupled to phospholipase C and adenylyl cyclase. *J Biol Chem* **272**, 31 969–73.

Communi D, Janssens R, Robaye B, Zeelis N, and Boeynaems JM (2000). Rapid upregulation of P2Y messengers during granulocytic differentiation of HL-60 cells. *FEBS Lett* **475**, 39–42.

Communi D, Motte S, Boeynaems JM, and Pirotton S (1996*a*). Pharmacological characterization of the human P2Y$_4$ receptor. *Eur J Pharmacol* **317**, 383–9.

Communi D, Parmentier M, and Boeynaems JM (1996*b*). Cloning, functional expression and tissue distribution of the human P2Y$_6$ receptor. *Biochem Biophys Res Comm* **222**, 303–8.

Communi D, Pirotton S, Parmentier M, and Boeynaems JM (1995). Cloning and functional expression of a human uridine nucleotide receptor. *J Biol Chem* **270**, 30 849–52.

Communi D, Robaye B, and Boeynaems JM (1999). Pharmacological characterization of the human P2Y$_{11}$ receptor. *Br J Pharmacol* **128**, 1199–206.

Communi D, Suarez-Huerta N, Dussossoy D, Savi P, and Boeynaems JM (2001). Cotranscription and intergenic splicing of human P2Y$_{11}$ and SSF1 genes. *J Biol Chem* **276**, 16 561–6.

Conigrave AD, Lee JY, van der Weyden L, Jiang L, Ward P, Tasevski V *et al.* (1998). Pharmacological profile of a novel cyclic AMP-linked P2 receptor on undifferentiated HL-60 leukemia cells. *Br J Pharmacol* **124**, 1580–5.

Cressman VL, Lazarowski E, Homolya L, Boucher RC, Koller BH, and Grubb BR (1999). Effect of loss of P2Y$_2$ receptor gene expression on nucleotide regulation of murine epithelial Cl$^-$ transport. *J Biol Chem* **274**, 26 461–8.

Dranoff JA, O'Neill AF, Franco AM, Cai SY, Connolly GC, Ballatori N *et al.* (2000). A primitive ATP receptor from the little skate *Raja erinacea*. *J Biol Chem* **275**, 307 010–6.

Dubyak GR (1991). Signal transduction by P2-purinergic receptors for extracellular ATP. *Am J Respir Cell Mol Biol* **4**, 295–300.

Dubyak GR and El-Moatassim C (1993). Signal transduction via P2-purinergic receptors for extracellular ATP and other nucleotides. *Am J Physiol* 265, C577–606.

Erb L, Lustig KD, Sullivan DM, Turner JT, and Weisman GA (1993). Functional expression and photoaffinity labelling of a cloned P2U purinergic receptor. *Proc Nat Acad Sci USA* 90, 10 449–53.

Fabre JE, Nguyen M, Latour A, Keifer JA, Audoly LP, Coffman TM, and Koller BH (1999). Decreased platelet aggregation, increased bleeding time and resistance to thromboembolism in P2Y$_1$-deficient mice. *Nature Med* 5, 1199–202.

Filippov AK, Brown DA, and Barnard EA (2000). The P2Y$_1$ receptor closes the N-type Ca^{2+} channel in neurones, with both adenosine triphosphates and diphosphates as potent agonists. *Br J Pharmacol* 129, 1063–6.

Filippov AK, Webb TE, Barnard EA, and Brown DA (1997). Inhibition by heterologously-expressed P2Y$_2$ nucleotide receptors of N-type calcium currents in rat sympathetic neurones. *Br J Pharmacol* 121, 849–51.

Filippov AK, Webb TE, Barnard EA, and Brown DA (1998). P2Y$_2$ nucleotide receptors expressed heterologously in sympathetic neurons inhibit both N-type Ca^{2+} and M-type K$^+$ currents. *J Neurosci* 18, 5170–9.

Filippov AK, Webb TE, Barnard EA, and Brown DA (1999). Dual coupling of heterologously-expressed rat P2Y$_6$ nucleotide receptors to N-type Ca^{2+} and M-type K$^+$ currents in rat sympathetic neurones. *Br J Pharmacol* 126, 1009–17.

Fredholm BB, Abbracchio MP, Burnstock G, Dubyak GR, Harden TK, Jacobson KA *et al.* (1997). Towards a revised nomenclature for P1 and P2 receptors. *Trends Pharmacol Sci* 18, 79–82.

Gordon JL (1986) Extracellular ATP: effects, sources and fate. *Biochem J* 233, 309–19.

Harada H, Chan CM, Loesch A, Unwin R, and Burnstock G (2000). Induction of proliferation and apoptotic cell death via P2Y and P2X receptors, respectively, in rat glomerular mesangial cells. *Kidney Int* 57, 949–58.

Herzog H, Darby K, Hort YJ, and Shine J (1996). Intron 17 of the human retinoblastoma susceptibility gene encodes an actively transcribed G protein-coupled receptor gene. *Genome Res* 6, 858–61.

Hiruma H and Bourque CW (1995). P2 purinoceptor-mediated depolarization of rat supraoptic neurosecretory cells in vitro. *J Physiol* 489, 805–11.

Hoebertz A, Meghji S, Burnstock G, and Arnett TR (2001). Extracellular ADP is a powerful osteolytic agent: evidence for signalling through the P2Y$_1$ receptor on bone cells. *FASEB J* 15, 1339–148.

Hollopeter G, Jantzen HM, Vincent D, Li G, England L, Ramakrishnan V *et al.* (2001). Identification of the platelet ADP receptor targeted by antithrombotic drugs. *Nature* 409, 202–7.

Homolya L, Watt WC, Lazarowski ER, Koller BH, and Boucher RC (1999). Nucleotide-regulated calcium signalling in lung fibroblasts and epithelial cells from normal and P2Y$_2$ receptor (-/-) mice. *J Biol Chem* 274, 26 454–60.

Hourani SMO (2001). Discovery and recognition of purine receptor subtypes on platelets. *Drug Devel Res* 52, 140–9.

Humphries RG, Tomlinson W, Clegg JA, Ingall AH, Kindon ND, and Leff P (1995). Pharmacological profile of the novel P2T-purinoceptor antagonist, FPL 67085, *in vitro* and in the anaesthetized rat *in vivo*. *Br J Pharmacol* 115, 1110–16.

Jacobson KA, King BF, and Burnstock G (2000). Pharmacological characterization of P2 (nucleotide) receptors. *Celltransmissions* 16, 3–16.

Janssens R, Boeynaems JM, Godart M, and Communi D (1997). Cloning of a human heptahelical receptor closely related to the P2Y$_5$ receptor. *Biochem Biophys Res Comm* 236, 106–12.

Janssens R, Paindavoine P, Parmentier M, and Boeynaems JM (1999). Human P2Y$_2$ receptor polymorphism: identification and pharmacological characterization of two allelic variants. *Br J Pharmacol* 127, 709–16.

Jin J, Daniel JL, and Kunapuli SP (1998). Molecular basis for ADP-induced platelet activation. II. The P2Y$_1$ receptor mediates ADP-induced intracellular calcium mobilization and shape change in platelets. *J Biol Chem* **273**, 2030–4.

Kennedy C, Qi AD, Herold CL, Harden TK, and Nicholas RA (2000). ATP, an agonist at the rat P2Y$_4$ receptor, is an antagonist at the human P2Y$_4$ receptor. *Mol Pharmacol* **57**, 926–31.

King BF, Burnstock G, Boyer JL, Boeynaems, JM, Weisman GA, Kennedy C *et al.* (2001). Nucleotide receptors: P2Y receptors. In *The IUPHAR Compendium of Receptor Characterization and Classification*, Edn 2, IUPHAR Media Publications.

King BF, Neary JT, Zhu Q, Wang S, and Burnstock G (1996). P2 purinoceptors in rat cortical astrocytes: expression, calcium-imaging and signalling studies. *Neurosci* **74**, 1187–96.

King BF and Townsend-Nicholson A (2000). Recombinant P2Y receptors: the UCL experience. *J Auton Nerv Syst* **81**, 164–170.

King BF, Townsend-Nicholson A, and Burnstock G (1998). Metabotropic receptors for ATP and UTP: exploring the correspondence between native and recombinant nucleotide receptors. *Trends Pharmacol Sci* **19**, 506–14.

Lazarowski ER, Paradiso AM, Watt WC, Harden TK, and Boucher RC (1997). UDP activates a mucosal-restricted receptor on human nasal epithelial cells that is distinct from the P2Y$_2$ receptor. *Proc Nat Acad Sci USA* **94**, 2599–603.

Lazarowski ER, Rochelle LG, O'Neal WK, Ribeiro CM, Grubb BR, Zhang V *et al.* (2001). Cloning and functional characterization of two murine uridine nucleotide receptors reveal a potential target for correcting ion transport deficiency in cystic fibrosis gallbladder. *J Pharmacol Exp Ther* **297**, 43–9.

Leon C, Hechler B, Freund M, Eckly A, Vial C, Ohlmann P *et al.* (1999). Defective platelet aggregation and increased resistance to thrombosis in purinergic P2Y$_1$ receptor-null mice. *J Clin Invest* **104**, 1731–7.

Li Q, Olesky M, Palmer RK, Harden TK, and Nicholas RA (1998). Evidence that the p2y3 receptor is the avian homologue of the mammalian P2Y$_6$ receptor. *Mol Pharmacol* **54**, 541–6.

Li Q, Schachter JB, Harden TK, and Nicholas RA (1997). The 6H1 orphan receptor, claimed to be the p2y5 receptor, does not mediate nucleotide-promoted second messenger responses. *Biochem Biophys Res Comm* **236**, 455–60.

Lustig KD, Shiau AK, Brake AJ, and Julius D (1993). Expression cloning of an ATP receptor from mouse neuroblastoma cells. *Proc Nat Acad Sci USA* **90**, 5113–17.

Lustig KD, Weisman GA, Turner JT, Garrad R, Shiau AK, and Erb L (1996). P2U purinoceptors: cDNA cloning, signal transduction mechanisms and structure-function analysis. *Ciba Found Symp* **198**, 193–204.

Maier R, Glatz A, Mosbacher J, and Bilbe G (1997). Cloning of P2Y$_6$ cDNAs and identification of a pseudogene: comparison of P2Y receptor subtype expression in bone and brain tissues. *Biochem Biophys Res Comm* **240**, 298–302.

Malmsjo M, Hou M, Harden TK, Pendergast W, Pantev E, Edvinsson L, and Erlinge D (2000). Characterization of contractile P2 receptors in human coronary arteries by use of the stable pyrimidines uridine 5′-O-thiodiphosphate and uridine 5′-O-3-thiotriphosphate. *J Pharmacol Exp Ther* **293**, 755–60.

Marchese A, George SR, Kolakowski Jr. LF, Lynch KR, and O'Dowd BF (1999). Novel GPCRs and their endogenous ligands: expanding the boundaries of physiology and pharmacology. *Trends Pharmacol Sci* **20**, 370–5.

Mendoza-Fernandez V, Andrew RD, and Barajas-Lopez C (2000). ATP inhibits glutamate synaptic release by acting at P2Y receptors in pyramidal neurons of hippocampal slices. *J Pharmacol Exp Ther* **293**, 172–9.

Meyer MP, Clarke JD, Patel K, Townsend-Nicholson A, and Burnstock G (1999). Selective expression of purinoceptor cP2Y$_1$ suggests a role for nucleotide signalling in development of the chick embryo. *Dev Dyn* **214**, 152–8.

Mironov SL (1994). Metabotropic ATP receptor in hippocampal and thalamic neurones: pharmacology and modulation of Ca^{2+} mobilizing mechanisms. *Neuropharmacology* 33, 1–13.

Moore D, Chambers J, Waldvogel H, Faull R, and Emson P (2000). Regional and cellular distribution of the $P2Y_1$ purinergic receptor in the human brain: striking neuronal localisation. *J Comp Neurol* 421, 3743–84.

Moran-Jimenez MJ and Matute C (2000). Immunohistochemical localization of the $P2Y_1$ purinergic receptor in neurons and glial cells of the central nervous system. *Mol Brain Res* 78, 50–8.

Mosbacher J, Maier R, Fakler B, Glatz A, Crespo J, and Bilbe G (1998). P2Y receptor subtypes differently couple to inwardly-rectifying potassium channels. *FEBS Lett* 436, 104–10.

Nandanan E, Jang SY, Moro S, Kim HO, Siddiqui MA, Russ P *et al.* (2000). Synthesis, biological activity, and molecular modeling of ribose-modified deoxyadenosine bisphosphate analogues as $P2Y_1$ receptor ligands. *J Med Chem* 43, 829–42.

Neary JT (2000). Trophic actions of extracellular ATP: gene expression profiling by DNA array analysis. *J Auton Nerv Syst* 81, 200–4.

Neary JT, Kang Y, Bu Y, Yu E, Akong K, and Peters CM (1999). Mitogenic signaling by ATP/P2Y purinergic receptors in astrocytes: involvement of a calcium-independent protein kinase C, extra-cellular signal-regulated protein kinase pathway distinct from the phosphatidylinositol-specific phospholipase C/calcium pathway. *J Neurosci* 19, 4211–20.

Neary JT, McCarthy M, Kang Y, and Zuniga S (1998). Mitogenic signaling from P1 and P2 purinergic receptors to mitogen-activated protein kinase in human fetal astrocyte cultures. *Neurosci Lett* 242, 159–62.

Nguyen T, Erb L, Weisman GA, Marchese A, Heng HH, Garrad RC *et al.* (1995). Cloning, expression, and chromosomal localization of the human uridine nucleotide receptor gene. *J Biol Chem* 270, 30 845–8.

O'Connor SE, Dainty IA, and Leff P (1991). Further subclassification of ATP receptors based on agonist studies. *Trends Pharmacol Sci* 12, 137–41.

Owman C, Nilsson C, and Lolait SJ (1996). Cloning of cDNA encoding a putative chemoattractant receptor. *Genomics* 37, 187–94.

Parr CE, Sullivan DM, Paradiso AM, Lazarowski ER, Burch LH, Olsen JC *et al.* (1994). Cloning and expression of a human P2U nucleotide receptor, a target for cystic fibrosis pharmacotherapy. *Proc Nat Acad Sci USA* 91, 3275–9.

Pendergast W, Yerxa BR, Douglass JG 3rd, Shaver SR, Dougherty RW, Redick CC *et al.* (2001). Synthesis and P2Y receptor activity of a series of uridine dinucleoside 5′-polyphosphates. *Bioorg Med Chem Lett* 11, 157–60.

Pidlaoan LV, Jin J, Sandhu AK, Athwal RS, and Kunapuli SP (1997). Colocalization of $P2Y_2$ and $P2Y_6$ receptor genes at human chromosome 11q13.3–14.1. *Somat Cell Mol Genet* 23, 291–6.

Pintor J and Peral A (2001). Therapeutic potential of nucleotides in the eye. *Drug Devel Res* 52, 190–5.

Qi AD, Kennedy C, Harden TK, and Nicholas RA (2001). Differential coupling of the human $P2Y_{11}$ receptor to phospholipase C and adenylyl cyclase. *Br J Pharmacol* 132, 318–26.

Ralevic V and Burnstock G (1998). Receptors for purines and pyrimidines. *Pharmacol Rev* 50, 413–92.

Ralevic V, Thomas T, Burnstock G, and Spyer KM (1999). Characterization of P2 receptors modulating neural activity in rat rostral ventrolateral medulla. *Neurosci* 94, 867–78.

Robaye B, Boeynaems JM, and Communi D (1997). Slow desensitization of the human $P2Y_6$ receptor. *Eur J Pharmacol* 329, 231–6.

Scarborough RM, Laibelman AM, Clizbe LA, Fretto LJ, Conley PB, Reynolds EE *et al.* (2001). Novel tricyclic benzothiazolo[2,3-c]thiadiazene antagonists of the platelet ADP receptor ($P2Y_{12}$). *Bio-organ Med Lett* in press.

Schachter JB, Boyer JL, Li Q, Nicholas RA, and Harden TK (1997). Fidelity in functional coupling of the rat $P2Y_1$ receptor to phospholipase C. *Br J Pharmacol* 122, 1021–4.

Schäfer R and Reiser G (1999). ATPαS is a ligand for P2Y receptors in synaptosomal membranes: solubilization of [^{35}S]ATPαS binding proteins associated with G-proteins. *Neurochem Int* **34**, 303–17.

Shaffer C, Jacobus K, Yerxa B, Johnson F, Griffin W, Evans R, and Edgar P (1998). INS365, a novel P2Y$_2$ receptor agonist and ion channel modulator for the treatment of cystic fibrosis: results from initial phase I study. *J. Pediat Pulmon* **S17**, 254.

Simon J, Vigne P, Eklund KM, Michel AD, Carruthers AM, Humphrey PP *et al.* (2001). Activity of adenosine diphosphates and triphosphates on a P2Y$_T$-type receptor in brain capillary endothelial cells. *Br J Pharmacol* **132**, 173–82.

Simon J, Webb TE, and Barnard EA (1997). Distribution of [^{35}S]-dATPαS binding sites in the adult rat neuraxis. *Neuropharmacol* **36**, 1243–51.

Simon J, Webb TE, King BF, Burnstock G, and Barnard EA (1995). Characterisation of a recombinant P2Y purinoceptor. *Eur J Pharmacol* **291**, 281–9.

Somers GR, Hammet FM, Trute L, Southey MC, and Venter DJ (1998). Expression of the P2Y$_6$ purinergic receptor in human T cells infiltrating inflammatory bowel disease. *Lab Invest* **78**, 1375–83.

Somers GR, Hammet F, Woollatt E, Richards RI, Southey MC, and Venter DJ (1997). Chromosomal localization of the human P2Y$_6$ purinoceptor gene and phylogenetic analysis of the P2Y purinoceptor family. *Genomics* **44**, 127–30.

Sromek SM and Harden TK (1998). Agonist-induced internalization of the P2Y$_2$ receptor. *Mol Pharmacol* **54**, 485–94.

Suarez-Huerta N, Boeynaems JM, and Communi D (2000). Cloning, genomic organization, and tissue distribution of human SSF-1. *Biochem Biophys Res Comm* **275**, 37–42.

Suh BC, Kim TD, Lee IS, and Kim KT (2000). Differential regulation of P2Y$_{11}$ receptor-mediated signalling to phospholipase C and adenylyl cyclase by protein kinase C in HL-60 promyelocytes. *Br J Pharmacol* **131**, 489–97.

Vöhringer C, Schäfer R, and Reiser G (2000). A chimeric rat brain P2Y$_1$ receptor tagged with green-fluorescent protein: high-affinity ligand recognition of adenosine diphosphates and triphosphates and selectivity identical to that of the wild-type receptor. *Biochem Pharmacol* **59**, 791–800.

Von Kügelgen I, Koch H, and Starke K (1997). P2-receptor-mediated inhibition of serotonin release in the rat brain cortex. *Neuropharmacol* **36**, 1221–7.

Von Kügelgen I, Späth L, and Starke K (1994). Evidence for P2-purinoceptor-mediated inhibition of noradrenaline release in rat brain cortex. *Br J Pharmacol* **113**, 815–22.

Webb TE, Feolde E, Vigne P, Neary JT, Runberg A, Frelin C, and Barnard EA (1996a). The P2Y purinoceptor in rat brain microvascular endothelial cells couple to inhibition of adenylate cyclase. *Br J Pharmacol* **119**, 1385–92.

Webb TE, Henderson D, King BF, Wang S, Simon J, Bateson AN *et al.* (1996b). A novel G protein-coupled P2 purinoceptor (P2Y$_3$) activated preferentially by nucleoside diphosphates. *Mol Pharmacol* **50**, 258–65.

Webb TE, Henderson D, Roberts JA, and Barnard EA (1998). Molecular cloning and characterization of the rat P2Y$_4$ receptor. *J Neurochem* **71**, 1348–57.

Webb TE, Kaplan MG, and Barnard EA (1996). Identification of 6H1 as a P2Y purinoceptor: P2Y$_5$. *Biochem Biophys Res Comm* **219**, 105–10.

Webb TE, Simon J, Krishek BJ, Bateson AN, Smart TG, King BF *et al.* (1993). Cloning and functional expression of a brain G-protein-coupled ATP receptor. *FEBS Lett* **324**, 219–25.

Webb TE, Simon J, Bateson AN, and Barnard EA (1994). Transient expression of the recombinant chick brain P2Y$_1$ purinoceptor and localization of the corresponding mRNA. *Cellul Molec Biol* **40**, 437–42.

Williams M and Jarvis MF (2000). Purinergic and pyrimidinergic receptors as potential drug targets. *Biochem Pharmacol* **59**, 1173–85.

Yerxa B (2001). Therapeutic use of nucleotides in respiratory and ophthalmic disease. *Drug Dev Res* **52**, 196–201.

Yokomizo T, Izumi T, Chang K, Takuwa Y, and Shimizu T (1997). A G-protein-coupled receptor for leukotriene B_4 that mediates chemotaxis. *Nature* 387, 620–4.

Zambon AC, Brunton LL, Barrett KE, Hughes RJ, Torres B, and Insel P (2001). Cloning expression, signaling mechanisms and membrane targeting of $P2Y_{11}$ receptors in cultured MDCK-D1 cells. *Mol Pharmacol* in press.

Zambon AC, Hughes RJ, Meszaros JG, Wu JJ, Torres B, Brunton LL, and Insel PA (2000). $P2Y_2$ receptor of MDCK cells: cloning, expression and cell-specific signalling. *Am J Physiol* 279, F1045–52.

Zhang YX, Yamashita H, Ohshita T, Sawamoto N, and Nakamura S (1995). ATP increases extracellular dopamine level through stimulation of P2Y purinoceptors in the rat striatum. *Brain Res* 691, 205–12.

Chapter 22

Serotonin receptors

Claire Roberts, Gary W. Price, and
Derek N. Middlemiss

22.1 Introduction

The pharmacology of serotonin (5-hydroxytryptamine; 5-HT) was reviewed in 1986 (Bradley
et al.) acknowledging the existence of at least three 5-HT receptor families: 5-HT$_{1-3}$. In
the 15 years following this classification molecular cloning has confirmed the existence of
multiple 5-HT receptors subtypes. The 5-HT family is now divided into seven main classes,
5-HT$_{1-7}$, comprising a total of 14 receptor subtypes. Apart from the 5-HT$_3$ receptor subtype,
which is a ligand gated channel, the 5-HT family are G protein-coupled receptors (GPCRs).
Mammalian G protein coupled 5-HT receptors represent the results of more than 750 million
years of molecular evolution. The evolutionary relationships between the subtypes has been
determined by a phylogenetic tree analysis (Peroutka 1994). This has demonstrated that
the primordial G protein coupled 5-HT receptor differentiated into three major branches:
5-HT$_1$, 5-HT$_2$, and 5-HT$_6$ families (Table 22.1). This chapter will review the molecular
characterization, localization, signalling mechanisms, pharmacology, and disease relevance
of the 5-HT GPCR family.

22.2 Molecular characterization

22.2.1 The 5-HT$_1$ receptor family

The 5-HT$_1$ receptor family includes 5-HT$_{1A}$, 5-HT$_{1B}$, 5-HT$_{1D}$, 5-ht$_{1E}$, 5-ht$_{1F}$, 5-ht$_5$ and
5-HT$_7$ receptor subtypes. Convention dictates that lower cases denote receptors which
await confirmation of function in native tissue. The 5-HT$_{1A}$ receptor was the first
5-HT receptor to be successfully cloned. Both the human and rat 5-HT$_{1A}$ receptors
were identified by screening a genomic library for homologous sequences to the β-
adrenoreceptor (Koblinka *et al.* 1987; Fargin *et al.* 1988; Albert *et al.* 1990). The
human 5-HT$_{1A}$ receptor has 422 amino acids and the gene is intronless, with sites
for glycosylation and phosphorylation and localized on chromosome 5q11.2–q13. Polymorphisms have been reported in the human 5-HT$_{1A}$ receptor by several groups. Warren
et al. (1993*a*) identified a silent substitution which was found in 22 per cent of African
Americans. Nakhai *et al.* (1995) reported two polymorphisms that altered amino acid
composition in the extracellular terminal domain (Gly22Ser and Iso28Val), present in
American and Finnish Caucasians and native American Indians. Kawanishi *et al.* (1998*a*)

found an Arg substitution for Leu at amino acid 219. In addition Kawanishi *et al.* (1998*b*) identified five mutations in the 5′ untranslated region and four mutations in the coding region, two of which were silent (Gly294Ala and Cys549Thr), Pro16Leu and Gly272Asp. In the polymorphisms reported to date no disease association has been identified.

During 1991 and 1992 two related intronless human genes were isolated on the basis of their sequence homology with an orphan receptor (dog RDC4) which had 5-HT$_1$ receptor characteristics. When expressed in cell lines they both demonstrated pharmacology of the 5-HT$_{1D}$ and not the rodent 5-HT$_{1B}$ receptor subtype and so were designated 5-HT$_{1D\alpha}$ and 5-HT$_{1D\beta}$ receptor subtypes. However, when the rodent 5-HT$_{1B}$ receptor was eventually cloned (Voigt *et al.* 1991; Adham *et al.* 1992) it was found to have greater than 95 per cent homology with the human 5-HT$_{1D\beta}$ receptor (Jin *et al.* 1992). The discovery of a rat gene orthologous to the human 5-HT$_{1D\alpha}$ receptor, encoding a receptor with 5-HT$_{1D}$ binding properties, together with the earlier findings led to a reassessment of the nomenclature for 5-HT receptors (Hartig *et al.* 1996). Despite differing pharmacology the human 5-HT$_{1D\beta}$ receptor is a species equivalent of the rodent 5-HT$_{1B}$ receptor and the human 5-HT$_{1D\alpha}$ receptor is a species equivalent of the rodent 5-HT$_{1D}$ receptor. To take account of the fact that the pharmacology of these receptors differ across species prefixes can be used to denote species. The genes encoding the human 5-HT$_{1B}$ and 5-HT$_{1D}$ receptors are located on different chromosomes, 6q13 and 1p34.3–36.3 respectively (Saudou and Hen 1994). Polymorphism of the h5-HT$_{1B}$ receptor has been identified in 40 unrelated Caucasians of European descent (Sidenberg *et al.* 1993) but the importance of this polymorphism remains uncertain. In addition, three groups have identified a silent Gly861Cys substitution (Hunag *et al.* 1999; Lappalainen *et al.* 1998; Maassen Van Den Brink *et al.* 1998) and Lappalainen *et al.* (1998), concluded that, in Finns, this polymorphism was linked to alcoholism. Two other substitutions were reported, Cys129Thr (Hunag *et al.* 1999) and Thr261Gly (Maassen Van Den Brink *et al.* 1998) but their disease association, if any, remains to be elucidated.

The human 5-ht$_{1E}$ receptor gene is intronless, encodes a protein of 365 amino acids (McAllister *et al.* 1992; Zgombick *et al.* 1992; Gudermann *et al.* 1993) and locates to human chromosome 6q14–q15 (Levy *et al.* 1992*b*). The 5-HT$_{1F}$ receptor is another intronless gene encoding a 366 amino acid protein (Amlaiky *et al.* 1992; Adham *et al.* 1993*b*) and is located on chromosome 3q11 (Saudou and Hen 1994).

Two 5-ht$_5$ genes have been identified in mouse and rat, designated 5-ht$_{5A}$ and 5-ht$_{5B}$ (Hen 1992; Matthes *et al.* 1993; Erlander *et al.* 1993; Wisden *et al.* 1993). Both genes are present in human tissue however the 5-ht$_{5B}$ gene contains a stop codon which results in a non-functional G protein (Rees *et al.* 1994). Unlike the rest of the 5-HT$_1$-family, both 5-ht$_5$ genes contain an intron in the third cytoplasmic loop (Matthes *et al.* 1993). Due to the position of this intron any splice variants are likely to be non-functional. The 5-ht$_{5A}$ gene was located on human chromosome 7q36 while the 5-ht$_{5B}$ gene was found on human chromosome 2q11–13. The 5-ht$_{5A}$ gene encodes a 357 amino acid protein (Plassat *et al.* 1992; Erlander *et al.* 1993; Rees *et al.* 1994) while the 5-ht$_{5B}$ gene encodes a 370–371 amino acid protein (Erlander *et al.* 1993; Matthes *et al.* 1993; Wisden *et al.* 1993), both containing glycosylation and phosphorylation sites.

The mammalian 5-HT$_7$ receptor is predicted to be 445–448 amino acids in length (Nelson *et al.* 1995; Lovenberg *et al.* 1993*a,b*; Heidmann *et al.* 1997; Jasper *et al.* 1997). It is located on human chromosome 10q21–q24 (Gelernter *et al.* 1995) and contains two introns (Ruat *et al.* 1993*a,b*; Erdmann *et al.* 1996; Heidmann *et al.* 1997). The existence of introns has led

Table 22.1 Serotonergic receptors

	5-HT$_{1A}$	5-HT$_{1B}$	5-HT$_{1D}$	5-HT$_{1E}$	5-HT$_{1F}$	5-HT$_{5A}$	5-HT$_{5B}$	5-HT$_7$
Alternative names	—	5-HT$_{1D\beta}$	5-HT$_{1D\alpha}$	5-HT$_{1E\alpha}$	5-HT$_{1F\beta}$	5-HT$_{1\alpha}$	5-HT$_{1D\beta}$	5-HT$_{1-like}$, 5-HT$_\gamma$
Structural information (Accession no.)	h421 aa (P08908) r 422 aa (P19327) m 421 aa (Q64264)	h390 aa (P28222) r 386 aa (P28564) m 386 aa (P28334)	h377 aa (P28221) r 374 aa (P28565) m 374 aa (Q61224)	h365 aa (P28566)	h366 aa (P30939) r 366 aa (P30940) m 366 aa (Q02284)	h357 aa (P47898) r 357 aa (P5364) m 357 aa (P30966)	r 370 aa (P35365) m 370 aa (P31387)	h445 aa (P50407)[AS] r 448aa(P32305)[AS] m 448 aa (P32304)
Chromosomal location	5q11.2–q13	6q13	1p34.3–36.3	6q14–15	3p11	7q35–36	2q11–q13	10q23.3–24.3
Selective agonists	8-OH-DPAT, U92016A	—	L775606	—	LY334370 LY344864	—	—	—
Selective antagonists	WAY100635, NAD299	GR127935, SB216641, SB224289, SB236057	GR127935, BRL15572, ketanserin, ritanserin	—	—	Methiothepin	Methiothepin	SB258719, SB266970, clozapine
Radioligands	[^{3}H]-WAY100635, [^{3}H]-8-OH-DPAT	[^{3}H]-sumatriptan, [^{3}H]-GR127935	[^{3}H]-sumatriptan, [^{3}H]-GR127935	[^{3}H]-5-HT	[^{3}H]-LY334370, [^{125}I]-I-LSD	[^{3}H]-5-CT, [^{125}I]-I-LSD	[^{3}H]-5-CT, [^{125}I]-I-LSD	[^{3}H]-5-CT, [^{3}H]-SB269970
G protein coupling	G$_i$/G$_O$	G$_i$/G$_O$	G$_i$/G$_O$	G$_i$/G$_O$	G$_i$/G$_O$	G$_i$/G$_O$	G$_i$/G$_O$	G$_s$
Expression profile	Cerebral cortex, hippocampus, septum, amygdala, raphe, myenteric plexus	Striatum, substantia nigra, globus pallidus, hippocampus, raphe, superior colliculi, cerebellum, vascular smooth muscle, autonomic terminals	Striatum, substantia nigra, globus pallidus, locus coeruleus, hippocampus, dorsal raphe nucleus, accumbens, cerebellum, vascular smooth muscle, autonomic and trigeminal nerve terminals	Cerebral cortex, caudate, amygdala, hippocampus, hypothalamus	Hippocampus, caudate, claustrum, cerebral cortex, raphe nucleus thalamus, cerebellum, hypothalamus, spinal cord, uterus, mesentry	Cerebral cortex, SCN, raphe, caudate, habenula, cerebellum, hippocampus, spinal cord, olfactory bulb	Cerebral cortex, olfactory bulb, hippocampus, cerebellum, habenula	Cerebral cortex, globus pallidus, amygdala, sunbstantia nigra, hypothalamus thalamus, raphe, SCN, superior colliculus, hippocampus, gastrointestinal and vascular smooth muscle, sympathetic ganglia

Table 22.1 (Continued)

	5-HT$_{1A}$	5-HT$_{1B}$	5-HT$_{1D}$	5-HT$_{1E}$	5-HT$_{1F}$	5-HT$_{5A}$	5-HT$_{5B}$	5-HT$_7$
Physiological function	Somatodendritic autoreceptor in hippocampus and raphe, somatodendritic at cholinergic terminals of myenteric plexus	Presynaptic autoreceptor and heteroreceptor, contraction of vascular smooth muscle	Somatodendritic autoreceptor in hippocampus and raphe, vasoconstriction in vascular smooth muscle, inhibition in autonomic neurones	Unclear	Trigeminal neuroinhibition	Unclear	Unclear	Smooth muscle relaxation, circadian phase shifts
Knockout phenotype	Anxious, less reactive, decreased depressive related behaviours, less aggressive	Aggressive, more reactive, abnormal exploratory behaviour, altered sleep patterns, decreased startle response, increased body weight	—	—	—	Abnormal baseline startle, anxiety related behaviour, increased exploratory activity	—	—
Disease relevance	Depression, schizophrenia, cognitive decline, attention deficit hyperactivity disorder, neurodegeneration	Depression, migraine, Parkinson's disease, schizophrenia, cognitive decline	Depression, migraine, schizophrenia, cognitive decline	Unclear	Migraine	Unclear but possibly cognitive decline, schizophrenia, anxiety, depression	Anxiety, depression, sleep disorders	

	5-HT$_{2A}$	5-HT$_{2B}$	5-HT$_{2C}$	5-HT$_4$	5-HT$_6$
Alternative names	D, 5-HT$_2$	5-HT$_{2F}$	5-HT$_{2C}$	—	—
Structural information (Accession no.)	h 471aa (P28223) r 471 aa (P14844) m 471 aa (P35363)	h 481aa (P41595) r 479 aa (P30994) m 504 aa (Q012152)	h 377 aa (P28221) r 377 aa (P49145) m 374 aa (P28565)	h 387aa(Y09756)[AS] r 387 aa (U20906)[AS] m 387aa (Y09587)[AS]	h 440 aa (P50406) r 436 aa (P31388) m 440 aa (NP_067333)
Chromosomal location	13q14–q21	2q36.3–2q37.1	Xq24	5q31–33	1p35–36
G protein coupling	G$_q$/G$_{11}$	G$_q$/G$_{11}$	G$_q$/G$_{11}$	G$_s$	G$_s$
Selective agonists	(partially selective) DOI, DOB	BW 723C86	m-CPP, Ro600175	BIMU8, RS67506, ML10302, SC53116	LSD
Selective antagonists	Ketanserin, spiperone MDL100,907	SB204741, SB200646, SB206553	SB200646, SB206553, SB242084, SB243213	GR113808, SB204070, RS100235	Ro046790, Ro630563, SB271046, SB357134
Radioligands	[^{3}H]-ketanserin	[^{3}H]-5-HT	[^{3}H]-mesulergine	[^{3}H]-GR113808 [^{125}I]-SB207710	[^{3}H]-LSD [^{125}I]-SB258585 [^{3}H]-Ro630563
Expression profile	Cerebral cortex, caudate, nucleus accumbens, hippocampus, caudate, olfactory bulb, gastrointestinal tract, vascular and bronchial smooth muscle, vascular endothelium, platelets	Cerebellum, lateral septum, hypothalamus amygdala, smooth muscle of ileum, stomach, uterus, vasculature, endothelium	Cerebral cortex, choroid plexus, nucleus accumbens, amygdala, substantia nigra, medulla pons, striatum hippocampus, hypothalamus, habenula	Cerebral cortex, nucleus accumbens, striatum, thalamus, hippocampus, olfactory bulb, substantia nigra, brainstem, parasympathetic neurones, oesophogeal and vascular smooth muscle	cerebral cortex, striatum, olfactory tubercule nucleus accumbens, hippocampus, superior cervical ganglia
Physiological function	Vascular smooth muscle constriction, platelet activation, possibly involved in neuroinhibition	Endothelium dependent vasorelaxation via NO production, vascular smooth muscle constriction	Possibly involved in modulation of transferrin production and regulation of CSF volume	Smooth muscle relaxation, myenteric cholinergic neuroexcitation, increased EEG amplitude, increased myocyte contractility	Possible modulation of CNS acetycholine release
Knockout phenotype	—	Embryonic and neonatal death due to heart defects	Spontaneous convulsions, weight gain, cognitive impairment	—	—
Disease relevance	Depression, shizophrenia	Anxiety, eating disorders	Anxiety, schizophrenia, depression, sleep disorders, Parkinson's disesase	Cognitive decline, schizophrenia, Parkinson's disease, anxiety, oesophageal reflux	Cognitive decline

to the generation of at least four splice variants of the receptor (5-HT_{7a}, 5-HT_{7b}, 5-HT_{7c}, and 5-HT_{7d}). Only the first three isoforms are expressed in the rat while 5-HT_{7a}, 5-HT_{7b}, and 5-HT_{7d} have been identified in human tissue.

22.2.2 The 5-HT$_2$ receptor family

The 5-HT$_2$ family includes 5-HT_{2A}, 5-HT_{2B}, and 5-HT_{2C} receptor subtypes. The human 5-HT_{2A} gene encodes a 471 amino acid protein, contains two introns and is located on human chromosome 13q12–q21. The amino acid sequence has potential sites for gylycosylation, phosphorylation, and palmitoylation (Saltzman *et al.* 1991). At least two allelic forms have been identified in Caucasians but they have proven to be silent (Warren *et al.* 1993*b*). The human 5-HT_{2B} gene encodes a 481 amino acid protein, contains two introns and is located on human chromosome 2q36.3–2q37.1. The 5-HT_{2C} gene encodes a 458 amino acid protein and contains three introns in the coding sequence and is an X-linked gene found on human chromosome Xq24 (Saltzman *et al.* 1991). Splice variants have been isolated in rat, mouse, and human brain but their functional significance remains to be determined (Canton *et al.* 1996). In addition, 5-HT_{2C} mRNA has been reported to undergo post-transcriptional editing to yield multiple isoforms which may have different pharmacology (Burns *et al.* 1977).

22.2.3 The 5-HT$_6$ receptor family

The 5-HT$_6$ family includes 5-HT$_4$ and 5-HT$_6$ receptor subtypes. The genomic structure of the 5-HT$_4$ gene has yet to be published but it appears to be highly fragmented and is comprised of at least five introns (Bockaert *et al.* 1998). Therefore, it is not surprising that multiple splice variants have been identified in human, mouse, and rat, designated 5-HT_{4a}, 5-HT_{4b}, 5-HT_{4c}, and 5-HT_{4d} of 387, 406, 380, and 360 amino acids long respectively (Blondel *et al.* 1998*a*; Bockaert *et al.* 1998; Medhurst *et al.* 2001). There are reports of differential expression of the splice variants with 5-HT_{4a}, 5-HT_{4b}, and 5-HT_{4c} isoforms mainly found in the brain (with some in the atrium and gut) while 5-HT_{4d} is only detected in the gut (Blondel *et al.* 1998*a*). The pharmacology between the splice variants appears to be similar, although some full agonists are only partial agonists at 5-HT_{4c} receptors (Blondel *et al.* 1998*a*). In addition the 5-HT_{4c} receptor is the only isoform to display constitutive activation of adenylyl cyclase when expressed in cell lines (Blondel *et al.* 1998*a,b*). All the splice variants contain glycosylation and phosphorylation sites. The human gene is located on chromosome 5q31–q33 (Claeysen *et al.* 1997*a–c*; Cichol *et al.* 1998).

The rat and human 5-HT$_6$ receptors are comprised of 436–440 amino acids (Ruat *et al.* 1993*a,b*; Kohen *et al.* 1996), a glycosylation site in the N-terminal extracellular domain and a number of phosphorylation sites in the third cytoplasmic loop and the C-terminal. The receptor phosphorylation is catalyzed by cAMP-dependent protein kinase and is responsible for agonist-induced desensitization reported by Sleight *et al.* (1997). The gene has two introns however functional splice variants have yet to be reported (Monsma *et al.* 1993; Ruat *et al.* 1993*a,b*; Kohen *et al.* 1996; Olsen *et al.* 1999). The human gene is located on chromosome p35–36. Several groups have reported a Cys267Thr polymorphism of this gene that has been linked to bipolar affective disorder (Vogt *et al.* 2000), Alzheimer's disease (Tsai *et al.* 1999) and schizophrenia (Tsai *et al.* 1999; Yu *et al.* 1999). However, three groups failed to replicate this polymorphic association with schizophrenia (Chiu *et al.* 2001; Masellis *et al.* 2001; Shinkai *et al.* 1999).

22.3 Cellular and subcellular localization

22.3.1 The 5-HT$_1$ receptor family

The distribution of the 5-HT$_{1A}$ receptor has been mapped extensively by receptor auto-radiography using ligands such as [^{3}H]5-HT, [^{3}H]8-OH-DPAT, [^{3}H]ipsapirone, [^{125}I][BH-8-MeO-N-PAT, [^{125}I]p-MPPI, and [^{3}H]WAY100635 (Pazos and Palacios 1985; Weissmann-Nanopoulus et al. 1985; Hoyer et al. 1986; Verge et al. 1986; Radja et al. 1991; Khawaja, 1995; Kung et al. 1995). High levels were found in limbic areas such as the hippocampus, septum, amygdala, and cortex, and also the raphe nuclei (Radja et al. 1991). Immunohistochemical studies revealed a similar 5-HT$_{1A}$ receptor protein distribution but, in addition, identified 5-HT$_{1A}$ receptors in the habenula (El Mestikawy et al. 1990). The distribution of 5-HT$_{1A}$ receptor mRNA in raphe is similar to that of the radioligand binding sites (Chalmers and Watson 1991; Pompeiano et al. 1992). Selective 5-HT lesioning studies, using the neurotoxin 5,7-dihydroxytryptamine, produced a decrease in 5-HT$_{1A}$ binding sites (Verge et al. 1986) and mRNA (Miquel et al. 1992) but only in the raphe, confirming their somatodendritic location in the raphe nuclei and a low density, if any, on presynaptic terminals in projection areas.

Autoradiographical studies have identified 5-HT$_{1B}$ receptors in the basal ganglia (caudate, globus pallidus, substantia nigra, ventral pallidum, and entopeduncular nucleus) and regions such as the subiculum, superior colliculi and cerebellum (Bruinvels et al. 1993; Pazos and Palacios 1985; Verge et al. 1986). Some areas with a high level of 5-HT$_{1B}$ receptor binding (e.g. the striatum), also express a high level of 5-HT$_{1B}$ receptor mRNA suggesting that these receptors are present on indigenous neurones. However, other areas with a high level of receptor binding (globus pallidus and ventral pallidum) have little detectable mRNA and are therefore likely to be present on afferent terminals. In addition, 5-HT$_{1B}$ receptor mRNA was identified in raphe nuclei (Boschert et al. 1994; Bruinvels et al. 1994a,b; Doucet et al. 1995; Voigt et al. 1991). Evidence from 5-HT neuronal lesioning experiments using radioligand binding have generated contradictory results, with some groups demonstrating upregulation and others downregulation of 5-HT$_{1B}$ receptors in the same brain region (Bruinvels et al. 1994a,b; Middlemiss and Hutson 1990). However, in situ hybridization studies have shown a consistent decrease in 5-HT$_{1B}$ receptor mRNA in the dorsal and medial raphe nuclei following 5-HT neuronal lesioning (Doucet et al. 1995). Together these studies support both the pre- and post-synaptic location of 5-HT$_{1B}$ receptors.

Due to the relatively low abundance of 5-HT$_{1D}$ receptors in the brain and the lack of a suitable high affinity and selective radioligand, it has been difficult to ascertain its distribution in the brain. However, it appears that the 5-HT$_{1D}$ receptor distribution is similar to that of the 5-HT$_{1B}$ receptor, with detectable levels of receptor binding sites in the basal ganglia (Bruinvels et al. 1993; Castro et al. 1997). In situ hybridization studies have demonstrated that 5-HT$_{1D}$ mRNA is present in the caudate, nucleus accumbens, hippocampus, olfactory tubercule, cortex, cerebellum, dorsal raphe, and locus coeruleus (LC) (Hamblin et al. 1992; Bach et al. 1993; Bruinvels et al. 1994a,b; Bonaventure et al. 1998).

Autoradiographical studies, determining the distribution of non-5-HT$_{1A/1B/1D/2C}$ [^{3}H]5-HT binding sites, indicate that 5-ht$_{1E}$ protein is present in cortex, caudate putamen, claustrum, hippocampus, and amygdala of human, mouse, rat, and guinea-pig brain (Miller and Teitler 1992; Barone et al. 1993; Bruinvels et al. 1994c). In addition, mRNA has been identified in cortical areas, caudate, putamen, amygdala, and hypothalamic regions of the human and monkey (Bruinvels *et al.* 1994*a,b*). Although this pattern is very similar to that of the 5-HT$_{1B}$ and 5-HT$_{1D}$ receptors, as yet there is no evidence for a raphe location. Together with the lack of change in 5-ht$_{1E}$ binding sites following 5-HT neuronal lesions it would appear to be consistent with a post-synaptic location for the 5-ht$_{1E}$ receptor (Barone *et al.* 1993).

Autoradiographical studies of 5-ht$_{1F}$ receptor binding sites have identified high levels in hippocampus, caudate, and claustrum (Bruinvels *et al.* 1994*a,b*; Lucaites *et al.* 1996). The receptor is also found in low levels in parts of the basal ganglia but is notably absent from the substantia nigra (Lucaites *et al.* 1996; Waeber and Moskowitz 1995*a,b*). *In situ* hybridization studies have demonstrated that 5-ht$_{1F}$ mRNA is also present in the hippocampus and caudate and, in addition, in the dorsal raphe, cortex, thalamus, pons, cerebellum, and hypothalamus (Adham *et al.* 1993, 1997; Bruinvels *et al.* 1994*a,b*; Lovenberg *et al.* 1993*a*).

Comparative autoradiography with wild type and 5-ht$_{5A}$ receptor knock-out mice indicates that 5-ht$_{5A}$ receptors are concentrated in the olfactory bulb, neocortex, hippocampus, caudate, and medial habenula (Grailhe *et al.* 1999). These results were confirmed by immunohistochemical studies (Oliver *et al.* 1999) which, in addition, provided evidence for the existence of 5-ht$_{5A}$ receptors in the raphe and suprachiasmatic nucleus. *In situ* hybridization studies have identified 5-ht$_{5A}$ receptor mRNA in the cortex, hippocampus, granule cells of the cerebellum, medial habenula, amygdala, septum, thalamic nuclei, and olfactory bulb (Plassat *et al.* 1992; Erlander *et al.* 1993).

Autoradiographical studies have identified 5-HT$_7$ receptors in cortex, septum, globus pallidus, thalamus, hypothalamus, amygdala, substantia nigra, periaquaductal grey, and superior colliculus (Lucas and Hen 1995; To *et al.* 1995; Waeber and Moskowitz 1995*a*; Gustafson *et al.* 1996). Immunohistochemical analysis (Oliver *et al.* 1999) confirmed the results of the autoradiography studies and in addition provided evidence for the existence of 5-HT$_7$ receptors in the raphe and SCN. *In situ* hybridization studies have identified 5-HT$_7$ mRNA in the hippocampus, hypothalamus, thalamus, amygdala, cortex, superior colliculus, and dorsal and median raphe nuclei (Lovenberg *et al.* 1993*b*; Ruat *et al.* 1993*a*; Shen *et al.* 1993; Lucas and Hen 1995; To *et al.* 1995; Gustafson *et al.* 1996; Heidmann *et al.* 1998).

22.3.2 The 5-HT$_2$ receptor family

High levels of 5-HT$_{2A}$ binding sites have been found in cortical areas, caudate, nucleus accumbens, olfactory tubercule, and hippocampus (Pazos *et al.* 1985, 1987; Lopez-Gimenez *et al.* 1997). Generally, there is close agreement between the distribution of 5-HT$_{2A}$ binding sites, immunoreactivity and mRNA suggesting that this subtype has a post-synaptic location (Mengod *et al.* 1990*a*; Morilak *et al.* 1993, 1994; Pompeiano *et al.* 1994). In areas such as the cortex, 5-HT$_{2A}$ receptors have been localized on GABAergic interneurones (Francis *et al.* 1992; Morilak *et al.* 1993, 1994; Burnet *et al.* 1995) and also glutamatergic projections (Burnet *et al.* 1995; Wright *et al.* 1995). The presence of 5-HT$_{2B}$ receptors in the brain has been controversial. However, low levels of mRNA for the receptor have been detected in both human and mouse brain (Loric *et al.* 1992; Kursar *et al.* 1994; Bonhaus *et al.* 1995) but it awaits confirmation in the rat. 5-HT$_{2B}$ immunoreactivity has been reported in rat brain, in cerebellum, lateral septum, dorsal hypothalamus, and amygdala (Duxon *et al.* 1997). 5-HT$_{2C}$ receptor localization is restricted to the CNS, unlike 5-HT$_{2A}$ and 5-HT$_{2B}$ receptors. Autoradiographical studies have identified this receptor in choroid plexus, cortex, nucleus accumbens, hippocampus, amygdala, caudate, and substantia nigra (Palacios *et al.*

1991; Radja *et al.* 1991). The distribution of mRNA is very similar except there are high levels in the habenular nucleus; where binding site levels are very low (Mengod *et al.* 1990*b*). Therefore, 5-HT$_{2C}$ receptors in addition to having a post-synaptic location may also be presynaptic.

22.3.3 The 5-HT$_6$ receptor family

High levels of 5-HT$_4$ binding sites have been identified in nigrostriatal and mesolimbic systems of all species studied (Mengod *et al.* 1996; Waeber *et al.* 1993). A similar distribution of 5-HT$_4$ mRNA has also been described in the rat CNS and in peripheral tissues such as the gut (Gerald *et al.* 1995; Claeysen *et al.* 1996; Mengod *et al.* 1996; Blondel *et al.* 1998*a*), although there are conflicting reports as to whether the splice variants of this receptor are differentially expressed (Claeysen *et al.* 1996; Gerald *et al.* 1995). Levels of 5-HT$_6$ receptor mRNA have been detected in stomach, adrenal glands and CNS (Ruat *et al.* 1993*a,b*; Monsma *et al.* 1993). In the human, rat and guinea-pig brain 5-HT$_6$ is found in striatum, olfactory tubecules, nucleus accumbens and hippocampus (Monsma *et al.* 1993; Ruat *et al.* 1993*a,b*; Ward *et al.* 1995; Gerard *et al.* 1996; Kohen *et al.* 1996). 5-HT$_6$ immunoreactivity in rat brain is consistent with that of the mRNA and is associated with dendritic processes making synapses with unlabelled axon terminals (Gerard *et al.* 1997). This is consistent with the 5-HT$_6$ receptors being post-synaptic to 5-HT neurones and is supported by the lack of decrease in mRNA following serotonergic lesioning (Gerard *et al.* 1997).

22.4 Pharmacology

22.4.1 Agonists

The 5-HT$_1$ receptor family

Several agonists show some selectivity for the 5-HT$_{1A}$ receptor. 8-OH-DPAT is a full agonist, and buspirone, gepirone, and MDL72832 partial agonists. There are a large number of agonists with high affinity for the 5-HT$_{1B}$ receptor but most are not selective for this receptor (e.g. RU24969, 5-CT, sumatriptan, and CP93129). L775606 (Longmore *et al.* 2000), RU25969, GR127935 as well as 5-CT, and sumatriptan have high affinity for the 5-HT$_{1D}$ receptor. Currently there are no selective ligands for the 5-ht$_{1E}$ receptor. This receptor has been characterized by a high affinity for 5-HT but low affinity for 5-CT. Recently an agonist BRL54443 was reported as having high affinity for both 5-ht$_{1E}$ and 5-ht$_{1F}$ receptors (Kune and Watts 2001). The 5-ht$_{1F}$ receptor is also characterized by high affinity for 5-HT and low affinity for 5-CT but is discriminated by its high affinity for sumatriptan. The agonists LY344864 and LY334370 were recently reported to have selectivity for 5-ht$_{1F}$ receptors but they only have 10-fold selectivity over 5-HT$_{1A}$ receptors (Phebus *et al.* 1997). There are no known selective ligands for 5-ht$_5$ receptors although its pharmacology is similar to that of the 5-HT$_{1D}$ receptor with a high affinity for 5-CT, LSD, methiothepin, and sumatriptan. The 5-HT$_7$ receptor is also characterized by high affinity to 5-CT, 5-HT, and atypical antipsychotics (e.g. clozapine) but there are currently no identified selecive agonists.

The 5-HT$_2$ receptor family

The 5-HT$_{2A}$ receptor has low affinity for 5-HT but high affinity for the agonists 2,5-dimethoxy-4-iodophenylisopropylamine (DOI), (-)2,5-dimethoxy-4-iodoamphetamine

(DOB), and α-methyl-5-HT. Unfortunately these three agonists have little selectivity over the other 5-HT$_2$ receptor subtypes. The pharmacology of the 5-HT$_{2B}$ receptor is similar to the other 5-HT$_2$ subtypes, however, the agonists α-methyl-5-HT and BW723C86 (Kennett *et al.* 1996*a*) do have some selectivity for the 5-HT$_{2B}$ subtype. α-Methyl-5-HT and mCPP have affinity for the 5-HT$_{2C}$ receptor as well as the remaining 5-HT$_2$ subtypes and Martin *et al.* (1995) later reported that RO600175 was a selective agonist.

The 5-HT$_6$ receptor family

Agonists such as BIMU8, zacopride, renzapride, and cisapride have affinity at the 5-HT$_4$ receptor but also possess some 5-HT$_3$ receptor activity. Less progress has been made in the design of selective 5-HT$_6$ agonists with only LSD having high affinity. However, Glennon *et al.* (2000) have described a substituted tryptamine that has high affinity for 5-HT$_6$ receptors, with 10–30-fold selectivity over 5-HT$_{1A}$, 5-HT$_{1D}$, and 5-HT$_7$ receptors.

22.4.2 **Antagonists**

The 5-HT$_1$ receptor family

The synthesis of selective 5-HT$_{1A}$ receptor antagonists has proven difficult. Several apparent antagonists, such as NAN190, BMY73778, MDL73005, SDZ216525, and WAY100135, demonstrated agonist properties in brain regions of high receptor reserve (Fletcher *et al.* 1993). However, selective 5-HT$_{1A}$ antagonists such as WAY100635 (Fletcher *et al.* 1996) and NAD299 (Johansson *et al.* 1997) have been identified. GR127935 (Skingle *et al.* 1996) is a mixed 5-HT$_{1B/1D}$, while SB216641 (Price *et al.* 1997), SB224289 (Gaster *et al.* 1998) and SB236057 (Roberts *et al.* 2000) are selective and potent 5-HT$_{1B}$ receptor antagonists. In contrast, ketanserin and ritanserin have around 30-fold selectivity (Pauwels *et al.* 1996) while BRL15572 (Price *et al.* 1997) has 60-fold selectivity for 5-HT$_{1D}$ over 5-HT$_{1B}$ receptors. SB258719 (Thomas *et al.* 1998) and SB269970 (Hagan *et al.* 2000) are selective 5HT$_7$ receptor antagonists.

The 5-HT$_2$ receptor family

The receptor antagonists, ketanserin and spiperone, are 20-fold selective for 5-HT$_{2A}$ over 5-HT$_{2B/2C}$ but have appreciable affinity for other monoamine receptors. The receptor antagonist, MDL100907, is however a reasonable selective 5-HT$_{2A}$ receptor antagonist. 5-HT$_{2B}$ receptors have lower affinity for the antagonist ritanserin and higher affinity for yohimbine. Recent reports have identified SB200646 and SB206553, which have high affinity and selectivity for 5-HT$_{2B/2C}$ receptors (Kennett *et al.* 1996*b*). SB204741 has 100-fold selectivity for 5-HT$_{2B}$ over all other receptors (Baxter *et al.* 1995) and SB242084 and SB243213 are selective 5-HT$_{2C}$ receptor antagonists (Kennett *et al.* 1997*a*; Bromidge *et al.* 2000).

The 5-HT$_6$ receptor family

There are many selective and potent 5-HT$_4$ receptor antagonists available, such as GR113808 (Gale *et al.* 1994), SB204070 (Wardle *et al.* 1994), and RS100235 (Sellers *et al.* 2000). Up until recently there have been no selective antagonists at the 5-HT$_6$ receptor with several antipsychotic drugs having high affinity but lacking selectivity (Roth *et al.* 1994). However, the selective receptor antagonists, Ro046790, Ro630563 (Sleight *et al.* 1998), SB271046 (Routledge *et al.* 2000), and SB357134 (Bromidge *et al.* 2001) have recently been reported.

22.5 Signal transduction and receptor modulation

22.5.1 The 5-HT$_1$ receptor family

The 5-HT$_{1A}$ receptor negatively couples to adenylyl cyclase via G$_i$ proteins in both rat and guinea-pig hippocampal tissue and cell lines (Saudou and Hen 1994). There are also reports of positive coupling of 5-HT$_{1A}$ receptors in hippocampal tissue (Shenker *et al.* 1983) but this response has recently been attributed to 5-HT$_7$ receptor activation (Thomas *et al.* 1999). However, 5-HT$_{1A}$ activation is able to enhance the effects of 5-HT$_7$ receptor activation (Thomas *et al.* 1999). This was hypothesized to be an action of $\beta\gamma$ subunits released on 5-HT$_{1A}$ receptor activation, potentiating the G$_s$ stimulation of adenylyl cyclase produced by 5-HT$_7$ receptor agonists. 5-HT$_{1A}$ receptors have also been shown to decrease intracellular calcium, activate phospholipase C and increase intracellular calcium in cloned systems but these observations await confirmation in native tissues (Boess and Martin 1994; Albert *et al.* 1996).

Both rat and human cloned 5-HT$_{1B}$ receptors have been shown to negatively couple to adenylyl cyclase (Adham *et al.* 1992; Weinshank *et al.* 1992). This was also demonstrated in rat and calf substantia nigra (Bouhelal *et al.* 1988; Schoeffter and Hoyer 1989). Cloned 5-HT$_{1D}$ receptors have been demonstrated to negatively couple to adenylyl cyclase (Hamblin and Metcalf 1991; Weinshank *et al.* 1992). In addition, there are reports that 5-HT$_{1D}$ receptor activation will increase cAMP accummulation in cloned cell lines (Watson *et al.* 1996). To date there are no second messenger responses associated with the 5-HT$_{1D}$ receptor which has been demonstrated in native tissues.

There is very little functional data available for the 5-ht$_{1E}$ receptor. However, in human cloned cell lines the 5-ht$_{1E}$ receptor has been shown to weakly inhibit forskolin-stimulated adenylyl cyclase (Levy *et al.* 1992a; McAllister *et al.* 1992; Zgombick *et al.* 1992; Adham *et al.* 1994). Both the human and mouse cloned 5-ht$_{1F}$ receptors couple to the inhibition of forskolin-stimulated adenylyl cyclase (Amlaiky *et al.* 1992; Adham *et al.* 1993a; Lovenberg *et al.* 1993a,b) but to date no second messenger response has been reported in native tissues.

Although demonstrated to be G protein coupled (Plassat *et al.* 1992; Matthes *et al.* 1993; Wisden *et al.* 1993) initial investigations were unable to show 5-ht$_{5A}$ or 5-ht$_{5B}$ receptor modulation of adenylyl cyclase or phospholipase c in cloned cell lines (Plassat *et al.* 1992; Erlander *et al.* 1993; Matthes *et al.* 1993; Wisden *et al.* 1993). Recent reports were more successful and demonstrated that 5-ht$_{5A}$ receptors, expressed in either HEK-293 (Hurley *et al.* 1998; Franken *et al.* 1998) or C6 glioma cell lines (Thomas *et al.* 2000), were negatively coupled to adenylyl cyclase. We now await selective 5-ht$_5$ receptor ligands to probe second messenger systems in native tissue.

Both recombinant and native 5-HT$_7$ receptors have been demonstrated to stimulate adenylyl cyclase (Bard *et al.* 1993; Lovenberg *et al.* 1993a,b; Plassat *et al.* 1993; Ruat *et al.* 1993a,b; Shen *et al.* 1993; Tsou *et al.* 1994; Heidmann *et al.* 1997, 1998; Stam *et al.* 1997). It is thought that this occurs through coupling to G$_s$ (Obosi *et al.* 1997). In recombinant systems it has been shown that activation of the 5-HT$_7$ receptor can activate G$_s$ insensitive isoforms of adenylyl cyclase, ACI and ACVIII, as well as the G$_s$ sensitive isoform, ACV (Baker *et al.* 1998). These effects now await confirmation in native tissue.

22.5.2 The 5-HT$_2$ receptor family

All three 5-HT$_{2A}$, 5-HT$_{2B}$, and 5-HT$_{2C}$ receptor subtypes positively couple to phospholipase C, leading to accumulation of inositol phosphates and intracellular calcium in both cloned and native tissue systems (Boess and Martin 1994). 5-HT$_{2A}$ receptor have been demonstrated in brain (Conn and Sanders-Bush 1984; Godfrey *et al.* 1988), 5-HT$_{2B}$ receptor effects in rat fundus (Baxter *et al.* 1995) and 5-HT$_{2C}$ receptor effects in choroid plexus (Sanders-Bush *et al.* 1988).

22.5.3 The 5-HT$_6$ receptor family

Both native and cloned 5-HT$_4$ receptors are positively coupled to adenylyl cyclase (Gerald *et al.* 1995; Claeysen *et al.* 1996; Van den Wyngaert *et al.* 1997). The splice variants generated at this receptor are in the C-terminal domain which may infer that there might be differences in their transduction systems and hence functional response. Recombinant 5-HT$_6$ receptors have been demonstrated to enhance adenylyl cyclase activity (Monsma *et al.* 1993; Routledge *et al.* 2000; Ruat *et al.* 1993*a,b*; Kohen *et al.* 1996; Boess *et al.* 1997). Studies in pig striatum (Schoeffter and Waeber 1994) and mouse cultured striatal cells (Sebben *et al.* 1994) have also suggested that the 5-HT$_6$ receptor is positively coupled to adenylyl cyclase.

22.6 Physiology and disease relevance

22.6.1 Depression

The rationale for 5-HT receptor modulation in the treatment of depression is strong, with a number of receptor subtypes being pursued as drug targets. 5-HT$_{1A}$ receptors are known to modulate 5-HT release and cell firing in the raphe nuclei, that is, function as soma-todendritic autoreceptors. Therefore, 5-HT$_{1A}$ receptor agonists, for example, buspirone, will decrease 5-HT release and many studies support both an anxiolytic and antidepressant action for these compounds (Traeber and Glasser 1987; Charney *et al.* 1990; Handley 1995). 5-HT$_{1A}$ receptor agonists were also demonstrated to increase NA release in the LC, which was attenuated by WAY100635 (Suzuki *et al.* 1995; Hajos-Korcsok and Sharp 1996). The LC projects to the medial pontine reticular formation, which is responsible for control of REM sleep. Therefore, as expected, 5-HT$_{1A}$ receptor agonists also modulated sleep patterns, with agonists reducing slow wave and REM sleep while increasing wakefulness (Monti and Monti 2000). Antidepressants also decrease REM sleep and this may indicate an antidepressant action of 5-HT$_{1A}$ receptor agonists. 5-HT$_{1A}$ receptor antagonists have no effect on 5-HT release, unless the endogenous 5-HT tone is enhanced. However, 5-HT$_{1A}$ receptor antag-onists have been reported to potentiate the increase in 5-HT release observed with SSRIs, monoamine oxidase inhibitors and certain tricyclic antidepressant drugs (Invernizzi *et al.* 1992; Hjorth 1993; Gartside *et al.* 1995; Artigas *et al.* 1996; Romero *et al.* 1996; Sharp *et al.* 1997). In addition, it has been demonstrated that the therapeutic effect of antidepressants in the clinic was improved in the presence of pindolol, a 5-HT$_{1A}$ partial agonist (Artigas *et al.* 1996). Therefore, 5-HT receptor antagonists may be beneficial in the treatment of depression, in the presence of 5-HT uptake blockade.

The 5-HT$_{1B}$ receptor is now recognized as a terminal 5-HT autoreceptor (Schlicker *et al.* 1997), regulating 5-HT release. 5-HT$_{1B}$ receptors have also been localized in cell body regions such as the raphe (Davidson and Stamford 1995; Starkey and Skingle 1994) where they also regulate 5-HT release. 5-HT$_{1B}$ receptor agonists have been reported to decrease while ant-agonists increase 5-HT release, although responses may be brain region dependent (Roberts *et al.* 1998). The increase in 5-HT release observed after acute 5-HT$_{1B}$ receptor antagonist administration led to the hypothesis that such a drug may be a fast acting antidepressant

(Roberts *et al.* 2000). In addition, 5-HT$_{1B}$ receptor antagonists have been demonstrated to potentiate the effects of SSRIs (Huson *et al.* 1995; Roberts *et al.* 1999) and 5-HT$_{1A}$ receptor antagonists (Invernizzi *et al.* 1996; Roberts *et al.* 1999) on 5-HT release, indicating that combinations of these compounds may also be beneficial in the treatment of depression.

5-HT$_{1D}$ receptors have been identified as autoreceptors in the dorsal raphe (Davidson and Stamford 1995; Starkey and Skingle 1994; Pinyero *et al.* 1996) but not the median raphe (Hopwood and Stamford 2001) or terminal brain regions (Schlicker *et al.* 1997). Although the availability of a suitable *in vivo* tools has limited the investigations into the importance of this receptor, it is possible that, as a consequence of its autoreceptor activity, 5-HT$_{1D}$ receptor antagonists may be antidepressant. In support of this, Roberts *et al.* (1999) demonstrated that, to maximize 5-HT release in terminal brain regions, 5-HT$_{1D}$ as well as 5-HT$_{1B}$ and 5-HT$_{1A}$ receptors must be blocked simultaneously.

Finally, 5-HT$_{2A}$ receptor function is reduced in untreated depressives and chronic anti-depressant treatment increases 5-HT$_{2A}$ receptor density (Zanardi *et al.* 2001). In addition, stimulation of the 5-HT$_{2A}$ receptor causes activation of a biochemical cascade leading to altered expression of brain-derived neurotrophic factor (BDNF; Vaidya *et al.* 1997). Repeated treatment with antidepressants also increases BDNF expression, and therefore 5-HT$_{2A}$ activation may contribute to the therapeutic effect of antidepressants (Duman *et al.* 1997). It should however be noted that direct acting 5-HT$_{2A}$ receptor agonists are hallucinogenic (Glennon *et al.* 1978).

22.6.2 Arousal, attention, and sleep

There are a number of reports that implicate 5-HT$_7$ receptors in the regulation of circadian rhythms (Medanic and Gillette 1992; Edgar *et al.* 1993; Lovenberg *et al.* 1993b; Prosser *et al.* 1993; Ying and Rusak 1997). Initial data demonstrated that the 5-HT$_{1A}$ receptor agonists, 5-CT and 8-OH-DPAT, were able to phase advance circadian cell firing in the suprachiasmatic nucleus. However, it was subsequently shown that 5-HT$_{1A}$ receptor selective antagonists (WAY100635, pindolol, p-MPPI) had very limited efficacy at attenuating this response. Both 5-CT and 8-OH-DPAT also have activity at 5-HT$_7$ receptors and it has been demonstrated that ritanserin and clozapine, two 5-HT$_7$ receptor antagonists, attenuated the agonist-induced phase advance (Lovenberg *et al.* 1993b; Ying and Rusak 1997). The 5-HT$_7$ receptor is therefore a strong contender for modulating circadian activity. As mentioned earlier, circadian rhythms are also disrupted during depression and so drugs acting at this receptor are also potential antidepressants. In support of this, downregulation of 5-HT$_7$ receptors was reported in the hypothalamus following chronic SSRI treatment (Sleight *et al.* 1995; Gobbi *et al.* 1996; Atkinson *et al.* 2001). Blockade of 5-HT$_{2C}$ receptors in humans has also been shown to increase slow wave sleep (Sharpley *et al.* 1994) and implies a potential use for treating sleep disorders and/or depression.

22.6.3 Alzhimer's disease and cognition

5-HT$_{1A}$ receptors modulate the release of non-serotonergic neurotransmitters. 5-HT$_{1A}$ receptor agonists and antagonists applied to the cortex decrease or increase glutamate release in the striatum respectively (Dijk *et al.* 1995). This has led to the suggestion that 5-HT$_{1A}$ receptor antagonists may enhance cognition by removal of an inhibitory input to cortical pyramidal cells and compensates for a loss of excitatory cholinergic input (Bowen *et al.* 1992; Carli *et al.* 1995, 1997; Harder *et al.* 1996). 5-HT$_{1A}$ receptor agonists also increase acetylcholine (ACh) release in the cortex and hippocampus, which would suggest that the agonists and not the antagonists would be beneficial in the treatment of cognitive dysfunction (Bianchi *et al.* 1990; Izumi *et al.* 1994; Consolo *et al.* 1996). In a similar manner, 5-HT$_{1B}$ receptors have been localized on non-5-HT neurones, acting as heteroreceptors. This explains the ability of 5-HT$_{1B}$ agonists to decrease ACh release in hippocampus (Maura and Raiteri 1986), increase ACh release in the frontal cortex (Consolo *et al.* 1996) and increase DA release in the frontal cortex (Lyer and Bradberry 1996). Evidence is also available demonstrating the 5-HT$_{1B}$ receptor modulation of glutamatergic (Boeijinga and Boddeke 1993, 1996; Tanaka and North 1993) and GABAergic (Waeber *et al.* 1990*a*,*b*) release. 5-HT$_{1D}$ heteroreceptors have also been reported to inhibit glutamate release from cerebellum (Maura and Raiteri 1996) and cortex (Maura *et al.* 1998), GABA from cortex (Feuerstein *et al.* 1996) and ACh from hippocampus (Harel-Dupas *et al.* 1991). However, these studies need to be verified with more selective tools before they can be assessed for relevance in diseases such as Alzheimer's disease and schizophrenia.

The physiological function of 5-ht$_5$ receptors is still unclear. However, the limbic distribution (dentate gyrus, CA1, CA3, amygdala, entorhinal cortex) suggest a possible role in learning and memory, as well as emotional behaviour (Plassat *et al.* 1992; Erlander *et al.* 1993; Pasqualetti *et al.* 1998).

5-HT$_4$ receptor agonists were demonstrated to increase ACh release in cortex and hippocampus (Consolo *et al.* 1994), an effect that was attenuated by 5-HT$_4$ receptor antagonists. Therefore, this approach has been suggested to facilitate both long and short term cognitive performance (Fontana *et al.* 1997; Galeotti *et al.* 1997, 1998; Letty *et al.* 1997; Marchetti-Gauthier *et al.* 1997; Menses and Hong 1997; Terry *et al.* 1998). However, 5-HT$_4$ receptors are unlikely to be expressed on cholinergic neurones (Vilaro *et al.* 1996) and it has been suggested that an additional mechanism, such as induction of long-term potentiation may explain the 5-HT$_4$ facilitation of cognition (Bliss and Collingridge 1993). Prior to the discovery of selective ligands, antisense studies indicated that 5-HT$_6$ receptors modulated ACh in the brain, visualized as increased yawns and stretches which were attenuated by muscarinic receptor antagonist, atropine (Bourson *et al.* 1995). These observations have since been repeated with the selective 5-HT$_6$ receptor antagonist, Ro046790 (Sleight *et al.* 1998). Recently, the selective 5-HT$_6$ receptor antagonist, SB274106, was reported to improve retention in the water maze test (Rogers *et al.* in press). These independent observations suggest that 5-HT$_6$ receptor antagonism is another possible 5-HT mechanism for enhancing cognition.

22.6.4 Parkinson's disease

Activation of 5-HT$_{1B}$ receptors on the nigrostriatal GABAergic pathway leads to activation of the nigrostriatal DA pathway (Higgins *et al.* 1991; Oberlander 1983) suggesting that 5-HT$_{1B}$ heteroreceptors may be targets for Parkinson's disease. 5-HT$_{2C}$ receptor antagonists were also recently reported to increase both DA and NA release (Millan *et al.* 1998; Di Matteo *et al.* 1998) and so influence mesolimbic and mesocortical DA and NA function. This would implicate the importance of this receptor subtype in both schizophrenia and Parkinson's disease.

5-HT$_4$ receptors can also modulate DA release. Benloucif *et al.* (1993) and Gale *et al.* (1994) demonstrated that 5-HT$_4$ receptor agonists could increase DA release in the striatum. These receptors are not located on DA neurones but on neurones that have their cell bodies in the striatum, for example, GABAergic neurones (Gerfen 1992; Kawaguchi *et al.* 1995), as 6-hydroxydopamine lesions of the nigrostriatal dopamine pathway did not alter the density of 5-HT$_4$ receptor binding (Patel *et al.* 1995). Therefore, 5-HT$_4$ receptor ligands could be beneficial in the treatment of schizophrenia or Parkinson's disease.

22.6.5 Ischemia

5-HT$_{1A}$ receptor agonists, such as BAYx3702 and 8-OH-DPAT, have been reported to be neuroprotective in rat models of cerebral ischaemia and traumatic brain injury (Osborne *et al.* 2000; Suchanek *et al.* 1998) and NMDA-induced excitotoxicity (Oosterink *et al.* 1998).

22.6.6 Schizophrenia

Increasing attention has been directed towards the role of 5-HT in schizophrenia. Post-mortem studies in schizophrenic patients has revealed an increase in 5-HT$_{1A}$ receptor density in the prefrontal cortex (Bantick *et al.* 2001). These 5-HT$_{1A}$ receptors are located on pyramidal cells and so could reflect an abnormal glutamatergic network. In addition, 5-HT$_{1A}$ receptor agonists increase dopamine release in the prefrontal cortex which is thought to improve negative symptoms of schizophrenia (Ichikawa and Meltzer 1999; Ichikawa *et al.* 2001; Sakaue *et al.* 2000). 5-HT receptor agonists are also anticataleptic (Bantick *et al.* 2001). Therefore, there is a strong rationale that ligands acting at 5-HT$_{1A}$ receptors may be beneficial in the treatment of schizophrenia (Bantick *et al.* 2001).

Currently there is considerable interest in the role of 5-HT$_{2A}$ receptors in antipsychotic drug action. Many antipsychotics (e.g. clozapine and olanzapine) have high affinity for the 5-HT$_{2A}$ as well as dopamine receptors (Leysen *et al.* 1993). In addition, selective 5-HT$_{2A}$ receptor antagonists are active in animal models that predict atypical antipsychotic action (Kehne *et al.* 1996) and have been demonstrated to increase dopamine release in the prefrontal cortex. Therefore, it is predicted that a 5-HT$_{2A}$ receptor antagonist would be antipsychotic with enhanced efficacy versus negative symptoms with reduced propensity for EPS. However, recent clinical studies with the 5-HT$_{2A}$ receptor antagonist, MDL100907, demonstrated that a non-dopaminergic compound has reduced EPS potential but failed to show any advantage over haloperidol on positive and negative symptoms (Potkin *et al.* 2001).

22.6.7 Migraine

5-HT$_{1B}$ receptors have been localized on meningeal blood vessels and receptor agonists act to constrict these vessels (Hamel 1999). This action was thought to be important in the treatment of migraine, as during an attack the cranial vessels are dilated (Pascul 1998). Clinically effective antimigraine drugs possess 5-HT$_{1B}$ as well as 5-HT$_{1D}$ and 5-HT$_{1F}$ receptor agonism (Hamel 1999). However, more recent studies preclude a role for vascular 5-HT$_{1B}$ receptor vasoconstriction in the treatment of migraine (Hargreaves and Shepheard 1999). The selective 5-ht$_{1F}$ agonist, LY334370, did not modulate 5-HT release but was active in the rat dural extravasation model, indicating a possible use in the treatment of migraine (Johnson *et al.* 1997; Phebus *et al.* 1997). In support of this, clinically effective antimigraine drugs also display 5-HT$_{1F}$ agonist activity (Hamel 1999).

22.6.8 **Anxiety**

5-HT$_5$ receptor mRNA expression in the habenula (Plassat *et al.* 1992; Matthes *et al.* 1993) suggest a possible role in acquisition of adaptive behaviour while stressed (Branchek and Zgombick 1997). In addition, 5-ht$_{5A}$ knock-out mice displayed enhanced exploratory activity, suggesting that this receptor might modulate exploratory behaviour (Grailhe *et al.* 1999). The distribution of 5-HT$_7$ mRNA and protein suggest that it might have an autoreceptor role, regulating 5-HT release. No evidence has been found to support an inhibitory role on 5-HT release in terminal or cell body brain regions (Roberts *et al.* 2001*a*). However, recent reports have identified a 5-HT$_7$ receptor-mediated stimulation of 5-HT release (Roberts *et al.* 2001*b*). Therefore, 5-HT$_7$ receptor antagonists will decrease 5-HT release and so are potentially anxiolytic in nature.

5-HT$_{2B}$ receptor agonists have been reported to have anxiolytic properties in a rat social interaction test (Kennett *et al.* 1996*a*,*b*) and 5-HT$_{2C}$ receptor antagonists may also be efficacious anxiolytics. 5-HT$_{2C}$ antagonists were anxiolytic in their own right (Kennett *et al.* 1996*a*,*b*, 1997*a*,*b*) and also reversed anxiogenic effect of mCPP (Kennett *et al.* 1994, 1996*a*,*b*). There are also reports in the literature suggesting anxiolytic like actions of 5-HT$_4$ receptor agonists (Kennett *et al.* 1997*a*,*b*; Silvestre *et al.* 1996). This was supported by the observation that 5-HT$_4$ receptor activation increased 5-HT release in the hippocampus (Ge and Barnes 1996) and substantia nigra (Thorre *et al.* 1998).

22.6.9 **Epilepsy**

5-HT$_7$ receptor activation inhibits the slow hyperpolarization in CA3 hippocampal neurones (Beck and Bacon 1998*a*,*b*). Together with the report that 5-HT$_7$ receptor antagonists might attenuate audiogenic seizures (Bourson *et al.* 1998) these observations indicate a possible role for ligands acting at this receptor in the treatment of epilepsy.

22.6.10 **Satiety and feeding**

5-HT$_{2C}$ receptor antagonists attenuate mCPP-induced hypophagia (Kennett *et al.* 1997*a*,*b*) which demonstrate the importance of this receptor in weight maintenance. This was reinforced when 5-HT$_{2C}$ knock-out mice were reported to be over weight (Tecott *et al.* 1995), although it should be noted that chronic administration of 5-HT$_{2C}$ receptor antagonists does not induce weight gain in rats (Kennett *et al.* 1997*a*).

22.6.11 **Irritable bowel syndrome**

Stimulation of 5-HT$_4$ receptors in the myenteric plexus (Craig and Clarke 1990) oesophagus (Baxter *et al.* 1991), and colon (Craig and Clarke 1990) lead to muscle contraction through facilitation of Ach release (Tonini *et al.* 1989; Kilbinger and Wolf 1992). This evokes secretions and peristaltic reflux. 5-HT$_4$ receptor antagonists such as cisapride have been demonstrated to be beneficial in the treatment of irritable bowel syndrome (De Ponti and Tonini 2001), although they have recently been plagued with cardiovascular side effect liabilities.

22.7 **Concluding remarks**

Since 1986 the number of recognized 5-HT subtypes in the CNS has increased to 14, with additional complexity achieved by either alternative gene splicing, post transcriptional RNA editing, or the presence of naturally occurring polymorphic receptor variants. CNS distribution and function of these receptors has relied on the discovery of selective agonist and antagonist tool compounds. Where selective tools are unavailable function has been eluded through molecular biology techniques, such as receptor knock-out or overexpression. The clinical utility has already been established for some 5-HT ligands such as buspirone (anxiety) and sumatriptan (migraine), however 5-HT receptors still hold much promise for the future as targets for the treatment of a diverse range of diseases.

References

Adham N, Romanienko P, Hartig P *et al.* (1992). The rat 5-HT$_{1B}$ receptor is the species homologue of the human 5-HT$_{1D\beta}$ receptor. *Mol Pharmacol* **41**, 1–7.

Adham N, Kao H-T, Schechter LE, Bard J, Olsen M, Urqurt D *et al.* (1993). Cloning of another human serotonin receptor (5-HT$_{1F}$): a fifth 5-HT$_1$ receptor subtype coupled to the inhibition of adenylate cyclase. *Proc Natl Acad Sci USA* **90**, 408–12.

Adham N, Vaysse PJ, Weinshank RL *et al.* (1994). The cloned human 5-ht$_{1E}$ receptor couples to inhibition and activation of adenylyl cyclase via two distinct pathways in transfected BS-C-1 cells. *Neuropharmacology* **33**, 403–10.

Adham N, Bard JA, Zgombick JM, Durkin MM, Kucharewicz S, Weinshank RL, and Branchek TA (1997). Cloning and characterization of the guinea pig 5-HT$_{1F}$ receptor subtype—a comparison of the pharmacological profile to the human species homologue. *Neuropharmacology* **36**, 569–76.

Albert PR, Zhou QY, Van Tol HHM *et al.* (1990). Cloning, functional expression and messenger RNA tissue distribution of the rat 5-HT$_{1A}$ receptor gene. *J Biol Chem* **265**, 5825–32.

Albert PR, Lembo P, Storring JM *et al.* (1996). The 5-HT$_{1A}$ receptor: signaling, desensitisation and gene transcription. *Neuropsychopharmacology* **14**, 19–25.

Amlaiky N, Ramboz S, Boschert U *et al.* (1992). Isolation of a mouse 5-HT$_{1E}$-like serotonin receptor expressed predominantly in hippocampus. *J Biol Chem* **267**, 19761–4.

Artigas F, Romero L, De Montigny C *et al.* (1996). Acceleration of the effect of selected antidepressant drugs in major depression by 5-HT1A antagonists. *TINS* **19**, 378–83.

Atkinson PJ, Duxon MS, Price GW, Hastie PG, and Thomas DR (2001). Paroxetine down-regulates 5-HT$_7$ receptors in rat hypothalamus but not hippocampus following chronic administration. *Br J Pharmacol* (in press).

Bach AW, Unger L, Sprengel R, Mengod G, Palacios J, Seeburg PH *et al.* (1993). Structure, functional expression and spatial distribution of a cloned cDNA encoding a rat 5-HT$_{1D}$-like receptor *J Recept Res* **13**, 479–502.

Baker LP, Neilson MD, Impey S *et al.* (1998). Stimulation of type 1 and type 8 Ca^{2+}/calmodulin-sensitive adenylyl cyclases by the G$_s$-coupled 5-HT subtype 5-HT$_7$ receptor. *J Biol Chem* **273**, 17 469–76.

Bantik RA, Deakin JF, and Grasby PM (2001). The 5-HT$_{1A}$ receptor in schizophrenia: a promising target for novel atypical neuroleptics? *J Psychopharmacol* **15**, 37–46.

Bard JA, Zgombick J, Adham N *et al.* (1993). Cloning of a novel human serotonin receptor (5-HT$_7$) positively linked to adenylate cyclase. *J Biol Chem* **268**, 23 422–6.

Barone P, Millet S, Moret C *et al.* (1993). Quantitative autoradiography of 5-HT$_{1E}$ binding sites in rodent brains: effect of lesion of serotonergic neurones. *Eur J Pharmacol* **249**, 221–30.

Baxter GS, Craig DA, and Clarke DE (1991). 5-HT$_4$ receptors mediate relaxation of the rat oesophageal tunica muscularis mucosae. *Naunyn Schmiedeberg's Arch Pharmacol* **343**, 439–46.

Baxter G, Kennett G, Blaney F *et al.* (1995). 5-HT$_2$ receptor subtypes: a family re-united? *TIPS* **16**, 105–10.

Beck SG and Bacon WL (1998*a*). 5-HT₇ receptor mediated inhibition of sAHP in CA3 hippocampal pyramidal cells. *4th IUPHAR Satellite meeting on Serotonin*, Rotterdam, S1.3.

Beck SG and Bacon WL (1998*b*). 5-HT₇ receptor mediated inhibition of sAHP in CA3 hippocampal pyramidal cells. *Soc Neurosci Abstr* **24**, 435.4.

Benloucif S, Keegan MJ, and Galloway MP (1993). Serotonin-facilitated dopamine release *in vivo* pharmacological characterisation. *J Pharmacol Exp Ther* **265**, 373–7.

Bianchi C, Siniscalchi A, and Beani L (1990). 5-HT₁A agonists increase and 5-HT₃ agonists decrease acetylcholine efflux from the cerebral cortex of freely-moving guinea-pigs. *Br J Pharamcol* **101**, 448–52.

Bliss TVP and Collingridge GL (1993). A synaptic model of memory-long-term potentiation in the hippocampus. *Nature* **361**, 31–9.

Blondel O, Gastineau M, Dahmoune Y *et al.* (1998). Cloning, expression and pharmacology of four human 5-HT₄ receptor isoforms produced by alternative splicing in the carboxyl terminus. *J Neurochem* **70**, 2252–61.

Bockaert J, Claeysen S, and Dumuis A (1998). Molecular biology, function and pharmacological role of 5-HT₄ receptors. *Naunyn-Schmiedeberg's Arch Pharmacol* **358**, 1.4.

Boeijinga PH and Boddeke HW (1993). Serotonergic modulation of neurotransmission in the rat subicular cortex *in vitro*: a role for 5-HT₁B receptors. *Naunyn-Schmiedeberg's Arch Pharmacol* **348**, 553–7.

Boeijinga PH and Boddeke HW (1996). Activation of 5-HT₁B receptors suppresses low but not high frequency synaptic transmission in the rat subicular cortex *in vitro*. *Brain Res* **721**, 59 –65.

Boess FG, Monsma FJ, Carolo C *et al.* (1997). Functional and radioligand binding characterisation of rat 5-HT₆ receptors stably expressed in HEK293 cells. *Neuropharmacology* **36**, 713–20.

Boess FG and Martin IL (1994). Molecular biology of 5-HT receptors. *Neuropharmacology* **33**, 275–317.

Bonaventure P, Langlois X, and Leysen JE (1998). Co-localization of 5-HT₁B and 5-HT₁D receptor mRNA in serotonergic cell bodies in guinea pig dorsal raphe nucleus—a double labeling *in situ* hybridization histochemistry study. *Neurosci Lett* **254**, 113–16.

Bonhaus DW, Bach C, De Souza A *et al.* (1995). The pharmacology and distribution of human 5-HT₂B receptor gene products: comparison with 5-HT₂A and 5-HT₂C receptors. *Br J Pharmacol* **115**, 622–8.

Boschert U, Amara DA, Segu L, and Hen R (1994). The mouse 5-HT₁B receptor is localised predominantly on axon terminals. *Neuroscience* **58**, 167–82.

Bouhelal R, Smounya L, and Bockaert J (1988). 5-HT₁B receptors are negatively coupled to adenylate cyclase in rat substantia nigra. *Eur J Pharmacol* **151**, 189–96.

Bourson A, Borroni E, Austin RH *et al.* (1995). Determination of the role of the 5-HT₆ receptor in the rat brain: a study using antisense oligonucleotides. *J Pharmacol Exp Ther* **274**, 173–80.

Bourson A, Boess FG, Bos M *et al.* (1998). Involvement of 5-HT₆ receptors in nigrostriatal function in rodents. *Br J Pharmacol* **125**, 1562–6.

Bowen DM, Francis PT, Pangalos MN *et al.* (1992). Treatment strategies for Alzheimer's disease. *Lancet* **339**, 132–3.

Bradley PB, Engel G, Feniuk W *et al.* (1986). Proposals for the classification and nomenclature of functional receptors for 5-HT. *Neuropharmacology* **25**, 563–76.

Branchek TA and Zgombick JM (1997). Molecular biology and potential functional role of 5-ht₅, 5-HT₆ and 5-HT₇ receptors. In (eds. HG Baumgarten and M Gothert), *Handbook of experimental pharmacology. Serotonin Neurons and 5-HT receptors in the CNS*, Springer-Verlag, Berlin, pp. 475–97.

Bromidge SM, Dabbs S, Davies DT *et al.* (2000). Biarylcarbamoylindolines are novel and selective 5-HT₍₂C₎ receptor inverse agonists: identification of 5-methyl-1-[[2-[(2-methyl-3-pyridyl)oxy]-5-pyridyl]carbamoyl]-6-trifluoromethylindoline (SB-243213) as a potential antidepressant/anxiolytic agent. *J Med Chem* **43**, 1123–34.

Bromidge SM, Clarke SE, Gager T *et al.* (2001). Phenyl benzenesulfonamides are novel and selective 5-HT$_6$ antagonists: identification of N-(2,5-dibromo-3-fluorophenyl)-4-methoxy-3-piperazin-1-ylbenzenesulfonamide (SB-357134). *Bioorganic Med Chem Lett* 11, 55–8.

Bruinvels AT, Palacios JM, and Hoyer D (1993). Autoradiographic characterisation and localisation of 5-HT$_{1D}$ compared to 5-HT$_{1B}$ binding sites in rat brain. *Naunyn Schmiedeberg's Arch Pharmacol* 347, 569–82.

Bruinvels AT, Landwehrmeyer B, Gustafson EL, Durkin MM, Mengod G, Branchek TA *et al.* (1994a). Localisation of 5-HT$_{1B}$, 5-HT$_{1D\alpha}$, 5-HT$_{1E}$ and 5-HT$_{1F}$ mRNA in rodent and primate brain. *Neuropharmacology* 33, 367–86.

Bruinvels AT, Landwehrmeyer B, Probst A, Palacios JM, and Hoyer D (1994b). A comparative autoradiographic study of 5-HT$_{1D}$ binding sites in human and guinea-pig brain using different radioligands. *Brain Res Mol Brain Res* 21, 19–29.

Bruinvels AT, Palacios JM, and Hoyer D (1994c). 5-HT$_1$ recognition sites in rat brain: heterogeneity of non-5-HT$_{1A/1C}$ binding sites revealed by quantitative receptor autoradiography. *Neuroscience* 53, 465–73.

Burnet PWJ, Eastwood SL, Lacey K *et al.* (1995). The distribution of 5-HT$_{1A}$ and 5-HT$_{2A}$ receptor mRNA in human brain. *Brain Res* 676, 157–68.

Burns CM, Chu H, Rueter SM *et al.* (1997). Regulation of 5-HT$_{2C}$ receptor G-protein coupling by RNA editing. *Nature* 387, 303–8.

Canton H, Emerson RB, Barker EL *et al.* (1996). Identification, molecular cloning and distribution of a short variant of the 5-HT$_{2C}$ receptor produced by alternative splicing. *Mol Pharmacol* 50, 799–807.

Carli M, Luschi R, and Samarin R (1995). (S)-WAY 100 135, a 5-HT$_{1A}$ receptor anatagonist, prevents the impairment of spatial learning caused by intrahippocampal scopolamine. *Eur J Pharmacol* 283, 133–9.

Carli M, Bonalumi P, and Samarin R (1997). WAY 100 635, a 5-HT$_{1A}$ anatagonist, prevents the impairment of spatial learning caused by intrahippocampal administration of scopolamine or 7-chlorokyneurenic acid. *Eur J Pharmacol* 774, 167–74.

Carson MJ, Thomas EA, Danielson PE, and Sutcliffe JG (1996). The 5-ht$_{5A}$ serotonin receptor is expressed predominantly by astrocytes in which it inhibits cAMP accumulation: a mechanism for neuronal suppression of reactive astrocytes. *Glia* 17, 317–26.

Castro ME, Pascual J, Romon T, Delarco C, Delolmo E, and Pazos A (1997). Differential distribution of [^{3}H]sumatriptan binding sites (5-HT$_{1B}$, 5-HT$_{1D}$ and 5-HT$_{1F}$) in human brain: focus on brain stem and spinal cord. *Neuropharmacology* 36, 535–42.

Chalmers DT and Watson SJ (1991). Comparative anatomical distribution of 5-HT$_{1A}$ receptor mRNA and 5-HT$_{1A}$ binding in rat brain—a combined *in situ* hybridisation/*in vitro* receptor autoradiographic study. *Brain Res* 561, 51–60.

Charney DS, Krystal J, Delgado PL *et al.* (1990). Serotonin specific drugs for anxiety and depressive disorders. *Ann Rev Med* 41, 437–46.

Chiu HJ, Wang YC, Liou JH, Chao CH, Lee H, Tsai KY *et al.* (2001). 5-HT$_6$ receptor polymorphism in schizophrenia: frequency, age at onset and cognitive function. *Neuropsychobiology* 43, 113–16.

Choi DS, Ward SJ, Messaddeq N *et al.* (1997). 5-HT$_{2B}$ receptor-mediated serotonin morphogenetic functions in mouse cranial neural crest and myocardiac cells. *Development* 124, 1745–55.

Cichol S, Kesper K, Propping P *et al.* (1998). Assignment of the human serotonin 4 receptor gene (HTR$_4$) to the long arm of chromosome 5 (5q31–q33). *Mol Memr Biol* 15, 75–8.

Claeyson S, Sebben M, and Journot L (1996). Cloning, expression and pharmacology of the mouse 5-HT$_{4L}$ receptor. *FEBS Lett* 398, 19–25.

Claeyson S, Faye P, Sebben M *et al.* (1997a). Assignment of 5-HT receptor (HTR$_4$) to human chromosome 5 bands q31–q33 by *in situ* hybridisation. *Cytogenet Cell Genet* 78, 133–4.

Claeyson S, Faye P, Sebben M *et al.* (1997*b*). Cloning and expression of human 5-HT$_{4S}$ receptors. Effect of receptor density on their coupling to adenylyl cyclase. *NeuroReport* **8**, 3189–96.

Claeyson S, Faye P, Sebben M, Taviaux S, Bockaert J, and Dumuis A (1997*c*). 5-HT$_4$ receptors: from gene to function. In *Proceedings of the Advances in Serotonin Receptor research.* San Francisco, USA, 8–10, October.

Conn PJ and Sanders-Bush E (1984). Selective 5-HT$_2$ antagonists inhibit serotonin stimulated phosphatidyl inositol metabolism in cerebral cortex. *Neuropharmacology* **23**, 993–6.

Consolo S, Arnaboldi S, Giorgi S *et al.* (1994). 5-HT$_4$ receptor stimulation facilitates acetylcholine release in rat frontal cortex. *NeuroReport* **5**, 1230–2.

Consolo S, Arnaboldi S, Ramponi S *et al.* (1996). Endogenous serotonin facilitates *in vivo* acetylcholine release in rat frontal cortex through 5-HT$_{1B}$ receptors. *J Pharmacol Exp Ther* **277**, 823–30.

Craig DA and Clarke DE (1990). Pharmacological characterisation of a neuronal receptor for 5-HT in guinea-pig ileum with properties similar to the 5-HT$_4$ receptor. *J Pharmacol Exp Ther* **252**, 1378–86.

Davidson C and Stamford JA (1995). Evidence that 5-HT release in rat dorsal raphe nucleus is controlled by 5-HT$_{1A}$, 5-HT$_{1B}$ and 5-HT$_{1D}$ autoreceptors. *Br J Pharmacol* **114**, 1107–9.

De Ponti F and Tonini M (2001). Irritable bowel syndrome: new agents targeting serotonin receptor subtypes. *Drugs* **61**, 317–32.

Dijk SN, Francis PT, Stratmann GC *et al.* (1995). NMDA induced glutamte and aspartate release from rat cortical pyramidal neurones: evidence for modulation by a 5-HT$_{1A}$ antagonist. *Br J Pharmacol* **115**, 1169–74.

Di Matteo V, Di Giovanni G, Di Mascio M, and Esposito E (1998). Selective blockade of 5-HT$_{2C/2B}$ receptors enhances dopamine release in the rat nucleus accumbens. *Neuropharmacology* **37**, 265–72.

Doucet E, Pohl M, Fattaccini CM, Adrien J, El Mestikawy S, and Hamon M (1995). *In situ* hybridisation evidence for the synthesis of 5-HT$_{1B}$ receptors in serotonergic neurons of anterior raphe nuclei in the rat brain. *Synapse* **19**, 18–28.

Duman RS, Heninger GR, and Nestler EJ (1997). A molecular and celular theory of depression. *Arch Gen Psychiatry* **54**, 597–606.

Duxon MS, Flanigan TP, Reavely AC *et al.* (1997). Evidence for the expression of the 5-HT$_{2B}$ receptor protein in the rat central nervous system. *Neuroscience* **76**, 323–9.

Edgar DM, Miller JD, Prosser RA *et al.* (1993). Serotonin and the mammalian circadian system. II. Phase shifting rat behaviour rhythms with serotonergic agonists. *J Biol Rhythms* **8**, 17–31.

El Mestikawy S, Riad M, Laporte AM *et al.* (1990). Production of specific anti-rat 5-HT$_{1A}$ receptor antibodies in rabbits injected with a synthetic peptide. *Neurosci Lett* **118**, 189–92.

Erdmann J, Nothen MM, Shimron-Abarbanell D *et al.* (1996). The human 5-HT$_7$ receptor gene: genomic organisation and systematic mutation screening in schizophrenia and bipolar affective disorder. *Mol Psychiatry* **1**, 392–7.

Erlander MG, Lovenberg TW, Baron BM *et al.* (1993). Two members of a distinct subfamily of 5-hydroxytryptamine receptors differentially expressed in rat brain. *Proc Natl Acad Sci USA* **90**, 3452–6.

Fargin A, Raymond JR, Lohse MJ *et al.* (1988). The genomic clone G-21 which resembles a β-adrenergic receptor sequence encodes the 5-HT$_{1A}$ receptor. *Nature* **335**, 358–60.

Francis PT, Pangalos MN, Pearson RC *et al.* (1992). 5-HT$_{1A}$ but not 5-HT$_2$ receptors are enriched on neocortical pyramidal neurones destroyed by intrastriatal volkensin. *J Pharmacol Exp Ther* **261**, 1273–81.

Franken BJB, Jurak M, Vanhauwe JFM, Luyten WHML, and Leysen JE (1998). The human 5-ht$_{5A}$ receptor couples to G$_i$/G$_o$ proteins and inhibits adenylate cyclase in HEK 293 cells. *Eur J Pharmacol* **361**, 299–309.

Feuerstein TJ, Huring H, Van Velthoven V *et al.* (1996). 5-HT$_{1D}$-like receptors inhibit the release of endogenously formed [^{3}H]GABA in human, but not in rabbit, neocortex. *Neurosci Lett* **209**, 210–14.

Fletcher A, Cliffe IA, and Dourish CT (1993). Silent 5-HT$_{1A}$ receptor antagonists: utility as research tools and therapeutic agents. *TIPS* 14, 41–8.

Fletcher A, Forster EA, Bill DJ *et al.* (1996). Electrophysiological, biochemical, neurohormonal and behavioural studies with WAY 100635, a potent, selective and silent 5-HT$_{1A}$ receptor antagonist. *Behar Brain Res* 73, 337–53.

Fontana DJ, Daniekls SE, Wong EHK *et al.* (1997). The effects of novel, selective 5-HT$_4$ receptor ligands in rats spatial navigation. *Neuropharmacology* 4/5, 689–96.

Forbes IT, Dabbs S, Duckworth DM *et al.* (1998). (R)-3,N-Dimethyl-N-[1-methyl-3-(4-methyl-piperidin-1-yl)propyl] benzenesulphonamide): the first selective 5-HT$_7$ receptor antagonist. *J Med Chem* 41, 655–7.

Gale JD, Grossman CJ, Whithead JWF *et al.* (1994). GR 113 808—A novel, selective antagonist with high affinity at the 5-HT$_4$ receptor. *Br J Pharmacol* 111, 332–8.

Galeotti N, Ghelardini C, Teodori E *et al.* (1997). Antiamnesic activity of metoclopramide, cisapride and SR-17 in the mouse passive avoidance test. *Pharmacol Res* 36, 59–67.

Galeotti N, Ghelardini C, and Bartolini A (1998). Role of 5-HT$_4$ receptors in the mouse passive avoidance test. *J Pharmacol Exp Ther* 286, 1115–21.

Gartside SE, Umbers V, Hajos M *et al.* (1995). Interaction between a selective 5-HT$_{1A}$ receptor antagonist and an SSRI *in vivo*: effects on cell firing and extracellular 5-HT. *Br J Pharmacol* 115, 1064–70.

Gaster L, Blaney FE, Duckworth M *et al.* (1998). The selective 5-HT$_{1B}$ receptor inverse agonist, SB-224289, 1'-methyl-5-[2'-methyl-4'-(5-methyl-1,2,4- oxadiazol-3-yl)biphenyl-4-carbonyl]-2,3,6,7-tetrahydrospiro[furo[2,3- f]indole-3,4'-piperidine], potently blocks terminal 5-HT autoreceptor function both *in vitro* and *in vivo*. *J Med Chem* 44, 1218–35.

Ge J and Barnes NM (1996). 5-HT$_4$ receptor mediated modulation of 5-HT release in the rat hippocampus *in vivo*. *Br J Pharmacol* 117, 1474–80.

Gelernter J, Rao PA, Pauls DL *et al.* (1995). Assignment of the 5-H$_7$ receptor gene (HTR7) to chromosome 10$_q$ and exclusion of genetic linkage with Tourette syndrome. *Genomics* 26, 207–9.

Gerald C, Adham A, Kao HT *et al* (1995). The 5-HT$_4$ receptor: molecular cloning and pharmacological characterisation of two splice variants. *EMBO J* 14, 2806–15.

Gerard C, El Mestikawy S, Lebrand C *et al.* (1996). Quantitative RT-PCR distribution of 5-HT$_6$ receptor mRNA in the central nervous system of control or 5,7-dihydroxytryptamine treated rats. *Synapse* 23, 164–73.

Gerard C, Martres M-P, Lefevre K *et al.* (1997). Immuno-localisation of 5-HT$_6$ receptor-like material in the rat central nervous system. *Brain Res* 746, 207–19.

Gerfen CR (1992). The neostriatal mosaic: multiple levels of compartmental organisation. *TINS* 15, 133–9.

Glennon RA, Liebowitz SM, and Mack EC (1978). Serotonin receptor binding affinities of several hallucinogenic phenylalkylamine and N,N-dimethyltryptamine analogues. *J Med Chem* 21, 822–5.

Glennon RA, Lee M, Rangisetty JB *et al.* (2000). 2-Substituted tryptamines: agents with selectivity for 5-HT$_6$ serotonin receptors. *J Med Chem* 43, 1011–18.

Gobbi M, Parotti L, and Mennini T (1996). Are 5-HT$_7$ receptors involved in [^{3}H]5-HT binding to 5-HT$_{1nonA-nonB}$ receptors in rat hypothalamus? *Mol Pharmacol* 49, 556–9.

Godfrey PP, McClue SJ, Young MM *et al.* (1988). 5-HT-stimulated inositol phospholipid hydrolysis in the mouse cortex has pharmacological characteristics compatible with mediation via 5-HT$_2$ receptors but this response does not reflect altered 5-HT$_2$ function after 5,7-dihyroxytryptamine lesioning or repeated antidepressant treatments. *J Neurochem* 50, 730–8.

Grailhe R, Waeber C, Dulawa SC, Hornung JP, Zhuang X, Brunner D *et al.* (1999). Increased exploratory activity and altered response to LSD in mice lacking the 5-HT$_{5A}$ receptor. *Neuron* 22, 581–91.

Gudermann T, Levy FO, Birnbaumer M *et al.* (1993). Human S31 serotonin receptor encodes a 5-HT$_{1E}$-like serotonin receptor. *Mol Pharmacol* 43, 412–18.

Gustafson EL, Durkin MM, Bard JA, Zgombick J, and Branchek TA (1996). A receptor autoradiographic and *in situ* hybridisation analysis of the distribution of the 5-HT$_7$ receptor in rat brain. *Br J Pharmacol* 117, 657–66.

Hagan JJ, Price GW, Jeffrey P *et al.* (2000). Characterization of SB-269970-A, a selective 5-HT$_{(7)}$ receptor antagonist. *Br J Pharmacol* 130, 539–48.

Hamblin MW and Metcalf MA (1991). Primary structure and functional characterisation of a human 5-HT$_{1D}$-type serotonin receptor. *Mol Pharmacol* 40, 143–8.

Hamblin MW, McGuffin RW, Metcalf MA, Dorsa DM, and Merchan KM (1992). Distinct 5-HT$_{1B}$ and 5-HT$_{1D}$ serotonin receptors in rat: structural and pharmacological comparison of the two cloned receptors. *Mol Cell Neurosci* 3, 578–87.

Hamel E (1999). The biology of serotonin receptors: focus on migraine pathophysiology and treatment. *Canadian J Neurological Sci* 26 (suppl 3), S2–S6.

Handley SL (1995). 5-HT pathways in anxiety and its treatment. *Pharmacol Ther* 66, 103–48.

Harder JA, Maclean CJ, Alder JT *et al.* (1996). The 5-HT$_{1A}$ antagonist, WAY 100635, ameliorates the cognitive impairment induced by fornix transection in the marmoset. *Psychopharmacol* 127, 245–54.

Harel-Dupas C, Cloez I, and Fillion G (1991). The inhibitory effect of trifluoromethylphenylpiperazine on [^{3}H]acetylcholine release in guinea-pig hippocampal synaptosomes is mediated by a 5-HT$_1$ receptor distinct from 1A, 1B and 1C subtypes. *J Neurochem* 56, 221–7.

Hargreaves RJ and Shepheard SL (1999). Pathophysiology of migraine–new insights. *Canadian J Neurological Sci* 26, S12–S19.

Hartig PR, Hoyer D, Humphrey PPA *et al.* (1996). Alignment of receptor nomenclature with the human genome: classification of 5-HT$_{1B}$ and 5-HT$_{1D}$ receptor subtypes. *TIPS* 17, 103–5.

Hajos-Korcsok E and Sharp T (1996). 8-OH-DPAT-induced release of hippocampal noradrenaline *in vivo*: evidence for a role of both 5-HT$_{1A}$ and D$_1$ receptors. *Eur J Pharmacol* 314, 285–91.

Heidmann DEA, Metcalf MA, Kohen R *et al.* (1997). Four 5-HT$_7$ receptor isoforms in human and rat produced by alternative splicing: species differences due to altered intron-exon organisation. *J Neurochem* 68, 1272–381.

Heidmann DEA, Szot P, Kohlen R, and Hamblin MW (1998). Function and distribution of three rat 5-HT$_7$ receptor isoforms produced by alternative splicing. *Neuropharmacology* 37, 1621–32.

Hen R (1992). Of mice and flies—commonalities among 5-HT receptors. *TIPS* 13, 160–5.

Higgins GA, Jordoan CC, and Skingle M (1991). Evidence that the unilateral activation of 5-HT$_{1D}$ receptors in the substantia nigra of the guinea-pig elicits contralateral rotation. *Br J Pharmacol* 102, 305–10.

Hjorth S (1993). Serotonin 5-HT$_{1A}$ autoreceptor blockade potentiates the ability of the 5-HT reuptake inhibitor citalopram to increase nerve terminal output of 5-HT *in vivo*: a microdialysis study. *J Neurochem* 60, 776–9.

Hopwood SE and Stamford JA (2001). Multiple 5-HT$_1$ autoreceptor subtypes govern serotonin release in dorsal and median raphe nuclei. *Neuropharmacology* 40, 508–19.

Hoyer D, Pazos A, Probst A, and Palacios JM (1986). Serotonin receptor in human brain. I. Characterisation and autoradiographic localisation of 5-HT$_{1A}$ recognition sites, apparent absence of 5-HT$_{1B}$ recognition sites. *Brain Res* 376, 85–96.

Huang YY, Grailhe R, Arango V, Hen R, and Mann JJ (1999). Relationship pf psychopathology to the human 5-HT$_{1B}$ genotype and receptor binding kinetics in postmortem brain tissue. *Neuropsychopharmacology* 21, 238–46.

Hurley PT, McMahon RA, Fanning P, O'Boyle KM, Rogers M, and Martin F (1998). Functional coupling of a recombinant human 5-HT$_{5A}$ receptor to G-proteins in HEK-293 cells. *Br J Pharmacol* 124, 1238–44.

Huson PH, Bristow LJ, Cunningham JR et al. (1995). The effects of GR 127935, a putative 5-HT$_{1D}$ receptor antagonist, on brain 5-HT metabolism, extracellular 5-HT concentration and behaviour in the guinea-pig. Neuropharmacol 34, 383–92.

Ichikawa J, Ishii H, Bonaccorso S, Fowler WL, O'Laughlin IA, and Meltzer HY (2001). 5-HT$_{2A}$ and D$_2$ receptor blockade increases cortical DA release via 5-HT$_{1A}$ receptor activation: a possible mechanism of atypical antipsychotic-induced cortical dopamine release. J Neurochem 76, 1521–31.

Ichikawa J and Meltzer HY (1999). R(+)-8-OH-DPAT, a 5-HT$_{1A}$ receptor agonist, potentiated S(−)-sulpiride-induced dopamine release in rat medial prefrontal cortex and nucleus accumbens but not striatum. J Pharmacol Exp Ther 291, 1227–32.

Invernizzi R, Belli S, and Samanin R (1992). Citalopram's ability to increase the extracellular concentrations of serotonin in the dorsal raphe prevents the drugs effect in he frontal cortex. Brain Res 584, 322–4.

Invernizzi R, Bramante M, and Samanin R (1996). Role of 5-HT$_{1A}$ receptors in the effects of acute and chronic fluoxetine on extracellular serotonin in the frontal cortex. Pharmacol Biochem Behav 54, 143–7.

Izumi J, Washizuka M, Miura N et al. (1994). Hippocampal 5-HT$_{1A}$ receptor enhances acetylcholine release in conscious rats. J Neurochem 62, 1804–8.

Jasper JR, Kosaka A, To ZP, Chang DJ, and Eglen RM (1997). Cloning, expression and pharmacology of a truncated splice variant of the human 5-HT$_7$ receptor (h5-HT$_{7(b)}$). Br J Pharacmol 122, 126–32.

Jin H, Oksenberg D, Askenazi A et al. (1992). Characterisation of the human 5-HT$_{1B}$ receptor. J Biol Chem 267, 5735–8.

Johansson L, Sohn D, Thorberg SO et al. (1997). The pharmacological characterisation of a novel selective 5-HT$_{1A}$ receptor antagonist, NAD-299. J Pharamcol Exp Ther 283, 216–25.

Johnson KW, Schaus JM, Durkin MM et al. (1997). 5-HT$_{1F}$ receptor agonists inhibit neurogenic dural inflammation in guinea-pigs. NeuroReport 8, 2237–40.

Kawaguchi Y, Wilson CJ, Augood SJ et al. (1995). Striatal interneurones: chemical, physiological and morphological characterisation. TINS 18, 527–35.

Kawanishi C, Hanihara T, Shimoda Y et al. (1998a). Lack of association between neuroleptic malignant syndrome and polymorphisms in the 5-HT$_{1A}$ and 5-HT$_{2A}$ receptor genes. Am J Psychiatry 155, 1275–7.

Kawanishi Y, Harada S, Tachikawa H, Okubo T, and Shiraishi H (1998b). Novel mutations in the promoter and coding region of the human 5-HT$_{1A}$ receptor gene and association analysis in schizophrenia. Am J Med Gen 81, 434–9.

Kehne JH, Baron BM, Carr AA et al. (1996). Preclinical characterisation of the putative atypical antipsychotic MDL 100907 as a potent 5-HT$_{2A}$ antagonists with a favorable CNS safety profile. J Pharmacol Exp Ther 277, 968–81.

Kennett GA, Wood MD, Grewal GS et al. (1994). In vivo properties of SB-200646A, a 5-HT$_{2C/2B}$ receptor antagonist. Br J Pharmacol 111, 797–802.

Kennett GA, Bright F, Trail B et al. (1996a) Effects of the 5-HT$_{2B}$ receptor agonist, BW 723C86, on three rat models of anxiety. Br J Pharmacol 117, 1443–8.

Kennett GA, Wood MD, Bright F et al. (1996b). In vitro and in vivo profile of SB-206553, a potent 5-HT$_{2C/2B}$ receptor antagonist with anxiolytic-like properties. Br J Pharmacol 117, 427–34.

Kennett GA, Wood MD, Bright F et al. (1997a). SB-242084, a selective and brain penetrant 5-HT$_{2C}$ receptor antagonist. Neuropharmacology 36, 609–20.

Kennett GA, Bright F, Trail B et al. (1997b). Anxiolytic-like actions of the selective 5-HT$_4$ receptor antagonits, SB-204070A and SB-207266A in rats. Neuropharmacology 4/5, 707–12.

Khawaja X (1995). Quantitative autoradiographic characterisation of the binding of [^{3}H]WAY 100635, a selective 5-HT$_{1A}$ receptor antagonist. Brain Res 673, 217–25.

Kilbinger H and Wolf D (1992). Effect of 5-HT$_4$ receptor stimulation on basal and electrically evoked release of acetylcholine from guinea-pig myenteric plexus. *Nauyn Schmiedeberg's Arch Pharmacol* **345**, 270–75.

Koblinka BK, Frielle T, Colins S *et al.* (1987). An intronless gene encoding a potential member of the family of receptors coupled to guanine nucleotide regulatory proteins. *Nature* **329**, 75–7.

Kohen R, Metcalf MA, Khan N *et al.* (1996). Cloning, characterisation and chromosomal localisation of a human 5-HT$_6$ serotonin receptor. *J Neurochem* **66**, 47–56.

Kung MP, Frederick D, Mu M *et al.* (1995). 4-(2′-methoxyphenyl)-1-[2′-(n-2″-pyrinyl)-p-iodobenzamido]-ethyl-piperazine ([^{125}I]p-MPPI) as a new selective radioligand of 5-HT$_{1A}$ sites in rat brain: *in vitro* binding and autoradiographic studies. *J Pharmacol Exp Ther* **272**, 429–37.

Kursar JD, Nelson DL, Wainscott DB *et al.* (1994). Molecular cloning, functional expression and mRNA distribution of the human 5-HT$_{2B}$ receptor. *Mol Pharmacol* **46**, 227–34.

Lappalainen J, Long JC, Eggert M *et al.* (1998). Linkage of antisocial alcoholism to the 5-HT$_{1B}$ receptor gene in two populations. *Arch Gen Psychiatry* **55**, 989–94.

Letty S, Child R, Dumais A *et al.* (1997). 5-HT$_4$ receptors improve social olfactory memory in rat. *Neuropharmacol* **4/5**, 681–8.

Levy FO, Gudermann T, Birnbaumer M *et al.* (1992*a*) Molecular cloning of a human gene (S31) encoding a novel serotonin receptor mediating inhibition of adenylyl cyclase. *FEBS Lett* **296**, 201–6.

Levy FO, Gudermann T, Perezreyes E *et al.* (1992*b*) Molecular cloning of a human serotonin receptor (S12) with a pharmacological profile resembling that of the 5-HT$_{1D}$ subtype. *J Biol Chem* **267**, 7553–62.

Leysen JE, Janssen PMF, Schotte A *et al.* (1993). Interaction of antipsychotic drugs with neurotransmitter receptor sites *in vitro* and *in vivo* in relation to pharmacology and clinical effects-role of 5-HT$_2$ receptors. *Psychopharmacology* **112**, S40–S54.

Longmore J, Maguire JJ, MacLeod A, Street L, Schofield WN, and Hill RG (2000). Comparison of the vasoconstrictor effects of the selective 5-HT$_{1D}$-receptor agonist L-775 606 with the mixed 5-HT$_{1B/1D}$-receptor agonist sumatriptan and 5-HT in human isolated coronary artery. *Br J Clin Pharmacol* **49**, 126–31.

Lopez-Gimenez JF, Mengod G, Palacios JM *et al.* (1997). Selectivite visualisation of rat brain 5-HT$_{2A}$ receptors by autoradiography with [^{3}H]MDL 100907. *Naunyn-Schmiedeberg's Arch Pharmacol* **356**, 446–54.

Loric S, Launay JM, Colas JF, and Maroteaux L (1992). New mouse 5-HT$_2$-like receptor: expression in brain, heart and intestine. *FEBS Lett* **312**, 203–7.

Lovenberg TW, Erlander MG, Baron MG, Racke M, Slone AL, and Siegel BW (1993*a*). Molecular cloning and functional expression of 5-HT$_{1E}$-like rat and human 5-HT receptor genes. *Proc Natl Acad Sci USA* **90**, 2184–8.

Lovenberg TW, Baron BM, De Lecea L, Miller JD, Prosser RA, and Rea MA (1993*b*). A novel adenylyl cyclase activating serotonin receptor (5-HT$_7$) implicated in the regulation of mammalian circadian rhythms. *Neuron* **11**, 449–58.

Lucaites VL, Krushinski JH, Schaus JM *et al.* (1996). Autoradiographic localisation of the serotonin$_{1F}$ receptor in rat brain using [^{3}H]LY 334370, a selective 5-HT$_{1F}$ receptor ligand. *Soc Neurosci Abstr* **22**, 258.

Lucas J and Hen R (1995). New players in the 5-HT receptor field: genes and knockouts. *TIPS* **16**, 246–52.

Lyer RN and Bradberry CW (1996). Serotonin-mediated increase in prefrontal cortex dopamine release: pharmacological characterisation. *J Pharm Exp Ther* **277**, 40–7.

Maassen Van Den Brink A, Vergrouwe MN, Ophoff RA, Saxena PR, Ferrari MD, and Frants RR (1998). 5-HT$_{1B}$ receptor polymorphism and clinical response to sumatriptan. *Headache* **38**, 288–91.

Marchetti-Gauthier E, Roman FS, Dumais A *et al.* (1997). BIMUI increases associative memory in rats by activating 5-HT$_4$ receptors. *Neuropharmacology* **4/5**, 697–706.

Martin JR *et al.* (1995). 5-HT$_{2C}$ receptor agonists and antagonists in animal models of anxiety. *Eur Neuropharmacol* 5, 209.

Masellis M, Basile VS, Meltzer HY *et al.* (2001). Lack of association between T-C 267 5-HT$_6$ receptor gene (HTR6) polymorphism and prediction of response to clozapine in schizophrenia. *Schizophrenia Res* 47, 49–58.

Matthes H, Boshert U, Amlaiky N *et al.* (1993). Mouse 5-HT$_{5A}$ and 5-HT$_{5B}$ receptors define a new family of serotonin receptors: cloning, functional expression and chromosomal localisation. *Mol Pharmacol* 43, 313–19.

Maura G and Raiteri M (1986). Cholinergic terminal in rat hippocampus possess 5-HT$_{1B}$ receptors mediating inhibition of acetylcholine release. *Eur J Pharmacol* 129, 333–7.

Maura G and Raiteri M (1996). Serotonin 5-HT$_{1D}$ and 5-HT$_{1A}$ receptors respectively mediate inhibition of glutamate release and inhibtiion of cyclic GMP producation in rat cerebellum *in vitro*. *J Neurochem* 66, 203–9.

Maura G, Marcoli M, Tortarolo M *et al.* (1998). Glutamate release in human cerebral cortex and its modulation by 5-HT acting at h5-HT$_{1D}$ receptors. *Br J Pharmacol* 123, 45–50.

McAllister G, Charlesworth A, Snodin C *et al.* (1992). Molecular cloning of a serotonin receptor from human brain (5-HT$_{1E}$): a fifth 5-HT$_1$-like subtype. *Proc Natl Acad Sci USA* 89, 5517–21.

Medanic M and Gillette MU (1992). Serotonin regulates the phase of the rat superchiasmatic circadian pacemaker *in vitro* only during the subjective day. *J Physiol* 450, 629–42.

Medhurst AD, Lezoualc'h F, Fischmeister R, Middlemiss DN, and Sanger GJ (2001). Quantitative mRNA analysis of five C-terminal splice variants of the human 5-HT$_{(4)}$ receptor in the central nervous system by TaqMan real time RT-PCR. *Brain Res Mol Brain Res* 90, 125–34.

Mengod G, Pompeiano M, Martinez Mir MI *et al.* (1990*a*). Localisation of the mRNA for the 5-HT$_2$ receptor by *in situ* hybridisation histochemistry. Correlation with the distribution of receptor sites. *Brain Res* 524, 139–43.

Mengod G, Nguyen H, Le H *et al.* (1990*b*). The distribution and cellular localisation of the 5-HT$_{1C}$ receptor mRNA in the rodent brain examined by *in situ* hybridisation histochemistry. Comparison with receptor binding distribution. *Neuroscience* 35, 577–91.

Mengod G, Vilaro MT, Raurich A *et al.* (1996). 5-HT receptors in mammalian brain: receptor autoradiography and *in situ* hybridisation studies of new ligands and newly identified receptors. *Histochem J* 28, 747–58.

Menses A and Hong E (1997). Effects of 5-HT$_4$ receptor agonists and antagonists in learning. *Pharmacol Biochem Behav* 56, 347–51.

Middlemiss DN and Hutson PH (1990). The 5-HT$_{1B}$ receptor. *Ann NY Acad Sci* 60, 132–47.

Millan MJ, Dekeyne A, and Gobert A (1998). 5-HT$_{2C}$ receptors tonically inhibit dopamine and noradrenaline, but not 5-HT, release in the frontal cortex *in vivo*. *Neuropharmacology* 37, 953–5.

Miller KJ and Teitler M (1992). Quantitative autoradiography of 5-CT sensitive (5-HT$_{1D}$) and 5-CT insensitive (5-HT$_{1E}$) serotonin receptors in human brain. *Neurosci Lett* 136, 223–6.

Miquel MC, Doucet E, Riad M, Adrien J, Verge D, and Hamon M (1992). Effect of the selective lesion of serotonergic neurons on the regional distribution of 5-HT$_{1A}$ receptor mRNA in the rat brain. *Molec Brain Res* 14, 357–62.

Monsma FJ, Shen Y, Ward RP *et al.* (1993). Cloning and expression of a novel serotonin receptor with high affinity for tricyclic psychotropic drugs. *Mol Pharmacol* 43, 320–7.

Monti JM and Monti D (2000). Role of dorsal raphe nucleus serotonin 5-HT$_{1A}$ receptor in the regulation of REM sleep. *Life Sci* 66, 1999–2012.

Morilak DA, Garlow SJ, and Ciaranello RD (1993). Immunocytochemical localisation and description of neurons expressing 5-HT$_2$ receptors in the rat brain. *Neuroscience* 54, 701–17.

Morilak DA, Somogyi P, Lujan-Miras R *et al.* (1994). Neurons expressing 5-HT$_2$ receptors in the rat brain: neurochemical identification of cell types by immunocytochemistry. *Neuropsychopharmacology* 11, 157–66.

Nakhai B, Nielsen DA, Linnoila M, and Goldman D (1995). Two naturally occuring amino acid substitutions in the human 5-HT$_{1A}$ receptor: glycine 22 to serine 22 and isoleucine 28 to valine 28. *Biochem Biophys Res Commun* **210**, 530–6.

Nebigil C, Choi DS, Launay JM *et al.* (1998). Mouse 5-HT$_{2B}$ receptors mediate serotonin embryonic functions. *4th IUPHAR Satellite Meeting of the Serotonin Club* Rotterdam, Abstr. S2.3.

Nelson CS, Cone RD, Robbins LS *et al.* (1995). Cloning and expression of a 5-HT$_7$ receptor from *Xenopus* laevis. *Recept Channel* **3**, 61–70.

Oberlander C (1983). Effects of a potent 5-HT agonist, RU 24969, on the mesocortico-limbic and nigrostriatal dopamine systems. *Br J Pharmacol* **80**, 675P.

Obosi LA, Hen R, Beadle DJ, Bermudez I, and King LA (1997). Mutational analysis of the mouse 5-HT$_7$ receptor: importance of the third intracellular loop for receptor-G-protein interaction. *FEBS Lett* **412**, 321–24.

Oliver KR, Kinsey AM, Wainwright A, McAllister G, and Sirinathsinghji D (1999). Localisation of 5-HT$_7$ and 5-HT$_{5A}$ receptor immunoreactivity in the rat brain. *Br Neurosci Assoc Abstr* **15**, 25.08.

Olsen MA, Nawoschik SP, Schurman BR, Schmitt HL, Burno M, Smith DL, and Schechter LE (1999). Identification of a human 5-HT$_6$ receptor variant prduced by alternative splicing. *Mol Brain Res* **64**, 255–63.

Oosterink BJ, Korte SM, Nyakas C, Korf J, and Luiten PG (1998). Neuroprotection against N-methyl-D-aspartate-induced excitotoxicity in rat magnocellular nucleus basalis by the 5-HT$_{1A}$ receptor agonist 8-OH-DPAT. *Eur J Pharmacol* **358**, 147–52.

Osborne NN, Wood JP, Melena J, Chao HM, Nash MS, Bron AJ, and Chidlow G (2000). 5-Hydroxytryptamine$_{1A}$ agonists: potential use in glaucoma. Evidence from animal studies. *Eye* **14**, 454–63.

Palacios JM, Waeber C, Mengod G *et al.* (1991). Autoradiography of 5-HT receptors: a critical appraisal. *Neurochem Int* **18**, 17–25.

Pascual J (1998). Mechanism of action of zolmitriptan. *Neurologia* **13** (suppl 2), 9–15.

Pasqualetti M, Ori , Nardi I, Castagna M, Cassano GB, and Marazziti D (1998). Distribution of the 5-HT$_{5A}$ serotonin receptor mRNA in the human brain. *Mol Brain Res* **56**, 1–8.

Patel S, Roberts J, Moorman *et al.* (1995). Localisation of 5-HT$_4$ receptors in the striatonigral pathway in rat brain. *Neuroscience* **69**, 1159–67.

Pauwels PJ, Palmier C, Wurch T *et al.* (1996). Pharmacology of cloned human 5-HT$_{1D}$ receptor-mediated functional responses in stably transfected C6-glial cell lines: further evidence differentiating human 5-HT$_{1D}$ and 5-HT$_{1B}$ receptors. *Nauyn-Schmiedeberg's Arch Pharmacol* **353**, 144–56.

Pazos A and Palacios JM (1985). Quantitative autoradiographic mapping of serotonin receptors in the rat brain. I. Serotonin-1 receptors. *Brain Res* **346**, 205–30.

Peroutka SJ (1994). Molecular biology of 5-HT receptors. *Synapse* **18**, 241–60.

Phebus LA, Johnson KW, Zgombick JM *et al.* (1997). Characterisation of LY 344864 as a pharmacological tool to study 5-HT$_{1F}$ receptors: binding affinities, brain penetration and activity in the neurogenic inflammation model of migraine. *Life Sci* **61**, 2117–26.

Pineyro G, Castanon N, Hen R, and Blier P (1995). regulation of [^{3}H]5-HT release in raphe, frontal cortex and hippocampus of 5-HT$_{1B}$ knockout mice. *Neuroreport* **7**, 353–9.

Plassat J-L, Boshert U, Amlaiky N, and Hen R (1992). The mouse 5-HT$_5$ receptor reveals a remarkable hetergeneity within the 5-HT$_{1D}$ receptor family. *EMBO J* **11**, 4779–86.

Plassat J-L, Amlaiky N, and Hen R (1993). Molecular cloning of a mammalian serotonin receptor that activates adenylate cyclase. *Mol Pharmacol* **44**, 229–36.

Pompeiano M, Palacios JM, and Mengod G (1992). Distribution and cellular localisation of mRNA coding for 5-HT$_{1A}$ receptor in the rat brain: correlation with receptor binding. *J Neurosci* **12**, 440–53.

Pompeiano M, Palacios JM, and Mengod G (1994). Distribution of the serotonin 5-HT$_2$ receptor family messenger RNAs: comparison between 5-HT$_{2A}$ and 5-HT$_{2C}$ receptors. *Mol Brain Res* **23**, 163–78.

Potkin SG, Shipley J, Bera RB, Carreon D, Fallon J, Alva G, and Keator D (2001). Clinical and PET effects of M100907, a selective 5-HT$_{2A}$ receptor antagonist. *Schizophrenia Res Abstr VIIIth Int Congress Schizophrenia Res* pp. 242.

Price GW, Burton MJ, Collin LJ *et al.* (1997). SB-216641 and BRL-15572 compounds to pharmacologically discriminate h5-HT$_{1B}$ and h5-HT$_{1D}$ receptors. *Naunyn-Schmiedeberg's Arch Pharmacol* 356, 312–20.

Prosser RA, Dean RR, Edgar DM *et al.* (1993). Serotonin and the mammalian circadian system: I. *In vitro* phase shifts by serotonergic agonists and antagonists. *J Biol Rhythms* 8, 1–16.

Radja F, Laporte A-M, Daval G *et al.* (1991). Autoradiography of serotonin receptor subtypes in the central nervous system. *Neurochem Int* 18, 1–15.

Rees S, Den Daas I, Foord S, Goodson S, Bull D, Kilpatrick G, and Lee M (1994). Cloning and characterisation of the human 5-HT$_{5A}$ serotonin receptor. *FEBS Lett* 355, 242–6.

Roberts C, Belenguer A, Middlemiss DN, and Routledge C (1998). Differential effects of 5-HT$_{1B/1D}$ receptor antagonists in dorsal and median raphe innervated brain regions. *Eur J Pharmacol* 346, 175–80.

Roberts C, Boyd DF, Middlemiss DN, and Routledge C (1999). Enhancement of 5-HT$_{1B}$ and 5-HT$_{1D}$ receptor antagonist effects on extracellular 5-HT following concurrent 5-HT$_{1A}$ or 5-HT re-uptake site blockade. *Neuropharmacology* 38, 1409–19.

Roberts C, Hatcher P, Hagan JJ *et al.* (2000). The effect of SB-236057-A, a selective 5-HT$_{1B}$ receptor inverse agonist, on *in vivo* extracellular 5-HT levels in the freely moving guinea-pig. *Naunyn-Schmiedeberg's Arch Pharmacol* 362, 177–83.

Roberts C, Allen L, Langmead C, Hagan JJ, Middlemiss DN, and Price GW (2001*a*). The effect of SB-269970, a 5-HT$_7$ receptor antagonist, on 5-HT release from serotonergic terminals and cell bodies. *Br J Pharmacol* 132, 1574–80.

Roberts C, Langmead CJ, Soffin EM, Davies CH, Lacroix L, and Heidbreder CA (2001*b*) The effect of SB-269970-A, a 5-HT$_7$ receptor antagonist, on 5-HT release and cell firing. In (eds, WT O'Connor, JP Lowry, JJ O'Connor, and RD O'Neill) *Monitoring Molecules in Neuroscience.* pp. 348–50.

Rogers DC and Hagan JJ (2001). 5-HT$_6$ receptor antagonists enhance retention of a water maze task in the rat. *Psychopharm* in press.

Romero L, Hervas I, and Artigas F (1996). The 5-HT$_{1A}$ antagonist, WAY 100635 selectively potentiates the presynaptic effects of serotonergic antidepressants in rat brain. *Neurosci Lett* 219, 123–26.

Roth BL, Craigo SC, Choudhary MS, Uluer A, Monsma FJ, Shen Y *et al.* (1994). Binding of typical and atypical antipsychotic agents to 5-hydroxytryptamine-6 and 5-hydroxytryptamine-7 receptors. *J Pharmacol Exp Ther* 268, 1403–10.

Routledge C, Bromidge SM, Moss SF, Price GW, Hirts W, Newman H *et al.* (2000). Characterisation of SB-271046: a potent and orally active 5-HT$_6$ receptor antagonist. *Br J Pharmacol* 130, 1606–12.

Ruat M, Traiffort E, Leurs R, Tardivel-Laconbe J, Diaz J, Arrang J-M, and Schwartz J-C (1993*a*). Molecular cloning, characterisation and localisation of a high affinity serotonin receptor (5-HT$_7$) activating cAMP formation. *Proc Natl Acad Sci USA* 90, 8547–51.

Ruat M, Traifford E, Arrang J-M *et al.* (1993*b*). A novel rat serotonin (5-HT$_6$) receptor: molecular cloning, localisation and stimulation of cAMP accumulation. *Biochem Biophys Res Comm* 193, 268–76.

Sakaue M, Somboonthum P, Nishihara B, Koyama Y, Hashimoto H, Baba A, and Matsuda T (2000). Postsynaptic 5-HT$_{1A}$ receptor activation increases *in vivo* dopamine release in rat prefrontal cortex. *Br J Pharmacol* 129, 1028–34.

Saltzman AG, Morse B, Whitman MM *et al.* (1991). Cloning of the human serotnin 5-HT$_2$ and 5-HT$_{1C}$ receptor subtypes. *Biochem Biophys Res Comm* 181, 1469–78.

Sanders-Bush E, Burris KD, and Knoth K (1988). Lysergic acid diethylamide and 2,5-dimethoxy-4-methylamphetamine are partial agonists at serotonin receptors linked to phosphoinositide hydrolysis. *J Pharmacol Exp Ther* 246, 924–28.

Saudou F and Hen R (1994). 5-HT receptor subtypes in vertebrates and invertebrates. *Neurochem Int* 25, 503–32.

Schlicker E, Fink K, Molderings GJ, Price GW, Duckworth M, Gaster L *et al.* (1997). Effects of selective h5-HT$_{1B}$ (SB-216641) and h5-HT$_{1D}$ (BRL-15572) receptor ligands on guinea-pig and human 5-HT auto- and heteroreceptors. *Naunyn-Schmiedeberg's Arch Pharmacol* 356, 321–7.

Schoeffter P and Hoyer D (1989). 5-HT, 5-HT$_{1B}$ and 5-HT$_{1D}$ receptors mediating inhibition of adenylate cyclase activity. Pharmacological comparison with special reference to the effects of yohimbine, rawolscine and some beta-adrenoceptor antagonists. *Naunyn-Scmiedeberg's Arch Pharmacol* 340, 285–92.

Schoeffter P and Waeber C (1994). 5-HT receptors with 5-HT$_6$ receptor-like profile stimulating adenylyl cyclase activity in pig caudate membranes. *Naunyn-Schmiedeberg's Arch Pharmacol* 350, 356–60.

Sebben M, Ansanay H, Bockaert J *et al.* (1994). 5-HT$_6$ receptors positively coupled to adenylyl cyclase in striatal neurons in culture. *NeuroReport* 5, 2553–7.

Sellers DJ, Chess-Williams R, and Chapple CR (2000). 5-Hydroxytryptamine-induced potentiation of cholinergic responses to electrical field stimulation in pig detrusor muscle. *BJU International* 86, 714–18.

Sharp T, Gartside SE, and Umbers V (1997). Effects of co-administration of a MAOI and a 5-HT$_{1A}$ receptor antagonist on 5-HT neuronal activity and release. *Eur J Pharmacol* 320, 15–19.

Sharpley AL, Elliott JM, Attenburrow MJ *et al.* (1994). Slow wave sleep in humans: role of 5-HT$_{2A}$ and 5-HT$_{2C}$ receptors. *Neuropharmacology* 33, 467–71.

Shen Y, Monsma FJ, Metcalf MA *et al.* (1993). Molecular cloning and expression of a 5-HT$_7$ serotonin receptor subtype. *J Biol Chem* 268, 18 200–4.

Shenker A, Maayani S, Weinstein H *et al.* (1983). Enhanced serotonin-stimulated adenylate cyclase activity in membranes from adult guinea-pig hippocampus. *Life Sci* 32, 2335–42.

Shinkai T, Ohmori O, Kojima H, Terao T, Suzuki T, and Abe K (1999). Association study of the 5-HT$_6$ recepetor gene in schizophrenia. *Am J Med Gen* 88, 120–2.

Sidenberg DG, Bassett AS, Demchychyn L, Niznik HB, Macciardi F, Kamble AB *et al.* (1993). New polymorphism for the human serotonin 1D receptor variant (5-HT$_{1D\beta}$) not linked to schizophrenia in five Canadian pedigrees. *Human Hered* 43, 315–18.

Silvestre JS, Fernandez AG, and Palacios JM (1996). Effects of 5-HT$_4$ receptor antagonists on rat behaviour in the elevated plus-maze test. *Eur J Pharmacol* 309, 219–22.

Skingle *et al.* (1996). *ref missing*.

Sleight AJ, Carolo C, Petit N *et al.* (1995). Identification of 5-HT$_7$ receptor binding sites in rat hypothalamus: sensitivity to chronic antidepressant treatment. *Mol Pharmacol* 47, 99–103.

Sleight AJ, Boess FG, Bourson A *et al.* (1997). 5-HT$_6$ and 5-HT$_7$ receptors: molecular biology, functional correlates and possible therapeutic indications. *Drug News and Perspectives* 10, 214–24.

Sleight AJ, Boess FG, Bos M *et al.* (1998). Characterisation of Ro 04-6790 and Ro 63-0563: potent and selective antagonists at human and rat 5-HT$_6$ receptors. *Br J Pharmacol* 124, 556–62.

Stam NJ, Roesink C, Dijcks F *et al.* (1997). Human serotonin 5-HT$_7$ receptor: cloning and pharmacological characterisation of two receptor variants. *FEBS Lett* 413, 489–94.

Starkey SJ and Skingle M (1994). 5-HT$_{1D}$ as well as 5-HT$_{1A}$ autoreceptors modulate 5-HT release in the guinea-pig dorsal raphe nucleus. *Neuropharmacology* 33, 393–402.

Suchanek B, Struppeck H, and Fahrig T (1998). The 5-HT$_{1A}$ receptor agonist BAY × 3702 prevents staurosporine-induced apoptosis. *Eur J Pharmacol* 355, 95–101.

Suzuki M, Matsuda T, Asano S *et al.* (1995). Increase of noradrenaline release in the hypothalamus of freely moving rat by postsynaptic 5-HT$_{1A}$ receptor activation. *Br J Pharmacol* 115, 703–11.

Tanaka E and North RA (1993). Actions of 5-HT on neurones of the rat cingulate cortex. *J Neurophysiol* 69, 1749–57.

Tecott LH, Sun LM, Akana SF *et al.* (1995). Eating disorder and epilepsy in mice lacking 5-HT$_{2C}$ serotonin receptors. *Nature* 374, 542–6.

Terry AV, Buccafusco JL, Jackson WJ *et al.* (1998). Enhanced delayed matching performance in younger and older macaques administered the 5-HT$_4$ receptor antagonist, RS17017. *Psychopharmacology* **135**, 407–15.

Thomas DR, Gittins SA, Collin LL *et al.* (1998). Functional characterisation of the human cloned 5-HT$_7$ receptor (long form) — antagonist profile of SB-258719. *Br J Pharmacol* **124**, 1300–6.

Thomas DR, Middlemiss DN, Taylor SG, Nelson P, and Brown AM (1999). 5-CT stimulation of adenylyl cyclase activity in guinea-pig hippocampus: evidence for involvement of 5-HT$_7$ and 5-HT$_{1A}$ receptors. *Br J Pharmacol* **128**, 158–64.

Thomas EA, Matli JR, Hu JL, Carson MJ, and Sutcliffe JG (2000). Pertusis toxin treatment prevents 5-HT$_{5A}$ receptor-mediated inhibition of cyclic AMP accumulation in rat C6 glioma cells. *J Neurosci Res* **61**, 75–81.

Thorre K, Ebinger G, and Michotte Y (1998). 5-HT$_4$ receptor involvement in the serotonin-enhanced dopamine efflux from the substantia nigra of the freely-moving rat: a microdialysis study. *Brain Res* **796**, 117–24.

To ZP, Bonhaus DW, Eglen RM, and Jakeman LB (1995). Characterisation and distribution of putative 5-HT$_7$ receptors in guinea-pig brain. *Br J Pharmacol* **115**, 107–16.

Tonini M, Galligan JJ, and North RA (1989). Effects of cisapride on cholinergic neurotransmission and propulsive motility in the guinea-pig ileum. *Gastroenterology* **96**, 1257–64.

Traeber J and Glasser T (1987). 5-HT$_{1A}$-related anxiolytics. *TIPS* **8**, 432–7.

Tsai SJ, Liu HC, Liu TY, Wang YC, and Hong CJ (1999*a*). Association analysis of the 5-HT$_6$ receptor polymorphism C267T in Alzheimer's disease. *Neurosci Lett* **276**, 138–9.

Tsai SJ, Chiu HJ, Wang YC, and Hong CJ (1999*b*). Association study of 5-HT$_6$ receptor variant (C267T) with schizophrenia and aggressive behaviour. *Neurosci Lett* **271**, 135–7.

Tsou A-P, Kosaka A, Bach C *et al.* (1994). Cloning and expression of a 5-HT$_7$ receptor positively cuopled to adenylyl cuclase. *J Neurochem* **63**, 456–64.

Vaidya VA, Marek GJ, Aghajanian GK *et al.* (1997). 5-HT$_{2A}$ receptor-mediated regulation of brain derived neurotrophic factor mRNA in the hippocampus and the neocortex. *J Neurosci* **17**, 2785–95.

Van den Wyngaert I, Gommeren W, Verhasselt P *et al.* (1997). Cloning and expression of a human serotonin 5-HT$_4$ receptor cDNA. *J Neurochem* **69**, 1810–19.

Verge D, Daval G, Marcinkiewicz M *et al.* (1986). Quantitative autoradiography of multiple 5-HT$_1$ receptor subtypes in the brain of control or 5,7-dihydroxytryptamine treated rats. *J Neurosci* **6**, 3473–82.

Vilaro MT, Cortes R, Gerald C *et al.* (1996). Localisation of 5-HT$_4$ receptor mRNA in rat brain by *in situ* hybridisation histochemistry. *Mol Brain Res* **43**, 356–60.

Vogt IR, Shimron-Abarbanell D, Neidt H *et al.* (2000). Investigation of the 5-HT$_6$ receptor gene in bipolar affective disorder. *Am J Med Gen* **96**, 217–21.

Voigt MM, Laurie DJ, Seeburg PH, and Bach A (1991). Molecular cloning and characterisation of a rat brain cDNA encoding a 5-HT$_{1B}$ receptor. *EMBO J* **10**, 4017–23.

Waeber C, Sebben M, Grossman C *et al.* (1993). [^{3}H]GR113808 labels 5-HT(4) receptors in the human and guinea-pig brain. *Neuroreport* **4**, 1239–42.

Waeber C and Moskowitz MA (1995*a*). Autoradiographic visualisation of [^{3}H]5-carboxamidotrypt-amine binding sites I the guinea-pig and rat brain. *Eur J Pharmacol* **283**, 31–46.

Waeber C and Moskowitz MA (1995*b*). [^{3}H]sumatriptan labels both 5-HT$_{1D}$ and 5-HT$_{1F}$ receptor binding sites in the guinea-pig brain: an autoradiographic study. *Naunyn-Schmiedeberg's Arch Pharmacol* **352**, 263–75.

Waeber C, Schoeffter P, Hoyer D *et al.* (1990*a*). The serotonin 5-HT$_{1D}$ receptor: a progress review. *Neurochem Res* **15**, 567–82.

Waeber C, Zhang LA, and Palacios JM (1990*b*). 5-HT$_{1D}$ receptors in the guinea-pig brain: pre and postsynaptic localisations in the striatonigral pathway. *Brain Res* **528**, 197–206.

Wardle KA, Ellis ES, Baxter GS *et al.* (1994). The effects of SB 204070, a highly potent and selective 5-HT$_4$ receptor antagonist, on guinea-pig distal colon. *Br J Pharmacol* 112, 789–94.

Ward RP, Hamblin MW, Lachowicz JE *et al.* (1995). Localisation of serotonin subtype 6 receptor messenger RNA in the rat brain by *in situ* hybridisation histochemistry. *Neuroscience* 64, 1105–11.

Warren JTJ, Peacock ML, and Fink JK (1993a). An Rsal polymorphism in the human serotonin gene (HTR$_{1A}$): Detection by DGGE and RFLP analysis. *Hum Mol Genet* 1, 778.

Warren JTJ, Peacock ML, Rodriguez LC, and Fink JK (1993b). An Mspl polymorphism in the human serotonin receptor gene (HTR$_2$): Detection by DGGE and RFLP analysis. *Hum Mol Genet* 2, 338.

Watson JM, Burton MJ, Price GW, Jones BJ, and Middlemiss DN (1996). GR 127935 acts as a partial agonist at recombinant human 5-HT$_{1D\alpha}$ and 5-HT$_{1D\beta}$ receptors. *Eur J Pharmacol* 314, 365–72.

Weinshank RL, Zgombick JM, Macchi MJ *et al.* (1992). Human 5-HT$_{1D}$ receptor is encoded by a subfamily of two distinct genes: 5-HT$_{1D\alpha}$ and 5-HT$_{1D\beta}$. *Proc Natl Acad Sci USA* 89, 3630–4.

Weissmann-Nanopoulus D, Mach E, Margre J *et al.* (1985). Evidence for the localisation of 5-HT$_{1A}$ binding sites on serotonin containing neurones in the raphe dorsalis and raphe centralis nuclei of the rat brain. *Neurochem Int* 7, 1061–72.

Wisden W, Parker EM, Mahle CD *et al.* (1993). Cloning and characterisation of the rat 5-HT$_{5B}$ receptor: evidence that the 5-HT$_{5B}$ receptor couples to a G-protein in mammalian cell membranes. *FEBS Lett* 333, 25–31.

Wright DE, Seroogy KB, Lundgren KH *et al.* (1995). Comparative localisation of 5-HT 1A, 1C and 2 receptor subtype mRNAs in rat brain. *J Comp Neurol* 351, 357–73.

Ying S-W and Rusak B (1997). 5-HT$_7$ receptors mediate serotonergic effects on light sensitive suprachiasmatic nucleus neurons. *Brain Res* 755, 246–54.

Yu YW, Tsai SJ, Lin CH, Hsu CP, Yang KH, and Hong CJ (1999). 5-HT$_6$ receptor variant (C267T) and clinical response to clozapine. *Neuroreport* 10, 1231–3.

Zanardi R, Artigas F, Moresco R *et al.* (2001). Increased 5-HT$_2$ receptor binding in the frontal cortex of depressed patients responding to paroxetine treatment: a positron emission tomography scan study. *J Clin Psychopharmacol* 21, 53–8.

Zgombick JM, Schechter LE, Macchi M *et al.* (1992). Human gene S31 encodes the pharmacologically defined serotonin 5-HT$_{1E}$ receptor. *Mol Pharmacol* 42, 180–5.

Chapter 23

Somatostatin receptors

Inger-Sofie Selmer, Caroline Nunn,
Jason P. Hannon, Wasyl Feniuk,
Patrick P. A. Humphrey, and Daniel Hoyer

23.1 Introduction

Somatostatin (Somatotropin-release inhibiting factor, SRIF) produced by neuronal, neuro-endocrine, inflammatory and immune cells as well as tumours, produces its many (patho)physiological functions through activation of specific cell membrane receptors. There are two bioactive forms of SRIF; $SRIF_{14}$ and the N-terminal extended form, $SRIF_{28}$ (Fig. 23.1; see Schindler *et al.* 1996). Both peptides act as neuromodulators or hormones, depending on the site of action and/or the target cell type. SRIF produces its effects via activation of specific membrane bound receptors (see Bell and Reisine 1993) which have high affinity for the endogenous peptides, $SRIF_{14}$ and $SRIF_{28}$, and the related brain specific cortistat-ins (CST; Fig. 23.1). CST_{14} has eleven amino acids in common with $SRIF_{14}$ and displaces iodine-labelled $SRIF_{14}$ and analogues as potently as SRIF itself (De Lecea *et al.* 1996). CST mRNA is present in the brain, but the distribution of the corresponding peptide remains to be elucidated as no selective antibodies are available.

23.2 Molecular characterization

For simplicity SRIF and CST receptors will be referred to as SRIF receptors, although these may be common to both SRIF and CST. Indeed, the cloned and native SRIF receptors display high affinity for both families of peptides. The SRIF receptors were initially subdivided into two main classes, $SRIF_1$ and $SRIF_2$, based on binding studies with iodinated analogues of SRIF. $SRIF_1$ receptors demonstrate differential structural and pharmacological activity with high (nM) affinity for the short chain synthetic SRIF analogues (Fig. 23.1 and Table 23.1) including SMS 201–995 (octreotide, Sandostatin®) and MK-678 (seglitide) (see Hoyer *et al.* 2000). $SRIF_2$ receptors have low (μM) affinity for octreotide and MK-678, but high affinity for CGP 23996 (see Hoyer *et al.* 1995, 2000). Conclusive evidence for the existence of several SRIF receptor types came from cloning of five intronless SRIF receptor genes and, as such, SRIF receptors have been reclassified as sst_1–sst_5 (Table 23.2). In humans, sst_1–sst_5 receptors are encoded by five non-allelic genes, which have been identified on chromosomes 14, 17, 22, 20, and 16, respectively (see Hoyer *et al.* 2000). The sst_2 receptor exists in two splice variant forms, $sst_{2(a)}$ and $sst_{2(b)}$, in rat and mouse (Schindler *et al.* 1996). The amino acid sequence of the cloned SRIF receptors range in size from 364 (sst_5) to 418 (sst_2) amino

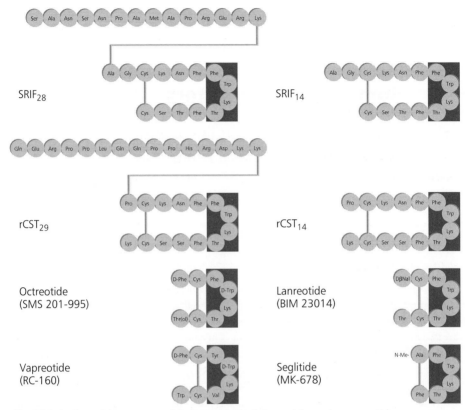

Fig. 23.1 Amino acid sequences of human somatostatin and its analogue peptides.

acids and their sequence homology varies from 39–57 per cent between receptors. There is greater sequence homology in the transmembrane domains compared with the extracellular N- and intracellular C-terminal domains. The highest sequence homology is evident within the SRIF₁ (sst₂, sst₃, sst₅) and SRIF₂ (sst₁, sst₄) classes (Hoyer *et al.* 1994), consistent with their distinctive pharmacological profiles.

23.3 Cellular and subcellular localization

Perhaps one of the surprising findings obtained following the molecular cloning of the SRIF receptors was the lack of marked tissue-specificity for any SRIF receptor. Rather, the receptors have a widespread and overlapping expression pattern, with the exception of the sst₅ receptor which has a limited distribution in the brain (see Selmer *et al.* 2000). Regional specific expression patterns of SRIF receptors in the mammalian central nervous system (CNS) have been examined by *in situ* hybridization. Thus, while sst₁ and sst₂ receptor mRNA exhibits a widespread distribution that encompasses expression in the cerebral cortex, the hypothalamus and to a lesser extent the cerebellum and the spinal cord there is still debate regarding the exact expression profile of sst₃. One consistent finding is that sst₃ receptor mRNA is highly expressed in the cerebellum whereas the sst₄ receptor has been detected at high levels in the hippocampus. In contrast, sst₅ receptor mRNA is most highly expressed

Table 23.1 Ligand binding profile on human recombinant SRIF receptors

	sst_1	sst_4	sst_2	sst_3	sst_5
SRIF-14	9.17, 8.95	8.90, 8.87	9.97, 10.10	9.1. 9.71	9.88, 9.82
SRIF-28	9.13, 9.38	8.62, 9.06	8.87, 9.99	9.2, 9.94	9.28, 10.15
rCST-14	9.01, 9.07	8.63, 8.57	9.49, 9.04	9.09, 9.27	9.06, 9.24
hCST-17	9.48	9.26	9.33	9.88	10.21
CGP-23996	8.67, 8.45	8.67, 8.62	9.30, 9.06	8.40, 9.15	8.47, 8.26
Seglitide	5.08, 4.87	5.46, 5.04	10.47, 9.82	7.50, 7.68	7.26, 10.22
Octreotide	6.20, 6.23	5.61, 6.02	9.37, 9.16	8.10, 8.44	7.68, 8.96
BIM-23027	< 5, 5.52	5.47, 5.51	9.77, 10.48	7.70, 7.98	7.45, 10.02
L-362855	6.35, 6.08	7.27, 7.08	9.00, 8.69	8.20, 8.29	9.55, 8.72
BIM-23056	6.67, 6.56	7.10, 7.04	6.98, 6.33	7.30, 7.20	7.83, 7.77
L-797,591	8.85, 7.21	6.77, 6.47	5.73, 6.34	5.65, 5.96	5.44, 6.34
L-779,976	5.56	6.51	10.30	6.14	5.37
L-796,778	5.90, 5.51	5.06, 5.36	< 5.00, 5.71	7.62, 8.20	5.92, 7.92
L-803,087	6.70	9.15	5.32	5.89	5.41
L-817,818	8.48	7.08	7.28	7.19	9.40

First figure for each data set are pIC_{50} values from Feniuk and coll. (unpublished observations) using $[^{125}I] - Tyr^{11}$ SRIF-14 on membranes from CHO-K1 transfected cells and second figure for each data set are pK_d values from Siehler et al. (1999) using $[^{125}I]$-CGP23996 on membranes from CCL39 transfected cells. The data from the bottom five (Merck) compounds are taken from Rohrer et al. (1998) using $[^{125}I]$-LTT SRIF-28 binding in membranes from CHO-K1 cells or from our laboratories.

in the hypothalamus and preoptic area (Hoyer et al. 1994; Raulf et al. 1994). All five SRIF receptors are expressed at varying levels in the pituitary (see Selmer et al. 2000).

A large number of autoradiographic studies have investigated the distribution of SRIF receptor protein throughout the brain (see Schindler et al. 1996). However, the majority of these studies utilize non-selective radioligands, with affinity for several of the SRIF receptor subtypes. Thus, based on this type of analysis, the distribution of the individual receptor subtypes remains uncertain. Exceptions are the studies using the sst_2 receptor selective radioligands $[^{125}I]$BIM-23027 and $[^{125}I]Tyr^3$-octreotide to investigate the receptor distribution in the rat (Holloway et al. 1996; Thoss et al. 1996) and human brain (Thoss et al. 1997; Schindler et al. 1998). In agreement with the mRNA studies, the radioligand studies identified a widespread distribution pattern for the sst_2 receptor. Recently, receptor specific antibodies raised against individual SRIF receptors have been used to re-evaluate the accuracy of autoradiographic distribution data and specifically to determine receptor protein localization in the rat brain (see Selmer et al. 2000). Surprisingly, work with the sst_1 receptor antibody revealed that the expression of this receptor protein was limited to within the hypothalamus. In contrast, the splice-variant forms of the sst_2 receptor are widely distributed throughout the rat brain, in a partially overlapping pattern. Furthermore, studies

Table 23.2 Somatostatin receptors

	sst$_1$	sst$_2$	sst$_3$	sst$_4$	sst$_5$
Alternate names	SRIF$_2$, SSTR$_1$, SFIF$_{2A}$	SRIF$_1$, SSTR$_2$, SFIF$_{1A}$	SSTR$_3$ SFIF$_{1C}$	SSTR$_4$, SSTR$_5$ SFIF$_{2B}$	SSTR$_5$, SSTR$_4$ SFIF$_{1B}$
Structural information (Accession no.)	h 391 aa (P30872) r 391 aa (P28646) m 391 aa (P30873)	h 369 aa (P30874)^{AS} r 369 aa (P30680)^{AS} m 369 aa (P30875)^{AS}	h 418aa (P32745) r 428 aa (P30936) m 428 aa (P30935)	h 388aa (P31391) r 384 aa (P30937) m 346 aa (P30934)	h 364aa (P35346) r 363 aa (P30938) m 363 aa (P08858)
Chromosome location	14q13	17q24	22q13.1	20p11.2	16p13.3
Selective agonists	CH275, L797591	BIM23027, L7799 76	L796778	NNC2629100, L803087	L362855, L817818
Selective antagonists		CYN154806	ODN8		BIM23056
Radioligands	[^{125}I] – [Tyr^{11}]SRIF-14 [^{125}I]-CGP23996	[^{125}I]-BIM-23027 [^{125}I] – [Tyr^3]octreotide [^{125}I]-CGP23996	[^{125}I] – [Tyr^{11}]SRIF-14 [^{125}I]-CGP23996	[^{125}I] – [Tyr^{11}]SRIF-14 [^{125}I]-CGP23996	[^{125}I] – [Tyr^{11}]SRIF-14 [^{125}I]-CGP23996
G protein coupling	G$_i$/G$_o$	G$_i$/G$_o$/G$_q$	G$_i$/G$_o$/G$_q$	G$_i$/G$_o$	G$_i$/G$_o$/G$_q$
Expression profile	Throughout the brain including cerebral cortex, hypothalamus, cerebellum, spinal cord, pituitary, islets, stomach, liver, kidney	Throughout brain, pituitary, islets, stomach, kidneys	Throughout brain and high in cerebellum, pituitary, islets stomach	Throughout brain and high in hippocampus, pituitary, islets, stomach lung, eye, placenta	Pituitary, islets, stomach, pancreas

Physiological function	Regulation of neurotransmission, inhibition of gastrointestinal motility, gastric acid flow, intestinal absorption, pancreatic enzyme secretion, growth hormone release
Knockout phenotype	Altered growth hormone regulation, high basal gastric acid secretion, increased anxiety-related behaviour, decreased locomotor activity in stress-inducing situations
Disease relevance	Somatostatin receptors have been implicated in a many disease states (although not by receptor subtype) including stroke, epilepsy, pain, depression, meningitis, metabolic encephalopathy, Parkinson's disease, neurodegeneration, diabetes

have identified the sst$_3$ receptor protein within neuronal cilia in a number of brain regions; however, the functional significance of this localization remains unclear. High levels of the sst$_4$ receptor in the hippocampus were confirmed in antibody studies. Whether mRNA or antibody studies are used it is clear that SRIF receptors can co-localize at the cellular level (see Lanneau *et al.* 2000; Selmer *et al.* 2000).

23.4 **Pharmacology**

Only recently have non-peptide compounds, which act as SRIF receptor agonists and ant-agonists, been identified (Table 23.1; see Rohrer *et al.* 1998; Selmer *et al.* 2000). So far, these ligands have been studied almost exclusively in recombinant systems and as such, the pharmacology of SRIF receptors has been largely studied using radioligand-binding and second messenger analysis (see Table 23.1; Siehler *et al.* 1999; Siehler and Hoyer 1999a–c). Furthermore, the recombinant nature of these experiments means that the data reported may not accurately reflect the physiological functions of these receptors. In this respect, the potency of the ligands utilized may be different in native systems where receptor density is likely to be lower and/or their coupling mechanism dissimilar. To date, extension of these investigations to the *in vivo* situation has been limited and, as such, the therapeutic utility of these compounds not extensively evaluated.

23.4.1 **Agonists**

Considerable effort has been directed towards the identification of SRIF analogues with increased selectivity for distinct SRIF receptors. However at present, these studies have only had a limited success and the characterization of the pharmacological properties of the endogenous SRIF receptors have been hampered by a lack of potent receptor-selective agonists and antagonists. In addition, the peptide nature of some of these compounds has reduced their usefulness for *in vivo* studies. CH-275 binds to the sst$_1$ receptor with high affinity and selectivity (see Hoyer *et al.* 2000). Selective agonists for sst$_2$ include BIM-23027 and L-779,976. To date, *in vivo* analysis of agonists has been relatively restricted. However, one notable exception is the sst$_2$ receptor agonist, L-054,552, which has a potent inhibitory effect on growth hormone release when intravenously infused into rats (Yang *et al.* 1998). BIM-23056 demonstrates high affinity and some agonist selectivity for the sst$_3$ receptor, however, its affinity would appear to be species-dependent (Bruns *et al.* 1996; Reisine and Bell 1995). Moreover, in CHO-K1 cells expressing the human sst$_5$ receptor, this compound acts as a potent full agonist for inhibition of cAMP formation (Carruthers *et al.* 1999). The compounds BIM-23268, BIM-23313 and L-362,855 appear to demonstrate specificity and high affinity for the sst$_5$ receptor (Coy and Taylor 1999; Shimon *et al.* 1997).

23.4.2 **Antagonists**

The limited number of antagonists available, which have been shown to bind to SRIF receptors with high affinity and block receptor effector coupling, include Cyanamid 154806 which binds predominantly to the human sst$_2$ receptor (Bass *et al.* 1996; Feniuk *et al.* 2000). An analogue of Cyanamid 154806, ODN-8, binds selectively and with an affinity similar to that of SRIF$_{28}$ to the recombinant sst$_3$ receptor (Reubi *et al.* 2001), and inhibits SRIF-induced inhibition of forskolin-stimulated cAMP accumulation in CCL39 cells (Reubi *et al.* 2001). BIM-23056, which is reported to be an agonist at both the sst$_3$ and sst$_5$ receptors, has also been shown to block human sst$_5$ receptor signalling (Wilkinson *et al.* 1997). Thus, antagonism for this receptor appears dependent on tissue type, receptor coupling and/or receptor density when tested in recombinant systems.

23.5 **Signal transduction and receptor modulation**

Mammalian SRIF receptors couple to intracellular signal transduction cascades via both pertussis toxin (PTX)-sensitive (G$_i$ and G$_o$) and PTX-insensitive G proteins (e.g. G$_q$, G$_{14}$, G$_{16}$; see Florio and Schettini 1996). However, the signalling mechanisms identified for the SRIF receptors have been shown to be dependent upon both the species of the receptor expressed and the cell types investigated. Furthermore, it should be noted that multiple subunit isoforms of heterotrimeric G proteins are thought to couple SRIF receptors specifically to divergent intracellular targets.

Each of the human SRIF receptors is able to inhibit adenylyl cyclase (AC) activity in a PTX-sensitive manner (see Hoyer *et al.* 2000). A number of intracellular effectors are known to couple through the inhibition of AC, but with respect to SRIF receptors, the protein kinase (PK) A system has been the most widely studied (see Florio and Schettini 1996). One of the downstream targets for PKA is the cAMP response element-binding-protein, which is activated following stimulation of recombinant sst$_{2(a)}$ receptors in GH$_4$ cells (Tentler *et al.* 1997).

All five human SRIF receptor subtypes have been shown to couple to phospholipase (PL) C at least partially via a PTX-insensitive mechanism, when expressed in COS-7 cells (Akbar *et al.* 1994), resulting in Ca^{2+} mobilization. However, in intestinal smooth muscle cells, endogenously expressed sst$_3$ receptors induce activation of PLC$_{\beta 3}$ and Ca^{2+}-release via activation of G$_{\beta\gamma}$-subunits of G$_{i1}$/G$_o$ (Murthy *et al.* 1996). It remains to be established whether other SRIF receptors modulate PLC activity in native cells.

Early studies of the electrophysiological actions of SRIF remain somewhat paradoxical, however, a number of more recent studies have been carried out in an attempt to clarify the neuronal modulations of SRIF (see Selmer *et al.* 2000). In acutely dissociated hippocampal pyramidal cells, SRIF inhibits a high-voltage activated Ca^{2+} current (Ishibashi and Akaike 1995). The high-voltage activated inhibition was voltage- and concentration-dependent and found to be N-type channel specific. Following the inhibition of Ca^{2+} currents, SRIF has been found to inhibit the release of noradrenaline from chick sympathetic neurones, possibly mediated by the sst$_3$ receptor (Boehm and Betz 1997). However, involvement of other endogenous SRIF receptors in modulation of Ca^{2+} currents remains to be identified. In noradrenergic neurones in the rat locus coeruleus, SRIF inhibits spontaneous firing by opening an inwardly rectifying K$^+$ channel (Chessell *et al.* 1996). Similar activation of K$^+$ currents by SRIF has been identified in the rat anterior cingulate cortical neurones (Hicks *et al.* 1998) and in hippocampal pyramidal neurones (Schweitzer *et al.* 1998), where the current was identified as the K$^+$ M current, I$_M$. A second voltage-insensitive outward K$^+$ current is also augmented by SRIF, I$_{K(L)}$ (Schweitzer *et al.* 1998), which predominates at rest. So far, the receptor subtype(s) responsible for the modulation of K$^+$ currents remain undetermined.

Activation of the mitogen-activated protein (MAP) kinase pathway by sst$_{1-5}$ receptors has been observed in diverse cellular systems and the involvement of this pathway on proliferation has been well characterized. Phosphorylation, and hence activation, of the extracellular signal-regulated kinase (ERK) has been determined following activation of recombinant SRIF receptors (Sellers *et al.* 1999). Prolonged activation of ERK results in translocation into the nucleus and activation of transcription factors resulting in enhanced cell growth. This prolonged activation of ERK is necessary for SRIF to promote growth through the human sst_4 receptor by $G_{i/o}$ (Sellers 1999). The proliferative effects of SRIF were abolished by a PKC inhibitor, although the presence of a dominant-negative Ras had no effect (Sellers 1999), indicating that the activation of PKC is distinct from that initiated by activation of receptor protein tyrosine kinases. Moreover, opposing effects, upon proliferation, have been observed following the activation of rat recombinant $sst_{2(a)}$ and $sst_{2(b)}$ receptors (Sellers *et al.* 2000). Only the $sst_{2(b)}$ receptor stimulated basal proliferation, while the $sst_{2(a)}$ receptor was without effect. However, the $sst_{2(a)}$ receptor did inhibit basic fibroblast growth factor mediated proliferation (Sellers *et al.* 2000). Since the difference between the two receptor isoforms is solely restricted to their respective C-termini, it implies that this region of the receptor may determine which downstream signalling pathways are stimulated upon receptor activation (Sellers *et al.* 2000). Due to such subtleties it is difficult to predict what effects native SRIF receptors might have upon proliferation in any given situation.

23.6 **Physiology and disease relevance**

In view of the biological actions of SRIF, an important question remains to be determined, namely, whether the effects of SRIF, for a given response, are mediated by a single receptor subtype or whether multiple receptor subtypes are involved. To date, the SRIF receptor subtype specifically involved in any given effect, is rarely defined. What is known is that activation/inhibition of SRIF receptors peripherally and centrally mediate a wide range of physiological effects. Thus, SRIF agonists play an important inhibitory role in the secretion of a number of gastroenterological and pancreatic peptides, inhibit growth hormone (GH), glucagon and insulin secretion.

One of the best characterized physiological effects of SRIF is its ability to lower circulating levels of GH (see Gillies 1997). Within the hypothalamus, the SRIF secreting neurones in the periventricular nucleus (PeVN) control the neurohumoral regulation of GH secretion. The majority of the neurones in the PeVN terminate in the median eminence and the secreted SRIF is then transported to the anterior pituitary gland. Others connect possibly via interneurones with other hypothalamic nuclei, including the arcuate nucleus (ArcN) where the growth hormone releasing hormone (GHRH) is synthesized. It is well established that SRIF acts on the somatotroph cells in the anterior pituitary to inhibit the pulsatile secretion of GH, an action that is counterbalanced by GHRH.

The receptor type(s) responsible for the inhibitory effects of SRIF on GH secretion remain equivocal. Nevertheless, the involvement of the sst_2 receptor in GH regulation has been demonstrated both *in vitro* and *in vivo* (Hoyer *et al.* 1994) and in sst_2 receptor KO mice (Zheng *et al.* 1997). In these KO mice, which appear normal in size, SRIF did not inhibit the release of GHRH from the ArcN. In the rat immunopositive $sst_{2(a)}$ receptor neurones have been identified in the ArcN (Schindler *et al.* 1997). As the levels of circulating GH were not measured in the KO mice, the overall importance of the sst_2 receptor in controlling GH levels remains unknown. However, infusion in rat of the sst_2 receptor specific agonist L-054,522, causes a decrease in circulating GH levels (Yang *et al.* 1998). Surprisingly sst_2 receptor antisense studies in rat showed no changes in the circulating levels of GH (Lanneau *et al.* 2000) following infusion of antisense to the sst_2 receptor. Although, there is much evidence for the involvement of the sst_2 receptors in mediating GH control, recent studies have indicated that additional SRIF receptor subtypes may also play a role. Thus, infusion of an antisense oligonucleotide for the sst_1 receptor, in rats led to a marked decrease in circulating plasma GH levels (Lanneau *et al.* 2000). The sst_1 receptor co-localizes with SRIF in nerve fibres in the PeVN and the external median eminence (Helboe *et al.* 1998), suggesting that it may be involved in an autocrine negative feedback loop in the median eminence and the hypothalamus. Moreover, the sst_5 receptor has been suggested to play a role in GH control, as this receptor is expressed in somatotrophs of the anterior pituitary (O'Carroll and Krempels 1995) and hence, may mediate inhibition of GH release from the pituitary.

In disease states SRIF agonists have been shown to have benefit in pain, epileptic seizures, vascular remodelling after angioplasty/transplantation, and tumour cell growth. Furthermore, alterations of SRIF levels in cerebrospinal fluid (CSF) and/or brain accompany many CNS disorders (Epelbaum 1986; Epelbaum *et al.* 1994; Patel 1992). For instance, increased SRIF levels in CSF were found in meningitis, metabolic encephalopathy, cerebral tumours, spinal cord disease, epilepsy, and other inflammatory disorders. In contrast, decreased CSF/SRIF levels accompany multiple sclerosis and depression (unipolar and bipolar). In several cases, especially Parkinson's disease, the CSF findings have been confirmed post-mortem in cerebral cortex or hippocampus of patients. In depression, the severity of the disease has been correlated with the extent to which CSF/brain SRIF levels are decreased. Various studies have reported changes in SRIF concentration in schizophrenic brains, but the findings are equivocal with increases, decreases, and no change described (see Beal *et al.* 1983; Rubinow *et al.* 1983; Nemeroff *et al.* 1986; Vecsei and Widerlöv 1990). However, in the brain, the specific function attributable to SRIF receptor activation is still largely unknown, so the relevance of reductions in striatal sst_2 receptor levels in conditions such as Huntington's disease is still under evaluation.

23.6.1 Arousal, attention, and sleep

Pharmacological studies have shown that SRIF has significant effects on behaviour and state of arousal (Epelbaum *et al.* 1994). Thus, intracerebroventricular administration of SRIF, and its analogues, has been shown to cause hypermotility in rats. Electroencephalographic (EEG) analysis of SRIF-treated rats also revealed that the animals spent significantly more time under paradoxical (Rapid Eye Movement) sleep than control rats. Somewhat surprisingly, CST and SRIF have opposite effects on sleep (De Lecea *et al.* 1996), suggesting that different (combinations of) receptors might be involved; however, so far no specific receptor for CST has been identified and since both the long and short forms of SRIF and CST have equal high affinity to all five SRIF receptors (Siehler *et al.* 1998, 1999), the reason for these discrepancies remains elusive.

23.6.2 Neuroprotection

SRIF may play a role in neuroprotection: intracerebroventricular administration of SRIF following middle cerebral artery occlusion reduced infarct size. However, the mechanism of action of this neuroprotective effect is not known and the receptors involved remain to be defined.

23.6.3 Epilepsy

There is convincing evidence that SRIF plays a favourable role in seizures, especially in the type observed in various animal models for epilepsy. In particular, it has been shown that compounds with preferential affinity for sst$_2$ receptors, when applied to the hippocampus, have inhibitory effects on the seizure development (see Vezzani and Hoyer 1999).

23.6.4 **Pain and analgesia**

There is also circumstantial evidence that SRIF plays a role in pain. Thus, it is known that acromegalic patients treated with octreotide report less migraine attacks, and intrathecal SRIF or analogues have been used in the management of intractable pain. Indeed, the distribution of SRIF receptors in pain associated-nuclei in the PAG and spinal cord, is in support of such reports (Schindler *et al.* 1998).

23.6.5 **Alzheimer's disease and cognition**

Decreased SRIF levels in the CSF are consistently reported in a variety of dementias including multi infarct dementia with delirium, senile dementia, Alzheimer's disease, Parkinson's disease with dementia, ACTH-dependent Cushing's syndrome, and occasionally Huntington's disease. SRIF-containing neurones in the cortex are selectively depleted in Alzheimer's disease patients, and studies in rats have linked the absence of SRIF with memory and learning deficits. CST, has also been shown to be involved in memory processes. Intracerebroventricular administration of CST impairs short-term memory in passive avoidance. In Alzheimer's disease, the severity of the disease has been correlated with the extent to which CSF/brain SRIF levels are decreased.

High levels of the sst$_4$ receptor in the hippocampus were confirmed in antibody studies, indicating a possible function of this receptor in learning and memory.

23.6.6 **Peripheral indications**

Specific SRIF receptor agonists such as the SRIF analogues octreotide (Sandostatin®) and lanreotide, have been available for more than a decade, for the treatment of acromegaly, gastroenterological and pancreatic tumours and other functional gastrointestinal disorders. The sst$_2$ receptor is the most frequently expressed SRIF receptor type in human tumours (Taylor *et al.* 1994). As such, it is thought that the effect of octreotide and the related SRIF analogues in potently inhibiting the growth of SRIF receptor-expressing tumours, is mediated through the sst$_2$ receptor. However, in isolated vascular smooth muscle cells, anti-proliferative SRIF effects are mediated through the sst$_5$ receptor (Lauder *et al.* 1997).

23.7 **Concluding remarks**

Extensive physiological exploration of somatostatinergic systems has been hindered by a lack of *in vivo* models. There is now however, extensive opportunity in the use of molecular genetic technology to generate mouse models with altered SRIF function. To date, phenotypic characteristics have been described for the sst$_2$ receptor KO mouse only (Zheng *et al.* 1997; Martinez *et al.* 1998). The development of models in which the other SRIF receptor genes or SRIF itself have been ablated will aid our understanding of the role of the individual SRIF receptor subtypes. Recently, two novel aspects of SRIF receptor pharmacology have been identified. First, studies have demonstrated that SRIF receptor subtypes have the ability to form agonist dependent homo- and heterodimers *in vitro* (Rocheville *et al.* 2000a). Presumably, such dimerization results in a change in receptor conformation which alters its pharmacological properties, including ligand binding affinity and receptor internalization (Rocheville *et al.* 2000*a*). Moreover, the human sst$_5$ receptor has been found to dimerize with the human dopamine D$_2$ receptor and suggested to form a 'new' receptor with a distinct and novel functional activity (Rocheville *et al.* 2000*b* and see Chapters 21, 27, 28). Second, yeast-two hybrid screens have identified proteins that interact with SRIF receptors through their C-terminal sequences (Schwärzler *et al.* 2000; Zitzer *et al.* 1999*a,b*). With regard to the sst$_2$ receptor, the interactions are through a PDZ domain to proteins that might facilitate inter-action with the cytoskeleton, which could be important for immobilizing the sst$_2$ receptor to specific sites in the membrane (Zitzer *et al.* 1999*a,b*). In contrast, the protein identified to interact with the sst$_1$ receptor seems to be involved in targeting the receptor to the cell surface (Schwärzler *et al.* 2000). Future studies are required to confirm how SRIF receptors interact with each other and with other proteins. These findings add a new dimension to the way we perceive receptor signalling today but will, no doubt, assist in the elucidation of the SRIF receptor mediated responses in physiological processes in the brain.

Acknowledgements

Some of the work which led to these ideas were supported by EC Contract QLG3-CT-1999-00908 and Swiss grant BBW 00-0427.

References

Akbar M, Okajima F, Tomura H, Majid MA, Yamada Y, Seino S, and Kondo Y (1994). Phospholi-pase C activation and Ca^{2+} mobilization by cloned human somatostatin receptor subtypes 1–5, in transfected COS-7 cells. *FEBS Lett* **348**, 192–6.

Bass RT, Buckwalter BL, Patel BP, Pausch MH, Price LA, Strnad J, and Hadcock, JR (1996). Identification and characterization of novel somatostatin antagonists. *Mol Pharmacol* **50**, 709–15.

Beal MF, Mazurek MF, Svendsen CN, Bird ED, and Martin JB (1986). Widespread reduction of somatostatin-like immunoreactivity in the cerebral cortex in Alzheimer's disease. *Ann Neurol* **20**, 489–95.

Bell, GI and Reisine T (1993). Molecular biology of somatostatin receptors. *Trends Neurosci* **16**, 34–8.

Boehm S and Betz H (1997). Somatostatin inhibits excitatory transmission at rat hippocampal synapses via presynaptic receptors. *J Neurosci* **17**, 4066–75.

Bruns C, Raulf F, Hoyer D, Schloos J, Luebbert H, and Weckbecker G (1996). Binding properties of somatostatin receptor subtypes. *Metab Clin Exp* **44**, 17–20.

Carruthers AM, Warner AJ, Michel AD, Feniuk W, and Humphrey PPA (1999). Activation of adenylate cyclase by human recombinant sst$_5$ receptors expressed in CHO-K1 cells and involvement of G.alpha.s proteins. *Br J Pharmacol* **126**, 1221–9.

Chessell IP, Black MD, Feniuk W, and Humphrey PPA (1996). Operational characteristics of soma-tostatin receptors mediating inhibitory actions on rat locus coeruleus neurones. *Br J Pharmacol* **117**, 1673–8.

Coy DH and Taylor JE (1999). Development of somatostatin agonists with high affinity and specificity for the human and rat type 5 receptor subtype. *Pept Proc Am Pept Symp*, 15, 559–60.

De Lecea L, Criado JR, Prospero-Garcia O, Gautvik KM, Schweitzer P, Danielson PE *et al.* (1996). A cortical neuropeptide with neuronal depressant and sleep-modulating properties. *Nature* **381**, 242–5.

Epelbaum J (1986). Somatostatin in the central nervous system: physiology and pathological modifications. *Prog Neurobiol* **27**, 63–100.

Epelbaum J, Dournaud P, Fodor M, and Viollet C (1994). The neurobiology of somatostatin. *Crit Rev Neurobiol* **8**, 25–44.

Feniuk W, Jarvie E, Luo J, and Humphrey PPA (2000). Selective somatostatin sst$_2$ receptor blockade with the novel cyclic octapeptide, CYN-154806. *Neuropharmacology* **39**, 1443–50.

Florio T and Schettini G (1996). Multiple intracellular effectors modulate physiological functions of the cloned somatostatin receptors. *J Mol Endocrinol* **17**, 89–100.

Gillies G (1997). Somatostatin: the neuroendocrine story. *TIPS* **18**, 87–95.

Helboe L, Stidsen CE, and Møller M (1998). Immunohistochemical and cytochemical localisation of the somatostatin receptor subtype sst$_1$ in the somatostatinergic parvocellular neuronal system of the rat hypothalamus. *J Neurosci* **18**, 4938–45.

Hicks GA, Feniuk W, and Humphrey PPA (1998). Outward current produced by somatotstatin (SRIF) in rat anterior cingulate pyramidal cells *in vitro*. *Br J Pharmacol* **124**, 252–8.

Holloway S, Feniuk W, Kidd EJ, and Humphrey PPA (1996). A quantitative autoradiographical study on the distribution of somatostatin sst$_2$ receptors in the rat central nervous system using [^{125}I]BIM-23027. *Neuropharmacology* **35**, 1109–20.

Hoyer D, Lubbert H, and Bruns C (1994). Molecular pharmacology of somatostatin receptors. *Naunyn-Schmiedeberg's Arch Pharmacol* **350**, 441–53.

Hoyer D, Bell GI, Berelowitz M, Epelbaum J, Feniuk W, Humphrey PPA *et al.* (1995). Classification and nomenclature of somatostatin receptors. *Trends Pharmacol Sci* **16**, 86–8.

Hoyer D, Epelbaum J, Feniuk W, Humphrey PPA, Meyerhof W, Patel Y *et al.* (2000). Somatostatin receptors. IUPHAR Media, London, pp. 354–64.

Ishibashi H and Akaike N (1995). Somatostatin modulates high-voltage-activated Ca^{2+} channels in freshly dissociated rat hippocampal-neurones. *J Neruophysiol* **74**, 1028–36.

Lanneau C, Bluet-Pajot MT, Zizzari P, Csaba Z, Dournaud P, Helboe L *et al.* (2000). Involvement of the sst$_1$ somatostatin receptor subtype in the intrahypothalamic neuronal network regulating growth hormone secretion: an *in vitro* and *in vivo* antisense study. *Endocrinology* **141**, 967–79.

Lauder J, Sellers LA, Fan T-PD, Feniuk W, and Humphrey PPA (1997). Somatostatin sst5 inhibition of receptor mediated regeneration of rat aortic vascular smooth muscle cells. *Br J Pharmacol* **122**, 663–70.

Martinez V, Curi AP, Torkian B, Schaeffer JM, Wilkinson HA, Walsh JH *et al.* (1998). High basal gastric acid secretion in somatostatin receptor subtype 2 knockout mice. *Gastroenterology* **114**, 1125–32.

Murthy KS, Coy DH, and Makhlouf GM (1996). Somatostatin receptor-mediated signaling in smooth muscle. Activation of phospholipase C-β3 by G$\beta\gamma$ and inhibition of adenylyl cyclase by Gα_{i1} and Gα_o. *J Biol Chem* **271**, 23 458–63.

Nemeroff CB, Youngblood WW, Manberg PJ, Prange AJ, and Kizer JS (1983). Regional brain concentration of neuropeptides in Huntington's chorea and schizophrenia. *Science* **221**, 972–5.

O'Carroll A-M and Krempels K (1995). Widespread distribution of somatostatin receptor messenger ribonucleic acids in the rat pituitary. *Endocrinology* **136**, 5224–7.

Patel YC (1992). General aspects of the biology and function of somatostatin. In (eds C Weil, EE Müller, MO Thorner), *Somatostatin*, Springer-Verlag, Berlin, pp. 1–16.

Raulf F, Perez J, Hoyer D, and Bruns C (1994). Differential expression of five somatostatin receptor subtypes, SSTR1-5, in the CNS and peripheral tissue. *Digestion* **55**, 346–53.

Reisine T and Bell GI (1995). Molecular biology of somatostatin receptors. *Endocrinol Rev* **16**, 427–42.

Reubi JC, Schaer J-C, Wenger S, Hoeger C, Erchegyi J, Waser B *et al.* (2001). SST3-selective potent peptidic somatostatin receptor antagonist. *PNAS* **97**, 13 973–8.

Rocheville M, Lange DC, Kumar U, Sasi R, Patel RC, and Patel YC (2000*a*). Subtypes of the somatostatin receptor assemble as functional homo- and heterodimers. *J Biol Chem* **275**, 7862–9.

Rocheville M, Lange DC, Kumar U, Patel SC, Patel RC, and Patel YC (2000*b*). Receptors for dopamine and somatostatin: formation of hetero-oligomers with enhanced functional activity. *Science (Washington, D.C.)* **288**, 154–7.

Rohrer SP, Birzin ET, Mosley RT, Berk SC, Hutchins SM, Shen DM *et al.* (1998). Rapid identification of subtype-selective agonists of the somatostatin receptor through combinatorial chemistry. *Science (Washington, D.C.)* **282**, 737–40.

Rubinow DR, Gold PW, Post RM, and Ballenger JC (1983). CSF Somatostatin in affective illness. *Archs Gen Psychiat* **40**, 409–12.

Schindler M, Emson PC, and Humphrey PPA (1996). Somatostatin receptors in the central nervous system. *Prog Neurobiol* **50**, 9–47.

Schindler M, Sellers LA, Humphrey PPA, and Emson PC (1997). Immunohistochemical localisation of the somatostatin $SST_{2(A)}$ receptor in the rat brain and spinal cord. *Neuroscience* **76**, 225–40.

Schindler M, Holloway S, Hathway G, Woolf CJ, Humphrey PPA, and Emson PC (1998). Identification of somatostatin sst_{2a} receptor expressing neurones in central regions involved in nociception. *Brain Res* **798**, 25–35.

Schwärzler A, Kreienkamp H-J, and Richter D (2000). Interaction of the somatostatin receptor subtype 1 with the human homolog of the Shk1 kinase-binding protein from yeast. *J Biol Chem* **275**, 9557–62.

Schweitzer P, Madamba S, and Siggins GR (1998). Somatostatin increases a voltage-insensitive K^+ conductance in rat CA1 hippocampal neurones. *J Neurophysiol* **79**, 1230–8.

Sellers LA (1999). Prolonged activation of extracellular signal-regulated kinase by a protein kinase C-dependent and N17Ras-insensitive mechanism mediates the proliferative response of $G_{i/o}$-coupled somatostatin sst_4 receptors. *J Biol Chem* **274**, 24 280–8.

Sellers LA, Feniuk W, Humphrey PPA and Lauder H (1999). Activated G protein-coupled receptor induces tyrosine phosphorylation of STAT3 and agonist-selective serine phosphorylation via sustained stimulation of mitogen-activated protein kinase. *J Biol Chem* **274**, 16 423–30.

Sellers LA, Alderton F, Carruthers AM, Schindler M, and Humphrey, PPA (2000). Receptor isoforms mediate opposing proliferative effects through Gβγ-activated p38 or Akt pathways. *Mol Cell Biol* **20**, 5974–85.

Selmer I-S, Schindler M, Allen JP, Humphrey PPA, and Emson PC (2000). Advances in understanding neuronal somatostatin receptors. *Reg Pep* **90**, 1–18.

Shimon I, Taylor JE, Dong JZ, Bitonte RA, Kim S, Morgan B *et al.* (1997). Somatostatin receptor subtype specificity in human fetal pituitary cultures. Differential role of SSTR2 and SSTR5 for growth hormone, thyroid-stimulating hormone, and prolactin regulation. *J Clin Invest* **99**, 789–98.

Siehler S, Seuwen K, and Hoyer D (1998) $[^{125}I]Tyr^{10}$Cortistatin$_{14}$ labels all five somatostatin receptors. *Naunyn Schmiedeberg's Arch Pharmacol* **357**, 483–9.

Siehler S, Seuwen K, and Hoyer D (1999). Characterization of human recombinant somatostatin receptors. 1. Radioligand binding studies. *Naunyn-Schmiedeberg's Arch Pharmacol* **360**, 488–99.

Siehler S and Hoyer D (1999*a*). Characterization of human recombinant somatostatin receptors. 2. Modulation of GTPgammaS binding. *Naunyn-Schmiedeberg's Arch Pharmacol* **360**, 500–9.

Siehler S, and Hoyer D (1999*b*). Characterization of human recombinant somatostatin receptors. 3. Modulation of adenylate cyclase activity. *Naunyn-Schmiedeberg's Arch Pharmacol* **360**, 510–21.

Siehler S and Hoyer D (1999*c*). Characterization of human recombinant somatostatin receptors. 4. Modulation of phospholipase C activity. *Naunyn-Schmiedeberg's Arch Pharmacol* **360**, 522–32.

Taylor JE, Theveniau MA, Bashirzadeh R, Reisine T and Eden PA (1994). Detection of somatostatin receptor subtype 2 (SSTR2) in established tumors and tumor cell lines: evidence for SSTR2 heterogeneity. *Peptides* **15**, 1229–36.

Tentler JJ, Hadcock JR, and Gutierrez-Hartmann, A (1997). Somatostatin acts by inhibiting the cyclic 3′, 5′-adenosine monophosphate (cAMP)/protein kinase A pathway, cAMP response element-binding protein (CREB) phosphorylation, and CREB transcription potency. *Mol Endocrinol* **11**, 859–66.

Thoss VS, Duc D, and Hoyer D (1996). Somatostatin receptors in the developing rat brain. *Eur J Pharmacol* **297**, 145–55.

Thoss VS, Piwko C, Probst A, and Hoyer D (1997). Somatostatin receptors in the human brain and pituitary: an autoradiographic study. *Naunyn Schmiedeberg's Arch Pharmacol* 355, 168–76.

Vecsei L and Widerlöv E (1990). Preclinical and clinical studies with somatostatin related to the central nervous system. *Prog Neuro-Psychopharmacol Biol Psych* 14, 473–502.

Vezzani A and Hoyer D (1999). Brain somatostatin: a candidate inhibitory role in seizures and epileptogenesis, *Eur J Neuroscience* 11, 3767–76.

Wilkinson GF, Feniuk W, and Humphrey PP (1997). Characterization of human recombinant somatostatin sst$_5$ receptors mediating activation of phosphoinositide metabolism. *Br J Pharmacol* 121, 91–6.

Yang L, Berk SC, Rohrer SP, Mosley RT, Guo L, Underwood D *et al.* (1998). Synthesis and biological activities of potent peptidomimetics selective for somatostatin receptor subtype 2. *Proc Natl Acad Sci USA* 95, 10 836–41.

Zheng H, Bailey A, Jiang MH, Honda K, Chen HY, Trumbauer ME *et al.* (1997). Somatostatin receptor subtype 2 knockout mice are refractory to growth hormone-negative feedback on arcuate neurons. *Mol Endocrinol* 11, 1709–17.

Zitzer H, Hönck H-H, Bächner D, Richter D, and Kreienkamp H-J (1999*a*). Somatostatin receptor interacting protein defines a novel family of multidomain proteins present in human and rodent brain. *J Biol Chem*, 274, 32 997–3001.

Zitzer H, Richter D, and Kreienkamp H-J (1999*b*). Agonist-dependent interaction of the rat somatostatin receptor subtype 2 with cortactin-binding protein 1. *J Biol Chem* 274, 18 153–6.

Chapter 24

Tachykinin receptors

Andrew D. Medhurst and Douglas W. P. Hay

24.1 Introduction

The mammalian tachykinins, also known as neurokinins, are a family of small peptides that share the common COOH-terminal sequence, Phe-X-Gly-Leu-Met-NH_2, where X is a Val or Phe residue, and NH_2 represents amidation of the COOH-terminus which is important for biological activity (Maggio 1988; Maggi *et al.* 1993; Otsuka and Yoshioka 1993). The name 'tachykinin'—from the Greek words tachys ('swift') and kinesis ('movement')—originally referred to the fast onset of action of these peptides compared to the slower acting peptide bradykinin (Maggio and Mantyh, 1994). Tachykinin peptides exert their major biological effects through activation of 7TM spanning tachykinin receptors (Simon *et al.* 1991; Strader *et al.* 1994).

Substance P (SP) was the first tachykinin to be identified (von Euler and Gaddum 1931), with its primary sequence of 11 amino acids determined 40 years later (Chang *et al.* 1971). Neurokinin A (NKA), also known as substance K, neurokinin α or neuromedin L (Kimura *et al.* 1983; Minamino *et al.* 1984; Nawa *et al.* 1984), and neurokinin B (NKB), also called neurokinin β or neuromedin K (Kangawa *et al.* 1983; Kimura *et al.* 1983), are the other major members of the tachykinin family, each consisting of 10 amino acid residues. Subsequently, biologically active N-terminally extended forms of NKA have been reported, known as neuropeptide K (NPK; Tatemoto *et al.* 1985) and neuropeptide γ (NPγ; Kage *et al.* 1988), which are 36 and 21 amino acids in length, respectively. In addition, a short eight amino acid form of NKA known as NKA(3–10) has been described, although there is limited understanding of the biological relevance of this peptide (MacDonald *et al.* 1989).

Mammalian tachykinin peptides are the product of two distinct but similar genes: the preprotachykinin A or I (PPT-A or PPT-I) gene, and the preprotachykinin B or II (PPT-B or PPT-II) gene (Nawa *et al.* 1983; Helke *et al.* 1990; Guard and Watson 1991; Nakanishi 1991). Alternative splicing of primary RNA transcripts of the PPT-A gene (consisting of 7 exons) results in the production of three SP-encoding mRNAs (Krause *et al.* 1987), known as α-PPT-A (exons 1–5 and 7), β-PPT-A (all 7 exons) and γ-PPT-A (exons 1–3 and 5–7). β-PPT-A and γ-PPT-A mRNAs also encode NKA, whilst NPK and NPγ are produced from β-PPT-A and γ-PPT-A respectively. The PPT-B gene encodes two precursor mRNAs, α-PPT-B and β-PPT-B, which both result in NKB formation. Each tachykinin is liberated from its corresponding preprotachykinin precursor (MacDonald *et al.* 1989) by cleavage between pairs of basic N-terminal residues, a process catalyzed by serine endoproteases called convertases

(Steiner *et al.* 1992). COOH-terminal amidation of each tachykinin then occurs, which is essential for biological activity.

The tachykinin peptides have a widespread distribution, being principally localized in neurones of both the central nervous system (CNS) and peripheral tissues. In the CNS, SP and NKA are often co-localized in the same neurones, while the distribution of NKB is distinct (Helke *et al.* 1990). For example, in spinal cord, mRNAs derived from PPT-A are prominent in laminae I and II, while PPT-B-derived mRNA is prominent in lamina III (Helke *et al.* 1990). In addition, NKB is thought to predominate in interneurones, while NKA and SP are present in interneurones and primary afferent terminals (Ogawa *et al.* 1985; Houghton *et al.* 2000). Reverse phase high performance liquid chromatography (HPLC) measurements show that SP is the most abundant tachykinin in many regions of the rat brain, while NKB is the least abundant. Despite differences in relative abundance, the highest levels of SP, NKA, and NKB, were all detected in substantia nigra, striatum, cerebral cortex, and hippocampus (Kanazawa *et al.* 1984; Arai and Emson 1986). In the periphery, the tachykinins have a prominent location in peripheral endings of capsaicin-sensitive primary afferent neurones (unmyelinated C-fibres) that innervate many tissues including airways, gastrointestinal and urinary tracts, skin and the eye (Taniguchi *et al.* 1986; Tateishi *et al.* 1990; Maggi *et al.* 1993).

24.2 Molecular characterization

Three major pharmacologically distinct receptors for the tachykinin peptides have so far been identified by molecular cloning and sequence analysis techniques (Maggi *et al.* 1993, Table 24.1). Further subdivisions of each tachykinin receptor have also been proposed, primarily based on differences in pharmacology of selective receptor antagonists (rather than differences at the molecular level), although these could often be attributed to species variants (Hall *et al.* 1993*a*).

The first tachykinin receptor to be cloned was the bovine stomach NK_2 receptor, using an expression cloning strategy and electrophysiological measurements in *Xenopus* oocytes (Masu *et al.* 1987). Within three years, functional clones had been isolated for the rat NK_1 (Yokota *et al.* 1989), NK_2 (Sasai and Nakanishi 1989), and NK_3 receptors (Shigemoto *et al.* 1990). Subsequently, the human NK_1 receptor cDNA was cloned from lung tissue and a lymphoblast cell line (Hopkins *et al.* 1991; Takeda *et al.* 1991), the human NK_2 receptor cDNA from trachea and jejunum (Gerard *et al.* 1991; Graham *et al.* 1991; Kris *et al.* 1991), and the human NK_3 receptor cDNA from brain (Buell *et al.* 1992; Huang *et al.* 1992). All three tachykinin receptor subtypes are members of the Type I subfamily of GPCRs (Simon *et al.* 1991; Strader *et al.* 1994). Unlike many other GPCRs, the human tachykinin receptors are encoded by genes containing introns which divide the protein coding regions into five exons (Takahashi *et al.* 1992). The NK_1, NK_2 and NK_3 receptor genes extend over 60, 12, and 45 kilobases (kb) respectively, although the four introns in each occur at equivalent positions, suggesting a common ancestral evolution (Takahashi *et al.* 1992).

At the amino acid level, tachykinin receptors exhibit significant homology to each other with highest similarity present in the TM domains. For example, the human NK_3 receptor has 74 and 68 per cent homology with human NK_1 and NK_2 receptors, respectively (Buell *et al.* 1992). Homology between species variants of each receptor are even higher. For example, the human NK_1 receptor is about 95 per cent homologous to the rat NK_1 receptor (Quartara and Maggi 1997) and the human and rabbit NK_3 receptors share 91 per cent homology (Medhurst *et al.* 1999). Certain specific sequence characteristics are observed in

Table 24.1 Tachykinin receptors

	NK$_1$	NK$_2$	NK$_3$
Alternative names	SP-P	SP-E	SP-N
Preferred endogenous ligand	Substance P	Substance K, neurokinin A	Neuromedin K, neurokinin B
Structural information (Accession no.)	h 407 aa (P25103) r 407 aa (M31477) m 407 aa (X62934)	h 398 aa (M57414) r 390 aa (M31838) m 384 aa (X62933)	h 468 aa (M89473) r 452 aa (J05189) m 385 aa (P47937)
Chromosomal location	2	10q11–21	4q25
Selective agonists	substance P methyl ester, GR73632, [Pro9]-SP, septide	[β-Ala8]-neurokinin A(4–10), GR64349	Senktide, [Me-Phe7]-NKB
Selective antagonists	CP99994, CP96345, RP67580, SR140333, PD154075, LY303870, RPR100893, CGP94823, GW597599	SR48968, GR159897, MDL29913, MEN10207, L659877, MEN11420	SR142801, SB223412, SB222200, PD157672, GR138676, PD168073
Radioligands	[^{3}H]-substance P [^{125}I]-L703,776	[^{125}I]-neurokinin A [^{3}H]-SR48968	[^{125}I]-Me-Phe7NKB [^{3}H]-senktide
G protein coupling	G$_q$/G$_{11}$	G$_q$/G$_{11}$	G$_q$/G$_{11}$
Expression profile	Striatum, hypothalamus, hippocampus, dentate gyrus, locus coeruleus, brain stem, olfactory bulb, spinal cord, intestine, bladder, salivary glands	Foetal brain, septum, thalamic nuclei, lung, intestine, bladder, stomach, adrenal gland	Cerebral cortex, amygdala, striatum, substantia nigra, hippocampus, habenula, hypothalamus, spinal cord, intestine, bladder, eye
Physiological function	Modulation of brainstem emetic nuclei	Modulation of bronchial airways, bladder and gut motility	Control of firing in dopamine sensitve neurones, regulation of noradrenergic, dopaminergic, cholinergic and serotonergic pathways
Knockout phenotype	No abnormalities decreased stress-induced vocalization, hypoalgesia	—	—
Disease relevance	Emesis, pain, depression, neurogenic inflammation	Irritable bowel syndrome cardiac pulmonary disease, asthma, urinary incontinence, anxiety	Irritable bowel syndrome COPD, urinary incontinence schizophrenia, pain

the three receptors. For example, one His residue is present in both TM domains V and VI of all tachykinin receptors. Moreover, TM domain II of the NK_2 and NK_3 receptor contains an Asp residue (as in other GPCRs), but this is replaced by a Glu in the NK_1 receptor. No other Asp or Glu residues are present in TM domains of tachykinin receptors (Nakanishi 1991). In contrast, the number of Ser and Thr residues in the third cytoplasmic loop and COOH-terminal differ in the three receptors, which may explain the differences observed in desensitization after receptor activation ($NK_1 > NK_3 > NK_2$; Shigemoto et al. 1990). Furthermore, short homologous sequences on the cytoplasmic side of the third intracellular loop between TM domains V and VI maybe important in coupling to G proteins (Nakanishi, 1991).

Many studies have investigated the potential peptide ligand-binding domains in the interactions of the tachykinins to their receptors, typically using site-directed mutagenesis and chimeric receptor approaches (Strader et al. 1994). Several features of the primary tachykinin receptor sequences are important in determining specific interactions with tachykinin peptides and G proteins. Since the conserved COOH-terminus of SP, NKA, and NKB (i.e. Phe-X-Gly-Leu-Met-NH_2) is required for agonist activity at the three tachykinin receptor subtypes, the remaining divergent residues of the N-terminus are thought to define peptide selectivity for their receptors. For example, TM domains V and VII have been proposed to be responsible for fundamental recognition of the common COOH-terminal tachykinin sequence (Yokota et al. 1992). In contrast to the classical type I receptors (e.g. β_2-adrenoceptor where ligand binds to the TM domains), substitution or deletion of the first, second, or third extracellular segments of the NK_1 receptor decreases binding affinity of tachykinin peptides, suggesting that tachykinins interact with extracellular parts of the receptor (Fong et al. 1992a,b). It has also been demonstrated that non-peptide NK_1 receptor antagonists bind to different epitopes on the NK_1 receptor than does SP (Gether et al. 1993b). Recent site-directed mutagenesis studies of aromatic residues in the transmembrane segments of the NK_2 receptor revealed different, but partially overlapping, residues that were important in binding of different peptide and non-peptide NK_2 receptor antagonists (Renzetti et al. 1999). Chimeric receptor and site-directed mutagenesis studies have identified several regions of the NK_3 receptor that are important in determining agonist and antagonist pharmacology. For example, TM domains III and IV are important for senktide recognition, while the third extracellular loop and TM domain VII are important for determining the affinity of NKB (Gether et al. 1993a). In addition, the presence of Met134 and Ala146 in the human NK_3 receptor instead of Val and Gly, respectively, in corresponding positions of the rat NK_3 receptor, confer higher affinity of NK_3 receptor antagonists for the human than for the rat NK_3 receptor (Wu et al. 1994; Chung et al. 1995).

Species differences in tachykinin receptor antagonist pharmacology is a classical characteristic of NK_1 (Snider et al. 1991), NK_2 (Advenier et al. 1992; Chung et al. 1994) and NK_3 receptors (Nguyen et al. 1994; Wu et al. 1994). These differences in pharmacology have often been linked to specific amino acid changes in the receptors between species. For example, human Ile290 instead of rat Ser290, is critical for high affinity binding of CP96345 for human compared to rat NK_1 receptors (Fong et al. 1992c; Sachais et al. 1993). The putative existence of NK_3 receptor subtypes was proposed in 1994 after the observation that two weak NK_3 receptor antagonists, SR48968 and [Trp^7, βAla^8]-NKA(4–10), had different pharmacological profiles in the isolated guinea-pig ileum and rat portal vein assays (Nguyen et al. 1994). However, these differences are most likely due to species differences, as there was little evidence for the existence of NK_3 receptors in the same species or even in the same tissue between species. Preliminary *in vitro* pharmacological evidence was also generated for the putative existence of two NK_3 receptor subtypes within the same species and tissue type, namely the rabbit iris sphincter muscle (Medhurst *et al.* 1997*a,b*). However, molecular cloning studies isolated only a single NK_3 receptor subtype in rabbit iris, with comparable pharmacology to the human and native rabbit iris NK_3 receptors. Further molecular studies failed to identify other NK_3 receptor subtypes, and additional pharmacological analysis of the rabbit recombinant NK_3 receptor suggested the existence of two affinity states of this receptor, providing an alternative explanation for the two putative NK_3 receptor subtypes proposed previously (Medhurst *et al.* 1999).

Another putative tachykinin receptor subtype, the 'septide-sensitive' receptor, has been shown to be pharmacologically distinct from the NK_1 receptor, although there is no molecular evidence to confirm that this is a genuine subtype (Glowinski 1995; Maggi and Schwartz 1997). The pharmacological evidence for this receptor is primarily the low affinity of the tachykinin agonist septide for NK_1 receptor binding sites, despite its potent functional activity in NK_1 receptor-mediated assays (Pradier *et al.* 1994). Furthermore, NK_1 receptor antagonists are more potent at blocking responses induced by septide than by standard NK_1 receptor agonists (Petitet *et al.* 1992). Recent studies have shown that a mutation of Gly166 at the extracellular surface of TM domain IV in the NK_1 receptor induces a selective and significant increase in this receptor's apparent low affinity for septide and NKA (Ciucci *et al.* 1997). In addition, further experiments have identified a region of the NK_1 receptor located at the end of the second extracellular loop that maybe part of a common high-affinity binding domain for both NKA and septide, distinct from the SP binding site (Wijkhuisen *et al.* 1999). Alternatively, NK_1 receptors may exist in two different states with similar affinity for SP and NK_1 antagonists, but with a high and a low affinity for NKA and septide (Ciucci *et al.* 1998). Recent evidence suggests that septide and SP act at the same immunohistologically identified receptor and are equipotent inducers of NK_1 receptor endocytosis (Jenkinson *et al.* 1999).

In 1992, an orphan 7TM GPCR was cloned and described as an atypical opioid receptor, based exclusively on radioligand binding and sequence homology (Xie *et al.* 1992). Interestingly it resembled the NK_3 receptor (80 per cent homology), but was shown not to bind [^{3}H]-eledoisin (Xie *et al.* 1992). More recently, this novel receptor was characterized electrophysiologically, and shown to bind both [^{125}I]-[MePhe^7]-NKB and [^{3}H]-senktide, resulting in it being named by two independent groups as a putative NK_4 receptor (Donaldson *et al.* 1996), or a putative NK_3 receptor homologue (Krause *et al.* 1997), despite each group obtaining the clone from the same source (Xie *et al.* 1992). The main differences between the two receptors was the selective inhibition of NK_4, but not NK_3 receptor activation, by dynorphin, blockade of NK_3 but not NK_4 receptor-mediated responses by [Trp^7,βAla^8]-NKA(4–10), and activation of NK_4 but not NK_3 receptors by SP (Donaldson *et al.* 1996). In addition, poor competitor activity of NKB for [^{3}H]-senktide binding was observed for the NK_3 receptor homologue, while weaker stimulation of IP$_3$ and arachidonic acid production by all agonists was seen with the NK_3 receptor homologue, compared to the classical NK_3 receptor (Krause *et al.* 1997). In the last four years, other groups have failed to isolate this NK_4/NK_3 receptor homologue gene from human cDNA libraries or genomic DNA using a number of molecular approaches (Sarau *et al.* 2000), suggesting that this gene, only isolated from one source, may not be a human gene.

24.3 Cellular and subcellular localization

Receptor autoradiography and RNA expression studies have revealed distinct tissue distribution patterns for NK_1, NK_2, and NK_3 receptors (Tsuchida *et al.* 1990; Maggi *et al.* 1993). NK_1 and NK_3 receptors are the major tachykinin receptor subtypes expressed in the CNS, being localized both discretely and often differentially in mammalian CNS (Buck *et al.* 1986; Tsuchida *et al.* 1990), including human brain (Aubry *et al.* 1994; Mileusnic *et al.* 1999). NK_1 receptors are highly expressed in striatum, hypothalamus, dentate gyrus, locus coeruleus, olfactory bulb, brainstem, and in the dorsal horn of the spinal cord (Saffroy *et al.* 1988; Tsuchida *et al.* 1990; Guard and Watson 1991).

The distribution of NK_3 receptors is widespread in rat brain, especially in cerebral cortex, amygdala, substantia nigra, hypothalamus, hippocampus, and medial habenula (Laufer *et al.* 1986; Saffroy *et al.* 1988; Dam *et al.* 1990). Early evidence for the presence of NK_3 receptors in human CNS was conflicting with Dietl and Palacios (1991) failing to detect binding sites using the non-optimal radioligand $[^{125}I]$-eledoisin, but Buell *et al.* (1992) demonstrating widespread expression of NK_3 receptor mRNA in human brain tissues. More recently, however, direct evidence was reported for the expression of NK_3 receptors in human brain using autoradiography and antibody techniques (Mileusnic *et al.* 1999), with intense staining being observed in cortical layer I, on pyramidal neurones and astrocytes in the neurophil and white matter, and also surrounding blood vessels. NK_3 receptors are also expressed specifically in laminae I–III of the substantia gelatinosa in spinal cord (Shughrue *et al.* 1996). It is noteworthy that species differences have been reported regarding the relative expression of NK_1 and NK_3 receptors in particular brain regions including basal ganglia. For example, in human striatum NK_1 appears to predominate over NK_3 receptors, whereas in rat striatum there is a high level of NK_3 receptor expression compared to NK_1 receptors (Dietl and Palacios 1991; Bannon and Whitty 1995; Mileusnic *et al.* 1999).

The presence of NK_2 receptors in mammalian CNS is controversial, being reported as undetectable (Tsuchida *et al.* 1990) or very discretely localized to the septum (Steinberg *et al.* 1998) and some thalamic nuclei (Hagan *et al.* 1993). In the CNS, NK_2 receptors may well be more important in development given their expression in foetal brain, but in adult rat brain, NK_2 receptor mRNA accounts for only around 1 per cent of the total tachykinin receptor mRNA (Poosch *et al.* 1991). Most NK_2 receptors are expressed in the periphery, with particular abundance in urinary bladder, large intestine, lung, stomach, and adrenal gland. NK_1 receptors are also present in the periphery, especially in the urinary bladder, salivary glands and intestine, while NK_3 receptors have only been detected at low levels in a few peripheral tissues such as the gut and bladder, and at higher levels in the eye (Tsuchida *et al.* 1990). This may be due to the difficulties of observing discrete neuronal localization of NK_3 receptors in larger tissues, especially given that functional receptors have been demonstrated in several peripheral tissues. Peripheral tachykinin receptor function is beyond the scope of this chapter but is reviewed in detail elsewhere (Hall 1994).

24.4 Pharmacology

Prior to the availability of recombinant tachykinin receptors and selective antagonists, early receptor classification was based on the rank order of potency of the tachykinin peptides in selective pharmacological tissue preparations thought to contain single tachykinin receptor populations (Maggi *et al.* 1993; Otsuka and Yoshioka 1993; Regoli *et al.* 1994). Hence, what was referred to initially as the SP-P receptor in dog carotid artery was most sensitive to SP, the SP-E subtype in rabbit pulmonary artery was most sensitive to NKA, and the SP-N receptor in rat portal vein was most sensitive to NKB (Drapeau *et al.* 1987). These subtypes were later renamed NK_1 (SP-P), NK_2 (SP-E), and NK_3 (SP-N) receptors (Henry 1987), referred to as the 'Montreal Nomenclature'. Despite recent discussions over potentially confusing nomenclature (particularly tachykinin vs neurokinin), the use of the terms tachykinins and tachykinin receptors, and the labels NK_1, NK_2, and NK_3 receptors is still recommended (Maggi 2000, 2001).

24.4.1 Agonists

Although the three major tachykinin receptor subtypes have a preferred endogenous ligand (i.e. $NK_1/SP, NK_2/NKA$, and NK_3/NKB), all natural tachykinins can act as full agonists at each receptor subtype, but with different potencies, as demonstrated in early functional isolated tissue assays (Drapeau *et al.* 1987). Radioligand binding experiments using transfected cells expressing each of the recombinant rat (Ingi *et al.* 1991) and human (Sarau *et al.* 1997) tachykinin receptor subtypes have also shown that the rank order of potency of the tachykinins at the three tachykinin receptors is generally SP $\gg$ NKA $\geq$ NKB for the NK_1 receptor, NKA $\gg$ NKB > SP for the NK_2 receptor, and NKB $\gg$ NKA > SP for the NK_3 receptor, with each agonist exhibiting around 50–100-fold higher affinity for their preferred receptor over the other receptor subtypes. It is notable that a number of mismatches have been demonstrated where the preferred endogenous tachykinin has been observed in the absence of its expected receptor, and vice versa. For example, NKA is present in various CNS tissues where very few, if any, NK_2 receptors are detected, while NKB has been undetectable in guinea-pig ileum where functional NK_3 receptors exist, providing evidence for extensive cross talk between endogenous tachykinins and their different tachykinin receptor subtypes (Maggi *et al.* 1993).

Because of the limited selectivity of naturally occurring tachykinins for each receptor subtype, the development of selective ligands was essential to aid investigation of the physiological and pathophysiological roles of NK_1, NK_2, and NK_3 receptors. Examples of selective agonists and antagonists for each tachykinin receptor are listed in Table 24.1. Receptor selective agonists were developed primarily through various modifications of the peptide sequences of the naturally occurring endogenous tachykinins, and were initially characterized in functional isolated tissue preparations (Drapeau *et al.* 1987; Regoli *et al.* 1988). SP methyl ester, [Pro^9]-SP and [Sar^{9}Met(O$_2$)^{11}]-SP are examples of potent and selective NK_1 receptor agonists with pD_2 values of 9.5–10.5 in dog carotid artery isolated tissue assay (Regoli *et al.* 1988). A number of potent and selective NK_2 receptor agonists are widely used, including GR64349 (Hagan *et al.* 1991) and a number of N-terminal truncated derivatives of NKA such as [β-Ala^8]NKA(4–10) and [Nle^{10}]NKA(4–10) (Maggi 1995). Senktide (succinyl-[Asp^6, MePhe^8]SP$_{6-11}$), a conformationally restricted analogue of SP, was the first selective NK_3 receptor agonist to be described with high potency (EC$_{50}$ = 0.5 nM) for the NK_3 receptor in guinea-pig ileum and over 60 000-fold selectivity for NK_3 receptors over NK_1 and NK_2 receptors (Wormser *et al.* 1986). Subsequently, [MePhe^7]-NKB, one of a series of neurokinin analogues, was shown to be the most potent stimulant of receptors for NKB (Drapeau *et al.* 1987). Other tachykinin derivatives have been reported to be selective for NK_3 receptors, including [Pro^7]-NKB, but they are not as potent or selective as senktide and [MePhe^7]-NKB (Lavielle *et al.* 1990). To date, senktide and [MePhe^7]-NKB are the most widely used NK$_3$ receptor agonists and their high potency and selectivity has resulted in them being suitable as radioligands, that is, [^{125}I]-[MePhe^7]-NKB (Sadowski *et al.* 1993), [^{125}I]-Bolton Hunter senktide (Laufer *et al.* 1986), and [^{3}H]-senktide (Guard *et al.* 1989; Dam *et al.* 1990).

24.4.2 Antagonists

Tachykinin peptide analogues were also the starting point for the development of tachykinin receptor antagonists, produced either by insertion of D-amino acids into the peptide (e.g. spantide) or N-terminally truncating various peptides (e.g. [D-Pro^4, D-Trp^{7,9,10}]-SP). However, these compounds exhibited very little selectivity for tachykinin receptor subtypes and had limited potency (Swain 1996). Second generation peptide antagonists were then developed with greater selectivity and improved potency, but still with the general limitations of peptide analogues for *in vivo* studies. For example, a number of conformationally constrained tachykinin peptide analogues, including compounds such as GR82334, were progressed as selective NK$_1$ antagonists. Similarly for NK$_2$ receptors, truncated as well as substituted cyclic hexapeptide derivatives of NKA were synthesized with some selectivity over NK$_1$ and NK$_3$ receptors, including L659877 and the more potent compounds MEN10207 and MDL29913, although the use of these compounds was somewhat complicated due to species differences with NK$_2$ receptors (Maggi 1995; Swain 1996). The first published NK$_3$ receptor antagonists were peptide derivatives of NKA or NKB, with pA$_2$ values no greater than 5.8 in the guinea-pig ileum (Hashimoto *et al.* 1987). Compounds such as [Trp^7, β-Ala^8]-NKA(4–10) were later identified with a pA$_2$ of 7.5 in rat portal vein (Drapeau *et al.* 1990). However, despite moderate selectivity over NK$_1$ and NK$_2$ receptors, this compound showed weak agonist effects at all three tachykinin receptor subtypes, and was later found to have limited NK$_3$ receptor antagonist activity in the guinea-pig ileum (Nguyen *et al.* 1994). Using an alternative approach, GR138676 was identified by conformationally restricting the NKB sequence (Stables *et al.* 1994). Despite high affinity in rat portal vein (pK$_B$ = 8.2) this compound had no selectivity for NK$_3$ over NK$_1$ receptors (pK$_B$ = 8.3 for NK$_1$ receptor). A series of dipeptide-derived NK$_3$ receptor antagonists was developed by Parke-Davis, including PD157672 (Boc(S)Phe(R)αMePheNH (CH$_2$)$_7$NHCONH$_2$), which exhibited nanomolar affinity for NK$_3$ receptors (Boden *et al.* 1996).

A major breakthrough occurred when the first non-peptide tachykinin antagonist, CP96345 was reported to be selective for NK$_1$ receptors (Snider *et al.* 1991). Despite its high potency and selectivity, it was later shown to have micromolar affinity for certain ion channels. A similar analogue to this compound, CP99994, showed 1000-fold selectivity for NK$_1$ over the other tachykinin receptors, possessed limited affinity for calcium channels, and excellent *in vivo* activity after intravenous administration (McLean *et al.* 1993). Subsequent random screening of chemical collections and lead optimization by several pharmaceutical companies resulted in the generation of several chemically distinct selective NK$_1$ receptor antagonists including LY303870, RP67580, CGP94823, SR140333, and RPR100893 (see Maggi 1995; Swain 1996) and more recently MK869 (Kramer *et al.* 1998). SR48968 was the first non-peptide NK$_2$ receptor antagonist to be described (Emonds-Alt *et al.* 1992) with high potency for NK$_2$ receptors, high selectivity over the other tachykinin receptors and prolonged duration of action *in vivo* (e.g. inhibition of NK$_2$ receptor-mediated bronchoconstriction). Interestingly there has been less extensive chemical diversity in the NK$_2$ receptor antagonist area compared to NK$_1$ receptor antagonists, although GR159897, another non-peptide

NK$_2$ receptor antagonist, has nanomolar affinity and potent *in vivo* efficacy following intravenous or oral administration (Ball *et al.* 1994; Renzetti *et al.* 1999). The first potent and selective non-peptide NK$_3$ receptor antagonist reported was SR142801 (Emonds-Alt *et al.* 1995), a compound with high selectivity for NK$_3$ receptors over NK$_1$ and NK$_2$ receptors and nanomolar affinity for human and guinea-pig NK$_3$ receptors. Since its discovery, SR142801 has been widely used as an experimental tool compound and as a comparator molecule in profiling novel classes of non-peptide NK$_3$ receptor antagonists. Further optimization of various chemical groups of the dipeptide-derived antagonists by Parke-Davis was also carried out to obtain true non-peptide NK$_3$ receptor antagonists. For example, PD161182 has around 10 nM affinity for the human NK$_3$ receptor (Boden *et al.* 1996), while later compounds such as PD168073 were even more potent (Jordan *et al.* 1998). In 1996, the 2-phenyl-4-quinolinecarboxamide series was identified by SmithKline Beecham as a novel class of potent and selective non-peptide NK$_3$ receptor antagonists, including SB223412 which had greater selectivity and similar potency to SR142801 (Giardina *et al.* 1996, 1999). In addition, this compound showed reasonable efficacy *in vivo*. For example, mouse behavioural responses to senktide (i.e. rapid head shakes and tail whips), thought to be mediated by 5-HT release (Stoessl *et al.* 1990), were inhibited by oral administration of SB223412 (5–20 mg/kg; Sarau *et al.* 1997). Other members of this series, including SB222200, are potentially more useful for CNS studies given their improved brain penetrance (Giardina and Raveglia 1997; Sarau *et al.* 2000*a*).

24.5 Signal transduction and receptor modulation

All tachykinin receptors operate primarily through the phosphoinositide second messenger system (Guard and Watson 1991). Following tachykinin agonist binding, stimulation of each receptor subtype is thought to result in a conformational change in associated G protein heterotrimer to which GDP binds. Dissociation of the G protein complex into Gαq and G$\beta\gamma$ subunits then occurs, facilitated by substitution of GDP with GTP. Consequently, the Gαq subunit activates phospholipase C (PLC) and then reforms into the G protein heterotrimer via intrinsic GTPase activity involving hydrolysis of GTP back to GDP (Simon *et al.* 1991). The activated PLC then catalyzes hydrolysis of phosphatidylinositol 4,5-bisphosphate (PIP$_2$) into inositol 1,4,5-trisphosphate (IP$_3$) and diacylglycerol (DAG). IP$_3$ can then mobilize intracellular Ca^{2+} via stimulation of the IP$_3$ receptor located on the endoplasmic reticular membrane, resulting in effector responses such as smooth muscle contraction, while DAG can stimulate protein kinase C (PKC) to phosphorylate other target proteins (Guard and Watson 1991; Simon *et al.* 1991).

Tachykinin induced IP$_3$ production has been demonstrated in several tissues including rat salivary gland (NK$_1$; Hanley *et al.* 1980), hamster bladder (NK$_2$; Bristow *et al.* 1987) and guinea-pig ileum (NK$_1$ and NK$_3$; Guard and Watson 1991), as well as in mammalian cells stably transfected with each of the three recombinant tachykinin receptors (Nakajima *et al.* 1992). There is also evidence that tachykinins can stimulate other second messenger pathways. For example, Nakajima and co-workers (1992) demonstrated that tachykinins induce cAMP formation in cells expressing recombinant tachykinin receptors at around 10-fold higher concentrations than those for inducing IP$_3$ formation, although there is little evidence for direct tachykinin receptor coupling to Gs proteins. Stimulation of arachidonic acid release by tachykinin receptor activation has also been demonstrated in transfected cells (Oury-Donat *et al.* 1995). However, this promiscuous signalling may be a result of assaying cells that are overexpressing high levels of receptors, and has rarely been demonstrated in tissues. For instance, in primary cultures of spinal neurones, tachykinins alter IP_3 formation but not cAMP levels (Parsons *et al.* 1995). Interestingly, in iris sphincter muscle, tachykinins can induce either IP_3 generation or cAMP formation depending on the species investigated (Abdel-Latif 1989).

Recent studies have revealed that tachykinin receptors undergo endocytosis (internalization) into cytoplasmic organelles following activation by agonists (Garland *et al.* 1996; Southwell *et al.* 1998). Subsequent phosphorylation of activated receptor by G protein receptor kinase occurs, resulting in desensitization by uncoupling the receptor from the heterotrimeric G protein. This internalization process removes the receptors from the cell surface, away from access to activating ligand, as part of the desensitization mechanisms. The receptors are then recycled to the cell surface where they rapidly regain their ability to bind ligand and undergo endocytosis again (Garland *et al.* 1996; Southwell *et al.* 1998).

24.6 Physiology and disease relevance

An important role for tachykinins in several physiological functions is suggested by both the ubiquitous distribution of these neuropeptides and their receptors, and the widespread activity of tachykinin receptor antagonists (reviewed by Maggi *et al.* 1993; Maggi 1995; Pritchard and Boden 1995; Longmore *et al.* 1997; Quartara and Maggi 1998). Due to the earlier availability of selective non-peptide antagonists for NK_1 and NK_2 receptors, more is known about the relevant disease states and the possible therapeutic implications of blocking these receptors than for NK_3 receptors (Watling 1992).

24.6.1 Pain and analgesia

Preclinical evidence, both from studies with selective NK_1 antagonists (Longmore *et al.* 1997; Quartara and Maggi 1998) and more recently from NK_1 knock-out mice (de Felipe *et al.* 1998; Rupniak *et al.* 2000), suggests the involvement of NK_1 receptors in several physiological paradigms including neurogenic inflammation and pain. For example, in both inflammatory and neuropathic pain models, several distinct chemical classes of NK_1 antagonist have been shown to be active anti-hyperalgesics, although not in every model tested (Urban and Fox 2000), while NK_1 receptor knock-out mice were hypoalgesic (de Felipe *et al.* 1998). However, the recent unexpected failure of NK_1 receptor antagonists in the treatment of neuropathic pain, dental pain, osteoarthritis, and migraine has raised much debate (Rupniak and Kramer 1999; Urban and Fox 2000). It has been questioned as to whether the most suitable compounds were tested in clinical trials (e.g. CP99994, LY303870), given the known species differences in antagonist affinities, doubts over CNS penetration and lack of efficacy in pain models in species other than rats. There was also discussion as to whether the clinical models (e.g. dental pain) were the correct ones in which to test the analgesic efficacy of NK_1 antagonists and also the suggestion that more trials in other pain paradigms were necessary for full evaluation of the potential analgesic properties of NK_1 receptor antagonists. The NK_2 receptor antagonist SR48968 (Saredutant) is also listed as being progressed in clinical trials for CNS indications such as pain (as well as anxiety and depression), although there is limited and conflicting published preclinical data to suggest activity in this area. The presence of NK_3 receptors in the dorsal horn of the spinal cord implies a potential role for NK_3 receptors in modulation of nociceptive activity, together with evidence for Ca^{2+}-dependent release of NKB from rat spinal cord after a depolarization stimulus (Lindefors *et al.* 1985). Recently, it has been demonstrated that SP release from capsaicin sensitive afferent terminals in spinal cord is enhanced by NK_3 receptor agonists, an effect that can be blocked by NK_3, but not NK_1 or NK_2 receptor antagonists (Schmid *et al.* 1998). Other studies have tried to address the role of NK_3 receptors in pain models. Intrathecal NKB was reported to elicit antinociceptive effects mediated by NK_3 receptors indirectly via endogenous opiate release (Couture *et al.* 1993), while upregulation of NK_1 and NK_3 receptor mRNA has been shown in rat spinal cord during formalin-induced nociception (McCarson and Krause 1994). Further evidence for a functional role of NK_3 receptors in spinal cord has been obtained showing that selective antagonists reduce thermal hyperalgesia in monoarthritic rats (Zaratin *et al.* 2000). Interestingly, NK_3 receptors have also been implicated in the central sensitization of spinal withdrawal reflex in the decerebrate, spinalized rabbit, a mechanism thought to contribute to the development of hyperalgesia and allodynia after injury (Houghton *et al.* 2000).

24.6.2 Depression

One of the most exciting recent developments in the tachykinin receptor field was the unexpected discovery of the potential for NK_1 receptor antagonists to be novel antidepressants. This finding arose from testing the effects of the NK_1 antagonist MK869 on behavioural parameters in less traditional species such as the guinea pig and gerbil, instead of rats, again highlighting the well documented species differences observed with tachykinin receptor pharmacology. It was found that MK869 suppressed not only NK_1 receptor agonist induced vocalization, but isolation-induced vocalization in guinea pigs (Kramer *et al.* 1998). This antagonist also showed activity in NK_1 mediated foot tapping behaviour in gerbils (Kramer *et al.* 1998). A more recent paper demonstrated in homozygous NK_1 receptor knock-out mice that separation-induced vocalization was 80 per cent less than in wild-type pups (Rupniak *et al.* 2000), providing more evidence for the potential involvement of NK_1 receptors in depression and anxiety states. A large number of NK_1 receptor antagonists have progressed to clinical trials in recent years including GW597599, CP96345, MK869, L758298, NKP608, CP122721 and SR140333. There is clinical support for the antidepressant (Kramer *et al.* 1998; Stout *et al.* 2001) activity of NK_1 receptor antagonists, although these compounds lacked efficacy in acute pain. The potential of NK_1 receptor antagonists as antidepressants was the subject of a very recent review (Stout *et al.* 2001) in which it was highlighted that the present evidence is suggestive, but not conclusive, for a role of NK_1 receptors in depression, and that further clinical trials are warranted in this therapeutic area.

24.6.3 Anxiety

Despite the limited expression of NK_2 receptors in the CNS, some studies have shown behavioural responses to centrally administered NK_2 receptor agonists (Hagan *et al.* 1993). Some evidence also exists for activity of NK_2 receptor antagonists in rodent (Hagan and McLean 1993; Stratton *et al.* 1993; Walsh *et al.* 1995) and primate (Walsh *et al.* 1995) models of anxiety, but in recent years there have been no reports with other antagonists to support these findings. Studies on individual neurones in brain slices have demonstrated that the NK_3 agonist senktide induced increases in firing rate in primarily dopamine sensitive neurones (Keegan *et al.* 1992; Norris *et al.* 1993; Boden and Woodruff 1994; Seabrook *et al.* 1995), but also in noradrenergic neurones of the locus coeruleus (Jung *et al.* 1996). In some cases, these neuronal responses have been shown to be susceptible to NK$_3$ receptor blockade (Pritchard and Boden 1995). Infusion of senktide using microdialysis probes into specific brain regions in animals has been shown to induce activation of noradrenergic (Jung *et al.* 1996), dopaminergic (Stoessl *et al.* 1991), cholinergic (Marco *et al.* 1998) and serotonergic pathways (Humpel *et al.* 1991). The major responses to central administration of selective NK$_3$ receptor agonists *in vivo* are behavioural, and are not observed with NK$_1$ or NK$_2$ receptor agonists. Intracerebroventricular administration of senktide results in behavioural responses characteristic of activation of dopaminergic pathways (Stoessl *et al.* 1991) and a 5-HT-like behavioural syndrome mediated by several 5-HT receptors. These behaviours include wet dog shakes in guinea pigs (Piot *et al.* 1995), rats and mice (Stoessl *et al.* 1990; Picard *et al.* 1994), forepaw treading and hindlimb splaying (Wormser *et al.* 1986; Stoessl *et al.* 1990), and increased locomotor activity in guinea pigs (Johnston and Chahl 1993). A recent report suggests that NK$_3$ receptor agonists may actually be anxiolytic in mice, since senktide increased the frequency of entry and time spent in open arms of the elevated plus-maze (Ribeiro and De Lima 1998). In addition, an anxiogenic-like effect was also observed with the NK$_3$ receptor antagonist SR142801 (Ribeiro and De Lima 1998), while intrastriatally-administered senktide induced turning behaviour in gerbils, an effect which is inhibited by SR142801 (Emonds-Alt *et al.* 1995). Indeed, Osanetant (SR142801) is listed as being in clinical trials for anxiety (as well as schizophrenia and depression), despite a certain lack of published preclinical data to support its efficacy in this indication.

24.6.4 **Emesis**

NK$_1$ receptor antagonists have been reported to attenuate both acute and delayed emesis induced by cisplatin in ferrets via activity on brainstem emetic nuclei and several NK$_1$ receptor antagonists (including CP122721 and MK869) are clinically effective against post-operative and cisplatin-induced vomiting (Kris *et al.* 1997; Rupniak and Kramer 1999).

24.6.5 **Peripheral indications**

A large proportion of the preclinical studies with NK$_2$ receptor antagonists have been in peripheral models investigating the modulation of airway, gut, and bladder function (Maggi *et al.* 1993; Maggi 1995). The focus of clinical studies with selective NK$_2$ receptor antagonists (e.g. SR48968, UK224671, and MEN11420) has also been predominantly towards peripheral applications such as anti-asthmatic/anti-spasmodic actions of NK$_2$ receptor antagonists against NKA-induced bronchoconstriction (Van Schoor *et al.* 1998), or targeting urinary incontinence and irritable bowel syndrome. Clinical studies for the treatment of peripheral indications with selective NK$_3$ receptor antagonists are in their infancy, although SB223412 is currently being pursued for chronic obstructive pulmonary disease.

24.7 **Concluding remarks**

Since the molecular cloning of tachykinin receptor subtypes and the development of potent and selective antagonists for each receptor, progress in understanding the contribution of the tachykininergic system to pathophysiology has increased significantly in recent years. There have been a number of clinical trials that have highlighted both successes (e.g. anti-emetic activity of NK$_1$ antagonists) and disappointments (lack of analgesia with NK$_1$ antagonists), which highlights the potential limited predictability value of some preclinical models. The future value of pursuing tachykinin receptors as clinical targets now depends on further clinical data with compounds with appropriate pharmacodynamic profiles, including improved CNS penetration. This, plus identifying better, more predictive preclinical models will assist greatly in clarifying the therapeutic potential of selective tachykinin receptor antagonists for peripheral and CNS diseases.

References

Abdel-Latif AA (1989). Calcium-mobilizing receptors, polyphosphoinositides, generation of second messengers and contraction in the mammalian iris smooth muscle: historical perspectives and current status. *Life Sci* 45, 757–86.

Advenier C, Rouissi N, Nguyen QT, Emonds-Alt X, Breliere JC, Neliat G, *et al.* (1992). Neurokinin A (NK2) receptor revisited with SR 48968, a potent non-peptide antagonist. *Biochem Biophys Res Commun* 184, 1418–24.

Arai H and Emson PC (1986). Regional distribution of neuropeptide K and other tachykinins (neurokinin A, neurokinin B and substance P) in rat central nervous system. *Brain Res* 399, 240–9.

Aubry JM, Lundstrom K, Kawashima E, Ayala G, Schulz P, Bartanusz V *et al.* (1994). NK1 receptor expression by cholinergic interneurones in human striatum. *Neuroreport* 5, 1597–600.

Ball DJ, Beresford IJM, Wren GPA, Pendry YD, Sheldrick RLG, Walsh DM *et al.* (1994). *In vitro* and *in vivo* pharmacology of the non-peptide antagonist at tachykinin NK2 receptors GR 159,897. *Br J Pharmacol* 112, 48P.

Bannon MJ and Whitty CJ (1995). Neurokinin receptor gene expression in substantia nigra: localization, regulation, and potential physiological significance. *Can J Phys Pharmacol* 73, 866–70.

Boden P and Woodruff GN (1994). Presence of NK3-sensitive neurones in different proportions in the medial habenula of guinea-pig, rat and gerbil. *Br J Pharmacol* 112, 717–19.

Boden P, Eden JM, Hodgson J, Horwell DC, Hughes J, McKnight AT *et al.* (1996). Use of a dipeptide chemical library in the development of non-peptide tachykinin NK3 receptor selective antagonists. *J Med Chem* 39, 1664–75.

Bristow DR, Curtis NR, Suman-Chauhan N, Watling KJ, and Williams BJ (1987). Effects of tachykinins on inositol phospholipid hydrolysis in slices of hamster urinary bladder. *Br J Pharmacol* 90, 211–17.

Buck SH, Helke CJ, Burcher E, Shults CW, and O'Donohue TL (1986). Pharmacologic characterization and autoradiographic distribution of binding sites for iodinated tachykinins in the rat central nervous system. *Peptides* 7, 1109–20.

Buell G, Schulz MF, Arkinstall SJ, Maury K, Missotten M, Adami N *et al.* (1992). Molecular characterisation, expression and localisation of human neurokinin-3 receptor. *FEBS Lett* 299, 90–5.

Chang MM, Leeman SE, and Niall HD (1971). Amino-acid sequence of substance P. *Nat—New Biol* 232, 86–7.

Chung FZ, Wu LH, Vartanian MA, Watling KJ, Guard S, Woodruff GN *et al.* (1994). The non-peptide tachykinin NK2 receptor antagonist SR 48968 interacts with human, but not rat, cloned tachykinin NK3 receptors. *Biochem Biophys Res Commun* 198, 967–72.

Chung FZ, Wu LH, Tian Y, Vartanian MA, Lee H, Bikker J *et al.* (1995). Two classes of structurally different antagonists display similar species preference for the human tachykinin neurokinin3 receptor. *Mol Pharmacol* 48, 711–16.

Ciucci A, Palma C, Riitano D, Manzini S, and Werge TM (1997). Gly166 in the NK1 receptor regulates tachykinin selectivity and receptor conformation. *FEBS Lett* 416, 335–8.

Ciucci A, Palma C, Manzini S, and Werge TM (1998). Point mutation increases a form of the NK1 receptor with high affinity for neurokinin A and B and septide. *Br J Pharmacol* 125, 393–401.

Couture R, Boucher S, Picard P, and Regoli D (1993). Receptor characterization of the spinal action of neurokinins on nociception: a three receptor hypothesis. *Regul Pept* 46, 426–9.

Dam TV, Escher E, and Quirion R (1990). Visualization of neurokinin-3 receptor sites in rat brain using the highly selective ligand [3H]senktide. *Brain Res* **506**, 175–9.

de Felipe C, Herrero JF, O'Brien JA, Palmer JA, Doyle CA, and Smith AJH (1998). Altered nociception, analgesia and aggression in mice lacking the receptor for substance P. *Nature* **392**, 394–7.

Dietl MM and Palacios JM (1991). Phylogeny of tachykinin receptor localization in the vertebrate central nervous system: apparent absence of neurokinin-2 and neurokinin-3 binding sites in the human brain. *Brain Res* **539**, 211–22.

Donaldson LF, Haskell CA, and Hanley MR (1996). Functional characterization by heterologous expression of a novel cloned tachykinin peptide receptor. *Biochem J* **320**, 1–5.

Drapeau G, D'Orleans-Juste P, Dion S, Rhaleb NE, Rouissi NE, and Regoli D (1987). Selective agonists for substance P and neurokinin receptors. *Neuropeptides* **10**, 43–54.

Drapeau G, Rouissi N, Nantel F, Rhaleb NE, Tousignant C, and Regoli D (1990). Antagonists for the neurokinin NK-3 receptor evaluated in selective receptor systems. *Regul Pept* **31**, 125–35.

Emonds-Alt X, Vilain P, Goulaouic P, Proietto V, Van Broeck D, Advenier C et al. (1992). A potent and selective non-peptide antagonist of the neurokinin A (NK2) receptor. *Life Sci Pharmac Lett* **50**, PL101–6.

Emonds-Alt X, Bichon D, Ducoux JP, Heaulme M, Miloux B, Poncelet M et al. (1995). SR 142801, the first potent non-peptide antagonist of the tachykinin NK3 receptor. *Life Sci* **56**, L27–L32.

Fong TM, Huang RR, and Strader CD (1992*a*). Localization of agonist and antagonist binding domains of the human neurokinin-1 receptor. *J Biol Chem* **267**, 25 664–7.

Fong TM, Yu H, Huang RR, and Strader CD (1992*b*). The extracellular domain of the neurokinin-1 receptor is required for high-affinity binding of peptides. *Biochemistry* **31**, 11 806–11.

Fong TM, Yu H, and Strader CD (1992*c*). Molecular basis for the species selectivity of the neurokinin-1 receptor antagonists CP-96,345 and RP67580. *J Biol Chem* **267**, 25 668–71.

Garland AM, Grady EF, Lovett M, Vigna SR, Frucht MM, Krause JE et al. (1996). Mechanisms of desensitization and resensitization of G protein-coupled neurokinin1 and neurokinin2 receptors. *Mol Pharmacol* **49**, 438–46.

Gerard NP and Gerard C (1991). Molecular cloning of the human neurokinin-2 receptor cDNA by polymerase chain reaction and isolation of the gene. *Ann N Y Acad Sci* **632**, 389–90.

Gether U, Johansen TE, and Schwartz TW (1993*a*). Chimeric NK1 (substance P)/NK3 (neurokinin B) receptors. Identification of domains determining the binding specificity of tachykinin agonists. *J Biol Chem* **268**, 7893–8.

Gether U, Johansen TE, Snider RM, Lowe JA, Nakanishi S, and Schwartz TW (1993*b*). Different binding epitopes on the NK1 receptor for substance P and non-peptide antagonist. *Nature* **362**, 345–8.

Giardina GA and Raveglia LF (1997). Neurokinin-3 receptor antagonists. *Exp Opin Ther Pat* **7**, 307–23.

Giardina GA, Raveglia LF, Grugni M, Sarau HM, Farina C, Medhurst AD et al. (1999). Discovery of a novel class of selective non-peptide antagonists for the human neurokinin-3 receptor. 2. Identification of (S)-N-(1-phenylpropyl)-3-hydroxy-2-phenylquinoline-4-carboxamide (SB 223412). *J Med Chem* **42**, 1053–65.

Giardina GAM, Sarau HM, Farina C, Medhurst AD, Grugni M, Foley JJ et al. (1996). 2-phenyl-4-quinolinecarboxamides: a novel class of potent and selective non-peptide competitive antagonists for the human neurokinin-3 receptor. *J Med Chem* **39**, 2281–4.

Glowinski J (1995). The 'septide-sensitive' tachykinin receptor: still an enigma. *Trends Pharmacol Sci* **16**, 365–7.

Graham A, Hopkins B, Powell SJ, Danks P, and Briggs I (1991). Isolation and characterisation of the human lung NK-2 receptor gene using rapid amplification of cDNA ends. *Biochem Biophys Res Commun* **177**, 8–16.

Guard S, Watson SP, Maggio JE, Too HP, and Watling KJ (1990). Pharmacological analysis of [3H]-senktide binding to NK3 tachykinin receptors in guinea-pig ileum longitudinal muscle-myenteric plexus and cerebral cortex membranes. *Br J Pharmacol* **99**, 767–73.

Guard S and Watson SP (1991). Tachykinin receptor types: Classification and transmembrane signalling mechanisms. *Neurochem Int* 18, 149–65.

Hagan RM and McLean S (1993). Vineyard peptide conference bears fruit. *Trends Pharmac Sci* 14, 315–18.

Hagan RM, Beresford IJM, Stables J, Dupere J, Stubbs CM, Elliott PJ *et al.* (1993). Characterisation, CNS distribution and function of NK2 receptors studied using potent NK2 receptor antagonists. *Regul Peptides* 46, 9–19.

Hagan RM, Ireland SJ, Jordan CC, Beresford IJM, Deal MJ, and Ward P (1991). Receptor-selective, peptidase-resistant agonists at NK1 and NK2 receptors: new tools for investigating neurokinin function. *Neuropeptides* 19, 127–35.

Hall JM (1994). Receptor function in the periphery. In *The Tachykinin Receptors* (ed., SH Buck), Humana press Inc, Totawa, NJ, pp. 515–79.

Hanley MR, Lee CM, Michell RH, and Jones LM (1980). Similar effects of substance P and related peptides on salivation and on phosphatidylinositol turnover in rat salivary glands. *Mol Pharmacol* 18, 78–83.

Hashimoto T, Uchida Y, Naminohira S, and Sakai T (1987). Tachykinin antagonist I: Specific, competitive and tissue-selective neurokinin B antagonists on contractile activity in smooth muscles. *Jpn J Pharmacol* 45, 570–3.

Helke CJ, Krause JE, Mantyh PW, Couture R, and Bannon MJ (1990). Diversity in mammalian tachykinin peptidergic neurons: multiple peptides, receptors, and regulatory mechanisms. *FASEB J* 4, 1606–15.

Henry JL (1987). Discussion of nomenclature for tachykinins and tachykinin receptors. In *Substance P and Neurokinins* (eds., JL Henry, R Couture, AC Cuello, G Pelletier, R Quirion, and D Regoli), Springer-Verlag, New York, pp. xvii–xviii.

Hopkins B, Powell SJ, Danks P, Briggs I, and Graham A (1991). Isolation and characterisation of the human lung NK-1 receptor cDNA. *Biochem Biophys Res Commun* 180, 1110–17.

Houghton AK, Ogilvie J, and Clarke RW (2000). The involvement of tachykinin NK2 and NK3 receptors in central sensitization of a spinal withdrawal reflex in the decerebrated, spinalized rabbit. *Neuropharmacology* 39, 133–40.

Huang RR, Cheung AH, Mazina KE, Strader CD, and Fong TM (1992). cDNA sequence and heterologous expression of the human neurokinin-3 receptor. *Biochem Biophys Res Commun* 184, 966–72.

Humpel C, Saria A, and Regoli D (1991). Injection of tachykinins and selective neurokinin receptor ligands into the substantia nigra reticulata increases striatal dopamine and 5-hydroxytryptamine metabolism. *Eur J Pharmacol* 195, 107–14.

Ingi T, Kitajima Y, Minamitake Y, and Nakanishi, S (1991). Characterization of ligand-binding properties and selectivities of three rat tachykinin receptors by transfection and functional expression of their cloned cDNAs in mammalian cells. *J Pharmacol Exp Ther* 259, 968–75.

Jenkinson KM, Southwell BR, and Furness JB (1999). Two affinities for a single antagonist at the neuronal NK1 tachykinin receptor: evidence from quantitation of receptor endocytosis. *Br J Pharmacol* 126, 131–6.

Johnston PA and Chahl LA (1993). Neurokinin-1, -2 and -3 receptors are candidates for a role in the opiate withdrawal response in guinea-pigs. *Regul Pept* 46, 376–8.

Jordan RE, Smart D, Grimson P, Suman-Chauhan N, and McKnight AT (1998). Activation of the cloned human NK3 receptor in Chinese Hamster Ovary cells characterized by the cellular acidification response using the Cytosensor microphysiometer. *Br J Pharmacol* 125, 761–6.

Jung M, Michaud JC, Steinberg R, Barnouin MC, Hayar A, Mons G *et al.* (1996). Electrophysiological, behavioural and biochemical evidence for activation of brain noradrenergic systems following neurokinin NK3 receptor stimulation. *Neuroscience* 74, 403–14.

Kage R, McGregor GP, Thim L, and Conlon JM (1988). Neuropeptide-gamma: a peptide isolated from rabbit intestine that is derived from gamma-preprotachykinin. *J Neurochem* 50, 1412–17.

Kanazawa I, Ogawa T, Kimura S, and Munekata E (1984). Regional distribution of substance P, neurokinin alpha and neurokinin beta in rat central nervous system. *Neurosci Res* 2, 111–20.

Kangawa K, Minamino N, Fukuda A, and Matsuo H (1983). Neuromedin K: a novel mammalian tachykinin identified in porcine spinal cord. *Biochem Biophys Res Commun* 114, 533–40.

Keegan KD, Woodruff GN, and Pinnock RD (1992). The selective NK3 receptor agonist senktide excites a subpopulation of dopamine-sensitive neurones in the rat substantia nigra pars compacta *in vitro*. *Br J Pharmacol* 105, 3–5.

Kimura S, Okada M, Sugita I, Kanazawa I, and Munekata E (1983). Novel neuropeptides, neurokinin alpha and beta, isolated from porcine spinal cord. *Proc Jpn Acad* 59, 101–4.

Kramer MS, Cutler N, Feighner J, Shrivastava R, Carman J, Sramek JJ *et al.* (1998). Distinct mechanism for antidepressant activity by blockade of central substance P receptors. *Science* 281, 1640–5.

Krause JE, Chirgwin JM, Carter MS, Xu ZS, and Hershey AD (1987). Three rat preprotachykinin mRNAs encode the neuropeptides substance P and neurokinin A. *Proc Natl Acad Sci USA* 84, 881–5.

Krause JE, Staveteig PT, Mentzer JN, Schmidt SK, Tucker JB, Brodbeck RM *et al.* (1997). Functional expression of a novel human neurokinin-3 receptor homolog that binds [3H]senktide and [125I-MePhe7]neurokinin B, and is responsive to tachykinin peptide agonists. *Proc Natl Acad Sci USA* 94, 310–15.

Kris MG, Radford JE, Pizzo BA, Inabinet R, Hesketh A, and Hesketh PJ (1997). Use of an NK1 receptor antagonist to prevent delayed emesis after cisplatin. *J Natl Cancer Inst* 89, 817–18.

Kris RM, South V, Saltzman A, Felder S, Ricca GA, Jaye M *et al.* (1991). Cloning and expression of the human substance K receptor and analysis of its role in mitogenesis. *Cell Growth Differ* 2, 15–22.

Laufer R, Gilon C, Chorev M, and Selinger Z (1986). Characterization of a neurokinin B receptor site in rat brain using a highly selective radioligand. *J Biol Chem* 261, 10257–63.

Lavielle S, Chassaing G, Loeuillet D, Convert O, Torrens Y, Beaujouan JC *et al.* (1990). Selective agonists of tachykinin binding sites. *Fundam Clin Pharmacol* 4, 257–68.

Lindefors N, Brodin E, Theodorsson-Norheim E, and Ungerstedt U (1985). Calcium-dependent potassium-stimulated release of neurokinin A and neurokinin B from rat brain regions *in vitro*. *Neuropeptides* 6, 453–61.

Longmore J, Hill RG, and Hargreaves RJ (1997). Neurokinin-receptor antagonists: pharmacological tools and therapeutic drugs. *Can J Phys Pharmacol* 75, 612–21.

MacDonald MR, Takeda J, Rice CM, and Krause JE (1989). Multiple tachykinins are produced and secreted upon post-translational processing of the three substance P precursor proteins, alpha-, beta-, and gamma-preprotachykinin. Expression of the preprotachykinins in AtT-20 cells infected with vaccinia virus recombinants. *J Biol Chem* 264, 15 578–92.

Maggi CA (1995). The mammalian tachykinin receptors. *Gen Pharmacol* 26, 911–44.

Maggi CA (2000). The troubled story of tachykinins and neurokinins. *Trends Pharmacol Sci* 21, 173–5.

Maggi CA (2001). The troubled story of tachykinins and neurokinins: an update. *Trends Pharmacol Sci* 22, 16.

Maggi CA, Patacchini R, Rovero P, and Giachetti A (1993). Tachykinin receptors and tachykinin receptor antagonists. *J Auton Pharmacol* 13, 23–93.

Maggi CA and Schwartz TW (1997). The dual nature of the tachykinin NK1 receptor. *Trends Pharmacol Sci* 18, 351–5.

Maggio JE (1988). Tachykinins. *Annu Rev Neurosci* 11, 13–28.

Maggio JE and Mantyh PW (1994). Historical perspectives. History of tachykinin peptides. In *The Tachykinin Receptors* (ed., SH Buck), Humana Press Inc, Totowa, NJ.

Marco N, Thirion A, Mons G, Bougault I, Le Fur G, Soubrie P, and Steinberg R (1998). Activation of dopaminergic and cholinergic neurotransmission by tachykinin NK3 receptor stimulation: an *in vivo* microdialysis approach in guinea pig. *Neuropeptides* 32, 481–8.

Masu Y, Nakayama K, Tamaki H, Harada Y, Kuno M, and Nakanishi S (1987). cDNA cloning of bovine substance-K receptor through oocyte expression system. *Nature* 329, 836–8.

McCarson KE and Krause JE (1994). NK-1 and NK-3 type tachykinin receptor mRNA expression in the rat spinal cord dorsal horn is increased during adjuvant or formalin-induced nociception. *J Neurosci* 14, 712–20.

McLean S, Ganong A, Seymour PA, Snider RM, Desai MC, Rosen T *et al.* (1993). Pharmacology of CP-99 994; a nonpeptide antagonist of the tachykinin neurokinin-1 receptor. *J Pharmacol Exp Ther* 267, 472–9.

Medhurst AD, Hirst W, Jerman JC, Meakin J, Roberts JC, Testa T, and Smart D (1999). Molecular and pharmacological characterisation of a functional tachykinin NK$_3$ receptor cloned from the rabbit iris sphincter muscle. *Br J Pharmacol* 128, 627–36.

Medhurst AD, Hay DWP, Parsons AA, Martin LD, and Griswold DE (1997). *In vitro* and *in vivo* characterisation of NK$_3$ receptors in the rabbit eye using selective non-peptide NK$_3$ receptor antagonists. *Br J Pharmacol* 122, 469–76.

Medhurst AD, Parsons AA, Roberts JC, and Hay DWP (1997). Characterisation of NK$_3$ receptors in rabbit isolated iris sphincter muscle. *Br J Pharmacol* 120, 93–101.

Mileusnic D, Lee JM, Magnuson DJ, Hejna MJ, Krause JE, Lorens JB, and Lorens SA (1999). Neurokinin-3 receptor distribution in rat and human brain: an immunohistochemical study. *Neuroscience* 89, 1269–90.

Minamino N, Kangawa K, Fukuda A, and Matsuo H (1984). Neuromedin L: a novel mammalian tachykinin identified in porcine spinal cord. *Neuropeptides* 4, 157–66.

Nakajima Y, Tsuchida K, Negishi M, Ito S, and Nakanishi S (1992). Direct linkage of three tachykinin receptors to stimulation of both phosphatidylinositol hydrolysis and cyclic AMP cascades in transfected Chinese hamster ovary cells. *J Biol Chem* 267, 2437–42.

Nakanishi S (1991). Mammalian tachykinin receptors. *Annu Rev Neurosci* 14, 123–36.

Nawa H, Hirose T, Takashima H, Inayama S, and Nakanishi S (1983). Nucleotide sequences of cloned cDNAs for two types of bovine brain substance P precursor. *Nature* 306, 32–6.

Nawa H, Doteuchi M, Igano K, Inouye K, and Nakanishi S (1984). Substance K: a novel mammalian tachykinin that differs from substance P in its pharmacological profile. *Life Sci* 34, 1153–60.

Nguyen QT, Jukic D, Chretien L, Gobeil F, Boussougou M, and Regoli D (1994). Two NK-3 receptor subtypes: demonstration by biological and binding assays. *Neuropeptides* 27, 157–61.

Norris SK, Boden PR, and Woodruff GN (1993). Agonists selective for tachykinin NK1 and NK3 receptors excite subpopulations of neurons in the rat medial habenula nucleus *in vitro*. *Eur J Pharmacol* 234, 223–8.

Ogawa T, Kanazawa I, and Kimura S (1985). Regional distribution of substance P, neurokinin alpha and neurokinin beta in rat spinal cord, nerve roots and dorsal root ganglia, and the effects of dorsal root section or spinal transection. *Brain Res* 359, 152–7.

Otsuka M and Yoshioka K (1993). Neurotransmitter functions of mammalian tachykinins. *Physiol Rev* 73, 229–308.

Oury-Donat F, Carayon P, Thurneyssen O, Pailhon V, Emonds-Alt X, Soubrie P, and Le Fur G (1995). Functional characterization of the nonpeptide neurokinin3 (NK3) receptor antagonist, SR142801 on the human NK3 receptor expressed in Chinese hamster ovary cells. *J Pharmacol Exp Ther* 274, 148–54.

Parsons AM, el-Fakahany EE, and Seybold VS (1995). Tachykinins alter inositol phosphate formation, but not cyclic AMP levels, in primary cultures of neonatal rat spinal neurons through activation of neurokinin receptors. *Neuroscience* 68, 855–65.

Petitet F, Saffroy M, Torrens Y, Lavielle S, Chassaing G, Loeuillet D *et al.* (1992). Possible existence of a new tachykinin receptor subtype in the guinea pig ileum. *Peptides* 13, 383–8.

Picard P, Regoli D, and Couture R (1994). Cardiovascular and behavioural effects of centrally administered tachykinins in the rat: characterization of receptors with selective antagonists. *Br J Pharmacol* 112, 240–9.

Piot O, Betschart J, Grall I, Ravard S, Garret C, and Blanchard JC (1995). Comparative behavioural profile of centrally administered tachykinin NK1, NK2 and NK3 receptor agonists in the guinea-pig. *Br J Pharmacol* 116, 2496–502.

Poosch MS, Goebel DJ, and Bannon MJ (1991). Distribution of neurokinin receptor gene expression in the rat brain. *Soc Neurosci (Abstr)* 17, 806.

Pradier L, Menager J, Le Guern J, Bock MD, Heuillet E, Fardin V *et al.* (1994). Septide: an agonist for the NK1 receptor acting at a site distinct from substance P. *Mol Pharmacol* 45, 287–93.

Pritchard MC and Boden P (1995). Tachykinin NK3 receptors: biology and development of selective peptide and non-peptide ligands. *Drugs Fut* 20, 1163–73.

Quartara L and Maggi CA (1997). The tachykinin NK1 receptor. Part I: ligands and mechanisms of cellular activation. *Neuropeptides* 31, 537–63.

Quartara L. and Maggi CA (1998). The tachykinin NK1 receptor. Part II: Distribution and pathophysiological roles. *Neuropeptides* 32, 1–49.

Regoli D, Drapeau G, Dion S, and Couture R (1988). New selective agonists for neurokinin receptors: pharmacological tools for receptor characterization. *Trends Pharmacol Sci* 9, 290–5.

Regoli D, Nguyen QT, and Jukic D (1994). Neurokinin receptor subtypes characterized by biological assays. *Life Sci* 54, 2035–47.

Renzetti AR, Catalioto RM, Criscuoli M, Cucchi P, Ferrer C, Giolittl A *et al.* (1999). Relevance of aromatic residues in transmembrane segments V to VII for binding of peptide and nonpeptide antagonists to the human tachykinin NK2 receptor. *J Pharmacol Exp Ther* 290, 487–95.

Ribeiro SJ and De Lima TC (1998). Naloxone-induced changes in tachykinin NK3 receptor modulation of experimental anxiety in mice. *Neurosci Lett* 258, 155–8.

Rupniak NM and Kramer MS (1999). Discovery of the anti-depressant and anti-emetic efficacy of substance P (NK1) antagonists. *Trends Pharmacol Sci* 20, 485–90.

Rupniak NMJ, Carlson EC, Harrison T, Oates B, Seward E, Owen S *et al.* (2000). Pharmacological blockade or genetic deletion of substance P (NK1) receptors attenuates neonatal vocalisation in guinea-pigs and mice. *Neuropharmacol* 39, 1413–21.

Sachais BS, Snider RM, Lowe JA, and Krause JE (1993). Molecular basis for the species selectivity of the substance P antagonist CP-96,345. *J Biol Chem* 268, 2319–23.

Sadowski S, Huang RR, Fong TM, Marko O, and Cascieri MA (1993). Characterization of the binding of [125I-iodo-histidyl, methyl-Phe7] neurokinin B to the neurokinin-3 receptor. *Neuropeptides* 24, 317–19.

Saffroy M, Beaujouan JC, Torrens Y, Besseyre J, Bergstrom L, and Glowinski J (1988). Localization of tachykinin binding sites (NK1, NK2, NK3 ligands) in the rat brain. *Peptides* 9, 227–41.

Sarau HM, Griswold DE, Potts W, Foley JJ, Schmidt DB, Martin LD *et al.* (1997). Non-peptide tachykinin receptor antagonists. I. Pharmacological and pharmacokinetic characterisation of SB 223412, a novel, potent and selective NK3 receptor antagonist. *J Pharmacol Exp Ther* 281, 1303–11.

Sarau HM, Griswold DE, Potts W, Foley JJ, Schmidt DB, Webb EF *et al.* (2000*a*). Nonpeptide tachykinin receptor antagonists. II. Pharmacological and pharmacokinetic profile of SB 222200, a CNS-penetrant, potent and selective NK-3 receptor antagonist. *J Pharmacol Exp Ther* 295, 373–81.

Sarau HM, Mooney JL, Schmidt DB, Foley JJ, Buckley PT, Giardina GA *et al.* (2000*b*). Evidence that the proposed novel human 'neurokinin-4' receptor is pharmacologically similar to the human neurokinin-3 receptor but is not of human origin. *Mol Pharmacol* 58, 552–9.

Sasai Y and Nakanishi S (1989). Molecular characterization of rat substance K receptor and its mRNAs. *Biochem Biophys Res Commun* 165, 695–702.

Schmid G, Carita F, Bonanno G, and Raiteri M (1998). NK-3 receptors mediate enhancement of substance P release from capsaicin-sensitive spinal cord afferent terminals. *Br J Pharmacol* **125**, 621–6.

Seabrook GR, Bowery BJ, and Hill RG (1995). Pharmacology of tachykinin receptors on neurones in the ventral tegmental area of rat brain slices. *Eur J Pharmacol* **273**, 113–19.

Shigemoto R, Yokota Y, Tsuchida K, and Nakanishi S (1990). Cloning and expression of a rat neuromedin K receptor cDNA. *J Biol Chem* **265**, 623–8.

Shughrue PJ, Lane MV, and Merchenthaler I (1996). *In situ* hybridization analysis of the distribution of neurokinin-3 mRNA in the rat central nervous system. *J Comp Neurol* **372**, 395–414.

Simon MI, Strathmann MP, and Gautam N (1991). Diversity of G proteins in signal transduction. *Science* **252**, 802–8.

Snider RM, Constantine JW, Lowe JA, Longo KP, Lebel WS, Woody HA *et al.* (1991). A potent nonpeptide antagonist of the substance P (NK1) receptor. *Science* **251**, 435–7.

Southwell BR, Seybold VS, Woodman HL, Jenkinson KM, and Furness JB (1998). Quantitation of neurokinin 1 receptor internalization and recycling in guinea-pig myenteric neurons. *Neuroscience* **87**, 925–31.

Stables JM, Beresford IJ, Arkinstall S, Ireland SJ, Walsh DM, Seale PW *et al.* (1994). GR138676, a novel peptidic tachykinin antagonist which is potent at NK3 receptors. *Neuropeptides* **27**, 333–41.

Steinberg R, Marco N, Voutsinos B, Bensaid M, Rodier D, Souilhac J *et al.* (1998). Expression and presence of septal neurokinin-2 receptors controlling hippocampal acetylcholine release during sensory stimulation in rat. *Eur J Neurosci* **10**, 2337–45.

Steiner DF, Smeekens SP, Ohagi S, and Chan SJ (1992). The new enzymology of precursor processing endoproteases. *J Biol Chem* **267**, 23 435–8.

Stoessl AJ, Dourish CT, and Iversen SD (1990). Pharmacological characterization of the behavioural syndrome induced by the NK-3 tachykinin agonist senktide in rodents: evidence for mediation by endogenous 5-HT. *Brain Res* **517**, 111–16.

Stoessl AJ, Szczutkowski E, Glenn B, and Watson I (1991). Behavioural effects of selective tachykinin agonists in midbrain dopamine regions. *Brain Res* **565**, 254–62.

Stout SC, Owens MJ, and Nemeroff CB (2001). Neurokinin1 receptor antagonists as potential antidepressants. *Annu Rev Pharmacol Toxicol* **41**, 877–906.

Strader CD, Fong TM, Tota MR, Underwood D, and Dixon RA (1994). Structure and function of G protein-coupled receptors. *Annu Rev Biochem* **63**, 101–32.

Stratton SC, Beresford IJM, Harvey FJ, Turpin MJ, Hagan RM, and Tyers MB (1993). Anxiolytic activity of tachykinin NK2 receptor antagonists in the mouse light dark box. *Eur J Pharmacol* **250**, R11–2.

Swain CJ (1996). Neurokinin receptor antagonists. *Exp Opin Ther Patents* **6**, 367–78.

Takahashi K, Tanaka A, Hara M, and Nakanishi S (1992). The primary structure and gene organization of human substance P and neuromedin K receptors. *Eur J Biochem* **204**, 1025–33.

Takeda Y, Chou KB, Takeda J, Sachais BS, and Krause JE (1991). Molecular cloning, structural characterization and functional expression of the human substance P receptor. *Biochem Biophys Res Commun* **179**, 1232–40.

Taniguchi T, Fujiwara M, Masuo Y, and Kanazawa I (1986). Levels of neurokinin A, neurokinin B and substance P in rabbit iris sphincter muscle. *Jpn J Pharmacol* **42**, 590–3.

Tateishi K, Kishimoto S, Kobayashi H, Kobuke K, and Matsuoka Y (1990). Distribution and localization of neurokinin A-like immunoreactivity and neurokinin B-like immunoreactivity in rat peripheral tissue. *Regul Pept* **30**, 193–200.

Tatemoto K, Lundberg JM, Jornvall H, and Mutt V (1985). Neuropeptide K: isolation, structure and biological activities of a novel brain tachykinin. *Biochem Biophys Res Commun* **128**, 947–53.

Tsuchida K, Shigemoto R, Yokota Y, and Nakanishi S (1990). Tissue distribution and quantitation of the mRNAs for three rat tachykinin receptors. *Eur J Biochem* **193**, 751–7.

Urban LA and Fox AJ (2000). NK1 receptor antagonists—are they really without effect in the pain clinic? *Trends Pharmacol Sci* 21, 462–4.

Van Schoor J, Joos GF, Chasson BL, Brouard RJ, and Pauwels RA (1998). The effect of the NK2 tachykinin receptor antagonist SR 48968 (saredutant) on neurokinin A-induced bronchoconstriction in asthmatics. *Eur Respir J* 12, 17–23.

von Euler US and Gaddum JH (1931). An unidentified depressor substance in certain tissue extracts. *J Physiol* 72, 74–87.

Walsh DM, Stratton SC, Harvey FJ, Beresford IJ, and Hagan RM (1995). The anxiolytic-like activity of GR159897, a non-peptide NK2 receptor antagonist, in rodent and primate models of anxiety. *Psychopharmacology* 121, 186–91.

Watling KJ (1992). Nonpeptide antagonists herald new era in tachykinin research. *Trends Pharmacol Sci* 13, 266–9.

Wijkhuisen A, Sagot MA, Frobert Y, Creminon C, Grassi J, Boquet D, and Couraud JY (1999). Identification in the NK1 tachykinin receptor of a domain involved in recognition of neurokinin A and septide but not of substance P. *FEBS Lett* 447, 155–9.

Wormser U, Laufer R, Hart Y, Chorev M, Gilon C, and Selinger Z (1986). Highly selective agonists for substance P receptor subtypes. *EMBO J* 5, 2805–8.

Wu LH, Vartanian MA, Oxender DL, and Chung FZ (1994). Identification of methionine134 and alanine146 in the second transmembrane segment of the human tachykinin NK3 receptor as reduces involved in species-selective binding to SR 48968. *Biochem Biophys Res Commun* 198, 961–6.

Xie GX, Miyajima A, and Goldstein A (1992). Expression cloning of cDNA encoding a seven-helix receptor from human placenta with affinity for opioid ligands. *Proc Natl Acad Sci USA* 89, 4124–8.

Yokota Y, Sasai Y, Tanaka K, Fujiwara T, Tsuchida K, Shigemoto R *et al.* (1989). Molecular characterization of a functional cDNA for rat substance P receptor. *J Biol Chem* 264, 17 649–52.

Yokota Y, Akazawa C, Ohkubo H, and Nakanishi S (1992). Delineation of structural domains involved in the subtype specificity of tachykinin receptors through chimeric formation of substance P/substance K receptors. *EMBO J* 11, 3585–91.

Zaratin P, Angelici O, Clarke GD, Schmid G, Raiteri M, Carita F, and Bonanno G (2000). NK3 receptor blockade prevents hyperalgesia and the associated spinal cord substance P release in monoarthritic rats. *Neuropharmacology* 39, 141–9.

Part 3

Family 2 GPCRs

Chapter 25

Corticotropin releasing factor receptors

Marilyn H. Perrin and Wylie W. Vale

25.1 Introduction

Throughout life, organisms experience myriad environmental and internal perturbations perceived as 'stressors'. The diverse strategies used in response to particular stressors are viewed as successful when they restore homeostasis and unsuccessful when they fail to do so and culminate in a disease state. True homeostasis is often not achieved; rather, a state of allostasis exists in which a less than ideal short-term accommodation to the stress exists and results in long-term deleterious effects.

The primary response to a stressful event is activation of the hypothalamic–pituitary–adrenal axis, resulting in the synthesis and release of corticotropin releasing factor, (CRF) from the hypothalamus followed by an increased release of pro-opiomelanocortin peptides ACTH, MSH's, and endorphins from the pituitary (Orth 1992). The isolation of CRF, a 41 amino acid peptide, was based on its function as the major secretagogue of ACTH (Vale *et al.* 1981). Subsequently, three more mammalian CRF-related ligands, urocortin, urocortin II, and urocortin III have been cloned. The cascade of physiological responses to CRF is initiated at the cellular level by binding to receptors that are integral membrane proteins.

25.2 Molecular characterization

There are two major types of CRF receptors (see Table 25.1). The first CRF receptor, CRFR1, was cloned both in human, from a pituitary tumour removed from a patient with Cushing's disease (Chen *et al.* 1993) and from brain (Vita *et al.* 1993), as well as in rodent (Vita *et al.* 1993; Perrin *et al.* 1993; Chang *et al.* 1993). Subsequently, a second receptor, CRFR2, encoded by a distinct gene was cloned (Perrin *et al.* 1995; Kishimoto *et al.* 1995; Lovenberg *et al.* 1995*a*; Stenzel *et al.* 1995). The existence of conserved genomic sequences corresponding to CRF receptor proteins in species ranging from drosophila (Adams *et al.* 2000) to human (Polymeropoulos *et al.* 1995; Meyer *et al.* 1997) suggests crucial evolutionarily-conserved roles for the receptors. The gene encoding CRFR1, on human chromosome 17 (Polymeropoulos *et al.* 1995), and that for CRFR2, on human chromosome 7 (Meyer *et al.* 1997), are both multi-exonic (Tsai-Morris *et al.* 1996; Kostich *et al.* 1998; Liaw *et al.* 1996) and encode differently-spliced G protein coupled, 7-transmembrane domain proteins. CRFR1 and CRFR2 are ~70 per cent homologous at the amino acid level and belong to the type 2 receptor family which includes receptors for parathyroid hormone (PTH) (Juppner 1994),

Table 25.1 CRF receptors

	CRF$_1$	CRF$_2$
Alternative names	CRF-RA, CRF-RB, PC-CRFR	CRF-Rβ, HM-CRFR, CRF-Rα, CRF-R2γ
Structural information (Accession no.)	h 415 aa (P34998)^{AS} r 415 aa (P35353) m 415 aa (P35347)	h 411aa (Q133324)^{AS} r 411aa (P47866)^{AS} m 411 aa (Q60748)
Chromosomal location	17q21–22	7p21–p15
Selective agonists	(Partially selective) CRF, sauvagine, urocortin, urotensin I	Urocortin II, urocortin III
Selective antagonists	NB127914, CP154526, antalarmin, CRA1000	(Partially selective) alpha helical CRF (9–41), [DPhe^{12},NLe^{21,38}]r/h CRF (12–41)
Radioligands	[^{125}I]-Tyr-oCRF, [^{125}I]-Tyr-sauvagine, [^{3}H]-urocortin, [^{125}I]-astressin	[^{125}I]-Tyr-sauvagine, [^{3}H]-urocortin
G protein coupling	G$_s$, G$_q$	G$_s$
Expression profile	Cerebral cortex,hippocampus, amygdala, cerebellum, medial septum, pituitary, olfactory bulb, gonad, adrenal, thymus, spleen, placenta	Hypothalamus, lateral septum, choroid plexus, olfactory bulb, heart, lung, cerebral arterioles, skeletal muscle, gastrointestinal tract, blood vessel, epididymus
Physiological function	Coordination of homeostasis and the HPA axis, stimulation of ACTH and corticosterone secretion from pituitary gland	
Knockout phenotype	Decreased anxiety like behaviour, impaired stress response, increased CRF levels in paraventricular nucleus, decreased corticosterone levels	Increased anxiety like behaviour, hypersensitivity to stress, increased CRF levels in amygdala
Disease association	Anxiety, depression, feeding disorders, cognitive decline	

vasoactive intestinal peptide (Sreedharan *et al.* 1991), secretin (Ishihara *et al.* 1991), growth hormone releasing factor (Mayo 1992), calcitonin (Jelinek *et al.* 1993), glucagon (Jelinek *et al.* 1993), and insect diuretic hormone (Reagan 1994).

The type 1 CRF receptor, CRFR1, has been cloned in mammals (Chen *et al.* 1993; Vita *et al.* 1993; Perrin *et al.* 1993; Palchaudhuri *et al.* 1998; Myers *et al.* 1998), bird (Yu *et al.* 1996), catfish (Arai *et al.* 2001), salmon (Pohl *et al.* 2001), and Xenopus (Dautzenberg *et al.* 1997). The human gene for CRFR1 contains 14 exons (Sakai *et al.* 1998) and has been found to encode four receptor splice variants, CRFR1a, CRFR1b, CRFR1c, and CRFR1d. The rat gene contains 13 exons and lacks exon 6 of the human gene (Tsai-Morris *et al.* 1996). The protein corresponding to CRFR1a contains 415 amino acids. Alternative splicing of exon 6, in the human, produces CRFR1b, whose corresponding protein contains an additional 29 amino acids in the first intracellular loop; thus far, this variant has been cloned only from a human Cushing's tumour (Chen *et al.* 1993). The protein corresponding to CRFR1c lacks amino acids 40–80 in the first extracellular domain (ECD-1) as a result of a deletion of the third exon; this variant has been cloned only from human brain (Ross *et al.* 1994). A deletion in exon 12 produces the variant protein, CRFR1d, in which 14 amino acids are missing in the seventh transmembrane domain; this variant was cloned from human pregnant term myometrium and fetal membranes (Grammatopoulos *et al.* 1999). Another variant of CRFR1, cloned from sheep, is a protein of 392 amino acids in which there is an alternative sequence in the intracellular C-terminus (Myers *et al.* 1998).

The type 2 CRF receptor, CRFR2, has been cloned in mammals (Perrin *et al.* 1995; Kishimoto *et al.* 1995; Lovenberg *et al.* 1995a; Kostich *et al.* 1998; Liaw *et al.* 1996; Valdenaire *et al.* 1997; Palchaudhuri *et al.* 1999; Miyata *et al.* 1999), catfish (Arai *et al.* 2001), salmon (Pohl *et al.* 2001) and Xenopus (Dautzenberg *et al.* 1997); homologous partial sequences have been reported in zebrafish and chicken (gene bank numbers: AW128724 and BG711051). The gene for the human CRFR2 encodes three alternatively spliced receptors: CRFR2α, CRFR2β and CRFR2γ. There are three rodent variants CRFR2α, CRFR2a-tr, and CRFR2β. From the genomic structure for the human type 2 receptor (Kostich *et al.* 1998), it was found that the exons encoding CRFR2β and CRFR2γ are located 5′ of the exon encoding CRFR2α and may help explain the initial failure to find evidence for the existence of CRFR2β in human (Liaw *et al.* 1996). In the rodent, CRFR2α, CRFRγ and CRFR2a-tr contain 411, 431, and 236 amino acids, respectively. The human CRFR2γ contains 397 amino acids. The subtype CRFR2α is the only form found in Xenopus and catfish.

A third receptor, CRFR3, cloned from catfish (Arai *et al.* 2001) is ~90 and ~80 per cent homologous to mouse CRFR1 and CRFR2 respectively. Thus far, the mammalian orthologue has not been identified.

All CRF receptor proteins have putative N-linked glycosylation sites and six conserved cysteine residues in the ECD-1. There are two other conserved cysteines, one each in ECD-2 and ECD-3. There are putative protein kinase-A and -C (PKA and PKC) phosphorylation sites in the intracellular loops and C-terminal tail. A study of CRFR1 expressed in COS cells showed that the receptor was phosphorylated in response to CRF whereas activation of PKA or calmodulin-dependent kinase did not cause phosphorylation (Hauger *et al.* 2000).

The molecular weights of native receptors in rat and mouse brain determined by Western blots were 80–76 kD and 83–79 kD and differed from those for rat and mouse pituitary which were found to be 72–59 kD (Radulovic *et al.* 1998). By chemical cross-linking the receptor in bovine anterior pituitary was found to be ~70 kD (Nishimura *et al.* 1987) and by photoaffinity labelling, the cloned CRFR1 was found to be ~66 kD (Bonk and Ruhmann 2000). The differences in the molecular weights of the native receptors were attributed to differential glycosylation (Grigoriadis and DeSouza 1989).

Mutational analyses have been used to identify regions of the receptor that are involved in binding and activation. In one approach, an interchange between different regions of CRFR1 and CRFR2 showed that ECD-2 and -3 were important for the selectivity of CRFR1 in binding of, and subsequent activation, by CRF. Specifically, amino acids Asp266, Leu267, and Val268 were shown to govern the selectivity (Liaw *et al.* 1997a). The receptor region important for the selectivity of the Xenopus receptor was between amino acids 70 and 89 in ECD-1 (Dautzenberg *et al.* 1998). In contrast, the ECD-1 was not important for the binding selectivity of Xenopus CRFR2 or human CRFR2α (Dautzenberg *et al.* 1999).

Another approach employed chimeric receptors in which regions of CRF receptors were interchanged with corresponding regions of other receptors. Using CRF/GRF receptor chimeras it was found that the ECD-1 of CRFR1 was important for binding of the peptide antagonist astressin and the agonist urocortin (Perrin *et al.* 1998). Another chimera, created by replacing the ECD-1 of an activin receptor (a single transmembrane domain serine/threonine kinase) with the ECD-1 of the CRFR1 displayed nanomolar binding of astressin and urocortin but not of either CRF or sauvagine (Perrin *et al.* 1998). Chimeric receptors between CRF and glucagon or PACAP receptors, showed that ECD-4 of CRFR1 was also involved in binding and signalling of ovine CRF (Sydow *et al.* 1999).

Mutations of the conserved extracellular cysteines showed that C30S, C54S, and C30S/C54S mutations did not affect CRF binding or cAMP accumulation, but C44S, C68S, C87S, C102S, C188S, C258S, C30S/C44S, C30S/C68S, C54S/C68S, C87S/C102S, and C188S/C258S mutants showed no CRF binding and impaired signalling (Qi *et al.* 1997). Using a soluble protein corresponding to the ECD-1 of CRFR1 that was capable of binding astressin and urocortin, a determination of the disulfide pattern showed that the Cys30 and Cys54, Cys44 and Cys87, and Cys68 and Cys102 form disulphide bonds (Perrin *et al.* 2001). This arrangement differed from that derived from the mutational analysis (Qi *et al.* 1997), but was similar to that proposed for the PTH receptor (Grauschopf *et al.* 2000) and the GLP-1 receptor (Wilmen *et al.* 1996).

The regions of the receptors that are involved in the binding of agonists differ from those that affect antagonist binding. For example, His 199 in TMD-3 and Met 276 in TMD-5 were important for recognition of a non-peptide antagonist (Liaw *et al.* 1997*b*). A detailed study of human CRFR2α showed that arginine occurs at position 185, as it does in every other species' receptors, rather than His, as originally reported (Liaw *et al.* 1996), and that this position played an important role in ligand specificity (Dautzenberg *et al.* 2000).

In another approach to studying the determinants for signal transduction, a chimeric receptor was created in which a portion of CRF peptide was tethered to the receptor (Nielsen *et al.* 2000). One of the chimeras, in which the N-terminal half of the peptide, CRF(1–16), replaced the first 119 amino acids (in the ECD-1) of CRFR1 displayed continual cAMP signalling. The accumulation of intracellular cAMP was blocked by a non-peptide antagonist, antalarmin, but not by the peptide antagonist, astressin. The sempiternal signalling of this chimeric receptor was also blocked by the Leu8Ala mutation in the CRF moiety consistent with the reduced biological activity of [Ala^8]CRF. Thus, neither the ECD-1 of CRFR1 nor the C-terminal half of the CRF appeared to be required for signalling. The data suggest the following model for CRFR1 activation: the C-terminal half of CRF serves to bind the peptide to the receptor's ECD-1, positioning the peptide in suitable spatial propinquity so that its N-terminal half may interact with the receptor's other extracellular and transmembrane domains.

25.2.1 CRF receptor regulation

Cellular responses to CRF ligands may be influenced at the receptor level by stability of its message, by regulation of its rate of transcription, or by internalization or desensitization resulting from interaction with intracellular proteins such as arrestins or kinases.

Icv infusion of CRF induced CRFR1 mRNA expression in the paraventricular nucleus of the hypothalamus, whereas there was no change in mRNA levels for CRFR2 (Mansi *et al.* 1996). Salt loading increased CRFR1 levels and binding in hypothalamus (Imaki *et al.*

2001). In CATH.a cells, pretreatment with CRF, TPA or blockage of calcium channels also decreased CRFR1 message (Iredale *et al.* 1996, 1997*a*). In cultured rat anterior pituitary cells, the CRFR1 message was decreased by treatment with CRF, vasopressin, forskolin, TPA, and glucocorticoids (Pozzoli *et al.* 1996). In the rat pituitary, salt loading produced an increase in mRNA for CRFR1 levels (Imaki *et al.* 2001); endotoxin decreased those levels (Aubry *et al.* 1997). In the AtT-20 cell line, pretreatment with CRF increased, whereas treatment with glucocorticoids decreased, CRFR1 mRNA levels (Iredale *et al.* 1996, 1997), respectively while dexamethasone decreased both the rate of gene transcription as well as the stability of the message (Iredale *et al.* 1997*b*). The CRFR1 expressed in mouse spleen was upregulated following inflammatory challenge with LPS (Radulovic *et al.* 1999).

In Y-79 cells that endogenously express CRFR1, as well as in Ltk⁻ cells stably expressing CRFR1, pretreatment with CRF induced a desensitization in the cAMP response and a concomitant downregulation of the number of binding sites (Dieterich *et al.* 1996; Hauger *et al.* 1997). In a study using Y-79 cells, inhibition of G protein-coupled receptor kinase, GRK3, blocked phosphorylation of the receptor, so that it is possible that this kinase may be involved in the desensitization of receptor (Dautzenberg *et al.* 2001).

Using ribonuclease protection assays and *in situ* hybridization techniques it was found that injection of LPS, corticosterone, or physical restraint resulted in a decrease in CRFR2β mRNA in the rat heart (Kageyama *et al.* 2000; Asaba *et al.* 2000). In A7R5 cells, CRFR2β mRNA was also decreased by urocortin, dexamethasone, and cytokines (Kageyama *et al.* 2000). Sub-lethal hypoxia upregulated CRFR1 mRNA in fetal hippocampal neurones (Wang *et al.* 2000).

25.3 Cellular and subcellular localization

The sites of expression of the two CRF receptors are distinct and, in general, non-overlapping (Van Pett *et al.* 2000). This observation suggests that the two receptors may subserve distinct physiological functions. The type 1 receptor has been identified and its expression described in the central nervous system (CNS) (Van Pett *et al.* 2000), in human neuroblastoma cells (Schoeffter *et al.* 1999) and peripherally in the gonads, placenta, adrenal, thymus, and spleen (Muramatsu *et al.* 2001; Baigent and Lowry 2000; Radulovic *et al.* 1999; Florio *et al.* 2000). A comparison between the distribution of receptor mRNA's for CRFR1 in rat and mouse showed that both rodents expressed similar distributions with predominant expression in cerebral cortex, sensory relay nuclei, and in the cerebellum and its major afferents, while in the pituitary, expression was scattered over the intermediate lobe and in distinct areas of the anterior pituitary (Van Pett *et al.* 2000; Potter *et al.* 1994). Semiquantitative RT-PCR showed that in the rat, the expression levels of CRFR1 are: brain > thymus > spleen (Baigent and Lowry 2000; Radulovic *et al.* 1999). The localization of CRF to synovial tissue in various forms of arthritis, but not in normal tissue (Murphy *et al.* 2001) combined with the autoradiographic localization of CRF receptors in the macrophage-rich regions of marginal and red pulp zones of the spleen (Grigoriadis *et al.* 1993), suggest a paracrine action of locally produced CRF. An increase in the levels of an antibody to CRFR1 was seen in mouse spleen neutrophils following LPS-induced inflammation (Radulovic *et al.* 1999).

In comparison to CRFR1, the expression of CRFR2 in the rodent displays a more restricted distribution including the olfactory bulb, lateral septal nucleus, bed nucleus of the stria terminalis, ventromedial hypothalamic nucleus, medial and posterior cortical nuclei of the amygdala, ventral hypothalamus, mesencephalic raphe nuclei and areas of the nucleus of the solitary tract and area postrema (Van Pett *et al.* 2000). In the rodent, CRFR2α was found in the CNS CRFR2β was found in non-neuronal tissues like the cerebral arterioles and choroid plexus (Lovenberg *et al.* 1995*b*) and in peripheral tissues, namely heart, lung, skeletal muscle, GI tract, epididymus, and blood vessels (Perrin *et al.* 1995; Kimura *et al.* 2002). Semiquantitative RT-PCR showed that in the rat, comparable levels of CRFR2 are seen in heart and brain with lower levels in the thymus and even lower levels in spleen (Baigent and Lowry 2000). CRFR2a-tr was localized to the lateral septum, amygdala, VMH, and frontal cortex (Miyata *et al.* 2001). Recently, sites of expression of CRFR2 have been shown to coincide with the distribution of urocortin III terminal fields in the rat brain (Li *et al.* 2002).

In the pituitary, CRFR2 was expressed mainly in the posterior lobe (Van Pett *et al.* 2000). In transgenic mice lacking a functional CRFR1, the anterior pituitary displayed expression of CRFR2, but there was no CRF-stimulated ACTH release, so that the expression of the type 2 receptor is probably restricted to blood vessels (Muller *et al.* 2001).

The sites of expression of the type 2 receptors differ between rodent and primate. In the rhesus monkey, the sites of expression CRFR1 and CRFR2 were probed both by binding and *in situ* hybridization. Both receptors were identified in the pituitary and throughout the neocortex (Sanchez *et al.* 1999). The human variants, CRFR2β and CRFR2γ were found to be expressed in the CNS (Kostich *et al.* 1998) whereas the human CRFR2α was found in the periphery (Valdenaire *et al.* 1997).

In normal cycling human ovaries, mRNA's for CRFR1 and CRFR2α were found to be greater in the regressing corpus luteum than in the functional corpus luteum (Muramatsu *et al.* 2001). In another study, CRFR1a was found in human placental trophoblast, amnion/chorion and decidua and CRFR2β was expressed in trophoblast and fetal membranes (Florio *et al.* 2000). Human umbilical vein endothelial cells express CRFR2 (Simoncini *et al.* 1999).

The human retinoblastoma cell line Y-79 expresses CRFR1 (Bonk and Ruhmann 2000) and the human neuronal cell line, SH-SY5Y, expresses both CRFR1 and a small amount of CRFR2 (Schoeffter *et al.* 1999). A mouse neuronal cell line, CATH.a, also expresses CRFR1. The pituitary corticotrope AtT-20 cell line expresses CRFR1a (Vita *et al.* 1993) and a rat aorta-derived cell line, A7R5, expresses CRFR2β (Kageyama *et al.* 2000).

25.4 **Pharmacology**

25.4.1 **Agonists**

Originally, the CRF peptide family included sauvagine, found in one species of frog, urotensins, found in many species of fish and insect diuretic hormones (Lovejoy and Balment 1999). The cloning of orthologous CRF-like peptides in both fish (Okawara *et al.* 1988) and frog (Stenzel-Poore *et al.* 1992*a*) indicated that more than one family member existed in these species and raised the possibility of multiple related peptides in mammals. This stimulated a search for other mammalian CRF family members resulting in the cloning of another mammalian CRF family member, urocortin (Vaughan *et al.* 1995; Donaldson *et al.* 1996). Continuing efforts were rewarded by the cloning of two more family members, Ucn II (Reyes *et al.* 2001) and Ucn III (Lewis *et al.* 2001), also known as stresscopin-related peptide and stresscopin (Hsu and Hsueb 2001). Urocortin is 63 and 45 per cent identical to urotensin and CRF, respectively. Urocortin II is 36 and 40 per cent and urocortin III is 30 and 20 per cent identical to CRF and urocortin, respectively. Both urocortin II and III are expressed centrally and in many peripheral tissues (Reyes *et al.* 2001; Lewis *et al.* 2001; Hsu and Hsueb 2001; Li *et al.* 2002). From an evolutionary viewpoint, CRF appears to form one lineage while urotensin, sauvagine, and the urocortins form another (Lovejoy and Balment 1999).

Radioreceptor assays using the two receptors expressed in different cell types have disclosed striking differences in the receptors' affinity and selectivity for the CRF family ligands. Both types of mammalian CRF receptors bind CRF, urotensin, and sauvagine with similar affinities, but bind urocortin with higher affinity. The type 2 receptor is highly selective for urocortins II and III, binding them with 100–1000 × greater affinity than does the type 1 receptor (Reyes *et al.* 2001; Lewis *et al.* 2001). Indeed, even at a dose of 1 μM Ucn III does not displace labelled sauvagine bound to CRFR1. Compared to CRFR1a, the variants CRFR1b and CRFR1c have reduced affinities for CRF (Ross *et al.* 1994; Xiong *et al.* 1995), whereas the deletion producing CRFR1d does not affect binding (Grammatopoulos *et al.* 1999). The ligand specificities of both chicken and Xenopus CRFR1 differ from those of the mammalian receptor. The affinity of chicken CRFR1 for ovine CRF is lower than for CRF (Yu *et al.* 1996) whereas the mammalian CRFR1 displayed the same affinity for the two analogues; Xenopus CRFR1 bound urotensin with 10 × higher affinity compared to the binding of CRF and sauvagine (Dautzenberg *et al.* 1997).

The CRFR2 splice variants show only moderate differences in binding affinities for different ligands. The human CRFR2α, and CRFR2β displayed the same pharmacological profile for the series of CRF ligands: urocortin > sauvagine > urotensin 1 > r/hCRF > alpha-helical CRF(9–41) > oCRF (Ardati *et al.* 1999) whereas, rodent CRFR2α had ∼2-fold lower binding affinity for urocortins II and III compared to CRFR2β (Lewis *et al.* 2001). The variant rat CRFR2a-tr bound CRF with the same affinity as the full-length CRFR2α but did not bind sauvagine or urocortin (Miyata *et al.* 1999).

The specificity of the catfish CRFR3 is very different from that of the catfish CRFR1 even though the proteins are 85 per cent homologous: the affinity of CRF for cfCRFR3 is 4–5 × greater than that of urotensin or sauvagine, whereas the ligands' affinities are nearly equal for cfCRFR1 (Arai *et al.* 2001).

25.4.2 Antagonists

There are many peptide antagonists such as astressin (Gulyas *et al.* 1995) that are equally potent on both types of receptors. However, there are also a series of small organic non-peptide CRF mimetics (Chen *et al.* 1996; Arvanitis *et al.* 1999) that are specific antagonists for the type 1 receptor (McCarthy *et al.* 1999; Webster *et al.* 1996). These include NB127914, CP154526, antalarmin, and CRA1000. Peptide antagonists that are more potent against the type 2 receptor (Perrin *et al.* 1999; Higelin *et al.* 2001) also exist and include alpha-helical CRF(9–41) and anti-sauvagine-30 (Ruhmann *et al.* 1998).

25.5 Signal transduction and receptor modulation

The signalling pathways activated by CRF receptors have been studied in cells expressing either endogenous or cloned receptors. The receptors are coupled to GTP-binding proteins, which activate either adenylate cyclase or phosholipase C, producing an increase in intracellular cAMP or IP$_3$. There are also data showing activation of pathways involving MAP kinase and for coupling of CRF receptors to calcium channels.

In general, the relative potencies of CRF-family ligands for stimulating cAMP accumulation paralleled the relative affinities in the radioreceptor assay. For example, the splice variant CRFR1b, showed ~3-fold lower potency and lower efficacy in CRF-stimulated cAMP accumulation and no response in total IP$_3$ in transiently transfected cells (Xiong et al. 1995). The reduced signalling was attributed to impaired coupling to G proteins (Nabhan et al. 1995). The variant CRFR1c displayed reduced potency in stimulating cAMP (Ross et al. 1994). The new ligands, urocortins II and III displayed the same high selectivity for stimulating cAMP when tested either on the cloned receptors or on the endogenous CRFR1 expressed in rat anterior pituitary cells and AtT-20 cells and the endogenous CRFR2 expressed in A7R5 cells (Reyes et al. 2001; Lewis et al. 2001).

In some cases, however, there was discordance between the binding affinities and the signalling potencies. The variant receptor CRFR1d showed poor coupling to all the G proteins, reduced potency in activating cAMP and no ability to stimulate IP$_3$ accumulation (Grammatopoulos et al. 1999). The CRFR2a-tr did not transduce any CRF-stimulated cAMP accumulation. In spite of the compromised signalling of CRFR1d and CRFR2α, they may still have physiological relevance since they displayed normal binding affinities and may act as 'binding proteins', thereby modulating CRF effects by altering its effective concentration (Miyata et al. 1999).

Activation by CRF induced Ca^{2+} uptake in AtT-20 cells (Tojo and Abou-Samra 1993) via cAMP-dependent and cAMP-independent mechanisms and also in A431 cells (Kiang et al. 1998; Kiang 1997). In contrast, in rat astrocytes, the increased Ca^{2+} influx was CRF receptor-mediated but did not involve cAMP (Takuma et al. 1994). In rat pancreatic beta cells, which express both types of receptors, application of CRF resulted in depolarization of the cell membrane and enhanced Ca^{2+} currents through L-type channels via activation of PKA signalling pathway (Kanno et al. 1999). In acutely dispersed neurones from the central nucleus of the amygdala, CRF increased dihydropyridine- and neurotoxin-resistant Ca^{2+} currents that were blocked by the CRF receptor antagonist, alpha-helical CRF(9–41) (Yu and Shinnick-Gallagher 1998). In human and rodent melanoma cells, CRF induced a rapid increase in cytosolic Ca^{2+} that was blocked by a CRF antagonist and by the presence of EGTA in the media. The rapid increase (1–3 s) may mean that the Ca^{2+} influx was not mediated through cAMP or IP$_3$ pathways (Fazal et al. 1998).

In another study, CRF-induced neurite outgrowth in immortalized noradrenergic neuronal CATH.a cells was blocked both by inhibitors of PKA and of extracellular signal-regulated protein kinase (ERK), which is a member of the MAP kinase family (Cibelli et al. 2001). In CHO cells stably expressing either CRFR1 or CRFR2α, sauvagine increased activation of p42/p44 MAP kinase independent of cAMP, suggesting that CRF regulates gene transcription by a combination of PKA and MAP kinase (Rossant et al. 1999). In human cells from pregnant myometrium, urocortin but not CRF induced phosphorylation and activation of MAP kinase (Grammatopoulos et al. 2000). Further, the cardioprotective effects of urocortin appeared to involve interaction with the MAP kinase pathway (Brar et al. 2000).

25.6 Physiology and disease relevance

A consequence of the broad distribution of CRF ligands and their receptors is the diverse physiological responses that are mediated by both central and peripheral receptors. The locus ceruleus, paraventricular nucleus of the hypothalamus, the bed nucleus of the stria terminalis, and the central nucleus of the amygdala are some of the areas of the brain that are involved in the central effects of CRF (Koob and Heinrichs 1999). Indeed, the CRF system is a major integrator of behavioural, autonomic, and endocrine responses to stress as underscored by its effects on learning and memory (Croiset *et al.* 2000), appetite and weight control (Heinrichs and Richard 1999). In addition, CRF has effects on affective disorders such as melancholic depression (Arborelius *et al.* 1999; Steckler *et al.* 1999) and on drug dependence as suggested by CRFR1 antagonist attenuation of opiate withdrawal symptoms in rats (Lu *et al.* 2000).

25.6.1 Anxiety

Changes in CRF receptor expression have been measured in response to stressors. Thus, following immobilization stress, an increase in CRFR1 expression levels was seen (Radulovic *et al.* 2000). The antidepressant, amitriptyline decreased CRFR1 gene expression in the amygdala and chronic administration of the anxiolytic benzodiazipine, alprazolam, decreased CRFR1 mRNA in the locus ceruleus (Skelton *et al.* 2000). This might suggest a role for CRFR1 in anxiety states. However, pharmacological manipulation has generated results that suggest the effects of CRFRs on anxiety are more complex. In this respect, icv injections of selective CRFR1 non-peptide antagonists produce anxiolytic and antidepressant effects in rats (Harro *et al.* 2001). In another study antisense nucleotides derived from the sequence of CRFR2, when injected into the lateral septum, reduced fear-induced freezing in rats (Ho *et al.* 2001). Paradoxically, another study using CRFR2-sequence specific antisense nucleotides, infused into the lateral ventricle of the rat, produced anxiogenic effects. However, in this study antisense blockade of CRFR1 confirmed the anxiolytic effects of blocking this receptor subtype (Liebsch *et al.* 1999). Pharmacological approaches to studying CRFR2 involvement in anxiety demonstrate anxiolytic activity of antagonists. Thus, blockade of rat CRFR2 by antisauvagine-30 produced anxiolytic responses in three different anxiety test models (Takahashi *et al.* 2001) and in models of panic behaviour in rats, urocortin was more potent than CRF when injected into the basolateral nucleus of the amygdala (Sajdyk *et al.* 1999).

Experiments using transgenic mice, in which the receptors have been rendered non-functional, complement the studies mentioned above in which receptor systems are intact. In this respect, within the CRF field receptor-deficient mice studies have been performed together with studies using transgenic mice that either lack (Muglia *et al.* 1997; Swiergiel and Dunn 1999; Dunn and Swiergiel 1999) or overexpress CRF itself (Heinrichs *et al.* 1996; Stenzel-Poore *et al.* 1992*b*, 1994; Beckmann *et al.* 2001).

Analysis of transgenic mice in which there was a targeted disruption of CRFR1 demonstrated the absence of a functional receptor in that no CRF-stimulated response in cultured pituitary cells taken from these animals could be evoked (Smith *et al.* 1998). The mice were viable, fertile and had reduced basal corticosterone levels. The basal plasma levels of ACTH were similar to controls (Smith *et al.* 1998; Timpl *et al.* 1998) but given the low levels of corticosterone, the ACTH levels were actually suppressed. The pituitaries contained a normal number of corticotropes but there was significant decrease in the size of the corticosterone-producing zona fasciculata area of the adrenal gland. Progeny of homozygous females died within 48 h because of neonatal respiratory insufficiency as a result of limited maternal corticosterone; the lethality was reversed by *in utero* treatment with corticosterone from embryonic day 12 to post-natal day 14. There was increased expression of CRF and of CRF mRNA in the paraventricular nucleus (PVN) of the hypothalamus but no change in expression of arginine vasopressin in the PVN. There was no change in expression of CRF in other CRF-expressing regions of the brain and no changes in either distribution or levels of CRFR2 expression. When challenged, the CRFR1-deficient mice exhibited reduced ACTH and corticosterone responses to restraint and forced-swim stress (Smith *et al.* 1998; Timpl *et al* 1998). They exhibited reduced sensitivity to well-established anxiogenic situations. Thus, in the light–dark emergence task and the elevated-plus maze, the mutant mice entered the aversive environment of the open areas with shorter latency and spent more time there compared to the controls (Smith *et al.* 1998). A similar increase in exploratory behaviour was observed in an alcohol-withdrawal test (Timpl *et al.* 1998). Icv injection of CRF increased locomotion in wild type but had no effect on homozygous mice (Contarino *et al.* 2000). In summary, CRFR1 appears to play a key, but not exclusive role, in mediating the anxiety-related behavioural and endocrine responses to stress; the reduced stress response of these mice may justify their description as 'mellow mice'.

Targeted disruption of the type 2 receptor produced no obvious differences in pituitary or adrenal structure. The mice had normal basal ACTH and corticosterone levels. In response to restraint stress, disruption of CRFR2 produced no difference in the levels of ACTH or corticosterone, but there was a difference in the temporal profile of the responses: specifically, the mutants displayed an earlier response and a faster decline compared to the wild type (Bale *et al.* 2000; Coste *et al.* 2000; Kishimoto *et al.* 2000). There was increased expression of urocortin in the Edinger-Westphal and of CRF in the central nucleus of the amygdala with no change in CRF expression in the PVN or of CRFR1 in the brain and pituitary (Bale *et al.* 2000). In the elevated-plus maze and the open-field test, the mutants displayed anxiety-like behaviour compared to the controls, but in the light–dark box paradigm their behaviour was like that of the wild type (Bale *et al.* 2000). Two further studies contest some of these findings in that no significant differences between wild-type and mutants were found in the elevated-plus maze and open field test in one (Coste *et al.* 2000) and in the other, mutant mice were found to have increased anxiety-like behaviour in the elevated-plus maze but were indistinguishable from controls in the open-field test (Kishimoto *et al.* 2000). Thus, it appears that the role of CRFR2 in anxiety is complicated and the behavioural responses to various stress paradigms are probably not only species dependent, but also gender and tissue-site dependent. In summary, these data suggest that the CRFR2 and its activating ligands may be involved in mediating anxiogenic behaviours and in modulating the anxiolytic roles of CRFR1.

A logical extension to these findings is to generate an animal with simultaneous disruption of both CRF receptors. In these doubly-mutant mice, there were no differences in the basal ACTH levels compared to control mice, but the basal corticosterone levels were lower in the double mutant compared to CRFR1 mutants (Bale *et al.* 2002; Preil *et al.* 2001). The doubly-mutant mice showed reduced HPA responses to restraint stress compared to controls, with levels of ACTH and corticosterone levels in males lower than those seen in the CRFR1-deficient mice. The adrenal glands of the double mutants and CRFR1 mutants were similarly reduced in size. Interestingly, there appeared to be a sexual dichotomy in the anxiety-like behaviour of the mutant mice (Bale *et al.* 2002). The doubly mutant females displayed decreased levels of anxiety-like behaviours compared to the wild type whereas the corresponding males did not. Another intriguing finding was anxiety-like behaviour in male pups (regardless of their genotype) born to mothers who were either homozygous or heterozygous for mutation of CRFR2. These data suggest a possible interaction between the two types of receptors and a role for CRFR2 in mediating the response of the HPA to stress. Since the type 2 receptor is not expressed in the pituitary, the effect of CRFR2 may be at the level of CNS CRF cell bodies.

25.6.2 Satiety and feeding

The regulation of feeding, appetite, energy balance, and weight maintenance involves complex interactions between CRF and other peptides and proteins such as NPY, MSH, MCH, agouti-related protein, and leptin. Urocortin, urocortin II, and CRF inhibited food intake and weight gain through central actions at the PVN, locus ceruleus, BNST, and the central nucleus of the amygdala (Reyes *et al.* 2001; Koob and Heinrichs 1999; Spina *et al.* 1996). A study in lean mice showed that urocortin was more potent than CRF in decreasing food intake and body weight gain (Asakawa *et al.* 1999). The specific CRFR2 antagonist, antisauvagine-30, blocked the CRF- and Ucn-stimulated inhibition of feeding and weight gain but not the ligands' effects on the metabolism (Cullen *et al.* 2001). In another study with marsupials, a CRFR1-specific non-peptide antagonist was not able to block the peripherally induced CRF inhibition of food intake, suggesting that CRFR2 may mediate the peripheral effects also (Hope *et al.* 2000). However, it was also found in rats that the intraperitoneal injection of a CRFR1-specific antagonist was able to block an emotional stress-induced inhibition of feeding suggesting that CRFR1 may also play a role in this response (Hotta *et al.* 1999).

In another study, icv injection of urocortin resulted in transient decrease in food intake which was mirrored by a loss in body weight in control mice whereas, a similar decrease in food consumption was accompanied by a sustained reduction in body weight in the CRFR1-deficient mouse (Bradbury *et al.* 2000). CRFR2-deficient mice exhibited normal basal feeding and weight gain, but had decreased food ingestion following food deprivation with no change in body weight (Bale *et al.* 2000; Kishimoto *et al.* 2000). Icv injection of urocortin resulted in decreased feeding in both CRFR2 mutant and control mice, but the mutants displayed a lack of continued suppression of feeding compared to the wild type (Coste *et al.* 2000).

25.6.3 Alzheimer's disease and cognition

CRFR1 has been implicated in learning because the enhancement of learning following hippocampal injection of CRF was inhibited by astressin but not by antisauvagine-30 (Radulovic *et al.* 1999). Furthermore, in a test of spatial recognition memory, CRFR1-deficient mice displayed deficiency compared to wild-type controls (Contarino *et al.* 1999). In fear related memory in rodents, the type 1 receptor is believed to mediate memory retrieval but not acquisition or retention (Kikusui *et al.* 2000). A role for CRFR2 in mnemonic processing has also been suggested based on the observation that injection of CRF into the lateral intermediate septum caused an impairment in learning in the rat that was inhibited by antisauvagine-30 (Radulovic *et al.* 1999).

25.6.4 Peripheral indications

Clearly CRF and its related ligands have been localized outside the CNS where they have been found to affect the skin (Slominski *et al.* 2000), gastrointestinal tract (Muramatsu *et al.* 2000) as well as the cardiovascular (Nishikimi *et al.* 2000), reproductive (Muramatsu *et al.*

2001), and immune systems (Bamberger and Bamberger 2000). Thus, for example, CRFR1-deficient mice displayed a considerable ACTH response to turpentine-induced inflammation associated with, and perhaps mediated by, an exaggerated IL-6 response and also with possible direct actions of cytokines on corticotropes (Turnbull *et al.* 1999).

In the gastrointestinal tract, stress produced a decrease in gastric emptying and an increase in colonic motility (Lenz *et al.* 1988; Sheldon *et al.* 1990; Coskun *et al.* 1997). Both central and peripheral sites of action have been implicated and direct effects on the colon have also been observed (Mancinelli *et al.* 1998). These effects are stimulated by CRF and inhibited by specific CRF antagonists, confirming that CRF receptors are involved (Martinez *et al.* 1998; Maillot *et al.* 2000; Tache *et al.* 2001). The type 1 receptor appears to be involved in the lower colonic response and may play a role in irritable bowel syndrome (Tache *et al.* 2001). The CRF- and urocortin-induced gastric emptying were not reversed by a CRFR1-selective non-peptide antagonist supporting the hypothesis that gastric emptying involves type 2 CRF receptors (Nozu *et al.* 1999). In both lean and ob/ob mice, gastric emptying was shown to be closely related to food intake; intraperitoneal injection of urocortin was more potent than CRF at inhibiting both gastric emptying and food intake (Asakawa *et al.* 1999).

In the cardiovascular system, the hypertension that was elicited by central administration of CRF was mediated by CRFR1 whereas the hypotension that resulted from peripheral CRF stimulation was not blocked by a non-peptide CRFR1 specific antagonist but was blocked by alpha-helical CRF(9–41), suggesting that the peripheral effects were mediated by type 2 receptors (Briscoe *et al.* 2000). In another study it was found that the expression of urocortin was higher and that of CRFR2β lower in rat left ventricle hypertrophy compared with the expressions in normal left ventricle (Nishikimi *et al.* 2000). In the isolated heart, Ucn was more potent than CRF in producing vasodilation and positive ionotropic effects, suggesting that the type 2 receptor was the one that was involved (Terui *et al.* 2001). Confirming that there are direct actions of CRF ligands on the heart, urocortin prevented ischaemia/hypoxia induced cell death in neonatal rat cardiomyocytes and protected hearts *ex vivo* from ischaemic damage (Brar *et al.* 2000). Additional studies have revealed that cultured rat cardiomyocytes exhibit a pharmacological profile that is consistent with expression of CRFR2 (Heldwein *et al.* 1996). Assessment of the cardiac functions in the CRFR2-deficient mice showed increased mean arterial and diastolic pressure with a normal resting heart rate. Systemic administration of urocortin did not produce the expected hypotension in the mutant mice (Coste *et al.* 2000). In summary, these data suggest that the CRFR2 and its activating ligands may be involved in maintenance of normal cardiac function.

25.7 Concluding remarks

The cast of characters in the CRF story currently includes four mammalian CRF-related ligands and two receptor genes. All the data support the conclusion that CRFR1 is the pituitary receptor responsible for the HPA response to stress. Centrally, it is this receptor that appears to mediate anxiogenic behavioural responses to some stresses. The other experiments showing effects on learning, memory, sleep, and substance-abuse are consistent with the broad expression of CRFR1 in many brain areas. The responses observed in the CRFR1-deficient mice may have relevance to pathological conditions such as depression, panic attacks, and anorexia nervosa. In the periphery, for example, the skin and gastrointestinal tract, where there is expression of both CRFR1 and CRF family ligands, they may act in a paracrine manner and constitute a local stress-response system. The role of CRFR2 in behavioural responses to stress may be to modulate some of the actions of CRFR1, but the picture is complicated. The role of CRFR2 in the periphery is somewhat clearer: in the heart, CRFR2 plays a role in maintaining blood pressure; in the GI tract, the receptor appears to mediate gastric responses to stress.

When urocortin was cloned, CRFR2 was proposed to be the cognate receptor for Ucn because its affinity for the receptor was greater than that of CRF. Now, the observations that Ucns II and III are expressed in the periphery, that in the brain there is overlap of Ucn III terminal fields with CRFR2 expression and that Ucns II and III are highly selective for CRFR2, support the proposition that CRFR2 is the cognate receptor for the urocortins.

Many questions have now been answered as a result of the success in cloning new ligands and receptors. However, new questions have been raised. For example, what receptors and ligands are responsible for the actions of CRF on the locus ceruleus? Thus far, there is not a complete correspondence between the known receptors and ligands. There remains the question of the interaction between the two receptor subtypes. There was a provocative hint provided by the CRFR2(-/-) transgenic mice in which the HPA axis appeared to be influenced by the absence of a functional type 2 receptor and also from the study on the CRFR1(-/-)/CRFR2(-/-) mutant mice in which, again, the type 2 receptor appeared to play a modulatory role on the actions of the type 1 receptor. An important answer to the question of the interaction of the two signalling pathways will be derived from an analysis of the physiology of a mouse in which all receptors and ligands are functionally absent. The development of CRFR2-specific non-peptide antagonists will also provide a means to study the relative importance of each of the receptors. It remains to be seen if there are other mammalian CRF family members as well as other receptors. Additionally, the other intracellular components that interact with the receptor to modulate its signal transduction await discovery. Finally, we may look forward to a more detailed molecular description of the receptor–ligand interaction when a high-resolution X-ray structure of the ECD-1-ligand complex is obtained.

Is it possible to live without CRF and its receptors? Unlikely. Our complicated environment will continue to demand appropriate metabolic, autonomic, and behavioural responses to stress. The discovery of so many unexpected roles for CRF has led to a greater understanding of normal and pathophysiology. It is exciting to contemplate a brave new world in which knowledge of the CRF system is used to develop pharmaceuticals targeted to the treatment of conditions such as heart disease, obesity, and depression.

References

Adams MD, Celniker SE, Holt RA et al. (2000). The genome sequence of Drosophila melanogaster. *Science* 287, 2185–95.

Arai M, Assil IQ, and Abou-Samra AB (2001). Characterization of three corticotropin-releasing factor receptors in catfish: a novel third receptor is predominantly expressed in pituitary and urophysis. *Endocrinology* 142, 446–54.

Arborelius L, Owens MJ, Plotsky PM, and Nemeroff CB (1999). The role of corticotropin-releasing factor in depression and anxiety disorders. *J Endocrinol* 160, 1–12.

Ardati A, Goetschy V, Gottowick J et al. (1999). Human CRF2 alpha and beta splice variants: pharmacological characterization using radioligand binding and a luciferase gene expression assay. *Neuropharmacology* 38, 441–8.

Arvanitis AG, Gilligan PJ, Chorvat RJ et al. (1999). Non-peptide corticotropin-releasing hormone antagonists: syntheses and structure-activity relationships of 2-anilinopyrimidines and -triazines. J Med Chem 42, 805–18.

Asaba K, Makino S, Nishiyama M, and Hashimoto K (2000). Regulation of type-2 corticotropin-releasing hormone receptor mRNA in rat heart by glucocorticoids and urocortin. J Cardiovasc Pharmacol 36, 493–7.

Asakawa A, Inui A, Ueno N, Makino S, Fujino MA, and Kasuga M (1999). Urocortin reduces food intake and gastric emptying in lean and ob/ob obese mice. Gastroenterology 116, 1287–92.

Aubry JM, Turnbull AV, Pozzoli G, Rivier C, and Vale W (1997). Endotoxin decreases corticotropin-releasing factor receptor 1 messenger ribonucleic acid levels in the rat pituitary. Endocrinology 138, 1621–6.

Baigent SM and Lowry PJ (2000). mRNA expression profiles for corticotrophin-releasing factor (CRF), urocortin, CRF receptors and CRF-binding protein in peripheral rat tissues. J Mol Endocrinol 25, 43–52.

Bale TL, Contarino A, Smith GW et al. (2000). Mice deficient for corticotropin-releasing hormone receptor-2 display anxiety-like behaviour and are hypersensitive to stress. Nat Genet 24, 410–4.

Bale TL, Picetti R, Contarino A, Koob GF, Vale WW, and Lee KF (2002). Mice deficient for both corticotropin-releasing factor receptor 1 (CRFR1) and CRFR2 have an impaired stress response and display sexually dichotomous anxiety-like behavior. J Neurosci 22, 193–9.

Bamberger CM and Bamberger AM (2000). The peripheral CRH/urocortin system. Ann N Y Acad Sci 917, 290–6.

Beckmann N, Gentsch C, Baumann D et al. (2001). Non-invasive, quantitative assessment of the anatomical phenotype of corticotropin-releasing factor-overexpressing mice by MRI. NMR Biomed 14, 210–6.

Bonk I and Ruhmann A (2000). Novel high-affinity photoactivatable antagonists of corticotropin-releasing factor (CRF) photoaffinity labeling studies on CRF receptor, type 1 (CRFR1). Eur J Biochem 267, 3017–24.

Bradbury MJ, McBurnie MI, Denton DA, Lee KF, and Vale WW (2000). Modulation of urocortin-induced hypophagia and weight loss by corticotropin-releasing factor receptor 1 deficiency in mice. Endocrinology 141, 2715–24.

Brar BK, Jonassen AK, Stephanou A et al. (2000). Urocortin protects against ischemic and reperfusion injury via a MAPK-dependent pathway. J Biol Chem 275, 8508–14.

Briscoe RJ, Cabrera CL, Baird TJ, Rice KC, and Woods JH (2000). Antalarmin blockade of corticotropin releasing hormone-induced hypertension in rats. Brain Res 881, 204–7.

Chang C-P, Pearse IRV, O'Connell S, and Rosenfeld MG (1993). Identification of a seven transmembrane helix receptor for corticotropin-releasing factor and sauvagine in mammalian brain. Neuron 11, 1187–95.

Chen C, Dagnino JR, DeSouza EB et al. (1996). Design and synthesis of a series of non-peptide high-affinity human corticotropin-releasing factor1 receptor antagonists,. J Med Chem 39, 4358–60.

Chen R, Lewis KA, Perrin MH, and Vale WW (1993). Expression cloning of a human corticotropin-releasing factor receptor. Proc Natl Acad Sci USA 90, 8967–71.

Cibelli G, Corsi P, Diana G, Vitiello F, and Thiel G (2001). Corticotropin-releasing factor triggers neurite outgrowth of a catecholaminergic immortalized neuron via cAMP and MAP kinase signalling pathways. Eur J Neurosci 13, 1339–48.

Contarino A, Dellu F, Koob GF et al. (1999). Reduced anxiety-like and cognitive performance in mice lacking the corticotropin-releasing factor receptor 1. Brain Res 835, 1–9.

Contarino A, Dellu F, Koob GF et al. (2000). Dissociation of locomotor activation and suppression of food intake induced by CRF in CRFR1-deficient mice. Endocrinology 141, 2698–702.

Coskun T, Bozkurt A, Alican I, Ozkutlu U, Kurtel H, and Yegen BC (1997). Pathways mediating CRF-induced inhibition of gastric emptying in rats. Regul Pept 69, 113–20.

Coste SC, Kesterson RA, Heldwein KA *et al.* (2000). Abnormal adaptations to stress and impaired cardiovascular function in mice lacking corticotropin-releasing hormone receptor-2. *Nat Genet* **24**, 403–9.

Croiset G, Nijsen MJ, and Kamphuis PJ (2000). Role of corticotropin-releasing factor, vasopressin and the autonomic nervous system in learning and memory. *Eur J Pharmacol* **405**, 225–34.

Cullen MJ, Ling N, Foster AC, and Pelleymounter MA (2001). Urocortin, corticotropin releasing factor-2 receptors and energy balance. *Endocrinology* **142**, 992–9.

Dautzenberg FM, Braun S, and Hauger RL (2001). GRK3 mediates desensitization of CRF(1) receptors: a potential mechanism regulating stress adaptation. *Am J Physiol Regul Integr Comp Physiol* **280**, R935–R946.

Dautzenberg FM, Dietrich K, Palchaudhuri MR, and Spiess J (1997). Identification of two corticotropin-releasing factor receptors from Xenopus laevis with high ligand selectivity: unusual pharmacology of the type 1 receptor. *J Neurochem* **69**, 1640–9.

Dautzenberg FM, Huber G, Higelin J, Py-Lang G, and Kilpatrick GJ (2000). Evidence for the abundant expression of arginine 185 containing human CRF(2alpha) receptors and the role of position 185 for receptor-ligand selectivity. *Neuropharmacology* **39**, 1368–76.

Dautzenberg FM, Kilpatrick GJ, Wille S, and Hauger RL (1999). The ligand-selective domains of corticotropin-releasing factor type 1 and type 2 receptor reside in different extracellular domains: generation of chimeric receptors with a novel ligand-selective profile. *J Neurochem* **73**, 821–9.

Dautzenberg FM, Wille S, Lohmann R, and Spiess J (1998). Mapping of the ligand-selective domain of the Xenopus laevis corticotropin-releasing factor receptor 1: implications for the ligand- binding site. *Proc Natl Acad Sci USA* **95**, 4941–6.

Dieterich KD, Grigoriadis DE, and De Souza EB (1996). Homologous desensitization of human corticotropin-releasing factor1 receptor in stable transfected mouse fibroblast cells. *Brain Res* **710**, 287–92.

Donaldson CJ, Sutton SW, Perrin MH *et al.* (1996). Cloning and characterization of human urocortin. *Endocrinology* **137**, 3896.

Dunn AJ and Swiergiel AH (1999). Behavioral responses to stress are intact in CRF-deficient mice. *Brain Res* **845**, 14–20.

Fazal N, Slominski A, Choudhry MA, Wei ET, and Sayeed MM (1998). Effect of CRF and related peptides on calcium signaling in human and rodent melanoma cells. *FEBS Lett* **435**, 187–90.

Florio P, Franchini A, Reis FM, Pezzani I, Ottaviani E, and Petraglia F (2000). Human placenta, chorion, amnion and decidua express different variants of corticotropin-releasing factor receptor messenger RNA. *Placenta* **21**, 32–7.

Grammatopoulos DK, Dai Y, Randeva HS *et al.* (1999). A novel spliced variant of the type 1 corticotropin-releasing hormone receptor with a deletion in the seventh transmembrane domain present in the human pregnant term myometrium and fetal membranes. *Mol Endocrinol* **13**, 2189–202.

Grammatopoulos DK, Randeva HS, Levine MA, Katsanou ES, and Hillhouse EW (2000). Urocortin, but not corticotropin-releasing hormone (CRH), activates the mitogen-activated protein kinase signal transduction pathway in human pregnant myometrium: an effect mediated via R1alpha and R2beta CRH receptor subtypes and stimulation of Gq-proteins. *Mol Endocrinol* **14**, 2076–91.

Grauschopf U, Lilie H, Honold K *et al.* (2000). The N-terminal fragment of human parathyroid hormone receptor 1 constitutes a hormone binding domain and reveals a distinct disulfide pattern. *Biochemistry* **39**, 8878–87.

Grigoriadis DE and De Souza EB (1989). Heterogeneity between brain and pituitary corticotropin-releasing factor receptors is due to differential glycosylation. *Endocrinology* **125**, 1877–88.

Grigoriadis DE, Heroux JA, and De Souza EB (1993). Characterization and regulation of corticotropin-releasing factor receptors in the central nervous, endocrine and immune systems. *Ciba Found Symp* **172**, 85–101.

Gulyas J, Rivier C, Perrin M *et al.* (1995). Potent, structurally constrained agonists and competitive antagonists of corticotropin-releasing factor. *Proc Natl Acad Sci USA* **92**, 10575–9.

Harro J, Tonissaar M, and Eller M (2001). The effects of CRA 1000, a non-peptide antagonist of corticotropin-releasing factor receptor type 1, on adaptive behaviour in the rat. *Neuropeptides* **35**, 100–109.

Hauger RL, Dautzenberg FM, Flaccus A, Liepold T, and Spiess J (1997). Regulation of corticotropin-releasing factor receptor function in human Y-79 retinoblastoma cells: rapid and reversible homologous desensitization but prolonged recovery. *J Neurochem* **68**, 2308–16.

Hauger RL, Smith RD, Braun S, Dautzenberg FM, and Catt KJ (2000). Rapid agonist-induced phosphorylation of the human CRF receptor, type 1: a potential mechanism for homologous desensitization. *Biochem Biophys Res Commun* **268**, 572–6.

Heinrichs SC and Richard D (1999). The role of corticotropin-releasing factor and urocortin in the modulation of ingestive behavior. *Neuropeptides* **33**, 350–9.

Heinrichs SC, Stenzel-Poore MP, Gold LH *et al.* (1996). Learning impairment in transgenic mice with central overexpression of corticotropin-releasing factor. *Neuroscience* **74**, 303–11.

Heldwein KA, Redick DL, Rittenberg MB, Claycomb WC, and Stenzel-Poore MP (1996). Corticotropin-releasing hormone receptor expression and functional coupling in neonatal cardiac myocytes and AT-1 cells. *Endocrinology* **137**, 3631–9.

Higelin J, Py-Lang G, Paternoster C, Ellis GJ, Patel A, and Dautzenberg FM (2001). 125I-Antisauvagine-30: a novel and specific high-affinity radioligand for the characterization of corticotropin-releasing factor type 2 receptors. *Neuropharmacology* **40**, 114–22.

Ho SP, Takahashi LK, Livanov V *et al.* (2001). Attenuation of fear conditioning by antisense inhibition of brain corticotropin releasing factor-2 receptor. *Brain Res Mol Brain Res* **89**, 29–40.

Hope PJ, Turnbull H, Farr S *et al.* (2000). Peripheral administration of CRF and urocortin: effects on food intake and the HPA axis in the marsupial Sminthopsis crassicaudata. *Peptides* **21**, 669–77.

Hotta M, Shibasaki T, Arai K, and Demura H (1999). Corticotropin-releasing factor receptor type 1 mediates emotional stress-induced inhibition of food intake and behavioral changes in rats. *Brain Res* **823**, 221–5.

Hsu SY and Hsueh AJ (2001). Human stresscopin and stresscopin-related peptide are selective ligands for the type 2 corticotropin-releasing hormone receptor. *Nat Med* **7**, 605–11.

Imaki T, Katsumata H, Miyata M, Naruse M, Imaki J, and Minami S (2001). Expression of corticotropin releasing factor (CRF), urocortin and CRF type 1 receptors in hypothalamic-hypophyseal systems under osmotic stimulation. *J Neuroendocrinol* **13**, 328–38.

Iredale PA, Bundey R, and Duman RS (1997*a*). Phorbol ester and calcium regulation of corticotrophin-releasing factor receptor 1 expression in a neuronal cell line. *J Neurochem* **69**, 1912–9.

Iredale PA and Duman RS (1997*b*). Glucocorticoid regulation of corticotropin-releasing factor1 receptor expression in pituitary-derived AtT-20 cells. *Mol Pharmacol* **51**, 794–9.

Iredale PA, Terwilliger R, Widnell KL, Nestler EJ, and Duman RS (1996). Differential regulation of corticotropin-releasing factor1 receptor expression by stress and agonist treatments in brain and cultured cells. *Mol Pharmacol* **50**, 1103–10.

Ishihara T, Nakamura S, Kaziro Y, Takahashi T, Takahashi K, and Nagata S (1991). Molecular cloning and expression of a cDNA encoding the secretin receptor. *EMBO J* **10**, 1635–41.

Jelinek LJ, Lok S, Rosenberg GB *et al.* (1993). Expression cloning and signaling properties of the rat glucagon receptor. *Science* **259**, 1614–16.

Juppner H (1994). Molecular cloning and characterization of a parathyroid hormone/parathyroid hormone-related peptide receptor: a member of an ancient family of G protein-coupled receptors. *Curr Opin Nephrol Hypertens* **3**, 371–8.

Kageyama K, Gaudriault GE, Bradbury MJ, and Vale WW (2000). Regulation of corticotropin-releasing factor receptor type 2 beta messenger ribonucleic acid in the rat cardiovascular system by urocortin, glucocorticoids, and cytokines. *Endocrinology* **141**, 2285–93.

Kanno T, Suga S, Nakano K, Kamimura N, and Wakui M (1999). Corticotropin-releasing factor modulation of Ca2+ influx in rat pancreatic beta-cells. *Diabetes* **48**, 1741–6.

Kiang JG (1997). Corticotropin-releasing factor-like peptides increase cytosolic [Ca2+] in human epidermoid A-431 cells. *Eur J Pharmacol* **329**, 237–44.

Kiang JG, Ding XZ, Gist ID, Jones RR, and Tsokos GC (1998). Corticotropin-releasing factor induces phosphorylation of phospholipase C-gamma at tyrosine residues via its receptor 2beta in human epidermoid A-431 cells. *Eur J Pharmacol* **363**, 203–10.

Kikusui T, Takeuchi Y, and Mori Y (2000). Involvement of corticotropin-releasing factor in the retrieval process of fear-conditioned ultrasonic vocalization in rats. *Physiol Behav* **71**, 323–8.

Kimura Y, Takahashi K, Totsune K *et al.* (2002). Expression of urocortin and corticotropin-releasing factor receptor subtypes in the human heart. *J Clin Endocrinol Metab* **87**, 340–6.

Kishimoto T, Pearse II RV, Lin CR, and Rosenfeld MG (1995). A sauvagine/corticotropin-releasing factor receptor expressed in heart and skeletal muscle. *Proc Natl Acad Sci USA* **92**, 1108–12.

Kishimoto T, Radulovic J, Radulovic M *et al.* (2000). Deletion of crhr2 reveals an anxiolytic role for corticotropin- releasing hormone receptor-2. *Nat Genet* **24**, 415–9.

Koob GF and Heinrichs SC (1999). A role for corticotropin releasing factor and urocortin in behavioral responses to stressors. *Brain Res* **848**, 141–52.

Kostich WA, Chen A, Sperle K, and Largent BL (1998). Molecular identification and analysis of a novel human corticotropin- releasing factor (CRF) receptor: the CRF2gamma receptor. *Mol Endocrinol* **12**, 1077–85.

Lenz HJ, Raedler A, Greten H, Vale WW, and Rivier JE (1988). Stress-induced gastrointestinal secretory and motor responses in rats are mediated by endogenous corticotropin-releasing factor. *Gastroenterology* **95**, 1510–7.

Lewis K, Li C, Perrin MH *et al.* (2001). Identification of urocortin III, an additional member of the corticotropin-releasing factor (CRF) family with high affinity for the CRF2 receptor. *Proc Natl Acad Sci USA* **98**, 7570–5.

Li C, Vaughan, J, Sawchenko PE, and Vale WW (2002). Urocortin III-immunoreactive projections in rat brain: Partial overlap with sites of type 2 corticotropin-releasing factor receptor expression. *J Neuroscience* **22**, 991–1001.

Liaw CW, Grigoriadis DE, Lorang MT, DeSouza EB, and Maki RA (1997a). Localization of agonist-and antagonist-binding domains of human corticotropin-releasing factor receptors. *Mol Endocrinology* **11**, 2048–53.

Liaw CW, Grigoriadis DE, Lovenberg TW, De Souza EB, and Maki RA (1997b). Localization of ligand-binding domains of human corticotropin-releasing factor receptor: a chimeric receptor approach. *Mol Endocrinol* **11**, 980–5.

Liaw CW, Lovenberg TW, Oltersdorf T, Grigoriadis DE, and De Souza EB (1996). Cloning and characterization of the human corticotropin-releasing factor-2 receptor complementary deoxyribonucleic acid. *Endocrinology* **137**, 72–7.

Liebsch G, Landgraf R, Engelmann M, Lorscher P, and Holsboer F (1999). Differential behavioural effects of chronic infusion of CRH 1 and CRH 2 receptor antisense oligonucleotides into the rat brain. *J Psychiatr Res* **33**, 153–63.

Lovejoy DA and Balment RJ (1999). Evolution and physiology of the corticotropin-releasing factor (CRF) family of neuropeptides in vertebrates. *Gen Comp Endocrinol* **115**, 1–22.

Lovenberg TW, Chalmers DT, Liu C, and De Souza EB (1995a). CRF2 alpha and CRF2 beta receptor mRNAs are differentially distributed between the rat central nervous system and peripheral tissues. *Endocrinology* **136**, 4139–42.

Lovenberg TW, Liaw CW, Grigoriadis DE *et al.* (1995b). Cloning and characterization of a functionally distinct corticotropin-releasing factor receptor subtype from rat brain. *Proc Natl Acad Sci USA* **92**, 836–40.

Lu L, Liu D, Ceng X, and Ma L (2000). Differential roles of corticotropin-releasing factor receptor subtypes 1 and 2 in opiate withdrawal and in relapse to opiate dependence. *Eur J Neurosci* 12, 4398–404.

Maillot C, Million M, Wei JY, Gauthier A, and Tache Y (2000). Peripheral corticotropin-releasing factor and stress-stimulated colonic motor activity involve type 1 receptor in rats. *Gastroenterology* 119, 1569–79.

Mancinelli R, Azzena GB, Diana M, Forgione A, and Fratta W (1998). *In vitro* excitatory actions of corticotropin-releasing factor on rat colonic motility. *J Auton Pharmacol* 18, 319–24.

Mansi JA, Rivest S, and Drolet G (1996). Regulation of corticotropin-releasing factor type 1 (CRF1) receptor messenger ribonucleic acid in the paraventricular nucleus of rat hypothalamus by exogenous CRF. *Endocrinology* 137, 4619–29.

Martinez V, Barquist E, Rivier J, and Tache Y (1998). Central CRF inhibits gastric emptying of a nutrient solid meal in rats: the role of CRF2 receptors. *Am J Physiol* 274, G965–70.

Mayo KE (1992). Molecular cloning and expression of a pituitary specific receptor for growth hormone releasing hormone. *Molec Endocrinol* 6, 1734–44.

McCarthy JR, Heinrichs SC, and Grigoriadis DE (1999). Recent advances with the CRF1 receptor: design of small molecule inhibitors, receptor subtypes and clinical indications. *Curr Pharm Des* 5, 289–315.

Meyer AH, Ullmer C, Schmuck K *et al.* (1997). Localization of the human CRF2 receptor to 7p21–p15 by radiation hybrid mapping and FISH analysis. *Genomics* 40, 189–90.

Miyata I, Shiota C, Chaki S, Okuyama S, and Inagami T (2001). Localization and characterization of a short isoform of the corticotropin-releasing factor receptor type 2alpha (CRF(2)alpha-tr) in the rat brain. *Biochem Biophys Res Commun* 280, 553–7.

Miyata I, Shiota C, Ikeda Y *et al.* (1999). Cloning and characterization of a short variant of the corticotropin-releasing factor receptor subtype from rat amygdala. *Biochem Biophys Res Commun* 256, 692–6.

Muglia LJ, Jacobson L, Weninger SC *et al.* (1997). Impaired diurnal adrenal rhythmicity restored by constant infusion of corticotropin-releasing hormone in corticotropin-releasing hormone-deficient mice. *J Clin Invest* 99, 2923–9.

Muller MB, Preil J, Renner U *et al.* (2001). Expression of CRHR1 and CRHR2 in mouse pituitary and adrenal gland: implications for HPA system regulation. *Endocrinology* 142, 4150–3.

Muramatsu Y, Fukushima K, Iino K *et al.* (2000). Urocortin and corticotropin-releasing factor receptor expression in the human colonic mucosa. *Peptides* 21, 1799–809.

Muramatsu Y, Sugino N, Suzuki T *et al.* (2001). Urocortin and corticotropin-releasing factor receptor expression in normal cycling human ovaries. *J Clin Endocrinol Metab* 86, 1362–9.

Murphy EP, McEvoy A, Conneely OM, Bresnihan B, and FitzGerald O (2001). Involvement of the nuclear orphan receptor NURR1 in the regulation of corticotropin-releasing hormone expression and actions in human inflammatory arthritis. *Arthritis Rheum* 44, 782–93.

Myers DA, Trinh JV, and Myers TR (1998). Structure and function of the ovine type 1 corticotropin releasing factor receptor (CRF1) and a carboxyl-terminal variant. *Mol Cell Endocrinol* 144, 21–35.

Nabhan C, Xiong Y, Xie LY, and Abou-Samra AB (1995). The alternatively spliced type II corticotropin-releasing factor receptor, stably expressed in LLCPK-1 cells, is not well coupled to the G protein(s). *Biochem Biophys Res Commun* 212, 1015–21.

Nielsen SM, Nielsen LZ, Hjorth SA, Perrin MH, and Vale WW (2000). Constitutive activation of tethered-peptide/corticotropin-releasing factor receptor chimeras. *Proc Natl Acad Sci USA* 97, 10 277–81.

Nishikimi T, Miyata A, Horio T *et al.* (2000). Urocortin, a member of the corticotropin-releasing factor family, in normal and diseased heart. *Am J Physiol Heart Circ Physiol* 279, H3031–9.

Nishimura E, Billestrup N, Perrin M, and Vale W (1987). Identification and characterization of a pituitary corticotropin releasing factor binding protein by chemical cross-linking. *J Biol Chem* **262**, 12 893–6.

Nozu T, Martinez V, Rivier J, and Tache Y (1999). Peripheral urocortin delays gastric emptying: role of CRF receptor 2. *Am J Physiol* **276**, G867–74.

Okawara Y, Morley SD, Burzio LO, Zwiers H, Lederis K, and Richter D (1988). Cloning and sequence analysis of cDNA for corticotropin-releasing factor precursor from the teleost fish Catostomus commersoni. *Proc Natl Acad Sci USA* **85**, 8439–43.

Orth DN (1992). Corticotropin-releasing hormone in humans. *Endocr Rev* **13**, 164–91.

Palchaudhuri MR, Hauger RL, Wille S, Fuchs E, and Dautzenberg FM (1999). Isolation and pharmacological characterization of two functional splice variants of corticotropin-releasing factor type 2 receptor from Tupaia belangeri. *J Neuroendocrinol* **11**, 419–28.

Palchaudhuri MR, Wille S, Mevenkamp G, Spiess J, Fuchs E, and Dautzenberg FM (1998). Corticotropin-releasing factor receptor type 1 from Tupaia belangeri—cloning, functional expression and tissue distribution. *Eur J Biochem* **258**, 78–84.

Perrin M, Donaldson C, Chen R *et al.* (1995). Identification of a second corticotropin-releasing factor receptor gene and characterization of a cDNA expressed in heart. *Proc Natl Acad Sci USA* **92**, 2969–73.

Perrin MH, Donaldson CJ, Chen R, Lewis KA, and Vale WW (1993). Cloning and functional expression of a rat brain corticotropin releasing factor (CRF) receptor. *Endocrinology* **133**, 3058–61.

Perrin MH, Fischer WH, Kunitake KS *et al.* (2001). Expression, purification, and characterization of a soluble form of the first extracellular domain of the human type 1 corticotropin releasing factor receptor. *J Biol Chem* **276**, 31 528–34.

Perrin MH, Sutton S, Bain DB, Berggren TW, and Vale WW (1998). The first extracellular domain of corticotropin releasing factor-R1 contains major binding determinants for urocortin and astressin. *Endocrinology* **139**, 566–70.

Perrin MH, Sutton SW, Cervini LA, Rivier JE, and Vale WW (1999). Comparison of an agonist, urocortin, and an antagonist, astressin, as radioligands for characterization of corticotropin-releasing factor receptors. *J Pharmacol Exp Ther* **288**, 729–34.

Pohl S, Darlison MG, Clarke WC, Lederis K, and Richter D (2001). Cloning and functional pharmacology of two corticotropin-releasing factor receptors from a teleost fish. *Eur J Pharmacol* **430**, 193–202.

Polymeropoulos MH, Torres R, Yanovski JA, Chandrasekharappa SC, and Ledbetter DH (1995). The human corticotropin-releasing factor receptor (CRHR) gene maps to chromosome 17q12–q22. *Genomics* **28**, 123–4.

Potter E, Sutton S, Donaldson C *et al.* (1994). Distribution of corticotropin-releasing factor receptor mRNA expression in the rat brain and pituitary. *Proc Natl Acad Sci USA* **91**, 8777–81.

Pozzoli G, Bilezikjian LM, Perrin MH, Blount AL, and Vale WW (1996). Corticotropin-releasing factor (CRF) and glucocorticoids modulate the expression of type 1 CRF receptor messenger ribonucleic acid in rat anterior pituitary cell cultures. *Endocrinology* **137**, 65–71.

Preil J, Muller MB, Gesing A *et al.* (2001). Regulation of the hypothalamic-pituitary-adrenocortical system in mice deficient for CRH receptors 1 and 2. *Endocrinology* **142**, 4946–55.

Qi LJ, Leung AT, Xiong Y, Marx KA, and Abou-Samra AB (1997). Extracellular cysteines of the corticotropin-releasing factor receptor are critical for ligand interaction. *Biochemistry* **36**, 12 442–8.

Radulovic J, Ruhmann A, Liepold T, and Spiess J (1999). Modulation of learning and anxiety by corticotropin-releasing factor (CRF) and stress: differential roles of CRF receptors 1 and 2. *J Neurosci* **19**, 5016–25.

Radulovic J, Sydow S, and Spiess J (1998). Characterization of native corticotropin-releasing factor receptor type 1 (CRFR1) in the rat and mouse central nervous system. *J Neurosci Res* **54**, 507–21.

Radulovic M, Dautzenberg FM, Sydow S, Radulovic J, and Spiess J (1999). Corticotropin-releasing factor receptor 1 in mouse spleen: expression after immune stimulation and identification of receptor-bearing cells. *J Immunol* **162**, 3013–21.

Radulovic M, Weber C, and Spiess J (2000). The effect of acute immobilization stress on the abundance of corticotropin-releasing factor receptor in lymphoid organs. *J Neuroimmunol* **103**, 153–64.

Reagan JD (1994). Expression cloning of an insect diuretic hormone receptor. A member of the calcitonin/secretin receptor family. *J Biol Chem* **269**, 9–12.

Reyes TM, Lewis K, Perrin MH *et al.* (2001). Urocortin II: A member of the corticotropin-releasing factor (CRF) neuropeptide family that is selectively bound by type 2 CRF receptors. *Proc Natl Acad Sci USA* **98**, 2843–48.

Ross PC, Kostas CM, and Ramabhadran TV (1994). A variant of the human corticotropin-releasing factor (CRF) receptor: cloning, expression and pharmacology. *Biochem Biophys Res Commun* **205**, 1836–42.

Rossant CJ, Pinnock RD, Hughes J, Hall MD, and McNulty S (1999). Corticotropin-releasing factor type 1 and type 2alpha receptors regulate phosphorylation of calcium/cyclic adenosine $3',5'$-monophosphate response element-binding protein and activation of p42/p44 mitogen-activated protein kinase. *Endocrinology* **140**, 1525–36.

Ruhmann A, Bonk I, Lin CR, Rosenfeld MG, and Spiess J (1998). Structural requirements for peptidic antagonists of the corticotropin-releasing factor receptor (CRFR): development of CRFR2beta-selective antisauvagine-30. *Proc Natl Acad Sci USA* **95**, 15 264–9.

Sajdyk TJ, Schober DA, Gehlert DR, and Shekhar A (1999). Role of corticotropin-releasing factor and urocortin within the basolateral amygdala of rats in anxiety and panic responses. *Behav Brain Res* **100**, 207–15.

Sakai K, Nagafuchi S, Horiba N, Yamada M, and Suda T (1998). The genomic organization of the human corticotropin-releasing factor type-1 receptor.(Abst. # P3-571). Endocrine Soc Meeting, New Orleans.

Sanchez MM, Young LJ, Plotsky PM, and Insel TR (1999). Autoradiographic and in situ hybridization localization of corticotropin-releasing factor 1 and 2 receptors in nonhuman primate brain. *J Comp Neurol* **408**, 365–77.

Schoeffter P, Feuerbach D, Bobirnac I, Gazi L, and Longato R (1999). Functional, endogenously expressed corticotropin-releasing factor receptor type 1 (CRF1) and CRF1 receptor mRNA expression in human neuroblastoma SH-SY5Y cells. *Fundam Clin Pharmacol* **13**, 484–9.

Sheldon RJ, Qi JA, Porreca F, and Fisher LA (1990). Gastrointestinal motor effects of corticotropin-releasing factor in mice. *Regul Pept* **28**, 137–51.

Simoncini T, Apa R, Reis FM *et al.* (1999). Human umbilical vein endothelial cells: a new source and potential target for corticotropin-releasing factor. *J Clin Endocrinol Metab* **84**, 2802–6.

Skelton KH, Nemeroff CB, Knight DL, and Owens MJ (2000). Chronic administration of the triazolobenzodiazepine alprazolam produces opposite effects on corticotropin-releasing factor and urocortin neuronal systems. *J Neurosci* **20**, 1240–8.

Slominski A, Wortsman J, Luger T, Paus R, and Solomon S (2000). Corticotropin releasing hormone and proopiomelanocortin involvement in the cutaneous response to stress. *Physiol Rev* **80**, 979–1020.

Smith GW, Aubry J-M, Dellu F *et al.* (1998). Corticotropin releasing factor receptor 1 deficient mice display decreased anxiety, an impaired stress response and aberrant development of the neuroendocrine system. *Neuron* **20**, 1–20.

Spina M, Merlo-Pich E, Chan RK *et al.* (1996). Appetite-suppressing effects of urocortin, a CRF-related neuropeptide. *Science* **273**, 1561–4.

Sreedharan S, Robichon A, Peterson K, and Goetzl E (1991). Cloning and expression of the human vasoactive intestinal peptide receptor. *Proc Nat Acad Sci USA* **88**, 4986–990.

Steckler T, Holsboer F, and Reul JM (1999). Glucocorticoids and depression. *Baillieres Best Pract Res Clin Endocrinol Metab* **13**, 597–614.

Stenzel P, Kesterson R, Yeung W, Cone RD, Rittenberg MB, and Stenzel-Poore MP (1995). Identification of a novel murine receptor for corticotropin-releasing hormone expressed in the heart. *Mol Endocrinol* 9, 637–45.

Stenzel-Poore MP, Heldwein KA, Stenzel P, Lee S, and Vale WW (1992*a*). Characterization of the genomic corticotropin-releasing factor (CRF) gene from Xenopus laevis: two members of the CRF family exist in amphibians. *Mol Endocrinol* 6, 1716–24.

Stenzel-Poore MP, Cameron VA, Vaughan J, Sawchenko PE, and Vale W (1992*b*). Development of Cushing's syndrome in corticotropin-releasing factor transgenic mice.

Stenzel-Poore MP, Heinrichs SC, Rivest S, Koob GF, and Vale WW (1994). Overproduction of corticotropin-releasing factor in transgenic mice: a genetic model of anxiogenic behavior. *J Neurosci* 14, 2579–84.

Swiergiel AH and Dunn AJ (1999). CRF-deficient mice respond like wild-type mice to hypophagic stimuli. *Pharmacol Biochem Behav* 64, 59–64.

Sydow S, Flaccus A, Fischer A, and Spiess J (1999). The role of the fourth extracellular domain of the rat corticotropin- releasing factor receptor type 1 in ligand binding. *Eur J Biochem* 259, 55–62.

Tache Y, Martinez V, Million M, and Wang L (2001). Stress and the gastrointestinal tract III. Stress-related alterations of gut motor function: role of brain corticotropin-releasing factor receptors. *Am J Physiol Gastrointest Liver Physiol* 280, G173–7.

Takahashi LK, Ho SP, Livanov V, Graciani N, and Arneric SP (2001). Antagonism of CRF(2) receptors produces anxiolytic behavior in animal models of anxiety. *Brain Res* 902, 135–42.

Takuma K, Matsuda T, Yoshikawa T *et al.* (1994). Corticotropin-releasing factor stimulates Ca2+ influx in cultured rat astrocytes. *Biochem Biophys Res Commun* 199, 1103–7.

Terui K, Higashiyama A, Horiba N, Furukawa KI, Motomura S, and Suda T (2001). Coronary vasodilation and positive inotropism by urocortin in the isolated rat heart. *J Endocrinol* 169, 177–83.

Timpl P, Spanagel R, Sillaber I *et al.* (1998). Impaired stress response and reduced anxiety in mice lacking a functional corticotropin-releasing hormone receptor 1. *Nat Genet* 19, 162–6.

Tojo K and Abou-Samra AB (1993). Corticotropin-releasing factor (CRF) stimulates 45Ca2+ uptake in the mouse corticotroph cell line AtT-20. *Life Sci* 52, 621–30.

Tsai-Morris CH, Buczko E, Geng Y, Gamboa-Pinto A, and Dufau ML (1996). The genomic structure of the rat corticotropin releasing factor receptor. *J Biol Chem* 271, 14 519–25.

Turnbull AV, Smith GW, Lee S, Vale WW, Lee KF, and Rivier C (1999). CRF type I receptor-deficient mice exhibit a pronounced pituitary–adrenal response to local inflammation. *Endocrinology* 140, 1013–7.

Valdenaire O, Giller T, Breu V, Gottowik J, and Kilpatrick G (1997). A new functional isoform of the human CRF2 receptor for corticotropin-releasing factor. *Biochim Biophys Acta* 1352, 129–32.

Vale W, Spiess J, Rivier C, and Rivier J (1981). Characterization of a 41-residue ovine hypothalamic peptide that stimulates secretion of corticotropin and b-endorphin. *Science* 213, 1394–7.

Van Pett K, Viau V, Bittencourt JC *et al.* (2000). Distribution of mRNAs encoding CRF receptors in brain and pituitary of rat and mouse. *J Comp Neurol* 428, 191–212.

Vaughan J, Donaldson C, Bittencourt J *et al.* (1995). Urocortin, a mammalian neuropeptide related to fish urotensin I and to corticotropin-releasing factor. *Nature* 378, 287–92.

Vita N, Laurent P, Lefort S *et al.* (1993). Primary structure and functional expression of mouse pituitary and human brain corticotrophin releasing factor receptors. *FEBS Lett* 335, 1–5.

Wang W, Ross GM, Riopelle RJ, and Dow KE (2000). Sublethal hypoxia up-regulates corticotropin releasing factor receptor type 1 in fetal hippocampal neurons. *Neuroreport* 11, 3123–6.

Webster WL, Lewis DB, Torpy DJ, Zachman EK, Rice KC, and Chrousos GP (1996). *In vivo* and *in vitro* characterization of antalarmin, a nonpeptide corticotropin-releasing hormone (CRH) receptor antagonist: suppression of pituitary ACTH release and peripheral inflammation. *Endocrinology* 137, 5747–50.

Wilmen A, Goke B, and Goke R (1996). The isolated N-terminal extracellular domain of the glucagon-like peptide-1 (GLP)-1 receptor has intrinsic binding activity. *FEBS Lett* **398**, 43–7.

Xiong Y, Xie LY, and Abou-Samra AB (1995). Signaling properties of mouse and human corticotropin-releasing factor (CRF) receptors: decreased coupling efficiency of human type II CRF receptor. *Endocrinology* **136**, 1828–34.

Yu B and Shinnick-Gallagher P (1998). Corticotropin-releasing factor increases dihydropyridine- and neurotoxin-resistant calcium currents in neurons of the central amygdala. *J Pharmacol Exp Ther* **284**, 170–9.

Yu J, Xie LY, and Abou-Samra A-B (1996). Molecular cloning of a type A chicken corticotropin-releasing factor receptor with high affinity for urotensin I. *Endocrinology* **137**, 192–7.

Chapter 26

VIP and PACAP receptors

Seiji Shioda and James A. Waschek

26.1 Introduction

Vasoactive intestinal peptide (VIP) and pituitary adenylate cyclase-activating polypeptide (PACAP), like other neuropeptides in this family (glucagon, glucagon-like peptide, secretin, and growth hormone-releasing hormone), act on heterotrimeric G protein coupled receptors. VIP and PACAP receptors have been identified in numerous tissues, and three different types have been discerned. The official names of these were recently established by a recent IUPHAR nomenclature committee (Harmar *et al.* 1998, see Table 26.1). Two of these, the VPAC$_1$ (VPAC1-R) and VPAC$_2$ (VPAC2-R) receptors, bind PACAP38, PACAP27, and VIP with similar high affinity. The third type, the PAC$_1$ receptor (PAC1-R), is a PACAP-specific receptor. This binds PACAP38 and PACAP27 with high affinity, and interacts with VIP at only high VIP concentrations. The PAC1-R is also unique because it exists in at least ten different isoforms derived by alternate mRNA splicing. These splice variants regulate receptor ligand affinity, specificity and coupling to adenylate cyclase (AC) and/or phospholipase C (PLC) pathways (Spengler *et al.* 1993; Harmar *et al.* 1998). The PAC1-R, VPAC1-R, and VPAC2-R have been identified in the nervous system, endocrine glands (pituitary, thyroid, gonads, and adrenal), the gastrointestinal tract, liver, pancreas, respiratory system, cardiovascular system, immune system, bones, and tumour cells. Although a very large number of functions have been described, it is likely that new roles for PACAP and VIP will continue to be discovered.

26.2 Molecular characterization

The VPAC1-R and VPAC2-R were referred to by several names (Table 26.1) until an official nomenclature was established (Harmar *et al.* 1998). The VPAC1-R was identified from rat lung (Ishihara *et al.* 1992) and binds both VIP and PACAP with a similar high affinity. The VPAC2-R was first cloned from rat olfactory bulb (Lutz *et al.* 1993) and later published independently by Usdin *et al.* (1994). Similar to VPAC1-R, VPAC2-R binds VIP and PACAP with equal high affinity. While the N-terminal domains of the secretin/VIP receptor family are highly divergent, they have substantial similarity at specific amino acid positions, including six cysteine residues postulated to confer a general conformation to the extracellular ligand-binding domains of these receptors (Mayo 1992, see Chapter 1).

The PAC1-R has greater similarity to the VPAC1-R (51 per cent) than to other members of the family (Ishihara *et al.* 1992; Christophe 1993). The PAC1-R is also similar to the receptors for calcitonin (31 per cent), parathyroid hormone (37 per cent), glucagon (38 per cent), glucagon-like peptide (37 per cent), GHRH (41 per cent), and secretin (47 per cent). Ten

Table 26.1 VIP/PACAP receptors

	VPAC₁	VPAC₂	PAC₁
Alternative names	VIP, VIP₁, VIP/PAC₁, PVR₂, VIP₁/PACAP₂, PACAP-II	VIP₂, VIP/PAC₂, PVR3, VIP₂/PACAP₃, PACAP-3	PAC₁, PVR₁
Structural information (Accession no)	h 457 aa (L13288) r 459 aa (M86835) m 150 aa (P97751)	h 438 aa (L36566) r 437 aa (Z25885) m 437 aa (D28132)	h 468 aa (D17516)[AS] r 523 aa (L16680)[AS] m 496 aa (D82935)[AS]
Chromosomal location	3p22	7q36.3	7p14
Selective agonists	[Lys¹⁵, Arg¹⁶, Leu²⁷]VIP(1–7)GRF(8–27)-NH₂	Ro251553; Ro251392	Maxadilan
Selective antagonists	[Ac-His¹, D-Phe², Lys¹⁵, Arg¹⁶]VIP(3–7)GRF(8–27)-NH₂	PG9946	M65 (maxadilan derivative)
Radioligands	¹²⁵I-VIP	¹²⁵I-VIP	¹²⁵I-PACAP-27
G protein coupling	G_s	G_s	G_s
Expression profile	Cerebral cortex, hippocampus, olfactory bulb, cerebellum, pineal gland	Suprachiasmatic nucleus, amygdala, thalamus, olfactory bulb, hippocampus, dentate gyrus, spinal cord, DRG, pancreas, skeletal muscle, heart, kidney, adipose tissue, testis, stomach	Widespread throughout the brain, particularly cerebral cortex, hippocampus, olfactory bulb, area postrema, thalamus, pineal gland, cerebellum, retina
Physiological function	Regulation of circadian rhythms, reaction to nerve injury, pain neurotransmission, immunomodulation	Regulation of circadian rhythms, reaction to nerve injury, pain neurotransmission, immunomodulation	Regulation of circadian rhythms, reaction to nerve injury, pain neurotransmission, immunomodulation, contextual fear conditioning
Knockout phenotype	Altered circadian rhythms	—	Altered circadian rhythms
Disease relevance	No clear subtype specific role but potentially useful for sleep disorders, neurodegeneration, multiple sclerosis, stroke, pain, epilepsy, schizophrenia, Alzheimer's disease, depression, inflammatory diseases		

subtypes of the PAC1-R that result from alternative splicing have been described. Among the subtypes, six differ from one another by the absence or presence of two cassettes named 'hip' and 'hop', inserted at the presumed end of the third intracellular loop of the receptor (Spengler *et al.* 1993). The resulting variants have been named PAC1-R-s (short receptor without either cassette), PAC1-R-hip, PAC1-R-hop1, PAC1-R-hop2, PAC1-R-hiphop1, and PAC1-R-hiphop2 (Fig. 26.1). Both the short form and the hop variants potently activate AC and PLC in transfected cells, whereas the hip variant does not couple to PLC. The hip-hop variants display an intermediate signal transduction pattern with an altered ability to activate PLC. More recently, splice variants at the N-terminal extracellular domain of the PAC1-R have been cloned. The first of these reported was characterized by a 21 amino acid deletion in the N-terminal extracellular domain (PAC1-R-vs1; Pantaloni *et al.* 1996). In PLC functional assays, the potency of PACAP38 was approximately 10-fold greater than that of PACAP27 in cells expressing PAC1-R-s, whereas PACAP38 was only 3-fold more potent than PACAP27 in PAC1-R-vs expressing cells (Fig. 26.1). In contrast, the PKA stimulating potencies of PACAP38 and PACAP27 were similar in cells expressing either receptor (Pantaloni *et al.* 1996). The extracellular 21 amino acid domain may thus modulate receptor activity to PACAPs in regard to stimulation of the PLC pathway. Dautzenberg *et al.* (1999) reported the existence of an additional variant with a 57 amino acid deletion in the N-terminal extra-cellular domain (PAC1-R-vs2; Fig. 26.1). The pharmacology of this variant was compared

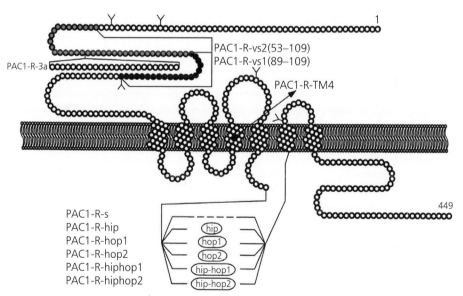

Fig. 26.1 Schematic illustration of the structure of the PAC1-receptor, which is specific for PACAP. Ten variants of the PAC1-R that result from alternative splicing have been identified. Six of them differ from one another by the presence or absence of the hip, hop1, or hop2 cassette insertion at the C-terminal region of the third intracellular domain. The subtypes PAC1-R-vs1 and PAC1-R-vs2 lack 21 or 57 amino acids respectively in the N-terminal extracellular domain. The subtype PAC1-R-3a inserts 24 amino acids in the N-terminal extracellular domain. The subtype PAC1-R-TM4 differs from the other PAC1-R subtypes by two amino acid substitutions and deletions in the 4th transmembrane domain, as shown by the arrow.

with the PAC1-R and 21 amino acid deleted form, PAC1-Rvs. As expected, PAC1-R preferentially bound PACAP38 and PACAP27 over VIP with affinities in the low nanomolar range. Surprisingly however, PAC1-Rvs bound all three ligands with high affinity and coupled efficiently to cAMP production (Dautzenberg et al. 1999). This suggests that PAC1-R-vs may function like VPAC1-R or VPAC2-R. In contrast, PAC1-R-vs2 bound all three ligands relatively poorly, although PACAP38 and PACAP27 exhibited lower IC_{50} than VIP. Yet another splice form in the N-terminal extracellular domain was recently characterized (Daniel et al. 2001). This variant PAC1-R(3a) contains a 24 amino acid insertion (Fig. 26.1) and was found to confer an increase in binding affinity to PACAP38 compared to PACAP27, but resulted in less potent activation of PKA and PLC pathways by both peptides. PAC1-R-TM4 differs to PAC1-R in discrete sequences within TM domains II and IV (Chatterjee et al. 1996; Fig. 26.1). Unlike the other PAC1-R variants, PAC1-R-TM4 activated neither AC nor PLC in response to PACAP in transient and stable expression systems. PACAP stimulated an increase in $[Ca^{2+}]_i$ in PAC1-R-TM4 expressing cells by activating L-type Ca^{2+} channels, a response not elicited by stimulation with VIP. mRNA co-expression of PAC1-R-TM4 and other PAC1-R splice variants has been observed in cerebellum, cerebral cortex, and brainstem. In contrast, the mRNA for PAC1-R-TM4 was not detected in the spinal cord where transcripts encoding other splice variants of PAC1-R were clearly expressed (Chatterjee et al. 1996).

26.3 Cellular and subcellular localization

In brain, the topographical distribution pattern of VIP and PACAP binding sites is quite different in the hippocampal formation, lateral hypothalamic area, substantia nigra pars compacta, pontine reticular nucleus and cerebellum (Masuo et al. 1992). Studies looking at VPAC1-R and VPAC2-R subtypes have largely been carried out at the mRNA level and have also revealed striking differences in expression profiles. VPAC1-R mRNA is widely distributed in the CNS, most abundantly in the cerebral cortex, hippocampus, and mitral cell layer of the olfactory bulb, while lower levels are present in some subcortical areas and in the cerebellar cortex (Ishihara et al. 1992). In contrast, VPAC2-R mRNA is expressed in the suprachiasmatic nucleus, the central nucleus of the amygdala and the thalamus and at low levels in the olfactory bulb and in the CA1-3 of the hippocampus and dentate gyrus, brainstem, spinal cord and, dorsal root ganglion (Usdin et al. 1994; Sheward et al. 1995). VCAP2-R mRNA is also expressed in a number of peripheral tissues including pancreas, skeletal muscle, heart, kidney, adipose tissue, testis, and stomach (Usdin et al. 1994; Adamou et al. 1995; Krempels et al. 1995; Wei and Mojsov 1996).

The distribution of the PAC1-R mRNA has been studied in the adult rat brain using in situ hybridization and immunohistochemistry (Shioda et al. 1994, 1997a; Hashimoto et al. 1996). PAC1-R mRNA is expressed in almost all major regions of the brain and at particularly high levels in parts of the rat cerebral cortex, hippocampal region, olfactory bulb, area postrema and supraoptic nuclei (Hashimoto et al. 1996; Shioda et al. 1997a; Zhou et al. 1999, 2000b). At the cellular level, PAC1-R mRNA is highly expressed in granule cells of the cerebellar cortex (Hashimoto et al. 1996; Shioda et al. 1997a) and in arginine-vasopressin (AVP)-containing, but not oxytocin-containing, neurones of the hypothalamic supraoptic nucleus (Shioda et al. 1997b). Similar to findings in the rat, PAC1-R gene expression is widespread in the human brain (Ogi et al. 1993). Two PAC1-R splice variants, PAC1-R-s and

PAC1-R-hop, are widely expressed in the adult rat brain (Spengler *et al.* 1993; Zhou *et al.* 1999, 2000*b*).

Although a high density of PACAP binding sites is present in the pineal gland (Masuo *et al.* 1992; Shimonneaux *et al.* 1998), only VPAC1-R gene transcripts have been detected in this structure (Usdin *et al.* 1994; Hashimoto *et al.* 1996), indicating that the PACAP binding sites probably correspond to VPAC1-R. PAC1-R transcripts are also expressed in rat retina, in ganglion and amacrine cells (Seki *et al.* 1997), and in murine superior cervical ganglion neurones where neither VPAC1-R nor VPAC2-R are thought to be present (Moller *et al.* 1997; Nogi *et al.* 1997). Type I astrocytes in culture express PAC1-R, VPAC1-R, and VPAC2-R mRNA, as determined by RT-PCR (Ashur-Fabian *et al.* 1997; Grimaldi and Cavallaro 1999). There is also evidence that microglial cells express PAC1-R and VPAC1-R (Ichinose *et al.* 1998; Kim *et al.* 2000) and that purified oligodendrocyte progenitor cells express PAC1-R mRNA, and have functional PACAP-preferring receptors (Lee *et al.* 2001). Gene expression for VPAC1-R, VPAC2-R, and PAC1-R has also been detected in the developing brain, with two PAC1-R splice variants, PAC1-R-s and PAC1-R-hop, detected in the neural plate of embryonic day (E)9 rats by both RT-PCR and *in situ* RT-PCR (Zhou *et al.* 2000*a*). PAC1-R exhibits early and widespread expression in the embryonic brain, even at the very earliest stages of neurogenesis (Sheward *et al.* 1996; Sheward *et al.* 1998; Waschek *et al.* 1998). At later stages, PAC1-R gene expression remains at particularly high levels in the ventricular zone and other areas of the brain undergoing active proliferation. Widespread PACAP gene expression has also been detected in early post-mitotic neurones in the neural tube (Waschek *et al.* 1998), suggesting that PACAP may signal to proliferating neural precursors to regulate their proliferation or development. The ontogenetic expression patterns of VPAC1-R and VPAC2-R differ from that of the PAC1-R: they are expressed at very low levels in the rodent brain at early developmental stages (Waschek *et al.* 1996; Pei 1997; Basille *et al.* 2000). VIP gene expression is also detected in the embryonic brain, but is much more restricted than that of PACAP (Waschek *et al.* 1996).

At the ultrastructural level, PAC1-R immunoreactivity appears to be concentrated predominantly in neuronal perikarya and dendrites (Shioda *et al.* 1997*a*,*b*). In granule cells of the hippocampal dentate gyrus, PAC1-R immunoreactivity was observed in cytoplasmic regions in the vicinity of rough endoplasmic reticulum (rER), the cytoplasmic face of the nuclear membrane, the plasma membrane, and in cytoplasmic matrices. Labelling of the Golgi apparatus has not been detected. PAC1-R were also densely associated with post-synaptic membranes at axo-somatic and axo-dendritic synapses (Shioda *et al.* 1997*a*) and in the hypothalamus, PAC1-R immunoreactivity was detected in post-synaptic membranes as well as on the rER and cytoplasmic matrices of magnocellular neurosecretory neurones (Shioda *et al.* 1997*b*).

26.4 **Pharmacology**

26.4.1 **Agonists**

The VIP/GRF hybrid [Lys^{15}, Arg^{16}, Leu^{27}]VIP(1–7)GRF(8–27)-NH2 is a selective VPAC1-R agonist (K_i 1 nM) which does not activate GRF receptors (Gourlet *et al.* 1997*a*). [Arg^{16}] chicken secretin is as a highly selective VPAC1-R agonist (K_i 1 nM) in those tissues which do not express the secretin receptor (Gourlet *et al.* 1997*a*). Two cyclic peptides with high affinity and selectivity for VPAC2-R have been reported: Ro251553 (K_i < VIP; Gourlet *et al.* 1997*b*) and Ro251392 (K_i 10 nM; Xia *et al.* 1997).

PACAP27 and PACAP38 are commonly recognized as the natural endogenous agonists for PAC1-R. In addition, other peptides can activate PAC1-Rs. Maxadilan is a potent 61 amino acid vasodilator peptide isolated from salivary gland lysates of the sand fly *Lutzomyia longipalipis* (Lerner *et al.* 1991) and is a specific PAC1-R agonist (K_i 3 nM; Moro and Lerner 1997; Soares *et al.* 1998; Uchida *et al.* 1998; Tatsuno *et al.* 2001). It is of some interest that maxadilan and PACAP do not share any significant primary sequence homology, but yet activate PAC1-R with a similar high affinity with little effect on VPAC1-R or VPAC2-R ($K_i > 1000$ nM; Moro and Lerner 1997).

26.4.2 Antagonists

[Acetyl-His^1, D-Phe^2, Lys^{15}, Arg^{16}]VIP(3–7)GRF(8–27)-NH2, is a selective antagonist of rat and human VPAC1-R (IC_{50} 1–10 nM; Gourlet *et al.* 1997*c*). The PACAP fragment PACAP (6–38) is a potent antagonist of PAC1-R (K_i 14 nM), does not interact with VPAC1-R ($K_i > 3000$ nM) but does have significant affinity for VPAC2-R (K_i 170 nM) (Robberecht *et al.* 1992; Dickinson *et al.* 1997). PACAP(6–38) has been used for in *in vivo* and *in vitro* investigations at 10-30 nM (Lai *et al.* 1997) and for other *in vitro* studies at 1 μM (Liu and Madsen 1997). Deletion of the 19 amino acids of maxadilan between positions 24 and 42 results in a peptide (Max.d.4, also called M65), which functions as a specific antagonist at PAC1-R (K_i 6 nM; Moro *et al.* 1999).

26.5 Signal transduction and receptor modulation

VPAC1-R and VPAC2-R are primarily coupled via G_s to adenylyl cylclase. VIP and PACAP potently increase $[Ca^{2+}]_i$ in GH4C1 pituitary cells which natively express VPAC2-R (Rawlings *et al.* 1995). This rise in $[Ca^{2+}]_i$ is due to activation of a membrane Ca^{2+} conductance secondary to stimulation of cAMP production. However, Inagaki *et al.* (1994) have demonstrated that *Xenopus* oocytes injected with the mouse VPAC2-R mRNA respond to VIP and PACAP with an increase in Ca^{2+}-dependent Cl^- conductance, presumably secondary to stimulation of PLC activation. Recent studies show that VIP and PACAP potently stimulate insulin release from HIT-T15 pancreatic tumour cells (Straub and Sharp 1996). The mechanism is most likely similar to the one described for GH4C1 pituitary cells (Rawlings *et al.* 1995).

The most abundant PACAP-containing neurones and PACAP binding sites in the nervous system are found in the hypothalamus (Arimura 1992; Arimura and Shioda 1995; Vaudry *et al.* 2000*a*). PACAP in nanomolar concentrations increased $[Ca^{2+}]_i$ in AVP-containing neurones *in vitro*, and the effect of PACAP was blocked by an inhibitor of protein kinase A (PKA; Shioda *et al.* 1997*b*). Both PACAP (1 nM) and noradrenalin (NA; 1 μM) induced large increases in the $[Ca^{2+}]_i$ in isolated AVP-containing neurones (Shioda *et al.* 1997*b*). PACAP at 0.1 nM and NA at 0.1 μM had little effects, if any, on $[Ca^{2+}]_i$. However, when 0.1 nM PACAP and 0.1 μM NA were added together, they evoked a large increase in $[Ca^{2+}]_i$ in the AVP-containing neurones (Shioda *et al.* 1998*b*). PACAP and NA may thus act in synergy to evoke calcium signalling and secretion in AVP-containing neurones. This synergism may be mediated by an interaction of the cAMP-PKA pathway with a yet unidentified factor 'X' linked to L-type Ca^{2+} channels (Shioda *et al.* 2000). In parasympathetic neurones, the potentiation of acetylcholine-evoked currents by PACAP and VIP may be mediated by a membrane-delimited signal transduction cascade involving the pertussis toxin (PTX)-sensitive Go protein (Liu *et al.* 2000).

In rat hippocampal neuronal cultures, PACAP38 (0.1 nM) increased $[Ca^{2+}]_i$ in a sub-population of neurones (Tatsuno *et al.* 1992). Forskolin and dibutyryl cAMP did not mimic the PACAP38-induced increase in $[Ca^{2+}]_i$. PACAP may thus increase $[Ca^{2+}]_i$ in rat hippocampal neurones by mobilizing Ca^{2+} from intracellular stores, an action not linked to activation of the cAMP/PKA signalling pathway. In the ischaemic hippocampus, including the CA1 region, c-Jun N-terminal kinase (JNK), p38, and extracellular signal-regulated kinase (ERK) activities are all increased during ischaemia-reperfusion (Ozawa *et al.* 1999). However, no significant in JNK and p38 activity was detected in the CA1 region in PACAP-treated animals (Shioda *et al.* 1998a).

In cultured cerebellar granule cells, activation of PAC1-R induced a dose-dependent stimulation of cAMP formation and polyphospho-inositide hydrolysis (Gonzalez *et al.* 1994; Basille *et al.* 1995; Favit *et al.* 1995; Villalba *et al.* 1997). The effect of PACAP on cerebellar granule cell survival appears to be mediated through activation of the AC pathway, leading to phosphorylation of the ERK-type of MAP kinase (Villalba *et al.* 1997). However, other data suggest that PACAP prevents cerebellar granule neurones from apoptotic cell death through a PKA- and PKC-dependent inhibition of caspase-3 activity that does not require ERK1/2 activation (Vaudry *et al.* 2000b).

PACAP actions have been examined on very early neuroepithelial cells isolated from neural fold (head fold) at the primitive streak stage of embryonic day (E) 9 rats (Zhou *et al.* 1999, 2000a). Two splice variants (PAC1-R-s and PAC1-R-hop) were detected in this region by RT-PCR and *in situ* RT-PCR methods (Zhou *et al.* 1999, 2000a). We (S.S. and colleagues) have shown that PKA is the main signalling cascade (and that PKC is the minor signalling cascade) in PACAP-responsive E9 neuroepithelial cells (Zhou *et al.* 2001). On the other hand, VIP showed no effects. These studies showed that PACAP, via one or both PAC1-R splice forms, activates both PKA and PKC signalling cascades in PACAP-responsive neuroepithelial cells.

26.6 Physiology and disease relevance

There is extensive literature which addresses the potential involvement of VIP and PACAP receptors in physiology and human disease (for general reviews, see Fahrenkrug 1993; Waschek 1994; Vaudry *et al.* 2000a). We discuss here only a subset of topics, some of which have received considerable study. Not discussed here but reviewed elsewhere is the significance of VIP and PACAP receptors in gastrointestinal physiology (Ulrich *et al.* 1998), lung function (Groneberg *et al.* 2001) and human tumours (Lelièvre *et al.* 2002).

26.6.1 Arousal, attention, and sleep

Both VIP and PACAP appear to play important, but distinct roles in circadian rhythm modulation. Intense PACAP immunoreactivity is found in retinal afferents in the SCN (Hannibal *et al.* 1997), suggesting that PACAP modulates the ability of light to reset the circadian clock. In fact, injection of PACAP in the vicinity of the SCN has been found to reset the circadian clock in a manner similar to light (Chen *et al.* 1999; Gillete and Tischekau 1999; Harrington *et al.* 1999). VIP, on the other hand, is expressed specifically by neurones within the retinorecipient portion of the SCN. VIP immunoreactivity increases in the SCN during the dark period and decreased during the subsequent light period (Takahashi *et al.* 1989;

Morin *et al.* 1991; Shinohara *et al.* 1993). VIP mRNA also shows a clear diurnal rhythm in the adult rat SCN with lowest levels in the light phase and a peak in the dark phase. VIP neurones in the SCN may be involved as secondary mediators of photic information to the rest of the pacemaker, and perhaps to other parts of the brain (Shinohara *et al.* 1993). Daily variations in the density of the receptor mRNAs have also been observed. PAC1-R mRNA peaks at both noon and midnight in the SCN (Cagampang *et al.* 1998). Similar biphasic SCN variations of VPAC2-R mRNA levels were reported (Cagampang *et al.* 1998; Shinohara *et al.* 1999).

Analyses of transgenic and knock-out mice have begun to reveal the specific roles of PAC1-R and VPAC2-R in circadian rhythms. In PAC1-R knock-out mice, light exposure in the late night induced a small phase delay rather than the usual advance, and light exposure in the early night induced a phase delay than was greater than in wild-type mice (Hannibal *et al.* 2001). These data strongly support the idea that PACAP plays a role on the ability of light to reset the circadian clock. On the other hand, transgenic mice overexpressing VPAC2-R in the SCN resynchronized their clocks after a 8 h advance more quickly than wild-type mice and when kept in constant darkness exhibited a significantly shorter period of running activity (Shen *et al.* 2000). These findings strongly suggest that both VIP and PACAP regulate the circadian pacemaker.

26.6.2 Ischaemia

VIP and PACAP are induced in several types of neurones after different types of experimental injury, suggesting that these peptides are involved in the reaction to nerve injury (Zigmond *et al.* 1997; Waschek 2002). Moreover, PAC1-R was found to be upregulated in the cortex and caudate putamen after transient focal cerebral ischaemia in mice (Gillardon *et al.* 1998). VIP and PACAP are reported to be neuroprotective in several *in vivo* models of brain injury (Table 26.2). For example, neurones in the CA1 region of the hippocampus are vulnerable to global forebrain ischaemia, and this model has been widely used for evaluating neuro-protective agents (Pulsinelli and Brierly 1979; Dohi *et al.* 1998). Intracerebroventricular infusion of PACAP38 into ischaemic animals prevented the otherwise total loss of pyramidal cells and their dendritic processes throughout the CA1 (Banks *et al.* 1996; Uchida *et al.* 1996; Somogyvari-Vigh *et al.* 1998; Reglodi *et al.* 2000). PACAP38 is neuroprotective at concen-trations as low as 0.1 pM *in vitro*, and it can cross the blood–brain barrier by a saturable mechanism (Banks *et al.* 1996). It has been reported that PACAP38 is effective even when administered intravenously (Uchida *et al.* 1996). Thus, PACAP may act on the hippocampal

Table 26.2 Protection against neuronal injury by PACAP or VIP*

Observed action	Injury model
Decreased hippocampal neuron loss (PACAP)	Global forebrain ischemia (Uchida *et al.* 1996)
Reduction of infarct size (PACAP)	Transient forebrain ischemia (Reglodi *et al.* 2000)
Decreased forebrain cholinergic neurone loss (PACAP)	Fimbria fornix transection (Takei *et al.* 2000)
Reduction of lesion size (VIP)	Local chemical excitotoxic injury (Gressens *et al.* 1997)

*Administered neuropeptide (PACAP or VIP) indicated after observed action.

neurones by crossing the blood–brain barrier. Although PAC₁-R gene expression is detected in rat CA1 pyramidal and nonpyramidal neurones (Shioda *et al.* 1997*a*), only a small amount of PAC1-R immunoreactivity is detected in the CA1 region in the normal rat brain (Arimura and Shioda 1995; Shioda *et al.* 1997*a*).

26.6.3 Neurodegeneration

A number of lines of *in vitro* and *in vivo* evidence suggest that VIP and PACAP mediate effects that may be useful in promoting neuronal survival and indeed neuronal regeneration in the CNS. PACAP has been reported to prevent programmed cell death in cultured cerebellar granule cells (Cavallaro *et al.* 1996; Chang *et al.* 1996; Campard *et al.* 1997; Villalba *et al.* 1997; Vaudry *et al.* 2000*b*) and basal forebrain cholinergic neurones (Takei *et al.* 2000). VIP and PACAP have also been shown to exhibit a variety of trophic or growth factor-like actions on other populations of cultured neurones and glia subtypes (see Waschek 1995; Waschek 2002). For example, in cultured rat astrocytes, PACAP mobilizes intracellular free calcium (Tatsuno and Arimura 1994), increases IL-6 secretion (Gottschall *et al.* 1994) and regulates glial glutamate transport and metabolism through activation of PKA and PKC (Figiel and Engele 2000). Early studies have also shown that PACAP stimulated the outgrowth of neurites from PC12 cells and enhanced the survival of sympathetic ganglion cells (Deutsch and Sun 1992; Deutsch *et al.* 1992). More recently, it has been shown that PACAP induces cell cycle withdrawal and promotes the transition from proliferation to differentiation in cultured cortical precursor cells from E13 rats (Lu and DiCicco-Bloom 1997). This antimitotic action of PACAP appears to be mediated by an increase in activity of the cyclin-dependent kinase inhibitor p57^{KIP2} (Carey *et al.* 2002). PACAP potently increased cAMP levels in cultured hindbrain neuroepithelial cells from E10.5 mice and downregulated the expression of the *sonic hedgehog*-dependent and PKA-dependent target gene *gli*-1 (Waschek *et al.* 1998). More recently, it was reported that PACAP increased mitosis in cultured embryonic superior cervical ganglion (SCG) precursors and potently enhanced precursor survival (DiCicco-Bloom *et al.* 2000). At the same time, PACAP promoted neuronal differentiation, increased neurite outgrowth and enhanced expression of the neurotrophin receptors trkC and trkA. Most or all of the physiological effects of PACAP on SCG precursors were shown to be mediated by the PAC1-R, via increased intracellular second messengers, including cAMP and PI as well as Ca^{2+} (DiCicco-Bloom *et al.* 2000). Early studies demonstrated that high concentrations of VIP stimulate mitosis, promote neurite outgrowth and enhance survival of sympathetic neuroblasts in culture (Pincus *et al.* 1990). However, these VIP effects have now been attributed to action on the PAC1-R (DiCicco-Bloom *et al.* 2000). In other studies, lower concentrations of VIP increased neuronal survival during a critical period of development in dissociated spinal cord cultures (Brenneman *et al.* 1985; Brenneman and Eiden 1986), and VIP administration reduced the size of excitotoxin-induced lesions in brains of neonatal mice (Gressens *et al.* 1997). VIP and PACAP were also reported to increase BDNF mRNA in cultured cortical neurones through glutamate action on NMDA receptors (Pellegri *et al.* 1998).

26.6.4 Alzheimer's disease and cognition

Alzheimer's disease is characterized by the death of neurones, likely resulting from the accumulation of the β amyloid proteins produced by a specific processing of the amyloid precursor protein. Stearyl-Nle17-VIP (SNV) is an agonist of VIP exhibiting a 100-fold greater potency than the parent VIP molecule. Daily injection of SNV to mouse pups deficient in apolipopro-tein E (ApoE), a model for Alzheimer's disease, was associated with marked improvements in the time of acquisition of behavioural milestones (Gozes *et al.* 1997). Peptide-treated animals developed as fast as control animals and exhibited improved cognitive function after cessation of peptide treatment. In contrast, treatment with PACAP produced only limited amelioration (Gozes *et al.* 1997). Although the mechanism of action is unclear in these studies, this data suggests that VIP agonism may have therapeutic benefit in the treatment of cognitive decline.

26.6.5 Pain and analgesia

The potential regulatory roles of VIP and PACAP in pain neurotransmission were recently reviewed (Dickinson and Fleetwood-Walker 1999). Small fibres immunoreactive for VIP and PACAP are present in the dorsal horn of the spinal cord, as are gene transcripts for VPAC1-R, VPAC2-R, and PAC1-R. PACAP is synthesized in a subpopulation of neurones of sensory ganglia (Mulder *et al.* 1994). Exogenous PACAP reduces the instances of flinching behaviour in the formalin test, indicating that PACAP may possess antinociceptive properties (Yamamoto and Tatsuno 1995; Zhang *et al.* 1996). Other reports suggest that PACAP may play a facilitatory role in pain transmission (Narita *et al.* 1996; Dickinson *et al.* 1997). Gene expression of PAC1-R, VPAC1-R, and VPAC2-R has been detected in neurons in the dorsal horn of the spinal cord. Several observations suggest PAC1-Rs represent an excellent target for the treatment of pain. PAC1-Rs are located in terminals of c-fibres where they inhibit synaptic transmission in the dorsal horn of the spinal cord. Moreover, PAC1-R has been shown to modulate primary sensory responses in the thalamic nucleus. It was reported that PAC1-R KO mice have a substantial decrease in nociceptive response to the late, but not early, phase of inflammatory pain, suggesting that PAC1-R is involved in the mediation of nociceptive responses during chronic conditions (Jongsma *et al.* 2001). VIP and PACAP can prevent or suppress tissue injury and inflammation (reviewed in Vaudry *et al.* 2000*a*) but the specific roles of VPAC-Rs and PAC1-R in pain and inflammation await further exploration.

26.6.6 Epilepsy

Subcutaneous administration of kainic acid (KA) causes a progressive development of seizures and increased expression of the PACAP gene in the paraventricular nucleus (PVN) of rats (Nomura *et al.* 2000). PACAP immunoreactivity gradually increased after KA administration. The induction of PACAP gene expression following KA-induced seizure was significantly reduced by pretreatment with diazepam or MK-801 (nonselective N-methly-D-aspartate receptor antagonist). These results suggest that PACAP may have a hypophysiotropic role during KA-induced seizures. AVP has been shown to reduce the tonic-clonic seizures' latency as well as the duration of the seizures brainstem generalization on the third and fifth post-partum days in rats (Chepurnova *et al.* 2001). PACAP may thus be involved in mechanisms of experimental seizures through its effect upon AVP neurosecretion.

26.6.7 Schizophrenia

The PACAP gene (ADCYAP1) is located in chromosome 18p11, a locus of that is linked to bipolar disorders and schizophrenia. However, scanning of the coding region of the PACAP gene for mutations in patients with schizophrenia or affective disorders did not provide evidence for this in one study (Ishiguro *et al.* 2001).

26.6.8 Anxiety

It has recently been shown that PAC1-R may play a crucial role in contextual fear conditioning. Mice with a deletion of the PAC1-R gene exhibited elevated locomotor activity and strongly reduced anxiety-like behaviour (Otto *et al.* 2001*a*). Otto *et al.* (2001*b*) have shown that mice with forebrain-specific disruption of the PAC1-R gene have an impairment in mossy fibre long-term potentiation, and an impairment of contextual fear conditioning. Recently, PACAP gene knock-out mice have been shown to display remarkable behavioural changes, such as hyperactive and explosive jumping behaviours in the open field, increased exploratory behaviour, and less anxiety in the elevated-plus maze, emergence, and novel-object tests (Hashimoto *et al.* 2001). PACAP may thus play a previously unrecognized role in the regulation of psychomotor behaviours.

26.6.9 Inflammation

VIP and PACAP have been shown to exert several other actions that may be important after nerve injury. For example, studies have indicated that PACAP regulates the release of cytokines from microglia (Kim *et al.* 2000). In addition to actions on brain cells, numerous studies indicate that PACAP and VIP exert a variety of inhibitory actions on lymphocytes, macrophages, and other immune cells that potentially act at the nerve injury site. These immunomodulatory actions are mediated primarily by receptors that recognize both VIP and PACAP with high affinity (reviewed in Vaudry *et al.* 2000*a*). These actions include (but are not limited to) regulation of chemotaxis and cytokine release. Thus, VIP and PACAP may act on numerous distinct targets to produce a coordinated response after nerve injury. The immunomodulatory actions of VIP and PACAP may also have significance in primary neuroinflammatory diseases such as multiple sclerosis, and in other neurodegenerative diseases with a deleterious inflammatory component. Finally, PACAP may also be important in promoting remyelination in neuroinflammatory diseases because it has recently been shown to promote oligodendrocyte progenitor proliferation *in vitro*, and regulate myelinogenesis in cerebellar slices (Lee *et al.* 2001).

VIP and PACAP receptors are also expressed on B and T lymphocytes, macrophages, neutrophils and many other types of inflammatory cells. Readers are referred to reviews which discuss the numerous reported immunomodulatory actions of these peptides (Ganea 1996, Goetzl *et al.* 1998, Pozo *et al.* 2000). Recently, immunomodulatory actions of VIP and PACAP have been addressed in receptor knock-out and transgenic mouse models. VPAC2-R knock-out mice exhibited enhanced delayed-type hypersensitivity, but depressed immediate-type hypersensitivity (Goetzl *et al.* 2001), whereas forced cell-specific transgenic expression of VPAC2-R in CD4+ T lymphocytes resulted in increased IgE antibody responses and depressed delayed-type hypersensitivity (Voice *et al.* 2001). The potential protective role of PAC1-R in endotoxin-induced shock was recently examined (Martinez *et al.* 2002). Mice which lacked this gene were found to exhibit higher and more rapid mortality in response to lipopolysaccharide-induced shock than wild-type mice. The investigators showed that this result may be partially due to a deficiency of PACAP induction of IL-6 in PAC1-R knock-out mice.

26.7 **Concluding remarks**

VIP and PACAP are now regarded as hormones, transmitters, modulators, and tropic factors which act in both the central and peripheral nervous systems through specific high-affinity receptors. VIP and PACAP receptors are differentially expressed in discrete areas of the nervous system as well as other systems, including reproductive, digestive, respiratory, circulatory, and immune systems. Increased understanding of the multiple VIP and PACAP receptor subtypes and their signalling pathways may lead to the development of new therapeutic agents acting on selected target organs and tissues.

Acknowledgements

We thank Dr Toshiteru Kikuta, U.S.-Japan Biomedical Research Laboratories, for illustrating Fig. 26.1. This work was supported by grants from the Ministry of Education, Science, Sports and Culture of Japan (to S.S.), and by National Institutes of Health grants HD04612, HD06576, and HD34475 (to J.W.).

References

Adamou JE, Aiyar N, Van Horn S, and Elshourbagy NA (1995). Cloning and functional characterization of the human vasoactive intestinal peptide (VIP)-2 receptor. *Biochem Biophys Res Commun* **209**, 385–92.

Arimura A (1992). Pituitary adenylate cyclase-activating polypeptide (PACAP): discovery and current status of research. *Regul Pept* **37**, 287–303.

Arimura A and Shioda S (1995). Pituitary adenylate cyclase activating polypeptide (PACAP) and its receptors: neuroendocrine and endocrine interaction. *Front Neuroendocrinol* **16**, 53–88.

Ashur-Fabian O, Giladi E, Brenneman DE, and Gozes I (1997). Identification of VIP/PACAP receptors on rat astrocytes using antisense oligonucleotides. *J Mol Neurosci* **9**, 11–22.

Banks WA, Uchida D, Arimura A, and Somogyvari-Vigh A, and Shioda S (1996). Transport of adenylate cyclase-activating polypeptide across the blood-brain barrier and the prevention of ischemia-induced death of hippocampal neurons. *Ann NY Acad Sci* **805**, 270–9.

Basille M, Gonzalez BJ, Desrues L, Demas M, Fournier A, and Vaudry H (1995). Pituitary adenylate cyclase-activating polypeptide (PACAP) stimulates adenylyl cyclase and phospholipase C activity in rat cerebellar neuroblasts. *J Neurochem* **65**, 1318–24.

Basille M, Vaudry D, Coulouam Y, Jegou S, Lihnman I, Fournier A *et al.* (2000). Comparative distribution of pituitary adenylate cyclase-activating polypeptide (PACAP) binding sites and PACAP receptor mRNAs in the rat brain during development. *J Comp Neurol* **425**, 495–509.

Brenneman D, Eiden LE, and Siegel RE (1985). Neuroprotective action of VIP on spinal cord cultures. *Peptides* **6**, 35–9.

Brenneman D and Eiden LD (1986). Vasoactive intestinal peptide and electrical activity influence neuronal survival. *Proc Natl Acad Sci USA* **83**, 1159–62.

Cagampang FR, Piggins HD, Sheward WJ, Harmar AJ, and Coen CW (1998). Circadian changes in PACAP type 1(PAC1) receptor mRNA in the rat suprachiasmatic and supraoptic nuclei. *Brain Res* **813**, 218–22.

Campard PK, Crochemore C, Rene F, Monnier D, Koch B, and Loeffler JP (1997). PACAP type I receptor activation promotes cerebellar neuron survival through the cAMP/PKA signaling pathway. *DNA Cell Biol* **16**, 323–33.

Carey RG, Li B, and DiCicco-Bloom E (2002). Pituitary adenylate cyclase activating polypeptide anti-mitogenic signaling in cerebral cortical progenitors is regulated by p57Kip2-dependent CDK2 activity. *J Neurosci* 22, 1583–91.

Cavallaro S, Copani A, D'Agata V, Musco S, Petralia S, Ventra C *et al.* (1996). Pituitary adenylate cyclase-activating polypeptide prevents apoptosis in cultured cerebellar granule neurons. *Mol Pharmacol* 50, 60–6.

Chang JY, Korolev VV, and Wang J-Z (1996). Cyclic AMP and pituitary adenylate cyclase activating polypeptide (PACAP) prevent programmed cell death of cultured rat cerebellar granule cells. *Neurosci Lett* 206, 181–4.

Chatterjee TK, Sharma RV, and Fisher RA (1996). Molecular cloning of a novel variant of the pituitary adenylate cyclase-activating polypeptide (PACAP) receptor that stimulates calcium influx by activation of L-type calcium channels. *J Biol Chem* 271, 32 226–32.

Chen D, Buchanan GF, Ding JM, Hannibal J, and Gillette MU (1999). Pituitary adenylate cyclase-activating peptide: A pivotal modulator of glutamatergic regulation of the suprachiasmatic circadian clock. *Proc Natl Acad Sci* 96, 13 468–73.

Chepurnova NE, Ponomarenko AA, and Chepurnov SA (2001). Peptidergic mechanisms of hypothermia-induced seizures in rats during early ontogenesis. *Ross Fiziol Zh Im I M Sechenova* 87, 217–26.

Christophe J (1993). Type I receptors for PACAP (a neuropeptide even more important than VIP?). *Biochim Biophys Acta* 1154, 183–99.

Daniel PB, Kieffer TJ, Leech CA, and Habener JF (2001). Novel alternatively spliced exon in the extra-cellular ligand-binding domain of the pituitary adenylate cyclase-activating polypeptide (PACAP) type 1 receptor (PAC1R) selectively increases ligand affinity and alters signal transduction coupling during spermatogenesis. *J Biol Chem* 276, 12 938–44.

Dautzenberg FM, Mevenkamp G, Wille S, and Hauger RL (1999). N-terminal splice variants of the type I PACAP receptor: isolation, characterization and ligand binding/selectivity determinants. *J Neuroendocrinol* 11, 941–94.

Deutsch PJ and Sun Y (1992). The 38-amino acid form of pituitary adenylate cyclase-activating poly-peptide stimulates dual signaling cascades in PC12 cells and promotes neurite out growth. *J Biol Chem* 267, 5108–13.

Deutsch PJ, Schadlow V, and Barzilai N (1992). 38-Amino acid form of pituitary adenylate cyclase activating peptide induces process outgrowth in human neuroblastoma cells. *J Neurosci Res* 35, 312–20.

DiCicco-Bloom E, Deutsch PJ, Maltzman J, Zhang J, Pintar JE, Zheng J *et al.* (2000). Autocrine expression and ontogenetic functions of the PACAP ligand/receptor system during sympathetic development. *Dev Biol* 219, 197–213.

Dickinson T and Fleetwood-Walker SM (1999). VIP and PACAP: very important in pain? *Trends Pharmacol Sci* 20, 324–9.

Dickinson T, Fleetwood-Walker SM, Mitchell R, and Lutz EM (1997). Evidence for roles of vasoactive intestinal polypeptide (VIP) and pituitary adenylate cyclase activating polypeptide (PACAP) recep-tors in modulating the responses of rat dorsal horn neurons to sensory inputs. *Neuropeptides* 31, 175–85.

Dohi K, Shioda S, Mizushima H, Homma H, Ozawa H, Nakai Y, and Matsumoto K (1998). Delayed neuronal cell death and microglial cell reactivity in the CA1 region of the rat hippocampus in the cardiac arrest model. *Med Electron Microsc* 31, 85–9.

Fahrenkrug J (1993). Transmitter role of vasoactive intestinal peptide. *Pharmacol Toxicol* 72, 354–63.

Favit A, Scapagnini U, and Canonico PL (1995). Pituitary adenylate cyclase-activating poly-peptide activates different signal transducing mechanism in cultured cerebellar granule cells. *Neuroendocrinology* 61, 377–82.

Figiel M and Engele J (2000). Pituitary adenylate cyclase-activating polypeptide (PACAP): a neuron-derived peptide regulating glial glutamate transport and metabolism. *J Neurosci* 20, 3596–605.

Ganea D (1996). Regulatory effects of vasoactive intestinal peptide on cytokine production in central and peripheral lymphoid organs. *Adv Neuroimmunol* 6, 61–7.

Gillardon F, Hata R, and Hossmann KA (1998). Delayed up-regulation of Zac1 and PACAP type I receptor after transient focal cerebral ischemia in mice. *Brain Res Mol Brain Res* 61, 207–10.

Gillete MU and Tischkau SA (1999). Suprachiasmatic nucleus: the brain's circadian clock. *Recent Prog Horm Res* 54, 33–58.

Goetzl EJ, Pankhaniya RR, Gaufo GO, Mu Y, Xia M, and Sreedharan SP (1998). Selectivity of effects of vasoactive intestinal peptide on macrophages and lymphocytes in compartmental immune responses. *Ann N Y Acad Sci* 840, 540–50.

Goetzl EJ, Voice JK, Shen S, Dorsam G, Kong Y, West KM *et al.* (2001). Enhanced delayed-type hyper-sensitivity and diminished immediate-type hypersensitivity in mice lacking the inducible VPAC(2) receptor for vasoactive intestinal peptide. *Proc Natl Acad Sci USA* 98, 13 854–59.

Gonzalez BJ, Leroux P, BasilleM, Bodenant C, and Vaudry H (1994). Somatostatin and pituitary adenylate cyclase-activating polypeptide (PACAP): Two neuropeptides potentially involved in the development of the rat cerebellum. *Ann Endocrinol* 55, 243–7.

Gottschall PE, Tatsuno I, and Arimura A (1994). Regulation of interleukin-6 (IL-6) secretion in primary cultured rat astrocytes: synergism of interleukin-(IL-1) and pituitary adenylate cyclase-activating polypeptide (PACAP). *Brain Res* 637, 197–203.

Gourlet P, Vandermeers A, Vertongen P, Rathe J, De Neef P, Cnudde J *et al.* (1997a). Development of high affinity selective VIP1 receptors agonists. *Peptides* 18, 1539–45.

Gourlet P, Vertongen P, Vandermeers A, Vandermeers-Piret MC, Rathe J, De Neef P *et al.* (1997b). The long-lasting vasoactive intestinal polypeptide agonist RO 25-1553 is highly selective of the VIP2 receptor subclass. *Peptides* 18, 403–8.

Gourlet P, De Neef P, Cnudde J, Waelbroeck M, and Robberecht P (1997c). *In vitro* properties of a high affinity selective antagonist of the VIP1 receptor. *Peptides* 18, 1555–60.

Gozes I, Bachar M, Bardea A, Davidson A, Rubinraut S, Fridkin M, and Giladi E (1997). Protection against developmental retardation in apolipoprotein E-deficient mice by a fatty neuropeptide: implications for early treatment of Alzheimer's disease. *J Neurobiol* 33, 329–42.

Gressens P, Marret S, Hill JM, Brenneman DE, Gozes I, Fridkin M, and Evrard P (1997). Vasoactive intestinal peptide prevents excitotoxic cell death in the murine developing brain. *J Clin Invest* 100, 390–7.

Grimaldi M and Cavallaro S (1999). Functional and molecular diversity of PACAP/VIP receptors in cortical neurons and type I astrocytes. *Eur J Neurosci* 11, 2767–72.

Groneberg DA, Springer J, and Fischer A (2001). Vasoactive intestinal polypeptide as mediator of asthma. *Pulm Pharmacol Ther* 14, 391–401.

Hannibal J, Ding JM, and Chen D (1997). Pituitary adenylate cyclase-activating peptide (PACAP) in the retinohypothalamic tract: a potential daytime regulator of the biological clock. *J Neurosci* 178, 207–17.

Hannibal J, Jamen F, Nielsen HS, Journot L, Brabet P, and Fahrenkrug J (2001). Dissociation between light-induced phase shift of the circadian rhythm and clock gene expression in mice lacking the pituitary adenylate cyclase activating polypeptide type 1 receptor. *J Neurosci* 21, 4883–90.

Harmar AJ, Arimura A, Gozes I, Journot L, Laburthe M, Pisegna JR *et al.* (1998). International Union of Pharmacology. XVIII. Nomenclature of receptors for vasoactive intestinal peptide and pituitary adenylate cyclase-activating polypeptide. *Pharmacol Rev* 50, 265–70.

Harrington ME, Hoque S, Hall A, Golombek D, and Biello S (1999). Pituitary adenylate cyclase-activating peptide phase shifts circadian rhythms in a manner similar to light. *J Neurosci* 19, 6637–42.

Hashimoto H, Nogi H, Mori K, Ohishi H, Shigemoto R, Yamamoto K *et al.* (1996). Distribution of the mRNA for a pituitary adenylate cyclase-activating polypeptide receptor in the rat brain: an *in situ* hybridization study. *J Comp Neurol* **371**, 567–77.

Hashimoto H, Shintani N, Tanaka K, Mori W, Hirose M, Matusda T *et al.* (2001). Altered psychomotor behaviors in mice lacking pituitary adenylate syclase-activating polypeptide (PACAP). *Proc Natl Acad Sci USA* **98**, 13 355–60.

Ichinose M, Asai M, and Sawada M (1998). Activation of outward current by pituitary adenylate cyclase-activating polypeptide in mouse microglial cells. *J Neurosci Res* **51**, 382–90.

Inagaki N, Yoshida H, Mizuta M, Mizuno N, Fujii Y, Gonoi T *et al.* (1994). Cloning and functional characterization of a third pituitary adenylate cyclase-activating polypeptide receptors subtype expressed in insulin-secreting cells. *Proc Natl Acad Sci USA* **91**, 2679–83.

Ishiguro H, Ohtuki T, Okubo Y, Kurumai A, and Arinami T (2001). Association analysis of the pituitary adenylate cyclase activating peptide gene (PACAP) on chromosome 18p11 with schizophrenia bipolar disorders. *J Neural Transm* **108**, 849–54.

Ishihara T, Shigemoto R, Mori K, Takahashi T, Yakahashi K, and Nagata S *et al.* (1992). Molecular and tissue distribution of a novel receptor for vasoactive intestinal polypeptide. *Neuron* **8**, 811–19.

Jongsma H, Pettersson LM, Zhang Yz, Reimer MK, Kanje M, Waldenstrom A *et al.* (2001). Markedly induced chronic nociceptive response in mice lacking the PAC1 receptor. *NeuroReport* **12**, 2215–19.

Kim WK, Kan Y, Ganea D, Hart RP, Gozes I, and Jonakait GM (2000). Vasoactive intestinal peptide and pituitary adenylyl cyclase-activating polypeptide inhibit tumor necrosis factor-alpha production in injured spinal cord and in activated microglia via a cAMP-dependent pathway. *J Neurosci* **20**, 3622–30.

Krempels K, Usdin TB, Harta G, and Mezey E (1995). PACAP acts through VIP type 2 receptors in the rat testis. *Neuropeptides* **29**, 315–20.

Lai CC, Wu SY, Lin HH, and Dun NJ (1997). Excitatory action of pituitary adenylae cyclase activating polypeptide on rat sympathetic preganglionic neuros *in vivo* and *in vitro*. *Brain Res* **748**, 189–94.

Lee M, Lelievre V, Zhao P, Torres M, Rodriguez W, Byun JY *et al.* (2001). Pituitary adenylyl cyclase-activating polypeptide stimulates DNA synthesis but delays maturation of oligodendrocyte progenitors. *J Neurosci* **21**, 3849–59.

Lelièvre V, Pineau N, and Waschek J (2002). The biological significance of PACAP and PACAP receptors in human tumors. In (ed. K Vaudry) *Pituitary Adenylate Cyclase-Activating Polypeptide*, Kluwer Academic Publishers, Norwell, MA, *in press*.

Lerner EA, Ribeiro JM, Nelson RJ, and Lerner MR (1991). Isolation of maxadilan, a protein vasodilatory peptide from the salivary galnds of the sand fly *Lutzomyia longipalpis*. *J Biol Chem* **266**, 11 234–6.

Liu DM, Cuevas J, and Adams DJ. (2000). VIP and PACAP potentiation of nicotinic ACh-evoked currents in rat parasympathetic neurons is mediated by G-protein activation. *Eur J Neurosci* **12**, 2243–51.

Liu GJ and Madsen BW (1997). PACAP38 modulates activity of NMDA receptors in cultured chick cortical neurons. *J Neurophysiol* **78**, 2231–4.

Lu N and DiCicco-Bloom E (1997). Pituitary adenylate cyclase-activating polypeptide is an autocrine inhibitor of mitosis in cultured cortical precursor cells. *Proc Natl Acad Sci USA* **94**, 3357–62.

Lutz EM, Sheward WJ, West KM, Morrow JA, Fink G, and Harmar AJ (1993). The VIP_2 receptor: Molecular characterization of a cDNA encoding a novel receptor for vasoactive intestinal peptide. *FEBS Lett* **334**, 3–8.

Martinez C, Abad C, Delgado M, Arranz A, Juarranz MG, Rodriguez-Henche N *et al.* (2002). Anti-inflammatory role in septic shock of pituitary adenylate cyclase-activating polypeptide receptor. *Proc Natl Acad Sci USA* **99**, 1053–8.

Masuo Y, Ohtaki T, Masuda Y, Tsuda M, and Fujino M (1992). Binding sites for pituitary adenylate cyclase activating polypeptide (PACAP): comparison with vasoactive intestinal polypeptide (VIP) binding site localization in rat brain sections. *Brain Res* **575**, 113–23.

Mayo KE (1992). Molecular cloning and expression of a pituitary-specific receptor for growth hormone-releasing hormone. *Mol Endocrinol* **6**, 1734–44.

Moller K, Reimer M, Ekblad E, Hannibal J, Fahrenkrug J, Kanje M, and Sundler F (1997). The effects of axotomy and preganglionic denervation on the expression of pituitary adenylate cyclase-activating peptide (PACAP), galanin and PACAP type 1 receptors in the rat superior cervical ganglion. *Brain Res* **775**, 166–82.

Morin AJ, Denoroy L, and Jouvet M (1991). Daily variations in concentration of vasoactive intestinal polypeptide immunoreactivity in discrete brain areas of the rat. *Brain Res* **538**, 136–40.

Moro O and Lerner EA. (1997). Maxadilan, the vasodilator from sand flies, is a specific pituitary adenylate cyclase-activating peptide type I receptor agonist. *J Biol Chem* **272**, 966–70.

Moro O, Wakita K, Ohmura M, Denda S, Lerner EA, and Tajima M (1999). Functional characterization of structural alterations in the sequence of the vasodilatory peptide maxadilan yields a pituitary adenylate cyclase-activating peptide type 1 receptor-specific antagonist. *J Biol Chem* **274**, 23 103–10.

Mulder H, Uddman R, Moller K, Zhang YZ, Ekblad E, Alumets J, and Sundler F (1994). Pituitary adenylate cyclase-activating polypeptide expression in sensory neurons. *Neuroscience* **63**, 307–12.

Narita M, Dun SL, Dun NJ, and Tseng LF (1996). Hyperalgesia induced by pituitary adenylate cyclase-activating polypeptide (PACAP) in the mouse spinal cord. *Eur J Pharmacol* **311**, 121–6.

Nogi H, Hashimoto H, Hagihara N, Shimada S, Yamamoto K, Matsuda T *et al.* (1997). Distribution of mRNAs for pituitary adenylate cyclase-activating polypeptide (PACAP), PACAP receptor, vasoactive intestinal polypeptide (VIP), and VIP receptors in the rat superior cervical ganglion. *Neurosci Lett* **227**, 37–40.

Nomura M, Ueta Y, Hannibal J, Serino R, Yamamoto Y, Shibuya I *et al.* (2000). Induction of pituitary adenylate cyclase-activating polypeptide mRNA in the medial parvocellular part of the paraventricular nucleus of rats following kainic-acid-induced seizure. *Neuroendocrinology* **71**, 318–26.

Ogi K, Miyamoto Y, Masuda Y, Habata Y, Hosoya M, Ohtaki T *et al.* (1993). Molecular cloning and functional expression of a cDNA encoding a human pituitary adenylate cyclase activating polypeptide receptor. *Biochem Biophys Res Commun* **196**, 1511–21.

Otto C, Martin M, Wolfer DP, Lipp HP, Maldonado R, Konnerth A, and Schutz G (2001a). Altered emotional behavior in PACAP-type-I-receptor-deficient mice. *Mol Brain Res* **92**, 78–84.

Otto C, Kovalchuk Y, Wolfer DP, Gass P, Martin M, Zuschratter W *et al.* (2001b). Impairment of mossy fiber long-term potentiation and associative learning in pituitary adenylate cyclase-activating polypeptide type 1 receptor-deficient mice. *J Neurosci* **21**, 5520–7.

Ozawa H, Shioda S, Dohi K, Matsumoto H, Mizushima H, Zhou C-J *et al.* (1999). Delayed neuronal cell death in the rat hippocampus is mediated by the mitogen-activated protein kinase signal transduction pathway. *Neurosci Lett* **262**, 57–60.

Pantaloni C, Brabet P, Bilanges B, Dumuis A, Houssami S, Spengler D *et al.* (1996). Alternative splicing in the N-terminal extracellular domain of the pituitary adenylate cyclase-activating polypeptide (PACAP) receptor modulates receptor selectivity and relative potencies of PACAP-27 and PACAP-38 in phospholipase C activation. *J Biol Chem* **271**, 22 146–221.

Pei L (1997). Genomic structure and embryonic expression of the rat type 1 vasoactive intestinal polypeptide receptor gene. *Regul Pept* **71**, 153–61.

Pellegri G, Magistretti PJ, and Martin JL (1998). VIP and PACAP potentiate the action of glutamate on BDNF expression in mouse cortical neurons. *Eur J Neurosci* **10**, 272–80.

Pincus DW, DiCicco-Bloom E, and Bkack IB (1990). Active intestnal peptide regulates, differentiation and survival of cultures sympathetic neuroblasts. *Nature* **343**, 564–7.

Pozo D, Delgado M, Martinez M, Guerrero JM, Leceta J, Gomariz RP, and Calvo JR (2000). Immunobiology of vasoactive intestinal peptide (VIP). *Immunol Today* **21**, 7–11.

Pulsinelli WA and Brierly JB (1979). A new model of bilateral hemispheric ischemia in the unanesthetized rat. *Stroke* **10**, 267–72.

Rawlings SR, Piuz I, Schlegel W, Bockaert J, and Journot L (1995). Differential expression of pituitary adenylate cyclase-activating polypeptide/vasoactive intestinal polypeptide receptor subtypes in clonal pituitary somatotrophs and gonadotrophs. *Endocrinology* **136**, 2088–98.

Reglodi D, Somogyvari-Vigh A, Vigh S, Kozicz T, and Arimura A (2000). Delayed systemic administration of PACAP38 is neuroprotective in transient middle cerebral artery occlusion in the rat. *Stroke* **31**, 1411–7.

Robberecht P, Gourlet P, and De Neef P (1992). Structural requirements for the occupancy of pituitary adenylate cyclase activating polypeptide (PACAP) receptors and adenylte cyclase activation in human neuroblastoma NB-OK-1 cell membranes. *Eur J Biochem* **207**, 239–46.

Seki T, Shioda S, Ogino D, Nakai Y, Arimura A, and Koide R (1997). Distribution and ultrastructural localization of a receptor for pituitary adenylate cyclase activating polypeptide and its mRNA in the rat retina. *Neurosci Lett* **238**, 127–30.

Shen S, Spratt C, Sheward WJ, Kallo I, West K, Morrison CF *et al.* (2000). Overexpression of the human VPAC2 receptor in the suprachiasmatic nucleus alters the circadian phenotype of mice. *Proc Natl Acad Sci* **97**, 11 575–80.

Sheward WJ, Lutz EM, and Harmar AJ (1995). The distribution of vasoactive intestinal peptide 2 receptor messenger RNA in the rat brain and pituitary gland as assessed by *in situ* hybridization. *Neuroscience* **67**, 409–18.

Sheward WJ, Lutz EM, and Harmar AJ (1996). Expression of pituitary adenylate cyclase activating polypeptide receptors in the early mouse embryo as assessed by reverse transcription polymerase chain reaction and in situ hybridisation. *Neurosci Lett* **216**, 45–8.

Sheward WJ, Lutz EM, Copp AJ, and Harmar AJ (1998). Expression of PACAP, and PACAP type 1 (PAC1) receptor mRNA during development of the mouse embryo. *Dev Brain Res* **109**, 245–53.

Shimonneaux V, Kienlen-Campard P, Loeffer JP, Basille M, Gonzalez BJ, Vaudry H *et al.* (1998). Pharmacological, molecular and functional characterization of VIP/PACAP receptors in the rat pineal gland. *Neuroscience* **85**, 887–96.

Shinohara K, Tominaga K, Isobe Y, and Inoue SIT (1993). Photic regulation of peptides located in the ventrolateral subdivision of the suprachiasmatic nucelus of the rat: daily variations of vasoactive intestinal polypeptide, gastrin releasing peptide and neuropeptide. *J Neurosci* **13**, 793–800.

Shinohara K, Funahashi T, and Kimura F (1999). Temporal profiles of vasoactive intestinal polypeptide precursor mRNA and its receptor mRNA in the rat suprachiasmatic nucleus. *Mol Brain Res* **63**, 262–7.

Shioda S, Shuto Y, Somogyvari-Vigh A, Legradi G, Onda H, and Arimura A (1994). Distribution of pituitary adenylate cyclase activating polypeptide (PACAP) receptor in rat brain. *Soc Neurosci Abstr* **20**, 516, 221.14.

Shioda S, Shuto Y, Somogyvari-Vigh A, Legradi G, Onda H, Coy D *et al.* (1997*a*). Localization and gene expression of the receptor for pituitary adenylate cyclase-activating polypeptide in the rat brain. *Neurosci Res* **28**, 345–54.

Shioda S, Yada T, Nakajo S, Nakaya K, Nakai Y, and Arimura A (1997*b*). Pituitary adenylate cyclase-activating polypeptide (PACAP): a novel regulator of vasopressin-containing neurons. *Brain Res* **765** 81–90.

Shioda S, Ozawa H, Dohi K, Mizushima H, Matsumoto K, Nakajo S *et al.* (1998*a*). PACAP protects hippocampal neurons against apoptosis. Involvement of JNK/SAPK signaling pathway. *Ann NY Acad Sci* **865**, 111–7.

Shioda S, Yada T, Nakajo S, Nakai Y, and Arimura A (1998*b*). PACAP increases cytosolic calcium in vasopressin neurons: synergism with noradrenaline. *Ann NY Acad Sci* **865**, 427–30.

Shioda S, Yada T, Muroya S, Uramura S, Nakajo S, Ohtaki H *et al.* (2000). Functional significance of co-localization of PACAP and catecholamine in nerve terminals. *Ann NY Acad Sci* **921**, 211–7.

Soares MBP, Titus RG, Shoemaker CB, David JR, and Bozza M (1998). The vasoactive peptide maxadilan from sand fly saliva inhibits TNF-α and induces IL-6 by mouse macrophages through interaction with the pituitary adenylate cyclase-activating polypeptide (PACAP) receptor. *J Immunol* 160, 1811–16.

Somogyvari-Vigh A, Svoboda-Teet J, Vigh S, and Arimura A (1998). Is an intravenous infusion of PACAP38 for prevention of neuronal death induced by global ischemia? The possible presence of a binding protein for PACAP38 in blood. *Ann NY Acad Sci* 865, 595–600.

Spengler D, Waeber C, Pantaloni C, Holsboer F, Bockaert J, Seeburg FH, and Journot L (1993). Differential signal transduction by five splice variants of the PACAP receptor. *Nature* 365, 170–5.

Straub SG and Sharp GWG. (1996). A wortmannin-sensitive signal transduction pathway is involved in the stimulation of insulin release by vasoactive intestinal polypeptide and pituitary adenylate cyclase -activating polypetide. *J Biol Chem* 271, 1660–8.

Takahashi Y, Okamura H, Yanaihara N, Hamada S, Fujita S, and Ibata Y (1989). The influence of light stimulus on VIP-like immunoreactive neurons in the suprachiasmatic nucleus. *Brain Res* 497, 374–7.

Takei N, Torres E, Yuhara A, Jongsma H, Otto C, Korhonen L *et al.* (2000). Pituitary adenylate cyclase-activating polypeptide promotes the survival of basal forebrain cholinergic neurons *in vitro* and *in vivo* comparison with effects of nerve growth factor. *Eur J Neurosci* 12, 2273–80.

Tatsuno I, Yada T, Vigh S, Hidaka H, and Arimura A (1992). Pituitary adenylate cyclase activating polypeptide and vasoactive intestinal peptide increase cytosolic free calcium concentration in cultured rat hippocampal neurons. *Endocrinology* 131, 73–81.

Tatsuno I and Arimura A (1994). Pituitary adenylate cyclase-activating polypeptide (PACAP) mobilizes intracellular free calcium in cultured type-2, but not type-1, astrocytes. *Brain Res* 662, 1–10.

Tatsuno I, Uchida D, Tanaka T, Saeki N, Hirai A, Saito Y *et al.* (2001). Maxadilan specifically interacts with PAC1 receptor, which is a dominant form of PACAP/VIP family receptors in cultured rat cortical neurons. *Brain Res* 889, 138–48.

Uchida A, Arimura A, Somogyvári-Vigh A, Shioda S, and Banks WA (1996). Prevention of ischemia-induced death of hippocampal neurons by pituitary adenylate cyclase activating polypeptide. *Brain Res* 736, 280–6.

Uchida A, Tatsuno I, Tanaka T, Hirai A, Saito Y, Moro O, and Tajima M (1998). Maxadilan is a specific agonist and its deleted peptide (M65) is a specific antagonist for PACAP type 1 receptor. *Ann NY Acad Sci* 865, 253–8.

Ulrich CD 2nd, Holtmann M, and Miller LJ (1998). Secretin and vasoactive intestinal peptide receptors: members of a unique family of G protein-coupled receptors. *Gastroenterology* 114, 382–97.

Usdin TB, Bonner TI, and Mezey E (1994). Two receptors for vasoactive intestinal polypeptide with similar specificity and complementary distributions. *Endocrinology* 135, 2662–80.

Vaudry D, Gonzalez BJ, Basille M, Yon L, Fournier A, and Vaudry H (2000*a*). Pituitary adenylate cyclase-activating polypeptide and its receptors: from structure to functions. *Pharmacol Rev* 52, 269–324.

Vaudry D, Gonzalez BJ, Basille M, Pamantung TF, Fontaine M, Fournier A, and Vaudry H (2000*b*). The neuroprotective effect of pituitary adenylate cyclase-activating polypeptide on cerebellar granule cells is mediated through inhibition of the CED3-related cysteine protease caspase-3/CPP32. *Proc Natl Acad Sci USA* 97, 13 390–5.

Voice JK, Dorsam G, Lee H, Kong Y, and Goetzl EJ (2001). Allergic diathesis in transgenic mice with constitutive T cell expression of inducible vasoactive intestinal peptide receptor. *FASEB J* 15, 2489–96.

Villalba M, Bockaert J, and Journot L (1997). Pituitary adenylate cyclase-activating polypeptide (PACAP) protects cerebellar granule neuron from apoptosis by activating the mitogen-activated protein kinase (MAP kinase) pathway. *J Neurosci* 17, 83–90.

Waschek JA (1994). Environmental and Hormonal Control of Vasoactive Intestinal Peptide (VIP) Gene Expression and Changes in Pathological Conditions. In (ed. AJ Turner) *Frontiers in Molecular Neurobiology* Vol. 1: *Neuropeptide Gene Expression*, Portland Press, London.

Waschek JA (1995). Vasoactive intestinal peptide: an important trophic factor and developmental regulator? *Dev Neurosci* 17, 1–7.

Waschek JA (2002). Multiple actions of pituitary adenylyl cyclase activating peptide (PACAP) in nervous system development and regeneration. *Dev. Neuroscience* (in press).

Waschek JA, Ellison J, Bravo DT, and Handley V (1996). Embryonic expression of vasoactive intestinal peptide (VIP) and VIP receptor genes. *J Neurochem* **66**, 1762-5.

Waschek JA, Casillas RA, Nguyen TB, DiCicco-Bloom EM, Carpenter EM, and Rodriguez WI (1998). Neural tube expression of pituitary adenylate cyclase-activating peptide (PACAP) and receptor: Potential role in patterning and neurogenesis. *Proc Natl Acad Sci USA* **95**, 9602–7.

Wei Y and Mojsov S (1996). Tissue specific expression of different human receptor types for pituitary adenylate cyclase-activating polypeptide and vasoactive intestinal polypeptide: Implication for their role in human physiology. *J Neuroendocrinol* **8**, 811–17.

Xia M, Sreedharan SP, Bolin DR, Gaufo GO, and Goetzl EJ (1997). Novel cyclic peptide agonist of high potency and selectivity for the type II vasoactive intestinal peptide receptor. *J Pharmacol Exp Ther* **281**, 629–33.

Yamamoto T and Tatsuno I (1995). Antinociceptive effect of intrathecally administered pituitary adenylate cyclase activating polypeptide (PACAP) on the rat formalin test. *Neurosci Lett* **16**, 32–5.

Zhang Y, Malmberg AB, Sjolund B, and Yaksh TL (1996). The effect of pituitary adenylate cyclase activating peptide (PACAP) on the nociceptive formalin test. *Neurosci Lett* **207**, 187–90.

Zhou C-J, Shioda S, Shibanuma M, Nakajo S, Funahashi H, Nakai Y *et al.* (1999). Pituitary adenylate cyclase-activating polypeptide receptors during development: expression in the rat embryo at primitive streak stage. *Neuroscience* **93**, 375–91.

Zhou C-J, Kikuyama S, Nakajo S, Hirabayashi T, Mizusima H, and Shioda S (2000*a*). Splice variants of PAC1 receptor during early neural development of rats. *Peptides* **21**, 1177–83.

Zhou C-J, Kikuyama S, Shibanuma M, Hirabayashi T, Nakajo S, Arimura A, and Shioda S (2000*b*). Cellular distribution of the splice variants of the receptor for pituitary adenylate cyclase-activating polypeptide (PAC1-R) in the rat brain by *in situ* RT-PCR. *Mol Brain Res* **75**, 150–8.

Zhou C-J, Yada T, Kohno D, Kikuyama S, Suzuki R, Mizushima H, and Shioda S (2001). PACAP activates PKA, PKC, and Ca signaling cascades in rat neuroepithelial cells. *Peptides* **22**, 1111–7.

Zigmond RE (1997). LIF, NGF, and the Cell Body Response to Axotomy. *New Scientist* **3**, 176–85.

Part 4

Family 3 GPCRs

Chapter 27

Calcium-sensing receptors

Donald T. Ward and Daniela Riccardi

27.1 Introduction

Regulation of blood calcium levels involves the integration of effects of three calcium homeostatic hormones, parathyroid hormone (PTH) released from the parathyroid glands, $1,25(OH)_2$ vitamin D_3 produced ultimately in the renal proximal tubule and calcitonin secreted from thyroidal C-cells. Of the three, PTH exerts perhaps the greatest influence on calcium homeostasis and its release or suppression of release is primarily dependent upon the accurate sensing of circulating ionic Ca^{2+} levels by the parathyroid glands. PTH secretion from the parathyroid glands rectifies hypocalcaemia by increasing renal Ca^{2+} reabsorption, bone resorption and $1,25(OH)_2D_3$ production, while hypercalcaemia is resolved by inhibition of PTH secretion. The protein that senses this hypercalcaemia is the extracellular Ca^{2+}-sensing receptor (CaR; Brown et al. 1993). Similarly, in the kidney, urinary excretion of Ca^{2+} is also dependent on circulating ionized Ca^{2+} levels and the relationship between the two mimics that of $[Ca^{2+}]_o$ versus PTH secretion and again the CaR is responsible for regulating this (Riccardi et al. 1995). In the kidney, CaR is expressed in tubular epithelium, such as in the thick ascending limb and collecting duct cells where it regulates salt and water reabsorption respectively (Hebert et al. 1997; Sands et al. 1997). CaR has since been found in other calcium homeostatic tissues such as gut and bone, but also in many other tissues and cells not directly involved in Ca^{2+} homeostasis such as brain, pancreas and skin (reviewed by Brown and MacLeod 2001).

Although CaRs share the defining properties of GPCRs, they also differ from Type I and Type II receptors in three major respects: (1) the primary physiological agonist of parathyroid CaR is an elemental ion and not a modified amino acid or other organic molecule (2) Ca^{2+} binding occurs in the CaRs unusually large extracellular domain rather than the transmembrane region, and (3) CaRs are expressed as homodimers, that is, two CaR molecules bound together, and not monomers. In this review we will describe some features of CaR structure and discuss the role of extracellular cysteine residues in this area.

27.2 Molecular characterization

The CaR gene is located on chromosome 3 in humans (Chou et al. 1992) and chromosome 11 in rats (Janicic et al. 1995) and encodes seven exons of which six (exons 2–7) contribute to the sequence of the full-length protein (Pearce et al. 1996; Table 27.1). Alternative splicing of the CaR gene can occur either in the coding regions, where in osteoblasts it produces a protein with apparently severely impaired function (Oda et al. 1998), or, in the 5'-untranslated region

Table 27.1 Ca^{2+}-sensing receptors

Alternative names	Extracellular Ca^{2+}/polyvalent cation-sensing receptor, calcium receptor (CaR or CaSR), BoPCaR (bovine parathyroid CaR), RaKCaR (r kidney), HuPCaR (h parathyroid)
Structural information (Accession no.)	h 1078 aa (U20759)^{AS} r 1079 aa (U10354) m 1079 aa (Q9QY96)^{AS}
Chromosomal location	3q13.3–24
Selective agonists	Divalent metals: Ca^{2+}, Mg^{2+}, trivalent metals: Gd^{3+}, polyvalent cations: Aminoglycoside antibiotics e.g. neomycin, amyloid β peptide, spermine, protamine
Allosteric modulators	Calcimimetics: NPSR467, NPSR568 Amino Acids: L-Histidine, L-Phenylalanine, L-Tyrosine Ionic strength: NaCl Calcilytics: NPS2143 (allosteric antagonist)
G protein coupling	Cr$_{q/11}$/Cr$_i$
Expression profile	Ca^{2+} homeostatic system; parathyroid, kidney, bone and bone-derived cells, thyroidal C cells, intestine Non-calcium homeostatic system; brain (highest in subfornical organ and the olfactory bulbs), pancreas, lens, mammary epithelium, skin keratnocytes and fibroblasts, bone marrow, blood monocytes, liver, ovarian/leydig cells, perivascular arteries
Physiological functions	Regulation of PTH secretion and renal Ca^{2+} excretion
Knockout phenotype	Elevated circulating PTH levels with parathyroid hyperplasia, reduced radiodensity of bones and bowing of long bones
Disease association	Inactivating mutations; FHH, NSHPT Activating mutations; autosomal dominant hyperparathyroidism (ADH)

(Garrett *et al.* 1995*b*) which could potentially allow for differential expression or regulation. Two cDNA clones for CaR with different 5′-untranslated regions have been isolated from parathyroid adenomas (Garrett *et al.* 1995*b*) and these encode two 5′-untranslated exons (exons 1A and 1B; Chikatsu *et al.* 2000). Chikatsu *et al.* (2000) reported that exon 1A expression is reduced in adenomatous parathyroid gland whereas exon 1B expression is unchanged from normal gland. They also found that the CaR gene has two promoters, the downstream one being GC-rich and the upstream one containing TATA and CAAT boxes.

An interesting and unusual feature of Type III (or Family C) GPCRs is that they exist in the plasma membrane as disulfide-linked homodimers (Bai *et al.* 1998; Ward *et al.* 1998; Romano *et al.* 1996). In addition to their non-covalent intracellular G protein coupling, they are also bound covalently to an adjacent receptor. Disulphide-linked CaR dimerization can be demonstrated by several methods including immunoblotting under nonreducing/reducing conditions, estimation of native mass by sucrose density gradient ultracentrifugation and use of covalent cross-linking reagents (Bai *et al.* 1998*a*; Ward *et al.* 1998). To establish that the intermolecular association is actually homodimeric (i.e. two CaRs bound together rather than one CaR bound to a different protein(s) of equivalent molecular mass), Bai *et al.* (1998*a*) successfully coprecipitated flagged and non-flagged CaRs from membranes of HEK cells cotransfected with both FLAG epitope-tagged and wild-type CaRs. To dissociate the monomers of a CaR homodimer, one must both denature it using an ionic detergent (e.g. SDS) but also reduce it using a reducing agent such as β-mercaptoethanol (Ward *et al.* 1998). The ionic detergent disrupts the tertiary structure of the protein breaking any ionic bonds that may exist between the receptors, while the reducing agent breaks the disulphide linkages including those responsible for dimerization. Based on the high degree of conservation between the extracellular cysteine residues of CaRs and mGluRs and the fact that the N-terminal 15 kDa extracellular portion of mGluR5 is the region responsible for the disulphide-linkage of the mGluR5 dimer (Romano *et al.* 1996), it was proposed that extracellular cysteine residues also mediate CaR dimerization (Ward *et al.* 1998). This has been clearly demonstrated by Goldsmith *et al.* (1999) who, expressing only the extracellular domain of CaR, found that it still forms disulphide-linked dimers. Several attempts have been made to determine which of the 21 extracellular cysteine residues present in CaR are responsible for intermolecular dimerization and currently the most likely candidates appear to be Cys-129 and Cys-131 (Ray *et al.* 1999; Zhang *et al.* 2001). Thus, while wild-type CaRs expressed in HEK cells migrate in non-reducing SDS-PAGE gels as a dimer, mutant CaRs in which Cys-129 and Cys-131 have been replaced by serine residues migrate mostly as a monomer (Ray *et al.* 1999; Zhang *et al.* 2001). Since treatment with reducing agents is insufficient to disassociate the CaR dimer in the absence of an ionic detergent (Ward *et al.* 1998), it has been postulated that the CaR dimer may also be held together by intramembrane ionic bonds (Bai *et al.* 1998*a*). This concept has been recently examined using the C129S/C131S mutant CaR. The mutant was expressed in HEK cells with either a Flag-tag or with its extreme C-terminus truncated and it was found that an anti-Flag antibody coprecipitated both the Flag-tagged receptor and the non-tagged truncated receptor (Zhang *et al.* 2001). This indicates that even without the two cysteine residues most likely responsible for intermolecular disulphide bonding, the CaR still remains a homodimer probably via ionic interactions.

Another interesting feature of the singly mutated CaRs C129S and C131S and the doubly mutated C129S/C131S is that they all exhibit greater sensitivity to $[Ca^{2+}]_o$ than wild-type CaRs (Zhang *et al.* 2001). This suggests that not only are these residues of structural importance, but they may also be functionally important as well. In addition, activating mutations of the CaR that are found in people with autosomal dominant hypocalcaemia include the mutations E127A and F128L which lie in the immediate vicinity of Cys-129 and Cys-131 and mimic the functional responses of the C129S and C131S CaR mutations.

Regarding CaR dimerization, it is important to recognize that the dimer pre-exists agonist treatment and thus does not serve the same activating function as for the receptor tyrosine kinases. While the Triton X-100-solubilized CaR dimer is more resistant to SDS/β-mercaptoethanol-induced monomerization following exposure to Ca^{2+} (Ward *et al.* 1998), there is currently no indication that Ca^{2+} treatment affects the dimeric property *in vivo*. Thus the functional advantage of dimerization remains unclear. Interestingly, when two inactivating CaR mutants, one with a mutation in the extracellular domain (ECD) and the other with a mutation in the intracellular domain are co-expressed in HEK cells, there is a partial recovery of signalling function (Bai *et al.* 1999). Although, this double mutant receptor does not signal exactly as for wild-type CaR, it is still more active than either of the mutant receptors expressed individually. In contrast, when two mutant CaRs both containing mutations in their ECDs, or, both containing mutations in their transmembrane/intracellular domains were co-expressed, no recovery of function occurred (Bai *et al.* 1999).

Despite the fact that mGluR1a and mGluR5 do not heterodimerize with each other when co-transfected in HEK cells (Romano *et al.* 1996), there is some evidence suggesting that both mGluR1a and mGluR5 can heterodimerise with CaR (Breitwieser *et al.* 2001). Although this remains controversial, the idea of CaR/mGluR dimers would be interesting since in neuronal cells where they are naturally co-expressed, their physical interaction could allow for integration of their signals.

Since only 2 of the possible 21 extracellular cysteine residues are apparently required for intermolecular disulphide bond formation (Ray *et al.* 1999; Zhang *et al.* 2001), the majority of the remaining cysteine residues are likely required for intramolecular binding, contributing towards extracellular secondary structure of the receptor. Furthermore, as most of the extracellular cysteine residues are either located in a juxta-membrane cysteine rich region, or spread out throughout the N-terminal portion of the ECD, an obvious arrangement of intramolecular disulphide bonds would be covalent association of the N-terminal region to the cysteine-rich region to 'anchor' the ECD. However, specific protease cleavage of the ECD between the cysteine-rich region and the N-terminal region, fully dissociates these two domains even under non-reducing conditions (Hu *et al.* 2001) suggesting that the juxta-membrane region is not disulphide linked to the large N terminus.

27.3 Cellular and subcellular localization

The CaR was first found in the CNS by Ruat *et al* (1995) who cloned it from a rat striatal cDNA library. By *in situ* hybridization, CaR mRNA has since been localized to multiple cell-types widely scattered throughout the brain with the highest levels of expression in the olfactory bulb, hippocampus, amygdala, zona incerta, hypothalamus, koliker-fuse nucleus, Purkinje cells, and the sub-fornical organ (SFO; Rogers *et al.* 1997). Neuronal CaR expression is highest of all in the SFO or 'thirst-centre', where it could potentially integrate ionic and water homeostatic mechanisms. In all of these CNS regions, the function of CaR remains unclear but it has been associated with regulation of local ionic homeostasis, long-term potentiation, cognitive function, regulation of neurotransmitter release, and neuronal excitability.

Expression of CaR in the rat hippocampus rises at a time when long-term potentiation (a putative *in vitro* analogue of memory) can first be induced in hippocampus and persists at high levels during the time when brain development is proceeding most rapidly (Chattopadhyay *et al.* 1997). CaR expression has been reported in immature oligodendrocytes and its activation most likely explains the effect of CaR agonists on inducing cellular proliferation and increasing the opening probability of an outward K^+ channel (Chattopadhyay *et al.* 1998*b*). Similarly, CaR expression has been reported in human astrocytomas, meningiomas and primary embryonic astrocytes and the amounts of PTH-related protein these cells secrete is elevated by CaR agonist treatment (Chattopadhyay *et al.* 2000). Thus, CaR also appears to affect the function of non-neuronal cells within the brain.

CaR is also found in the peripheral nervous system where it has been suggested to play a role in small artery contractility (Bukoski *et al.* 1997). The perivascular nerves that innervate rat mesenteric arteries have been shown to express CaR protein and in the presence of elevated

$[Ca^{2+}]_o$ there is a dose-dependent, endothelium-independent vasorelaxation of arteries pre-constricted with norepinephrine (Bukoski *et al.* 1997), possibly occurring via the release of anandamide (Ishioka and Bukoski 1999). In a different study, CaR has also been found in the smooth muscle cells of the spiral modiolar artery, a vessel that supplies blood to the cochlea however, in these vessels, increasing $[Ca^{2+}]_o$ causes vasoconstriction (Wonneberger *et al.* 2000). As described previously, the CaR was first discovered in the parathyroid glands and kidney and subsequently in a wide variety of tissues including all of those associated with calcium homeostasis. In kidney, CaR is found on the apical side of proximal tubule (PT) cells, the basolateral side of cortical thick ascending limb cells (cTAL), both sides of distal convoluted tubule (DCT) cells and the apical side of inner medullary collecting duct (IMCD) cells. The intestine is another important tissue for calcium homeostasis since it is the site of Ca^{2+} entry into the body, occurring under the control of $1, 25(OH)_2D_3$. Here, CaR has been identified in both small and large intestine and it has been suggested to modulate absorption (Chattopadhyay *et al.* 1998*a*). Finally, CaR expression has been reported in a number of bone-forming, osteoblast-derived cell lines. Although the role of the receptor in these cells is still controversial, CaR agonists have been shown to induce either chemotaxis and/or proliferation (Godwin *et al.* 1997).

27.4 **Pharmacology**

Primarily, the CaR responds to divalent metals such as Ca^{2+} and Mg^{2+}, trivalent metals including Gd^{3+} and polyvalent cations such as the aminoglycoside antibiotics and amyloid β-peptide (Brown *et al.* 1993; Ye *et al.* 1997, and see Table 27.2). Currently there are no CaR-specific agonists as such, however a group of phenylalkylamines including NPSR467 and R568 can act specifically on CaR as allosteric activators and thus are considered as 'calcimimetics' (Nemeth *et al.* 1998). That is, they are capable of increasing the sensitivity of CaR to challenge with $[Ca^{2+}]_o$ but in the absence of Ca^{2+}_o they are without effect. Calcimimetics are undergoing clinical trials for the treatment of primary and secondary hyperparathyroidism, as they will stimulate the CaR to inhibit PTH production and secretion. Another phenylalkylamine, NPS2143, is reported to be an allosteric antagonist of the CaR, a so-called 'calcilytic' (Gowen *et al.* 2000). These agents are currently being investigated as a possible treatment for osteoporosis since by antagonizing the CaR-mediated inhibition of PTH secretion, they appear to cause pulsatile PTH release, which in turn promotes bone formation (Gowen *et al.* 2000).

In addition to the phenylalkylamines acting as allosteric modulators of CaR function, there is evidence that certain L-amino acids stereoselectively increase the sensitivity of the receptor to $[Ca^{2+}]_o$ (Conigrave *et al.* 2000). The amino acids that exert the greatest shift in Ca^{2+}_o sensitivity are histidine (basic), phenylalanine, tyrosine and tryptophan (all aromatic), whereas arginine (basic) and isoleucine (neutral) have little effect and leucine (neutral) is totally without effect. This phenomenon potentially enables the integration of signals from distinct classes of mineral ions and amino acids (Conigrave *et al.* 2000) and thus the CaR could actually represent a nutrient sensor in cell types not involved in Ca^{2+}-homeostasis. It is also possible that amino acid-induced CaR activation is somehow involved in conditions of aromatic amino acid-induced neurotoxicity such as hepatic encephalopathy and in particular phenylketonuria (Conigrave *et al.* 2000), especially given that CaR is highly expressed in myelin-synthesizing oligodendrocytes (Chattopadhyay *et al.* 1998*b*).

Table 27.2 Differential pharmacology of CaR endogenously expressed in particular cell types

Cell/Tissue	Readout	Agonists	Ref
Bovine Parathyroid	↑ Ca^{2+}_i	EC_{50}'s for CaR agonists: 3 mM Ca^{2+}; 10–15 mM Mg^{2+}; 35 µM Gd^{3+}; 70 µM neomycin and 500 µM spermine. No response to Pb^{2+}, Ni^{2+}, Cd^{2+} or Al^{3+} (<500 µM) Stereoselective response to calcimimetics	Brown et al. 1993 Brown et al. 2001
Human keratinocytes	IP_3 production	Increased sensitivity to Ca^{2+}_o (0.07–0.25 mM) Stereoselective response to calcimimetics	Oda et al. 1998
Murine Leydig cells	↑ Ca^{2+}_i	Reduced sensitivity to Ca^{2+}_o (2.5–15 mM) and no response to Mg^{2+} (0.8–15 mM). Sensitive to Ni^{2+} (0.5–5 mM) Calcimimetics not tested	Adebanjo et al. 1998
Sheep thyroid parafollicular cells	Ionic conductances ↑ Ca^{2+}_i	No response to Mg^{2+} (up to 10 mM) Calcimimetics not tested	McGhee et al. 1997
Rabbit thick ascending limb	↑ Ca^{2+}_i	No response to 500 µM Gd^{3+}, 5 mM Mg^{2+} or 1 mM neomycin Calcimimetics not tested	Desfleurs et al. 1999
Liver	↑ Ca^{2+}_i	Greatly reduced sensitivity to Gd^{3+} (EC_{50} > 2 mM) and to spermine (EC_{50} ~ 5 mM) Stereoselective response to calcimimetics	Canaff et al. 2001
Pancreatic β cells (endocrine)	Insulin release, ↑ Ca^{2+}_i	Not sensitive to neomycin, Mg^{2+} or Gd^{3+} Stereoselective response to calcimimetics	Straub et al. 2000
Pancreatic acinar cells (exocrine)	↑ Ca^{2+}_i	Not sensitive to neomycin (up to 1 mM) Reduced sensitivity to Gd^{3+} (>1 mM) Calcimimetics not tested	Bruce et al. 1999
AT-3 prostate carcinoma cells	Cell death	Reduced sensitivity to Ca^{2+}_o (EC_{50}. 6.1 mM) and to Mg^{2+}_o (EC_{50}. 23.4 mM) Calcimimetics not tested	Lin et al. 1998

An interesting feature arising from the many studies of CaR pharmacology is that in some cell types, the receptor exhibits differential sensitivities to the various agonists. For example, for CaR overexpressed in *Xenopus* oocytes or HEK cells, the EC_{50} for Ca^{2+}_o is approximately 4 mM (Brown *et al.* 1993; Bai *et al.* 1998). However, in the body this would be considered supra-hypercalcaemic since the IC_{50} of Ca^{2+}_o-induced PTH release suppression is $\sim$1.25 mM (Brown *et al.* 1993). This value is ideal for maintaining a free $[Ca^{2+}]_o$ in the 1.1–1.3 mM range. However, in other regions of the body, cells are exposed to considerably higher concentration of $[Ca^{2+}]_o$ such as renal inner medullary collecting ducts where $[Ca^{2+}]_o$ can reach 5 mM and of course bone where $[Ca^{2+}]_o$ can reach 20–30 mM. To date, only one isoform of CaR has been found and thus it seems likely that CaR somehow adapts *in situ* to sense significantly differing levels of free Ca^{2+}. How this is achieved exactly is unclear, though interaction with other regulatory molecules such as caveolin-1 could play a part (Kifor *et al.* 1998). Table 27.2 lists a number of studies in which CaR pharmacology has been shown to be discrepant from its classic pharmacology as determined using bovine parathyroid cells.

27.5 Signal transduction and receptor modulation

CaRs couple both with $G_{q/11}\alpha$ and $G_i\alpha$ proteins (Brown *et al.* 1993; Tamir *et al.* 1996; Arthur *et al.* 1997) and accordingly CaR activation can cause both IP_3 formation and suppression of cAMP generation. In the presence of normal extracellular Mg^{2+} levels ($\sim$1 mM), there is a basal inhibition of this CaR/$G\alpha$ subunit axis due to direct binding of cytosolic Mg^{2+} to the $G\alpha$ subunits. However, reduction of $[Mg^{2+}]_o$ to below 0.5 mM removes this Mg^{2+}_i-dependent inhibition causing elevation in IP3 formation (due to G_q activation) and decreased cAMP formation (due to G_i activation) (Quitterer *et al.* 2000). In parathyroid cells, this results in suppressed PTH secretion and this could explain why, paradoxically, patients with severe hypomagnesemia exhibit blunted PTH secretion (Quitterer *et al.* 2000).

CaR agonists activate phospholipases C, A_2 and D in both CaR-transfected HEK cells and bovine parathyroid cells (Kifor *et al.* 1997) resulting in the formation of inositol trisphosphate, arachidonic acid, and phosphatidic acid respectively. CaR activation can also result in activation of serine/threonine kinases such as PKC (Kifor *et al.* 1997; Bai *et al.* 1998b). Indeed, there is evidence that the activity of the CaR itself may be modulated by PKC. Bai *et al.* (1998b) showed that phosphorylation of putative PKC phosphorylation sites in the intracellular domain of CaR (Brown *et al.* 1993; Riccardi *et al.* 1995) significantly decreases the Ca^{2+} sensitivity of the receptor.

CaRs have also been shown to induce protein tyrosine phosphorylation (Filvaroff *et al.* 1994) and may employ this to mediate certain responses such as CaR-mediated differentiation in mouse keratinocytes (Filvaroff *et al.* 1994) and rat fibroblasts (McNeil *et al.* 1998). More recently, a series of studies have reported that CaR stimulation can activate the extracellular signal-regulated kinases (ERKs) in a variety of cell types (McNeil *et al.* 1998; Yamaguchi *et al.* 2000; Kifor *et al.* 2001) and JNK activity in MDCK cells (Arthur *et al.* 2000).

27.6 Physiology and disease relevance

Additional physiological and disease roles to these previously described (i.e cellular proliferation, vasorelaxation and vasoconstriction) may well exist for the CaR (and see Table 27.2).

27.6.1 Alzheimer's disease and cognition

The activity of nonselective cation channels (NCC) in cultured rat hippocampal neurones was increased by polycationic CaR agonists (Ye et al. 1996a). This effect is likely to be mediated via the CaR since increased $[Ca^{2+}]_o$ also raised the opening probability for a similar NCC in HEK293 cells transfected with the cloned human CaR but not in non-transfected HEK cells (Ye et al. 1996b). Ye et al. (1997) also demonstrated that the amyloid β peptides (Aβ), produced in excess in Alzheimer's disease, cause CaR-mediated activation of a Ca^{2+}-permeable, NCC in both CaR-transfected HEK cells and in cultured rat hippocampal pyramidal neurones. The same effect was also seen in hippocampal pyramidal neurones collected from wild-type mice but not in those collected from mice with a targeted disruption of the CaR gene (CaR knockouts) (Ye et al. 1997). Sustained elevation of $[Ca^{2+}]_i$ produced by Aβ-induced CaR activation could contribute to the neuronal degeneration seen in Alzheimer's.

27.6.2 Hypercalcaemia and hyperparathyroidism

In the parathyroid gland, CaR not only suppresses PTH formation and release but also appears to inhibit proliferation of the secretory cells themselves. This is demonstrated by the fact that both mice that are homozygous for the CaR knockout and patients who are homozygous for inactivating mutations of CaR (neonatal severe hyperparathyroidism (NSHPT)) exhibit hyperplasia of their parathyroids. The condition in which a person is heterozygous for an inactivating mutation of CaR is called familial hypocalciuric hypercalcaemia (FHH) and fortunately is much more benign than the more dangerous NSHPT (Pollak et al. 1996). An inherited disorder resulting from activating mutations of CaR is autosomal dominant hypocalcaemia. Study of these conditions, which arise mostly from single point mutations in the CaR sequence, have revealed significant structure/function information regarding the likely Ca^{2+}-binding sites and the significance of CaR dimerization. In addition to suppressing PTH secretion, hypercalcaemia also stimulates release of calcitonin (CT) from thyroidal C cells through stimulation of a CaR (Garrett et al. 1995a). CT acts on osteoclasts to inhibit bone resorption, so reducing the release of calcium into the bloodstream and counteracting the hypercalcaemia (Austin et al. 1981).

27.6.3 Other peripheral indications

It still remains unclear exactly how CaR expression is regulated, and a number of studies have reported alterations in CaR expression under certain experimental and developmental conditions. For example, CaR expression is upregulated in the neonatal period of rat kidney development (Chattopadhyay et al. 1997) and in the parathyroid glands of sheep exposed to experimental burn injury (Murphey et al. 2000), possibly resulting from elevated levels of inflammatory cytokines. Meanwhile CaR becomes markedly downregulated in primary cultures of bovine parathyroid cells (Brown et al. 1995; Mithal et al. 1995) though less so in human parathyroid cells (Rousanne et al. 1998). Parathyroid CaR expression is also diminished in patients with primary hyperparathyroidism (Gogusev et al. 1997) or severe secondary hyperparathyroidism that can result from chronic haemodialysis in patients with renal failure in end-stage renal disease (Kifor et al. 1996). In addition, CaR protein levels are reduced in the kidneys of rats made diabetic with streptozotocin (Ward et al. 2001) and in the proximal tubules of rats exposed to high P_i diet or acute PTH treatment (Riccardi et al. 2000).

There is also some evidence that CaR becomes downregulated in the parathyroids of rats treated with a high P_i diet (Brown *et al.* 1999). The molecular mechanisms controlling either the regulation of CaR gene expression or indeed of CaR protein degradation have not yet been identified. In cTAL (Hebert *et al.* 1997) and DCT (Blankenship *et al.* 2001) cells of the kidney, CaR is apparently involved in preventing excessive Ca^{2+} reabsorption, whereas in IMCD cells it inhibits excess vasopressin-elicited water reabsorption, presumably to protect against stone formation (Sands *et al.* 1997). Finally, Ca^{2+}, Ba^{2+} and the calcimimetic compound NPS568 all induced potent stimulation of adrenocorticotropic hormone (ACTH) secretion in AtT-20 cells (Ferry *et al.* 1997) suggesting a possible role for CaR in controlling secretion of ACTH, a hormone implicated in several types of stress.

27.7 Concluding remarks

The last decade has seen fundamental change in our understanding of cell responsiveness to extracellular calcium and much of this has been due to the discovery of the cell surface CaR. Since not every effect of $[Ca^{2+}]_o$ is necessarily due to CaR activation it remains to be determined the precise physiological functions of the CaR at each of its many sites of expression including the CNS. Furthermore, since in some CaR-expressing tissues $[Ca^{2+}]_o$ remains constant, it is likely that Ca^{2+} is not even the exclusive physiological agonist of the receptor and that endogenous polycations and amino acids also exert physiological, or indeed pathological, effects via CaR activation. Thus, despite having been discovered only relatively recently, the CaR is a G protein-coupled receptor that senses some of nature's most fundamental ions and molecules essential for the welfare of vertebrates.

References

Adebanjo OA, Igietseme J, Huang CL, and Zaidi M (1998). The effect of extracellularly applied divalent cations on cytosolic Ca^{2+} in murine leydig cells: evidence for a Ca^{2+}-sensing receptor. *J Physiol* 513, 399–410.

Arthur JM, Collinsworth GP, Gettys TW, Quarles LD, and Raymond JR (1997). Specific coupling of a cation-sensing receptor to G protein α-subunits in MDCK cells. *American J Physiol* 273, F129–35.

Arthur JM, Lawrence MS, Payne CR, Rane MJ, and McLeish KR (2000). The calcium-sensing receptor stimulates JNK in MDCK cells. *Biochem Biophys Res Commun* 275, 538–41.

Austin LA and Heath H 3rd (1981). Calcitonin: physiology and pathophysiology. *N Eng J Med* 304, 269–78.

Bai M, Trivedi S, and Brown EM (1998). Dimerization of the extracellular calcium-sensing receptor (CaR) on the cell surface of CaR-transfected HEK293 cells. *J Biol Chem* 273, 23 605–10.

Bai M, Trivedi S, Lane CR, Yang Y, Quinn SJ, and Brown EM (1998). Protein kinase C phosphorylation of threonine at position 888 in Ca^{2+}_o-sensing receptor (CaR) inhibits coupling to Ca^{2+} store release. *J Biol Chem* 273, 21 267–75.

Bai M, Trivedi S, Kifor O, Quinn SJ, and Brown EM (1999). Intermolecular interactions between dimeric calcium-sensing receptor monomers are important for its normal function. *Proc Natl Acad Sci (USA)* 96, 2834–39.

Blankenship KA, Williams JJ, Lawrence MS, McLeish KR, Dean WL, and Arthur JM (2001). The calcium-sensing receptor regulates calcium absorption in MDCK cells by inhibition of PMCA. *Am J Physiol* 280, F815–22.

Breitwieser GE, Gama L, and Wilt S (2001). Heterodimerization of calcium sensing receptors with group 1 metabotropic glutamate receptors in hippocampal and cerebellar neurons. *Fed Am Soc Exp Biol J* 15, A1114.

Brown AJ, Zhong M, Ritter C, Brown EM, and Slatopolsky E (1995). Loss of calcium responsiveness in cultured bovine parathyroid cells is associated with decreased calcium receptor expression. *Biochem Biophys Res Commun* 212, 861–67.

Brown AJ, Ritter CS, Finch JL, and Slatopolsky EA (1999). Decreased calcium-sensing receptor expression in hyperplastic parathyroid glands of uremic rats: role of dietary phosphate. *Kidney Int* 55, 1284–92.

Brown EM, Gamba G, Riccardi D, *et al.* (1993). Cloning and characterization of an extracellular Ca^{2+}-sensing receptor from bovine parathyroid. *Nature* 366, 575–80.

Brown EM and MacLeod RJ (2001). Extracellular Calcium sensing and extracellular calcium signaling. *Physiol Rev* 81, 239–97.

Bruce JI, Yang X, Ferguson CJ, *et al.* (1999). Molecular and functional identification of a Ca^{2+} (polyvalent cation)-sensing receptor in rat pancreas. *J Biol Chem* 274, 20 561–68.

Bukoski RD, Bian K, Wang Y, and Mupanomunda M (1997). Perivascular sensory nerve Ca^{2+} receptor and Ca^{2+}-induced relaxation of isolated arteries. *Hypertension* 30, 1431–9.

Canaff L, Petit JL, Kisiel M, Watson PH, Gascon-Barre M, and Hendy GN (2000). Extracellular calcium-sensing receptor is expressed in rat hepatocytes. Coupling to intracellular calcium mobilization and stimulation of bile flow. *J Biol Chem* 276, 4070–9.

Chattopadhyay N, Legradi G, Bai M, *et al.* (1997). Calcium-sensing receptor in the rat hippocampus: a developmental study. *Brain Res Dev Brain Res* 100, 13–21.

Chattopadhyay N, Cheng I, Rogers K, *et al.* (1998a). Identification and localization of extracellular Ca^{2+}-sensing receptor in rat intestine. *Am J Physiol* 274, G122–30.

Chattopadhyay N, Ye CP, Yamaguchi T, *et al.* (1998b). Extracellular calcium-sensing receptor in rat oligodendrocytes: expression and potential role in regulation of cellular proliferation and an outward K^+ channel. *Glia* 24, 449–58.

Chattopadhyay N, Evliyaoglu C, Heese O, *et al.* (2000) Regulation of secretion of PTHrP by (Ca^{2+})-sensing receptor in human astrocytes, astrocytomas, and meningiomas. *American J Physiol* 279, C691–99.

Chikatsu N, Fukumoto S, Takeuchi Y, *et al.* (2000) Cloning and characterization of two promoters for the human calcium-sensing receptor (CaSR) and changes of CaSR expression in parathyroid adenomas. *J Biol Chem* 275, 7553–57.

Chou YH, Brown EM, Levi T, *et al.* (1992). The gene responsible for familial hypocalciuric hypercalcemia maps to chromosome 3q in four unrelated families. *Nat Genet* 4, 295–300.

Conigrave AD, Quinn SJ, and Brown EM (2000). L-amino acid sensing by the extracellular Ca^{2+}-sensing receptor. *Proc Natl Acad Sci (USA)* 97, 4814–19.

Desfleurs E, Wittner M, Pajaud S, Nitschke R, Rajerison RM, and Di Stefano A (1999). The Ca^{2+}-sensing receptor in the rabbit cortical thick ascending limb (CTAL) is functionally not coupled to phospholipase C. *Pflugers Archive* 437, 716–23.

Ferry S, Chatel B, Dodd RH, *et al.* (1997). Effects of divalent cations and of a calcimimetic on adrenocorticotropic hormone release in pituitary tumor cells. *Biochem Biophys Res Commun* 238, 866–73.

Filvaroff E, Calautti E, Reiss M, and Dotto GP (1994). Functional evidence for an extracellular calcium receptor mechanism triggering tyrosine kinase activation associated with mouse keratinocyte differentiation. *J Biol Chem* 269, 21 735–40.

Garrett JE, Tamir H, Kifor O, *et al.* (1995a). Calcitonin-secreting cells of the thyroid express an extracellular calcium receptor gene. *Endocrinology* 136, 5202–11.

Garrett JE, Capuano IV, Hammerland LG, *et al.* (1995b). Molecular cloning and functional expression of human parathyroid calcium receptor cDNAs. *J Biol Chem* 270, 12 919–25.

Godwin SL and Soltoff SP (1997). Extracellular calcium and platelet-derived growth factor promote receptor-mediated chemotaxis in osteoblasts through different signaling pathways. *J Biol Chem* 272, 11 307–12.

Gogusev J, Duchambon P, Hory B, *et al.* (1997). Depressed expression of calcium receptor in parathyroid gland tissue of patients with hyperparathyroidism. *Kidney Int* 51, 328–36.

Goldsmith PK, Fan GF, Ray K, *et al.* (1999). Expression, purification, and biochemical characterization of the amino-terminal extracellular domain of the human calcium receptor. *J Biol Chem* 274, 11 303–09.

Gowen M, Stroup GB, Dodds RA, *et al.* (2000). Antagonizing the parathyroid calcium receptor stimulates parathyroid hormone secretion and bone formation in osteopenic rats. *J Clin Invest* 105, 1595–604.

Hebert SC, Brown EM, and Harris HW (1997). Role of calcium-sensing receptor in divalent mineral homeostasis. *J Exp Med* 200, 295–302.

Hu J, Reyes-Cruz G, Goldsmith PK, and Spiegel AM (2001). The Venus's-flytrap and cysteine-rich domains of the human Ca^{2+} receptor are not linked by disulfide bonds. *J Biol Chem*, 276, 6901–04.

Ishioka N and Bukoski RD (1999). A role for N-arachidonylethanolamine (anandamide) as the mediator of sensory nerve-dependent Ca^{2+}-induced relaxation. *J Pharmacol Exp Ther* 289, 245–50.

Janicic N, Soliman E, Pausova Z *et al.* (1995). Mapping of the calcium-sensing receptor gene (CASR) to human chromosome 3q13.3–21 by fluorescence *in situ* hybridization, and localization to rat chromosome 11 and mouse chromosome 16. *Mammalian Genome* 6, 798–801.

Kifor O, Moore FD Jr, Wang P *et al.* (1996). Reduced immunostaining for the extracellular Ca^{2+}-sensing receptor in primary and uremic secondary hyperparathyroidism. *J Clin Endocrinol Metabol* 81, 1598–606.

Kifor O, Diaz R, Butters R, and Brown EM (1997). The Ca^{2+}-sensing receptor (CaR) activates phospholipases C, A2, and D in bovine parathyroid and CaR-transfected, human embryonic kidney (HEK293) cells. *J Bone Miner Res* 12, 715–25.

Kifor O, Diaz R, Butters R, Kifor I, and Brown EM (1998). The calcium-sensing receptor is localized in caveolin-rich plasma membrane domains of bovine parathyroid cells. *J Biol Chem* 273, 21 708–13.

Kifor O, MacLeod RJ, Diaz R, *et al.* (2001). Regulation of MAP kinase by calcium-sensing receptor in bovine parathyroid and CaR-transfected HEK293 cells. *Am J Physiol* 280, F291–F302.

Lin KI, Chattopadhyay N, Bai M, *et al.* (1998). Elevated extracellular calcium can prevent apoptosis via the calcium-sensing receptor. *Biochem Biophys Res Commun* 249, 325–31.

McGehee DS, Aldersberg M, Liu KP, Hsuing S, Heath MJ, and Tamir H. (1997). Mechanism of extracellular Ca^{2+} receptor-stimulated hormone release from sheep thyroid parafollicular cells. *J Physiol* 502, 31–44.

McNeil SE, Hobson SA, Nipper V, and Rodland KD (1998). Functional calcium-sensing receptors in rat fibroblasts are required for activation of SRC kinase and mitogen-activated protein kinase in response to extracellular calcium. *J Biol Chem* 273, 1114–20.

Mithal A, Kifor O, Kifor I, *et al.* (1995). The reduced responsiveness of cultured bovine parathyroid cells to extracellular Ca^{2+} is associated with marked reduction in the expression of extracellular $Ca(^{2+})$-sensing receptor messenger ribonucleic acid and protein. *Endocrinology* 136, 3087–92.

Murphey ED, Chattopadhyay N, Bai M, *et al.* (2000). Up-regulation of the parathyroid calcium-sensing receptor after burn injury in sheep: a potential contributory factor to postburn hypocalcemia. *Crit Care Med* 28, 3885–90.

Nemeth EF, Steffey ME, Hammerland LG, *et al.* (1998). Calcimimetics with potent and selective activity on the parathyroid calcium receptor. *Proc Natl Acad Sci (USA)* 95, 4040–5.

Oda Y, Tu CL, Pillai S, and Bikle DD (1998). The calcium sensing receptor and its alternatively spliced form in keratinocyte differentiation. *J Biol Chem* 273, 23 344–52.

Pearce SH, Bai M, Quinn SJ, Kifor O, Brown EM, and Thakker RV (1996). Functional characterization of calcium-sensing receptor mutations expressed in human embryonic kidney cells. *J Clin Invest* 98, 1860–6.

Pollak MR, Seidman CE, and Brown EM (1996). Three inherited disorders of calcium sensing. *Medicine (Baltimore)* 75, 115–23.

Quitterer U, Hoffmann M, Freichel M, and Lohse M (2000). Paradoxical block of parathormone secretion is mediated by increased activity of G-alpha subunits. *J Biol Chem* **276**, 6763–9.

Ray K, Hauschild BC, Steinbach PJ, Goldsmith PK, Hauache O, and Spiegel AM (1999). Identification of the cysteine residues in the amino-terminal extracellular domain of the human Ca^{2+} receptor critical for dimerization. Implications for function of monomeric Ca^{2+} receptor. *J Biol Chem* **274**, 27 642–50.

Riccardi D, Park J, Lee W, Gamba G, Brown EM, and Hebert SC (1995). Cloning and functional expression of a rat kidney extracellular calcium/polyvalent cation-sensing receptor. *Proc Natl Acad Sci (USA)* **92**, 131–5.

Riccardi D, Traebert M, Ward DT, *et al.* (2000). Dietary phosphate and PTH alter the expression of the calcium-sensing receptor and the Na^+-dependent P_i transporter (NaPi-2) in the rat proximal tubule. *Pflügers Arch Eur J Physiol* **441**, 379–87.

Rogers KV, Dunn CK, Hebert SC, and Brown EM (1997). Localization of calcium receptor mRNA in the adult rat central nervous system by in situ hybridization. *Brain Res* **744**, 47–56.

Romano C, Yang WL, and O'Malley KL (1996). Metabotropic glutamate receptor 5 is a disulfide-linked dimer. *J Biol Chem* **271**, 28 612–16.

Roussanne MC, Gogusev J, Hory B, *et al.* (1998). Persistence of Ca^{2+}-sensing receptor expression in functionally active, long-term human parathyroid cell cultures. *J Bone Miner Res* **13**, 354–62.

Ruat M, Snowman AM, Hester LD, and Snyder SH (1996). Cloned and expressed rat Ca^{2+}-sensing receptor. *J Biol Chem* **271**, 5972–5.

Sands JM, Naruse M, Baum M, *et al.* (1997). Apical extracellular calcium/polyvalent cation-sensing receptor regulates vasopressin-elicited water permeability in rat kidney inner medullary collecting duct. *J Clin Invest* **99**, 1399–405.

Straub SG, Kornreich B, Oswald RE, Nemeth EF, and Sharp GW (2000). The calcimimetic R-467 potentiates insulin secretion in pancreatic beta cells by activation of a nonspecific cation channel. *J Biol Chem* **275**, 18 777–84.

Tamir H, Liu KP, Adlersberg M, Hsiung SC, and Gershon MD (1996). Acidification of serotonin-containing secretory vesicles induced by a plasma membrane calcium receptor. *J Biol Chem* **271**, 6441–50.

Ward DT, Brown EM, and Harris HW (1998). Disulfide bonds in the extracellular calcium-polyvalent cation-sensing receptor correlate with dimer formation and its response to divalent cations in vitro. *J Biol Chem* **273**, 14 476–83.

Ward DT, Yau SK, Mee AP, *et al.* (2001). Functional, molecular and biochemical characterization of Streptozotocin-induced diabetes. *J Am Soc Nephrol* **12**, 779–90.

Wonneberger K, Scofield MA, and Wangemann P (2000). Evidence for a calcium-sensing receptor in the vascular smooth muscle cells of the spiral modiolar artery. *J Membr Biol* **175**, 203–12.

Yamaguchi T, Chattopadhyay N, Kifor O, Sanders JL, and Brown EM (2000). Activation of p42/44 and p38 mitogen-activated protein kinases by extracellular calcium-sensing receptor agonists induces mitogenic responses in the mouse osteoblastic MC3T3-E1 cell line. *Biochem Biophys Res Commun* **279**, 363–8.

Ye C, Kanazirska M, Quinn S, Brown EM, and Vassilev PM (1996*a*). Modulation by polycationic Ca^{2+}-sensing receptor agonists of nonselective cation channels in rat hippocampal neurons. *Biochem Biophys Res Commun* **224**, 271–80.

Ye C, Rogers K, Bai M, Quinn SJ, Brown EM, and Vassilev PM (1996*b*). Agonists of the $Ca(^{2+})$-sensing receptor (CaR) activate nonselective cation channels in HEK293 cells stably transfected with the human CaR. *Biochem Biophys Res Commun* **226**, 572–79.

Ye C, Ho-Pao CL, Kanazirska M, *et al.* (1997). Amyloid-beta proteins activate Ca^{2+}-permeable channels through calcium-sensing receptors. *J Neurosci Res* **47**, 547–54.

Zhang Z, Sun S, Quinn SJ, Brown EM, and Bai M (2000). The extracellular calcium-sensing receptor dimerizes through multiple types of intermolecular interactions. *J Biol Chem* **276**, 5316–22.

Chapter 28

GABA$_B$ receptors

Norman G. Bowery

28.1 Introduction

The GABA$_B$ receptor was originally described based on pharmacological observations. In studies focusing on GABA control of transmitter release it was noted that the GABA receptor responsible for modulating the evoked release of transmitters, in a variety of isolated tissue preparations, appeared to have different pharmacological characteristics from that of the receptor responsible for the Cl$^-$ dependent pre- and post-synaptic action of GABA. The modulatory action of GABA could not be blocked by bicuculline, was not mimicked by isoguvacine, was activated by only high concentrations of the normally potent agonist, muscimol and was not dependent on [Cl$^-$] but was dependent on [Ca^{2+}] or [Mg^{2+}]. Most striking of all was that the therapeutic agent, baclofen (β-chlorophenyl GABA), which had already been introduced into therapeutics as an antispastic agent on the basis that it would mimic the effect of endogenous GABA (Bein 1972; Keberle and Faigle 1972), was stereospecifically active at this novel receptor. Membrane receptor binding studies using radiolabelled baclofen and GABA provided the crucial evidence showing the existence of distinct binding sites on central neurones. It was then that the term 'GABA$_B$ receptor' was coined to distinguish the receptor from the bicuculline-sensitive receptor, which was in turn designated 'GABA$_A$' (Hill and Bowery 1981).

In previous chapters of this volume it is apparent that G protein coupled receptors can be expressed in cell membranes in different ways. They may simply reside as a single protein unit after initial transcription in the endoplasmic reticulum or in some instances the functional receptor may be expressed as a combination of two identical protein molecules linked to form a homodimer. The discovery of the structure of the metabotropic GABA$_B$ receptor, in 1998, introduced even further complexity to the situation as the functional GABA$_B$ receptor was revealed to exist as a dimer but the two protein components, designated GABA$_{B1}$ and GABA$_{B2}$ (White *et al.* 1998; Kaupmann *et al.* 1998a; Jones *et al.* 1998; Kuner *et al.* 1999; Ng *et al.* 1999), were structurally different and thus, the native receptor exists as a unique heterodimer.

28.2 Molecular characterization

Detection of this structural dimer was triggered by the original cloning of the receptor subunit, GABA$_{B1}$, by Bettler and colleagues (Kaupmann *et al.* 1997). This receptor was shown to be a large molecular weight, seven transmembrane domain monomer. However, it soon became apparent that the binding affinity of agonists at this single receptor subunit, expressed in COS-1 cells, was up to three orders of magnitude lower than for wild-type receptors in brain homogenates. This was not altered if other cell types (e.g. CHO cells) were used for the expression studies. Separate studies also demonstrated that the GABA$_{B1}$ protein is not normally expressed in the plasma membrane of such cells but remains in the endoplasmic reticulum (Couve *et al.* 1998). This led to the possibility than an unknown trafficking protein such as a 'RAMP' (receptor activity modifying protein), which had previously been described in connection with the membrane expression of CGRP receptors (McLatchie *et al.* 1998), might be necessary to obtain the full expression of GABA$_{B1}$. Studies reported by four independent groups in December 1998 and January 1999 provided the initial evidence that the GABA$_B$ receptor exists as a heterodimer with a second receptor subunit apparently linked to GABA$_{B1}$ through coiled-coil domains at the C-terminal (Jones *et al.* 1998; White *et al.* 1998; Kaupmann *et al.* 1998*a*; Kuner *et al.* 1999) in a stoichiometric 1 : 1 ratio (Fig. 28.1). This second subunit has been designated GABA$_{B2}$ and has many of the structural features of GABA$_{B1}$ including a large molecular weight (110 kDa), 7 transmembrane domains, a long extracellular chain at the N-terminus with 54 per cent similarity (35 per cent homology). However, whether it is really a receptor or a trafficking protein is unclear as no binding sites have been demonstrated on GABA$_{B2}$ and large majority of functional studies using this subunit have been negative. However, the protein has been reported to be expressed in the absence of GABA$_{B1}$ to produce a functional site by some (Kaupmann *et al.* 1998*a*) and to couple to adenylate cyclase in cultured Chinese Hamster Ovary (CHO) cells by

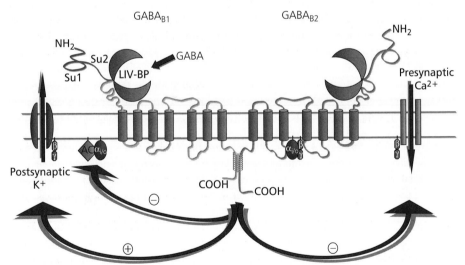

Fig. 28.1 Schematic representation of GABA$_B$ receptor system and its downstream signalling pathways. The GABA$_{B1a}$ subunit containing two Sushi repeats (Su1 and Su2) is depicted to interact via its C-terminal coiled-coil domain with the coiled-coil domain of the GABA$_{B2}$ subunit. The N-terminal domain of the GABA$_{B1}$ subunit contains a region with clear homology to the leucine-isoleucine-valine bacterial periplasmic binding protein (LIV-BP), and is predicted to form two lobes able to trap the agonist in a manner reminiscent of a Venus Flytrap. Evidence that the N-terminal region of GABA$_{B2}$ is able to bind GABA remains in question. The GABA$_B$ receptor couples through G$_{i/o}$ to inhibit adenylyl cyclase. The βγ subunit of the G protein activates inwardly rectifying potassium channels and inhibits voltage gated calcium channels.

others (Martin *et al.* 1999). Whether this would occur in native tissues has been questioned. Certainly the coupling of GABA_B2 with GABA_B1 enables the receptor to be fully expressed in the plasma membrane of cells allowing the high affinity binding of radiolabelled ligands to occur at GABA_B1 (White *et al.* 1998; Kaupmann *et al.* 1998a). Moreover, it appears that GABA_B1 expressed as part of the heterodimer has the same affinity for receptor agonists as the native receptor when assessed by displacement of bound radiolabelled antagonist (White *et al.* 1998).

At least six isoforms (1a–1f) of subunit GABA_B1 (Fig. 28.2) and three forms of subunit GABA_B2 have been reported (Kaupmann *et al.* 1997; Isomoto *et al.* 1998; Pfaff *et al.* 1999; Calver *et al.* 2000; Schwarz *et al.* 2000; Wei *et al.* 2001) but whether these all make functional forms of the receptor is unclear. At present it is known that if 1a, 1b, or 1c is expressed with subunit 2 and they can all produce a functional receptor (Kaupmann *et al.* 1998a; Pfaff *et al.* 1999). The ligand binding domain of the heterodimer (Galvez *et al.* 1999, 2000; Malitschek *et al.* 1999; Pin *et al.* 2000) appears to reside in the N-terminal segment of GABA_B1 with no evidence for any ligand binding to GABA_B2. Current evidence would suggest that even though GABA_B2 has a large N-terminus comparable in length to that in GABA_B1 with considerable homology between them, GABA_B2 is more important for effector coupling and signalling of the receptor (Jones *et al.* 2000; Calver *et al.* 2001; Robbins *et al.* 2001; Galvez *et al.* 2001). On this basis one could predict that G protein coupling is only mediated through the GABA_B2 subunit and this seems to be the current view. Thus, the ligand binds to the GABA_B1 subunit and this, presumably, produces a conformational change in the subunit which in turn activates GABA_B2 to enable the G protein coupled signalling mechanism to proceed (Calver *et al.* 2001; Robbins *et al.*, 2001; Galvez *et al.* 2001). In support of this concept it has been shown that GABA_B2 must remain linked to GABA_B1 after it has been transported to the cell membrane and only GABA_B2 can fulfil this role. While many other

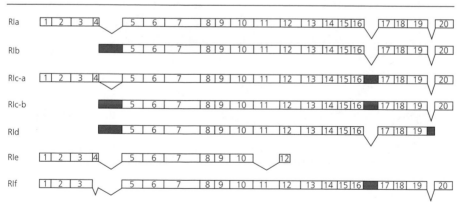

Fig. 28.2 Schematic summary of the isoforms of rat GABA_BR1. These include GABA_BR1a (R1a), GABA_BR1b (R1b), GABA_BR1c-a form (R1c-a), GABA_BR1c-b (R1c-b), GABA_BR1d (R1d), GABA_BR1e (R1e) and GABA_BR1f (R1f). Rat GABA_BR1 contains 20 exons indicated by numbered unfilled boxes 1–20. Alternative spliced exons are indicated by black boxes. Wedged lines indicate intronic sequences where alternative splicing of GABA_BR1 occurs. The R1f isoform results from omission of exon 4 and insertion of 93 bp between exon 16 and 17. (Taken from Wei *et al.* (2001) with permission.)

proteins have been shown to be capable of associating with high affinity to GABA$_{B1}$ and in some cases enable it to be transported to the cell membrane, none have been able to produce a functional receptor (White *et al.* 2000; Nehring *et al.* 2000; Couve *et al.* 2001). Conversely, the direct interaction of GABA$_{B1/2}$ protein with transcription factors such as ATF-4 may serve to regulate activity dependent gene expression (Nehring *et al.* 2000).

The coupling of GABA$_{B1}$ and GABA$_{B2}$ within cells seems to be crucial for the correct assembly of the heterodimer in the surface membrane. The interaction of the C-terminal coiled-coil domains of the subunits appears to mask the action of the retention motif present in the C-terminal of GABA$_{B1}$ (Margeta-Mitrovic *et al.* 2000) and these authors suggest that this provides a 'trafficking checkpoint' to ensure organized assembly of the functional receptor. Expression of the coupled heterodimer in cell membranes can still arise when the C-terminal in GABA$_{B1}$ has been removed or even when the C-terminals in GABA$_{B1}$ and GABA$_{B2}$ have both been removed (Calver *et al.* 2001; Pagano *et al.* 2001). But in neither case is the receptor functional although these observations indicate that the subunits can be linked through other points on their structures and not just the C-termini to allow expression of the dimer. The coiled-coil structures are clearly not essential for the heterodimerization to occur. Other proteins which can associate with GABA$_{B1}$, such as mGlu4 (Sullivan *et al.* 2000), and which facilitate expression presumably also mask the retention signal in GABA$_{B1}$ but, unlike GABA$_{B2}$ are unable to provide effector coupling within the cell membrane. Whether there are other association proteins which remain to be discovered and which can form functional heterodimers is an intriguing possibility. A current summary of some of the characteristics of the subunit relationships are highlighted in Table 28.1.

A recent and important step towards understanding the role of GABA$_{B1}$ has been made by Prosser *et al.* (2001) with the production of a GABA$_{B1}$ subunit knock-out mouse. It appears from their *in vitro* and *in vivo* studies with these mice that the GABA$_{B1}$ subunit is essential for all GABA$_{B}$-mediated effects. The knock-out animals die prematurely around 4 weeks post-natal suggesting that the GABA$_{B}$ receptor system provides a major inhibitory role within the CNS during the animal's early development.

28.3 Cellular and subcellular localization

Mammalian brain regions with the highest density of GABA$_{B}$ binding sites are the thalamic nuclei, the molecular layer of the cerebellum, the cerebral cortex, the interpeduncular nucleus, and the dorsal horn of the spinal cord (Bowery *et al.* 1987; Chu *et al.* 1990).

Distribution studies of mRNA for two of the splice variants of the GABA$_{B1}$ subunit, GABA$_{B(1a)}$ and GABA$_{B(1b)}$ using *in situ* hybridization techniques have revealed that they appear to be differentially distributed (Liang *et al.* 2000). Initial studies suggested that the individual isoforms may even be associated with separate brain structures. Studies in rat and human cerebellum and spinal cord indicate that GABA$_{B(1a)}$ is associated with pre-synaptic receptors whereas GABA$_{B(1b)}$ may be responsible for post-synaptic receptor formation in the cerebellum (Kaupmann *et al.* 1998*b*; Billinton *et al.* 1999; Bischoff *et al.* 1999; Princivalle *et al.* 2000; Towers *et al.* 2000). However, other studies elsewhere in the brain have revealed that the converse can also occur, that is 1b pre-synaptic and 1a post-synaptic (Benke *et al.* 1999; Princivalle *et al.* 2001; Charles *et al.* 2001). In the dorsal horn of the rat spinal cord the density of GABA$_{B(1a)}$ mRNA is low whereas in the dorsal root ganglia, which contain the cell bodies of the primary afferent fibres, >90 per cent of the GABA$_{B}$ subunit mRNA.is GABA$_{B(1a)}$ with GABA$_{B(1b)}$ mRNA comprising less than 10 per cent of the total

Table 28.1 GABA$_B$ receptor subunits

	GABA$_{B1}$ subunit	GABA$_{B2}$ subunit
Alternative names		
Structural information (Accession no.)	h 961 aa (AJ012185)^{AS} r 960 aa (Y10369)^{AS} m 960 (Q9WV18)AS	h 941 aa (AJ012188) r 941 aa (AJ011318)
Chromosomal location	6p21.3	9q22.1
Selective agonists	(–) baclofen, 3-aminopropylphosphinic acid (3-APPA), 3 aminopropyl (methyl)phosphinic acid [3-APMPA], CGP44532	No ligand binding domain identified on this subunit
Selective antagonists	CGP35348, CGP36742, CGP55845A, CGP62349, SCH50911	No ligand binding domain identified on this subunit
Radioligands	[^{3}H]-GABA, [^{3}H]-baclofen, [^{3}H]-CGP54626, [^{3}H]-CGP62349, [^{125}I]-CGP64213, [^{125}I]-CGP71872	
G protein coupling	Coupling is via the GABA$_{B2}$ sub unit	G$_i$/G$_o$
Expression profile	Pre- and post-synaptic locations on CNS neurones, peripheral autonomic nerve fibres and sensory fibres, ubiquitiously throughout areas of the CNS including, thalamic nuclei, cerebellar molecular layer, cerebral cortex, interpeduncular nucleus, superior colliculus, dorsal horn and large motor neurons of the spinal cord	Similar to GABA$_{B1}$ subunit but lower levels in the striatum and some periperal tissues such as placenta, heart, lung and gut
Physiological function	Auto- and hetero-receptor mediated inhibition of transmitter release, slow inhibitory postsynaptic potentials	
Knockout phenotype	Prone to epileptic seizures, increased pain threshold, cognitive impairments and death dependent on strain of mouse	—
Disease relevance	Spasticity, analgesia, reversal of drug craving, absence seizures, cognitive decline, neuroprotection, depression and anxiety, asthma, chronic hiccups	

GABA$_{B1}$ mRNA (Towers *et al.* 2000). In support of this conclusion we have recently shown, using an immunocytochemical technique, that the level of GABA$_{B(1a)}$ protein appears to be higher than that of GABA$_{B(1b)}$ in the dorsal horn of the spinal cord (Princivalle and Bowery, unpublished observations). A similar conclusion has arisen from observations in rat and human cerebellum where GABA$_{B(1a)}$ mRNA was detected over the granule cells which send their excitatory fibres into the molecular layer to innervate the Purkinje cell dendrites (Kaupmann *et al.* 1998*b*; Billinton *et al.* 1999; Bischoff *et al.* 1999). The terminals of these fibres express GABA$_B$ sites to modulate the output of the excitatory transmitter. In contrast, GABA$_{B(1b)}$ mRNA was observed to be associated with the Purkinje cell bodies which express GABA$_B$ receptors on their dendrites in the molecular layer. These would be post-synaptic to GABAergic Stellate cells. However, we and others have also observed the contrary arrangement in, for example, the thalamo-cortical circuitry where GABA$_{B(1a)}$ subunits appear to be post-synaptic on cell bodies (Princivalle *et al.* 2001). As a consequence it seems unlikely that functional role and cellular location can be assigned in a general manner to specific GABA$_B$ receptor subunit splice variants (Poorrkhalkali *et al.* 2000; Princivalle *et al.* 2001; Charles *et al.* 2001).

The regional distributions of GABA$_{B1}$ and GABA$_{B2}$ protein subunits are in broad alignment with that of the native receptor (e.g. Margeta-Mitrovic *et al.* 1999; Durkin *et al.* 1999; Charles *et al.* 2001) although in some brain areas, such as the caudate putamen, GABA$_{B1}$ and the native receptor are present whereas GABA$_{B2}$ appears to be absent (Durkin *et al.* 1999; Clark *et al.* 2000). In addition it has been noted that there is only a low level of GABA$_{B2}$ mRNA relative to GABA$_{B1}$ mRNA in the hypothalamus (Jones *et al.* 1998; Clark *et al.* 2000). This would surely support the existence of an additional, as yet undiscovered, subunit.

Functional GABA$_B$ receptors are not restricted to the CNS but appear to be significant in the periphery as well. In the enteric nervous system GABA has long been considered to play a crucial role in modulating autonomic inputs to the intestine (see Ong and Kerr 1990) and clearly GABA$_B$ receptors are implicated in mediating numerous effects in peripheral organs. For example, a particular focus which has attracted considerable attention in the recent past is the influence of the GABA$_B$ system on vagal innervation of the oesophageal sphincter. GABA$_B$ receptor agonists inhibit relaxations of the lower oesophageal sphincter in dogs, ferrets, and humans and inhibit oesophageal reflux by an inhibitory action on the vagus (Lehmann *et al.* 1999; Smid and Blackshaw 2000; Lidums *et al.* 2000) and it is suggested that this might provide a potential treatment for reflux disease in man. The intestine is not the only peripheral organ where GABA$_B$ receptors have been detected. Original studies monitoring functional responses in mammalian isolated tissues indicated a wide peripheral distribution (see Bowery 1993). More recent studies using Northern blot analysis have also shown the presence of RNA's for GABA$_{B(1a)}$ and GABA$_{B(1b)}$ throughout the periphery of the rat (Castelli *et al.* 1999). Detection of the receptor protein by immunoblotting has also shown a wide distribution of the receptor subunits throughout the periphery (Calver *et al.* 2000). However, the GABA$_{B2}$ subunit was not always present, for example, in the uterus and spleen where the GABA$_{B1}$ subunits could be detected (Calver *et al.* 2000). In rat heart myocytes Western blotting has enabled the presence of GABA$_{B1}$ and GABA$_{B2}$ proteins to be demonstrated (Lorente *et al.* 2000). This supports the observation that baclofen has an action on inwardly rectifying K$^+$ currents in these cells. Photoaffinity labelling studies by Belley *et al.* (1999) suggest that GABA$_{B(1a)}$ and GABA$_{B(1b)}$ are differentially distributed in the periphery as well as in the CNS. These authors observed that GABA$_{B(1a)}$ was not present in the rat kidney or liver whereas GABA$_{B(1b)}$ was not in adrenal gland tissue, pituitary, spleen, or prostate but was in the kidney and liver whilst GABA$_{B(1a)}$ was present in the adrenals, pituitary, spleen, and prostate.

28.4 **Pharmacology**

Although GABA$_B$ receptors are structurally heterogeneous the current evidence for receptor subtypes which are functionally distinct is limited. Transmitter release studies have suggested differences between heteroreceptors and autoreceptors (Gemignani *et al.* 1994; Ong *et al.* 1998; Bonanno *et al.* 1998; Phelan 1999) and the dual action of GABA$_B$ agonists on adenylate cyclase in brain slices would support the concept of receptor subtypes (Cunningham and Enna 1996). Also comparative data obtained with native GABA$_B$ receptors and recombinant GABA$_{B(1b)}$/GABA$_{B2}$/Gqi5 receptors expressed in CHO cells indicate that the latter receptor, when compared with the native receptor, was insensitive to the antagonists phaclofen, saclofen, and CGP35348 (Wood *et al.* 2000). However, comparison of the two isoforms GABA$_{B(1a)}$ and GABA$_{B(1b)}$ expressed with GABA$_{B2}$ in CHO cells indicated that they were pharmacologically indistinguishable (Green *et al.* 2000). Electrophysiological studies in mammalian brain suggest that there may be subtle distinctions between pre- and post-synaptic receptors (Dutar and Nicoll 1988*b*; Harrison 1990; Colmers and Williams 1988; Thompson and Gahwiler 1992; Deisz *et al.* 1997; Chan *et al.* 1998) but, unfortunately, the only receptor ligands currently available do not appear to reliably distinguish between receptor subtypes. Some antagonists do appear to be selective in certain systems such as synaptosomes or brain slices (e.g. Gemignani *et al.* 1994; Cunningham and Enna 1996) but these same compounds have not been reported to produce the same separation of effects in other neuronal systems

28.4.1 **Agonists**

The observation that β-[4-chorophenyl] GABA (baclofen), is a stereospecifically active agonist at the GABA$_B$ receptor (Bowery *et al.* 1980, 1981) provided part of the original evidence for the existence of a distinct receptor. Since then 3-aminopropyl phosphinic acid (2APPA) and its methyl homologue(AMPPA, SKF97541) have emerged and are reported to be 3–7 fold more potent than the active isomer of baclofen. A variety of other phosphinic based agonists have been produced (Froestl *et al.* 1995*a*) with varying potencies but overall there are very few selective compounds with greater efficacy or affinity than baclofen for GABA$_B$ sites.

28.4.2 **Antagonists**

The design and development of selective GABA$_B$ receptor antagonists with increasing receptor affinity and potency has been an important process in establishing the significance and structure of the GABA$_B$ receptor. The original selective antagonists, phaclofen and 2-hydroxy saclofen were first described by Kerr, Ong and colleagues (1987, 1988). Although these compounds have low potency with affinities of 100 μM and 12 μM respectively for GABA$_B$ binding sites in rat brain membranes they were the first selective compounds and it was the use of phaclofen which provided the first evidence for the synaptic involvement of GABA$_B$ receptors (Dutar and Nicoll 1988*a*). Subsequent discoveries of antagonists derived largely from the group at Novartis were led by Froestl and Mickle. They developed the first antagonist which was able to cross the blood–brain barrier after intraperitoneal injection, CGP35348 (Olpe *et al.* 1990) and subsequently CGP36742 which was shown to be centrally-active after oral administration in rats (Olpe *et al.* 1993). However, both of these compounds and others in the same series only have potencies similar to that of saclofen even though they are selective for the GABA$_B$ receptor. The only other low affinity compound of note which exhibits significant CNS activity after peripheral administration is SCH50911 (Bolser *et al.* 1995) which, unlike all of the CGP antagonist series, is not phosphorus containing.

The most crucial breakthrough in the discovery of antagonists came with the development, by Froestl and colleagues, of compounds with affinities about 10 000 times higher than any previous antagonist. This major advance stemmed from the substitution of a dichloroben-zene moiety into the existing molecules. This produced a profusion of compounds with affinities in the nanomolar or even subnanomolar range (Froestl *et al.* 1995b). Perhaps the most notable compounds among these are CGP55845, CGP54626, and CGP62349 although many more were produced. This series eventually led to the development of the iodinated high affinity antagonist ^{125}I-CGP64213 which was used in the elucidation of the structure of GABA$_{B1}$ (Kaupmann *et al.* 1997).

28.4.3 Allosteric modulators

Ca^{2+} has been proposed to act as a positive modulator that increases the potency of GABA at GABA$_B$ receptors (Galvez *et al.* 2000), in a manner similar to that reported for mGluRs (Saunders *et al.* 1998). Small molecules have also recently been reported as positive allosteric modulators of the GABA$_B$ receptor (Urwyler *et al.* 2000). Such compounds have no effect on receptor activity in the absence of GABA, but do potentiate the effect of GABA when it is bound to the receptor.

28.5 Signal transduction and receptor modulation

The effector mechanisms to which neuronal GABA$_B$ receptors are coupled are adenylate cyclase (Karbon *et al.* 1984; Hill *et al.* 1984; Hill 1985; Xu and Wojcik 1986) and Ca^{2+} and K$^+$ channels (Inoue *et al.* 1985; Andrade *et al.* 1986; Dolphin *et al.* 1990; Bindokas and Ishida 1991; Gage 1992). While it can be demonstrated that the majority of these effects are mediated via G proteins, in particular Gi2α as described in rat cerebral cortex (Odagaki *et al.* 2000; Odagaki and Koyama 2001), not all effects of GABA$_B$ receptor activation appear to be coupled. Pertussis toxin insensitive effects of baclofen have been noted, for example, in rat spinal cord and magnocellular neurones of the paraventricular and supraoptic nuclei (Noguchi and Yamashita 1999; Moran *et al.* 2000; Cui *et al.* 2000). It has also been reported that pre-synaptic as compared to post-synaptic GABA$_B$ receptor mechanisms are insensitive to pertussis toxin (Harrison *et al.* 1990).

It is well established that GABA$_B$ agonists inhibit basal and forskolin-stimulated neuronal adenylate cyclase activity in brain slices (e.g. Xu and Wocjik 1986). While this effect is unrelated to the channel events it is similarly mediated via G proteins and produces a reduced level of intracellular cAMP. Enhancement of cAMP formation produced by G$_s$ coupled receptor agonists such as isoprenaline, is also a well-documented response to GABA$_B$ receptor agonists in brain slice preparations but not in neuronal membranes. Both of these actions have also been observed *in vivo*. Using a microdialysis technique in freely moving rats, Hashimoto and Kuriyama (1997) were able to show that baclofen and GABA could reduce the increase in cAMP generated by infusion of forskolin in the cerebral cortex. This effect was blocked by the GABA$_B$ antagonist CGP54626. Baclofen also potentiated the generation of cAMP by isoprenaline in this preparation. Interestingly, a direct GABA$_B$-mediated increase in basal adenylate cyclase activity has been demonstrated in membranes of rat olfactory bulb (Olianas and Onali 1999) and this effect was blocked by pertussis toxin suggesting an involvement of G$_i$/G$_o$ rather than G$_s$ protein. Using recombinant receptors, it appears that both GABA$_{B1}$ and GABA$_{B2}$ can individually modulate adenylate cyclase activity in cell lines (Kaupmann et al. 1998a; Martin et al. 1999) but there is currently no information about the nature of the G protein coupling to the heterodimer. However, the recent evidence that GABA$_{B2}$ is responsible for signalling would suggest that functional G protein coupling would be more likely to be through this subunit rather than GABA$_{B1}$.

GABA$_B$ receptor activation decreases Ca^{2+} conductance but increases the membrane conductance to K$^+$ ions. The decrease in Ca^{2+} conductance appears to be primarily associated with pre-synaptic sites (e.g. Chen and Van der Pol 1998; Takahashi et al. 1998; Barral et al. 2000) suppressing 'P/Q' and 'N' type channels (e.g. Santos et al. 1995; Lambert and Wilson 1996; Bussieres and El Manira 1999; Barral et al. 2000) although facilitation of an 'L' type channel in non-mammalian retina has been described (Shen and Slaughter 1999) and inhibition of N and P/Q type channels at postsynaptic sites has also been reported (Harayama et al. 1998). Modulation of K$^+$ conductances appears to be primarily linked with post-synaptic GABA$_B$ sites (e.g. Luscher et al. 1997) and possibly more than one type of K$^+$ channel (Wagner and Dekin 1993). Conversely a K$^+$(A) current appears to be coupled to GABA$_B$ receptors on pre-synaptic terminals in hippocampal cultures (Saint et al. 1990). But overall, changes in membrane K$^+$ flux appear to make the major contribution to postsynaptic GABA$_B$ receptor activation.

While suppression of Ca^{2+} influx is probably the most frequently observed mechanism associated with pre-synaptic GABA$_B$ sites, (e.g. Doze et al. 1995; Wu and Saggau 1995; Isaacson 1998), a process independent of Ca^{2+} or K$^+$ channels has been noted in rodent CA1 hippocampal pyramidal cells by Jarolimek and Misgeld (1997) who suggest that activation of protein kinase C (PKC) may be responsible. GABA$_B$ receptor activation has previously been shown to induce a rapid increase in PKC activity in rat hippocampal slices but this was only apparent in the early post-natal period of life (Tremblay et al. 1995). In additon, low threshold Ca^{2+} T-currents, which are inactivated at normal resting membrane potentials, may be involved in the response to GABA$_B$ receptor activation at least within the thalamus (Scott et al. 1990). GABA$_B$ receptor activation produces a post-synaptic hyperpolarization of long duration which initiates Ca^{2+} spiking activity in thalamocortical cells (Crunelli and Leresche 1991) and may be implicated in the generation of spike and wave discharges associated with absence seizures.

28.6 Physiology and disease relevance

Numerous effects have been attributed to the action of GABA$_B$ receptor agonists and antagonists (Table 28.2). Some of these include central muscle relaxant action, antitussive action, bronchiolar relaxation, urinary bladder inhibition, gastric motility increase, epileptogenesis, suppression of self-administered cocaine, nicotine and opiates, antinociception, yawning, hypotension, brown fat thermogenesis, cognitive impairment, reduction in release of hormones such as corticotrophin releasing hormone, prolactin releasing hormone, luteinizing hormone and melanocyte stimulating hormone, and reduced gastric acid secretion.

Table 28.2 Therapeutic potential of GABA$_B$ receptor ligands

Effect	Agonist(Ag)/ Antagonist(Ant)	Site of action	Possible mechanism(s) of action	Effect of Ag/Ant on response
Anti-spastic	Ag	Spinal cord Excitatory presynaptic Terminals	Suppression of release of excitatory transmitter on to motoneurones. Possible post-synaptic effect on reflex neurones	Ant blocks effect
Analgesic	Ag	Spinal cord C and A delta sensory primary afferent terminals CNS higher centres	Suppression of release of transmitter from small diameter primary afferent terminals	Ant blocks Ag effect and enhances nociception produced by inflammation
Reversal of durg craving	Ag	Mesolimbic region of Cl, probably ventral tegmental area	Unknown	Ant blocks Ag effect
Anti-asthma	Ag	Bronchii	Inhibition of Neural release of acetylcholine and substance P on to airway smooth muscle Decrease in hypersensitivity reaction	
Anti-absence	Ant	Thalamus-Thalamocortical neurone	Suppression of GABA-mediated Ca++ Spikes	Ag exacerbates seizure activity
Cognitive enhancement	Ant	Cerebral cortex/Hippocampus?		Ag produces cognitive impairment in rats
Anti-depressant	Ant		Modulation of Catecholamine Release	Ag has opposite action
Neuroprotection	Ant		Enhancement of neurotrophin release?	
Anxiolytic	Ant		Blockade of autoreceptors on GABA terminals?	

28.6.1 **Spasticity**

The centrally mediated muscle relaxant action of baclofen is the most well recognized of its effects and the mechanism underlying this appears to derive from its ability to reduce the release of excitatory neurotransmitter on to motoneurones in the ventral horn of the spinal cord. However, this has been questioned in a recent article in which it is suggested that the clinical anti-spastic effect derives from a post-, rather than a pre-, synaptic action on motoneurones (Orsnes *et al.* 2000*b*). But irrespective of its site of action its efficacy has made it the drug of choice in the treatment of spasticity irrespective of its cause. Unfortunately, baclofen is not without significant side effects in some patients making it poorly tolerated after systemic administration. This can be avoided by intrathecal infusion of very low amounts of the drug. This mode of treatment has proved to be very successful in spasticity associated with tardive dystonia, brain and spinal cord injury, cerebral palsy, tetanus, multiple sclerosis and stiff-man syndrome (Penn *et al.* 1989; Ochs *et al.* 1989, 1999; Penn and Mangieri 1993; Becker *et al.* 1995; Seitz *et al.* 1995; Albright *et al.* 1996; Paret *et al.* 1996; Dressnandt and Conrad 1996; Azouvi *et al.* 1996; Armstrong *et al.* 1997; Dressler *et al.* 1997; Meythaler *et al.* 1997; Francois *et al.* 1997, 2001; Auer *et al.* 1999; Trampitsch *et al.* 2000; Orsnes *et al.* 2000*a*).

28.6.2 **Pain and analgesia**

Although pain relief has been noted with baclofen in trigeminal neuralgia in man (Fromm 1994) as well as in a rodent model (IdänpäänHeikklä and Guilbaud 1999), its usefulness as an analgesic has always been questioned (see Hansson and Kinnman 1996). Nevertheless, more recent clinical observations have indicated that baclofen can reduce pain due to stroke or spinal cord injury and musculoskeletal pain when administered by intrathecal infusion (Taira *et al.* 1995; Loubser and Akman 1996; Becker *et al.* 2000). But, despite these reports, GABA_B-mediated clinical analgesia has still to be established possibly because of receptor desensitization or receptor inactivation following systemic administration of baclofen.

In animal acute pain models baclofen has long been known to have an antinociceptive action (Cutting and Jordan 1975; Levy and Proudfit 1979; Serrano *et al.* 1992; Przesmycki *et al.* 1998). Even in chronic neuropathic pain models in rats, baclofen clearly exhibits an antinociceptive or anti-allodynic response (Smith *et al.* 1994; Wiesenfeld Hallin *et al.* 1997; Cui *et al.* 1998). The locus of this action is probably, in part, within higher centres of the brain (Liebman and Pastor 1980; Thomas *et al.* 1995) but an action within the spinal cord also makes an important contribution (Sawynok and Dickson 1985; Hammond and Washington 1993; Dirig and Yaksh 1995; Thomas *et al.* 1996). Even a single intrathecal injection of GABA given within one week of nerve injury in the rat could permanently reverse neuropathic pain (Eaton *et al.* 1999). This led the authors to suggest that altered spinal GABA levels may contribute to the induction phase of the chronic pain. In the dorsal horn of the rat spinal cord GABA_B receptors appear to be located on small diameter afferent fibre terminals (Price *et al.* 1987) where their activation decreases the evoked release of sensory transmitters such as substance P and glutamate (see Malcangio and Bowery 1996; Iyadomi *et al.* 2000; Ataka *et al.* 2000; Riley *et al.* 2001). This suppression of transmitter release appears to contribute to the antinociceptive action of baclofen after systemic or intrathecal administration. An additional mechanism has been proposed by Voisin and Nagy (2001) who suggest that modulation of voltage-dependent nifedipine sensitive calcium channels in dorsal horn neurones may contribute to the antinociceptive effect of baclofen.

A study supporting the analgesic potential of GABA$_B$ agonists stems from examining the effect of the GABA transport inhibitor tiagabine as an antinociceptive compound in rodents (Ipponi *et al.* 1999). It was effective in a variety of paradigms and its action was associated with an increase in the extracellular concentration of GABA within the thalamus. Moreover, the antinociceptive effect was blocked by the GABA$_B$ antagonist, CGP35348. In a preliminary report by Hering-Hanit (1999) it has also been shown that baclofen can be beneficial in the prophylactic treatment of migraine to suppress headache pain.

28.6.3 Drug dependence

An important and recent observation with baclofen indicates that it may be a very effective treatment for cocaine, nicotine, and opiate addiction by reducing the craving for the drug. In rats, baclofen, administered at doses of 1–5 mg/kg, suppressed the self-administration of cocaine without affecting responding for food reinforcement (Roberts and Andrews 1997; Shoaib *et al.* 1998; Campbell *et al.* 1999; but see Munzar *et al.* 2000). Moreover the selective GABA$_B$ receptor agonist, CGP 44532 (0.063–0.5 mg/kg) mimicked the action of baclofen and failed to disrupt responding for food (Brebner *et al.* 1999, 2000). This was also observed in rats following injection of baclofen directly into the ventral tegmental area (VTA) and this was obtained at doses (56 ng in each side) much lower than required to produce the same effect after injection into the nucleus accumbens or striatum. Preliminary clinical observations with baclofen indicate that the anti-craving effect may well extend to man (Ling *et al.* 1998). This is an important observation which could have major consequences in the future therapy of drug addiction as comparable results have also been obtained in rats self-administering heroin, alcohol, and nicotine (Xi and Stein 1999; Lobina *et al.* 2000; Corrigall *et al.* 2000). Elevating endogenous GABA levels in the mesolimbic area of the brain by administration of vigabatrin or the uptake inhibitor, NO 711, attenuated heroin self-administration in rats (Xi and Stein 2000) and prevented the cocaine-induced increase in dopamine in this region as well as the self-administration of cocaine (Ashby *et al.* 1999). Both of these studies implicate GABA$_B$ receptor activation in the underlying mechanism.

An interesting clinical study recently performed by Myrick *et al.* (2001) in 30 cocaine-dependent adults has shown that the craving for cocaine could be significantly reduced by the antiepileptic/analgesic agent gabapentin (1.2 g p.o./day). While the basis for this is potentially manifold and may be linked with enhanced release of GABA the results interestingly coincide with a report by Ng *et al.* (2001) which identifies gabapentin as an agonist at the GABA$_{B(1a)}$/GABA$_{B2}$ (but not the GABA$_{B(1b)}$/GABA$_{B2}$) receptor expressed in oocytes. This would be an attractive mechanism to explain the clinical observations. However, there are other studies that fail to support this view. For example, Lanneau *et al.* (2001) report that gabapentin was completely inactive at hippocampal pre- and post-synaptic GABA$_B$ receptors as well as recombinant GABA$_B$ heterodimers expressing either GABA$_{B(1a)}$ or GABA$_{B(1b)}$ receptor subunits in combination with GABA$_{B2}$ receptor subunits. Furthermore, even though gabapentin and baclofen produce the same long-lasting reduction in paired-pulse inhibition in rat dentate gyrus, the action of baclofen could be blocked by GABA$_B$ receptor antagonism whereas the effect of gabapentin was not affected (Stringer and Lorenzo 1999). Similarly, while the anti-hyperalgesic effects of baclofen and CGP35024, a baclofen mimetic, could be blocked by the GABA$_B$ receptor antagonist CGP56433A in a rat model of chronic neuropathy, the effect of gabapentin was unaltered (Patel *et al.* 2001).

As further studies emerge in this area it will be intriguing to know if gabapentin has only limited/specific effects related to GABA$_B$ sites.

28.6.4 Epilepsy

Perhaps the most established and consistent effect of GABA$_B$ antagonists is their ability to suppress absence seizures in a variety of animal models. Marescaux *et al.* (1992) first showed that GABA$_B$ antagonists administered systemically or directly into the thalamus can prevent the spike and wave discharges manifest in the EEG of genetic absence epilepsy rats (GAERS). Similar observations have been made in the Lethargic mouse (Hosford *et al.* 1992) and also in rats injected with γ-hydroxybutyric acid (or γ-butyrolactone, GHB) which produces seizure activity reminiscent of absence epilepsy (Snead 1992). In all cases GABA$_B$ antagonists dose-dependently reduced the seizure activity. In GAERS the spontaneous absence seizures can be blocked by bilateral administration of pertussis toxin implicating the involvement of G$_i$/G$_o$ protein coupling in the generation of seizures (Bowery *et al.* 1999). These and other data have prompted the suggestion that GABA$_B$ receptor activation may be involved in the generation of the absence syndrome within the thalamus possibly through Ca^{2+} spike generation (Crunelli and Leresche 1991; but see Charpier *et al.* 1999). It is also of interest, that the production of absence-like seizures by GHB in rats appears to be due, at least in part, to a weak partial agonist action at GABA$_B$ receptors (Bernasconi *et al.* 1999; Lingenhoehl *et al.* 1999) although GHB acts additionally via receptors distinct from GABA$_B$ sites (Snead 2000).

Somewhat surprisingly, GABA$_B$ antagonists, at much higher doses, can produce convulsive activity in rats (Vergnes *et al.* 1997) but the mechanism(s) underlying this is unknown although GABA$_B$ receptor agonists will reverse the effect. However, not every antagonist appears to produce the same effect. For example, we failed to observe any convulsant activity with SCH50911 at doses 10–100-fold higher than the dose which completely blocks absence seizures in the GAERS (Richards and Bowery 1996).

28.6.5 Alzheimer's disease and cognition

GABA$_B$ receptor antagonists improve cognitive performance in a variety of animal paradigms (Mondadori *et al.* 1993; Carletti *et al.* 1993; Getova *et al.* 1997; Getova and Bowery1998; Yu *et al.* 1997; Nakagawa and Takashima 1997; Staubli *et al.* 1999; Genkova-Papazova *et al.* 2000; Farr *et al.* 2000; but see Brucato *et al.* 1996). Perhaps not surprisingly, therefore, GABA$_B$ agonists impair learning behaviour in animal models (Tong and Hasselmo 1996; Arolfo *et al.* 1998; McNamara and Skelton 1996; Nakagawa *et al.* 1995) and this induced amnesia appears to be mediated via G protein linked receptors as the impairment produced by baclofen in mice can be blocked by pertussis toxin administered intracerebroventricularly (Galeotti *et al.* 1998).

28.6.6 Neurodegeneration

GABA$_B$ antagonists and agonists might both have the potential to produce neuroprotection. Lal *et al.* (1995) suggest that the agonist, baclofen, could be cytoprotective in a cerebral ischaemia model in gerbils but very large doses, well in excess of that producing muscle relaxation, were required. Beskid *et al.* (1999) have also reported that baclofen can attentuate the neurotoxic effect of quinolinic acid on cells in the CA1 region of the rat hippocampus.

By contrast studies on mouse cultured striatal neurones have shown that GABA$_B$ receptor activation enhances the neurotoxic effects of NMDA (Lafoncazal *et al.* 1999) supporting the view that GABA$_B$ antagonists are more likely to be neuroprotective. In line with this Heese *et al.* (2000) have shown that low doses of GABA$_B$ antagonists can increase the tissue levels of NGF and BDNF in rats with a single systemic dose of CGP36742, CGP56433A, or CGP56999A increasing the growth factor levels in hippocampus, neocortex and spinal cord by 2–3-fold within 6 h. This could have a major influence on the neurodegenerative process.

28.6.7 Depression

The potential significance GABA$_B$ receptor mechanisms in depression was first suggested by Lloyd *et al.* (1985) but this was challenged by other groups. More recently, however, further suggestions that GABA$_B$ antagonists, for example, CGP36742, are effective in animal models of depression have emerged (Nakagawa *et al.* 1999). This might be supported in due course by the observations of Heese *et al.* (2000) who showed that GABA$_B$ antagonists produce a rapid increase in nerve growth factors. As antidepressants produce the same response but only after 2–3 weeks this might be linked.

28.6.8 Peripheral indications

Baclofen, has been shown to be very effective in the clinical treatment of otherwise intractable hiccups (Guelaud *et al.* 1995; Marino 1998; Nickerson *et al.* 1997; Kumar and Dromerick 1998). This effect is believed to stem from an inhibition of the hiccup reflex arc and possibly involves GABAergic inputs from the nucleus raphe magnus (Oshima *et al.* 1998). Baclofen also has an antitussive action in low oral doses in man (Dicpinigaitis and Dobkin 1997) which confirms earlier reports of an antitussive action in the cat and guinea pig (Bolser *et al.* 1994). Baclofen exerts a significant inhibitory effect on the growth of murine mammary cancer and a correlation between glandular GABA levels and mammary pathology, also in humans, has been reported (Opolski *et al.* 2000) suggesting that the GABA$_B$ system may provide a future focus for treatment.

28.7 Concluding remarks

While the existence of the G protein coupled GABA$_B$ receptor has been known for over 20 years there is no doubt that the recent cloning and identification of its unique heterodimeric structure has stimulated a large volume of research. The most recent of this has been the introduction of GABA$_{B1}$ receptor knock-out mice (Prosser *et al.* 2001). However, even though we now know more about the formation and characteristics of the receptor there is still much to ascertain. For example, uncertainty about the existence of functional receptor subtypes still remains and information about the potential clinical significance of GABA$_B$ receptor antagonism is still lacking. No doubt these and other queries will soon be resolved.

References

Albright AL, Barry MJ, Fasick P, Barron W, and Shultz B (1996). Continuous intrathecal baclofen infusion for symptomatic generalized dystonia. *Neurosurgery* 38, 934–8.

Andrade R, Malenka RC, and Nicoll RA (1986). A G protein couples serotonin and GABA$_B$ receptors to the same channels in hippocampus. *Science* 234, 1261–5.

Armstrong RW, Steinbrok P, Cochrane DD, Kube SD, and Fife SE (1997). Intrathecally administered baclofen for treatment of children with spasticity of cerebral origin. *Neurosurg* 87, 409–14.

Arolfo MP, Zanudio MA, and Ramirez OA (1998). Baclofen infused in rat hippocampal formation impairs spatial learning. *Hippocampus* 8, 109–13.

Ashby CR, Rohatgi R, Ngosuwan J, Borda T, Gerasimov MR, Morgan AE *et al.* (1999). Implication of the GABA(B) receptor in gamma vinyl-GABA's inhibition of cocaine-induced increases in nucleus accumbens dopamine. *Synapse* 31, 151–3.

Ataka T, Kumamoto E, Shimoji K, and Yoshimura M (2000). Baclofen inhibits more effectively C-efferent than A delta-afferent glutamatergic transmission in substantia gelatinosa neurons of adult rat spinal cord slices. *Pain* 86, 273–82.

Auer C, Siebner HR, Dressnandt J, and Conrad B (1999). Intrathecal baclofen increases corticospinal output to hand muscles in multiple sclerosis. *Neurology* 1298–99.

Azouvi P, Mane M, Thiebaut JB, Denys P, Remyneris O, and Bussel B (1996). Intrathecal baclofen administration for control of severe spinal spasticity: functional improvement and long-term follow-up. *Arch Phys Med Rehab* 77, 35–8.

Barral J, Toro S, Galarraga E, and Bargas J (2000). GABAergic presynaptic inhibition of rat neostriatal afferents is mediated by Q-type Ca2+ channels. *Neurosci Lett* 283, 33–6.

Becker R, Benes L, Sure U, Hellwig D, and Bertalanffy H (2000). Intrathecal baclofen alleviates autonomic dysfunction in severe brain injury. *J Clin Neurosci* 7, 316–19.

Becker WJ, Harris CJ, Long ML, Ablett DP, Klein GM, and DeForge DA (1995). Long-term intrathecal baclofen therapy in patients with intractable spasticity. *Can J Neurol Sci* 22, 208–17.

Bein HJ (1972). Pharmacological differentiations of muscle relaxants. In (ed. Birkmayer W) *Spasticity: A Topical Survey*. Hans Huber, Vienna.

Belley M, Sullivan R, Reeves A, Evans J, O'Neill G, and Ng GYK (1999). Synthesis of the nanomolar photoaffinity GABA (B) receptor ligand CGP 71872 reveals diversity in the tissue distribution of GABA(B) receptor forms. *Bioorg Med Chem* 7, 2697–704.

Benke D, Honer M, Michel C, Bettler B, and Mohler H (1999). Gamma-aminobutyric acid type B receptor splice variant proteins GBR1a and GBR1b are both associated with GBR2 *in situ* and display differential regional and subcellular distribution. *J Biol Chem* 274, 27 323–30.

Bernasconi R, Mathivet P, Bischoff S, and Marescaux C (1999). Gamma-hydroxybutyric acid: an endogenous neuromodulator with abuse potential? *Trends in Pharmacol Sci* 20, 135–41.

Beskid M, Rozycka Z, and Taraszewska A (1999). Quinolinic acid and GABA_B receptor ligand: effect on pyramidal neurons of the CA1 sector of rat's dorsal hippocampus following peripheral administration. *Folia Neuropathol* 37, 99–106.

Billinton A, Upton N, and Bowery NG (1999). GABA_B receptor isoforms GBR1a and GBR1b, appear to be associated with pre- and post-synaptic elements respectively in rat and human cerebellum. *Br J Pharmacol* 126, 1387–92.

Bindokas VP and Ishida AT (1991). (-)-baclofen and gamma-aminobutyric acid inhibit calcium currents in isolated retinal ganglion cells. *Proc Natl Acad Sci USA* 88, 10 759–63.

Bischoff S, Leonhard S, Reymann N, Schuler V, Shigemoto R, Kaupmann K, and Bettler B (1999). Spatial distribution of GABA(B) R1 receptor mRNA and binding sites in the rat brain. *J Comp Neurol* 412, 1–6.

Bolser DC, Blythin DJ, Chapman RW, Egan RW, Hey JA, Rizzo C, Kuo SC *et al.* (1995). The pharmacology of SCH 50911: a novel, orally- active GABA-beta receptor antagonist. *J Pharmacol Exp Ther* 274, 1393–8.

Bolser DC, DeGennaro FC, O'Reilly S, Chapman RW, Kreutner W, Egan RW, and Hey JA (1994). Peripheral and central sites of action of GABA-B agonists to inhibit the cough reflex in the cat and guinea pig. *Br J Pharmacol* 113, 1344–8.

Bonanno G, Fassio A, Sala R, Schmid G, and Raiteri M (1998). GABA$_B$ receptors as potential targets for drugs able to prevent excessive excitatory amino acid transmission in the spinal cord. *Eur J Pharmacol* **362**, 143–8.

Bowery NG, Parry K, Boehrer A, Mathivet P, Marescaux C, and Bernasconi R (1999). Pertussis toxin decreases absence seizures and GABA(B) receptor binding in thalamus of a genetically prone rat (GAERS). *Neuropharmacology* **38**, 1691–7.

Bowery NG (1993). GABA$_B$ receptor pharmacology. *Annu Rev Pharmacol Toxicol* **33**, 109–47.

Bowery NG, Doble A, Hill DR, Hudson AL, Shaw JS, Turnbull MJ, and Warrington R (1981). Bicuculline-insensitive GABA receptors on peripheral autonomic nerve terminals. *Eur J Pharmacol* **71**, 53–70.

Bowery NG, Hill DR, Hudson AL, Doble A, Middlemiss DN, Shaw JS, and Turnbull MJ (1980). (-) Baclofen decreases neurotransmitter release in the mammalian CNS by an action at a novel GABA receptor. *Nature* **283**, 92–4.

Bowery NG, Hudson AL, and Price GW (1987). GABA$_A$ and GABA$_B$ receptor site distribution in the rat central nervous system. *Neuroscience* **20**, 365–83.

Brebner K, Phelan R, and Roberts DCS (2000). Intra-VTA baclofen attenuates cocaine self-administration on a progressive ratio schedule of reinforcement. *Pharmacol Biochem Behav* **66**, 857–62.

Brebner K, Froestl W, Andrews M, Phelan R, and Roberts DCS (1999). The GABA$_B$ agonist CGP44532 decreases cocaine self-administration in rats: demonstration using a progressive ratio and a discrete trials procedure. *Neuropharmacology* **38**, 1797–804.

Brucato FH, Levin ED, Mott DD, Lewis DV, Wilson WA, and Swartzwelder HS (1996). Hippocampal long-term potentiation and spatial learning in the rat: Effects of GABA$_B$ receptor blockade. *Neuroscience* **74**, 331–9.

Bussieres N and El Manira A (1999). GABA(B) receptor activation inhibits N- and P/Q-type calcium channels in cultured lamprey sensory neurons. *Brain Res* **847**, 175–85.

Calver AR, Medhurst AD, Robbins MJ, Charles KJ, Evans ML, Harrison DC *et al.* (2000). The expression of GABA(B1) and GABA(B2) receptor subunits in the CNS differs from that in peripheral tissues. *Neuroscience* **100**, 155–70.

Calver AR, Robbins MJ, Cosio C, Rice SQJ, Babbs AJ, Hirst WD *et al.* (2001). The C-terminal domains of the GABA(B) receptor subunits mediate intracellular trafficking but are not required for receptor signaling. *Neuroscience* **21**, 1203–10.

Campbell UC, Lac ST, and Carroll ME (1999). Effects of baclofen on maintenance and reinstatement of intravenous cocaine self-administration in rats. *Psychopharmacology* **143**, 209–14.

Carletti R, Libri V, and Bowery NG (1993). The GABA$_B$ antagonist CGP 36742 enhances spatial learning performance and antagonises baclofen-induced amnesia in mice. *Br J Pharmacol* **109**, 74P.

Castelli MP, Ingianni A, Stefanini E, and Gessa GL (1999). Distribution of GABA(B) receptor mRNAs in the rat brain and peripheral organs. *Life Sci* **64**, 1321–8.

Chan PKY, Leung CKS, and Yung WH (1998). Differential expression of pre- and postsynaptic GABA$_B$ receptors in rat substantia nigra pars reticulata neurones. *Eur J Pharmacol* **349**, 187–97.

Charles KJ, Evans ML, Robbins MJ, Calver AR, Leslie RA, and Pangalos MN (2001). Comparative immunohistochemical localisation of GABA$_{B1a}$, GABA$_{B1b}$ and GABA$_{B2}$ subunits in rat brain, spinal cord and dorsal root ganglion. *Neuroscience* **106**, 447–67.

Charpier S, Leresche N, Deniau JM, Mahon S, Hughes SW, and Crunelli V (1999). On the putative contribution of GABA(B) receptors to the electrical events occurring during spontaneous spike and wave discharges. *Neuropharmacology* **38**, 1699–706.

Chen G and Van der Pol AN (1998). Presynaptic GABA$_B$ autoreceptor modulation of P/Q-type calcium channels and GABA release in rat suprachiasmatic nucleus neurons. *J Neurosci* **18**, 1913–22.

Chu DCM, Albin RL, Young AB, and Penney JB (1990). Distribution and kinetics of GABA$_B$ binding sites in rat central nervous system: a quantitative autoradiographic study. *Neuroscience* **34**, 341–57.

Clark JA, Mezey E, Lam AS, and Bonner TI (2000). Distribution of the GABA(B) receptor subunit gb2 in rat CNS. *Brain Res* 860, 41–52.

Colmers WF and Williams JT (1988). Pertussis toxin pretreatment discriminates between pre- and postsynaptic actions of baclofen in rat dorsal aphe nucleus *in vitro*. *Neurosci Lett* 93, 300–6.

Corrigall WA, Coen KM, Adamson KL, Chow BLC, and Zhang J (2000). Response of nicotine self-administration in the rat to manipulations of mu-opioid and gamma-aminobutyric acid receptors in the ventral tegmental area. *Pschopharmacology* 149, 107–14.

Couve A, Kittler JT, Uren JM, Calver AR, Pangalos MN, Walsh FS, and Moss SJ (2001). Association of GABA(B) receptors and members of the 14-3-3 family of signaling proteins. *Mol Cell Neurosci* 17, 317–28.

Crunelli V and Leresche N (1991). A role for the GABA$_B$ receptors in excitation and inhibition of thalamocortical cells. *Trends Neurosci* 14, 16–21.

Cui LN, Coderre E, and Renaud LP (2000). GABA(B) presynaptically modulates suprachiasmatic input to hypothalamic paraventricular magnocellular neurons. *Am J Physio-Regul Integr Comp Physiol* 278, R1210–16.

Cui JG, Meyerson BA, Sollevi A, and Linderoth B (1998). Effect of spinal cord stimulation on tactile hypersensitivity in mononeuropathic rats is potentiated by simultaneous GABA$_B$ and adenosine receptor activation. *Neurosci Lett* 247, 183–6.

Cunningham MD and Enna SJ (1996). Evidence for pharmacologically distinct GABA$_B$ receptors associated with cAMP production in rat brain. *Brain Res* 720, 220–4.

Cutting DA and Jordan CC (1975). Alternative approaches to analgesia: baclofen as a model compound. *Br J Pharmacol* 54, 171–9.

Deisz RA, Billard JM, and Zieglgänsberger W (1997). Presynaptic and postsynaptic GABA$_B$ receptors of neocortical neurons of the rat in vitro: Differences in pharmacology and ionic mechanisms. *Synapse* 25, 62–72.

Dicpinigaitis PV and Dobkin JB (1997). Antitussive effect of the GABA-agonist baclofen. *Chest* 111, 996–9.

Dirig DM and Yaksh TL (1995). Intrathecal baclofen and muscimol, but not midazolam, are antinociceptive using the rat-formalin model. *J Pharmacol Exp Ther* 275, 219–27.

Dolphin AC, Huston E, and Scott RH (1990). GABA$_B$-mediated inhibition of calcium currents: a possible role in presynaptic inhibition. (eds. NG Bowery, H Bittiger, and H-R Olpe), *GABA$_B$ receptors in mammalian function*. J. Wiley, Chichester, pp. 259–71.

Doze VA, Cohen GA, and Madison DV (1995). Calcium channel involvement in GABA$_B$ receptor-mediated inhibition of GABA release in area CA1 of the rat hippocampus. *J Neurophysiol* 74, 43–53.

Dressler D, Oeljeschlager RO, and Ruther E (1997). Severe tardive dystonia: Treatment with continuous intrathecal baclofen administration. *Movement Disorders* 12, 585–7.

Dressnandt J and Conrad B (1996). Lasting reduction of severe spasticity after ending chronic treatment with intrathecal baclofen. *J Neurol Neurosurg Psychiatry* 2, 168–73.

Durkin MM, Gunwaldsen CA, Borowsky B, Jones KA, and Branchek TA (1999). An in situ hybridization study of the distribution of the GABA(B2) protein mRNA in the rat CNS. *Mol Brain Res* 71, 185–200.

Dutar P and Nicoll RA (1988a). A physiological role for GABA$_B$ receptors in the central nervous system. *Nature* 332, 156–8.

Dutar P and Nicoll RA (1988b). Pre- and postsynaptic GABA$_B$ receptors in the hippocampus have different pharmacological properties. *Neuron* 1, 585–91.

Eaton MJ, Martinez MA, and Karmally S (1999). A single intrathecal injection of GABA permanently reverses neuropathic pain after nerve injury. *Brain Res* 835, 334–9.

Farr SA, Uezu K, Creonte TA, Flood JF, and Morley JE (2000). Modulation of memory processing in the cingulate cortex of mice. *Pharmacol Biochem Behav* 65, 363–8.

Francois B, Clavel M, Desachy A, Vignon P, Salle JY, and Gastinne H (1997). Continuous intrathecal baclofen in tetanus—An alternative management. *Presse Medicale* 26, 1045–7.

Francois B, Vacher P, Roustan J, Salle JY, Vidal J, Moreau JJ, and Vignon P (2001). Intrathecal baclofen after traumatic brain injury: Early treatment using a new technique to prevent spasticity. *J Trauma-Injury Infect Crit Care* 50, 158–61.

Froestl W, Mickel SJ, Hall RG, Von Sprecher G, Strub D, Baumann PA *et al.* (1995a). Phosphinic acid analogues of GABA. 1. New potent and selective GABA$_B$ agonists. *J Med Chem* 38, 3297–312.

Froestl W, Mickel SJ, Von Sprecher G *et al.* (1995b). Phosphinic acid analogues of GABA. 2. Selective, orally active GABA$_B$ antagonists. *J Med Chem* 3, 3313–31.

Fromm GH (1994). Baclofen as an adjuvant analgesic. *J Pain Symptom Manag* 9, 500–9.

Gage PW (1992). Activation and modulation of neuronal K$^+$ channels by GABA. *Trends Neurosci* 15, 46–51.

Galeotti N, Ghelardini C, and Bartolini A (1998). Effect of pertussis toxin on baclofen- and diphenhydramine-induced amnesia. *Psychopharmacology* 136, 328–34.

Galvez T, Parmentier ML, Joly C, Malitschek B, Kaupmann K, Kuhn R *et al.* (1999). Mutagenesis and modeling of the GABA(B) receptor extracellular domain support a Venus flytrap mechanism for ligand binding. *J Biol Chem* 33, 13 362–9.

Galvez T, Prezeau L, Milioti G, Franek M, Joly C, Froestl W *et al.* (2000). Mapping the agonist-binding site of GABA(B) type 1 subunit sheds light on the activation process of GABA(B) receptors. *J Biol Chem* 275, 41 166–74.

Galvez T, Duthey B, Kniazeff J, Blahos J, Rovelli G, Bettler B *et al.* (2001). Allosteric interactions between GB1 and GB2 subunits are required for optimal GABA$_B$ receptor function. *EMBO J* 20(9), 2152–9.

Gemignani A, Paudice P, Bonanno G, and Raiteri M (1994). Pharmacological discrimination between gamma-aminobutyric acid type B receptors regulating cholecystokinin and somatostatin release from rat neocortex synaptosomes. *Mol Pharmacol* 46, 558–62.

Genkova-Papazova MG, Petkova B, Shishkova N, and Lazarova-Bakarova M (2000). The GABA-B antagonist CGP 36742 prevents PTZ-kindling-provoked amnesia in rats. *Eur Neuropsychopharmacol* 10, 273–8.

Getova D, Bowery NG, and Spassov V (1997). Effects of GABA$_B$ receptor antagonists on learning and memory retention in a rat model of absence epilepsy. *Eur J Pharmacol* 320, 9–13.

Getova D and Bowery NG (1998). The modulatory effects of high affinity GABA$_B$ receptor antagonists in an active avoidance learning paradigm in rats. *Psychpharmacology* 137, 369–73.

Green A, Walls S, Wise A, Green RH, Martin AK, and Marshall FH (2000). Characterization of [^{3}H]-CGP54626A binding to heterodimeric GABA$_B$ receptors stably expressed in mammalian cells. *Br J Pharmacol* 131, 1766–74.

Guelaud C, Similowski T, Bizec JL, Cabane J, Whitelaw WA, and Derenne JP (1995). Baclofen therapy for chronic hiccup. *Eur Respir J* 8, 235–7.

Hammond DL and Washington JD (1993). Antagonism of L-baclofen-induced antinociception by CGP 35348 in the spinal cord of the rat. *Eur J Pharmacol* 234, 255–62.

Hansson P and Kinnman E (1996). Unmasking mechanisms of peripheral neuropathic pain in a clinical perspective. *Pain Rev* 3, 272–92.

Harayama N, Shibuya I, Tanaka K, Kabashima N, Ueta Y, and Yamashita H (1998). Inhibition of N- and P/Q-type calcium channels by postsynaptic GABA$_B$ receptor activation in rat supraoptic neurones. *J Physiol* 509, 371–83.

Harrison NL, Lambert NA, and Lovinger DM (1990). Presynaptic GABA$_B$ receptors on rat hippocampal neurons. (eds. NG Bowery, H Bittiger, and H-R Olpe) *GABA$_B$ receptors in mammalian function*, Wiley, Chichester, pp. 208–21.

Hashimoto T and Kuriyama K (1997). *In vivo* evidence that GABA$_B$ receptors are negatively coupled to adenylate cyclase in rat striatum. *J Neurochem* 69, 365–70.

Heese K, Otten U, Mathivet P, Raiteri M, Marescaux C, and Bernasconis R (2000). GABA$_B$ receptor antagonists elevate both mRNA and protein levels of the neurotrophins nerve growth factor(NGF) and brain-derived neurotrophic factor(BDNF) but not neurotrophin-3(NT-3) in brain and spinal cord of rats. *Neuropharmacology* **39**, 449–62.

Hering-Hanit R (1999). Baclofen for prevention of migraine. *Cephalalgia* **19**, 589–91.

Hill DR and Bowery NG (1981). ^{3}H-baclofen and ^{3}H-GABA bind to bicuculine-insensitive GABA_B sites in rat brain. *Nature* **290**, 149–52.

Hill DR (1985). GABAB receptor modulation of adenylate cyclase activity in rat brain slices. *Br J Pharmacol* **84**, 249–57.

Hill DR, Bowery NG, and Hudson AL (1984). Inhibition of GABAB receptor binding by guanyl nucleotides. *J Neurochem* **42**, 652–7.

Hosford DA, Clark S, Cao Z, Wilson WA, Jr., Lin F, Morrisett RA, and Huin A (1992). The role of GABA_B receptor activation in absence seizures of lethargic (*lh/lh*) mice. *Science* **257**, 398–401.

Idänpään Heikkilä JJ and Guilbaud G (1999). Pharmacological studies on a rat model of trigeminal neuropathic pain: baclofen, but not carbamazepine, morphine or tricyclic antidepressants, attenuate the allodynia-like behaviours. *Pain* **79**, 281–90.

Inoue M, Matsuo T, and Ogata N (1985). Possible involvement of K^+ conductance in the action of gamma-aminobutyric acid in the guinea-pig hippocampus. *Br J Pharmacol* **86**, 515–24.

Ipponi A, Lamberti C, Medica A, Bartolini A, and MalmbergAiello P (1999). Tiagabine antinociception in rodents depends on GABA_B receptor activation: parallel antinociception testing and medial thalamus GABA microdialysis. *Eur J Pharmacol* **368**, 205–11.

Isaacson JS (1998). GABA_B receptor-mediated modulation of presynaptic currents and excitatory transmission at a fast central synapse. *J Neurophysiol* **80**, 1571–6.

Isomoto S, Kaibara M, Sakurai-Yamashita Y, Nagayama Y, Uezono Y, Yano K, and Taniyama K (1998). Cloning and tissue distribution of novel splice variants of the rat GABA(B) receptor. *Biochem Biophys Res Commun* **253**, 10–15.

Iyadomi M, Iyadomi I, Kumamoto E, Tomokuni K, and Yoshimura M (2000). Presynaptic inhibition by baclofen of miniature EPSC and IPSCs in substantia gelatinosa neurons of the adult rat spinal dorsal horn. *Pain* **85**, 385–93.

Jarolimek W and Misgeld U (1997). GABA_B receptor-mediated inhibition of tetrodotoxin-resistant GABA release in rodent hippocampal CA1 pyramidal cells. *J Neuroscience* **17**, 1025–32.

Jones KA, Tamm JA, Craig DA, Yao WJ, and Panico R (2000). Signal transduction by GABAB receptor heterodimers. *Neuropsychopharmacology* **23**, S41–S49.

Jones KA, Borowsky B, Tamm JA, Craig DA, Durkin MM, Dai M *et al.* (1998). GABA_B receptors function as a heteromeric assembly of the subunits GABA_BR1 and GABA_BR2. *Nature* **396**, 674–9.

Karbon EW, Duman RS, and Enna SJ (1984). GABA_B receptors and norepinephrine-stimulated cAMP production in rat brain cortex. *Brain Res* **306**, 327–32.

Kaupmann K, Huggel K, Heid J, Flor PJ, Bischoff S, Mickel SJ *et al.* (1997). Expression cloning of GABA_B receptors uncovers similarity to metabotropic glutamate receptors. *Nature* **386**, 239–46.

Kaupmann K, Malitschek B, Schuler V, Heid J, Froestl W, Beck P *et al.* (1998*a*) GABA_B-receptor subtypes assemble into functional heteromeric complexes. *Nature* **396**, 683–7.

Kaupmann K, Schuler V, Mosbacher J, Bischoff S, Bittiger H, Heid J *et al.* (1998*b*) Human gamma-aminobutyric acid type B receptors are differentially expressed and regulate inwardly rectifying K+ channels. *Proc Natl Acad Sci USA* **95**, 14 991–6.

Keberle H and Faigle JW (1972). Synthesis and structure-activity relationship of the gamma-aminobutyric acid derivatives. In (ed. W. Birkmayer) *Spasticity: A Topical Survey*, Hans Huber, Vienna.

Kerr DIB, Ong J, Johnston GAR., Abbenante J, and Prager RH (1988). 2-Hydroxy-saclofen: an improved antagonist at central and peripheral GABA_B receptors. *Neurosci Lett* **92**, 92–6.

Kerr DIB, Ong J, Prager RH, Gynther BD, and Curtis DR (1987). Phaclofen: a peripheral and central baclofen antagonist. *Brain Res* **405**, 150–4.

Kumar A and Dromerick AW (1998). Intractable hiccups during stroke rehabilitation. *Arch Physical Med and Rehab* 79, 697–9.

Kuner R, Köhr G, Grünewald S, Eisenhardt G, Bach A, and Kornau HC (1999). Role of heteromer formation in GABA$_B$ receptor function. *Science* 283, 74–7.

Lafoncazal M, Viennois G, Kuhn R, Malitschek B, Pin JP, Shigemoto R, and Bockaert J (1999). mGluR7-like receptor and GABA$_B$ receptor activation enhance neurotoxic effects of N-methyl-D-aspartate in cultured mouse striatal GABAergic neurones. *Neuropharmacology* 38, 1631–40.

Lal S, Shuaib A, and Ijaz S (1995). Baclofen is cytoprotective to cerebral ischemia in gerbils. *Neurochem Res* 20, 115–19.

Lambert NA and Wilson WA (1996). High-threshold Ca^{2+} currents in rat hippocampal interneurones and their selective inhibition by activation of GABA$_B$ receptors. *J Physiol* 492, 115–27.

Lanneau C, Green A, Hirst WD, Wise A, Brown JT, Donnier E *et al.* (2001). Gabapentin is not a GABA(B) receptor agonist. *Neuropharmacology* 41(8), 965–75.

Lehmann A, Antonsson M, Bremner-Danielsen M, Flardh M, Hansson-Branden L, and Karrberg L (1999). Activation of the GABA(B) receptor inhibits transient lower esophageal sphincter relaxations in dogs. *Gastroenterology* 117, 1147–54.

Levy RA and Proudfit HK (1979). Analgesia produced by microinjection of baclofen and morphine at brain stem sites. *Eur J Pharmacol* 57, 43–55.

Liang FY, Hatanaka Y, Saito H, Yamamori T, and Hashikawa T (2000). Differential expression of gamma-aminobutyric acid type B receptor-1a and -1b mRNA variants in GABA and non-GABAergic neurons of the rat brain. *J Comp Neurol* 416, 475–95.

Lidums I, Lehmann A, Checklin H, Dent J, and Holloway RH (2000). Control of transient lower esophageal sphincter relaxations and reflux by the GABA$_B$ agonist baclofen in normal subjects. *Gastroenterology* 118, 7–13.

Liebman JM and Pastor G (1980). Antinociceptive effects of baclofen and muscimol upon intraventricular administration. *Eur J Pharmacol* 61, 225–30.

Ling W, Shoptaw S, and Majewska D (1998). Baclofen as a cocaine anti-craving medication: a preliminary clinical study. *Neuropsychopharmacology* 18, 403–4.

Lingenhoehl K, Brom R, Heid J, Beck P, Froestl W, Kaupmann K *et al.* (1999). Gamma-Hydroxybutyrate is a weak agonist at recombinant GABA$_B$ receptors. *Neuropharmacology* 38, 1667–73.

Lloyd KG, Thuret F, and Pilc A (1985). Upregulation of gamma-aminobutyric acid(GABA$_B$) binding sites in rat frontal cortex: a common action of repeated administration of different classes of antidepressant and electroshock. *J Pharmacol Exp Ther* 235, 191–9.

Lobina C, Agabio R, Reali R, Gessa GL, and Colombo G (1999). Contribution of GABA(A) and GABA(B) receptors to the discriminative stimulus produced by gamma-hydroxybutyric acid. *Pharmacol Biochem Behav* 64, 363–5.

Lorente P, Lacampagne A, Pouzeratte Y, Richards S, Malitschek B, Kuhn R *et al.* (2000). Gamma-aminobutyric acid type B receptors are expressed and functional in mammalian cardiomyocytes. *Proc Natl Acad Sci USA* 97, 8664–9.

Loubser PG and Akman NM (1996). Effects of intrathecal baclofen on chronic spinal cord injury pain. *J Pain Symptom Manag* 12, 241–7.

Luscher C, Jan LY, Stoffel M, Malenka RC, and Nicoll RA (1997). G protein-coupled inwardly rectifying K+ channels (GIRKs) mediate postsynaptic but not presynaptic transmitter actions in hippocampal neurons. *Neuron* 19, 687–95.

Malcangio M and Bowery NG (1996). GABA and its receptors in the spinal cord. *Trends Pharmacol Sci* 17, 457–62.

Malitschek B, Schweizer C, Keir M, Heid J, Froestl W, Mosbacher J *et al.* (1999). The N-terminal domain of gamma-aminobutyric acid (B) receptors is sufficient to specify agonist and antagonist binding. *Mol Pharmacol* 56, 448–54.

Marescaux C, Vergnes M, and Bernasconi R (1992). GABA$_B$ receptor antagonists: potential new anti-absence drugs. *J Neural Transm* 35, 179–88.

Margeta-Mitrovic M, Jan YN, and Jan LY (2000). A trafficking checkpoint controls GABA(B) receptor heterodimerization. *Neuron* 27, 97–106.

Margeta-Mitrovic M, Mitrovic I, Riley RC, Jan LY, and Basbaum AI (1999). Immunohistochemical localization of GABA(B) receptors in the rat central nervous system. *J Comp Neurol* 405, 299–321.

Marino RA (1998). Baclofen therapy for intractable hiccups in pancreatic carcinoma. *Am J Gastroenterol* 93, 2000.

Martin SC, Russek SJ, and Farb DH (1999). Molecular identification of the human GABA$_B$ R2: cell surface expression and coupling to adenylyl cyclase in the absence of GABA$_B$ R1. *Mol Cell Neurosci* 13, 180–91.

McLatchie LM, Fraser NJ, Main MJ, Wise A, Brown J, Thompson N *et al.* (1998). RAMPs regulate the transport and ligand specificity of the calcitonin-receptor-like receptor. *Nature* 393, 333–9.

McNamara RK and Skelton RW (1996). Baclofen, a selective GABA$_B$ receptor agonist, dose-dependently impairs spatial learning in rats. *Pharmacol Biochem Behav* 53, 303–8.

Meythaler JM, McCary A, and Hadley MN (1997). Prospective assessment of continuous intrathecal infusion of baclofen for spasticity caused by acquired brain injury: a preliminary report. *J Neurosurg* 87, 415–19.

Mondadori C, Jaekel J, and Preiswerk G (1993). CGP 36742: The first orally active GABAB blocker improves the cognitive performance of mice, rats, and rhesus monkeys. *Behav Neural Biol* 60, 62–8.

Moran JM, Enna SJ, and McCarson KE (2000). Functional G-protein coupling of GABA(B) receptors in brain but not spinal cord of the rat. *FASEB J* 14, 213.

Munzar P, Kutkat SW, Miller CR, and Goldberg SR (2000). Failure of baclofen to modulate discriminative-stimulus effects of cocaine or methamphetamine in rats. *Eur J Pharmacol* 408, 169–74.

Myrick H, Henderson S, Brady KT, and Malcolm R (2001). Gabapentin in the treatment of cocaine dependence: a case series. *J Clin Psychiatry* 62, 19–23.

Nakagawa Y and Takashima T (1997). The GABA$_B$ receptor antagonist CGP36742 attenuates the baclofen- and scopolamine-induced deficit in Morris water maze task in rats. *Brain Res* 766, 101–6.

Nakagawa Y, Ishibashi Y, Yoshii T, and Tagashira E (1995). Involvement of cholinergic systems in the deficit of place learning in Morris water maze task induced by baclofen in rats *Brain Res* 683, 209–14.

Nakagawa Y, Sasaki A, and Takashima T (1999). The GABA$_B$ receptor antagonist CGP36742 improves learned helplessness in rats. *Eur J Pharmacol* 381, 1–7.

Nehring RB, Horikawa HPM, El Far O, Kneussel M, Brandstatter JH, Stamm S *et al.* (2000). The metabotropic GABA(B) receptor directly interacts with the activating transcription factor 4. *J Biol Chem* 275, 35 185–91.

Ng GYK, Bertrand S, Sullivan R, Ethier N, Wang J, Yergey J *et al.* (2001). Gamma-Aminobutyric acid type B receptors with specific heterodimer composition and postsynaptic actions in hippocampal neurons are targets of anticonvulsant gabapentin action. *Mol Pharmacol* 59, 144–52.

Ng GYK, Clark J, Coulombe N, Ethier N, Hebert TE, Sullivan R *et al.* (1999). Identification of a GABA$_B$ receptor subunit, gb2, required for functional GABA$_B$ receptor activity. *J Biol Chem* 274, 7607–10.

Nickerson RB, Atchison JW, Van Hoose JD, and Hayes D (1997). Hiccups associated with lateral medullary syndrome. A case report. *Am J Physical Med Rehab* 76, 144–6.

Noguchi J and Yamashita H (1999). Baclofen inhibits postsynaptic voltage-dependent calcium currents of supraoptic nucleus neurons isolated from young rats. *Biomed Res-Tokyo* 20, 239–47.

Ochs G, Naumann C, Dimitrijevic M, and Sindou M (1999). Intrathecal baclofen therapy for spinal origin spasticity: Spinal cord injury, spinal cord disease, and multiple sclerosis. *Neuromodulation* 2, 108–19.

Ochs G, Struppler A, Meyerson BA, Linderoth G, and Gybels J (1989). Intrathecal baclofen for long term treatment of spasticity: an multicentre study. *J Neurol Neurosurg Psychiatry* 52, 933–9.

Odagaki Y and Koyama T (2001). Identification of G alpha subtype(s) involved in gamma-aminobutyric acid(B) receptor-mediated high-affinity guanosine triphosphatase activity in rat cerebral cortical membranes. *Neurosci Lett* **297**, 137–41.

Odagaki Y, Nishi N, and Koyama T (2000). Functional coupling of GABA(B) receptors with G proteins that are sensitive to N-ethylmaleimide treatment, suramin, and benzalkonium chloride in rat cerebral cortical membrane. *J Neural Transm* **107**, 1101–16.

Olianas MC and Onali P (1999). GABA(B) receptor-mediated stimulation of adenylyl cyclase activity in membranes of rat olfactory bulb. *Br J Pharmacol* **126**, 657–64.

Olpe H-R, Karlsson G, Pozza MF, Brugger F, Steinmann M, Van Riezen H *et al.* (1990). CGP 35348: a centrally active blocker of GABAB receptors. *Eur J Pharmacol* **187**, 27–38.

Olpe H-R, Steinmann MW, Ferrat T, Pozza MF, Greiner K, Brugger F *et al.* (1993). The actions of orally active GABA$_B$ receptor antagonists on GABAergic transmission in vivo and in vitro. *Eur J Pharmacol* **233**, 179–186.

Ong J and Kerr DI (1990). GABA-receptors in peripheral tissues. *Life Sci* **46**, 1489–501.

Ong J, Marino V, Parker DAS, and Kerr DIB (1998). Differential effects of phosphonic analogues of GABA on GABA$_B$ autoreceptors in rat neocortical slices. *Naunyn-Schmiedeberg's Arch Pharmacol* **357**, 408–12.

Opolski A, Mazurkiewicz M, Wietrayk J, Kleinrok Z, and Radzikowski C (2000). The role of GABA-ergic system in human mammary gland pathology and in growth of transplantable murine mammary cancer. *J Exp Clin Cancer Res* **19**, 383–90.

Orsnes GB, Sorensen PS, Larsen TK, and Ravnborg M (2000*a*). Effect of baclofen on gait in spastic MS patients. *Acta Neurol Scand* **101**, 244–8.

Orsnes G, Crone C, Krarup C, Petersen N, and Nielsen J (2000*b*). The effect of baclofen on the transmission in spinal pathways in spastic multiple sclerosis patients. *Clin Neurophysiol*, **111**, 1372–9.

Oshima T, Sakamoto M, Tatsuta H, and Arita H (1998). GABAergic inhibition of hiccup-like reflex induced by electrical stimulation in medulla of cats. *Neurosci Res* **30**, 287–93.

Pagano A, Rovelli G, Mosbacher J, Lohmann T, Duthey B, Stauffer D *et al.* (2001). C-terminal inter-action is essential for surface trafficking but not for heteromeric assembly of GABA(B) receptors. *J Neurosci* **21**, 1189–202.

Paret G, Tirosh R, Benzeev B, Vardi A, Brandt N, and Barzilay Z (1996). Intrathecal baclofen for severe torsion dystonia in a child. *Acta Paediatr* **85**, 635–7.

Patel S, Naeem S, Kesingland A, Froestl W, Capogna M, Urban L, and Fox A (2001). The effects of GABA(B) agonists and gabapentin on mechanical hyperalgesia in models of neuropathic and inflammatory pain in the rat. *Pain* **90**, 217–26.

Penn RD, Savoy SM, Corcos D, Latash M, Gottlieb G, and Kroin J (1989). Intrathecal baclofen for severe spinal spasiticity. *N Eng J Med* **320**, 1517–21.

Penn RD and Mangieri EA (1993). Stiff-man syndrome treated with intrathecal baclofen. *Neurology* **43**, 2412.

Pfaff T, Malitschek B, Kaupmann K, Prezeau L, Pin JP, Bettler B, and Karschin A (1999). Alternative splicing generates a novel isoform of the rat metabotropic GABA(B) R1 receptor. *Eur J Neurosci* **11**, 2874–82.

Phelan KD (1999). N-Ethylmaleimide selectively blocks presynaptic GABA$_B$ autoreceptor but not heteroreceptor-mediated inhibition in adult rat striatal slices. *Brain Res* **847**, 308–13.

Pin JP, Galvez T, Franek M, Bertrand HO, Prezeau L, Blahos J *et al.* (2000). The characterization of GABA binding site in GABA-(B)R1 receptor shed some light functioning of the heteromeric GABA-B receptor. *Eur J Neurosci* **12**, 67.

Poorrkhalkali N, Juneblad K, Jonsson AC, Lindberg M, Karlsson O, Wallbrandt P *et al.* (2000). Immunocytochemical distribution of the GABA(B) receptor splice variants GABA(B) R1a and R1b in the rat CNS and dorsal root ganglia. *Anat Embryol* **201**, 1–13.

Price GW, Kelly JS, and Bowery NG (1987). The location of GABAB receptor binding sites in mammalian spinal cord. *Synapse* 1, 530–8.

Princivalle A, Spreafico R, Bowery N, and de Curtis M (2000). Layer-specific immunocytochemical localization of GABA(B) R1a and GABA(B) R1b receptors in the rat piriform cortex. *Eur J Neurosci* 12, 1516–20.

Princivalle AP, Pangalos MN, Bowery NG, and Spreafico R (2001). Distribution of GABA(B(1a)), GABA(B(1b)) and GABA(B2) receptor protein in cerebral cortex and thalamus of adult rats. *Neuroreport* 12, 591–5.

Prosser HM, Gill CH, Hirst WD, Grau E, Robbins M, Calver A *et al.* (2001). Epileptogenesis and enhanced prepulse inhibition in GABA$_{B1}$-deficient mice. *Mol Cell Neurosci* 10, 1–10.

Przesmycki K, Dzieciuch JA, Czuczwar SJ, and Kleinrok Z (1998). An isobolographic analysis of drug interaction between intrathecal clonidine and baclofen in the formalin test in rats. *Neuropharmacology* 37, 207–14.

Richards DA and Bowery NG (1996). Anti-seizure effects of the GABA$_B$ antagonist, SCH-50911, in the genetic absence epilepsy rat from Strasbourg (GAERS). *Pharmacol Comm* 8, 227–30.

Riley RC, Trafton JA, Chi SI, and Basbaum AI (2001). Presynaptic regulation of spinal cord tachykinin signaling via GABA(B) but not GABA(A) receptor activation. *Neuroscience* 103, 725–37.

Robbins MJ, Calver AR, Fillipov AK, Couve A, Moss SJ, and Pangalos MN (2001). The GABA$_{B2}$ subunit is essential for G protein coupling of the GABA$_B$ receptor heterodimer. *J Neurosci* 21, 8043–52.

Roberts DCS and Andrews MM (1997). Baclofen suppression of cocaine self-administration: Demonstration using a discrete trials procedure. *Psychopharmacology* 131, 271–7.

Saint DA, Thomas T, and Gage PW (1990). GABAB agonists modulate a transient potassium current in cultured mammalian hippocampal neurons. *Neurosci Lett* 118, 9–13.

Santos AE, Carvalho CM, Macedo TA, and Carvalho AP (1995). Regulation of intracellular [Ca^{2+}] and GABA release by presynaptic GABA$_B$ receptors in rat cerebrocortical synaptosomes. *Neurochem Int* 27, 397–406.

Saunders R, Nahorski SR, and Challiss RAJ (1998). A modulatory effect of extracellular Ca^{2+} on type 1a metabotropic glutamate receptor-mediated signalling. *Neuropharmacology* 37, 273–6.

Sawynok J and Dickson C (1985). D-Baclofen is an antagonist at baclofen receptors mediating antinociception in the spinal cord. *Pharmacology* 31, 248–59.

Schwarz DA, Barry G, Eliasof SD, Petroski RE, Conlon PJ, and Maki RA (2000). Characterization of gamma-aminobutyric acid receptor GABA(B(1e)), a GABA(B(1)) splice variant encoding a truncated receptor. *J Biol Chem* 275, 32 174–81.

Scott RH, Wootton JF, and Dolphin AC (1990). Modulation of neuronal T-type calcium channel currents by photoactivation of intracellular guanosine 5'-O(3-thio) triphosphate. *Neuroscience* 38, 285–94.

Seitz RJ, Blank B, Kiwit JCW, and Benecke R (1995). Stiff-person syndrome with anti-glutamic acid decarboxylase autoantibodies-complete remission of symtoms after intrathecal baclofen administration. *J Neurology* 242, 618–22.

Serrano I, Ruiz RM, Serrano JS, and Fernandez A (1992). GABAergic and cholinergic mediation in the antinociceptive action of homotaurine. *General Pharmacology* 23, 421–6.

Shen W and Slaughter MM (1999). Metabotropic GABA$_B$ receptors facilitate L-type and inhibit N-type calcium channels in single salamander retinal neurons. *J Physiol* 516, 711–18.

Shoaib M, Swanner LS, Beyer CE, Goldberg SR, and Schindler CW (1998). The GABA$_B$ agonist baclofen modifies cocaine self-administration in rats. *Behav Pharmacol* 9, 195–206.

Smid SD and Blackshaw LA (2000). Vagal neurotransmission to the ferret lower oesophageal sphincter: inhibition via GABA(B) receptors. *Br J Pharmacol* 131, 624–30.

Smith GD, Harrison SM, Birch PJ, Elliott PJ, Malcangio M, and Bowery NG (1994). Increased sensitivity to the antinociceptive activity of (+/−)-baclofen in an animal model of chronic neuropathic, but not chronic inflammatory hyperalgesia. *Neuropharmacology* 33, 1103–8.

Snead OC (2000). Evidence for G protein-coupled gamma-hydroxybutyric acid receptor. *J Neurochem* 75, 1986–96.

Snead OC III. (1992). Evidence for GABA$_B$-mediated mechanisms in experimental generalized absence seizures. *Eur J Pharmacol* 213, 343–9.

Staubli U, Scafidi J, and Chun D. (1999). GABA(B), receptor antagonism: Facilitatory effects on memory parallel those on LTP induced by TBS but not HFS. *J Neuroscience* 19, 4609–15.

Stringer JL and Lorenzo N (1999). The reduction in paired-pulse inhibition in the rat hippocampus by gabapentin is independent of GABA(B) receptor activation. *Epilepsy Res* 33, 169–76.

Sullivan R, Chateauneuf A, Coulombe N, Kolakowski LF, Johnson MP, Hebert TE *et al.* (2000). Coexpression of full-length gamma-aminobutyric acid (B) (GABA(B)) receptors with truncated receptors and metabotropic glutamate receptor 4 supports the GAGA(B) heterdimer as the functional receptor. *J Pharmacol Exp Ther* 293, 460–7.

Taira T, Kawamura H, Tanikawa T, Iseki H, Kawabatake H, and Takakura K (1995). A new approach to control central deafferentation pain: spinal intrathecal baclofen. *Stereotac Funct Neurosurg* 65, 101–5.

Takahashi T, Kajikawa Y, and Tsujimoto T (1998). G-protein-coupled modulation of presynaptic calcium currents and transmitter release by a GABA(B) receptor. *J Neurosci* 18, 3138–46.

Thomas DA, McGowan MK, and Hammond DL (1995). Microinjection of baclofen in the ventromedial medulla of rats: antinociception at low doses and hyperalgesia at high doses. *J Pharmacol Exp Ther* 275, 274–84.

Thomas DA, Navarrete IM, Graham BA, McGowan MK, and Hammond DL (1996). Antinociception produced by systemic R(+)-baclofen hydrochloride is attenuated by CGP 35348 administered to the spinal cord or ventromedial medulla of rats. *Brain Res* 718, 129–37.

Thompson SM and Gahwiler BH (1992). Comparison of the actions of baclofen at pre- and postsynaptic receptors in the rat hippocampus in vitro. *J Physiol* 451, 329–45.

Tong AC and Hasselmo ME (1996). Effects of long term baclofen treatment on recognition memory and novelty detection. *Behav Brain Res* 74, 145–52.

Towers S, Princivalle A, Edmunds M, Bettler B, Urban L, Castro-Lopes J, and Bowery NG (2000). GABA$_B$ receptor protein and mRNA distribution in rat spinal cord and dorsal root ganglia. *Eur J Neuroxci* 12, 3201–10.

Tremblay E, Ben-Ari Y, and Roisin MP (1995). Different GABAB-mediated effects on protein kinase C activity and immunoreactivity in neonatal and adult rat hippocampal slices. *J Neurochem* 65, 863–70.

Trampitsch E, Krumpholz R, Likar R, Oher M, and Gulle D (2000). Continuous intrathecal administration of baclofen in severe tetanus. *Anasthesiol Intensivmed NotfMed Schmerzther* 35, 532–3.

Urwyler S, Mosbacher J, Lingenhoehl K, Heid J, Hofstetter K, Froestl W, Bettler B, and Kaupmann K (2001). Positive allosteric modulation of native and recombinant γ-aminobutyric acid$_B$ receptors by 2,6-Di-text-butyl-4-(3-hydroxy-2,2-dimethyl-propyl)-phenol (CGP 7930) and its aldehyde analog CGP 13501. *Mol Pharmacol* 60, 963–71.

Vergnes M, Boehrer A, Simler S, Bernasconi R, and Marescaux C (1997). Opposite effects of GABA$_B$ receptor antagonists on absences and convulsive seizures. *Eur J Pharmacol* 332, 245–55.

Voisin DL and Nagy F (2001). Sustained L-type calcium currents in dissociated deep dorsal horn neurone of the rat: characteristics and modulation. *Neuroscience* 102, 461–72.

Wagner PG and Dekin MS (1993). GABA$_B$ receptors are coupled to a barium-insensitive outward rectifying potassium conductance in premotor respiratory neurons. *J Neurophysiol* 69, 286–9.

Wei K, Eubanks JH, Francis J, Jia Z, and Carter Snead III O (2001). Cloning and tissue distribution of a novel isoform of the rat GABABR1 receptor subunit. *Neuroreport* 12, 833–7.

White JH, McIllhinney RA, Wise A, Ciruela F, Chan WY, Emson PC *et al.* (2000). The GABA(B) receptor interacts directly with the related transcription factors CREB2 and ATFx. *Proc Natl Acad Sci USA* 97, 13 967–72.

White JH, Wise A, Main M, Green A, Fraser NJ, Disney GH *et al.* (1998). Heterodimerization is required for the formation of a functional GABA$_B$ receptor. *Nature* **396**, 679–82.

Wiesenfeld Hallin Z, Aldskogius H, Grant G, Hao JX, Hokfelt T, and Xu XJ (1997). Central inhibitory dysfunctions: mechanisms and clinical implications. *Behav Brain Sci* **20**, 420–30.

Wood MD, Murkitt KL, Rice SQ, Testa T, Punia PK, Stammers J *et al.* (2000). The human GABA$_{B1b}$ and GABA$_{B2}$ heterodimeric recombinant receptor shows low sensitivity to phaclofen and saclofen. *Br J Pharmacol* **131**, 1050–4.

Wu LG and Saggau P (1995). GABAB receptor-mediated presynaptic inhibition in guinea-pig hippocampus is caused by reduction of presynaptic Ca2+ influx. *J Physiol* **485**, 649–57.

Xi ZX and Stein EA (2000). Increased mesolimbic GABA concentration blocks heroin self-administration in the rat. *J Pharmacol Exp Ther* **294**, 613–19.

Xi ZX and Stein EA (1999). Baclofen inhibits heroin self-administration behavior and mesolimbic dopamine release. *J Pharmacol Exp Ther* **290**, 1369–74.

Xu J and Wojcik WJ (1986). Gamma aminobutyric acid B receptor-mediated inhibition of adenylate cyclase in cultured cerebellar granule cells: blockade by islet-activating protein. *J Pharmacol Exp Ther* **239**, 568–73.

Yu ZF, Cheng GJ, and Hu BR (1997). Mechanism of colchicine impairment on learning and memory, and protective effect of CGP36742 in mice. *Brain Res* **750**, 53–8.

Chapter 29

Metabotropic glutamate receptors

Jean-Philippe Pin and Joël Bockaert

29.1 Introduction

Glutamate, like other neurotransmitters (acetylcholine, ATP, serotonine, glycine, GABA), acts on two main types of membrane receptors: ligand-gated channels, also called ionotropic receptors (iGluRs), and G protein coupled receptors also called metabotropic glutamate receptors (mGluRs). In contrast to the iGluRs that are primarily responsible for the fast excitatory synaptic transmission, mGluRs are generally involved in the tuning not only of glutamatergic synapses, but also GABA-ergic, dopaminergic and serotoninergic synapses. As such these receptors represent an excellent target for the development of drugs with various therapeutic application including anxiety, pain, ischaemia, Parkinson disease, epilepsy, and schizophrenia. MGluRs have also been identified outside the brain, in bones, skeletal muscles, taste buds, sensory and enteric neurones . . . indicating that new roles for mGluRs have still to be discovered.

29.2 Molecular characterization

Thus far eight genes encoding mGluRs have been identified in mammalian genomes. No additional sequences clearly related to mGluRs have been identified from the latest public human genome databanks. Sequence similarity searches revealed that these receptors are related to fish olfactory receptors, some putative pheromone and taste receptors, the CaSR and the $GABA_B$ receptor subunits (Bockaert and Pin 1999). Additional membrane proteins also share sequence similarity with the 7TM domains of mGluRs. These are the Drosophila Bride of sevenless (BOSS) and some retinoic-acid induced 7TM proteins called RAIG or GPRC5 (Cheng and Lotan 1998; Robbins *et al.* 2000; Bräuner-Osborne and Krogsgaard-Larsen 2000).

The vertebrate mGluR sequences can be classified into three groups according to their sequence similarity (Fig. 29.1). Two mGlu-like sequences can be identified in the Drosophila genome (DmGluA and DmGluB), and three in C. elegans. Further comparison of the putative glutamate binding pocket of these distant receptors allow DmGluA and one C. elegans sequence to be classified in the group-II mGluRs. Similarly, the same analysis allows the other C. elegans sequences to be classified as group-I and -III mGluRs, respectively (Fig. 29.1). These observations suggest that all 3 mGluR groups are already defined in C. elegans and that only one is conserved in Drosophila.

In mammals, several splice variants of mGluRs have been identified for mGlu1R (Houamed *et al.* 1991; Masu *et al.* 1991; Tanabe *et al.* 1992; Desai *et al.* 1995; Laurie *et al.* 1996;

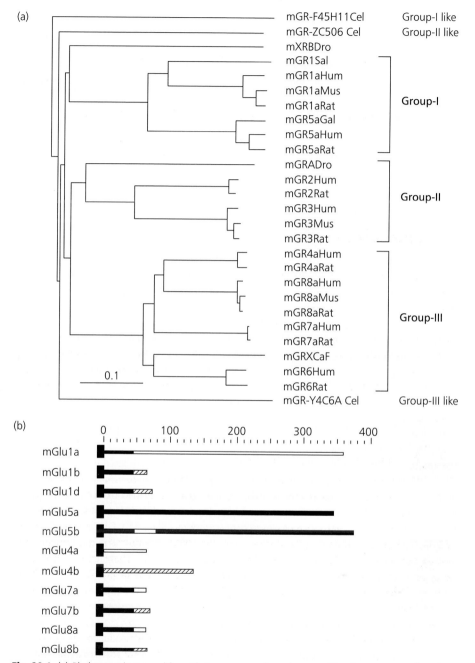

Fig. 29.1 (a) Phylogenetic tree with mGluR sequences from human (Hum), rat, mouse (Mus), chick (Gal), salmon (Sal), cat fish (CaF), Drosophila melanogaster (Dro) and Caenorhabditis elegans (Cel). The alignment and the calculation of the phylogenetic distances were generated with Clustal-W (1.6) using the default parameters. The tree was drawn using TreeView. The groups indicated are defined with bootstrap value higher than 990. (b) Scheme of some described splice variants. Only the C-terminal intracellular tail is shown, residue number 1 (scale bar on the top) being the Pro residue at the end of TM-VII conserved in all sequences. Sequences conserved identical between the variants generated from the same gene are indicated in black.

Stephan *et al.* 1996; Mary *et al.* 1997), mGlu5R (Abe *et al.* 1992; Minakami *et al.* 1993; Joly *et al.* 1995; Lin *et al.* 1997), mGlu4R (Tanabe *et al.* 1992; Flor *et al.* 1995; Makoff *et al.* 1996*a*; Thomsen *et al.* 1997), mGlu7R (Okamoto *et al.* 1994; Makoff *et al.* 1996*b*; Flor *et al.* 1997; Corti *et al.* 1998) and mGlu8R (Duvoisin *et al.* 1995; Corti *et al.* 1998; Wu *et al.* 1998; Malherbe *et al.* 1999) (Table 29.1 and Fig. 29.1). All the variants identified at the protein level and found in both rat and human tissues, differ in their C-terminal intracellular sequences. Further mRNA transcripts likely generated by alternative splicing have been identified (mGlu1cR (Pin *et al.* 1992), 1e (Pin and Duvoisin 1995), 1f (Soloviev *et al.* 1999), 1g (Makoff *et al.* 1997); mGlu8cR (Malherbe *et al.* 1999), taste-mGlu4R (Chaudhari *et al.* 2000)). However, the clear demonstration that these mRNAs do not correspond to miss-splicing events remains to be demonstrated. The taste-mGlu4R variant has been identified in taste buds, where it represents a small percentage of the total mGlu4 mRNA (Chaudhari *et al.* 2000). Because this receptor lacks most of the ligand binding domain (see below), additional experiments are required to demonstrate that this isoform plays an important functional role in umami taste detection.

Sequence analysis of all the mGluRs revealed a large N-terminal extracellular domain (ECD) of about 600 residues, a 7TM region, and an intracellular C-terminal tail variable in length. The extracellular sequence of mGluRs consists of a domain similar to bacterial periplasmic amino acid binding proteins (PBP) (O'Hara *et al.* 1993) linked to the transmembrane region by a cysteine-rich region (about 90 residues with 9 conserved cysteines). Accordingly, O'Hara proposed that the glutamate binding site was located within the ECD (O'Hara *et al.* 1993). This is in marked contrast to the ligand binding pocket of the rhodopsin-like receptors that is located within the 7TM region (Bockaert and Pin 1999). Since 1993, many studies have supported O'Hara's hypothesis (Takahashi *et al.* 1993; Okamoto *et al.* 1998; Parmentier *et al.* 1998; Han and Hampson 1999; Peltekova *et al.* 2000). In parallel, 3D models of the PBP-like domain of several mGluRs (O'Hara *et al.* 1993; Costantino and Pellicciari 1996; Hampson *et al.* 1999; Bessis *et al.* 2000; Parmentier *et al.* 2000) were described and residues critical for glutamate binding identified. These models have been recently validated by the determination of the mGlu1 ECD structure with and without bound glutamate (Fig. 29.2; (Kunishima *et al.* 2000)). This part of the ECD of the mGluRs is composed of two lobes, one large (lobe-I) and one small (lobe-II; Fig. 29.2) that are interconnected by three linkers forming a hinge-like structure allowing both open or closed conformations (Fig. 29.2). Glutamate was found to interact with lobe-I, within the crevice separating the two lobes. The α-carboxylic and α-amino groups form hydrogen bonds with Ser165 and Thr188, whereas the γ-carboxylic group makes an ionic interaction with Lys409, and indirectly interacts with Arg78 via a water molecule. In the closed form conformation of the binding domain, additional interactions occur between glutamate and residues from lobe-II. A cation–pi interaction and an ionic interaction occurs between the α-amino-group of glutamate and Tyr236 and Asp318 respectively, and an additional ionic interaction occurs between the γ-carboxylic group and Arg323 (Fig. 29.2). As often observed for similar proteins, this suggests that glutamate binding stabilizes the closed state of the protein. Interestingly, all these key residues are conserved in all mGluRs suggesting that glutamate binds similarly in all these receptors. This is in agreement with the proposed pharmacophore models for group-I, -II and -III mGluRs (Bessis *et al.* 1999; Costantino *et al.* 1999; Jullian *et al.* 1999). Indeed, this mode of glutamate binding observed in the mGlu1-ECD structure has also been proposed for the mGlu4R by homology modelling and ligand docking (Bessis *et al.* 2000). As such, a restricted number of residues within the binding pocket may be involved in ligand selectivity.

The 7TM domains of mGluRs share no significant sequence similarity with that of other GPCRs (10–12 per cent sequence identity, none of the rhodopsin-like receptor signatures are found in the 7TM region of mGluRs). However, several arguments suggest that the general structure of this domain is similar to that of rhodopsin. First, the two cysteines forming a disulphide bond between the top of the third TM domain and extracellular loop 2 (e2) in all rhodopsin-like receptors, are conserved in all mGlu-like receptors. Second, 3D models based on the rhodopsin structure have received recent validation. In these models, residues involved in the selective recognition of agents acting in this domain of mGluRs (see below) form a binding pocket in such models, in which these ligands can be docked (Pagano *et al.* 2000). Finally, the same molecular determinants controlling G protein coupling have been identified in rhodopsin-like receptors and mGluRs (Gomeza *et al.* 1996*a,b*; Blahos II *et al.* 1998; Francesconi and Duvoisin 1998; Blahos *et al.* 2001). Indeed, the C-terminal tail of the G protein α subunit has also been proposed to interact in a cavity formed by the second and third intracellular loops of mGluRs.

How can agonist binding within the ECD induce a conformational change in the 7TM region that is required for G protein activation? It was previously proposed that the agonist-stabilized closed state of the ECD activates the 7TM region by interacting with the extracellular loops (Pin and Bockaert 1995). However, recent data suggests another more likely scenario. A few years ago, mGlu5R was shown to be a dimer stabilized by disulphide bonds (Romano *et al.* 1996), a fact confirmed for many other receptors of this family, including the CaSR (Bai *et al.* 1999; Ray *et al.* 1999; see Chapter x.y). However, although the cysteines involved were identified and shown to cross-link the two protomers (Ray and Hauschild 2000; Tsuji *et al.* 2000), they were shown not to be necessary for dimer formation (Tsuji *et al.* 2000; Romano *et al.* 2001). Nevertheless, the ECDs of mGluRs, when produced as soluble proteins spontaneously form dimers (Okamoto *et al.* 1998; Han and Hampson 1999; Peltekova *et al.* 2000; Tsuji *et al.* 2000) and indeed, X-rays revealed the presence of dimeric ECDs in the crystals. In the presence of glutamate, not only was a closed form of the protomer observed, but also a big change in the general conformation of the dimer (Kunishima *et al.* 2000). Such was the change, that the C-terminal ends of each ECD were 88 Å apart in the absence of glutamate, but closed to 63 Å in the presence of agonist. Because the C-terminal ends of each ECD are connected to the first TM domain by the cysteine-rich region, one can imagine that the changes in the conformation of the dimeric ECDs can be transduced to the dimer of the 7TM domains. This model is in good agreement with recent data obtained on the GABA$_B$ receptor (Galvez *et al.* 2001). This receptor is a heterodimer formed from GABA$_{B1}$ and GABA$_{B2}$ subunits, each structurally related to mGluRs (Billinton *et al.* 2001, see Chapter 28). Recently, we have also shown that the heterodimeric ECD was necessary, first to maintain the receptor in an inactive state (Fig. 29.3), and second to allow GABA to activate the receptor (Galvez *et al.* 2001).

29.3 Cellular and subcellular localization

29.3.1 Group-I mGluRs

In situ hybridization and immunostaining experiments have revealed widespread expression of mGlu1R and mGlu5R in the major targets of putative glutamatergic pathways of the brain. These two receptor types are generally expressed in different cell populations. In

Table 29.1 Metabotropic glutamate receptors

	Group-I		Group-II		Group-III			
	mGlu₁	**mGlu₅**	**mGlu₂**	**mGlu₃**	**mGlu₄**	**mGlu₆**	**mGlu₇**	**mGlu₈**
Alternative names	mGluR₁	mGluR₅	mGluR₂	mGluR₃	mGluR₄	mGluR₆	mGluR₇	mGluR₈
Structural information (Accession no.)	h 1194 aa (NM_000838)AS r 1199 aa (X57569)AS m 1199 aa (AF320126)	h 1212 aa (XM_011959)AS r 1203 aa (NM_017012)AS m 58 aa (Q9WUK8)	h (XM_003207) r (M92075)	h (XM_004604) r (M92076) m (AF170698)	h (XM_004326) r (M92077)	h (XM_011318) r (D13963)	h (AF083081) r (D16817)	h(XM_011620) r (NM_022202) m (NM_008174)
Chromosomal location	6q24	11cen–q12.1	3p24.3	7q21.1–2	6p21.3	5q35	3p26.1–p25.1	7q31.3–q32.1
Selective agonists	Quisqualate, DHPG	Quisqualate, DHPG, CHPG, cycloquis	LY379268, LY354740, DCG-IV, APDC, NAAG	LY379268, LY354740, DCG-IV, APDC, NAAG	L-AP4,L-SOP, PPG, ACPT-I+ ACPT-III	L-AP4, L-SOP, PPG, ACPT-I +ACPT-III, H-AMPA	L-AP4, L-SOP, PPG, ACPT-I+ACPT-III	L-AP4, L-SOP, PPG, ACPT-I +ACPT-III, S-3,4-DCPG
Selective competitive antagonists	LY393053 LY393675, CMPG	LY393053, LY393675, LY344545	LY341495, MCCG-I, E-Glu	LY341495, MCCG-I, E-Glu	LY341495, DCG-IV, AP4, MPPG	LY341495, DCG-IV,MAP4, MPPG	LY341495, DCG-IV, MAP4, MPPG	LY341495, DCG-IV, MAP4, MPPG
Selective non-competitive antagonists	CPCCOEt, BAY367620	MPEP	RO64-5229	—	—	—	—	—
Selective inverse agonists	BAY367620	MPEP	—	—	—	—	—	—

Radioligands	[3H]-glutamate, [3H]-quisqualate	[3H]-quisqualate, [3H]-MMPEP	[3H]-LY354740, [3H]-LY341495, [3H]-DCG-IV	[3H]-glutamate, [3H]-LY354740, [3H]-LY341495, [3H]-DCG-IV	[3H]-L-AP4	[3H]-LY341495	[3H]-LY341495	[3H]-LAP4, [3H]-LCPPG, [3H]-LY341495
G-coupling	$G_q/G_o/(G_s)$	G_q, G_o, (G_s)	G_i/G_o	G_i/G_o	G_i/G_o	G_i/G_o	G_i/G_o	G_i/G_o
Expression profile	Cerebral cortex, thalamus, septum, globus pallidus, cerebellum, substantia nigra, hippocampus, dentate gyrus, superior colliculus, olfactory bulb, spinal cord, retina, bone	Cerebral cortex, hippocampus, dentate gyrus, spinal cord, septum, striatum, nucleus accumbens, inferior colliculus, olfactory bulb, retina, liver	Cerebral cortex, dentate gyrus, thalamus, striatum, cerebellum, olfactory bulb, retina	Cerebral cortex, thalamus, dentate gyrus, striatum, retina	Cerebral cortex, cerebellum, globus pallidus, substantia nigra, entopeduncular nucleus, striatum, thalamus, hippocampus, olfactory bulb, taste buds	Retina	Cerebral cortex, globus pallidus, superior colliculus, hippocampus, locus coeruleus, olfactory bulb, spinal cord, retina	Cerebral cortex, hippocampus, thalamus, mammillary body, olfactory bulb, retina
Physiological function	Increasing neuronal excitability, pain sensitivity	Increasing neuronal excitability pain sensitivity	Tuning of fast excitatory synaptic transmission	Tuning of fast excitatory synaptic transmission, release of trophic factors from astrocytes	Tuning of fast excitatory synaptic transmission, control of glutamate and GABA release, detection of taste	Tuning of fast excitatory synaptic transmission	Inhibition of synaptic transmission in the spinal cord, tuning of fast excitator ysynaptic transmission	Tuning of fast excitatory synaptic transmission
Knockout phenotype	Reduced pain sensitivity, cerebellar ataxia, loss of cerebellar LTD, spatial memory loss	Impaired learning, impaired CA1 LTP	Impaired LTD in CA3, normal learning	—	Absence seizures, spacial learning and memory impairment	Deficit of ON response, impaired behavioural suppression by light	Epilepsy and premature death, deficit in fear response	—
Disease relevance	Pain, neurodegeneration, Parkinson's disease, cerebellar ataxia,	Pain, neurodegeneration, Parkinson's disease, synaptic plasticity	Schizophrenia, anxiety, stroke, neurodegeneration, Parkinson's disease, addiction	Schizophrenia, neurodegeneration, stroke, addiction	Schizophrenia, absence epilepsy, neurodegeneration, stroke, addiction	—	Schizophrenia, epilepsy, neurodegeneration, stroke, pain, addiction	Schizophrenia, neurodegeneration, stroke, addiction

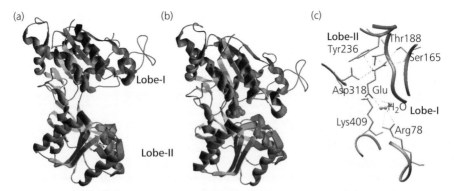

Fig. 29.2 Structure of the mGlu1R binding domain. (a) Ribbon view of the open conformation without glutamate. (b) Ribbon view of the closed conformation with glutamate. (c) Detailed view of the glutamate binding site.

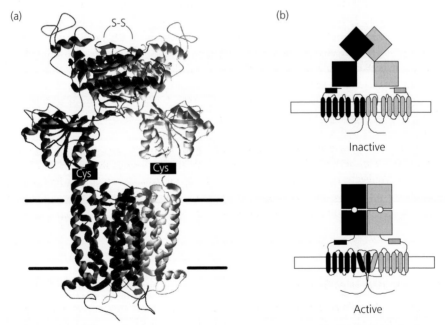

Fig. 29.3 (a) Ribbon representation of a mGluR dimer. One protomer is shown in black, the other in light grey. The dimeric form of the ECD is drawn according to the coordinates deposited in the protein data bank for the unliganded form (pdb accession number 1EWK). The 7TM regions correspond to a 3D model of the mGlu1R constructed according to the rhodopsin structure (pdb accession number 1F88), each 7TM region being in contact via their TM6. (b) Proposed activation mechanism.

the central nervous system (CNS), mGlu1R is found only in neurones. The most intense levels of expression are found in Purkinje neurones of the cerebellum, mitral and tufted cells of the olfactory bulb, and neurones of the hippocampus (interneurones, and pyramidal neurones of CA3), lateral septum, thalamus, globus pallidus, and substantia nigra. Less intense expression was found in dentate gyrus, striatum, cingulate and entorhinal cortices, and superior colliculus. In contrast, mGlu5R is found on both astrocytes and neurones, with most intense expression in the olfactory bulb (granule cells), cerebral cortex, hippocampus (dentate gyrus and pyramidal neurones of the CA1-3 area) lateral septum, striatum, nucleus accumbens and inferior colliculus. In the cerebellum, mGlu5R is expressed only in a sub-population of Golgi neurones. Both mGlu1R and mGlu5R are also expressed in the retina as well as in the spinal cord.

At the subcellular level, both mGlu1a and mGlu5 proteins were found in post-synaptic elements on the side of the post-synaptic density (Lujàn *et al.* 1996, 1997) and were rarely detected in pre-synaptic elements (Fotuhi *et al.* 1993; Romano *et al.* 1995). However, elec-trophysiological as well as biochemical studies suggest a pre-synaptic action of a group-I mGluR, likely mGlu5R (Gereau and Conn 1995; Manzoni and Bockaert 1995; Lu *et al.* 1997; Rodriguez-Moreno *et al.* 1998; Sistiaga *et al.* 1998).

The specific localization of the short mGlu1R variants (mGlu1bR and mGlu1dR) has not been studied extensively. However, Western blotting and PCR experiments revealed that the major mGlu1R isoform in most regions outside of the cerebellum is a short vari-ant (Fotuhi *et al.* 1993; Hampson *et al.* 1994; Ferraguti *et al.* 1998; Mateos *et al.* 1998). In the hippocampus, mGlu1aR is expressed in non-pyramidal cells in all areas, whereas the 1b variant is expressed in the CA3 pyramidal neurones and the granule cells of the dendate gyrus (Ferraguti *et al.* 1998). When examined at the electron microscopic level, mGlu1bR was found at the same perisynaptic location as mGlu1aR (Lujàn *et al.* 1996; Mateos *et al.* 2000). The two mGlu5 variants are expressed in the same brain area (Joly *et al.* 1995), with the 5a isoform being predominant in the developing brain (Minakami *et al.* 1995).

Group-I mGluRs have also been detected outside the brain (for a review see Skerry and Genever 2001), in peripheral unmyelinated sensory afferent terminals (both mGlu1 and 5 receptors; Walker *et al.* 2000*b*; Bhave *et al.* 2001); in melanocytes (mGlu5R; Frati 2000); in osteoblast (mGlu1bR; Gu and Publicover 2000); in heart cells (both mGlu1a and mGlu5 receptors; Gill 1999), and hepatocytes (mGlu5R; Storto *et al.* 2000).

29.3.2 Group-II mGluRs

MGlu2R is widely distributed in the brain, with most prominent expression observed in the Golgi cells of the cerebellum, mitral and granule cells of the accessory olfactory bulb, in the dentate gyrus and in some neurones of the cortex, thalamus, and striatum (Ohishi *et al.* 1993, 1994; Neki *et al.* 1996). In contrast to mGlu2Rs which are found exclusively in neurones, mGlu3Rs are also found in non-neuronal cells (mostly astrocytes). The mGlu3Rs are also found in many neurones with a most prominent expression in the reticular nucleus in the thalamus, dentate gyrus, striatum and cortex (Tanabe *et al.* 1993). Both mGlu2 and mGlu3 receptors are found in several types of cells in the retina (Koulen *et al.* 1996).

In neurones, both mGlu2R and mGlu3R are found in axonal and somato-dendritic areas. At the subcellular level, these receptors are found either pre-synaptically, as well as in the somato-dendritic area of many neurones, distant from the post-synaptic density (Lujan *et al.* 1997). Strong labelling with mGlu2-3R antibodies is also observed in glia, in agreement with the expression of mGlu3 mRNA in these cells.

29.3.3 **Group-III mGluRs**

Among the group-III mGluRs, mGlu6R has a limited distribution. So far, this receptor has been found exclusively in single cell type of the retina, the ON bipolar cell. This receptor is responsible for the hyperpolarization of these cells in the dark, when glutamate is released from the rod or cone photoreceptor cells. As such, this glutamatergic synapse is the only one that functions almost exclusively via a post-synaptic mGluR. Note that this makes this glutamatergic synapse inhibitory in nature, rather than excitatory.

The other group-III mGluRs have a wider distribution in the CNS each being expressed in specific neuronal types. The mGlu4Rs are mostly found in parallel fibre terminals of the cerebellum (Kinoshita *et al.* 1996*b*) and internal granules cells of the olfactory bulb (Tanabe *et al.* 1993). They are also found in the hippocampus (Shigemoto *et al.* 1997), in globus pallidus, substantia nigra pars reticulata and entopeduncular nucleus, and to a lesser extent in the striatum, retina (Akazawa *et al.* 1994; Koulen *et al.* 1996), neocortex, and thalamus (Bradley *et al.* 1999). The mGlu4R has also been found outside the CNS, in taste buds where it could play a role in the detection of the umami taste (Chaudhari *et al.* 1996, 2000). The mGlu7aRs were seen in the olfactory bulb, anterior olfactory nucleus, piriform and entorhinal cortices, periamygdaloid cortex, amygdalohippocampal area, hippocampus, layer I of the neocortical regions, globus pallidus, superficial layers of the superior colliculus, locus coeruleus, and superficial layers of the medullary and spinal dorsal horns (Kinoshita *et al.* 1998). The distribution of mGluR7bR was more restricted. The mGlu7R is also found in spinal cord in the terminals of primary afferent fibres (Ohishi *et al.* 1995; Li *et al.* 1997) and in the retina (Akazawa *et al.* 1994; Brandstatter *et al.* 1996). The mGlu8Rs are found in the rhinencephal, in the terminals of the main olfactory bulb neurones (Kinoshita *et al.* 1996*a*), in the retina (Koulen *et al.* 1996). mGlu8 mRNA are found primarily in the olfactory bulb, pontine gray, lateral reticular nucleus of the thalamus, and piriform cortex. Less abundant expression was detected in cerebral cortex, hippocampus, cerebellum, and mammillary body (Duvoisin *et al.* 1995; Saugstad *et al.* 1997).

At the subcellular level, mGlu4, 7 and 8 receptors are mostly located in pre-synaptic terminals and in many cases act as autoreceptors. However, mGlu4aRs have also been found in the soma and dendrites of pyramidal and granule cells as well as interneurones in the hippocampus (Bradley *et al.* 1996). MGlu4a, 7a and 7b receptors are also found in terminals of inhibitory GABA-ergic neurones (Bradley *et al.* 1996; Kinoshita *et al.* 1998).

At the electron microscopic level, these three group-III mGluRs are located in the synaptic grid where they likely modulate the release process. In addition, these three receptor types are found exclusively in synapses making contact with a specific target cell (Shigemoto *et al.* 1996, 1997). For example, in the hippocampus, mGlu7R is mostly found in pre-synaptic grids facing GABAergic interneurones, and to a much lesser extent in those facing glutamatergic pyramidal cells even when these two types of synapses are located within the same nerve terminal (Shigemoto *et al.* 1996). Similarly, in the retina, mGlu7R is located in only one of the two active zones of the cone bipolar cell ribbon synapses (Brandstatter *et al.* 1996).

Little is known about the molecular mechanisms that control mGluR localization in neurones. Several studies identified the intracellular C-terminal tails of these receptors as an important determinant for their localization. For example, deletion of this domain plus the last transmembrane domain of mGlu1R prevents the targeting of the receptor outside the soma *in vivo* (Conquet *et al.* 1994). Expression of recombinant wild-type and chimeric receptors in neurones revealed that *axon exclusion of mGlu2R* versus axon targeting of mGlu7R is mediated by their 60 amino acid C-terminal cytoplasmic domains (Stowell and Craig 1999). This suggests that intracellular proteins interacting with mGluRs may control their specific localization in neurones. In agreement with this idea, PICK1 that interacts with mGlu7R (Dev *et al.* 2000) is involved in the pre-synaptic clustering of this mGluR (Boudin *et al.* 2000), but not in its dendrite/axon sorting. Moreover, Homer proteins that interact with group-I mGluRs (Xiao *et al.* 2000) control the axon/dendrite (Ango *et al.* 2000) as well as the plasma membrane insertion (Roche *et al.* 1999; Ciruela *et al.* 2000) of these receptors.

29.4 Pharmacology

Several reviews have been published recently that described in detail the activity of most known mGluR ligands (Pin *et al.* 1999; Schoepp *et al.* 1999; Bräuner-Osborne *et al.* 2000). In the following passage we will highlight the most important of these compounds.

29.4.1 Agonists

Glutamate is commonly recognized as the natural endogenous agonist for all mGluRs. The affinity values for glutamate, as determined from binding studies, are quite variable depending on the receptor subtype ranging from $0.3–0.7\,\mu M$ for mGlu1, 3 and 5 receptors, to $869\,\mu M$ for mGlu7R. Glutamate affinity is 1.6, 2.5, 5 and $12.3\,\mu M$ for mGlu4, 2, 8, and 6 receptors, respectively (Thomsen *et al.* 1993; Eriksen and Thomsen 1995; Laurie *et al.* 1995; Cartmell *et al.* 1998; Johnson *et al.* 1999; Mutel *et al.* 2000; Naples and Hampson 2000; Peltekova *et al.* 2000; Schweitzer *et al.* 2000; Wright *et al.* 2000). In addition to glutamate, other endogenous compounds activate mGluRs. For example, the N-acetylated dipeptide NAAG activates group-II mGluRs (Wroblewska *et al.* 1997; Schweitzer *et al.* 2000), and L-SOP is a selective agonist of all group-III mGluRs, being even 10 times more potent than glutamate (Schweitzer *et al.* 2000). Sulphur containing amino acids also activate mGluRs with a certain group selectivity (Eriksen and Thomsen 1995; Johansen *et al.* 1995; Kingston *et al.* 1998*a*; Parmentier *et al.* 2000; Schweitzer *et al.* 2000). These observations suggest that endogenous molecules other than glutamate may act on mGluRs *in vivo*. In agreement with this observation, many PBP-like proteins can bind a variety of natural ligands (Speca *et al.* 1999). In addition, we have recently reported that the binding pocket of mGluRs is highly conserved during evolution from Drosophila to mammals (Parmentier *et al.* 2000). Such a high degree of conservation of both the binding pocket and pharmacological profile of a receptor during evolution has never been observed for other receptors. This supports the idea that recognition of multiple natural ligands has important biological meaning that still remains to be elucidated.

Group-I mGluR agonists

So far, all selective mGluR agonists are glutamate derivatives, and rare are those that can discriminate mGluRs from the same subgroup. For the group-I, quisqualate remains the most potent agonist known (Affinity around 10–30 nM), although not selective because of its agonist activity at non-NMDA-iGlu receptors. As such DHPG remains the only potent $(1–4\,\mu M)$ and selective group-I agonist. Homo-quisqualate has been shown to fully activate mGlu5R $(36\,\mu M)$ and to antagonize mGlu1R, but this compound is also active on group-II mGluRs (Bräuner-Osborne and Krogsgaard-Larsen 1998). Recently, Z-cyclobutane-quisqualate was shown to be a potent (11 μM) mGlu5R agonist devoid of activity at mGlu1, 2 and 4a receptors (Littman *et al.* 1999). Several derivatives of phenylglycine have been reported to have selective action on a single group-I mGluR, either mGlu1R or mGlu5R. These include CHPG which activates mGlu5R at high concentration (Doherty *et al.* 1997), CBPG, 4CPG, and 4C3HPG which act as mGlu5R partial agonists and mGlu1R antagonists (Brabet *et al.* 1995; Pellicciari *et al.* 1996) and CMPG (LY367585) which antagonizes mGlu1R and is without effect on mGlu5R (Clark *et al.* 1997). However, most of these compounds exert a complex action on these receptors (Littman and Robinson 1994; Brabet *et al.* 1995) and their binding properties are not related with their action on these receptors (Mutel *et al.* 2000). It is possible that these compounds not only bind at the glutamate binding site but also at a regulatory site which may actually be different between mGlu1 and mGlu5 receptors (Mutel *et al.* 2000).

Group-II mGluR agonists

For the group-II mGluRs, LY354740 (affinity around 10–50 nM) and its analog LY379268 (0.2 nM), are still the most potent, selective and systemically active agonists (Schoepp *et al.* 1997; Monn *et al.* 1999). Even more potent derivatives have been described (Nakazato *et al.* 2000). The other commonly used group-II agonist, L-CCG-I, is much less selective, being active on group-I and -III mGluRs with a 10–50 time lower potency (Brabet *et al.* 1998; Wright *et al.* 2000). DCG-IV (0.1–0.5 μM) is a possible alternative to the LY compounds (Hayashi *et al.* 1993), although one should take care of its antagonist activity at all group-III receptors (10 μM) (Brabet *et al.* 1998; Wright *et al.* 2000) and agonist activity at NMDA receptors (Hayashi *et al.* 1993). Recently, NAAG has been proposed as a selective mGlu3 over mGlu2 receptor (100 μM) (Wroblewska *et al.* 1997). However, further analysis of its activity revealed that it is also active at mGlu2R albeit with a 10-fold lower affinity (Schweitzer *et al.* 2000).

Group-III mGluR agonists

For group-III mGluRs, L-AP4, L-SOP, PPG, and ACPT-I and (+)-III are selective agonists but act on all receptors from this group, although they are all less active at the mGlu7R. PPG is 10–30 times more potent on mGlu8R (2 μM) than on the other group-III mGluRs (Gasparini *et al.* 1999a; Wright *et al.* 2000), and S-Homo-AMPA(100 μM) is a selective mGlu6R agonist (Ahmadian *et al.* 1997; Wright *et al.* 2000). Although being the most potent, L-AP4, L-SOP and PPG also act on the so-called Ca^{2+}/Cl^--dependent glutamate transport system (Pin *et al.* 1984) such that care should be taken when using these compounds on neurones or brain slices. Recently, DCPG has been reported to be a selective mGlu8R agonist (EC50 = 30 nM), being 10–100 times more potent on this receptor subtype compared to the other mGluRs (Thomas *et al.* 2001).

29.4.2 **Antagonists**

Group-I mGluR antagonists

There is a definite lack of potent and selective group-I competitive antagonists. The first described was the phenylglycine derivative MCPG (which antagonizes both mGlu1 and mGlu5 receptors, as well as the group-II and some group-III mGluRs), and CMPG, 4CPG and 4C3HPG which antagonize mGlu1R and have either no effect or act as partial agonists at mGlu5R. However, as described above, these compounds have complex action on these receptors that cannot be explained by action at a single site. The best competitive antagonists of group-I mGluRs are LY393053 (1 μM; Chen *et al.* 2000) and the non-selective compounds LY341495 (7–8 μM; Kingston *et al.* 1998*b*) and LY344545 (5.5 μM). This latter compound displays a slightly higher affinity at mGlu5R than other mGluRs (Doherty *et al.* 2000).

Group-II mGluR antagonists

The most potent group-II mGluR competitive antagonist is the LY341495 (10 nM), a derivative of L-CCG-I, which is 100 times more potent at group-II than group-I mGluRs, but is also active in the nanomolar range at most group-III mGluRs. Other group-II antagonists are E-Glu (4–40 μM) (Cartmell *et al.* 1998; De Colle *et al.* 2000; Parmentier *et al.* 2000; Schweitzer *et al.* 2000) and PCCG-IV (Thomsen *et al.* 1996; Parmentier *et al.* 2000). E-Glu is much less potent on the group-III mGlu4 and mGlu8 receptors, but its activity at cloned group-I mGluRs remains to be identified.

Group-III mGluR antagonists

There are still no good group-III antagonists. The rank order of potency is LY341495 > DCG-IV > MAP4 > MPPG. However, LY341495, MAP4, and MPPG are also active at group-II mGluRs, and DCG-IV is a potent group-II agonist.

29.4.3 Non-competitive antagonists and inverse agonists

CPCCOEt was the first antagonist described as selective for a single mGluR (mGlu1R) that did not have a glutamate-like structure. Later, this compound was shown to be a non-competitive antagonist interacting within the 7TMD region of mGlu1R in a cavity lined by TMD-III, -VI and -VII (Litschig *et al.* 1999; Ott *et al.* 2000). More recently, another mGlu1 selective non-competitive antagonist, BAY367620, with a higher affinity has been described and shown also to interact in the 7TMD region of this receptor (Carroll *et al.* 2001). In parallel, selective mGlu5 antagonists also acting in the 7TMD region of this receptor have been identified (Varney *et al.* 1999). Based on the structure of these molecules, the most potent, selective and systemically active mGlu5 antagonist has been synthesized, MPEP (Gasparini *et al.* 1999*b*; Pagano *et al.* 2000). Recently, a non-competitive selective mGlu2 antagonist (RO64-5229) was also reported (Adam *et al.* 1999).

A few years ago, we reported that mGlu1 and mGlu5 receptors display constitutive activity when expressed in different cell lines (Joly *et al.* 1995; Prézeau *et al.* 1996). Because none of the competitive antagonists were able to inhibit this constitutive activity, we speculated that the constitutive activity comes from a spontaneous oscillation of the 7TMD region of these receptors between active and inactive conformations. In agreement with this proposal, we recently found that the mGlu1R selective antagonist BAY367620, and the mGlu5R selective antagonist MPEP, which both bind in the 7TMD region of these receptors, were able to inhibit the constitutive activity (Pagano *et al.* 2000; Carroll *et al.* 2001). These two compounds are therefore the first mGlu1R and mGlu5R selective inverse agonists. As observed for other GPCRs, not all antagonists acting in the 7TMD region have inverse agonists properties. Indeed, CPCCOEt has almost no inverse agonist activity at mGlu1R (Litschig *et al.* 1999; Carroll *et al.* 2001).

Using such inverse agonists, we recently demonstrated that the constitutive activity of mGlu1 and mGlu5 receptors is tightly regulated by the interacting proteins Homer in cultured neurones. Homer1b, 1c, 2 and 3 proteins directly interact with the C-terminal tail of these receptors, are constitutively expressed in neurones and dimerize via their coiled-coil domain (Xiao *et al.* 2000, see Chapter x.y??). Homer3 was found to inhibit the constitutive activity of mGlu1 and mGlu5 receptors in cerebellar granule neurones (Ango *et al.* 2001). Of interest, a short version of the Homer1 protein, Homer1a is produced after simulation of these neurones like an immediate early gene (Xiao *et al.* 2000). This short Homer1a protein lacks the dimerization motif, and as such, competes with the dimeric Homer3, and prevents its inhibition of the mGluR constitutive activity. Accordingly, group-I mGluRs can be activated either by glutamate (fast and transient activation), or by Homer1a (slowly generating and long lasting activation) as a result of increased neuronal activity (Ango *et al.* 2001).

29.4.4 Allosteric modulators

Ca^{2+} has been proposed to activate mGlu1, mGlu5, and mGlu3 receptors and Ser166 of mGlu1R has been shown to be involved in this effect (Kubo *et al.* 1998). However, other authors described Ca^{2+} as a positive modulator of these mGluRs, that increases the potency of glutamate (Saunders *et al.* 1998), in agreement with what has been reported for the related $GABA_B$ receptor (Galvez *et al.* 2000).

Artificial molecules have also been identified as positive allosteric modulators of the CaSR (Hammerland *et al.* 1998) and the $GABA_B$ receptor (Urwyler *et al.* 2000) and recently similar allosteric modulators have been described for mGlu1R (Knoflach *et al.* 2001). Such molecules have no activity when applied alone, but potentiate (increase of potency and efficacy) the action of agonists. Such compounds may be extremely useful as therapeutics since in contrast to agonists, they will neither induce receptor desensitization nor constant receptor activation. Such molecules will only result in receptor activation in the presence of low concentrations of endogenous agonist thus potentiating agonist function.

29.5 Signal transduction and receptor modulation

29.5.1 Group-I mGluRs

Group-I mGluRs are coupled to the G_q-type of G protein and as such activate PLC and Ca^{2+} release from intracellular stores. Coupling of group-I mGluRs to this transduction pathway is sensitive to PKC-dependent phosphorylation (Francesconi and Duvoisin 1998; Gereau and Heinemann 1998). By interacting with the C-terminal tail of mGlu1a and mGlu5 receptors, Homer proteins affect the coupling efficacy of these receptors to Ca^{2+} release. Three different Homer proteins have been characterized, each existing as different isoforms generated by alternative splicing (Brakeman *et al.* 1997; Kato *et al.* 1998; Xiao *et al.* 1998, 2000; Soloviev *et al.* 2000). These proteins can dimerize and interact with either group-I mGluRs, IP3 receptors or ryanodine channels and as such directly connect mGluRs to these channels (Tu *et al.* 1998). This observation is in agreement with the tight coupling of mGluRs to ryanodine channels observed in cultured cerebellar granule neurones (Chavis *et al.* 1996; Fagni *et al.* 2000). As mentioned earlier, a short Homer1 variant is expressed in neurones like an immediate early gene upon neuronal activation (Brakeman *et al.* 1997). This variant lacks the coiled-coil domain responsible for Homer dimerization and as such disrupts the physical link between mGlu and IP3 receptors established by constitutively expressed dimeric Homer. This slows down the kinetics of the Ca^{2+} signal generated by the group-I mGluRs (Xiao *et al.* 1998) and facilitates modulation of Ca^{2+} and K^+ channels by group I mGluRs (Kammermeier *et al.* 2000).

Group-I mGluRs also couple to G_s proteins in heterologous expression systems but the physiological consequences of this coupling in neurones remains to be identified (Aramori and Nakanishi 1992; Joly et al. 1995; Francesconi and Duvoisin 1998). Of interest, this coupling to G_s is not sensitive to PKC phosphorylation in contrast to the coupling to G_q (Francesconi and Duvoisin 2000). Accordingly, group-I mGluRs may activate different transduction pathways depending on their phosphorylation state. In agreement with this proposal, group-I mGluRs can either potentiate or inhibit neurotransmitter release depending on the PKC activity (Herrero et al. 1998; Rodriguez-Moreno et al. 1998). In addition, depending on the kinetics of receptor activation, group-I mGluRs can either hyperpolarize or depolarize neurones (Fiorillo and Williams 1998). Coupling of group-I mGluRs to PTX-sensitive G proteins G_i and G_o has also been observed in many cells including neurones. Activation of these G proteins can also activate PLC and modulate the activity of a variety of channels.

Group-I mGluRs can also regulate the activity of a number of ion channels in neurones, including K^+, Ca^{2+} and cationic non-selective channels as well as the NMDA receptors (Conn and Pin 1997). Accordingly, activation of Ca^{2+}-activated K^+-channels hyperpolarizes neurones, whereas the activation of cationic non-selective channels or the inhibition of K^+-channels induces a slowly generated depolarization, or changes the firing pattern of neurones. Group-I mGluRs are among the few receptors which have been shown to potentiate the activation of L-type Ca^{2+} channels, an effect involving ryanodine channels and likely the Homer proteins (Chavis et al. 1996; Fagni et al. 2000). Another interesting observation, is the cross-talk between NMDA receptors and group-I mGluRs, especially when one considers that these two receptors are both involved in synaptic plasticity. MGlu5R not only potentiates NMDA receptor currents (Pin and Duvoisin 1995; Conn and Pin 1997; Lan et al. 2001), but activation of the NMDA receptor also reverses desensitization of mGlu5R (Alagarsamy et al. 1999). This positive feedback mechanism likely results from co-localization of these two receptors in the same protein complex (Husi et al. 2000) and may play a pivotal role in the induction of long-term potentiation. Recently, coupling of group-I mGluRs to cationic non-selective channels has been shown to be G protein independent, and as such represents one of the few examples of G protein independent cascade activated by heptahelical receptors. Indeed, coupling of mGluRs to these channels involves activation of Src-like protein tyrosine kinases (Heuss et al. 1999).

Finally, Group-I mGluRs can activate other intracellular pathways such as the MAPK pathway (Ferraguti et al. 1999; Karim et al. 2001). This effect likely results from the activation of G_i/G_o G proteins.

29.5.2 Group-II and Group-III mGluRs

Both group-II and group-III mGluRs couple to G_i and G_o types of G proteins and as such can inhibit adenylyl cyclase and modulate ion channel activity (Pin and Duvoisin 1995; Conn and Pin 1997). All these receptors have been shown to inhibit various types of Ca^{2+}-channels in different neurones (Anwyl 1999). Although these receptors are coupled to the same type of G proteins some specificity in the regulation of Ca^{2+}-channels have been reported. For example, whereas mGlu2R inhibits N- and L-type Ca^{2+}-channels in cerebellar granule neurones (Chavis et al. 1994, 1995), the group-III mGlu7R inhibits P/Q-type channel selectively (Perroy et al. 2000). In both cases, a PTX sensitive G protein (likely G_o) is involved, but the downstream transduction cascades are different since mGlu7R activation of PLC and PKC appears to play a critical role for its action on the P/Q-type channel. These effects on

Ca^{2+}-channels are likely to be responsible, at least in part, for the pre-synaptic inhibition of transmitter release at synapses. However, other actions are likely to be involved in this effect since the receptors appear to directly inhibit the release process (Gereau and Conn 1995; Chavis *et al.* 1998). Indeed, the localization of group-III mGluRs within the synaptic grid makes them likely to directly control the release process.

Group-II and -III mGluRs also regulate the activity of K^+-channels (Anwyl 1999). The transduction mechanism of mGlu6R is specific to this receptor subtype expressed on ON bipolar neurones. This receptor appears to activate a transduction cascade similar to that activated by rhodopsin in the photoreceptor cells: activation of a PDE, leading to the closure of a cGMP activated cationic channel (for reviews see (Nakanishi 1994; Conn and Pin 1997)).

29.6 Physiology and disease relevance

29.6.1 Anxiety

The recent development of systemically active mGluR ligands has allowed the examination of their action in several behavioural tests. This has led to the demonstration of the anxiolytic action of the mGlu2R agonist LY354740 (Helton *et al.* 1998), and the mGlu5R selective antagonist MPEP (Spooren *et al.* 2000c; Tatarczynska *et al.* 2001).

29.6.2 Pain and analgesia

Several observations suggest mGluRs represent an excellent target for the treatment of pain. MGlu7R is located on the terminals of c-fibres where they inhibit synaptic transmission, and both group-I and -II mGluRs have been found in the dorsal horn in the spinal cord. Moreover, mGluRs have been shown to modulate the primary sensory responses in the thalamic nucleus (Eaton *et al.* 1993). In agreement with the possible involvement of mGlu1Rs in nociception, pain sensitivity is reduced in mGlu1R knock-out mice (Corsi *et al.* 1996), and inhibition of group-I mGluRs in the spinal cord has been reported to attenuate pain in several disease models (Fisher and Coderre 1996; Fisher *et al.* 1998; Fundytus *et al.* 1998, 2001). However, the most exciting observation is that both group-I mGluRs are found in the sensory terminals of the skin where they likely play a critical role in inflammatory as well as acute pain sensitivity (Walker *et al.* 2000a,b; Bhave *et al.* 2001).

29.6.3 Epilepsy

By tuning fast excitatory synaptic transmission, mGluRs are possible targets for the treatment of epilepsy. Among the different mGluR subtypes, mGlu7R probably appears the best target. This receptor is located in the pre-synaptic grid of glutamatergic synapses where, due to its low affinity for glutamate, it may protect the synapse from an overflow of glutamate in the synaptic cleft. In agreement with this hypothesis, mGlu7R knock-out mice are susceptible to epileptic seizures leading to premature death of the animals (Masugi *et al.* 1999). Moreover, the potent group-III mGluR agonist PPG has anti-convulsive activity *in vivo* (Gasparini *et al.* 1999a). Another group-III mGluR, mGlu4R, controls glutamate and GABA release, especially within the thalamocortical circuitry. Alterations in the regulation of these synapses have been reported in several animal models of absence seizure, suggesting that mGluRs, and especially mGlu4R may play a role in these pathologies. In agreement with this hypothesis, the GABA$_B$ agonist baclofen no longer induces absence seizures in mGlu4R knock-out mice (Snead *et al.* 2000).

29.6.4 Schizophrenia

Schizophrenia is assumed to be the consequence of an imbalance of the glutamatergic and dopaminergic systems. Indeed, PCP or 'angel dust', is known to block NMDA receptors as well as dopamine transporters and induces schizophrenic-like symptoms in both humans and rodents. By regulating glutamate release, group-II and -III mGluR agonists may have beneficial effects in schizophrenic patients. In agreement with this proposal, the group-II agonist LY354740 has been shown to inhibit some of the PCP-induced motor activity in rats (Moghaddam and Adams 1998). However, the behavioural effects of the group-II agonists differ in some respects with those of the antipsychotic agent clozapine (Cartmell *et al.* 1999, 2000*a,b*; Schreiber *et al.* 2000). These effects are likely the results from the activation of the mGlu2R subtype, since they were not seen in mGlu2 knock-out mice (Spooren *et al.* 2000*b*).

29.6.5 Ischaemia

Due to their pre-synaptic inhibitory action at many glutamatergic terminals activation of either group-II or -III mGluRs would be expected to be neuroprotective and such effects have been observed in many situations *vitro* and *in vivo* (for reviews see Pin and Duvoisin 1995; Conn and Pin 1997). The use of knock-out mice has helped identify mGlu4R as being responsible for the L-AP4 induced protection from excitotoxic insults (Bruno *et al.* 2000*a*). In addition, group-II mGluR agonists stimulate the release of a trophic factor (TGFβ) from astrocytes (likely by activating mGlu3Rs expressed in these cells), leading to a long term decrease in the sensitivity of neurones to excitotoxic insults (Bruno *et al.* 1997). More recently, *in vivo* experiments have been conducted with the group-II agonists, including LY354740, and revealed an important protection from ischaemic damage in gerbil (Bond *et al.* 1998, 2000), but not in a rat model of permanent ischaemia (Lam *et al.* 1998). Although inhibition of glutamate release may be neuroprotective, inhibition of GABA release may have the opposite effect. As such, the action of group-II and group-III mGluR agonists on excitotoxic insults may depend on the neuronal sub-populations involved. Indeed, activation of mGlu7Rs has been found to be protective in cultured cerebellar granule neurones (most of these cells being glutamatergic) (Lafon-Cazal *et al.* 1999*a*) but to potentiate toxicity in striatal GABAergic cultures (Lafon-Cazal *et al.* 1999*b*).

By increasing neuronal excitability, the activation of group-I mGluRs likely facilitates the excitotoxic action of glutamate. In agreement with this proposal, group-I mGluR antagonists are neuroprotective in cultures (for reviews see Pin and Duvoisin 1995; Conn and Pin 1997). Although mGlu1R knock-out mice are equally sensitive to ischaemia compared to control mice (Ferraguti *et al.* 1997), the selective mGlu5R antagonist MPEP is protective both *in vitro* and *in vivo* (Bruno *et al.* 2000*b*).

29.6.6 Parkinson's disease

The loss of dopaminergic neurones from the substantia nigra pars compacta observed in Parkinson's patients is at the origin of the abnormal functioning of a neuronal network controlling movements. Indeed, over activity and burst firing of neurones from the subthalamic nucleus (STN) may play an important role in the abnormal movements of Parkinson's patients (Rouse *et al.* 2000). By decreasing glutamate release from cortico-striatal (Lovinger and McCool 1995) and STN terminals (Bradley *et al.* 2000), group-II mGluR agonists may be beneficial for these patients. In addition, decreasing the excitability of neurones from both the STN and the subtantia nigra pars reticulata, using group-I antagonists may also attenuate the consequences of the loss of dopaminergic neurones (Awad *et al.* 2000). Indeed, the mGlu2R agonist LY354740 reverses the movement abnormalities in two rat models of Parkinson's disease (Konieczny *et al.* 1998; Bradley *et al.* 2000). However, the mGlu5R antagonist MPEP had no effect in these models (Spooren *et al.* 2000*a*).

29.6.7 Alzheimer's disease and cognition

Alzheimer's disease is characterized by the death of neurones, likely resulting at least in part from accumulation of β-amyloid (Aβ) peptides produced by the amyloidogenic processing of the amyloid precursor protein (APP). It is known that increased cleavage of APP by the enzyme α-secretase diminishes the level of amyloidogenic Aβ peptides produced. It is of interest therefore that activation of the PLC pathway by various GPCRs, including the group-I mGluRs stimulate the neuronal secretion of APP, cleaved by α-secretase (Lee *et al.* 1995), so decreasing the production of toxic Aβ peptides. It has therefore been proposed that decreased glutamatergic neurotransmission in the brain of Alzheimer's patients may result in cognitive decline, not only as a result of alterations in LTP and LTD function, but also by promoting the formation of Aβ containing senile plaques.

29.6.8 Drug dependence

In humans addicted to nicotine, alcohol, psychostimulants (cocaine, amphetamine), or opiates (morphine, heroin) drug withdrawal results in a number of symptoms which include anxiety and irritability. In rodents, these symptoms are more difficult to analyse. One model used, is the increase in startle response five days following drug withdrawal. Using this model, it has been shown that re-administration of nicotine but also of LY354740 greatly reduces this withdrawal-induced behaviour in nicotine addicted rats (Helton *et al.* 1997). Similarly, group-II mGluR agonists attenuate morphine withdrawal symptoms (Fundytus and Coderre 1997; Fundytus *et al.* 1997). Moreover, it has been reported that the efficacy of pre-synaptic mGluRs was augmented in both the nucleus accumbens (Martin *et al.* 1999) and the ventro basal nucleus (Manzoni and Williams 1999) following chronic morphine treatment. Group-II mGluR agonists may therefore be useful in helping substance abuse patients.

29.6.9 Cerebellar ataxia

Recently, autoantibodies against mGlu1Rs have been identified in patients suffering from paraneoplastic cerebellar ataxia, and injection of such antibodies in the cerebellar subarachnoid space of mice impaired motor coordination (Sillevis Smitt *et al.* 2000). These antibodies were found to inhibit mGlu1R activation, and as such affect the normal functioning of the cerebellum. Indeed, mGlu1Rs are known to control the synaptic efficacy at the parallel fibre—Purkinje neurones synapses (Aiba *et al.* 1994; Conquet *et al.* 1994; Shigemoto *et al.* 1994; Ichise *et al.* 2000).

29.7 Concluding remarks

Discovered 16 years ago (Sladeczek *et al.* 1985; Nicoletti *et al.* 1986), the metabotropic glutamate receptors are now among the best characterized of all receptors. Their original structure makes them excellent receptor models for understanding the general mechanisms of GPCR activation. Indeed the importance of dimerization in their activation questions whether or not all GPCRs may actually function as dimers. Several observations also make the mGluRs excellent targets for the development of new therapeutic drugs. First, the high resolution structure of their binding domain can be used to design new agonists and antagonists. Second, the original structural organization of these proteins is such that other sites can be targeted, allowing the development of positive and negative highly specific allosteric modulators. Third, the various mGluRs are expressed in specific areas of the brain where they modulate fast excitatory synaptic transmission. Finally, the recent demonstration that mGluRs are expressed outside the brain, in organs such as liver, heart and bones, opens the field to many other areas, and offers new possibilities for the development of therapeutic mGluR ligands unable to cross the blood–brain barrier.

Acknowledgements

The authors wish to thank Drs O. Manzoni, Ph. Rondard, L. Prézeau, L. Fagni (Montpellier, France) and F. Acher (Paris, France) for critical reading of the manuscript. This work was supported by grants from the CNRS (J.B.), Bayer company (France and Germany) (J.B.), the Fondation pour la Recherche Médicale (J.B.), GIP-HMR (J.B.), the 'Action Incitative Physique et Chimie du Vivant' (PCV00-134) from the CNRS (J.P.P.), the programme 'Molécules et Cibles Thérapeutiques' from INSERM and CNRS (J.P.P.) and Retina France (J.P.P.).

References

Abe T, Sugihara H, Nawa H, Shigemoto R, Mizuno N, and Nakanishi S (1992). Molecular characterization of a novel metabotropic glutamate receptor mGluR5 coupled to inositol phosphate/Ca2+ signal transduction. *J Biol Chem* **267**, 13 361–8.

Adam G, Kolczewski S, Ohresser S, Wichmann J, Woltering T, and Mutel V (1999). Synthesis, structure activity relationship and receptor pharmacology of new non competitive, non-amino acid, mGlu2 receptor selective antagonist. *Neuropharmacology* **38**, A1.

Ahmadian H, Nielsen B, Bräuner-Osborne H, Johansen TN, Stensbøl TB, Sløk FA *et al.* (1997). (S)-Homo-AMPA, a specific agonist at the mGlu6 subtype of metabotropic glutamic acid receptors. *J Med Chem* **40**, 3700–5.

Aiba A, Kano M, Chen C, Stanton ME, Fox GD, Herrup K *et al.* (1994). Deficient cerebellar long-term depression and impaired motor learning in mGluR1 mutant mice. *Cell* **79**, 377–88.

Akazawa C, Ohishi H, Nakajima Y, Okamoto N, Shigemoto R, Nakanishi S *et al.* (1994). Expression of mRNAs of L-AP4-sensitive metabotropic glutamate receptors (mGluR4, mGluR6, mGluR7) in the rat retina. *Neurosci Lett* **171**, 52–4.

Alagarsamy S, Marino MJ, Rouse ST, Gereau RWt, Heinemann SF, and Conn PJ (1999). Activation of NMDA receptors reverses desensitization of mGluR5 in native and recombinant systems. *Nat Neurosci* **2**, 234–40.

Ango F, Pin J-P, Tu JC, Xiao B, Worley PF, Bockaert J *et al.* (2000). Dendritic and axonal targeting of type 5 metabotropic glutamate receptor (mGluR5) is regulated by homer1 proteins and neuronal excitation. *J Neurosci* **20**, 8710–9716.

Ango F, Prézeau L, Muller T, Worley PF, Pin JP, Bockaert J *et al.* (2001). Agonist-independent activation of mGluRs by the intracellular interacting protein, Homer. *Nature* **411**, 962–5.

Anwyl R (1999). Metabotropic glutamate receptors: electrophysiological properties and role in plasticity. *Brain Res Brain Res Rev* **29**, 83–120.

Aramori I and Nakanishi S (1992). Signal transduction and pharmacological characteristics of a metabotropic glutamate receptor, mGluR1, in transfected CHO cells. *Neuron* 8, 757–65.

Awad H, Hubert GW, Smith Y, Levey AI, and Conn PJ (2000). Activation of metabotropic glutamate receptor 5 has direct excitatory effects and potentiates NMDA receptor currents in neurones of the subthalamic nucleus. *J Neurosci* 20, 7871–9.

Bai M, Trivedi S, Kifor O, Quinn SJ, and Brown EM (1999). Intermolecular interactions between dimeric calcium-sensing receptor monomers are important for its normal function. *Proc Natl Acad Sci USA* 96, 2834–9.

Bessis A-S, Bertrand H-O, Galvez T, De Colle C, Pin J-P, and Acher F (2000). 3D-model of the extracellular domain of the type 4a metabotropic glutamate receptor: new insights into the activation process. *Prot Sci* 9, 2200–9.

Bessis A-S, Jullian N, Coudert E, Pin J-P, and Acher F (1999). Extended glutamate activates metabotropic receptor types 1, 2 and 4: selective features at mGluR4 binding site. *Neuropharmacology* 38, 1543–51.

Bhave G, Karim F, Carlton SM, and Gereau IV RW (2001). Peripheral group I metabotropic glutamate receptors modulate nociception in mice. *Nat Neurosci* 4, 417–23.

Billinton A, Ige AO, Bolam JP, White JH, Marshall FH, and Emson PC (2001). Advances in the molecular understanding of GABA(B) receptors. *Trends Neurosci* 24, 277–82.

Blahos II J, Mary S, Perroy J, De Colle C, Brabet I, Bockaert J *et al.* (1998). Extreme C-terminus of G-protein α-subunits contains a site that discriminates between Gi-coupled metabotropic glutamate receptors. *J Biol Chem* 273, 25765–9.

Blahos J, Fischer T, Brabet I, Stauffer D, Rovelli G, Bockaert J *et al.* (2001). A novel site on the Gα protein that recognizes heptahelical receptors. *J Biol Chem* 276, 3262–9.

Bockaert J and Pin J-P (1999). Molecular tinkering of G-protein coupled receptors: an evolutionary success. *EMBO J* 18, 1723–9.

Bond A, Jones NM, Hicks CA, Whiffin GM, Ward MA, O'Neill MF *et al.* (2000). Neuroprotective effects of LY379268, a selective mGlu2/3 receptor agonist: investigations into possible mechanism of action *in vivo*. *J Pharmacol Exp Ther* 294, 800–9.

Bond A, Oneill MJ, Hicks CA, Monn JA, and Lodge D (1998). Neuroprotective effects of a systemically active Group II metabotropic glutamate receptor agonist LY354740 in a gerbil model of global ischaemia. *Neuroreport* 9, 1191–3.

Boudin H, Doan A, Xia J, Shigemoto R, Huganir RL, Worley PF *et al.* (2000). Presynaptic clustering of mGluR7a requires the PICK1 PDZ domain binding site. *Neuron* 28, 485–97.

Brabet I, Mary S, Bockaert J, and Pin J-P (1995). Phenylglycine derivatives discriminate between mGluR1 and mGluR5 mediated responses. *Neuropharmacology* 34, 895–903.

Brabet I, Parmentier M-L, De Colle C, Bockaert J, Acher F, and Pin J-P (1998). Comparative effect of L-CCG-I, DCG-IV and γ-carboxy-L-glutamate on all cloned metabotropic glutamate receptor subtypes. *Neuropharmacology* 37, 1043–51.

Bradley SR, Levey AI, Hersch SM, and Conn PJ (1996). Immunocytochemical localization of group III metabotropic glutamate receptors in the hippocampus with subtype-specific antibodies. *J Neurosci* 16, 2044–56.

Bradley SR, Marino MJ, Wittmann M, Rouse ST, Awad H, Levey AI *et al.* (2000). Activation of group II metabotropic glutamate receptors inhibits synaptic excitation of the substantia Nigra pars reticulata. *J Neurosci* 20, 3085–94.

Bradley SR, Standaert DG, Rhodes KJ, Rees HD, Testa CM, Levey AI *et al.* (1999). Immunohistochemical localization of subtype 4a metabotropic glutamate receptors in the rat and mouse basal ganglia. *J Comp Neurol* 407, 33–46.

Brakeman PR, Lanahan AA, O'Brien RJ, Roche K, Barnes CA, Huganir RL *et al.* (1997). Homer: a protein that selectively binds metabotropic glutamate receptors. *Nature* 286, 284–8.

Brandstatter JH, Koulen P, Kuhn R, Vanderputten H, and Wassle H (1996). Compartmental localization of a metabotropic glutamate receptor (mGluR7): Two different active sites at a retinal synapse. *J Neurosci* **16**, 4749–56.

Bräuner-Osborne H, Egebjerg J, Nielsen EO, Madsen U, and Krogsgaard-Larsen P (2000). Ligands for glutamate receptors: design and therapeutic prospects. *J Med Chem* **43**, 2609–45.

Bräuner-Osborne H and Krogsgaard-Larsen P (1998). Pharmacology of (S)-homoquisqualic acid and (S)-2-amino-5-phosphonopentanoic acid [(S)-AP5] at cloned metabotropic glutamate receptors. *Br J Pharmacol* **123**, 269–74.

Bräuner-Osborne H and Krogsgaard-Larsen P (2000). Sequence and expression pattern of a novel human orphan G-protein-coupled receptor, GPRC5B, a family C receptor with a short amino-terminal domain. *Genomics* **65**, 121–8.

Bruno V, Battaglia G, Ksiazek I, van der Putten H, Catania MV, Giuffrida R *et al.* (2000*a*). Selective activation of mGlu4 metabotropic glutamate receptors is protective against excitotoxic neuronal death. *J Neurosci* **20**, 6413–20.

Bruno V, Ksiazek I, Battaglia G, Lukic S, Leonhardt T, Sauer D *et al.* (2000*b*). Selective blockade of metabotropic glutamate receptor subtype 5 is neuroprotective. *Neuropharmacology* **39**, 2223–30.

Bruno V, Sureda FX, Storto M, Casabona G, Caruso A, Knopfel T *et al.* (1997). The neuroprotective activity of group-II metabotropic glutamate receptors requires new protein synthesis and involves a glial-neuronal signaling. *J Neurosci* **17**, 1891–7.

Carroll FY, Stolle A, Beart PM, Voerste A, Brabet I, Mauler F *et al.* (2001). BAY36-7620: a potent non-competitive mGlu1 receptor antagonist with inverse agonist activity. *Mol Pharmacol* **59**, 965–73.

Cartmell J, Adam G, Chaboz S, Henningsen R, Kemp JA, Klingelschmidt A *et al.* (1998). Characterization of [H-3]-(2S,2'R,3'R)-2-(2',3'-dicarboxycyclopropyl)glycine ([H-3]-DCG IV) binding to metabotropic mGlu(2) receptor-transfected cell membranes. *Br J Pharmacol* **123**, 497–504.

Cartmell J, Monn JA, and Schoepp DD (1999). The metabotropic glutamate 2/3 receptor agonists LY354740 and LY379268 selectively attenuate phencyclidine versus d-amphetamine motor behaviors in rats. *J Pharmacol Exp Ther* **291**, 161–70.

Cartmell J, Monn JA, and Schoepp DD (2000*a*). Attenuation of specific PCP-evoked behaviors by the potent mGlu2/3 receptor agonist, LY379268 and comparison with the atypical antipsychotic, clozapine. *Psychopharmacology (Berl)* **148**, 423–9.

Cartmell J, Monn JA, and Schoepp DD (2000*b*). Tolerance to the motor impairment, but not to the reversal of PCP-induced motor activities by oral administration of the mGlu2/3 receptor agonist, LY379268. *Naunyn Schmiedeberg's Arch Pharmacol* **361**, 39–46.

Chaudhari N, Landin AM, and Roper SD (2000). A metabotropic glutamate receptor variant functions as a taste receptor. *Nat Neurosci* **3**, 113–19.

Chaudhari N, Yang H, Lamp C, Delay E, Cartford C, Than T *et al.* (1996). The taste of monosodium glutamate: membrane receptors in taste buds. *J Neurosci* **16**, 3817–26.

Chavis P, Fagni L, Bockaert J, and Lansman JB (1995). Modulation of calcium channels by metabotropic glutamate receptors in cerebellar granule cells. *Neuropharmacology* **34**, 929–37.

Chavis P, Fagni L, Lansman JB, and Bockaert J (1996). Functional coupling between ryanodine receptors and L-type calcium channels in neurons. *Nature* **382**, 719–22.

Chavis P, Mollard P, Bockaert J, and Manzoni O (1998). Visualization of cyclic AMP-regulated presynaptic activity at cerebellar granule cells. *Neuron* **20**, 773–81.

Chavis P, Shinozaki H, Bockaert J, and Fagni L (1994). The metabotropic glutamate receptor types 2/3 inhibit L-type calcium channels via a Pertussis toxin-sensitive G-protein in cultured cerebellar granule cells. *J Neurosci* **14**, 7067–76.

Chen Y, Bacon G, Sher E, Clark BP, Kallman MJ, Wright RA *et al.* (2000). Evaluation of the activity of a novel metabotropic glutamate receptor antagonist (+/−)-2-amino-2-(3-cis and trans-carboxycyclobutyl-3-(9-thioxanthyl)propionic acid) in the *in vitro* neonatal spinal cord and in an *in vivo* pain model. *Neuroscience* **95**, 787–93.

Cheng Y and Lotan R (1998). Molecular cloning and characterization of a novel retinoic acid- inducible gene that encodes a putative G protein-coupled receptor. *J Biol Chem* **273**, 35008–15.

Ciruela F, Soloviev MM, Chan WY, and McIlhinney RA (2000). Homer-1c/Vesl-1L modulates the cell surface targeting of metabotropic glutamate receptor type 1 alpha: evidence for an anchoring function. *Mol Cell Neurosci* **15**, 36–50.

Clark BP, Baker SR, Goldsworthy J, Harris JR, and Kingston AE (1997). (+)-2-methyl-4-carboxyphenylglycine (LY367385) selectively antagonises metabotropic glutamate mGluR1 receptors. *Bioorg Med Chem Lett* **7**, 2777–80.

Conn P and Pin J-P (1997). Pharmacology and functions of metabotropic glutamate receptors. *Ann Rev Pharmacol Toxicol* **37**, 205–37.

Conquet F, Bashir ZI, Davies CH, Daniel H, Ferraguti F, Bordi F et al. (1994). Motor deficit and impairment of synaptic plasticity in mice lacking mGluR1. *Nature* **372**, 237–43.

Corsi M, Quartaroli M, Maraia G, Chiamulera C, Ugolini A, Conquet F et al. (1996). PLC-coupled-mGluRs and their possible role in pain. *Neuropharmacology* **35**, A9.

Corti C, Restituito S, Rimland JM, Brabet I, Corsi M, Pin J-P et al. (1998). Cloning and characterization of alternative mRNA forms for the rat metabotropic glutamate receptors mGluR7 and mGluR8. *Eur J Neurosci* **10**, 3629–41.

Costantino G, Macchiarulo A, and Pellicciari R (1999). Pharmacophore models of group I and group II metabotropic glutamate receptor agonists. Analysis of conformational, steric, and topological parameters affecting potency and selectivity. *J Med Chem* **42**, 2816–27.

Costantino G and Pellicciari R (1996). Homology modeling of metabotropic glutamate receptors. (mGluRs) structural motifs affecting binding modes and pharmacological profile of mGluR1 agonists and competitive antagonists. *J Med Chem* **39**, 3998–4006.

De Colle C, Bessis A-S, Bockaert J, Acher F, and Pin J-P (2000). Pharmacological characterization of the rat metabotropic glutamate receptor type 8a revealed strong similarities and slight differences with the type 4a receptor. *Eur J Pharmacol* **394**, 17–26.

Desai MA, Burnett JP, Mayne NG, and Schoepp DD (1995). Cloning and expression of a human metabotropic glutamate receptor 1 alpha: Enhanced coupling on co-transfection with a glutamate transporter. *Mol Pharmacol* **48**, 648–57.

Dev KK, Nakajima Y, Kitano J, Braithwaite SP, Henley JM, and Nakanishi S (2000). PICK1 interacts with and regulates PKC phosphorylation of mGLUR7. *J Neurosci* **20**, 7252–7.

Doherty AJ, Palmer MJ, Bortolotto ZA, Hargreaves A, Kingston AE, Ornstein PL et al. (2000). A novel, competitive mGlu(5) receptor antagonist (LY344545) blocks DHPG-induced potentiation of NMDA responses but not the induction of LTP in rat hippocampal slices. *Br J Pharmacol* **131**, 239–44.

Doherty AJ, Palmer MJ, Henley JM, Collingridge GL, and Jane DE (1997). (RS)-2-chloro-5-hydroxyphenylglycine (CHPG) activates mGlu(5), but not mGlu(1), receptors expressed in CHO cells and potentiates NMDA responses in the hippocampus. *Neuropharmacology* **36**, 265–7.

Duvoisin RM, Zhang C, and Ramonell K (1995). A novel metabotropic glutamate receptor expressed in the retina and olfactory bulb. *J Neurosci* **15**, 3075–83.

Eaton SA, Birse EF, Wharton B, Sunter DC, Udvarhelyi PM, Watkins JC et al. (1993). Mediation of thalamic sensory responses. *In vivo* by ACPD-activated excitatory amino acid receptors. *Eur J Neurosci* **5**, 186–9.

Eriksen L and Thomsen C (1995). [H-3]-L-2-amino-4-phosphonobutyrate labels a metabotropic glutamate receptor, mGluR4a. *Br J Pharmacol* **116**, 3279–87.

Fagni L, Chavis P, Ango F, and Bockaert J (2000). Complex interactions between mGluRs, intracellular Ca2+ stores and ion channels. *Trends Neurosci* **23**, 80–8.

Ferraguti F, Baldani-Guerra B, Corsi M, Nakanishi S, and Corti C (1999). Activation of the extracellular signal-regulated kinase 2 by metabotropic glutamate receptors. *Eur J Neurosci* **11**, 2073–82.

Ferraguti F, Conquet F, Corti C, Grandes P, Kuhn R, and Knopfel T (1998). Immunohistochemical localization of the mGluR1beta metabotropic glutamate receptor in the adult rodent forebrain: evidence for a differential distribution of mGluR1 splice variants. *J Comp Neurol* **400**, 391–407.

Ferraguti F, Pietra C, Valerio E, Corti C, Chiamulera C, and Conquet F (1997). Evidence against a permissive role of the metabotropic glutamate receptor 1 in acute excitotoxicity. *Neurosci* **79**, 1–5.

Fiorillo CD and Williams JT (1998). Glutamate mediates an inhibitory postsynaptic potential in dopamine neurones. *Nature* **394**, 78–82.

Fisher K and Coderre TJ (1996). The contribution of metabotropic glutamate receptors (mGluRs) to formalin-induced nociception. *Pain* **68**, 255–63.

Fisher K, Fundytus ME, Cahill CM, and Coderre TJ (1998). Intrathecal administration of the mGluR compound, (S)-4CPG, attenuates hyperalgesia and allodynia associated with sciatic nerve constriction injury in rats. *Pain* **77**, 59–66.

Flor PJ, Lukic S, Rüegg D, Leonhardt T, Knöpfel T, and Kuhn R (1995). Molecular cloning, functional expression and pharmacological characterization of the human metabotropic glutamate receptor type 4. *Neuropharmacology* **34**, 149–55.

Flor PJ, Vanderputten H, Ruegg D, Lukic S, Leonhardt T, Bence M et al. (1997). A novel splice variant of a metabotropic glutamate receptor, human mGluR7b. *Neuropharmacology* **36**, 153–9.

Fotuhi M, Sharp AH, Glatt CE, Hwang PM, Von Krosigk M, Snyder SH et al. (1993). Differential localization of phosphoinositide-linked metabotropic glutamate receptor (mGluR1) and the inositol 1,4,5-triphosphate receptor in rat brain. *J Neurosci* **13**, 2001–12.

Francesconi A and Duvoisin RM (1998). Role of the second and third intracellular loops of metabotropic glutamate receptors in mediating dual signal transduction activation. *J Biol Chem* **273**, 5615–24.

Francesconi A and Duvoisin RM (2000). Opposing effects of protein kinase C and protein kinase A on metabotropic glutamate receptor signaling: Selective desensitization of the inositol trisphosphate/Ca^{2+} pathway by phosphorylation of the receptor-G protein-coupling domain. *Proc Natl Acad Sci USA* **97**, 6185–90.

Frati Cea (2000). Expression of functional mGlu5 metabotropic glutamate receptors in human melanocytes. *J Cell Physiol* **183**, 364–72.

Fundytus ME and Coderre TJ (1997). Attenuation of precipitated morphine withdrawal symptoms by acute i.c.v. administration of a group II mGluR agonist. *Br J Pharmacol* **121**, 511–4.

Fundytus ME, Fisher K, Dray A, Henry JL, and Coderre TJ (1998). *In vivo* antinociceptive activity of anti-rat mGluR1 and mGluR5 antibodies in rats. *Neuroreport* **9**, 731–5.

Fundytus ME, Ritchie J, and Coderre TJ (1997). Attenuation of morphine withdrawal symptoms by subtype-selective metabotropic glutamate receptor antagonists. *Br J Pharmacol* **120**, 1015–20.

Fundytus ME, Yashpal K, Chabot JG, Osborne MG, Lefebvre CD, Dray A et al. (2001). Knockdown of spinal metabotropic glutamate receptor 1 (mGluR(1)) alleviates pain and restores opioid efficacy after nerve injury in rats. *Br J Pharmacol* **132**, 354–67.

Galvez T, Duthey B, Kniazeff J, Blahos J, Rovelli G, Bettler B et al. (2001). Allosteric interactions between GB1 and GB2 subunits are required for optimal GABAB receptor function. *EMBO J* **20**, 2152–9.

Galvez T, Urwyler S, Prézeau L, Mosbacher J, Joly C, Malitschek B et al. (2000). Ca^{2+}-requirement for high affinity GABA binding at $GABA_B$ receptors: involvement of serine 269 of the $GABA_BR1$ subunit. *Mol Pharmacol* **57**, 419–26.

Gasparini F, Bruno V, Battaglia G, Lukic S, Leonhardt T, Inderbitzin W et al. (1999a). (R,S)-4-phosphonophenylglycine, a potent and selective group III metabotropic glutamate receptor agonist, is anticonvulsive and neuroprotective *in vivo*. *J Pharmacol Exp Ther* **289**, 1678–87.

Gasparini F, Lingenhohl K, Stoehr N, Flor PJ, Heinrich M, Vranesic I et al. (1999b). 2-Methyl-6-(phenylethynyl)-pyridine (MPEP), a potent, selective and systemically active mGlu5 receptor antagonist. *Neuropharmacology* **38**, 1493–503.

Gereau RW and Conn PJ (1995). Multiple presynaptic metabotropic glutamate receptors modulate excitatory and inhibitory synaptic transmission in hippocampal area CA1. *J Neurosci* 15, 6879–89.

Gereau RWt and Heinemann SF (1998). Role of protein kinase C phosphorylation in rapid desensitization of metabotropic glutamate receptor 5. *Neuron* 20, 143–51.

Gill SSea (1999). Immunochemical localization of the metabotropic glutamate receptors in the rat heart. *Brain Res Bull* 48, 143–6.

Gomeza J, Joly C, Kuhn R, Knöpfel T, Bockaert J, and Pin J-P (1996a). The second intracellular loop of mGluR1 cooperates with the other intracellular domains to control coupling to G-protein. *J Biol Chem* 271, 2199–205.

Gomeza J, Mary S, Brabet I, Parmentier M-L, Restituito S, Bockaert J et al. (1996b). Coupling of mGluR2 and mGluR4 to Gα15, Gα16 and chimeric Gαq/i proteins: characterization of new antagonists. *Mol Pharmacol* 50, 923–30.

Gu Y and Publicover SJ (2000). Expression of functional metabotropic glutamate receptors in primary cultured rat osteoblasts. CROSS-TALK WITH N-METHYL-D-ASPARTATE RECEPTORS. *J Biol Chem* 275, 34 252–9.

Hammerland L, Garrett J, Hung B, Levinthal C, and Nemeth E (1998). Allosteric activation of the Ca2+ receptor expressed in Xenopus laevis oocytes by NPS 467 or NPS 568. *Mol Pharmacol* 53, 1083–8.

Hampson DR, Huang XP, Pekhletski R, Peltekova V, Hornby G, Thomsen C et al. (1999). Probing the ligand-binding domain of the mGluR4 subtype of metabotropic glutamate receptor. *J Biol Chem* 274, 33 488–95.

Hampson DR, Theriault E, Huang XP, Kristensen P, Pickering DS, Franck JE et al. (1994). Characterization of two alternatively spliced forms of a metabotropic glutamate receptor in the central nervous system of the rat. *Neuroscience* 60, 325–36.

Han G and Hampson DR (1999). Ligand binding to the amino terminal domain of the mGlurR4 subtype of metabotropic glutamate receptor. *J Biol Chem* 274, 10 008–13.

Hayashi Y, Momiyama A, Takahashi T, Ohishi H, Ogawa-Meguro R, Shigemoto R et al. (1993). Role of a metabotropic glutamate receptor in synaptic modulation in the accessory olfactory bulb. *Nature* 366, 687–90.

Helton DR, Tizzano JP, Monn JA, Schoepp DD, and Kallman (1997). LY354740: a metabotropic glutamate receptor agonist which ameliorates symptoms of nicotine withdrawal in rats. *Neuropharmacology* 36, 1511–16.

Helton DR, Tizzano JP, Monn JA, Schoepp DD, and Kallman MJ (1998). Anxiolytic and side-effect profile of LY354740: A potent, highly selective, orally active agonist for group II metabotropic glutamate receptors. *J Pharmacol Exp Ther* 284, 651–60.

Herrero I, Miras-Portugal MT, and Sanchez-Prieto J (1998). Functional switch from facilitation to inhibition in the control of glutamate release by metabotropic glutamate receptors. *J Biol Chem* 273, 1951–8.

Heuss C, Scanziani M, Gähwiler BH, and Gerber U (1999). G-protein-independent signaling mediated by metabotropic glutamate receptors. *Nat Neurosci* 2, 1070–7.

Houamed KM, Kuijper JL, Gilbert TL, Haldeman BA, O'Hara PJ, Mulvihill ER et al. (1991). Cloning, expression, and gene structure of a G-protein-coupled glutamate receptor from rat brain. *Science* 252, 1318–21.

Husi H, Ward MA, Choudhary JS, Blackstock WP, and Grant SG (2000). Proteomic analysis of NMDA receptor-adhesion protein signaling complexes. *Nat Neurosci* 3, 661–9.

Ichise T, Kano M, Hashimoto K, Yanagihara D, Nakao K, Shigemoto R et al. (2000). mGluR1 in cerebellar Purkinje cells essential for long-term depression, synapse elimination, and motor coordination. *Science* 288, 1832–5.

Johansen PA, Chase LA, Sinor AD, Koerner JF, Johnson RL, and Robinson MB (1995). Type 4a metabotropic glutamate receptor: Identification of new potent agonists and differentiation from the L-(+)-2-amino-4-phosphonobutanoic acid-sensitive receptor in the lateral perforant pathway in rats. *Mol Pharmacol* **48**, 140–9.

Johnson BG, Wright RA, Arnold MB, Wheeler WJ, Ornstein PL, and Schoepp DD (1999). [3H]-LY341495 as a novel antagonist radioligand for group II metabotropic glutamate (mGlu) receptors: characterization of binding to membranes of mGlu receptor subtype expressing cells. *Neuropharmacology* **38**, 1519–29.

Joly C, Gomeza J, Brabet I, Curry K, Bockaert J, and Pin J-P (1995). Molecular, functional and pharmacological characterization of the metabotropic glutamate receptor type 5 splice variants: comparison with mGluR1. *J Neurosci* **15**, 3970–81.

Jullian N, Brabet I, Pin J-P, and Acher FC (1999). Agonist selectivity of mGluR1 and mGluR2 metabotropic receptors: a different environment but similar recognition of an extended glutamate conformation. *J Med Chem* **42**, 1546–55.

Kammermeier PJ, Xiao B, Tu JC, Worley PF, and Ikeda SR (2000). Homer proteins regulate coupling of group I metabotropic glutamate receptors to N-type calcium and M-type potassium channels. *J Neurosci* **20**, 7238–45.

Karim F, Wang C-C, and Gereau IV RW (2001). Metabotropic glutamate receptor subtypes 1 and 5 are activators of extracellular signal-regulated kinase signaling required for inflammatory pain in mice. *J Neurosci* **21**, 3771–9.

Kato A, Ozawa F, Saitoh Y, Fukazawa Y, Sugiyama H, and Inokuchi K (1998). Novel members of the Vesl/Homer family of PDZ proteins that bind metabotropic glutamate receptors. *J Biol Chem* **273**, 23 969–75.

Kingston A, Lowndes J, Evans N, Clark B, Tomlinson R, Burnett J *et al.* (1998a). Sulphur-containing amino acids are agonists for group 1 metabotropic receptors expressed in clonal RGT cell lines. *Neuropharmacology* **37**, 277–87.

Kingston AE, Ornstein PL, Wright RA, Johnson BG, Mayne NG, Burnett JP *et al.* (1998b). LY341495 is a nanomolar potent and selective antagonist of group II metabotropic glutamate receptors. *Neuropharmacology* **37**, 1–12.

Kinoshita A, Ohishi H, Neki A, Nomura S, Shigemoto R, Takada M *et al.* (1996a). Presynaptic localization of a metabotropic glutamate receptor, mGluR8, in the rhinencephalic areas: a light and electron microscope study in the rat. *Neurosci Lett* **207**, 61–4.

Kinoshita A, Ohishi H, Nomura S, Shigemoto R, Nakanishi S, and Mizuno N (1996b). Presynaptic localization of a metabotropic glutamate receptor, mGluR4a, in the cerebellar cortex: a light and electron microscope study in the rat. *Neurosci Lett* **207**, 199–202.

Kinoshita A, Shigemoto R, Ohishi H, van der Putten H, and Mizuno N (1998). Immunohistochemical localization of metabotropic glutamate receptors, mGluR7a and mGluR7b, in the central nervous system of the adult rat and mouse: a light and electron microscopic study. *J Comp Neurol* **393**, 332–52.

Knoflach F, Mutel V, Jolidon S, Kew JN, Malherbe P, Vieira E *et al.* (2001). Positive allosteric modulators of metabotropic glutamate 1 receptor: Characterization, mechanism of action, and binding site. *Proc Natl Acad Sci USA* **98**, 13 402–7.

Konieczny J, Ossowska K, Wolfarth S, and Pilc A (1998). LY354740, a group II metabotropic glutamate receptor agonist with potential antiparkinsonian properties in rats. *Naunyn Schmiedeberg's Arch Pharmacol* **358**, 500–2.

Koulen P, Malitschek B, Kuhn R, Wassle H, and Brandstatter JH (1996). Group II and group III metabotropic glutamate receptors in the rat retina: distributions and developmental expression patterns. *Eur J Neurosci* **8**, 2177–87.

Kubo Y, Miyashita T, and Murata Y (1998). Structural basis for a Ca^{2+}-sensing function of the metabotropic glutamate receptors. *Science* **279**, 1722–5.

Kunishima N, Shimada Y, Tsuji Y, Sato T, Yamamoto M, Kumasaka T *et al.* (2000). Structural basis of glutamate recognition by a dimeric metabotropic glutamate receptor. *Nature* **407**, 971–7.

Lafon-Cazal M, Fagni L, Guiraud M-J, Mary S, Lerner-Natoli M, Pin J-P *et al.* (1999*a*). mGluR7 like metabotropic glutamate receptors inhibit NMDA-mediated excitotoxicity in cultured cerebellar granule neurones. *Eur J Neurosci* 11, 663–72.

Lafon-Cazal M, Viennois G, Kuhn R, Malitschek B, Pin J-P, Shigemoto R *et al.* (1999*b*). mGluR7-like receptor and GABA$_B$ receptor activation enhance neurotoxic effects of NMDA in striatal GABAergic neurones. *Neuropharmacology* 38, 1631–40.

Lam AG, Soriano MA, Monn JA, Schoepp DD, Lodge D, and McCulloch J (1998). Effects of the selective metabotropic glutamate agonist LY354740 in a rat model of permanent ischaemia. *Neurosci Lett* 254, 121–3.

Lan JY, Skeberdis VA, Jover T, Grooms SY, Lin Y, Araneda RC *et al.* (2001). Protein kinase C modulates NMDA receptor trafficking and gating. *Nat Neurosci* 4, 382–90.

Laurie DJ, Boddeke HWGM, Hiltscher R, and Sommer B (1996). HmGlu1d, a novel splice variant of the human type-I metabotropic glutamate receptor. *Eur J Pharmacol* 296, R1–R3.

Laurie DJ, Danzeisen M, Boddeke HWGM, and Sommer B (1995). Ligand binding profile of the rat metabotropic glutamate receptor mGluR3 expressed in a transfected cell line. *Naunyn-Schmiedeberg's Arch Pharmacol* 351, 565–8.

Lee RK, Wurtman RJ, Cox AJ, and Nitsch RM (1995). Amyloid precursor protein processing is stimulated by metabotropic glutamate receptors. *Proc Natl Acad Sci USA* 92, 8083–7.

Li H, Ohishi H, Kinoshita A, Shigemoto R, Nomura S, and Mizuno N (1997). Localization of a metabotropic glutamate receptor, mGluR7, in axon terminals of presumed nociceptive, primary afferent fibers in the superficial layers of the spinal dorsal horn: an electron microscope study in the rat. *Neurosci Lett* 223, 153–6.

Lin FF, Varney M, Sacaan AI, Jachec C, Daggett LP, Rao S *et al.* (1997). Cloning and stable expression of the mGluR1b subtype of human metabotropic receptors and pharmacological comparison with the mGluR5a subtype. *Neuropharmacol* 36, 917–31.

Litschig S, Gasparini F, Rueegg D, Munier N, Flor PJ, Vranesic I-T *et al.* (1999). CPCCOEt, a noncompetitive mGluR1 antagonist, inhibits receptor signaling without affecting glutamate binding. *Mol Pharmacol* 55, 453–61.

Littman L and Robinson MB (1994). The effects of L-glutamate and trans-($\pm$)-1-amino-1,3-cyclopentanedicarboxylate on phosphoinositide hydrolysis can be pharmacologically differentiated. *J Neurochem* 63, 1291–302.

Littman L, Tokar C, Venkatraman S, Roon RJ, Koerner JF, Robinson MB *et al.* (1999). Cyclobutane Quisqualic acid analogues as selective mluR5a metabotropic gluutamic acid receptor ligands. *J Med Chem* 42, 1639–47.

Lovinger DM and McCool BA (1995). Metabotropic glutamate receptor-mediated presynaptic depression at corticostriatal synapses involves mGluR2 or 3. *J Neurophysiol* 73, 1076–83.

Lu YM, Jia Z, Janus C, Henderson JT, Gerlai R, Wojtowicz JM *et al.* (1997). Mice lacking metabotropic glutamate receptor 5 show impaired learning and reduced CA1 long-term potentiation (LTP) but normal CA3 LTP. *J Neurosci* 17, 5196–205.

Lujàn R, Nusser Z, Roberts JDB, Shigemoto R, and Somogyi P (1996). Perisynaptic location of metabotropic glutamate receptors mGluR1 and mGluR5 on dendrites and dendritic spines in the rat hippocampus. *Eur J Neurosci* 8, 1488–500.

Lujan R, Roberts JD, Shigemoto R, Ohishi H, and Somogyi P (1997). Differential plasma membrane distribution of metabotropic glutamate receptors mGluR1 alpha, mGluR2 and mGluR5, relative to neurotransmitter release sites. *J Chem Neuroanat* 13, 219–41.

Makoff A, Lelchuk R, Oxer M, Harrington K, and Emson P (1996*a*). Molecular characterization and localization of human metabotropic glutamate receptor type 4. *Mol Brain Res* 37, 239–48.

Makoff A, Pilling C, Harrington K, and Emson P (1996*b*). Human metabotropic glutamate receptor type 7: molecular cloning and mRNA distribution in the CNS. *Mol Brain Res* 40, 165–70.

Makoff AJ, Phillips T, Pilling C, and Emson P (1997). Expression of a novel splice variant of human mGluR1 in the cerebellum. *Neuroreport* **8**, 2943–7.

Malherbe P, Kratzeisen C, Lundstrom K, Richards JG, Faull RLM, and Mutel V (1999). Cloning and functional expression of alternative spliced variants of the human metabotropic glutamate receptor 8. *Mol Brain Res* **67**, 201–10.

Manzoni O and Bockaert J (1995). Metabotropic glutamate receptors inhibiting excitatory synapses in the CA1 area of the rat hippocampus. *Eur J Neurosci* **7**, 2518–23.

Manzoni OJ and Williams JT (1999). Presynaptic regulation of glutamate release in the ventral tegmental area during morphine withdrawal. *J Neurosci* **19**, 6629–36.

Martin G, Przewlocki R, and Siggins GR (1999). Chronic morphine treatment selectively augments metabotropic glutamate receptor-induced inhibition of N-methyl-D-aspartate receptor-mediated neurotransmission in nucleus accumbens. *J Pharmacol Exp Ther* **288**, 30–5.

Mary S, Stephan D, Gomeza J, Bockaert J, Pruss R, and Pin J-P (1997). The rat mGluR1d splice variant is devoid of significant constitutive activity like the other short isoforms of mGluR1. *Eur J Pharmacol* **335**, 65–72.

Masu M, Tanabe Y, Tsuchida K, Shigemoto R, and Nakanishi S (1991). Sequence and expression of a metabotropic glutamate receptor. *Nature* **349**, 760–65.

Masugi M, Yokoi M, Shigemoto R, Muguruma K, Watanabe Y, Sansig G *et al.* (1999). Metabotropic glutamate receptor subtype 7 ablation causes deficit in fear response and conditioned taste aversion. *J Neurosci* **19**, 955–63.

Mateos JM, Azkue J, Benitez R, Sarria R, Losada J, Conquet F *et al.* (1998). Immunocytochemical localization of the mGluR1b metabotropic glutamate receptor in the rat hypothalamus. *J Comp Neurol* **390**, 225–33.

Mateos JM, Benitez R, Elezgarai I, Azkue JJ, Lazaro E, Osorio A *et al.* (2000). Immunolocalization of the mGluR1b splice variant of the metabotropic glutamate receptor 1 at parallel fiber-Purkinje cell synapses in the rat cerebellar cortex. *J Neurochem* **74**, 1301–9.

Minakami R, Iida K, Hirakawa N, and Sugiyama H (1995). The expression of two splice variants of metabotropic glutamate receptor subtype 5 in the rat brain and neuronal cells during development. *J Neurochem* **65**, 1536–42.

Minakami R, Katsuki F, and Sugiyama H (1993). A variant of metabotropic glutamate receptor subtype 5: an evolutionarily conserved insertion with no termination codon. *Biochem Biophys Res Com* **194**, 622–7.

Moghaddam B and Adams BW (1998). Reversal of phencyclidine effects by a group II metabotropic glutamate receptor agonist in rats. *Science* **281**, 1349–52.

Monn JA, Valli MJ, Massey SM, Hansen MM, Kress TJ, Wepsiec JP *et al.* (1999). Synthesis, pharmacological characterization, and molecular modeling of heterobicyclic amino acids related to (+)-2-aminobicyclo[3.1.0] hexane-2,6-dicarboxylic acid (LY354740): identification of two new potent, selective, and systemically active agonists for group II metabotropic glutamate receptors. *J Med Chem* **42**, 1027–40.

Mutel V, Ellis GJ, Adam G, Chaboz S, Nilly A, Messer J *et al.* (2000). Characterization of [(3)H]Quisqualate binding to recombinant rat metabotropic glutamate 1a and 5a receptors and to rat and human brain sections. *J Neurochem* **75**, 2590–601.

Nakanishi S (1994). Metabotropic glutamate receptors: synaptic transmission, modulation, and plasticity. *Neuron* **13**, 1031–37.

Nakazato A, Kumagai T, Sakagami K, Yoshikawa R, Suzuki Y, Chaki S *et al.* (2000). Synthesis, SARs, and pharmacological characterization of 2-amino-3 or 6-fluorobicyclo[3.1.0]hexane-2,6-dicarboxylic acid derivatives as potent, selective, and orally active group II metabotropic glutamate receptor agonists. *J Med Chem* **43**, 4893–909.

Naples MA and Hampson DR (2000). Pharmacological profiles of the metabotropic glutamate receptor ligands [3H]L-AP4 and [3H]CPPG. *Neuropharmacology* **39**.

Neki A, Ohishi H, Kaneko T, Shigemoto R, Nakanishi S, and Mizuno N (1996). Pre- and postsynaptic localization of a metabotropic glutamate receptor, mGluR2, in the rat brain: An immunohistochemical study with a monoclonal antibody. *Neurosci Lett* **202**, 197–200.

Nicoletti F, Wroblewski JT, Novelli A, Alho H, Guidotti A, and Costa E (1986). The activation of inositol phospholipid metabolism as a signal-transduction system for excitatory amino acids in primary cultures of cerebellar granule cells. *J Neurosci* **6**, 1905–11.

O'Hara PJ, Sheppard PO, Thøgersen H, Venezia D, Haldeman BA, McGrane V *et al.* (1993). The ligand-binding domain in metabotropic glutamate receptors is related to bacterial periplasmic binding proteins. *Neuron* **11**, 41–52.

Ohishi H, Nomura S, Ding YQ, Shigemoto R, Wada E, Kinoshita A *et al.* (1995). Presynaptic localization of a metabotropic glutamate receptor, mGluR7, in the primary afferent neurones: An immunohistochemical study in the rat. *Neurosci Lett* **202**, 85–8.

Ohishi H, Ogawa-Meguro R, Shigemoto R, Kaneko T, Nakanishi S, and Mizuno N (1994). Immunohistochemical localization of metabotropic glutamate receptors, mGluR2 and mGluR3, in rat cerebellar cortex. *Neuron* **13**, 55–66.

Ohishi H, Shigemoto R, Nakanishi S, and Mizuno N (1993). Distribution of the messenger RNA for a metabotropic glutamate receptor, mGluR2, in the central nervous system of the rat. *Neurosci* **53**, 1009–18.

Okamoto N, Hori S, Akazawa C, Hayashi Y, Shigemoto R, Mizuno N *et al.* (1994). Molecular characterization of a new metabotropic glutamate receptor mGluR7 coupled to inhibitory cyclic AMP signal transduction. *J Biol Chem* **269**, 1231–6.

Okamoto T, Sekiyama N, Otsu M, Shimada Y, Sato A, Nakanishi S *et al.* (1998). Expression and purification of the extracellular ligand binding region of metabotropic glutamate receptor subtype 1. *J Biol Chem* **273**, 13089–96.

Ott D, Floersheim P, Inderbitzin W, Stoehr N, Francotte E, Lecis G *et al.* (2000). Chiral resolution, pharmacological characterization, and receptor docking of the noncompetitive mGlu1 receptor antagonist (±)-2-Hydroxyimino-1a,2-dihydro-1H-7-oxacyclopropa[b]naphthalene-7a-carboxylic acid ethyl ester. *J Med Chem* **43**, 4428–36.

Pagano A, Rüegg D, Litschig S, Stoehr N, Stierlin C, Heinrich M *et al.* (2000). The non-competitive antagonists 2-Methyl-6-(phenylethynyl)pyridine and 7-hydroxyiminocyclopropan[b]chromen-1a-carboxylic acid ethyl ester interact with overlapping binding pockets in the transmembrane region of group I metabotropic glutamate receptors. *J Biol Chem* **275**, 33750–8.

Parmentier M-L, Galvez T, Acher F, Peyre B, Pellicciari R, Grau Y *et al.* (2000). Conservation of the ligand recognition selectivity between the Drosophila metabotropic glutamate type A receptor and its mammalian homolog. *Neuropharmacology* **39**, 1119–31.

Parmentier ML, Joly C, Restituito S, Bockaert J, Grau Y, and Pin J-P (1998). The G-protein coupling profile of metabotropic glutamate receptors, as determined with exogenous G-proteins, is independent of their ligand recognition domain. *Mol Pharmacol* **53**, 778–86.

Pellicciari R, Raimondo M, Marinozzi M, Natalini B, Costantino G, and Thomsen C (1996). (S)-(+)-2-(3′-carboxybicyclo[1.1.1]pentyl)glycine, a structurally new group I metabotropic glutamate receptor antagonist. *J Med Chem* **39**, 2874–6.

Peltekova V, Han G, Soleymanlou N, and Hampson DR (2000). Constraints on proper folding of the amino terminal domains of group III metabotropic glutamate receptors. *Brain Res Mol Brain Res* **76**, 180–90.

Perroy J, Prezeau L, De Waard M, Shigemoto R, Bockaert J, and Fagni L (2000). Selective blockade of P/Q-type calcium channels by the metabotropic glutamate receptor type 7 involves a phospholipase C pathway in neurones. *J Neurosci* **20**, 7896–904.

Pin J-P and Bockaert J (1995). Get receptive to metabotropic glutamate receptors. *Curr Op Neur* **5**, 342–9.

Pin J-P, Bockaert J, and Récasens M (1984). The Ca^{2+}/Cl^--dependent L[3H]-glutamate binding: a new receptor or a particular transport process? *FEBS Lett* **175**, 31–6.

Pin J-P, De Colle C, Bessis A-S, and Acher F (1999). New perspectives for the development of selective metabotropic glutamate receptor ligands. *Eur J Pharmacol* **375**, 277–94.

Pin J-P and Duvoisin R (1995). The metabotropic glutamate receptors: Structure and functions. *Neuropharmacology* **34**, 1–26.

Pin J-P, Waeber C, Prézeau L, Bockaert J, and Heinemann SF (1992). Alternative splicing generates metabotropic glutamate receptors inducing different patterns of calcium release in *Xenopus* oocytes. *Proc Natl Acad Sci USA* **89**, 10 331–5.

Prézeau L, Gomeza J, Ahern S, Mary S, Galvez T, Bockaert J *et al.* (1996). Changes of the C-terminal domain of mGluR1 by alternative splicing generate receptors with different agonit independent activity. *Mol Pharmacol* **49**, 422–9.

Ray K and Hauschild BC (2000). Cys-140 Is critical for metabotropic glutamate receptor-1 (mGluR-1) dimerization. *J Biol Chem* **275**, 34245–51.

Ray K, Hauschild BC, Steinbach PJ, Goldsmith PK, Hauache O, and Spiegel AM (1999). Identification of the cysteine residues in the amino-terminal extracellular domain of the human Ca(2+) receptor critical for dimerization. Implications for function of monomeric Ca(2+) receptor. *J Biol Chem* **274**, 27642–50.

Robbins MJ, Michalovich D, Hill J, Calver AR, Medhurst AD, Gloger I *et al.* (2000). Molecular cloning and characterization of two novel retinoic acid-inducible orphan G-protein-coupled receptors (GPRC5B and GPRC5C). *Genomics* **67**, 8–18.

Roche K, Tu JC, Petralia RS, Xiao B, Wenthold RJ, and Worley PF (1999). Homer 1b regulates the trafficking of group I metabotropic glutamate receptors. *J Biol Chem* **274**, 25 953–7.

Rodriguez-Moreno A, Sistiaga A, Lerma J, and Sanchez-Prieto J (1998). Switch from facilitation to inhibition of excitatory synaptic transmission by group I mGluR desensitization. *Neuron* **21**, 1477–86.

Romano C, Miller JK, Hyrc K, Dikranian S, Mennerick S, Takeuchi Y *et al.* (2001). Covalent and non-covalent interactions mediate metabotropic glutamate receptor mGlu5 dimerization. *Mol Pharmacol* (In press).

Romano C, Sesma MA, MacDonald C, O'Malley K, van den Pol AN, and Olney JW (1995). Distribution of metabotropic glutamate receptor mGluR5 immunoreactivity in rat brain. *J Comp Neurol* **355**, 455–69.

Romano C, Yang W-L, and O'Malley KL (1996). Metabotropic glutamate receptor 5 is a disulphide-linked dimer. *J Biol Chem* **271**, 28 612–16.

Rouse ST, Marino MJ, Bradley SR, Awad H, Wittmann M, and Conn PJ (2000). Distribution and roles of metabotropic glutamate receptors in the basal ganglia motor circuit: implications for treatment of Parkinson's disease and related disorders. *Pharmacol Ther* **88**, 427–35.

Saugstad JA, Kinzie JM, Shinohara MM, Segerson TP, and Westbrook GL (1997). Cloning and expression of rat metabotropic glutamate receptor 8 reveals a distinct pharmacological profile. *Mol Pharmacol* **51**, 119–25.

Saunders R, Nahorski SR, and Challiss RA (1998). A modulatory effect of extracellular Ca^{2+} on type 1α metabotropic glutamate receptor-mediated signalling. *Neuropharmacology* **37**, 273–6.

Schoepp DD, Jane DE, and Monn JA (1999). Pharmacological agents acting at subtypes of metabotropic glutamate receptors. *Neuropharmacology* **38**, 1431–76.

Schoepp DD, Johnson BG, Wright RA, Salhoff CR, Mayne NG, Wu S *et al.* (1997). LY354740 is a potent and highly selective group II metabotropic glutamate receptor agonist in cells expressing human glutamate receptors. *Neuropharmacology* **36**, 1–11.

Schreiber R, Lowe D, Voerste A, and De Vry J (2000). LY354740 affects startle responding but not sensorimotor gating or discriminative effects of phencyclidine. *Eur J Pharmacol* **388**, R3–4.

Schweitzer C, Kratzeisen C, Adam G, Lundstrom K, Malherbe P, Ohresser S *et al.* (2000). Characterization of [(3)H]-LY354740 binding to rat mGlu2 and mGlu3 receptors expressed in CHO cells using semliki forest virus vectors. *Neuropharmacology* **39**, 1700–6.

Shigemoto R, Abe T, Nomura S, Nakanishi S, and Hirano T (1994). Antibodies inactivating mGluR1 metabotropic glutamate receptor block long-term depression in cultured Purkinje cells. *Neuron* **12**, 1245–55.

Shigemoto R, Kinoshita A, Wada E, Nomura S, Ohishi H, Takada M *et al.* (1997). Differential presynaptic localization of metabotropic glutamate receptor subtypes in the rat hippocampus. *J Neurosci* **17**, 7503–22.

Shigemoto R, Kulik A, Roberts JDB, Ohishi H, Nusser Z, Kaneko T *et al.* (1996). Target-cell-specific concentration of a metabotropic glutamate receptor in the presynaptic active zone. *Nature* **381**, 523–5.

Sillevis Smitt P, Kinoshita A, De Leeuw B, Moll W, Coesmans M, Jaarsma D *et al.* (2000). Paraneoplastic cerebellar ataxia due to autoantibodies against a glutamate receptor. *N Engl J Med* **342**, 21–7.

Sistiaga A, Herrero I, Conquet F, and Sanchez-Prieto J (1998). The metabotropic glutamate receptor 1 is not involved in the facilitation of glutamate release in cerebrocortical nerve terminals. *Neuropharmacology* **37**, 1485–92.

Skerry TM and Genever PG (2001). Glutamate signalling in non-neuronal tissues. *Trends Neurosci* **22**, 174–81.

Sladeczek F, Pin J-P, Récasens M, Bockaert J, and Weiss S (1985). Glutamate stimulates inositol phosphate formation in striatal neurones. *Nature* **317**, 717–19.

Snead OC, 3rd, Banerjee PK, Burnham M, and Hampson D (2000). Modulation of absence seizures by the GABA(A) receptor: a critical role for metabotropic glutamate receptor 4 (mGluR4). *J Neurosci* **20**, 6218–24.

Soloviev MM, Ciruela F, Chan WY, and McIlhinney RA (1999). Identification, cloning and analysis of expression of a new alternatively spliced form of the metabotropic glutamate receptor mGluR1 mRNA1. *Biochim Biophys Acta* **1446**, 161–6.

Soloviev MM, Ciruela F, Chan WY, and McIlhinney RA (2000). Molecular characterisation of two structurally distinct groups of human homers, generated by extensive alternative splicing. *J Mol Biol* **295**, 1185–200.

Speca DJ, Lin DM, Sorensen PW, Isacoff EY, Ngai J, and Dittman AH (1999). Functional identification of a goldfish odorant receptor. *Neuron* **23**, 487–98.

Spooren WP, Gasparini F, Bergmann R, and Kuhn R (2000*a*). Effects of the prototypical mGlu(5) receptor antagonist 2-methyl-6-(phenylethynyl)-pyridine on rotarod, locomotor activity and rotational responses in unilateral 6-OHDA-lesioned rats. *Eur J Pharmacol* **406**, 403–10.

Spooren WP, Gasparini F, van der Putten H, Koller M, Nakanishi S, and Kuhn R (2000*b*). Lack of effect of LY314582 (a group 2 metabotropic glutamate receptor agonist) on phencyclidine-induced locomotor activity in metabotropic glutamate receptor 2 knock-out mice. *Eur J Pharmacol* **397**, R1–2.

Spooren WP, Vassout A, Neijt HC, Kuhn R, Gasparini F, Roux S *et al.* (2000*c*). Anxiolytic-like effects of the prototypical metabotropic glutamate receptor 5 antagonist 2-methyl-6-(phenylethynyl)pyridine in rodents. *J Pharmacol Exp Ther* **295**, 1267–75.

Stephan D, Bon C, Holzwarth JA, Galvan M, and Pruss RM (1996). Human metabotropic glutamate receptor 1: mRNA distribution, chromosome localization and functional expression of two splice variants. *Neuropharmacology* **35**, 1649–60.

Storto M, de Grazia U, Knopfel T, Canonico PL, Copani A, Richelmi P *et al.* (2000). Selective blockade of mGlu5 metabotropic glutamate receptors protects rat hepatocytes against hypoxic damage. *Hepatology* **31**, 649–55.

Stowell JN and Craig AM (1999). Axon/dendrite targeting of metabotropic glutamate receptor by their cytoplasmic carboxy-terminal domains. *Neuron* **22**, 525–36.

Takahashi K, Tsuchida K, Tanabe Y, Masu M, and Nakanishi S (1993). Role of the large extracellular domain of metabotropic glutamate receptors in agonist selectivity determination. *J Biol Chem* **268**, 19 341–5.

Tanabe Y, Masu M, Ishii T, Shigemoto R, and Nakanishi S (1992). A family of metabotropic glutamate receptors. *Neuron* **8**, 169–79.

Tanabe Y, Nomura A, Masu M, Shigemoto R, Mizuno N, and Nakanishi S (1993). Signal transduction, pharmacological properties, and expression patterns of two rat metabotropic glutamate receptors, mGluR3 and mGluR4. *J Neurosci* **13**, 1372–8.

Tatarczynska E, Klodzinska A, Chojnacka-Wojcik E, Palucha A, Gasparini F, Kuhn R *et al.* (2001). Potential anxiolytic- and antidepressant-like effects of MPEP, a potent, selective and systemically active mGlu5 receptor antagonist. *Br J Pharmacol* **132**, 1423–30.

Thomas NK, Wright RA, Howson PA, Kingston AE, Schoepp DD, and Jane DE (2001). (S)-3,4-DCPG, a potent and selective mGlu8a receptor agonist, activates metabotropic glutamate receptors on primary afferent terminals in the neonatal rat spinal cord. *Neuropharmacology* **40**, 311–8.

Thomsen C, Bruno V, Nicoletti F, Marinozzi M, and Pellicciari R (1996). (2S,1′S,2′S,3′R)-2-(2′-carboxy-3′-phenylcyclopropyl)glycine, a potent and selective antagonist of type 2 metabotropic glutamate receptors. *Mol Pharmacol* **50**, 6–9.

Thomsen C, Mulvihill ER, Haldeman B, Pickering DS, Hampson DR, and Suzdak PD (1993). A pharmacological characterization of the mGluR1α subtype of the metabotropic glutamate receptor expressed in a cloned baby hamster kidney cell line. *Brain Res* **619**, 22–8.

Thomsen C, Pekhletski R, Haldeman B, Gilbert TA, O'Hara P, and Hampson DR (1997). Cloning and characterization of a metabotropic glutamate receptor, mGluR4b. *Neuropharmacol* **36**, 21–30.

Tsuji Y, Shimada Y, Takeshita T, Kajimura N, Nomura S, Sekiyama N *et al.* (2000). Cryptic dimer interface and domain organization of the extracellular region of metabotropic glutamate receptor subtype 1. *J Biol Chem* **275**, 28 144–51.

Tu JC, Xiao B, Yuan JP, Lanahan AA, Leoffert K, Li M *et al.*(1998). Homer Binds a Novel Proline-Rich Motif and Links Group 1 Metabotropic Glutamate Receptors with IP3 Receptors. *Neuron* **21**, 717–26.

Urwyler S, Bettler B, Froestl W, Gama AL, Heid J, Hofstetter K *et al.* (2000). Positive allosteric modulation of native and recombinant GABAB receptor activity. *Soc Neurosci Abstr* **26**, 1660.

Varney MA, Cosford ND, Jachec C, Rao SP, Sacaan A, Lin FF *et al.* (1999). SIB-1757 and SIB-1893: selective, noncompetitive antagonists of metabotropic glutamate receptor type 5. *J Pharmacol Exp Ther* **290**, 170–81.

Walker K, Bowes M, Panesar M, Davis A, Gentry C, Kesingland A *et al.* (2000*a*). Metabotropic glutamate receptor subtype 5 (mGlu5) and nociceptive function. I. Selective blockade of mGlu5 receptors in models of acute, persistent and chronic pain. *Neuropharmacology* **40**, 1–9.

Walker K, Reeve A, Bowes M, Winter J, Wotherspoon G, Davis A *et al.* (2000*b*). mGlu5 receptors and nociceptive function II. mGlu5 receptors functionally expressed on peripheral sensory neurones mediate inflammatory hyperalgesia. *Neuropharmacology* **40**, 10–19.

Wright RA, Arnold MB, Wheeler WJ, Ornstein PL, and Schoepp DD (2000). Binding of [3H](2S,1′S,2′S)-2-(9-xanthylmethyl)-2-(2′-carboxycyclopropyl) glycine ([3H]LY341495) to cell membranes expressing recombinant human group III metabotropic glutamate receptor subtypes. *Naunyn Schmiedeberg's Arch Pharmacol* **362**, 546–54.

Wroblewska B, Wroblewski JT, Pshenichkin S, Surin A, Sullivan SE, and Neale JH (1997). N-acetylaspartylglutamate selectively activates mGluR3 receptors in transfected cells. *J Neurochem* **69**, 174–81.

Wu S, Wright RA, Rockey PK, Burgett SG, Arnold JS, Rosteck PR, Jr *et al.* (1998). Group III human metabotropic glutamate receptors 4, 7 and 8: molecular cloning, functional expression, and comparison of pharmacological properties in RGT cells. *Brain Res Mol Brain Res* **53**, 88–97.

Xiao B, Tu JC, Petralia RS, Yuan JP, Doan A, Breder CD *et al.* (1998). Homer regulates the association of group 1 metabotropic glutamate receptors with multivalent complexes of Homer-related, synaptic proteins. *Neuron* 21, 707–16.

Xiao B, Tu JC, and Worley PF (2000). Homer: a link between neural activity and glutamate receptor function. *Curr Op Neurobiol* 10, 370–4.

Index

DATE DUE

Palci # 214911 due 5/16/04			

Demco